Botanica Marina

Volume 34, 1991

Walter de Gruyter · Berlin · New York

Editor-in-Chief

Dr G. T. Boalch
The Laboratory
Citadel Hill
Plymouth PL1 2PB
Great Britain
Telephone (0752) 22 27 72
Telefax (0752) 22 68 65

Editorial Office

Botanica Marina
Genthiner Str. 13
D-1000 Berlin 30
F. R. of Germany
Telephone: (030) 260 05-133
Telefax: (030) 260 05-251
Telex: 184 027

Title No. 3 10 900 676 6 ISSN 0006-8055

Typesetting and Printing Arthur Collignon GmbH, Berlin
Binding Buchbinderei Spiller, Berlin

Editorial Assistant and Technical Coordinator Asta Wonneberger

Printed in Germany

Walter de Gruyter & Co., Genthiner Straße 13, D-1000 Berlin 30, F.R. of Germany
Tel. (030) 260 05-133, Telefax (030) 260 05-251, Telex 184 027
Walter de Gruyter, Inc., 200 Saw Mill River Road, Hawthorne, N.Y. 10532, U.S.A.
Tel. (914) 747-0110, Telefax (914) 747-1326, Telex 646 677

Walter de Gruyter · Berlin · New York

Contents

Obituary

Original Papers

Author Index

Subject Index

Taxonomical Index

Botanica Marina

Volume 34, 1991

Walter de Gruyter · Berlin · New York

Botanica Marina Volume 34, Fasc. 1
January 1991, pages 1 – 61

Editor-in-Chief
Dr G. T. Boalch
The Laboratory
Citadel Hill
Plymouth PL1 2PB
Great Britain
Telephone (0752) 22 27 72
Telefax (0752) 22 68 65

Editorial Office
Botanica Marina
Genthiner Str. 13
D-1000 Berlin 30
F. R. of Germany
Telephone: (030) 260 05-133
Telefax: (030) 260 05-251
Telex: 184 027

Title No. 3 10 900 676 7 ISSN 0006-8055

Typesetting and Printing Arthur Collignon GmbH, Berlin
Binding Buchbinderei Spiller, Berlin

Editorial Assistant and Technical Coordinator Asta Wonneberger

Printed in Germany

Walter de Gruyter & Co., Genthiner Straße 13, D-1000 Berlin 30, F.R. of Germany
Tel. (030) 260 05-133, Telefax (030) 260 05-251, Telex 184 027
Walter de Gruyter, Inc., 200 Saw Mill River Road, Hawthorne, N.Y. 10532, U.S.A.
Tel. (914) 747-0110, Telefax (914) 747-1326, Telex 646 677

Walter de Gruyter · Berlin · New York

Botanica Marina
Vol. 34, pp. 1–61, 1991

Illustrated Key to the Filamentous Higher Marine Fungi

J. Kohlmeyer and B. Volkmann-Kohlmeyer

Institute of Marine Sciences, University of North Carolina at Chapel Hill, Morehead City, North Carolina 28557, U.S.A.

(Accepted 31 July 1990)

Abstract

This treatise deals with all filamentous higher marine fungi described up to the present time, namely, 255 Ascomycotina, 6 Basidiomycotina and 60 anamorphic fungi. A dichotomous key leads to the species and varieties, mainly by using characters of the propagules. All (except 5) taxa are illustrated by original camera lucida drawings of ascospores, basidiospores and conidia, more than half based on type material, the remainder on authentic specimens. The annotated list of species includes concise characterizations of the propagules. *Quintaria* gen. nov. (*Q. lignatilis* comb. nov.) and *Corollospora novofusca* sp. nov. are described, and five new combinations in the genera *Coronopapilla, Halosarpheia, Lindra* and *Lulworthia* are proposed.

Introduction

The last complete key to the filamentous higher marine fungi, including 209 species, was published over a decade ago (Kohlmeyer and Kohlmeyer 1979). In the present treatise we recognize 321 taxa, representing 255 ascomycetes, 6 basidiomycetes, and 60 anamorphic ('imperfect') fungi. The monograph is divided into three major parts, namely, the key, the plates, and the list of species.

The dichotomous **Key** is mostly based on the propagules (ascospores, basidiospores, conidia), and leads to specific taxa; the figure number and letter are added in italics. In some cases, the experienced marine mycologist can use 'short-cuts' to reach certain morphological, taxonomical or ecological groups or genera. For instance, species characterized by apothecia or hysterothecia are included in leads 1–8′; algicolous species (except those in *Corollospora* and *Lulworthia*) are between the leads 50 and 93′; well-circumscribed genera with five or more taxa are usually contained in 'keys within the key' (*Chadefaudia*: 57′–66′; *Corollospora*: 9–24′; *Haloguignardia*: 57–61′; *Koralionastes*: 35–39′; *Lulworthia*: 25–34′; *Pontogeneïa*: 87′–93′; *Spathulospora*: 52–56′). Those species that are often collected with immature or partly ripened propagules have been included in two locations of the key, to permit identification of such material.

The **Plates** are originals, containing camera lucida drawings of all but 5 species for which we could not obtain type or authentical material. We examined type specimens of all available species and used other collections only when types could not be located, or where the material was not preserved well enough to show critical features. Whenever feasible, members of the same genus, or morphologically similar species were grouped together and drawn at the same magnification. Species received after completion of Figures 1–22 are illustrated in Figure 23.

The alphabetical **List of Species** includes concise descriptions of major characters used in identification, namely, ascospores, basidiospores and conidia. Besides dimensions of propagules, the characterization consists of color (if other than hyaline) and septation (if variable). If not indicated otherwise, the sizes exclude gelatinous sheaths or appendages. The kind of type or authentical material and its location are included in parentheses. A synonym is listed for a particular fungus if it appeared under this name in Kohlmeyer and Kohlmeyer (1979). In many cases, taxonomical or other annotations are added in a separate paragraph. A concluding list contains rejected names, doubtful or excluded species. The three parts of this treatise, viz., key, plates and descriptions of species, are an integrity and should not be applied one without the others. In particular, we urge users of this key not to base identifications on illustrations alone, but to consider all features included in the **Key** and **List** and, if in doubt, also its known geographical distribution (see Kohlmeyer and Kohlmeyer 1979, chapter 23).

Key to the Filamentous Higher Marine Fungi

I. Key to Subdivisions of Eumycota

1 Reproduction by sexual spores 2
1′ Reproduction by asexual propagules 3
2 (1) Spores produced in asci .... **Ascomycotina**, p. 2
2′(1) Spores produced on basidia .... **Basidiomycotina**, p. 13
3 (1′) Smut spores, originating inside the host .... **Basidiomycotina**, p. 13
3′(1′) Smut spores absent; conidia or chlamydospores formed .... **Anamorphs**, p. 13

II. Key to Ascomycotina

1 Ascomata opening with a cleft-like ostiole (hysterothecial), or discoid and hymenium completely exposed at maturity (apothecial) 2
1′ Ascomata without opening (cleistothecial), or with round ostiole (perithecial) 9
2 (1) Ascomata embedded or superficial; not discoid, ostiole cleft-like 3
2′(1) Ascomata superficial, discoid; hymenium exposed 7
3 (2) Ascomata lirelliform, embedded in calcareous substrates ..... ***Halographis runica*, *Fig. 2CC***
3′(2) Ascomata not lirelliform, in or on wood ... 4
4 (3′) Ascomata superficial, boat-shaped; ascospores with transverse septa only .... ***Lophiostoma mangrovei*, *Fig. 13H***
4′(3′) Ascomata immersed; ascospores muriform 5
5 (4′) Ascospores with 13 or more transsepta, up to 96 μm long ...... ***Aigialus grandis*, *Fig. 16A***
5′(4′) Ascospores with less than 13 transsepta, less than 75 μm long 6
6 (5′) Ascospores 16–27 μm diam, more than 8 transsepta ...... ***Aigialus parvus*, *Fig. 16B***
6′(5′) Ascospores 10–16 μm diam, less than 8 transsepta ... ***Aigialus mangrovei*, *Fig. 16C***
7 (2′) On decaying algae; ascospores non-septate ... . ***Laetinaevia marina*, *Fig. 1T***
7′(2′) On wood; ascospores 1 septate 8
8 (7′) In temperate waters; ascospores with bristle-like setae at both ends .... ***Banhegyia setispora*, *Fig. 8K***
8′(7′) In tropical waters; ascospores without setae .... ***Dactylospora haliotrepha*, *Fig. 3E***
9 (1′) Ascospores with hyaline, ribbon-like appendages all around the middle and at each apex, developing by the peeling-off of the exospore; most species with a long spine-like appendage at each end (*Corollospora* spp.) 10
9′(1′) Ascospores without or with different appendages 25
10 (9) Ascospores 1 septate 11
10′(9) Ascospores multi-septate 13
11 (10) Ascospores brown .... ***Corollospora cinnamomea*, *Fig. 10C***
11′(10) Ascospores hyaline 12
12 (11′) Ascospore diameter usually 7 μm or less .... ***Corollospora gracilis*, *Fig. 10A***
12′(11′) Ascospore diameter 8 μm or more .... ***Corollospora maritima*, *Fig. 10B***
13 (10′) Ascospores brown, muriform 14
13′(10′) Ascospores hyaline, with cross septa only 15
14 (13) Ascospore length (excluding apical spines) up to 220 μm; spines up to 65 μm long .. .. ***Corollospora fusca*, *Fig. 12B***
14′(13) Ascospore length less than 100 μm; spines up to 86 μm long .... ***Corollospora novofusca*, *Fig. 12A***
15 (13′) Ascospores exclusively 3 septate 16
15′(13′) Ascospores with 3 or more septa 18
16 (15) Ascospore length 35 μm or more .... ***Corollospora angusta*, *Fig. 10D***
16′(15) Ascospore length usually less than 35 μm ... 17
17 (16′) Ascospore diameter 7–12 μm; spines 10–14 μm long .... ***Corollospora intermedia*, *Fig. 10E***
17′(16′) Ascospore diameter up to 7.5 μm; spines up to 9.5 μm long ... ***Corollospora sp.*, *Fig. 10G***
18 (15′) Ascospore diameter 13–26 μm .... ***Corollospora colossa*, *Fig. 12C***

18′(15′) Ascospore diameter normally less than 13 µm ... 19
19(18′) Ascospores with long apical spines ... 20
19′(18′) Ascospores apically rounded, without spines ... 22
20(19) Ascospore diameter 10 µm or more, spines up to 24 µm long ... ***Corollospora lacera****, Fig. 11B*
20′(19) Ascospore diameter 10 µm or less, spines up to 13 µm long ... 21
21(20′) Ascospores normally 3, rarely 4 or 5 septate ... ***Corollospora armoricana****, Fig. 10F*
21′(20′) Ascospores normally 5, rarely 3 or 6–8 septate ... ***Corollospora quinqueseptata****, Fig. 11A*
22(19′) Ascospores 5 septate ... ***Corollospora luteola****, Fig. 11C*
22′(19′) Ascospores with more septa ... 23
23(22′) Ascospores normally more than 12 septate ... ***Corollospora filiformis****, Fig. 11F*
23′(22′) Ascospores normally less than 12 septate ... 24
24(23′) Ascospores attenuated toward both ends; ascomata smooth and metallic ... ***Corollospora pseudopulchella****, Fig. 11D*
24′(23′) Ascospores rounded, not distinctly tapering toward the ends; ascomata rough and tuberculate ... ***Corollospora pulchella****, Fig. 11E*
25(9′) Ascospores filiform (at least 15 times longer than wide), tapering at both ends into mucilage-containing, cell-like processes; use stains, *e.g.*, hematoxylin (*Lulworthia* spp.) ... 26
25′(9′) Ascospores different; if filiform, without apical chambers ... 35
26(25) Ascomata on/in corals or coralline algae ... 27
26′(25) Ascomata on/in other substrates ... 29
27(26) Ascospores 200 µm or longer, immersed in coralline algae ***Lulworthia kniepii****, Fig. 18K*
27′(26) Ascospores shorter than 200 µm ... 28
28(27′) Ascospores 5.5–8.5 µm diam; superficial on corals ... ***Lulworthia calcicola****, Fig. 18E*
28′(27′) Ascospores 4–6 µm diam; immersed in coralline algae ... ***Lulworthia curalii****, Fig. 18M*
29(26′) Ascospores septate ... 30
29′(26′) Ascospores without septa ... 32
30(29) Ascomata on sand grains; ascospores with more than 30 septa ... ***Lulworthia lignoarenaria****, Fig. 18L*
30′(29) Ascospores with less than 30 septa ... 31
31(30′) Ascospores 9–12 septate, longer than 150 µm ... ***Lulworthia lindroidea****, Fig. 18J*
31′(30′) Ascospores 1 septate, shorter than 150 µm ... ***Lulworthia uniseptata****, Fig. 18G*
32(29′) Ascomata on sand grains; ascospores 140–205 µm long ... ***Lulworthia crassa****, Fig. 18F*
32′(29′) Ascomata mostly on wood (also on algae or seagrasses) ... 33
33(32′) Ascospores longer than 500 µm ... ***Lulworthia grandispora****, Fig. 18H*
33′(32′) Ascospores shorter than 500 µm ... 34
34(33′) Ascospores normally 70–110 µm long ... ***Lulworthia fucicola****, Fig. 18I*
34′(33′) Ascospores longer than 110 µm ... ***Lulworthia*** *spp.*
35(25′) On corals; ascomata surrounded or covered by crustose sponges (*Koralionastes* spp.) ... 36
35′(25′) In other habitats and substrates ... 40
36(35) Ascospore diameter 50 µm or more ... 37
36′(35) Ascospore diameter less than 50 µm ... 38
37(36) Ascospores mostly 2 septate, up to 170 µm long ... ***Koralionastes giganteus****, Fig. 15A*
37′(36) Ascospores mostly 4 septate, up to 131 µm long ... ***Koralionastes ovalis****, Fig. 15B*
38(36′) Ascospores mostly 4 and 5 septate, up to 130 µm long, an Australian species ... ***Koralionastes violaceus****, Fig. 15E*
38′(36′) Ascospores mostly 6–8 septate, up to 127 µm long, Caribbean species ... 39
39(38′) Ascospore diameter 18–28 µm; mostly 8 septate ... ***Koralionastes angustus****, Fig. 15D*
39′(38′) Ascospore diameter 27–45 µm; mostly 6 septate ... ***Koralionastes ellipticus****, Fig. 15C*
40(35′) Ascospores filiform, diameter 10 µm or less ... 41
40′(35′) Ascospores not filiform ... 49
41(40) Ascospores 6 or more septate, without mucilaginous appendages (*Lindra* spp.) ... 42
41′(40) Ascospores non-septate, with a mucilaginous cap at one or both ends (*Linocarpon* spp.) ... 47
42(41) Ascospore diameter 8 µm or more, length up to 520 µm ... ***Lindra crassa****, Fig. 17C*
42′(41) Ascospore diameter less than 8 µm, maximum length usually less than 400 µm ... 43
43(42′) Ascomata forming subicula on hard surfaces (*e.g.*, sand) ... ***Lindra obtusa****, Fig. 17F*
43′(42′) Ascomata without subicula, not attached to hard surfaces, usually immersed in the substrate ... 44
44(43′) In wood ... 45
44′(43′) In algae or sea grasses ... 46
45(44) Ascospores shorter than 200 µm; tropical species ... ***Lindra hawaiiensis****, Fig. 17E*

45′(44) Ascospores longer than 200 µm; temperate species .... ***Lindra inflata**, Fig. 17B*

46(44′) Ascospores usually less than 230 µm long, mean around 210 µm .... ***Lindra marinera**, Fig. 17A*

46′(44′) Ascospores usually 230 µm or longer, mean around 270 µm .... ***Lindra thalassiae**, Fig. 17D*

47(41′) Ascospores with appendage at each end ... . ***Linocarpon** cfr. **pandani**, Fig. 17I*

47′(41′) Ascospores with one cap at the base ... 48

48(47′) Ascospores arranged spirally in the ascus, distinctly curved, up to 4 µm diam .... ***Linocarpon nypae**, Fig. 17H*

48′(47′) Ascospores arranged in a straight row in the ascus, straight or slightly curved, up to 3.6 µm diam .... ***Linocarpon appendiculatum**, Fig. 17G*

49(40′) In or on macroalgae 50

49′(40′) In or on other substrates 94

50(49) Ascospores one-celled at maturity 51

50′(49) Ascospores several-celled at maturity ... 69

51(50) Ascospores with distinct appendages or mucilaginous sheaths, or with chambered, deciduous or non-deciduous apices 52

51′(50) Ascospores without appendages, sheaths or chambered apices 67

52(51) Hyphae absent; crustose thallus bearing flask-shaped antheridia; ascomata superficial on *Ballia* spp., Rhodophyta (*Spathulospora* spp.) 53

52′(51) Hyphae present; antheridia absent; on other hosts 57

53(52) Ascomata without hairs; in antarctic and subantarctic waters .... ***Spathulospora antarctica**, Fig. 1P*

53′(52) Ascomata with hairs; in Australia or New Zealand 54

54(53′) Ascospores longer than 65 µm 55

54′(53′) Ascospores usually shorter than 65 µm 56

55(54) Ascospores less than 14 µm diam, spathulate to spoon-shaped at the apices; few antheridia on long hairs .... ***Spathulospora phycophila**, Fig. 1L*

55′(54) Ascospores more than 14 µm diam, with a conical appendage at each end; many antheridia on short stalks .... ***Spathulospora adelpha**, Fig. 1M*

56(54′) Ascomata enclosed by long hairs; ascospore tips spathulate; gelatinous appendages subterminal .... ***Spathulospora lanata**, Fig. 1N*

56′(54′) Ascomata bare, except for a few short hairs around the ostiole; ascospore tips rounded, with terminal appendages .... ***Spathulospora calva**, Fig. 1O*

57(52′) In Phaeophyta; conical ascospore appendages chambered or striate (*Haloguignardia* spp.) 58

57′(52′) In Rhodophyta; gelatinous, rounded appendages without structure (*Chadefaudia* spp.) 62

58(57) In *Cystoseira* or *Halidrys* 59

58′(57) In *Sargassum* spp. 60

59(58) Ascospores longer than 60 µm; Mediterranean species .... ***Haloguignardia cystoseirae**, Fig. 1K*

59′(58) Ascospores shorter than 60 µm; Californian species .... ***Haloguignardia irritans**, Fig. 1I*

60(58′) Ascospore appendages persistent .... ***Haloguignardia tumefaciens**, Fig. 1H*

60′(58′) Ascospore appendages deciduous 61

61(60′) Galls with fingerlike projections; ascospores, including appendages, longer than 35 µm; Pacific Ocean .... ***Haloguignardia decidua**, Fig. 1G*

61′(60′) Galls without projections; ascospores, including appendages, commonly shorter than 35 µm; Atlantic Ocean .... ***Haloguignardia oceanica**, Fig. 1J*

62(57′) Ascospores longer than 25 µm (excluding appendages) .... ***Chadefaudia balliae**, Fig. 1F*

62′(57′) Ascospores shorter than 25 µm 63

63(62′) Ascomata superficial, but often covered by epiphytes 64

63′(62′) Ascomata immersed in the host tissue .. 65

64(63) Symbiont with epiphytic algae; ascoma diameter generally less than 300 µm; upper peridium less than 40 µm thick; ostiolar projection into ascoma cavity absent .... ***Chadefaudia corallinarum**, Fig. 1A*

64′(63) Parasite; ascoma diameter generally over 300 µm; ostiolar canal forming tubelike projection into ascoma cavity .... ***Chadefaudia polyporolithi**, Fig. 1B*

65(63′) In *Rhodymenia*; ascospores up to 16 µm long (excluding appendages) .... ***Chadefaudia marina**, Fig. 1D*

65′(63′) In other hosts 66

66(65′) Ascospores cylindrical, curved, 1–4 µm diam; in *Schizymenia* .... ***Chadefaudia schizymeniae**, Fig. 1C*

66′(65′) Ascospores ellipsoidal, not curved, 4–7 µm diam; in other hosts .... ***Chadefaudia gymnogongri**, Fig. 1E*

67 (51′) In Rhodophyta; New Zealand .... ***Polystigma apophlaeae**, Fig. 1Q*

67′(51′) In Chlorophyta; other geographic regions 68

68 (67′) Ascospore diameter up to 5 µm; in *Prasiola* .... ***Mastodia tessellata**, Fig. 1S*

68′(67′) Ascospore diameter up to 7 µm; in *Blidingia* .... ***Turgidosculum ulvae**, Fig. 1R*

69 (50′) Ascospores always 1 septate 70

69′(50′) Ascospores with one septum or more .. 85

70 (69) In Rhodophyta 71

70′(69) In Phaeophyta 74

71 (70) Systemic symbiont of *Apophlaea* in New Zealand ... ***Mycosphaerella apophlaeae**, Fig. 2S*

71′(70) In other hosts 72

72 (71′) In cystocarps and tetrasporic pustules of *Chondrus crispus* .... ***Lautitia danica**, Fig. 2P*

72′(71′) In other hosts 73

73 (72′) In *Gloiopeltis*; Japan .... ***Didymella gloiopeltidis**, Fig. 2X*

73′(72′) In *Rhodymenia*; France .... ***Didymella magnei**, Fig. 2Y*

74 (70′) Ascospore diameter over 10 µm 75

74′(70′) Ascospore diameter less than 10 µm ... 78

75 (74) Ascospores with gelatinous caps at each end; in *Cystophora* .... ***Massarina cystophorae**, Fig. 6J*

75′(74) Ascospores without caps; in other hosts .. 76

76 (75′) On *Cystoseira* (Fucales); ascomata stalked ..... ***Thalassoascus cystoseirae**, Fig. 3G*

76′(75′) On other hosts; ascomata sessile 77

77 (76′) On *Lessonia* (Laminariales); ascospore diameter 26 µm or less .... ***Thalassoascus lessoniae**, Fig. 3F*

77′(76′) On *Aglaozonia* and *Zanardinia* (Cutleriales); ascospore diameter over 26 µm .... ***Thalassoascus tregoubovii***

78 (74′) Systemic symbiont of *Ascophyllum* and *Pelvetia* .... ***Mycosphaerella ascophylli**, Fig. 2T*

78′(74′) Not systemic, in or on other hosts ... 79

79 (78′) In or on *Laminaria* 80

79′(78′) In or on other hosts 83

80 (79) Parasitic, causing black spots in the stems of the host .... ***Phycomelaina laminariae**, Fig. 3N*

80′(79) Saprobic or epiphytic 81

81 (80′) Saprobic, immersed in dead tissues .... ***Nectriella laminariae**, Fig. 2AA*

81′(80′) Symbiotic, lichenoid with epiphytic *Ectocarpus* or cyanobacteria, on the surface ... 82

82 (81′) Ascospores with apical caplike appendages; spore diameter more than 6 µm ... . ***Pharcidia laminariicola**, Fig. 2Q*

82′(81′) Ascospores without appendages; diameter less than 6 µm .... ***Pharcidia rachiana**, Fig. 2R*

83 (79′) Symbiotic (epiphyte on *Pelvetia*), lichenoid with cyanobacteria, ascospore length 16 µm or less .. ***Pyrenocollema pelvetiae**, Fig. 2DD*

83′(79′) Perthophytic in senescent algal tissues .. 84

84 (83′) Ascospores longer than 15 µm, without appendages .... ***Didymella fucicola**, Fig. 2Z*

84′(83′) Ascospores shorter than 15 µm, with a whorl of threadlike appendages below each apex ...... ***Crinigera maritima**, Fig. 8H*

85 (69′) Ascospores muriform, brown ... ***Pleospora pelvetiae*** and ***P. gracilariae**, Fig. 16F* and *23F*

85′(69′) Ascospores with cross septa only, hyaline 86

86 (85′) Ascospores filamentous; broad at one end, tapering to the very narrow other end ... . ***Trailia ascophylli**, Fig. 13F*

86′(85′) Ascospores ellipsoidal, or if elongate, diameter more or less equal throughout 87

87 (86′) In *Ascophyllum*, *Fucus* and *Pelvetia*; ascus opening with an operculum .... ***Orcadia ascophylli**, Fig. 13E*

87′(86′) In other hosts, asci without opercula (*Pontogeneia* spp.) 88

88 (87′) Ascospores with one septum near each apex, diameter up to 10 µm; on *Egregia*, associated with epiphytic *Ectocarpus* (?) .... ***Pontogeneia erikae**, Fig. 15H*

88′(87′) Ascospores with 2 or more septa, the first one developed in the center, diameter 10 µm or more 89

89 (88′) In Chlorophyta; ascospore diameter 20 µm or more 90

89′(88′) In Phaeophyta; ascospore diameter less than 20 µm 91

90 (89) Ascospores 3 septate; in *Codium* .... ***Pontogeneia codiicola**, Fig. 15J*

90′(89) Ascospores 2–5 septate; in *Valoniopsis* .. .. ***Pontogeneia valoniopsidis***

91 (89′) Ascospores longer than 200 µm; in *Halopteris* 92

91′(89′) Ascospores shorter than 200 µm; in other host genera 93

92 (91) Ascospores with 10 or more septa .... ***Pontogeneia cubensis**, Fig. 15 I*

92′(91) Ascospores with 5 or less septa .... ***Pontogeneia enormis***

93 (91′) Ascospores normally 3–5 septate, 85 µm long or shorter; in *Castagnea* . . . . ***Pontogeneia calospora**, Fig. 15G*

93′(91′) Ascospores normally 6–9 septate, longer than 95 µm; in *Padina* . . . . ***Pontogeneia padinae**, Fig. 15F*

94 (49′) Ascospores non-septate at maturity . . 95

94′(49′) Ascospores with one septum or more at maturity 112

95 (94) Ascospores elongate (50 µm or longer) . . 96

95′(94) Ascospores subglobose, ellipsoidal or shorter than 50 µm 99

96 (95) Ascospore diameter over 16 µm, both ends with a tubelike, mucus-filled appendage . . . . ***Kohlmeyeriella tubulata**, Fig. 2A*

96′(95) Ascospores without tubes, diameter less than 16 µm (*Bathyascus* spp.) 97

97 (96′) Ascospores less than 80 µm long; a deep-sea species . . ***Bathyascus vermisporus**, Fig. 18C*

97′(96′) Ascospores longer than 80 µm; shallow-water species 98

98 (97′) Ascospores longer than 200 µm, diameter over 8 µm . . . . ***Bathyascus grandisporus**, Fig. 18D*

98′(97′) Ascospores shorter than 200 µm, diameter less than 8 µm . . . . ***Bathyascus avicenniae**, Fig. 18A*

99 (95′) Ascospores hyaline 100

99′(95′) Ascospores yellow-brown or brown . . . 109

100 (99) Ascospores without appendages . . . 101

100′(99) Ascospores with appendages 104

101 (100) Ascospores elongate-fusiform . . . . ***Halonectria milfordensis**, Fig. 2G*

101′(100) Ascospores subglobose or ellipsoidal 102

102 (101′) Ascospores more than 12 µm diam . . . . ***Thalassogena sphaerica**, Fig. 2C*

102′(101′) Ascospores less than 12 µm diam . . . 103

103 (102′) In wood . . . . ***Payosphaeria minuta**, Fig. 23A*

103′(102′) Lichenized, on calcareous beach rock . . . . ***Verrucaria cribbii**, Fig. 1U*

104 (100′) Ascospores with longitudinal striations, a moustache-shaped appendage at one end; on ferns . . . . ***Phomatospora acrostichi**, Fig. 2B*

104′(100′) Ascospores without striations, or if present, on other substrates 105

105 (104′) Ascospore with one turban-like appendage, unfurling into a long ribbon . . . . ***Moana turbinulata**, Fig. 2D*

105′(104′) Ascospores different 106

106 (105′) Ascospores with rigid, acuminate appendages, evenly distributed over the surface . . . . ***Amylocarpus encephaloides**, Fig. 2J*

106′(105′) Ascospore ornamentations different 107

107 (106′) Ascospores with a tuft of bristlelike appendages at each end and four tufts around the equator . . . . ***Nautosphaeria cristaminuta**, Fig. 8M*

107′(106′) Ascospores different, appendages only apical or subapical 108

108 (107′) Ascospores with longitudinal striations; tropical species . . . . ***Lanspora coronata**, Fig. 8L*

108′(107′) Ascospores without striations; temperate species . . . ***Eiona tunicata**, Fig. 2I*

109 (99′) Ascospores lenticular with a longitudinal band . . . . . ***Gymnascella littoralis**, Fig. 2H*

109′(99′) Ascospores different 110

110 (109′) Ascospores with a long tail-like appendage . . . . . . ***Adomia avicenniae**, Fig. 2E*

110′(109′) Ascospores without appendages . . 111

111 (110′) Ascospores dark brown, with longitudinal germ slit . ***Hypoxylon oceanicum**, Fig. 4L*

111′(110′) Ascospores yellowish to yellow-brown, without slit . . . . ***Rhizophila marina**, Fig. 2F*

112 (94′) Ascospores with one septum at maturity 113

112′(94′) Ascospores with one and more septa at maturity 201

113 (112) Ascospores septate near the base; large cell dark, small cell light-colored 114

113′(112) Ascospores septate near the middle or, if asymmetrical, concolorous 116

114 (113) Ascospores longer than 70 µm . . . . ***Manglicola guatemalensis**, Fig. 4K*

114′(113) Ascospores less than 40 µm long 115

115 (114′) Ascospores with subapical germ pore; basal cell broadly cylindrical, up to 7–8 µm diam . . . . . . ***Zopfiella latipes**, Fig. 4N*

115′(114′) Ascospores with apical germ pore; basal cell elongate cylindrical, up to 4 µm diam . . . . ***Zopfiella marina**, Fig. 4M*

116 (113′) Ascospores hyaline, light grayish, yellow, or yellowish brown at maturity 117

116′(113′) Ascospores dark-colored (brown or blackish) at maturity 192

117 (116) Ascospores without appendages or gelatinous or mucilaginous sheaths 118

117′(116) Ascospores with appendages or sheaths 136

118 (117) Ascospores thick-walled 119

118′(117) Ascospores thin-walled 120

119(118) Ascoma without ostiole; ascospores broad ellipsoidal, up to 21 μm long .... ***Heleococcum japonense***, *Fig. 3Q*

119′(118) Ascoma ostiolate; ascospores elongate, up to 37 μm long .... ***Aniptodera chesapeakensis***, *Fig. 3 O*

120(118′) On animals 121

120′(118′) In plant material 122

121(120) In shells of mollusca and balanidae .... ***Arthopyrenia halodytes***, *Fig. 2BB*

121′(120) On chitin of insects .... ***Laboulbenia marina***, *Fig. 3K*

122(120′) Ascomata orange-yellowish, ostiolate 123

122′(120′) Ascoma color different, ostiolate or non-ostiolate 124

123(122) Ascospores distinctly ridged longitudinally .... ***Hydronectria tethys***, *Fig. 3D*

123′(122) Ascospores without longitudinal ridges .. .. ***Hydronectria tethys var. glabra***, *Fig. 3C*

124(122′) Ascospore diameter over 20 μm .... ***Nais glitra***, *Fig. 3A*

124′(122′) Ascospore diameter below 20 μm ... 125

125(124′) Ascospores longer than 60 μm .... ***Bathyascus tropicalis***, *Fig. 18B*

125′(124′) Ascospores shorter than 60 μm 126

126(125′) In bark of living pneumatophores of *Avicennia* .... ***Mycosphaerella pneumatophorae***, *Fig. 2U*

126′(125′) In other substrates 127

127(126′) Asci fissitunicate, thick-walled 128

127′(126′) Asci not fissitunicate, thin-walled .. 131

128(127) Ascospore diameter over 9 μm .... ***Acrocordiopsis patilii***, *Fig. 2L*

128′(127) Ascospore diameter less than 9 μm 129

129(128′) In Plumbaginaceae .... ***Mycosphaerella staticicola***, *Fig. 2W*

129′(128′) In Chenopodiaceae 130

130(129′) Ascomata 100 μm or more diam; ascospores 18 μm long or longer .... ***Mycosphaerella suaedae-australis***

130′(129′) Ascomata less than 100 μm diam; ascospores up to 18 μm long (in fresh material in a gelatinous sheath) .... ***Mycosphaerella salicorniae***, *Fig. 2V*

131(127′) Ascospores apricot-colored in masses ... . ***Swampomyces armeniacus***, *Fig. 2K*

131′(127′) Ascospores hyaline in masses 132

132(131′) Ascospore diameter over 9 μm ... 133

132′(131′) Ascospore diameter less than 9 μm 135

133(132) Ascospores cylindrical, 34–54 μm long .. .. ***Aniptodera longispora***, *Fig. 23E*

133′(132) Ascospores ellipsoidal, mostly shorter 134

134(133′) Ascomata up to 250 μm high; peridium thin (up to 15 μm); temperate species . ... ***Nais inornata***, *Fig. 3B*

134′(133′) Ascomata over 300 μm high; peridium thick (30 μm or more); tropical species .... ***Lignincola tropica***, *Fig. 2O*

135(132′) Asci with apical pore, remaining in ascoma in a fascicle; catenophyses absent .... ***Lignincola longirostris***, *Fig. 2N*

135′(132′) Asci without apical pore, detached and released from ascoma; catenophyses present .... ***Lignincola laevis***, *Fig. 2M*

136(117′) Ascospores without appendages, but enclosed in uniform gelatinous sheath 137

136′(117′) Ascospores with appendages; if covered by a sheath, also additional fibrous appendages present 143

137(136) Ascomata over 1 mm diam, conical, superficial; ascospores become 3 septate at maturity .. ***Ascocratera manglicola***, *Fig. 13M*

137′(136) Ascomata smaller; ascospores never 3 septate 138

138(137′) Ascospores warty .... ***Belizeana tuberculata***, *Fig. 3H*

138′(137′) Ascospores smooth 139

139(138′) In *Salicornia* .... ***Mycospharella salicorniae***, *Fig. 2V*

139′(138′) In other hosts or substrates 140

140(139′) In ferns .... ***Massarina acrostichi***, *Fig. 13P*

140′(139′) In wood 141

141(140′) Ascospores shorter than 20 μm; temperate species .... ***Biflua physasca***, *Fig. 3 I*

141′(140′) Ascospores longer than 20 μm; tropical species 142

142(141′) Asci obclavate, with long oculus; in *Rhizophora* .... ***Massarina lacertensis***, *Fig. 13Q*

142′(141′) Asci clavate, without oculus; in *Avicennia* .... ***Didymella avicenniae***, *Fig. 3J*

143(136′) Ascospores with one caplike appendage 144

143′(136′) Ascospores with more than one appendage 147

144(143) Appendage filamentous, longer than 30 μm; a deep-sea species .... ***Oceanitis scuticella***, *Fig. 7O*

144′(143) Appendages subglobose or lenticular; shallow-water species 145

145(144′) Asci bitunicate; 1 lenticular, lateral cap on the ascospore septum (hematoxylin!) .... ***Paraliomyces lentiferus***, *Fig. 3M*

145′(144′) Asci unitunicate; ascospore with 1 deciduous apical or subapical cap 146

146(145′) Ascospores up to 21 µm long; no additional septa formed during germination ...... ***Ophiodeira monosemeia***, *Fig. 7I*

146′(145′) Ascospores up to 68 µm long; 1–3 additional septa may be formed during germination .... ***Halosphaeria cucullata***, *Fig. 3L*

147(143′) Ascospores with terminal or subterminal radiating deciduous setae (thin bristles) 148

147′(143′) Ascospore appendages different 152

148(147) Setae on one end of the ascospore ... 149

148′(147) Setae on each end of the ascospore 150

149(148) Exclusively on bark of *Rhizophora*; up to 7 ascospore appendages .... ***Etheirophora blepharospora***, *Fig. 8G*

149′(148) In other hosts; more than 7 appendages . ... ***Etheirophora unijubata***, *Fig. 8E*

150(148′) Ascospores shorter than 15 µm, appendages subpolar .... ***Dryosphaera navigans***, *Fig. 8I*

150′(148′) Ascospores longer than 15 µm, appendages polar 151

151(150′) Ascomata superficial on bark, with brown setae around the ostiole; a temperate species .... ***Capronia ciliomaris***, *Fig. 8J*

151′(150′) Ascomata immersed in wood, without setae; a tropical species .... ***Etheirophora bijubata***, *Fig. 8F*

152(147′) Ascospores completely or partially enclosed in a sheath; also fibrillar or strand-like appendages present ... 153

152′(147′) Ascospores different 155

153(152) Sheath eccentric, strand-like appendages lateral ...... ***Nimbospora effusa***, *Fig. 9D*

153′(152) Sheath surrounding the spore completely, equatorial appendages equidistantly around the septum 154

154(153′) With equatorial and polar ascospore appendages .... ***Nimbospora octonae***, *Fig. 9E*

154′(153′) With equatorial appendages only .... ***Nimbospora bipolaris***, *Fig. 9F*

155(152′) Ascospore appendages only terminal or subterminal (an additional gelatinous cover around the spore may occur) 156

155′(152′) Ascospore appendages terminally and laterally attached 182

156(155) Ascospores with 3–4 thin, more or less rigid appendages at each apex 157

156′(155) Appendages different 159

157(156) Ascospore diameter less than 6 µm .... ***Arenariomyces parvulus***, *Fig. 8D*

157′(156) Ascospore diameter more than 7 µm ... 158

158(157′) Ascomata normally smaller than 200 µm, usually on sand or other hard substrates; ascospores consistently with 3 appendages at each end .... ***Arenariomyces trifurcatus***, *Fig. 8A*

158′(157′) Ascomata larger than 200 µm, partly immersed in wood; ascospores with 3 or 4 appendages at each end .... ***Arenariomyces majusculus***, *Fig. 8B*

159(156′) Mature ascospores enclosed in an exosporic sheath that is pierced at each apex by an elongated outward growing, tapering appendage 160

159′(156′) Mature ascospores without sheath .. 162

160(159) Appendages tube-like .... ***Ceriosporopsis capillacea***, *Fig. 6A*

160′(159) Appendages solid 161

161(160′) Peridium thin; ascoma without persistent sterile elements .... ***Ceriosporopsis halima***, *Fig. 6E*

161′(160′) Peridium thick; centrum of ascoma containing persistent, wide filaments, similar to pseudoparaphyses or catenophyses .. .. ***Ceriosporopsis cambrensis*** and ***Ceriosporopsis sundica***, *Fig. 6F*

162(159′) Ascospores with 2–9 appendages (more or less identical) at each end 163

162′(159′) Ascospore appendages different, not more than 1 at each end 166

163(162) 2 subterminal appendages at each ascospore apex, these pairs arranged at right angles to one another (anomalous spores with 3–4 appendages on both ends may occur) ..... ***Halosphaeria quadricornuta***, *Fig. 5A*

163′(162) 3–9 appendages at each ascospore apex 164

164(163′) 3–4 (rarely 5) subterminal ascospore appendages; species of tropical and subtropical waters .... ***Halosphaeria salina***, *Fig. 5B*

164′(163′) 4 or more terminal ascospore appendages; species of temperate zones ... 165

165(164′) Generally 4 radiating appendages at each ascospore apex .... ***Remispora quadriremis***, *Fig. 5H*

165′(164′) Generally 6 appendages at each apex . . . . ***Remispora stellata***, *Fig. 5I*

166(162′) Appendages initially enclosing the ascospores completely, finally detached around the septum; polymorphous, yoke-shaped, veil-like, or forming spines 167

166′(162′) Base of appendages at or near the apices of the mature ascospores only . . . 170

167(166) Ascospores thick-walled, rhomboid; distinct striae in the base of appendages (hematoxylin!) . . ***Remispora pilleata***, *Fig. 5K*

167′(166) Ascospores thin-walled, ovoid or ellipsoidal 168

168(167′) Ascospores with a whorl of subpolar spines and an apical tail-like appendage at each end . . . . ***Remispora spinibarbata***, *Fig. 5E*

168′(167′) Ascospores without spines, a yoke-like appendage covering each apex . . . 169

169(168′) Ascospore appendages with a spoon-shaped, smooth tip; tropical species . . . . ***Remispora crispa***, *Fig. 5J*

169′(168′) Ascospore appendages without such tips; temperate species . . . . ***Remispora maritima***, *Fig. 5F*

170(166′) Ascospores with a tubelike wall extension at each apex; unfurling long filaments extending from the tubes; spores longer than 48 µm . . . . ***Cucullosporella mangrovei***, *Fig. 7P*

170′(166′) Ascospores without apical tubes, usually shorter than 48 µm 171

171(170′) Immature ascospores enclosed in a gelatinous sheath; each apex of mature spores is surrounded by a large, subglobose, subgelatinous cap with delicate radiating striae (stain!) . . . . . ***Remispora galerita***, *Fig. 5G*

171′(170′) Immature ascospores not completely enclosed in a sheath; appendages apical or subapical, mostly unfurling into long threads 172

172(171′) Appendages semiglobose or lenticular, apical 173

172′(171′) Appendages hamate, lateral 177

173(172) Ascospore diameter 18 µm or more . 174

173′(172) Ascospore diameter usually less than 18 µm 175

174(173) Catenophyses absent; ascospore appendages not unfurling; temperate species . . . . ***Gnomonia salina***, *Fig. 7H*

174′(173) Catenophyses present; appendages unfurling into long filaments; tropical species . . . . ***Halosarpheia fibrosa***, *Fig. 7C*

175(173′) Ascospore diameter more than 13 µm; temperate species . . . . ***Halosarpheia trullifera***, *Fig. 7E*

175′(173′) Ascospore diameter less than 13 µm; tropical species 176

176(175′) Ascospores longer than 15 µm . . . . ***Halosarpheia marina***, *Fig. 7D*

176′(175′) Ascospores shorter than 15 µm . . . . ***Halosarpheia sp.***, *Fig. 23B*

177(172′) Base of ascospore appendages spoon-shaped . . . . ***Haligena salina***, *Fig. 6K*

177′(172′) Base of appendages not spoon-shaped 178

178(177′) Ascospore diameter 11 µm or less 179

178′(177′) Ascospore diameter over 11 µm . . 180

179(178) A freshwater species . . . . ***Halosarpheia retorquens***, *Fig. 7G*

179′(178) A marine species . . . . ***Halosarpheia viscosa***, *Fig. 7F*

180(178′) Ascospores longer than 47 µm . . . . ***Halosarpheia ratnagiriensis***, *Fig. 7A*

180′(178′) Ascospores 47 µm or shorter 181

181(180′) Ascospore diameter 14–22 µm . . . . ***Halosarpheia abonnis***, *Fig. 7B*

181′(180′) Ascospore diameter 12–14 µm . . . . ***Aniptodera mangrovei***, *Fig. 3P*

182(155′) Appendages surrounding the ascospore completely at maturity 183

182′(155′) Appendages not enclosing the ascospore completely at maturity 186

183(182) Ascospores surrounded by an inconspicuous halo of fine hairlike appendages . . . . ***Capillataspora corticola***, *Fig. 23C*

183′(182) Ascospore appendages different, lobe- or tube-like 184

184(183′) Ascospore appendages inconspicuous, irregular, extended into lobes around the septum and into 1 process at each end of the ascospore; core of apical process different from outer part (staining dark with hematoxylin), without distinct tubes or ring . . . . ***Ceriosporopsis circumvestita***, *Fig. 6C*

184′(183′) Ascospore appendages consisting of a distinct tube at each apex and a ring around the septum 185

185(184′) Polar appendages up to 15 µm long, releasing mucilage . . . . ***Ceriosporopsis tubulifera***, *Fig. 6D*

185′(184′) Polar appendages up to 54 µm long, not releasing mucilage . . . . ***Ceriosporopsis caduca***, *Fig. 6B*

186(182′) A tubular annulus around septum . . . . ***Ondiniella torquata***, *Fig. 6G*

186′(182′) Equatorial appendages different . . 187

187 (186′) 2–3 lunate, rigid appendages, attached with the middle part to the ascospore over the septum; in addition a small apical cap, which may become reversed, at each end . . . . ***Halosphaeriopsis mediosetigera**, Fig. 9A*

187′(186′) Ascospore appendages different 188

188 (187′) Ascomata whitish, with horn-like appendages . . ***Marisolaris ansata**, Fig. 9C*

188′(187′) Ascomata without appendages . . . 189

189 (188′) Ascospore diameter less than 8 μm; 6–8 awl-like appendages around the septum . . . . . . ***Ocostaspora apilongissima**, Fig. 5C*

189′(188′) Ascospore diameter usually 8 μm or more; less than 6 equatorial appendages . . 190

190 (189′) Ascospore appendages spoon-shaped at the bases, their tips without refractive bodies or small caps; without persistent paraphysoid chains in the ascomatal center . . . . ***Halosphaeria appendiculata**, Fig. 5D*

190′(189′) Base of ascospore appendage cylindrical, its tip with refractive body and covered with a small cap; persistent paraphysoid chains in the ascomatal center 191

191 (190′) Apical appendages up to 187 μm long; ascospores enclosed in a delicate, irregular gelatinous sheath . . . . ***Marinospora longissima**, Fig. 6I*

191′(190′) Apical appendages up to 27 μm long; ascospores without sheath . . . . ***Marinospora calyptrata**, Fig. 6H*

192 (116′) Ascomata in valsoid groups with one common pore; ascospores surrounded by a uniform gelatinous sheath, which may be deliquescing; wall smooth . . . . ***Helicascus kanaloanus**, Fig. 4H*

192′(116′) Ascomata single, not joined together; ascospores without gelatinous sheath; but appendages may be present; wall smooth or sculptured 193

193 (192′) Ascospores initially surrounded by a delicate sheath that fragments into lobe-like polar and equatorial appendages . . . . ***Groenhiella bivestia**, Fig. 9B*

193′(192′) Ascospores without appendages 194

194 (193′) Ascospores verrucose to verruculose 195

194′(193′) Ascospores smooth or striate, not verrucose 197

195 (194) Ascomata subglobose, superficial, over 800 μm diam; Ascospores 24 μm long or more . . . . . . ***Caryosporella rhizophorae**, Fig. 4F*

195′(194) Ascomata not subglobose, less than 800 μm diam; ascospores less than 24 μm long 196

196 (195′) Ascospores with a distinct dark band around the septum; temperate species . . . . ***Didymosphaeria lignomaris**, Fig. 4D*

196′(195′) Ascospores without distinct band; tropical species . . . . ***Verruculina enalia**, Fig. 4C*

197 (194′) Ascospores distinctly striate; on *Rhizophora* . . . ***Lineolata rhizophorae**, Fig. 4E*

197′(194′) Ascospores not striate; on other substrates 198

198 (197′) Ascospores shorter than 22 μm; diameter less than 10 μm . . . . ***Kirschsteiniothelia maritima**, Fig. 4B*

198′(197′) Ascospores longer than 22 μm; diameter over 10 μm 199

199 (198′) Ascospores shorter than 35 μm; diameter less than 16 μm; in *Atriplex* . . . . ***Bicrouania maritima**, Fig. 4A*

199′(198′) Ascospores longer than 35 μm; diameter 16 μm or more; in *Posidonia* 200

200 (199′) Ascospores shorter than 61 μm; dark band around the septum; without distinct germ pores . . . . ***Halotthia posidoniae**, Fig. 4G*

200′(199′) Ascospores longer than 65 μm; without band around the septum; with hyaline, apical germ pores . . . . ***Pontoporeia biturbinata**, Fig. 4J*

201 (112′) Ascospores with transverse septa only 202

201′(112′) Ascospores muriform (with transverse and longitudinal septa) 254

202 (201) Ascospores without appendages or gelatinous sheaths (since sheaths are often difficult to observe, some species are included in both leads) 203

202′(201) Ascospores with appendages or gelatinous sheaths 228

203 (202) Ascospores hyaline at maturity 204

203′(202) Ascospores colored at maturity (light yellow, yellow, brown, or olivaceous), at least some of the cells 215

204 (203) Ascospores attenuate at one end . . . . ***Buergenerula spartinae**, Fig. 13A*

204′(203) Ascospores never attenuate at one end 205

205 (204′) Ascospores distoseptate (narrow lumina); asci 4 spored . . . . ***Luttrellia estuarina**, Fig. 15K*

205′(204′) Ascospores with large lumina; asci 8 spored 206

206 (205′) Ascospores 45 μm and longer 207

206′(205′) Ascospores shorter than 45 μm .. 210
207 (206) Ascospores with 5 septa .... ***Quintaria lignatilis***, *Fig. 13G*
207′(206) Ascospores 1–3 septate 208
208 (207′) Ascoma conical, superficial, over 1 mm diam; ascospores 15 μm diam or more ..... ***Ascocratera manglicola***, *Fig. 13M*
208′(207′) Ascoma different; ascospores less than 15 μm diam 209
209 (208′) Upper cell of ascospores tapering .... ***Wettsteinia marina***, *Fig. 14A*
209′(208′) Upper cell of ascospores obtusely rounded .... ***Phaeosphaeria macrosporidium***, *Fig. 14B*
210 (206′) Asci not fissitunicate, thin-walled ... 211
210′(206′) Asci fissitunicate, thick-walled ... 213
211 (210) Temperate species, not in mangroves .... ***Sphaerulina oraemaris***, *Fig. 13B*
211′(210) Tropical species, in mangroves 212
212 (211′) Ascospore diameter up to 6 μm .... ***Hypophloeda rhizospora***, *Fig. 23D*
212′(211′) Ascospore diameter 6 μm or more .... .. ***Marinosphaera mangrovei***, *Fig. 13C*
213 (210′) Ascospores shorter than 28 μm .... ***Leptosphaeria australiensis***, *Fig. 13 I*
213′(210′) Ascospores 28 μm and longer 214
214 (213′) Ascospores up to 12 μm diam; temperate species .... ***Leptosphaeria pelagica***, *Fig. 13K*
214′(213′) Ascospores up to 15 μm diam; tropical species .... ***Massarina thalassiae***, *Fig. 13O*
215 (203′) Apical cells of ascospores hyaline or lighter colored than the brown middle cells 216
215′(203′) Ascospores concolorous throughout 220
216 (215) Ascospores unequally biconical, with 7–8 tubercules around the equator ... ... ***Biconiosporella corniculata***, *Fig. 9K*
216′(215) Ascospores ellipsoidal to fusoid, without tubercules around the middle 217
217 (216′) Asci not fissitunicate, 8-spored; pseudoparaphyses absent .... ***Savoryella lignicola***, *Fig. 4R*
217′(216′) Asci fissitunicate, 8- or 4-spored; pseudoparaphyses present 218
218 (217′) Ascospores longer than 60 μm, diameter over 20 μm .... ***Passeriniella savoryellopsis***, *Fig. 4O*
218′(217′) Ascospores shorter than 60 μm, diameter less than 20 μm 219
219 (218′) Ascospore length 36 μm or less; asci typically 8-spored, longer than 110 μm; temperate species .... ***Passeriniella obiones***, *Fig. 4P*
219′(218′) Ascospore length 36 μm or more; asci typically 2-spored, shorter than 110 μm; tropical species .... ***Savoryella paucispora***, *Fig. 4Q*
220 (215′) Ascospores 1–3 septate 221
220′(215′) Ascospores have 5 and more septa ... 226
221 (220) Ascospores 1–3 septate 222
221′(220) Ascospores consistently 3 septate ... 223
222 (221) Ascospores shorter than 35 μm; temperate species .... ***Leptosphaeria oraemaris***, *Fig. 14G*
222′(221) Ascospores longer than 35 μm; tropical species ...... ***Coronopapilla mangrovei***, *Fig. 4 I*
223 (221′) Ascomata thick-walled, carbonaceous; asci longer than 150 μm .... ***Trematosphaeria mangrovei***, *Fig. 14H*
223′(221′) Ascomata thin-walled, coriaceous; asci shorter than 150 μm 224
224 (223′) Ascospores shorter than 20 μm, diameter less than 7 μm .... ***Leptosphaeria peruviana***, *Fig. 14F*
224′(223′) Ascospores longer than 20 μm, diameter over 7 μm 225
225 (224′) A freshwater species on *Typha* spp. .... ***Phaeosphaeria typharum***, *Fig. 14C*
225′(224′) A marine species on *Spartina* spp. .... ***Phaeosphaeria spartinaecola***, *Fig. 14D*
226 (220′) Ascospores with 2–4 septa near each apex, non-septate in the center; with hyaline, refructive apices .... ***Biatriospora marina***, *Fig. 15L*
226′(220′) Ascospores evenly septate, without hyaline apices 227
227 (226′) Ascospores shorter than 60 μm .... ***Phaeosphaeria spartinae***, *Fig. 14K*
227′(226′) Ascospores longer than 60 μm .... ***Phaeosphaeria gessneri***, *Fig. 14L*
228 (202′) Mature ascospores completely enclosed in a uniform gelatinous sheath; apical cells may be free in *Carbosphaerella* (use India ink!) 229
228′(202′) Mature ascospores without gelatinous sheath, but with other appendages ... 240
229 (228) Asci thin-walled, early deliquescent; without pseudoparaphyses 230
229′(228) Asci thick-walled, not deliquescent; pseudoparaphyses present 231
230 (229) Ascospore diameter 12–20 μm; central cells dark .... ***Carbosphaerella leptosphaerioides***, *Fig. 9 I*
230′(229) Ascospore diameter 7–12 μm; hyaline throughout .... ***Appendichordella amicta***, *Fig. 13L*

231 (229′) Ascospores with 1–3 septa 232
231′(229′) Ascospores with 3 and more septa .. 237
232 (231) Ascospores yellow-brown ....
...... ***Phaeosphaeria halima**, Fig. 14E*
232′(231) Ascospores hyaline 233
233 (232′) Ascospore length 45 µm or more ... 234
233′(232′) Ascospores shorter than 45 µm 235
234 (233) Ascomata conical, superficial, over 1 mm diam ....
..... ***Ascocratera manglicola**, Fig. 13M*
234′(233) Ascomata ± immersed under a clypeus, smaller ....
...... ***Massarina velatospora**, Fig. 13N*
235 (233′) Ascospores shorter than 26 µm; in *Avicennia* spp. ....
..... ***Leptosphaeria avicenniae**, Fig. 13J*
235′(233′) Ascospores longer than 26 µm; in other hosts 236
236 (235′) Ascomata under a black clypeus; tropical species ....
.... ***Massarina thalassiae**, Fig. 13 O*
236′(235′) Ascomata not clypeate; temperate species .. ***Leptosphaeria pelagica**, Fig. 13K*
237 (231′) Ascospores with distinct longitudinal striations, central cells darker than end cells
.... ***Trematosphaeria striatispora**, Fig. 14 I*
237′(231′) Ascospores without striations, cells uniformly colored 238
238 (237′) Ascospores 3–5 septate; third cell from apex largest ....
.. ***Phaeosphaeria neomaritima**, Fig. 14J*
238′(237′) Ascospores usually have 6 or more septa; fourth or fifth cell largest .. 239
239 (238′) Ascospores usually 5–7 septate (rarely 8 septa); diameter 12 µm or more; wall smooth; mostly in *Ammophila* ....
.. ***Amarenomyces ammophilae**, Fig. 14M*
239′(238′) Ascospores 7–11 septate; diameter rarely over 10 µm; wall verrucose in age; in *Juncus*, *Spartina*, *Phragmites*, and *Typha* ...
...... ***Chaetomastia typhicola**, Fig. 14N*
240 (228′) Ascospore appendages only lateral or apical 241
240′(228′) Apical as well as lateral appendages (tufts of bristles) on the ascospore ....
.... 253
241 (240) Lateral spore appendages only, cilia-like .
.... ***Chaetosphaeria chaetosa**, Fig. 8N*
241′(240) Apical or subapical spore appendages only, not cilia-like 242
242 (241′) Ascospore appendages on one end only
.... 243
242′(241′) Ascospore appendages on both ends ..
.... 246
243 (242) Ascospores with 3–4 radiating, acuminate appendages ....
.... ***Torpedospora radiata**, Fig. 8P*
243′(242) Ascospores with one cap or sheath-like appendage 244
244 (243′) Ascospores 1 septate (upon germination up to 4 septa); diameter 6 µm or more; appendage subglobose, terminal ......
.... ***Halosphaeria cucullata**, Fig. 3L*
244′(243′) Ascospores 3 or more septate; diameter 5 µm or less; appendage an elongate irregular sheath around the apex or along the upper side 245
245 (244′) Ascospores 3, rarely up to 5 septate ....
...... ***Halosarpheia unicaudata**, Fig. 7N*
245′(244′) Ascospores 5–11 septate ....
..... ***Halosarpheia cincinnatula**, Fig. 7M*
246 (242′) Ascospores with 3 or more radiating, acuminate, terminal or subterminal appendages around each apex 247
246′(242′) Ascospore appendages single, different
.... 248
247 (246) Ascospores longer than 25 µm, 3 appendages at each apex ....
...... ***Arenariomyces triseptatus**, Fig. 8C*
247′(246) Ascospores shorter than 25 µm, normally 5 appendages below each apex ....
...... ***Torpedospora ambispinosa**, Fig. 8O*
248 (246′) Ascospores 22 µm or shorter; on animal substrates 249
248′(246′) Ascospores longer than 22 µm; on plant material 250
249 (248) In the carapace of crabs ....
.... ***Trichomaris invadens**, Fig. 6M*
249′(248) On hydrorhiza and hydrocaulon of hydrozoa ... ***Abyssomyces hydrozoicus**, Fig. 13D*
250 (248′) Ascospores normally 3 septate ... 251
250′(248′) Ascospores always with 4 or more septa
.... 252
251 (250) Ascospore diameter less than 9 µm; appendages not enveloping the spore completely
Halosarpheia bentotensis, Fig. 7J
251′(250) Ascospore diameter 9 µm or more; appendages very long, enclosing the spore completely within the ascus ....
.... ***Haligena elaterophora**, Fig. 6L*
252 (250′) Ascospores more than 10 µm diam, predominantly 5 septa ....
...... ***Halosarpheia spartinae**, Fig. 7K*
252′(250′) Ascospores less than 10 µm diam, predominantly 11 septa ....
...... ***Halosarpheia viscidula**, Fig. 7L*
253 (240′) Mature ascospores typically 3 septate; asci less than 100 µm long ....
.... ***Nereiospora cristata**, Fig. 9H*

253′(240′) Mature ascospores typically 5 septate; asci longer than 100 μm .... ***Nereiospora comata***, *Fig. 9G*

254(201′) Ascospores hyaline, longer than 100 μm .... ***Lautospora gigantea***, *Fig. 16J*

254′(201′) Ascospores brown, shorter than 100 μm 255

255(254′) Ascospores with striate gelatinous sheath or subconical appendages 256

255′(254′) Ascospores without sheaths or appendages; if a simple sheath present, not striate 257

256(255) Ascospores with gelatinous sheath without striae; subconical appendages at each apex; asci thick-walled, persistent .... ***Pleospora gaudefroyi***, *Fig. 16I*

256′(255) Ascospores with a gelatinous, striate sheath, without appendages; asci thin-walled, deliquescing .... ***Carbosphaerella pleosporoides***, *Fig. 9J*

257(255′) In *Avicennia* .... ***Pleospora avicenniae***, *Fig. 16H*

257′(255′) In other hosts 258

258(257′) In algae .... ***Pleospora gracilariae***, *Fig. 23F* and ***Pleospora pelvetiae***, *Fig. 16F*

258′(257′) In other hosts 259

259(258′) In *Triglochin*; ascospores typically with 7 transsepta .... ***Pleospora triglochinicola***, *Fig. 16D*

259′(258′) In *Spartina*; ascospores with 5 or up to 9 septa 260

260(259′) Ascospores typically with 5 transverse septa; up to 38 μm long .... ***Pleospora spartinae***, *Fig. 16E*

260′(259′) Ascospores with 7–9 transverse septa; up to 52 μm long .... ***Pleospora pelagica***, *Fig. 16G*

III. Key to Basidiomycotina

1 Parasites in algae or sea grasses 2

1′ Saprobes on the surface of wood or *Spartina* . 3

2(1) Smut in the stem of *Ruppia* (Monocotyledoneae) .... ***Melanotaenium ruppiae***

2′(1) Causing leasions in *Dilsea* (Rhodophyta) ... . ***Mycaureola dilseae***, *Fig. 19E*

3(1′) Basidiospores consisting of 4 radiating arms; formed on the surface of a flat, cushion-like basidioma 4

3′(1′) Basidiospores subglobose; in a globular or funnel-shaped basidioma 5

4(3) Apical arms of basidiospores up to 23 μm long, up to 8 μm diam .... ***Digitatispora lignicola***, *Fig. 19B*

4′(3) Apical arms of basidiospores up to 41(–45) μm long, up to 4 μm diam .... ***Digitatispora marina***, *Fig. 19A*

5(3′) Basidioma subglobose, puffball-like, irregularly dehiscing at maturity; basidiospores with 5 filamentous appendages .... ***Nia vibrissa***, *Fig. 19C*

5′(3′) Basidioma funnel-shaped, releasing the non-appendaged basidiospores apically as a ball through a wide aperture .... ***Halocyphina villosa***, *Fig. 19D*

IV. Key to Anamorphs

1 Conidia borne on or in fruit bodies, *e.g.* acervuli, pycnidia, and sporodochia (Coelomycetes) 2

1′ Conidia not produced on or in fruit bodies, but on hyphae (Hyphomycetes) 24

2(1) Conidia on sporodochia, on the surface of the substrate 3

2′(1) Conidia enclosed in pycnidia or acervuli . 4

3(2) Sporodochia brown, on wood in the deep sea .... ***Allescheriella bathygena***, *Fig. 20A*

3′(2) Sporodochia brightly colored, on marsh plants (*Salicornia* spp.) .... ***Tubercularia pulverulenta***, *Fig. 20C*

4(2′) Acervuli with thin upper wall, rupturing irregularly; hyperparasitic in galls formed by *Haloguignardia* spp. in Phaeophyta .... ***Gloeosporidina cecidii***, *Fig. 19K*

4′(2′) Pycnidia with thick upper wall, ostiolate; not hyperparasitic 5

5(4′) Pycnidia with a long beak, almost equal in diameter with the venter; on wood .... anamorph of ***Halonectria milfordensis***, *Fig. 19O*

5′(4′) Pycnidia without beak or with a short papilla; in wood, algae, mangroves, or marsh plants 6

6(5′) Conidia always nonseptate 7

6′(5′) Conidia zero- to multiseptate 14

7(6) Conidia with apical setae; produced in cupulate, setose acervuli .... ***Dinemasporium marinum***, *Fig. 19Q*

7′(6) Conidia without setae; produced in pycnidia without setae 8

8(7′) Composite pycnidia, 1.5–1.7 mm diam, with a central ostiole; under the bark of *Rhizophora* .... ***Cytospora rhizophorae***, *Fig. 19L*

8′(7′) Pycnidia different, smaller 9

9(8′) Pycnidia on the bark of pneumatophores and trunks of *Avicennia* .... ***Rhabdospora avicenniae***, *Fig. 19P*

9′(8′) Pycnidia on other substrates 10
10(9′) Conidia brownish . . . . ***Coniothyrium obiones**, Fig. 19N*
10′(9′) Conidia hyaline 11
11(10′) Conidia produced endogenously in phialides (oil immersion!) . . . . ***Phialophorophoma litoralis**, Fig. 19F*
11′(10′) Conidia produced exogenously on conidiophores 12
12(11′) On algae . . . ***Phoma laminariae**, Fig. 19G*
12′(11′) On other substrates 13
13(12′) On *Suaeda* . . . . . ***Phoma suaedae**, Fig. 19H*
13′(12′) On other substrates . . . . ***Phoma** and **Macrophoma** spp., Fig. 19J* and *I*
14(6′) Conidia muriform 15
14′(6′) Conidia with transverse septa only . . . 17
15(14) Conidia predominantly with 3 transverse septa, usually not longer than 20 µm; without gelatinous sheaths or appendages . . . . ***Camarosporium roumeguerii**, Fig. 19W*
15′(14) Conidia with 5 or more transverse septa; with distinctive gelatinous sheaths or appendages 16
16(15′) Conidia with a gelatinous cap-like appendage at each end; mostly on Gramineae in dunes . . ***Amarenographium metableticum**, Fig. 19U*
16′(15′) Conidia completely surrounded by a gelatinous sheath; on salt-marsh *Salicornia* spp. . . . ***Camarosporium palliatum**, Fig. 19V*
17(14′) Conidia filiform or elongate cylindrical, zero- to multiseptate 18
17′(14′) Conidia ellipsoidal or ovoid, always 1 septum when mature 20
18(17) Conidia shorter than 35 µm, 0–3 septa; in *Pelvetia* . . . . ***Stagonospora haliclysta**, Fig. 19Z*
18′(17) Conidia longer than 35 µm, with 5 or more septa; in higher plants (Gramineae) . . 19
19(18′) Conidia without appendages . . . . anamorph of ***Pleospora spartinae**, Fig. 19X*
19′(18′) Conidia with apical gelatinous caps . . . . ***Stagonospora**, sp., Fig. 19Y*
20(17′) Conidia with appendages, in *Rhizophora* . . . . . . ***Robillarda rhizophorae**, Fig. 19M*
20′(17′) Conidia without appendages, in other substrates 21
21(20′) Conidia hyaline, in *Laminaria* . . . . ***Phoma laminariae**, Fig. 19G*
21′(20′) Conidia colored; in wood or phanerogams 22
22(21′) Conidia shorter than 9 µm; in wood . . . . ***Diplodia oraemaris**, Fig. 19T*
22′(21′) Conidia 9 µm or longer; in marsh plants 23
23(22′) In *Salicornia*; conidia up to 20 µm long, up to 7 µm diam . . . . ***Ascochyta salicorniae**, Fig. 19R*
23′(22′) In *Halimione*; conidia up to 12 µm long, up to 5 µm diam . . . . ***Ascochyta obiones**, Fig. 19S*
24(1′) Conidia filamentous or longer than 100 µm 25
24′(1′) Conidia not filamentous, shorter than 100 µm 28
25(24) Conidia brown, apical cell over 10 µm diam . . . . ***Sporidesmium salinum**, Fig. 22A*
25′(24) Conidia hyaline, less than 10 µm diam . . . 26
26(25′) Apical cells devoid of cytoplasm 27
26′(25′) Apical cells with cytoplasm . . . . ***Anguillospora marina**, Fig. 22B*
27(26) Conidial development sympodial; no teleomorph known . . . . ***Sigmoidea marina***, Fig. 22C
27′(26) Conidial development not sympodial; anamorph of *Corollospora luteola* . . . . ***Sigmoidea luteola**, Fig. 22D*
28(24′) Conidia 1 celled, but single cells may form long chains that break up eventually . . . 29
28′(24′) Conidia or chlamydospores several-celled; or 1 celled conidia remaining attached to, and distributed with, a central sporogenous cell 32
29(28) Conidia hyaline 30
29′(28) Conidia dark colored 31
30(29) Conidia with radiating projections (staurospores) . . . . . ***Clavatospora stellatacula**, Fig. 20B*
30′(29) Conidia simple . . . . ***Botryophialophora marina**, Fig. 20D*
31(29′) Conidia 16 µm diam or more; deep-sea species . . . . ***Periconia abyssa**, Fig. 20E*
31′(29′) Conidia generally 13 µm diam or less; littoral species . . . . . . ***Periconia prolifica**, Fig. 20F*
32(28′) Conidia with 3 or more arms, often with a large bulbous cell 33
32′(28′) Conidia without radiating arms or appendages 37
33(32) Conidia hyaline, without bulbous cell . . . 34
33′(32) Conidia dark, with bulbous cell 35
34(33) Conidia 2–4(–7) septa in the main axis . . . . ***Varicosporina prolifera**, Fig. 22F*
34′(33) Conidia 1–2(–4) septa in the main axis . . . . . . ***Varicosporina ramulosa**, Fig. 22E*
35(33′) Conidia 1 celled, often distributed in aggregates of 5–9 cells, attached to the central sporogenous cell . . . . ***Asteromyces cruciatus**, Fig. 20G*

35′(33′) Conidia composed of 1 basal, bulbous, dark cell with 1 or 2 crowns of radiating, nondeciduous arms 36

36(35′) Basal cell hyaline to light olive, 6–20 µm diam .... ***Clavatospora bulbosa***, *Fig. 22H*

36′(35′) Basal cell dark brown, 24–42 µm diam .... ***Orbimyces spectabilis***, Fig. 22G

37(32′) Conidia 1 or 2 celled, rarely 3 celled .... ***Humicola alopallonella***, *Fig. 20M*

37′(32′) Conidia or chlamydospores multicellular, with more than 2 cells 38

38(37′) Conidia with transverse septa only .. 39

38′(37′) Conidia muriform, or irregular conglomerates of cells 55

39(38) Symbiotic fungus with chlamydospores, forming chains inside the walls of living Chlorophyta (*Cladophora*) .... ***Blodgettia confervoides***, *Fig. 20K*

39′(38) Saprobic fungi with true conidia, on decaying algae, sea grasses, marsh plants, or wood 40

40(39′) Conidia borne in chains on the conidiophore 41

40′(39′) Conidia borne singly on the conidiophore 42

41(40) Conidia 0–2 (some 3 or 4) septate .... ***Cladosporium algarum***, *Fig. 20H*

41′(40) Conidia 1–9 (sometimes 10 or 11) septate, predominantly with 3–5 septa .... ***Dendryphiella salina***, *Fig. 20I*

42(40′) Conidia more or less straight 43

42′(40′) Conidia distinctly coiled or curved (helicoid) 48

43(42) Apical cell not broader than basal cell .. 44

43′(42) Apical cell broader than basal cell 46

44(43) Conidia thick-walled (distoseptate), with 6–12 pseudosepta; apical cells light-colored, separated from adjoining cells by dark septa .... ***Exserohilum sp.***, *Fig. 20L*

44′(43) Conidia thin-walled, usually with less than 6 true septa, concolorous throughout ... 45

45(44′) Conidia 1–3 septate; shorter than 20 µm .. .. ***Dendryphiella arenaria***, *Fig. 20J*

45′(44′) Conidia usually 3–5 septate; rarely with fewer septa or up to 11 septa; usually longer than 20 µm (up to 75 µm) .... ***Dendryphiella salina***, *Fig. 20I*

46(43′) Conidia light brown, slightly constricted at the septa, apical cell up to 14 (rarely 17) µm diam .... ***Trichocladium lignincola***, *Fig. 20O*

46′(43′) Conidia dark brown, strongly constricted at the septa, apical cell up to 20 µm diam or more 47

47(46′) Cells subglobose, reddish-brown .... ***Trichocladium constrictum***, *Fig. 20P*

47′(46′) Cells compressed, not reddish .... ***Trichocladium achrasporum***, *Fig. 20N*

48(42′) Apical cell considerably broader than basal cell (*Cirrenalia*) 49

48′(42′) Apical cell not conspicuously broader than basal cell 54

49(48) Mature conidia black, shiny, fist-shaped, not constricted at the obscured septa .... ***Cirrenalia pygmea***, *Fig. 21F*

49′(48) Mature conidia brown, all septa distinct, constricted at septa 50

50(49′) Conidia 6–12 septate, slightly constricted at septa .... ***Cirrenalia tropicalis***, *Fig. 21E*

50′(49′) Conidia mostly with less than 6 septa, distinctly constricted at septa 51

51(50′) Conidia reddish-brown; diameter of coil 12–24 µm .... ***Cirrenalia macrocephala***, *Fig. 21C*

51′(50′) Conidia without reddish tint; diameter of coil up to 30 µm or more 52

52(51′) Conidia 3 septate, rarely 2 or 4 septate, apical cell often sausage-shaped; temperate species ...... ***Cirrenalia fusca***, *Fig. 21B*

52′(51′) Conidia 3–5 (sometimes 6) septate, apical cell subglobose to ellipsoidal; tropical species 53

53(52′) Conidia $\frac{1}{2}$–$1\frac{1}{4}$ times contorted, apical cell 16–20 µm diam .... ***Cirrenalia pseudomacrocephala***, *Fig. 21D*

53′(52′) Conidia 1/2 time contorted, apical cell up to 16 µm diam .... ***Cirrenalia basiminuta***, *Fig. 21A*

54(48′) Conidial filament turns to produce a terminal, regular spiral; no additional complex spores formed .... ***Zalerion maritimum***, *Fig. 21H*

54′(48′) Conidial filament turns to a lateral, variable spiral; additional complex spores composed of up to several hundred cells formed in the substrate .... ***Zalerion varium***, *Fig. 21I*

55(38′) Conidia borne acropetally in chains .... ***Alternaria spp.***, *Fig. 20R*

55′(38′) Conidia single on the conidiophore 56

56(55′) Conidia black at maturity, septa obscured .... ***Monodictys pelagica***, *Fig. 21G*

56′(55′) Conidia fuscous, gray or brown at maturity, septa distinct 57

57(56′) Conidia consisting of 3–8 parallel branches arising from a single cell .... ***Dictyosporium pelagicum***, *Fig. 20Q*

57′(56′) Cells of the conidium not in parallel branches 58

58(57′) Conidiophores united into sporodochia . . . . *Epicoccum spp.*, *Fig. 21J*

58′(57′) Conidia not on sporodochia 59

59(58′) Conidia composed of clusters of subglobose cells on algae or wood . . . . *Cumulospora marina*, *Fig. 21K*

59′(58′) Conidia muriform, anamorphs of *Pleospora* spp. 60

60(59′) Conidia trigonous, on *Triglochin* . . . . *Stemphylium triglochinicola*, *Fig. 21L*

60′(59′) Conidia ellipsoidal, on algae . . . . *Stemphylium gracilariae*, *Fig. 23G*

List of Species

1. Ascomycotina

Abyssomyces hydrozoicus Kohlm., *Ber. Dtsch. Bot. Ges. 83*: 505–506, 1970. Ascospores 18–20 × 3.5–4 µm. (Isotype, IMS, J.K. 2754) Fig. 13D

Acrocordiopsis patilii Borse *et* Hyde, *Mycotaxon 34*: 535, 1989. Ascospores 16–25 × 10–16 µm. (Holotype, IMI 297769) Fig. 2L

Two characters of *A. patilii* not included in the original diagnosis by Borse and Hyde (1989) need to be added. The type material shows ascomata to be seated on a thin black stroma that covers the wood surface. Also, next to the large ascomata are similar smaller fruiting bodies, containing hyaline, one-celled spores, most probably spermatia of the same species.

Adomia avicenniae Schatz, *Trans. Br. Mycol. Soc. 84*: 555, 1985. Ascospores (excl. appendages) 20–33 × 8–12 µm; light brown. (Holotype, IMS 275) Fig. 2E

Aigialus grandis Kohlm. *et* Schatz, *Trans. Br. Mycol. Soc. 85*: 699–703, 1985. Ascospores 67–96(–101) × 18–29 broad × 14–20 µm thick, 14–17(–18) transsepta, 1–3 longisepta per segment, except in the end cells, subapical cells with or without longisepta; yellow-brown. (Holotype, IMS, J.K. 4332) Fig. 16A

Aigialus mangrovei Borse (as '*mangrovis*'), *Trans. Br. Mycol. Soc. 88*: 424, 1987. Ascospores 35–55 × 10–16 µm, 6–7 transsepta, 1–2 longisepta per segment, except in the end cells; yellow-brown. (IMS, J.K. 4860) Fig. 16C

Aigialus parvus Schatz *et* Kohlm., *Trans. Br. Mycol. Soc. 85*: 704–705, 1985. Ascospores (44–)49–71 (–74) × 19–27 broad, 16–22 µm thick, (9–) 10–11(–12) transsepta, 1–3 longisepta per segment, except in the end cells; yellow-brown. (Holotype, IMS 287a) Fig. 16B

Aigialus rhizophorae Borse, *Trans. Br. Mycol. Soc. 88*: 424–425, 1987. Ascospores 70–95 × 18–25 µm, 13–15(–17) transsepta, 1–4 longisepta, except in the end cells; yellow-brown.

This species from India is not illustrated. It is very similar to the Caribbean *A. grandis* and is most probably a synonym. Borse (1987) separates the two species by absence (*A. grandis*) or presence (*A. rhizophorae*) of a vertical septum in the subapical cell. But this does not appear to be a constant character, because ascospores in a type slide of *A. rhizophorae* (IMI 304216) show frequently subapical cells without longisepta. On the other hand, *A. grandis* may have a vertical septum in the subapical cell (see *A. grandis*). Also, the shape and size of ascospores in both species are more or less identical. Moreover, Borse (1987) collected *A. grandis* on the same host and in the same location as *A. rhizophorae*. New collections from India need to be examined in detail to clarify if *A. rhizophorae* is a valid species.

Amarenomyces ammophilae (Lasch) Eriksson, *Opera Bot. 60*: 124, 1981 [= *Phaeosphaeria ammophilae* (Lasch) Kohlm. *et* Kohlm., the teleomorph of *Amarenographium metableticum*]. Ascospores (35–)38–51(–55) × 12–16 µm, 5–7(–8) septate; yellow-brown. (Isotype (?), B, Rabenhorst "Klotzschii Herbarium ..." 1340) Fig. 14M

Shoemaker and Babcock (1989) give a detailed description and discussion of this species.

Amylocarpus encephaloides Currey, *Proc. R. Soc. Lond. 9*: 119–123, 1857–1859. Ascospores 8–16 µm diam. (IMS, J.K. 1895) Fig. 2J

Aniptodera chesapeakensis Shearer *et* Miller, *Mycologia 69*: 894, 1977. Ascospores 21–37 × 7–15 µm. (IMS, J.K. 3825) Fig. 3O

Aniptodera chesapeakensis is the type species of the genus. Shearer (1989) recently described three freshwater species in *Aniptodera*. Although the original protologue did not include ascospore appendages, Shearer and Crane (1980) later identified an isolate from Virginia with appendaged ascospores as *A. chesapeakensis*, implying that this species can occur in appendaged and unappendaged forms. We do not agree with this conclusion because we believe that the presence or absence of appendages are stable characters. There are other cases of closely related, but definitely distinct taxa, one with (*Moana turbinulata*) and the other without (*Thalassogena sphaerica*) appendages.

Aniptodera longispora Hyde, *Bot. Mar. 33*: 335, 1990. Ascospores (32–)34–54(–56) × 9–13.5 µm. (IMS, J.K. 5338) Fig. 23E
The ascospore dimensions are based on our own collections.

Aniptodera mangrovei Hyde *in* Hyde, Farrant *et* Jones (as *'mangrovii'*), *Can. J. Bot. 64*: 2989, 1986. Ascospores 37–45 × 12–14 µm. (Holotype, IMI 297766) Fig. 3P
It is doubtful that this species with appendaged ascospores is correctly placed in *Aniptodera*.

Appendichordella amicta (Kohlm.) Johnson, Jones *et* Moss, *Can. J. Bot. 65*: 941, 1987 (= *Sphaerulina amicta* Kohlm.). Ascospores (15–)18–27 × 7–12 µm. (Isotype, IMS, J.K. 543) Fig. 13L

Arenariomyces majusculus Kohlm. *et* Volkm.-Kohlm., *Mycol. Res. 92*: 411, 1989. Ascospores 28–39 × 10–14 µm, 1 septate (rarely 2 or 3 septate), 3 or 4 subterminal appendages. (Holotype, IMS, J.K. 5097) Fig. 8B

Arenariomyces parvulus Koch, *Nordic J. Bot. 6*: 497, 1986. Ascospores 16–25 × 3–6 µm (see Kohlmeyer and Volkmann-Kohlmeyer 1989). (IMS, J.K. 4891) Fig. 8D

Arenariomyces trifurcatus Höhnk, *Veroeff. Inst. Meeresforsch. Bremerhaven 3*: 30, 1954. Ascospores 24–38 × 7–16 µm. (IMS, J.K. 4839) Fig. 8A

Arenariomyces triseptatus Kohlm., *Mar. Ecol. (P.S.Z.N.I) 5*: 333, 1984. Ascospores 27–34 × 6–8 µm. (Isotype, IMS, J.K. 4426) · Fig. 8C

Arthopyrenia halodytes (Nyl.) Arn., *Ber. Bayer. Bot. Ges. 1* (Anhang): 121, 1891. Ascospores 12–23 (–27) x 5–9 µm. (IMS, J.K. 2254) Fig. 2BB
Cannon et al. (1985) list this species with synonyms [*Pharcidia balani* (Winter) Bauch, etc.]. Santesson (1984) treats the fungus on barnacles as a separate species, namely, *A. sublitoralis* (Leight.) Arnold.

Ascocratera manglicola Kohlm., *Can. J. Bot. 64*: 3037, 1986. Ascospores 46–62(–66) × 15–22 µm; often found in the immature 1 septate state. (Paratype, IMS, J.K. 4408) Fig. 13M

Banhegyia setispora Zeller *et* Tóth, *Sydowia 14*: 327–328, 1960. Ascospores 15–27 × 6–10 µm; hyaline to light brown. (IMS, J.K. 2258) Fig. 8K

Bathyascus avicenniae Kohlm., *Bot. Mar. 23*: 530, 1980. Ascospores 90–145 × 2.5–4 µm. (Paratype, IMS, J.K. 3928) Fig. 18A

Bathyascus grandisporus Hyde *et* Jones, *Bot. Mar. 30*: 413, 1987. Ascospores 205–300 × 9–15 µm. (Holotype, IMI 308587) Fig. 18D

Bathyascus tropicalis Kohlm., *Bot. Mar. 23*: 532, 1980. Ascospores 70–100 × 8–10 µm. (Isotype, IMS, J.K. 3946) Fig. 18B

Bathyascus vermisporus Kohlm., *Rev. Mycol. 41*: 191, 1977. Ascospores 50–72 × 4–6 µm. (Paratype, IMS, J.K. 1821) Fig. 18C

Belizeana tuberculata Kohlm. *et* Volkm.-Kohlm., *Bot. Mar. 30*: 196, 1987. Ascospores 17–25 × 10–15 µm. (Holotype, IMS, J.K. 4398a) Fig. 3H

Biatriospora marina Hyde *et* Borse, *Mycotaxon 26*: 263, 1986. Ascospores 55–82 × 16–25 µm, 2–4 septate at each end; brown to dark brown. (IMS, J.K. 4963) Fig. 15L

Biconiosporella corniculata Schaumann, *Veroeff. Inst. Meeresforsch. Bremerhaven 14*: 24–25, 1972. Ascospores 26–43 × 16–28 µm; central cell dark brown to black with subhyaline protuberances. (IMS, J.K. 3117) Fig. 9K
The tropical fungus illustrated by Hyde (1988b) and identified as *B. corniculata* has longer ascospores and is possibly a different species.

Bicrouania maritima (Crouan *et* Crouan) Kohlm. *et* Volkm.-Kohlm., *Mycol. Res. 94*: 685, 1990 [= *Didymosphaeria maritima* (Crouan *et* Crouan) Sacc.]. Ascospores 24–32 × 12–15 µm; reddish-brown. (Isotype, IMS, J.K. 3097) Fig. 4A

Biflua physasca Koch *et* Jones, *Can. J. Bot. 67*: 1183, 1989. Ascospores 10–16 × 6–9 µm. (Holotype, CP, Koch 564/10) Fig. 3I

Buergenerula spartinae Kohlm. *et* Gessner, *Can. J. Bot. 54*: 1764, 1976. Ascospores 37–66 × 10–14 µm, 3 (rarely 4) septate. (Isotype, IMS, J.K. 3498) Fig. 13A

Capillataspora corticola Hyde, *Can. J. Bot. 67*: 2522, 1989. Ascospores 18–27 × 8–14 µm. (Holotype, IMI 326618) Fig. 23C

We were unable to detect in the type material the 'halo of fine, hairlike appendages' mentioned by Hyde.

Capronia ciliomaris (Kohlm.) Müller, Petrini, Fisher, Samuels *et* Rossmann, *Trans. Br. Mycol. Soc. 88*: 73, 1987 (= *Herpotrichiella ciliomaris* Kohlm.). Ascospores (16–)20–28 × 8–10 μm. (IMS, J. K. 883) Fig. 8J

The necessity of the recent transfer of this species to *Capronia* can be questioned, and the authors do not discuss this decision. It could now be argued that the species is sufficiently different from *Capronia*, *Herpotrichiella* or any other genus with otherwise exclusively terrestrial species, to justify the creation of a new genus. More and more obligate marine species that were once tentatively assigned to 'terrestrial' genera have to be removed and transferred to newly described genera because of revised generic circumscriptions. Representatives of *Capronia* have black ascomata, whereas *C. ciliomaris* is whitish to rarely dark blue. It is the only species in *Capronia* with hyaline appendaged ascospores and the only marine species occurring on submersed bark (not wood, as Müller *et al.* 1987 write). Also, the asci are surrounded by short sterile hairs (paraphyses or apical paraphyses, Kohlmeyer 1960) that do not occur in species of *Capronia*.

Carbosphaerella leptosphaerioides I. Schmidt, *Nat. Naturschutz Mecklenburg 7*: 9–10, 1969. Ascospores (25–)27–42 × (12–)16–24 μm; central cells dark brown. (IMS, J. K. 2472) Fig. 9I

Carbosphaerella pleosporoides I. Schmidt, *Feddes Repert. 80*: 108, 1969. Ascospores 22–30 x 12–20 μm; central cells dark brown. (Paratype, IMS, I. Schmidt) Fig. 9J

Caryosporella rhizophorae Kohlm., *Proc. Ind. Acad. Sci. (Plant Sci.) 94*: 356, 1985. Ascospores (22–)24–30 × 10–12 μm; dark brown. (Isotype, IMS, J. K. 4532b) Fig. 4F

Ceriosporopsis caduca Jones *et* Zainal, *Mycotaxon 32*: 327, 1988. Ascospores (10–)21–34 × 10–17 μm. (Holotype, IMI 322552) Fig. 6B

Ceriosporopsis cambrensis Wilson, *Trans. Br. Mycol. Soc. 37*: 276, 1954. Ascospores 20–24 × 9–12 μm (as measured in holotype material, IMI 53595).

Apparently, ascospore sizes given by Wilson (1954), 29–31.5 × 10.5–14.5 μm, are in error, because we found considerably lower dimensions in the type. This species is not illustrated because appendages are not well preserved in the type material, but see *Ceriosporopsis sundica* that is a possible synonym.

Ceriosporopsis capillacea Kohlm., *Can. J. Bot. 59*: 1314, 1981. Ascospores (26–)30–38(–42) × 10–12 μm. (Isotype, IMS, J. K. 4066b) Fig. 6A

Ceriosporopsis circumvestita (Kohlm.) Kohlm., *Can. J. Bot. 50*: 1953, 1972. Ascospores 16–25 × 9–13 μm. (IMS, J. K. 807) Fig. 6C

Ceriosporopsis halima Linder *in* Barghoorn *et* Linder, *Farlowia 1*: 409, 1944. Ascospores 18–27(–35) × 6–12 μm. (IMS, J. K. 5089) Fig. 6E

Ceriosporopsis sundica Koch *et* Jones, *Nordic J. Bot. 6*: 339, 1986. Ascospores 19–24 × 8–10 μm. (Holotype, CP, Koch 402/3) Fig. 6F

This species appears to be identical with *C. cambrensis*. Koch and Jones (1986) distinguish the two species mainly by ascospore dimensions that, however, are actually almost identical (see remarks under *C. cambrensis*). We found interthecial filaments similar to catenophyses, about 70–160 × 3–12 μm, in material of *C. sundica* from Denmark (Herb. J. Koch 594 and 770, kindly supplied by Dr. Koch). These filaments are not included in the protologue for *C. sundica* (Koch and Jones 1986), but occur also in *C. cambrensis* (Wilson 1954).

Ceriosporopsis tubulifera (Kohlm.) Kirk *in* Kohlm., *Can. J. Bot. 50*: 1953, 1972. Ascospores 14–23 × 7–11 μm. (IMS, J. K. 1898) Fig. 6D

Chadefaudia balliae Kohlm., *Mycologia 65*: 244–245, 1973. Ascospores 29–38 × 14–22 μm. (Paratype, IMS, J. K. 3001) Fig. 1F

Chadefaudia corallinarum (Crouan *et* Crouan) Müller *et* von Arx, *in Ainsworth, Sparrow and Sussman, The Fungi 4A*: 116, 1973. Ascospores (9–)12–19 × 4–8 μm. (IMS, J. K. 5237) Fig. 1A

Chadefaudia gymnogongri (J. Feldmann) Kohlm., *Bot. Mar. 16*: 202, 1973. Ascospores 14–20 × 4–7 μm. (IMS, G. R. South NZ 873) Fig. 1E

Chadefaudia marina G. Feldmann, *Rev. Gén. Bot. 64*: 150, 1957. Ascospores 12–16 × 4–6 μm. (IMS, J. K. 3106) Fig. 1D

Chadefaudia polyporolithi (Bonar) Kohlm., *Bot. Mar. 16*: 205, 1973. Ascospores 16–24 × 4–8 μm. (Holotype, UC 1272162) Fig. 1B

Chadefaudia schizymeniae Stegenga *et* Kemperman, *Bot. Mar. 27*: 443, 1984. Ascospores 12–18 × 1–4 μm. (Holotype, CBS 9369) Fig. 1C

Chaetomastia typhicola (Karsten) Barr, *Mycotaxon 34*: 514, 1989 (= *Leptosphaeria typhicola* Karsten). Ascospores 34–52(–62) × 7–10(–13) μm, 7–11 septate; hyaline to light- or golden-brown. (IMS, J. K. 6000) Fig. 14N

Chaetosphaeria chaetosa Kohlm., *Nova Hedwigia 6*: 307–308, 1963. Ascospores 24–36 × 6–12 µm. (Isotype, IMS, J. K. 870) Fig. 8N

Corollospora angusta Nakagiri *et* Tokura, *Trans. Mycol. Soc. Jpn. 28*: 417, 1987. Ascospores 35–57 × 3–7.5 µm, 3(–5) septate, spines 3–8 µm long. (Holotype, Nakagiri, TKB-F-5053) Fig. 10D

Corollospora armoricana Kohlm. *et* Volkm.-Kohlm., *Can. J. Bot. 67*: 1281, 1989. Ascospores 29–44 × 6.5–8.5 µm, 3 septate (rarely 4 or 5 septate), spines 9–13 µm long. (Holotype, IMS, J. K. 5135a) Fig. 10F

Corollospora cinnamomea Koch, *Nordic J. Bot. 6*: 498, 1986. Ascospores 18–25 × 6–9 µm (Koch 1986), 16.5–24.5 × 8–12 µm (Kohlmeyer and Volkmann-Kohlmeyer 1987b), spines 10–15 µm long; dark brown. (IMS, J. K. 4930) Fig. 10C

Corollospora colossa Nakagiri *et* Tokura, *Trans. Mycol. Soc. Jpn. 28*: 418, 1987. Ascospores 60–108 × 13–26 µm, (6–)7(–8) septate, without spines. (Holotype, Nakagiri, TKB-F-5054) Fig. 12C

Corollospora filiformis Nakagiri *in* Nakagiri *et* Tokura, *Trans. Mycol. Soc. Jpn. 28*: 422, 1987. Ascospores (73–)87–120 × 5–8(–10)µm, (9–)13(–17) septate. (Holotype, Nakagiri, TKB-F-5055) Fig. 11F

Corollospora fusca Nakagiri *et* Tokura, *Trans. Mycol. Soc. Jpn. 28*: 424, 1987. Ascospores 63–220 × 20–38 µm, (5–)12–21 transsepta, spines 28–65 µm long; dark brown. (Holotype, Nakagiri, TKB-F-5056) Fig. 12B

Corollospora gracilis Nakagiri *et* Tokura, *Trans. Mycol. Soc. Jpn. 28*: 426, 1987. Ascospores 26–45 × 3–7 µm, spines 6–12 µm long. (Holotype, Nakagiri, TKB-F-5057) Fig. 10A

Corollospora intermedia I. Schmidt, *Nat. Naturschutz Mecklenburg 7*: 6, 1969 (anamorph: *Varicosporina prolifera*). Ascospores 25–34(–36) × 7–12 µm, 3 septate (rarely more or less septa), spines 10–14 µm long. (IMS, Schaumann 768) Fig. 10E

Corollospora lacera (Linder *in* Barghoorn *et* Linder) Kohlm., *Ber. Dtsch. Bot. Ges. 75*: 126, 1962. Ascospores 39–60(–63) × 10–16(–19) µm, 5 (rarely 4 or 6) septate, spines 10–14(–28) µm long. (IMS, J. K. 5285) Fig. 11B

Corollospora luteola Nakagiri *et* Tubaki, *Trans. Mycol. Soc. Jpn. 23*: 102–103, 1982 (anamorph: *Sigmoidea luteola*). Ascospores 50–85 × 5–8 µm, 5 septate. (Holotype, Nakagiri, TKB-F-5002) Fig. 11C

Corollospora maritima Werdermann, *Notizbl. Bot. Gart. u. Mus. Berlin-Dahlem* 8: 248, 1922. Ascospores 22–33 × 8–10 µm. (IMS, I. Schmidt, Baltic) Fig. 10B

This cosmopolitan species appears to be composed of physiological races with different temperature requirements (Bebout *et al.* 1987). Morphologically these races cannot be distinguished with certainty, although slight differences in ascospore dimension in material from different geographical areas may occur. The type material of *C. maritima* derived from the North Sea (Büsum). Because ascospore diameters in the original diagnosis (Werdermann 1922) appear too small, we base our sizes on new collections from temperate European waters (Brittany and Baltic Sea).

Corollospora novofusca Kohlm. *et* Volkm.-Kohlm., sp. nov. Ascospores 74–99 × 24–34 µm, (12–)13 transsepta, spines 52–86 µm long; dark brown. (Holotype, IMS, J. K. 5102) Fig. 12A

The complete description of this species is given in the Appendix.

Corollospora pseudopulchella Nakagiri *et* Tokura, *Trans. Mycol. Soc. Jpn. 28*: 428, 1987. Ascospores 65–98 × 8–12 µm, 7–11 septate. (Holotype, Nakagiri, TKB-F-5058) Fig. 11D

Corollospora pulchella Kohlm., Schmidt *et* Nair, *Ber. Dtsch. Bot. Ges. 80*: 98–99, 1967 (anamorph: *Clavatospora bulbosa*). Ascospores 52–102(–112) × 7–12(–16) µm, 7 (rarely 9–13) septate. (IMS, J. K. 4888) Fig. 11E

Corollospora quinqueseptata Nakagiri *in* Nakagiri *et* Tokura, *Trans. Mycol. Soc. Jpn. 28*: 430, 1987. Ascospores (38–)41–59 × 8–10 µm, (3–)5(–8) septate, spines 5–12 µm long. (Holotype, Nakagiri, TKB-F-5059) Fig. 11A

Corollospora sp. Ascospores 26–34 × 6–7.5 µm, 3 septate, spines 7.5–9.5 µm long. (IMS, Honrubia AR 9521, J. K. 5186) Fig. 10G

Ascomata of this species from the Mediterranean coast of Spain (Arenales del Sol, Alicante) developed on sand grains attached to leaves of *Cymodocea nodosa* (Ucria) Aschers. (Roldán and Honrubia 1989). It remains undescribed until sufficient new material becomes available.

Coronopapilla mangrovei (Hyde) Kohlm. *et* Volkm.-Kohlm. **comb. nov.** (Basionym: *Caryospora mangrovei* Hyde, *Trans. Mycol. Soc. Jpn. 30*: 336, 1989; Synonym: *Coronopapilla avellina* Kohlm. *et* Volkm.-Kohlm., *Mycol. Res.* 94:687, 1990). Ascospores 36–60 × 16–24 µm, 1–3 septate; hazel brown. (Holotype of *C. avellina*, IMS, J. K. 5144) Fig. 4I

We have examined the holotype material of *Caryospora mangrovei* (IMI 327272) from Brunei, compared it with *Coronopapilla avellina* from Belize (both had been in press at the same time) and conclude that the two are conspecific. We do not agree with Hyde (1989 b) that the species belongs to *Caryospora* De Not. (*sensu* Barr 1979) and transfer it herewith to the monospecific genus *Coronopapilla* Kohlm. *et* Volkm.-Kohlm. *Caryospora* is characterized by erumpent or superficial ascomata with flattened bases that leave circular black lines in the substrate when removed and that are not immersed under a stroma and clypeus; the peridium is carbonaceous, and the ascospores are surrounded by a gelationous sheath that develops into surface ornamentations (Barr 1979). In contrast, ascomata of *Coronopapilla* are submerged under a stroma that merges into thick clypei around the ostioles, the bases are not flattened and the peridium is hyaline and soft, not carbonaceous, the ascospores have no sheath, nor signs of surface ornamentations (Kohlmeyer and Volkmann-Kohlmeyer 1990). We believe that *Coronopapilla* is best placed in the Didymosphaeriaceae, rather than the Massariaceae where *Caryospora* is listed (Barr 1987).

Hyde (1989 b) considers the ascospores of *C. mangrovei* to be one-septate with pseudosepta often forming near the poles, whereas we maintain that most mature spores are three-septate with true septa near the apices. The faint apical ring of the ascus, mentioned by Hyde (1989 b) can be observed with Nomarski interference contrast optics in the type of *C. mangrovei*, whereas it was not visible in the material from Belize (Kohlmeyer and Volkmann-Kohlmeyer 1990).

The type material of *Caryospora mangrovei* contains conidiomata (or spermogonia?), not mentioned in the description. They could represent an anamorph of the ascomycete because they are immersed next to the ascomata under the stroma. The conidia (or spermatia?) are one-celled, hyaline, ellipsoidal or subglobose, 1.5 – 2 µm in diameter, and contain a large oil globule.

Crinigera maritima I. Schmidt, *Nat. Naturschutz Mecklenburg 7*: 11, 1969; *Mycotaxon 24:* 420, 1985; *emend.* Koch *et* Jones, *Can. J. Bot. 67:* 1184, 1989. Ascospores 10 – 14 × 4 – 5 µm. (Holotype, JE, I. Schmidt 211) Fig. 8H

Koch and Jones (1989) reexamined Schmidt's type material and concluded that the original description was based on two different species, viz. *C. maritima* (as found in the holotype on *Fucus vesiculosus*) and *Dryosphaera navigans* Koch *et* Jones (as found in the paratype on wood). We also examined the type specimens from Herb. JE and concur with Koch and Jones' conclusions, except for their statement that the material does not show whether *C. maritima* is an ascomycete or a fungus imperfectus. We found that the ascospores in the slides prepared by Schmidt (No. 211 on *Fucus*) are often arranged in clusters of eight. There is no indication of conidiophores, therefore, it is clear that *C. maritima* is an ascomycete, not a coelomycete.

Cucullosporella mangrovei (Hyde *et* Jones) Hyde *et* Jones *in* Jones *et* Hyde, *Mycotaxon 37*: 200, 1990. Ascospores 49 – 64(– 69) × (14 –)16 – 24 µm; the delicate wall striation is visible only at the highest magnification under oil. (Paratype, IMS, J. K. 4661) Fig. 7P

Dactylospora haliotrepha (Kohlm. *et* Kohlm.) Hafellner, *Beih. Nova Hedwigia 62*: 111, 1979 [= *Kymadiscus haliotrephus* (Kohlm. *et* Kohlm.) Kohlm. *et* Kohlm.]. Ascospores (15 –)18 – 28(– 32) × 8 – 12(– 14) µm; grayish-green to brownish. (IMS, J. K. 5175) Fig. 3E

Didymella avicenniae Patil *et* Borse, *Trans. Mycol. Soc. Jpn. 26*: 271, 1985. Ascospores 25 – 32 × 10 – 15 µm. (IMS, Herb. B. D. Borse) Fig. 3J

Didymella fucicola (Sutherland) Kohlm., *Phytopath. Z. 63*: 342, 1968. Ascospores 16 – 23 × 6 – 8 µm. (IMS, J. K. 3120) Fig. 2Z

Didymella gloeopeltidis (Miyabe *et* Tokida) Kohlm. *et* Kohlm., *Marine Mycology*, Academic Press, p. 382, 1979. Ascospores (14 –)17 – 22 × 4 – 6 µm. (IMS, J. K. 2727) Fig. 2X

Didymella magnei G. Feldmann, Rev. *Gén. Bot. 65*: 414 – 415, 1958. Ascospores (8 –)12 – 16 × 2 – 4 µm. (Type slide, Herb. G. Feldmann) Fig. 2Y

Didymosphaeria lignomaris (Strongman *et* Miller, *in* Strongman, Miller *et* Whitney, *Proc. N. S. Inst. Sci. 35*: 102, 1985. Ascospores 18 – 23 × 9 – 11 µm, as measured in type material (length not 24.7 – 26 µm, as listed by Strongman and Miller); brown. (Holotype, DAOM 193904) Fig. 4D

The generic placement of this species is doubtful, but it appears not to belong to *Didymosphaeria* (Kohlmeyer and Volkmann-Kohlmeyer 1990).

Dryosphaera navigans Koch *et* Jones, *Can. J. Bot. 67*: 1185, 1989. Ascospores 8 – 12 × 3 – 5 µm. (Paratype, JE, I. Schmidt 214) Fig. 8I

As Koch and Jones (1989) have shown, combined features of this species and *Crinigera maritima* were used by Schmidt (1969) in the description of *C. maritima*. Tropical ascomycetes, resembling *D. na-*

vigans or *C. maritima* are most likely different species.

Eiona tunicata Kohlm., *Ber. Dtsch. Bot. Ges. 81*: 58, 1968. Ascospores 7.5–12 × 3.5–5 μm. (Isotype, IMS, J. K. 5190) Fig. 2I

Etheirophora bijubata Kohlm. *et* Volkm.-Kohlm., *Mycol. Res. 92*: 414–415, 1989. Ascospores 16–22 × 6–8 μm. (Holotype, IMS, J. K. 5098) Fig. 8F

Etheirophora blepharospora (Kohlm. *et* Kohlm.) Kohlm. *et* Volkm.-Kohlm. (= *Keissleriella blepharospora* Kohlm. *et* Kohlm.), *Mycol. Res. 92*: 415, 1989. Ascospores 12–21 × 6–8 μm. (IMS, J. K. 4402) Fig. 8G

Etheirophora unijubata Kohlm. *et* Volkm.-Kohlm., *Mycol. Res. 92*: 415–416, 1989. Ascospores 18–24 × 5.5–7 μm. (Holotype, IMS, J. K. 5110) Fig. 8E

Gnomonia salina Jones, *Trans. Brit. Mycol. Soc. 45*: 107, 1962. Ascospores 36–72 × 20–32 μm. (Holotype, IMI 80274) Fig. 7H

This is an ill-defined species that does not fit in *Gnomonia*. It has not been studied again after the original description, and the type material is insufficient to observe critical details. Possibly, *G. salina* belongs to Halosphaeriaceae.

Groenhiella bivestia Koch, Jones *et* Moss, *Bot. Mar. 26*: 265–266, 1983. Ascospores 12–16 × 6–8.5 μm; first hyaline, later yellow-brown. (Isotype, IMI 260599) Fig. 9B

Gymnascella littoralis (Orr) Currah, *Mycotaxon 24*: 87, 1985 (= *Plunkettomyces littoralis* Orr). Ascospores 4–5 μm diam in face view; yellow- to orange-brown. (Isotype, NY) Fig. 2H

Haligena elaterophora Kohlm., *Nova Hedwigia 3*: 87–88, 1961. Ascospores 24–54 × 9–18(–20) μm, 3 (rarely up to 5) septate. (IMS, J. K. 1922) Fig. 6L

Haligena salina Farrant *et* Jones, *Bot. J. Linn. Soc. 93*: 406, 1986. Ascospores (14–)16–18(–20) × 8–10 μm. (Holotype, IMI 297765) Fig. 6K

Halographis runica Kohlm. *et* Volkm.-Kohlm., *Can. J. Bot. 66*: 1138–1139, 1988. Ascospores 17–29 × 6–9 μm. (Holotype, IMS, J. K. 4970a) Fig. 2CC

Haloguignardia cystoseirae Kohlm. *et* Demoulin, *Bot. Mar. 24*: 9–10, 1981. Ascospores 32–40 × 9–12 μm (excl. appendages). (Isotype, IMS, J. K. 4024) Fig. 1K

Haloguignardia decidua Cribb *et* Cribb, *Univ. Queensl. Pap.*, Dept. Bot. *3*: 97, 1956. Ascospores 38–48 × 6–9 μm (incl. appendages). (Holotype, Herb. Cribb) Fig. 1G

Haloguignardia irritans (Setchell *et* Estee *in* Estee) Cribb *et* Cribb, *Univ. Queensl. Pap.*, Dept. Bot. *3*: 98, 1956. Ascospores (30–)34–50 × (10–)12–15 μm (incl. appendages). (IMS, J. K. 2606) Fig. 1I

Haloguignardia oceanica (Ferdinandsen *et* Winge) Kohlm., *Mar. Biol. 8*: 344, 1971. Ascospores 20–32 (–36) × 9–14 μm (excl. appendages). (Holotype C) Fig. 1J

Haloguignardia tumefaciens (Cribb *et* Herbert) Cribb *et* Cribb, *Univ. Queensl. Pap.*, Dept. Bot. *3*: 98, 1956. Ascospores (30–)35–70 × (10–)15–24(–26) μm (incl. appendages). (Holotype, Herb. Cribb) Fig. 1H

Halonectria milfordensis Jones, *Trans. Brit. Mycol. Soc. 48*: 287, 1965 (see also the anamorph). Ascospores 16–30 × 2–3.5 μm. (Holotype, IMI 86722) Fig. 2G

Halosarpheia abonnis Kohlm., *Mar. Ecol.* (P. S. Z. N. I) *5*: 339, 1984. Ascospores 33–47 × 14–22 μm. (Paratype, IMS, J. K. 4355) Fig. 7B

Halosarpheia bentotensis Koch, *Nordic J. Bot. 2*: 165, 1982. Ascospores 28–36 × 6–8 μm, 3(–4) septate. (Holotype, CP, Koch 377) Fig. 7J

Halosarpheia cincinnatula Shearer *et* Crane, *Bot. Mar. 23*: 613, 1980. Ascospores 34–59 × 4–5 μm, 5–11 septate. (Holotype, ILLS 39006) Fig. 7M

Halosarpheia fibrosa Kohlm. *et* Kohlm., *Trans. Br. Mycol. Soc. 68*: 208, 1977. Ascospores 32–44 × 18–24 μm. (Paratype, IMS, J. K. 3745) Fig. 7C

Halosarpheia marina (Cribb *et* Cribb) Kohlm., *Mar. Ecol. (P. S. Z. N. I) 5*: 345, 1984 (= *Gnomonia marina* Cribb *et* Cribb). Ascospores 18–23(–26) × 9–12 μm. (IMS, J. K. 5107) Fig. 7D

Halosarpheia ratnagiriensis Patil *et* Borse, *Ind. Bot. Rep. 1*: 102, 1982. Ascospores 48–64(–72) × 22–28 μm. (IMS, J. K. 4470) Fig. 7A

Halosarpheia retorquens Shearer *et* Crane, *Bot. Mar. 23*: 608–609, 1980. Ascospores 20–34 × 7–11 μm. (Holotype, ILLS 38994) Fig. 7G

This species was originally described only from freshwater habitats, but it was fruiting in culture in 100 percent seawater (Shearer and Crane 1980). We include it here because it has been reported frequently from marine habitats; however, the identity of these collections with *H. retorquens* is not certain. *Halosarpheia retorquens* is very close to *H. viscosa*, and they may belong to a larger group of closely related taxa that cannot be separated at this time.

Halosarpheia spartinae (Jones) Shearer *et* Crane, *Bot. Mar. 23*: 608, 1980 (= *Haligena spartinae* Jones). Ascospores 40–76(–90) × 14–22 µm, 4–7(–9) septate. (IMS, R. D. Brooks 24) Fig. 7K

Halosarpheia trullifera (Kohlm.) Jones, Moss *et* Cuomo, *Trans. Br. Mycol. Soc. 80*: 200, 1983 [= *Halosphaeria trullifera* (Kohlm.) Kohlm.]. Ascospores 23–32(–36) × 14–18(–20) µm. (IMS, J. K. 4833) Fig. 7E

Halosarpheia unicaudata (Jones *et* LeCampion-Alsumard) Johnson, Jones *et* Moss *ex* Kohlm. *et* Volkm.-Kohlm. ***comb. nov.*** (basionym: *Haligena unicaudata* Jones *et* LeCampion-Alsumard, *Nova Hedwigia 19*: 574, 1970). Ascospores 36–60 × 2.5–5 µm., 3 (rarely up to 5) septate. (IMS, J. K. 4109) Fig. 7N

According to the International Code of Botanical Nomenclature (Article 33.2) the combination made by Johnson *et al.* (1987) was invalid because the basionym was not fully cited and we validate it herewith.

Halosarpheia viscidula (Kohlm. *et* Kohlm.) Shearer *et* Crane, *Bot. Mar. 23*: 608, 1980 (= *Haligena viscidula* Kohlm. *et* Kohlm.). Ascospores (38–)44–79(–89) × 3–6.5 µm, 5–11(–16) septate. (Isotype, IMS, J. K. 1773b) Fig. 7L

Halosarpheia viscosa (I. Schmidt) Shearer *et* Crane *ex* Kohlm. *et* Volkm.-Kohlm. ***comb. nov.*** (basionym: *Halosphaeria viscosa* I. Schmidt, *Mycotaxon 24*: 420, 1985); *non Halosphaeria viscosa* I. Schmidt, *Nat. Naturschutz Mecklenburg 12*: 70, 1974 (no type designated); *nec non Halosarpheia viscosa* (I. Schmidt) Shearer *et* Crane, *Bot. Mar. 23*: 608, 1980 (invalid new combination). Ascospores 21–28(–31) × 8–11 µm [our measurements from the type slides No. 147 and 148 from Herb. JE; they do not agree with the sizes published by Schmidt (1974): 16.5–23 × 6.5–12.5 µm]. (Paratype, JE, I. Schmidt 147) Fig. 7F

The species from Maryland, identified by Shearer and Crane (1980) as *H. viscosa* has smaller spores than the type.

Halosarpheia sp. (Leong, Tan, Hyde *et* Jones, *Can. J. Bot.*, manuscript). Ascospores 10–14 × 4–6.5 µm. (Holotype, IMI 327473) Fig. 23B

Halosphaeria appendiculata Linder *in* Barghoorn *et* Linder, *Farlowia 1*: 412, 1944. Ascospores (16–)18–29 × (6–)8–12 µm. (IMS, J. K. 4887) Fig. 5D

Halosphaeria cucullata (Kohlm.) Kohlm., *Can. J. Bot. 50*: 1956, 1972 (anamorph: *Periconia prolifica*). Ascospores 20–68 × 6–12 µm, 1 septate, becoming 2–4 septate upon germination. (Isotype, IMS, J. K. 1686) Fig. 3L

Halosphaeria quadricornuta Cribb *et* Cribb, *Univ. Queensl. Pap.*, Dept. Bot. *3*: 99, 1956. Ascospores 20–35 × (6–)8–12 µm. (IMS, J. K. 5148) Fig. 5A

This species is frequently listed as *Antennospora quadricornuta* (Cribb *et* Cribb) Johnson (1958). The reasons for keeping it in *Halosphaeria* were discussed by Kohlmeyer and Volkmann-Kohlmeyer (1987 a).

Halosphaeria salina (Meyers) Kohlm., *Can. J. Bot. 50*: 1957, 1972. Ascospores 19–28 × 8–14 µm. (IMS, J. K. 5087) Fig. 5B

Ascomata developing on incubated wood often contain spores with abnormal shapes and appendages.

Halosphaeriopsis mediosetigera (Cribb *et* Cribb) Johnson, *J. Elisha Mitchell Sci. Soc. 74*: 44, 1958 (anamorph: *Trichocladium achrasporum*). Ascospores 24–44 × 8–17(–20) µm. (IMS, J. K. 5233) Fig. 9A

Halotthia posidoniae (Durieu *et* Montagne *in* Montagne) Kohlm., *Nova Hedwigia 6*: 9, 1963. Ascospores 37–60 × 16–26 µm; dark brown. (IMS, J. K. 477) Fig. 4G

Heleococcum japonense Tubaki, *Trans. Mycol. Soc. Jpn. 8*: 5, 1967. Ascospores 18–21 × 10–14 µm. (Holotype, Inst. Ferment. Osaka H-12121[8643]) Fig. 3Q

Ascospores from the type material have irregular ridges. Fresh material is required to determine if these are remnants of the ascus or features of the spore wall.

Helicascus kanaloanus Kohlm., *Can. J. Bot. 47*: 1471, 1969. Ascospores 36–51 × 18–26 µm; brown. (Paratype, IMS, J. K. 2565) Fig. 4H

Ascospore sizes given are based on the type material from Hawaii. There are other species with similar ascospores but unilocular ascomata that can be confused with *H. kanaloanus*.

Hydronectria tethys Kohlm. *et* Kohlm., *Nova Hedwigia 9*: 95, 1965. Ascospores 17–26 × 8–13 µm. (Isotype, IMS, J. K. 1709e) Fig. 3D

Hydronectria tethys var. *glabra* Kohlm., *Mar. Ecol. (P. S. Z. N. I) 5*: 351, 1984. Ascospores 17–24(–27) × 8–10 µm. (Isotype, IMS, J. K. 4463b) Fig. 3C

Hypophloeda rhizospora Hyde *et* Jones, *Trans. Mycol. Soc. Jpn. 30*: 62, 1989. Ascospores 21.5–36 × 3.5–6 µm. (Holotype, IMI 315891) Fig. 23D

Hypoxylon oceanicum Schatz, *Mycotaxon 33*: 413, 1988. Ascospores 19–25 × 8–11 µm; brown. (Paratype, IMS, J. K. 4010) Fig. 4L

This species will be transferred to another genus (E. B. G. Jones, personal communication).

Kirschsteiniothelia maritima (Linder) Hawksworth, *Bot. J. Linn. Soc. 91*: 193, 1985 (= *Microthelia linderi*

Kohlm.). Ascospores 14–21 × 5–8 µm; brown. (IMS, J. K. 1665) Fig. 4B

Kohlmeyeriella tubulata (Kohlm.) Jones, Johnson *et* Moss, *Bot. J. Linn. Soc. 87*: 210, 1983 (= *Corollospora tubulata* Kohlm.). Ascospores 137–152 × 17.5–18.5 µm. (Isotype, IMS, J. K. 5136) Fig. 2A

Koralionastes angustus Kohlm. *et* Volkm.-Kohlm., *Mycologia 79*: 768, 1987. Ascospores 81–122 × 18–28 µm, 6–10 (mostly 8) septate. (Holotype, IMS, J. K. 4966a) Fig. 15D

Koralionastes ellipticus Kohlm. *et* Volkm.-Kohlm., *Mycologia 79*: 765–768, 1987. Ascospores 81–127 × 27–45 µm., 1–8 (mostly 6) septate. (Holotype, IMS, J. K. 4965c) Fig. 15C

Koralionastes giganteus Kohlm. *et* Volkm.-Kohlm., *Can. J. Bot. 68*: 1554, 1990. Ascospores 126–170 × 68–89 µm, 2–4 (mostly 2) septate. (Holotype, IMS, J. K. 5022a) Fig. 15A

Koralionastes ovalis Kohlm. *et* Volkm.-Kohlm., *Mycologia 79*: 765, 1987. Ascospores 94–131 × 50–77 µm, 1–6 (mostly 4) septate. (Holotype, IMS, J. K. 4965a) Fig. 15B

Koralionastes violaceus Kohlm. *et* Volkm.-Kohlm., *Can. J. Bot. 68*: 1556, 1990. Ascospores 85–130 × 25–34 µm, 3–6 (mostly 4 and 5) septate. (Isotype, IMS, J. K. 5227b) Fig. 15E

Laboulbenia marina Picard, *C. R. Séances Soc. Biol. 65*: 484–485, 1908. Ascospores 26–35 × 4 µm. (R. K. Benjamin 2537) Fig. 3K

Laetinaevia marina (Boyd) Spooner *apud* Kirk *et* Spooner, *Kew Bull. 38*: 568, 1984 (= *Orbilia marina* Boyd *in* Smith). Ascospores 8–10 × 4–5 µm. (K, Herb. D. A. Boyd) Fig. 1T

Lanspora coronata Hyde *et* Jones, *Can. J. Bot. 64*: 1581, 1986. Ascospores 10–15 × 2.5–4 µm. (IMS, J. K. 4839) Fig. 8L

Lautitia danica (Berlese) Schatz, *Can. J. Bot. 62*: 31, 1984 [= *Didymosphaeria danica* (Berlese) Wilson *et* Knoyle]. Ascospores (25–)33–40(–44) × 5–8 µm. (IMS, J. K. 1988) Fig. 2P

Lautospora gigantea Hyde *et* Jones, *Bot. Mar. 32*: 479–481, 1989. Ascospores 140–195 × 36–45 µm. (Holotype, IMI 327273) Fig. 16J

Leptosphaeria australiensis (Cribb *et* Cribb) G. C. Hughes, *Syesis 2*: 132, 1969. Ascospores 19–27 × 6–9 µm. (IMS, J. K. 5124) Fig. 13I

Leptosphaeria avicenniae Kohlm. *et* Kohlm., *Nova Hedwigia 9*: 98, 1965. Ascospores 18–25 × 6–8 µm. (IMS, J. K. 4377) Fig. 13J

Leptosphaeria oraemaris Linder *in* Barghoorn *et* Linder, *Farlowia 1*: 413, 1944. Ascospores 17–29(–32) × 4–8 µm, 1–3 septate; brown. (IMS, J. K. 1647) Fig. 14G

Leptosphaeria pelagica Jones, *Trans. Br. Mycol. Soc. 45*: 105, 1962. Ascospores 28–44 × 8–12 µm, 3 (rarely 4) septate. (Isotype, IMI 87372) Fig. 13K

We have examined type material of *L. pelagica* and found a sheath around the ascospores that was not included in the original protologue. Comparison with *L. contecta* Kohlm. (Kohlmeyer 1963) showed that the two species are identical, and *L. pelagica* has priority.

Leptosphaeria peruviana Spegazzini, *An. Soc. Cient. Argent. 12*: 179, 1881. Ascospores 12–16 × 4–5.5 µm, 3 septate; olive-brown. (Holotype, LPS, Spegazzini 2172) Fig. 14F

Ligníncola laevis Höhnk, *Veroeff. Inst. Meeresforsch. Bremerhaven 3*: 216, 1955. Ascospores (12–)16–24 × 5–8 µm. (IMS, J. K. 4835) Fig. 2M

Ligníncola longirostris (Cribb *et* Cribb) Kohlm., *Mar. Ecol. (P. S. Z. N. I) 5*: 353, 1984 (= *Gnomonia longirostris* Cribb *et* Cribb). Ascospores 13–20 × 4–8 µm. (IMS, J. K. 4971) Fig. 2N

Ligníncola tropica Kohlm., *Mar. Ecol. (P. S. Z. N. I) 5*: 355, 1984. Ascospores 22–36 × 12–16 µm. (Isotype, IMS, J. K. 4431) Fig. 2O

Lindra crassa (Kohlm.) Kohlm. *et* Volkm.-Kohlm. ***comb. nov.*** (basionym: *Lindra thalassiae* Orpurt, Meyers, Boral *et* Simms var. *crassa* Kohlm., *Mar. Ecol. (P. S. Z. N. I) 5*: 358–359, 1984). Ascospores 320–520 × 8–10 µm, 15–23 septate. (Paratype, IMS, J. K. 4424) Fig. 17C

Originally, *L. crassa* was described as a variety of *L. thalassiae*, however, the two fungi differ sufficiently from each other by the length and diameter of their ascospores and separation on the species level is warranted.

Lindra hawaiiensis Kohlm. *et* Volkm.-Kohlm., *Can. J. Bot. 65*: 574, 1987. Ascospores (101–)110–179 (–188) × 3.5–5.5 µm, 8–17 septate. (Holotype, IMS, J. K. 4485a) Fig. 17E

Lindra inflata Wilson, *Trans. Br. Mycol. Soc. 39*: 411, 1956. Ascospores 210–415 × 4–6 µm, 30–50 septate. (Holotype, IMI 62909) Fig. 17B

Lindra marinera Meyers, *Mycologia 61*: 488, 1969. Ascospores 152–230 × 4–7 µm, 6–17 septate. (IMS, J. K. 5089) Fig. 17A

Lindra obtusa Nakagiri *et* Tubaki, *Mycologia 75*: 488, 1983 (anamorph: *Anguillospora marina*). Ascospores 182–250(–313) × 2.5–3.5 µm, 6–16(–21) septate. (Holotype, Nakagiri, TKB-F-5008-1) Fig. 17F

Lindra thalassiae Orpurt, Meyers, Boral *et* Simms, *Bull. Mar. Sci. Gulf Caribb. 14*: 406, 1964. Ascospores (220–)230–390 × 3–6 µm, 14–26 septate. (IMS, J. K. 5090) Fig. 17D

Lineolata rhizophorae (Kohlm. *et* Kohlm.) Kohlm. *et* Volkm.-Kohlm., *Mycol. Res. 94*: 688, 1990 (= *Didymosphaeria rhizophorae* Kohlm. *et* Kohlm.). Ascospores 23–33 × 9–12 µm; dark brown. (Isotype, IMS, J. K. 2390) Fig. 4E

Linocarpon appendiculatum Hyde, *Trans. Mycol. Soc. Jpn. 29*: 339–340, 1988. Ascospores 75–120 × 2.2–3.6 µm. (Holotype, IMI 326619) Fig. 17G

Linocarpon nypae (P. Henn.) Hyde, *Trans. Mycol. Soc. Jpn. 29*: 346, 1988 (as *'nipae'*). Ascospores 91–124 × 3.5–4 µm. (Herb. K. D. Hyde 471) Fig. 17H

Linocarpon cfr. *pandani* (H. Sydow *et* P. Sydow) H. Sydow *et* P. Sydow, *Ann. Mycol. 15*: 210, 1917. Ascospores 99–122 × 2.2–3.4 µm. (Herb. K. D. Hyde 391) Fig. 17I

A fungus on *Nypa fruticans* (Thunb.) Wurmb. from Brunei was tentatively identified by Hyde (1988 a) as *L. pandani*. As Hyde pointed out, further studies are needed to establish if the intertidal fungus on *Nypa* is identical with *L. pandani*, that was reported from *Pandanus* in terrestrial habitats.

Lophiostoma mangrovei Kohlm. *et* Vittal (as *'mangrovis'*), *Mycologia 78*: 487, 1986. Ascospores (36–)38–55 × 7–11 µm. Often found in the immature 1 septate stage. (IMS, J. K. 5074) Fig. 13H

Lulworthia calcicola Kohlm. *et* Volkm.-Kohlm., *Mycologia 81*: 289, 1989. Ascospores 80–150(–155) × 5.5–8.5 µm. (Holotype, IMS, J. K. 5028a) Fig. 18E

Lulworthia crassa Nakagiri, *Trans. Mycol. Soc. Jpn. 25*: 378–379, 1984. Ascospores 140–205 × 5–8 µm. (Holotype, Nakagiri, TKB-F-5049) Fig. 18F

Lulworthia curalii (Kohlm.) Kohlm. *et* Volkm.-Kohlm. ***comb. nov.*** (basionym: *Lulworthia kniepii* Kohlm. var. *curalii* Kohlm., *Mar. Ecol. (P. S. Z. N. I) 5*: 361, 1984). Ascospores 120–180 × 4–5(–6) µm. (Holotype, IMS, J. K. 4379) Fig. 18M

The original material of *L. curalii* came from a calcified red alga in Belize. In the meantime we have found the fungus in an empty shell of the conch *Strombus gigas* L., also on a coral reef of Belize. The ascospores of *L. curalii* differ distinctly from those of *L. kniepii* and separation of the two taxa on the species level is warranted.

Lulworthia fucicola Sutherland, *Trans. Br. Mycol. Soc. 5*: 259, 1916. Ascospores 70–110(–126) x 4–6 µm. (IMS, J. K. 1938) Fig. 18I

Lulworthia grandispora Meyers, *Mycologia 49*: 513, 1957. Ascospores 500–756 × 3–5 µm. (IMS, J. K. 5058) Fig. 18H

Lulworthia kniepii Kohlm., *Nova Hedwigia 6*: 140–141, 1963. Ascospores 200–270 × 2.5–6 µm. (IMS, J. K. 3622) Fig. 18K

Lulworthia lignoarenaria Koch *et* Jones, *Mycotaxon 20*: 389, 1984. Ascospores 409–472 × 6–8 µm, 31–38 septate. (Isotype, IMI 281466, Koch 398) Fig. 18L

Lulworthia lindroidea Kohlm., *Bot. Mar. 23*: 537–539, 1980. Ascospores 170–240 × 4 µm, 9–12 septate. (Isotype, IMS, J. K. 3932) Fig. 18J

Lulworthia uniseptata Nakagiri, *Trans. Mycol. Soc. Jpn. 25*: 382–384, 1984 (anamorph: *Zalerion maritimum*). Ascospores 94–148 × 2.5–5 µm, 1 septate. (Holotype, Nakagiri, TKB-F-5050) Fig. 18G

Lulworthia uniseptata has not been recorded as yet in the American and European habitats of *Z. maritimum*.

Lulworthia spp.

This group includes *Lulworthia* spp. (ascospore lengths between 150 and 500 µm) that cannot be classified at the present time (see also Kohlmeyer and Kohlmeyer 1979). Ascospores of this group resemble those of *L. grandispora* (Fig. 18H).

Luttrellia estuarina Shearer, *Mycologia 70*: 692, 1978. Ascospores 42–55 × 16–22 µm, 5 (rarely to 9) septate. (Paratype, ILLS 36983) Fig. 15K

Manglicola guatemalensis Kohlm. *et* Kohlm., *Mycologia 63*: 841, 1971. Ascospores 80–109 × 18–34 µm; large cell chestnut-brown, small cell light brown. (Isotype, IMS, J. K. 2700a) Fig. 4K

The species was known only from the type locality, the Pacific Coast of Guatemala, until Hyde (1988 b) found it in Brunei.

Marinosphaera mangrovei Hyde, *Can. J. Bot. 67*: 3080, 1989. Ascospores 26–31 × 6–8.5 µm (Holotype, IMI 325412) Fig. 13C

Marinospora calyptrata (Kohlm.) Cavaliere, *Nova Hedwigia 11*: 548, 1966 (= *Ceriosporopsis calyptrata* Kohlm.). Ascospores 20–36 × (7–)9–19 µm; apical and equatorial appendages up to 27 µm long. (IMS, J. K. 4149) Fig. 6H

Marinospora longissima (Kohlm.) Cavaliere, *Nova Hedwigia 11*: 548, 1966 (= *Ceriosporopsis longissima* Kohlm.). Ascospores 20–34 × 8–16.5(–18) µm; apical appendages up to 187 µm long. (IMS, Herb. P. Kirk R-626) Fig. 6I

Marinospora longissima was considered conspecific with *M. calyptrata* (Kohlmeyer and Kohlmeyer 1979), but Jones and Moss (1980) opined that the two species could be separated by the presence of a mucilaginous envelope in *M. longissima* and consistently longer equatorial appendages in *M. calyptrata.*

Marisolaris ansata Koch *et* Jones, *Can. J. Bot. 67*: 1183, 1989. Ascospores 6–9 × 3–4 µm. (Holotype, CP, Koch 532) Fig. 9C

Massarina acrostichi Hyde, *Mycol. Res. 93*: 437, 1989. Ascospores 24–33 × 6–10 µm, 1 septate (possibly 3 septate ?); hyaline becoming light brown with age (?). (Holotype, IMI 327274) Fig. 13P

Although Hyde (1989) describes ascopores as "(1)–3 septate", spores in holotype material from IMI have only one true septum. Two additional pseudosepta, formed by the cytoplasm or cell inclusions may occur. Some ascospores in slide KDH 0105 (paratype) kindly made available by Dr. Hyde appear to have one or two additional septa, but the question of spore septation must remain open until fresh mature material can be examined.

Massarina cystophorae (Cribb *et* Herbert) Kohlm. *et* Kohlm., *Marine Mycology*, Academic Press, p. 427, 1979. Ascospores 50–65(–73) × 15–23(–25) µm. (IMS, M. J. Parsons, A625) Fig. 6J

Massarina lacertensis Kohlm. *et* Volkm.-Kohlm., *Austr. J. Mar. Freshw. Res. 42*, 1991 (in press). Ascospores 25.5–34 × 9–12 µm. (Isotype, IMS, J. K. 5301b) Fig. 13Q

Massarina thalassiae Kohlm. *et* Volkm.-Kohlm., *Can. J. Bot. 65*: 575, 1987. Ascospores 28–44(–47) × 10–15 µm, 1–3 septate. (Paratype, IMS, J. K. 4227) Fig. 13O

Massarina velatospora Hyde *et* Borse (as *'velataspora'*), *Mycotaxon 27*: 161, 1986. Ascospores 45–56 × 14–19 µm, 1–3 septate. (IMS, J. K. 5064) Fig. 13N

Mastodia tessellata (Hooker f. *et* Harvey) Hooker f. *et* Harvey *ex* Hooker, Flora Antarctica, Part II, p. 499, 1847. Ascospores (8–)11–18 × 3–5 µm. (Type of *Leptogiopsis complicatula* Nylander, H, Nyl. 41012) Fig. 1S

This species was included in Kohlmeyer and Kohlmeyer (1979) as *Turgidosculum complicatulum* (Nylander) Kohlm. *et* Kohlm. and a list of synonyms was given. Eriksson (1981) maintains that the valid name is *Mastodia tessellata. Kohlmeyera complicatula* (Nyl.) Schatz (1980) and *Plagiostoma prasiolae* (Winter) Clauzade, Diederich *et* Roux (Clauzade *et* al. 1989) are additional synonyms.

Moana turbinulata Kohlm. *et* Volkm.-Kohlm., *Mycol. Res. 92*: 418, 1989. Ascospores 12–16 µm diam. (Holotype, IMS, J. K. 5103) Fig. 2D

Mycosphaerella apophlaeae Kohlm. *in* Kohlm. *et* Demoulin, *Bot. Mar. 24*: 13, 1981. Ascospores 15–20 × 4–5 µm. (Isotype, IMS, J. K. 4028) Fig. 2S

According to Kohlmeyer and Demoulin (1981) and Kohlmeyer and Hawkes (1983) *M. apophlaeae* can be considered as mycobiont in a mycophycobiosis with *Apophlaea lyallii* Hooker f. *et* Harvey and *A. sinclairii* Harvey *in* Hooker *et* Harvey.

Mycosphaerella ascophylli Cotton, *Trans. Br. Mycol. Soc. 3*: 96, 1907. Ascospores 15–22(–25) × 4–6 µm. (IMI 37292) Fig. 2T

Mycosphaerella pneumatophorae Kohlm., *Ber. Dtsch. Bot. Ges. 79*: 32, 1966. Ascospores 14–18(–21) × 6–8.5 µm. (Paratype, IMS, J. K. 2431) Fig. 2U

Mycosphaerella salicorniae (Auerswald) Petrak, *Hedwigia 74*: 35, 1934. Ascospores (8–)10–18 × 2–6 µm. (IMS, J. K. 1994) Fig. 2V

Mycosphaerella staticicola (Patouillard) Dias, *Mem. Soc. Broteriana 21*: 72, 1970–1971. Ascospores 12–15 × 4–6 µm. (Holotype, FH) Fig. 2W

Mycosphaerella suaedae-australis Hansford, *Proc. Linn. Soc. N. S. W. 79*: 122–123, 1954. Ascospores 18–20 × 3–3.5 µm. The anamorph is *Septoria suaedae-australis* Hansford.

This species is not illustrated because type material is not available at present for loan, according to the curator of the Waite Agricultural Research Institute, Adelaide, Australia and new collections have not been made.

Nais glitra Crane *et* Shearer, *Trans. Br. Mycol. Soc. 86*: 509, 1986. Ascospores 42–59 × 21–31 µm. (Holotype, ILLS 44900) Fig. 3A

Nais inornata Kohlm., *Nova Hedwigia 4*: 409, 1962. Ascospores 22–30 × 12–16 µm. (Isotype, IMS, J. K. 516) Fig. 3B

Nautosphaeria cristaminuta Jones, *Trans. Br. Mycol. Soc. 47*: 97, 1964. Ascospores 13–18 × 7–11 µm. (IMS, J. K. 862) Fig. 8M

Nectriella laminariae O. Eriksson, *Sven. Bot. Tidskr. 58*: 233, 1964. Ascospores 13–20 × 7–9 µm. (Isotype, UME 25269) Fig. 2AA

Nereiospora comata (Kohlm.) Jones, Johnson *et* Moss, *Bot. J. Linn. Soc. 87*: 206, 1983 [= *Corollospora comata* (Kohlm.) Kohlm.]. Ascospores (32–)35–54 × 12–17 μm, 5 (rarely 3, 4 or 6) septate; central cells fuscous. (IMS, ex Herb. A. Munk) Fig. 9G

Nereiospora cristata (Kohlm.) Jones, Johnson *et* Moss, *Bot. J. Linn. Soc. 87*: 206, 1983 [= *Corollospora cristata* (Kohlm.) Kohlm.]. Ascospores 24–38(–41) × 8–16 μm, 3 (rarely 2, 4 or 5) septate; central cells fuscous. (IMS, J.K. 3887) Fig. 9H

Nimbospora bipolaris Hyde *et* Jones, *Can. J. Bot. 63*: 611, 1985. Ascospores 18–27 × 8–12 μm. (IMS, J.K. 4844) Fig. 9F

Nimbospora effusa Koch, *Nordic J. Bot. 2*: 166, 1982. Ascospores 17–24 × 8–10 μm. (IMS, J.K. 4483) Fig. 9D

Nimbospora octonae Kohlm., *Can. J. Bot. 63*: 1122, 1985. Ascospores 22–29(–31) × 12–16(–18) μm. (Holotype, IMS, J.K. 4474) Fig. 9E

Oceanitis scuticella Kohlm., *Rev. Mycol. 41*: 194, 1977. Ascospores 60–80 × 4–6 μm. (Isotype, IMS, J.K. 2971) Fig. 7O

Ocostaspora apilongissima Jones, Johnson *et* Moss, *Bot. Mar. 26*: 354, 1983. Ascospores 15–22 × 5–7 μm. (IMS, J.K. 1903) Fig. 5C

Ondiniella torquata (Kohlm.) Jones, Johnson *et* Moss, *Bot. Mar. 27*: 136, 1984 (= *Halosphaeria torquata* Kohlm.). Ascospores 20–30 × 10–16 μm (excl. appendages). (IMS, J.K. 4434) Fig. 6G

Ophiodeira monosemeia Kohlm. *et* Volkm.-Kohlm., *Can. J. Bot. 66*: 2062–2063, 1988. Ascospores 16–21 × 6–8 μm. (Holotype, IMS, J.K. 5060a) Fig. 7I

Orcadia ascophylli Sutherland, *Trans. Br. Mycol. Soc. 5*: 151, 1915. Ascospores 34–56 × 4–7 μm. (IMS, J.K. 5373) Fig. 13E

Paraliomyces lentiferus Kohlm., *Nova Hedwigia 1*: 81–82, 1959. Ascospores 17–26(–28) × 8–12 μm. (IMS, J.K. 1956) Fig. 3M

Passeriniella obiones (Crouan *et* Crouan) Hyde *et* Mouzouras, *Trans. Br. Mycol. Soc. 91*: 179, 1988 [= *Leptosphaeria obiones* (Crouan *et* Crouan) Saccardo]. Ascospores 25–36(–38) × 9–14 μm; central cells brown. (Holotype, CO) Fig. 4P

The genus *Passeriniella* was described by Berlese (1892) with two species, *P. dichroa* (Pass.) Berl. and *P. incarcerata* Berl., without designation of a type species. Usually the first-named species (*P. dichroa*) could be considered as *species typica*. We have tried in vain to locate type material of this species and received negative answers from FH, IMI, K, NY, PAD and PAV. Because Berlese's diagnosis and illustrations of *P. dichroa* are insufficient to characterize it (what is the structure of ascomata, are asci uni- or bitunicate, are apical apparatuses present?) this must be considered a *nomen dubium*. We propose that *P. incarcerata* should serve as type species of the genus.

Berlese's (1892) original text reads: "*Passeriniella incarcerata* (B. et C.) Berl. *Sphaeria incarcerata* B. et C. North Amer. Fungi n. 960, *Leptosphaeria incarc.* Sacc. Syll. Pyr. II, p. 86 ...Ex specim. orig. a claro Cooke benevole communic. ...". Obviously, Berlese intended to transfer *Sphaeria incarcerata* to *Passeriniella* and based his diagnosis and illustrations on part of the type material obtained through Cooke. Unfortunately, the type specimen of *Sphaeria incarcerata* contains two different ascomycetes as we detected upon examination of the original material from K (Fig. 24). One species has large ascomata with a cleft-like ostiole and cylindrical, 5 septate, hyaline ascospores (Fig. 24, area marked A), whereas the second ascomycete has smaller ascomata with round ostioles, 3 septate ascospores with brown middle cells and light-colored apical cells (Fig. 24, area marked B). Evidently, the first species is the "real" *Sphaeria incarcerata* Berkeley *et* Curtis, overlooked by Berlese. He mistook the second species with 3 septate spores for *S. incarcerata* and named it *Passeriniella incarcerata*. The case is further complicated by the fact that the earliest available name for the fungus he called *P. incarcerata* is *Pleospora obiones* Crouan *et* Crouan 1867 (Gessner and Kohlmeyer 1976), therefore, *Passeriniella incarcerata* is a synonym. Because *P. obiones* does not belong to *Pleospora*, and is identical with *P. incarcerata*, the new combination *Passeriniella obiones* was necessary (Hyde and Mouzouras 1988).

Passeriniella savoryellopsis Hyde *et* Mouzouras, *Trans. Br. Mycol. Soc. 91*: 179, 1988. Ascospores 64–88 × 24–28 μm; central cells brown. (IMS, J.K. 5167) Fig. 4O

Most probably, this species does not belong in the genus *Passeriniella* because hamathecia and asci of the type species, *P. incarcerata* (= *P. obiones*), are different.

Payosphaeria minuta Leong *in* Leong, Tan, Hyde *et* Jones, *Bot. Mar. 33*: 511–512, 1990. Ascospores 6–11 × 4–7 μm. (Holotype, IMI 327472) Fig. 23A

Phaeosphaeria gessneri Shoemaker *et* Babcock, *Can. J. Bot. 67*: 1567, 1989. Ascospores 68–85 × (10–)15–20 μm, 5(–7) septate; yellowish-brown. (Isotype, DAOM 147881) Fig. 14L

Phaeosphaeria halima (Johnson) Shoemaker *et* Babcock, *Can. J. Bot. 67*: 1514, 1989 (= *Leptosphaeria halima* Johnson). Ascospores 12–18 × 5–8 µm, 3 septate; yellow-brown. (IMS, J. K. 4783) Fig. 14E

Phaeosphaeria macrosporidium (Jones) Shoemaker *et* Babcock, *Can. J. Bot. 67*: 1532, 1989 (= *Leptosphaeria macrosporidium* Jones). Ascospores 45–68(–72) × 10–14 µm (sizes from Jones 1962 and our examination of the type), 1–3 septate; hyaline to pale yellow. (Holotype, IMI 80279) Fig. 14B

Ascospores of this species are hyaline, at the most pale yellow, therefore, the placement in *Phaeosphaeria* appears to be doubtful.

Phaeosphaeria neomaritima (Gessner *et* Kohlm.) Shoemaker *et* Babcock, *Can. J. Bot. 67*: 1572, 1989 (= *Leptosphaeria neomaritima* Gessner *et* Kohlm.). Ascospores (30–)32–45 × (6–)8–14 µm, 3–5 septate; yellowish to yellow-brown. (IMS, J. K. 3384) Fig. 14J

Phaeosphaeria spartinae (Ellis *et* Everhart) Shoemaker *et* Babcock, *Can. J. Bot. 67*: 1573, 1989 (= *Leptosphaeria spartinae* Ellis *et* Everhart). Ascospores 35–40(–52) × 9–11(–14) µm, 5 septate; yellow-brown. (Holotype, NY) Fig. 14K

Until recently, this species was included in the marine mycological literature as a synonym of *Leptosphaeria albopunctata* (see remarks on *Phaeosphaeria albopunctata* under Excluded or Doubtful Species). The name *P. spartinae* applies to the fungus on *Spartina* in saltwater, whereas *P. albopunctata* occurs on *Phragmites* in freshwater.

Phaeosphaeria spartinaecola Leuchtmann *in* Leuchtmann *et* Newell, *Mycotaxon 39*, 1991 (in press). Ascospores 23–35 × 9–13 µm, 3 septate; brown to yellow-brown. (IMS, J. K. 4740) Fig. 14D

See comments for *P. typharum* under which name this fungus on *Spartina* was listed until now in the marine mycological literature.

Phaeosphaeria typharum (Desmazières) Holm, *Symb. Bot. Ups. 14*: 126, 1957. Ascospores (21–)24–30(–35) × 8–12 µm, 3 septate; reddish-brown. (BPI, Rabenhorst Herb. Mycol. II, 731) Fig. 14C

This fungus was originally described from *Typha* spp. in freshwater habitats and occurs widely on these hosts (Shoemaker and Babcock 1989). The name *P. typharum* has been applied in the marine mycological literature to a fungus on *Spartina* spp. It appears clear now that the freshwater and the marine fungus are actually different species and the latter is *P. spartinaecola*.. Ascospores in *P. typharum* have a sheath, which, according to Shoemaker and Babcock (1989), is not visible in the type material.

Pharcidia laminariicola Kohlm., *Bot. Mar. 16*: 209, 1973. Ascospores 18–25 × 6–8 µm. (Isotype, IMS, J. K. 3118) Fig. 2Q

Pharcidia rhachiana Kohlm., *Bot. Mar. 16*: 210, 1973. Ascospores 12–20 × 4–5 µm. (Isotype, IMS, J. K. 3119) Fig. 2R

Phomatospora acrostichi Hyde, *Trans. Br. Mycol. Soc. 90*: 135, 1988. Ascospores 6–7 × 2–3 µm. (Holotype, IMI 315912) Fig. 2B

Phycomelaina laminariae (Rostrup) Kohlm., *Phytopathol. Z. 63*: 350, 1968. Ascospores (18–)20–25(–28) × 6–8 µm. (IMS, J. K. 2501) Fig. 3N

Pleospora avicenniae Borse, *Curr. Sci. India 56*: 1109, 1987. Ascospores 28–40 × 12–15 µm, 5–9 transsepta, 2–3 longisepta per segment; yellow-brown. (Holotype, IMI 304217) Fig. 16H

Pleospora gaudefroyi Patouillard, *Tabulae Analyticae Fungorum*, Paris, 2 ser. p. 40, No. 602, 1886. Ascospores (20–)30–46(–56) × (8–)14–20(–25) µm, 7–9 transsepta, 1–3 longisepta per segment; yellow- to golden-brown. (IMS, J. K. 3520) Fig. 16I

Pleospora gracilariae Simmons *et* Schatz *in* Simmons, *Mem. N. Y. Bot. Gard. 49*: 305, 1989 (anamorph: *Stemphylium gracilariae*). Ascospores 28–29.5 × 13–13.5 µm, 6–7 transsepta, 1 (rarely 2) longisepta in each segment, ochraceous. (Type culture EGS 37-073) Fig. 23F

The ascospores have a distinct gelatinous sheath that was not included in the original protologue. The sheath is visible under Nomarski differential interference optics and can also be demonstrated by using India ink. This species is difficult to distinguish from *P. pelvetiae*.

Pleospora pelagica Johnson, *Mycologia 48*: 504, 1956. Ascospores 35–52 × 10–15 µm, 7–9 transsepta, 1 longiseptum in 1 to 8 segments; yellow-brown. (IMS, J. K. 3680) Fig. 16G

Pleospora pelvetiae Sutherland, *New Phytol. 14*: 41–42, 1915. Ascospores 25–35 × 12–17 µm, 6–7 transsepta, several longisepta in each segment; yellowish-brown to brown; ascospores have a gelatinous deliquescing sheath. (Neotype, PC, Desmazières, *Pl. Crypt. France II*, Fasc. 36, No. 1772) Fig. 16F

Pleospora spartinae (Webster *et* Lucas) Apinis *et* Chesters, *Trans. Br. Mycol. Soc. 47*: 432, 1964 (see anamorph). Ascospores 24–38 × 10–13 µm, 5 transsepta, one longiseptum in the third or fourth cell from the top; yellowish-brown. (IMS, Herb. R. V. Gessner) Fig. 16E

Pleospora triglochinicola Webster, *Trans. Br. Mycol. Soc. 53*: 481, 1969 (anamorph: *Stemphylium triglochinicola*). Ascospores 45–65 × 16–25 μm, typically with 7 transsepta, 1–3 longisepta in most segments; golden- to olive-brown. (IMS, Herb. A. Munk) Fig. 16D

Polystigma apophlaeae Kohlm., *Bot. Mar. 24*: 13, 1981. Ascospores 15–18 × 5–6 μm. (Isotype, IMS, J.K. 4027) Fig. 1Q

Pontogeneia calospora (Patouillard) Kohlm., *Bot. Jahrb. 96*: 205, 1975. Ascospores (54–)62–85 × 8–12(–14) μm, 3–6 septate. (Holotype, P) Fig. 15G

Pontogeneia codiicola (Dawson) Kohlm. *et* Kohlm., *Marine Mycology*, Academic Press, p. 350, 1979. Ascospores 52–79 × 20–28 μm, 3 septate. (Holotype, AHFH, Dawson 4766b) Fig. 15J

Pontogeneia cubensis (Hariot *et* Patouillard) Kohlm., *Bot. Jahrb. 96*: 207, 1975. Ascospores 215–325 × 14–17(–21) μm, (10–)12–13 septate. (Holotype, PC) Fig. 15 I

Pontogeneia enormis (Patouillard *et* Hariot) Kohlm., *Bot. Jahrb. 96*: 208, 1975. Ascospores 280–350 × 12–14 μm, 4–5 septate.

This species is known only from the original description in Spain on *Halopteris scoparia*. It is not illustrated because we were unable to locate the type material.

Pontogeneia erikae Kohlm. *in* Kohlm. *et* Demoulin, *Bot. Mar. 24*: 16, 1981. Ascospores 18–26(–28) × 8–10 μm, 2 septate near the apices. (Isotype, IMS, J.K. 4015) Fig. 15H

Pontogeneia padinae Kohlm., *Bot. Jahrb. 96*: 201, 1975. Ascospores 99–167(–172) × 10–12 μm, 6–9(–10) septate. (Isotype, IMS, J.K. 3541) Fig. 15F

Pontogeneia valoniopsidis (Cribb *et* Cribb) Kohlm., *Bot. Jahrb. 96*: 209, 1975. Ascospores 72–100 × 21–30 μm, 2–5 septate.

This species is not illustrated because no type or authentic specimens have become available.

Pontoporeia biturbinata (Durieu *et* Montagne) Kohlm., *Nova Hedwigia 6*: 5–6, 1963. Ascospores 66–90 × 32–44 μm, dark- to blackish-brown. (IMS, J.K. 819) Fig. 4J

Pyrenocollema pelvetiae (Sutherland) Hawksworth, *Bot. J. Linn. Soc. 96*: 10, 1988 [= *Leiophloea pelvetiae* (Sutherland) Kohlm. *et* Kohlm.]. Ascospores 12–16 × 5–6.5 μm. (Holotype of *Plowrightia pelvetiae* Fragoso, MA) Fig. 2DD

Quintaria lignatilis (Kohlm.) Kohlm. *et* Volkm.-Kohlm., *Bot. Mar. 34*: 35, 1991 (= *Trematosphaeria lignatilis* Kohlm.). Ascospores 51–80 × 14–20(–22) μm, 5 septate, hyaline. (IMS, J.K. 5074) Fig. 13G

See Appendix for the description of *Quintaria* and the proposed new combination.

Remispora crispa Kohlm., *Can. J. Bot. 59*: 1317, 1981. Ascospores 22–34 × 8–12 μm. (Isotype, IMS, J.K. 4066) Fig. 5J

Remispora galerita Tubaki, *Publ. Seto Mar. Biol. Lab. 15*: 362, 1968 [= *Halosphaeria galerita* (Tubaki) I. Schmidt]. Ascospores (16-)20–28(–35) × 7–12 (–15) μm. (IMS, J.K. 2686) Fig. 5G

Remispora maritima Linder *in* Barghoorn *et* Linder, *Farlowia 1*: 410, 1944. Ascospores 18–30(–32) × 8–13 μm. (IMS, J.K. 3317) Fig. 5F

Remispora pilleata Kohlm., *Nova Hedwigia 6*: 319, 1963. Ascospores 24–34(–36) × 12–20 μm. (Isotype, IMS, J.K. 914) Fig. 5K

Remispora quadriremis (Höhnk) Kohlm., *Nova Hedwigia 2*: 332, 1960. Ascospores 18–20(–34) × 8–12(–16) μm. (IMS, ex Herb. I. Schmidt) Fig. 5H

Remispora spinibarbata Koch, *Nordic J. Bot. 8*: 517, 1989. Ascospores 18–26 × 8–12 μm. (Holotype, CP, Koch 581/10) Fig. 5E

Remispora stellata Kohlm., *Nova Hedwigia 2*: 334, 1960. Ascospores 24–30 × 8–12 μm. (Holotype, B, J.K. 156) Fig. 5I

Rhizophila marina Hyde *et* Jones, *Mycotaxon 34*: 527, 1989. Ascospores 20–28 × 7–10 μm; yellowish to yellow-brown. (IMS, J.K. 4458) Fig. 2F

Savoryella lignicola Jones *et* Eaton, *Trans. Br. Mycol. Soc. 52*: 162, 1969. Ascospores 24–34(–43) × 8–13 μm; central cells brown. (Holotype, IMI 129784) Fig. 4R

Savoryella paucispora (Cribb *et* Cribb) Koch, *Nordic J. Bot. 2*: 169, 1982 (= *Leptosphaeria paucispora* Cribb *et* Cribb). Ascospores 36–50 × 13–16 μm; central cells brown. (IMS, J.K. 4478) Fig. 4Q

Spathulospora adelpha Kohlm., *Mycologia 65*: 615–617, 1973. Ascospores 70–104 × 16–23 μm. (Paratype, IMS, J.K. 2981) Fig. 1M

Spathulospora antarctica Kohlm., *Mycologia 65*: 619–620, 1973. Ascospores 40–92 × 4–12 μm. (IMS, J.K. 5375) Fig. 1P

Spathulospora calva Kohlm., *Mycologia 65*: 622–623, 1973. Ascospores 45–56 × 14–20 µm. (IMS, J. K. 4233) Fig. 1O

Spathulospora lanata Kohlm., *Mycologia 65*: 625–626, 1973. Ascospores 39–62(–74) × 8–12(–16) µm. (IMS, J. K. 3488) Fig. 1N

Spathulospora phycophila Cavaliere *et* Johnson, *Mycologia 57*: 927–928, 1965. Ascospores 80–110 × 10–13 µm. (Isotype, BPI) Fig. 1L

Sphaerulina oraemaris Linder *in* Barghoorn *et* Linder, *Farlowia 1*: 413, 1944. Ascospores 26–32 × 5–8 µm. (Holotype, FH) Fig. 13B

Swampomyces armeniacus Kohlm. *et* Volkm.-Kohlm., *Bot Mar. 30*: 200, 1987. Ascospores 13–20 × 6–9 µm; hyaline to yellowish. (Holotype, IMS, J. K. 4387) Fig. 2K

Thalassoascus cystoseirae (Ollivier) Kohlm., *Mycologia 73*: 837, 1981 [= *Melanopsamma tregoubovii* (Oll.) Oll. var. *cystoseirae* Oll.]. Ascospores 29–48 × 19–34 µm. (IMS, J. K. 835) Fig. 3G

Thalassoascus lessoniae Kohlm., *Mycologia 73*: 837, 1981. Ascospores (30–)32–40 × 18–26 µm. (Isotype, IMS, J. K. 4080) Fig. 3F

Thalassoascus tregoubovii Ollivier, *C. R. Hebd. Séances Acad. Sci. 182*: 1348–1349, 1926. Ascospores 40–45 × 28–30 µm.

This species from *Zanardinia* and *Aglaozonia* is not illustrated because we have been unable to locate type or authentical material. A similar fungus with stalked ascomata on *Cystoseira* is treated as a separate species, viz. *T. cystoseirae.*

Thalassogena sphaerica Kohlm. *et* Volkm.-Kohlm., *Syst. Ascom. 6*: 225, 1987. Ascospores 12–16 µm diam, subglobose to ellipsoidal. (Holotype, IMS, J. K. 5044a) Fig. 2C

Ascospores in the type material were most spherical, but later collections contained both, subglobose and ellipsoidal, spores. Sometimes asci may produce less than 8 ascospores which can be up to 22 µm long.

Torpedospora ambispinosa Kohlm., *Nova Hedwigia 2*: 336, 1960. Ascospores 17–24 × 6–10 µm. (Isotype, IMS, J. K. 178) Fig. 8O

Torpedospora radiata Meyers, *Mycologia 49*: 496, 1957. Ascospores 30–52 × 4–9 µm. (IMS, J. K. 5089) Fig. 8P

Trailia ascophylli Sutherland, *Trans. Br. Mycol. Soc. 5*: 149, 1915. Ascospores 90–110 µm long, 3–4 µm diam at the upper end. (IMS, J. K. 5373) Fig. 13F

Trematosphaeria mangrovei Kohlm., *Mycopath. Mycol. Appl. 34*: 1–2, 1968 (as *'mangrovis'*). Ascospores 30–41 × 10–13(–16) µm, 3 septate; fuscous. (Isotype, IMS, J. K. 1845a) Fig. 14H

Trematosphaeria striatispora Hyde (as *'striataspora'*), *Bot. J. Linn. Soc. 98*: 142, 1988. Ascospores 31–38 × 6–9 µm, 3 (rarely up to 6) septate; brown. (Holotype, IMI 312390) Fig. 14I

Trichomaris invadens Hibbits, Hughes *et* Sparks, *Can. J. Bot. 59*: 2123–2125, 1981. Ascospores 15–22 × 4.5–5 µm, 1–3 septate. (IMS, J. K. 5159) Fig. 6M

Turgidosculum ulvae (Reed) Kohlm. *et* Kohlm., *Bot. Jahrb. 92*: 429, 1972. Ascospores (8–)10–14 × 4–7 µm. (Lectotype, FH) Fig. 1R

Verrucaria cribbii Rogers, *Austr. Syst. Bot. 1*: 181, 1988. Ascospores 7–10 × 5–6 µm. (IMS, J. K. 5345) Fig. 1U

Verruculina enalia (Kohlm.) Kohlm. *et* Volkm.-Kohlm., *Mycol. Res. 94*: 689, 1990 (= *Didymosphaeria enalia* Kohlm.). Ascospores 16–23 × 7–11 µm; dark brown. (Paratype, IMS, J. K. 1802) Fig. 4C

Wettsteinia marina (Ellis *et* Everhart) Shoemaker *et* Babcock, *Can. J. Bot. 67*: 1596, 1989 (= *Leptosphaeria marina* Ellis *et* Everhart). Ascospores 50–65(–70) × 10–14 µm, 1–3 septate; hyaline to yellow. (IMS, J. K. 4762) Fig. 14A

A possible synonym is *Phaeosphaeria macrosporidium.*

Zopfiella latipes (Lundqvist) Malloch *et* Cain, *Can. J. Bot. 49*: 876, 1971. Ascospores with subapical germ pore, upper cell 16–22(–25) × 10–13(–15) µm, olivaceous to brown; basal cell up to 7–8 µm diam, hyaline. (ILLS 35139) Fig. 4N

Zopfiella marina Furuya *et* Udagawa, *J. Jpn. Bot. 50*: 249, 1975. Ascospores with apical germ pore, upper cell 14–20 × 10–14 µm, olivaceous to dark brown; basal cell up to 4 µm diam, hyaline. (Type culture, IMS, Udagawa NHL2731) Fig. 4M

2. Basidiomycotina

Digitatispora lignicola Jones, *Mycotaxon 27*: 155, 1986. Basidiospores tetraradiate, apical arms 7–23 × 3–8 µm. (Holotype, IMI 292870) Fig. 19B

Digitatispora marina Doguet, *C. R. Hebd. Séances Acad. Sci. 254*: 4338, 1962. Basidiospores tetraradiate, apical arms 26–41(–45) × 2–4 µm. (IMS, J. K. 905) Fig. 19A

Halocyphina villosa Kohlm. *et* Kohlm., *Nova Hedwigia 9*: 100, 1965. Basidiospores 8 – 10 × 7 – 10 µm. (IMS, J. K. 5067) Fig. 19D

Melanotaenium ruppiae G. Feldmann, *Rev. Gén. Bot. 66*: 36, 1959. Chlamydospores 12 – 25 × 12 – 13 µm; dark fuscous.

The species is not illustrated because type or authentic material has not become available.

Mycaureola dilseae Maire *et* Chemin, *C. R. Hebd. Séances Acad. Sci. 175*: 321, 1922. Basidiospores 105 – 118 × 2.5 – 3 µm. (IMS, J. K. 5158) Fig. 19E

Porter and Farnham (1986) reexamined this species that was originally described as an ascomycete and determined that it belongs to the basidiomycotina.

Nia vibrissa Moore *et* Meyers, *Mycologia 51*: 874, 1959. Basidiospores 8 – 15 × 6 – 11 µm. (IMS, J. K. 5071) Fig. 19C

3. Anamorphs

Allescheriella bathygena Kohlm., *Rev. Mycol. 41*: 199 – 201, 1977. Conidia 16.5 – 24(– 26) × 6.5 – 9 µm; light brown. (Isotype, IMS, J. K. 2469) Fig. 20A

Alternaria spp. Conidia muriform, in chains; brown. (IMS, J. K. 5184) Fig. 20R

No obligate marine species is known, but conidia of this genus are occasionally found on detritus along the shore.

Amarenographium metableticum (Trail) Eriksson, *Mycotaxon 15*: 199, 1982 (= *Camarosporium metableticum* Trail, teleomorph: *Amarenomyces ammophilae*). Conidia 22 – 30 × 9 – 13 µm (*fide* Eriksson 1982), 22 – 42(– 44) × (8 –)10 – 17 µm (*fide* Nag Raj 1989), 3 – 7(– 9) transsepta, 0 – 2 longisepta per segment; yellowish to fuscous. (B, Shear N. Y. Fungi 379) Fig. 19U

Nag Raj (1989) found additional microconidia (3.5 – 6.5 × 2.5 – 3 µm) restricted to the neck and ostiolar region of the conidiomata.

Anguillospora marina Nakagiri *et* Tubaki, *Mycologia 75*: 488, 1983 (teleomorph: *Lindra obtusa*). Conidia 150 – 255(– 312) × 2.5 – 4 µm, 9 – 13(– 19) septate. (Holotype, Nakagiri, TKB-F-5008-3) Fig. 22B

Ascochyta obiones (Jaap) Buchanan, *Mycol. Pap. 156*: 28, 1987 [= *Ascochytula obiones* (Jaap) Diedicke]. Conidia 9 – 12 × 4 – 5 µm; pale yellowish to brownish. (IMS, J. K. 2903) Fig. 19S

Ascochyta salicorniae P. Magnus *in* Jaap, *Schr. Naturwiss. Ver. Schleswig-Holstein 12*: 345, 1902. Conidia 10 – 20 × 4 – 7 µm; hyaline to yellowish or light brown. (IMS, J. K. 3342) Fig. 19R

Asteromyces cruciatus Moreau *et* Moreau *ex* Hennebert, *Can. J. Bot. 40*: 1213, 1962. Conidia 9 – 20 × 4 – 9 µm, dispersed in nature as aggregates of up to 13 cells; brown. (IMS, J. K. 3678) Fig. 20G

Blodgettia confervoides Harvey, *Smithson. Contrib. Knowl. 10*: 48, 1858. Chlamydospores 35 – 150 × 8 – 33 µm, 1 – 7 septate; hyaline to yellow-brown. (IMS, J. K. 3853) Fig. 20K

Hawksworth (1987) discusses the nomenclatural status of the species, maintaining that *B. confervoides* is the valid name that should be applied, versus *B. bornetii* Wright.

Botryophialophora marina Linder *in* Barghoorn et Linder, *Farlowia 1*: 404, 1944. Conidia 2 – 3.5 µm diam. (Holotype, FH) Fig. 20D

Camarosporium palliatum Kohlm. *et* Kohlm., *Marine Mycology*, Academic Press, pp. 519 – 520, 1979. Conidia 20 – 34 × 9 – 20 µm, (3 –)5(– 6) transsepta, 1 – 3 longisepta; yellowish-brown to dark brown. (Paratype, IMS, J. K. 3521) Fig. 19V

Camarosporium roumeguerii Sacc., *Michelia 2*: 112 – 113, 1880. Conidia 10 – 20(– 22) × 7 – 13 µm, 3 transsepta, 1 – 2 longisepta; golden- to olive-brown. (IMS, J. K. 3896) Fig. 19W

Cirrenalia basiminuta Raghu-Kumar *et* Zainal *in* Raghu-Kumar, Zainal *et* Jones, *Mycotaxon 31*: 163, 1988. Conidia 3 – 5 septate, ½ times contorted, apical cell 9 – 16 µm diam; pale brown. (Isotype, IMS, Zainal and Jones) Fig. 21A

Cirrenalia fusca I. Schmidt, *Feddes Repert. 80*: 110, 1969. Conidia (2 –)3(– 4) septate, ½ – 1½ times contorted, apical cell 11 – 22 µm diam; brown to dark brown. (Paratype, IMS, I. Schmidt) Fig. 21B

Cirrenalia macrocephala (Kohlm.) Meyers *et* Moore, *Am. J. Bot. 47*: 347, 1960. Conidia 2 – 7 septate, ¼ – 1 time contorted, apical cell 6 – 14(– 17) µm diam; reddish fuscous. (IMS, J. K. 5188) Fig. 21C

Cirrenalia pseudomacrocephala Kohlm., *Mycologia 60*: 266, 1968. Conidia 3 – 5(– 6) septate, ½ – 1¼ times contorted, apical cell 16 – 20 µm diam; fuscous to grayish-brown. (Isotype, IMS, J. K. 2447b) Fig. 21D

Cirrenalia pygmea Kohlm., *Ber. Dtsch. Bot. Ges. 79*: 35, 1966. Conidia 3 – 4 septate, ½ – 1 time contorted, apical cell 16 – 23 µm diam; black or fuscous. (IMS, J. K. 1889) Fig. 21F

Cirrenalia tropicalis Kohlm., *Mycologia 60*: 267, 1968. Conidia 6 – 12 septate, 1 – 1½ times contorted, apical cell 10 – 20 µm diam; umber to reddish-brown. (Isotype, IMS, J. K. 1888b) Fig. 21E

Cladosporium algarum Cooke *et* Massee *in* Cooke, *Grevillea 16*: 80, 1888. Conidia 8–24 × 4–8 µm, 0–2(–4) septate; olive-brown. (Lectotype, K) Fig. 20H

Clavatospora bulbosa (Anast.) Nakagiri *et* Tubaki, *Bot. Mar. 28*: 489, 1985 (= *Clavariopsis bulbosa* Anast., teleomorph: *Corollospora pulchella*). Conidia tetraradiate, divergent upper arms 20–70 × 4–6(–8) µm; light brown. (IMS, J. K. 2577) Fig. 22H

Clavatospora stellatacula Kirk *ex* Marvanova *et* Nilsson, *Trans. Br. Mycol. Soc. 57*: 531, 1971. Conidia 7–9 × 6–7 µm. (Isotype, IMS, from culture Kirk F83) Fig. 20B

Coniothyrium obiones Jaap, *Schr. Naturwiss. Ver. Schleswig-Holstein 14*: 29, 1907. Conidia 4–8 × 4–6 µm; olivaceous to light brown. (BPI, Jaap, Fungi sel. exs. 98) Fig. 19N

Cumulospora marina I. Schmidt, *Mycotaxon 24*: 420, 1985 (= *Vesicularia marina* Schmidt). Conidia 42–85 × 42–85 µm, outer cells 15–30 µm diam; gray to brownish. (Paratype, JE, I. Schmidt 169) Fig. 21K

Because the generic name *Vesicularia* Schmidt was preoccupied by earlier homonyms, Schmidt (1985) proposed *Cumulospora*. *Basramyces marinus* Abdullah, Abdulkadder *et* Goos (Abdullah *et al.* 1989) is a synonym.

Cytospora rhizophorae Kohlm. *et* Kohlm., *Mycologia 63*: 847–848, 1971. Conidia 3–6 × 1–1.5 µm, exuded cirrhi orange-colored. (Isotype, IMS, J. K. 2693) Fig. 19L

Dendryphiella arenaria Nicot, *Rev. Mycol. 23*: 93, 1958. Conidia 9–20 × 4–6 µm, 1–3 septate; pale golden-brown. (IMS, J. K. 1999) Fig. 20J

Dendryphiella salina (Sutherland) Pugh *et* Nicot, *Trans. Br. Mycol. Soc. 47*: 266, 1964. Conidia 14–75 × 6–10 µm, 1–9(–11) septate; pale brown to light olivaceous. (IMS, J. K. 2953) Fig. 20I

Dictyosporium pelagicum (Linder) G. C. Hughes *ex* Johnson et Sparrow, *Fungi in Oceans and Estuaries*, Cramer, p. 391, 1961. Conidia 12–66 × 9–28(–36) µm; dark brown to black. (Holotype, FH) Fig. 20Q

Dinemasporium marinum Nilsson, *Bot. Not. 110*: 321, 1957. Conidia 11–15 × 2.5–3.5(–6.5) µm; hyaline to very slightly greenish. (Holotype, UPS 11450) Fig. 19Q

Diplodia oraemaris Linder *in* Barghoorn *et* Linder, *Farlowia 1*: 403, 1944. Conidia 6–8 × 3.5–4.5 µm; yellowish-brown. (IMS, J. K. 172) Fig. 19T

Epicoccum spp. Sporodochial; conidia muriform, verrucose; dark golden-brown. (IMS, J. K. 1857) Fig. 21J

No obligate marine species of *Epicoccum* has been described; representatives of this genus occur on washed-up algae and other detritus on beaches.

Exserohilum spp. Conidia cylindrical to ellipsoidal, with a protruberant basal scar, with pseudosepta, thick-walled; brown; apical cells somewhat lighter colored, separated from adjoining cells by dark-colored septa. (IMS, J. K. 5374) Fig. 20L

Traditionally, marine collections of fungi with conidia like the one illustrated in Fig. 20L have been identified as *Drechslera halodes* (Drechsler) Subram. *et* Jain or *Exserohilum halodes* (Drechsler) Leonard *et* Suggs (Kohlmeyer and Kohlmeyer 1979). Leonard (1976) considers *E. halodes* to be synonymous with *E. rostratum* (Drechsler) Leonard *et* Suggs. For the time being we prefer to list species similar to *E. halodes* from marine habitats as *Exserohilum* sp. until thorough studies have clarified the identity of these collections.

Gloeosporidina cecidii (Kohlm.) Sutton, *The Coelomycetes*, Commonwealth Mycol. Inst. Kew, p. 517, 1980 (= *Sphaceloma cecidii* Kohlm.). Conidia 3–5 × 2–2.5 µm (excluding appendages). (Isotype, IMS, J. K. 3022) Fig. 19K

Halonectria milfordensis Jones (anamorph), *Trans. Br. Mycol. Soc. 48*: 287, 1965. Conidia 10–16 × 0.5 µm. (IMS, J. K. 1903) Fig. 19O

Humicola alopallonella Meyers *et* Moore, *Am. J. Bot. 47*: 346, 1960. Conidia 10–22(–38) × 8–18 µm, 1 or 2 celled; fuscous. (Holotype, FH) Fig. 20M

The illustrations and sizes of *H. alopallonella* are based on type material, i. e., conidia developed in culture. When this fungus is collected in nature, the conidia are usually smaller and measure about 11–20 × 5–9 µm.

Macrophoma spp. See remarks under *Phoma*. (IMS, J. K. 4605) Fig. 19I

Monodictys pelagica (Johnson) Jones, *Trans. Br. Mycol. Soc. 46*: 138, 1963. Conidia 15–41(–44) × 12–37 µm, muriform; black. (IMS, J. K. 1919) Fig. 21G

Orbimyces spectabilis Linder *in* Barghoorn *et* Linder, *Farlowia 1*: 404–405, 1944. Conidial basal cell 24–42 × 18–37 µm; dark brown. (IMS, J. K. 5288) Fig. 22G

Periconia abyssa Kohlm., *Rev. Mycol. 41*: 202, 1977. Conidia 16–20 µm diam; light brown. (Isotype, IMS, J. K. 3811) Fig. 20E

Periconia prolifica Anastasiou, *Nova Hedwigia 6*: 260, 1963 (teleomorph: *Halosphaeria cucullata*). Conidia 6–13(–20) µm; light to dark brown. (IMS, J.K. 4838) Fig. 20F

Phialophorophoma litoralis Linder *in* Barghoorn *et* Linder, *Farlowia 1*: 403, 1944. Conidia 2.5–4.5 × 1.5–2 µm. (Holotype, FH, Barghoorn 24) Fig. 19F

Phoma laminariae Cooke *et* Massee *in* Cooke, *Grevillea 18*: 53, 1889. Conidia 6–10(–12) × 3–4 µm, 0–1 septate. (Holotype, K) Fig. 19G

Phoma suaedae Jaap, *Schr. Naturwiss. Ver. Schleswig-Holstein 14*: 27, 1907. Conidia 5–8(–10) × 3–5 µm. (IMS, J.K. 3168) Fig. 19H

Phoma spp. (IMS, J.K. 5069) Fig. 19J

Coelomycetes resembling *Phoma* and *Macrophoma* are frequently encountered in the marine and estuarine environment on a variety of substrates. Some of these may be anamorphs of ascomycetes, others may be spermogonia. No effort has been made so far to separate and describe the different species.

Pleospora spartinae (Webster *et* Lucas) Apinis *et* Chesters (anamorph). Conidia 36–72 × 3–6 µm, 5–7 septate; yellow to pale brown. (IMS, J.K. 4738) Fig. 19X

Rhabdospora avicenniae Kohlm. *et* Kohlm., *Mycologia 63*: 851, 1971. Conidia 9–12 × 1.5–2 µm. (Isotype, IMS, J.K. 2740) Fig. 19P

Robillarda rhizophorae Kohlm., *Can. J. Bot. 47*: 1483, 1969. Conidia 10–14 × 3–4.5 µm. (Isotype, IMS, J.K. 2551) Fig. 19M

Sigmoidea luteola Nakagiri *et* Tubaki, *Trans. Mycol. Soc. Jpn. 23*: 102–103, 1982 (teleomorph: *Corollospora luteola*). Conidia 106–222 × 4–8 µm, 7–13(–18) septate. (Holotype, Nakagiri, TKB-F-5002) Fig. 22D

Sigmoidea marina Haythorn, Jones *et* Harrison, *Trans. Br. Mycol. Soc. 74*: 615, 1980. Conidia (80–)110–180(–231) × 2–6 µm, 7–11 septate. (Holotype, IMI 239282) Fig. 22C

Sporidesmium salinum Jones, *Trans. Br. Mycol. Soc. 46*: 135, 1963. Conidia 120–296 × 11–34 µm, 2–8 septate; brown. (Isotype, IMS, from Herb. E.B.G. Jones) Fig. 22A

Stagonospora haliclysta Kohlm., *Bot. Mar. 16*: 213, 1973. Conidia 20–28 × 3.5–4.5 µm. (Isotype, IMS, J.K. 3151) Fig. 19Z

Stagonospora sp., Kohlmeyer *et* Kohlmeyer, *Marine Mycology*, Academic Press, pp. 540–541, 1979. Conidia 38–61(–76) × (4–)6–7 µm, (5–)7 septate; hyaline to light brown. (IMS, J.K. 3386) Fig. 19Y

This species from *Spartina* spp. has remained unnamed thus far (see Kohlmeyer and Kohlmeyer 1979).

Stemphylium gracilariae Simmons, *Mem. N.Y. Bot. Gard. 49*: 305–306, 1989 (teleomorph: *Pleospora gracilariae*). Conidia 22–31 × 13–20 µm, 3–4 transsepta, 1–2 longisepta in each segment; medium golden-brown. (Type culture EGS 37-073) Fig. 23G

Stemphylium triglochinicola Sutton *et* Pirozynski, *Trans. Br. Mycol. Soc. 46*: 519, 1963 (teleomorph: *Pleospora triglochinicola*). Conidia 40–82 × 18–46 µm, up to twelve transsepta, 1–4 longisepta in each segment; greenish-brown to black. (Paratype, IMI 95209) Fig. 21L

Trichocladium achrasporum (Meyers *et* Moore) Dixon *in* Shearer *et* Crane, *Mycologia 63*: 244, 1971 (teleomorph: *Halosphaeriopsis mediosetigera*). Conidia (15–)20–34(–45) × (8–)10–24 µm, 2–5 septate; dark brown. (Holotype, FH) Fig. 20N

Trichocladium constrictum Schmidt, *Nat. Naturschutz Mecklenburg 12*: 114, 1974; *Mycotaxon 24*: 419–420, 1985. Conidia 25–43 µm long, apical cell 13–20 µm diam, 2–4 septate; reddish-brown. (Holotype, JE, Schmidt 41) Fig. 20P

Trichocladium lignincola Schmidt, *Nat. Naturschutz Mecklenburg 12*: 116, 1974; *Mycotaxon 24*: 420, 1985. Conidia 25–32 × 12–14(–17) µm, 2–5(–6) septate; light to middle brown. (Holotype, JE, Schmidt 238) Fig. 20O

Tubercularia pulverulenta Spegazzini, *An. Soc. Cient. Argent. 13*: 32–33, 1882. Conidia 3–4 × 1.5–2 µm. (IMS, J.K. 3514) Fig. 20C

Varicosporina prolifera Nakagiri, *Trans. Mycol. Soc. Jpn. 27*: 198, 1986 (teleomorph: *Corollospora intermedia*). Conidia with 2–4(–7) septa in the main axis, 1–3(–4) in the first branch, 1–3 in the second branch, 1–2 in the third branch. (Holotype, Nakagiri, TKB-F-5095) Fig. 22F

Varicosporina ramulosa Meyers *et* Kohlm., *Can. J. Bot. 43*: 916, 1965. Conidia with 1–2(–4) septa in the main axis, 1–3 in the first branch, 1–2 in the second branch. (Paratype, IMS, J.K. 3995) Fig. 22E

This species forms sclerocarps, *i.e.* sterile survival structures similar to ascomata of *Corollospora* spp. (Kohlmeyer and Charles 1981).

Zalerion maritimum (Linder) Anastasiou, *Can. J. Bot. 41*: 1136, 1963 [according to Nakagiri (1984) an anamorph of *Lulworthia uniseptata*]. Conidia 4–27(–35) septate, 1–5(–6) times coiled, conidial filament (4–)6–12 µm diam; fuscous or almost black. (Nakagiri, TBK-F-5051) Fig. 21H

Zalerion varium Anastasiou, *Can. J. Bot. 41*: 1136, 1963. Conidial coil irregular, 15–65 × 14–56 µm diam, composed of about 10–30 cells; brown to dark brown. (IMS, J. K. 4081) Fig. 21I

4. Rejected Names, Doubtful and Excluded Species

Ceriosporopsis hamata Höhnk, *Veroeff. Inst. Meeresforsch. Bremerhaven 3*: 211, 1955 [*Halosphaeria hamata* (Höhnk) Kohlm., Can. J. Bot. 50: 1956, 1972; = *Remispora hamata* (Höhnk) Kohlm., *Ber. Dtsch. Bot. Ges.* 74: 305, 1961].

This species most probably belongs to *Halosarpheia*, a genus with 12 described and probably numerous unnamed species. Because type material of *C. hamata* does not exist, Shearer and Crane (1980) lectotypified it by the original protologue and illustrations. However, we feel that the original description is not sufficient to differentiate it from other taxa and it must be considered a *nomen dubium*.

Cremasteria cymatilis Meyers *et* Moore, *Am. J. Bot. 47*: 348, 1960.

The doubtful nature of this species was discussed in detail by Kohlmeyer and Kohlmeyer (1979). We consider *C. cymatilis* a *nomen dubium* because similar catenulate chlamydospores occur as anamorphs in many marine ascomycetes.

Cylindrodendrum album var. *paralion* Summerbell, Seifert *et* MacKinnon, *Can. J. Bot. 67*: 573, 1989.

For the time being, we exclude this species that has been isolated from incubated *Fucus distichus* L. until its occurrence in the marine environment has been confirmed.

Halosarpheia indica Borse, *nomen nudum*.

This name was listed by Hyde et al. (1986), but has not been published so far.

Massariosphaeria typhicola (Karsten) Leuchtmann, *Sydowia* 37: 168, 1984 (1985) (= *Leptosphaeria typhicola* Karsten).

A fungus closely resembling *M. typhicola* occurs on *Spartina* and *Juncus* (Kohlmeyer and Kohlmeyer 1979), however, the identity of this saltmarsh species with the fungus from freshwater habitats on *Juncus, Phalaris* and *Typha* needs to be confirmed.

Ophiobolus australiensis Johnson *et* Sparrow, *Fungi in Oceans and Estuaries*, Cramer, p. 419, 1961 (= *O. littoralis* Cribb *et* Cribb).

This fungus has not been collected again after its original description from Australia. We have seen only a squash mount where the ascospores are reminescent of *Bathyascus* spp. We consider *O. australiensis* a *nomen dubium* until new collections provide material for detailed examinations.

Papulaspora halima Anastasiou, *Nova Hedwigia 6*: 266, 1963. Papulaspores 35–870 µm diam, forming irregular, subglobose accumulations of cells; black.

This member of the "Mycelia sterilia" is uncharacteristic and therefore poorly circumscribed.

Phaeosphaeria albopunctata (Westendorp) Shoemaker *et* Babcock, *Can. J. Bot. 67*: 1566, 1989 [= *Leptosphaeria albopunctata* (Westendorp) Sacc.]. (IMS, J. K. 4803)

This species is clearly a fungus from freshwater habitats on *Phragmites communis* Trin. The name *Leptosphaeria albopunctata* was first used by Ellis and Everhart (1892) for a species on *Spartina* and *Juncus*. In the marine mycological literature Johnson (1956) applied the name to a species with yellow, 5–7 septate ascospores on *Spartina alterniflora*. The *Spartina* fungus is listed now as *Phaeosphaeria spartinae*.

Phoma marina Lind, *Danish Fungi as Represented in the Herbarium of E. Rostrup*, Nordisk Forlag, Copenhagen, p. 214, 1913.

This fungus represents most probably the spermatial state of *Lautitia danica*.

Podospora inquinata Udagawa *et* Ueda, *Mycotaxon 22*: 400, 1985.

This ascomycete was isolated from marine sediment and could be a facultative marine species. It is excluded from our list because its actual growth in the marine environment has not been demonstrated as yet.

Septoria ascophylli Melnik *et* Petrov, *Nov. Sist. Niz. Rast.* p. 211, 1966.

This fungus is most probably the spermatial state of *Mycosphaerella ascophylli*.

Septoria thalassica Spegazzini, *An. Mus. Nac. Buenos Aires 20*: 387, 1910.

The hosts, *Distichlis* spp., occur above the high-water mark. Therefore, we exclude this species that cannot be considered marine in the strict sense.

Sphaerulina albispiculata Tubaki, *Publ. Seto Mar. Biol. Lab. 15*: 366–367, 1968.

As discussed by Kohlmeyer and Kohlmeyer (1979), the taxonomic position of this species is not clear and it has not been verified up to date that *S. albispiculata* is a marine species *sensu stricto*.

Splanchnonema britzelmayriana (Rehm) Boise, *Mycotaxon 22*: 480, 1985 (= *Melanomma megalosporum* var. *britzelmayrianum* Rehm).
For the time being, we exclude this species from the list of marine fungi because its occurrence in marine habitats is doubtful. If fungi similar to *S. britzelmayriana* should be collected in marine localities, they need to be carefully compared with type material of the species.

Xylaria psamathos Boise, *Mycologia 74*: 467, 1982.
Although *X. psamathos* occurs on buried culms of *Spartina patens* (Ait.) Muhl., a saltwater marsh plant, it cannot be considered a marine fungus *sensu stricto*.

5. Appendix

Corollospora novofusca Kohlm. *et* Volkm.-Kohlm. ***sp. nov.***

Etym: From the Latin *novus* = new and the epithet *fusca*, to indicate resemblance with *C. fusca*.

Ascomata *200–350 μm diametro, subglobosa, superficiales, ostiolata, breve papillata, subiculata, carbonacea, obscure cano-nigra, singularia, in granis arenariis;* canalis ostioli *pseudoparenchyma dilute brunnea impletus;* peridium *22–26 μm crassum, brunneum; paraphyses absentes, centrum ascomatis immaturi cellulis pseudoparenchymaticis, hyalinis, leptodermis, foveolatis, deliquescentibus;* asci *leptodermi, unitunicati, deliquescentes ante maturitatem ascosporarum;* ascosporae *(spinis exclusis) 74–99 × 24–34 μm ($\bar{x}$ = 89 × 27 μm; n = 76), fusiformes, muriformes, transverse (12–)13 septatae, ad septa non vel leniter constrictae, atro-brunneae, striatae longitrorsum in superficie; appendices polares 52–(68)–86 μm longae, spiniformes, dilute brunneae; appendices equatoriales et apicales hyalines, fragmentis exosporis evolutae.*

Substratum: Fructus *Casuarinae* sp.
Distributio: Oceanus Pacificus (Hawaii)
Holotypus: J. K. 5102 (IMS)

Ascomata 200–350 μm diam, subglobose, superficial, ostiolate, short papillate, subiculate, carbonaceous, dull greyish-black, single, attached to grains of sand. *Ostiolar canal* filled with a light brown pseudoparenchyma. *Peridium* 22–26 μm thick, brown. *Pseudoparenchyma* of thin-walled polygonal cells filling the centrum of young ascomata, deliquescing and compressed by ascospores accumulating in the venter; pit-connections connect plasmatic strands between neighboring cells. *Asci* thin-walled, unitunicate, deliquescing before ascospore maturation. *Ascospores* (excluding polar spines) 74–99 × 24–34 μm ($\bar{x}$ = 89 × 27 μm; n = 76), fusiform, muriform, transverse (12–)13 septate, not or slightly constricted at the septa, dark brown, surface longitudinally striate; polar appendages 52–(68)–86 μm long, thorn-like, light brown; frill-like equatorial and tube-like apical appendages hyaline, developing by the fragmentation and peeling off of the exospore (Fig. 12A). Mode of life: Saprobic.

Substrate: Fruits of *Casuarina* sp.
Range: Pacific Ocean (known only from Kauai and Maui, Hawaii)
Material Examined: Fruits of *Casuarina* sp., among detritus from the intertidal area, Haena Beach, Hanalei, Kauai (Hawaii), 22°13′18″N, 159°34′16″W, 14 Mar. 1987, incubated with moist sand for 9 mo, together with *Arenariomyces parvulus*, J. K. 5102 (HOLOTYPE), also growing in pure culture; ascospores in seafoam, Hanalei Bay, Hanalei, Kauai (Hawaii), 22°12′12″N, 159°30′49″W, 16 June 1968, together with *C. maritima*, *C. pulchella*, *A. trifurcatus* and *Varicosporina ramulosa*, J. K. 2586 (PARATYPE); in seafoam, mouth of Wainiha River, Wainiha, Kauai (Hawaii), 22°12′42″N, 159°32′36″W, 3 Nov. 1983, water temperature 26 °C, salinity 34‰, together with *A. trifurcatus*, J. K. 4489 (PARATYPE); in seafoam, Hana Bay State Park, Maui (Hawaii), 20°45′30″N, 155°59′W, 9 Nov. 1983, water temperature 26 °C, salinity 34‰, together with *A. trifurcatus* and *C. maritima*, J. K. 4491 (PARATYPE). All types are lodged at Herb. IMS.

Corollospora novofusca is very similar to *C. fusca*, but can be distinguished from the latter by shorter means of ascospore length, longer apical thorns and the almost constant presence of 13 transsepta. In contrast to the original description, we found the apical thorns of *C. fusca* to be light brown. So far, *C. fusca* has been reported only from Japan, whereas *C. novofusca* occurred only in collections from Hawaiian islands. The new species was listed in an earlier publication as *C. fusca* (Kohlmeyer and Volkmann-Kohlmeyer 1989).

Quintaria Kohlm. *et* Volkm.-Kohlm. **gen. nov.**

Etym.: From the Latin, *quintarius* = of five, in reference to the five-septate ascospores.

Genus Melanommatalium, Platystomatacearum. Ascomata *obpyriformia, aliquantum lateraliter compressa, omnino immersa, ostiolata, papillata, carbonacea ad subcoriacea, nigra;* peridium *bistratosum, stratum externum hyphoideum, stratum internum texturam angularem formans;* canalis ostioli *reticulo hypharum, in matrice gelatinosa incluso, paries canalis materia nigra incrustatus;* hamathecium *pseudoparaphysoideum, trabeculatum;* asci *octospori, cylindrici, pedunculati, pachydermi, fissitunicati, disco refractivo apicale, non-amyloidei;* ascosporae *biseriatae, fusiformes, quinqueseptatae, ad septum medium leviter constrictae, hyalinae. Saprophytica in cortice et ligno.*

Typus generis: ***Quintaria lignatilis*** (Kohlm.) Kohlm. *et* Volkm.-Kohlm. **comb. nov.**

Basionym: *Trematosphaeria lignatilis* Kohlm., *Mar. Ecol. (P. S. Z. N. I) 5*: 365–366, 1984.

Ascomata obpyriform, somewhat compressed laterally, completely immersed, ostiolate, papillate, carbonaceous to subcoriaceous, black. *Peridium* two-layered, outer stratum of hyphoid cells, inner stratum of flattened cells, forming a *textura angularis. Ostiolar canal* filled with a network of thin hyphae, embedded in a gelatinous matrix, black incrustations lining the sides. *Pseudoparaphyses* trabeculate above the asci, embedded in gelatinous matrix. *Asci* eight-spored, cylindrical, pedunculate, thick-walled, fissitunicate, with a refractive apical plate, IKI negative. *Ascospores* biseriate, fusiform, five-septate, hyaline. Mode of life: Saprobic in wood and bark.

Papers by Barr (1987) and Boise (1985) that appeared after publication of *T. lignatilis* convinced us that this species should be removed from *Trematosphaeria* and assigned to a new genus. *Quintaria* is differentiated from *Trematosphaeria* by completely immersed ascomata with rounded bases, by black incrustations lining the sides of the ostiolar canal, by a non-amyloid apical plate of the ascus, and by hyaline ascospores.

Acknowledgements

We are grateful for support by the U. S. National Science Foundation (Grant BSR-8815719) and for loan of type collections by curators of herbaria AHFH, B, BM, BPI, C, CBS, CO, CP, DAOM, FH, H, ILLS, IMI, JE, K, LPS, MA, NY, P, PC, UC, UME, and UPS. The following colleagues kindly made their own collections available: R. K. Benjamin, B. D. Borse, R. D. Brooks, A. B. Cribb, O. Eriksson, G. Feldmann, R. V. Gessner, M. Honrubia, K. D. Hyde, E. B. G. Jones, P. W. Kirk, J. Koch, A. Munk, A. Nakagiri, M. J. Parsons, D. Porter, R. Ricker, K. Schaumann, I. Schmidt, R. A. Shoemaker, E. G. Simmons, G. R. South, K. Tubaki, and S. Udagawa. We acknowledge gratefully information on nomenclatural and other matters by J. R. Boise, D. L. Hawksworth, R. P. Korf, and B. Sutton. W. Burk at the UNC Botany Library and the staff of Duke University Biological Library assisted us in the literature search. We are grateful to B. B. Bright for library service and typing of the manuscript.

References

Abdullah, S. K., M. A. Abdulkadder and R. D. Goos. 1989. *Basramyces marinus* nom. nov. (hyphomycete) from southern marshes of Iraq. *Int. J. Mycol. Lichenol. 4*: 181–186.

Barr, M. E. 1979. On the Massariaceae in North America. *Mycotaxon 9*: 17–37.

Barr, M. E. 1987. *Prodromus to Class Loculoascomycetes.* Amherst, Massachusetts, publ. by the author. pp. 168.

Bebout, B., S. Schatz, J. Kohlmeyer and M. Haibach. 1987. Temperature-dependent growth in isolates of *Corollospora maritima* Werderm. (Ascomycetes) from different geographical regions. *J. Exp. Mar. Biol. Ecol. 106*: 203–210.

Berlese, A. N. 1892. *Icones Fungorum 1, Fasc. 3.* Abellini. pp. 51–90.

Boise, J. 1985. An amended description of *Trematosphaeria. Mycologia 77*: 230–237.

Borse, B. D. 1987. New species of *Aigialus* from India. *Trans. Br. Mycol. Soc. 88*: 424–426.

Borse, B. D. and K. D. Hyde. 1989. Marine Fungi from India. III. *Acrocordiopsis patilii* gen. et. sp. nov. from mangrove wood. *Mycotaxon 34*: 535–540.

Cannon, P. F., D. L. Hawksworth and M. A. Sherwood-Pike. 1985. *The British Ascomycotina. An Annotated Checklist.* CMI, Kew. pp. 302.

Clauzade, G., P. Diederich and C. Roux. 1989. Nelikenigintaj fungoj likenlogaj. Illustrita determinlibro. *Bull. Soc. Linn. Provence No. Spéc. 1*: 142 pp.

Ellis, J. B. and B. M. Everhart. 1892. *North American Pyrenomycetes.* Ellis and Everhart, Newfield, New Jersey. pp. 793.

Eriksson, O. 1981. The families of bitunicate ascomycetes. *Opera Bot. 60*: 1–220.

Eriksson, O. 1982. Notes on ascomycetes and coelomycetes from northwest Europe. *Mycotaxon 15*: 189–202.

Gessner, R. V. and J. Kohlmeyer. 1976. Geographical distribution and taxonomy of fungi from salt marsh *Spartina. Can. J. Bot. 54*: 2032–2037.

Hawksworth, D. L. 1987. Observations on three algicolous microfungi. *Notes RBG Edinb. 44*: 549–560.

Hyde, K. D. 1988 a. The genus *Linocarpon* from the mangrove palm *Nypa fruticans. Trans. Mycol. Soc. Jpn. 29*: 339–350.

Hyde, K. D. 1988 b. Studies on the tropical marine fungi of Brunei. *Bot. J. Linn. Soc. 98*: 135–151.

Hyde, K. D. 1989a. Intertidal fungi from the mangrove fern, *Acrostichum speciosum*, including *Massarina acrostichi* sp. nov. *Mycol. Res. 93*: 435–438.

Hyde, K. D. 1989b. *Caryospora mangrovei* sp. nov. and notes on marine fungi from Thailand. *Trans. Mycol. Soc. Jpn. 30*: 333–341.

Hyde, K. D., C. A. Farrant and E. B. G. Jones. 1986. Marine fungi from Seychelles. III. *Aniptodera mangrovii* sp. nov. from mangrove wood. *Can. J. Bot. 64*: 2989–2992.

Hyde, K. D. and R. Mouzouras. 1988. *Passeriniella savoryellopsis* sp. nov., a new ascomycete from intertidal mangrove wood. *Trans. Br. Mycol. Soc. 91*: 179–185.

Johnson, R. G., E. B. G. Jones and S. T. Moss. 1987. Taxonomic studies of the Halosphaeriaceae: *Ceriosporopsis, Haligena,* and *Appendichordella* gen. nov. *Can. J. Bot. 65*: 931–942.

Johnson, T. W., Jr. 1956. Marine fungi. I. *Leptosphaeria* and *Pleospora. Mycologia 48*: 495–505.

Johnson, T. W., Jr. 1958. Some lignicolous marine fungi from the North Carolina coast. *J. Elisha Mitchell Sci. Soc. 74*: 42–48.

Jones, E. B. G. 1962. Marine fungi. *Trans. Br. Mycol. Soc. 45*: 93–114.

Jones, E. B. G. and S. T. Moss. 1980. Further observations on the taxonomy of the Halosphaeriaceae. *Bot. Mar. 23*: 483–500.

Koch, J. and E. B. G. Jones. 1986. *Ceriosporopsis sundica*, a new lignicolous marine ascomycete from Denmark. *Nordic J. Bot. 6*: 339–344.

Koch, J. and E. B. G. Jones. 1989. The identity of *Crinigera maritima* and three new genera of marine cleistothecial ascomycetes. *Can. J. Bot. 67*: 1183–1197.

Kohlmeyer, J. 1960. Wood-inhabiting marine fungi from the Pacific Northwest and California. *Nova Hedwigia 2*: 293–343.

Kohlmeyer, J. 1963. Fungi marini novi vel critici. *Nova Hedwigia 6*: 297–329.

Kohlmeyer, J. and T. M. Charles. 1981. Sclerocarps: undescribed propagules in a sand-inhabiting marine fungus. *Can. J. Bot. 59*: 1787–1791.

Kohlmeyer, J. and V. Demoulin. 1981. Parasitic and symbiotic fungi on marine algae. *Bot. Mar. 24*: 9–18.

Kohlmeyer, J. and M. W. Hawkes. 1983. A suspected case of mycophycobiosis between *Mycosphaerella apophlaeae* (Ascomycetes) and *Apophlaea* spp. (Rhodophyta). *J. Phycol. 19*: 257–260.

Kohlmeyer, J. and E. Kohlmeyer. 1979. *Marine Mycology. The Higher Fungi.* Academic Press, New York and London. pp. 690.

Kohlmeyer, J. and B. Volkmann-Kohlmeyer. 1987 a. Marine fungi from Aldabra, the Galapagos, and other tropical islands. *Can. J. Bot. 65*: 571–582.

Kohlmeyer, J. and B. Volkmann-Kohlmeyer. 1987 b. Reflections on the genus *Corollospora* (ascomycetes). *Trans. Br. Mycol. Soc. 88*: 181–188.

Kohlmeyer, J. and B. Volkmann-Kohlmeyer. 1989. Hawaiian marine fungi, including two new genera of Ascomycotina. *Mycol. Res. 92*: 410–421.

Kohlmeyer, J. and B. Volkmann-Kohlmeyer. 1990. Revision of marine species of *Didymosphaeria* (Ascomycotina). *Mycol. Res. 94*: 685–690.

Leonard, K. J. 1976. Synonymy of *Exserohilum halodes* with *E. rostratum*, and induction of the ascigerous state, *Setosphaeria rostrata. Mycologia 68*: 402–411.

Müller, E., P. J. Fisher, O. Petrini, A. Y. Rossman and G. J. Samuels. 1987. Taxonomy and anamorphs of the Herpotrichiellaceae with notes on generic synonymy. *Trans. Br. Mycol. Soc. 88*: 63–74.

Nag Raj, T. R. 1989. Genera coelomycetum. XXVI. *Amarenographium, Callistospora, Hyalothyridium, Orphanocoela* anam.-gen. nov., *Scolecosporiella,* and *Urohendersoniella. Can. J. Bot. 67*: 3169–3186.

Nakagiri, A. 1984. Two new species of *Lulworthia* and evaluation of genera-delimiting characters between *Lulworthia* and *Lindra* (Halosphaeriaceae). *Trans. Mycol. Soc. Jpn. 25*: 377–388.

Porter, D. and W. F. Farnham. 1986. *Mycaureola dilseae*, a marine basidiomycete parasite of the red alga, *Dilsea carnosa. Trans. Br. Mycol. Soc. 87*: 575–582.

Roldán, A. and M. Honrubia. 1989. Hongos marinos saprófitos en la provincia de Alicante. *Anales Jardín Bot. Madrid 46*: 207–214.

Santesson, R. 1984. *The Lichens of Sweden and Norway.* Swedish Museum of Natural History, Stockholm, and Uppsala. pp. 333.

Schatz, S. 1980. Taxonomic revision of two pyrenomycetes associated with littoral-marine green algae. *Mycologia 72*: 110–117.

Schmidt, I. 1969. *Corollospora intermedia*, nov. spec., *Carbosphaerella leptosphaerioides*, nov. spec. und *Crinigera maritima*, nov. gen., nov. spec., 3 neue marine Pilzarten von der Ostseeküste. *Nat. Naturschutz Mecklenburg 7*: 5–14.

Schmidt, I. 1974. Untersuchungen über höhere Meerespilze an der Ostseeküste der DDR. *Nat. Naturschutz Mecklenburg 12*: 1–148.

Schmidt, I. 1985. Types and type collections of new higher marine and freshwater fungi from the Baltic coast. *Mycotaxon 24*: 419–421.

Shearer, C. A. 1989. *Aniptodera* (Halosphaeriaceae) from wood in freshwater habitats. *Mycologia 81*: 139–146.

Shearer, C. A. and J. L. Crane. 1980. Fungi of the Chesapeake Bay and its tributaries. VIII. Ascomycetes with unfurling appendages. *Bot. Mar. 23*: 607–615.

Shoemaker, R. A. and C. E. Babcock. 1989. *Phaeosphaeria. Can. J. Bot. 67*: 1500–1599.

Werdermann, E. 1922. *Corollospora maritima* Werd., ein salzliebender Pilz von der Nordseeküste. *Notizbl. Bot. Gart. u. Mus., Berlin-Dahlem 8*: 248–250.

Wilson, I. M. 1954. *Ceriosporopsis halima* Linder and *Ceriosporopsis cambrensis* sp. nov.: two Pyrenomycetes on wood. *Trans. Br. Mycol. Soc. 37*: 272–285.

Note added in proof:

Nectriella laminariae Eriksson was recently transferred to *Pronectria* [*P. laminariae* (O. Eriks.) Lowen, *Mycotaxon 39*: 461, 1990]

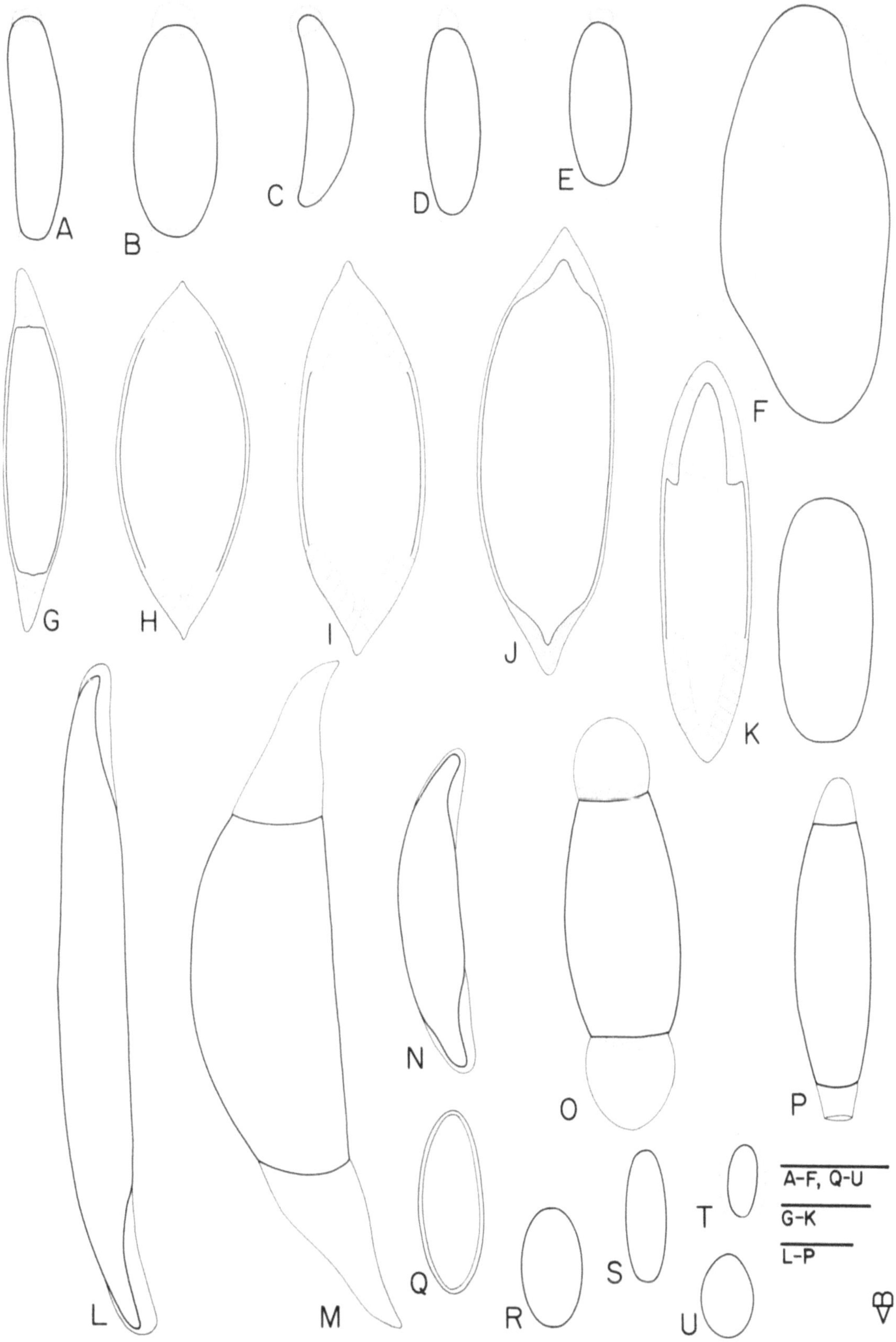

Fig. 1. Ascospores. **A.** *Chadefaudia corallinarum.* **B.** *C. polyporolithi.* **C.** *C. schizymeniae.* **D.** *C. marina.* **E.** *C. gymnogongri.* **F.** *C. balliae.* **G.** *Haloguignardia decidua.* **H.** *H. tumefaciens.* **I.** *H. irritans.* **J.** *H. oceanica.* **K.** *H. cystoseirae* (appendages fallen off in the right spore). **L.** *Spathulospora phycophila.* **M.** *S. adelpha.* **N.** *S. lanata.* **O.** *S. calva.* **P.** *S. antarctica.* **Q.** *Polystigma apophlaeae.* **R.** *Turgidosculum ulvae.* **S.** *Mastodia tessellata.* **T.** *Laetinaevia marina.* **U.** *Verrucaria cribbii.* (All scales = 10 μm)

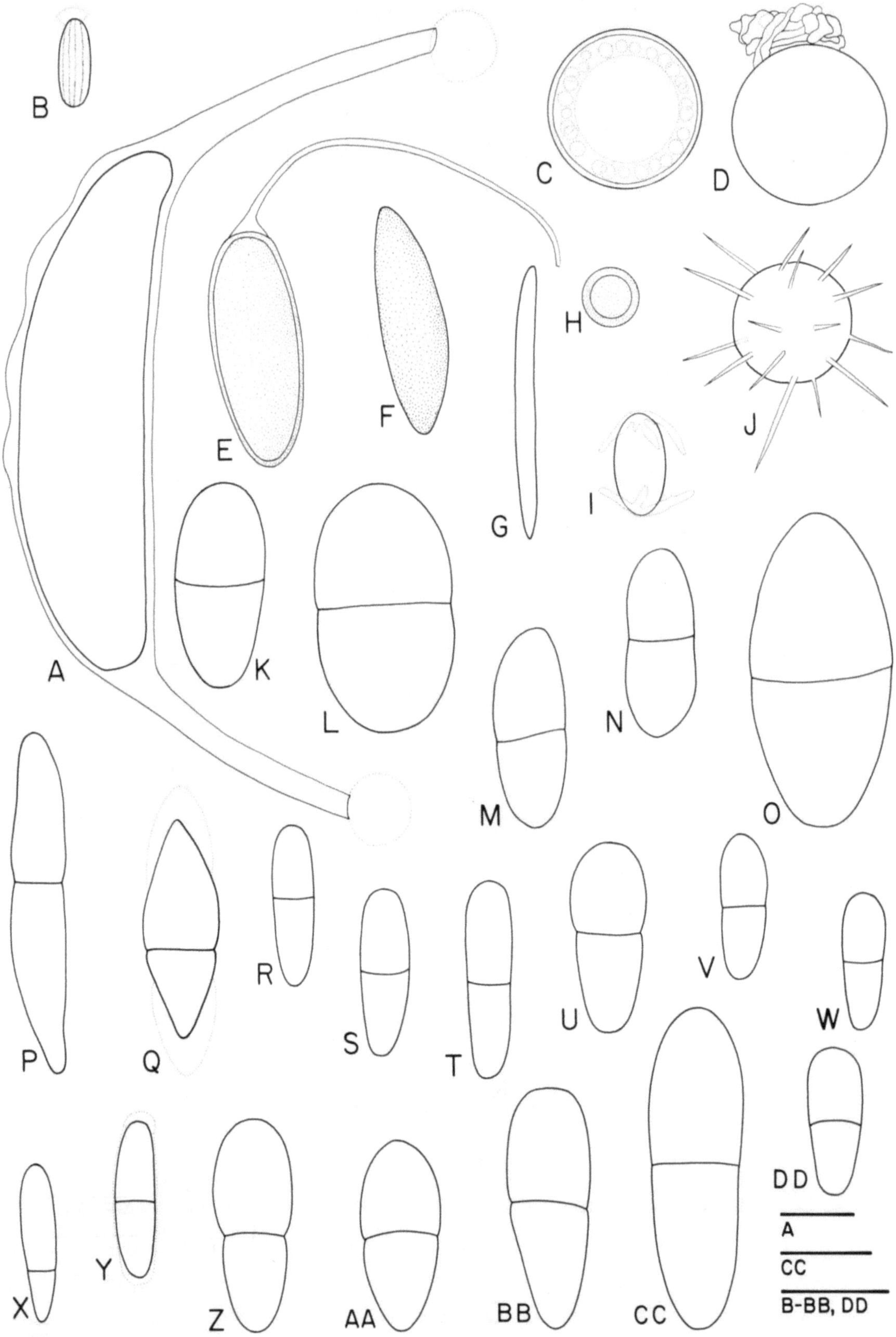

Fig. 2. Ascospores. **A.** *Kohlmeyeriella tubulata.* **B.** *Phomatospora acrostichi.* **C.** *Thalassogena sphaerica.* **D.** *Moana turbinulata.* **E.** *Adomia avicenniae.* **F.** *Rhizophila marina.* **G.** *Halonectria milfordensis.* **H.** *Gymnascella littoralis.* **I.** *Eiona tunicata.* **J.** *Amylocarpus encephaloides.* **K.** *Swampomyces armeniacus.* **L.** *Acrocordiopsis patilii.* **M.** *Lignincola laevis.* **N.** *L. longirostris.* **O.** *L. tropica.* **P.** *Lautitia danica.* **Q.** *Pharcidia laminariicola.* **R.** *P. rhachiana.* **S.** *Mycosphaerella apophlaeae.* **T.** *M. ascophylli.* **U.** *M. pneumatophorae.* **V.** *M. salicorniae.* **W.** *M. staticicola.* **X.** *Didymella gloiopeltidis.* **Y.** *D. magnei.* **Z.** *D. fucicola.* **AA.** *Nectriella laminariae.* **BB.** *Arthopyrenia halodytes.* **CC.** *Halographis runica.* **DD.** *Pyrenocollema pelvetiae.* (All scales = 10 μm)

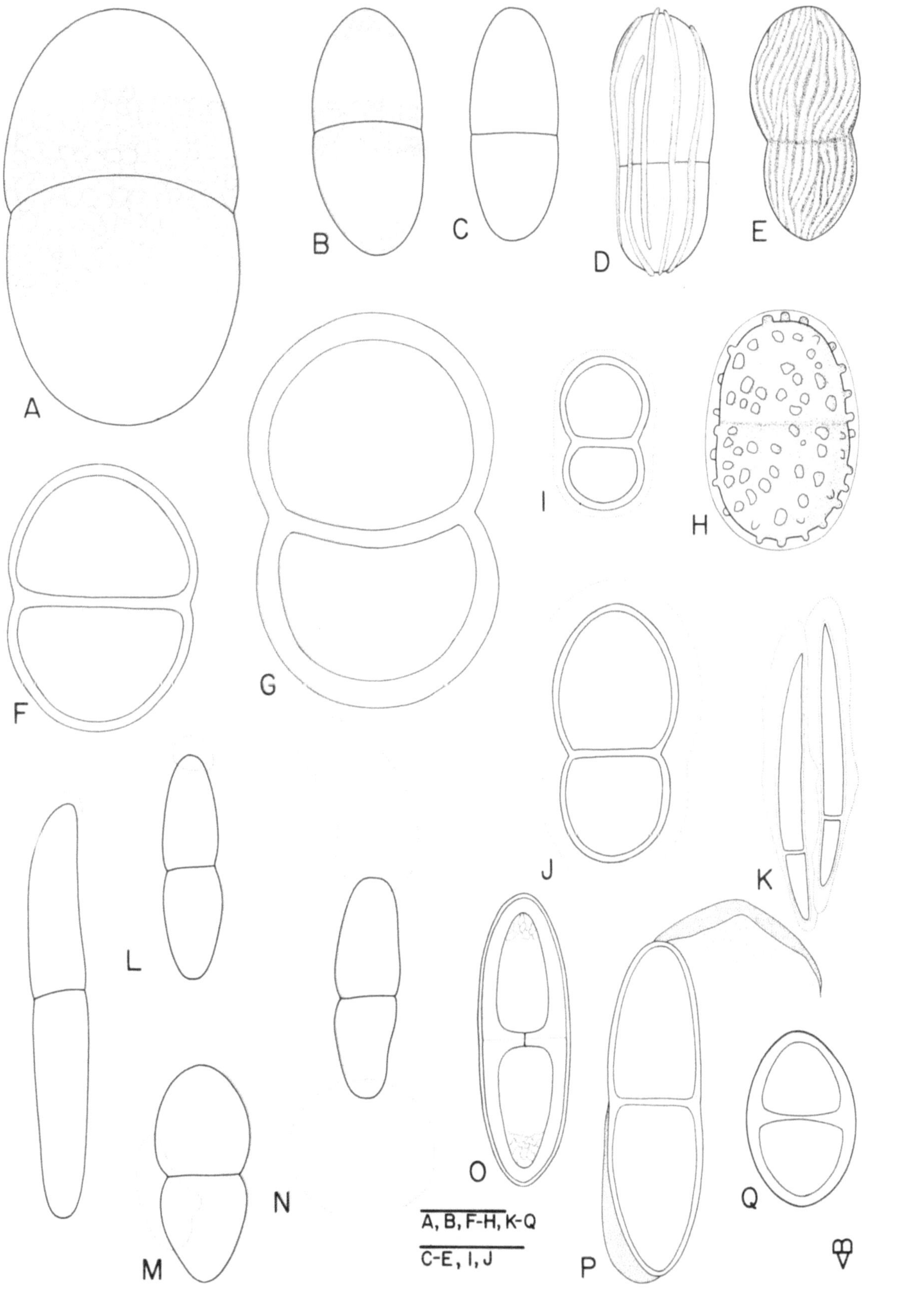

Fig. 3. Ascospores. **A.** *Nais glitra.* **B.** *N. inornata.* **C.** *Hydronectria tethys* var. *glabra.* **D.** *H. tethys.* **E.** *Dactylospora haliotrepha.* **F.** *Thalassoascus lessoniae.* **G.** *T. cystoseirae.* **H.** *Belizeana tuberculata.* **I.** *Biflua physasca.* **J.** *Didymella avicenniae.* **K.** *Laboulbenia marina.* **L.** *Halosphaeria cucullata.* **M.** *Paraliomyces lentiferus.* **N.** *Phycomelaina laminariae.* **O.** *Aniptodera chesapeakensis.* **P.** *A. mangrovei.* **Q.** *Heleococcum japonense.* (Both scales = 10 μm)

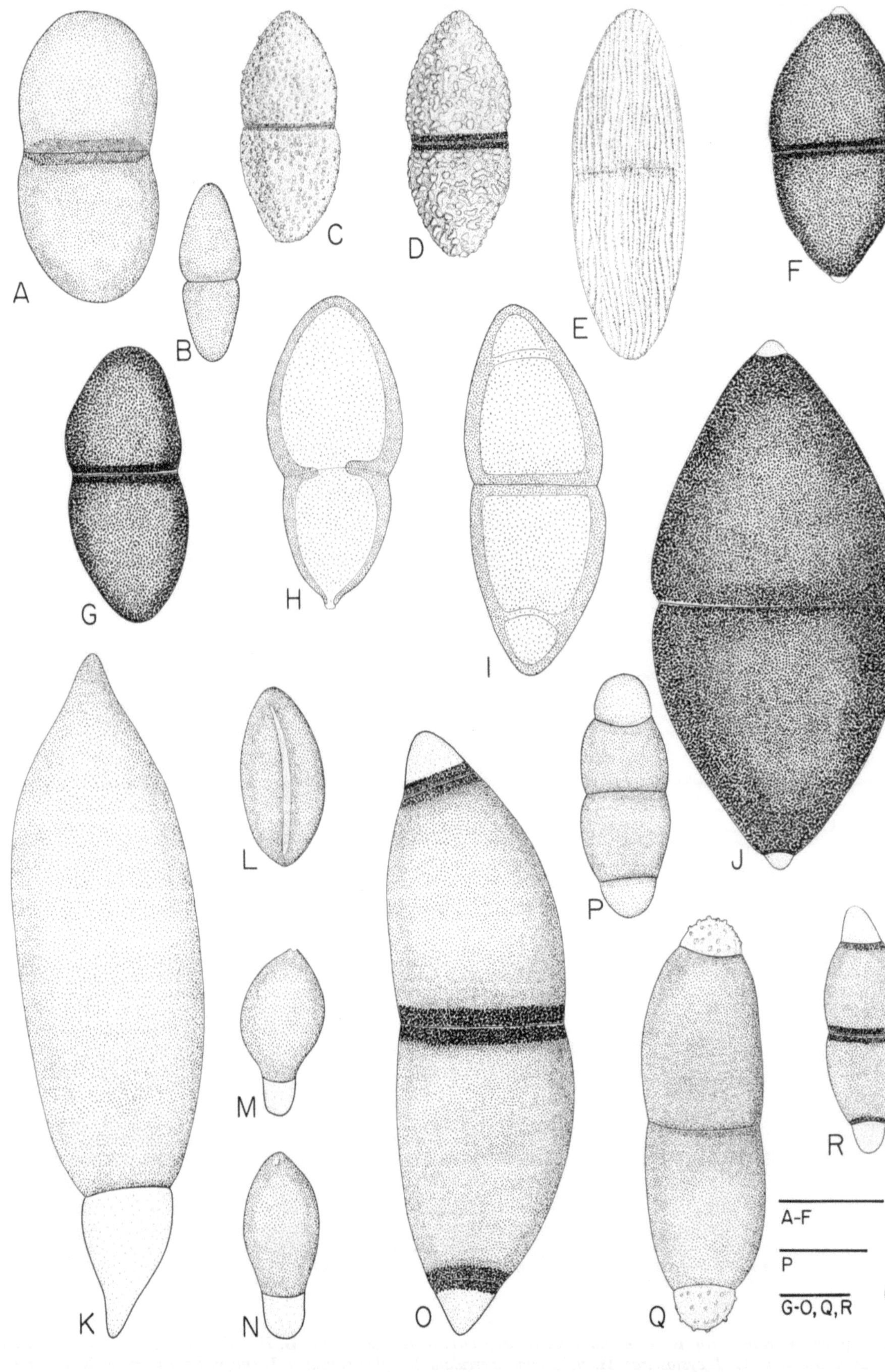

Fig. 4. Ascospores. **A.** *Bicrouania maritima*. **B.** *Kirschsteiniothelia maritima*. **C.** *Verruculina enalia*. **D.** *Didymosphaeria lignomaris*. **E.** *Lineolata rhizophorae*. **F.** *Caryosporella rhizophorae*. **G.** *Halotthia posidoniae*. **H.** *Helicascus kanaloanus*. **I.** *Coronopapilla mangrovei*. **J.** *Pontoporeia biturbinata*. **K.** *Manglicola guatemalensis*. **L.** *Hypoxylon oceanicum*. **M.** *Zopfiella marina*. **N.** *Z. latipes*. **O.** *Passeriniella savoryellopsis*. **P.** *P. obiones*. **Q.** *Savoryella paucispora*. **R.** *S. lignicola*. (All scales = 10 μm)

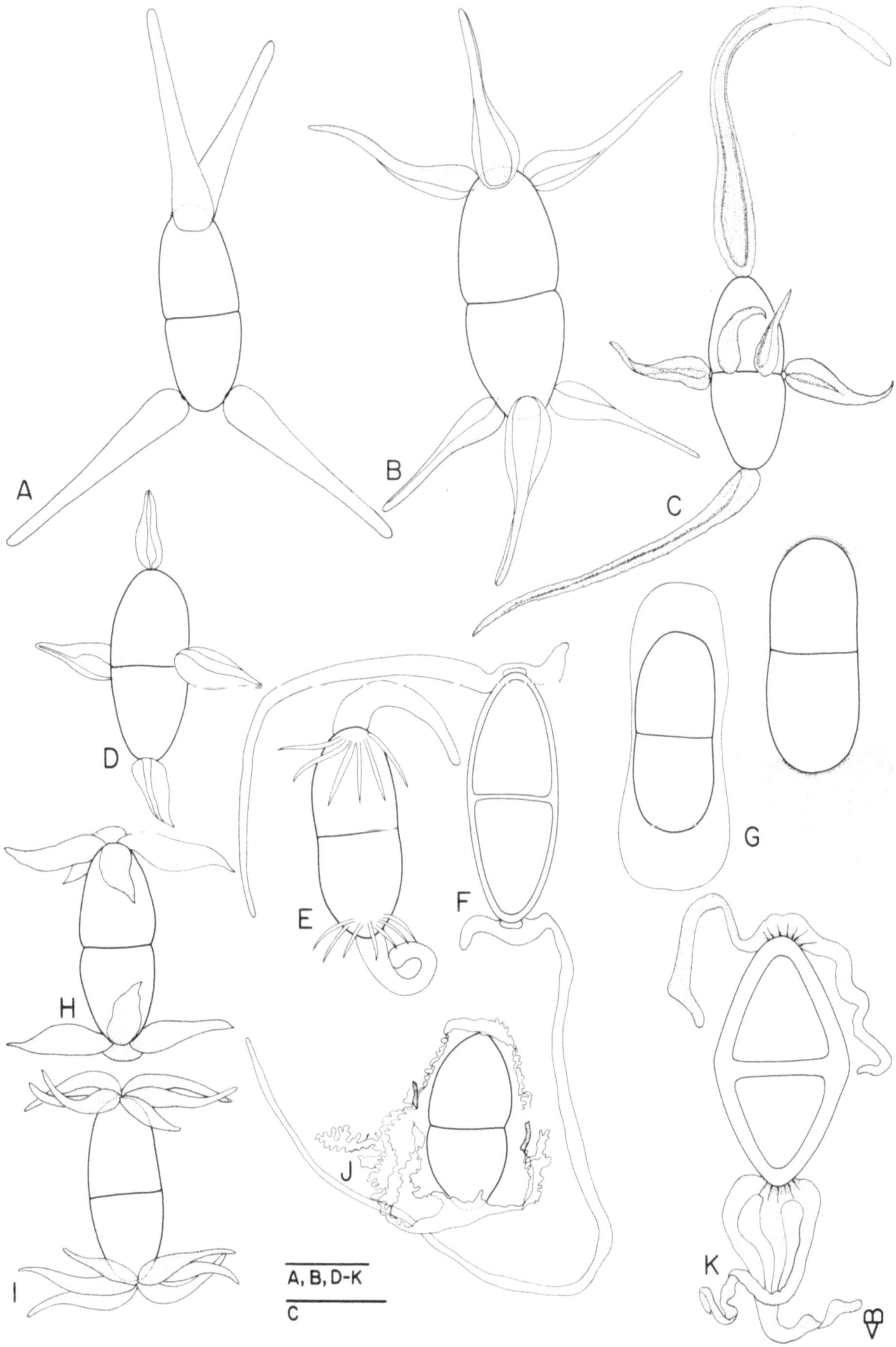

Fig. 5. Ascospores. **A.** *Halosphaeria quadricornuta.* **B.** *H. salina.* **C.** *Ocostaspora apilongissima.* **D.** *H. appendiculata.* **E.** *Remispora spinibarbata.* **F.** *R. maritima.* **G.** *R. galerita* (the right spore showing appendage striation). **H.** *R. quadriremis.* **I.** *R. stellata.* **J.** *R. crispa.* **K.** *R. pilleata.* (Both scales = 10 µm)

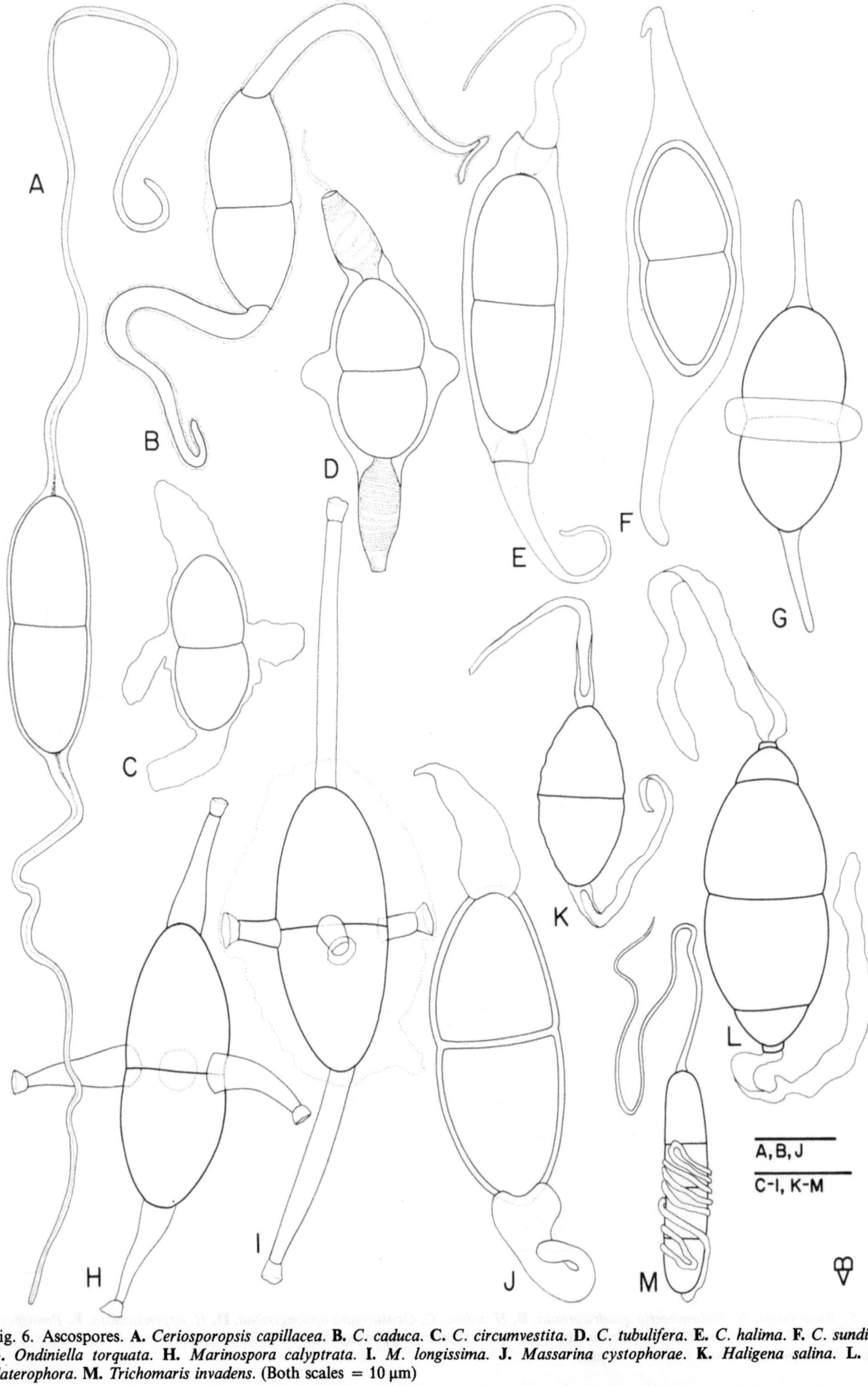

Fig. 6. Ascospores. **A.** *Ceriosporopsis capillacea.* **B.** *C. caduca.* **C.** *C. circumvestita.* **D.** *C. tubulifera.* **E.** *C. halima.* **F.** *C. sundica.* **G.** *Ondiniella torquata.* **H.** *Marinospora calyptrata.* **I.** *M. longissima.* **J.** *Massarina cystophorae.* **K.** *Haligena salina.* **L.** *H. elaterophora.* **M.** *Trichomaris invadens.* (Both scales = 10 µm)

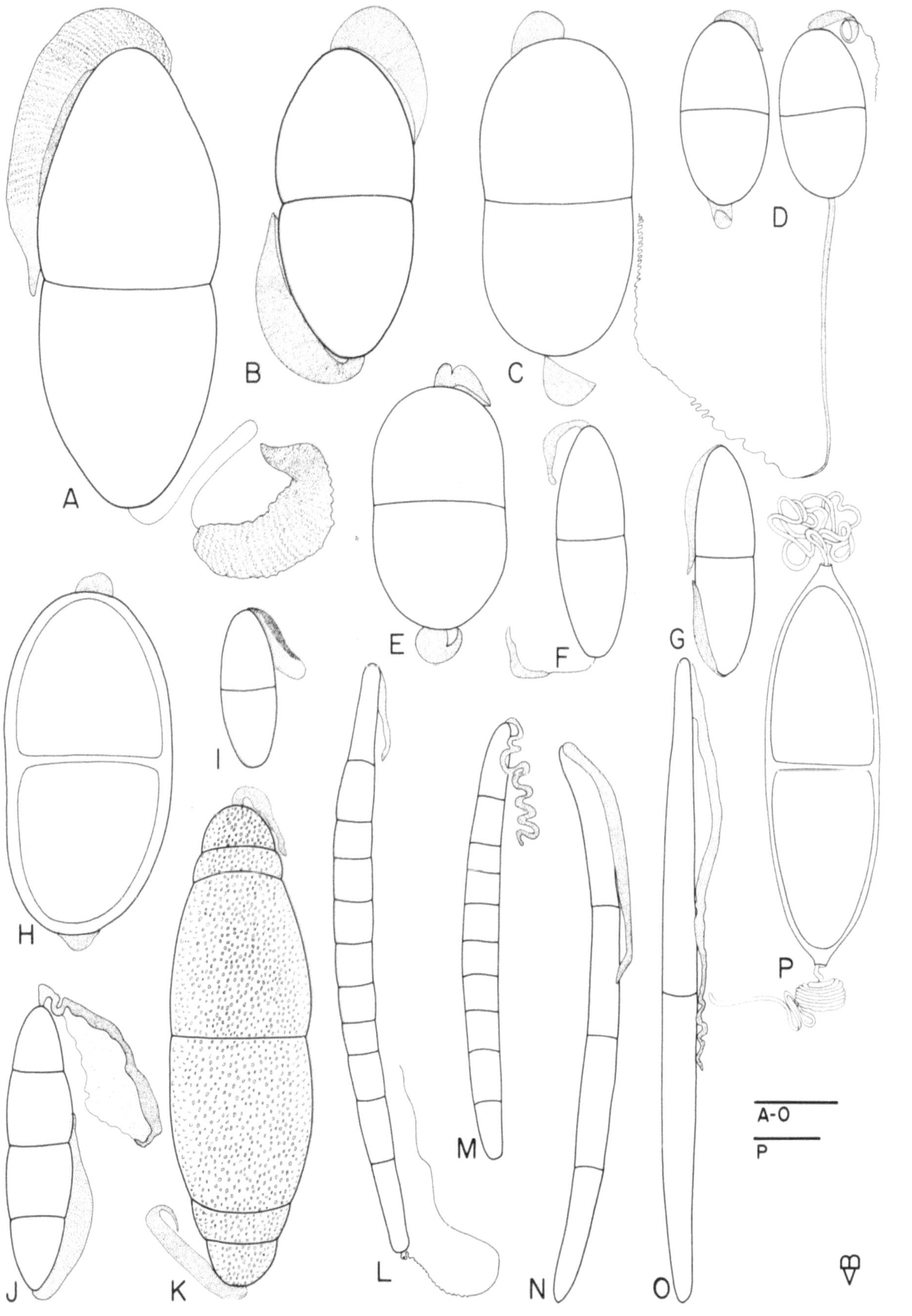

Fig. 7. Ascospores. **A.** *Halosarpheia ratnagiriensis.* **B.** *H. abonnis.* **C.** *H. fibrosa.* **D.** *H. marina.* **E.** *H. trullifera.* **F.** *H. viscosa.* **G.** *H. retorquens.* **H.** *Gnomonia salina.* **I.** *Ophiodeira monosemeia.* **J.** *H. bentotensis.* **K.** *H. spartinae.* **L.** *H. viscidula.* **M.** *H. cincinnatula.* **N.** *H. unicaudata.* **O.** *Oceanitis scuticella.* **P.** *Cucullosporella mangrovei.* (Both scales = 10 μm)

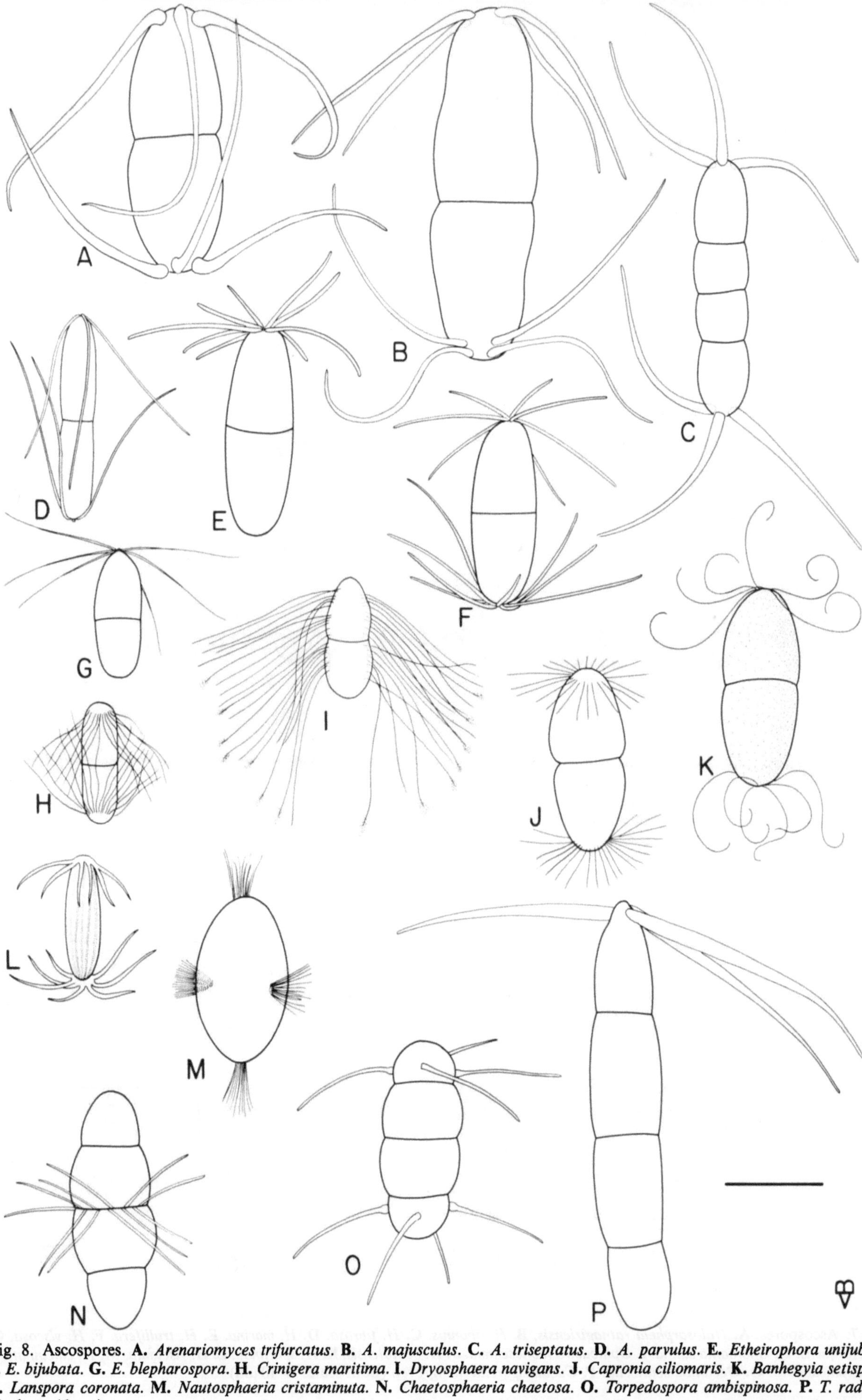

Fig. 8. Ascospores. **A.** *Arenariomyces trifurcatus.* **B.** *A. majusculus.* **C.** *A. triseptatus.* **D.** *A. parvulus.* **E.** *Etheirophora unijubata.* **F.** *E. bijubata.* **G.** *E. blepharospora.* **H.** *Crinigera maritima.* **I.** *Dryosphaera navigans.* **J.** *Capronia ciliomaris.* **K.** *Banhegyia setispora.* **L.** *Lanspora coronata.* **M.** *Nautosphaeria cristaminuta.* **N.** *Chaetosphaeria chaetosa.* **O.** *Torpedospora ambispinosa.* **P.** *T. radiata.* (Scale = 10 μm)

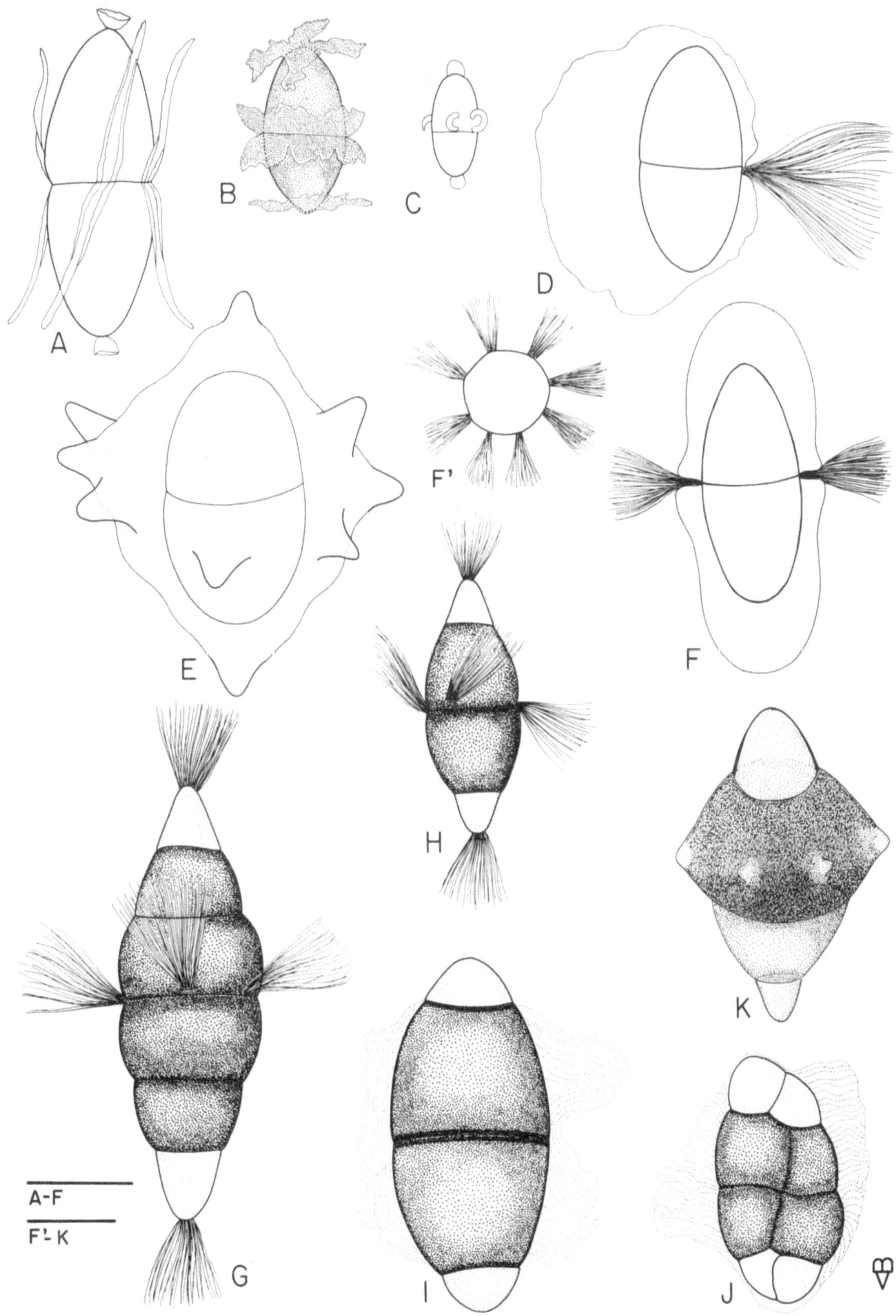

Fig. 9. Ascospores. **A.** *Halosphaeriopsis mediosetigera.* **B.** *Groenhiella bivestia.* **C.** *Marisolaris ansata.* **D.** *Nimbospora effusa.* **E.** *N. octonae.* **F.** *N. bipolaris* (F′ in apical view, at lower magnification). **G.** *Nereiospora comata.* **H.** *N. cristata.* **I.** *Carbosphaerella leptosphaerioides.* **J.** *C. pleosporoides.* **K.** *Biconiosporella corniculata.* (Both scales = 10 μm)

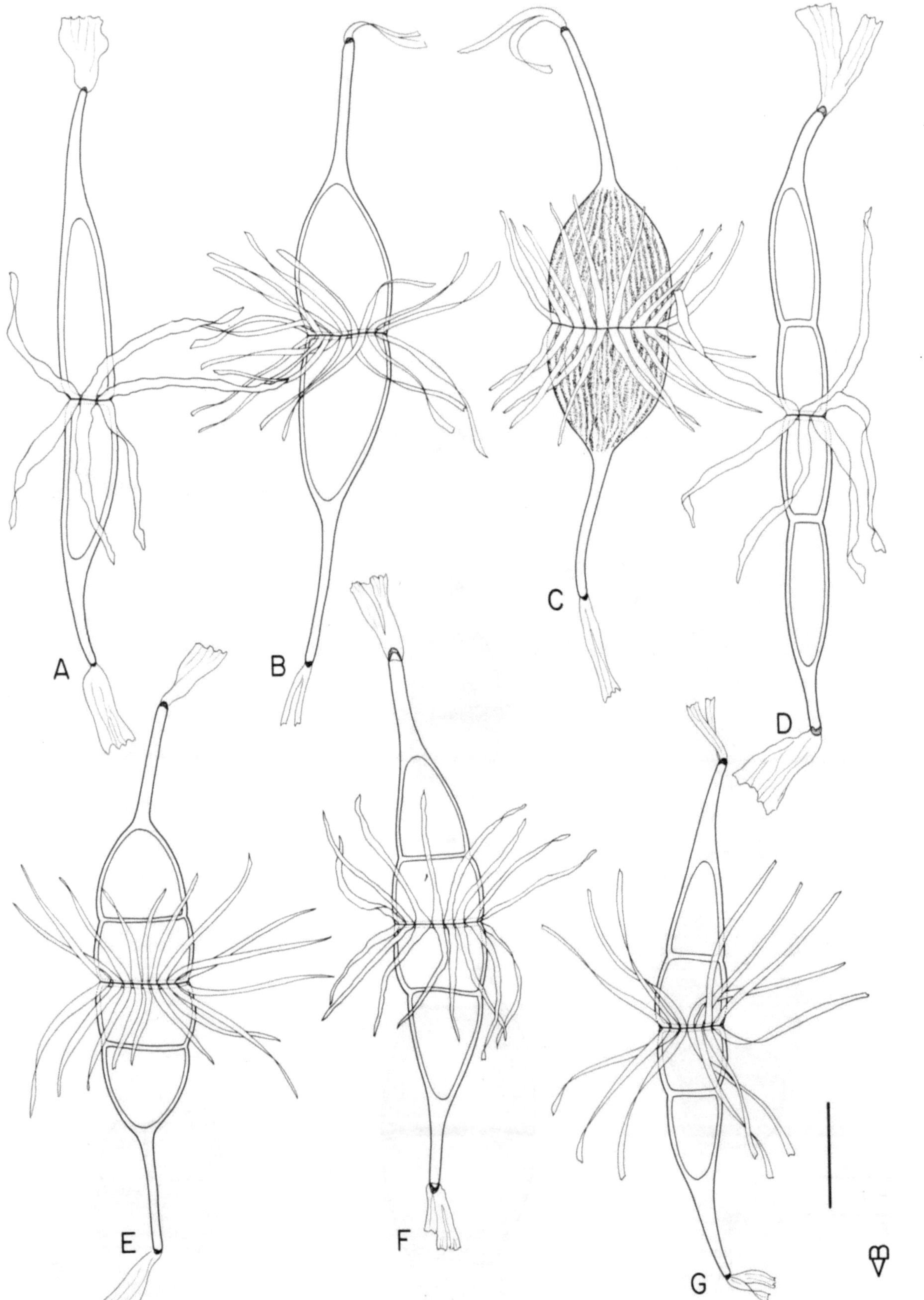

Fig. 10. Ascospores. **A.** *Corollospora gracilis*. **B.** *C. maritima*. **C.** *C. cinnamomea*. **D.** *C. angusta*. **E.** *C. intermedia*. **F.** *C. armoricana*. **G.** *Corollospora* sp. (Scale = 10 μm)

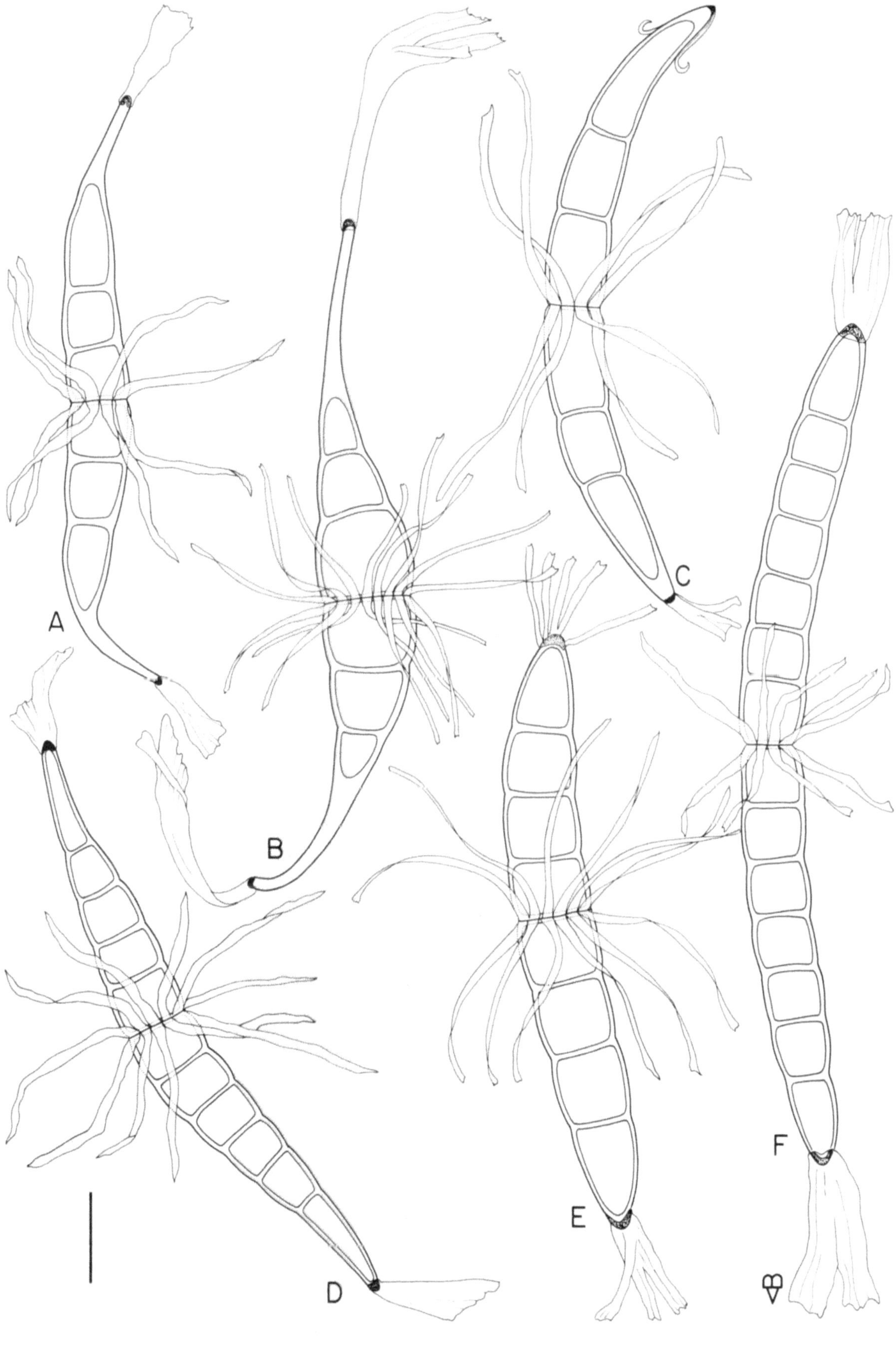

Fig. 11. Ascospores. **A.** *Corollospora quinqueseptata*. **B.** *C. lacera*. **C.** *C. luteola*. **D.** *C. pseudopulchella*. **E.** *C. pulchella*. **F.** *C. filiformis*. (Scale = 10 μm)

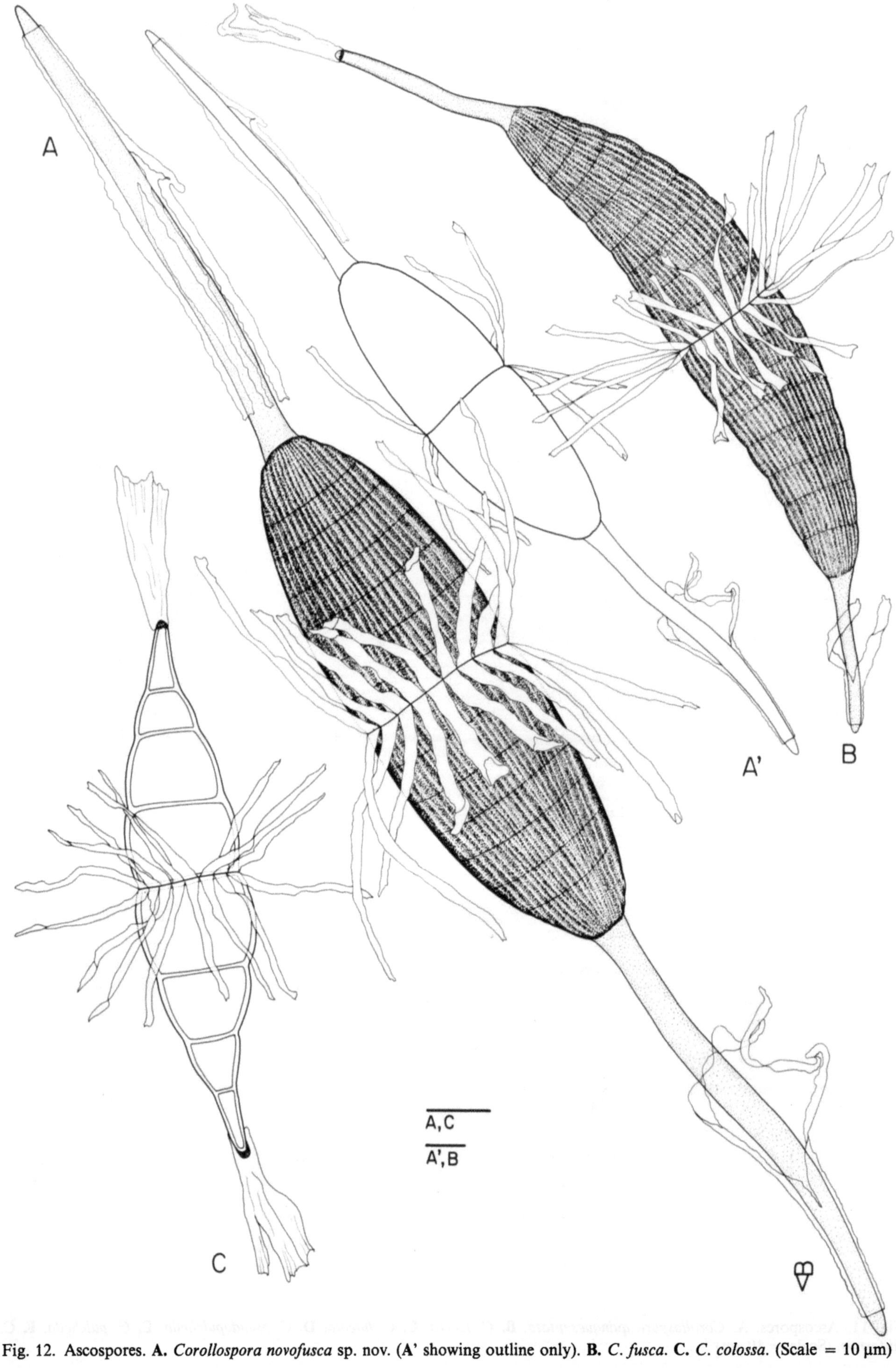

Fig. 12. Ascospores. **A.** *Corollospora novofusca* sp. nov. (**A'** showing outline only). **B.** *C. fusca.* **C.** *C. colossa.* (Scale = 10 µm)

Fig. 13. Ascospores. **A.** *Buergenerula spartinae.* **B.** *Sphaerulina oraemaris.* **C.** *Marinosphaera mangrovei.* **D.** *Abyssomyces hydrozoicus.* **E.** *Orcadia ascophylli.* **F.** *Trailia ascophylli.* **G.** *Quintaria lignatilis.* **H.** *Lophiostoma mangrovei.* **I.** *Leptosphaeria australiensis.* **J.** *L. avicenniae.* **K.** *L. pelagica.* **L.** *Appendichordella amicta.* **M.** *Ascocratera manglicola.* **N.** *Massarina velatospora.* **O.** *M. thalassiae.* **P.** *M. acrostichi.* **Q.** *M. lacertensis.* (Both scales = 10 μm)

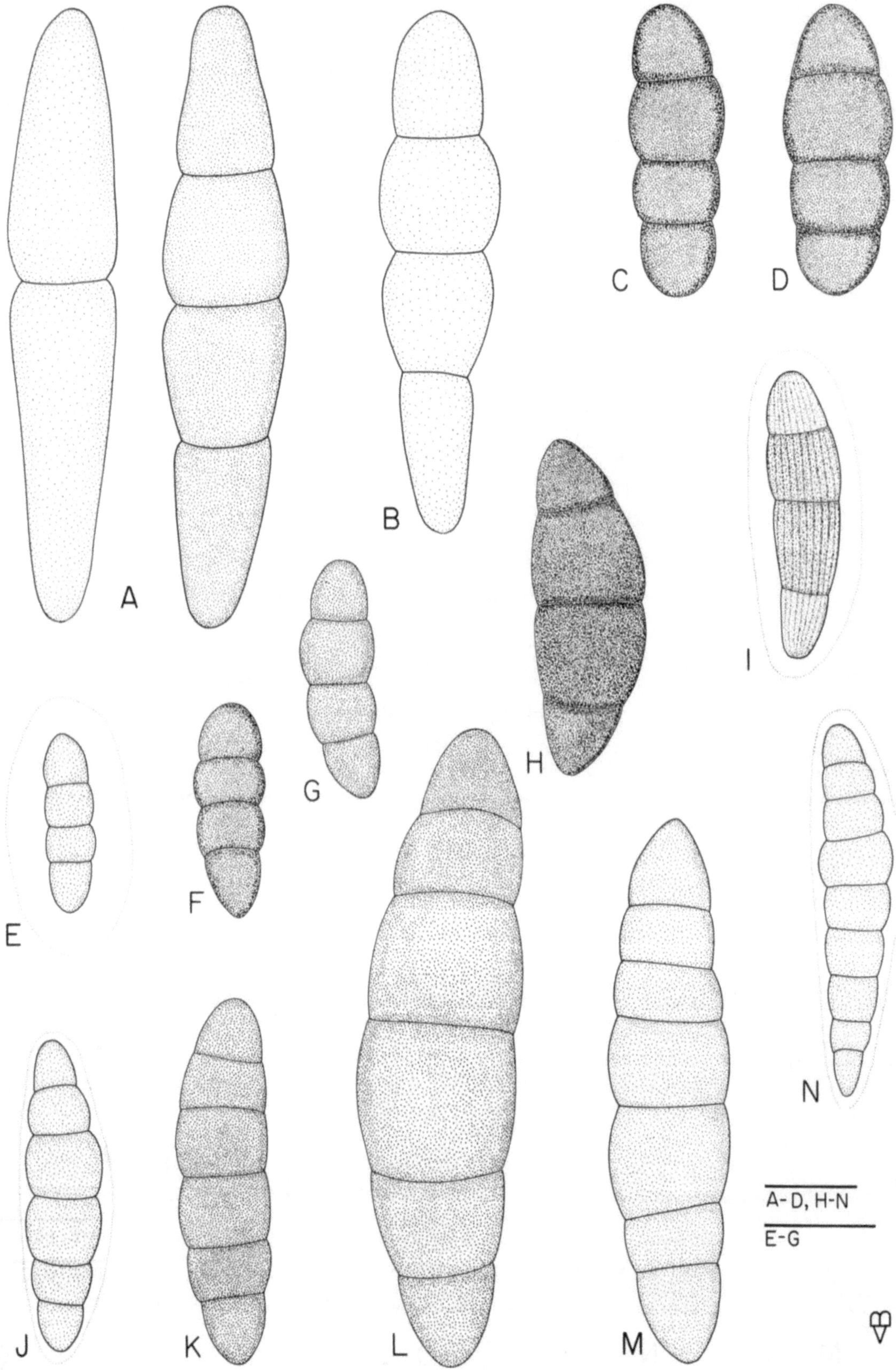

Fig. 14. Ascospores. **A.** *Wettsteinia marina.* **B.** *Phaeosphaeria macrosporidium.* **C.** *P. typharum.* **D.** *P. spartinaecola.* **E.** *P. halima.* **F.** *Leptosphaeria peruviana.* **G.** *L. oraemaris.* **H.** *Trematosphaeria mangrovei.* **I.** *T. striatispora.* **J.** *P. neomaritima.* **K.** *P. spartinae.* **L.** *P. gessneri.* **M.** *Amarenomyces ammophilae.* **N.** *Chaetomastia typhicola.* (Both scales = 10 µm)

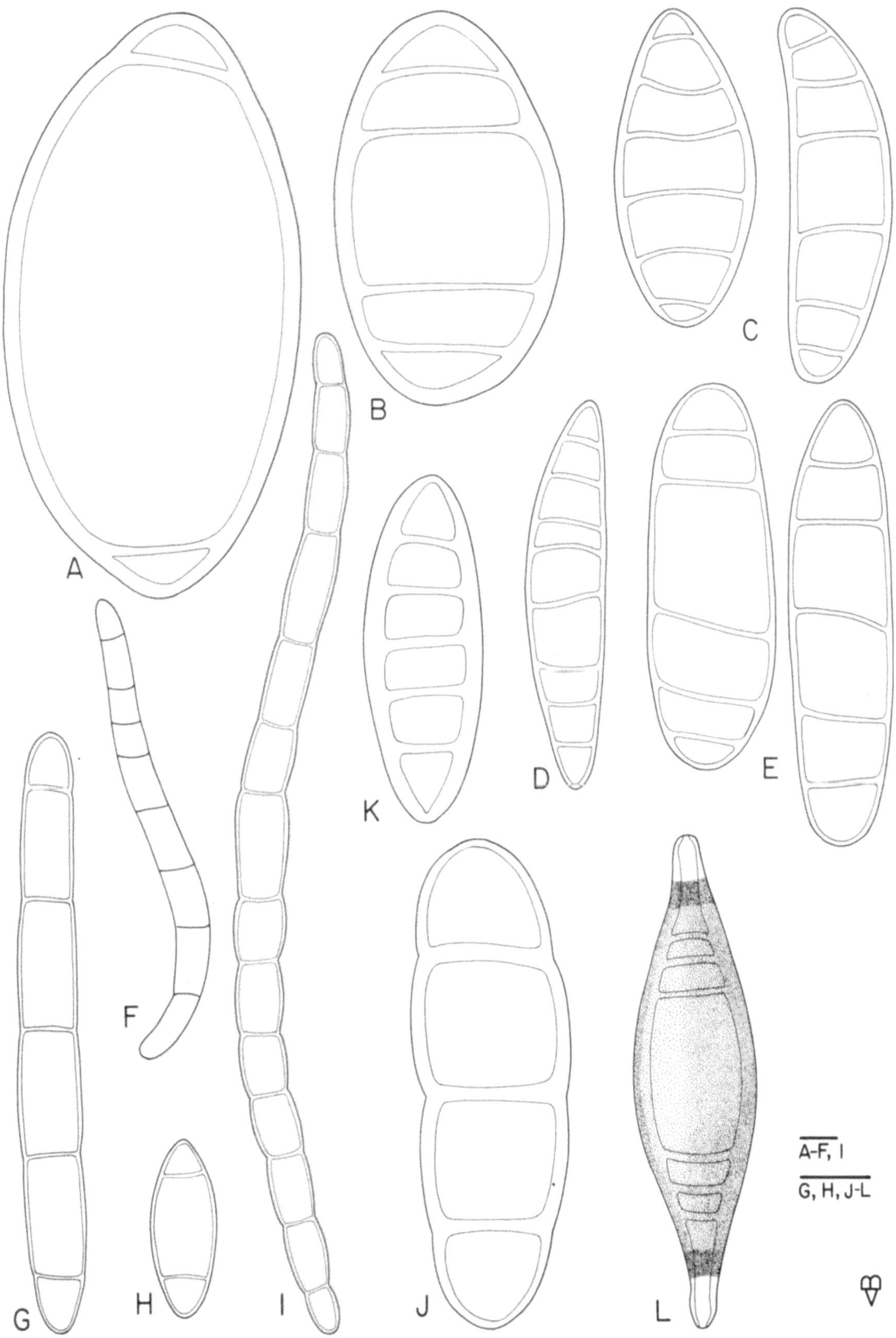

Fig. 15. Ascospores. **A.** *Koralionastes giganteus.* **B.** *K. ovalis.* **C.** *K. ellipticus.* **D.** *K. angustus.* **E.** *K. violaceus.* **F.** *Pontogeneia padinae.* **G.** *P. calospora.* **H.** *P. erikae.* **I.** *P. cubensis.* **J.** *P. codiicola.* **K.** *Luttrellia estuarina.* **L.** *Biatriospora marina.* (Both scales = 10 μm)

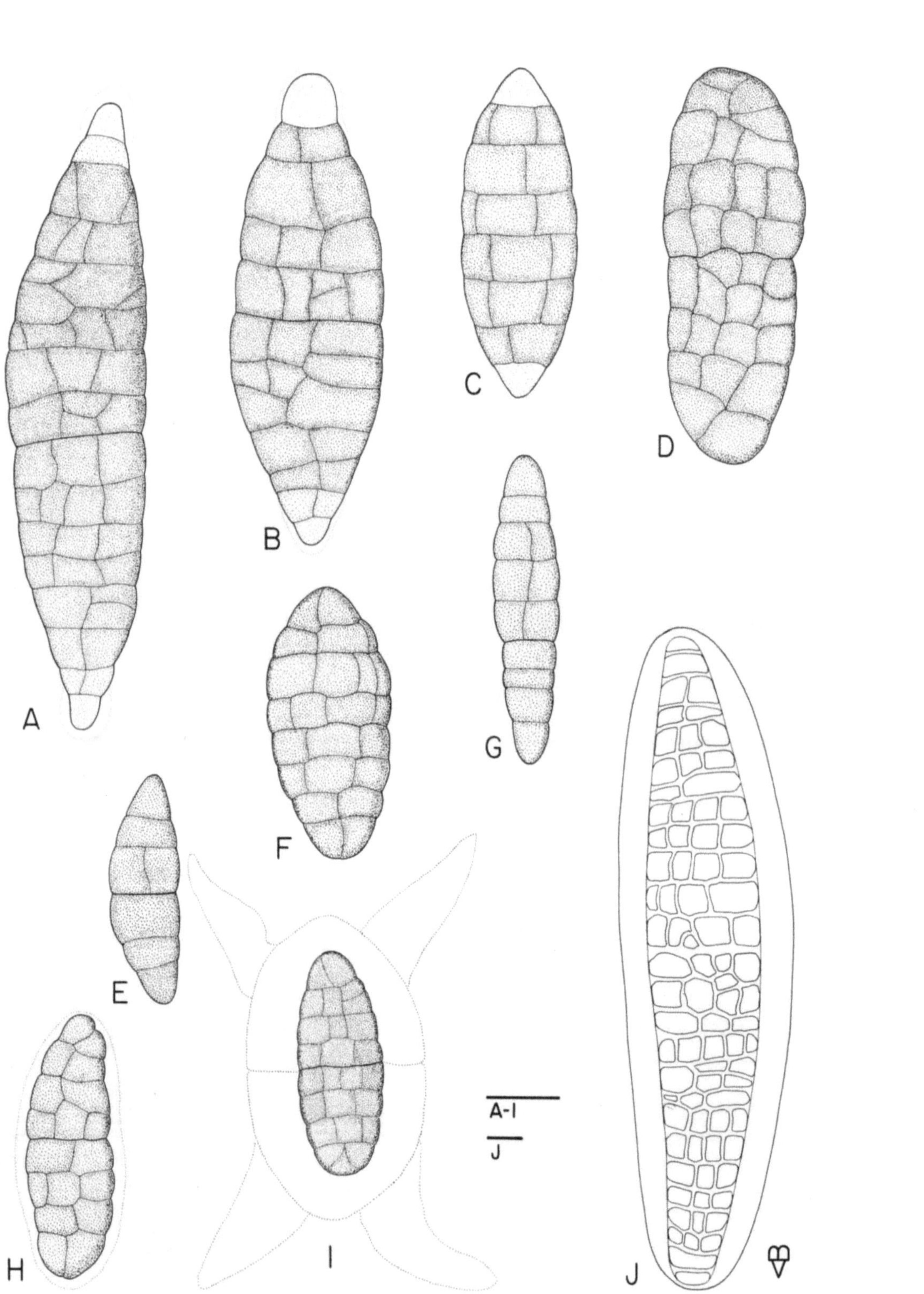

Fig. 16. Ascospores. **A.** *Aigialus grandis*. **B.** *A. parvus*. **C.** *A. mangrovei*. **D.** *Pleospora triglochinicola*. **E.** *P. spartinae*. **F.** *P. pelvetiae*. **G.** *P. pelagica*. **H.** *P. avicenniae*. **I.** *P. gaudefroyi*. **J.** *Lautospora gigantea*. (Both scales = 10 μm)

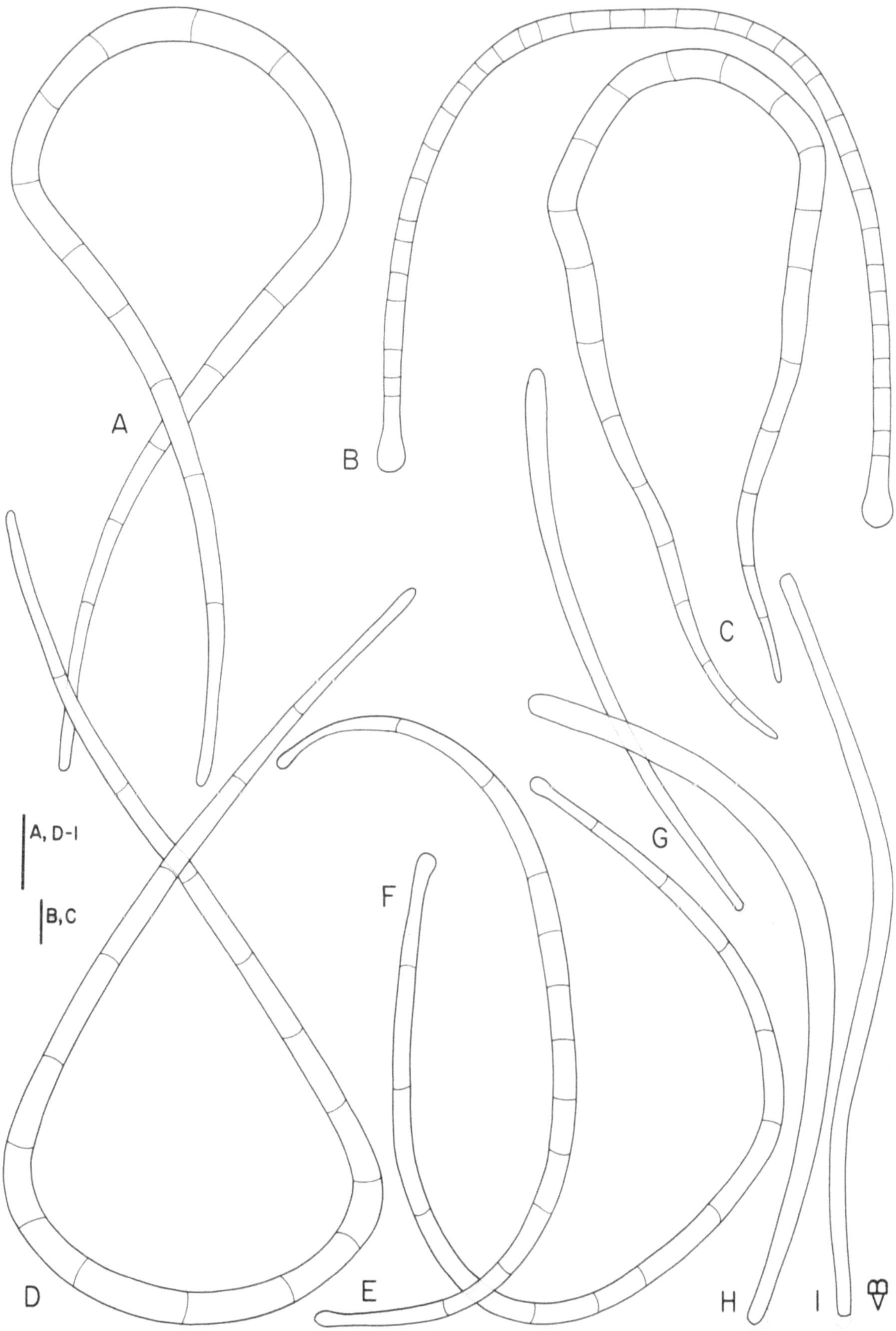

Fig. 17. Ascospores. **A.** *Lindra marinera.* **B.** *L. inflata.* **C.** *L. crassa.* **D.** *L. thalassiae.* **E.** *L. hawaiiensis.* **F.** *L. obtusa.* **G.** *Linocarpon appendiculatum.* **H.** *Linocarpon nypae.* **I.** *Linocarpon* cfr. *pandani.* (Both scales = 10 μm)

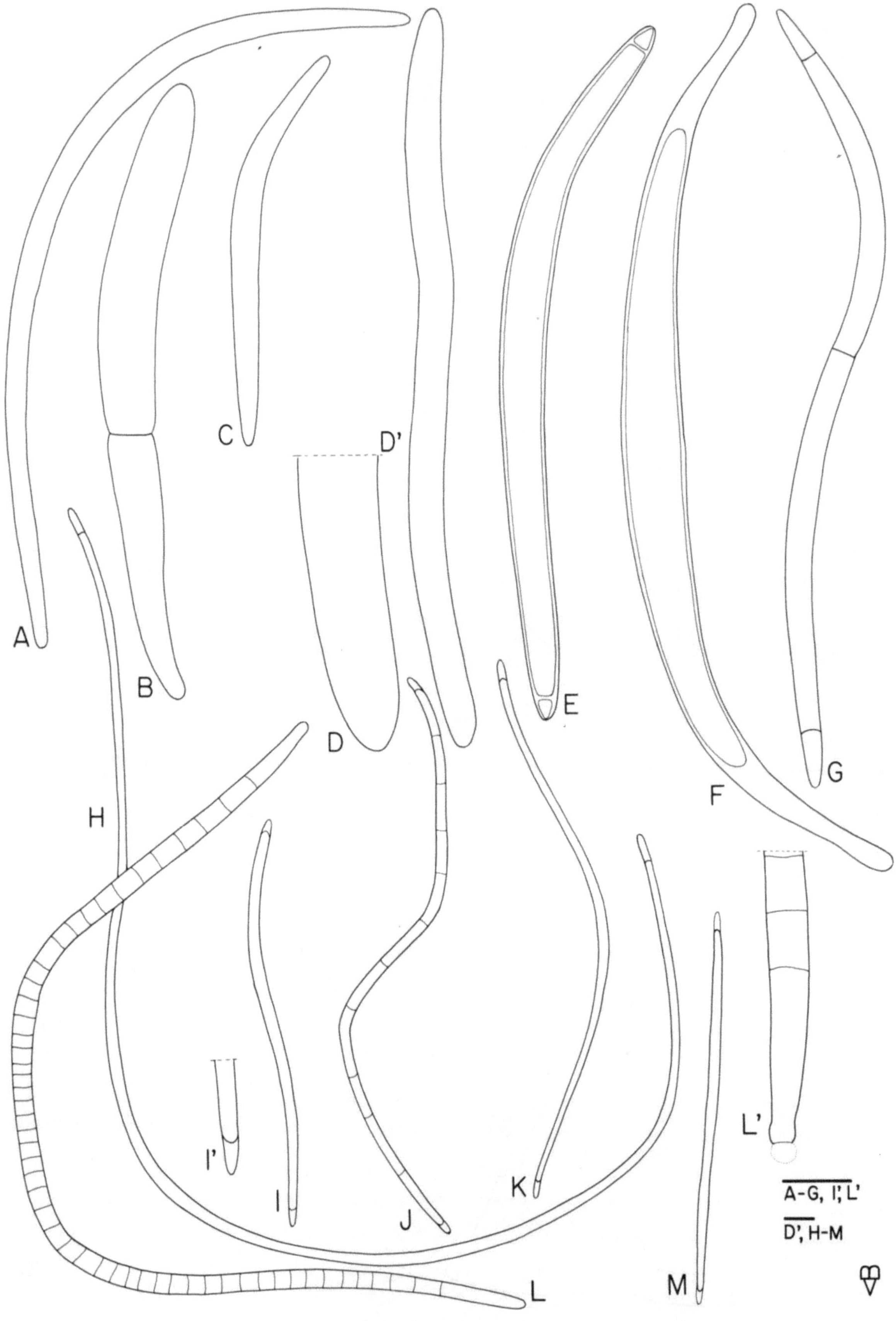

Fig. 18. Ascospores. **A.** *Bathyascus avicenniae*. **B.** *B. tropicalis*. **C.** *B. vermisporus*. **D.** *B. grandisporus* (**D'** at lower magnification). **E.** *Lulworthia calcicola*. **F.** *L. crassa*. **G.** *L. uniseptata*. **H.** *L. grandispora*. **I.** *L. fucicola* (**I'** at higher magnification). **J.** *L. lindroidea*. **K.** *L. kniepii*. **L.** *L. lignoarenaria* (**L'** at higher magnification). **M.** *L. curalii*. (Both scales = 10 μm)

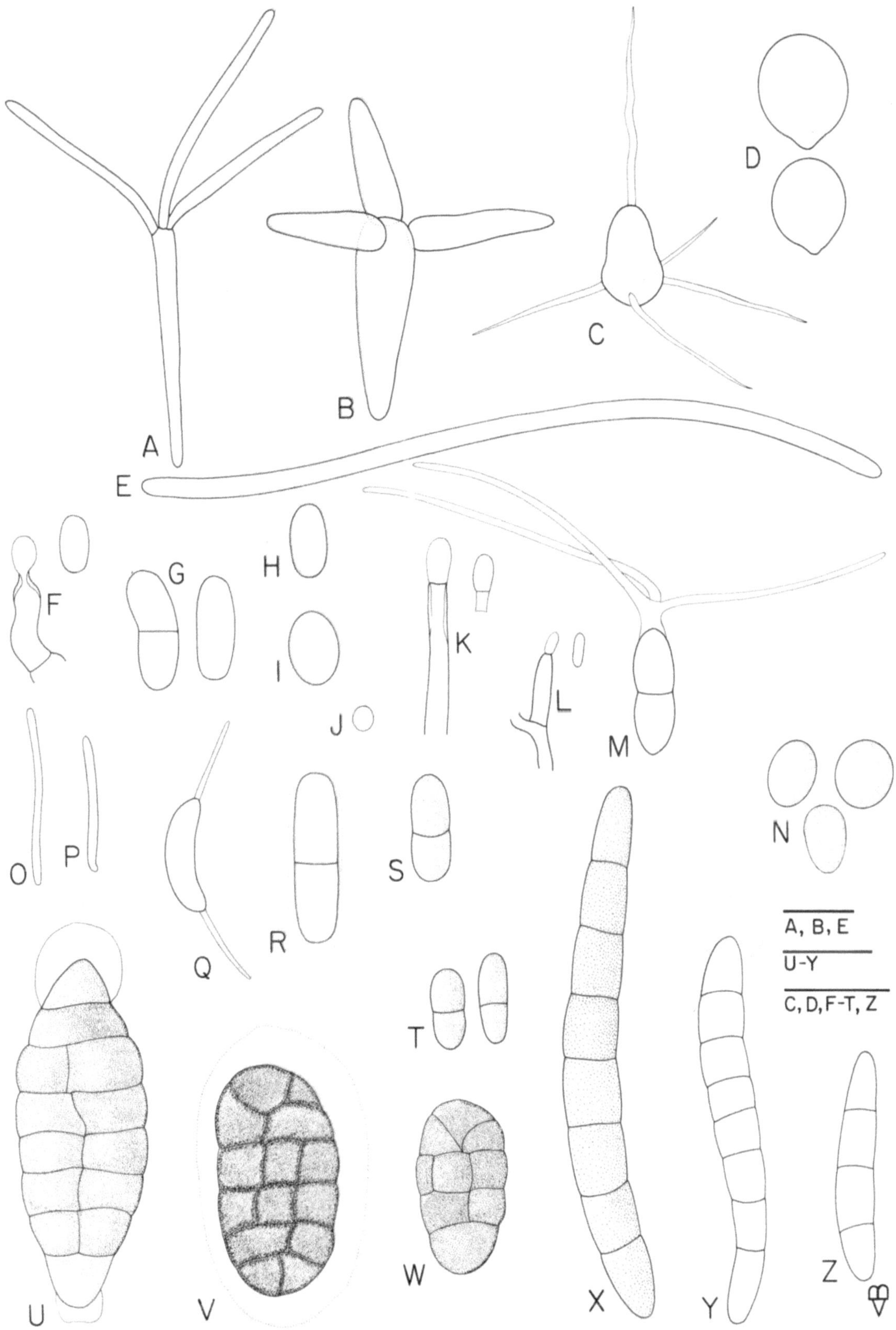

Fig. 19. Basidiospores. **A.** *Digitatispora marina.* **B.** *D. lignicola.* **C.** *Nia vibrissa.* **D.** *Halocyphina villosa.* **E.** *Mycaureola dilseae.* Conidia. **F.** *Phialophorophoma litoralis.* **G.** *Phoma laminariae.* **H.** *P. suaedae.* **I.** *Macrophoma* sp. **J.** *Phoma* sp. **K.** *Gloeosporidina cecidii.* **L.** *Cytospora rhizophorae.* **M.** *Robillarda rhizophorae.* **N.** *Coniothyrium obiones.* **O.** *Halonectria milfordensis*, anamorph. **P.** *Rhabdospora avicenniae.* **Q.** *Dinemasporium marinum.* **R.** *Ascochyta salicorniae.* **S.** *A. obiones.* **T.** *Diplodia oraemaris.* **U.** *Amarenographium metableticum.* **V.** *Camarosporium palliatum.* **W.** *C. roumeguerii.* **X.** *Pleospora spartinae*, anamorph. **Y.** *Stagonospora* sp. **Z.** *S. haliclysta.* (All scales = 10 μm)

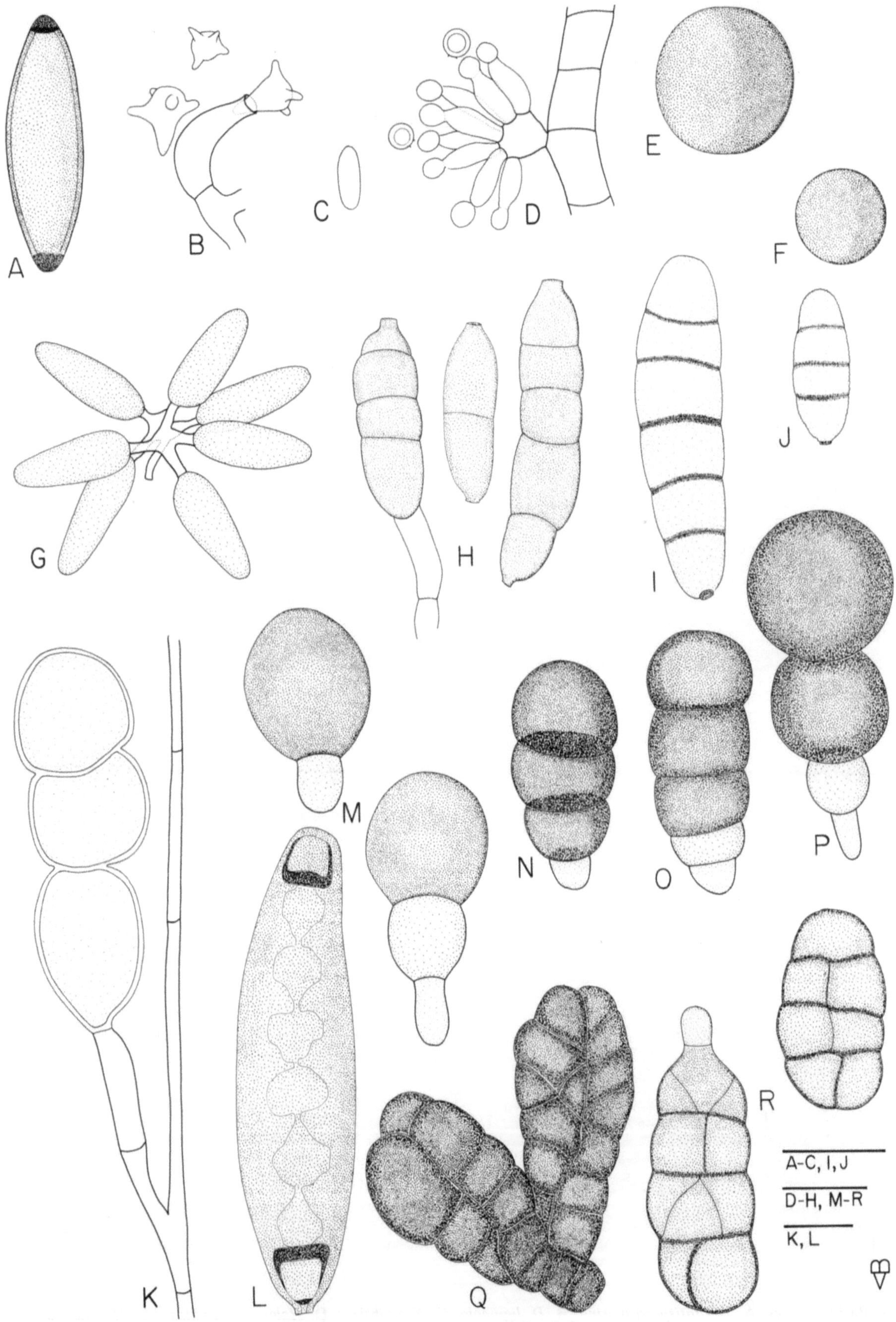

Fig. 20. Conidia. **A.** *Allescheriella bathygena.* **B.** *Clavatospora stellatacula.* **C.** *Tubercularia pulverulenta.* **D.** *Botryophialophora marina.* **E.** *Periconia abyssa.* **F.** *P. prolifica.* **G.** *Asteromyces cruciatus.* **H.** *Cladosporium algarum.* **I.** *Dendryphiella salina.* **J.** *D. arenaria.* **K.** *Blodgettia confervoides.* **L.** *Exserohilum* sp. **M.** *Humicola alopallonella.* **N.** *Trichocladium achrasporum.* **O.** *T. lignincola.* **P.** *T. constrictum.* **Q.** *Dictyosporium pelagicum.* **R.** *Alternaria* sp. (All scales = 10 µm)

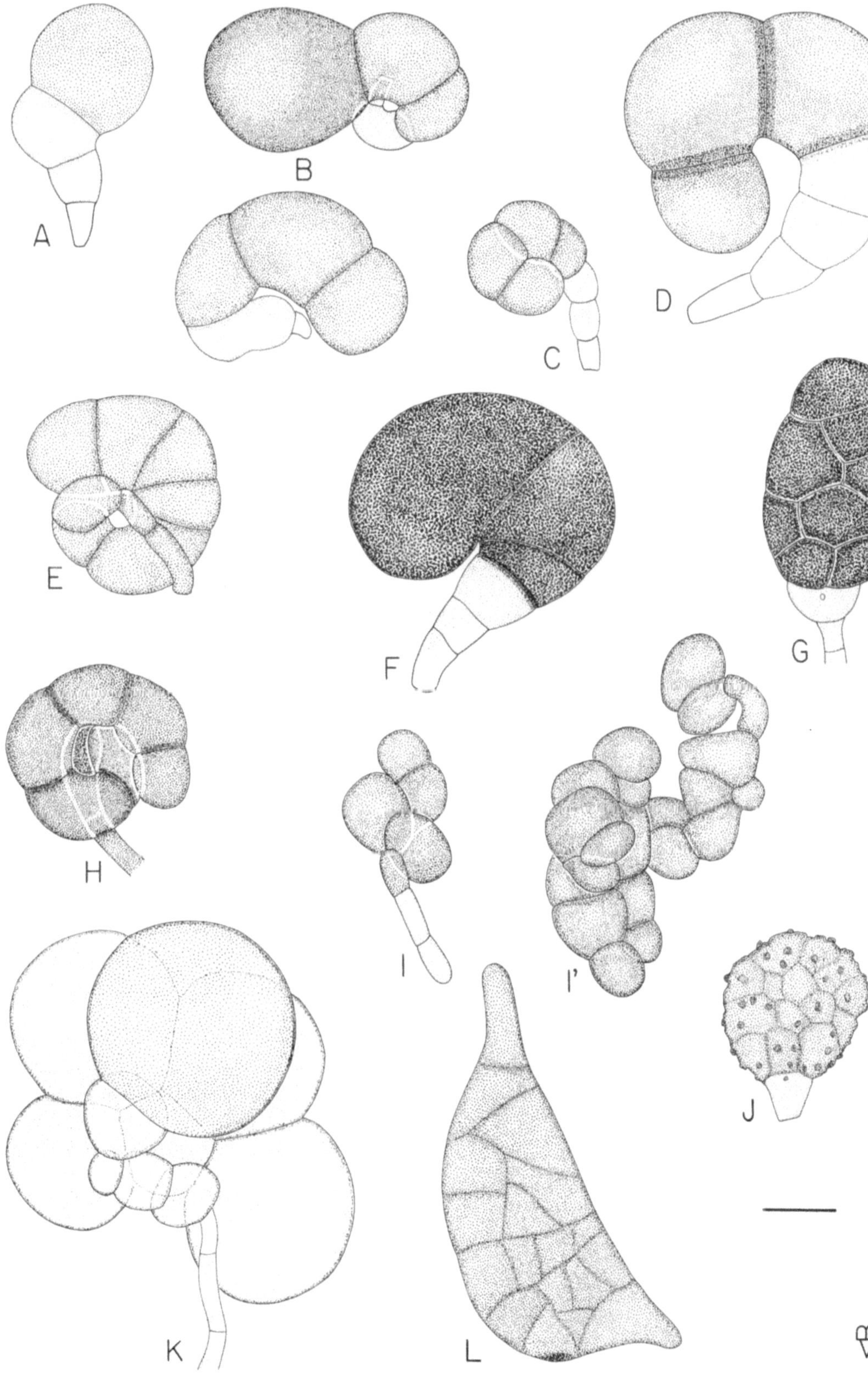

Fig. 21. Conidia. **A.** *Cirrenalia basiminuta.* **B.** *C. fusca.* **C.** *C. macrocephala.* **D.** *C. pseudomacrocephala.* **E.** *C. tropicalis.* **F.** *C. pygmea.* **G.** *Monodictys pelagica.* **H.** *Zalerion maritimum.* **I.** *Z. varium.* **J.** *Epicoccum* sp. **K.** *Cumulospora marina.* **L.** *Stemphylium triglochinicola.* (Scale = 10 μm)

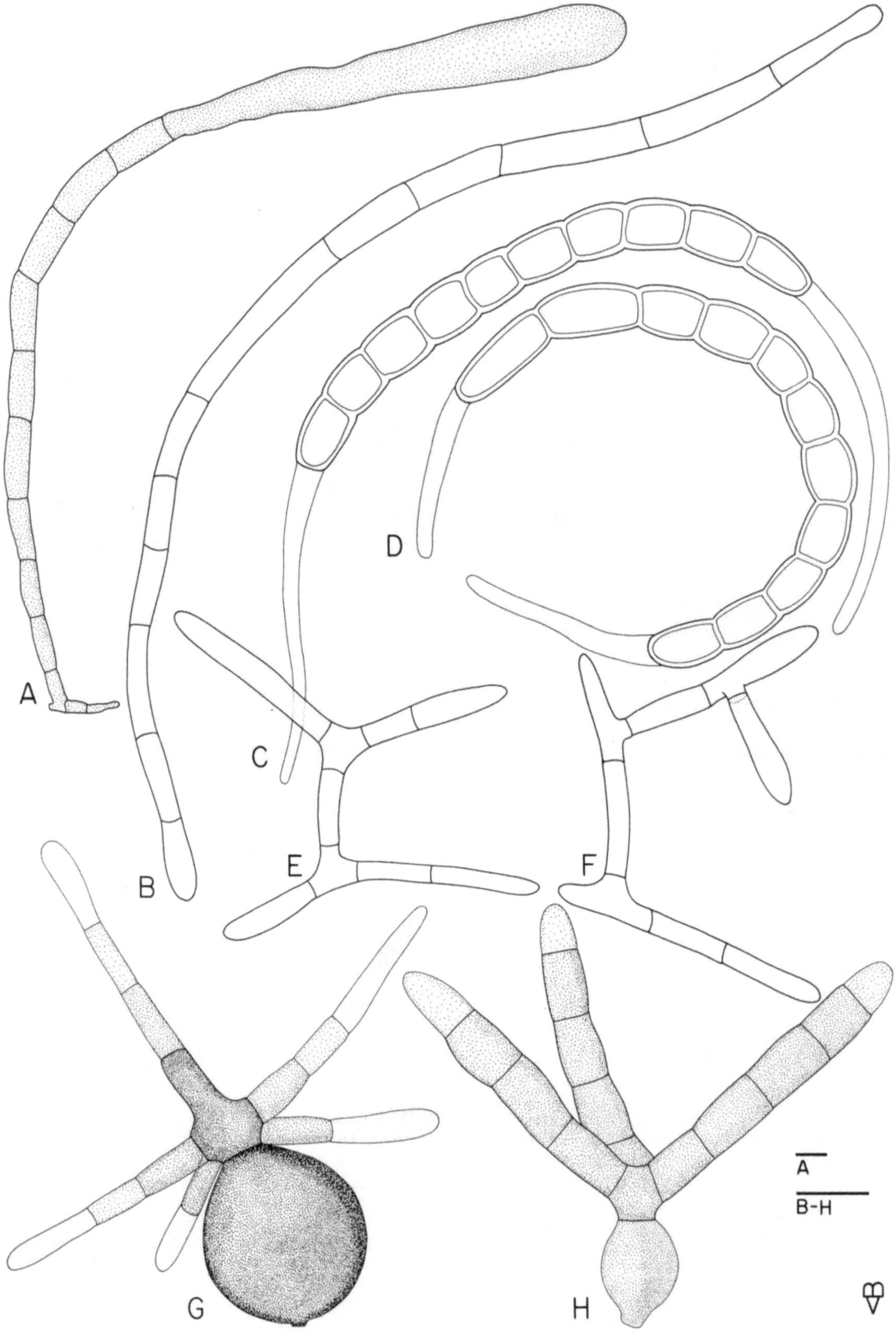

Fig. 22. Conidia. **A.** *Sporidesmium salinum.* **B.** *Anguillospora marina.* **C.** *Sigmoidea marina.* **D.** *S. luteola.* **E.** *Varicosporina ramulosa.* **F.** *V. prolifera.* **G.** *Orbimyces spectabilis.* **H.** *Clavatospora bulbosa.* (Both scales = 10 μm)

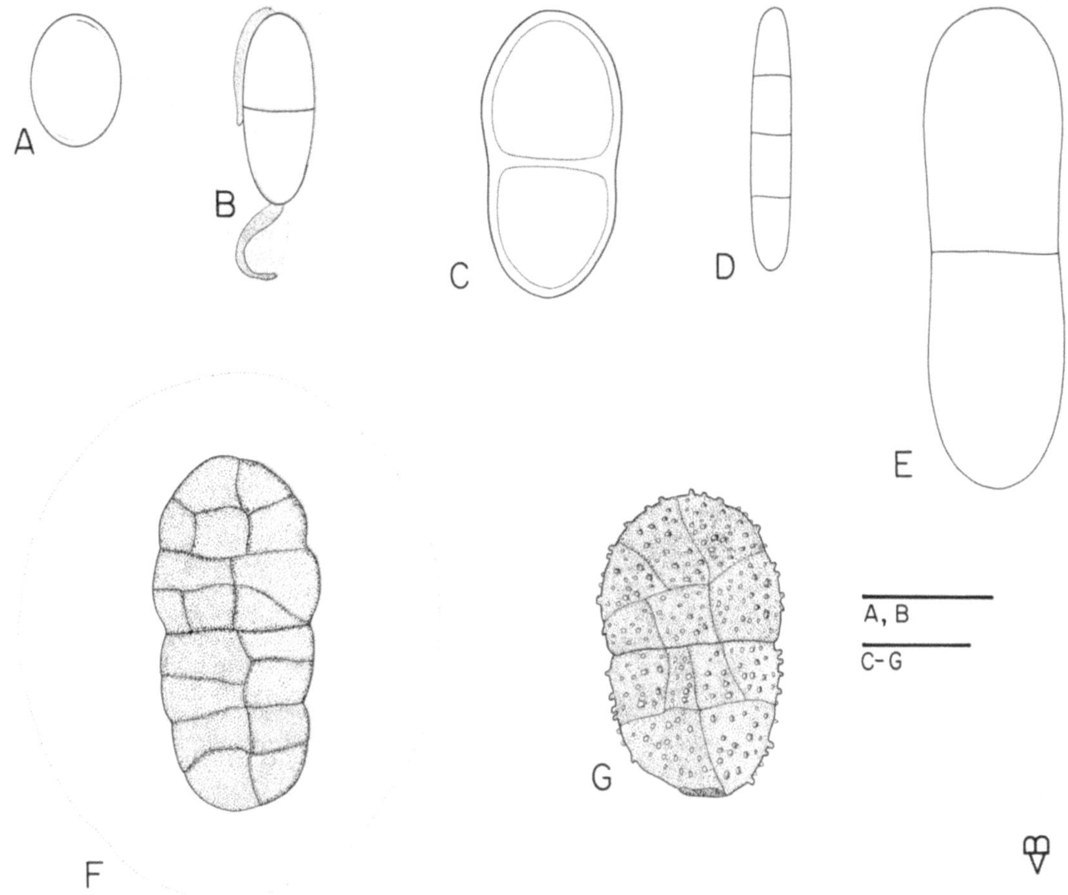

Fig. 23. Ascospores. **A.** *Payosphaeria minuta.* **B.** *Halosarpheia sp.* **C.** *Capillataspora corticola* (shown without appendages). **D.** *Hypophloeda rhizospora.* **E.** *Aniptodera longispora.* **F.** *Pleospora gracilariae.* Conidium. **G.** *Stemphylium gracilariae.* (Scales = 10 μm)

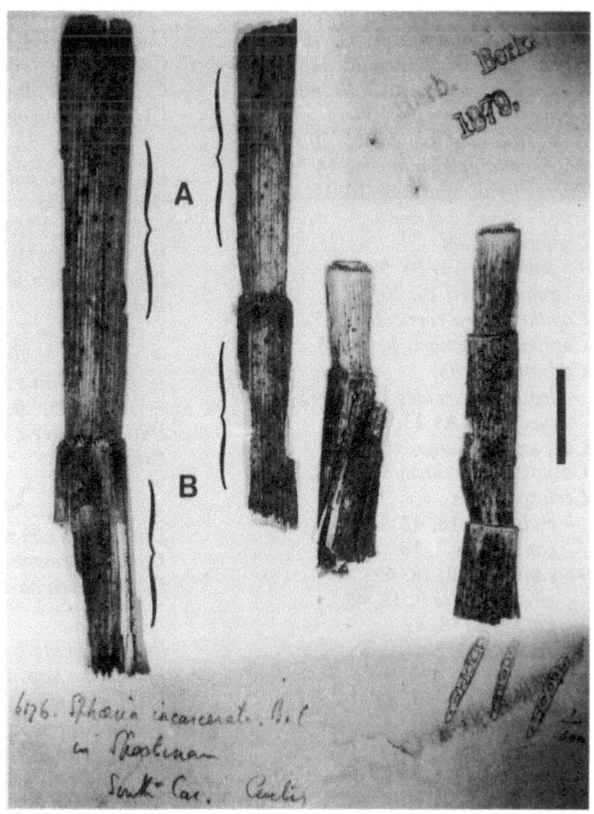

Fig. 24. Type material of *Sphaeria incarcerata* Berkeley et Curtis (A) and *Passeriniella incarcerata* Berlese (B) on culms of *Spartina* from South Carolina (Herb. K). (Scale = 15 mm)

Taxonomical Index

Botanica Marina
Vol. 34, pp. 63–67, 1991

OBITUARY

Prof. Dr William Randolph Taylor

21 December 1895 – 11 November 1990

Rich in years and accomplishments, Wm Randolph Taylor passed away peacefully in his home just five weeks from reaching the age of 95. Professor Taylor had a life and career that spanned the 20th Century. Perhaps no other contemporary phycologist served to bridge the past with the present. An early recollection as a young phycologist was meeting Frank S. Collins on the latter's arrival at Penn Station and taking him in a hansom cab to the Taylor home. Another indication of this link with the past is that he co-authored a paper in 1931 on Brazilian algae with Marshal A. Howe. During his career he carried on correspondence with such colleagues as Fritsch, Hamel, Pascher, Howe, Kylin, Børgesen, Skuja, Skottsberg, and many others around the world. He was a broadly trained and broadly oriented scientist, and his meticulous and thoughtful work gave him an international reputation of the highest rank.

Taylor was born in Philadelphia on 21 December 1895, the first of two sons of Dr William Long Taylor and Caroline (Sower) Taylor. His formal education was received in the Philadelphia area, receiving his B.S. (1916), M.S. (1917), and Ph. D. in botany (1920) from the University of Pennsylvania. He also became a faculty member at this same institution, becoming a professor of botany in 1927. One of his school friends was Ralph E. Cleland, later a cytogeneticist, who had an influence on Taylor's early common interest in cytology. His career showed his deep early involvement in microscopy, first with flowering plants and later with algae. His early algal papers were primarily with phytoplankton, Cyanobacteria, and desmids. His drawings of Newfoundland desmids (1934, 1935) demonstrate his conciseness and his skill. The move of his colleague B.M. Davis from Penn to the University of Michigan was significant, since Davis helped recruit Taylor to move to Ann Arbor.

As early as 1917 Taylor established life-long links with the Marine Biological Laboratory in Woods Hole, Massachusetts, where he conducted his research and also taught the algal course for many years. Taylor managed to spend part or all of his summers there from 1917 through 1989, except for 1918 when he was in the service of the U.S. Army in World War I. Taylor was more in the tradition of Harvey than the Agardhs in that he genuinely loved to be in the field collecting and he was happiest when he was aboard a ship or a boat. His field work was prodigious and wide-ranging. It involved arduous treks into the Selkirk Mountains of eastern British Columbia in the early 1920s and seaweed-collecting excursions along the coastline of New England and eastern Canada as well as island-hopping throughout the Caribbean. Thanks to support from the Carnegie Institution of Washington DC, the summers of 1924–25–26 were spent at the Dry Tortugas Laboratory in the Gulf of Mexico. This experience proved to be crucial in converting Taylor into a student of seaweeds. His initial goal was to compile a simple checklist of the algae, but the unexpected richness of the flora that he encountered, in large part by carrying out dredging, led him to publish his first book in 1928. Over the years he participated in many other expeditions, including two cruises made possible by Captain Allan Hancock, one in the Caribbean and northern South America and the other ranging from the islands off western Mexico to the Galápagos Islands. In 1946 he was a senior biologist in "Operation Crossroads" of the Department of Navy and conducted a botanical survey of Bikini and other Marshall Islands in the South Pacific prior to and immediately after the testing of atomic bombs.

He was also the recipient of many algal collections entrusted to him for identification. He published on the collections made on the cruises of the United States ships *Hassler* (1871–72), and *Albatross* (1887–88). He similarly worked up collections made by the 1937 Smithsonian-Hartford Expedition to the West Indies, the 1960 Smithsonian-Bredin Expedition to Yucatan and their 1957 Expedition to the Society and Tuamotu Islands. He published on collections gathered during the International Indian Ocean Expedition and those made by the *Te Vega* Expedition. He also received many algal collections made by Waldo L. Schmitt of the United States National Museum. He spent much time working in Bermuda, Jamaica, and Hawaii and also always managed to make collections when he attended Botanical Congresses or Seaweed Symposia. With advancing age he expressed regret that he had been unable to visit the shores of Australia, China, or Africa.

In the preface Taylor referred to his 1960 flora rather modestly as being "in the aggregate a pioneer work", and he stated that the book made no claims to monographic completeness for all groups. Yet after 30 years the flora remains unmatched in its scope and its synthesis of the far-flung literature sources. As he did for his "Northeast flora", Taylor compiled his flora from decades of unparalleled field work along with exhaustive laboratory work, supplemented by the scrutiny of types housed in European herbaria. So his taxonomic conclusions were arrived at with a mature judgment. Some genera have been subsequently subjected to more critical study, such as *Cladophora* by C. van den Hoek and *Udotea* by D. and M. Littler, and our understanding has been further refined. But it was Taylor's major contribution to set the stage for the next generation of investigators. His pioneering achievements will be long acknowledged and appreciated. In 1979 Taylor was the first recipient of the Gilbert Morgan Smith Gold Medal in Phycology awarded by the U.S. National Academy of Sciences for his "outstanding contribution to knowledge of marine algae of Florida, the Caribbean Sea, northwest Atlantic and tropical Pacific Ocean". His memory will be perpetuated by the generic names *Taylorophycus* E.Y. Dawson (1961) and *Tayloriella* H. Kylin (1938). Over the course of his career Taylor authored or coauthored the descriptions of numerous taxa: 1 family (Wurdemanniaceae), 5 genera, 250 species, 58 varieties, 27 formae, and he proposed 40 new combinations. Additional professional information can be obtained in his entry in Volume 6 of Stafleu and Cowan's *Taxonomic Literature*, 2nd ed.

In reading Taylor's (1969) account of the marine algae in the Lesser Antilles, one can gain a good insight into his ability to mix the hard facts of scientific data with the more colorful narrative from his own perspective as a traveler. In the section General Aspects of the Algal Vegetation Taylor focuses on each of the islands or island groups visited. He provides a very readable, personal account of his impressions, detailing which collecting sites are most accessible and which are not, what species were the most commonly encountered, and useful historical background of any prior work. He was clearly thinking of future workers who would be traveling these same routes, and he made their way a lot smoother. That characteristic was perhaps indicative of his lasting contribution. He was a pioneer in a sense, but more than that he set a very high standard of scholarship.

It was at the Marine Biological Laboratory that he met Jean Falconer Grant from Virginia, a student in his course, whom he married on 18 December 1926. Jean Taylor generously supported Randolph in his scientific endeavors in the field, at the University, and at Woods Hole for almost 64 years. Randolph expressed his deep gratitude to Jean with his simple dedications of his books to her. Taylor is survived by Jean, their sons, William R., Jr, of Langhorne, PA, and James Keith of Pottstown, PA, and grandchildren, W. R. Taylor III, Carol Ann Taylor, and Patti Jean Taylor. His final resting place is at the Church of the Messiah cemetery in Woods Hole close to the sea.

Michael J. Wynne

Bibliography
(restricted to algal references)

1921a A method of demonstrating the sheath structure of a desmid. *Trans. Amer. Micros. Soc. 40*: 94–95.
1921b Notes from the Woods Hole Laboratory – 1921. *Rhodora 23*: 249–256. (With I. F. Lewis).
1922a Notes on some algae from British Columbia. *Rhodora 24*: 101–111.
1922b Recent studies on Phaeophyceae and their bearing on classification. *Bot. Gaz. 74*: 431–441.
1924a Further notes on British Columbia algae. *Rhodora 26*: 160–166.
1924b Report on the marine algae of Dry Tortugas. *Carnegie Inst. Washington Yearbook 23*: 206–207.
1924c The flora of Penikese, fifty years after. (Marine Algae). *Rhodora 26*: 181–195, 211–219, 222–229, pls. 146, 147. (With I. F. Lewis).
1925a Observations upon the food of certain Tortugas fishes. *Carnegie Inst. Washington Yearbook 24*: 230–232. (With W. H. Longley and W. Schmitt).
1925b Second report on the marine algae of the Dry Tortugas. *Carnegie Inst. Washington Yearbook 24*: 239–240.
1926a Third report on the marine algae of the Dry Tortugas. *Carnegie Inst. Washington Yearbook 25*: 255–257.
1926b The marine flora of the Dry Tortugas. *Revue Algologique 2*: 113–135.
1927 Notes on some freshwater algae from Newfoundland. *Rhodora 29*: 160–164. (With J. M. Fogg, Jr).
1928a The marine algae of Florida, with special reference to the Dry Tortugas. *Carnegie Inst. Washington Publ. 379; Papers Tortugas Lab.* vol. *25*, pp. i–v, 1–219, 37 pls.
1928b The alpine algal vegetation of the mountains of British Columbia. *Proc. Acad. Nat. Sci. Philadelphia 80*: 45–115, pls. 9–13.
1928c Alpine algal flora of the mountains of British Columbia. *Ecology 9*: 343–348, pl. 18.
1928d Notes from the Woods Hole Laboratory – 1928. *Rhodora 30*: 193–198, pl. 176. (With I. F. Lewis).
1928e Observations on *Amoebidium parasiticum* Cienkowski. *J. Elisha Mitchell Sci. Soc. 44*: 126–132, pl. 7.
1929a A species of *Acrothrix* from the Massachusetts coast. *Amer. J. Bot. 15*: 577–583, pls. 42, 43. ["1928"].
1929b The phytoplankton of some Arizona pools and lakes. *Amer. J. Bot. 15*: 596–614, pls. 46, 47. ["1928"] (With H. S. Colton).
1929c General botanical microtechnique. *In*: (C. E. McClung, ed.) *Handbook of Microscopical Technique*. P. B. Hoeber, New York pp. 108–160, 164–180.
1929d Notes on the marine algae of Florida. *Bull. Torrey Bot. Club 56*: 199–210.
1929e Notes on algae from the tropical Atlantic Ocean. *Amer. J. Bot. 16*: 621–630, pl. 62.
1929f The marine algae of the southwestern peninsula of Hispaniola. *Amer. J. Bot. 16*: 651–662. (With C. H. Arndt).
1920a Algae collected by the "Hassler", "Albatross" and Schmitt Expeditions. I. Marine algae from Brazil. *Amer. J. Bot. 17*: 627–634, pl. 39.
1930b Algae from São Paulo, Brazil. *Amer. J. Bot. 13*: 635.
1931a A synopsis of the marine algae of Brazil. *Revue Algologique 5*: 279–313. ["1930"].
1931b Notes on new or little-known algae from Brazil. *Brittonia 1*: 7–33, pls. 1, 2. (With M. A. Howe).
1932 Notes on the genus *Anabaenopsis*. *Amer. J. Bot. 19*: 454–463, pls. 39, 40.
1933a Notes on algae from the tropical Atlantic Ocean, II. *Papers Michigan Acad. Sci., Arts & Lett. 17*: 395–407, pl. 36. ["1932"].
1933b Notes from the Woods Hole Laboratory, – 1932. *Rhodora 35*: 147–154, pl. 274. (With I. F. Lewis).
1933c Methods for collection and study of freshwater algae. Biological Conference Symposium. *J. Michigan Schoolmasters' Club 1933*: 114–125. *Univ. Michigan Official Publ. 35*: 132–143.
1934 The freshwater algae of Newfoundland, I. *Papers Michigan Acad. Sci., Arts & Lett. 19*: 217–278, pls, 45–57. ["1933"].
1935a The freshwater algae of Newfoundland, II. *Papers Michigan Acad. Sci., Arts & Lett. 20*: 185–230, pls. 33–49. ["1934"]
1935a The freshwater algae of Newfoundland, II. *Papers Michigan Acad. Sci., Arts & Lett. 20*: 185–230, pls. 33–49. ["1934"]
1935b Phytoplankton of Isle Royale. *Trans. Micros. Soc. 54*: 83–97.
1935c Alpine algae from the Santa Marta Mountains, Columbia. *Amer. J. Bot. 22*: 763–781, pls. 1–3.
1935d Marine algae from the Yucatan Peninsula. *Carnegie Inst. Washington Publ. 461*: 115–124.
1936a Notes on algae from the tropical Atlantic Ocean, III. *Papers Michigan Acad. Sci., Arts & Lett. 21*: 199–207. ["1935"]
1936b Phaeophycean life-histories in relation to classification. *Botanical Reviews 2*: 554–563.
1937a General botanical microtechnique. *In*: (C. E. McClung, ed.) *Handbook of Microscopical Technique*. 2nd edit. P. B. Hoeber, New York pp. 155–245.
1937b Notes on North Atlantic marine algae, I. *Papers Michigan Acad. Sci., Arts & Lett. 22*: 235–233, pls. 25–27. ["1936"]

1937c *Marine Algae of the Northeastern Coast of North America*. Univ. of Michigan Press, Ann Arbor, Michigan. pp. i–ix, 1–427, 60 pls.

1939a Algae collected by the "Hassler", "Albatross" and Schmitt Expeditions II. Marine algae from Uruguay, Argentina, the Falkland Islands, and the Strait of Magellan. *Papers Michigan Acad. Sci., Arts & Lett. 24*: 127–164, 7 pls. ["1938"]

1939b Freshwater algae from the Petén District of Guatemala. *Botaniska Notiser 1939*: 112–124, pl. 1.

1939c Algae from the Presidential Cruise of 1938. *Smithsonian Miscellaneous Collections 93 (9)*: 1–18, 2 pls.

1940a Marine algae from the Smithsonian-Hartford Expedition to the West Indies, 1937. *Contrib. U.S. Natl. Herb. 28*: 549–562, 2 pls.

1940b Marine algae from Long Island. *Torreya 40*: 185–195.

1941a Reappearance of rare New England marine algae. *Rhodora 43*: 72–74.

1941b Notes on the marine algae of Texas. *Papers Michigan Acad. Sci., Arts & Lett. 26*: 69–79. ["1940"]

1941c Tropical marine algae of the Arthur Schott Herbarium. *Publ. Field Mus. Nat. Hist., Bot., 20*: 87–104, 2 pls.

1942 Marine algae of the Allan Hancock Expedition to the Caribbean – 1937. *Allan Hancock Atlantic Exped., Rept. 2*: 1–193, pls. 1–20.

1943 Marine algae from Haiti collected by H. H. Bartlett in 1941. *Papers Michigan Acad. Arts, Sci. & Lett. 28*: 143–163, pls. 1–4. ["1942"]

1944 The collecting of seaweeds and freshwater algae. Instructions to Naturalists in the Armed Forces for Botanical Field Work, no. 1 Suppl. to Company D Newsletter. pp. 3–17.

1945a William Gilson Farlow, promoter of phycological research in America, 1844–1919. *Farlowia 2*: 53–70.

1945b Pacific marine algae of the Allan Hancock Expeditions to the Galapagos Islands. *Allan Hancock Pacific Exped. 12*: iv + 528 pp., 100 pls.

1945c The collecting of seaweeds and freshwater algae. Instructions to Naturalists in the Armed Forces for Botanical Field Work, no. 1, edit. 2, pp. 3–18.

1947 Algae collected by the "Hassler", "Albatross" and Schmitt Expeditions III. Marine algae from Peru and Chile. *Papers Michigan Acad. Sci., Arts & Lett. 31*: 57–90, 14 pls. ["1945"]

1950a Field preservation and shipping of biological specimens. *Turtox News 28*: 42–43.

1950b *Plants of Bikini and other Northern Marshall Islands. Univ. Michigan Sci. Ser. 18*: xv + 227 pp., frontisp. & 79 pls. Ann Arbor, Michigan.

1950c Reproduction of *Dudresnaya crassa* Howe. *Biol. Bull. 99*: 272–284.

1951a Structure and taxonomic position of *Trichogloea Herveyi*. *Hydrobiol. 3*: 113–121.

1951b Structure and reproduction of *Chrysophaeum Lewisii*. *Hydrobiol. 3*: 122–130.

1952a The algal genus *Chrysophaeum*. *Bull. Torrey Bot. Club 79*: 79.

1952b Reproduction of *Acrosymphyton caribaeum*. *Papers Michigan Acad. Sci., Arts, & Lett. 36*: 31–37, 3 pls. ["1950"]

1952c Survey of the marine algae of Bermuda. *Yearbook Amer. Philos. Soc. 1951*: 167–171.

1952d Notes on *Vaucheria longicaulis* Hoppaugh. *Madroño 11*: 274–277.

1952e Bermudian marine Vaucherias of the section Piloboloideae. *Papers Michigan Acad. Sci., Arts, & Lett. 37*: 75–85, 3 pls. ["1951"] (With A. J. Bernatowicz).

1953a Marine species of *Vaucheria* at Bermuda. *Bull. Mar. Sci. Gulf & Carib. 2*: 405–413, 2 pls. (With A. J. Bernatowicz).

1953b The relation of *Dichotomosiphon pusillus* to the genus *Boodleopsis*. *Papers Michigan Acad. Sci., Arts, & Lett.* 38: 97–107, 3 pls. ["1952"] (With A. B. Joly and A. J. Bernatowicz).

1954a Marine algal vegetation of the Marshall Islands in comparison with other tropical areas. *Proc. Seventh Internatl. Bot. Congr. (Stockholm)* 1950, pp. 826–827.

1954b Distribution of marine algae in the Gulf of Mexico. *Papers Michigan Acad. Sci., Arts & Lett. 39*: 85–109.

1954c The cryptogamic flora of the Arctic, II. Algae, nonplanktonic. *Botanical Review 20*: 363–399.

1954d Marine algal flora of the Caribbean and its extension in neighboring seas. *Huitème Congrès Internat. de Bot., Paris* 1954, Rapp. et Comm., sect. 17, pp. 149–150.

1954e Sketch of the character of the marine algal vegetation of the shores of the Gulf of Mexico. *In*: (P. S. Galtsoff, ed.) *Gulf of Mexico, its Origin, Waters and Marine Life*. pp. 177–192. U.S. Dept. Inter., Fish & Wildlife Serv., Fishery Bull. 55: i–xiv, 1–604.

1955a Notes on algae from the tropical Atlantic Ocean. IV. *Michigan Acad. Sci., Arts, & Lett. 40*: 67–76, 5 pls.

1955b Marine algal flora of the Caribbean and its extension into neighboring seas. *Essays in the Natural Sciences in Honor of Captain Allan Hancock*. Univ. Southern Calif. Press, Los Angeles. pp. 259–270.

1957a Allocution Présidentielle. *Huitème Congrès Internat. de Bot.*, Paris 1954. Comptes Rendus et Rapp. et Comm., sect. 17, pp. 1–4.

1957b *The Marine Algae of the Northeastern Coast of North America*. Revised Edit. Univ. of Michigan Press, Ann Arbor, Michigan. ix + 509 pp., 60 pls.

1959a Phycology in retrospect and anticipation. *In: Vistas in Botany*. Kew Bicentennary Volume. Pergamon Press, London, pp. 328–347.

1959b Associations algales des mangroves d'Amérique. *Colloques Internat. du Centre Nat. de la Recherche Sci., Dinard 81*: 143–152.

1960a Distribution of marine algae about Bermuda. Bermuda Biol. Stat. for Research. 50 pp. (With A. J. Bernatowicz).

1960b *Marine Algae of the Eastern Tropical and Subtropical Coasts of the Americas*. Univ. of Michigan Press, Ann Arbor, Michigan. ix. + 870 pp., 14 photos, 80 pls.

1961a Distribution in depth of marine algae in the Caribbean and adjacent seas. *Recent Advances in Botany. Vol. 1*. Univ. of Toronto Press, Toronto, pp. 193–197.

1961b Notes on three Bermudian marine algae. *Hydrobiol. 18*: 277–283.

1962a *Cladophoropsis philippinensis*, a new species from the western Pacific Ocean. *Bot. Mar. 3*: 56–59.

1962b Marine algae from the tropical Atlantic Ocean. V. Algae from the Lesser Antilles. *Contrib. U.S. Nat. Herb. 36*: 43–62, pls. 1–4.

1962c Observations on *Pseudobryopsis* and *Trichogloea* (Chlorophyceae-Bryopsidaceae) in America. *Brittonia 14*: 58–65.

1962d A note on *Bryopsis* in the West Indies. *Phycologia 2*: 24–28.

1962e Two undescribed species of *Halimeda. Bull. Torrey Bot. Club 89*: 172–177.
1962f *Marine Algae of the Northeastern Coast of North America*. Edit. 2 revised with additions and corrections. Univ. of Michigan Press, Ann Arbor, Michigan. ix + 509 pp., 60 pls.
1964a The genus *Turbinaria* in eastern seas. *J. Linn. Soc. Lond.*, (Bot.) *58*: 475–490.
1964b A valuable old collection of Florida marine algae. *Quart. J. Florida Acad. Sci. 27*: 1–8.
1965 A new *Turbinaria* from the Red Sea district. *Israel J. Bot. 14*: 97–100.
1966a An interesting *Caulerpa* from the Andaman Sea. *J. Phycol. 1*: 154–156. ["1965"]
1966b Records of Asian and western Pacific marine algae, particularly algae from Indonesia and the Philippines. *Pac. Sci. 20*: 342–359.
1966c Notes on Indo-Pacific Turbinarias. *Hydrobiol. 28*: 91–100.
1967a Algal research at the University of Michigan. *In: Research: Definitions and Reflections. Essays on the Occasion of the University of Michigan's Sesquicentennial.* University of Michigan Press, Ann Arbor, Michigan, pp. 137–145.
1967b Species of *Caulerpa* (Chlorophyceae) collected on the International Indian Ocean Expedition. *Blumea 15*: 45–53.
1967c A "Caulerpa" newly recorded for the West Indies. *Botaniste 50*: 467–470.
1967d Caulerpas of the Israel South Red Sea Expedition. Israel South Red Sea Exped., 1962, Reports, No. 24. *Sea Fish. Res. Stat. Haifa, Bull. 43*: 13–17.
1969a Distribution of marine algae about Bermuda. *Bermuda Biol. Stat. for Research, Special Publ. 1*: 1–42. (With A. J. Bernatowicz).
1969b Phycology. *In*: (J. Evan, ed.) *A Short History of Botany in the United States.* XI International Botanical Congress, August 1969, pp. 74–81.
1969c Notes on the distribution of West Indian marine algae particularly in the Lesser Antilles. With a bibliography of recent works on Eastern American tropical algae. *Contrib. Univ. Mich. Herb. 9*: 125–203.
1970 Marine algae of Dominica. *Smithsonian Contrib. Bot. 3*. 16 pp. (With C. F. Rhyne).
1971a A *Punctaria* new to New England waters. *Rhodora 73*: 293–295.
1971b A new *Naccaria* from the West Indies. *Hydrobiologia 38*: 207–212.
1971c Notes on algae from the tropical Atlantic Ocean – VI. *Br. Phycol. J. 6*: 145–156.
1972 Marine algae of the Smithsonian-Bredin Expedition to Yucatán – 1960. *Bull. Mar. Sci. 22*: 34–44.
1973a A new *Halimeda* (Chlorophyceae, Codiaceae) from the Philippines. *Pac. Sci. 27*: 34–36.
1973b Marine algae of the Smithsonian-Breden Expedition to the Society and Tuamotu Islands. *Pac. Sci. 27*: 37–43.
1973c The status of *Agardhiella tenera* and *Agardhiella baileyi* (Rhodophyta, Gigartinales). *Hydrobiol. 43*: 93–107. (With M. J. Wynne).
1974a A new species of *Botryocladia* from the West Indies. *Br. Phycol. J. 8*: 409–412. ["1973"] (With I. A. Abbott).
1974b Notes on algae from the tropical Atlantic Ocean – VII. *Revue Algologique*, n. s., *11*: 58–71.
1975a Marine algae of Great Swan Island. *Atoll Res. Bull.* No. *128*: 6–10.
1975b A pelagic *Sargassum* from the western Atlantic. *Contrib. Univ. Mich. Herb. 11*: 73–75.
1975c A noteworthy variant *Caulerpa. Contrib. Univ. Mich. Herb. 11*: 77–79.
1975d A new species of *Halimeda* from Malaysia. *Contrib. Univ. Mich. Herb. 11*: 81–83.
1976 A check-list of Venezuelan marine algae. *Bol. Soc. Venez. Cienc. Nat. 22*: 71–101.
1977a Notes on plants of the genus *Caulerpa* in the herbarium of Maxwell S. Doty at the University of Hawaii. *Atoll Res. Bull.* No. *208*, 17 pp.
1977b Marine algae of the Te Vega 1965 Expedition in the western Pacific Ocean. *Atoll Res. Bull.* No. *209*, 16 pp.
1980 Notes on marine algae from the tropical Atlantic Ocean – VIII. *Contrib. Univ. Mich. Herb. 14*: 205–207.
1989 A report on philatelic phycology. *Phycologia 28*: 395–396. (With M. J. Wynne).

Botanica Marina
Vol. 34, pp. 69–76, 1991

Fungal Colonization of Submerged *Bruguiera cylindrica* and *Rhizophora apiculata* Wood

W. F. Leong, T. K. Tan and E. B. G. Jones*

Department of Botany, National University of Singapore, Kent Ridge Crescent, Singapore
* *School of Biological Sciences, Portsmouth Polytechnic, King Henry Building, King Henry 1 Street, Portsmouth, Hants PO1 2DY, U.K.*

(Accepted 3 September 1990)

Abstract

The colonization of submerged wood of *Bruguiera cylindrica* and *Rhizophora apiculata* by fungi was followed over 60 weeks in Mandai mangrove, Singapore. Seawater temperature ranged from 26–32 °C, salinity from 21–32‰ and pH from 5.91–8.35. Thirty-two fungi, not all from the same wood species, representing 24 Ascomycotina, 7 Deuteromycotina and 1 Basidiomycotina were collected. Very frequent fungi (above 20%) were *Didymosphaeria enalia*, *Lulworthia* sp. 1, *Lignincola laevis* and *Aigialus parvus* on both timbers; *Payosphaeria minuta* on *B. cylindrica* and *Aigialus mangrovis* and *Nais inornata* on *R. apiculata*. Frequent fungi (10.0–20.0%) were *Halosarpheia marina* on both timbers; and *Nais inornata*, *Hypoxylon oceanicum*, Deuteromycete No. 456 and *Halosarpheia lotica* on *B. cylindrica*. The remaining fungi are classed as infrequent (less than 10%).

Introduction

There is considerable information on the microfungi found on mangrove material. Hyde and Jones (1988) list 90 fungi reported from 18 species of mangrove trees and their geographical location. Studies on their colonization and role in nature are in comparison, less numerous. Degradation studies have been undertaken by Fell and Master (1973: leaves) and Newell *et al.* (1986: leaves); and succession studies by Newell (1973, 1976: seedlings) and Lee and Baker (1973: roots). However, few studies exist on the ability of these fungi to degrade woody tissue (Leightley 1980, Mouzouras 1989) and on their succession on such material (Tan *et al.* 1989). This study reports on the colonization of submerged wood of *Bruguiera cylindrica* and *Rhizophora apiculata* by fungi.

Material and Methods

The study was carried out at Mandai mangrove, one of the most extensive along the Singapore coastline. Table I shows the physico-chemical parameters of the mangrove monitored at each sampling time from December 1985 to February 1987.

Table I. Physico-chemical parameters of Mandai mangrove swamp.

Date of retrieval	Parameter		
	Temperature of seawater	Salinity of seawater	pH of seawater
21. 1. 86	29 °C	29–30 ppt	7.55–7.88
25. 2. 86	26 °C	30–31 ppt	8.34–8.35
11. 4. 86	29 °C	21–23 ppt	7.83–7.90
13. 5. 86	30 °C	24–25 ppt	7.32–7.74
11. 6. 86	30 °C	30–31 ppt	7.41–7.61
24. 7. 86	29 °C	25–27 ppt	6.88–7.02
22. 8. 86	31 °C	26–27 ppt	7.28–7.68
22. 9. 86	32 °C	25–29 ppt	7.45–7.62
23. 10. 86	27 °C	30–32 ppt	7.57–8.35
25. 11. 86	28 °C	24–26 ppt	5.91–6.80
19. 12. 86	28 °C	25–26 ppt	6.77–7.25
3. 2. 87	27 °C	23–24 ppt	7.51–7.91

The mangrove vegetation at Mandai consists of *Avicennia alba* Blume, *A. lanata* Ridley, *Bruguiera cylindrica* (L.) Blume, *Rhizophora apiculata* Blume and *Sonneratia alba* J. Smith, and throughout the period of study, remained undisturbed. Mangrove roots were submerged twice a day during high tides, each time for an average of 6 to 8 hours.

Wood samples for submersion were prepared by splitting young *B. cylindrica* and *R. apiculata* stems (of 5 – 7 cm girth and 8 cm length) into quarters so that each sample contained a surface with intact bark and 2 surfaces with exposed woody tissue. These were drilled centrally and strung into ladders, with the blocks separated 2 – 3 cm apart by knots on the nylon string. For each wood species, 6 replicate strings each carrying 12 blocks were prepared. These were sterilized by autoclaving at 121 °C for 20 minutes, and later attached to the prop roots of mangrove trees standing 5 – 10 m apart at the mid-tide level where they were submerged twice a day.

At each sampling, one block was removed from each of the replicate strings of both timbers. In the laboratory, fouling organisms were removed, the blocks washed in autoclaved seawater and incubated at 28 – 30 °C under diurnal light conditions in plastic boxes lined with a layer of moist, sterile filter paper. The test blocks were examined for the presence of fungi immediately and over a 6-month incubation period. Fungi were recorded on an occurence basis with 6 replicates per timber per sampling time. Voucher slides of all the fungi collected were prepared and are held in the Department of Botany, National University of Singapore.

Results

Seawater temperature in Mandai mangrove ranged from 26 – 32 °C, salinity from 21 – 32‰ and pH from 5.91 – 8.35 (Table I).

Twenty-four species of fungi, not necessarily the same, were recorded on each timber species (Table II). The most numerous fungi were the Ascomycotina with 18 species, followed by 5 Deuteromycotina and 1 Basidiomycotina. The 32 species of fungi observed on both timbers, representing 24 Ascomycotina, 7 Deuteromycotina and 1 Basidiomycotina are listed in Table III which ranks them in order of their percentage occurrence. From the overall percentage occurrence, these fungi may be classed as 'very frequent' (above 20%), 'frequent' (10 – 20%) and 'infrequent' (less than 10%).

Didymosphaeria enalia was very frequent on *B. cylindrica and R. apiculata* (63.3% and 72.9% respectively). The other very frequently occurring species on both timbers were *Lulworthia* sp. 1 (58.3% and 30.0%), *Lignincola laevis* (50.0% and 47.1%) and *Aigialus parvus* (38.3% and 35.7%). *Payosphaeria minuta* (as ascomycete No. 25 in Tan *et al.* 1989, 33.3%) was very frequent on *B. cylindrica*; and *Aigialus mangrovis* (22.7%) and *Nais inornata* (21.4%) on *R. apiculata*. Frequent species were *Halosarpheia marina* on both timbers (18.3% and 10.0%), *Nais inornata* and *Hypoxylon oceanicum* (18.3% occurrence each), Deuteromycete No. 456 (15.0%) and *Halosarpheia lotica* (10.0%) on *B. cylindrica*. Infrequent fungi totalled 14 on *B. cylindrica* and 17 on *R. apiculata*.

Fungal colonization of the wood blocks took place within the first 6 – 18 weeks by 8 or 9 fungal species (Table III). This number increased with time of submersion, reaching 16 species (*B. cylindrica*) and 17 species (*R. apiculata*) during the 22 – 32 week period. The number of species increased slightly during the period 37 – 60 weeks but the number of wood blocks observed was also greater than that of the earlier 2 periods. The fungi observed can be grouped according to their appearance on the wood blocks into early (appeared during the first 18 weeks of submersion), intermediate (between 22 and 32 weeks) and late (after 32 weeks) colonizers (Table III). The 'cut-off' time for early, intermediate and late colonizers was selected in relation to the observed physical condition of the submerged wood blocks. During the early or first 18 weeks of submersion, the wood blocks remained hard and the bark layer intact in most of the retrieved

Table II. Occurrence of marine fungi on different mangrove wood.

Species of mangrove wood	No. of blocks examined*	No. of species observed				Total no. of fungal occurrences
		Asco-mycotina	Basidio-mycotina	Deutero-mycotina	Total	
Bruguiera cylindrica	60	18	1	5	24	234
Rhizophora apiculata	70	18	1	5	24	216
Total	130					

* 12 blocks of *B. cylindrica* and 2 of *R. apiculata* were lost during period of submersion

Table III. Colonization of *Bruguiera cylindrica* and *Rhizophora apiculata* wood blocks by marine fungi.

	Bruguiera cylindrica				*Rhizophora apiculata*			
	Early (6–18 weeks)[1]	Intermediate (22–32 weeks)[2]	Late (37–60 weeks)[3]	Overall % occurrence	Early (6–18 weeks)	Intermediate (22–32 weeks)	Late (37–60 weeks)	Overall % occurrence
Didymosphaeria enalia Kohlm.	33.3	66.7	83.3	63.3	38.9	77.8	88.2	72.9
Lulworthia sp. 1 (ascospores 240–320[350] × 2.5 μm)	44.4	77.8	54.2	58.3	33.3	61.1	11.8	30.0
Lignincola laevis Höhnk	72.2	50.0	33.3	50.0	77.8	72.2	17.6	47.1
Aigialus parvus Schatz *et* Kohlm.	–	33.3	70.8	38.3	5.6	27.8	55.9	35.7
Payosphaeria minuta Leong	11.1	61.1	29.2	33.3	–	11.1	2.9	4.3
Aigialus mangrovis Borse	–	–	16.8	6.7	–	22.2	35.3	22.7
Nais inornata Kohlm.	–	–	45.8	18.3	–	22.2	32.4	21.4
Halosarpheia marina (Cribb *et* Cribb) Kohlm.	–	27.8	25.0	18.3	5.6	16.7	8.8	10.0
Hypoxylon oceanicum Schatz	–	11.1	37.5	18.3	–	–	5.8	2.9
Deuteromycete No. 456	11.1	22.2	12.5	15.0	11.1	5.6	2.9	5.7
Halosarpheia lotica Crane *et* Shearer	–	5.6	20.8	10.0	–	–	–	–
Dactylospora haliotrepha (Kohlm. *et* Kohlm.) Hafellner	–	–	20.8	8.3	–	–	14.7	7.1
Halocyphina villosa Kohlm. *et* Kohlm.	11.1	–	8.3	6.7	–	16.7	2.9	5.7
Aniptodera chesapeakensis Shearer *et* Miller	–	–	16.8	6.7	–	5.6	2.9	2.9
Savoryella paucispora (Cribb *et* Cribb) Koch	5.6	11.6	4.2	6.7	–	5.6	–	1.4
Halosarpheia minuta Leong *et al.*	16.7	5.6	–	6.7	–	–	–	–
Cirrenalia pygmea Kohlm.	–	11.1	–	3.3	5.6	5.6	8.8	7.1
Deuteromycete No. 632	–	–	8.4	3.3	–	–	2.9	1.4
Antennospora quadricornuta (Cribb *et* Cribb) T. W. Johnson	–	5.6	4.2	3.3	–	–	–	–
Dictyosporium pelagicum (Linder) G. C. Hughes	5.6	–	–	1.7	–	–	–	–
Cirrenalia basiminuta Raghu-Kumar *et* Zainal	–	5.6	–	1.7	–	–	–	–
Savoryella lignicola Jones *et* Eaton	–	5.6	–	1.7	–	–	–	–
Halosarpheia cincinnatula Shearer *et* Crane	–	5.6	–	1.7	–	–	–	–
Nais glitra Shearer	–	–	4.2	1.7	–	–	–	–
Deuteromycete No. 140	–	–	–	–	5.6	11.2	–	4.3
Lulworthia medusa (Ellis *et* Everh.) Cribb *et* Cribb (ascospores 360–450 × 2.5–3.7 μm)	–	–	–	–	–	–	8.8	4.3
Halosarpheia retorquens Shearer *et* Crane	–	–	–	–	–	5.6	2.9	2.9
Clavariopsis bulbosa Anastasiou	–	–	–	–	–	5.6	2.9	2.9
Mycosphaerella pneumatophorae Kohlm.	–	–	–	–	–	–	5.8	2.9
Sphaerulina oraemaris Linder	–	–	–	–	–	5.6	–	1.4
Aigialus grandis Kohlm. *et* Schatz	–	–	–	–	–	–	2.9	1.4
Passeriniella savoryellopsis Hyde *et* Mouzouras	–	–	–	–	–	–	2.9	1.4
No. of species appearing	9	16	18		8	17	21	

1 & 2: 18 wood blocks each
3: 24 wood blocks for *B. cylindrica* and 34 wood blocks for *R. apiculata*

Frequency groupings: Very frequent (>20%), Frequent (10–20%), Infrequent (<10%)

Percentage occurrence: $\frac{\text{No. of collections of fungi}}{\text{No. of wood blocks}}$

blocks. With longer submersion (22 to 32 weeks), the bark fell off and the wood blocks began to show signs of decay, as shown by the softness of the woody tissues. Towards the later part of the study (after 32 and up to 60 weeks of submersion), the wood blocks were found to be extremely soft and fragile, some of which easily broke into pieces because of extensive decay.

Lignincola laevis was clearly an early colonizer, appearing on both timbers during 6–18 weeks with more than 70% frequency of occurrence (Table III). This high frequency decreased by the 37–60th week. Another early colonizer was *Didymosphaeria enalia* which differed in peaking later (37–60 weeks, 83.3% on *B. cylindrica*, 88.2% on *R. apiculata*). *Aigialus parvus* colonized the wood blocks at 22–32 weeks on *B. cylindrica* and reached its peak (70.8%) by 37–60 weeks. Likewise for *A. mangrovis* which appeared on *R. apiculata* during 22–32 weeks and reached its peak (35.3%) by 37–60 weeks. The late colonizers, appearing only during the 37–60th week when other species were on the decline, included *A. mangrovis* (16.8%) and *Nais inornata* (45.8%) on *B. cylindrica*, and *Dactylospora haliotrepha* on both timbers (20.8%, 14.7%).

Discussion

Various workers have compiled lists of fungi found growing on mangrove driftwood. Hyde and Jones (1988) collected 47 species in the Seychelles mangroves with *Halocyphina villosa* as the most common fungus. Fungi frequently collected were *Aniptodera mangrovii* Hyde, *Antennospora quadricornuta*, *Halosarpheia marina*, Ascomycete sp. 4 (described as *Rhizophila marina* Hyde *et* Jones) and *Lulworthia grandispora* Meyers. From Malaysia, the most common species were *Rosellinia* sp. (*Hypoxylon oceanicum*), *Savoryella paucispora*, *Halocyphina villosa* and *Trichocladium achrasporum* (Meyers *et* Moore) Dixon (Jones and Tan 1987); and *Halocyphina villosa*, *Hydronectria tethys* Kohlm. *et* Kohlm., *Lulworthia gandispora*, *Hypoxylon oceanicum*, *Halosarpheia marina*, *Didymosphaeria enalia*, *Savoryella lignicola*, *Lignincola laevis* and *Dactylospora haliotrepha* (Jones and Kuthubutheen 1989). In the Philippines, Jones *et al.* (1988) recorded as the most common species *Massarina velatospora* Hyde *et* Borse, *Savoryella lignicola*, *Zalerion varium* Anastasiou and *Rosellinia* sp. (*H. oceanium*), with the following listed as frequent: *Helicascus kanaloanus* Kohlm., *Halosarpheia marina*, *Trichocladium achrasporum* and *Dactylospora haliotrepha*. While these studies give us some insight into the fungi occurring on driftwood, they do not provide the absolute picture. Drift material is composed of wood of different origin (not necessarily arising in the mangroves), species and age of exposure to seawater. In their paper on fungi colonizing *Avicennia* wood, Tan *et al.* (1989) reported as very frequent, *Aigialus parvus*, *Halosarpheia marina*, *Didymosphaeria enalia*, *Halocyphina villosa*, *Lulworthia* sp. 1 and *Payosphaeria minuta* (as ascomycete No. 25).

The data presented in this paper provide information on the colonization of *Bruguiera* and *Rhizophora* wood by fungi, with supporting physico-chemical information. These data can be discussed from the point of:

(1) Frequency of occurrence of species,

(2) Colonization of wood blocks by fungi, and

(3) Comments on the taxonomy of some of the species found.

Frequency of occurrence of fungi on wood blocks

The mycota on *B. cylindrica* and *R. apiculata* were very similar with 24 species recorded for each, of which 16 were common to both timbers. The very frequent and frequent fungi are listed in Table IV which also compares their occurrence on submerged wood of *Avicennia alba* Blume and *A. lanata* Ridley (Tan *et al.* 1989). Eleven fungi were recorded on all four timbers. Nine of these were very frequent or frequent while *Aniptodera chesapeakensis* and *Dactylospora haliotrepha* were infrequent. The latter was a late colonizer of the test blocks (Table III) (which is in agreement with field observation of mangrove driftwood) and this accounts for its low frequency of occurrence. *Didymosphaeria enalia* and *Lulworthia* sp. 1 were very frequent on all four timbers, while *Aigialus parvus*, *Lignincola laevis* and *Payosphaeria minuta* were only very frequent on three timbers. The remaining 4 fungi were only very frequent on one or two of the timber species, indicating a preference for that wood (*e.g.* *Aigialus mangrovis*, *Nais inornata* for *Rhizophora apiculata*; *Halosarpheia marina* for *Avicennia* spp.) or were late colonizers of the wood (*e.g.* *Halocyphina villosa* on *Avicennia* spp., Tan *et al.* 1989).

Twenty four fungi occurred on three or less of the timbers, with only *Halosarpheia lotica*, *Hypoxylon oceanicum* and Deuteromycete No. 456 recorded as frequent; and these restricted to *Bruguiera cylindrica*. The low frequency of *H. oceanicum* may be regarded as a preference for *B. cylindrica*; however, a more plausible explanation is that this is a late colonizer of timber (Table V). Observations by Hyde (1988) and Jones and Kuthubutheen (1989) show that this species is common on mangrove wood, *e.g.* especially on the damaged prop roots of *Rhizophora mucronata*.

Table IV. Distribution of marine fungi on different mangrove wood.

Fungi	Wood species			
	Avicennia alba*	*Avicennia* lanata*	*Bruguiera cylindrica*	*Rhizophora apiculata*
Present on all 4 wood species				
Aigialus mangrovis	+ (I)	+ (I)	+ (I)	+ (VF)
Aigialus parvus	+ (VF)	+ (F)	+ (VF)	+ (VF)
Aniptodera chesapeakensis	+ (I)	+ (I)	+ (I)	+ (I)
Halosarpheia marina	+ (VF)	+ (VF)	+ (F)	+ (F)
Dactylospora haliotrepha	+ (I)	+ (I)	+ (I)	+ (I)
Didymosphaeria enalia	+ (VF)	+ (VF)	+ (VF)	+ (VF)
Halocyphina villosa	+ (VF)	+ (F)	+ (I)	+ (I)
Lignincola laevis	+ (F)	+ (VF)	+ (VF)	+ (VF)
Lulworthia sp. 1	+ (VF)	+ (VF)	+ (VF)	+ (VF)
Payosphaeria minuta	+ (VF)	+ (VF)	+ (VF)	+ (I)
Nais inornata	+ (I)	+ (I)	+ (F)	+ (VF)
Present on 3 wood species				
Antennospora quadricornuta	+ (I)	+ (I)	+ (I)	–
Halosarpheia minuta	+ (I)	+ (I)	+ (I)	–
Lulworthia medusa	+ (I)	+ (I)	–	+ (I)
Mycosphaerella pneumatophorae	+ (I)	+ (I)	–	+ (I)
Present on 2 wood species				
Cirrenalia pygmea	–	–	+ (I)	+ (I)
Cirrenalia basiminuta	+ (I)	–	+ (I)	–
Halosarpheia lotica	+ (I)	–	+ (F)	–
Halosarpheia retorquens	–	+ (I)	–	+ (I)
Humicola alopallonella	+ (I)	+ (I)	–	–
Nais glitra	–	+ (I)	+ (I)	–
Hypoxylon oceanicum	–	–	+ (F)	+ (I)
Savoryella paucispora	–	–	+ (I)	+ (I)
Deuteromycete No. 456	–	–	+ (F)	+ (I)
Deuteromycete No. 632	–	–	+ (I)	+ (I)
Present on 1 wood species				
Aigialus grandis	–	–	–	+ (I)
Aniptodera mangrovii	–	+ (I)	–	–
Clavariopsis bulbosa	–	–	–	+ (I)
Dictyosporium pelagicum	–	–	+ (I)	–
Halosarpheia cincinnatula	–	–	+ (I)	–
Passeriniella savoryellopsis	–	–	–	+ (I)
Lignincola longirostris	–	+ (I)	–	–
Savoryella lignicola	–	–	+ (I)	–
Sphaerulina oraemaris	–	–	–	+ (I)
Deuteromycete No. 140	–	–	–	+ (I)

Key
+ : present
– : absent
VF : very frequent >20% occurrence
F : frequent 10% to 20% occurrence
I : infrequent <10% occurrence
* : Data from Tan *et al.* (1989)

Various factors may account for the low frequency of occurrence of some of the fungi on the four timbers. Some fungi, like *Antennospora quadricornuta*, are usually found to be dominant on submerged substrata in the open sea (Zainal and Jones 1984, 1986). Others may prefer other substrata, *e.g. Mycosphaerella pneumatophorae* on pneumatorphores (Kohlmeyer 1966), or possibly wood exposed for longer periods, *e.g. Nais glitra, Aigialus grandis, Savoryella lignicola, S. paucispora, Passeriniella savoryellopsis* (Hyde and Mouzouras 1988). Jones and Eaton (1969) reported *Savoryella lignicola* as a primary colonizer of wood exposed in a brackish water cooling tower. Jones and Tan (1987) also reported *Savoryella lignicola* as frequent on mangrove driftwood collected at Geylang Patah, Malaysia, usually on well decayed wood. Clearly, further work is necessary to resolve the ecological niche of many of the fungi recorded infrequently in this study, for they have been noted previously from other mangrove collections (Hyde and

Jones 1988). Such factors as timber dimensions, position on the shore and degree of exposure and drying out of the timber need to be studied.

The species listed in Table III are all new records for *B. cylindrica* and *R. apiculata*, timber species not examined previously for the occurrence of marine fungi (Kohlmeyer 1984, Hyde and Jones 1988).

Colonization of wood blocks by fungi

During the first 6–18 weeks of submersion, fewer species were encountered (Table III) and this may be attributed to the presence of a bark layer and hardness of the wood. According to Kohlmeyer and Kohlmeyer (1979), only a few fungi are able to inhabit the intact bark surface because it is a complex tissue composed mainly of dead cork cells with high levels of suberin.

The test blocks offered two types of surfaces for colonization: two sides of debarked wood, while the remaining surface had intact bark. Fungi such as *Halosarpheia minuta* and *Lulworthia* sp. 1 colonized the phloem, parenchyma and cambial cells with the necks penetrating through the bark. *Didymosphaeria enalia* also showed the same pattern of colonization, but it was also found on debarked wood.

The majority of fungi observed in this study were found on the cut surfaces of the submerged wood blocks. Hyde and Jones (1988) also found the majority of mangrove-inhabiting fungi on exposed wood of dead or damaged mangrove roots and branches, and listed such fungi as *Aigialus grandis*, *A. parvus*, *Halocyphina villosa*, *Rosellinia* sp. (*Hypoxylon oceanicum*), *Aniptodera mangrovii* and *A. marina* (*Halosarpheia marina*). From the observations made in the present study, this list can be extended to include *Antennospora quadricornuta*, *Dactylospora haliotrepha*, *Didymosphaeria enalia*, *Passeriniella savoryellopsis*, *Lignincola laevis*, *L. longirostris* (Cribb *et* Cribb) Kohlm., *Lulworthia medusa* (Ellis *et* Everhart) Cribb *et* Cribb, *Savoryella lignicola*, *S. paucispora*, *Sphaerulina oraemaris* and *Nais glitra*, all of which formed superficial, semi-immersed to deep-seated fruiting bodies on cut or exposed wood.

With longer periods of submersion, there was some degree of rotting of the wood caused by the initial colonizers, and this would pre-condition the wood blocks and pave the way for other fungi to settle. Such pre-conditioning of wood is also known to favour the settlement of marine fouling organisms (Kohlmeyer 1984). Towards the end of this study, the wood blocks were found to be heavily infested with fouling organisms. These formed a physical barrier to the settlement and growth of marine fungi, and may have hindered the appearance of more species towards the end of the exposure perid. Jones *et al.* (1972) and Vrijmoed *et al.* (1986 b) previously reported that fewer fungal species were observed on wood heavily infested with fouling organisms compared to wood free of such organisms.

Studies on the colonization of timber submerged in the sea have shown the presence of a characteristic mycota which included *Periconia prolifica* Anastasiou, *Ceriosporopsis halima* Linder, *Antennospora quadricornuta*, *Cirrenalia macrocephala* (Kohlm.) Meyers *et* Moore and *Mondictys pelagica* (Johnson) Jones (Vrijmoed *et al.* 1986 a); and *Zalerion maritimum*, *Ceriosporopsis halima*, *Ceriosporopsis cambrensis* Wilson, *Corollospora maritima* Werdermann and *Halosphaeria appendiculata* Linder (Jones 1968, 1971, Curran 1975). Except for *Antennospora quadricornuta* and *Cirrenalia* spp., these 'open water' species were not observed on the retrieved mangrove wood. These differences could be attributed to the different substrata used and also to the different geographical location of the areas studied. However, the latter is considered unlikely with the studies of Kohlmeyer (1984), Zainal and Jones (1984, 1986), Hyde and Jones (1988) and Tan *et al.* (1989) demonstrating a distinctive mycota for wood exposed in mangroves and sawn wood submerged in the open sea.

A comparison of the fungi collected in this study with those listing fungi from driftwood collections suggests that the latter approach selects for more heavily decayed samples and therefore the later colonizers, *e.g.* *Aigialus grandis*, *A. mangrovis*, *Hypoxylon oceanicum* and *Halocyphina villosa*. It would also account for the absence or low frequency of the early colonizers seen in studies on other driftwood collections, *e.g.* *Lignincola laevis*, *Lulworthia* sp. 1 (Jones and Tan 1987; Hyde and Jones 1988; Jones *et al.* 1988) and *Payosphaeria minuta* (Tan *et al.* 1989).

In this study with submerged mangrove wood, the early, intermediate and late colonizers were mostly the Ascomycotina. This clearly shows that marine ascomycetes play an important role in the colonization of woody substrata in the mangrove habitat. Furthermore, Mouzouras (1989) has shown that many mangrove fungi are active in the decomposition of mangrove wood, *e.g.* white rot: *Halocyphina villosa*; soft rot: *Cirrenalia pygmea*, *C. tropicalis* Kohlm., *Lulworthia* sp. 1, *Savoryella lignicola* and *Trichocladium achrasporum*.

Taxonomy of fungi recorded

Lulworthia sp. 1 listed in Table III had ascospores that measured 240–320(350) × 2.5 µm while those

Table V. Ascospore measurements (µm) of median-sized *Lulworthia* species collected in the tropics.

	Group 1	Group 2	Group 3
Our collections	240 – 320(350) × 2.5 Singapore, mangroves	360 – 450 × 2.5 – 3.7 Singapore, mangroves	–
Hyde, 1988	220 – 335 Brunei, Thailand, mangroves	340 – 490 Brunei, Thailand, mangroves	–
Koch, 1986	280 × 3.5 Thailand, driftwood	440 – 469 × 3 Thailand, driftwood	–
Kohlmeyer and Volkmann-Kohlmeyer, 1987	253 – 341 Hawaii	310 – 429 Hawaii, Galapagos mangroves and Tobago	162 – 269 Hawaii, Galapagos mangroves and Tobago

of *Lulworthia medusa* were 360 – 450 × 2.5 – 3.7 µm. A number of *Lulworthia* species have been described, some are easily assigned, e. g. *L. crassa* Nakagiri, *L. fucicola* Suth., *L. grandispora*, *L. lignoarenaria* Koch *et* Jones, *L. uniseptata* Nakagiri. However, many of the tropical collections with medium-sized ascospores are difficult to assign to species, as has already been stated by Kohlmeyer (1984). From Table V, it can be seen that most collections fall into 2 size groups: (1) 220 – 350 × 2.5 – 3.5 µm; (2) 310 – 490 × 2.5 – 2.7 µm; while Kohlmeyer and Volkmann-Kohlmeyer (1987) recognize a third group. A number of workers have reported on these and clearly there is an urgent need to review the taxonomy of the genus. We have decided to refer to Group 2 collections under *Lulworthia medusa* for convenience while the Group 1 collections are not assigned to a species. *Lulworthia* sp. 1 has larger conical end chambers while those of *Lulworthia medusa* are much smaller (circa. 3.5 – 6.0 µm).

Payosphaeria minuta (as ascomycete No. 25 in Tan *et al.* 1989) was particularly common on *B. cylindrica* wood and is a new genus (Leong *et al.* 1990). It has hyaline, globose to pyriform ascocarps; septate paraphyses; asci that are 8-spored, cylindrical, short pedunculate, unitunicate; and round to oval, 1-celled, hyaline, non-appendaged ascospores.

Nais inornata was found frequently on *R. apiculata*; with ascospore measurements (16 – 20 × 4 – 6 µm) smaller than those of the type (22 – 30 × 11.5 – 15.5 µm). Other characters are in agreement with the type, *e. g.* ascocarp measurements (200 – 300 µm diameter), length of neck (200 – 270 × 30 – 40 µm) and oil globules around the central septum and at the apices of the spore. *Nais inornata* is regarded as a temperate water species and, other than a record by Suhirman and Jones (1983), it appears not to have been observed in the tropics (Kohlmeyer 1984, Hyde and Jones 1988).

Hypoxylon oceanicum has been reported by many workers from mangrove wood as a *Rosellinia* sp. (Jones and Tan 1987, Hyde 1988, Hyde and Jones 1988, Jones *et al.* 1988). Dr A. Whalley (personal communication) does not believe this species belongs in *Hypoxylon* and a new genus will be described to accomodate it.

Halosarpheia minuta (as *Halosarpheia* sp. 1 in Tan *et al.* 1989) is a new species being described by Leong, Tan, Hyde and Jones (in press). It has ascospores measuring 10 – 14 × 4.0 – 6.3 µm and most closely resembles *Halosarpheia viscosa* Schmidt but differs in ascocarp size and morphology, ascospore phenology and habitat preference. *Halosarpheia minuta* was found just beneath the bark of the wood and appears to be particularly common on *Avicennia* bark (Jones, unpublished records).

Three Deuteromycetes with insufficient material for identification are also listed in Table III as Deuteromycete No. 456, 140 and 632. Deuteromycete No. 456 forms white sporodochia on the wood and has small (4 – 6 µm long), clavate, hyaline conidia. This fungus is only known from 5 collections. Deuteromycete No. 140 forms solitary hyaline conidiophores, 40 × 2 µm, with a single terminal conidium that is spindle-shaped, unicellular, smooth, thin-walled and measuring 10 – 12 × 3 µm. Deuteromycete No. 632 forms semi-immersed, ellipsoidal, membranous, black pycnidia 200 × 135 µm with hyaline, globose, smooth, thin-walled spores 2 – 3 µm in diameter.

Acknowledgements

Ms W. F. Leong thanks the National University of Singapore for the award of a research scholarship for this study. Professor Jones is grateful to Portsmouth Polytechnic for granting study leave.

References

Curran, P. M. T. 1975. Lignicolous marine fungi from Ireland. *Nova Hedwigia 26*: 591–599.

Fell, J. W. and I. M. Master. 1973. Fungi associated with the degradation of mangrove (*Rhizophora mangle* L.) leaves in South Florida. *In*: (L. H. Stevenson and R. R. Colwell, eds) *Estuarine Microbial Ecology*. Univ. of South Carolina Press, Columbia. pp. 455–465.

Hyde, K. D. 1988. Studies on the tropical marine fungi of Brunei. *Bot. J. Linn. Soc. 98*: 135–151.

Hyde, K. D. and E. B. G. Jones. 1988. Marine mangrove fungi. *P. S. Z. N. I. Mar. Ecol. 9*: 15–33.

Hyde, K. D. and R. Mouzouras. 1988. *Passeriniella savoryellopsis* sp. nov., a new ascomycete from intertidal mangrove wood. *Trans. Br. mycol. Soc. 91*: 179–185.

Jones, E. B. G. 1968. The distribution of marine fungi on wood submerged in the sea. *In*: (A. H. Walters and J. J. Elphick, eds) *Biodeterioration of Materials*. Elsevier, Amsterdam. pp. 460–485

Jones, E. B. G. 1971. The ecology and rotting ability of marine fungi. *In*: (E. B. G. Jones and S. K. Eltringham, eds) *Marine Borers, Fungi and Fouling Organisms of Wood*. O. E. C. D., Paris. pp. 237–258.

Jones, E. B. G. and R. A. Eaton. 1969. *Savoryella lignicola* gen. et sp. nov. from water-cooling towers. *Trans. Br. mycol. Soc. 52*: 161–165.

Jones, E. B. G., H. Kühne, P. C. Trussell and R. D. Turner. 1972. Results of an international co-operative research programme on the biodeterioration of timber submerged in the sea. *Material & Organismen 7*: 93–118.

Jones, E. B. G. and A. J. Kuthubutheen. 1989. Malaysian mangrove fungi. *Sydowia 41*: 160–169.

Jones, E. B. G. and T. K. Tan. 1987. Observations on manglicolous fungi from Malaysia. *Trans. Br. mycol. Soc. 89*: 390–392.

Jones, E. B. G., F. R. Uyenco and M. P. Follosco. 1988. Fungi on driftwood collected from the Philippines. *Asian Marine Biol. 5*: 103–106.

Koch, J. 1986. Some lignicolous fungi from Thailand, including two new species. *Nord. J. Bot. 6*: 497–499.

Kohlmeyer, J. 1966. Neue Meerespilze on Mangroven. *Ber. Dtsch. Bot. Ges. 79*: 27–37.

Kohlmeyer, J. 1984. Tropical marine fungi. *P. S. Z. N. I. Mar. Ecol. 5*: 329–378.

Kohlmeyer, J. and E. Kohlmeyer. 1979. *Marine Mycology. The Higher Fungi*. Acad. Press, New York, 690 pp.

Kohlmeyer, J. and B. Volkmann-Kohlmeyer. 1987. Marine fungi from Aldabra, the Galapagos and other islands. *Can. J. Bot. 65*: 571–582.

Lee, B. K. H. and G. E. Baker. 1973. Fungi associated with the roots of red mangrove, *Rhizophora mangle*. *Mycologia 65*: 894–906.

Leightley, L. E. 1980. Wood decay activities of marine fungi. *Bot. Mar. 28*: 387–395.

Leong, W. F., T. K. Tan, K. D. Hyde and E. B. G. Jones. 1990. *Payosphaeria minuta* gen. et sp. nov., an ascomycete on mangrove wood. *Bot. Mar. 33*: 511–514.

Leong, W. F., T. K. Tan, K. D. Hyde and E. B. G. Jones. 1991. *Halosarpheia minuta* sp. nov., an ascomycete from submerged mangrove wood. *Can. J. Bot.* (in press).

Mouzouras, R. 1989. Decay of mangrove wood by marine fungi. *Bot. Mar. 32*: 65–69.

Newell, S. Y. 1973. Succession and role of fungi in the degradation of red mangrove seedlings. *In*: (L. H. Stevenson and R. R. Colwell, eds) *Estuarine Microbial Ecology*. Univ. of South Carolina Press, Columbia. pp. 467–480.

Newell, S. Y. 1976. Mangrove fungi: the succession in the mycoflora of red mangrove (*Rhizophora mangle* L.) seedlings. *In*: (E. B. G. Jones, ed) *Recent Advances in Aquatic Mycology*. Elek Science, London. pp. 51–91.

Newell, S. Y., R. D. Fallon and D. J. Miller. 1986. Measuring fungal biomass dynamics in standing-dead leaves of a saltmarsh vascular plant. *In*: (S. T. Moss, ed) *The Biology of Marine Fungi*. Cambridge Univ. Press, Cambridge. pp. 19–26.

Suhirman and E. B. G. Jones. 1983. Preliminary observations on lignicolous marine fungi in the Java Sea, Indonesia. *Annales Bogoriensis 8*: 35–49.

Tan, T. K., W. F. Leong and E. B. G. Jones. 1989. Succession of fungi on wood of *Avicennia alba* and *A. lanata* in Singapore. *Can. J. Bot. 67*: 2686–2691.

Vrijmoed, L. L. P., I. J. Hodgkiss and L. B. Thrower. 1986a. Occurrence of fungi on submerged pine and teak blocks in Hong Kong coastal waters. *Hydrobiologia 135*: 109–122.

Vrijmoed, L. L. P., I. J. Hodgkiss and L. B. Thrower. 1986b. Effects of surface fouling organisms on the occurrence of fungi on submerged pine blocks in Hong Kong coastal waters. *Hydrobiologia 135*: 123–130.

Zainal, A. and E. B. G. Jones. 1984. Observations on some lignicolous marine fungi from Kuwait. *Nova Hedwigia 39*: 569–583.

Zainal, A. and E. B. G. Jones. 1986. Occurrence and distribution of lignicolous marine fungi in Kuwait coastal waters. *In*: (S. Barry, D. R. Houghton, G. C. Llewellyn and C. E. O'Rear, eds) *Biodeterioration 6*. C. A. B. International Mycological Institute, The Biodeterioration Society, London. pp. 596–600.

Botanica Marina
Vol. 34, pp. 77–79, 1991

Phycochemical Studies on *Colpomenia sinuosa* (Scytosiphonales, Phaeophyta)

W. Shaikh, M. Shameel, V. U. Ahmad* and K. Usmanghani**

Department of Botany, University of Karachi, Karachi-75270, Pakistan
* *H. E. J. Research Institute of Chemistry, University of Karachi, Karachi-75270, Pakistan*
** *Department of Pharmacognosy, Faculty of Pharmacy, University of Karachi, Karachi-75270, Pakistan*

(Accepted 3 September 1990)

Abstract

Six fatty acids *i.e.* methyl myristate, methyl palmitate, methyl behenate, methyl lignocerate, methyl oleate and methyl tetradecatrienoate, a sterol, stigmasta-5,23-dien-3β-ol and a metabolite, *tris* β-butyrolactone have been isolated and identified from *Colpomenia sinuosa* (Roth) Derbès *et* Solier. The last two compounds are being reported for the first time from a brown seaweed. *Colpomenia sinuosa* has been found to differ remarkably from the closely related alga *Iyengeria stellata* in its natural products.

Introduction

Colpomenia sinuosa (Roth) Derbès *et* Solier is a globular, hollow, leathery, sessile brown alga, growing abundantly on rocky ledges in the upper and mid-littoral regions along the coast of Karachi during the months of December and February (Nizamuddin and Begum 1978). Very little information is available on the chemical constituents of this seaweed. Shiraiwa *et al.* (1975) reported alginate lyase acitvites in the extracts from several brown algae including *C. sinuosa*. Hussain (1975) identified two new sulphated polysaccharides *i.e.* glucuronomannogalactans from this alga, whereas Nisizawa (1978) reported on the distribution of aldolases in 27 species including *C. sinuosa*.

Dhargalkar *et al.* (1980) described the biochemical contents of *C. sinuosa* from the coast of Maharashtra. Zavodnic (1981) studied the phenolic constituents of this seaweed from the Adriatic Sea. Hussain *et al.* (1983) investigated the influence of seasonal changes on chemical composition of this alga from Alexandria. Nishibori and Namiki (1985) observed that an extract of *C. sinuosa* inhibited autoxidation of linoleic acid. In this paper we report on studies carried out on the fatty acid and sterol constituents of *C. sinuosa* and identify other related compounds.

Material and Methods

Large, healthy specimens of *Colpomenia sinuosa* were collected from the upper and mid-littoral zone from a large and shallow sandy flat at Buleji, near Goth Abdur Rehman during February 1987. The thalli, free from epiphytes, were thoroughly washed with fresh water, dried in the shade and milled.

A. *Isolation of fatty acids*

The dried seaweed (1.6 kg) was extracted three times by percolation at room temperature with $CHCl_3$: hexane (1 : 1) for 20 days. The combined extract was evaporated under vacuum and gave 50 g of thick dark-brownish residue. The procedures for saponification and esterification and the techniques for gas chromatography-mass spectometry (GC-MS) were the same as described by Shameel (1990).

B. *Isolation of sterol*

Dried material (1.2 kg) of *C. sinuosa* was extracted in a soxhlet with $CHCl_3$ (1 L) for 8 h. The extraction was carried out 3 times. The combined extracts were evaporated under reduced pressure, and yielded a thick dark-brownish syrupy residue (45 g). The residue was partitioned between EtOAc and H_2O (1 : 1), and this method was repeated several times. The total EtOAc component was evaporated under vacuum and 29.3 g of thick semi-solid residue was obtained.

The fraction was chromatographed on a silica gel 60 (230 – 400 mesh, E. Merck) column. The column was first eluted with hexane and thereafter Et_2O was added to increase the polarity. The mixture of the sterol-containing fraction was obtained from hexane: Et_2O (70 : 30). The sterol was purified by preparative layer chromatography (silica gel 70 – 230 mesh) using hexane : Et_2O : HOAc (1 : 1 : 0.8). The pure sterol was subjected to NMR and mass scanning.

C. *Isolation of a metabolite*

After evaporation on silica gel the $CHCl_3$: MeOH (3 : 1) extract (21 g) of *C. sinuosa* was loaded on to a silica gel column and eluted with $CHCl_3$: MeOH (9 : 1), which gave a crystalline compound. This compound was repeatedly purified on preparative TLC ($CHCl_3$: MeOH, 9 : 1) plates, and recrystallization was achieved with methanol (m. p. 252 °C).

Results and Discussion

The analysis of fatty acid methyl esters of *Colpomenia sinuosa* by GC-MS showed the presence of methyl myristate, methyl palmitate, methyl behenate, methyl lignocerate, methyl oleate and methyl tetradecatrienoate (Table I). The last mentioned is a tri-unsaturated and the last but one a mono-unsaturated fatty acid.

Of the four saturated fatty acids, only methyl palmitate was also found in *Iyengaria stellata* (Børg.) Børg. (Usmanghani *et al.* 1987 a), while methyl myristate, methyl behenate and methyl lignocerate could not be detected in the later. Similarly methyl pentadecylate, methyl margarate and methyl stearate have been reported from *I. stellata* but could not be detected in *C. sinuosa*. Furthermore, *I. stellata* did not contain saturated fatty acids higher than C 18:0, but much higher acids, *i. e.* C 22:0 and C 24:0 were found in *C. sinuosa*.

Amongst the unsaturated fatty acids methyl oleate is common both to *C. sinuosa* and *I. stellata*, while *I. stellata* also contains methyl palmitoleate (Usmanghani *et al.* 1987 a), which could not be detected in *C. sinuosa*. It appears that *I. stellata* possesses lower mono-unsaturated acids than those of *C. sinuosa*, while the later contains fatty acids which are more unsaturated than the former. Methyl tetradecatrienoate, a tri-unsaturated acid was found in *C. sinuosa* but could not be detected in *I. stellata*. This indicates that the two taxa differ remarkably from each other in their fatty acid composition.

The isolation and chromatographic procedures revealed the occurrence of a single sterol in *C. sinuosa*. This was subjected to NMR and mass spectrometry. The high resolution mass showed the molecular ion peak at m/z 412 indicating the molecular formula as $C_{29}H_{48}O$, which was confirmed by peak matching. The PMR assignments are quite consistent with the structure of stigmasta-5,23-dien-3β-ol, an unsaturated steroid alcohol.

Stigmasta-5,23-dien-3β-ol has already been shown to occur as a minor sterol in the sponge *Calyx nicaaensis* (Steiner *et al.* 1977). It is here reported for the first time from a brown seaweed. This sterol was not detected in *I. stellata*, the sterol fraction of which indicated the presence of cholesterol as a major sterol and 24-methyl cholesterol and ergosterol in minute quantities (Usmanghani *et al.* 1987 b). The last three natural products could not be detected in *C. sinuosa*.

The chloroform-methanol extract of *C. sinuosa* indicated yet another compound, probably a metabolite. The mass spectra showed a very weak molecular ion peak at m/z 258 corresponding with the molecular

Table I. Fatty acids from *Colpomenia sinuosa* analysed as methyl esters

Systematic name	Common name	Molecular formula	Mol. Wt.	Retention time (min.)	Relative percentage
A. *Saturated fatty acid methyl esters:*					
Methyl tetradecanoate	Methyl myristate	$C_{15}H_{30}O_2$	242	11′12″	12.6
Methyl hexadecanoate	Methyl palmitate	$C_{17}H_{34}O_2$	270	13′46″	33.07
Methyl docosanoate	Methyl behenate	$C_{23}H_{46}O_2$	354	18′26″	9.76
Methyl tetracosanoate	Methyl lignocerate	$C_{25}H_{50}O_2$	382	25′26″	3.46
B. *Unsaturated fatty acid methyl esters:*					
Methyl octadecenoate	Methyl oleate	$C_{19}H_{36}O_2$	296	15′52″	38.11
Methyl 2,4,5-tetra-decatrienoate	Methyl tetradeca-trienoate	$C_{15}H_{24}O_2$	236	7′28″	2.83

formula $C_{12}H_{18}O_6$, and provided evidence of a trimer of β-hydroxy butyric acid lactone. A signal observed at m/z 155 is due to the loss of $C_4H_7O_3$ from the molecular ion, while another fragment at m/z 86 is attributed to the monomer of β-butyrolactone. The mass arrangement, therefore, confirms the structure of tris β-butyrolactone. This may have been an artefact formed from 3-hydroxybutanoic acid during isolation. The compound isolated showed antibacterial activity against gram positive bacteria. It is being reported for the first time from a brown alga.

The presence of this trimer is another peculiarity of *C. sinuosa* as it has not been reported from *I. stellata*. This further strengthens the idea that *I. stellata* is a distinct taxon from *C. sinuosa*, with which it has been confused by certain taxonomists. *Iyengaria stellata* was initially reported from the coast of Karachi as *C. sinuosa* f. *tuberculata* Setchell *et* Gardner and *Colpomenia sinuosa* as *C. sinuosa* f. *expansa* Setchell *et* Gardner (Anand 1940, Salim 1965). Apart from morphological and anatomical differences (Nizamuddin and Begum 1978), these two also differ remarkably from one another phycochemically. It is, therefore, justified to keep them in different genera.

Acknowledgements

We would like to thank Dr (Mrs) Shaheen Bano, Research Associate, H. E. J. Research Institute of Chemistry, University of Karachi for her valuable help during the course of this study and Dr Gerald T. Boalch, The Laboratory, Citadel Hill, Plymouth for critical evaluation of the manuscript.

References

Anand, P. L. 1940. *Marine Algae from Karachi* I. *Chlorophyceae.* Punjab Univ. Bot. Publ., Lahore, 52 pp. + 6 pls.

Dhargalkar, V. K., T. G. Jagtap and A. G. Untawale. 1980. Biochemical constituents of seaweeds along the Maharashtra coast. *Indian J. Mar. Sci. 9*: 297–299.

Hussein, M. M.-D. 1975. Two new polysaccharides from *Colpomenia sinuosa. Phytochem. 14*: 1866–1868.

Hussein, M. H.-D., A. Abdel-Aziz and H. M. Salem. 1983. Chemical composition of *Colpomenia sinuosa* as influenced by seasonal variation. *Pak. J. Sci. Ind. Res. 26*: 152–154.

Nishibori, S. and K. Namiki. 1985. Antioxidative activity of seaweed lipids and their utilization in food. *Kaseig. Zasshi 36*: 845–850.

Nisizawa, K. 1978. Aldolases of multicellular marine algae. *In*: (J. A. Hellebust and J. S. Craigie eds.): *Handbook of Phycological Methods: Physiological and Biochemical Methods.* Cambridge Univ. Press, London. pp. 239–244.

Nizamuddin, M. and M. Begum. 1978. Studies on the marine algae (Punctariales) from Karachi. *Rev. Algol.* N. S. *13*: 315–326.

Salim, K. M. 1965. The distribution of marine algae along Karachi coast. *Bot. Mar. 8*: 183–198.

Shameel, M. 1990. Phycochemical studies on fatty acids from certain seaweeds. *Bot. Mar. 33*: 429–432.

Shiraiwa, Y., K. Abe, S. F. Saskai, T. Ikawa and K. Nisizawa. 1975. Alginate lyase activities in the extracts from several brown algae. *Bot. Mar. 18*: 97–104.

Steiner, E., C. Djerassi, E. Fattorusso, S. Magno, L. Mayol, C. Santacroce and D. Sica. 1977. Isolation, structure determination and synthesis of new acetylenic steroids from the sponge *Calyx nicaaensis. Helv. Chim. Acta 60*. 475–481.

Usmanghani, K., M. Shameel, S. Siddiqui and S. A. Siddiqui, 1987 a. Studies on the fatty acids of *Iyengaria stellata* (Scytosiphonales, Phaeophyta). *Bot. Mar. 30*: 305–307.

Usmanghani, K., M. Shameel, S. Siddiqui and M. Alam. 1987 b. Studies on the sterols of a brown seaweed *Iyengaria stellata* from Pakistan. *Pak. J. Bot. 19*: 249–252.

Zavodnic, C. 1981. Studies on phenolic content of some brown algae from Adriatic Sea. *Proc. Int. Seaweed Symp. 10*: 543–548.

Botanica Marina
Vol. 34, pp. 81–89, 1991

The Structure and Reproduction of a New Species *Helminthocladia nizamuddinii* (Nemaliales – Rhodophyta) from the Coast of Pakistan

S. Afaq-Husain and M. Shameel*

PCSIR Laboratories Complex, Off University Road, Karachi-75280, Pakistan

* *Department of Botany, University of Karachi, Karachi-75270, Pakistan*

(Accepted 22 September 1990)

Abstract

A new taxon, *Helminthocladia nizamuddinii* Afaq-Husain *et* Shameel sp. nov. is described. The vegetative, anatomical and reproductive structures, which are highly variable, have been investigated in detail. A variable placement of the cell wall of the first division of the zygote and up to four distal cells of gonimoblast filaments producing carposporangia are the characters which distinguish it from presently known species of *Helminthocladia*. Moreover, in the present species 2-celled carpogonial branches are also occasionally seen, involucral filaments are common but a fusion cell is not present.

Introduction

During a survey of the marine algal flora of the Karachi and Lasbela coasts several interesting red algae have been found (Shameel 1987, Shameel *et al.* 1989). One of these, which appeared to be a new species of *Helminthocladia* J. Agardh, is described here. Most of the early studies on *Helminthocladia* (*e.g.* Kylin 1938, Feldmann 1939, Levring 1953) were made on herbarium material and left several questions unanswered. Desikachary (1957) pointed out: "There is a definite need for studying as many species as possible of this genus with the help of fluid preserved material". The present study was made on specimens freshly collected *in situ* as well as freshly washed ashore and then preserved in formalin solution.

At present the Liagoraceae (Nemaliales – Rhodophyta) are little known from Pakistani waters (Afaq-Husain *et al.* 1991), and there is a need for a detailed study of the family from this area. This is the first report of *Helminthocladia* from this region.

Material and Methods

Fresh drift plants were collected from Manora, Sandspit, Hawksbay, Buleji and Paradise Point (all localities being within 2–5 km of each other) in the months of December–April 1986–1989. Many plants were collected *in situ* from Paradise Point as well as Buleji. Drift plants were also collected from the rocky ledge of Sandspit with their attachment discs intact. Some specimens were fixed in 4% formalin-seawater solution, others were mounted on herbarium sheets and stored in the Herbarium of PCSIR, Karachi (CLH) and Karachi University Herbarium (KUH). Materials were squashed in 1% aniline blue in 5% HCl after treatment with 10% HCl for 48 h. Permanent slides were prepared using equal volumes of glycerine, acetic acid and distilled water or in Karo, a thick syrup used as a baby food. Drawings were made with the help of a camera lucida.

Observations

External description

Thalli are brownish red, mucilaginous, slippery, tough, up to 70 cm long and caespitose. Three to five (–7) axes arise from a holdfast, which is discoid, soft, usually 2–5 (–10) mm broad, sometimes very minute. The stipe is conico-cylindrical, 1–2 mm broad at the base, becoming broader above. The main axes are commonly 2–5 mm broad, irregularly cylindrical or tereto-compressed, becoming as broad as 20 mm and giving rise to several palmate branches. Branching is slight to profuse, radial, irregular, up to 3–4 (–6) orders (Figs 29–31).

Branching of the thallus starts at a variable distance from the holdfast, occurring just above the holdfast or as much as 6 cm above. The main axes of all the plants examined and branches in general are narrow at the proximal end. Ultimate branches are thin, slender, highly flexible, 445–480 μm broad at obtuse apices which look conical to the naked eye. The thalli may also be beset, to a varying degree, with more or less fusiform ramuli which are 0.5–2.0 mm broad and up to 12 mm long, usually constricted at the bases and with obtuse apices.

Anatomical description

The thallus is multiaxial consisting of a compact medulla of thick and thin filaments running along the longitudinal axis of the thallus, and a cortex, 200–300 μm broad, made up of radially arranged filaments ending in assimilatory branches. The medullary filaments consist of large, elongate, cylindrical or fusiform cells, up to 134 μm long and 30–50 (–60) μm broad; sometimes a cell becomes as broad as 125 μm but not more than 25 μm at joints, resembling a large sack. The main filaments bear primary branches, 16–24 μm broad, running along the main filaments, which in turn produce secondary branches running along the central core as well as outwardly towards the periphery forming the cortex of the thallus. Their cells are 7–15 (–20) μm broad and 45–156 μm long, becoming shorter and broader distally.

The secondary branches of the cortex, on reaching the periphery, give rise to radially arranged assimilatory branch systems, which are subdi- or subtrichotomous due to the continuous formation of young branches from older ones (Fig. 1). Their cells are usually broader at the distal end and bear chloroplast in this portion (Figs 7, 8, 11). The terminal cells are more inflated than others, oblong or pyriform, 20–82 μm long, 14–40 μm broad and form the surface of the thallus. In surface view they appear round with a darkly stained body in the centre, the plastid which is stellate is clearly seen in the apical portion of the terminal cells (Fig. 8).

Some assimilatory branches end in cylindrical cells, which are usually narrow, 36–53 μm long and 10–24 μm broad (Fig. 21). The second cell is usually shorter in length than the first and third cell in the assimilatory branch and is 23–40 μm long and 12–26 μm broad. The third cell is 30–60 (–92) μm long and 7–36 μm broad. Red rectangular crystals of various sizes are also seen in assimilatory cells (Fig. 8). The rhizoid like filaments are 7 μm broad and usually arise from the lower cells of assimilatory branches but sometimes from the upper cells as well (Fig. 1). Assimilatory branches arise only from the distal part of the bearing cell, but medullary or rhizoidal filaments may arise from any part (Fig. 1).

Hairs

Three types of structures have been observed arising terminally from the apical cells of lateral assimilatory branches, which are identified as hairs:

(a) Single-celled hairs: up to 8 (–10) μm long, 7 (–8) μm broad, with dense cytoplasm and becoming darkly stained with aniline blue, usually one but sometimes two at a time on a cell (Fig. 7). Moreover, one to four empty and ruptured cells are frequently observed in a similar position, which appear to be the remnants of single-celled hairs (Fig. 15). Sometimes a new hair is observed beside a remnant (Fig. 8).

(b) Two-celled hairs: up to 20 μm long and 8 μm broad (Fig. 10) are less dense and less stained than the above cells.

(c) Long, hyaline, nonseptate hairs: up to 137 μm long and 4 μm broad, bearing a broad cylinder like base, 6–8 μm broad and with a mucilaginous ring in the lower half (Figs 9, 11).

Single-celled hairs have been observed frequently in all plants examined. The other two types are not so common and are also not found in every plant and only one hair was observed on a cell at a time. However a long hair is sometimes observed beside a remnant (Fig. 9). From observations it appears that:

(1) The initials and remnants of all three types of hairs are not distinguishable, and their positions are similar which indicate they are of similar origin.

(2) Single-celled hairs soon deteriorate but their walls remain on the cells for a longer time. A cell can produce up to four such hairs, one after the other.

(3) These cells sometimes may develop further into two-celled hairs (Fig. 10) or into long hyaline hairs (Fig. 11).

Adventitious branches

Several types of special adventitious branches, differing in appearance from assimilatory ones, develop from the second or third cells of a pre-existing assimilatory branch. Their apical cells are not enlarged or

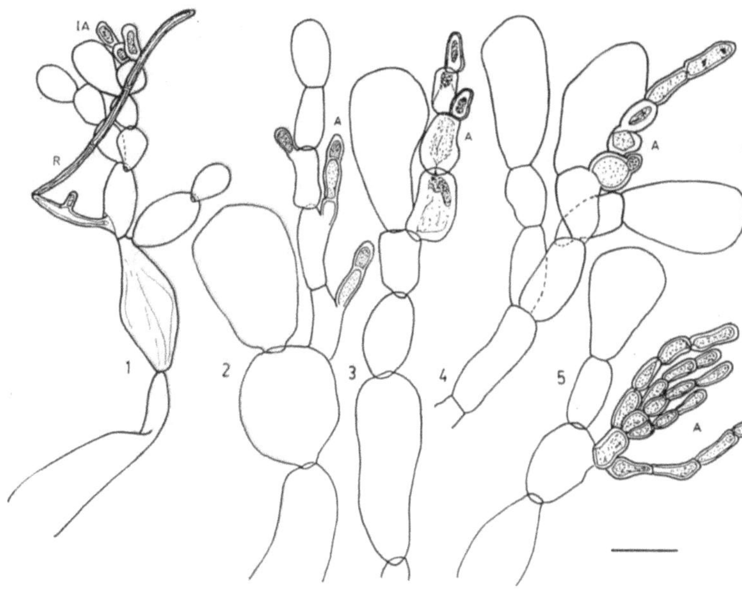

Figs 1–5. *Helminthocladia nizamuddinii* Afaq-Husain *et* Shameel
Fig. 1. A cortical branch system with sack like cells, a rhizoidal filament from upper cell and initials of adventitious branches. Figs 2–5. Different types of adventitious branches from pre-existing assimilatory branches. A = adventitious branch, IA = initials of adventitious branches, R = rhizoidal filament. Scale: Fig. 1 = 60 µm, Figs 2–5 = 30 µm.

pyriform (Figs 2–5). The mother branches of these adventitious branches consist of more or less fusiform cells with very narrow connections. Another type of branch, consisting of large sack like cells, up to 120 µm long and 90 µm broad, with narrow connections, has also been observed. The cells of these branches give rise to thin, elongated protuberances with dense cytoplasm, which develop into slender filaments with elongated cells (Fig. 6).

The plants are dioecious as well as monoecious, the two forms not being distinguishable in external appearance.

Spermatangia

Spermatangial branches, when young, appear to develop in panicules (Fig. 12) turning into heads embeded in mucilage on maturation. The spermatangial branches develop directly on the second cell of secondary assimilatory branches, which do not bear pyriform inflated cells. Special slender branches consisting of thin elongated cells also develop from the lower cells of the assimilatory branch system bearing spermatangial branches terminally (Fig. 12). The spermatangial branches are borne on apical parts of cells and are branched up to 5 times. The cells of spermatangial branches and their assimilatory mother branches become shorter towards their distal end. The ultimate cells of spermatangial branches bear 2–7 spermatangial mother cells, each bearing up to 4 spermatangia terminally. The cytoplasmic contents of a spermatangium rounds off forming a single spermatium. The spermatangial mother cell is oblong, 5–7 µm long and 3–4 µm broad. The spermatangium is round or slightly oblong, up to 4–5 µm in diameter. The Spermatium is round, 3–4 µm in diameter.

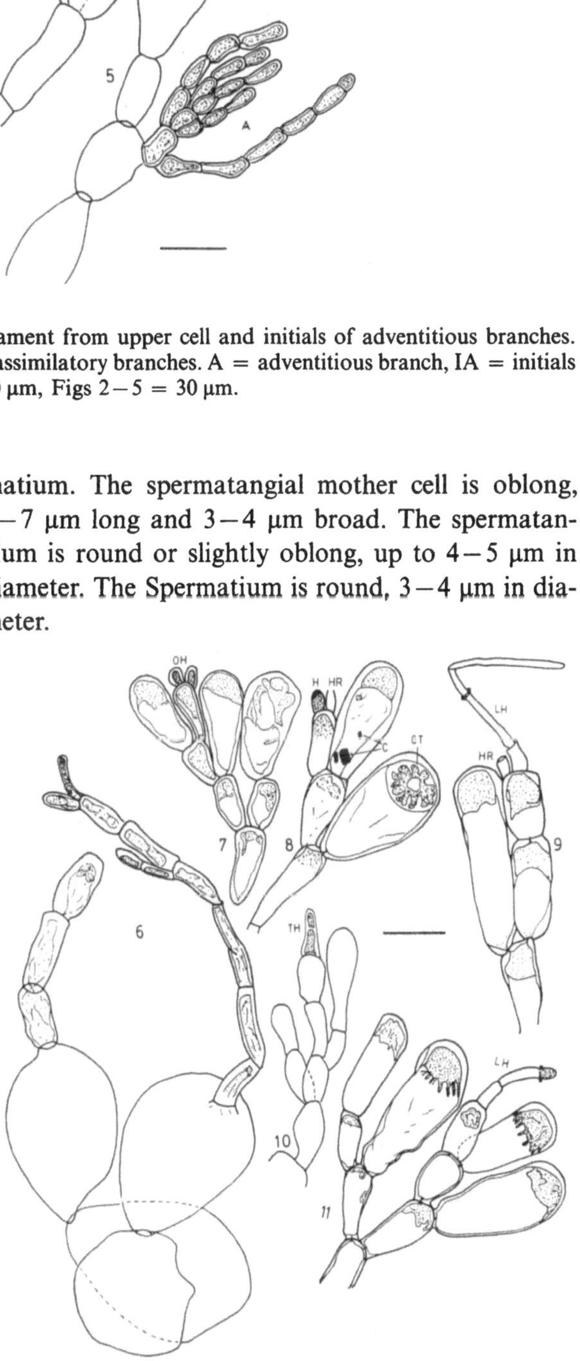

Figs 6–11. *Helminthocladia nizamuddinii*
Fig. 6. Sack like cells with slender branches at the top. Figs 7–11. Portions of assimilatory branches showing different types of hairs, hair remnants, stellate plastids in the apical parts and red crystals in the assimilatory cells. C = red crystals, CT = stellate plastid, H = hair initial, HR = hair remnant, LH = long hyaline hair, OH = one-celled hair, TH = two-celled hair. Scale: Figs 6–11 = 30 µm.

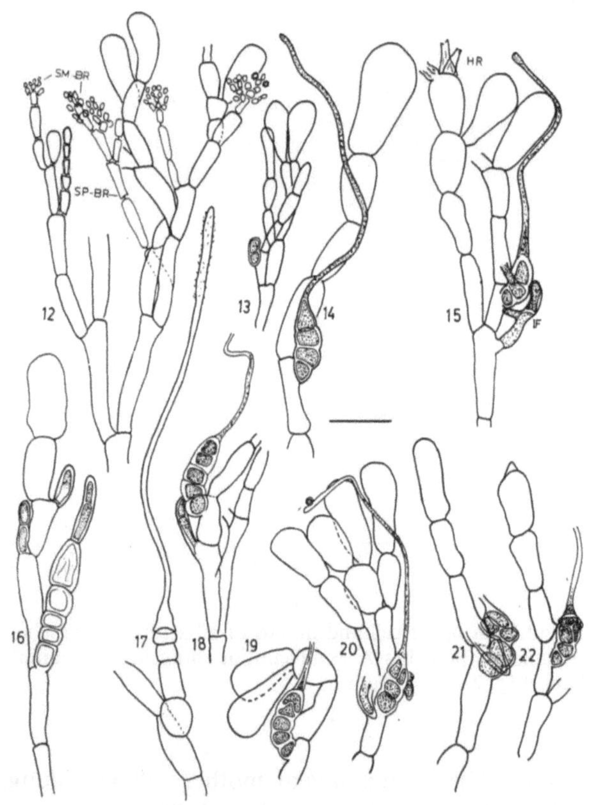

Figs 12–22. *H. nizamuddinii*
Fig. 12. Young spermatangial branches on special long slender branches which do not end in inflated terminal cells. Fig. 13. A young carpogonial branch (two-celled stage). Fig. 14. A four-celled carpogonial branch. Fig. 15. A three-celled carpogonial branch and development of involucral filaments before division in the carpogonium. Fig. 16. A five-celled (non-fertile) carpogonial branch with thick trichogyne probably turning into a vegetative filament. Fig. 17. A five-celled (non-fertile) carpogonial branch with a very long trichogyne having some mucilaginous matter attached on its apical portion. Figs 18–20. Carpogonial branches showing transverse (Fig. 18), conico-oblique (Fig. 19) and longitudinal (Fig. 20) division of the fertilized carpogonium. Figs 21–22. Carpogonial branches showing planes of first two divisions (transverse + longitudinal) in the fertilized carpogonium. HR = hair remnant, IF = involucral filamelnt, SM-BR = spermatangial branch, SP-BR = special branch bearing spermatangial branches. Scale: Figs 12–22 = 30 µm.

Carpogonia

The carpogonial branches consist of (2–) 3–4 cells, borne laterally on the upper half of the fourth or fifth cell of an assimilatory branch (Figs 14, 15). Sometimes more than one (up to 3) carpogonial branches develop on the same assimilatory branch, but if more than two they appear nonfunctional and look sterile. The carpogonium is elongate, conical, 12–14 µm long and 9–12 µm broad at the base. The trichogyne is 67–168 µm long, 1–4 spermatia have been seen attached to the tip of trichogyne (Fig. 20). No change has been noticed in the amount of cytoplasm and the size of the bearing cell, on which a carpogonial branch develops. The cells of carpogonial branch are cut off successively, the trichogyne is formed by the last cell of the carpogonial branch (Figs 13, 14).

Sterile carpogonial branches also occur to a varying degree in all the plants examined, and consist of 5–6 (–8) cells which are unstained with aniline blue like other cortical cells. These branches usually end in a small broadly conical cell or rarely terminate in a long trichogyne (Fig. 17) and sometimes turn into a vegetative filament (Fig. 16). Some plants are found containing many-celled (up to 16 or more) carpogonial branches in a greater frequency than normal and show a consistent, characteristic appearance.

Post fertilization changes in carpogonium

After fertilization the trichogyne remains partly or wholly intact with the dividing carpogonium for a long time (Figs 23–26), and no remarkable change in size of any cell of carpogonial branch or its bearing cell is observed. The first division in the fertilized carpogonium may be transverse, longitudinal or oblique, all three types have been clearly observed (Figs 18–20). The plane of second division is also not definite (Figs 21, 22). Both the daughter cells divide and redivide forming a rounded mass of cells (Fig. 23). Sometimes two such masses are observed (Figs 24, 25), which may be the result of independent behaviour of the two daughter cells of the first division product of the zygote. This mass increases in size and its cells give rise to gonimoblast filaments by transverse division forming a conico-round cystocarp, 100–220 µm broad (Figs 26, 27). The cells of the gonimoblast filaments are 7–15 (–20) µm long, (3–) 4–7 µm broad. The lower cells are comparatively shorter and broader becoming elongate and slender distally, and the apical cell is obtuse. Branching of gonimoblast filaments is copious but not dichotomous (Fig. 28). Any cell except the apical one puts forth a protuberance by the side of previous cell, which gives rise to a secondary branch, the branching is up to 3–4 orders.

One to four distal cells of the gonimoblast filaments become enlarged in size and rich in cytoplasm and act as carposporangia. The cytoplasmic content of each carposporangium rounds off forming a single carpospore sequentially. Carposporangia are 10.8–16.8 µm long, 7.2–9.6 µm broad; carpospores are oblong, 10–15 µm long, 7–9 µm broad. After formation of carpospores, the lower vegetative cell of the gonimoblast filament gives rise to a new branch from the apical part, which again produces 1–2 carpospores terminally in a similar fashion.

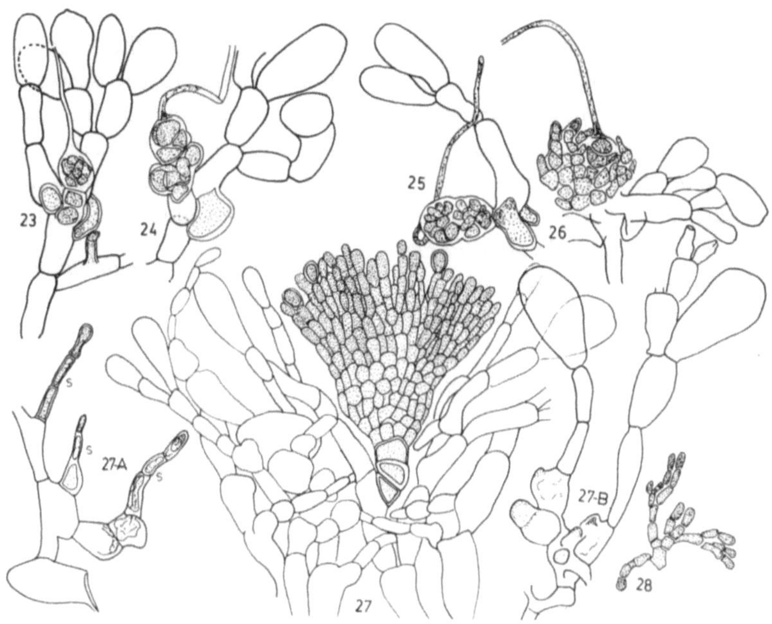

Figs 23–28. *H. nizamuddinii*
Fig. 23. Fertilized carpogonium giving rise to a single mass of cells of gonimoblast initial. Figs 24–25. Fertilized carpogonium giving rise to two masses of cells of gonimoblast initial probably from two daughter cells of the first division product, dividing separately. Fig. 26. Inception of gonimoblast filaments from cellular mass of the fertilized carpogonium. Fig. 27. A young cystocarp showing: lower three cells of the carpogonial branch forming a stalk like structure but do not fuse to form a fusion cell (all doted) and new branches forming loose network below the cystocarp (without dots). Fig. 27 A. Basal cells of assimilatory branches showing bulging. Fig. 27 B. Thin slender (S) branches arising from bulging cells. Fig. 28. Branching of a gonimoblast filament. Scale: Figs 23–28 = 30 µm.

Involucral filament

The involucral filaments develop from the cells above and below the bearing cell of the carpogonial branch as well as from other vegetative branches present near the carpogonial branch (Figs 15, 18, 20, 21, 23). They start growing towards the carpogonial branch and become few-celled before the first division occurs in the fertilized carpogonium (Fig. 15). Whether initiation of the involucral filaments takes place before or after fertilization, is not clear.

It has also been observed that the basal cells of a few cortical branches (which appear to be the part of the branch system on which the cystocarp is growing) near and around the cystocarp show irregular bulging or lobes (Figs 27 A) and produce filaments which either form a tangled mass of colourless cells below the cystocarp or develop upward towards the periphery consisting of thin elongated cells when young (Figs 27, 27 B).

General description

Helminthocladia nizamuddinii Afaq-Husain *et* Shameel *sp. nov.*: Thalli up to 70 cm long, branching radial of 5 (–6) orders (Fig. 29). Main axes cylindrico-terete, 2–5 mm diam. or compressed up to 20 mm broad. Large sack like cells in medullary filaments and various types of adventitious branches among assimilatory filaments are present. Three types of hairs (one-celled, two-celled and long hyaline) and one to four remnants of the hairs are seen on the apex of terminal cells of lateral assimilatory branches.

Plants mon- or dioecious; spermatangia 4–5 µm diam., borne in the form of small heads embedded in mucilage; spermatia round, 3–4 µm diam., carpogonial branch 2–4 celled; fusion cell not formed; involucral filaments present. First division in the fertilized carpogonium may be longitudinal, transverse or oblique, both the daughter cells partake in the formation of gonimoblast filaments, 1–4 terminal cells of gonimoblast filaments act as carposporangia from apex downwards, each producing one, oblong carpospore, 10–15 µm long, 7–9 µm broad. Basal cells of cortical branches near and around the cystocarp become irregularly lobed and produce new branches around cystocarp.

Diagnosis

Thalli ad 70 cm longi, rami radiali ad 5 (–6) ordini; axis principalis cylindrico teretes, 2–5 mm diametero vel compresso ad 20 mm lato; cellulis magnis sacciformi in filis medullaris et ramii adventitis varius typis inter assimilativus filamentis praesentis; pilis tri typis (uni-

cellularis, bicellularis et longis hyalinis) et 1–4 pila vestigia praesentia ab cellulae apicae ramulae lateralae; plant monoecia vel dioecia; spermatangia 4–5 μm diametero, protata in glomerula parva, in mucus inclosus; spermatia globosa, 3–4 mm diametero; rami carpogonii 2–4 cellulae; cellula conjunctionis nec formis; fili involucri praesenti; divisionus prima primarius longitudinalus, transversus vel obliques in fecundatio carpogonus; cellula filia bina participa in forma filia gonimoblasta; 1–4 cellula terminalia filia gonimoblasta facio carposporangio ad declivis; carpospori oblongi 10–15 μm longi, 7–9 μm lati; cellula basalia ramulae corticae prope et circa cystocarpa irregulare lobata et producta ramula nova ad circum cystocarpum.

Holotype

No. 258 PCSIR (Leg. S. Afaq-Husain 23-12-1988) Paradise Point, Karachi (Fig. 29).

Isotypes

Nos. 257, 259, 260, 261 PCSIR (Leg. S. Afaq-Husain 23-12-1988) Paradise Point, Karachi.

Other plants examined

Manora (Leg. Farida 5-2-1962, M. Nizamuddin 8-2-1963, 11-4-1987); Sandspit (Leg. S. Afaq-Husain 11-3-1989); Hawksbay (Leg. M. Nizamuddin 31-12-1963, S. Afaq-Husain 9-3-1986, 20-12-1987, 30-5-1988); Buleji (Leg. M. Shameel 23-1-1988, S. Afaq-Husain 6-2-1989); Paradise Point (Leg. S. Afaq-Husain 11-3-1989).

Habitat

Mostly grows on rocks covered by sand at low tide level of the intertidal zone, sometimes growing in tufts on oyster shells.

The new species, *H. nizamuddinii*, is named after Prof. Dr. Mohammad Nizamuddin, who made numerous contributions on the marine algae of Pakistan.

Discussion

The genus *Helminthocladia* is characterised by the presence of enlarged, inflated, obovate to clavate terminal cells of cortical filaments as well as by the development of gonimoblast filaments from both primary division products of the zygote (Papenfuss 1946, Kylin 1956, Desikachary 1957, Doty and Abbott 1961, Abbott and Hollenberg 1976). Moreover, Womersley (1965) emphasized the conical shape of the carpogonium and the formation of only upward involucral filaments (no descending rhizoidal filaments) as the only generic characters for separating *Helminthocladia* from related genera. Our specimens clearly show all these characters and their assignment to *Helminthocladia* is, therefore, justified.

Helminthocladia nizamuddinii differs from other known species in its large size, dense branching and several reproductive features. The thin, slender, highly flexible branches, which gradually taper to conical apices as well as those arising palmately from compressed axis or secondary branches (Figs 29–31) may also appear distinctive. It not only shows variations in habit but also exhibits variations in reproductive features:

(1) The carpogonial branch consists of two to four cells, two-celled forms are rare but three- and four-celled forms are equally common (Figs 14, 15).

(2) A few plants, not distinguishable from others, were found containing many-celled (up to 16 cells) sterile carpogonial branches in a greater frequency than normal. These branches bear a small trichogyne and do not resemble vegetative branches but have their own characteristic appearance.

(3) The plane of the first division of the zygote may be longitudinal, transverse or oblique (Figs 18–20).

(4) One to four distal cells of the gonimoblast filaments produce carpospores, two or three being most common.

Børgesen (1931) considered Indian plants from Okha Port, a place very close to Karachi, to be *Helminthocladia australis* Harvey on the basis of a study of herbarium material. Desikachary (1957) carried out a thorough investigation of *H. australis*, *H. calvadosii* (Lamouroux *et* Duby) Setchell and *H. papenfussii* Kylin on the basis of fluid preserved material. For identification he used three characters, viz.:

(1) the nature of the first division of the fertilized carpogonium or zygote whether longitudinal, transverse or oblique;

(2) presence or absence of a fusion cell formed by the enlargement of protoplasmic connections between the cells comprising carpogonial branch and the supporting cell and

(3) presence or absence of an involucre (sterile filaments) around the gonimoblasts.

He found that *H. calvadosii* showed longitudinal division, a fusion cell and involucral filaments, and these characters were also shown by all the Indian plants

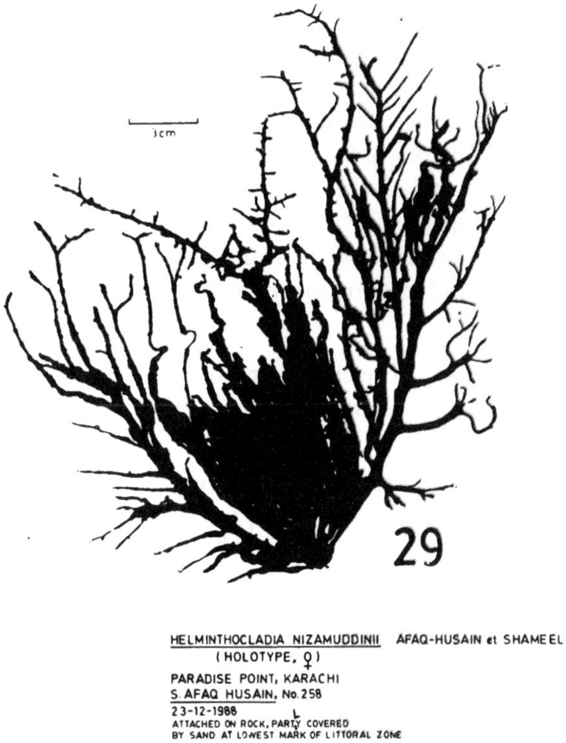

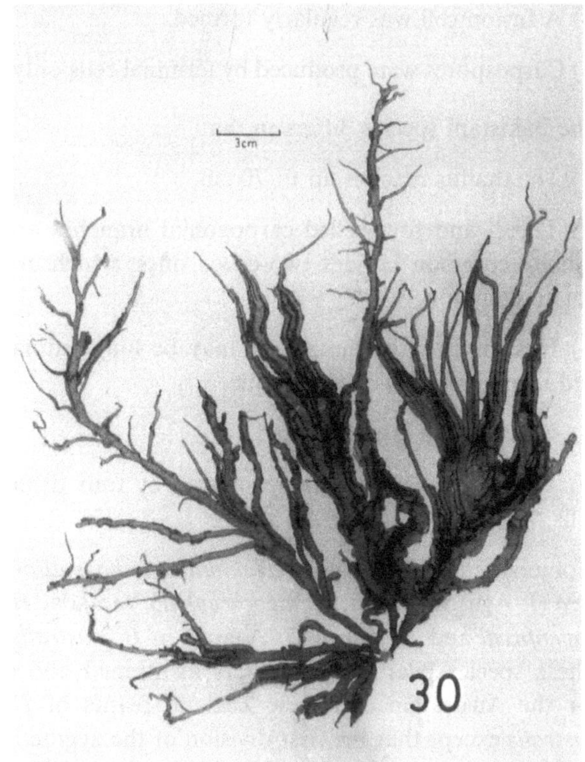

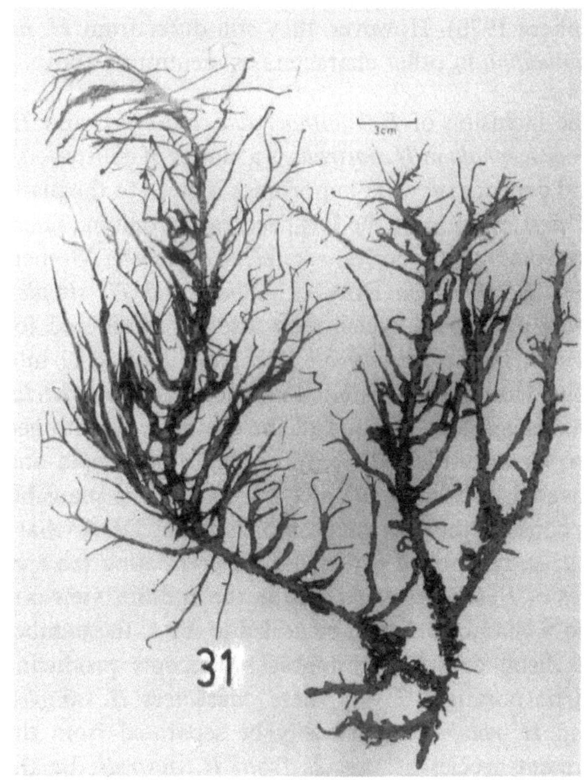

Figs 29–31. *H. nizamuddinii*
Fig. 29. The type specimen, preparation on a herbarium sheet. Fig. 30. Plant exhibiting palmate branching. Fig. 31. Plant showing much branched habit with dense ramuli.

he examined from Okha Port, therefore he considered the Indian plants to be *H. calvadosii* and not *H. australis*. This has also been confirmed by Krishnamurthy and Joshi (1970). However, there is some doubt about the involucral filaments in *H. calvadosii*. Rosenvinge (1909) and Hamel (1930) reported their presence, while Kylin (1930) and Doty and Abbott (1961) denied their occurrence, and Womersley (1965) emphasized a thorough reinvestigation of this aspect in the species.

Pakistani plants differ from those of India in size and branching as well as reproductive structures. Thalli of *H. calvadosii* are 50 cm tall, exhibit pinnate branching and a fusion cell in the developing carposporophyte, and only terminal cells of gonimoblast filaments produce carpospores (Desikachary 1957). In the present species thalli are 70 cm tall, branching is radial and irregular but not pinnate, there is no fusion cell and up to four distal cells of gonimoblasts produce carpospores.

The Pakistani species resembles *H. australis* in showing a large variation in habit. Levring (1953, 1955), Desikachary (1957) and Womersley (1965) studied specimens of *H. australis* from Australia and New Zealand and found that:

(1) The thallus height was about 35 cm.

(2) Carpogonial branches were generally three-celled, two-celled ones were not observed.

(3) The first division of the zygote was oblique.

(4) A fusion cell was regularly formed.

(5) Carpospores were produced by terminal cells only.

The Pakistani species differs in that:

(1) The thallus reaches up to 70 cm.

(2) Three- and four-celled carpogonial branches are equally common besides two-celled ones, which are not so common.

(3) First division of the zygote may be longitudinal and transverse as well as oblique.

(4) A fusion cell is not formed.

(5) The carposporangia are produced by four distal cells.

Womersley (1965) included *Helminthocladia californica* (J. Agardh) Kylin, *H. macrocephala* Yamada, *H. papenfussii* and *H. tumens* J. Agardh in *H. australis*. These species bear the characters mentioned above for the Australian and New Zealand plants of *H. australis* except that the first division of the zygote is longitudinal in *H. macrocephala* (Yamada 1941) as well as *H. californica* (Abbott 1965, Abbott and Hollenberg 1976). However, they still differ from *H. nizamuddinii* in other characters as mentioned above.

The inclusion of *Helminthocladia californica* and *H. macrocephala* in *H. australis* is a matter of controversy and depends on what importance is given to the plane of first division of the fertilized carpogonium. There are some other species, *e.g. H. beaugleholei* Womersley, *H. rhizoidea* Doty *et* Abbott and *H. simplex* Doty *et* Abbott, which may also be considered for merger in the *H. australis* complex, if we take into consideration only external characters. *Helminthocladia beaugleholei* exhibits all the characters mentioned for *H. australis*, but it differs in being simple and covered by ramuli (Womersley 1965). It is plausible to consider the characters emphasized by Desikachary (1957), mentioned earlier, for differentiating the species of *Helminthocladia*, and in the authors' view one more character should be added to it *i.e.* the number of distal cells of gonimoblast filaments producing carposporangia. Using these characters *H. simplex* and *H. rhizoidea* can easily be separated from the present species as well as from *H. australis* by the presence of longitudinal division and the absence of a fusion cell. Moreover, *H. simplex* will be further distinguished from others by the production of carposporangia from terminal and subterminal cells.

Helminthocladia agardhiana Dixon [as *H. hudsoni* (C. Agardh) J. Agardh by Feldmann 1939] is quite different from the present as well as other species in that it produces tetracarpospores instead of single carpospores. *Helminthocladia densa* (Harvey) Schmitz *et* Hauptfleisch and *H. dotyi* Womersley also differ from the present species in size, appearance, formation of very small terminal cells of the cortical branches, fusion cell and production of carpospores only by terminal cells.

There are a few more characters found in the present species, which have not yet been reported from others, viz.:

(1) large sack like cells in medullary filaments,

(2) long hyaline hairs (Figs 9, 11),

(3) variously lobed cells producing new filaments, which form a network below the cystocarp and radiate upward towards surface of the thallus (Figs 27 A).

It is, therefore, concluded that the Pakistani plants of *Helminthocladia* should be treated as a separate taxonomic unit.

No other order of red algae has seen such dramatic changes made in its concept and classification as Nemaliales since the classical publication of the treatise on red algae by Kylin (1956). A recently published cladistic analysis of the assemblage of Liagoraceae by Kraft (1989) brings us to the conclusion that this group of seaweeds cannot be divided into families or tribes along pre-existing lines. The Helminthocladiaceae, propounded and advocated by many (*e.g.* Abbott and Hollenberg 1976, Garbary *et al.* 1982), has been replaced by the Liagoraceae (Kraft 1989, Afaq-Husain *et al.* 1990), a name which has priority. Phycochemical investigations of *H. nizamuddinii* are under way, they might throw some light on its taxonomic position and phylogeny.

Acknowledgements

We are thankful to the Director, PCSIR Laboratories, Karachi and the Chairman, Department of Botany, University of Karachi for providing facilities for this research. We are also indebted to Mr Abdul Ghafoor and Dr Mohammad Qaiser, Professor of Botany, University of Karachi for the Latin diagnosis and to Dr Gerald T. Boalch, The Laboratory, Citadel Hill, Plymouth, G. Britain, for the critical evaluation of the manuscript.

References

Abbott, I. A. 1965. *Helminthora* and *Helminthocladia* from California. *Hydrobiol. 25*: 88–98.

Abbott, I. A. and G. J. Hollenberg. 1976. *Marine Algae of California*. Standford Univ. Press, California, 827 pp.

Afaq-Husain, S., M. Nizamuddin and M. Shameel. 1991. The structure and reproduction of a new taxon *Dermonema abbottiae* (Nemaliales – Rhodophyta) from the coast of Pakistan. *Pak. J. Sci. Ind. Res. 34*: (in press).

Børgesen, F. 1931. Some Indian Rhodophyceae especially from the shores of the presidency of Bombay – I. *Kew Bull. 1931*: 1–24.

Desikachary, T. V. 1957. *Helminthocladia* from India and New Zealand. *J. Ind. Bot. Soc. 36*: 441–456.

Doty, M. S. and I. A. Abbott. 1961. Studies in the Helminthocladiaceae (Rhodophyta): *Helminthocladia. Pac. Sci. 15*: 56–63.

Feldmann, J. 1939. Une Nemalionale a carpotetraspores: *Helminthocladia hudsoni* (C. Ag.) J. Ag. *Bull. Soc. Hist. Nat. Afr. Nord 30*: 87–94.

Garbary, D. J., G. I. Hansen and R. F. Scagel. 1982. The marine algae of British Columbia and northern Washington: Division Rhodophyta (red algae), class Florideophyceae, orders Acrochaetiales and Nemaliales. *Syesis 15* (Suppl. 1): 1–102.

Hamel, G. 1930. Floridees de France – VI. *Rev. Algol. 5*: 61–109.

Kraft, G. T. 1989. *Cylindraxis rotundatus* gen. et sp. nov. and its generic relationships within Liagoraceae (Nemaliales, Rhodophyta). *Phycologia 28*: 275–304.

Krishnamurthy, V. and H. V. Joshi. 1970. *A Check-list of Indian Marine Algae*. CSMCRI, Bhavnagar, 36 pp.

Kylin, H. 1930. Über die Entwicklungsgeschichte der Florideen. *Acta Univ. Lund.* Adv. 2, *26*: 1–104.

Kylin, H. 1938. Verzeichnis einiger Rhodophyceen von Südafrika. *Lunds. Univ. Arsskr.* n. f. Adv. 2, *34*: 1–26.

Kylin, H. 1956. *Die Gattungen der Rhodophyceen*. CWK Gleerups Forlag, Lund. 673 pp.

Levring, T. 1953. The marine algae of Australia. I. Rhodophyta: Goniotrichales, Bangiales and Nemaliales. *Ark. Bot.* Ser. 2, *2*: 457–566.

Levring, T. 1955. Contributions to the marine algae of New Zealand. I. Rhodophyta: Goniotrichales, Bangiales, Nemaliales and Bonnemaisoniales. *Ark. Bot.* Ser. 2, *3*: 407–432.

Martin, M. T. 1939. Some South African Rhodophyceae: I. *Helminthocladia papenfussii* Kylin. *J. Bot. Lond. 77*: 234–244.

Papenfuss, G. F. 1946. Structure and reproduction of *Trichogloea requienii*, with a comparison of the genera of Helminthocladiaceae. *Bull. Torrey Bot. Club 73*: 419–437.

Rosenvinge, L. K. 1909. The marine algae of Denmark. Part I. Introduction. Rhodophyceae (Bangiales and Nemaliales). *K. Dan. Vid. Delsk. Skr. 7*: 1–151.

Shameel, M. 1987. A preliminary survey of seaweeds from the coast of Lasbela, Pakistan. *Bot. Mar. 30*: 511–515.

Shameel, M., S. Afaq-Husain and S. Shahid-Husain. 1989. Addition to the knowledge of seaweeds from the coast of Lasbela, Pakistan. *Bot. Mar. 32*: 177–180.

Womersley, H. B. S. 1965. The Helminthocladiaceae (Rhodophyta) of southern Australia. *Aust. J. Bot. 13*: 451–487.

Yamada, Y. 1941. Notes on some Japanese algae – IX. *Sci. Pap. Inst. Algol. Res. Hokkaido Univ. 2*: 195–215.

Botanica Marina
Vol. 34, pp. 91–106, 1991

Cyst-theca Relationships in *Scrippsiella* (Dinophyceae) and Related Orthoperidinioid Genera

J. Lewis[1]

Biology Department, Royal Holloway & Bedford New College, Egham Hill, Egham, Surrey, TW20 OEX, U.K.

(Accepted 23 September 1990)

Abstract

Study of sediments from Loch Creran, a Scottish west coast sea-loch has yielded four new species of small calcareous dinoflagellate cysts. Single cyst incubation techniques and culture studies have shown all these to be members of the genus *Scrippsiella* Balech *ex* Loeblich. Descriptions are given of both the motile stage and cyst of these new species, (*S. crystallina, S. lachcrymosa, S. trifida*). These are compared with descriptions of *S. trochoidea* and *Pentapharsodinium dalei.* The archeopyle of *Scrippsiella* is theropylic and is interpreted as representing the loss of 2′–4′ and 1–3a paraplates, the operculum remaining attached to the cyst body by the first apical paraplate.

Introduction

The first description of a Scrippsielloid dinoflagellate was that of *Glenodinium trochoideum* by Stein (1883) (later transferred to the genus *Peridinium* by Lemmermann 1910). Another closely related species, *Peridinium faeroense*, was described by Paulsen in 1905. However, because of their small size, thecal details were not easily distinguished and much confusion existed about these two species, which were included in the orthoperidinium group (section conica) of Lebour (1925). The difficulty of accurate taxonomy combined with the superficial similarities of these species and their comparative ease of culture has led to much confusion in the literature concerning them (see for example discussion in Dale 1977 and Steidinger and Balech 1977). Detailed work by Balech (1974) led to his splitting of *Peridinium* into four genera (*Peridinium* sensu stricto, *Protoperidinium, Ensiculifera* and *Scrippsiella*) largely on the basis of cingular and sulcal plates. Following this scheme *P. trochoidea* (Stein) Lemmermann was transferred to *Scrippsiella.*

In 1977 Dale reported observations of the germination of *Scrippsiella trochoidea* (Stein) Loeblich from a partly calcareous cyst and *Peridinium faeroense* Paulsen from a wholly organic cyst. He went on in 1978 to fully endorse Balechs' splitting by emphasizing that as well as small differences in thecal morphology the four genera also have a range of cyst types. However, some inconsistencies remained and recently Indelicato and Loeblich (1986) have attempted to clarify these. In particular they strove to find "mutually exclusive criteria" to define genera. Cyst type and thecal morphology were important considerations in their reasoning but in addition they used newly described cingular plate homologies. They recognized three of the previously established genera (*Peridinium, Protoperidinium* and *Scrippsiella*); established a new genus (*Pentapharsodinium*) and discarded *Ensiculifera.* Species previously referred to *Ensiculifera* were transferred to *Pentapharsodinium* and *Scrippsiella.* Notably *Peridium faeroense*, as observed by Dale (1977), was described as the new species *Pentapharsodinium dalei* Indelicato *et* Loeblich.

As the small orthoperidinioid species have been investigated in more detail the number of new species has increased. In recent years five new *Scrippsiella*

[1] Present address: School of Biological & Health Sciences, The Polytechnic of Central London, 115 New Cavendish Street, London, W1M 8JS, U.K.

species have been described: *S. hexapraecingula* Horiguchi *et* Chihara (1983); *S. tinctoria* Indelicato *et* Loeblich (1985); *S. caponii* Horiguchi *et* Pienaar (1988 a); *S. arenicola* Horiguchi *et* Pienaar (1988 b) and *S. precaria* Montresor *et* Zingone (1988). Of these only the last species included a description of the cyst type. At present there are some ten species in the genus *Scrippsiella* of which four have known cysts. Cysts play an important role in the ecology of cyst forming species. Factors affecting this have been investigated for *S. trochoidea* by Binder and Anderson (1987).

Our knowledge of cyst-theca relationships in the dinoflagellates has grown considerably over the last twenty years since the first work of this kind by Wall and Dale (1966). In 1968 Wall and Dale described eight new types of calcareous dinoflagellate cysts from a variety of sites world wide. Two species were successfully germinated to give motile thecate cells and these were identified as *Peridinium trochoideum* (this was prior to its transfer to *Scrippsiella*) and *Scrippsiella sweeneyae* Balech *ex* Loeblich. Other species were also germinated but cells did not properly form thecae. This led to the conclusion that the Quaternary calcareous dinoflagellates were Peridinioid resting cysts. This study is based on four new species of small calcareous dinoflagellate cysts along with those of *S. trochoidea* and *Pentapharsodinium dalei*, all of which were found in Creran, a Scottish west coast sea-loch.

Material and Methods

Sediment was obtained from Loch Creran and Loch Melfort by Craib coring. Cysts used for raising cultures were taken from various depths from a core obtained on 15 September 1986, in Loch Creran which was extruded and sectioned into 14, 1 cm portions and kept at 4 °C until required. Cysts were isolated as described in Lewis *et al.* (1984) using sieve sizes of 85 μm and 20 μm. To obtain cultures, single cysts or small numbers of cysts (usually five) were placed in 5 cm petri-dishes in f/2 medium without silicate (Guillard and Ryther 1962) at 20 °C and where excystment was successful, cells were allowed to grow up. Thereafter cultures were maintained in conical flasks at 20 °C in a 24 hour light regime, using either f/2 or Erdshreiber medium (McLachlan 1973). Encystment of cultures was attempted by sub-culturing into petri-dishes containing f/2 medium without added nitrate. Cultures that survived the period of study were deposited with the Culture Collection of Algae and Protozoa at the Dunstaffnage Marine Laboratory, of these only *Scrippsiella trochoidea* now remains.

Cultures provided material for taxonomic investigations of the motile thecate cells by light and scanning microscopy. Thecal dissections were facilitated by the use of 5% sodium hypochlorite. The most effective method for determining thecal plate patterns under the light microscope was found to be Trypan blue staining (Taylor 1978) in combination with phase contrast microscopy. All light microscopy was carried out on an Olympus BH-2 microscope. Some cultures were measured on a Cambridge instrument Quantimet 970 image analysis system. Cysts for taxonomic investigation were obtained from sediments. The nature of the wall was tested with 30% hydrochloric acid. Trypan blue staining was also carried out on some acid-treated cysts in an attempt to discern paraplates (Dale 1983).

Scanning microscopy of cysts and thecate cells was carried out as described in Lewis *et al.* (1984) with the additional use of Cambridge S100 and S410 microscopes. Cysts were broken open by first air drying on to a cover slide glued to a stub then squashing with a sharpened mounted needle (Keupp personal communication).

Results

The history of separate phycological and palynological investigations of the two different life-cycle stages of dinoflagellates has left a legacy of dual nomenclature. The problem becomes most complex where the two systems meet in Quaternary and Recent cyst investigations and in cyst-theca relationship studies such as this. It is my belief and that of Dale (1976) that the situation should not be complicated unnecessarily. The Recent calcareous cysts have not been named by palynologists. As will be discussed later they have obvious affinities with some species from as far back as the Cretaceous (Keupp 1981) but one cannot be certain of the relationship of these fossil species with those of the present day. All cyst descriptions, therefore, will be made under the nomenclature used for the motile thecate cells which already exists for some of the species described. Describing cysts under a second name using palynological nomenclature will provide no extra scientific understanding of this group and so will not be carried out here.

The first observations presented will be those of the previously described species *Scrippsiella trochoidea* and *Pentapharsodinium dalei*. These highlight the major differences between these two genera. *Scrippsiella* having six cingular plates (5 + t), five sulcal plates and a partly calcareous cyst and *Pentapharsodinium* having five cingular plates (4 + t), four sulcal plates and a wholly organic cyst. Thereafter the four new species are described. All possess six cingular plates and partly calcareous cysts and therefore best conform to the genus *Scrippsiella.*

Order Peridiniales Haeckel 1894
Family Calciodinellaceae Deflandre 1947 emend Bujak and Davies 1983
Genus *Scrippsiella* Balech *ex* Loeblich 1965

Scrippsiella trochoidea (Stein) Loeblich Figs 1–8

Cyst size:— Length: 25–48 μm (10 measured); breadth: 25–44 μm (12 measured).

Thecate cell size:— length: 23–37 μm; breadth: 19–30 μm (36 measured).

Cysts

Cysts are round to oblate and contain a prominent red body (Fig. 1). They are covered with calcareous processes of variable length (Fig. 2). These processes have been shown by electron diffraction analysis to be calcite (Wall *et al.* 1970). The processes are composed of a polygonal base plate from which a three cornered spine emerges, short spines are therefore pyrimidal in shape (Fig. 3). When the processes are dissolved using dilute acid a clear inner organic wall is left with no apparent paratabulatory features. Scanning electron microscopic studies on the inner wall of broken open cysts revealed no indication of internal paratabulation as has been shown for other calcareous cysts (Keupp 1981). The archeopyle is a split towards one end of the cyst forming a cap shaped operculum which remains attached to the main body of the cyst by approximately one third of its circumference. In some specimens the opening is not a smooth split but appears to follow paraplate boundaries implying an apical archeopyle representing the loss of 2′–4′ and 1–3a but remaining attached by the first apical paraplate (Figs 4, 9).

Encystment was induced a number of times in subcultures from the main culture. In initial attempts the

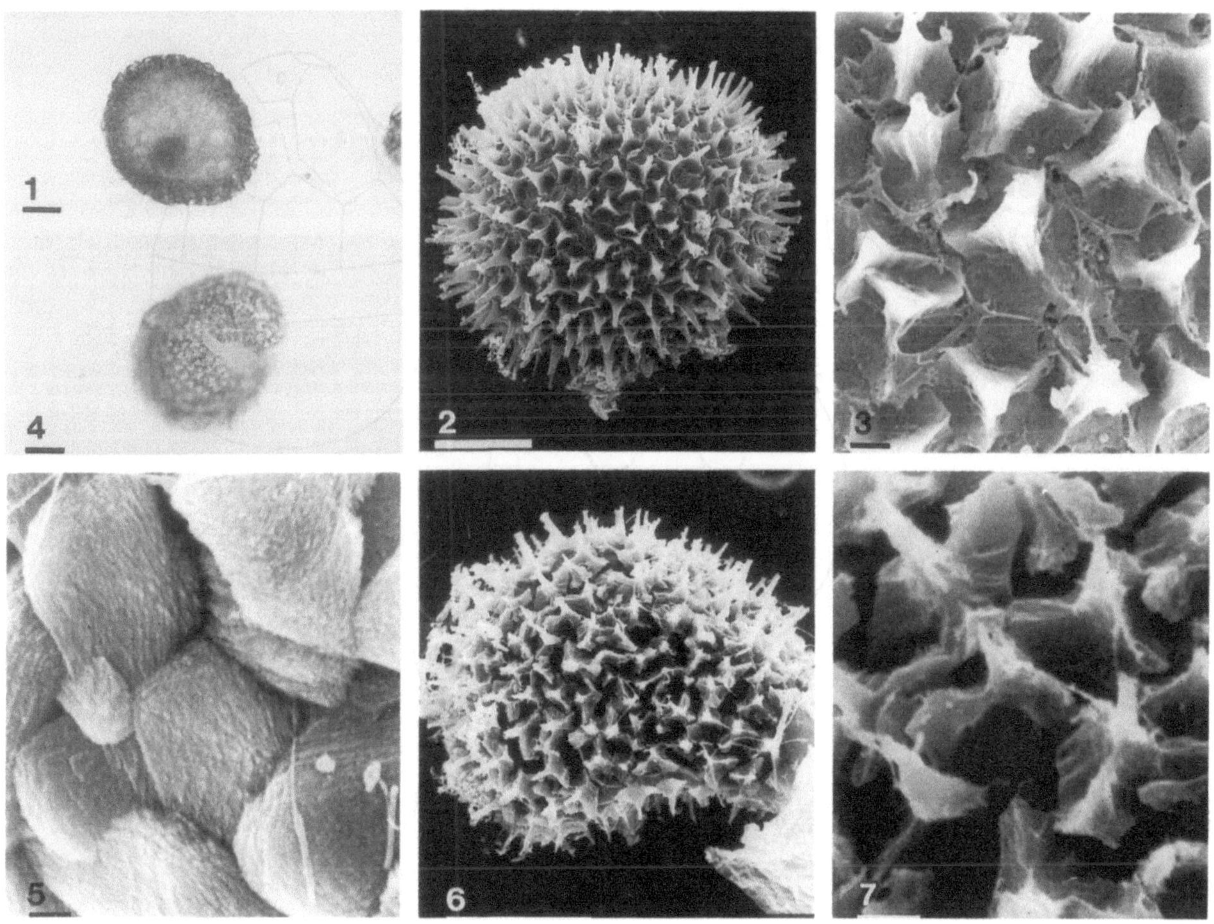

Figs 1–7. *Scrippsiella trochoidea.*
Figs 1–4. Light micrographs, all the rest scanning microgaphs. All thecate cells from culture material. Fig. 1. Ungerminated cyst from Loch Creran, showing red body; Fig. 2. Ungerminated cyst from Loch Creran showing calcareous ornament; Fig. 3. Detail of ornament of cyst in Fig. 2; Fig. 4. Germinated cyst showing notched archeopyle opening, from Loch Creran; Fig. 5. Detail of calcareous ornament of a cyst formed in laboratory culture; Fig. 6. Cyst formed in laboratory culture; Fig. 7. Detail of ornament of cyst in Fig. 6. (Scale bars 10 μm except Figs 3, 5, 7; 1 μm.)

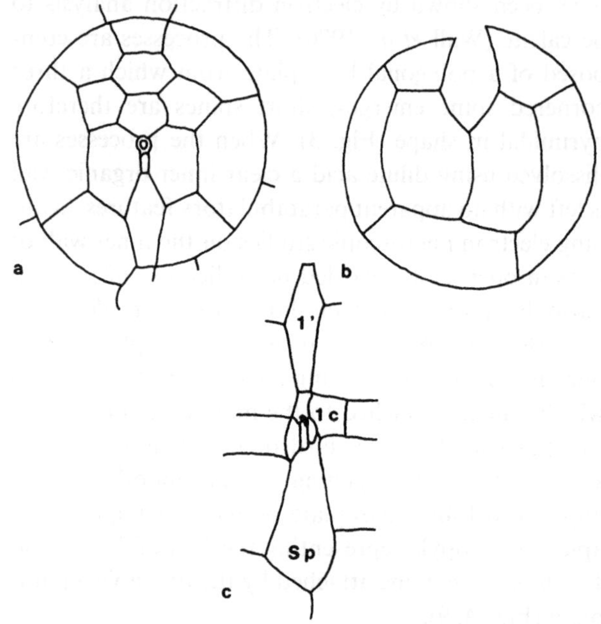

Fig. 8. Drawings to illustrate plate pattern of *Scrippsiella trochoidea*. (a) Epitheca; (b) Hypotheca; (c) Sulcus.

cysts produced had ornamentation that was different to that originally isolated (Fig. 5). Processes lacked spines but were perhaps analogous to the base plate of the wild type. Cysts of this kind have not been found in sediment samples. Cysts produced on a later occasion were the same as those originally isolated (Fig. 6) spine formation here has taken place before the 'base plates' are fully developed (Fig. 7). The culture of this species was started from five cysts so it is not clear whether this is an affect of culture conditions or shows some of the range of intraspecific variability that is possible. Similar differences in cultured cysts for this species were found by Wall *et al.* 1970.

Thecate cells

Thecate cells have a typical Scrippsielloid morphology with a central nucleus and numerous brownish chloroplasts. The thecal plates show no patterning and are covered with scattered simple trichocyst pores.

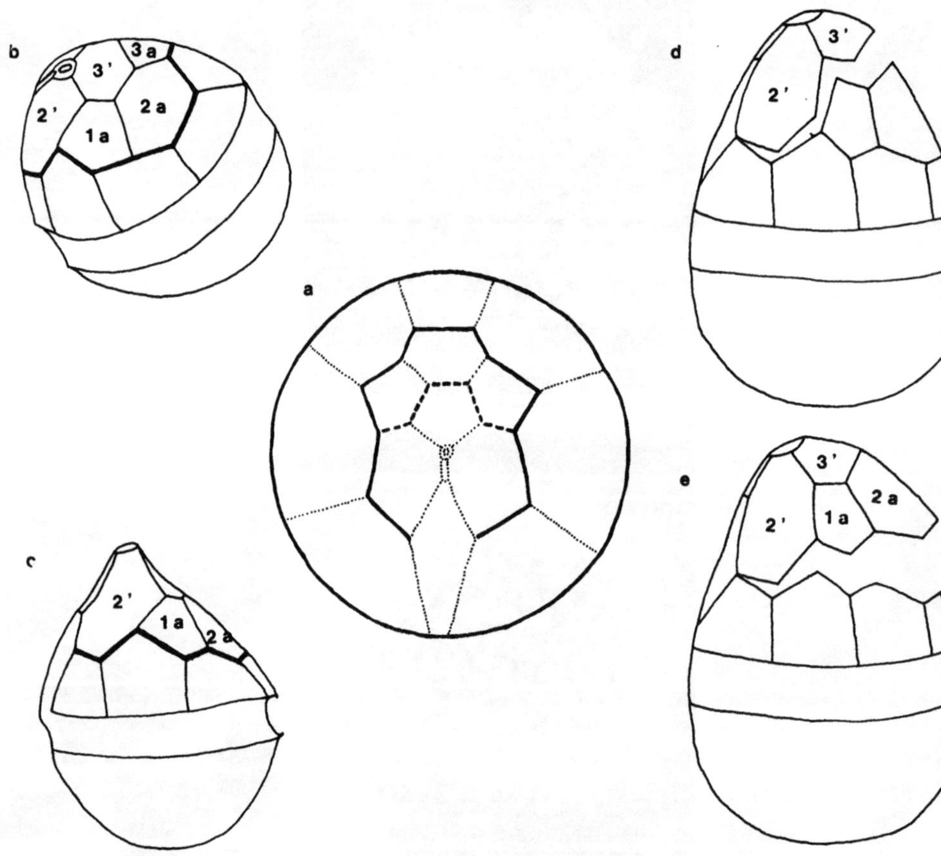

Fig. 9. Drawings to illustrate archeopyle formation in *Scrippsiella*.
(a) Thecal tabulation of epitheca in *Scrippsiella* with heavy lines and heavy dashed lines indicating possible archeopyle interpretations; (b) Epitheca of *S. rotunda* drawn from an S. E. M., heavy lines indicate archeopyle interpretation; (c) Epitheca of *S. trifida* drawn from an S. E. M., heavy lines indicate archeopyle interpretation; (d) Oval *Scrippsiella* cyst with projected paraplates, archeopyle represents the loss of the apical paraplates 2'–4'; (e) Oval *Scrippsiella* cyst with projected paraplates, archeopyle represents the loss of the apical and intercalary paraplates, 2'–4' and 1–3a.

The apical pore plate supports a circular smooth ridge and is surrounded by a collar, a long canal plate connects this to the first apical plate. The first apical plate is narrow and is slightly asymmetrical. The sulcus contains five plates, the Sp plate extending up and meeting the cingulum (Fig. 8c). The plate pattern is as follows: – P_o, x, 4′, 3a, 7″, 6c, 5s, 5‴, 2⁗ (Fig. 8). There is variation in the plate patterns encountered in cultures with extra intercalary or precingular plates being found. There also appears to be some variability in the sulcal pattern although not in sulcal plate number. The position of cingular sutures is also slightly variable with regard to the other plates.

Genus *Pentapharsodinium* Indelicato *et* Loeblich 1986

Pentapharsodinium dalei Indelicato *et* Loeblich.
Figs 10–16

Cyst size [diameter inc. spines 24–38 μm (10 measured)].

Thecate cell size length: 10–21 μm; breadth: 8–19 μm (7 measured).

Cysts

Cysts are spherical or slightly subspherical, contain a red body and are surrounded by spines (Figs 10, 11). The cyst wall is clear and wholly organic remaining unaffected by acid treatments. Examination by scanning electron microscopy shows the outer wall to be pitted. The spines are drawn up from this wall and are frequently flattened at the base. They are very variable in morphology and can be branched in a variety of ways and may have simple or complex tips (Figs 13–15, 16a). A single cyst frequently displays a variety of spine types. Despite careful observations no distinct archeopyle has been found in this study, Dale (1977) reported similar difficulty in discerning the archeopyle of this species and stated that excystment seemed to occur by a simple splitting of the cyst wall.

No encystment was achieved with this species but this may be due to a rather poor culture. The culture was started from a single cyst and never grew strongly.

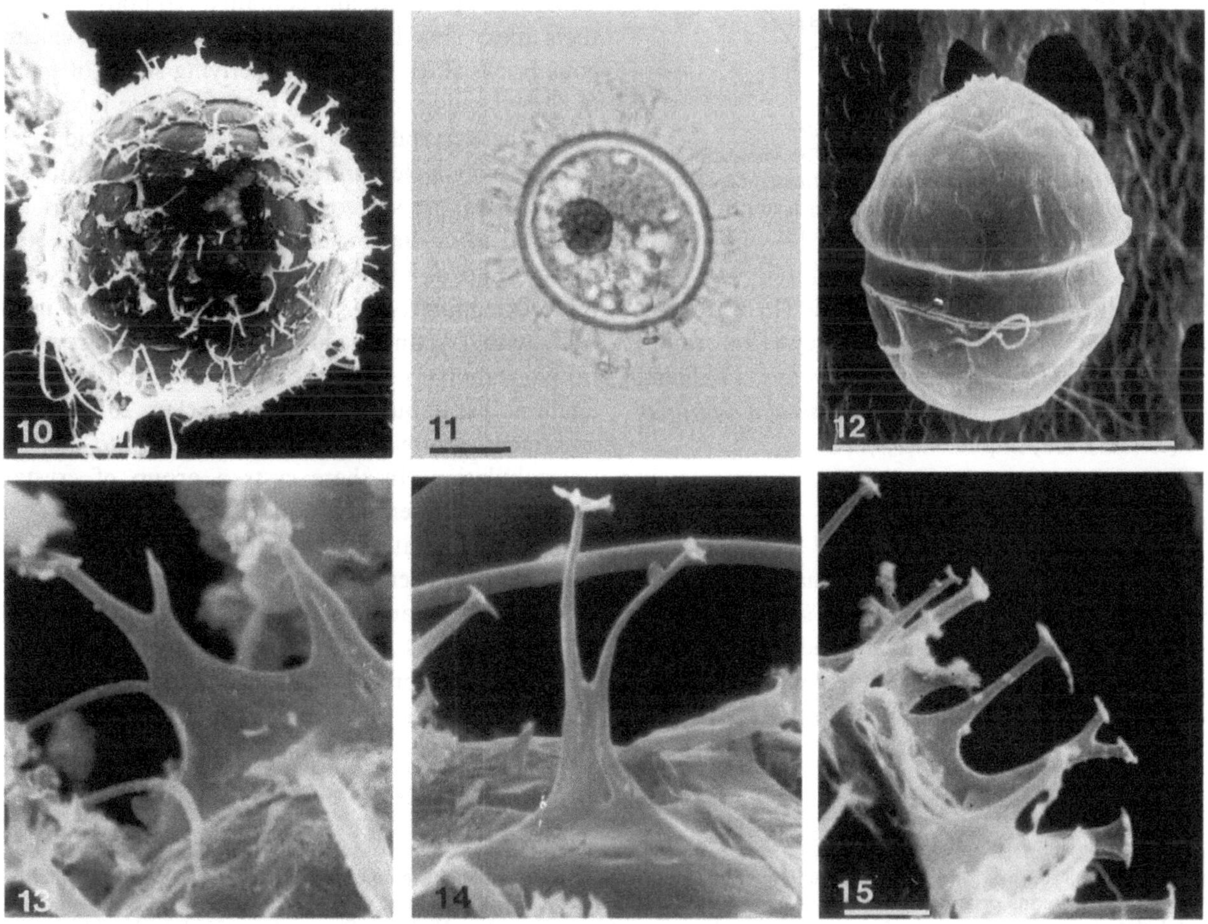

Figs 10–15. *Pentapharsodinium dalei.*
Fig. 11. Light micrograph, all the rest scanning micrographs. Fig. 10. Cyst from Loch Creran sediments; Fig. 11. Acid treated cyst, note spines still present, also shows red body, from Loch Creran; Fig. 12. Dorsal view of thecate cell, from culture, note arrangement of cingular plates; Figs 13–15. Different spine types found on *P. dalei* cysts, all cysts from Loch Creran. (Scale bars 10 μm except Figs 13–15; 1 μm.).

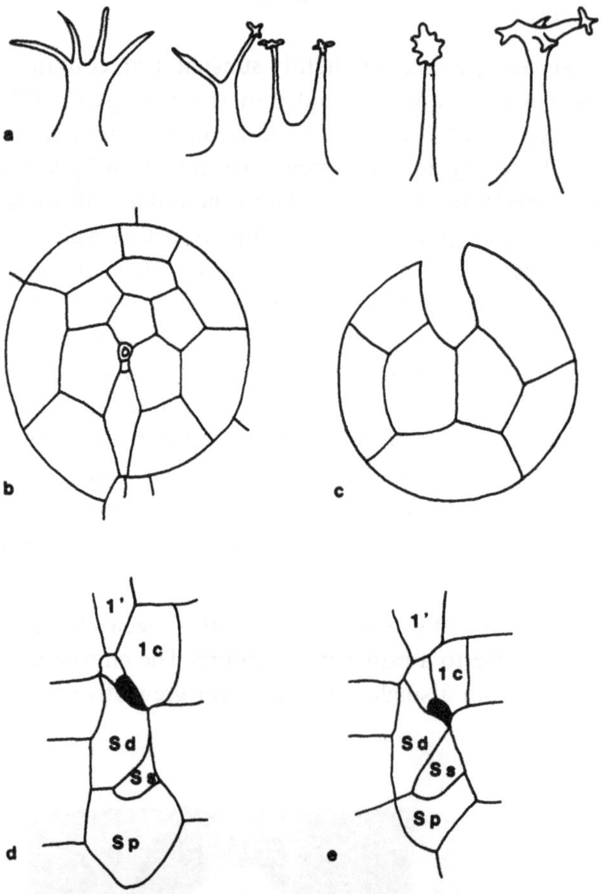

Fig. 16. Drawings of *Pentapharsodinium dalei*.
(a) Illustrations of different spine types to be found on cysts; (b) Epitheca (note cingular plates); (c) Hypotheca; (d–e) Sulcus showing slightly different plate arrangements, in neither does the Sp plate touch the cingulum.

This cyst type was described by Dale (1978) as corresponding to acritarchous tertiary cysts informally described as *Michrystridium bifurcatum* (Williams 1963, unpub. thesis) and *Impletosphaeridium bifurcatum* (Eaton 1969, unpub. thesis).

Thecate cells

Thecate cells are Scrippsielloid in shape but with a less pronounced apical horn and in some specimens a slightly reduced hypotheca. They differ from *Scrippsiella* in having only 5 cingular plates and 4 sulcal plates (Fig. 12). The plate formula is as follows:– P_o, x, 4′, 3a, 7″, 5c, 4s, 5‴, 2⁗ (Figs 12, 16b–e). The apical pore plate is surrounded by a small collar and a short canal plate connects this to the slightly asymmetric first apical plate. The Sp plate of the sulcus is short and does not touch the cingulum (Fig. 16d, e). There are some plate variations seen in cultured cells (see for example Fig. 16d, e). The most striking feature of this species is the thecal patterning consisting of a variable number of concentric ridges surrounding the trichocyst pores (Fig. 12). This patterning is carried over on to the sulcal plates.

Genus *Scrippsiella*

Scrippsiella crystallina Lewis sp. nov. Figs 17–29

Cyst size length: 50–68 µm; breadth: 36–48 µm (21 measured).

Thecate cell size length: 27–47 µm; breadth: 20–35 µm (40 measured).

Cysts

Cysts are egg-shaped and contain a red body towards their apex. They have an outer layer of thick calcareous bricks (Fig. 17). The underlying shape of many of the crystals is hexagonal with alternate faces concave. These crystals dissolve away by gentle acid treatment leaving a clear inner organic body with fibrous material on the surface (Fig. 19). These fibres are involved in crystal formation and it seems possible that they act as a template for the developing crystals (Gao Xiaoping *et al.* 1989). Staining of this inner body with Trypan blue after crystal removal revealed no paratabulation. Cysts were also broken open but again no paratabulation was found (Fig. 20). In some specimens, grouping of the cystals seems to indicate paratabulation. In these cysts a group of crystals appear to represent a paraplate and the gap between them the paraplate boundary (Fig. 17). However, no specimen has been found with clear, complete paratabulation. The archeopyle is a split around the apex of the cyst (Figs 21, 22). Sometimes this is notched and this is interpreted as an apical archeopyle representing the loss of 2′–4′ and 1–3a paraplates (Fig. 9).

Figs 17–28. *Scrippsiella crystallina* sp. nov.
Fig. 21. Light micrograph, all the rest scanning micrographs. Thecate cells from culture material. Fig. 17. Cyst from Loch Melfort with typical calcareous ornament, in this specimen the grouping of crystals is suggestive of a paracingulum and some antapical paratabulation; Fig. 18. Germinated cyst with unusual calcareous ornament; Fig. 19. Cyst after treatment with hydrochloric acid; Fig. 20. Cyst from Loch Creran, broken open to reveal the inside of the wall, no internal tabulation evident; Fig. 21. Hatched cyst from Loch Creran; Fig. 22. Hatched cyst showing archeopyle and organic wall beneath the calcareous ornament; Fig. 23. Cyst formed in laboratory culture; Fig. 24. Ventral view of thecate cell; Fig. 25. Apical view of thecate cell; Fig. 26. Antapical view of thecate cell; Fig. 27. Sulcus of thecate cell; Fig. 28. Cell from culture with no clear thecal plates, some indication on surface of girdle. (Scale bars 10 µm except Fig. 27; 1 µm.)

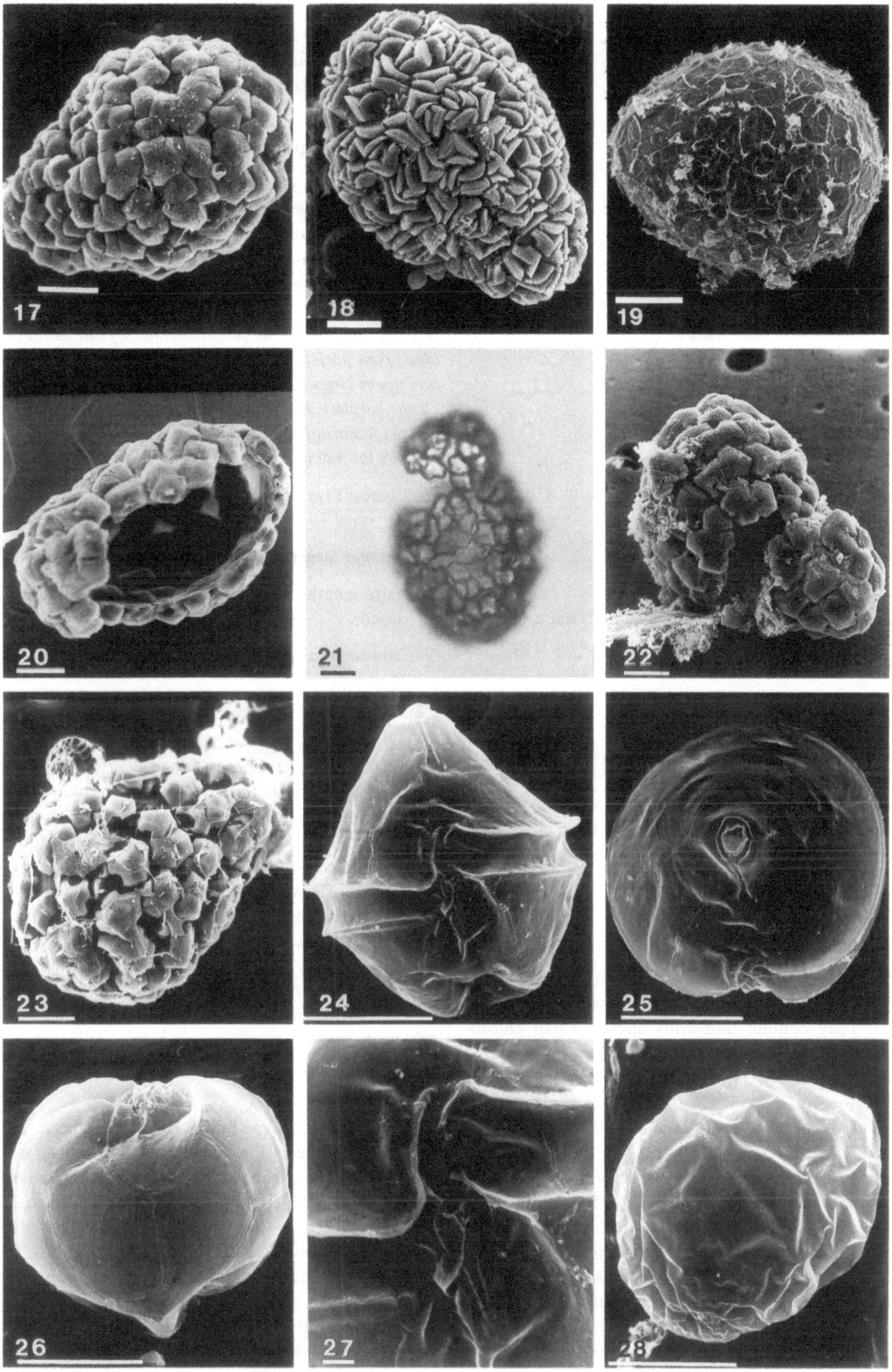
17
18
19
20
21
22
23
24
25
26
27
28

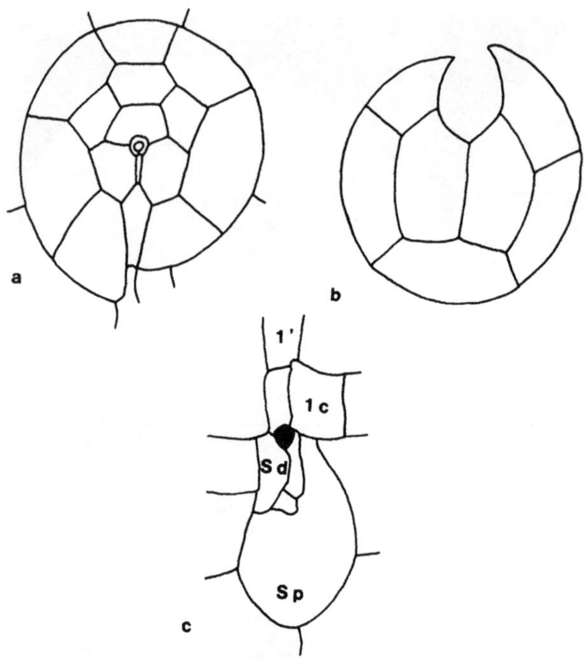

Fig. 29. Drawings to illustrate plate pattern of *S. crystallina* sp. nov.
(a) Epitheca; (b) Hypotheca; (c) Sulcus.

This culture was successfully encysted, the cysts that were formed were very similar to the wild type from which the culture was started (Fig. 23).

Another cyst type was found that closely resembles *S. crystallina* in form (Fig. 18). The arrangement of the calcerous ornament is different with crystals lying upright (not flat) but following the same underlying geometric design. This cyst type is treated as a morphotype of *S. crystallina* because of their great similarities. This morphotype was only rarely encountered and was not incubated in this study.

Thecate cells

Thecate cells have a conical epitheca and a rounded hypotheca sometimes with a flattened bottom (Fig. 24). They contain numerous discrete small chloroplasts. The wall is unornamented except for scattered simple trichocyst pores. Cells have a large nucleus contained in the centre of the cell towards the hypotheca. The plate formula is as follows: – P_o, x, 4′, 3a, 7″, 6c, 5s, 5‴, 2⁗ (Figs 25, 26, 29). The apical horn is pronounced and consists of a pore plate with a smooth circular plug surrounded by a low collar. There is a long thin canal plate connecting the pore plate to the symmetrical, wide, first apical plate. There are six cingular plates, the girdle is excavated and offset by one girdle width. The sulcus has five plates, the Sd plate covers the flagellar exits (Figs 27, 29c). The Sp plate is wide and touches the cingulum. The thecal patterning on this species was frequently underdeveloped (Fig. 28). A number of plate variations were observed including an extra precingular plate and an extra antapical plate.

Diagnosis: Scrippsiella crystallina *Lewis sp. nov.*

Cellulae conicas epithecas habent rotundatae ad hypothecam complanatam, proportione paene congruente. Cingulum excavatum unius cinguli latitudine deflexum. Prima lamina apicalis latior modo. Laminarum ornatus non amplior quam ex poris simplicibus. Laminarum formula haec: P_o, x, 4′, 3a, 7″, 6c, 5s, 5‴, 2⁗. 27–47 µm longitudine, 20–35 µm latitudine. Cystae sunt ovoideae cum pariete interiore pellucido animantis materiae, quem cingunt laterculi calcarei. Antequam extant cystae corpus rubrum continent. Archeopyle apicalis est et operculum affixum manet. 50–68 µm longitudine; 36–48 latitudine.

Holotypes: Figs 17, 24.

Scrippsiella lachrymosa Lewis sp. nov. Figs 30–36

Cyst size length: 34–44 µm; breadth: 18–32 µm (8 measured).

Thecate cell size length: 16–20 µm; breadth: 12–26 µm (8 measured).

Cysts

Elongate oval cysts with an inner clear organic wall are covered with thin, flattened calcareous plates (Fig. 30). In some specimens the crystals start to overgrow each other (Fig. 31). The red body is usually found in the middle of the cyst towards one wall (Fig. 32). The archeopyle is a smooth split towards the apex of the cyst remaining attached to the cyst. This is interpreted as representing the loss of paraplates 2′–4′ and 1–3a (Fig. 9). A more delicate, smaller and elongate cyst than that of *Scrippsiella crystallina.* This was a poorly growing culture and no encystment was achieved.

Thecate cells

Thecate cells have a fairly conical epitheca and rounded hypotheca (Fig. 33). The wall is smooth with scattered simple trichocyst pores. Narrow intercalary bands have been found. The nucleus tends to be centrally placed to one side of the cell. The plate formula is as foHows: – P_o, x, 4′, 3a, 7″, 6c, ?5s, 5‴, 2⁗ (Fig. 36). The apical horn is quite pronounced in some specimens and the apical pore plate supporting a circular smooth ridge is surrounded by a low collar which extends down the sides of the long narrow

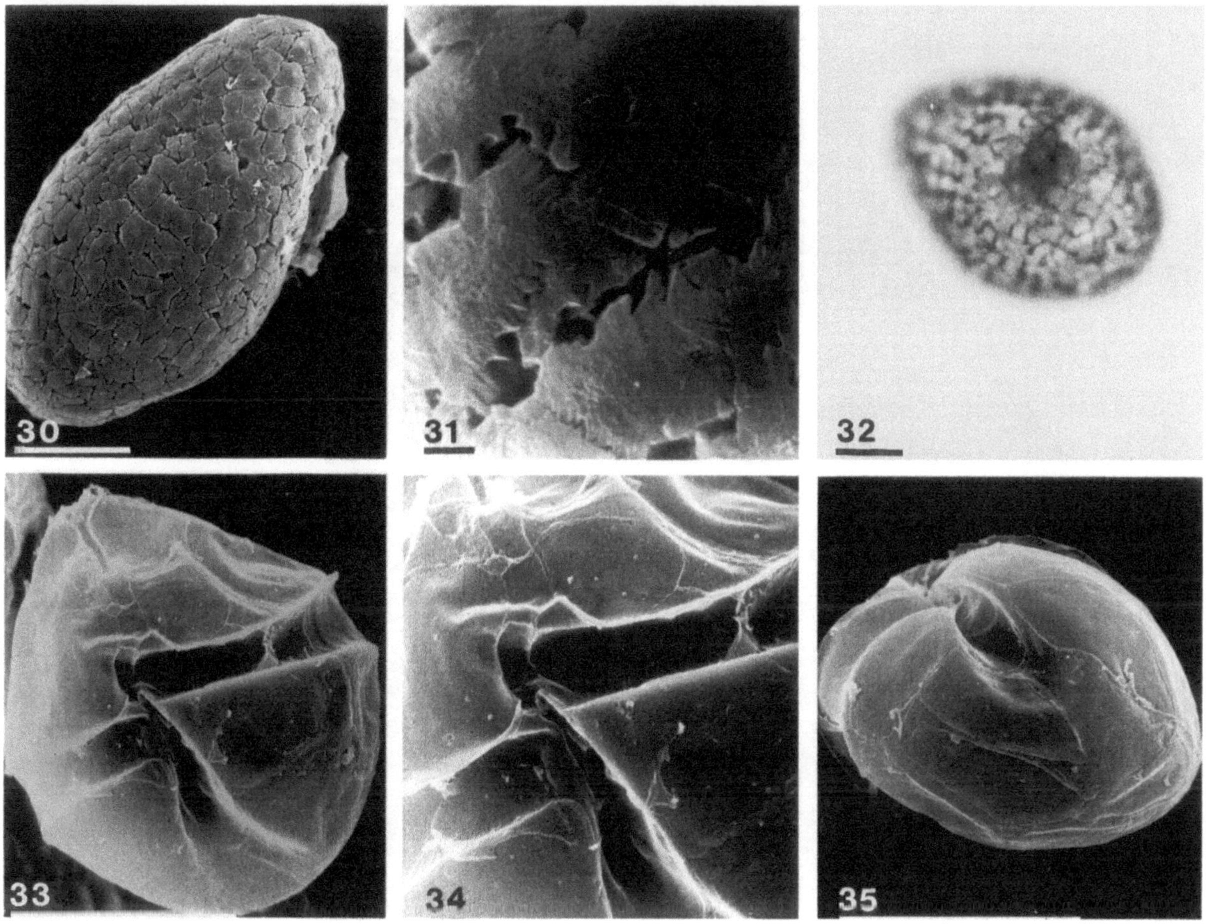

Figs 30–35. *Scrippsiella lachrymosa* sp. nov.
Fig. 32. Light micrograph, all the rest scanning micrographs. Thecate cells from culture material. Fig. 30. Cyst from Loch Creran showing calcareous ornament; Fig. 31. Detail of calcareous ornament on cyst showing plates starting to overgrow each other; Fig. 32. Cyst from Loch Creran showing red body; Fig. 33. Ventral view of thecate cell; Fig. 34. Sulcal area of thecate cell; Fig. 35. Antapical view of thecate cell. (Scale bars 10 μm except Figs 31, 34; 1 μm.)

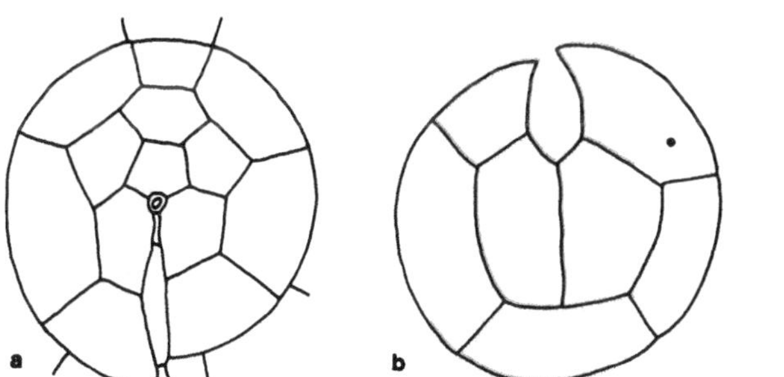

Fig. 36. Drawings to illustrate plate pattern of *S. lachrymosa* sp. nov.
(a) Epitheca; (b) Hypotheca.

canal plate (Fig. 33). This connects to a narrow first apical plate which is also asymmetrical. There are six cingular plates, the girdle is excavated and offset by one girdle width (Fig. 34). The sulcus is deeply excavated with a flap-like Sd plate. A complete sulcal description was not satisfactorily obtained. In one case, however, 5 sulcal plates were found. The antapical plates were not of even size (Fig. 35). As with *Scrippsiella crystallina* the thecal plate patterns were frequently poorly developed and a number of extra plates were found including an extra antapical plate. However, cells differ from those of *S. crystallina* as they are smaller in size and have a narrow first apical plate.

Diagnosis: Scrippsiella lachrymosa *Lewis sp. nov.*

Cellulae epithecam conicam et hypothecam rotundam habent. Cingulum medium excavatum unius cinguli latitudine deflexum. Cellulae primam laminam apicalem angustam habent. Laminarum formula haec: P_o, *x, 4′, 3a, 7″, 6c, ?s, 5‴, 2⁗. Laminarum ornatus non amplior quam ex poris simplicibus. Cysta ovoidea elongata quae antequam extat corpus rubrum continet. Paries interior, qui ex animata materia constat, complanatis laminis calcareis obtectus est. Archeopyle est apicalis et operculum affixum manet. 34–44 μm longitudine, 18–32 μm latitudine.*

Holotypes: Figs 30, 33.

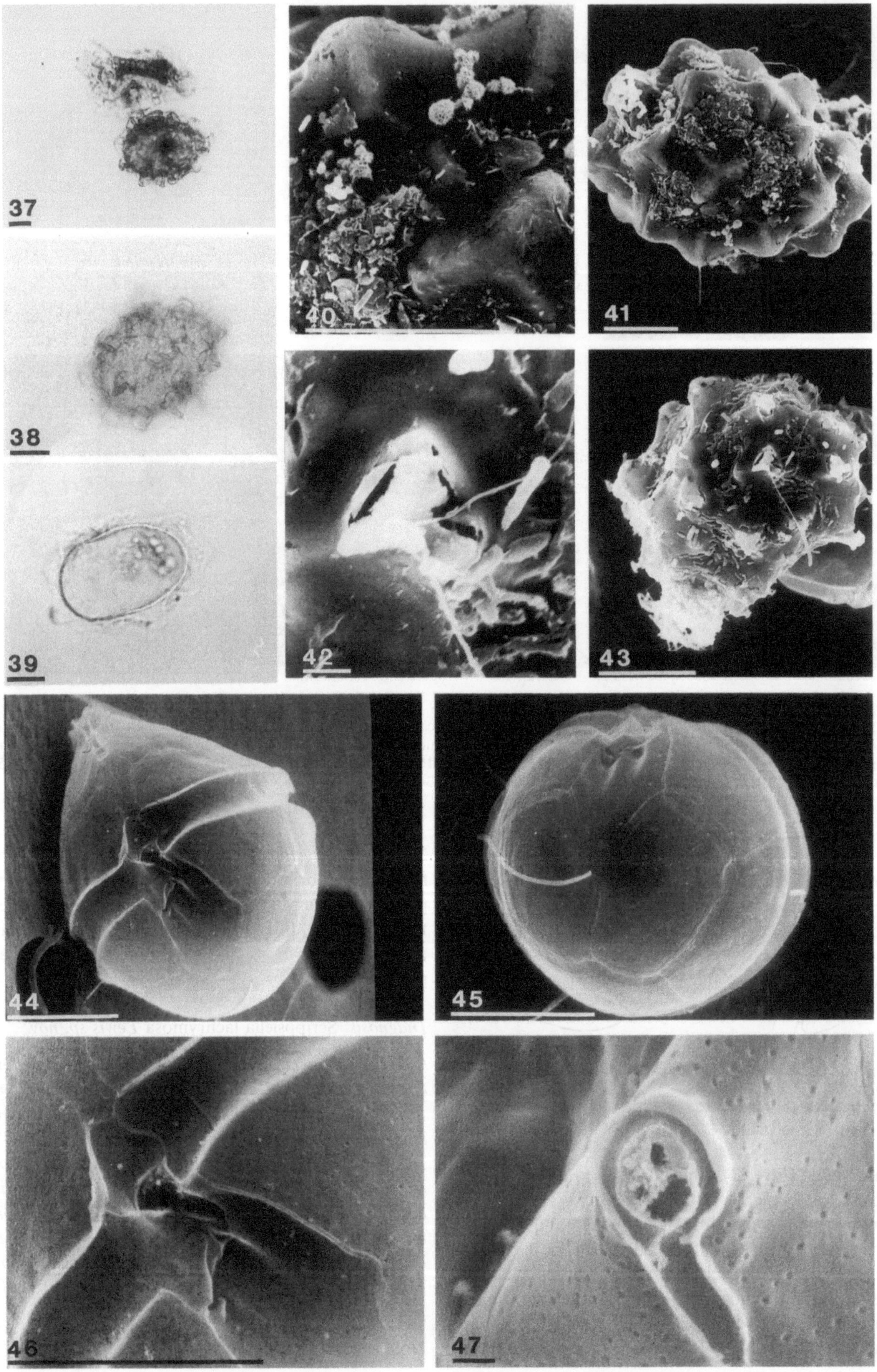
37
38
39
40
41
42
43
44
45
46
47

Scrippsiella trifida Lewis sp. nov. Figs 9c, 37–48

Cyst size length: 27–46 μm; breadth: 21–36 μm (11 measured).

Thecate cell size length: 20–37 μm; breadth: 14–36 μm (5 measured).

Cysts

The unusual oval shaped cysts contain a centrally placed red body (Fig. 37). The clear inner organic body has a number of trifurcate, recurving, calcareous spines (Figs 38, 40). These spines are surrounded by a thin clear outer membrane which undulates over the spines but remains close to the inner cyst body in places (Fig. 41). Addition of acid brings about dissolution of the spines but the inner and outer walls remain (Fig. 39). The archeopyle is an apical split which is sometimes notched, remaining attached to the main body of the cyst on one side. This is interpreted as an apical archeopyle representing the loss of 2′–4′ and 1–3a paraplates (Fig. 9). This species was successfully encysted several times to give cysts very reminiscent of the original but with slightly shorter appendages to the spines (Figs 42, 43).

Thecate cells

Thecate cells have a conical epitheca and a rounded hypotheca (Fig. 44). The theca is smooth with many scattered simple trichocyst pores. Thin sutural banding is evident on some specimens. The nucleus is centrally placed to one side of the cell. The plate formula is as follows:– P_o, x, 4′, 3a, 7″, 6c, 5s, 5‴, 2⁗ (Fig. 48). The apical pore plate supports a smooth circular ridge and is surrounded by a small wing which extends down the sides of the canal plate. The canal plate is quite long attaching to a wide, symmetrical, first apical plate. The girdle is excavated and displaced by half a girdle width. Only four sulcal plates are easily discernible in this species. This is because the flap-like Sd plate often obscures the Spa plate (Figs 46, 48c). The antapical plates are fairly even in size (Fig. 45). Some plate variations were seen in this species most notably in one specimen the absence of the third apical plate which was merged with the second apical plate (Fig. 47).

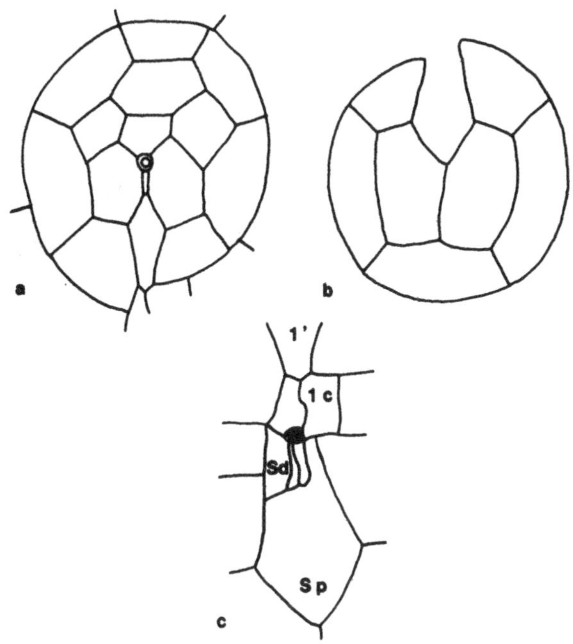

Fig. 48. Drawings to illustrate plate pattern of *S. trifida* sp. nov.
(a) Epitheca; (b) Hypotheca; (c) Sulcus.

Diagnosis: Scrippsiella trifida *Lewis sp. nov.*

Cellulae epithecam conicam et hypothecam rotundam habent. Cingulum excavatum et deflexum cinguli latitudine dimidiata. Laminarum formula haec: P_o, x, 4′, 3a, 7″, 6c, 5s, 5‴, 2⁗. Laminarum ornatus non amplior quam ex poris simplicibus. Cystae ovoideae, quarum centra, ex animata materia composita, pellucida sunt et spinis trifurcatis et recurvatis circumdata, quae ipsae in vicem externa membrana pellucida animatae materiae saepta est. Antequam extant cystae corpus rubrum continent. Archaeopyle est apicalis et operculum affixum manet. 27–46 μm longitudine; 21–36 μm latitudine.

Holotypes: Figs 41, 44.

Scrippsiella rotunda Lewis sp. nov. Figs 9b, 49–58

Cyst diameter: 25–33 μm (17 measured).

Thecate cell size length: 23–40 μm; breadth: 20–34 μm (32 measured).

Figs 37–47. *Scrippsiella trifida* sp. nov.
Figs 37–39. Light micrographs, all the rest scanning micrographs. Thecate cells from culture material. Fig. 37. Cyst from Loch Creran, median focus showing red body; Fig. 38. Cyst from Loch Creran, surface focus showing trifurcate recurving spines; Fig. 39. Cyst from Loch Creran after acid treatment, note two walls and lack of spines; Fig. 40. Detail of cyst wall shown in Fig. 41, note thin outer wall covering spines; Fig. 41. Cyst from Loch Creran; Fig. 42. Detail of outer wall of cyst formed in laboratory culture, the appendages to the spines are shorter than in the wild type and the thin outer wall has broken away around the spine; Fig. 43. Cyst formed in laboratory culture; Fig. 44. Ventral view of thecate cell; Fig. 45. Antapical view of thecate cell; Fig. 46. Sulcus of thecate cell; Fig. 47. Apical pore of thecate cell, note absence of a plate boundary between 2′ and 3′. (Scale bars 10 μm except Figs 42, 47; 1 μm.)

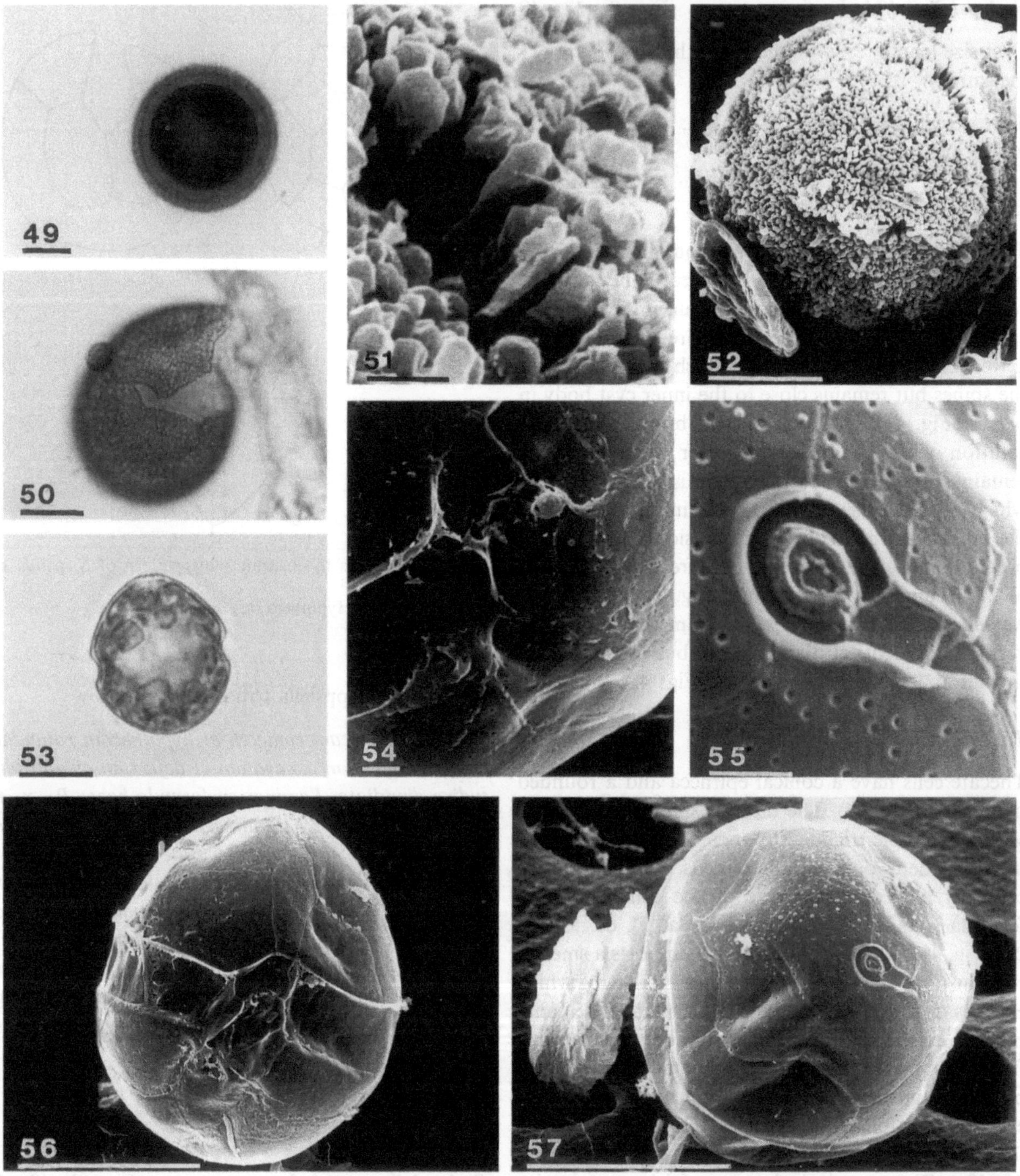

Figs 49–57. *Scrippsiella rotunda* sp. nov.
Figs 49, 50, 53. Light micrographs, all the rest scanning micrographs. Thecate cells from culture material. Fig. 49. Cyst from Loch Creran, median focus showing red body; Fig. 50. Germinated cyst from Loch Creran showing archeopyle; Fig. 51. Detail of cyst wall of cyst in Fig. 52, split in wall reveals close packing of upright crystals on inner organic wall; Fig. 52. Cyst from Loch Creran showing calcareous ornament; Fig. 53. Thecate cell showing rounded epicone; Fig. 54. Sulcus of thecate cell; Fig. 55. Apical pore of thecate cell; Fig. 56. Ventral view of thecate cell; Fig. 57. Apical view of thecate cell. (Scale bars 10 µm except Figs 51, 54, 55; 1 µm.)

Cysts

The cysts are thick walled, round to oval and brown in colour. The red body is difficult to discern because of the colour of the cyst (Fig. 49). The outer wall is composed of small (up to 2 µm) upright needle-shaped crystals packed together (Figs 51, 52). These are calcareous and when dissolved in acid a colourless inner organic body remains. Staining of this body with Trypan blue reveals no paratabulation. The archeopyle is somewhat angular in shape remaining attached

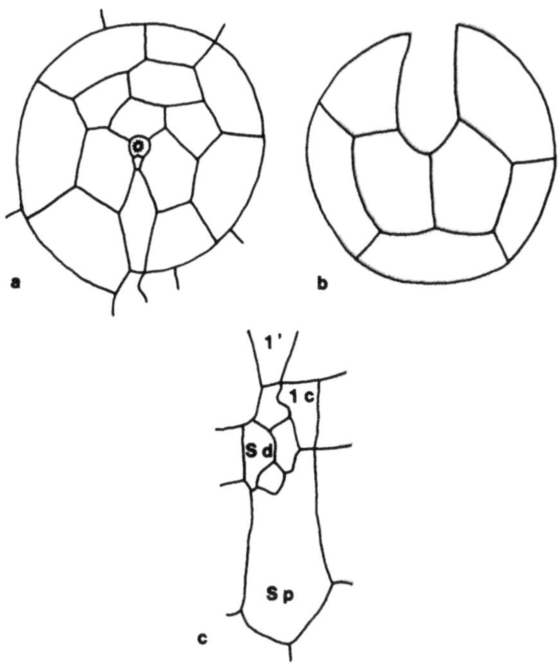

Fig. 58. Drawings to illustrate plate pattern of *S. rotunda* sp. nov.
(a) Epitheca; (b) Hypotheca; (c) Sulcus.

to the cyst body on one side. It is interpreted as representing the loss of the 2′–4′ and 1–3a paraplates (Figs 9, 50). Despite its thick wall this cyst is delicate especially when it has excysted, under examination it frequently squashed or broke into a number of pieces. This culture was not successfully encysted.

Thecate cells

This species has a rounded epitheca and hypotheca with no emphasized apical horn (Fig. 53). The cell has a fairly thick, smooth theca with scattered simple trichocyst pores (Fig. 56). The nucleus is very large (approximately one third of cell size) and centrally placed. There are many peripherally sited chloroplasts. The plate formula is as follows: – P_o, x, 4′, 3a, 7″. 6c, 5s, 5‴, 2⁗ (Figs 57, 58). The apical pore is slightly different from that of the other species (Fig. 55). The pore plate supports a toothed circular ridge and is surrounded by a narrow collar which extends along the sides of the canal plate. The connection to the first apical plate is made by a short canal plate. The cingulum is wide and generally poorly excavated and contains six plates. The apical edge has a narrow list which is absent on the antapical margin (Figs 54, 56). The sulcus is also fairly flat and contains five plates (Figs 54, 58c). The Sd plate forms a small flap. The Sp plate touches the first cingular plate and is somewhat rectangular in shape.

Diagnosis: Scrippsiella rotunda *Lewis sp. nov.*

Cellulae lato medio cingulo inexcavato rotundatae. Cellulae apex complanatus est, et pori lamina iugum eminens sustinet circulare et dentatum. Prima lamina apicalis latior est modo, et laminae ornatus non amplior est quam ex poris simplicibus. Laminae formula haec: P_o*, x, 4′, 3a, 7″, 6c, 5s, 5‴, 2⁗. 23–40 µm longitudine; 20–34 µm latitudine. Cystae sunt rotundae brunneoque colore. Corpus interius ex animata materia compositum conferto agmine crystallorum acicularium et calcareorum circumdatum est. Archeopyle est apicalis et operculum affixum manet. 25–33 µm transversim.*

Holotypes: Figs 52, 56.

Discussion

Table I summarises the differences between the species. All the new species are clearly *Scrippsiella* species and highlight Dale's contention (1978) that small thecal differences can manifest themselves in large differences in cyst types and that where possible both cysts and thecae should be included in species descriptions. Encystment of *S. trochoidea* cultures in this study and that of Wall *et al.* (1970) shows that in the laboratory a single species can produce some variety in calcareous ornament. However, the underlying cyst morphology (size, shape) remains the same. It is possible that this also occurs in nature, variations in calcium carbonate concentration causing different calcareous ornament (Keupp, pers. comm.). It is for these reasons that the second morphotype of *S. crystallina* was ascribed as such instead of being designated a new species. In the case of the four new species however, the base morphologies are sufficiently different to justify their description as new species.

With regard to other *Scrippsiella* species the new species differ in plate tabulation from *S. hexapraecingular* (syn. *S. gregaria* Loeblich *et al.* 1979) and *S. saladense* Balech in having seven not six precingular plates (Balech 1962, Horiguchi and Chihara 1983) and in plate arrangement from *S. subsalsa* (Ostenfeld) Steidinger *et* Balech, *S. arenicola* and *S. caponii* in having a contiguous set of intercalary plates instead of the third intercalary being separated from the first two (Steidinger and Balech 1977, Horiguchi and Pienaar 1988 a, b). *Scrippsiella mexicana* Indelicato *et* Loeblich (Syn. *Ensiculifera mexicana* Balech) has a distinctive first cingular plate with a projection into the epitheca. None of the species in this study were found to have such a projection. The cyst described for *S. mexicana* is also different from those of the new species (Wall *et al.* 1970, Wall and Dale 1968).

Table I. Comparison of the orthoperidinioid species described in this study.

	S. trochoidea	*P. dalei*	*S. lachrymosa*	*S. trifida*	*S. rotunda*	*S. crystallina*
Thecate cell length (μm)	23–37	10–21	16–30	20–37	23–40	27–47
Epitheca	Conical	Rounded	Conical/ Rounded	Conical	Rounded	Conical
Apical Horn	Upright collar, pore plate with smooth circular ridge	Upright collar, pore plate with smooth circular ridge	Upright collar, pore plate with smooth circular ridge	Upright collar, pore plate with smooth circular ridge	Flattened collar, pore plate with toothed circular ridge	Upright collar, pore plate with smooth circular ridge
l'plate	Narrow	Medium	Narrow	Medium	Wide	Wide
Cingular plates	6	5	6	6	6	6
Plate ornament	Simple pores	Concentric rings	Simple pores	Simple pores	Simple pores	Simple pores
Cyst shape	Spherical/ Subspherical	Spherical	Oval (elongate)	Oval	Spherical	Oval
Cyst ornament	Calcitic plates with spines	Organic branched spines	Calcareous flattened plates	Calcareous trifurcate spines with outer organic wall	Calcareous closely packed upright needles	Calcareous 'bricks'
Length of cyst (μm)	25–48	(Diam.) 24–38	34–44	27–46	(Diam.) 25–33	50–68

The cyst types of the new species are different from those described for *S. trochoidea*, *S. sweeneyae* and *S. precaria* (Wall and Dale 1968, Montresor and Zingone 1988). The remaining species in the genus is *S. tinctoria* described by Indelicato and Loeblich (1985). That species is most similar to *S. rotunda* described here. Both lack the typical Scrippsielloid conical apical horn and have a broad first apical plate. In size *S. tinctoria* is slightly smaller than *S. rotunda* (lengths: 19.2–28.8 μm compared to 23–40 μm). However, *S. rotunda* has a larger sulcus with a different arrangement of sulcal plates. In *S. tinctoria* one of the sulcal plates is described as being 'hidden' which is not the case for *S. rotunda* (Figs 54, 58c compared to figs 1, 7 of Indelicato and Loeblich 1985). *Scrippsiella rotunda* also appears to have a wider girdle than *S. tinctoria* and therefore I consider them to be different species. A feature of *S. tinctoria* that Indelicato and Loeblich (1985) regarded as important was the production of a yellow water soluble compound. A number of cultures raised during this study had this ability to a greater or lesser extent so this would not seem to be a good taxonomic feature.

Wall and Dale (1968) described a number of Quaternary calcareous dinoflagellate cysts. Apart from *S. trochoidea* the cysts they figured are different species to those discussed here. Dale (1983) reproduced photographs of a calcareous cyst ('so far unidentified') before and after acid treatment from Norwegian sediments (p. 87; figs 27, 28). That cyst appears to be identical to *S. crystallina*. Blanco (pers. comm.) has isolated five calcareous and organic orthoperidinioid cysts from sediments from north west Spain. Two of these appear conspecific with cysts described here. These are the cyst for *S. trochoidea* and possibly the cyst described here as *S. lachrymosa*. Both of these he attributes to *S. trochoidea* being uncertain of the taxonomic status of the cyst stage. I believe the evidence of the validity of cysts in dinoflagellate taxonomy is overwhelming and therefore it is more correct to separate these species. A large number of calcareous dinoflagellate cysts have been described from the Lower Cretaceous (for example see Keupp 1981). Some of these have obvious affinities with the new species described here particularly those within the fossil genera *Obliquepithonella* Keupp *ex* Lentin *et* Williams and *Orthopithonella* Keupp *ex* Lentin *et* Williams. However, it is difficult to be certain of any conspecifity when taxa are separated over such a long period of time.

It is clear that the calcareous dinoflagellates are a diverse group that merit attention. Dale (1986) has highlighted their potential importance in oceanic regions. They have probably been neglected because of their small size and the nature of the wall. In acidic palynological preparations any calcitic ornament would be lost leaving the inner organic wall or in the case of species such as *Scrippsiella trifida* the inner and outer organic wall. In this condition the species in the present study, if lacking an archeopyle, would not show any clear dinoflagellate features to allow their designation as such. Even with the archeopyle present their affinity might not be clear. The delicacy of this inner organic wall makes it probable that they would be destroyed completely during palynological processing. However, a calcitic cyst has been recognised from its organic wall in a study by Hultberg (1985) so it is clearly possible under some circumstances that such organic inner walls do appear in palynological preparations. This possibility should be carefully investigated when small organic cysts of this kind are encountered. Examination of the outside of the wall may give an indication of past crystalization either by remains of fibrous material or slight depressions in the wall where crystals had been present (Fig. 19). In the case of small cavate cysts (for example acid-treated *S. trifida*, Fig. 39) the space between the walls would need such examination. As Hultberg pointed out (1985) this could lead to a situation where one cyst species could be described under two names (one for the organic inner and the other for the whole calcareous cysts). In the case of the Recent calcareous cysts it is to be hoped this situation can be avoided as the possibility that one species could have three names (an additional one for the motile stage) could clearly lead to some confusion.

Many of the fossil calcareous cysts and those described by Wall and Dale (1968) have apical archeopyles. With the lack of paratabulation of the cysts described here it is difficult to decide which paraplates are involved in archeopyle formation. All the calcareous species in the present study have theropylic archeopyles with the operculum remaining attached to the body of the cyst by means of the first apical paraplate. Figure 9 shows two possible interpretations of the archeopyle, the first involving the loss of three apical paraplates ($[3A_{2-4'}]a$) and the second involving the loss of these and the intercalary paraplates ($[3A_{2-4'}\ 3I_{1-3a}]a$). Both styles of archeopyle have been found in calcareous dinoflagellate cysts (Keupp and Versteegh 1989). Wall and Dale (1968) were also uncertain about the involvement of the three intercalary paraplates in addition to the apical paraplates in the archeopyles of *S. trochoidea* and *S. sweeneyae*. As they noted in the case of a large archeopyle, involvement of all six paraplates seems likely. Archeopyles in the species described here can be somewhat variable. In many cases paratabulation is evident but in some cases archeopyles tend to roundness. Drawings taken from scanning micrographs of thecate cells (Figs 9b, c) show likely configurations of the archeopyle. If solely the three apical paraplates are involved the expectation would be for the operculum to be relatively small and the archeopyle opening to slope towards the apex of the cyst. This, however, does not appear to be the case (Figs 4, 21, 22, 50). The operculum is quite large and the archeopyle split is parallel to the equator of the cyst. Therefore, the best interpretation of the archeopyle in these species would seem to be that it represents the loss of both apical and intercalary paraplates.

The similarity of the motile thecate cells of these new species presents neontologists studying plankton samples with a problem. In seasonal studies of armoured dinoflagellates in Loch Creran *Scrippsiella* species were the most commonly occurring species from March to September (Lewis 1985). In these studies great variation of size and shape was noted in the *Scrippsiella* species but counting in chambers made precise taxonomic identification of these closely related species impossible and so they were grouped together. Cyst formation in both *S. trochoidea* and *S. crystallina* was noted sporadically throughout the times of *Scrippsiella* occurrence in the water column. The findings of Binder and Anderson (1987) from their laboratory study of *S. trochoidea* suggested that with the short dormancy period of the cyst, the fairly broad temperature ranges for germination and growth, rapid cycling of the *S. trochoidea* population between motile and resting stages was possible. This would seem to be the case for Loch Creran. The likelihood that such species are present in the water column for short periods of time (albeit overlapping with other species giving a broad background of elevated *Scrippsiella* numbers throughout the summer) and the closeness in morphology of the different species suggests that information on the range of species likely to be present in these situations would be most unambiguously obtained by study of sediment samples. The distribution of *S. crystallina* in Loch Creran (described as a calcareous *Scrippsiella* cyst, fig. 2B) was given in Lewis (1988).

It is important that the extent to which calcareous ornament varies with environmental factors or genetic factors should be determined. Such information may improve our understanding of the fossil record by giving indications of palaeoenvironment and also increasing our understanding of the species concept with regard to dinoflagellates. It is intended that such work will be carried out with this group of species to elucidate these mechanisms.

Acknowledgements

This work was carried out under N. E. R. C. Grant No. GR3/5511. Field sampling facilities were provided by the S. M. B. A. Support during manuscript preparation was provided by U. M. B. S. This paper was presented at the Fourth International Conference on Modern and Fossil Dinoflagellates. Thanks go to Dr S. Usher for Latin diagnoses, Mrs J. Young for photographic assistance and Professor J. D. Dodge for helpful discussion.

References

Balech, E. 1962. Dos Dinoflagelados de una laguna salobre de la Argentina. *Notas Del Museo XX*: 111–123.

Balech, E. 1974. El genero "Protoperidinium" Bergh, 1881 ("Peridinium" Ehrenberg, 1831, Partim). *Hidrobiologia IV*: 1–79.

Binder, B. and D. M. Anderson. 1987. Physiological and environmental control of germination in *Scrippsiella trochoidea* (Dinophyceae) resting cysts. *J. Phycol. 23*: 99–107.

Dale, B. 1976. Cyst formation, sedimentation, and preservation: factors affecting dinoflagellate assemblages in Recent sediments from Trondheimsfjord, Norway. *Rev. of Paleobotany & Palynology 22*: 39–60.

Dale, B. 1977. New observations on *Peridinium faeroense* Paulsen (1905), and classification of small orthoperidinioid dinoflagellates. *Br. phycol. J. 12*: 241–253.

Dale, B. 1978. Acritarchous cysts of *Peridinium faeroense* Paulsen: implications for dinoflagellate systematics. *Palynology 2*: 187–193.

Dale, B. 1983. Dinoflagellate resting cysts: "benthic plankton". *In*: (G. A. Fryxell, ed.) *Survival Strategies of the Algae*. Cambridge University Press, Cambridge. pp. 69–136.

Dale, B. 1986. Life cycle strategies of oceanic dinoflagellates. *Unesco Technical Papers in Marine Science 49*: 65–72.

Eaton, G. L. 1969. Dinoflagellate cysts and acritarchs from the Bracklesham Beds of the Isle of Wight. Ph. D. thesis Sheffield University.

Gao Xiaoping, J. D. Dodge and J. Lewis. 1989. An ultrastructural study of planozygotes and encystment of a marine dinoflagellate, *Scrippsiella* sp. *Br. Phycol. J. 24*: 153–165.

Guillard, R. R. L. and J. H. Ryther. 1962. Studies on marine planktonic diatoms I. *Cyclotella nana* Hustedt and *Detonula confervacea* (Cleve) Gran. *Can. J. Microbiol. 8*: 229–239.

Horiguchi, T. and M. Chihara. 1983. *Scrippsiella hexapraecingula* sp. nov. (Dinophyceae), a tide pool dinoflagellate from the Northwest Pacific. *Bot. Mag. Tokyo 96*: 351–358.

Horiguchi, T. and R. N. Pienaar. 1988 a. A redescription of the tidal pool dinoflagellate *Peridinium gregarium* based on re-examination of the type material. *Br. Phycol. J. 23*: 33–39.

Horiguchi, T. and R. N. Pienaar. 1988 b. Ultrastructure of a new sand-dwelling dinoflagellate, *Scrippsiella arenicola* sp. nov. *J. Phycol. 24*: 426–438.

Hultberg, S. U. 1985. *Pithonella organica* a new calcareous dinoflagellate with an inner organic wall. *Grana 24*: 115–120.

Indelicato, S. R. and A. R. Loeblich III. 1985. A description of the marine dinoflagellate, *Scrippsiella tinctoria* sp. nov. *Jap. J. Phycol. 33*: 127–134.

Indelicato, S. R. and A. R. Loeblich III. 1986. A revision of the marine peridinioid genera (Pyrrhophyta) utilizing hypothecal-cingular plate relationships as a taxonomic guideline. *Jap. J. Phycol. 34*: 153–162.

Keupp, H. 1981. Die kalkigen Dinoflagellaten – Zysten der borealen Unter-kreide (Unter-Hauterivium bis Unter-Albium). *Facies 5*: 1–190.

Keupp, H. and G. Versteegh. 1989. Ein neues systematisches Konzept für kalkige Dinoflagellaten – Zysten der Subfamilie Orthopithonelloideae Keupp 1987. *Berliner geowiss. Abh. A. 106*: 207–219.

Lebour, M. V. 1925. *The Dinoflagellates of Northern Seas*. Marine Biological Association, Plymouth. pp. 250.

Lemmermann, E. 1910. Klasses Peridiniales. *In*: (E. Lemmermann, ed.) *Kryptogamenflora der Mark Brandenburg und angrenzender Gebiete, III Algen* 1 *(Schizophyceen, Flagellaten, Peridineen)*, Gebrüder Borntraeger, Leipzig. pp. 563–682.

Lewis, J. M. 1985. The ecology and taxonomy of marine dinoflagellates in Scottish sea-lochs. Unpub. Ph. D. thesis. University of London.

Lewis, J. 1988. Cysts and sediments: *Gonyaulax polyedra* (*Lingulodinium machaerophorum*) in Loch Creran. *J. mar. biol. Ass. U. K. 68*: 701–714.

Lewis, J., J. D. Dodge and P. Tett. 1984. Cyst-theca relationships in some *Protoperidinium* species (Peridiniales) from Scottish sea lochs. *J. micropalaeontol. 3*: 25–34.

Loeblich, A. R. III, J. L. Sherley and R. J. Schmidt. 1979. Redescription of the thecal tabulation of *Scrippsiella gregaria* (Lombard and Capon) comb. nov. (Pyrrophyta) with light and scanning electron microscopy. *Proc. Biol. Soc. Wash. 92*: 45–50.

McLachlan, J. 1973. Growth media – marine. *In*: (J. R. Stein, ed.) *Handbook of Phycological Methods*. Cambridge University Press. Cambridge, pp. 25–51.

Montresor, M. and A. Zingone. 1988. *Scrippsiella precaria* sp. nov. (Dinophyceae), a marine dinoflagellate from the Gulf of Naples. *Phycologia 27*: 387–394.

Paulsen, O. 1905. On some Peridiniae and plankton diatoms. *Meddelelser fra Kommisionen for Havundersogelser, Serie Plankton B. I nr. 3*.

Steidinger, K. A. and E. Balech. 1977. *Scrippsiella subsala* (Ostenfeld) comb. nov. (Dinophyceae) with a discussion on *Scrippsiella*. *Phycologia 16*: 69–73.

Stein, F. R. 1883. *Der Organismus der Infusionsthiere nach Eigenen Forschungen in systematischen bei neufloge bearbeitet. III. Abteilung„I H"-Die Naturgeschichte der Arthrodelen Flagellaten*. Wilhelm Engelmann, Leipzig. pp. 1–30, pls. 25.

Taylor, F. J. R. 1978. Dinoflagellates. *In*: (A. Sournia, ed.) *Phytoplankton Manual. Monographs on Oceanographic Methodology 6*. U. N. E. S. C. O., Paris. pp. 143–147.

Wall, D. and B. Dale. 1966. "Living fossils" in Western Atlantic plankton. *Nature 211*: 1025–1026.

Wall, D. and B. Dale. 1968. Quaternary calcareous dinoflagellates (calciodinellideae) and their natural affinities. *J. of Paleontology 42*: 1395–1408.

Wall, D., R. R. L. Guillard, B. Dale, E. Swift and N. Watabe. 1970. Calcitic resting cysts in *Peridinium trochoideum* (Stein) Lemmermann, an autotrophic marine dinoflagellate. *Phycologia 9*: 151–156.

Williams, G. L. 1963. Organic-walled microplankton of the London Clay. Ph. D. thesis. Sheffield University.

Botanica Marina
Vol. 34, pp. 107 – 111, 1991

Phycochemical Studies on *Gracilaria foliifera* (Gigartinales, Rhodophyta)

A. Hayee-Memon, M. Shameel, M. Ahmad*, V. U. Ahmad**, and K. Usmanghani*

Department of Botany, University of Karachi, Karachi – 75270, Pakistan

* *Department of Pharmacognosy, Faculty of Pharmacy, University of Karachi, Karachi – 75270, Pakistan*

** *H. E. J. Research Institute of Chemistry, University of Karachi, Karachi – 75270, Pakistan*

(Accepted 23 September 1990)

Abstract

The red seaweed *Gracilaria foliifera* (Forsskål) Børgesen was examined for the presence of fatty acids, sterols, fatty alcohols and sugars. The fatty acids, identified by a combined GC-MS technique, showed the presence of five methyl esters as pentadecylate, palmitate, margarate, stearate and oleate. The sterol composition, analysed by mass spectrometry, revealed the presence of cholesterol, 22-dehydrocholesterol and desmosterol. An unsaturated fatty alcohol was isolated and its structure was determined by ^{13}C-NMR as 2-methyl-triacont-5,10,16,22,27-pentaen-15-ol. A sugar was also isolated and its structure was characterised through pmr, cmr and F. D. mass spectral techniques as 6-O-dodecyl-α-D-glucopyranosyl-(1→2)-β-D-fructofuranoside.

Introduction

Phycochemistry is the study of the natural products and the chemical constituents occurring in algae, from a biological point of view (Shameel 1990). The marine red alga *Gracilaria foliifera* (Forsskål) Børgesen is found abundantly around the seashore of Karachi and the neighbouring coastal areas (Shameel 1987, Shameel *et al.* 1989). In the last decade several studies have been made on the fatty acids, sterols, agar and polysaccharides, aminoacids, prostaglandins and other chemical constituents of various species of *Gracilaria* (Brasch *et al.* 1981, Friedlander *et al.* 1981, Marolia *et al.* 1982, Fusetani and Hashimoto 1984, Riguera *et al.* 1984, 1985, Helleur *et al.* 1985, Takagi *et al.* 1985). In the present study an attempt has been made to isolate and characterise fatty acid methyl esters, sterols, fatty alcohols and sugars from *Gracilaria foliifera*.

Material and Methods

Gracilaria foliifera was collected from mid- and lower littoral rocks of the seashore ledges at Buleji, near Karachi during October 1986. Healthy specimens were selected and thoroughly cleaned with fresh water to remove undesirable material attached to the thalli. They were then air dried under shade and milled.

A. Isolation of fatty acids

1. Extraction

The dried seaweed (250 g) was extracted with hexane : $CHCl_3$ (1 : 1) at room temperature for 15 days. Evaporation of the organic phase under reduced pressure yielded 2.5 g of thick syrupy dark-reddish-black material.

2. Saponification

Saponification of the extract (1.5 g) was carried out by refluxing at 100 °C for 6 hours with 15% KOH in 50% EtOH (25 mL). The resulting mixture was concentrated under vacuum and H_2O was added to make up the volume and finally ethyl acetate (EtOAc) was added. The partitioning procedure between H_2O and EtOAc was repeated 3 times. The total combined ethyl acetate fraction was acidified with 1N HCl (pH 4 – 5) and then re-extracted with EtOAc. The procedure was repeated several times to furnish a total of ½ L of EtOAc extract, which on evaporation under reduced pressure yielded 0.5 g of material.

3. Esterification

The extract so obtained was subjected to methylation with diazomethane. About 2 mL of diazomethane was added to 0.25 g fatty acid fraction and the reaction mixture was left at room temperature for 24 hours until dissolved, and the aliquotes were directly injected into a gas chromatograph-mass spectrometer (GC-MS).

4. Gas chromatograph-mass spectrometry

The GC-MS of the methylated fatty acid fraction was performed on a GC-Hewlett Packard with 11/73 DEC computer data system and a 1.2 m × 4 mm packed glass capillary column coated with gas chrome Q (100–120 mesh, OV 101 1%). The column temperature was programmed between 70–250 °C with a rate of increase of 80 °C per minute. The carrier gas (helium) flow rate was 32 mL/min and the injector temperature was 250 °C.

B. Isolation of sterols, fatty alcohol and sugar

1. Extraction

Dried material of *G. foliifera* (500 g) was percolated at room temperature in $CHCl_3$: MeOH (1 : 3) for 3 weeks. On evaporation of the organic layer the extract so obtained weighed 4 g and it was partitioned between EtOAc and H_2O and this procedure was repeated 3 times. The combined EtOAc extract (3 g) was absorbed onto silica gel (70–230 mesh) and eluted. The eluting solvents were hexane : ether, chloroform, chloroform : methanol in increasing order of polarity and the eluate was collected in 50 mL fractions. The fractions obtained from hexane : ether (7 : 3) contained a mixture of sterols, the hexane : ether (9 : 1) fraction gave an unsaturated fatty alcohol, whereas the chloroform : methanol (1 : 1) yielded a sugar derivative.

2. Chromatography

The sterol mixture was separated by preparation thin layer plate chromatography (silica gel 254 nm, Mikrokarten SI F, 5 × 10 cm, 0.2 mm, Riedel-de-Haen). The pure compounds so obtained were subjected to spectral characterisation. The ^{1}H-NMR spectra were obtained either in deuterochloroform ($CDCl_3$) or deuteropyridine (C_5D_5N) as a solvent and tetramethylsilane (TMS) as the internal reference on Bruker WM 400 instrument equipped with an Aspect 2000 computer. The ^{13}C-NMR spectra were measured at 75.43 MHz using a Bruker AM 300 spectrometer. Electron impact (EI) mass spectra were determined on a Finnigan MAT 312 double focusing mass spectrometer connected to a PDP 11/34 (DEC) computer system whereas field desorption (F. D.) mass spectra were measured on a MAT 312 (Germany) spectrometer connected to a PDP 11/34 (U. S. A.) computer system.

Results

Detailed analysis of the fatty acids found in *Gracilaria foliifera* are presented in Table I. The fatty acids ranged from C16 to C19 as saturated methyl esters of pentadecylate, palmitate, margarate, stearate and unsaturated methyl ester of oleate. The mixture of sterol fractions contained 3 intense spots by TLC and was purified by preparative TLC to give 3 fractions. The isolated pure sterols were subjected to analysis by mass spectrometry and the following spectral data have been obtained.

Cholesterol

Mass (EI) m/z 386 (M^+, $C_{27}H_{46}O$, 41%), 371 (M^+-CH_3, 33.2%), 368 (M^+-H_2O, 31%), 353 (M^+-CH_3 -H_2O, 37.3%), 301 (6%), 275 (13.5%), 273 (M^+-side chain, 16.8%), 255 (M^+-side chain -H_2O, 7.7%), 247 (5%), 231 (M^+-side chain – ring D cleavage, 18.4%), 229 (M^+-side chain -C_{16} -C_{17} -OH, 8.7%), 213 (M^+-side chain -H_2O-ring D cleavage, 15%), 129 (41%), 121 (15.6%), 107 (25.1%), 97 (47%), 83 (92%), 73 (100%).

22-Dehydrocholesterol

Mass (EI) m/z 384 (M^+, $C_{27}H_{44}O$, 44%), 300 (M^+-C_6H_{11} -H, 24%), 273 (M^+-C_8H_{15}, 14%), 271 (M^+-C_8H_{15} -2H, 16%), 255 (M^+-C_8H_{15} -H_2O, 8%), 213 (25%), 198 (11%), 157 (28.5%), 145 (46%), 95 (80%), 69 (100%). ^{1}H-NMR (400 MHz, $CDCl_3$), ppm 0.67 (3H, s, 18-Me), 0.83 (3H, d, J = 6.0 Hz, 26-Me), 0.85 (3H, d, J = 5.6 Hz, 27-Me), 1.00 (3H, s, 19-Me), 1.22 (3H, d, J = 7.0 Hz, 21-Me), 3.51 (1H, m, 3α-H), 4.22 (1H, t, J = 6.0 Hz, 6-H).

Desmosterol

Mass (EI) m/z 384 (M^+, $C_{27}H_{44}O$, 32%), 369 (M^+-CH_3, 14%), 366 (M^+-H_2O, 3%), 351 (M^+-CH_3 -H_2O, 3.5%), 299 (M^+-CH_3-C_2H_9 -H, 2%), 271 (M^+-C_8H_{15}, side chain -2H, 7%), 255 (M^+-C_8H_{15}-H_2O, 2%), 253 (M^+-side chain -2H -H_2O, 3%), 299 (13%), 161

Table I. Fatty acids of *Gracilaria foliifera* analysed as methyl esters.

Systematic name	Common name	Molecular formula	Mol. wt.	Retention time (Min)	Rel. % age	Mass and fragmentation pattern
A. *Saturated fatty acid methyl esters:*						
Methyl-*n*-pentadecanoate	Methyl pentadecylate	$C_{16}H_{32}O_2$	256	16′30″	6.69	GC-MS m/z 256 (M^+, $C_{16}H_{32}O_2$, 3%), 225 (M^+-31, 1%), 199 (4%), 185 (43%), 171 (17%), 157 (3%), 143 (7%), 129 (63%), 115 (22.5%), 101 (6%), 87 (23%), 73 (100%).
Methyl-*n*-hexadecanoate	Methyl palmitate	$C_{17}H_{34}O_2$	270	19′00″	26.52	GC-MS m/z 270 (M^+, $C_{17}H_{34}O_2$, 39%), 239 (M^+-31, 19%), 227 (22%), 213 (5%), 199 (6%), 185 (18%), 171 (18%), 157 (5%), 143 (75%), 129 (57%), 115 (22%), 101 (58%), 87 (76%), 74 (100%).
Methyl-*n*-heptadecanoate	Methyl margarate	$C_{18}H_{36}O_2$	284	19′30″	27.96	GC-MS m/z 284 (M^+, $C_{18}H_{36}O_2$, 40%), 253 (M^+-31, 1%), 241 (M^+-43, 10%), 227 (11%), 213 (42%), 199 (10%), 185 (22%), 171 (22%), 157 (25%), 143 (12%), 129 (89%), 115 (37%), 101 (30%), 87 (54%), 74 (100%).
Methyl-*n*-octadecanoate	Methyl stearate	$C_{19}H_{38}O_2$	298	22′00″	15.53	GC-MS m/z 298 (M^+, $C_{19}H_{38}O_2$, 56%), 284 (M^+-14, 1%), 267 (M^+-31, 19%), 255 (M^+-43, 24%), 241 (5%), 227 (7%), 213 (18%), 199 (24%), 185 (17%), 171 (6%), 157 (6%), 143 (75%), 129 (47%), 115 (16%), 101 (30%), 87 (81%), 74 (100%).
B. *Unsaturated fatty acid methyl ester:*						
Methyl-octadecenoate	Methyl oleate	$C_{19}H_{36}O_2$	296	26′00″	23.18	GC-MS m/z 296 (M^+, $C_{19}H_{36}O_2$, 1%), 264 (M^+ 32, 1%), 222 (5%), 183 (7%), 169 (5%), 155 (4%), 141 (6%), 127 (7%), 113 (97%), 99 (12%), 85 (42%), 71 (100%).

(20%), 121 (27%), 81 (63%), 55 (100%). ^{1}H-NMR (400 MHz, $CDCl_3$), ppm 0.55 (3H, s, 18-Me), 1.00 (3H, s, 19-Me), 1.24 (3H, d, J = 7.0 Hz, 21-Me), 1.57 (6H, s, 26, 27-Me), 3.56 (1H, m, 3α-H), 4.23 (2H, m, 6, 24-H).

The ^{1}H-NMR of the isolated sterols confirmed the presence of cholesterol [1], 22-dehydrocholesterol [2] (signal assignments such as 3.51 and 4.22 indicating OH and double bond) and desmosterol [3] (signal assignments such as 3.56 and 4.23 indicating OH and double bond). These are well known compounds. An unsaturated fatty alcohol was also isolated from the extract of *G. foliifera*, which exhibited the following spectral data:

Fatty alcohol

$C_{31}H_{54}O$, mol. wt. 442, ^{13}C-NMR (75 MHz, $CDCl_3$) ppm C-1′ (14.08), C-1 (19.30), C-2 (38.72), C-3 (29.67), C-4 (27.06), C-5 (130.90), C-6 (124.91), C-7 (29.43), C-8 (23.72), C-9 (30.01), C-10 (125.91), C-11 (130.90), C-12 (31.91), C-13 (28.90), C-14 (22.66), C-15 (68.15), C-16 (128.83), C-17 (135.80), C-18 (32.73), C-19 (22.66), C-20 (29.33), C-21 (30.34), C-22 (125.91), C-23 (130.90), C-24 (29.67), C-25 (34.20), C-26 (29.67), C-27 (128.83), C-28 (135.80), C-29 (22.95), C-30 (14.08).

The ^{13}C-NMR of the fatty alcohol furnished two signals at 19.30 and 14.08 ppm, indicating the presence of methyl groups. Whereas a peak at 68.15 showed -CH(OH)-linkage and peaks appearing between 22.66 and 38.70 ppm were representatives of the methylene groups. The unsaturation *i.e.* -C=C- signals were displayed from 124.91 to 135.80 ppm. The unsaturated fatty alcohol has been identified through proton magnetic resonance (pmr) and mass spectrometry (MS) as 2-methyl-triacont-5, 10, 16, 22, 27 -pentaen-15-ol [4]. The extract of *G. foliifera* also yielded a glycoside, through column chromatography, in sufficient quantity so as to scan the different spectra. The following spectral data have been obtained:

Glycoside

$C_{24}H_{44}O_{12}$, mol. wt. 524, ^{13}C-NMR (75.43 MHz, C_5D_5N) ppm C-1 (14.20), C-2 (22.86), C-3 (29.35), C-4 (25.18), C-5 (29.63), C-6 (32.05), C-7 (29.79), C-8 (34.19), C-9 (29.93), C-10 (34.41), C-11 (75.42), C-12

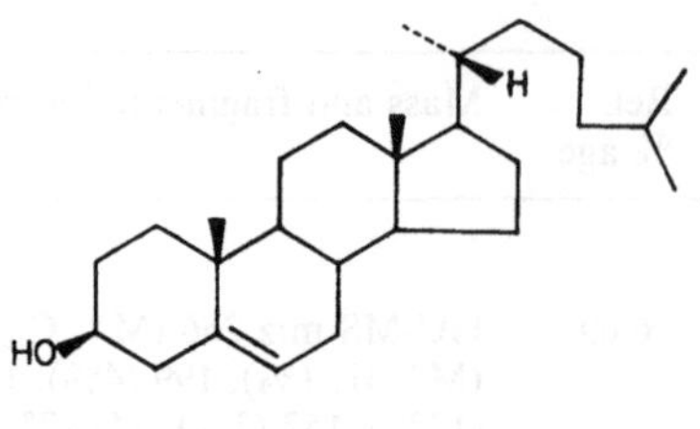

[1]

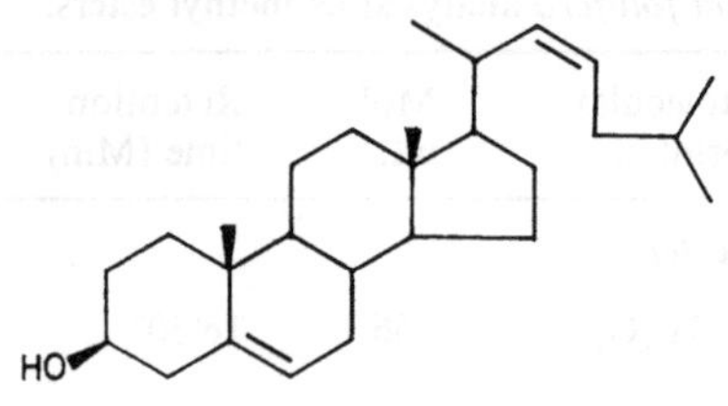

[2]

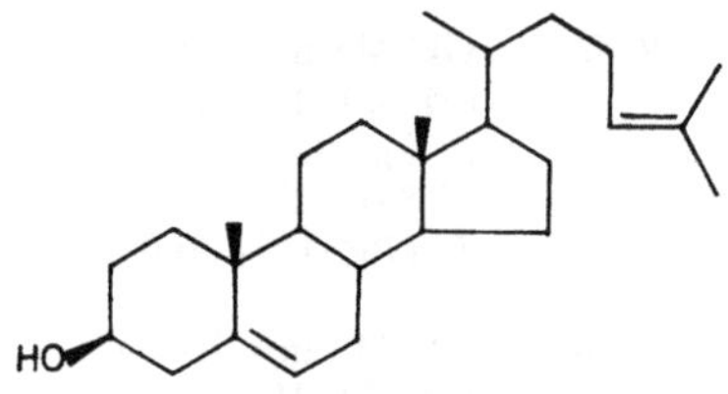

[3]

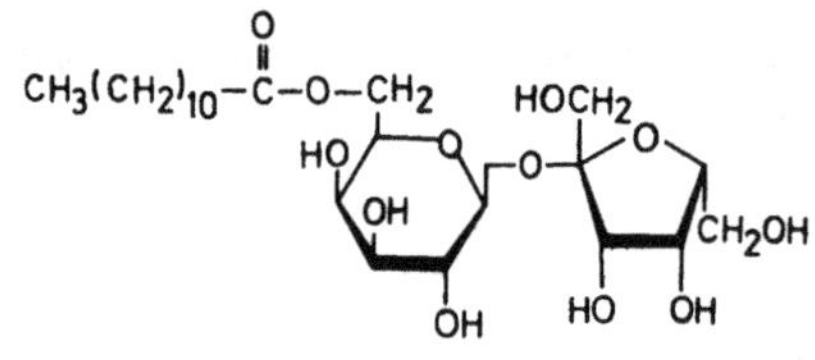

[5]

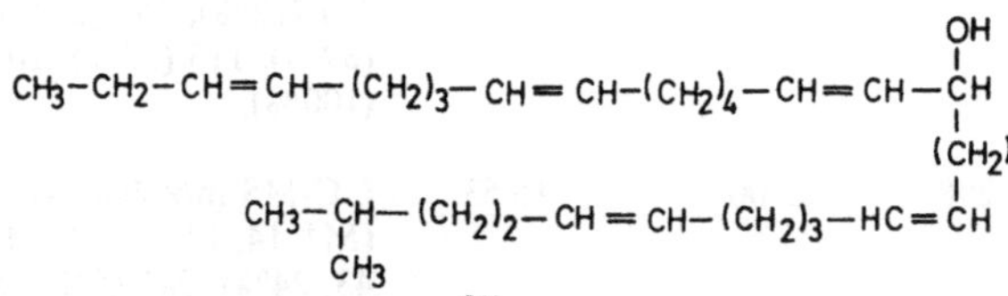

[4]

(173.41). Glucose: C-1′ (100.31), C-2′ (73.51), C-3′ (73.81), C-4′ (69.92), C-5′ (70.92), C-6′ (66.42). Fructose: C-1″ (64.64), C-2″ (102.33), C-3″ (86.13), C-4″ (76.00), C-5″ (75.03), C-6″ (63.48). Mass (FD and EI) m/z 524 (M^+, $C_{24}H_{44}O_{12}$, 32%), 522 (M^+-H_2, 26%), 342 (M^+-dodecoic acid, 4%), 199 (M^+-sucrose, 7%), 183 (M^+-sucrose with O atom, 4%).

The ^{13}C-NMR of the isolated glycoside is quite representative and showed two moieties, a sucrose and dodecoic acid. A fatty acid molecule, dodecoic acid, is attached at the C-6 of α-D-glucopyranosyl because the signal for C-6 of glucose is found at 66.42 ppm (downfield shifting due to -CO-O-). The ^{13}C-NMR values of glucose and fructose are quite similar to the reported values of α-D-glucopyranosyl-β-D-fructofuranoside (Breitmaier and Voelter 1978). The attachment between the fructose and glucose appears to be 1′−2″. On the basis of the spectral evidences the isolated sugar derivative has been identified as 6-O-dodecyl-α-D-glucopyranosyl-(1→2)-β-D-fructofuranoside [5].

Discussion

Five fatty acids have been isolated from *Gracilaria foliifera* (Table I). The frequency of occurrence of these acids in decreasing order of concentration was methyl margarate (27.96%), palmitate (26.52%), oleate (23.18%), stearate (15.53), whereas pentadecylate was the minor constituent (6.69%). Palmitic and oleic acids have previously been reported as major components of the fatty acids in *G. corticata* J. Ag. (Marolia *et al.* 1982), *G. lemanaeformis* and *G. verrucosa* (Huds.) Papenf. (Henriquez *et al.* 1972 a, Riguera *et al.* 1984). In several species of red algae from Karachi palmitic acid was the predominant fatty acid but oleic acid was absent (Qasim 1986, Shameel 1990). Although arachidonic acid (C20 : 4), a polyethenoid, has been reported in very high amounts (60%) from *G. verrucosa* (Takagi *et al.* 1985), it was absent in *G. foliifera*. Polyethenoic, as well as lower fatty acids, were absent from *G. corticata* (Marolia *et al.* 1982), and they were not found in *G. foliifera*.

Three sterols, cholesterol, 22-dehydrocholesterol and desmosterol were found in *G. foliifera*. Cholesterol has also been reported from *G. lemanaeformis*, *G. verrucosa* (Henriquez *et al.* 1972 a, b, Riguera *et al.* 1985) and *G. textorii* (Sur.) J. Ag. (Kanazawa *et al.* 1972). It appears to be of common occurrence in *Gracilaria*. A ketosteroid, cholest-4-en-3-one isolated from *G. textorii* (Kanazawa and Yoshioka 1971) and cholest-7-en-3β-ol reported from *G. foliifera* from Eastern Long Island (Goldberg *et al.* 1982), could not be detected in our algal specimens. This indicates that regional differences may occur in this seaweed. Although 22-dehydrocholesterol and desmosterol have been reported from other red algae (*e.g.* Kanazawa *et al.* 1972, Riguera *et al.* 1985), they are being reported for the first time from the genus *Gracilaria*.

The unsaturated fatty alcohol isolated from the hexane : ether (9 : 1) fraction of *G. foliifera* appeared to be triacontene, more specifically a pentaene. A pentatriacontane has been obtained in 60:40 petroleum ether: C_6H_6 eluate of *G. lemanaeformis* and *G. verrucosa* (Henriquez *et al.* 1972 a). Heptadecane was found to be the predominant hydrocarbon in *G. verrucosa* (Riguera *et al.* 1984). It appears that both short and long chain hydrocarbons and their derivatives are produced as secondary metabolites in various species of *Gracilaria*. The occurrence of this polyunsaturated alcohol in *G. foliifera* is not a unique feature, as certain polyunsaturated acids have been found in *G. verrucosa* (Takagi *et al.* 1985) and cis and trans-phytol were isolated from *G. andersonii* (Grun.) Kylin (Sims and Pettus Jr. 1976).

The glycoside isolated from the extract of *G. foliifera* was a sucrose derivative with a fatty acid molecule, dodecoic acid attached with it at C-6. Galactose derivatives, which are the major components of this seaweed, were not identified and no study could be made on agar-related polysaccharides. Very characteristic ^{13}C-NMR spectra have been obtained for agars of *G. foliifera*, *G. bursapastoris* (Gmel.) Silva (Bhattacharjee *et al.* 1979) and *G. verrucosa* (Friedlander *et al.* 1981), agaroses of *G. secundata* Harv. (Brasch *et al.* 1981) and polysaccharide pyrolysates, of *G. verrucosa* (Helleur *et al.* 1985). The analysis of polysaccharides in *Gracilaria* and other red algae may be useful chemotaxonomically. It might help in distinguishing between very closely related species such as *G. corticata* and *G. foliifera*.

Acknowledgements

We would like to thank Dr (Mrs) Shaheen Bano, Research Associate, H. E. J. Research Institute of Chemistry, University of Karachi, for her valuable help during the course of this study.

References

Bhattacharjee, S. S., W. Yaphe and G. K. Hamer. 1979. Study of agar and carrageenan by C-13 nuclear magnetic resonance spectroscopy. *Proc. Int. Seaweed Symp. 9*: 379–385.

Brasch, D. J., C.-T. Chuah and L. D. Milton. 1981. A carbon-13 NMR study of some agar related polysaccharides from New Zealand seaweeds. *Aust. J. Chem. 34*: 1095–1105.

Breitmaier, E. and W. Voelter. 1978. *^{13}C-NMR Spectroscopy*. Verlag Chemie, Weinheim. 344 pp.

Friedlander, M., Y. Lipkin and W. Yaphe. 1981. Composition of agars from *Gracilaria cf verrucosa* and *Pterocladia capillacea*. *Bot. Mar. 24*: 595–598.

Fusetani, N. and K. Hashimoto. 1984. Prostaglandin E_2: a candidate for causative agent of "Ogonori" poisoning. *Nip. Suis. Gakkai. 50*: 465–469.

Goldberg, A. S., C. Hubby, D. Cobb, P. Millard, N. Ferrara, G. Galdi, E. T. Premuzic and J. S. Gaffney. 1982. Sterol distribution in red algae from the waters of Eastern Long Island. *Bot. Mar. 25*: 351–355.

Helleur, R. J., E. R. Hayes, W. D. Jamieson and J. S. Craigie. 1985. Analysis of polysaccharide pyrolysate of red algae by capillary gas chromatography-mass spectrometry. *J. Anal. Appl. Pyrolysis 8*: 333–347.

Henriquez, P., R. Trucco and M. Silva. 1972 a. Chemical examination of Chilean marine algae. *Bot. Mar. 15*: 117–118.

Henriquez, P., R. Trucco, M. Silva and P. G. Sammes. 1972 b. Cholesterol in *Iridaea laminarioides* and *Gracilaria verrucosa* *Phytochem. 11*: 1171.

Kanazawa, A. and M. Yoshioka. 1971. Occurrence of cholest 4-en-3-one in the red alga *Gracilaria textorii*. *Proc. Int. Seaweed Symp. 7*: 502–505.

Kanazawa, A., M. Yoshioka and S. Teshima. 1972. Sterols in some red algae. *Kagosh. Daig. Suis. Gaku. Kiyo 21*: 103–107.

Marolia, V. J., S. Joshi and H. H. Mathur. 1982. Fatty acid composition of neutral lipids from some red algae. *Indian J. Mar. Sci. 11*: 102–103.

Qasim, R. 1986. Studies on fatty acid composition of eighteen species of seaweeds from the Karachi coast. *J. Chem. Soc. Pak. 8*: 223–230.

Riguera, R., L. Castedo, J. M. Quintela and R. Vilalta. 1984. Acids, esters and hydrocarbons in algae of the Galician coast. *Acta Cient. Compostel. 21*: 13–26.

Riguera, R., L. Castedo, J. M. Quintela and R. Vilalta. 1985. Sterols from red and brown algae from the Galician Coast. *An. Quim.* Ser. C, *81*: 113–115.

Shameel, M. 1987. A preliminary survey of seaweeds from the coast of Lasbella, Pakistan. *Bot. Mar. 30*: 511–515.

Shameel, M. 1990. Phycochemical studies on fatty acids from certain seaweeds. *Bot. Mar. 33*: 429–432.

Shameel, M., S. Afaq-Husain and S. Shahid-Husain. 1989. Addition to the knowledge of seaweeds from the coast of Lasbela, Pakistan. *Bot. Mar. 32*: 177–180.

Sims, J. J. and J. A. Pettus Jr. 1976. Isolation of free cis and trans-phytol from the red alga *Gracilaria andersoniana*. *Phytochem. 15*: 1076–1077.

Takagi, T., M. Asahi and Y. Itabashi. 1985. Fatty acid composition of twelve algae from Japanese waters. *Yukagaku 34*: 1008–1012.

Botanica Marina
Vol. 34, pp. 113–121, 1991

Vertical Migration of a Flagellate-dominated Bloom in a Shallow South African Estuary

T. Hilmer and G. C. Bate

Department of Botany, Institute for Coastal Research, University of Port Elizabeth, PO Box 1600, Port Elizabeth, 6000, Rep. of South Africa

(Accepted 23 September 1990)

Abstract

A bloom dominated by *Katodinium rotundatum* (Lohmann) Loeblich and *Micromonas pusilla* (Butcher) Parke *et* Manton was observed over a 24 hour period in the shallow upper reaches of the Sundays River estuary, South Africa. The bloom was displaced horizontally by the tides and displayed vertical migration, believed to be primarily due to *K. rotundatum*, with data for *M. pusilla* being inconclusive. The phasing of vertical migration with the ebb/flood 2-layer flow in the stratified water column retained the bloom within the upper estuary. The strong halocline posed no barrier to the migration of *K. rotundatum*, while calculations based on the swimming velocity of *M. pusilla* show it to have only been capable of moving into the halocline and not through it. Phototaxis did not appear to be involved in the migration phenomenon as the return to the surface was initiated 3 hours before sunrise.

Introduction

Investigations have been carried out on the spatial distribution of chlorophyll-*a* in the Sundays River estuary, which is generally a turbid and at times highly stratified estuary located in Algoa Bay, South Africa. The shallow upper reaches are characterized by high chlorophyll-*a* levels, and although this chlorophyll maximum is displaced horizontally with the tides and displays marked variations in concentration, it nonetheless was found to be a persistant feature in this part of the estuary.

Sampling for chlorophyll-*a* determinations was routinely conducted during daylight hours and on one occasion a diel session was included to observe the behaviour of the chlorophyll maximum over a 24 hour period. From subsequent microscopic investigations it was discovered that the horizontal displacement and vertical migration of a bloom dominated by the dinoflagellate *Katodinium rotundatum* (Lohmann) Loeblich and the prasinophyte *Micromonas pusilla* (Butcher) Parke *et* Manton had been monitored.

Material and Methods

Study area

The Sundays River estuary is located 35 km to the north-east of Port Elizabeth (Fig. 1). From the 800 m wide flood tidal delta it narrows to only 20 m at the tidal limit, which is approximately 21 km from the mouth. There is no clear freshwater/estuarine interface as the riverine input is brackish. The lower 10 km of the estuary is characterized by deep (5–7 m) scour holes associated with meanders, while the upper estuary is relatively straight and shallow (mean depth 2 m). These two sections also differ in their hydrodynamic features (Mackay and Schumann 1990) in that the shallow upper section displays a semi-closed circulation with a water residence time of between 3 and 7 tidal cycles. In the lower section it is only about 2 tidal cycles.

Sampling was carried out at approximately 4 hour intervals over a calm 24 hour period during a neap tide in May 1987. Nine stations (6–14 in Fig. 1) in the upper estuary were covered within 2 hours with

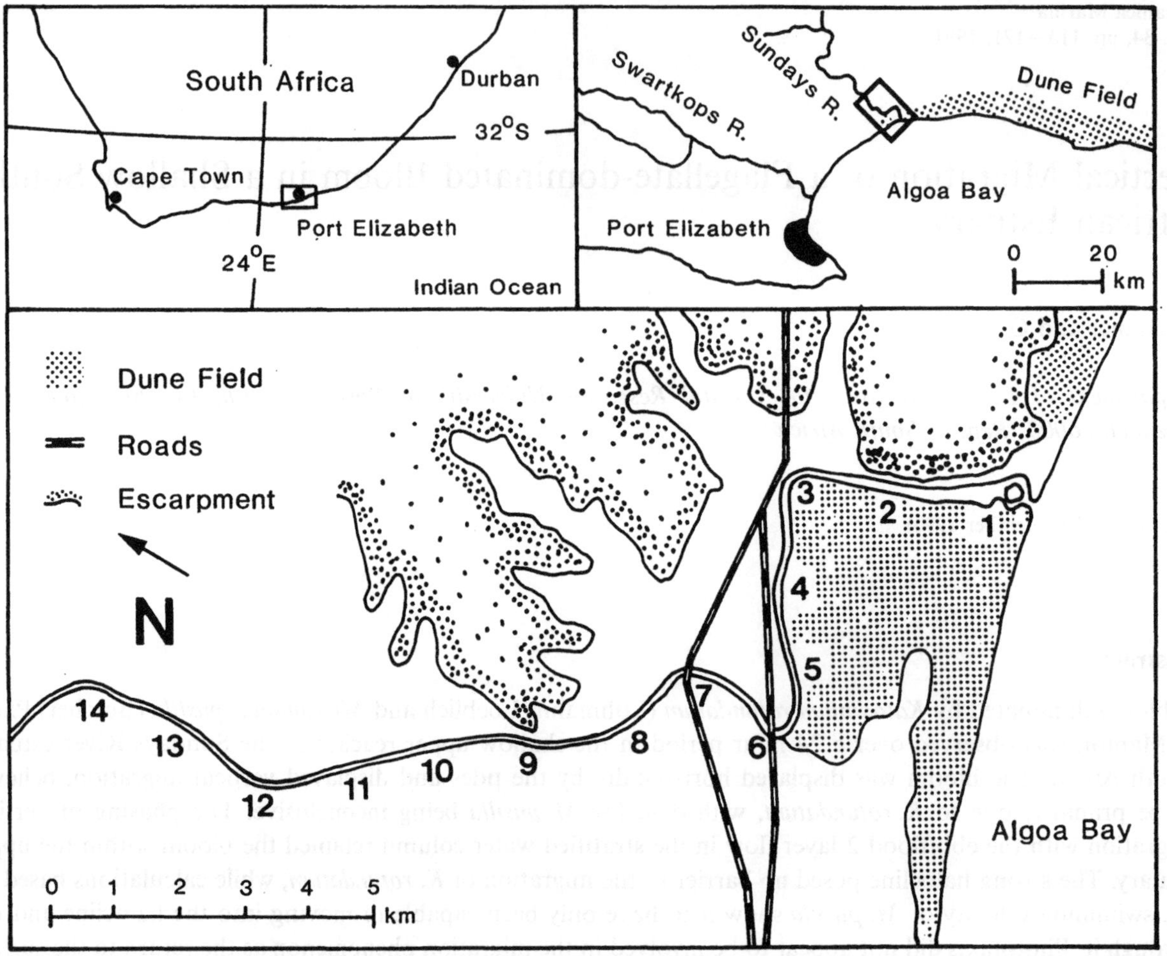

Fig. 1. Location map of the Sundays River estuary showing sampling stations.

the help of a ski-boat. Sampling was carried out in mid-channel and water samples were collected at 0.5 m depth intervals from the surface to the bottom using a narrow-necked 1 litre bottle sampler. Salinity was measured using a Braystoke series 600 CTD system. Temperature readings were not taken but values for May (early winter) are typically between 18 °C and 22 °C (Wooldridge and Bailey 1982). During this period there is little variation in temperature with depth (Wooldridge and Bailey 1982, Mackay and Schumann 1990).

Samples for chlorophyll-*a* were filtered on board by gravity filtration through Whatman GF/C filters (which were found to retain small cells as efficiently as 0.4 µm pore-size Nuclepore filters; Hilmer and Bate 1989). Chlorophyll-*a* was determined using high performance liquid chromatography as described in Hilmer and Bate (1989).

Sub-samples from 0.5 m below the surface were fixed with Lugol's solution for routine cell counting using settling chambers (Utermöhl 1958). No further samples for cell enumeration were taken as the main objective was to monitor phytoplankton chlorophyll-*a* distribution. Rose Bengal was added to stain organic matter and 25 mL samples were settled for at least 36 hours. Cells greater than 10 µm were counted at a magnification of × 630, and smaller cells at × 1250 (Zeiss inverted microscope, × 100 oil objective, × 10 eye-piece, × 1.25 nose-piece). At least 500 cells or 50 fields were counted per sample. Although the cell dimensions of *Micromonas pusilla* are small (about 1 µm diameter), the cells were very obvious when present by nature of their sheer numbers. Dominant species were identified from enrichment cultures of fresh, unstained samples (R. N. Pienaar, personal communication).

Results

The neap tide stage in the Sundays River estuary is characterized by a small tidal amplitude (Fig. 2) and reduced current speeds which allow a clear density and current interface to develop (Mackay and Schumann 1990). On the ebb tide, outflow is restricted to the surface layers (riverine inflow ≈ 1 m s^{-1}), while

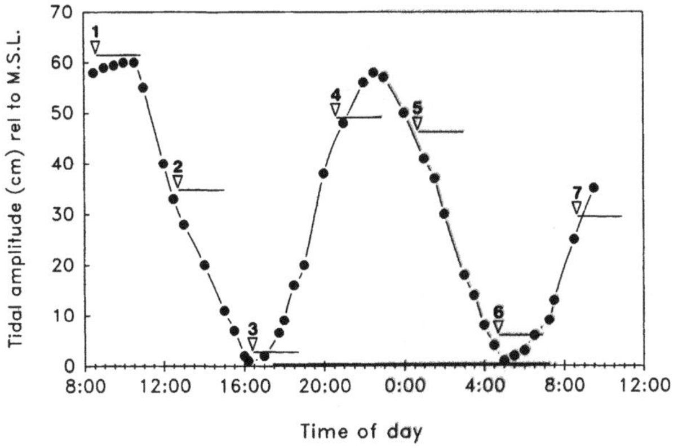

Fig. 2. Tidal elevation at Station 6 and start of sampling sessions (1 to 7). Thick horizontal black bar denotes period of darkness. Smaller bars indicate duration of sampling session.

on the flood tide currents tend to be confined to the bottom layers of inflowing seawater. Lateral displacement of currents and water masses due to ebb and flood only occurs in the wide mouth region and not in the upper estuary. The salinity distribution (Fig. 3a) reflects the stages of the tidal cycle quite clearly: the influx of less saline water over the halocline during low tide, while at high tide the more saline water from downstream wedges under the halocline, forcing it to the surface in this region.

At the start of the sampling session the chlorophyll-*a* distribution (Fig. 3b) indicated an extensive phytoplankton bloom above the halocline in the upper reaches of the estuary. The longitudinal extent of the bloom may be distorted due to the temporal lag in sampling per session, but this would not affect the point-source chlorophyll-*a* concentrations measured.

The bloom moved approximately 4 km downstream with the ebbing tide and at low tide was bounded at its seaward edge by the then vertical halocline. By nightfall, the incoming tide had pushed the bloom some way up the estuary, and it had been displaced downward in the water column. After the turn of the tide very little chlorophyll-*a* was present in the surface water, most being found below the halocline, which prevented any further downstream displacement on the second ebbing tide. By three hours before sunrise on the second low tide, the bloom had moved upward through the halocline and had returned to its former position above the halocline on the incoming tide. During the last four sampling periods a reduction in the size of the chlorophyll maximum was also evident.

The reduction in the size of the chlorophyll maximum could be ascribed to either lateral displacement of the chlorophyll maximum from the static mid-channel sampling stations, temporal sampling lag, grazing or simple dilution throughout the water column. To test the latter we calculated the volume of water contained within the estuarine section at each sampling time with data from 23 surveyed channel cross-sections (Mackay and Schumann 1990). The total amount of chlorophyll-*a* present at each sampling time was computed from the integrated vertical profiles and the volume of water spanning each sampling site (Table I). Assuming no destruction or new synthesis of chlorophyll-*a* took place, the possible fluctuations in the initial pigment concentration due to volume changes

Table I. Total volume of water and chlorophyll-*a* contained within the estuarine section at different tidal stages during sampling sessions (MSL = mean sea level).

Sampling time	Tidal height (cm) relative to MSL	Total volume (m^3)	Total chlorophyll-*a* (g)
8:40	57	1257080	22040
12:40	33	1096582	23721
16:20	0	885230	19672
20:40	44	1169467	14833
0:40	47	1189544	15454
4:40	3	903952	11059
8:40	25	1044328	11409

over the tidal cycle can be calculated and compared with the actual measured concentrations (Fig. 4). The observed levels of chlorophyll-*a* display a marked lower concentration after sunset and indicate that a decrease in chlorophyll-*a* took place after the onset of night which can not be accounted for by dilution over the tidal cycle. Advection to the lower reaches of the estuary can also be ruled out as the bloom was still wholly contained within the section under consideration at the first low tide (Fig. 3a, 16:20 h). Longitudinal distortion of the bloom size due to a temporal lag in sampling between stations would result in a 'longer' bloom on the outgoing tide and a 'shorter' one on the incoming one as sampling was always initiated from Station 14. This implies that there would be more apparent chlorophyll-*a* per total volume during outgoing than incoming tides. Sessions 1–4 (Fig. 3b) tend to show this pattern, but it is not carried through for Sessions 4–7, where little difference was observed between the tidal stages (Figs 3b, 4). The time lag during sampling therefore does not appear to be responsible for the overall decrease in chlorophyll-*a*. Lateral displacement of the bloom from the sampling stations due to tidal forces only is unlikely in the upper estuary. Strong winds may push a bloom to either bank only when it is close to the surface, as the effect of mixing due to wind was found to be limited to the top 50 cm of the water column

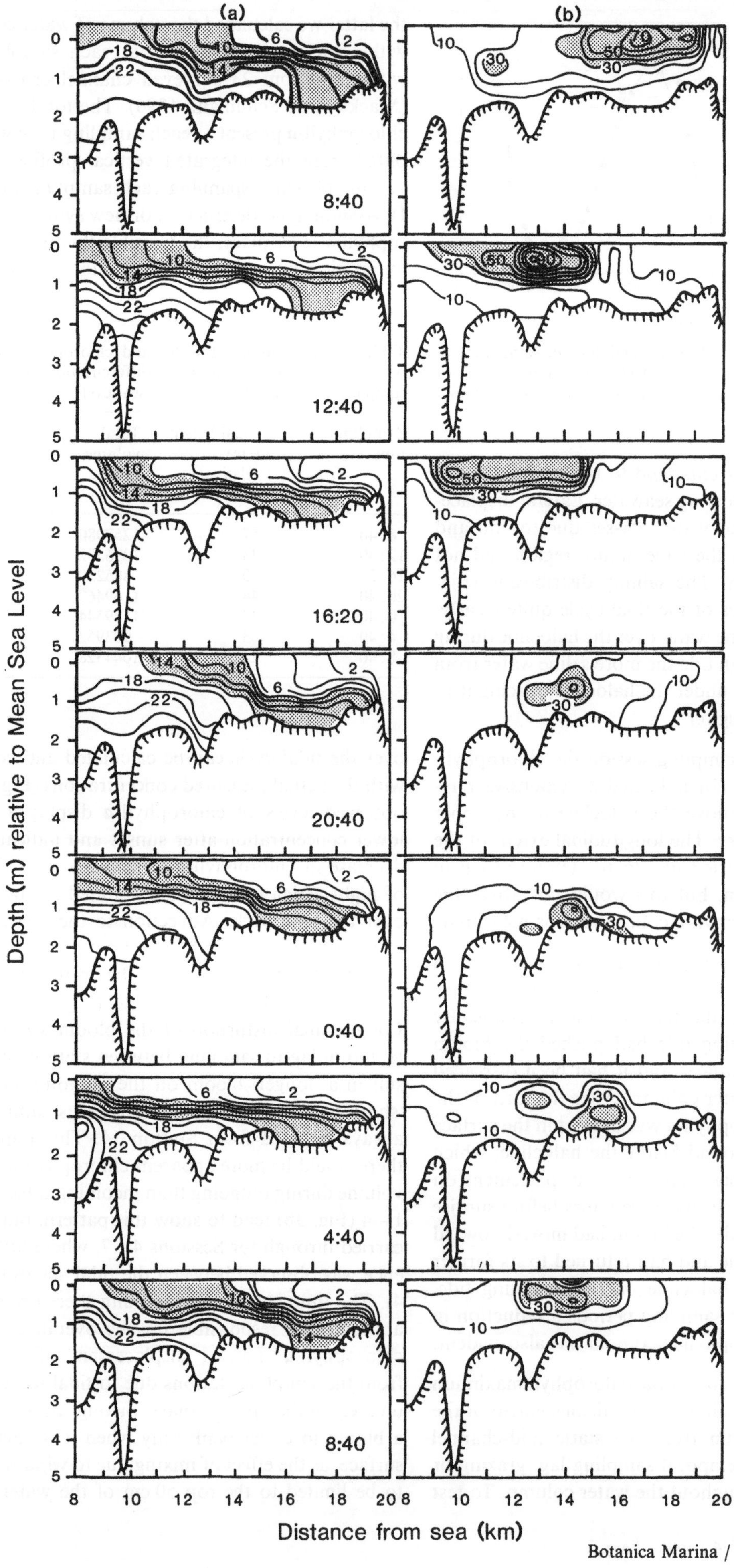

(a)
(b)
8:40
12:40
16:20
20:40
0:40
4:40
8:40
Depth (m) relative to Mean Sea Level
Distance from sea (km)

Table II. Results of regression analyses (r^2 values; n given in parentheses) displaying significant relationships between cell concentrations of the different groups of phytoplankton enumerated and chlorophyll-*a*. (* = $p < 0.05$; ** = $p < 0.025$).

	K. rotundatum	*M. pusilla*	diatoms	cyanophytes	flagellates[1]	others[2]
M. pusilla	0.3197 (51)**					
diatoms	0.0087 (51)	0.0016 (54)				
cyanophytes	0.0056 (43)	0.0027 (43)	0.1799 (43)			
flagellates[1]	0.0189 (38)	0.0175 (38)	0.0236 (38)	0.0890 (38)		
others[2]	0.1162 (46)	0.0075 (46)	0.2050 (46)	0.2495 (43)*	0.0548 (38)	
chlorophyll-*a*	0.7214 (51)**	0.5243 (54)**	0.0000 (61)	0.0002 (43)	0.0099 (38)	0.0495 (46)

[1] flagellated cells larger than 10 μm
[2] non-flagellated cells not falling into any of the above groups

Table III. Stepwise regression analysis of cell numbers in Table II against chlorophyll-*a*.

Step no.	Variable entered	Multiple r^2	Change in r^2
1	*K. rotundatum*	0.7214	0.7214
2	*M. pusilla*	0.8089	0.0875

(Mackay and Schumann 1990). Since no strong winds occurred during the sampling period, it is possible that the decrease in chlorophyll-*a* may have been caused by herbivory or by cells attaching themselves to the bottom substrate (Lombard and Capon 1971, Horstmann 1980). This would effectively remove them from the water column.

The horizontal distribution of cell numbers at a depth of 0.5 m is shown in Figure 5 (note logarithmic scales). The cell concentrations are representative of the maximum bloom concentrations as the 0.5 m depth contour corresponded to the near surface chlorophyll maximum during the day (Fig. 3). Diatom concentrations increased towards the head of the estuary and were present in similar quantities throughout the sampling period. Cyanophyte distribution was similar but numbers increased at low tide, indicating a riverine origin. The larger flagellates and non-flagellates displayed an uneven distribution. Apart from the 20:40 h session (Fig. 5), none of these groups showed any marked changes in numbers during the sampling period. Cell concentrations for the four groups were in the region of 10^6 cells l^{-1} and less, except for cyanophytes which peaked once at 1.65×10^7 cells l^{-1} (Fig. 5a, 16:20 h). The distribution of *Katodinium rotundatum* and *Micromonas pusilla* was very similar both in space and time. At 0:40 h *K. rotundatum* numbers had decreased from 10^7 to 10^6 cells l^{-1} and *M. pusilla* numbers from 10^8 to 10^7 cells l^{-1}. By the next morning these values had risen again.

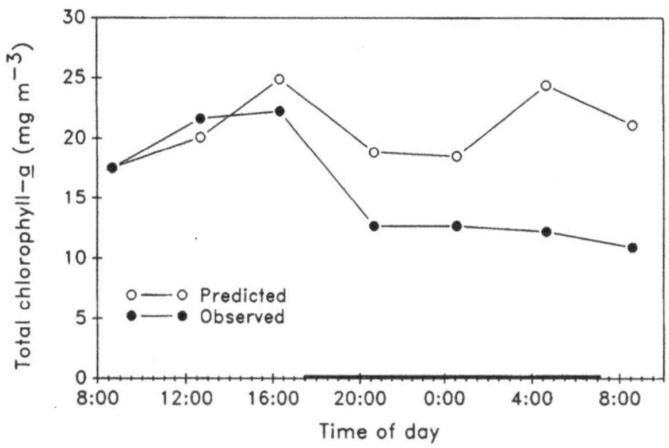

Fig. 4. Predicated and observed levels of chlorophyll-*a* over the tidal cycle used to explain loss of chlorophyll (see text for details). Predicted = initial total chlorophyll-*a* divided into estuarine volumes over tidal cycle (Table I). Observed = actual total chlorophyll-*a* at each sampling time divided into estuarine volumes over tidal cycle.

The relationships between the various groups of phytoplankton enumerated and chlorophyll-*a* are given in Table II. Cell concentrations of *K. rotundatum* and *M. pusilla* correlated significantly with each other and with chlorophyll-*a*, indicating their dominance within the bloom. A stepwise regression (Table III) showed that the dinoflagellate accounted for 72% and the picoflagellate for 9% of the variation in chlorophyll-*a*. The spatial displacement of these two species is thus closely linked to that of the chlorophyll maximum.

Discussion

There are two possible explanations for the distribution of chlorophyll-*a* observed. The first is that purely physical processes resulted in the vertical distribution, while the second invokes biological behaviourial patterns such as vertical migration. There is some evi-

Fig. 3. Contour diagrams of a) salinity (‰) and b) chlorophyll-*a* (μg l^{-1}) for the seven sampling sessions. Salinities between 8 and 16‰ and chlorophyll-*a* concentrations above 30 μg l^{-1} have been shaded to aid in visual interpretation of the diagrams.

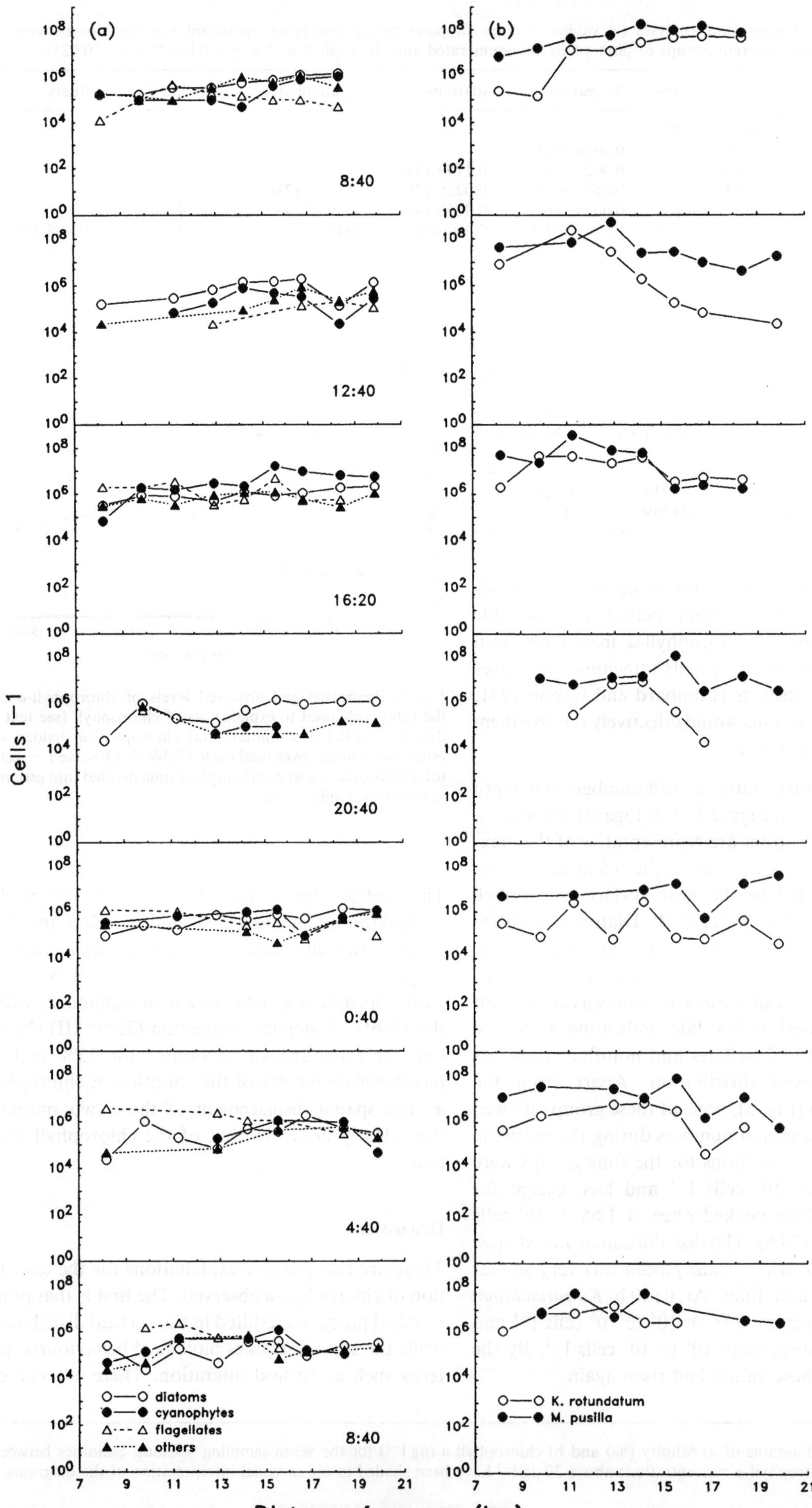
(a)
(b)
8:40
12:40
16:20
20:40
0:40
4:40
8:40
Cells l⁻¹
diatoms
cyanophytes
flagellates
others
K. rotundatum
M. pusilla
Distance from sea (km)

dence of physical vertical displacement of the phytoplankton: at 20:40 h the halocline was shifted downward by about 50 cm in the upper estuary (Fig. 3a) and a decrease in cell numbers of the phytoplankton at 0.5 m depth was observed (Fig. 5). During the subsequent sampling period, however, while *K. rotundatum* and *M. pusilla* numbers continued to decrease, those of the other species had risen again, which tends to indicate that this vertical mixing process was of short duration only. If vertical mixing was strong enough to advect the bloom from the surface to the bottom and back again, one would expect the pycnocline structure to be broken down in response to this turbulence, whereas it remained intact (Fig. 3a). Furthermore, one would also expect non-motile particles (*e.g.* diatoms) to follow a similar distribution pattern to that of the bloom if physical forces alone were responsible. The fact that the other groups of phytoplankton showed no distinct patterns of distribution over the sampling period, and were not correlated with changes in chlorophyll-*a* near the surface, indicates that processes other than entirely physical ones were operating with respect to *K. rotundatum* and *M. pusilla*. The timing of the vertical displacement phases is also similar to that of many dinoflagellates which do show vertical migration (Levandowski and Kaneta 1987).

Processes involving the vertical migration and diel behaviour of flagellates in the marine environment have received a fair amount of attention in the literature (*e.g.* Lombard and Capon 1971, Tyler and Seliger 1978, Horstmann 1980, Heaney and Eppley 1981, Yamochi and Abe 1984, Anderson and Stolzenbach 1985, Voltolina *et al.* 1985, Wangersky and Maass 1988). Features discussed have included possible phototactic and geotactic responses, physical and chemical barriers to migration, and behaviour in relation to hydrodynamic conditions. Many of the mechanisms involved appear to be species specific, for instance Kamykowski and Zentara (1977) observed that *Amphidinium carterae* Hulbert would pass through a thermocline while the latter posed a barrier to the downward migration of *Cachonina niei* Loeblich. Differing behaviour within a species have also been noted (Taylor *et al.* 1988).

The flagellate bloom described here displays some features which have not been observed in other studies, the most important being the synchronized vertical migration of the dinoflagellate and prasinophyte dominating the bloom. Vertical migration of *K. rotundatum* has been documented previously (Cohen 1985) and our data regarding the strong correlation between its cell numbers and the chlorophyll maximum support this. However, our results are not conclusive with respect to *M. pusilla*.

The swimming velocities of *K. rotundatum* and *M. pusilla* were found to differ by a factor of about 4 by Throndsen (1973), with *K. rotundatum* capable of migrating 15 m and *M. pusilla* 4 m in 12 hours. More exact measurements of swimming velocities of flagellates have been performed by Bauerfeind *et al.* (1986) using laser doppler spectroscopy. In general, they observed that velocities determined with their method agreed well with previous studies except for *M. pusilla*, which was found to be only capable of travelling about 0.75 m in 12 hours. Even so, this is sufficient to explain the decreases in cell numbers observed. Assuming the cells initiated their downward migration at the onset of darkness, they would have about 8 hours before reversing their direction. During this time they could cover 0.5 m, which would mean that they probably did not move below the halocline as did *K. rotundatum*, but remained within it. In this position the effect of downstream displacement by the ebbing tide would still be greatly reduced, but not as efficiently as if they had passed through the halocline. The small size of *M. pusilla*, however, makes it seem unlikely that it is capable of maintaining a vertical position in a moving water column. Another explanation for the decrease in near-surface numbers of *M. pusilla* is nocturnal grazing with subsequent new growth. This would require a 50–100% increase in numbers in about 8 hours in order to obtain the final cell concentrations observed. The increase in cell numbers at the end of the sampling period is not the result of concentration into a reduced estuarine volume at low tide since there was a continual decrease in total chlorophyll-*a* during the latter half of the sampling period (Fig. 4).

To our knowledge, no previous study has reported aspects of vertical migration by *M. pusilla* in the field. On the other hand, a *K. rotundatum* red tide bloom has been described by Cohen (1985) in the Potomac Estuary. The bloom persisted for 3 months during winter and attained chlorophyll-*a* concentrations of up to 200 $\mu g\ l^{-1}$ and cell concentrations of 1.7×10^8 cells l^{-1}. Cohen (1985) also showed evidence of vertical migration from the halocline upwards and not from below, suggesting that steep density gradients may inhibit positive phototaxis through the halocline.

Fig. 5. Cell concentrations at the 0.5 m depth contour for the seven sampling sessions. a) diatoms, cyanophytes, flagellates and others. b) *K. rotundatum* and *M. pusilla*.

It is quite obvious from our results that a strong halocline (10‰) posed no barrier to the vertical migration of *K. rotundatum* and this suggests that such physical, and perhaps chemical, gradients may only act as barriers to dinoflagellates under certain environmental conditions (*e.g.* water temperature: 4–6 °C Cohen, 1985; 18–22 °C this study). Furthermore, the vertical migration did not appear to involve phototaxis as both *K. rotundatum* and *M. pusilla* had begun to return to the surface at least 3 hours before sunrise, although light may ultimately be involved in the induction and timing of a diel rhythm of migration.

The combination of vertical migration and 2-layer water movement caused by wind stress or estuarine exchange has been shown to be capable of transporting populations horizontally and to accumulate and retain them positionally (Seliger *et al.* 1970, Tyler and Seliger 1981, Anderson and Stolzenbach 1985). A similar mechanism appears to be operative in the Sundays estuary.

The estuary usually varies between states of strong stratification at neap tide and complete mixing at spring tide (Mackay and Schumann 1990) and we postulate the following cycle of events to take place in the case of *K. rotundatum*: during the period of stratification, current velocities are reduced and the retention time of water within the upper estuary is in the region of 6–7 tidal cycles (Mackay and Schumann 1990). This stable environment would allow the dinoflagellate to bloom. Sources for the bloom may be remnants from previous blooms or resting stages in the sediment (Anderson *et al.* 1982, Cohen 1985). Light is not a limiting factor as the cells migrate to the upper layers during the day. Furthermore, the light compensation depth is usually greater than the depth of the estuary (Hilmer and Bate, in press). Nutrients are also not limiting due to run-off from agricultural irrigation schemes in the catchment area (Emmerson 1989). The degree of phasing of vertical migration with the tidal cycle would determine to a large extent the size of the bloom by reducing advective losses on the ebbing tide. The bloom would subsequently be dispersed and reduced due to mixing of the water column by higher current velocities associated with the spring tide period and due to grazing and also losses to the lower estuary resulting from the reduced water retention time (3 tidal cycles at spring).

Red tide blooms of the raphidophyte *Heterosigma akashiwo* Hada, which frequently causes serious damage to the aquaculture fisheries of Japan (Yamochi and Abe 1984), have also been noted in the Sundays River estuary (own unpublished observations). The identification of at least two red tide flagellates in one of our south coast estuaries necessitates further urgent investigations into their occurrence and mechanism of bloom formation in other systems. It is also important to establish their possible future effects on recreational activities, such as angling, and emergent shellfish industries along our coast.

Acknowledgements

We are deeply indebted to R. N. Pienaar, T. Horiguchi and S. D. Sym for the identification of *Katodinium rotundatum*, *Micromonas pusilla* and' *Heterosigma akashiwo*. Heather M. Mackay is thanked for making the survey data of the channel cross-sections available to us. The constructive comments of two unknown referees is appreciated. This work was supported by the South African National Committee for Oceanographic Research, the Department of Environment Affairs and Tourism, and the University of Port Elizabeth.

References

Anderson, D. M., D. G. Aubrey, M. A. Tyler and D. W. Coats. 1982. Vertical and horizontal distributions of dinoflagellate cysts in sediments. *Limnol. Oceanogr. 27*: 757–764.

Anderson, D. M. and K. D. Stolzenbach. 1985. Selective retention of two dinoflagellates in a well-mixed estuarine embankment: the importance of diel vertical migration and surface avoidance. *Mar. Ecol. Prog. Ser. 25*: 39–50.

Bauerfeind, E., M. Elbrächter, R. Steiner and J. Throndsen. 1986. Application of Laser Doppler Spectroscopy (LDS) in determining swimming velocities of motile phytoplankton. *Mar. Biol. 93*: 323–327.

Cohen, R. R. H. 1985. Physical processes and the ecology of a winter dinoflagellate bloom of *Katodinium rotundatum*. *Mar. Ecol. Prog. Ser. 26*: 135–144.

Emmerson, W. D. 1989. The nutrient status of the Sundays River estuary South Africa. *Wat. Res. 23*: 1059–1067.

Heaney, S. I. and R. W. Eppley. 1981. Light, temperature and nitrogen as interacting factors affecting diel vertical migrations of dinoflagellates in culture. *J. Plankton Res. 3*: 331–344.

Hilmer, T. and G. C. Bate. 1989. Filter types, filtration and post-filtration treatment in phytoplankton production studies. *J. Plankton Res. 11*: 49–63.

Hilmer, T. and G. C. Bate. 1991. Covariance analysis of chlorophyll-*a* distribution in the Sundays River estuary, eastern Cape. *Sthn Afr. J. Aquat. Sci.* (in press).

Horstmann, U. 1980. Observations on the peculiar diurnal migration of a red tide dinophyceae in tropical shallow waters. *J. Phycol. 16*: 481–485.

Kamykowski, D. and S. J. Zentara. 1977. The diurnal vertical migration of motile phytoplankton through temperature gradients. *Limnol. Oceanogr. 22*: 148–151.

Levandowski, M. and P. J. Kaneta. 1987. Behaviour in dinoflagellates. *In*: (F. J. H. Taylor, ed.) *The Biology of Dinoflagellates*. Botanical Monographs Vol. 21, Blackwell Scientific Publ., Oxford. pp. 360–397.

Lombard, E. H. and B. Capon. 1971. Observations on the tidepool ecology and behaviour of *Peridinium gregarium*. *J. Phycol. 7*: 184–187.

Mackay, H. M. and E. H. Schumann. 1990. Mixing and circulation in the Sundays River estuary, South Africa. *Estuar. coast. Shelf Sci. 31*: 203–216.

Seliger, H. H., J. H. Carpenter, M. Loftus and W. D. McElroy. 1970. Mechanisms for the accumulation of high concentrations of dinoflagellates in a bioluminescent bay. *Limnol. Oceanogr. 15*: 234–245.

Taylor, W. D., J. W. Barko and W. F. James. 1988. Contrasting diel patterns of vertical migration in the dinoflagellate *Ceratium hirundinella* in relation to phosphorus supply in a north temperate reservoir. *Can. J. Fish Aquat. Sci. 45*: 1093–1098.

Throndsen, J. 1973. Motility in some marine nanoplankton flagellates. *Norw. J. Zool. 21*: 193–200.

Tyler, M. A. and H. Seliger. 1978. Annual subsurface transport of a red tide dinoflagellate to its bloom area: Water circulation patterns and organism distributions in the Chesapeake Bay. *Limnol. Oceanogr. 23*: 227–246.

Tyler, M. A. and H. Seliger. 1981. Selection for a red tide organism: Physiological responses to the physical environment. *Limnol. Oceanogr. 26*: 310–324.

Utermöhl, H. 1958. Zur Vervollkommnung der quantitativen Phytoplankton-Methodik. *Mitt. Int. Ver. Theor. Angw. Limnol. 9*: 38 pp.

Voltolina, D., L. N. Brown and M. G. Robinson. 1985. Vertical variations of the chlorophyll maximum during a red tide in a shallow lagoon. *Estuar. coast. Shelf Sci. 21*: 817–822.

Wangersky, P. J. and R. L. Maass. 1988. Diurnal behaviour of cultures of *Dunaliella tertiolecta*. *J. Plankton Res. 10*: 327–329.

Wooldridge, T. and C. Bailey. 1982. Euryhaline zooplankton of the Sundays estuary and notes on trophic relationships. *S. Afr. J. Zool. 17*: 151–163.

Yamochi, S. and T. Abe. 1984. Mechanisms to initiate a *Heterosigma akashiwo* red tide in Osaka Bay. *Mar. Biol. 83*: 255–261.

Botanica Marina
Vol. 34, pp. 123–131, 1991

Ultrastructure of a Marine Dinoflagellate, *Peridinium quinquecorne* Abé (Peridiniales) from South Africa with Particular Reference to its Chrysophyte Endosymbiont

T. Horiguchi[1] and R. N. Pienaar

Department of Botany, University of the Witwatersrand, P.O. Wits 2050, Rep. of South Africa

(Accepted 23 September 1990)

Abstract

The micro-morphology and the ultrastructure of a marine dinoflagellate, *Peridinium quinquecorne* Abé were studied. The dinoflagellate shows a wide range of morphological variation, including cell shape and length of the antapical spines. The thecal plate arrangement is: pp, x, 3′, 2a, 7″, 5c, 5‴, 2⁗ 4s. Detailed study of the sulcus and the cingulum, which was not undertaken by the original author was carried out. The ultrastructural investigation revealed that all the chloroplasts belong to an endosymbiotic alga which is separated from the host cytoplasm by a single unit membrane. The host cytoplasm (dinokaryotic cytoplasm) revealed typical dinoflagellate organization. The eyespot is bounded by a triple membrane and is located in the host cytoplasm. The structure of the endosymbiont is simple and contains only a eukaryotic nucleus, mitochondria, dictyosomes and chloroplasts. The ultrastructural features of the chloroplasts suggest a chrysophyte affinity for the endosymbiont.

Introduction

Peridinium quinquecorne was originally described by Abé (1927) based on the two incomplete specimens found in summer plankton samples collected at Asamushi, Mutsu Bay, northern Japan. Later, however, Abé (1981) redescribed the organism in greater detail, although he was not able to analyze the details of the cingulum and the sulcus. The cytological features of the original material are not available from Abé's description, except for the brief mention of yellowish brown droplets.

Peridinium quinquecorne is an armoured dinoflagellate which is characterized by possessing four prominent antapical spines and a more or less angular shape (Abé 1981). During the course of our studies on the tidal pool dinoflagellates occurring along the east coast of South Africa, we found a spine-bearing dinoflagellate which possessed golden-brown chloroplasts and a prominent eyespot. Based on the morphological similarities, including the thecal plate arrangement and the presence of four antapical spines. this dinoflagellate was identified as *Peridinium quinquecorne* (Horiguchi and Pienaar 1986). In an earlier report we have mentioned that this species is essentially a colourless dinoflagellate and its chloroplasts belong to an eukaryotic endosymbiont which is separated from the host cytoplasm by a single unit membrane (Horiguchi and Pienaar 1986). Although we were unable to conduct an analysis of the photosynthetic pigments, the ultrastructure of the chloroplasts indicates the chrysophyte affinity of the endosymbiont.

The occurrence of a chrysophyte endosymbiont in dinoflagellates has been demonstrated in two other species; *Peridinium balticum* (Levander) Lemmermann (Tomas and Cox 1973) and *Glenodinium foliaceum* Stein (Jeffrey and Vesk 1976). This paper deals with further observations on the morphology and the ultrastructure of this dinoflagellate.

[1] Present address: Faculty of Education, Shinshu University, Nagano 380 Japan

Material and Methods

The specimens used in this study were collected from tidal pools at two localities along the Natal coast of South Africa, viz. Amanzimtoti (21 August 1985) and Palm Beach (28 September 1985).

For transmission electron microscopy, the natural samples were processed according to Pienaar and Aken (1985). Observations were made using a JEOL 100CX transmission electron microscope. For scanning electron microscopy, the method described by Horiguchi and Pienaar (1988) has been employed. Observations were made using a HITACHI S-570 scanning electron microscope.

Observations

Light and scanning electron microscopy

Peridinium quinquecorne is an armoured dinoflagellate, measuring 23.0 – 30.0 μm in length and 20.0 – 25.0 μm in width. The cell form exhibits a range of morphological variation. When seen in ventral view (Figs 1, 7) it is angular diamond-shaped with convex sides. Broad diamond-shaped cells (Fig. 2) or nearly ovoidal cell forms (Fig. 3) have also been observed within the same population. The cell is usually dorsiventrally compressed. The epitheca is conical and has a short apical horn, while the hypotheca is hexagonal with median convex sides half way between the cingulum and the antapex. Four antapical spines

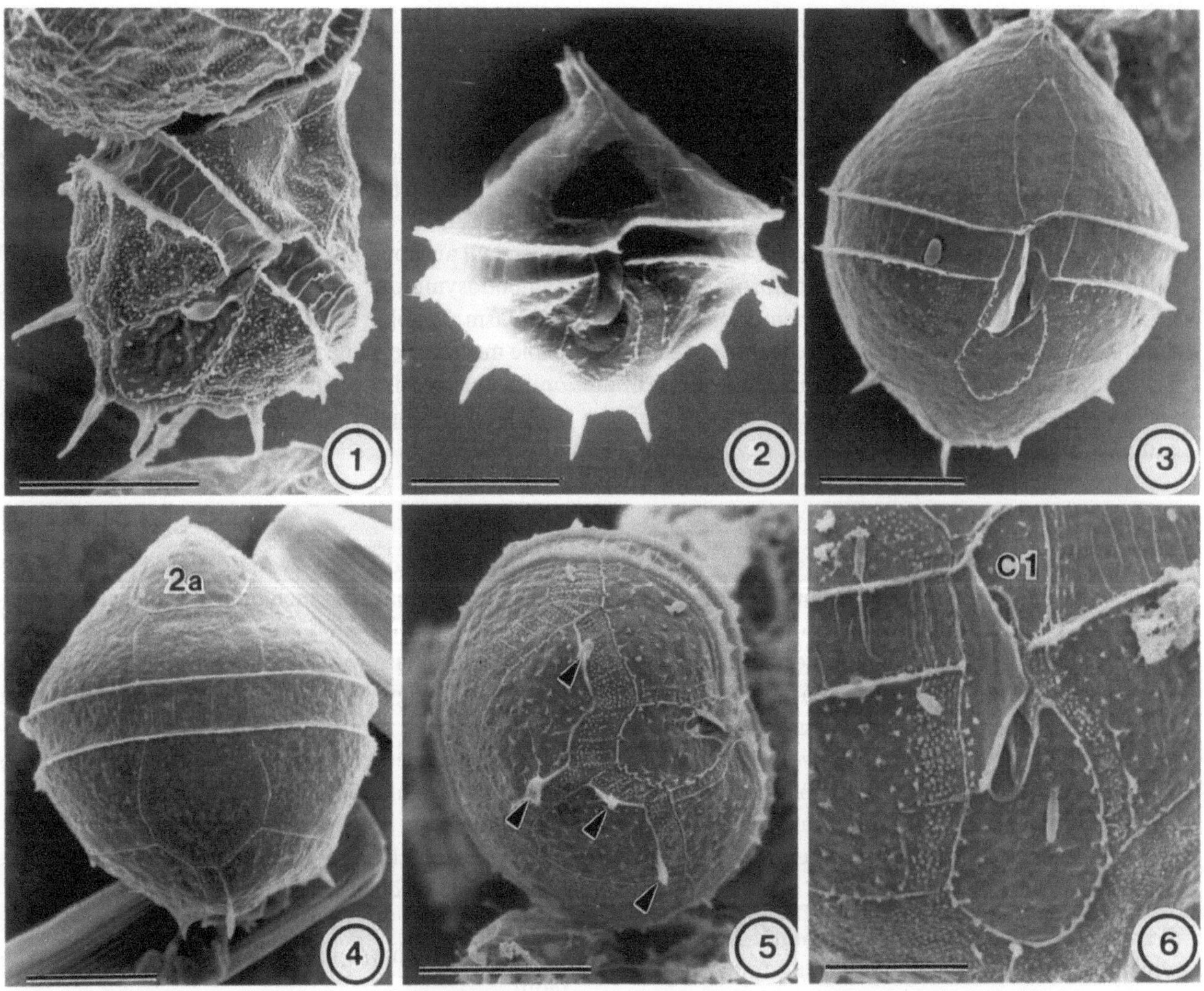

Figs 1 – 6. *Peridinium quinquecorne* Abé. Scanning electron micrographs.
Figs 1 – 3. Ventral view, showing morphological variations. Fig. 4. Dorsal view. Fig. 5. Antapical view. Fig. 6. Detail of the sulcus. (c1: first cingular plate = transitional plate) (Figs 1 – 5: scale = 10 μm, Fig. 6: scale = 5 μm)

Abbreviations used in the figures: C: chloroplast, CM: cytoplasmic membrane, D: dictyosomes, dm: dinoflagellate mitochodrion, DN: dinokaryotic nucleus, E: eyespot, em: endosymbiont mitochodrion, EN: eukaryotic nucleus, HV: hair-containing vesicle, M: membrane between the endosymbiont and the host cytoplasm, OM: outer membrane, OPM: outer plate membrane, P: pellicle, Pu: pusule, T: trichocyst, TP: thecal plate.

are always present on the hypotheca. The length of these spines varies from 1.0–5.0 μm. The cingulum is bordered by the narrow lists which are ornamented at regular intervals by ridges (Fig. 5). The cingulum is left-handed and is displaced about 1/3–1/2 of its own width. The sulcus becomes wider posteriorly, and extends to the antapex (Fig. 1), except when the intercalary bands are well developed (Figs 2, 3).

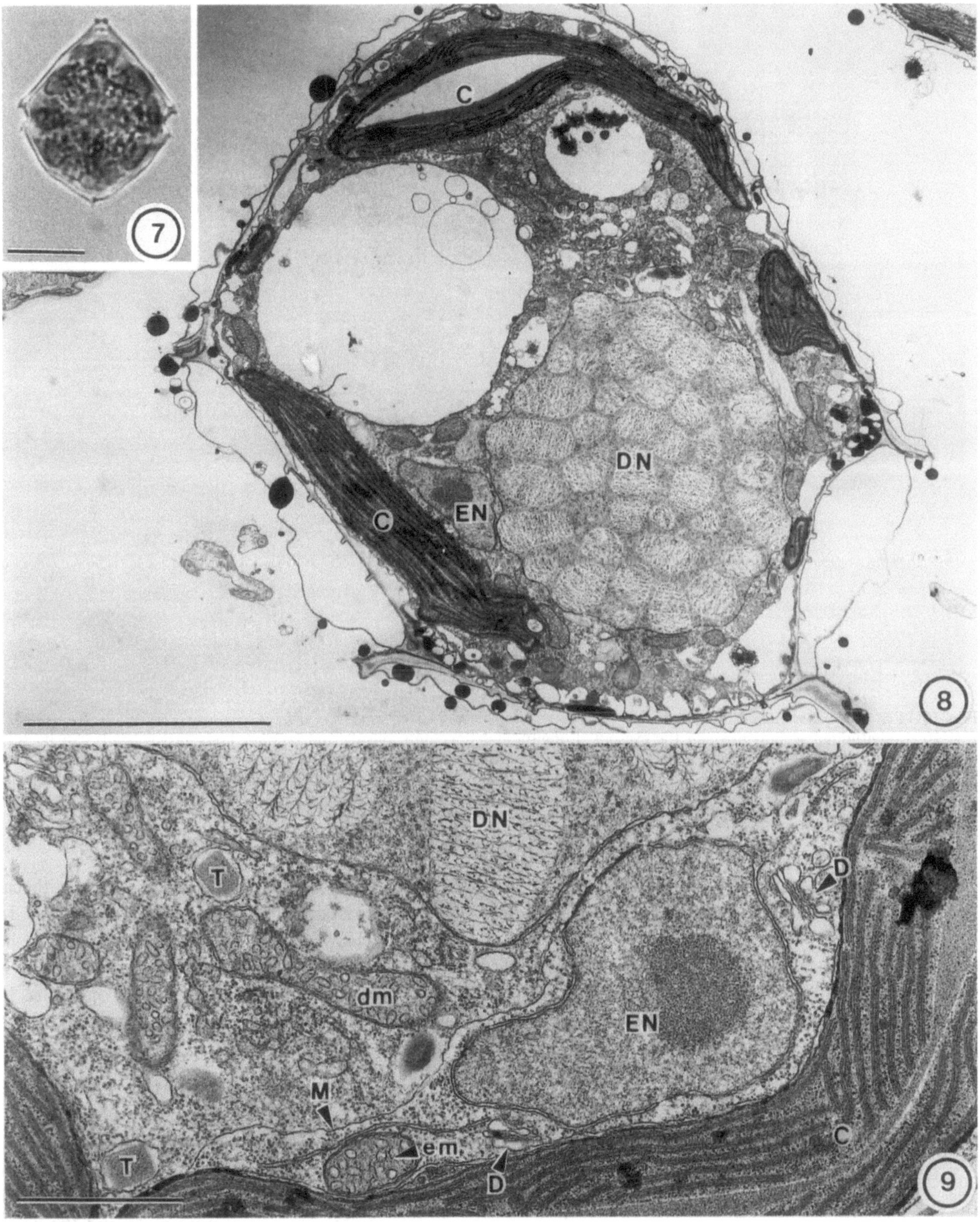

Figs 7–9. *Peridinium quinquecorne* Abé.
Fig. 7. Light micrograph, showing living specimen with short antapical spines (scale = 10 μm). Fig. 8. Section through the cell, showing general arrangement of the cellular organelles (scale = 5 μm). Fig. 9. A part of the cell, showing two nuclei, dinokaryotic and eukaryotic nuclei. Two nuclei are separated from each other by a single unit membrane (scale = 1 μm).

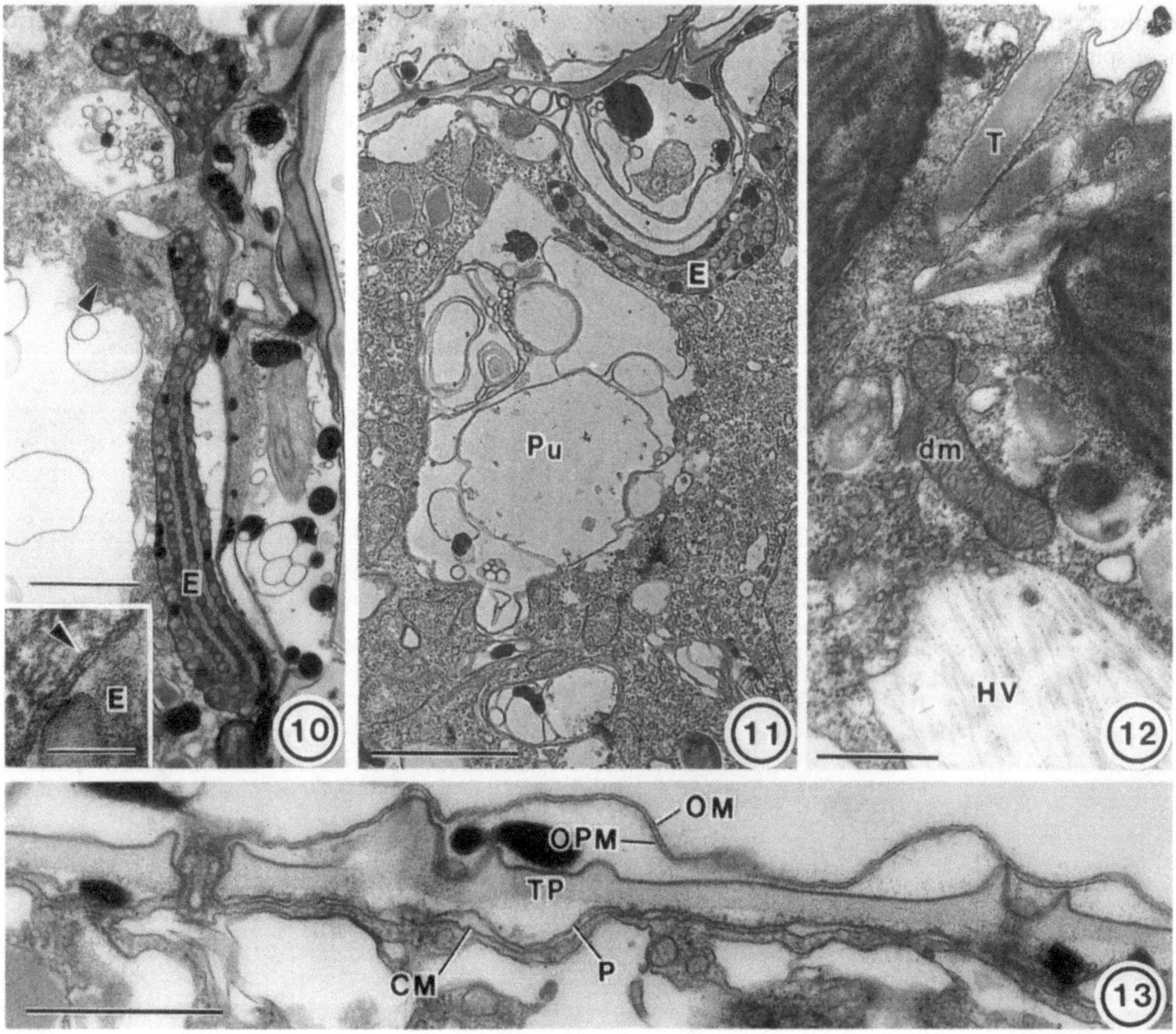

Figs 10–13. *Peridinium quinquecorne* Abé.
Fig. 10. Longitudinal section through the eyespot. Longitudinal flagellum, pusule and laminar body (arrowhead) can be seen (scale = 1 μm). Fig. 10 (inset). A part of the eyespot, showing a triple membrane boundary (arrowhead) (scale = 100 nm). Fig. 11. Transverse section through the eyespot and pusule (scale = 1 μm). Fig. 12. A part of the cell, showing trichocyst, mitochondria, starch grains and hair-containing vesicles (scale = 500 nm). Fig. 13. Detail of amphiesma (scale = 500 nm).

The dinokaryotic nucleus is spherical to ovoid in shape and is located near the centre of the left side of the cell (Fig. 24). The eukaryotic nucleus is located near the apex, just above the dinokaryotic nucleus (Fig. 24). The chloroplasts are golden-brown, ellipsoidal to strap-shaped and are peripherally arranged (Fig. 24). The number of chloroplasts is variable but the majority of cells usually possess about 10 chloroplasts. The eyespot is prominent, bright red and is located near the upper part of the sulcus. The upper part of the eyespot is hooked and this hooked part coincides with the junction between the cingulum and the sulcus (Fig. 24). A large vacuole is often present at right side of the cell (Fig. 24).

The ornamentation of the thecal plates is variable (Figs 1–6). With the exception of the cingular and the sulcal plates (Fig. 1), the surface of the thecal plates are covered with small wart-like projections. In other cases (Fig. 5), wart-like projections can only be seen on the intercalary bands. Each trichocyst pore is situated in the irregular-shaped small projections. In many cases, the intercalary bands are well developed (Figs 2–5).

The thecal plate arrangement is pp, x, 3′, 2a, 7″, 5c, 5‴, 2⁗ & 4s (Figs 21, 22). The apical pore plate (pp) is very small and is surrounded by the raised apex of the apical plates, resulting in the formation of the short apical horn. The canal plate (x) is also small and rectangular. The apical pore plate and the canal plate are surrounded by three apical plates. The 2a plate is hexagonal and is located just behind the apical pore plate. There is, however, no direct connection

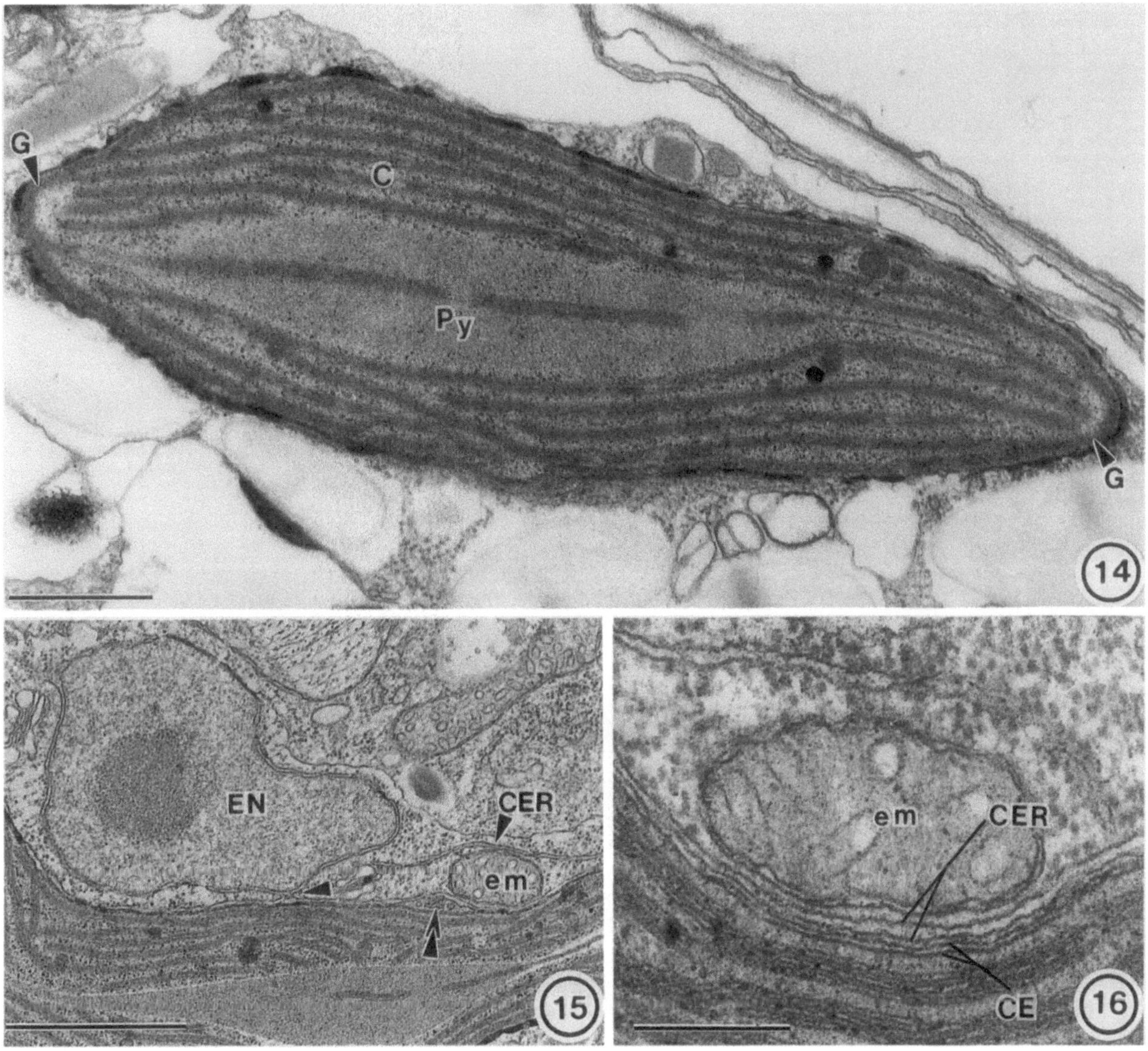

Figs 14–16. *Peridinium quinquecorne* Abé.
Fig. 14. Detail of the chloroplast (scale = 500 nm). Fig. 15. A part of the eukaryotic cytoplasm, showing the association between the nucleus and chloroplast endoplasmic reticulum (scale = 1 μm). Fig. 16. Detail of the mitochondrion and chloroplast boundary (scale = 200 nm).

between these two plates. The la plate is pentagonal in shape and touches the 1″ plate. The cingulum consists of five plates. The c1 plate is small and forms a part of the sulcus (transitional plate). The sulcus consists of four plates (Figs 3, 6, 23). The middle of the sulcus is occupied by an elongated ovoid plate, bearing the flagellar pores. The left edge of the right sulcus plate develops into the internal sulcal list which covers the flagellar pores. The left sulcal plate is very small. The posterior sulcal plate is scoop-shaped and is larger than the three other sulcal plates. The arrangement of the hypothecal plates is almost symmetrical (Fig. 22). Two of the antapical spines are located on plate 1⁗, while the other two are on plate 2⁗.

Transmission electron microscopy

As shown in Figure 8, the dinoflagellate contains two different types of nuclei, viz. the dinokaryotic nucleus and the eukaryotic nucleus. These two nuclei are separated from one another by a single unit membrane (Fig. 9). There are, therefore, two cytoplasmic regions; the dinokaryotic cytoplasm and the eukaryotic cytoplasm (terms from Tomas and Cox 1973).

The dinokaryotic cytoplasm includes the typical dinoflagellate organelles, including dinokaryotic nucleus with thick condensed chromosomes (Fig. 8), mitochondria with tubular cristae (Fig. 12), trichocysts (Fig. 12), hair-containing vesicles (Fig. 12), dictyosomes, eyespot (Fig. 10) and storage products, such as starch grains (Fig. 12) and lipid bodies.

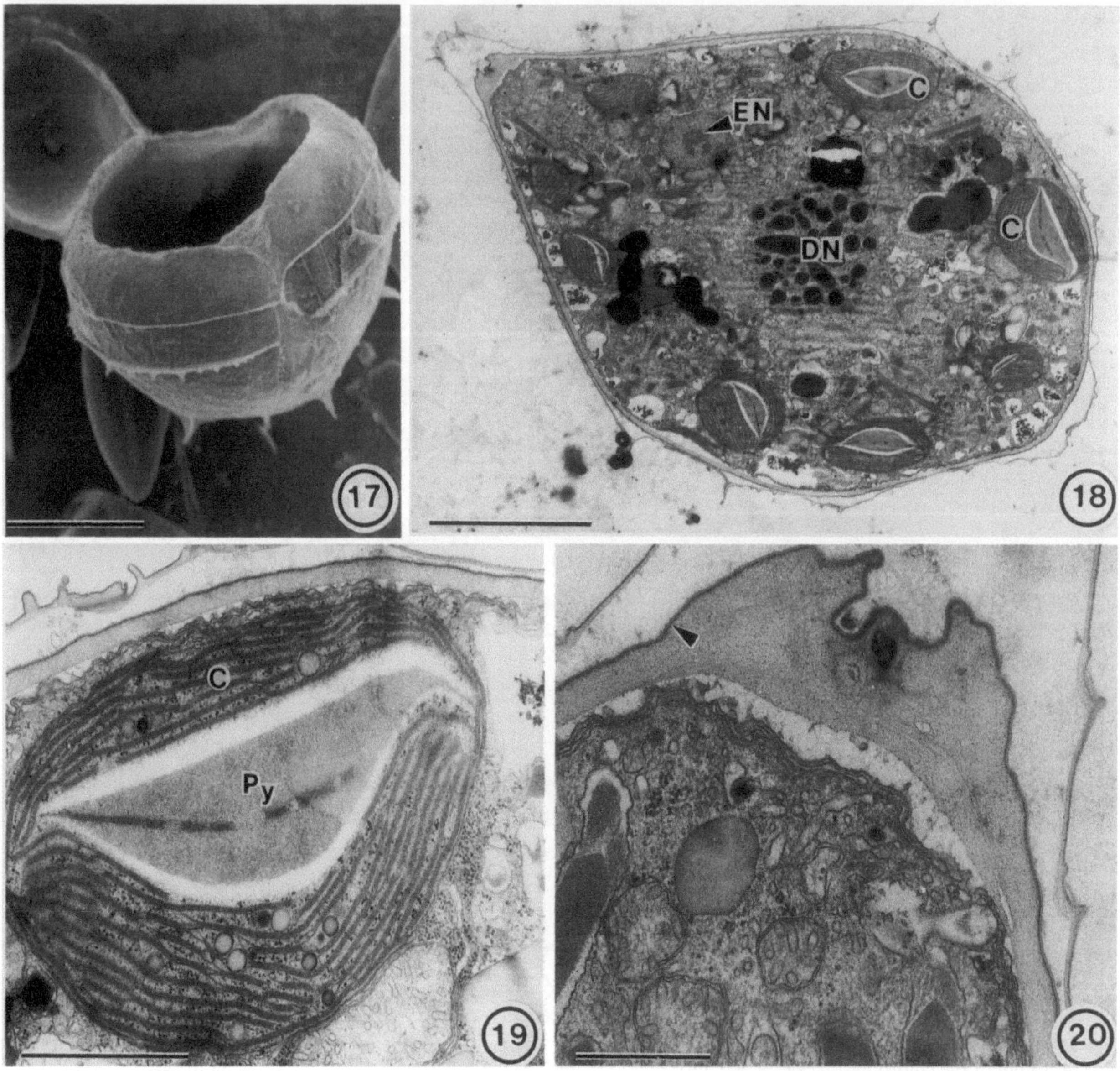

Figs 17–20. *Peridinium quinquecorne* Abé. (Non-motile cell)
Fig. 17. Scanning electron micrograph of the empty cell (scale = 10 μm). Fig. 18. Longitudinal section through non-motile cell, showing general arrangement of cellular organelles. Both eukaryotic and dinokaryotic nuclei can be seen (scale = 5 μm). Fig. 19. Detail of the chloroplast (scale = 5 μm). Fig. 20. Detail of the amphiesma (scale = 500 nm).

The eyespot is present in the dinokaryotic cytoplasm (Fig. 10). It is membrane-bound, consisting of many lipid globules, and is extraplastidial. A longitudinal section through the eyespot shows that these lipid globules are arranged in 2–4 rows. The eyespot is enclosed by a triple membrane (Fig. 10, inset). A laminar body has been observed near the eyespot (Fig. 10).

The pusule (Fig. 11) is located near the eyespot. It consists of a large collecting chamber and relatively few pusular vesicles which radiate from the collecting chamber. The collecting chamber includes vesicles of varying size.

The amphiesma is typical peridinioid dinoflagellates and consists of the outer membrane, the outer plate membrane, the thecal plates, the pellicle and the cytoplasmic membrane (Fig. 13).

The eukaryotic cytoplasm includes an eukaryotic nucleus (Figs 9, 15), mitochondria and endoplasmic reticulum. Small dictyosomes have been observed near the nucleus (Fig. 9). The mitochondria (Figs 9, 16) possess tubular cristae and mitochondrial profiles are often found near the chloroplast and are surrounded by endoplasmic reticulum (Fig. 9).

The chloroplasts are arranged in the peripheral region of the cell (Fig. 8). Almost all the chloroplasts have

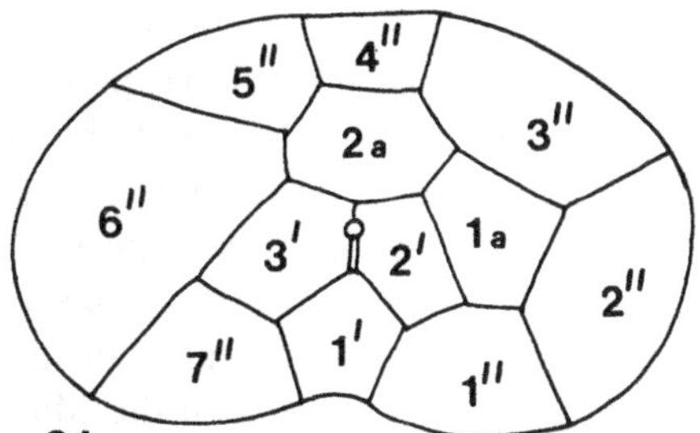

21

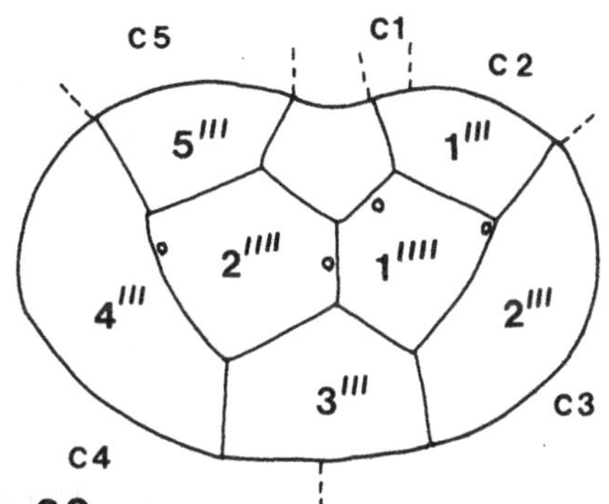

22

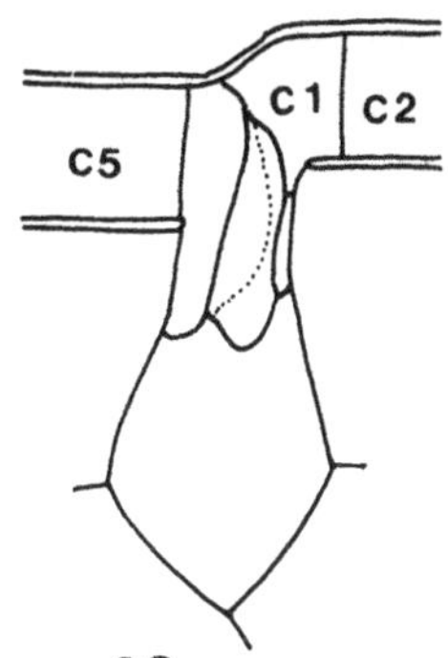

23

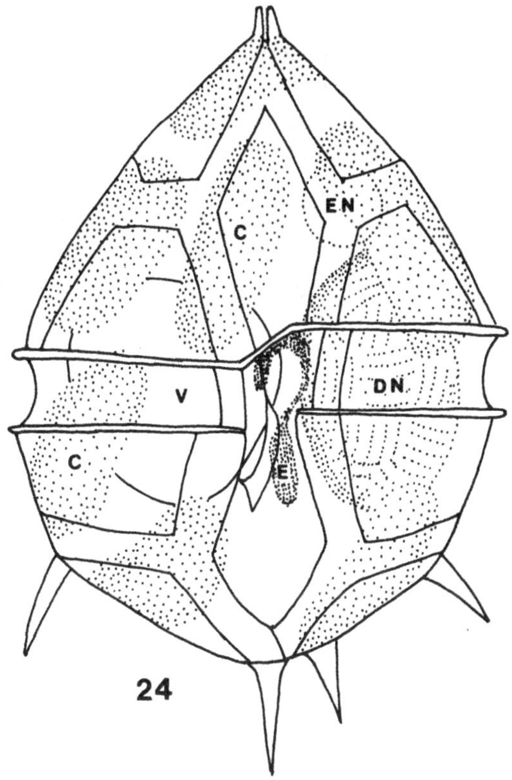

Figs 21–24. *Peridinium quinquecorne* Abé.
Fig. 21–23. Schematic diagram, showing thecal plate arrangement. Fig. 21. Apical view. Fig. 22. Antapical view. Fig. 23. Detail of the sulcus. Fig. 24. Ventral view of the cell.

an internal lenticular pyrenoid (Fig. 14). The pyrenoid matrix is always penetrated by thylakoids (Fig. 14). Girdle lamellae are always present (Fig. 14). Each lamella consists of three closely appressed thylakoids (Fig. 16). A few plastoglobuli have been found in the stroma (Fig. 14). The chloroplast is surrounded by two double unit membranes. The outer two form the chloroplast endoplasmic reticulum, while the inner two form the chloroplast envelope (Fig. 16). An association between the outer nuclear envelope and the chloroplast endoplasmic reticulum has been observed (Fig. 15).

Non-motile cell

In a sand sample from the same tidal pool in which the motile stage of *Peridinium quinquecorne* occurred, we found some non-motile stages of this dinoflagellate. When the collection was made at Amanzimtoti, the salinity of the pool was 43‰. Distilled water was added to the sample to reduce salinity to about 30‰ in order to induce germination. About three hours later, motile cells were released from the non-motile cells. In the most cases, only a single motile cell was released. The motile cell escapes from the non-motile cell through a large circular opening which is formed in the epitheca. There seems to be no relationship between the sutures of the thecal plates and the formation of this opening (Fig. 17). The non-motile cell has a thick cell wall which was, at least partially, resistant to cold concentrated sulfuric acid.

The components of both cytoplasmic regions of the non-motile cell are the same as those of the motile cell. The chloroplasts of the endosymbiont in the non-motile cell, however, seem to be reduced in size (Fig. 18), although their structure remains unchanged (Fig. 19). The electron-translucent region is conspicuous at the boundary between the pyrenoid matrix and the stroma in the chloroplasts of the non-motile cell. Such electron-translucent regions, however, have occasionally been observed in the chloroplast of the motile cells.

In the formation of the non-motile cell a thick wall is formed beneath the thecal plates (Fig. 20). The wall consists of two layers; an outer thin layer and the inner thickened layer. The outer thin layer is trilaminar (dark–light–dark) in appearance, while the inner layer consists of fine fibres which are randomly arranged. Beneath the cell wall, a single continuous membrane and many flattened vesicles of varying size have been observed (Figs 19, 20).

Discussion

According to Abé (1981), *Peridinium quinquecorne* exhibits a wide spectrum of variation in its body form, viz. from the angular specimen to the nearly ovoidal cell form. The length of the antapical spines is also variable. Our specimens from South Africa agree with these characteristics, including the arrangement of thecal plates, well developed intercalary bands, presence of four antapical spines and a variable cell shape.

The major differences between the type description and the South African specimens are the cell size and the habitat. However, although the average cell size is different, the size ranges of specimens from both localities do overlap. Based on the above mentioned similarities, we have identified our specimens as *Peridinium quinquecorne* Abé (Horiguchi and Pienaar 1986).

Halim (1967) described a dinoflagellate under the name of *Peridinium quinquecorne*. His description, however, does not agree with that of Abé (1981). The dinoflagellate described by Halim has some characteristic features which cannot be seen in the description given by Abé (1981), including a key-hole shaped apical pore plate complex (apical pore plate + canal plate) and possession of three apical intercalary plates with 2a and 3a being separated from one another by a direct connection between plates 3′ and 4″. Abe's *P. quinquecorne* has no such characters and, therefore, we believe that these two species are different entities. Halim's species probably belongs one of the following two species, *Peridinium sociale* (Henneguy ex Labbé) Biecheler or *Scrippsiella gregaria* (Lombard *et* Capon) Loeblich *et al.*

A red-tide forming dinoflagellate reported from the tropical shallow water, Maribago Bay (Horstmann 1980) was also identified as *Peridinium quinquecorne.* The size of cells in this population were closer to that of our specimens than that of the original description. Horstmann (1980) reported the presence of the conspicuous eyespot in his specimens. Although the general appearance of the specimens from Maribago Bay is similar to the South Africa specimens, there are a few differences.

1) Horstmann's figure (Fig. 1, 1980) shows that his species has a slightly right-handed rather than a left-handed cingulum,

2) the Maribago Bay specimens often possess five or more antapical spines,

3) Horstmann (1980) mentions that his species has a similar thecal plate formula to that of Halim's description.

First of all, we should point out that it is probable that his photograph (Fig. 1) is optically inverted, since the plate which seems to be the transitional plate is situated on the right side of the sulcus and this is quite an unusual situation among the dinoflagellates. Secondly, as previously mentioned, Halim's species is probably not *P. quinquecorne*. Therefore, if the species from Maribago Bay has the same thecal plate arrangement as that of Halim's species, then the Maribago Bay species is obviously not *P. quinquecorne.* Since Horstmann (1980) did not present a detailed description and figures to show the thecal plate arrangement, it is necessary to re-analyze the thecal plate arrangement of the Maribago Bay species and to illustrate them with adequate photographs in order to clarify whether this species is *P. quinquecorne.*

Balech (1974) transferred *Peridium quinquecorne* to the genus *Protoperidinium,* making the new combination, *Protoperidinium quinquecorne* (Abé) Balech. However, as we have demonstrated in this study, the species possesses five cingular plates. Therefore, it is not appropriate to place this species in the genus *Protoperidinium* which is characterized by possessing four cingular plates, including a transitional plate.

We have found and studied the non-motile stage of *Peridium quinquecorne*. The main difference between the motile and the non-motile form is the presence of a thick cell wall in the non-motile stage. A thick cell wall is formed just beneath the thecal plate layer and it is obvious that the trilaminar part of the wall (outer layer) corresponds to the pellicle layer of the motile cell. Although the sexuality of an endosymbiont-bearing dinoflagellate, *Peridinium balticum* has been demonstrated (Chesnick and Cox 1987), it it not yet known whether the thick-walled non-motile stage of *P. quinquecorne* is a zygotic stage or not.

Dodge (1983) compared the ultrastructural features of both host and endosymbiont of *Glenodinium foliaceum* and *Peridinium balticum*, the only two examples of dinoflagellates then known to possess a chrysophyte endosymbiont. Based on the chloroplast ultrastructure, photosynthetic pigment composition and configuration of the plastidial DNA, Dodge (1983) and Kite and Dodge (1985) concluded that the endosymbiont of *G. foliaceum* was of diatom origin.

When *Peridinium quinquecorne* is compared with these two examples, there are remarkable similarities between the three species, not only in the structure of the endosymbiont but also in the structure of the host cytoplasm. The major difference being that dictyosomes have been found in the endosymbiont of *P. balticum* and *P. quinquecorne*, while no such organelle has been found in *G. foliaceum*. The ultrastructural similarity of the endosymbionts suggests that the endosymbiont of *P. quinquecorne* is closely related to those of *G. foliaceum* and *P. balticum.*

The similarities of the host cytoplasm of these three species include the similarity of the eyespot structure, viz. extraplastidial and being surrounded by a triple membrane. According to Dodge (1984), this type of eyespot has so far only been found in *G. foliaceum* and *P. balticum*, both are a chrysophyte-bearing dinoflagellate. Our result added the third example of this type of eyespot and once again our organism is chrysophyte-bearing dinoflagellate. Therefore, it is possible that the endosymbiotic event had occurred only once before the speciation took place among these three species.

Another possibility is that those dinoflagellates were once photosynthetic or endosymbiont-bearing with an eyespot which is of the intraplastidial type as suggested by Dodge (1984). Once the photosynthetic dinoflagellate had acquired an endosymbiont, the host had lost its own chloroplasts or endosymbiont and only the eyespot was retained. If this is the case, we must assume that endosymbiotic event and subsequent loss of the original chloroplasts had occurred independently in these dinoflagellates. The electrophoretic data which compared several isoenzymes of *P. balticum* and *G. foliaceum* (Whitten and Hayhome 1986) revealed that these two taxa are genetically not closely related. In this connection, one must also take into account the fact that these three species have quite different arrangements of the thecal plates. It is, therefore, premature to draw a conclusion concerning this controversy. A more detailed comparative study of a biochemical (immunological, isoenzymatic etc.) or morphological nature (flagellar apparatuses, etc.) is required.

Acknowledgements

The authors wish to acknowledge the help of staff members of the University of Natal, Pietermaritzburg and University of Witwatersrand Electron Microscope Units and the CSIR Foundation for Research Development for financial assistance.

References

Abé, T. H. 1927. Notes on the protozoan fauna of Mutsu Bay. *Sci. Rep. Tohoku Imperial Univ.*, Ser. 4, Vol. *2*: 383–438.

Abé, T. H. 1981. Studies on the faimly Peridiniidae. An unfinished monograph of the armoured dinoflagellata. *Special Publications* from *the Seto Marine Biological Laboratory*, vol. *6*: 1–409.

Balech, E. 1974. El genero *"Protoperidinium"* Bergh, 1881 (*"Peridinium"* Ehrenberg, 1831, partim). *Revta Mus. argent. Cienc. nat. "B. Rivadavia", Hidrobiol. 4*: 1–79.

Chesnick, J. M. and E. R. Cox. 1987. Synchronized sexuality of an algal symbiont and its dinoflagellate host, *Peridinium balticum* (Levander) Lemmermann. *BioSystems 21*: 69–78.

Dodge, J. D. 1983. A re-examination of the relationship between unicellular host and eucaryotic endosymbiont with special reference to *Glenodinium foliaceum* Dinophyceae. *Endocytobiology 2*: 1015–1026.

Dodge, J. D. 1984. The functional and phylogenetic significance of dinoflagellate eyespots. *BioSystems 16*: 259–267.

Halim, Y. 1967. The phytoplankton of Venezuela. *Int. Rev. Hydrobiol. 52*: 701–755.

Horiguchi, T. and R. N. Pienaar. 1986. Ultrastructure of a marine dinoflagellate, *Peridinium quinquecorne* with special reference to its endosymbiotic alga. *Electron Micro. Soc. South. Afr. Proc. 16*: 107–108.

Horiguchi, T. and R. N. Pienaar. 1988. Ultrastructure of a new sand-dwelling dinoflagellate, *Scrippsiella arenicola* sp. nov. *J. Phycol. 24*: 426–438.

Horstmann, U. 1980. Observations on the peculiar diurnal migration of a red tide Dinophyceae in tropical shallow waters. *J. Phycol.* 16: 481–485.

Jeffrey, S. W. and M. Vesk. 1976. Further evidence for a membrane-bound endosymbiont within the dinoflagellate, *Peridinium foliaceum. J. Phycol. 12*: 450–455.

Kite, G. C. and J. D. Dodge. 1985. Structural organization of plastid DNA in two anomalously pigmented dinoflagellates. *J. Phycol. 21*: 50–56.

Pienaar, R. N. and M. E. Aken. 1985. The ultrastructure of *Pyramimonas pseudoparkeae* sp. nov. (Prasinophyceae) from South Africa. *J. Phycol. 6*: 79–86.

Tomas, R. N. and E. R. Cox. 1973. Observations on the symbiosis of *Peridinium balticum* and its intracellular alga. Ultrastructure. *J. Phycol. 9*: 304–323.

Whitten, D. J. and B. A. Hayhome. 1986. Comparative electrophoretic analysis of two binucleate dinoflagellates. *J. Phycol. 22*: 348–352.

Botanica Marina
Vol. 34, pp. 133–137, 1991

New Records of Benthic Marine Algae from Easter Island

M. E. Ramirez C. and D. G. Müller*

Botany Section, Museo Nacional de Historia Natural, Casilla 787, Santiago, Chile

* *Fakultät für Biologie der Universität, D-7750 Konstanz, Fed. Rep. of Germany*

(Accepted 29 September 1990)

Abstract

Six new records of marine algae for Easter Island are described: *Blastophysa rhizopus*, *Derbesia tenuissima*, *Caulerpa racemosa* var. *peltata*, *Eupogodon pilosus*, *Bangia atropurpurea*, and *Nemacystus novae-zelandiae*. These findings are in agreement with biogeographical distribution patterns recently established for this area. *Blastophysa* and *Derbesia* were not apparent in our field collections, but appeared in subsequent laboratory cultures.

Introduction

Easter Island (27°07′S; 109°22′W) is located in the South Pacific at a very isolated position. Santelices and Abbott (1987) compiled a comprehensive list of its marine benthic flora. They emphasized its richness and diversity compared with floras of other Pacific islands of similar size, and the low degree of endemism (14%). The Easter Island marine flora is characterized by a large number of species with a wide tropical and temperate distribution and is very different from that of the nearest Pacific South American coasts. The intertidal and shallow subtidal flora of Easter Island has been studied by several workers (Boergesen 1924, Etcheverry 1960, Abbott and Santelices 1985, Santelices and Abbott 1987), but seems to be still incompletely known. We report here on several taxa new to the area, which we recently encountered and identified in field collections and in culture studies.

Material and Methods

Our collections were made in September 1986, January 1987, and February 1988 at different localities on the Island (Fig. 1). Most of our collecting sites correspond to previously studied areas, but Hanga Kioe, Hanga Oteo, and Te Pahu had not been studied before. These habitats do not differ ecologically from the general character of the rocky coast of the Island.

For taxonomic studies we used mounted herbarium specimens, material preserved in 5% formalin or observations on living laboratory cultures. Sections were made by hand or with a freezing microtome. Permanent mounts were stained with aceto-carmine or aniline blue and mounted in syrup. The materials are deposited in the Herbarium of the Museo Nacional de Historia Natural of Santiago, Chile (SGO). Laboratory cultures were initiated by inoculating fragments of various macro-algae or inorganic substrates into polypropylene screw-capped tubes containing sterile culture medium. Laboratory cultures were kept at 20 °C under fluorescent light in enriched natural seawater (ES-enrichment, Starr and Zeikus 1987). The salinity of North Sea seawater (28‰) was corrected to 35‰ by addition of a commercial sea salt preparation.

The Species

Blastophysa rhizopus Reinke
Reinke, 1888, p. 240. Ballantine and Wynne 1986, 131–135, Iima and Tatewaki 1987, 241–250.

Specimens examined: SGO 111536, growing in crude cultures of *Zonaria stipitata* Tan. *et* Noz. from Pta. Cook. SGO 111537 and SGO 111538 on *Dasya* sp. from Anakena beach, collected in January 1987.

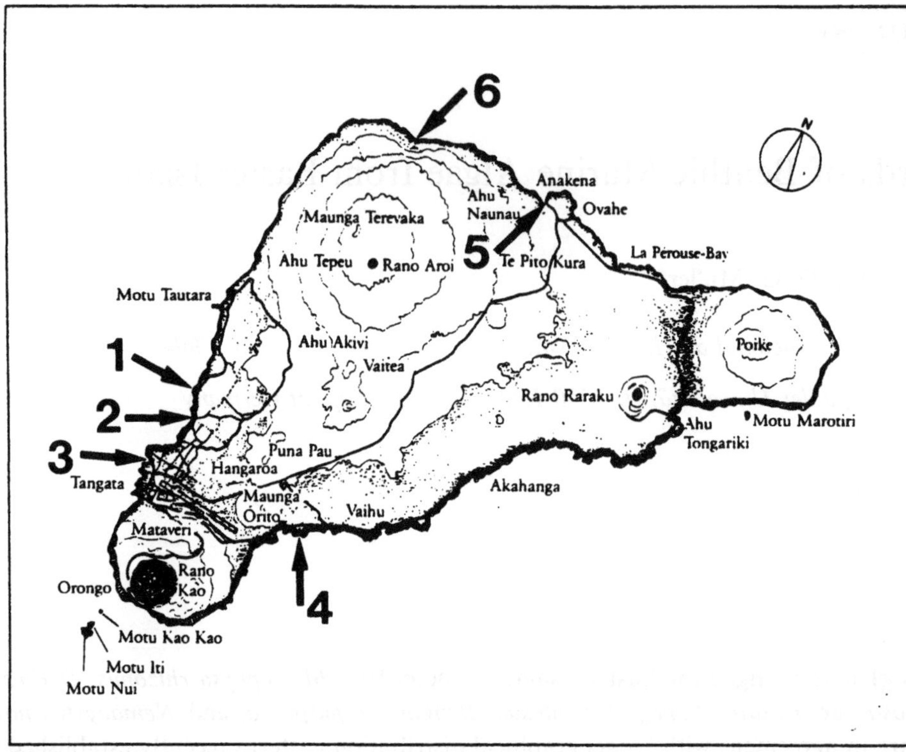

Fig. 1. Geographical map of Easter Island with collection sites. 1: Pta. Cook; 2: Hanga Kioe; 3: Hanga Piko; 4: Te Pahu; 5: Anakena; 6: Hanga Oteo.

Distribution: Reported from tropical and subtropical waters in the Atlantic and Pacific.

Blastophysa rhizopus is an endophytic green alga, which grows inside the cortical tissue of larger algae. The morphology of our specimens agrees well with previous descriptions. Plants form extensive colonies, in which neighbouring cells are connected by slender colourless filaments 5–10 μm in diameter. Pigmented cells are multinucleate, and vary in shape from spherical, lobulate to tubular. Their dimensions are 50–(150)–200 μm × 60–100 μm, and each forms 2 or three hairs. Zooidangia are rounded and zooids are released through an opening in a tubular filament.

Recent studies by Iima and Tatewaki (1987) have shown that *Blastophysa rhizopus* in Japan has a strong preference for endophytic growth in *Grateloupia turuturu* Yamada. It does not naturally penetrate other algae, and seems to use them as a favourable substrate only. There is no evidence for parasitism. We obtained excellent growth of *Blastophysa rhizopus* in unialgal clonal cultures.

Derbesia tenuissima (de Notaris) Crouan *et* Crouan
Boergesen 1925, p. 107, fig. 45; Womersley 1984, figs 99–100.

Specimens examined: SGO 111534, growing in crude cultures on shell fragments and small pebbles collected at Te Pahu in January 1987. SGO 111535 developed on *Padina australis* Hauck collected in February 1988 at Hanga Kioe.

Distribution: Mediterranean Sea, temperate eastern Atlantic and southern Australia (Womersley 1984).

Species of *Derbesia* grow epilithic or epiphytic. Our specimens reach a size of 6 cm and consist of erect cylindrical filaments arising from a branched rhizoidal base. Erect filaments are branched and have a diameter of 80–90 μm. Branches originate 1 to 3 mm apart and are frequently inserted without cross-walls. Chloroplasts are numerous, lenticular and each contains one pyrenoid. Sporangia are inserted laterally and separated by a double septum. They are pyriform, attenuated at the base, 115 μm in diameter and 270 μm long. Zoospores are rounded and 3–4 μm in diameter. In our cultures they developed into unicellular spherical or elongated coenocytic *Halicystis* plants, which are considered to represent the gametophyte phase.

Derbesia tenuissima is very similar to *Derbesia marina* (Lyngbye) Solier, but differs in its dimensions, chloroplast size and presence of a pyrenoid.

Caulerpa racemosa (Forsskal) J. Ag., var. *peltata* (Lamour.) Eubank
Eubank, 1946 p. 421, figs 2r–s Fig. 2

Specimens examined: SGO 109595 and 109596 collected February 1988 from Hanga Piko.

Distribution: Widely distributed in tropical and subtropical areas.

The plants of this species formed grass-like cushions in high intertidal pools and at subtidal levels at 2–4 m depth in quiet localities. Our specimens are 3–5 cm long, producing peltate and pedicellate apical ramuli flattened to a disk.

Most of our specimens agree well with those described from Hawaii by Eubank (1946), who placed it as a variety under *Caulerpa racemosa*. Some of our specimens, however, have up to 5 pedicellate fronds; others had erratic ramuli and deviations in shape, which brings them closer to other varieties of the *C. racemosa* complex.

Considerable variation in the apical assimilator ramuli of *Caulerpa* is frequently mentioned in the literature. Several authors suggested that this variability is due to environmental factors. Ohba and Enomoto (1978) studied morphological variability of *Caulerpa racemosa* var. *laetevirens* (Montagne) Weber van Bosse from southern Japan under different culture conditions. Plants kept at 20 to 22.5 °C and a high light intensity (5 8 kLux) formed *laetevirens* assimilator fronds, while at 20–30 °C and 1.5–5 kLux *peltata* fronds were formed. The variability of this character is thus attributed to environmental factors.

Caulerpa racemosa var. *peltata* differs from *C. webbiana* Mont., the other common species of Easter Island by its peltate and pedicellate apical assimilator ramuli. *Caulerpa webbiana* in contrast has assimilator fronds with branched whorls and mucronate ramuli.

Eupogodon pilosus (Weber van Bosse) Silva

Silva, 1987 p. 60.
Basionym: *Dasyopsis pilosa* Weber van Bosse
Weber van Bosse, 1923 p. 377, fig. 137. Fig. 3

Specimens examined: SGO 107356, Hanga Piko, September 1986 (tetrasporic only). SGO 109593, Hanga Kioe, February 1988 (all reproductive stages). SGO 109594, Hanga Oteo, February 1988 (tetrasporic).

Distribution: *Eupogodon pilosus* is known from New Guinea and Vietnam under *Dasyopsis* (Weber van Bosse 1923, Dawson 1954). Recently it has been reported from the Philippines by Silva *et al.* (1987), who transferred *Dasyopsis* to *Eupogodon*. It is also reported from Hawaii by Schlech and Abbott (1989).

This species grows fixed to rocks and stones in pools of the high and medium intertidal zone in 1–3 m depth in association with *Lobophora variegata* (Lamour.) Womersley, and *Stypopodium flabelliforma* Weber van Bosse. Thalli are terete, dark red, up to 12 cm high and attached by a small discoid holdfast of 2 mm diameter. Fronds are dichotomously branched and densely covered with long monosiphonous filaments, which give the plants a fluffy appearance. These filaments are pigmented and 5 mm or more long, consisting of cells 40–(58)–63 µm long. Tetrasporangia are tetrahedral, 19 µm in diameter and borne in stichidia, which arise singly on monosiphonous filaments. Spermatangial branches also originate on lateral filaments. Spermatia are colourless, rounded, 3 µm in diameter. Cystocarps are sessile, inserted laterally, 880 µm in diameter with a thick pericarp and an ostiole. Carpospores are pyriform, 14.5 × 6 µm.

Eupogodon pilosus differs from other species of the genus mainly in its dichotomous branching pattern and the lateral position of the cystocarp.

Bangia atropurpurea (Roth) C. Ag.
C. Agardh 1824, p. 76.

Specimens examined: SGO 107357, Hanga Piko September 1986.

Distribution: Cosmopolitan.

Our specimens are in good agreement with the descriptions of this species from all oceans.

Nemacystus novae-zelandiae Kylin
Kylin, 1940 p. 48, figs 26c, d, pl. 29; Lindauer, Chapman and Aiken 1961, p. 235. fig. 56; Womersley 1987 p. 130. Figs 4–8

Specimens examined: SGO 109936, Hanga Piko, February 1988, epiphytic on *Sargassum skottsbergii* Sjoestedt. Culture experiments were started with material from Pta. Cook, January 1987.

Distribution: South Pacific, to date only reported from New Zealand (Auckland) and southern Australia.

Our plants agree well with material from New Zealand described by Kylin (1940) and Australia (Womersley 1987), both of which lacked unilocular sporangia. Our field specimens were found with uniseriate plurilocular sporangia 35–50 µm × 5 µm. Immediately upon collection in 1988 dense crops of mature unilocular sporangia were seen, which released their contents during handling and microscopic observation. Their size was 25 × 35 µm, after release their

Fig. 2. *Caulerpa racemosa* var. *peltata*, rhizoids, creeping axis and upright parts of thallus, scale bar 3 cm. Fig. 3. *Eupogodon pilosus*. Specimen collected at Hanga Kioe, scale bar 5 cm. Figs 4–8. *Nemacystus novae-zelandiae*. Fig. 4. Field specimen growing on *Sargassum skottsbergii* (scale bar 3 cm). Fig. 5. Plurilocular sporangium (scale bar 50 μm). Fig. 6. Cross section showing medulla, assimilators and hairs (scale bar 200 μm). Fig. 7. Unilocular sporangium (scale bar 50 μm). Fig. 8. Macrothallus from culture, showing apical cell, central axis, cortication and insertion of laterals (scale bar 100 μm).

walls collapsed to inconspicuous tube-like structures 40 to 45 µm long. No development occurred in cultures initiated with uni- or plurizooids from field specimens. However, fragments of assimilator filaments from field specimens regenerated to microthalli, which gave rise directly to *Nemacystus* macrothalli with plurilocular sporangia.

Discussion

Our reports confirm the affinities and biogeographical patterns recently established for this region of the Pacific by Santelices and Abbott (1987). Four of the species are cosmopolitan and have been found in most temperate and tropical waters: *Blastophysa rhizopus*, *Derbesia tenuissima*, *Caulerpa racemosa*, and *Bangia atropurpurea*. *Eupogodon pilosus* and *Nemacystus novae-zelandiae* are representatives of the Indo-Pacific flora and have been reported from some localities in the West Pacific. *Nemacystus novae-zelandiae* seems to have a rather restricted distribution in the South Pacific. It is reported to date only from northern New Zealand and Australia. Our finding at Easter Island expands its presently known distribution eastward.

Several of the species we report here are macroscopic, and can be recorded with the standard methods of phytogeographic surveys: field collections followed by taxonomic examination of preserved material. Our results illustrate however, that smaller taxa may escape attention in a survey focusing on larger specimens. Occasionally, this deficit is mentioned in phytogeographic surveys (Taylor 1960, for *Blastophysa*).

Acknowledgements

The authors wish to thank director and staff of CONAF-Chile for logistic support on the Island, and Hugo Atan and Eduardo Villouta for contributions of various collections.

References

Abbott, I. A. and B. Santelices. 1985. The marine algae of Easter Island (Eastern Polynesia). *Proc. Fifth Coral Reef Congress 5*: 71–75.

Agardh, C. A. 1824. *Systema Algarum*. Lund xxxviii + 312 pp.

Ballantine, D. L. and M. J. Wynne. 1986. Notes on the marine algae of Puerto Rico I. Addition to the Flora. *Bot. Mar. 29*: 131–135.

Boergesen, F. 1924. Marine algae from Easter Island. *In*: (C. Skottsberg, ed.) *The Natural History of Juan Fernandez and Easter Island 2*: Almkvist & Wiksells, Uppsala, Sweden. pp. 247–309.

Boergesen, F. 1925. Marine algae from the Canary Islands. *Danske Vidensk. Selskab. Bid. Meddel. 5, 1*: 1–123.

Dawson, E. Y. 1954. Marine plants in the vicinity of Nha Trang, Viet Nam. *Pac. Sci. 8*: 373–471.

Etcheverry, H. 1960. Algas marinas de las islas oceanicas Chilenas. *Rev. Biol. Mar. 10*: 83–132, 6 pl.

Eubank, L. L. 1946. Hawaiian representatives of the genus *Caulerpa*. *Univ. Calif. Publ. Bot. 18*: 409–432.

Iima, M. and M. Tatewaki. 1987. On the life history and host-specificity of *Blastophysa rhizopus* (Codiales, Chaetosiphonaceae), and endophytic green algae from Muroran in laboratory cultures. *Jpn. J. Phycol. 35*: 241–250.

Kylin, H. 1940. Die Phaeophyceenordnung Chordariales. *Acta Univ. Lund* N. F. 2, *36*: 1–67, 8 pl.

Lindauer, V. W., V. J. Chapman and M. Aiken. 1961. The marine algae of New Zealand II Phaeophyceae. *Nova Hedwigia III*: 129–350.

Ohba, H. and S. Enomoto. 1978. Culture Studies on *Caulerpa* (Caulerpales – Chlorophyceae) II. Morphological variation of *Caulerpa racemosa* var. *laetevirens*, under various culture conditions. *Jpn. J. Phycol. 35*: 178–188.

Reinke, J. 1888. Einige neue braune und grüne Algen in der Kieler Bucht. *Ber. Dt. Bot. Ges. 6*: 240–241.

Santelices, B. and I. A. Abbott. 1987. Geographic and marine isolation: An assessment of the marine algae of Easter Island. *Pacif. Sci. 41*: 1–20.

Schlech, K. and I. A. Abbott. 1989. Species of Dasyaceae (Rhodophyta) from Hawaii. *Pacif. Sci. 43*: 332–351.

Silva, P. C., E. Menez and R. L. Moe. 1987. Catalog of the benthic marine algae of the Philippines; Nomenclatural notes. *Smithsonian Contrib. Mar. Sci. 27*: 127–131.

Starr, R. C. and J. A. Zeikus. 1987. UTEX, The culture collection of algae at the University of Texas at Austin. *J. Phycol. 23, Suppl.*: 1–47.

Taylor, W. R. 1960. *Marine Algae of the Eastern Tropical and Subtropical Coasts of the Americas*. Univ. of Michigan Press, Ann Arbor, Michigan, U. S. A. 870 pp.

Weber van Bosse, A. 1923. Liste des algues du Siboga II Rhodophyceae, 2e p. Ceramiales. *Siboga Expeditie Monogr. 59c*: 311–392 pls. 9 + 10.

Womersley, H. B. S. 1984. Part I; 1987. Part II: *The Marine Benthic Flora of Southern Australia*. South Australian Governm. Printer Div., Adelaide. 329 and 484 pp.

Botanica Marina
Vol. 34, pp. 139–144, 1991

Cytophotometric Estimation of Inter- and Intraspecific Nuclear DNA Content Variation in *Gracilaria* and *Gracilariopsis* (Gracilariales, Rhodophyta)

D. F. Kapraun and J. A. Dutcher

Department of Biological Sciences, University of North Carolina, Wilmington, North Carolina 28403, U.S.A.

(Accepted 29 September 1990)

Abstract

Microspectrophotometry with the DNA-localizing fluorochrome hydroethidine was used to estimate nuclear genome sizes in three geographic isolates of *Gracilaria tikvahiae*, *Gracilaria* sp. (cultivar G-6), *G. blodgettii* and *Gracilariopsis lemanieformis*. Relative fluorescence (I_f) values for the angiosperm *Antirrhinum majus* and the green alga *Cladophora albida* were plotted against their known DNA contents to derive a standard line. Mean 2 C nuclear DNA I_f values for *Gracilaria* and *Gracilariopsis* were extrapolated to obtain estimates of their genome sizes. Results indicate a narrow range of relatively small values (0.35–0.45 pg) in isolates examined. All *Gracilaria* species had similar nuclear genomes, within the range of 0.40–0.45 pg, while a smaller genome of 0.35 pg was indicated for *Gracilariopsis lemanieformis*. Variations in genome size and previously determined differences in nucleotide base pair compositions suggest genome repatterning processes which may have accompanied speciation in these taxa.

Introduction

Members of the Gracilariales (Fredericq and Hommersand 1989 a, 1989 b) are among the most economically important agarophytes (Dawes 1987) and support a significant industry based on both cultivated and wild plants (Hanisak 1987, Santelices and Doty 1989). Consequently, it seems remarkable that even the most basic information for the *Gracilaria* and *Gracilariopsis* genomes remains unknown. To date, genetic research has been limited to the pioneering studies of Mendelian inheritance by van der Meer and coworkers (see van der Meer 1987 and Santelices and Doty 1989 for reviews). Chromosome numbers have been published for only five (McLachlan *et al.* 1977, Bird and McLachlan 1982, Bird *et al.* 1982, Patwary and van der Meer 1984, Gargiulo *et al.* 1987) of the 100–160 recognized species of Gracilariales (Bird and McLachlan 1982, Kraft 1981, Guiry and Freamhainn 1986).

In contrast, information is now available for genome size, organization and complexity in several green marine macroalgae. Preliminary data indicate broad similarities with angiosperms. For example, flowering plants are characterized by nearly a thousand-fold range of variation in nuclear DNA contents, from less than 1 pg to several hundred picograms (Bennett 1972, Rees 1972, Rees and Jones 1972, Leutwiler *et al.* 1984, Ohri *et al.* 1986). Much of this variation is attributed to repeated sequences which compose 40–90% of their nuclear genomes (Hake and Walbot 1980, Flavell 1980, Wenzel and Hemleben 1982). Significantly, interspecific variation in their DNA contents often shows a discontinuous pattern, and in some cases, even a geometric progression (Rothfels *et al.* 1966, Price 1973, Sparrow and Nauman 1973, Ayonoadu 1974). Although multinucleate Chlorophyta investigated have relatively small genomes, ranging from 0.32–1.79 pg (Schnetter *et al.* 1981, Olsen *et al.* 1987, Bot *et al.* 1989 a, 1989 b, Kapraun and Shipley 1990), they are often characterized by large scale, discontinuous variation (Kapraun and Gargiulo 1987 a, 1987 b, Kapraun *et al.* 1988). In addition, reassociation kinetics have demonstrated three distinct components corresponding to fast and slow repeats, as well as single copy DNA (Bot *et al.* 1989a, 1989b).

It is not known to what extent these generalizations apply to members of the Rhodophyta as no published information is available for their genome size, organization or complexity. Consequently, we initiated an investigation to quantify and characterize the nuclear genomes of four species of Gracilariales from the western Atlantic. Analysis of their reassociation kinetics (Dutcher 1990), which permits characterization of genome organization and sequence complexity, will be published elsewhere. The present communication concerns cytophotometric estimation of their genome sizes using the DNA-localizing fluorochrome hydroethidine and microspectrophotometry (Kapraun *et al.* 1988, Kapraun and Bailey 1989, Kapraun and Shipley 1990).

Material and Methods

Source of specimens

Three red algal species were collected from southeastern North Carolina (Kapraun 1980): *Gracilariopsis lemanieformis* (Bory) Dawson, Acleto *et* Folvik from Kure Beach; *Gracilaria blodgettii* Harvey from Bogue Sound near Morehead City; *Gracilaria tikvahiae* McLachlan from a coastal jetty near Wrightsville Beach. Additional samples of *Gracilaria tikvahiae* were provided by C. Bird from the type locality (McLachlan 1979) and by K. Bird as a cultivar G-6 (Hanisak and Ryther 1984, Bird 1988) and field collected specimens from the Indian River, Ft. Pierce, Florida.

Fixation and karyotype analysis

Specimens were fixed in the laboratory in absolute ethanol-glacial acetic acid for 12–24 h and stored in 70% ethanol at 4 °C (Kapraun and Bailey 1989). Preserved material was soaked in distilled water to soften tissue and remove fixative prior to hydroethidine staining. Small sections from thalli containing mature medullary cells were squashed on coverslips and air dried 1–2 h. Coverslips were soaked in phosphate buffer (PBS) for 1 h, stained for 5–10 min in hydroethidine (Kapraun 1989), and destained in PBS for 24 h prior to viewing.

Determination of nuclear DNA

Microspectrophotometric data for an angiosperm, snapdragon (*Antirrhinum majus* L.) with a 2 C DNA content of 3.2 pg (Bennet and Smith 1976), and the green seaweed *Cladophora albida* (Huds.) Kuetz. with a 2 C DNA content of 0.7–0.8 pg (Bot *et al.* 1989b), were used to quantify I_f values for *Gracilaria* and *Gracilariopsis* specimens. Seeds were germinated in petri dishes and root tips 1 cm long with abundant root hairs were harvested, fixed and stained with hydroethidine as above. Field collected *C. albida* was cleaned and placed in enriched seawater medium (Freshwater and Kapraun 1986) in petri dishes lined with glass coverslips on to which released swarmers readily settled. Coverslips were transferred to fixative within 24 h of zoospore attachment and stored in 70% ethanol.

Frequency distributions for I_f values corresponding to 2 C and 4 C CNA contents were determined for *Antirrhinum majus* from root tip sqash preparations (Dhillon and Miksche 1982, Kurth and Gifford 1985). *Cladophora albida* zoospores, which were haploid, had I_f profiles corresponding to 1 C and 2 C DNA levels. Fluorescence values for 2 C nuclei were determined (Dhillon and Miksche 1982, Kurth and Gifford 1985) and plotted against their known DNA contents to derive a standard line. Nuclear DNA contents for *Gracilaria* and *Gracilariopsis* were extrapolated by plotting their I_f values of G_1- and G_2-phase nuclei along this I_f/DNA slope (Kapraun and Shipley 1990).

Cytophotometric measurements were made with a Kinetek photometer system in combination with an Olympus BH2-RFK microscope with dichroic mirror DM-580 and barrier filter 0-590, specific for hydroethidine emissions (Kapraun *et al.* 1988). The photometer system was equiped with a rotating array of perforated diaphragms, permitting selection of pinhole aperatures corresponding to the diameter of the nucleus being viewed. Consequently, error from cytoplasmic (extranuclear) fluorescence was greatly reduced (Hull *et al.* 1982, Goff and Coleman 1984a).

Results and Discussion

DNA cytofluorometry

Microspectrophotometry with DNA-localizing fluorochromes has been used previously for karyological studies of marine red algae (Goff and Coleman 1984a, 1984b, 1985, 1986, Kapraun and Bailey 1989). In the present investigation, hydroethidine staining for periods as brief as 2 min followed by destaining for 24 h at 4 °C resulted in reproducible, intense fluorescence in medullary cell nuclei with only slight cytoplasmic interference (Fig. 1). Stained interphase nuclei examined with episcopic UV illumination revealed interspecific variation in amounts of condensed chro-

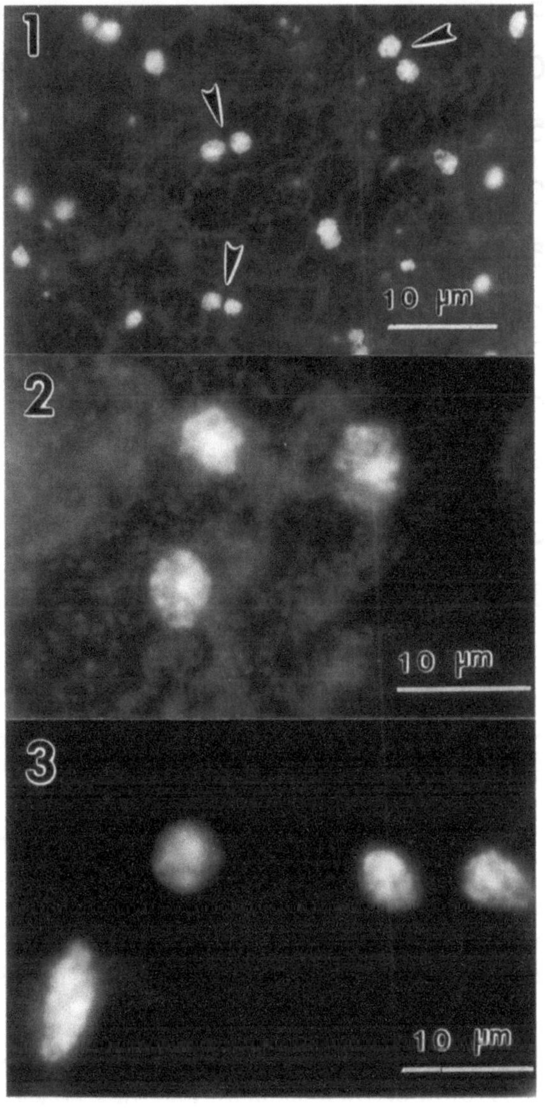

Figs 1–3. Haploid interphase nuclei in medullary cells following hydroethidine staining and visualized with episcopic UV illumination.
Fig. 1. *G. tikvahiae* NC, note pairs of telophase nuclei (arrows); Fig. 2. *Gracilariopsis lemanieformis*; Fig. 3. *Gracilaria tikvahiae* NC, note relative nuclear brightness associated with condensed chromatin.

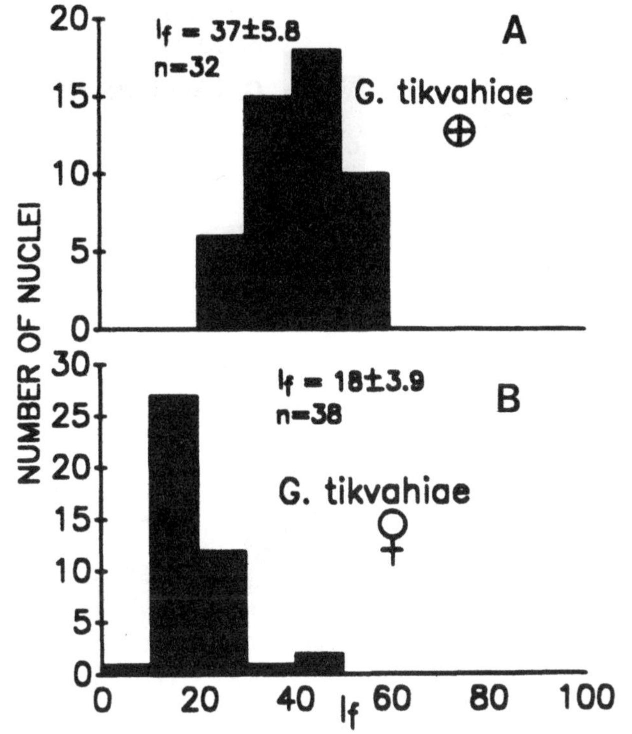

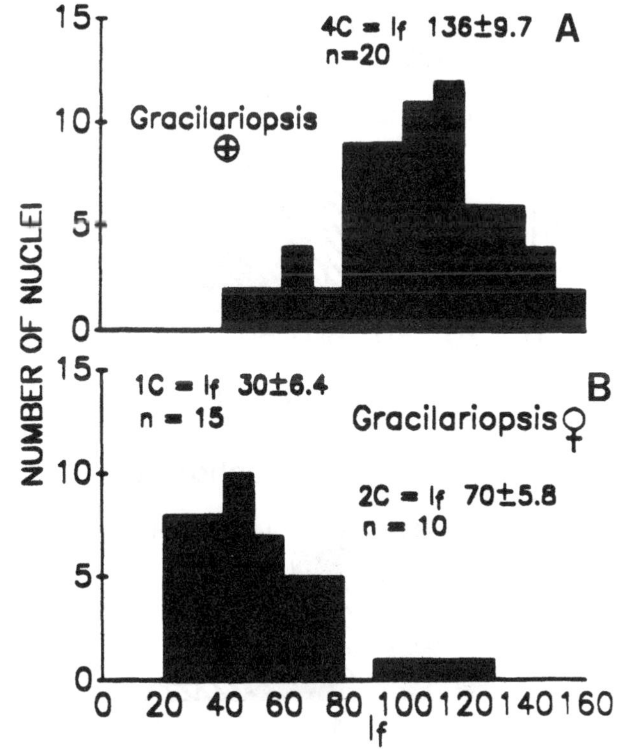

Figs 4–7. Comparison of frequency distributions of relative DNA values for nuclei after hydroethidine staining. n = number of nuclei, excluding S-phase nuclei, used to determine I_f values, I_f = fluorescence intensity mean ± SD.
Fig. 4. Fluoroscence intensity values (I_f) for nuclei of *Gracilaria tikvahiae*, (a) 4 C level nuclei in tetrasporophyte, (B) 2 C level nuclei in female gametophyte. Fig. 5. Fluorescence intensity values (I_f) for nuclei of *Gracilariopsis lemanieformis*, (A) 4 C level nuclei in tetrasporophyte (B) 2 C level nuclei in female gametophyte.

matin. For example, nuclei in *Gracilariopsis lemanieformis* had relatively small amounts of condensed chromatin (Fig. 2). In contrast, all regional isolates of *Gracilaria tikvahiae* were characterized by numerous, large chromatin bodies (Fig. 3).

Variations in nuclear DNA levels associated with ploidy level differences in gametophytic and tetrasporophytic phases of red algae have been demonstrated with micro-spectrophotometry (Goff and Coleman 1984 a, 1984 b, 1985, 1986, Kapraun and Bailey 1989). In the present study, mean DNA levels for 2 C nuclei closely approximate 50% of the 4 C values in *Gracilaria* (Fig. 4) and *Gracilariopsis* (Fig. 5).

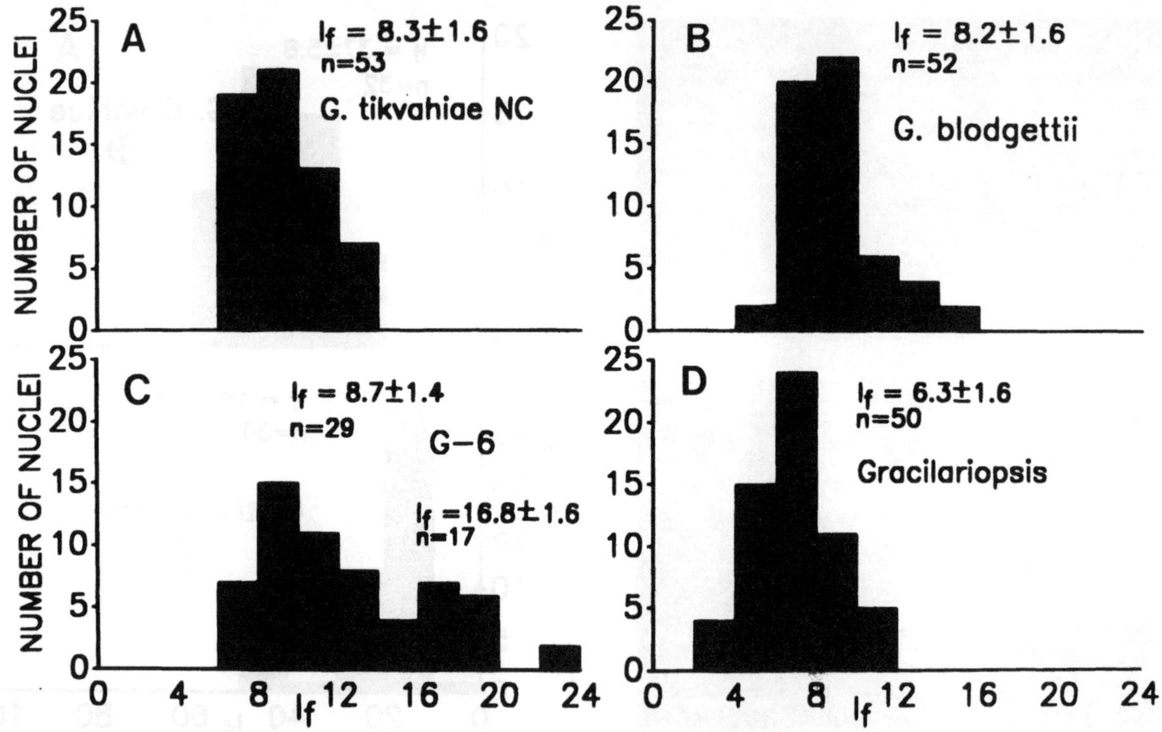

Fig. 6. Comparison of relative I_f values for nuclei of (A) *G. tikvahiae* NC, (B) *G. Blodgettii*, (C) cultivar G-6 and (D) *Gracilariopsis lemanieformis* after hydroethidine staining.

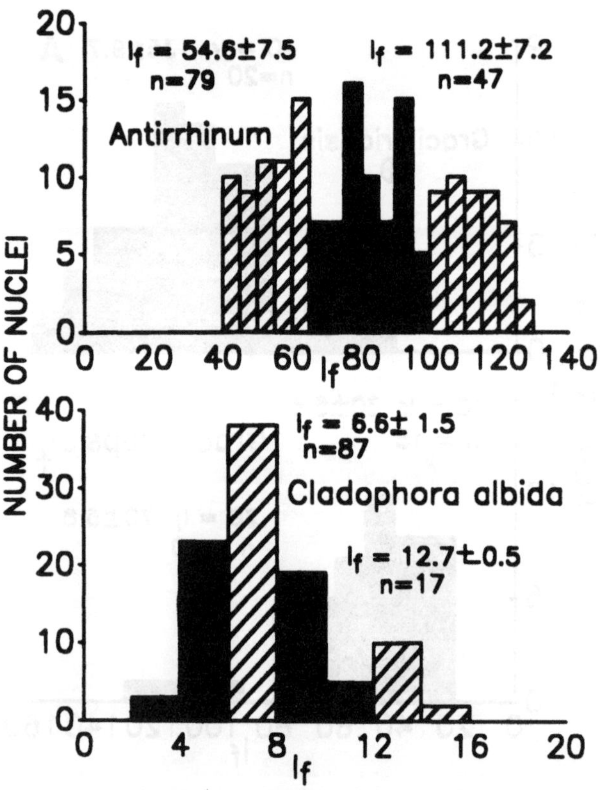

Fig. 7. Fluorescence intensity (I_f) profiles for 2 C and 4 C nuclei in (a) *Antirrhinum majus* and (b) *Cladophora albida* after hydroethidine staining. Diagonal lines indicate G_1^- and G_2^- phase nuclei (n) used to calculate C levels.

Results of microspectrophotometry indicate similar relative I_f values for *Gracilaria tikvahiae* (NC), *G. blodgettii* and cultivars assigned to *G. tikvahiae* (G-6), and slightly lower values for *Gracilariopsis lemanieformis* (Fig 6a–d). However, there was no significant differences in DNA contents among the isolates. Nuclear DNA contents were confirmed by quantifying I_f values with standards having known DNA contents (Schnetter *et al.* 1981, Hamada *et al.* 1985).

The *Antirrhinum* and *Cladophora* standards (Fig. 7) were used as previously described (Kapraun and Shipley 1990) to derive a standard line. In a typical set of observations, the observed I_f ratio for 2 C values of *C. albida* and *A. majus* (*e.g.* 13 : 55 = 0.24) closely approximated their DNA content (pg) ratio (0.8 : 3.2 = 0.25). Extrapolated mean 2 C nuclear DNA levels for all experiments indicate a genome size of 0.35–0.45 pg for geographic isolates of *Gracilaria tikvahiae*, *G. blodgettii*, and *Gracilaria* sp. cultivar G-6. *Gracilariopsis lemanieformis* is characterized by a smaller genome 0.35 pg (Table I).

Although no published data for other Rhodophyta are available for comparison, these values are similar to the nuclear DNA contents of 0.15–1.79 pg reported for multinucleate marine green algae (Schnetter *et al.* 1981, Olsen *et al.* 1987, Bot *et al.* 1989a, 1989b, Kapraun and Shipley 1990).

Efforts are currently underway to estimate genome sizes in additional populations of *Gracilaria* and *Gracilariopsis* from the tropical western Atlantic to determine if generalizations developed from our study of these four species have broader application among the 100–160 recognized species of Gracilariales.

Table I. Genome size (pg) for nuclei after hydroethidine staining for species of *Gracilaria* and *Gracilariopsis*. Data standardized to the 2 C DNA level of *Antirrhinum majus* (3.2 pg).

Species	Genome size (pg)			Number of nuclei
	1 C	2 C	4 C	
G. tikvahiae NC		0.42 ± .09		141
		0.42 ± .06		44
		* 0.52 ± .12		120
		$\bar{X}$ 0.45 ± .05		
G. tikvahiae NS		0.42 ± .05		89
		0.44 ± .07		22
		$\bar{X}$ 0.43 ± .01		
G. tikvahiae Fla		0.37 ± .04		38
		0.43 ± .06		31
			0.66 ± .05	36
		$\bar{X}$ 0.40 ± .04		
G. tikvahiae G-6		0.45 ± .09		124
	0.27 ± .05			77
		0.46 ± .05		44
		$\bar{X}$ 0.45 ± .01		
G. blodgettii		0.42 ± .16		132
		0.41 ± .09		86
		* 0.46 ± .09		97
		$\bar{X}$ 0.43 ± .02		
Gracilariopsis		0.36 ± .11		151
		0.41 ± .09		41
		* 0.29 ± .09		67
		$\bar{X}$ 0.35 ± .06		

* Data standardized to the 2 C DNA level of *Antirrhinum majus* (3.2 pg) and *Cladophora albida* (0.80 pg). Mean ± SD is given for each species.

Acknowledgements

The authors thank Dr Kimon Bird for critically reviewing the manuscript, and Dr Carolyn J. Bird for providing samples of *Gracilaria tikvahiae* from the type locality. Financial support for this research from the National Science Foundation-International Programs (NSF-INT-88-15226) and the North Carolina Center for Biotechnology is gratefully acknowledged. Contribution Number 18 to the Center for Marine Science Research, UNC-Wilmington.

References

Ayonoadu, U. W. U. 1974. Nuclear DNA variation in *Phaseolus*. *Chromosoma (Berl.) 48*: 41–49.

Bennett, M. D. 1972. Nuclear DNA content and minimum generation time in herbaeious plants. *Proc. Royal Soc. Lond. Ser B. 181*: 109–135.

Bennett, M. D. and J. B. Smith. 1976. Nuclear DNA amounts in Angiosperms. *Phil. Trans. R. Soc. (London) B. 274*: 227–274.

Bird, C. J. and J. McLachlan. 1982. Some underutilized taxonomic criteria in *Gracilaria* (Rhodophyta, Gigartinales). *Bot. Mar. 25*: 557–562.

Bird, C. J., J. P. van der Meer and J. McLachlan. 1982. A comment on *Gracilaria verrucosa* (Huds.) Papenf. (Rhodophyta: Gigartinales). *J. mar. biol. Ass. U.K. 62*: 453–459.

Bird, K. T. 1988. Agar production and quality from *Gracilaria* sp. strain G-16: Effects of environmental factors. *Bot. Mar. 31*: 33–39.

Bot, P. V. M., W. T. Stam, S. A. Boele-Bos, C. van den Hoek and W. van Delden. 1989 a. Biogeographic and phylogenetic studies in three North Atlantic species of *Cladophora* (Cladophorales, Chlorophyta) using DNA-DNA hybridization. *Phycologia 28*: 159–168.

Bot, P. V. M., R. W. Holton, W. T. Stam, C. van den Hoek. 1989 b. Molecular divergence between North Atlantic and Indo-West Pacific *Cladophora albida* (Cladophorales: Chlorophyta) isolates as indicated by DNA-DNA hybridization. *Mar. Biol. 102*: 307–313.

Dawes, C. J. 1987. The biology of commercially important tropical marine algae. *In*: (K. T. Bird and P. H. Benson, eds) *Seaweed Cultivation for Renewable Resources. Developments in Aquaculture and Fisheries Science*, Vol 16. Elsevier, Amsterdam. pp. 191–218.

Dhillon, S. S. and J. P. Miksche. 1982. DNA content and heterochromatin variations in various tissues of peanut (*Arachis hypogaea*). *Amer. J. Bot. 69*: 219–226.

Dutcher, J. A. 1990. A cytogenetic investigation of the marine red algae *Gracilaria* and *Gracilariopsis*. Unpubl. M.S. Thesis, University of North Carolina at Wilmington. 100 pp.

Flavell, R. 1980. The molecular characterization and organization of plant chromosomal DNA sequences. *Ann. Rev. Plant Physiol. 31*: 569–596.

Fredericq, S. and M. H. Hommersand. 1989a. Comparative morphology and taxonomic status of *Gracilariopsis* (Gracilariales, Rhodophyta). *J. Phycol. 25*: 228–241.

Fredericq, S. and M. H. Hommersand. 1989b. Proposal of the Gracilariales ord. nov. (Rhodophyta) based on an analysis of the reproductive development of *Gracilaria verrucosa. J. Phycol. 25*: 213–227.

Freshwater, D. W. and D. F. Kapraun. 1986. Field, culture and cytological studies of *Porphyra carolinensis* Coll et Cox (Bangiales, Rhodophyta) from North Carolina. *Jpn. J. Phycol. 43*: 251–262.

Gargiulo, G. M., F. DeMasi and G. Tripodi. 1987. Structure and reproduction of *Gracilaria longa* sp. nov. (Rhodophyta, Gigartinales) from the Mediterranean Sea. *Ital. J. Bot. 121*: 247–257.

Goff, L. J. and A. W. Coleman. 1984a. Elucidation of fertilization and development in a red alga by quantitative DNA microspectrophotometry. *Develop. Biol. 102*: 173–194.

Goff, L. J. and A. W. Coleman. 1984b. Transfer of nuclei from a parasite to its host. *Proc. Natl. Acad. Sci. USA 81*: 5420–5424.

Goff, L. J. and A. W. Coleman. 1985. The role of secondary pit connections in red algal parasitism. *J. Phycol. 21*: 483–508.

Goff, L. J. and A. W. Coleman. 1986. A novel pattern of apical intercellular nuclear transfer in the red alga *Polysiphonia. Amer. J. Bot. 73*: 1109–1130.

Guiry, M. D. and M. T. Freamhainn. 1986. Biosystematics of *Gracilaria foliifera* (Gigartinales, Rhodophyta). *Nord. J. Bot. 5*: 629–637.

Hake, S. and V. Walbot. 1980. The genome of *Zea mays*, its organization and homology to related grasses. *Chromosoma (Berl) 79*: 251–270.

Hamada, J., M. Saito and M. R. Ishida. 1985. Nuclear phase in vegetative and gamete cells of *Closterium ehrenbergii*: Fluorescence microspectrophotometry of DNA content. *Annu. Rep. Res. Reactor. Inst. Kyoto Univ. 18*: 56–61.

Hanisak, M. D. 1987. Cultivation of *Gracilaria* and other macroalgae in Florida for energy production. *In*: (K. T. Bird and P. H. Benson, eds) *Seaweed Cultivation for Renewable Resources. Developments in Aquaculture and Fisheries Science*, Vol 16. Elsevier, Amsterdam. pp. 191–218.

Hanisak, M. D. and J. H. Ryther. 1984. Cultivation biology of *Gracilaria tikvahiae* in the United States. *Hydrobiologia 116/117*: 295–298.

Hull, H. M., R. W. Howshaw and J. C. Wang. 1982. Cytofluorometric determination of nuclear DNA in living and preserved algae. *Stain Technol. 57*: 272–282.

Kapraun, D. F. 1980. *An Illustrated Guide to the Marine Algae of Coastal North Carolina. I. Rhodophyta*. Univ. North Carolina Press, Chapel Hill. pp. 206.

Kapraun, D. F. 1990. Parasexual fusion products in green algae: *Enteromorpha* and *Ulvaria. Proc. Intl. Seaweed Symposium 13*: 443–453.

Kapraun, D. F. and J. C. Bailey. 1989. Karyology and nuclear DNA content of *Gelidium pusillum* (Gelidiales, Rhodophyta) from North Carolina, USA. *Jpn. J. Phycol. 37*: 201–207.

Kapraun, D. F. and G. M. Gargiulo. 1987a. Karyological studies of three species of *Cladophora* (Cladophorales, Chlorophyta) from Bermuda. *Ital. J. Bot. 121*: 165–176.

Kapraun, D. F. and G. M. Gargiulo. 1987b. Karyological studies of four species of *Cladophora* (Cladophorales, Chlorophyta) from coastal North Carolina. *Ital. J. Bot. 121*: 1–26.

Kapraun, D. F. and M. J. Shipley. 1991. Karyology and nuclear DNA quantification in *Bryopsis* (Codiales, Chlorophyta) from North Carolina, USA. *Phycologia*. (in press).

Kapraun, D. F., M. G. Gargiulo and G. Tripodi. 1988. Nuclear DNA and karyotype variation in species of *Codium* (Codiales, Chlorophyta) from the North Atlantic. *Phycologia 27*: 273–282.

Kraft, G. T. 1981. Rhodophyta: Morphology and classification. *In*: (C. S. Lobban and M. J. Wynne, eds) *The Biology of Seaweeds*. Univ. Calif. Press, Berkeley. pp. 6–51.

Kurth, E. and E. M. Gifford Jr. 1985. Ontogenetic changes in DNA content in roots of the water fern *Azolla filiculoides. Amer. J. Bot. 72*: 1676–1683.

Leutwiler, L. S., B. R. Hough-Evans and E. M. Meyerowitz. 1984. The DNA of *Arabidopsis thaliana. Mol. Gen. Genet. 194*: 15–23.

McLachlan, J. 1979. *Gracilaria tikvahiae* sp. nov. (Rhodophyta, Gigartinales, Gracilariaceae), from the Northwestern Atlantic. *Phycologia 18*: 19–23.

McLachlan, J., J. P. van der Meer and N. L. Bird. 1977. Chromosome numbers of *Gracilaria foliifera* and *Gracilaria* (Rhodophyta) and attempted hybridization. *J. Mar. Biol. Assoc. U.K. 57*: 1137–1141.

Olsen, J. L., W. T. Stam, P. V. M. Bot and C. van den Hoek. 1987. scDNA-DNA hybridization studies in Pacific and Carribean isolates of *Dictyosphaeria cavernosa* (Chlorophyta) indicate a long divergence. *Helgoländer Wiss. Meeresunters. 41*: 377–383.

Ohri, D. and T. N. Khoshoo. 1986. Plant DNA: Contents and Systematics. *In*: (S. K. Dutta, ed) *DNA Systematics*. CRC Press, Boca Raton, Fla. pp. 2–19.

Patwary, M. U. and J. P. van der Meer. 1984. Growth experiments on autopolyploids of *Gracilaria tikvahiae* (Rhodophyceae). *Phycologia 23*: 21–27.

Price, H. J., A. H. Sparrow and A. F. Nauman. 1973. Evolutionary and developmental considerations of the variability of nuclear parameters in higher plants. I. Genome volume, interphase chromosome volume, and estimated DNA content of 236 gymnosperms. *Brookhaven Symp. Biol. 25*: 390–421.

Rees, H. 1972. DNA in higher plants. *In*: (H. H. Smith, ed.) *Evolution of Genetic Systems*. Gordon & Breach, New York. pp. 394–418.

Rees, H. and R. N. Jones. 1972. The origin of the wide species variation in nuclear DNA content. *Int. Rev. Cytol. 32*: 53–92.

Rothfels, K., E. Sexsmith, M. Heimburger and M. O. Krause. 1966. Chromosome size and DNA content of species of *Anemone* L. and related genera (Ranunculaceae). *Chromosoma 20*: 54–74.

Santelices, B. and M. S. Doty. 1989. A review of *Gracilaria* farming. *Aquaculture 78*: 95–133.

Schnetter, R., B. Mohr, G. Bula-Meyer and G. Seibold. 1981. Ecology, life history and nucleus DNA contents of *Derbesia tenuissima* from the Caribbean coast of Columbia. *In*: (T. Lewing, ed.) *Xth International Seaweed Symposium*. pp. 357–362.

Sparrow, A. H. and A. F. Nauman. 1973. Evolutionary changes in genome and chromosome sizes and in DNA content in the grasses. *Brookhaven Symp. Biol. 25*: 367–389.

van der Meer, J. P. 1987. Marine algal genetics and genomes. *Hydrobiologia. 151/152*: 49–56.

Wenzel, W. and V. Hemleben. 1982. A comparitive study of genomes of angiosperms. *Pl. Syst. Evol. 139*: 209–227.

Botanica Marina
Vol. 34, pp. 145–152, 1991

Life History of *Gymnogongrus furcellatus* (C. Ag.) J. Ag. (Rhodophyta, Phyllophoraceae) from Chile[1]

N. I. Lewis, M. Avila[2] and J. L. McLachlan

National Research Council, 1411 Oxford Street, Halifax NS, Canada B3H 3Z1

(Accepted 29 September 1990)

Abstract

The life history of *Gymnogongrus furcellatus*, a carrageenophyte, isolated from a population on the Pacific coast of central Chile, was completed in culture. Carpospores from field-collected, foliose gametophytes gave rise to a crustose tetrasporophytic phase, morphologically similar to plants of the genus *Erythrodermis*. These in turn formed tetrasporangia, the spores of which developed into the foliose, gametophytic phase that became fertile. The gametophyte of *G. furcellatus* is dioecious, and carposporophytes did not form in the absence of fertile male fronds. The life history of *G. furcellatus* was completed in about one year, and the rate of growth and development of fertility was affected by both temperature and photoperiod. The gametophytic phase grew well over a much broader temperature range than the sporophyte. Our results from culture are consistent with ecological observations on, and the distribution of, this species in Chile.

Introduction

Taxonomy of species that traditionally have been placed in the rhodophycean genera *Gymnogongrus* Martius and *Ahnfeltia* Fries is currently under review (Maggs and Pueschel 1989). Our study concerns the alga referred to as *Gymnogongrus furcellatus* (C. Agardh) J. Agardh which may properly fall within the proposed genus *Ahnfeltiopsis* (Maggs and Pueschel 1989). In this proposed genus, the cystocarp is internal, the tetraspores are cruciately arranged and intercalary and the cell-wall matrix is presumably carrageenan [certainly for *G. furcellatus* (McCandless *et al.* 1983)], rather than agar. For those species with internal carposporophytes, two types of life histories have been recognized (*cf.* Maggs 1988). This may be either a direct development of the carposopore into the gametophyte or, as in *G. furcellatus*, the carpospore forms a crustose tetrasporophyte, similar in morphology to species of *Erythrodermis* Batters (Batters 1900). Because the taxonomy of this group is as yet unresolved, we continue to refer to our alga as *Gymnogongrus furcellatus*.

When an alga is utilized commercially, clarification of its life history becomes important. *Gymnogongrus furcellatus* has been exploited for its cell-wall polysaccharide, which in the foliose gametophytes is iota-type carrageenan (McCandless *et al.* 1983). This species is an important source of iota carrageenan, and global resources of *Gymnogongrus* have been obtained almost exclusively from Chile as *G. furcellatus* (Santelices *et al.* 1989). The sporophytic stage of *G. furcellatus* produces a different type of carrageenan (McCandless *et al.* 1983). However, because of the obscure, small habit of this life-history phase, which remains virtually unknown in the field, there is no commercial interest in the sporophyte of *G. furcellatus*.

Of the three species of *Gymnogongrus* reported from Chile (Etcheverry 1986), only *G. furcellatus* is sufficiently abundant for exploitation (Santelices *et al.* 1989), and the largest standing stocks are present in

[1] Issued as NRCC no. 31951
[2] Present address: Instituto de Formento Pesquero, José Domingo, Cañas 2277, Casilla 1287, Santiago, Chile.

the intertidal zone along the central Chilean coast, where the foliose gametophytic fronds are collected generally by hand plucking. Habitats of *G. furcellatus* are moderate to low exposure with shifting sand.

Resources of *G. furcellatus* in Chile have been obtained exclusively from natural populations and this is apt to continue. Cultivation as an alternative or adjunct to harvesting of natural populations has been indicated as uneconomic (Santelices *et al.* 1989). At the same time Chilean beds of *G. furcellatus* have gone unmanaged, even though there have been pleas for implementation of management practices. This in part may result from lack of information on the biology of *G. furcellatus*, and if rational management is to be undertaken, detailing the life history of this alga becomes imperative.

Previously, Candia and Kim (1977) briefly reported on the culturing of *G. furcellatus*, although they were not successful in completing its life history. In the present instance, we have followed, *in vitro*, the life history of *G. furcellatus* from Chile and responses of both gametophytes and sporophytes to temperatures and photoperiods were recorded.

Material and Methods

Our culture of *Gymnogongrus furcellatus* was initiated from carpospores released from cystocarpic fronds collected 21-12-83 at Matanzas (33°56′S, 71°53′W) in central Chile. The carpospores, established in unialgal culture, developed into small crusts which were incubated for 4 months at 15 °C with a photon flux density of about 30 μmol m^{-2} s^{-1} provided by cool-white fluorescent lamps on a 12 : $\overline{12}$ light : dark cycle. The medium was modified "C" (McLachlan 1973). These crusts formed tetrasporangia, releasing spores that were used to initiate the life-history study.

Gametophytic fronds of 1-mm length were obtained from tetraspore-derived crusts and incubated with aeration in "C" medium. Conditions of temperature and light period are given in Table I.

Carpospores from cultured cystocarpic fronds were released over cover glasses at 15 °C and a 12 : $\overline{12}$, L : D cycle and incubated in small (60 × 20 mm) plastic dishes under conditions listed in Table II. As preliminary results suggested a possible short-day response in production of tetrasporangia, crustose plants from 'non-inducing' conditions, *i.e.* 20 °C and 12 : $\overline{12}$ L : D cycle, were incubated at 15 °C, with 16 : $\overline{8}$, L : D and 8 : $\overline{16}$, L : D cycles. Photon flux density was 30 and 60 μmol m^{-2} s^{-1} for the 16 : $\overline{8}$, L : D and 8 : $\overline{16}$, L : D periods respectively so that the total quantum dosage was the same for both light regimes.

Table I. Effects of temperature and photoperiod on development of cystocarps and spermatia of *G. furcellatus* in culture. Photon flux density = 30 μmol m^{-2} sec^{-1}.

Temp. (°C)	Photoperiod (h d^{-1})	Cystocarps	Spermatia
10	16	+	+
	8	+	+
15	16	+	+
	12	+	+
	8	+	+
20	16	–	–
	12	–	–

Table II. Effects of temperature and photoperiod on formation of tetrasporangia of *G. furcellatus* in culture and time after germination at which erect axes were first observed. Photon flux density = 30 μmol m^{-2} sec^{-1}.

Temp. (°C)	Photoperiod (h d^{-1})	Tetrasporangia (spores released)	Erect axis (wk)
10	16	+	9
	12	+	9
	8	+	18
15	16	+	5
	12	+	5
	8	+	9
20	16	–	none
	12	±	18
	8	±	18

Effect of temperature on the growth of gametophytes was determined by incubating 30 apices, excised in 10-mm lengths from clonal fronds, per 250-mL flask. These fronds had been incubated at 15 °C and 20 °C and in a 12 : $\overline{12}$, L : D cycle. The flasks, containing 100 mL of medium changed weekly, were incubated with aeration at temperatures from 0 °C to 24 °C, all at 40 μmol m^{-2} s^{-1} and a 16 : $\overline{8}$, L : D period. At the end of the 4-week experimental period, the length of the segments was measured and mean growth (increase in length) at each temperature calculated.

Growth of the crustose sporophytes, at the same temperatures used for the gametophytic apices, was determined following a 4-week incubation period. Carpospores were released over cover glasses (18 mm diam) and incubated in small plastic petri dishes on a 16 : $\overline{8}$, L : D cycle with a photon flux density of 40 μmol m^{-2} s^{-1}. Diameters of 100 plants at each temperature were determined from which the mean growth of the discs was calculated.

Tests for equality among means were done using one way ANOVA and for multiple comparisons among means the Student-Newman-Keuls (SNK) test was employed (Sokal and Rohlf 1969).

Results

Sporophytic crusts in culture, derived from carpospores of field-collected cystocarpic fronds, attained a maximum diameter of 5 to 6 mm (20–28 mm^2) but could become adjoined, forming large, continuous layers of sporophytic tissue. In younger crusts, the hypothallus was monostromatic, with the perithallus composed of up to 10 cells. Older crusts formed a sub-hypothallus, about 50 μm in thickness, consisting of 4 to 5 layers of dark-red cubical cells (Fig. 1).

In surface view, the tetrasporangia were noticeable as darkened spots, slightly elevated above the surface of the sporophytic crust. The intercalary tetrasporangia were formed as single chains in the perithalloid filaments, the tetraspores being cruciately arranged (Fig. 2). Released spores had a diameter of 8.5 ± 0.7 μm ($n = 17$) for a calculated volume of 24 μm^3.

Tetraspores were released in mass (Fig. 3) and became attached to the substratum within 24 hours. Germination resulted either directly in a crustose plant or first in a filamentous germling of 10 to 12 cells which subsequently gave rise to the crustose gametophyte (Fig. 4). After tetraspore-derived discs were 0.5 to 1.0 mm in diameter, an upright axis was initiated from the center of the disc, followed by initiation of fronds around the periphery. Accordingly the plant, or genet, of the mature gametophyte consists of a basal disc-like holdfast together with several terete to flattened dichotomously-branched fronds.

The gametophytic generation of *Gymnogongrus furcellatus* is dioecious and when vegetative the two sexes are indistinguishable. The male gametophytes were identifiable when spermatia were being actively released, which occurred from the upper region of the frond (Fig. 5). These gametes formed in the outer cortex (Fig. 6a) by oblique cleavage of spermatangia from superficial spermatangial mother cells (Fig. 6b). The carpogonial branches developed within the first mm of the apex of the female frond and were easily distinguishable in section (Fig. 7).

Cystocarps (Fig. 8) developed on the female frond, only when co-cultured with male fronds actively releasing spermatia. In the absence of male fronds, the females remained non-cystocarpic and there was no suggestion of apomixis in our isolates. Released carpospores were pale pink [diam 10.7 ± 1 μm, vol 38 μm^3 ($n = 30$)]. Germination and early development of carpospores were similar to that of the tetraspores noted above. With release of viable tetraspores, from the resultant tetrasporophytes, the life history of *G. furcellatus* was completed as illustrated in the diagramatic life cycle shown in Figure 10.

A minimum of about one year was required to complete the life history of *G. furcellatus*. Following tetraspore germination, a minimum of 10 months was required for the gametophytes to become cystocarpic and release viable carpospores. This occurred at 15 °C and a photoperiod of at least 12-hours light (Table I, Figure 10). Reducing the light period to 8 hours at 15 °C or decreasing the temperature to 10 °C, even with a long light period required a substantially longer time before carposporogenesis was apparent (Fig. 10). At 10° and 15 °C, both temperature and total irradiance affected the length of time required for carposporogenesis to occur. However, regardless of the light period, gametophytes failed to form cystocarps at 20 °C (Table I). Moreover, when fronds actively releasing spermatia at 15 °C were transferred to 20 °C and co-cultured with female gametophytes, cystocarps failed to develop and eventually spermatia ceased to be produced at the warmer temperature.

Four to 6 weeks after carpospore germination at 15 °C, tetrasporangia formed and viable tetraspores were released from crustose plants at both $12:\overline{12}$, L:D and $16:\overline{8}$, L:D periods. Tetraspores were released after 9-weeks incubation at the longer light period at 10 °C and at the shorter light period at 15 °C. A considerably longer interval was required for spore release to occur at 10 °C, short days. These results are summarized in Table II. Few tetrasporangia developed from sporophytic crusts incubated at 20 °C under $8:\overline{16}$, L:D and $12:\overline{12}$, L:D periods and none at a $16:\overline{8}$, L:D period. Although this suggested a possible 'short-day' response, crusts from 'non-inducing' conditions incubated with equivalent quantum doses produced spores within 1 month under both long and short days.

Under all conditions at which tetraspores were released, the spores were viable and eventually erect axes developed from the initially formed basal discs.

Gametophytic apices increased rapidly in length at temperatures from 10 °C to 20 °C, with maximum growth at 15 °C (Fig. 11). At 0 °C, 5 °C and 24 °C growth was significantly ($p \leq 0.05$) reduced compared with the mid-range temperatures (ANOVA; SNK). With increase in temperature, the fronds became more darkly pigmented and thicker with increased branching (Fig. 9).

Maximum increase in area of the sporophytic crusts occurred at 20 °C (Fig. 12). The increase in area of these crusts was about 1.5 times that of crusts incubated at 15 °C and this difference (ANOVA; SNK) was significant ($p < 0.05$). There was little growth of the crusts from 0 °C to 10 °C and at 24 °C (Fig. 12).

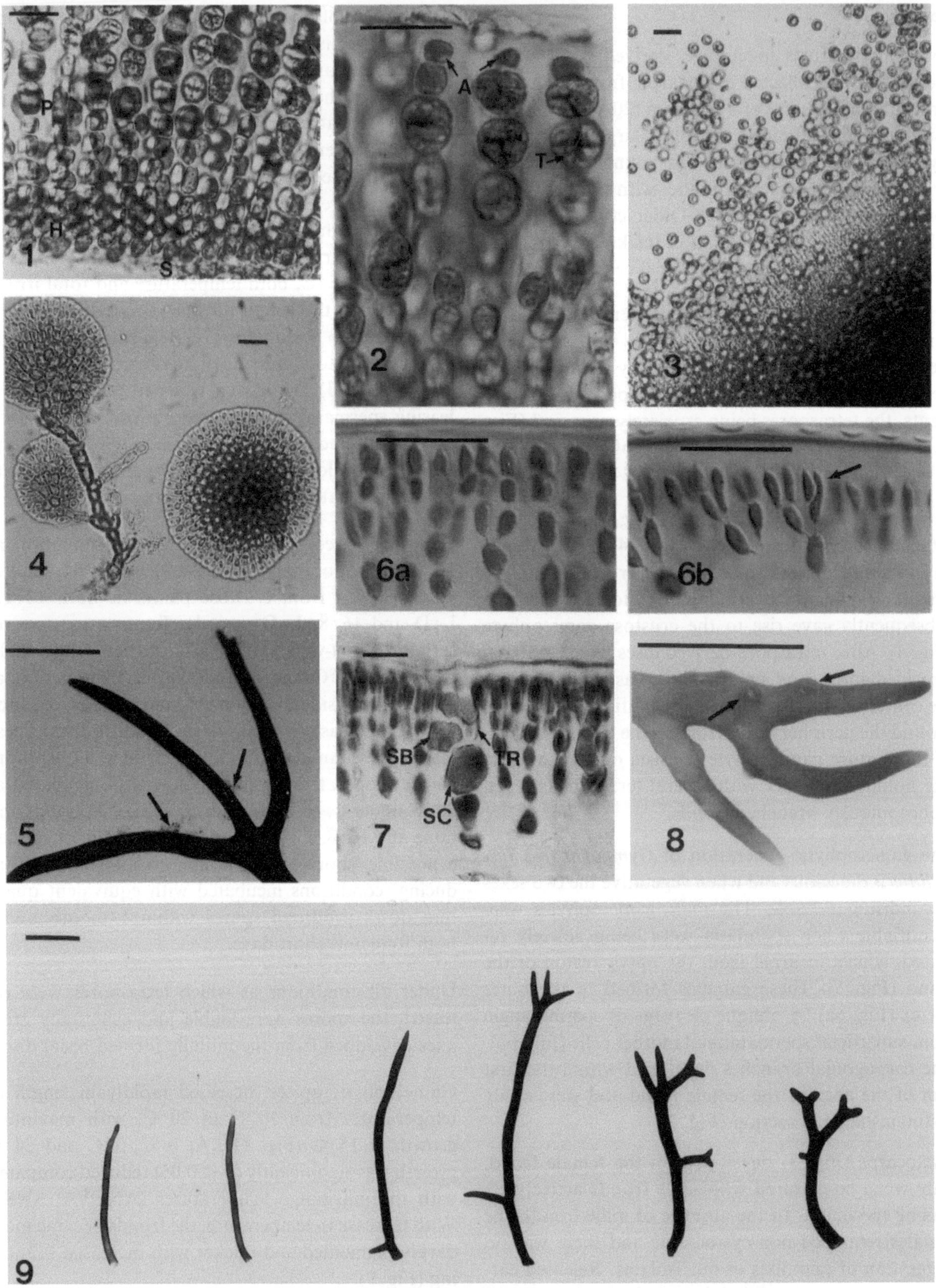

Figs 1–9.
Fig. 1. Vertical section of a tetrasporangial crust; the erect perithalloid filaments (P), the multi-layered hypothallus (H), and subhypothallus (S) are characteristic of older plants. Fig. 2. Section of a tetrasporic crust; chains of tetrasporangia (T), with cruciately arranged tetraspores, and apical sterile cells (A). Fig. 3. Surface view of a tetrasporangial crust; release en masse of tetraspores. Fig. 4. Tetraspore-derived sporelings that have developed either directly as crusts or first as uniseriate filaments, subsequently forming crustose plants. Fig. 5. Apical portion of a male gametophytic frond; release of spermatia (arrows). Fig. 6.

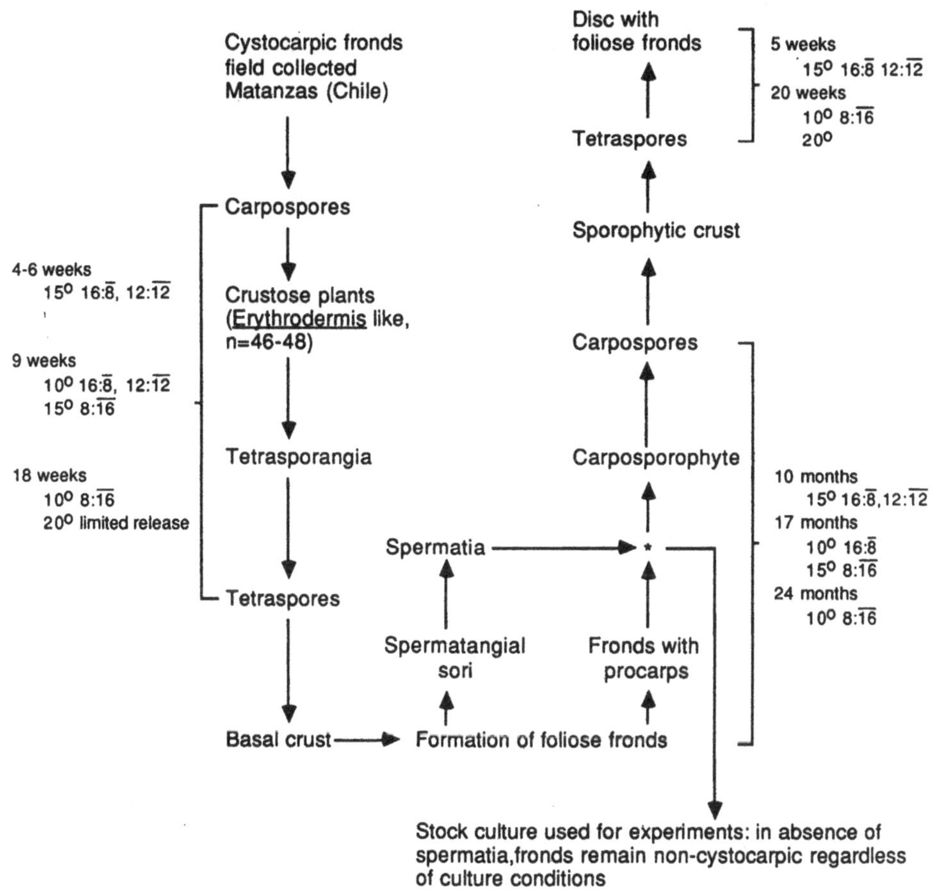

Fig. 10. Diagrammatic summary of the life history of *G. furcellatus* in culture.

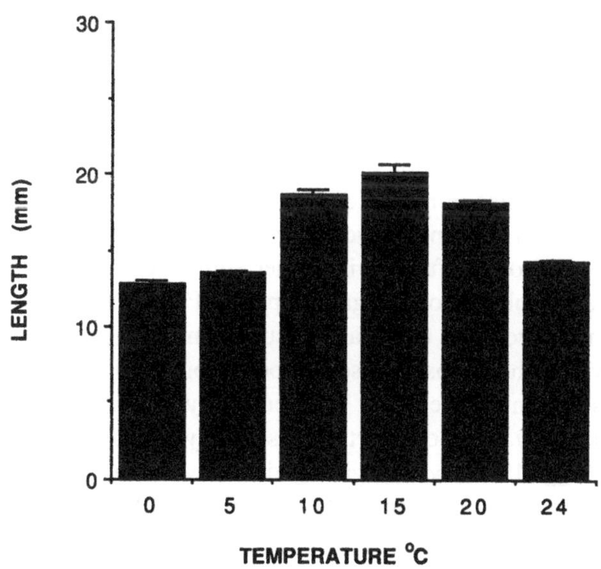

Fig. 11. Mean increase in length of gametophytic apices after 4-weeks incubation at temperatures indicated. Bars = SE.

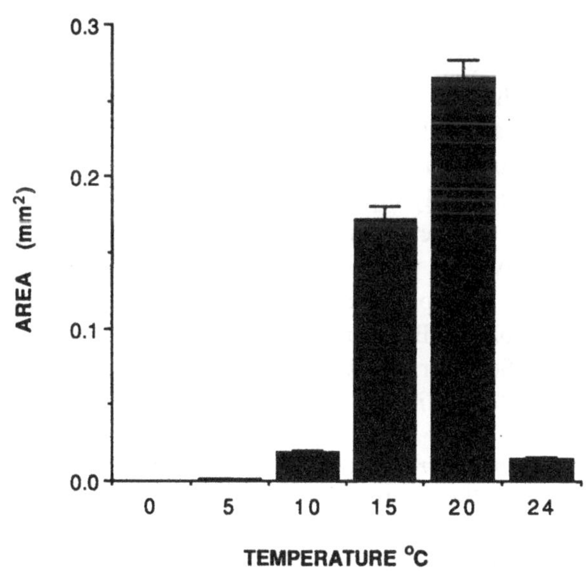

Fig. 12.. Mean increase in area of sporophytic crusts after 4-weeks incubation at temperatures indicated. Bars = SE.

Transverse section of a male gametophytic frond; (a) cortical cells, (b) spermatangia budding from spermatangial mother cells (arrow). Fig. 7. Transverse section of a female gametophytic frond; 4-celled carpogonial branch, showing supporting cell (SC), trichogyne (TR) and sterile branch (SB). The carpogonium is not in the focal plane. Fig. 8. Apical portion of a female gametophytic frond with mature cystocarps (arrows). Fig. 9. Apical segments of gametophytic fronds grown for 4 weeks at temperatures of (from left): 0 °C, 5 °C, 10 °C, 15 °C, 20 °C, 24 °C and 16 : $\overline{8}$ h photoperiod. Scale bars for Figs 1–6 and Fig. 7 = 20 μm; Scale bars for Figs 5, 8 & 9 = 5 mm.

At 0 °C, the plants died at the 3- to 4-cell stage, and at 24 °C, although the crusts remained viable, they became morphologically abnormal and badly contaminated with bacteria.

Discussion

More than 40 species of *Gymnogongrus* have been recognized world wide (P. A. Camus, pers. comm.), whereas only a few have been studied in culture (DeCew and West 1981, Maggs 1988). Consequently, apart from noting complexities of life histories in this genus, few generalizations are possible. The basic life history in the genus *Gymnogongrus* may be heteromorphic, where a foliose, dioecious gametophyte alternates with a crustose sporophyte having an *Erythrodermis*-like morphology and reproduction. The limited cytological evidence also suggests that the gametophyte is haploid and the sporophyte diploid (Maggs 1988).

Known life-histories in species of *Gymnogongrus* include intraspecific variations in which both a heteromorphic and apomictic life history have been reported, albeit from different populations (Ardré 1978, DeCew and West 1981). In the direct type, or apomixis, spermatangia are unknown, and carpospores from the foliose gametophyte give rise directly to the gametophytic phase [*e.g. G. devoniensis* (Greville) Schotter and *G. leptophyllus* J. Agardh]. In these instances, the carposporophyte may develop either from the auxiliary cell or from a medullary cell. This life-history variation is reported to cycle at a single ploidy level, presumably diploid (Maggs 1988). In another variation (Cordeiro-Marino and Candia 1981), only a single type of plant is present, which bears nemathecia of tetrasporangia referred to as the carpotetrasporophyte (Bold and Wynne 1985) or tetrasporoblast (Schotter 1968), and occasionally spermatangia [*e.g. G. griffithsiae* (Turner) Martius].

For *Gymnogongrus furcellatus*, we have demonstrated unequivocally that the isolate of this species from the population at Matanzas in central Chile has a heteromorphic life history. The gametophytes are dioecious and the sporophyte a crustose, *Erythrodermis*-like plant with intercalary tetrasporangia borne in single chains in the perithallial filaments. Although they failed to complete the life history of this species *in vitro*, Candia and Kim (1977) obtained foliose fronds from tetraspores of a crustose phase started from carpospores of *G. furcellatus*. These isolations were initiated from plants of a population from Concepción Bay, considerably farther to the south than the one we sampled. Consequently, there is evidence for a heteromorphic life history for this species along the Chilean coast.

The female gametophyte of *G. furcellatus* in culture formed the characteristic 4-celled procarp. The carposporophyte failed to develop in the absence of male fronds releasing spermatia, thus indicating that apomixis does not occur in the life history. Chromosome counts of 46–48 (Maggs 1988) for the sporophytic crust have been accepted by Maggs as circumstantial evidence that this phase is diploid. This number is considerably larger than those reported earlier for both *G. chiton* (Howe) Silva *et* DeCew (= *G. platyphyllus* Gardner) and *G. linearis* (Turner) J. Agardh (Magne 1964). In addition, the respective volumes of the tetra- and carpospores suggest that these spores contain different amounts of DNA related to their respective ploidy levels (*cf.* Goff and Coleman 1987).

In populations with a heteromorphic life history, both female and male gametophytes are usually present. However, for *G. furcellatus* we are unaware of records of male plants or fronds, which are difficult to recognize in the field and may be misinterpreted as being vegetative (*cf.* Santelices *et al.* 1989). Fertile male fronds developed readily in culture, under several conditions, and there is no apparent reason why they should not be present in field populations.

The sporophytic crust of *G. furcellatus* either has not been collected or reported in the Chilean flora (*cf.* Etcheverry 1986), and the occurrence of this life-history phase in natural populations remains to be confirmed. It is not surprising, perhaps, that the sporophyte has gone undetected as these crusts tend to be small and inconspicuous and can be easily overlooked or confused with other similar crusts or the basal holdfast of the gametophyte. Failure to record the sporophyte may have occurred in other instances where the foliose phase of species of *Gymnogongrus* is present in the flora. The genus *Erythrodermis*, presumably the sporophytic phase of *Gymnogongrus* (*cf.* Maggs 1988), was originally described from the flora of southern England (Batters 1900), although even there the plants have been reported as rare and sublittoral (Dixon and Irvine 1977). Thus, in addition to being inconspicuous, the sporophytic phase of *G. furcellatus* may not necessarily occur among the foliose plants but at another tidal level. However, sporophytic crusts of *Gymnogongrus*, when noted, have usually been present near the base of the gametophytic plants (Maggs 1988, DeCew and West 1981).

For some species of red algae, fertility depends on external stimuli such as photoperiod, temperature or both (*e.g.* Maggs and Guiry 1987). In other instances, plants have become fertile under various temperature and photo-regimes (*e.g.* Chen and McLachlan 1972), providing that conditions were favourable for growth. In *G. furcellatus* the formation of spermatangia and tetrasporangia, development of the carposporophyte, and production of viable tetra- and carpospores occurred over a range of temperatures and photoperiods. The temperatures conducive to fertility in our cultures occur within the distributional range (Etcheverry 1986, Santelices 1980, Dawson *et al.* 1964) of *G. furcellatus* along much of Pacific South America. Gametophytes, however, failed to become reproductive at 20 °C, so the northern distribution of this species may be limited by its reproductive thermal tolerances. The rates of cystocarp maturation and growth of gametophytic fronds were maximal between 10 °C and 15 °C and were similarly affected by photoperiod, suggesting that reproduction and growth in nature are closely coupled. Thus, considerable seasonal variation in development of cystocarps would be anticipated. This has been confirmed by the seasonal observations of Santelices *et al.* (1989). Masuda *et al.* (1979) and DeCew and West (1981), on the other hand, failed to obtain gametophytic reproduction in *Gymnogongrus flabelliformis* Harvey and *G. linearis* under a variety of temperatures and photoperiods. However, in a later report, Masuda (1981) completed the life history of the former species at 20 °C and 8:$\overline{16}$, L:D. A strict response to temperature or daylength or both for gametophytic fertility, apart from possible effects on growth, remains to be reported for species of *Gymnogongrus*.

Fertility responses of the sporophytic phase of *G. furcellatus* to temperature and photoperiod were similar to those of the gametophytic phase, except for reduced fertility occurring at 20 °C. As with gametophytes, a seasonal response in fertility of the sporophytic phase can be predicted, but must await field observations for confirmation. For *G. leptophyllus*, tetrasporogenesis failed to occur under any of the light or temperature conditions tested (DeCew and West 1981). Rather, abrasion of the sporophytic crusts induced sporogenesis, pointing to this factor as a strong ecological determinant of tetrasporogenesis in *G. leptophyllus*. Induction of reproduction in species of *Gymnogongrus* is likely a complex process, requiring further studies to clarify possible mechanisms involved.

Maximum growth of vegetative, gametophytic apices of *G. furcellatus* occurred within the normal ecological temperatures pertaining along the Chilean coast (10 °C – 15 °C). This is in agreement with Santelices *et al.* (1989) who cultured field-collected gametophytes of this alga. Our gametophytic apices also grew well at 20 °C but, as noted, failed to become reproductive at this temperature. Maximum increase in length of these apices (about 0.4 mm d^{-1}) was considerably faster than that observed for other species both in the field (Markham and Newroth 1972) and culture (DeCew and West 1981). Within the region of Chile where *G. furcellatus* occurs in abundance, it seems that factors other than temperature must markedly influence seasonal rates of growth of the gametophytes. This is likely light, a conclusion reached by Santelices *et al.* (1989) who reported a strong response in growth resulting from changes of quantum dosage and photoperiod. In their field observations, these authors noted, moreover, that growth was seasonal, maximal during spring and summer as has also been reported for *G. linearis* (Markham and Newroth 1972).

It is noteworthy that 0 °C was lethal to the sporophyte of *G. furcellatus*, whereas the gametophytic apices remained viable at this temperature. Growth of the sporophytic crusts of the alga, unlike the gametophytic apices (also Santelices *et al.* 1989), showed a direct response to temperature between 10 °C and 20 °C. Growth was considerably reduced at 10 °C and maximal at 20 °C, although, as in the gametophytes, reproduction failed to occur, or was very limited, at the higher temperature. At both 10 °C and 15 °C, the sporophytic crusts rapidly became fertile, indicating that for this life-history phase growth and reproduction may not be directly coupled. The crusts could be relatively short-lived under field conditions. As noted elsewhere (Santelices *et al.* 1989), fleshy crusts seem to be particularly susceptible to grazing, and probably to sand abrasion as well, and this may apply equally to the sporophytic phase of *G. furcellatus*. Furthermore, we have noted that after a year or so in culture, the sporophytic crusts become moribund, whereas the gametophytic phase seems to have an extended longevity (*i.e.* perennial). Rapid growth and reproduction would, consequently, tend to increase the probability of completion of the life history. In our experiments, a minimum of one year was necessary to complete the life history in culture. Under fluctuating environmental conditions possibly even longer periods are required as suggested by the experiments of Santelices *et al.* (1989). Nevertheless, more ecological information is needed on the sporophytic phase of the life history for all species of *Gymnogongrus*. It is interesting to note that *G. furcellatus* is one of the few species of red algae having a heteromorphic life history that are exploited commercially from field populations.

Acknowledgements

Comments and suggestions on drafts of this manuscript by C. A. Maggs, P. A. Camus, B. Santelices and C. J. Bird are gratefully acknowledged. M. Avila was supported by a travel grant from the International Development Research Centre (IDRC). This study was in part supported by a grant from the Natural Science and Engineering Research Council to JMcL.

References

Ardré, F. 1978. Sur les cycles morphologiques du *Gymnogongrus crenulatus* (Turn.) J. Ag. et du *Gymnogongrus devoniensis* (Grev.) Schott. (Gigartinales, Phyllophoracées) en culture. *Revue Algol.* n. s. *13*: 151–176.

Batters, E. A. L. 1900. New or critical British marine algae. *J. Bot.* (Lond.) *38*: 369–379.

Bold, H. C. and M. J. Wynne. 1985. *Introduction to the Algae*. Prentice-Hall Inc. Englewood Cliffs, NJ. 720 pp.

Candia, A. I. and D. H. Kim. 1977. Resultados preliminares de los estudios de ciclo de vida de *Gymnogongrus furcellatus* (C. Agardh) J. Agardh (Phyllophoraceae, Gigartinales). *Gayana*, Misc. Prim. *J. Nac. Acuicult.*, No. *5*: 77–78.

Chen, L. C. M. and J. McLachlan. 1972. The life history of *Chondrus crispus* in culture. *Can. J. Bot. 50*: 1055–1060.

Cordeiro-Marino, M. and A. Candia P. 1981. Life history of *Gymnogongrus griffithsiae* (Turner) Martius (Phyllophoraceae, Gigartinales). *Proc. Int. Seaweed Symp. 10*: 155–161.

Dawson, E. Y., C. Acleto and N. Foldvik. 1964. The seaweeds of Peru. Beih. *Nova Hedwigia 13*: 1–111.

DeCew, T. C. and J. A. West. 1981. Life histories in the Phyllophoraceae (Rhodophyta, Gigartinales) from the Pacific coast of North America. 1. *Gymnogongrus linearis* and *G. leptophyllus*. *J. Phycol. 17*: 240–250.

Dixon, P. S. and L. M. Irvine. 1977. *Seaweeds of the British Isles*. Vol. 1. *Rhodophyta*, Part 1. *Introduction, Nemaliales, Gigartinales*. British Museum (Natural History) London. 251 pp.

Etcheverry, H. D. 1986. *Algas Marinas bentónicas de Chile*. UNESCO Montevideo. 379 pp.

Goff, L. J. and A. W. Coleman. 1987. The solution to the cytological paradox of isomorphy. *J. Cell Biol. 104*: 739–748.

McCandless, E. L., J. A. West and M. D. Guiry. 1983. Carrageenan patterns in the Gigartinaceae. *Biochem. System. Ecol. 11*: 175–182.

McLachlan, J. 1973. Growth media-marine. *In*: (J. Stein, ed.) *Handbook of Phycological Methods*. Cambridge Univ. Press, Cambridge. pp. 25–51.

Maggs, C. A. 1988. Intraspecific life history variability in the Florideophycidae (Rhodophyta). *Bot. Mar. 31*: 465–490.

Maggs, C. A. and M. D. Guiry. 1987. Environmental control of macroalgal phenology. *In*: (R. M. M. Crawford, ed.) *Plant Life in Aquatic and Amphibious Habitats*. Blackwell Scientific Publ., Oxford. pp. 359–374.

Maggs, C. A. and C. M. Pueschel. 1989. Morphology and development of *Ahnfeltia plicata* (Rhodophyta): Proposal of Ahnfeltiales ord. nov. *J. Phycol. 25*: 333–351.

Magne, F. 1964. Recherches caryologiques chez les Floridées (Rhodophycées). *Cah. Biol. Mar. 5*: 461–671.

Markham, J. W. and P. R. Newroth. 1972. Observations on the ecology of *Gymnogongrus linearis* and related species. *Proc. Int. Seaweed Symp. 7*: 127–130.

Masuda, M. 1981. Further observations on the life history of *Gymnogongrus flabelliformis* Harvey (Rhodophyta) in culture. *J. Fac. Sci. Hokkaido Univ.* Ser. V, Bot. *12* (3): 159–164.

Masuda, M., T. C. DeCew and J. A. West. 1979. The tetrasporophyte of *Gymnogongrus flabelliformis* Harvey (Gigartinales, Phyllophoraceae). *Jpn. J. Phycol. (Sorui) 27*: 63–73.

Santelices, B. 1980. Phytogeographic characterization of the temperate coast of Pacific South America. *Phycologia 19*: 1–12.

Santelices, B., P. A. Camus and A. J. Hoffmann 1989. Ecological studies for managing and culturing *Gymnogongrus furcellatus* (Rhodophyta, Gigartinales) in central Chile. *J. Appl. Phycol. 1*: 171–181.

Schotter, G. 1968. Recherches sur les Phyllophoracées. *Bull. Inst. Océanogr. Monaco 67*: 1–99.

Sokal, R. R. and F. J. Rohlf. 1969. *Biometry*. W. H. Freeman & Co., San Francisco. 776 pp.

Botanica Marina
Vol. 34, pp. 153–160, 1991

Antithamnion algeriensis nov. sp. (Ceramiaceae, Rhodophyta) from Algeria (Mediterranean Sea)

M. Verlaque and H. Seridi*

Laboratoire de Biologie marine et d'Ecologie du Benthos, Faculté des Sciences de Luminy, 13009 Marseille cedex 9, France

* *Laboratoire de Biologie marine, Faculté des Sciences et techniques Houari Boumedienne, Algiers, Algeria*

(Accepted 30 September 1990)

Abstract

A new species of *Antithamnion* from Algeria, is described: *A. algeriensis*. It differs from the closely related species of *Antithamnion* (distichous whorl-branchlets, opposite and abaxial branchlets and gland cells) by the branching pattern of the axis, the arrangement of whorl-branchlets as well as the frequency, and the location and shape of the gland cells.

Introduction

Cormaci and Furnari (1987) recorded six members of the genus *Antithamnion* Nägeli (1847) *sensu* Wollaston (1968) in the Mediterranean Sea:

- *A. antillanum* Boergesen,
- *A. cruciatum* (C. Agardh) Nägeli,
- *A. heterocladum* Funk,
- *A. ogdeniae* Abbott,
- *A. piliferum* Cormaci *et* Furnari,
- *A. tenuissimum* (Hauck) Schiffner.

So far, the occurrence of only the last five species is established without any doubt. All Mediterranean citations of *A. antillanum* (Schiffner 1931, Boudouresque and Verlaque 1976, Barcelo 1987 and Boisset 1986, 1987) seem to correspond to *A. ogdeniae* (Verlaque 1990). Recently, a sixth taxon, *A. nipponicum* Yamada *et* Inagaki, of Japanese origin, has been introduced accidentally into the Thau lagoon (France) (Verlaque and Riouall 1989). This species belongs to an *Antithamnion* group possessing gland cells and distichous whorl-branchlets bearing opposite distichous proximal branchlets and abaxial distal branchlets. Members of this group were previously unknown in the Mediterranean and in the Atlantic. Research undertaken along the North-African coast has resulted in the discovery of a second member of this group. In qualitative and quantitative characteristics, this *Antithamnion* differs from all other taxa and is thus described as a new species: *Antithamnion algeriensis* nov. sp.

Material and Methods

The alga, isolated from samples of superficial communities from the coast near Algiers and fixed in 4% formalin in sea water, has been studied under a light microscope and drawn with a camera lucida.

Description

This species is epiphytic on a variety of algal hosts, is about 5–7 mm high, without cortication, pink in colour and has opposite and distichous determinate lateral branches (= whorl-branchlets) (Figs 1–6, 9). In the basal part of the thallus, pseudodichotomous branching of main axis occurs without loss of whorl-branchlets (= axial branching, Wollaston 1968) (Fig. 2). Each pair of whorl-branchlets is set at approximately 10° from the next pair, so that it lies slightly above or below the adjacent whorl-branchlet. As in

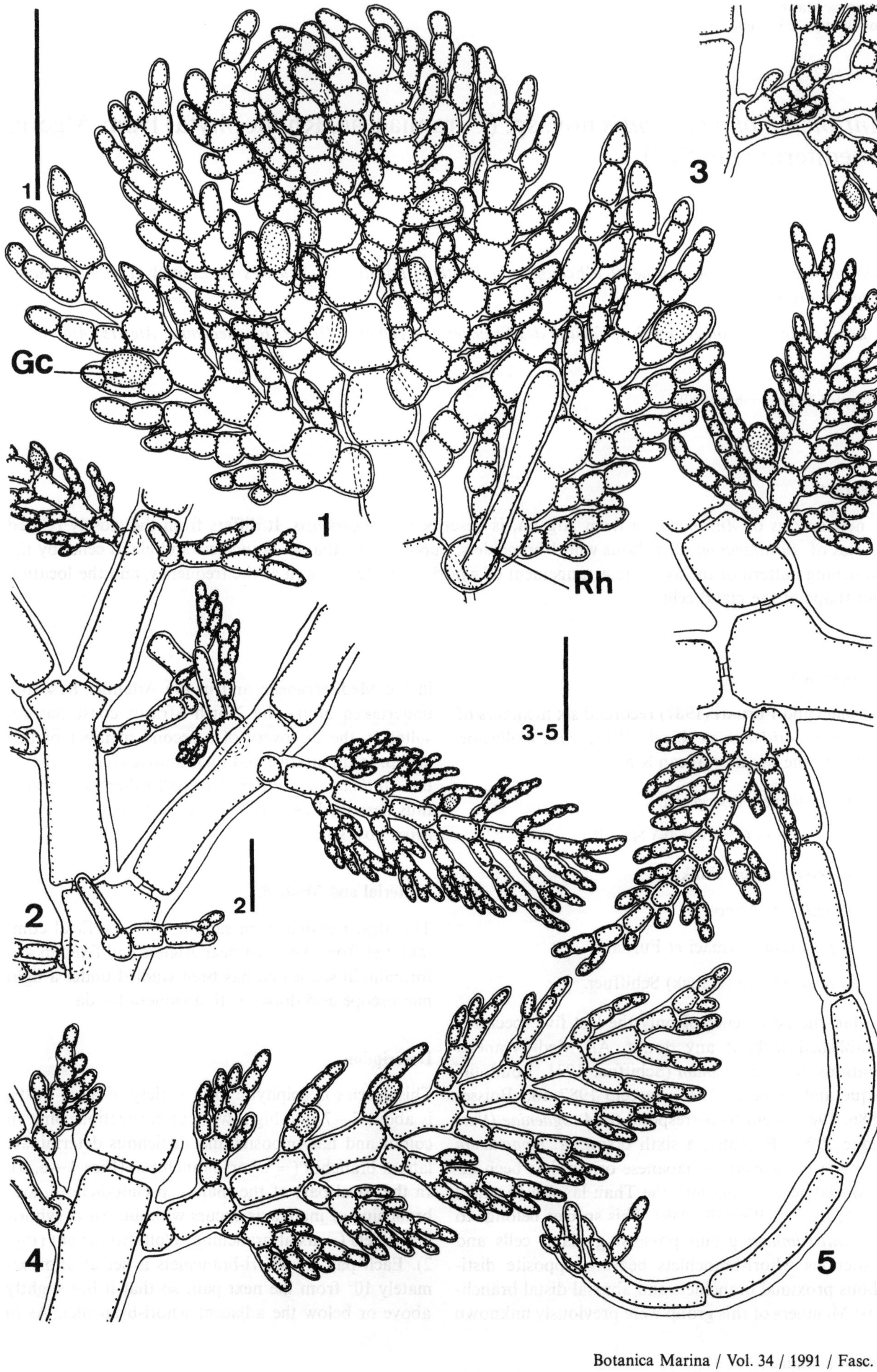
1
Gc
1
Rh
3
2
2
3-5
4
5

some other species of *Antithamnion* (L'Hardy-Halos 1970), the whorl-branchlets are produced on two zig-zag and inverse lines (= "phyllotaxie scorpiomère", Chadefaud 1954). Axial cells are 41 – 61 μm in diameter, and 2.6 – 4.2 times as long as broad. Whorl-branchlets, up to 320 μm long, possess a basal cell (= "cellule coxale", Chadefaud 1954) that has produced several orders of branching: primary (= rachis), secondary (= branchlets) and, for the most developed, tertiary (= pinnules) (Figs 1 – 2, 4 – 9). The rachis consist of 5 – 12 cells, 13 – 23 μm in diameter and 1.3 – 2.0 times as long as broad. Basal cells without indeterminate lateral branches are small and isodiametric: 14 – 27 μm in diameter. In the proximal part, whorl-branchlets bear opposite distichous, simple or pectinate or pinnate branchlets, but only abaxial branchlets are present near the distal region. Terminal cells are blunt. Apical cells of whorl-branchlets are 5.5 – 9 μm in diameter and 7.5 – 15 μm in length. Simple multicellular rhizoids, 11 – 26 μm in diameter, arise alone or in pairs from the basal cell of some whorl-branchlets and terminate in an irregular multicellular attachment process (Fig. 5). Elliptical gland-cells, 9 – 15 × 13 – 20 μm length, are abundant and occur on special short branchlets and pinnules, generally touching 2 basal cells and sometimes slightly touching a third cell (Figs 1 – 9). Indeterminate lateral branches are formed by the primary and secondary growth systems *sensu* Norris (1987). In the former, the branch (= "cladome-fils épi-axial", L'Hardy-Halos 1970) is produced by an axial cell without any whorl-branchlets, while a pair occurs on its basal cell (Figs 6 – 7). The phyllotactic order of whorl-branchlets is not disturbed and the mother axial cell is an "intertagme monomère" *sensu* Chadefaud 1967. Less frequently, a secondary indeterminate branch (= "cladome-fils épi-coxal", L'Hardy-Halos 1970) is formed by the basal cell of a whorl-branchlet (Figs 3, 8 – 9). After its development, its mother basal cell looks like an axial cell bearing one whorl-branchlet without a basal cell (Fig. 9). These features readily allow discrimination between these two types of branching. The intervals between two consecutive indeterminate branches are irregular (2 – 17 axial cells, usually 10 – 15). In the distal cells of whorl-branchlets, chloroplasts are angular or irregularly band-shaped, 4 – 7 × 3 – 4 μm (Fig. 10A), while they are elongated and band-shaped, 16 – 30 × 2 – 5 μm, in mature axial cells (Fig. 10B). Reproductive stages have not been observed.

Holotype and isotype collections: 0.5 m depth at Bordj-El-Kiffan (Algiers beach), Algeria on 30 April 1989 by H. Seridi.

Holotype F. 1306 (sterile) deposited in the phycological Herbarium of Muséum National d'Histoire Naturelle, Paris, France.

Isotypes F. 1307 (sterile) deposited in the phycological Herbarium of LBMEB, Faculté des Sciences de Luminy, Marseille, France.

Diagnosis: ***Antithamnion algeriensis*** sp. nov. (Figs 1 – 10)

Thallus minutus, 5 – 7 mm altus, epiphyticus, ecorticatus, roseolus, sparse ramosus et opposite distiche pinnatus. Axes primarii per multicellularia, uniseriata, rhizoidea affixi, 11 – 26 μm diam.; 1 – 2 rhizoidea a cellulis basalibus ramulorum verticillatorum emittentibus et discum irregulariter multicellularem desinens. Axium ramificatio pseudodichotoma prope basim, cum ramulis verticillatis permanentibus. Cellulae axiales usque ad 41 – 61 μm diam. et 2.6 – 4.2-plo longiores. Ramuli verticillati pinnate ramosi usque ad 320 μm. long. e circa 5 – 12 cellulis axialibus, 13 – 23 μm diam., 1.3 – 2-plo longiores, constituti. Ramulorum verticillatorum cellulae basales, sine indeterminato secondale ramo, isodiametrae et usque ad 14 27 μm diam. Ramuli oppositi et distichi, simplices aut pectinati aut pinnati in parte proximale ramulorum verticillatorum, sed ramuli secundi et abaxiales in parte distale. Cellulae terminales obtusae. Apices ramulorum verticillatorum, 5.5 – 9 μm diam. atque 7.5 – 15 μm long. Glandicellulae ellipsoideae, 9 – 15 diam. et usque ad 13 – 20 μm long., numerosae in ramulis propriis et duas cellulas basales firme aut duas et tertiam cellulam leviter contingentes. Rami indeterminati in incrementi primariis et secundariis systemibus producti. Rami indeterminati primarii pro ramulis verticillatis substituunt. Ramuli verticillati a rami indeterminati cellula basale emittenti. Rami indeterminati e cellula basale ramulorum verticillatorum secundale formati. Chloroplasti in ramulorum verticillatorum cellulis distalibus minuti, angulares usque ad 4 – 7 × 3 – 4 μm. In cellulis axialibus, chloroplasti aut fusiformes aut anguste vittaeformes usque ad 16 – 30 × 2 – 5 μm. Status reproductivi non observati.

Specimen typicum in loco Bjord-El-Kiffan (Alger-Plage) Algerie dicto (d. 30, m IV, 1989), ex acqua 0 – 0.5 metra profunda a H. Seridi collectum.

Fig. 1. – 5. *Antithamnion algeriensis.*
Fig. 1. Apex. Abbreviations. Gc: gland cell; Rh: rhizoid. Fig. 2. Pseudodichotomous branching (= axial branching) in basal part of thallus. Fig. 3. Young secondary indeterminate lateral branch. Fig. 4. Whorl-branchlet. Fig. 5. Part of thallus with a developed rhizoid. Scale lengths = 50 μm.

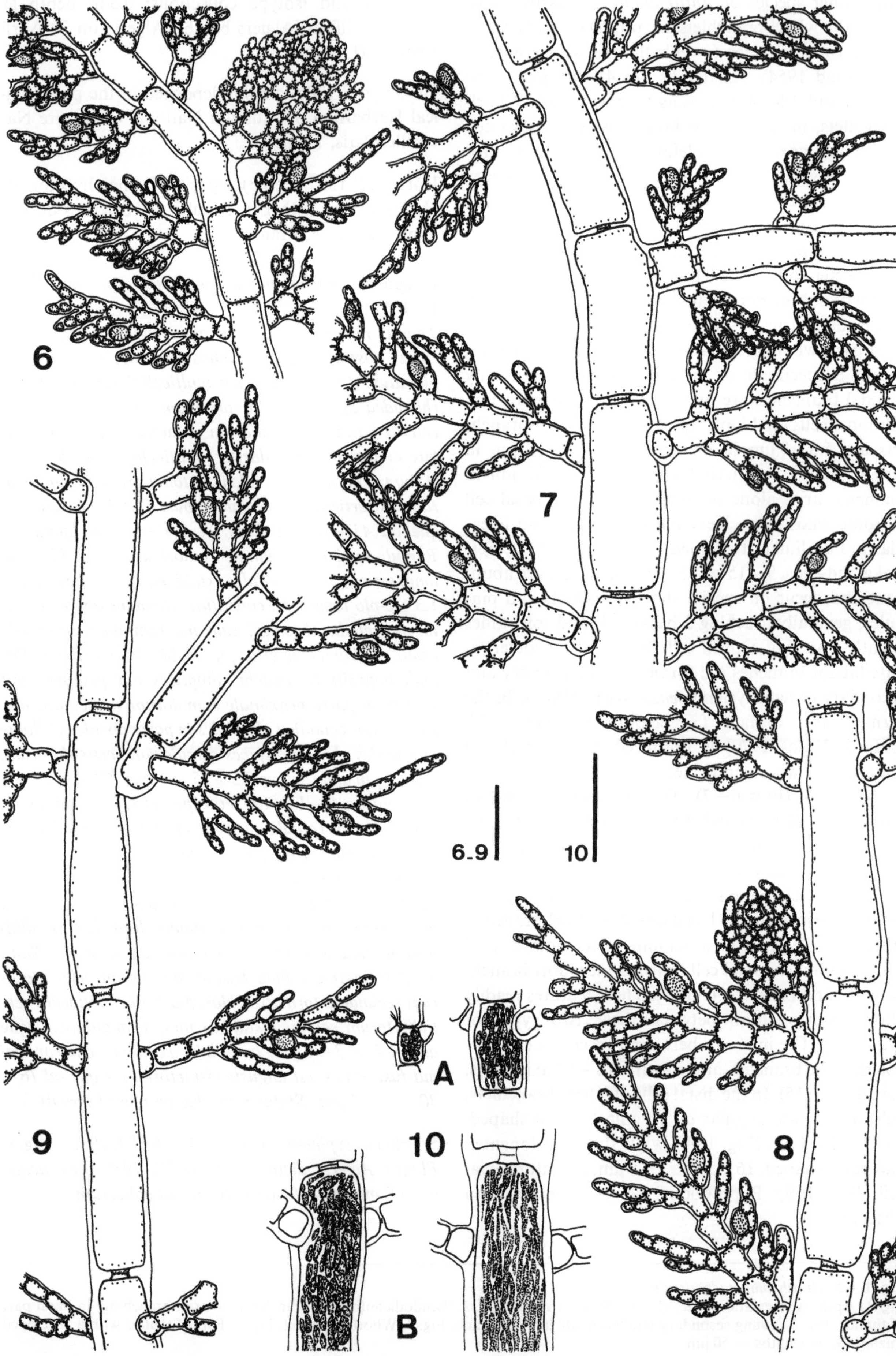
6
7
6-9
10
9
A
10
B
8

Systematic Position

Despite the lack of reproductive organs, the Algerian species must be related to the genus *Antithamnion* Nägeli (1847) *sensu* Wollaston (1968) because of the following combination of features:

- non-corticated thallus possessing erect branch tips and erect branches emerging from the prostrate system, both systems being similar in structure in that they produce two whorl-branchlets opposite one another on each axial cell,
- isodiametric basal cells of whorl-branchlets lacking any determinate branchlets,
- gland cells touching more than one cell in the special branchlet.

Hitherto, fifty species belonging to the genus *Antithamnion* have been described (Cormaci and Furnari 1989). Among them, the taxa with distichous whorl-branchlets bearing opposite distichous and abaxial secund branchlets are very few. According to the studies of Yamada and Inagaki (1935), Tokida and Inaba (1950), Dawson (1962), Wollaston (1968, 1971, 1976, 1984), Itono (1969, 1971, 1977), Kajimura (1987) and Norris (1987), only nine such algae are recorded, all of them originating from the Indo-Pacific area:

- *A. adenocladellum* Norris,
- *A. calloclados* Itono,
- *A. cristirhizophorum* Tokida *et* Inaba,
- *A. hubbsii* Dawson,
- "*A. hubbsii* Dawson" *sensu* Norris (1987) which differs from the original diagnosis by the lack of gland cells, the occurrence of pinnate proximal branchlets on whorl-branchlets and by its smaller measurements.
- *A. nipponicum* Yamada *et* Inagaki,
- *A. okiense* Kajimura,
- *A. pinnafolium* Wollaston,
- *A. pterocladellum* Norris (whorl-branchlets with distal branchlets sometimes alternate).

In Table I, the characteristics of these taxa are listed. *Antithamnion algeriensis* corresponds to none of these and must be considered a new species. This *Antithamnion* differs from all others in having the following features:

- distichous whorl-branchlets with blunt branchlet tips,
- all cells of rachis longer than broad,
- proximal branchlets of the whorl-branchlets with pinnules opposite and abaxial,
- elliptical gland cells abundant and touching 2 (3) cells.

Moreover, the primary indeterminate branching without any opposite whorl-branchlet, observed in *A. algeriensis* and in some species of *Antithamnion* with decussate whorl-branchlets or with alternate or secund branchlets [*A. cruciatum* (C. Agardh) Nägeli, *A. defectum* Kylin, *A. diminuatum* Wollaston, *A. eliseae* Norris, *A. tenuissimum* (Hauck) Schiffner] seems unusual in this group of species.

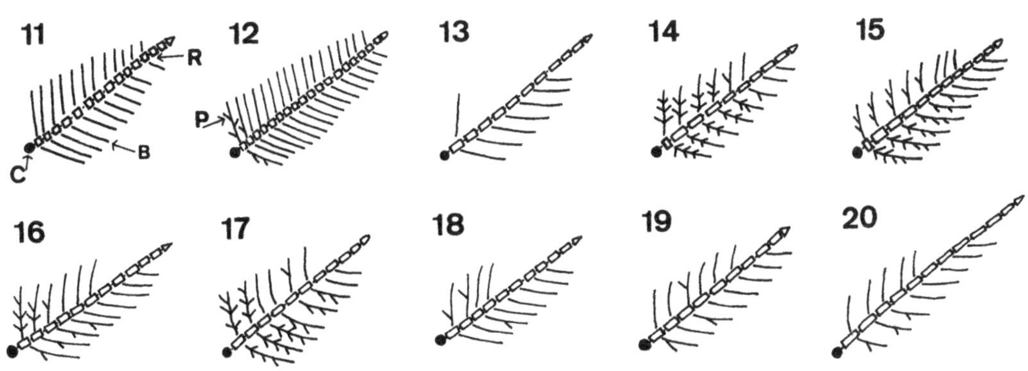

Figs 11–20. Diagrams of whorl-branchlets of the ten closely related species of *Antithamnion*; the gland cells are not figured in the diagrams. C: basal cell; R: rachis; B: branchlets; P: pinnules.
Fig. 11. *A. pterocladellum* Norris; Fig. 12. *A. pinnafolium* Wollaston; Fig. 13. *A. adenocladellum* Norris; Fig. 14. *A. calloclados* Itono; Fig. 15. *A. okiense* Kajimura; Fig. 16. "*A. hubbsii*" *sensu* Norris 1987; Fig. 17. *A. algeriensis* Verlaque *et* Seridi; Fig. 18. *A. cristirhizophorum* Tokida *et* Inaba; Fig. 19. *A. nipponicum* Yamada *et* Inagaki; Fig. 20. *A. hubbsii* Dawson.

Figs 5–10. *Antithamnion algeriensis*.
Fig. 6. Young primary indeterminate lateral branch. Fig. 7. Developed primary lateral branch. Fig. 8. Young secondary indeterminate lateral branch. Fig. 9. Developed secondary lateral branch. Fig. 10. Chloroplasts: A. Whorl-branchlet cells; B. Axial cells. Scale lengths = 50 µm.

Table I. Characteristics of the ten species of *Antithamnion* possessing distichous whorl-branchlets with opposite branchlets (in accordance with original diagnoses and figures).

Species and references	*A. pterocladellum* Norris 1987	*A. pinnafolium* Wollaston 1968	*A. adenocladellum* Norris 1987	*A. calloclados* Itono 1971, 1977	*A. okiense* Kajimura 1987
Localities	South-East Africa, Natal	South Australia	South-East Africa, Natal	South Japan Osumi	Sea of Japan Oki Island
Height, cm	unspecified	1.5	0.05*	1.0–2.0	6.0
Axis					
Ø, µm	20–40	64	70–90	60–90	60
Length/Ø	4–6	about 3	1–1.4	2.5–3.7	about 3
Whorl-branch.	Fig. 11	Fig. 12	Fig. 13	Fig. 14	Fig. 15
Length, µm	350	760–1000	unspecified	380–550	360
Arrangement	decussate becoming distichous	distichous	distichous	distichous	distichous
Basal cells, Ø, µm	12–18	25–50	12–30	45	(22) 30 (40)
Rachis,					
Number of cells	15–17	15–20	12–15	10–16	12
Length/Ø	about 1	about 1	>1	>1 except the first cell	>1 except the first cell
Terminal cells	acute	unspecified	acute	sub-acute	acute to blunt
Adaxial branchlets	well developed	well developed	rare, located on few proximal cells	like abaxial branchlets	less developed than abaxial branchlets
Pinnules (2nd order branchlets)	lacking	alternate and abaxial	lacking	opposite and abaxial	opposite and abaxial
Gland cells	abundant	abundant	abundant	very rare	abundant
Shape	elliptical	elliptical	elliptical	oblong	ellipsoid to mushroom-shaped
Length, µm	15–18	about 24	20–33	unspecified	20–40
Touching	2 (3) cells	3–4 cells	3 cells	3 cells	3–4 cells
Rhizoids	simple	branched	unspecified	simple	simple
Ø, µm	10–20	25–27	10	18–21	25–28
Primary ind. branches	replacing one whorl-branchlet	absent	replacing one whorl-branchlet	absent	absent
Secondary ind. branches	present	present	not observed	present	present

In the Mediterranean Sea, except for *A. nipponicum* which was recently discovered along the French coast, *A. algeriensis* has some tenuous similarities with *A. cruciatum* which sometimes may have pinnate whorl-branchlets but which differs in that its whorl-branchlets are always clearly decussate and never bipinnate.

There have been numerous studies of the marine vegetation of the Mediterranean Sea, in particular along the North-African coast, and it is surprising that so peculiar a species as *A. algeriensis* has not been detected previously. Its affinities with a strictly Indo-Pacific group of species and its discovery close to a great harbour (Algiers) suggest that it could be a foreign species recently introduced into the Mediterranean Sea, just like *A. nipponicum* in the Thau lagoon. If so, *A. algeriensis* could rapidly extend its distribution because newly introduced algae frequently assume an important vitality owing to the transit effect (selection of resistant strains) and the lack of usual predators and competitors (Farnham 1980). This discovery adds a seventh species of *Antithamnion* to the Mediterranean algal flora.

Acknowledgements

Sincere thanks are due to Dr Semroud for his help in exchanges between our two Laboratories, to Professors Boalch and Boudouresque for their critical reading of our manuscript, to Dr Crouzet for help with the Latin description and to Professor McKenna for reviewing the English text.

Table I. Continued

Species and references	"*A. hubbsii*" *sensu* Norris 1987	*A. algeriensis* sp. nov. F. 1306	*A. cristirhizo-phorum* Tokida et Inaba 1950	*A. nipponicum* Yamada et Inagaki 1935	*A. hubbsii* Dawson 1960 Wollaston 1976
Localities	South-East Africa, Natal	Mediterranean Sea, Algeria	Japan, South-East of Tokyo	North Japan Honshu Island	California
Height, cm	unspecified	0.5–0.7	0.5	0.6–1.5	1.5
Axis					
Ø, µm	33–58	41–61	60–150	75–90	50–78
Length/Ø	2–2.9	2.6–4.2	1.8–2.7	1.3–4.0	3.4–5
Whorl-branch.	Fig. 16	Fig. 17	Fig. 18	Fig. 19	Fig. 20
Length, µm	300	115–320	unspecified	335–443**	453–1000
Arrangement	distichous	distichous	distichous	distichous	distichous
Basal cells Ø, µm	15–20	14–27	64	18–30	30–46
Rachis,					
Number of cells	12–17	5–12	10–18	9–13	12–14
Length/Ø	>1	>1	>1	>1	>1
Terminal cells	acute	blunt	acute	acute	acute
Adaxial branch-lets	well developed	well developed	poorly developed	well developed	well developed
Pinnules (2nd order branchlets)	opposite and abaxial	opposite and abaxial	abaxial	abaxial	abaxial
Gland cells	not observed	abundant	abundant	abundant	abundant
Shape	–	elliptical	elliptical	elliptical	elliptical
Length, µm	–	13–18 (20)	22–40	12.5–18.5	23–31
Touching		2 (3) cells	3 (4) cells	2 cells	2 cells
Rhizoids	unspecified	simple	branched	simple	simple
Ø, µm	9–18	11–26	29–33	14–31	30–40
primary ind. branches	rare, replacing one whorl-branch.	cells without whorl-branchlets	absent	present unspecified	absent
Secondary ind. branches	present	present	present	present	present

*: probably a mistake between mm and cm;
**: up to 616 µm, in the specimens from the Thau lagoon (Verlaque and Riouall 1989);
Ø: diameter of cells (proximal cell for rhizoids);
Length/Ø: length/diameter ratios of cells;
Ind.: Indeterminate

References

Barcelo, M. C. 1987. Estudi de la flora bentonica marina del Pais Valencia. Thèse Pharmàcia, Barcelona, Esp., 485 pp.

Boisset, F. 1986. Contribucion al conocimiento algologico del litoral Levantino. I. *Acta Botanica Malacitana 11*: 3–8.

Boisset, F. 1987. Estudio del fitobentos esciafilo infralitoral de sustratos duros, en el litoral Valenciano (Espana), Flora y vegetation. Thèse Cien. biol., Univ. Valencia, Esp., 387 pp.

Boudouresque, C. F. and M. Verlaque. 1976. Sur quelques Rhodophycées intéressantes des côtes de Corse. *Soc. phycol. Fr. 21*: 56–64.

Chadefaud, M. 1954. Sur la morphologie de quelques Céramiacées. *Rev. algol. Fr. N. S. 1*: 71–87.

Chadefaud, M. 1967. Remarques sur la tagmatisation et la phyllotaxie des Floridées Rhodomélacées. *C. R. Acad. Sci. Fr. 264*: 2888–2890.

Cormaci, M. and G. Furnari. 1987. *Antithamnion piliferum* sp. nov. (Ceramiaceae, Rhodophyta) from eastern Sicily (Mediterranean Sea). *Cryptogamie, Algologie 8*: 223–232.

Cormaci, M. and G. Furnari. 1989. World distribution of the genus *Antithamnion* Naegeli (Rhodophyta, Ceramiaceae). *Sôrui, Jap. J. Phycol. 37*: 23–30.

Dawson, E. Y. 1962. Marine Red Algae of Pacific Mexico. Part 7. Ceramiales: Ceramiaceae, Delesseriaceae. *Allan Hancock Pac. Expeditions 26*: 1–205.

Farnham, W. F. 1980. Studies on aliens in the marine flora of Southern England. *In*: (J. H. Price, D. E. G. Irvine and W. F. Farnham, eds.) *The Shore Environment. 2. Ecosystems.* The Systematics Association, Spec. Vol. 17(a). Acad. Press, London. pp. 875–914.

Itono, H. 1969. The genus *Antithamnion* (Ceramiaceae) in Southern Japan and adjacent waters. I. *Mem. Fac. Fish. Kagoshima Univ. 18*: 29–45.

Itono, H. 1971. The genus *Antithamnion* (Ceramiaceae) in Southern Japan and adjacent waters. II. *Mem. Fac. Fish. Kagoshima Univ. 20*: 209–216.

Itono, H. 1977. Studies of the ceramiaceous algae (Rhodophyta) from southern parts of Japan. *Bibliotheca Phycologica 35*, J. Cramer, Vaduz, 499 pp.

Kajimura, M. 1987. *Cladophoropsis corallinicola* sp. nov. and *Antithamnion okiense* sp. nov.: two deep-water algae from the Sea of Japan. *Bot. Mar. 30*: 117–186.

L'Hardy-Halos, M. T. 1970. Recherches sur les Céramiacées (Rhodophycées, Céramiales) et leur morphogenèse. I. Structure de l'appareil végétatif et des organes reproducteurs. *Rev. gén. Bot. 77*: 211–287.

Nägeli, C. 1847. *Die neuern Algensysteme und Versuch zur Begründung eines eigenen Systems der Algen und Florideen.* F. Schulthess, Zürich. 275 pp., pl. I–X.

Norris, R. E. 1987. Species of *Antithamnion* (Rhodophyceae, Ceramiaceae) occuring on the southeast African coast (Natal). *J. Phycol. 23*: 18–36.

Schiffner, V. 1931. Neue und bemerkenswerte Meeresalgen. *Hedwigia 71*: 139–205.

Tokida, J. and T. Inaba. 1950. Contributions to the knowledge of the Pacific species of *Antithamnion* and related algae. *Pacific Sci. 4*: 118–134.

Verlaque, M. 1990. Végétation marine de la Corse (Méditerranée). VIII. Documents pour la flore des algues. *Vie et Milieu 40*: 79–92.

Verlaque, M. and R. Riouall. 1989. Introduction de *Polysiphonia nigrescens* et d'*Antithamnion nipponicum* sur le littoral méditerranéen français. *Cryptogamie, Algol. 10*: 313–323.

Wollaston, E. M. 1968. Morphology and taxonomy of southern australian genera of Crouaniae Schmitz (Ceramiaceae, Rhodophyta). *Aust. J. Bot. 16*: 217–417.

Wollaston, E. M. 1971. *Antithamnion* and related genera occuring on the Pacific coast of North America. *Syesis 4*: 73–92.

Wollaston, E. 1976. *Antithamnion* Nägeli 1847. *In*: (I. A. Abbott and G. J. Hollenberg, eds) *Marine Algae of California.* Stanford Univ. Press, Stanford Calif. pp. 572–577.

Wollaston, E. M. 1984. Species of Ceramiaceae (Rhodophyta) recorded from the International Indian Ocean Expedition, 1962. *Phycologia 23 (3)*: 281–299.

Yamada, Y. and K. Inagaki. 1935. On *Acrothamnion pulchellum* Yamada (non J. Agardh) from Japan. *Sci. Pap. Inst. Algol. Res. Fac. Sci. Hokkaido 1 (1)*: 37–40.

Botanica Marina
Vol. 34, pp. 161–166, 1991

The Effect of Light and Ammonium on Growth, Epiphytes and Chemical Constituents of *Gracilaria conferta* in Outdoor Cultures

M. Friedlander, M. D. Krom and A. Ben-Amotz

Israel Oceanographic & Limnological Research, Tel-Shikmona, P.O.B. 8030, Haifa 31080, Israel

(Accepted 15 December 1990)

Abstract

The relationship between *Gracilaria conferta* growth rate, its epiphytes' biomass and its chemical constituents was determined under various conditions of irradiance and pulse-fed ammonium concentration in an outdoor culture system. Under summer conditions, high pulse-fed ammonium concentration and relatively low irradiance promoted the growth rate of *Gracilaria conferta* and inhibited its epiphyte biomass, presenting a competitive relationship. Growth rate of *Gracilaria conferta* in summer showed an inverse relationship with ash free dry weight, C/N ratio and phycoerythrin to chlorophyll *a* ratio, and a positive relationship with ash, chlorophyll *a* and phycoerythrin content. Under winter and autumn conditions, the irradiance became the only limiting factor of growth rate of *Gracilaria conferta*. These results are related to the common model of relationship between environmental factors and the response of growing seaweeds.

Introduction

The growth rate and chemical constituents of *Gracilaria* spp. (Rhodophyta) in outdoor cultures is a function of irradiance, temperature, salinity, nitrogen and phosphate concentration, plant density, rate of water exchange, aeration, pH and inorganic carbon supply (DeBoer *et al.* 1978, Lapointe and Ryther 1979, Lapointe 1981, 1987, Rosenberg and Ramus 1982, DeBusk and Ryther 1984, Lapointe *et al.* 1984, Guerin and Bird 1987, Bird 1988, Daugherty and Bird 1988). Nitrogen pulse feeding in continuous flow outdoor seawater systems results in a growth rate similar to that obtained with a continuous supply of dissolved nitrogen (Ryther *et al.* 1981, Bird *et al.* 1982). A main interfering factor in the cultivation of *Gracilaria* spp. is the development of epiphytes which compete for light, nutrients and CO_2. The development of epiphytes can be limited by a pulse-feeding regime (Edelstein *et al.* 1976, Lapointe and Ryther 1978, Ryther *et al.* 1981, Bidwell *et al.* 1985). However, there is limited information about the interference of epiphytes with growth of seaweeds, and the effect of pulse-feeding on the combined growth of *Gracilaria* spp. with epiphytes (Neish *et al.* 1977, Ballantine 1979, Morgan *et al.* 1980).

The purpose of the work to be reported was:

(a) to determine quantitatively the relationship between *Gracilaria conferta* and its epiphytes under various conditions of irradiance and pulse-fed ammonium concentration in an outdoor culture system,

(b) to find the relationship between growth rate, epiphyte biomass and chemical constituents of the pulse-fed *Gracilaria conferta*.

Material and Methods

Gracilaria conferta (Schousboe) J. *et* G. Feldmann (identified By Dr C. Bird) was collected on a rocky coast at Sdot-Yam, Israel. One hundred grams of fresh vegetatively growing sporophytes were transferred to 36-litre outdoor fiberglass tanks, forming a density of 0.4 kg m^{-2}. Continous unfiltered seawater was supplied to each tank with a turnover of 24 volumes day^{-1}. Constant aeration kept the algae moving in the tank. A 4 × 4 factorial experiment with duplicate tanks per treatment was designed.

In the summer experiment (Exp. 1), four irradiances of 5, 25, 40 and 100% of full incident light (Io) were

Table I. Effect of irradiance and ammonium concentration on various dependent variables of *Gracilaria conferta* culture in the summer.

Dependent variable	Ammon. conc. (mM)	Week	Irradiance (Io) 0.05	0.25	0.40	1.00
A. Relative epiphyte biomass (%)	0.0	1	9	25	20	26
		2	19	49	39	50
		3	25	52	45	60
		4	33	52	55	70
	0.1	1	5	15	19	19
		2	12	29	42	39
		3	20	46	54	60
		4	29	62	65	79
	0.5	1	1	8	11	17
		2	1	16	22	33
		3	1	22	31	48
		4	1	27	40	62
	2.0	1	2	6	4	13
		2	5	9	7	24
		3	3	8	5	33
		4	2	6	4	44
B. Specific growth rate (doublings/day/10^2)	0.0		2.72	1.74	2.39	1.74
	0.1		4.13	3.35	2.72	3.08
	0.5		5.26	5.70	4.83	4.38
	2.0		5.48	5.59	5.74	4.70
C. Ash free dry weight (%)	0.0		15.0	17.4	15.9	14.1
	0.1		13.1	16.1	15.5	15.9
	0.5		10.5	13.1	14.7	13.9
	2.0		10.7	13.5	14.2	15.0
D. Ash content (%)	0.0		15.6	10.6	11.6	18.4
	0.1		17.3	12.6	15.2	11.3
	0.5		22.9	18.1	16.6	17.9
	2.0		25.1	18.9	17.0	16.2
E. C/N ratio	0.0		45.5	38.5	41.5	31.0
	0.1		43.8	44.9	47.9	28.8
	0.5		20.3	28.0	31.8	42.2
	2.0		18.0	21.5	20.7	18.2
F. Chlorophyll *a* conc. (μg/g FW)	0.0		44	19	43	14
	0.1		54	41	42	23
	0.5		100	56	52	27
	2.0		99	71	68	39
G. Phycoerythrin conc. (μg/g FW)	0.0		57	44	65	24
	0.1		80	61	58	61
	0.5		117	60	69	32
	2.0		104	90	66	27

used with different polypropylene screening. Four pulse-fed ammonium chloride additions, to give final concentrations of 0, 0.1, 0.5 and 2.0 mM, were supplied once a week for a period of 6 hours. Phosphate was added as NaH_2PO_4 at one tenth the concentration of ammonium. During the addition of nutrients, the seawater supply was stopped. The experiment was carried out for six weeks including two weeks of acclimation (July – August). Each week the plants were damp dried, weighed, and harvested back to their initial weight. Specific growth rate as doublings per day was calculated according to $\mu = 1/t\ (\ln B_t/B_o)$, where B_o is the initial biomass and B_t is the biomass after t days. The additional fresh material was cleaned of epiphytes and dried at 70 °C for 48 hours in an oven. Epiphytes were identified, weighed, and the weight as a percent of the total weight of the *Gracilaria conferta* was calculated. Irradiance was measured daily at noon at 10 cm depth in the tanks, and salinity was constant (40‰ ± 1). Similar experiments were conducted in the autumn (Exp. 2) and in the winter (Exp. 3 and Exp. 4). Conditions of the summer, autumn and winter growth experiments are reported in Table IV.

Chemical constituents were determined for samples from each tank, collected in the last week. Ash content was determined after ignition of the samples of the dry material at 500 °C for 24 hours in an oven. The C/N ratio was determined with a LECO CHN-600 Analyzer. Samples of fresh material were analyzed for chlorophyll *a* (Moran 1982) and for phycoerythrin (Moon and Dawes 1976). Ammonia-N was determined according to Solorzano (1969) and phosphate according to Strickland and Parsons (1972). Growth rate (doublings day^{-1}) and dry weight were averaged for the last 4 weeks and intercorrelated with the other measured parameters (Pearson correlation). The main effects of irradiance, ammonium concentration, time, and their interactions were determined by two-way and three-way ANOVA (Sokal and Rohlf 1981). Means were compared by Duncan's new multiple range test.

Results

The main interfering epiphytes in all four experiments were: *Enteromorpha compressa* Greville, *Cladophora pellucida* Kuetzing, *Ulva lactuca* Le Jolis. and *Ectocarpus confervoides* Le Jolis. Table I A (Exp. 1) shows the time dependent development of the relative epiphyte biomass as a function of irradiance and ammonium concentration. Only in four of the treatments was the relative epiphyte biomass inhibited (0.05, 0.25, 0.40 Io at 2.0 mM and 0.05 Io at 0.5 mM ammonium concentration) and showed non-significant differences between the relative epiphyte biomass at each week of the four-week experiment; all others increased with time from 10 to 75% of total biomass. Relative epiphyte biomass was positively related to irradiance and negatively to ammonium concentration. The growth rate of *Gracilaria conferta* fluctuated in most treatments; however, the average of four-week growth rates was calculated for comparative purposes (Table I B). The growth rate of *Gracilaria conferta* reached a steady state in the four treatments in which the relative epiphyte biomass was lowest, and ranged from 0.02 to 0.06 doublings day^{-1}. Growth rate was inhibited by high irradiance, reaching saturation at 0.05 Io, and was positively related to ammonium concentration, which generally reached saturation above 0.5 mM.

The main effects of irradiance (I), ammonium concentration (N) and the difference between four weekly measurements (T) on growth rate and relative epiphyte biomass were significant in Experiment 1, as shown by a three-way ANOVA test. Ammonium concentration accounted for most of the variation in growth (52.1%) as compared with irradiance (6.4%), with a low significant interaction, while both factors accounted almost equally for the variation in relative epiphyte biomass (25.9% and 28.7%) and showed low interaction. The difference between each of four week measurements accounted for 5.2% of the total variation in growth rate, and its interaction with ammonium concentration accounted for 11.1% of the total variation in growth rate. This time factor also had a significant effect on relative epiphyte biomass, representing their development, which accounted for a high percent (20.3%) of the total variation. Relative epiphyte biomass inversely correlated with growth rate ($r = -0.866$; $p < 0.001$).

At a low maximal underwater irradiance (184 μE m^{-2} s^{-1}) which was recorded in the second experiment made in the autumn (Exp. 2), the relative effect of irradiance variation on growth rate was much higher than that of ammonium concentrations (Table II B). Maximal growth rate coincided with maximal irradiance, and ranged from −0.006 to 0.07 doublings day^{-1}. The level of dissolved nitrogen in the ambient seawater (6 μM) was twice as high as that in the

Table II. Effect of irradiance and ammonium concentration on various dependent variables of *Gracilaria conferta* culture in the autumn.

Dependent variable	Ammon. conc. (mM)	Irradiance (Io)		
		0.06	0.53	1.00
A. Relative epiphyte biomass (%)	0.0	0.0	8.0	35.0
	0.1	0.0	9.0	34.1
	0.5	0.0	9.0	38.0
	2.0	0.0	6.0	40.0
B. Specific growth rate (doublings/day/10^2)	0.0	0.0	−0.6	6.1
	0.1	−0.2	0.0	6.0
	0.5	−0.1	1.0	6.8
	2.0	−0.1	0.8	6.0

Table III. Average weekly growth rate, final dry weight and relative epiphyte biomass of *Gracilaria conferta* as a function of irradiance in a 5 week winter experiment (Exp. 3). Each treatment was operated twice (n = 4).

Treatment	Irradiance (Io)	Specific growth rate (doublings/day/10^2)	Dry weight (%)	Epiphyte biomass (%)
High light	1.00	a 2.71 ± 0.40	a 12.75 ± 0.70	a 14.1 ± 5.1
Low light	0.48	b 1.61 ± 0.43	a 12.00 ± 0.23	b 2.6 ± 1.4

Weekly pulse feeding consisted of 2.0 mM ammonium and 0.2 mM phosphate.
Different letters express significant difference at the level of $p < 0.01$.

summer experiment. The winter experiment (Exp. 3), in which the growth rate reached 0.03 doublings day^{-1}, confirmed the positive relationship between growth rate and irradiance (Table III). In these low irradiance experiments, the relative epiphyte biomass was positively related to irradiance variations (Table II A, Table III) and to *Gracilaria conferta* growth rate ($r = 0.967$, $p = 0.001$). In the autumn experiment (Exp. 2), irradiance accounted for the major percentage of total variation of the relative epiphyte biomass. The culture conditions of the three experiments are presented in Table IV.

Ash free dry weight of *Gracilaria conferta* in the summer experiment (Exp. 1) was negatively related to ammonium concentration up to 0.5 mM and positively related to irradiance above 0.1 mM (Table I C), while ash content was positively related to ammonium concentration, which reached saturation above 0.5 mM, and negatively related to irradiance (Table I D). Ash free dry weight ranged from 10.5% to 17.4% and ash content from 11% to 25%. Ammonium concentration accounted for most of the variation in ash free dry weight (76.8%) as compared to irradiance and their interactions (10.5% and 9.0%, respectively), while the variation in ash content was accounted for gradually by ammonium concentration, irradiance and their interactions (36.7%, 23.9% and 22.1%, respectively). Growth rate was inversely correlated with ash free dry weight ($r = -0.681$, $p = 0.004$), and positively correlated with ash content ($r = 0.666$, $p = 0.005$). No response of dry weight to irradiance was shown in the winter experiment (Exp. 3, Table III).

In the summer experiment (Exp. 1), the C/N ratio, ranging from 18 to 48, showed an inverse relationship with ammonium concentration, which accounted for most of the variation in C/N ratio (75%), and a positive relationship with irradiance except for Io (Table I E). The C/N ratio was inversely correlated with growth rate ($r = -0.696$, $p = 0.01$).

Ash content (30.4 – 38.7%) and C/N ratio (8.8 – 9.8) increased significantly during ammonium starvation (Exp. 3, Table V), in which feeding was stopped but fresh seawater supplied about 0.1 mmole ammonia-N l^{-1} $week^{-1}$. In this winter experiment, the C/N ratio was only 20 – 50% of the summer C/N ratios (Exp. 1), while ash content was 50 – 170% higher in the winter experiment.

Table IV. Culture conditions of different *Gracilaria conferta* growth experiments.

Experiment	Irradiance (Io) ($\mu E\ m^{-2}\ s^{-1}$)	Day length (h)	Temperature (°C)	DIN (μM)
Summer (1)	1100	14	27.0	3
Autumn (2)	184	12	21.6	6
Winter (3)	138	10	15.3	0.4
Winter (4)	160	10	15.3	0.4

DIN: Dissolved inorganic nitrogen in the inlet water.

Table V. Ash content in dry weight and C/N ratio in *Gracilaria conferta* as a function of ammonium starvation time (n = 3) in the winter experiment (Exp. 3).

Variable	Days after feeding	
	0	9
Ash (%)	a 30.40 ± 1.14	b 38.73 ± 2.26
C/N ratio	a 8.77 ± 0.38	b 9.79 ± 0.87

Different letters in each variable represent significant differences ($p < 0.05$).
Irradiance 100% Io.

Table VI. Effect of irradiance and feeding duration on growth rate of *Gracilaria conferta* in a winter experiment (Exp. 4).

Feeding duration (h)	Irradiance (Io)			
	0.25	0.38	0.75	1.00
4	3.4	3.3	5.7	4.9
8	2.1	2.5	5.6	4.2
14	4.3	4.1	5.8	5.7
24	1.7	2.1	5.9	5.1

In Experiment 1, the concentrations of both pigments, chlorophyll *a* 14 – 100 μg g^{-1} f. w.) and phycoerythrin (24 – 117 μg g^{-1} f. w.), in *Gracilaria conferta* were negatively related to irradiance and showed generally positive relationships to ammonium concentration except for phycoerythrin in high light (Table I F, G). Irradiance accounted for most of the variation in both chlorophyll *a* (47.7%) and phycoerythrin (49.5%), while ammonium concentration accounted more for chlorophyll *a* (37.4%) than for phycoerythrin (11.1%), in which the interaction was relatively high (25.3%). Both pigments positively correlated with growth rate ($r = 0.772$; $p < 0.001$ and $r = 0.542$, $p = 0.037$, respectively), and positively intercorrelated ($r = 0.887$, $p < 0.001$). The ratio of phycoerythrin to chlorophyll *a* inversely correlated with growth rate ($r = -0.722$, $p = 0.002$).

Variation in pulse feeding duration accounted only for 13% of the total variation in growth rate, whereas different irradiances accounted for 81% of the total variation as expected in a winter experiment (Exp. 4, Table VI). Both main effects were significant; however, interaction was insignificant. Irradiance was again positively related to growth rate, up to 120 µE m^{-2} s^{-1}, while feeding duration caused maximal growth rate at 14 hours of feeding during the night. Comparison of short and long feeding durations in the winter experiment showed no significant effect on relative epiphyte biomass and dry weight; however, there was a significant increase of 230–460% in ammonium uptake.

Discussion

The lowest relative epiphyte biomass development coincided with the highest growth rate of *Gracilaria conferta* under high pulse-fed ammonium concentration and relatively low irradiances. The thick structured *Gracilaria conferta* thallus was light saturated, while the thin structured epiphytes were light limited. These data support the functional-form hypothesis of Littler (1980) and Ramus (1983). The thin structure of the main epiphytes might also be the reason for the growth inhibition caused by the high ammonium (> 0.5 mM) concentration, which was shown to reach a toxic level for *Ulva* sp. above 60 µM (Waite and Mitchell 1972). Biweekly pulse feeding (0.1–1 mM ammonium) of *Chondrus crispus* (L.) Stackhouse (Neish *et al.* 1977, Bidwell *et al.* 1985) and *Gracilaria tikvahiae* McLachlan (Ryther *et al.* 1981, Bird *et al.* 1982) supported maximum growth rate in nitrogen starved plants, inhibited epiphytes, and caused efficient utilization of ammonium. This is confirmed by the present weekly pulse-fed experiment, in which ammonium saturation supported maximum growth rate of *Gracilaria conferta* at 0.5 mM. It is hypothesized that epiphytes compete with *Gracilaria*, being involved in light decrease, nutrient and inorganic carbon consumption, and tissue damage of their host.

In the autumn growth experiment, nitrogen concentration in ambient seawater reached almost saturation (1 mmole L^{-1} $week^{-1}$). Additionally, the low saturating irradiance of the summer experiment is confirmed by the autumn and winter experiments in which irradiance as a limiting factor became the only significant variable, and by field studies with the local *Gracilaria* sp. (Friedlander and Zelikovitch 1984). Longer feeding hours significantly increased the rate of ammonium uptake by *Gracilaria conferta*, having almost no effect, however, on growth rate, relative epiphyte biomass and dry weight. A comparison of *Gracilaria conferta* maximal growth rate in the summer and autumn experiments showed no difference, presumably because temperatures were higher and lower than optimum. Only very low temperature or irradiance in the winter experiment showed a limited growth rate. This confirms the major contribution of temperature differences to growth rate variations in *Gracilaria* (Bird 1988). The inverse relationship between growth rate and ash free dry weight in the summer experiment suggests that the higher growth rate was related to more mineral and water absorption by the elongating algae, while at the lower growth rate, more organic synthesis took place.

The inverse relationship between growth rate and C/N ratio in the summer experiment, and the difference between the high C/N ratio under summer promoting conditions and the low C/N ratio under winter limiting conditions, confirm the assumption that a low growth rate usually coincides with nitrogen limitation (Lapointe and Duke 1984). While the increase in the C/N ratio through nitrogen starvation was expected, the huge difference between high summer and low winter C/N ratios was unexpected, and suggests nitrogen limitation in summer. The higher summer growth rate presumably needed more nitrogen which would have increased the epiphyte contamination.

Ash free dry weight and ash are mainly affected by changes in ammonium concentration and less by variation in irradiance under the growth conditions of the present study. This contradicts major significant light effects shown in previous studies with *Gracilaria tikvahiae* (Lapointe *et al.* 1984). It is suggested that this may be a result of low light saturation levels of the local *Gracilaria conferta*. Pigment concentration relationships with irradiance and ammonium concentration confirm similar relationships for *Gracilaria tikvahiae* (Ramus 1983, Lapointe and Duke 1984). Pigment accumulation accounts for the increasing protein fraction of growing seaweeds.

In this study using *Gracilaria conferta*, there are two unusual relationships: the weekly fluctuations in growth rate, and the extremely high C/N ratio in summer. These results do not fit the common model of relationships between environmental factors and the response of growing seaweeds. The common factor which might explain these unusual results is the competitive growth of epiphytes. Further work is needed to understand these relationships more completely.

Acknowledgements

This research was supported by Fertilizers & Chemicals Ltd., Israel. The authors wish to thank Ms R. Shalev and Ms S. Strimling for assistance in the experimental work.

References

Ballantine, D. L. 1979. The distribution of algal epiphytes on macrophyte hosts offshore from La Parguera, Puerto Rico. *Bot. Mar. 22*: 107–111.

Bidwell, R. G. S., J. McLachlan and N. D. H. Lloyd. 1985. Tank cultivation of Irish moss, *Chondrus crispus* Stackh. *Bot. Mar. 28*: 87–97.

Bird, K. T. 1988. Agar production and quality from *Gracilaria* sp. strain G-16; Efects of environmental factors. *Bot. Mar. 31*: 33–39.

Bird, K. T., C. Habig and T. DeBusk. 1982. Nitrogen allocation and storage patterns in *Gracilaria tikvahiae*. *J. Phycol. 18*: 344–348.

Daugherty, B. K. and K. T. Bird. 1988. Salinity and temperature effects on agar production from *Gracilaria verrucosa* strain G-16. *Aquaculture 75*: 105–113.

DeBoer, J. A., H. J. Guigly, T. L. Israel and C. F. D'Elia. 1978. Nutritional studies of red algae. I. Growth rates as a function of nitrogen source and concentration. *J. Phycol. 14*: 261–266.

DeBusk, T. A. and J. H. Ryther. 1984. Effects of seawater change, pH and carbon supply on the growth of *Gracilaria tikvahiae* in large scale cultures. *Bot. Mar. 27*: 357–362.

Edelstein, T., C. J. Bird and J. McLachlan. 1976. Studies on *Gracilaria*. 2. Growth under greenhouse conditions. *Can. J. Bot. 54*: 2275–2290.

Friedlander, M. and N. Zelikovitch. 1984. Growth rates, phycocolloid yield and quality of the red seaweeds, *Gracilaria* sp., *Pterocladia capillacea*, *Hypnea nusciformis* and *Hypnea cornuta*, in field studies in Israel. *Aquaculture 40*: 57–66.

Guerin, J. M. and K. T. Bird. 1987. Effects of aeration periods on the productivity and agar quality of *Gracilaria* sp. *Aquaculture 64*: 105–110.

Lapointe, B. E. 1981. The effects of nitrogen and light on growth, pigment content, and biochemical composition of *Gracilaria foliifera* v. *angustissima*. *J. Phycol. 17*: 90–95.

Lapointe, B. E. 1987. Phosphorus and nitrogen limited photosynthesis and growth of *Gracilaria tikvahiae* in the Florida Keys: an experimental field study. *Mar. Biol. 93*: 561–568.

Lapointe, B. E. and C. S. Duke. 1984. Biochemical strategies for growth of *Gracilaria tikvahiae* in relation to light intensity and nitrogen availability. *J. Phycol. 20*: 488–495.

Lapointe, B. E. and J. H. Ryther. 1978. Some aspects of the growth and yield of *Gracilaria tikvahiae* in culture. *Aquaculture 15*: 185–193.

Lapointe, B. E. and J. H. Rhyter. 1979. The effect of nitrogen and seawater rate on the growth and biochemical composition of *Gracilaria foliifera* v. *angustissima* in mass outdoor cultures. *Bot. Mar. 22*: 529–537.

Lapointe, B. E., K. R. Tenore and C. J. Dawes. 1984. Interaction between light and temperature on the physiological ecology of *Gracilaria tikvahiae*. 1. Growth, photosynthesis and respiration. *Mar. Biol. 80*: 161–170.

Littler, M. M. 1980. Morphological form and photosynthesis performances of marine macroalgae: test of a functional form hypothesis. *Bot. Mar. 23*: 161–165.

Moon, R. E. and C. J. Dawes. 1976. Pigment changes and photosynthetic rates under selected wavelengths in the growing tips of *Eucheuma isiforme* v. *denudatum* during vegetative growth. *Br. Phycol. J. 11*: 165–175.

Moran, R. 1982. Formulae for determination of chlorophyllous pigments extracted with N,N-dimethylformamide. *Plant Physiol. 69*: 1376–1381.

Morgan, K. C., P. F. Shacklock and F. J. Simpson. 1980. Some aspects of culture of *Palmaria palmata* in greenhouse tanks. *Bot. Mar. 23*: 765–770.

Neish, A. C., P. E. Shacklock, C. H. Fox and F. J. Simpson. 1977. The cultivation of *Chondrus crispus*. Factors affecting growth under greenhouse conditions. *Can. J. Bot. 55*: 2263–2271.

Ramus, J. 1983. A physiological test of the theory of complementary chromatic adaptation: brown, green and red seaweeds. *J. Phycol. 19*: 173–178.

Rosenberg, G. and J. Ramus. 1982. Ecological growth strategies in the seaweeds *Gracilaria foliifera* and *Ulva* sp.: soluble nitrogen and reserve carbohydrate. *Mar. Biol. 66*: 251–259.

Ryther, J. H., N. Corwin, T. A. DeBusk and L. D. Williams. 1981. Nitrogen uptake and storage by the red alga *Gracilaria tikvahiae*. *Aquaculture 26*: 107–115.

Sokal, R. R. and F. J. Rohlf. 1981. *Biometry. The Principles and Practice of Statistics in Biological Research*. 2nd ed. W. H. Freeman & Co., San Francisco.

Solorzano, L. 1969. Determination of ammonia in natural waters by the phenolhypochlorite method. *Limnol. Oceanogr. 14*: 799–801.

Strickland, J. D. H. and T. R. Parsons. 1972. *A Practical Handbook of Seawater Analysis*. Fish. Res. Bd. Can., Ottawa.

Waite, T. and R. Mitchell. 1972. The effect of nutrient fertilization on the benthic alga *Ulva lactuca*. *Bot. Mar. 15*: 151–156.

Botanica Marina
Vol. 34, pp. 167–175, 1991

Scrippsiella spinifera sp. nov. (Pyrrhophyta): A New Dinoflagellate from the Northern Adriatic Sea[1]

G. Honsell and M. Cabrini*

Dipartimento di Biologia, Università degli Studi di Trieste, via A. Valerio 30, 34127 Trieste, Italy

* *Laboratorio di Biologia Marina, Sorgenti di Aurisina, Strada Costiera 336, 34010 Trieste, Italy*

(Accepted 16 December 1990)

Abstract

Scrippsiella spinifera sp. nov. (Pyrrhophyta) is described using light and scanning electron microscopy. The species was first found in the Gulf of Trieste (Italy, Northern Adriatic Sea). The cells are 30–52 µm long and 21–36 µm wide. The epitheca is conical and the hypotheca trapezoidal with short spines at its terminal part. The plate tabulation does not differ from other *Scrippsiella* species. The thecal plates are perforated by numerous pores surrounded by concentric ridges. A taxonomical discussion and a comparison with the other species of the genus are made.

Introduction

Scrippsiella spinifera sp. nov. was first found in phytoplankton samples collected in the inner parts of the Gulf of Trieste (Italy, Northern Adriatic Sea) in September 1983. The area was dominated by a dense bloom which discolored the waters, forming yellow brown patches on the surface. This species represented the dominant organism of the phytoplankton community, which was composed of dinoflagellates and diatoms. Since that time it has been repeatedly encountered in the Gulf of Trieste, and along the northwestern coasts of the Adriatic Sea, indicating that it is a common component of the phytoplankton in summer and autumn. It has often been present in relatively high cell concentrations, especially in restricted inshore areas where there is little water exchange (Honsell and Cabrini 1990).

Material and Methods

Specimens of *Scrippsiella spinifera* sp. nov. were first collected in the Bay of Muggia in the Gulf of Trieste (Italy) during the last week of September 1983. The cells were fixed immediately with 2% neutralized formaldehyde and examined under Leitz Diavert and Leitz Ortholux light microscopes using phase contrast. The length, the transdiameter and the dorsoventral diameter of 100 cells were measured at 400 × magnification. The thecal plates were observed after squashing the cells by gentle pressure on the coverslip. A drop of sodium hypochlorite was placed beneath the coverslip to help separate the plates. Specimens for scanning electron microscopy were fixed in the same way, then washed in distilled water, dehydrated in a gradual series of ethanol, critical point dried from liquid CO_2 and sputter coated with gold-palladium. Observations were made using a Philips PSEM 500 scanning electron microscope.

Description

Scrippsiella spinifera sp. nov. is a small armored orthoperidinioid dinoflagellate with a cell length of 30–52 µm, a transdiameter of 21–36 µm and a dorso-ventral diameter of 16–33 µm (Figs 1–7). The cell body shows a slight dorso-ventral compression, which appears more pronounced in large cells: the dorso-ventral diameter is usually 1–2 µm shorter than the

[1] The work has been supported by grants from C.N.R. (Consiglio Nazionale delle Ricerche) and M.P.I. (Ministero della Pubblica Istruzione), Rome, Italy.

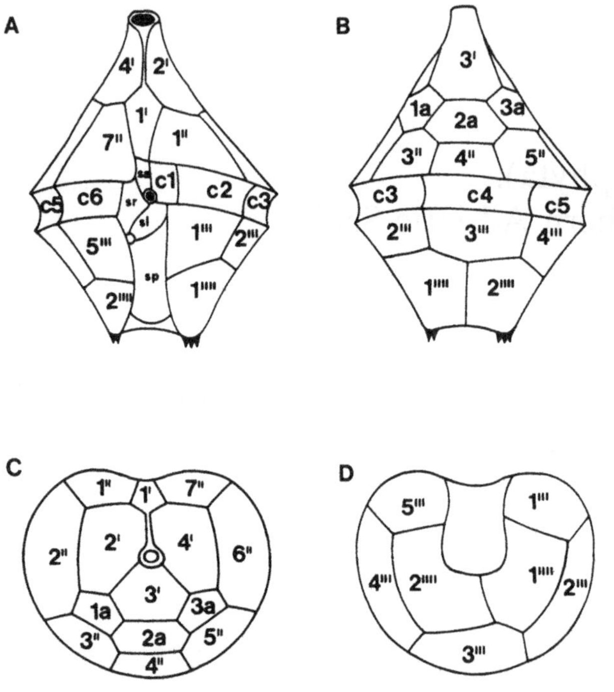

Fig. 1. *Scrippsiella spinifera* sp. nov. Plate tabulation: ventral (a), dorsal (b), apical (c) and antapical view (d) of the theca.

transdiameter. The epitheca is larger than the hypotheca; the mean ratio between epithecal and total cell length is about 0.60 (range 0.53–0.71). In ventral and dorsal view the epitheca appears conical, with a concavity near the apex, being particularly evident in cells of a small size (Figs 2–3). In large cells the epithecal sides tend to be slightly convex due to the presence of wide growth bands and the concavity near the apex is usually less pronounced (Figs 4–7). In lateral view the epitheca appears asymmetrical, being more compressed on its ventral side.

The hypotheca has a trapezoidal shape and it is deeply incised by the sulcus, which divides the antapical region into two distinct lobes. Each lobe bears on its terminal part two or three short spines, which are particularly evident in lateral view.

The girdle is postmedian and descends half its own height as it encircles the cell body. It is well excavated and bordered by narrow lists.

The sulcus is deeply incised and becomes deeper and wider, as it approaches the posterior end of the cell.

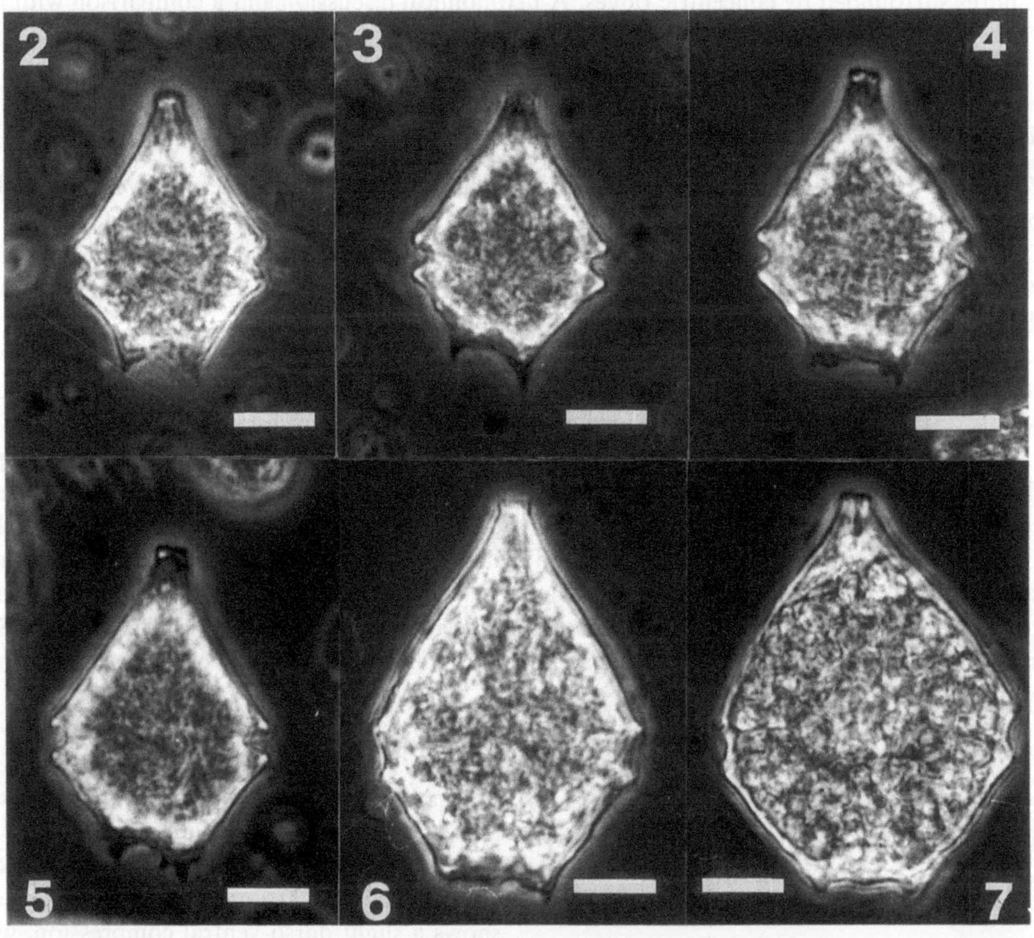

Figs 2–7. *Scrippsiella spinifera* sp. nov.: variations of size and morphology. Fixed cells. LM phase contrast. Scale bars = 10 µm. Fig. 2. Small cell. Fig. 3. Small cell. Fig. 4. Medium sized cell: the epithecal sides become slightly convex. Fig. 5. Medium sized cell. Fig. 6. Megacytic cell. Fig. 7. Megacytic cell. Many granules (starch?) occupy the cytoplasm.

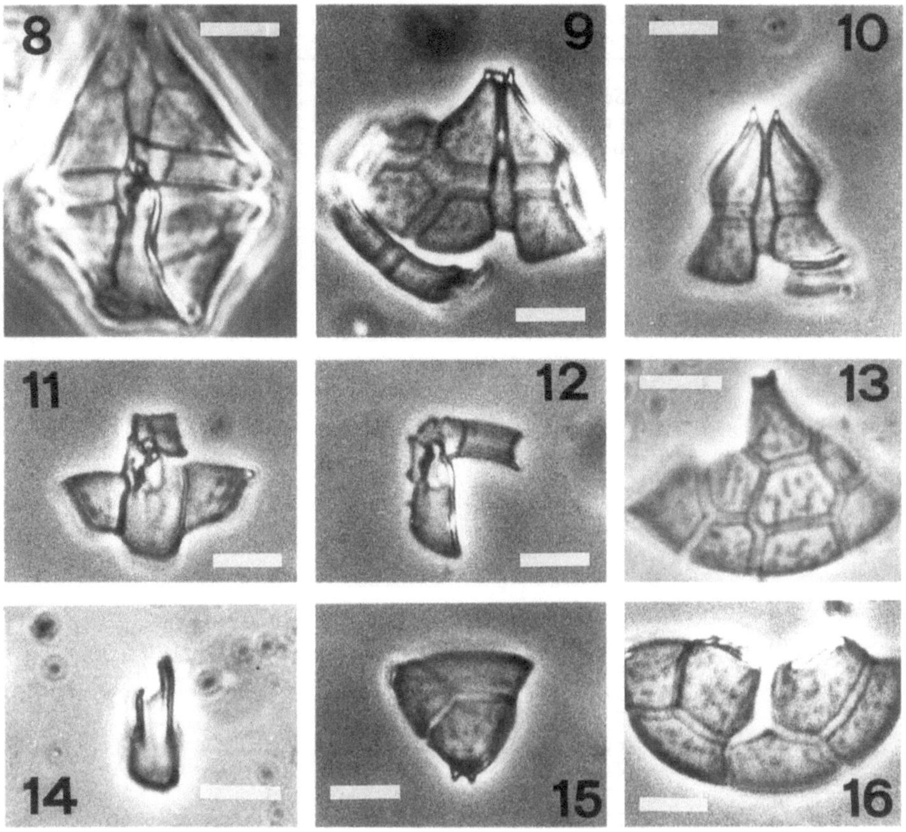

Figs 8–16. *Scrippsiella spinifera* sp. nov.: squashed thecae. LM phase contrast. Scale bars = 10 µm.
Fig. 8. Whole theca: ventral view. Fig. 9. Ventral view of epitheca with wide intercalary bands. Reversed image. Fig. 10. Ventral view of epitheca. Fig. 11. Sulcal and post-cingular plates. Fig. 12. Sulcal plates. Fig. 13. Dorsal view of epitheca. Fig. 14. Posterior sulcal plate. Fig. 15. Lateral view of hypothecal plates. The antapical plate shows two short spines. Fig. 16. Antapical view of hypotheca. A post-cingular plate is lacking.

The thecal plates (Figs 8–16) have pores irregularly distributed on their surface. These pores are usually arranged in groups of two, three or more. Observations by scanning electron microscopy (Figs 17–23) reveal finer details of the theca, and show that the pores are formed by two concentric ridges, with diameters of 0.6 and 0.3 µm respectively, surrounding an aperture 70 nm wide. The external ridges connect adjacent pores forming a continuous structure (Fig. 23). An almost continuous line of pores is present on the margins of the postcingular plates along the girdle. Pores are found also on cingular and sulcal plates. The plate formula is: pp, x, 4′, 3a, 7″, 6c, 5‴, 2⁗, 5s (Fig. 1).

Apical pore plates (Figs 9, 10, 22): The pore plate (pp) is circular in shape, with plates 2′, 3′ and 4′ forming a collar around it. The canal plate (x, or preapical plate) is very narrow and it is about one half the length of plates 2′ and 4′. It is not easily visible by SEM, as it is covered by the raised margins of these two plates (Figs 17, 18).

Apical plates: Plate 1′ is narrow and shows an ortho-arrangement. Plates 2′, 3′ and 4′ are hexagonal: 2′ and 4′ are similar in size and shape, while 3′ is somewhat smaller than 2′ and 4′.

There are three intercalary plates; 1a and 3a are pentagonal; 2a is hexagonal (Fig. 13) or, more rarely, pentagonal.

There are seven precingular plates and 2″ and 6″ are considerably larger than the other ones. Plate 1″ is smaller than 7″ due to the girdle displacement (Fig. 10). Its shape shows a certain variability (it may have four or five sides) as, on its left margin, it may form a suture only with plate 1′ or with this plate and the sulcal anterior plate (sa) (Figs 17, 20).

The cingulum consists of six plates of which five are large and similar in size (2c and 6c are somewhat smaller than 3c, 4c and 5c). Plate 1c (transitional) is considerably small and extends into the sulcal region, where it forms sutures with plates sa, sp, sl. It delimits, on the right, the flagellar pore (Fig. 20).

The sulcus is formed by five plates (Figs 11, 12, 14, 17, 20). The sulcal anterior plate (sa) is four-sided and narrow, and it is contiguous with plates 1′, 7″, 1c, sometimes 1″, and the flagellar pore. The sulcal

Table I. Comparison of the described *Scrippsiella* species.

Species	*Scrippsiella spinifera* sp. nov.	*Scrippsiella sweeneyae* Balech	*Scrippsiella saladense* Balech	*Scrippsiella subsalsa* (Ostenfeld) Steidinger *et* Balech	*Scrippsiella trochoidea* (Stein) Loeblich III
Synonyms					*Scrippsiella faeröense* Balech *et* Oliveira Soares
Cell size					
Length (µm)	30–52	24–32.5	30 –46.5	23.5–41	30–38
Width (µm)	21–36	19–24	23.5–40.5	21 –51	18–22.5
Plate tabulation					
Apical pore	pp x	pp x	pp x	pp x	pp x
Apical	4′	4′	4′	4′	4′
Intercalary	3 a	3 a (2 a)	3 a	3 a	3 a
Precingular	7″	7″ (4″ 5″ 6″)	6″ (7″)	7″	7″
Cingular	6 c	6 c	6 c	6 c	6 c
Sulcal	5 s	4 s	4 s	4 s	5 s
Postcingular	5‴	5‴	5‴	5‴	5‴
Antapical	2⁗	2⁗	2⁗	2⁗	2⁗
2 a morphology	hexa (penta)	penta (hexa)	penta	hexa	penta (hexa)
Thecal ornamentation	pores with concentric ridges			reticulation	pores
Other features	antapical spines			2 a and 3 a separated	
Organelles					
Chloroplasts	many greenish	many elliptical brown yellow or yellow green	present	many oval to rod shaped yellow brown	many elliptical greenish
Pyrenoid		present	present		present
Eyespot				present	
Other structures				orange accumulation globule	
Cysts					calcareous
Ecology	planktonic marine	planktonic marine	planktonic brackish	planktonic brackish marine	planktonic marine
References	present paper	Balech (1959)	Balech (1963)	Steidinger and Balech (1977)	Steidinger and Balech (1977) Balech and Oliveira Soares (1966) Loeblich (1976)

Scrippsiella hexapraecingula Horiguchi *et* Chihara	*Scrippsiella tinctoria* Indelicato *et* Loeblich	*Scrippsiella mexicana* Indelicato *et* Loeblich	*Scrippsiella precaria* Montresor *et* Zingone	*Scrippsiella caponii* Horiguchi *et* Pienaar	*Scrippsiella arenicola* Horiguchi *et* Pienaar
Scrippsiella gregaria (Lombard *et* Capon) Loeblich Sherley *et* Schmidt		*Ensiculifera mexicana* Balech		*Peridinium gregarium* Lombard *et* Capon	
25–32 20–28	19.2–28.8 17.3–24.3	42–45 35–36	15 –25 13.8–20	20–34 20–34	40.3–58.4 34.1–46.5
pp x 4′ 3 a (4 a–6 a) 6″ (7″) 6 c 5 s 5‴ 2⁗	pp x 4′ 3 a 7″ 6 c 5 s 5‴ 2⁗	pp x 4′ 3 a 7″ 6 c 5 s 5‴ 2⁗	pp x 4′ 3 a 7″ 6 c 5 s 5‴ 2⁗	pp x 4′ 3 a 7″ 6 c 5 s 5‴ 2⁗	pp x 4′ 3 a 7″ 6 c 4 s 5‴ 2⁗
penta (hexa)	hexa	hexa	rhombic	hexa	penta
pores			pores	striated plates	minute wart-like projections
		knife-like projection on 1 c plate		2 a and 3 a separated; antapical spines	2 a and 3 a separated
many oval to rod shaped yellow brown	not described	present	numerous drop shaped olive green	numerous elongate from light green to yellow brown	many, brown to reddish-brown rod shaped and radially arranged
present					present
present			present	present	
peduncle; apical stalk	yellow accumulation bodies (physodes)		yellow orange accumulation body		peduncle; apical stalk
		calcareous	calcareous		
in tide pools marine	planktonic marine	planktonic marine	planktonic marine	in tide pools marine	in tide pools marine
Horiguchi and Chihara (1983) Loeblich et al. (1979) Horiguchi and Poenaar (1988)	Indelicato and Loeblich (1985)	Balech (1967) Indelicato and Loeblich (1986)	Montresor and Zingone (1988)	Horiguchi and Pienaar (1988) Lombard and Capon (1971)	Horiguchi and Pienaar (1988)

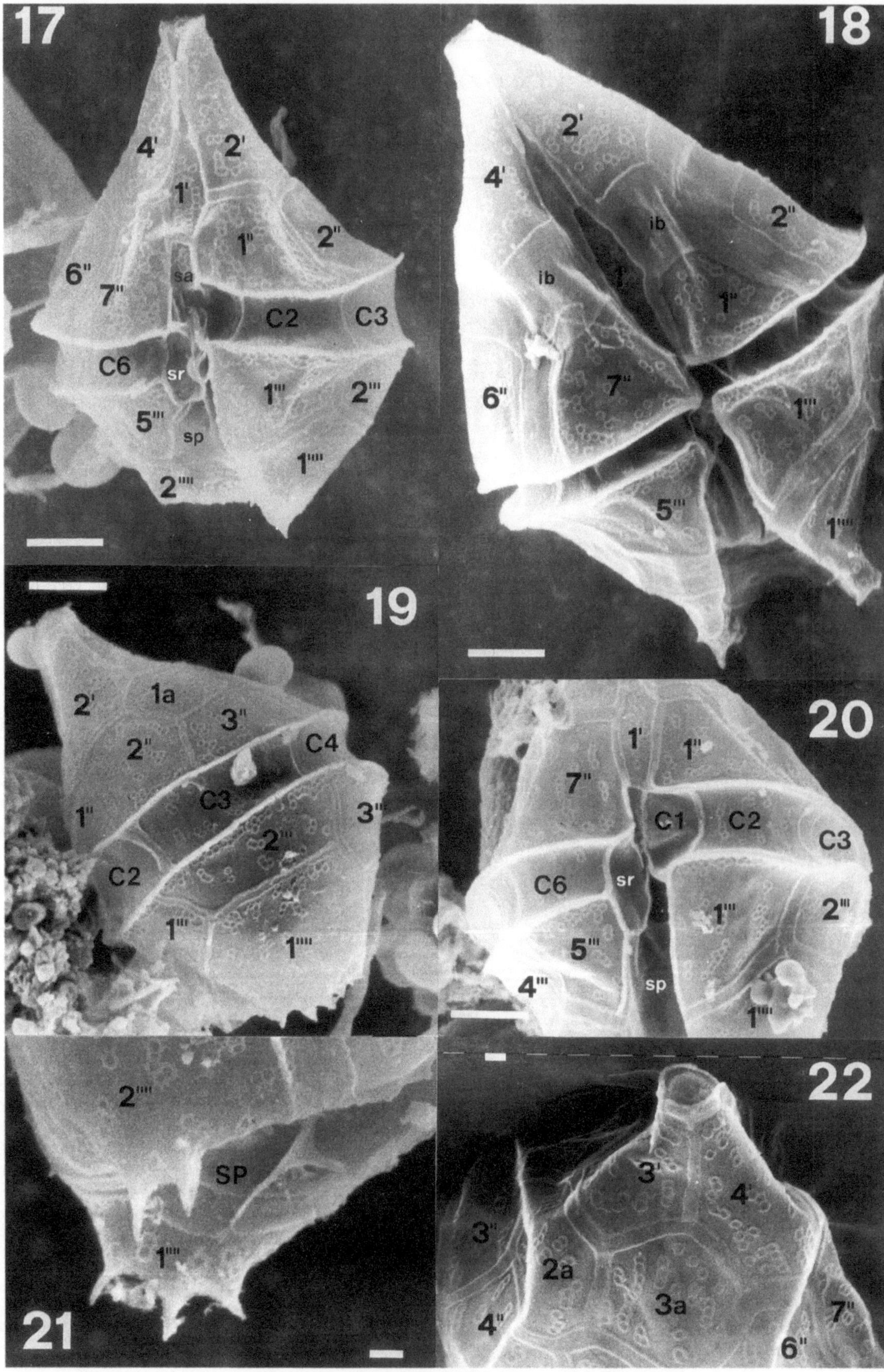

Figs 17–22. *Scrippsiella spinifera* sp. nov.: scanning electron microscopy. Fig. 17. Ventral view. Scale bar = 10 µm. Fig. 18. Ventral view of megacytic cell showing very wide intercalary bands (ib). Scale bar = 10 µm. Fig. 19. Lateral view. Scale bar = 10 µm. Fig. 20. Detail of sulcal region. Scale bar = 10 µm. Fig. 21. Lateral view of the terminal part of hypotheca showing antapical spines. Scale bar = 1 µm. Fig. 22. Apical pore. Scale bar = 1 µm.

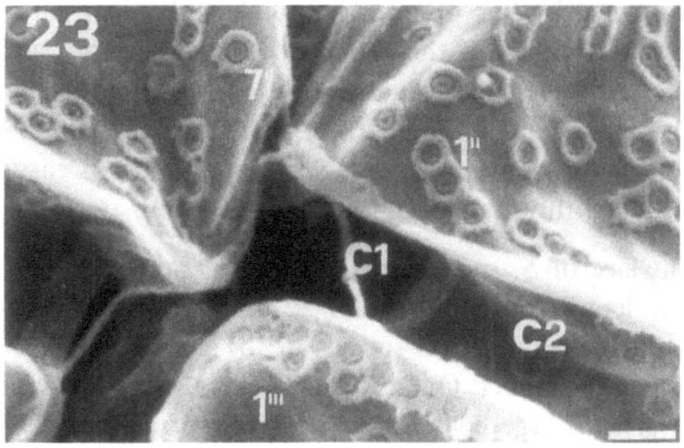

Fig. 23. *Scrippsiella spinifera* sp. nov.: scanning electron microscopy. Detail of ventral view of girdle showing thecal pores surrounded by concentric ridges. Scale bar = 1 µm.

posterior (sp) (Fig. 14) is the largest sulcal plate and extends into the hypotheca reaching the antapex. Its anterior end is deeply excavated and forms two lobes: the left is longer and articulates with plate 1c, while the right is much shorter and butts the right sulcal plate (rs). Three plates are inserted in the excavation of the sulcal posterior plate: the right sulcal (rs), the left sulcal (ls) and the internal sulcal plate (is). The right sulcal plate is rather small and elongated with rounded margins, and it forms sutures with plates 7″, 1c, 6c, ss and sp. Anteriorly, on the left, it delimits the flagellar pore. The left sulcal plate (ls) is not always easily visible by SEM, as it is frequently lost or dislodged during preparations. This plate is similar in size and shape to the right sulcal plate. The internal sulcal plate (is) is very small and it is inserted between sr, sl and sp. The area of the flagellar pore is very difficult to examine, as it is often covered by parts of the flagella or other material.

There are five post-cingular plates (Fig. 16) and they are symmetrically arranged. They show some differences in size, as plates 1‴ and 5‴ are smaller than 2‴, 3‴ and 4‴.

The arrangement of antapical plates is slightly asymmetrical (Figs 15, 16), as the suture between them is not perpendicular to plate 3‴, but oblique. They have short spines disposed in rows parallel to the right and left posterior margins of the sulcus. Usually three spines, about 1 µm long, arise from the surface of each plate (Figs 15, 16, 21).

Large cells with intercalary growth bands are frequently found (Fig. 18). The width of the bands may be considerable, reaching up to 7 µm. Their surface is different from that of normal thecal plates as it is smooth with no pores or ornamentation (Fig. 18). Growth seems to occur on the margins of both adjoining plates, as the suture lies in the middle of the band.

Numerous discoidal greenish chromatophores, with a mean diameter of 3 µm, are present in the peripheral cytoplasm. A large number of small granules (starch?) is frequently observed in the cytoplasm of large cells (Fig. 7). The nucleus is located in the equatorial region of the cell. Cell division occurs by ecdysis.

The species is marine and the type locality is the Gulf of Trieste (Northern Adriatic Sea, Italy) in summer and autumn. It may become abundant in inshore areas where there is little water exchange.

Diagnosis

Cellula 30–52 µm longa, 21–36 µm lata; epitheca conica cum latere concavo ad apicem; ipotheca trapezoidea cum brevibus spinis; epitheca ipotheca longior; formula laminarum: pp, x, 4′, 3a, 7″, 6c, 5‴, 2⁗, 5s dispositione orthoperidinioidea; laminae cum poris ab anulis concentricis circumcinctis. Cingulum descendens, per dimidiam partem latitudinis dispositum; sulcus magnus, latus versus antapicem; laminae 1⁗ et 2⁗ cum duobus vel tribus brevibus spinis secus margines posticos sulci; cromatophora plurima, lutea-viridia; nucleus positus in centro.

Habitat: in aqua marina, sinus tergestinus, aestivalis et autumnalis.

Holotype: Figure 1.

Type locality: Gulf of Trieste (Northern Adriatic Sea, Italy), salinity 32–37‰, in summer and autumn (temperature 18–26 °C).

Geographical distribution: North-western coasts of the Adriatic Sea.

Discussion

The taxonomy of orthoperidinioid dinoflagellates at generic rank is mainly based on cingular and sulcal plates, as the tabulation of these regions, which are connected with the most dynamic parts of the cell, is particularly conservative (Balech 1959, 1980). Recently Indelicato and Loeblich (1986) in a revision of the marine peridinioid genera consider the relationships between hypothecal and cingular plates as a further taxonomic criterion. *Scrippsiella spinifera* sp. nov. shows the typical tabulation of the genus with no significant variation. A comparison of the *Scrippsiella* species so far described is made in Table I[1]. The

[1] At the moment of publication of this paper four new species of *Scrippsiella* have been found along the west coast of Scotland (Lewis, personal communication).

general morphology of *S. spinifera* sp. nov. shows a certain similarity with *S. sweenyae* Balech (Balech 1959), *S. trochoidea* (Stein) Loeblich (= *S. faeroense* sensu Balech, 1967) and *S. mexicana* Indelicato *et* Loeblich (Balech 1967, Indelicato and Loeblich 1986), due to the presence of a pronounced apical horn. However the hypotheca, divided by the sulcus into two lobes bearing antapical spines, clearly distinguish this species. Spines at the end of the cell are unusual in *Scrippsiella*: they have been found previously only in *S. caponii* Horiguchi *et* Pieenar (Horiguchi and Pienaar 1988 a).

The apical pore structures (pore plate and canal plate) do not differ from those described by Dodge and Hermes (1981) for *Protoperidinium* and related genera (including *Scrippsiella*), although the canal plate is very narrow and it is often covered by the raised margins of plates 2′ and 4′. The intercalary plate 2a is usually hexagonal, as in *S. trochoidea*, *S. subsalsa* (Ostenfeld) Steidinger *et* Balech, *S. hexapraecingula* Horiguchi *et* Chihara, *S. tinctoria* Indelicato *et* Loeblich, *S. mexicana* and *S. caponii*.

The thecal pores observed in *S. spinifera* sp. nov. are unusual for dinoflagellates and they show a certain resemblance only with the pores surrounded by five concentric ridges described by Dale (1977) in *Peridinium faeroense* Paulsen, now *Pentapharsodinium dalei* Indelicato *et* Loeblich.

The girdle has the typical configuration with five plates and a small transitional one. The sulcus extends very far in the hypotheca reaching the antapex as in *S. hexapraecingula* (Horiguchi and Chihara 1983). It is formed by five plates. The number of sulcal plates in other species of the genus varies between four and five. However, it should be noticed that the internal or median sulcal plate (is or ms), which was not detected in some species, is very small and not easily visible without staining (von Stosch 1969).

The size of *S. spinifera* shows considerable variation (the length ranges between 30 and 52 µm), if compared with the other species of the genus which are generally smaller. Only *S. subsalsa* has a similar range of sizes. Large cells always possess very wide intercalary bands and a dense cytoplasm. In some *Peridinium* species it has been shown that intercalary growth is slight in vegetative thecae and greater in the thecae of plano- and hypnozygotes (Pfiester and Skvarla 1979, 1980, Pfiester and Anderson 1987). The presence of very wide intercalary bands could indicate a zygotic condition. If this is so it is possible that large individuals of *S. spinifera* represent zygotes and not vegetative cells. It would be of particular interest to study the life cycle of this species including also cyst morphology and structure as some species of *Scrippsiella* are known to produce calcareous cysts and the cyst type constitutes an important character in separating peridinioid dinoflagellates at the generic level (Indelicato and Loeblich 1986).

Acknowledgements

The authors wish to thank Prof. E. Balech for taxonomical discussion, Miss Sabrina Ivicevic for the drawings and Mr Tito Ubaldini for technical assistance in SEM preparations.

References

Balech, E. 1959. Two new genera of dinoflagellates from California. *Biol. Bull. 116*: 195–203.

Balech, E. 1963. Dos dinoflagelados de una laguna salobre de la Argentina. *Universidad Nacional de la Plata. Facultad de Ciencias Naturales y Museo. Notas del Museo XX*: 111–123.

Balech, E. 1967. Dinoflagelados nuevos o interesantes del Golfo de Mexico y Caribe. *Rev. Mus. Argent. Cienc. Nat. "Bernardino Rivadavia" Inst. Nac. Invest. Cienc. Nat., Hydrobiologia 2*: 77–126.

Balech, E. 1980. On thecal morphology of dinoflagellates with special emphasis on cingular and sulcal plates. *An. Centro Cienc. del Mar y Limnol. Univ. Nal. Auton. Mexico 7*: 57–68.

Balech, E. and L. de Oliveira Soares 1966. Dos dinoflagelados de la bahia de Guanabara y proximidades (Brasil). *Neotropica 12*: 103–109.

Dale, B. 1977. New observations on *Peridinium faeroense* Paulsen (1905) and classification of small orthoperidinioid dinoflagellates. *Br. Phycol. J. 12*: 241–253.

Dodge, J. E. and H. B. Hermes. 1981. A scanning electron microscopical study of the apical pores of marine dinoflagellates (Dinophyceae). *Phycologia 20*: 424–530.

Honsell, G. and M. Cabrini. 1990. Analisi quali-quantitative del fitoplancton. *In*: *Crociera oceanografica Po-Gargano (settembre 1989). Caratterizzazione fisico-chimica e trofica della fascia costiera con il rilevamento dei principali inquinanti.* Regione Emilia Romagna, Assessorato Ambiente e Difesa del Suolo, Bologna. pp. 21–24.

Horiguchi, T. and M. Chihara. 1983. *Scrippsiella hexapraecingula* sp. nov. (Dinophyceae), a tide pool dinoflagellate from the Northwest Pacific. *Bot. Mag.* Tokyo *96*: 351–358.

Horiguchi, T. and R. N. Pienaar. 1988 a. A redescription of the tidal pool dinoflagellate *Peridinium gregarium* based on reexamination of the type material. *Br. phycol. J. 23*: 33–39.

Horiguchi, T. and R. N. Pieenar. 1988 b. Ultrastructure of a new sand-dwelling dinoflagellate, *Scrippsiella arenicola* sp. nov. *J. Phycol. 24*: 426–438.

Indelicato, S. R. and A. R. Loeblich III. 1985. A description of dinoflagellate *Scrippsiella tinctoria* sp. nov. *Jpn. J. Phycol. 33*: 127–134.

Indelicato, S. R. and A. R. Loeblich III. 1986. A revision of marine peridinioid genera (Pyrrhophyta) utilizing hypothecal-cingular plate relationships as a taxonomic guideline. *Jpn. J. Phycol. 34*: 153–162.

Loeblich, A. R. III. 1976. Dinoflagellate evolution: speculation and evidence. *J. Protozool. 23*: 13–28.

Loeblich, A. R. III, J. L. Sherley and R. J. Schmidt. 1979. Redescription of the thecal tabulation of *Scrippsiella gregaria* (Lombard *et* Capon) *comb. nov.* (Pyrrhophyta) with light and scanning electron microscopy. *Proc. Biol. Soc. Wash. 92(1)*: 45–50.

Lombard, E. H. and B. Capon. 1971. *Peridinium gregarium*: a new species of dinoflagellate. *J. Phycol. 7*: 184–187.

Montresor, M. and A. Zingone. 1988. *Scrippsiella precaria* sp. nov. (Dinophyceae), a marine dinoflagellate from the Gulf of Naples. *Phycologia 27*: 387–394.

Pfiester, L. A. and D. M. Anderson. 1987. Dinoflagellate reproduction. *In*: (F. J. R. Taylor, ed.) *The Biology of Dinoflagellates.* Blackwell Scientific Publications, Oxford. pp. 611–648.

Pfiester, L. A. and J. J. Skvarla. 1979. Heterothallism and thecal development in the sexual life history of *Peridinium volzii* (Dinophyceae). *Phycologia 18*: 13–18.

Pfiester, L. A. and J. J. Skvarla. 1980. Comparative ultrastructure of vegetative and sexual thecae of *Peridinium limbatum* and *Peridinium cinctum* (Dinophyceae). *Amer. J. Bot. 67*: 955–958.

Steidinger, K. A. and E. Balech. 1977. *Scrippsiella subsalsa* (Ostenfeld) comb. nov. (Dinophyceae) with a discussion on *Scrippsiella. Phycologia 16*: 69–73.

Stosch, H. A. von. 1969. Dinoflagellaten aus der Nordsee I. Über *Cachonina niei* Loeblich 1968, *Gonyaulax grindley* Reinecke 1967, und eine Methode zur Darstellung von Peridineenpanzern. *Helgoland Wiss. Meeresunt. 19*: 558–568.

Botanica Marina
Vol. 34, pp. 177–185, 1991

The Vegetative Thallus of *Pterocladia capillacea* (Gelidiales, Rhodophyta)

I. An Ultrastructural and Cytochemical Study

N. Rascio, P. Mariani, F. Dalla Vecchia and R. Trevisan*

Dipartimento di Biologia, Università di Padova, Via Trieste 75, I-35121 Padua, Italy

* *Istituto di Biologia Agraria, Università di Potenza, Via Nazario Sauro 75, I-85100 Potenza, Italy*

(Accepted 16 December 1990)

Abstract

An ultrastructural study of the vegetative thallus of the red alga *Pterocladia capillacea* (Gmel.) Born. *et* Thur. has been carried out. Different kinds of cells have been recognized and described. The small cortical cells, rich in plastids, seem to be more specifically involved in photosynthesis, while the large medullary cells might essentially operate in nutrient storage. The distribution of mechanical elements has been well defined in different thallus regions. The cell wall organization and the localization of agar components in the extracellular compartment have been cytochemically analyzed.

Introduction

The Gelidiales are a small but economically important order of the Rhodophyta, comprising many species producing large quantities of the phycocolloid agar (Akatsuka and Iwamoto 1979, Dawes 1981).

The agarophytes of the genus *Pterocladia* (Friedlander *et al.* 1981, Berchez and Oliveira 1986 a, b) to which *Pterocladia capillacea* (Gmel.) Born. *et* Thur. belongs are particularly interesting, because of their high agar quality.

This perennial species is common along the Italian coasts, where it grows at the high infralitoral level on rocky substrates (Tolomio *et al.* 1986).

The present paper deals with an ultrastructural and cytochemical study of the vegetative thallus of this species, which is little known from a fine morphological point of view, in order to increase our knowledge as well as to obtain more information about some morphological features considered distinctive for the genus *Pterocladia*.

The morphology and organization of the thallus have been investigated by light and electron microscopy. The different cellular types, the cell wall structure and the phycocolloid distribution in the extracellular regions have been described.

Material and Methods

Vegetative (sterile) thalli (about 10 cm high) of *Pterocladia capillacea* (Gmel.) Born. *et* Thur. were collected in October along the coast of Capo Molino (Catania, Sicily).

Light and electron microscopy

One mm long segments from young (the middle region of a young lateral branch) and mature regions (the middle region of a main axis) of the thallus were fixed overnight in 6% glutaraldehyde in 0.1 M cacodylate buffer (pH 6.9), washed in buffer, then postfixed two hours in 1% osmium tetroxide in 0.1 M cacodylate buffer (pH 6.9), and dehydrated in a graded series of ethyl alcohol and propylene oxide. The samples were embedded in an Epon-Araldite mixture.

For light microscopy, thin sections (1 µm thick) were cut with an Ultracut Reichert-Jung ultramicrotome, stained with 1% toluidine blue and 1% Na tetraborate

(1 : 1 by volume) and observed and photographed under a Leitz ortholux microscope (LM).

For transmission electron microscopy ultra-thin sections, cut with the same ultramicrotome, were post-stained with uranyl acetate and lead citrate and examined with a Hitachi HS9 electron microscope (TEM) operating at 75 kV.

Cytochemical tests

Thin sections were stained with acidified toluidine blue, which is considered a diagnostic test for sulphated polysaccarides. A drop of 0.05% toluidine blue in 0.02 M phosphate buffer (pH 4.4) was placed on the sections for a few minutes (Feder and O'Brien 1968). The sections were then rinsed with distilled water, dried and observed under the LM. The sulphated polysaccarides were recognizable by their characteristic red metachromasia (O'Brian *et al.* 1964).

Ultra-thin sections mounted on gold grids were subjected to the periodic acid – thiosemicarbazide – silver proteinate (PATAg) procedure for ultrastructural localization of polysaccharides (Thiéry 1967). Sections were oxidized first with 1% periodic acid for 25 min and then were treated with thiosemicarbazide (1% in 10% acetic acid) for 15 h at room temperature. Successively they were stained with 1% aqueous silver proteinate for 30 min in the dark, rinsed with distilled water and observed under TEM without uranyl acetate and lead citrate contrast. Controls were carried out either by replacing periodic acid with hydrogen peroxide (Roland 1978), omitting the thiosemicarbazide reaction or without periodic acid oxidation (Courtoy and Simar 1974). This procedure reveals carbohydrates with molecules having vic-glycol groups.

Results and Discussion

Morphology of the thallus

The brownish-red thallus of *Pterocladia capillacea* (5–10 cm high) was bushy, forming dense wiry tufts. The thallus branching produced fronds regularly decreasing in length from the base to the apex. The younger branches appeared flattened, while older portions of the thallus exhibited a more swollen elliptical shape in transverse section. Frequently the thallus was covered or invaded by numerous small epiphytic organisms (see Figs 3, 13).

Cellular types

Under LM, a cortex, consisting of some layers of small cells, and a large medulla with more voluminous cells could be distinguished in a cross section of the young flattened region of the thallus (Fig. 1). Among these 'parenchymatous' cells (as defined by Akutsuka and Iwamoto 1979) numerous thick-walled elements, named 'hyphae' or 'intercellular fibres' by Feldman and Amel (1934, 1936) and 'rhizines' by Dawes (1981) were visible. This type of mechanical cell was particularly concentrated in the two lateral regions of the branch to form rod-like structures.

In the mature thallus the medullary region with larger cells (Fig. 2) and the cortical one with smaller cells (Fig. 3) were still recognizable. These living cells showed characteristic pit connections (Rascio *et al.* 1991) inserted in their cell walls. The hyphae, either single or in small groups, were now regularly distributed through the whole section, from the central region (Fig. 2) to the most peripheral cell layers (Fig. 3).

The hyphae have represented an important taxonomical character within the Gelidiceae. On the basis of their absence or presence, in fact, very similar genera have been distinguished (Feldmann and Hamel 1934).

According to the literature (Okamura 1934, Feldmann and Hamel 1936, Dawes 1981), hyphae are limited to the peripheral region of the thallus in *Gelidium*, while in *Pterocladia* they occur in the inner medullary region only.

Our observations have shown that in *Pterocladia capillacea* these mechanical elements have a wider distribution, going from the centre of the medulla to the outer cortical cell layers. Moreover the distribution of hyphae is variable and changes during the growth of the thallus. In fact, in the younger, flattened

Figs 1–3. Thin sections of the vegetative thallus of *Pterocladia capillacea* observed at LM. The young thallus (Fig. 1) shows small cortical and voluminous medullary cells. Hyphae are spread in the centre and form large clusters (arrows) in the lateral area of the section. In the mature thallus hyphae are also visible in the medullary region (Fig. 2) and in the cortical cell layers (Fig. 3) extending towards the outmost strata. Pit connections can be seen between adjacent cells. The thallus surface is invaded by epiphytes (arrow).

Abbreviations used in the figures: cc = cortex cells; cw = cell wall proper; ep = epiphytes; fs = floridean starch; hy = hypha; icw = inner cell wall region; im = intercellular matrix; ir = intercellular region; mc = medulla cell; ocw = outer cell wall region; p = plastid; pbs = phycobilisomes; pc = pit connection; pm = plasma membrane; pv = periplasmic vesicles.

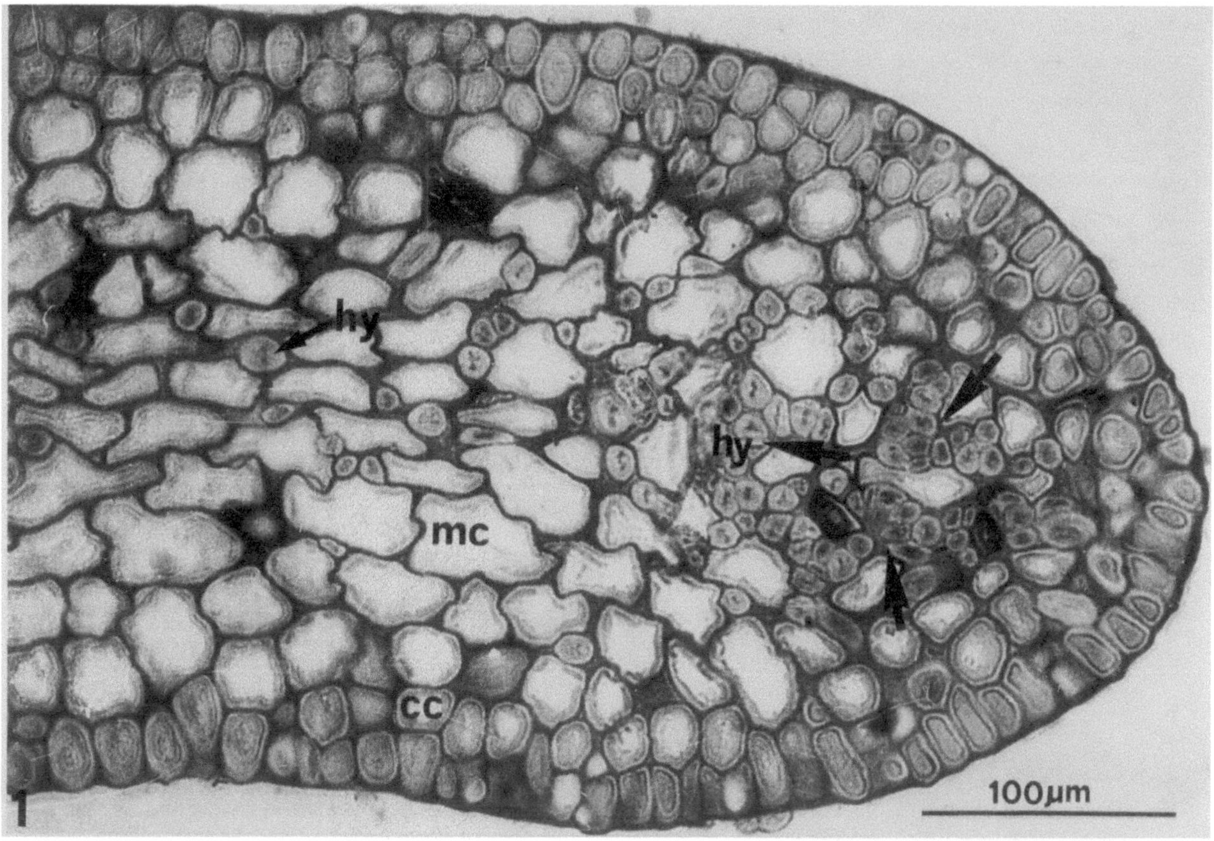
hy
hy
mc
cc
100µm
1

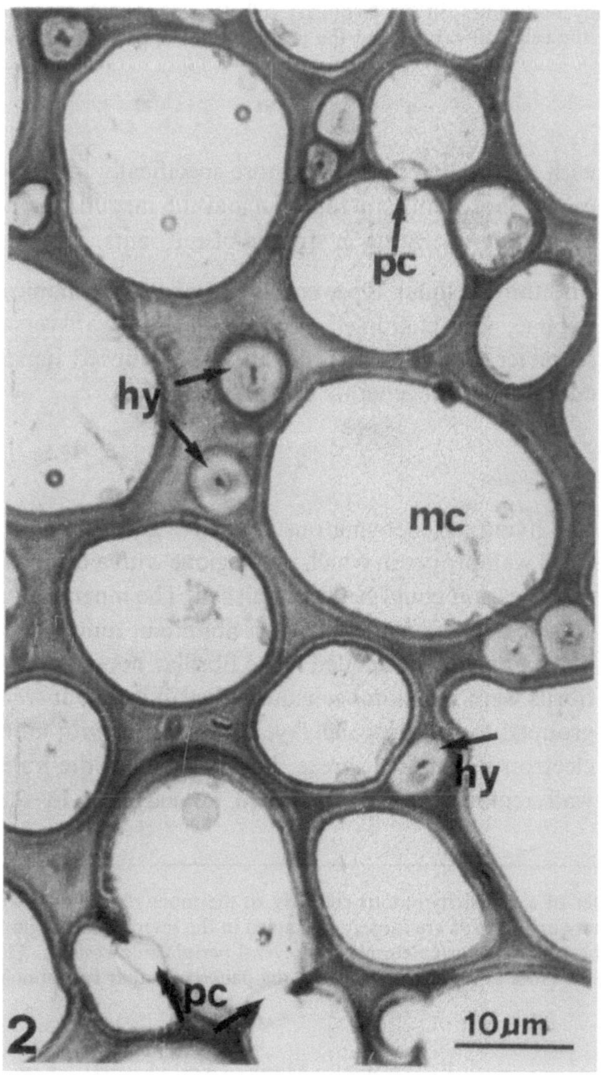
pc
hy
mc
hy
pc
10µm
2

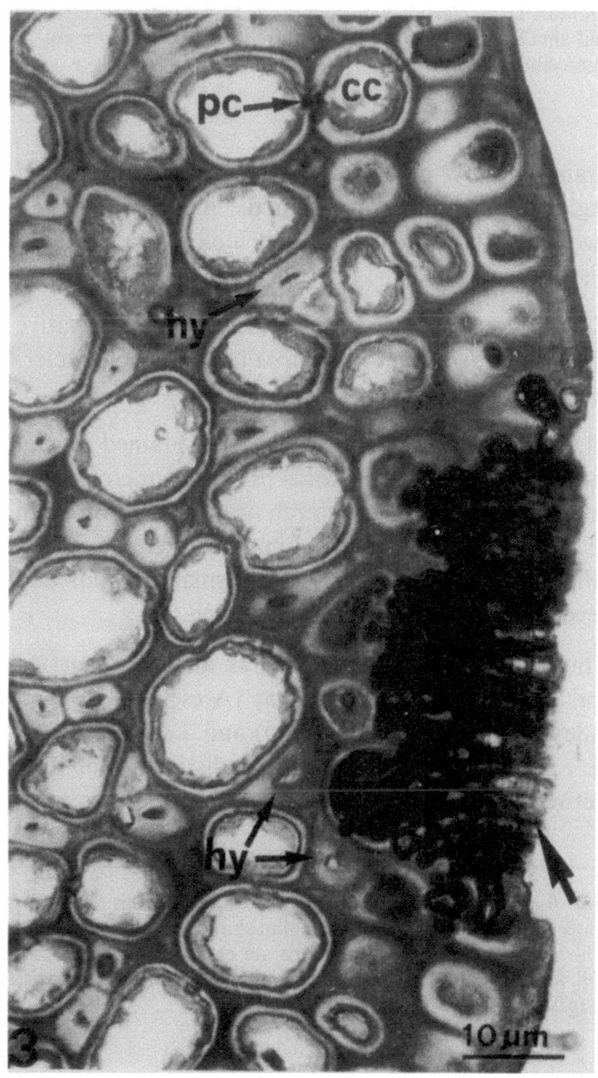
pc
cc
hy
hy
10µm
3

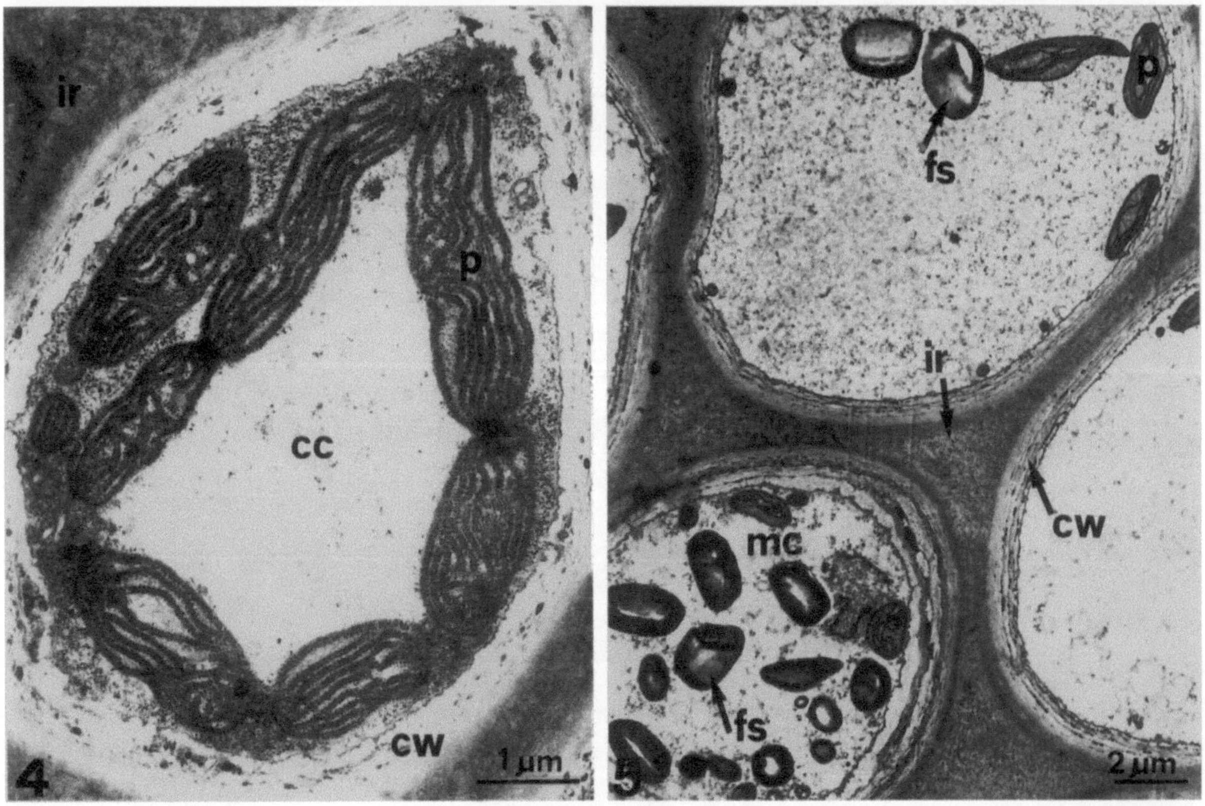

Figs 4, 5. TEM micrographs showing different cellular types in the thallus of *Pterocladia capillacea*. The cortical cell (Fig. 4) contains large plastids with numerous thylakoids provided with phycobilisomes. In the medullary cells (Fig. 5) plastids are rare and small and several floridean starch granules are present. Note the cell wall proper and the loose fibrillar appearance of the intercellular region.

branches most of them are grouped in two lateral regions, while successively, in the more mature and swollen parts of the thallus, they look homogeneously distributed.

The different cellular types present in the vegetative thallus of *Pterocladia capillacea* have been characterized by TEM.

The small peripheral cells (Fig. 4) contained several large plastids with thylakoids bearing numerous phycobilisomes. Floridean starch was rarely found in these cells. In the more voluminous medullary cells (Fig. 5), on the contrary, a few small plastids and several starch grains usually occurred.

Thus in *Pterocladia capillacea*, as already suggested for *Phymatolithon lenormandii* (Aresch. *in* J. Ag.) Adey (Millson and Moss 1985) and *Vidalia volubilis* (L.) Ag. (Andreoli *et al.* 1985), a distinction of roles between the two regions of the thallus seems to exist, with the peripheral region more specifically involved in the photosynthetic function and the medullary one essentially operating in storage of nutrients.

The third cellular type, consisting of the mechanical hyphae, was characterized by a narrow transversal diameter, a very thick cell wall and a reduced lumen containing a degenerate cytoplasm (Fig. 8).

Cell walls

The living parenchymatous cells were surrounded by a cell wall proper in which two regions with a different organization could be distinguished. The inner region (Figs 6, 7) was made up of an abundant amorphous matrix and an embedded loose fibrillar network. The fibrils were not homogeneously distributed, but were grouped to form parallel layers alternating with more electron transparent areas. In this manner, the inner wall region assumed a striped appearance. In this

Figs 6–9. TEM micrographs of cell walls. Note in the wall proper of a parenchymatous cell (Fig. 6) the inner region with the loose striped fibrils. The outer region shows a more close fibrillar network. Fibrils are loosely entangled in the intercellular region. The two different cell wall regions are evident in Figure 7, also showing the plasmalemma and several periplasmic vesicles. The cell wall of the hypha looks very homogeneous and compact (Fig. 8). Note at high magnification the packed fibrillar component (Fig. 9).

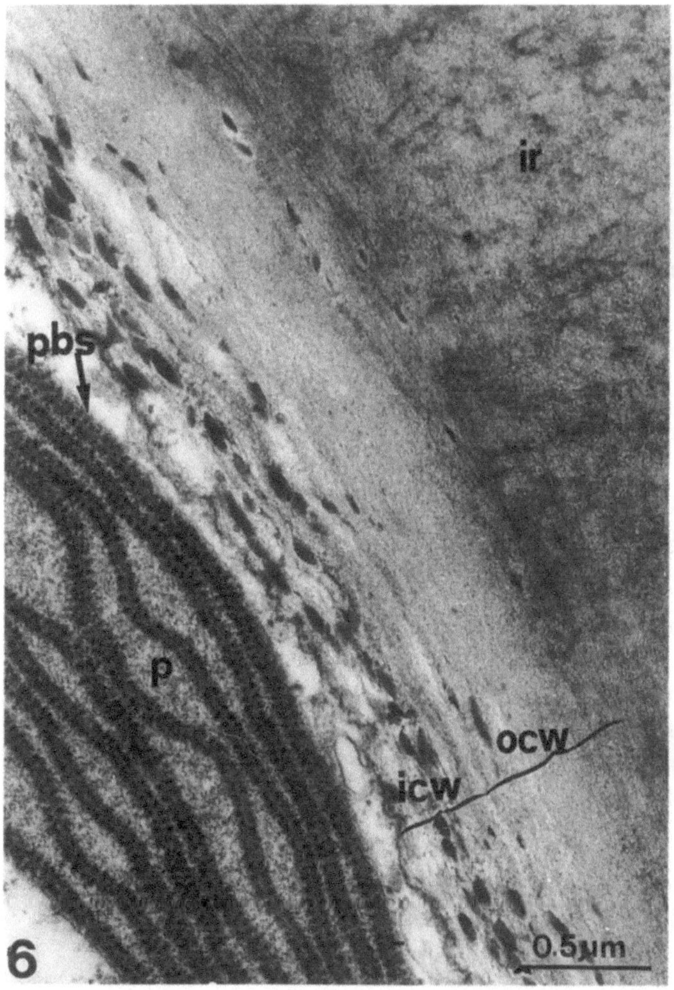
ir
pbs
p
ocw
icw
0.5µm
6

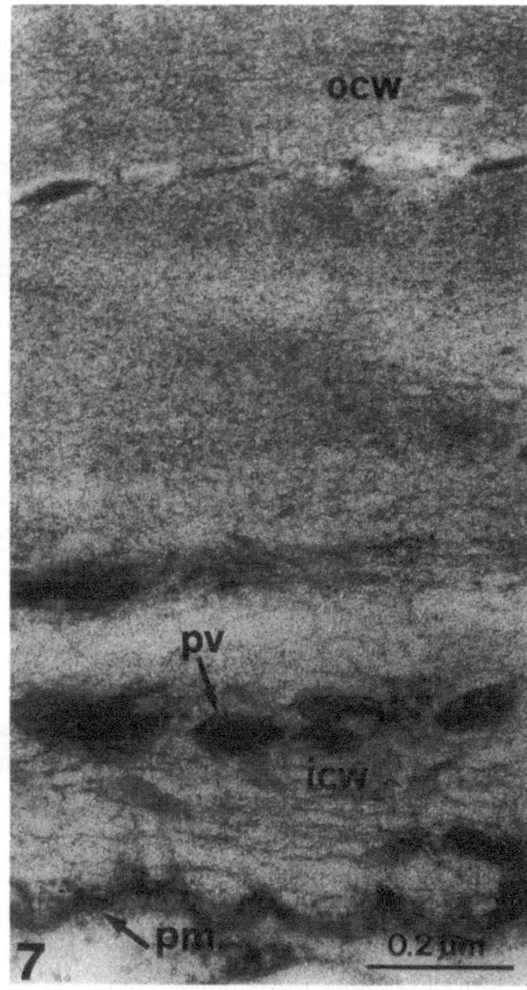
ocw
pv
icw
pm
0.2µm
7

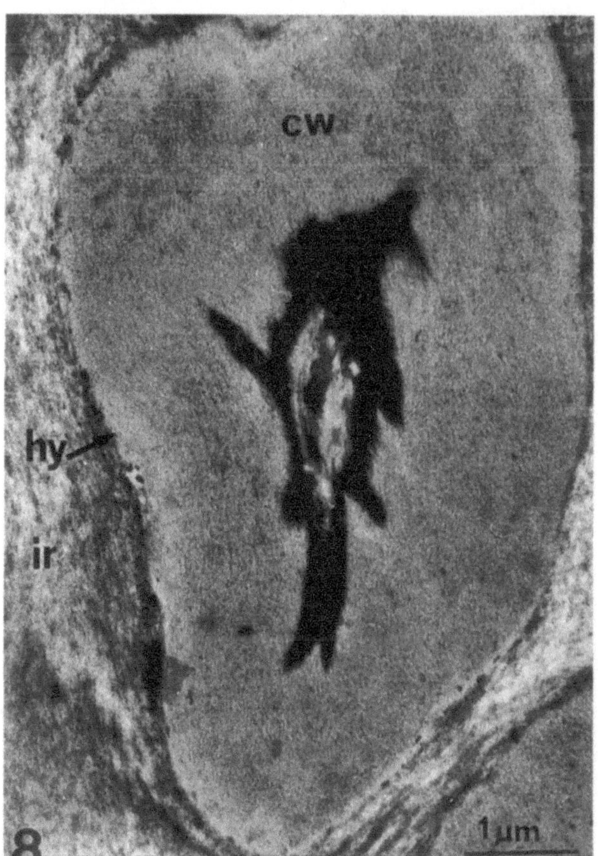
cw
hy
ir
1µm
8

0.2µm
9

region several electron dense periplasmic vesicles, like those already described in other red algae (Bisalputra *et al.* 1967, Young 1980, Verdus *et al.* 1986) were visible.

Nearer the external region of the cell wall, the fibrillar network became more closely arranged (Figs 6, 7), so that the outer wall region appeared as a rather homogeneous layer with fibrils prevailing over the amorphous matrix.

The organization of the cell wall of the hyphae was very different. In this kind of mechanical cell, the two distinct regions found in the parenchymatous ones did not occur, and the cell wall looked homogeneous and compact throughout its entire thickness (Fig. 8). At high magnifications (Fig. 9) a closely packed fibrillar component could be seen.

The parenchymatous cells as well as the thick walled hyphae were embedded in an intercellular matrix. As observed in other red algae (Lichtle 1975, Verdus *et al.* 1986, Mariani *et al.* 1990), also in these intercellular regions a fibrillar component, with fibrils entangled in a very loose network is present (Figs 4, 5, 6, 8).

According to Mariani *et al.* (1990), the presence of the fibrillar skeletal component in the intercellular areas might be related to a role in supporting the mass of amorphous matrix polysaccharides.

The fibrils of both the parenchymatous cell wall proper (Fig. 10) and the intercellular region (Figs 11, 12), were positive to the PATAg test and probably consisted of cellulose, as already assumed for red algae belonging to the Florideophycidae (Dawes *et al.* 1961, Bisalputra *et al.* 1967, Dodge 1973, Lichtle 1975, Tripodi and De Masi 1975, La Claire II and Dawes 1976, Duckett and Peel 1978, Young 1980, Brawley and Wetherbee 1981, Kloareg and Quatrano 1988).

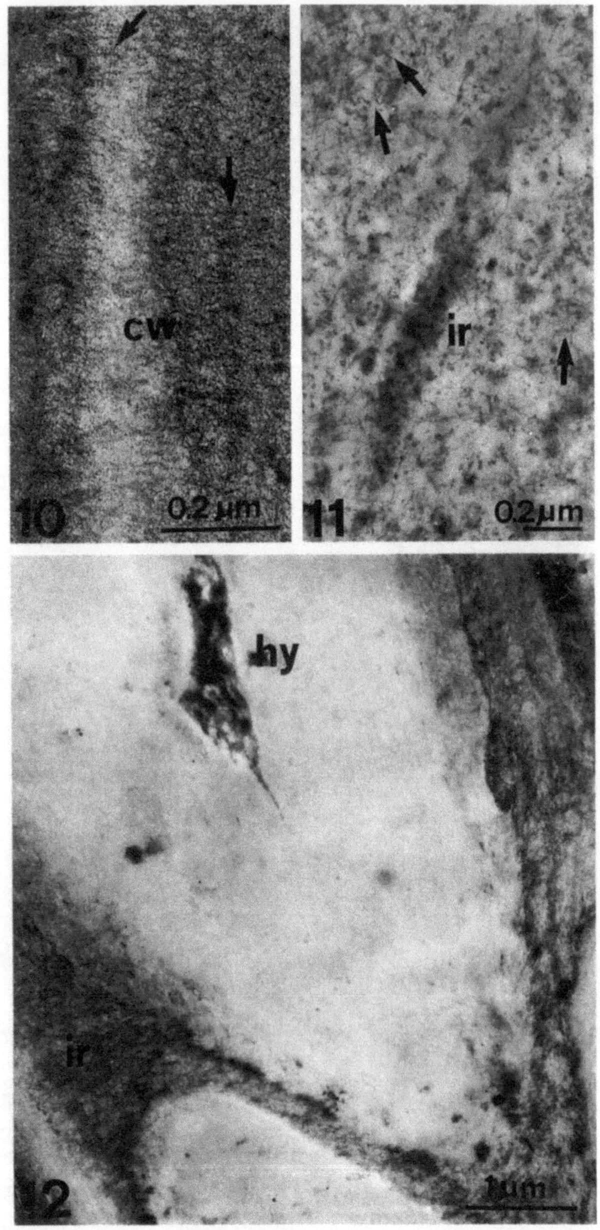

Figs 10–12. Responses of cell walls to PATAg staining technique for polysaccharides. The microfibrils of both the cell wall proper (arrows) (Fig. 10) and the intercellular region (arrows) (Fig. 11) are positive. The thick cell wall of the hypha (Fig. 12) has reacted faintly, while in the surrounding intercellular region a clear positive reaction can be seen.

The thick walls of the hyphae, which are considered to be essentially composed of cellulose (Akatsuka and Iwamoto 1979), reacted quite faintly to the PATAg staining (Fig. 12). This fact, however, might be explained by considering the very tight association of the cell wall fibrils in these mechanical elements. In fact, this kind of organization may prevent the reaction from taking place, by hiding the reactive groups of the cellulose molecules (Roland 1978).

In red algae, the agar is an amorphous constituent of the extracellular compartment, which becomes economically important in agarophyte species. The term agar refers to a complex mixture of related polysaccharides, consisting of a neutral component (agarose) and a sulphated fraction (sulphated agarose and sulphated galactan) which, in different species, can be present in different ratios, thus conferring peculiar characteristics to the commercial product (Friedlander *et al.* 1981, Kloareg and Quatrano 1988).

Using toluidine blue (pH 4.4) staining, the localization and distribution of the sulfated polysaccharides of agar were indicated in the thallus of *Pterocladia capillacea.*

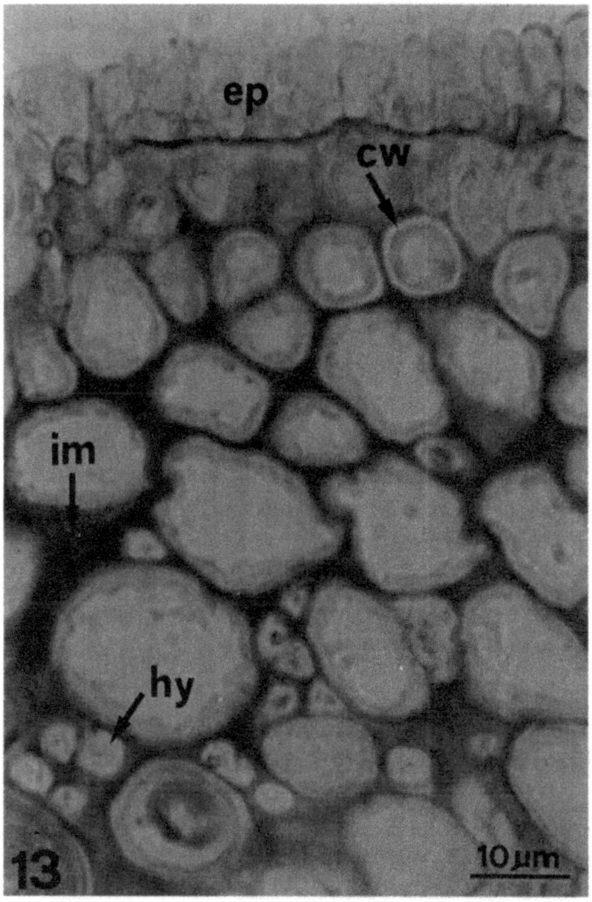

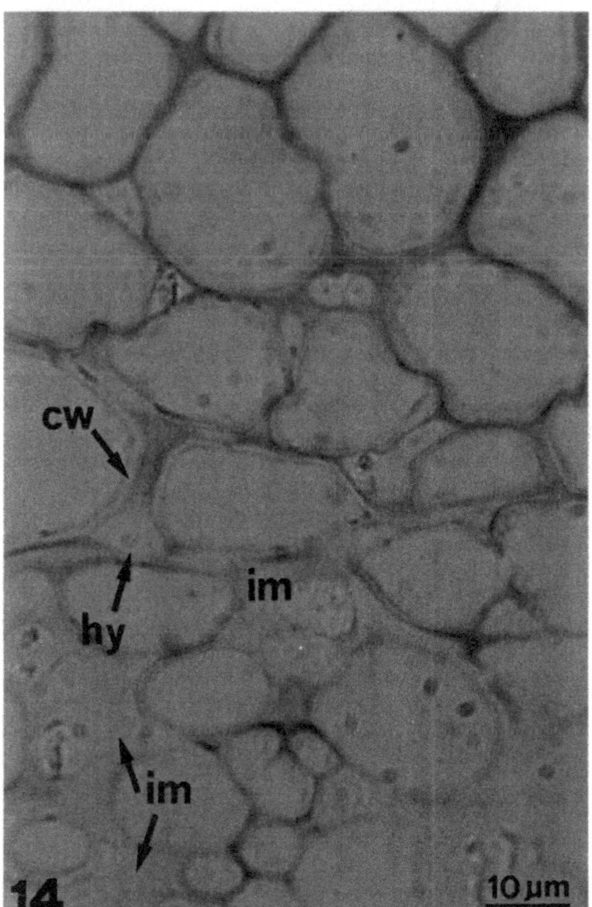

The specific red metachromasia (O'Brien *et al.* 1964) was observed in the intercellular matrix, while the cell wall proper of both parenchymatous cells and hyphae was negative or barely metachromatic (Figs 13, 14). The positive reaction was most obvious in the outer region (Fig. 13) and gradually decreased towards the inner medullary region of the thallus (Fig. 14).

The sulphated polysaccharide concentration thus appears to become lower going from the intercellular matrix to the cell wall proper and from the outside to the centre of the thallus.

The localization of the agar sulphate polysaccharides mainly in the intercellular matrix was generally observed in red algae (Gordon and McCandless 1973, La Claire II and Dawes 1976, Akatsuka and Iwamoto 1979, Verdus *et al.* 1986, Kloareg and Quatrano 1988). The distribution in the thallus regions, on the contrary, showed some differences, according to the organism considered. In *Eucheuma nudum* J. Agardh., for example, this component was distributed in all the regions of the thallus (Le Claire II and Dawes 1976), whereas in *Chondrus crispus* Stackhouse it mainly occurred in the cortical region, being also present in the medulla (Gordon and McCandless, 1973). In *Gelidium pacificum* Okamura, Akatsuka and Iwamoto (1979) found an intense metachromatic red staining of the intercellular matrix in the medullary region, but not in the cortical one. However, they justified this fact as due to the narrowness of the intercellular spaces among the cortical cells, and considered the agar sulphated polysaccarides as distributed in all the thallus regions.

In *Pterocladia capillacea* large masses of intercellular matrix were present in both the cortical and medullary regions. Thus the gradual reduction of metachromatic staining from the cortex towards the centre of the thallus could be imputed as a real decrease in the acidic component of the agar and to a plausible increase in its neutral fraction, the latter becoming the prevalent one in the large intercellular areas of the innermost medullary region. This might link up with chemical data showing that the agar of *Pterocladia*, with respect to that of other agarophytes, is characterized by a higher concentration of neutral agarose, which affects its physico-chemical properties and gives it its high commercial value (Friedlander *et al.* 1981).

Figs 13, 14. Responses of the intercellular material to toluidine blue staining for sulphated groups. The cell wall proper as well as the hyphae appear uncoloured, meanwhile the intercellular matrix shows a very intense purple metachromasia in the periphery of the thallus (Fig. 13). The red colouring is decidedly less intense in the intercellular matrix of the central region (Fig. 14).

Conclusions

The vegetative thallus of *Pterocladia capillacea* is characterized by the presence of two distinct regions, possibly playing different roles in the algal metabolism. A mainly photosynthetic function can be hypothesized for the small cortical cells, which are rich in plastids. The voluminous medullary cells, because of the scarcity of plastids and the abundance of starch grains, might play an essential role in storage of nutrients.

The thallus contains numerous hyphae, whose cellulosic cell walls show a remarkable compactness. The distribution of these mechanical elements has been extensively studied. In contrast to the reports in the literature it was found that the hyphae in *Pterocladia capillacea* are not limited to the inner medullary region but extend up to the peripheral cell layers of the thallus. Moreover their localization changes according to the thallus age, the young stage being characterized by large lateral groups, while the mature stage has a more homogeneous distribution. It might be suggested that this different distribution of mechanical elements is somehow related to the changes in thallus shape, from flattened to elliptic in section with aging.

The parenchymatous cells are surrounded by a cell wall proper, which probably consists, at least partially, of cellulose. Outside this cell wall an intercellular region is present, which shows a loose microfibrillar network that supports the amorphous matrix containing agar. The sulphate polysaccharide component of the agar does not appear to be homogeneously distributed in the thallus. In fact it is more abundant in the intercellular matrix of the peripheral region and gradually decreases towards the central area, where the neutral component of the phycocolloid, that this species produces at quantitatively and qualitatively high levels, seems to become prevalent.

Acknowledgements

The authors are very grateful to Mr Giorgio Varotto for the technical assistance. This work was supported by grants from C. N. R. and M. P. I.

References

Akatsuka, I. and K. Iwamoto. 1979. Histochemical localization of agar and cellulose in the tissue of *Gelidium pacificum* (Gelidiaceae, Rhodophyta). *Bot. Mar. 22*: 367–370.

Andreoli, C., C. Tolomio, N. Rascio and F. Dalla Vecchia. 1985. Ultrastructural observations by TEM and SEM on the vegetative thallus of the red alga *Vidalia volubilis* (L.) Ag. (Ceramiales). *Abstr. II Internat. Phycological Congress, Copenhagen, Denmark,* 6.

Berchez, F. A. DeS. and E. C. De Oliveira. 1986 a. Distribution of *Pterocladia capillacea* (Rhodophyta, Gelidiaceae) in sea coasts of Brasil. *Abstr. XIIth Internat. Seaweed Symp., Sao Paulo, Brasil, MS5–16*: 28.

Berchez, F. A. DeS. and E. C. De Oliveira. 1986 b. *Pterocladia capillacea* (Rhodophyta, Gelidiaceae) growth rate in vitro. *Abstr. XIIth Internat. Seaweed Symp., Sao Paulo, Brasil, P5–17*: 29.

Bisalputra, T., P. C. Rusanowski and W. S. Walker. 1967. Surface activity, cell wall, and fine structure of pit connection in the red alga *Laurencia spectabilis*. *J. Ultrastruct. Res. 20*: 277–289.

Brawley, S. H. and R. Wetherbee. 1981. Cytology and ultrastructure. *In*: (C. S. Lobban, M. J. Wynne, eds) *The Biology of Sea Weeds*. Blackwell Scientific Publications, Oxford. pp. 248–299.

Courtoy, R. and J. Simar. 1974. Importance of controls for the demonstration of carbohydrates in electron microscopy with the silver methenamine or the thyocarbohydrazide-silver proteinate methods. *J. Microscopy 100*: 199–211.

Dawes, C. J., F. M. Scott and E. Bowler. 1961. A light- and electron-microscopic survey of algal cell walls. I. Phaeophyta and Rhodophyta. *Am. J. Bot. 48*: 925–934.

Dawes, C. J. 1981. *Marine Botany*. Wiley Interscience Publication, New York. 629 pp.

Dodge, J. D. 1973. *The Fine Structure of Algal Cells*. Academic Press, London and New York. pp. 261.

Duckett, J. G. and M. C. Peel. 1978. The role of transmission electron microscopy in elucidating the taxonomy and phylogeny of the Rhodophyta. *In*: (D. E. G. Irvine and J. H. Price, eds) *Modern Approaches to the Taxonomy of Red and Brown Algae*. Academic Press, London and New York. pp. 157–204.

Feder, N. and T. P. O'Brien. 1968. Plant microtechnique: Some principles and new methods. *Am. J. Bot. 55*: 123–142.

Feldmann, J. and G. Hamel. 1934. Observations sur quelques Gélidiacéés. *Rev. Génér. de Bot. 46*: 1–84.

Feldmann, J. and G. Hamel. 1936. Floridéés de France, VII Gelidiales. *Rev. Algol. 9*: 95–140.

Friedlander, M., Y. Lipkin and W. Yaphe. 1981. Composition of agars from *Gracilaria* cf. *verrucosa* and *Pterocladia capillacea*. *Bot. Mar. 24*: 595–598.

Gordon, E. M. and E. L. McCandless. 1973. Ultrastructure and histochemistry of *Chondrus crispus* Stackhouse. *Proc. Nova Scot. Inst. Sci. 27* Supp: 111–133.

Kloareg, B. and R. S. Quatrano. 1988. Structure of the cell walls of marine algae and ecophysiological functions of the matrix polysaccharides. *Oceanogr. Mar. Biol. Annu. Rev. 26*: 259–315.

La Claire II, J. W. and C. J. Dawes. 1976. An autoradiographic and histochemical localization of sulfated polysaccharides in *Eucheuma nudum* (Rhodophyta). *J. Phycol. 12*: 368–375.

Lichtlé, C. 1975. Etude ultrastructurale de la paroi du *Polysiphonia elongata* (Harv.) Rhodophycée, Floridée, a l'aide d'action ménagée d'enzymes. *J. Microsc. Biol. Cell. 23*: 93–104.

Mariani, P., C. Tolomio, B. Baldan and P. Braghetta. 1990. Cell wall ultrastructure and cation localization in some benthic marine algae. *Phycologia 29*: 253–262.

Millson, C. and B. L. Moss. 1985. Ultrastructure of the vegetative thallus of *Phymatolithon lenormandii* (Aresch. *in* J. Ag.) Adey. *Bot. Mar. 28*: 123–132.

O'Brien, T. P., N. Feder and M. E. McCully. 1964. Polychromatic staining of plant cell walls by toluidine blue O. *Protoplasma 59*: 358–373.

Okamura, K. 1934. On *Gelidium* and *Pterocladia* of Japan. *J. Imp. Fisher Inst. 29*: 1–87.

Rascio, N., P. Mariani, F. Dalla Vecchia and R. Trevisan. 1991. The vegetative thallus of *Pterocladia capillacea* (Gelidiales, Rhodophyta): II. The pit connections. *Bot. Mar. 34*: 187–194.

Roland, J. C. 1978. General preparation and staining of thin sections. *In*: (J. L. Hall, ed.) *Electron Microscopy and Cytochemistry of Plant Cells*. Elsevier/North Holland Biochemical Press, Amsterdam, Oxford and New York. pp. 1–62.

Thiéry, J. P. 1967. Mise in évidence des polysaccharides sur coupes fines en microscopie électronique. *J. Microscopie 6*: 987–1017.

Tolomio, C., C. Andreoli, N. Rascio and L. Talarico. 1986. *Pterocladia pinnata* (Huds.) Papenf. (Gelidiales) in Italian seas. *Abstr. XIIth Internat. Seaweed Symp., Sao Paulo, Brasil, P1–177*: 109.

Tripodi, G. and F. De Masi. 1975. Cytological localization of polysaccharidic molecules in some red algae. *J. Submicrosc. Cytol. 7*: 197–209.

Verdus, M. C., D. Christiaen, T. Stadler and H. Morvan. 1986. Etude ultrastructurale et cytochimique de la paroi cellulaire chez *Gracilaria verrucosa* (Rhodophyceae). *Can. J. Bot. 64*: 96–101.

Young, D. N. 1980. Unusual cell wall ultrastructure in *Antithamnion* (Rhodophyta). *Br. Phycol. J. 15*: 119–124.

Botanica Marina
Vol. 34, pp. 187–194, 1991

The Vegetative Thallus of *Pterocladia capillacea* (Gelidiales, Rhodophyta)

II. Pit Connections

N. Rascio, P. Mariani, F. Dalla Vecchia and R. Trevisan*

Dipartimento di Biologia, Università di Padova, Via Trieste 75, I-35121 Padua, Italy

* *Istituto di Biologia Agraria, Università di Potenza, Via Nazario Sauro 75, I-85100 Potenza, Italy*

(Accepted 16 December 1990)

Abstract

Pit connections have been investigated in the vegetative thallus of the red alga *Pterocladia capillacea* (Gmel.) Born. *et* Thur. Both a polysaccharidic and a proteinaceous component have been found by cytochemical tests in these pit connections, which did not seem to be covered by a cap membrane.

Introduction

In all the red algae belonging to the Florideophycidae, as well as in the *Conchocelis* stage of Bangiophycidae, cytokinesis does not occur completely and in the wall between two adjacent cells the large persisting holes are occluded by peculiar plug-like structures, defined as pit connections or pit plugs (Pueschel and Cole 1982) and synapses by French authors (Peyrière 1977).

The structure and organization of the pit connections are generally considered to have a relevant taxonomical significance (Pueschel and Cole 1982, Gabrielson and Garbary 1986).

On the contrary, there are conflicting opinions on their chemical composition (Ramus 1971, Pueschel 1980) as well as on their function. According to some researchers (Wetherbee 1979, 1980, Wetherbee and Scott 1980, Wetherbee and Kraft 1981) the pit connections can play a role in intercellular transport. This assumption is not shared by other authors, who tend to criticize such an involvement of these structures (Goff 1979, Kugrens 1980), suggesting different roles for them (Broadwater and Scott 1982, Goff and Coleman 1985) as, for example, a mechanical one (Pueschel 1988).

We have studied the ultrastructure of pit connections of *Pterocladia pacillacea* (Gmel.) Born. *et* Thur. and compared it with the description of pit connections of other species belongig to the same order of Gelidiales. Cytochemical tests have been performed in an attempt to determine the chemical composition of these interesting structures.

Material and Methods

Vegetative thalli of *Pterocladia capillacea* (Gmel.) Born. *et* Thur. were collected in October along the coast of Capo Molino (Catania, Sicily).

Light and electron microscopy

Segments (1 mm long) from mature regions of the thallus were fixed overnight in 6% glutaraldehyde in 0.1 M cacodylate buffer (pH 6.9), washed in buffer, then postfixed for two hours in 1% osmium tetroxide in 0.1 M cacodylate buffer (pH 6.9), and dehydrated in a graded series of ethyl alcohol and propylene oxide. The samples were embedded in an Epon-Araldite mixture.

For light microscopy thin sections (1 µm thick) were cut with an Ultracut Reichert-Jung ultramicrotome, stained with 1% toluidine blue and 1% tetraborate (1 : 1 by volume) and observed and photographed under a Leitz ortholux micriscope (LM).

For transmission electron microscopy, ultra-thin sections cut with the same ultramicrotome were stained with uranyl acetate and lead citrate and examined with a Hitachi HS9 electron microscope (TEM) operating at 75 kV.

Cytochemical tests

Period acid – thiosemicarbazide – silver proteinate staining for polysaccharides. Utra-thin sections mounted on gold grids were subjected to the periodic acid – thiosemicarbazide – silver proteinate (PATAg) procedure for ultrastructural localization of polysaccharides (Thiéry 1967).

Sections were oxidized first with 1% periodic acid for 25 min and then were treated with thiosemicarbazide (1% in 10% acetic acid) for 15 h at room temperature. They were then stained with distilled water, and observed under the TEM without uranyl acetate and lead citrate contrast.

Controls were carried out either by replacing periodic acid with hydrogen peroxide (Roland 1978), omitting the thiosemicarbazide reaction or without periodic acid oxidation (Courtoy and Simar 1974). This procedure reveals carbohydrates with molecules having vic-glycol groups.

Phosphotungstic acid staining for proteins. According to the Silverman and Glick (1969) method, thallus samples were fixed overnight in 1% glutaraldehyde in 0.1 M potassium phosphate buffer (pH 7.4), washed in buffer, stained for 3 h in 5% phosphotungstic acid (PTA) in 6.25% Na_2SO_4 and washed for 3 h in 2% ammonium acetate (formic acid added to pH 2.0). After dehydration in graded ethanol with added formic acid (pH 2.0) and in propylene oxide, the samples were embedded in Epon-Araldite. The ultrathin sections were cut onto 0.01 N HCl.

For control ultra-thin sections were washed in droplets of 0.1 N NaOH (pH 13) to remove PTA before examination under TEM.

By this procedure proteins can be specifically revealed because PTA acts as an anionic stain for their positively charged groups. Lipids do not stain and cell membranes are demonstrable only as negative images (Silverman and Glick 1969).

Results and Discussion

Numerous pit connections were present in the wall proper of the living parenchymatous cells (Rascio *et al.* 1991) of *Pterocladia capillacea*. In the cortical region of the thallus they were smaller (about 1 μm in diameter), inserted within the thickness of the cell wall and clepsydra-like shaped (Figs 1, 3).

The pit connections between the cells of the medullary region were larger (with diameters of 4–5 μm) and lenticular in shape (Figs 2, 4).

Independently from their different shapes and dimensions, and probably related to both the different wall thicknesses and cell sizes, the pit connections showed the same pattern of organization (Figs 5, 6). At high magnification (Fig. 7) a single layer of plug cap and a well structured plug core were distinguishable from the outside to the inside. The plug core consisted of a peripheral granular region, with granules forming parallel rows perpendicular to the pit connection surface, a median region showing irregularly distributed granules and a central region with a fine granular background, clusters of granules and, sometimes, vesicles.

The organization of the pit connections in *Pterocladia capillacea* was similar to that already found in other species of Gelidiaceae (Tripodi 1971, Pueschel and Cole 1982, Pueschel 1989), thus confirming the importance of these structures as a taxonomical character.

Though the ultrastructural patterns of pit connections of red algae have been well defined by electron microscopy, their chemical composition is quite uncertain and it is still a matter for debate. According to Ramus (1971) the whole pit connections of *Griffithsia* consisted of a polysaccharide-protein complex. Investigations carried out by Pueschel (1980) on pit connections of different red algae, on the contrary, led him to suggest a proteinaceous nature for the plug core only, and a partial polysaccharidic composition for the outer plug cap.

In order to obtain information on the chemical composition of pit connections of *Pterocladia capillacea* some cytochemical analyses were performed.

The whole pit connection was found to be positive to the PATAg test for polysaccharide localization (Fig. 8a). The silver granules appeared widely distributed throughout the plug cap and the plug core, showing the presence of a polysaccharidic component in both these regions. All the controls, one of which is illustrated in Fig. 8b, were negative.

Our results are in contrast with the findings of both Lichtlé (1975) for the pit connections of *Polysiphonia* and Pueschel (1980) for the same structures of *Palmaria* and are, on the other hand, in agreement with the observations of Feldmann *et al.* (1977), who no-

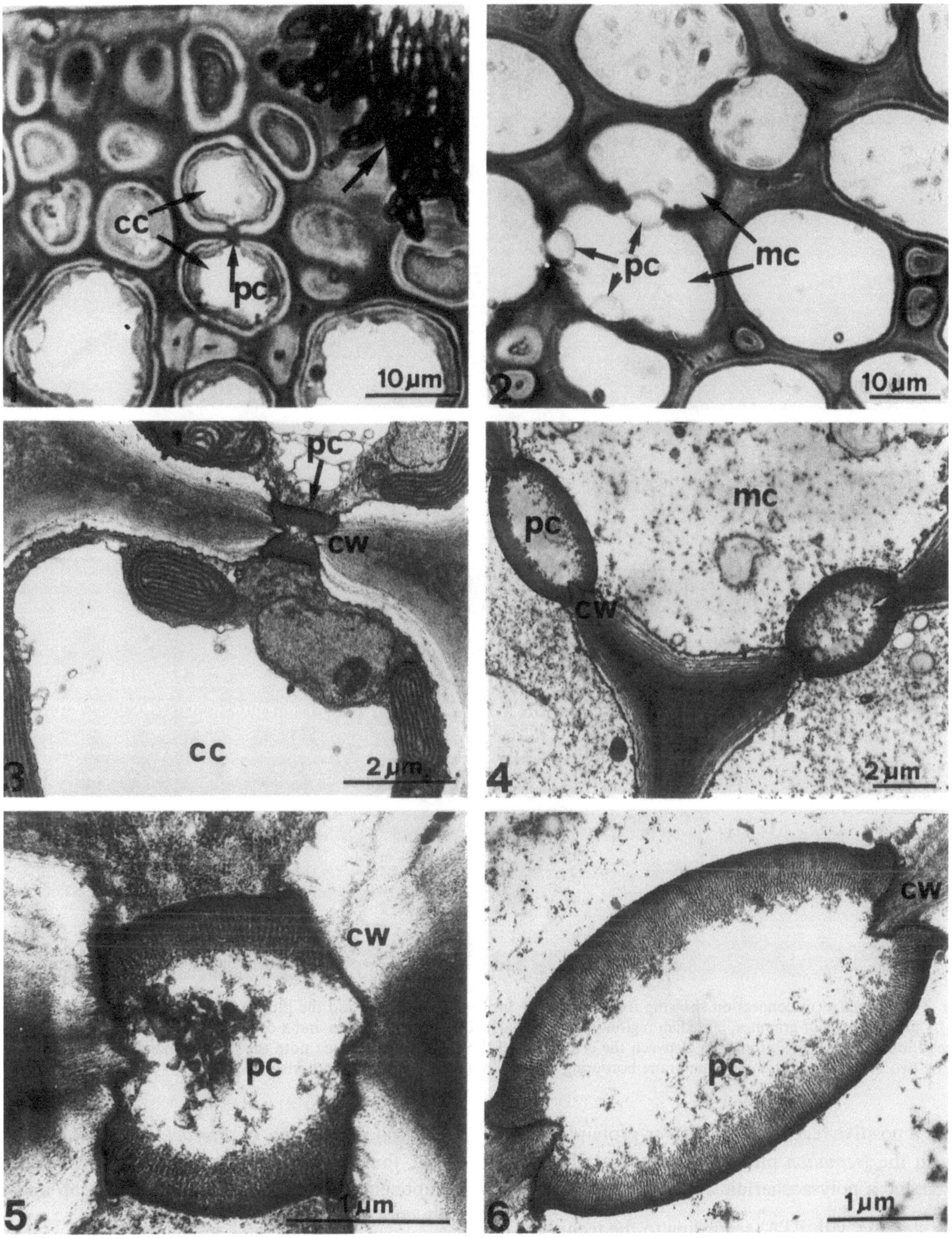

Figs 1–6. Pit connections in the thallus of *Pterocladia capillacea*. Note the different sizes and shapes of the pit connections between the cortical cells (Figs 1, 3) and the medullary ones (Figs 2, 4). Both kinds of pit connections exhibit the same organization (Figs 5, 6).

Abbreviations used in the figures: cc = cortical cells; cco = central plug core; pca = plug cap; cw = cell wall; mc = medulary cells; mco = median plug core; pc = pit connection; pco = peripheral plug core; pm = plasma membrane.

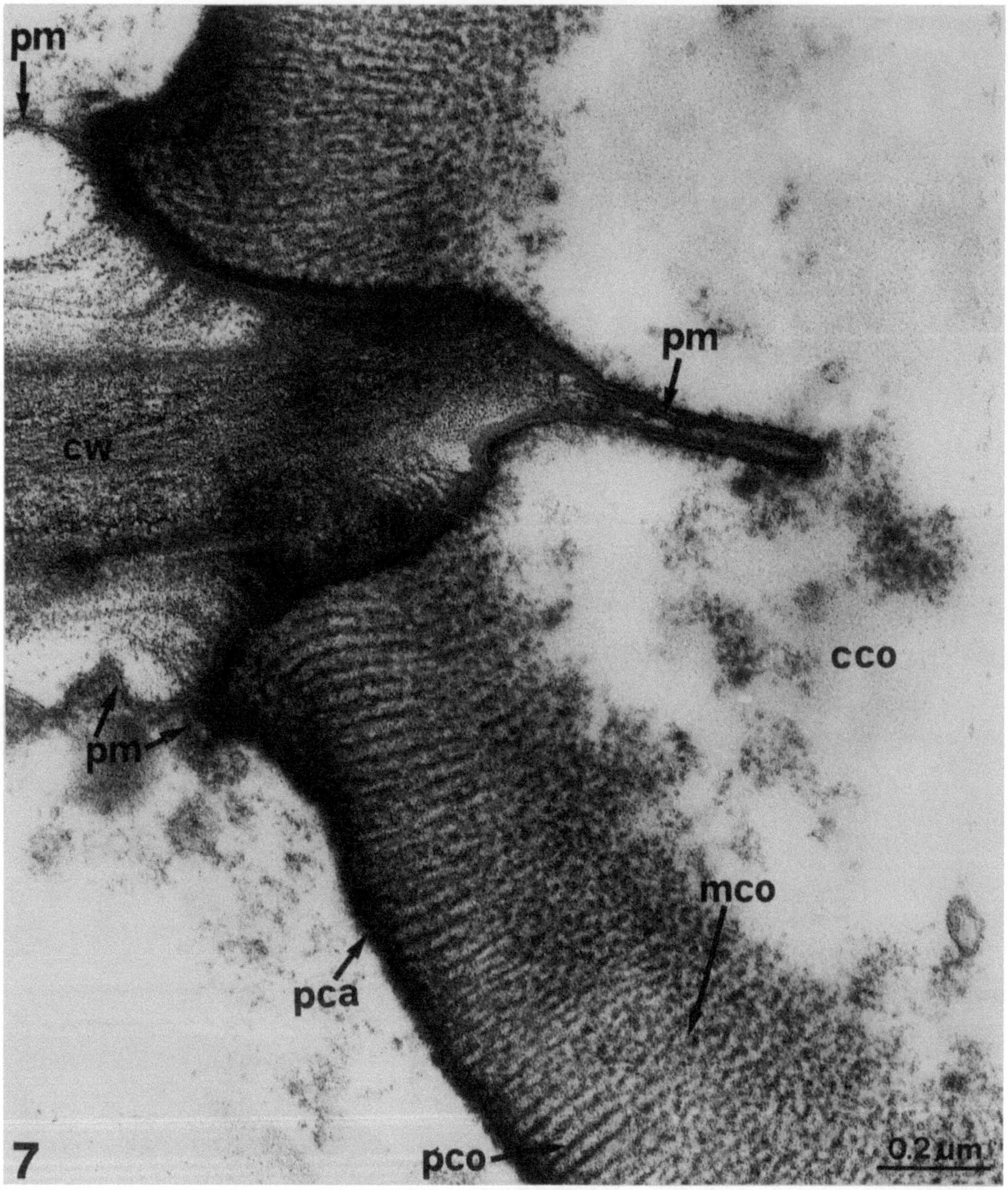

Fig. 7. Detail of a pit connection showing from outside to inside: the plug cap and the plug core with a peripheral region made up by parallel rows of granules, a median region with irregularly distributed granules and a central region with clusters of granules spread into a thin granular matrix. Between the cell wall and the pit connection side note the plasma membrane that continues in the two adjacent cells. A cap membrane bordering the pit connection surface is not visible.

ticed a positive reaction to the silver-proteinate staining in the *Nemalion* pit connections, for which they suggested a polysaccharidic composition.

The staining with PTA, according to the technique of Silverman and Glick (1969), a useful stain for proteins, also gave a positive reaction in all the regions of the pit connection of *Pterocladia capillacea* (Figs 9a, b), thus showing the presence of a proteinaceous component once again in both the plug cap and the plug core.

So, in agreement with the hypothesis by Ramus (1971), it seems that also in *Pterocladia capillacea* an

Fig. 8. Response of a pit connection to the PATAg test for polysaccharides. Note the positive reaction of all the regions with the reconstruction of the whole image (Fig. 8a). The control with no reaction (Fig. 8b) was obtained without adding thiosemicarbazide.

Fig. 9. Response of a pit connection to PTA staining technique for proteins. The positive reaction concerns the whole pit connection whose organization model is apparent (Fig. 9a). The positive response is evident if compared with the negative one of the control (Fig. 9b).

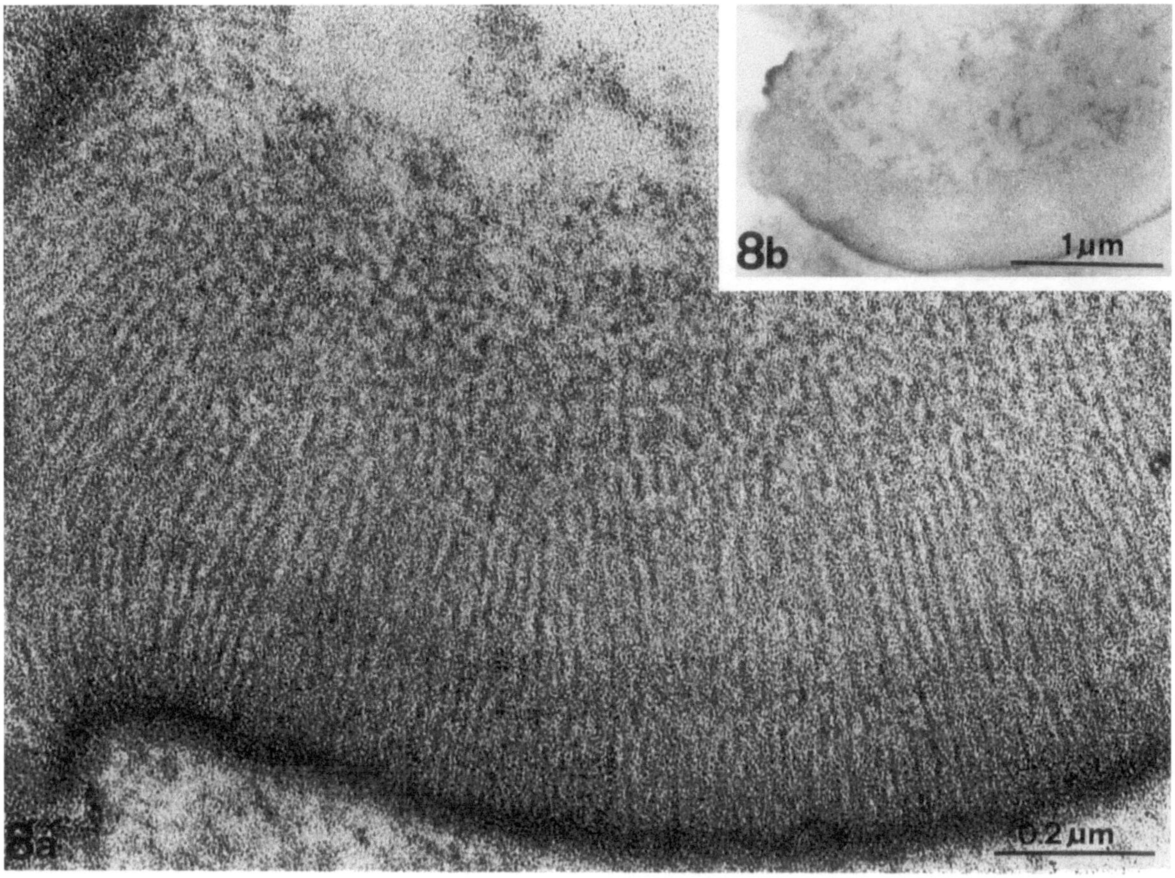
8b
1 µm
8a
0.2 µm

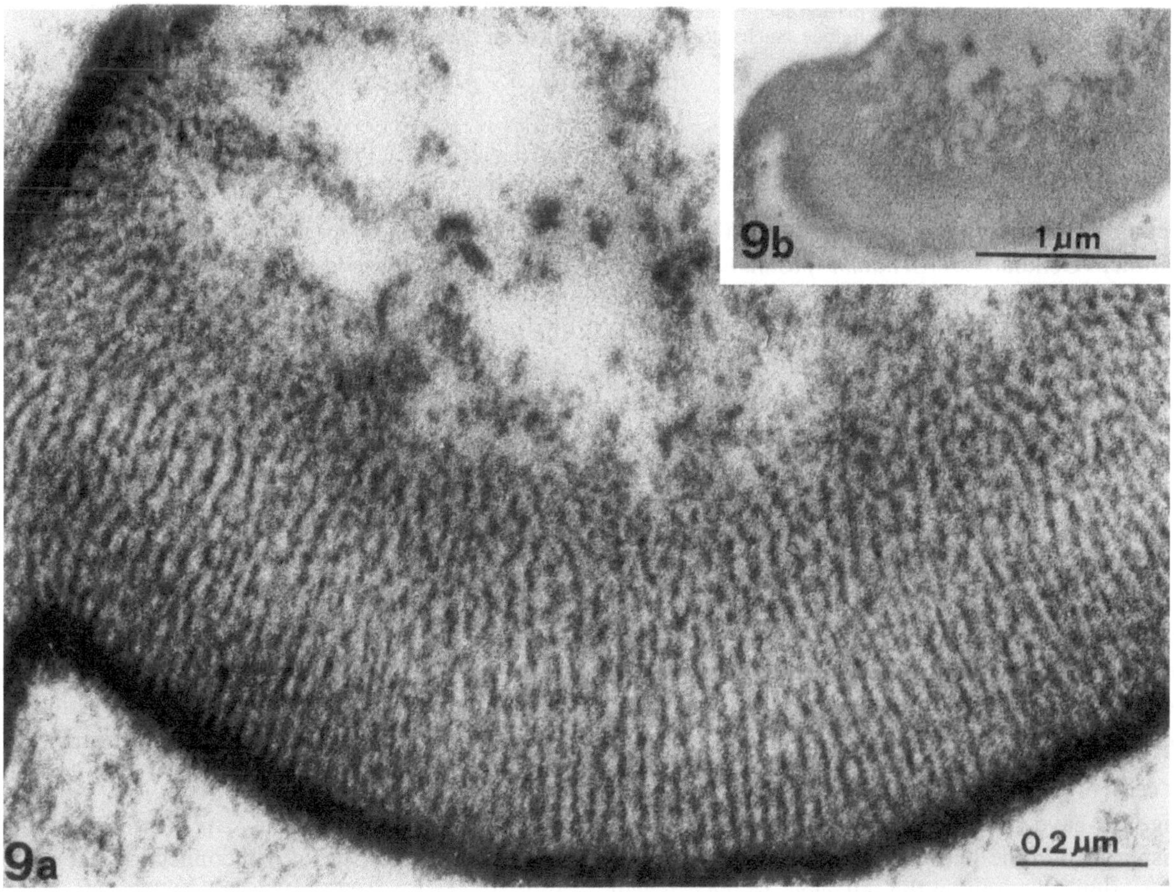
9b
1 µm
9a
0.2 µm

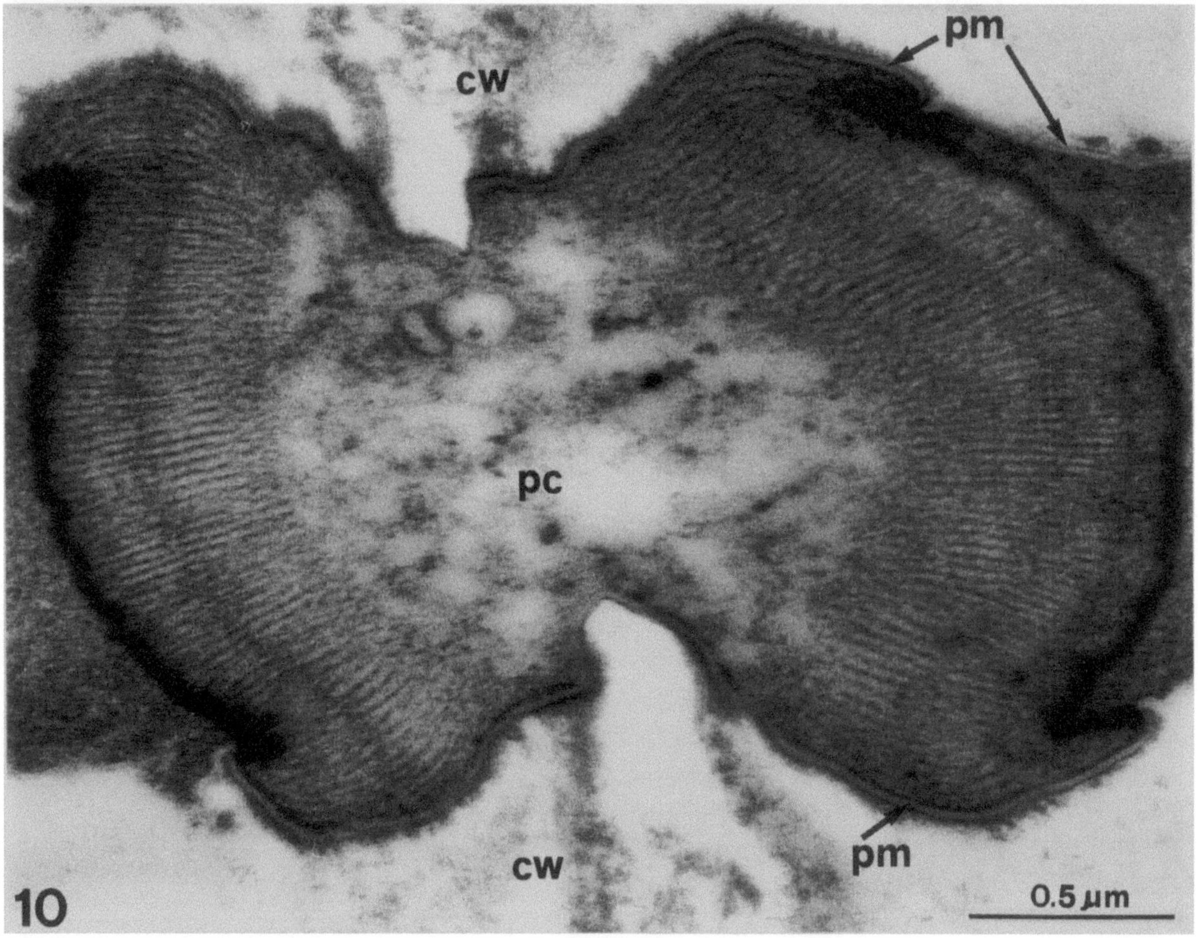

Fig. 10. The negative image of the plasma membrane between cell wall and pit connection sides as well as on the cell surface can be seen after staining with PTA technique. Such an image is not visible on the pit connection surfaces.

association of a proteinaceous and a polysaccharidic component contributes to the pit connection.

In contrast to what found by Pueschel and Cole (1982) in other Gelidiaceae [*Gelidium pusillum* (Stackh.) LeJol.; *Gelidium robustum* Gardn.; *Gelidium purpurascens* Gardn.], the pit connection of *Pterocladia capillacea* did not seem to be delimited by a cap membrane. The plasma membrane, distinguishable between the cell wall and the sides of the pit connection, could be seen to continue into the cell, tightly adherent to the wall surface, but did not appear to fork to cover the pit connection. The PTa test for proteins seemed to confirm the lack of the cap membrane. By this technique, which does not require the osmium postfixation, the lipids are not stained and the cell membranes are visible only as negative images against their positively stained surroundings (Silverman and Glick 1969).

In cells of *Pterocladia capillacea* the PTA staining (Fig. 10) gave rise to a clear negative image of the plasma membrane between the cell wall and the pit connection sides and, in the cell, close to the cell wall. Such a negative image of a membrane, on the contrary, was not shown on the pit connection surfaces.

It is interesting to point out that also in *Gelidium pusillum* indirect evidence for the presence of a cap membrane on pit connections had been produced by the negative image obtained by omission of osmium postfixation during the TEM routine techniques (Pueschel and Cole 1982).

Thus, in the two Gelidiaceae quite different results have been obtained with comparable methods, and this supports the idea that their pit connections, very similar in organization patterns, could be, on the contrary, different with regard to the presence of the cap membrane.

Sometimes the lack of such a cap membrane, observed in pit connections of some species, was related to an artifact due to the poor technique of sample preparation (Pueschel and Cole 1982).

Actually, the presence of a cap membrane is not a general characteristic of pit connections (Feldmann and Feldmann 1970, Peyrière 1977, Brawley and

Wetherbee 1981, Broadwater and Scott 1982, Tsekos and Schnepf 1985) and, at present, the absence of this component, noticed in several red algae, is regarded as a real feature (Pueschel 1987; Pueschel and Magne 1987). Therefore the idea that the pit connections of *Pterocladia capillacea* may actually lack the cap membrane seems plausible.

Conceding that perhaps transfer can occur even when a membrane is present (Ramm-Anderson and Wetherbee 1982), Wetherbee (1979) deems that pit connections lacking a cap membrane can assume a precise role in favouring the transport of nutritive substances between neighbouring cells, and to stress this function he defines this kind of pit connection 'transfer connections'. We agree with this hypothesis and suggest for the pit connections of *Pterocladia capillacea* an involvement in intercellular transport. It must be considered that a pit connection not covered by a cap membrane, is not an extracellular structure but, on the contrary, an intracellular component belonging to the symplatic environment.

Conclusions

The pit connections occurring in the cell wall of *Pterocladia capillacea* are found to be made up of an association of both a proteinaceous and a polysaccharidic component. Their organization is similar to that already described for other Gelidiaceae, thus confirming the taxonomical importance of these structures.

In *Pterocladia capillacea* the pit connections show a characteristic lack of the cap membrane. Consequently they are intracellular structures belonging to the symplastic route and might play an essential role in the mechanisms of cell to cell transport.

Acknowledgements

We wish to thank Mr Giorgio Varotto very much for his technical assistance. This work was supported by grants from C. N. R. and M. P. I.

References

Brawley, S. H. and R. Wetherbee. 1981. Cytology and ultrastructure. *In*: (C. S. Lobban and M. J. Wynne, eds) *The Biology of Seaweeds*. Blackwell Scientific Publications, Oxford. pp. 248–299.

Broadwater, S. T. and J. Scott. 1982. Ultrastructure of early development in the female reproductive system of *Polysiphonia harveyi* Bailey (Ceramiales, Rhodophyta). *J. Phycol. 18*: 427–441.

Courtoy, R. and J. Simar. 1974. Importance of controls for the demonstration of carbohydrates in electron microscopy with the silver methenamine or the thiocarbohydrazide-silver proteinate methods. *J. Microscopy 100*: 199–211.

Feldmann, J. and G. Feldmann. 1970. Sur l'ultrastructure des synapses des Algues rouges. *C. R. Acad. Sci. Paris, Série D 271*: 292–295.

Feldmann, J., G. Feldmann and G. Guglielmi. 1977. Nouvelles observations sur l'ultrastructure des synapses des Rhodophycées. *Rev. Algol. 7*: 11–33.

Gabrielson, P. W. and D. Garbary. 1986. Systematics of red algae (Rhodophyta). *CRC Critical Reviews in Plant Sciences 3*: 325–366.

Goff, L. J. 1979. The biology of *Harveyella mirabilis* (Cryptonemiales, Rhodophyceae) VII. Structure and proposed function of host penetrating cells. *J. Phycol. 15*: 87–100.

Goff, L. J. and A. W. Coleman. 1985. The role of secondary pit connections in red algal parasitism. *J. Phycol. 21*: 483–508.

Kugrens, P. 1980. Electron microscopic observations on the differentiation and release of spermatia in the marine red alga *Polysiphonia hendryi* (Ceramiales, Rhodomelaceae). *Am. J. Bot. 67*: 519–528.

Lichtlé, C. 1975. Etude ultrastructurale de la paroi du *Polysiphonia elongata* (Harv.) Rhodophycée, Floridée, à l'aide d'actions ménagées d'enzymes. *J. Microscopie Biol. Cell. 23*: 93–104.

Peyrière, M. 1977. Infrastructure des synapses du *Griffithsia flosculosa* (Ellis) Batters et de quelques autres Rhodophycées Floridées. *Rev. Algol. 12*: 31–43.

Pueschel, C. M. 1977. A freeze-etch study of the ultrastructure of red algal pit plugs. *Protoplasma 91*: 15–30.

Pueschel, C. M. 1980. A reappraisal of the cytochemical properties of rhodophycean pit plugs. *Phycologia 19*: 210–217.

Pueschel, C. M. 1987. Absence of cap membranes as a characteristic of pit plugs of some red algal orders. *J. Phycol. 23*: 150–156.

Pueschel, C. M. 1988. Secondary pit connections in *Hildenbrandia* (Rhodophyta, Hildenbrandiales). *Br. Phycol. J. 23*: 25–32.

Pueschel, C. M. 1989. An expanded survey of the ultrastructure of red algal pit plugs. *J. Phycol. 25*: 625–636.

Pueschel, C. M. and K. M. Cole. 1982. Rhodophycean pit plugs: An ultrastructural survey with taxonomic implications. *Am. J. Bot. 69*: 703–720.

Pueschel, C. M. and F. Magne. 1987. Pit plugs and other ultrastructural features of systematic value in *Rhodochaete parvula* (Rhodophyta, Rhodochaetales). *Cryptog. Algol. 8*: 201–209.

Ramm-Anderson, S. M. and R. Wetherbee. 1982. Structure and development of the carposporophyte of *Nemalion Helminthoides* (Nemalionales, Rhodophyta). *J. Phycol. 18*: 133–141.

Ramus, J. 1971. Properties of septal plugs from the red alga *Griffithsia pacifica*. *Phycologia 10*: 99–103.

Rascio, N., P. Mariani, F. Dalla Vecchia and R. Trevisan. 1991. The vegetative thallus of *Pterocladia capillacea* (Gelidiales, Rhodophyta): I. An ultrastructural and cytochemical study. *Bot. Mar. 34*: 177–185.

Roland, J. C. 1978. General preparation and staining of thin sections. *In*: (J. L. Hall, ed.) *Electron Microscopy and Cytochemistry of Plant Cells*. Elsevier/North Holland Biochemical Press, Amsterdam, Oxford and New York. pp. 1–62.

Silverman, L. and D. Glick. 1969. The reactivity and staining of tissue proteins with phosphotungstic acid. *J. Cell Biol. 40*: 761–767.

Thiéry, J. P. 1967. Mise en évidence des polysaccharides sur coupes fines en microscopie électronique. *J. Microscopie 6*: 987–1017.

Tripodi, G. 1971. Some observations on the ultrastructure of the red alga *Pterocladia capillacea* (Gmel.) Born. et Thur. *J. Submicr. Cytol. 3*: 63–70.

Tsekos, I. and E. Schnepf. 1985. Ultrastructure of the early stages of carposporophyte development in the red alga *Chondria tenuissima* (Rhodomelaceae, Ceramiales). *Pl. Syst. Evol. 151*: 1–18.

Wetherbee, R. 1979. "Transfer connections": Specialized pathways for nutrient translocation in a red alga? *Science 204*: 858–859.

Wetherbee, R. 1980. Post-fertilization development in the red alga *Polysiphonia* I. Proliferation of the carposporophyte. *J. Ultrastr. Res. 70*: 259–274.

Wetherbee, R. and G. T. Kraft. 1981. Morphological and fine structural features of pit connections of *Cryptonemia* sp. a highly differentiated red alga from Australia. *Protoplasma 106*: 167–172.

Wetherbee, R. and J. Scott. 1980. The fine structure and distribution of "transfer connections" in the red alga *Polysiphonia*. *Micron 11*: 509–510.

Botanica Marina
Vol. 34, pp. 195–199, 1991

A Preliminary List of Marine Algae from the Banc d'Arguin (Mauritania)

J. Marcot-Coqueugniot

Laboratoire d'Ecologie du Benthos, Faculté de Luminy, Université d'Aix-Marseille 2, 163 avenue de Luminy, 13288 Marseille Cedex 9, France

(Accepted 16 December 1990)

Abstract

Macroscopic marine algae were collected on the Banc d'Arguin (Mauritania). Among the determined taxa, 57 are new records from Mauritania; this represents an increase of 36% in the known Mauritanian marine algal flora. For 20 of these 57 taxa, this is the first record from the tropical west coast of the African continent.

Introduction

The macroscopic algal flora of Mauritania has not been extensively studied. The areas most investigated are those of the Cap Blanc Peninsula and those situated in the south, near Senegal (Bodard 1968, Dangeard 1951, Feldmann, G. 1964, Feldmann-Mazoyer 1942, Feldmann, J. 1937, Gayral 1960, Hariot 1911, Lawson and John 1977, 1982 and 1987, Price et al. 1986, 1988, Seoane-Camba 1960, Sourie 1954).

The coast between the northern and southern limits of Mauritania consists of sandy beaches and dunes with some rocky islands (Chickchitt, Iouik, Kiaone islands). Sand-banks not being very suitable for algal attachment, the macroscopic algal flora of Mauritania is relatively poor.

As part of a general study of the ecosystem of the Banc d'Arguin, algae were collected on three scientific expeditions: April 1987, October 1987 and May 1988.

Lawson and John (1977) reported the presence of 158 taxa of macroscopic marine algae from the Chlorophyta, Phaeophyta and Rhodophyta. In our check-list of taxa (of these same three taxonomic divisions) 57 taxa are recorded for the first time from Mauritania (without taking into account algae for which some doubt remains as to the identification as the samples were incomplete or sterile). This represents an increase of 36% in the known Mauritanian marine algal flora. This is the first time two of the genera have been recorded for Mauritania (*Myrionema* and *Dudresnaya*). For 20 of the 57 taxa, this is the first record from the tropical west coast of the African continent. Among these, some had previously been observed in the adjacent islands (Canarias, Madeira, Selvagens, Desertas, Cabo Verde).

In further publications we shall give a detailed study of the interesting taxa.

Sites at which collections have been made are referred to by numbers in the check-list (see Fig. 1).

II: Baie de l'Etoile; 7 April 1987.
V: Cap Tafarit; 10 April 1987.
VI: Ile Chickchitt; 10 April 1987.
VIII: Ile Petite Kiaone; 10 April 1987.
XIII: Ile Tidra; 11 April 1987, wrecks.
XIV: Offshore (see on the map); 16 October 1987.
XV: Ten Alloul; 20 October 1987, wrecks.
XV: Iouik; 21 October 1987.
XVII: Ile d'Arel; 23 October 1987.
XVIII: Ile d'Arel; 24 October 1987, wrecks.
XIX: Ile Niroumi; 26 October 1987, wrecks.
XX: Ile d'Arel; 15 May 1988.
XXI: Ile Petite Kiaone; 19 May 1988.
XXII: Iouik; 17 May 1988, wrecks.

Except for wrecks, the algae were collected between 2 and 6 meters deep. The high turbidity at the collecting sites is due to silt, sand particles and *Zostera* debris.

The Banc D'Arguin is relatively isolated from the cold mineral-rich water of the upwelling, but it does penetrate a little into the area in winter and early spring (Cuq and Campredon, in press). The Banc d'Arguin, isolated by sand-banks, consists of mudflats and channels of shallow water (2 to 6 meters). This, with high evaporation and the absence of any fresh water from land drainage, results in elevated water temperatures (from 23 °C in winter to 28 °C in summer) and high salinities (38 to 41%) (Sevrin-Reyssac 1980, 1983a, 1984) at our collecting sites.

* first record for Mauritania
** first record for the tropical west coast of the African continent

CHLOROPHYTA

** *Acetabularia parvula* Solms-Laubach, VI.
Acrochaete viridis (Reinke) Nielsen, XIII, XX.
Bryopsis (corymbosa?) J. Agardh, XIV.
** *Bryopsis penicillum* Meneghini, XVI, XVII, XVIII.
Bryopsis plumosa (Hudson) C. Agardh, VI.
Caulerpa cupressoides (West *ex* Wahl) C. Agardh var. *flabellata* Børgesen, XVII.
Caulerpa cupressoides (West *ex* Wahl) C. Agardh var. *lycopodium* (J. Agardh) Weber van Bosse f. *elegans* (Crouan) Weber van Bosse, XIV.
* *Caulerpa mexicana* Sonder *ex* Kützing, XVII, XX.
Chaetomorpha brachygona Harvey, XVII, XVIII, XIX, XX.
Chaetomorpha (nodosa?) Kützing, VI.
Cladophora prolifera (Roth) Kützing, XIX.
Cladophora (vagabunda?) (Linnaeus) Hoek, VI.
Cladophora sp., VI.
* *Codium isthmocladum* Vickers, XIII.
* *Codium taylori* Silva, VI: gametangia; XV.
Derbesia sp., II.
Enteromorpha clathrata (Roth) Greville, II, XVI.
* *Enteromorpha ramulosa* (J. E. Smith) Hooker, II.
** *Entocladia major* (Feldmann) Nielsen, XVI.
Entocladia sp., XIII, on *Codium isthmocladum*.
Pseudochlorodesmis furcellata (Zanardini) Børgesen, XX.
** *Pseudodictyon inflatum* Ercegović, XIII.
** *Ulvella lens* Crouan *et* Crouan. V, VI, VIII, XX, XXI.

PHAEOPHYTA

Colpomenia sinuosa (Mertens *ex* Roth) Derbès *et* Solier *in* Castagne, VI: sori, XXI.
Cystoseira discors C. Agardh emend. Sauvageau, VI.
Cystoseira humilis (Schousboe) Kützing, VI, receptacles.
Cystoseira mauritanica Sauvageau *ex* Hariot, V, VIII: male conceptacles; XXI.
Dictyopteris delicatula Lamouroux, XVII.
Dictyopteris membranacea (Stackhouse) Batters, VI, VIII, XVII: sporangia; XX.
** *Dictyota cervicornis* Kützing *f. curvula* W. Taylor, VI, XX.
Dictyota dichotoma (Hudson) Lamouroux, VI, XVI, XVIII, XIX: sporangia.
Dictyota spp., XIII, XVII.
* *Feldmannia indica* (Sonder) Womersley *et* Bailey, XIII: plurilocular sporangia.
Levringia atlantica (J. Feldmann) Kylin, VI.
Lobophora variegata (Lamouroux) Womersley, II, VI, XVII, XXI.
** *Myrionema sp.*, XIII.
** *Nemacystus erythraeus* (J. Agardh) Sauvageau (var. *hispanicus?*) Sauvageau, II, VI: polysporangia, monosporangia; VIII.
Padina tetrastromatica Hauck, VIII: female sori.
Padina vickersiae Hoyt *in* Howe, II: male sori; VI.
Ralfsia expansa (J. Agardh) J. Agardh, VI.
** *Rosenvingea sanctae-crucis* Børgesen, XVI: sori.
* *Sargassum filipendula* C. Agardh, II, XV: receptacles.
** *Sargassum ramifolium* Kützing, V; VIII, XXI: receptacles.
Sargassum vulgare C. Agardh, XV: receptacles; XVII.
Spatoglossum schroederi (C. Agardh) Kützing, VIII: sporangia.
Sphacelaria rigidula Kützing, VI, VIII, XXI: propagules.
Sphacelaria tribuloides Meneghini, VIII: propagules.
Sphacelaria sp., VI, XVIII, XX.
** *Sporochnus bolleanus* Montagne, XIII.

RHODOPHYTA

* *Acanthophora spicifera* (Vahl) Børgesen, XVI, XVII, XVIII, XIX.
* *Acrochaetium daviesii* (Dillwyn) Nägeli, V, VIII, XIII, XVI, XVII, XX: monosporangia.
* *Acrochaetium hallandicum* (Kylin) Hamel, XXI: monosporangia.
* *Acrochaetium microscopicum* (Nägeli *in* Kützing) Nägeli, VI, XV, XXI: monosporangia.
* *Acrochaetium parvulum* (Kylin) Hoyt, VIII: monosporangia.
* *Acrochaetium (savianum?)* (Meneghini) Nägeli, XIX, XX.
Acrochaetium (secundatum?) (Lyngbye) Nägeli, XIII.
Acrochaetium seriatum Børgesen, XVII: monosporangia.
** *Acrosorium venulosum* (Zanardini) Kylin, VIII.

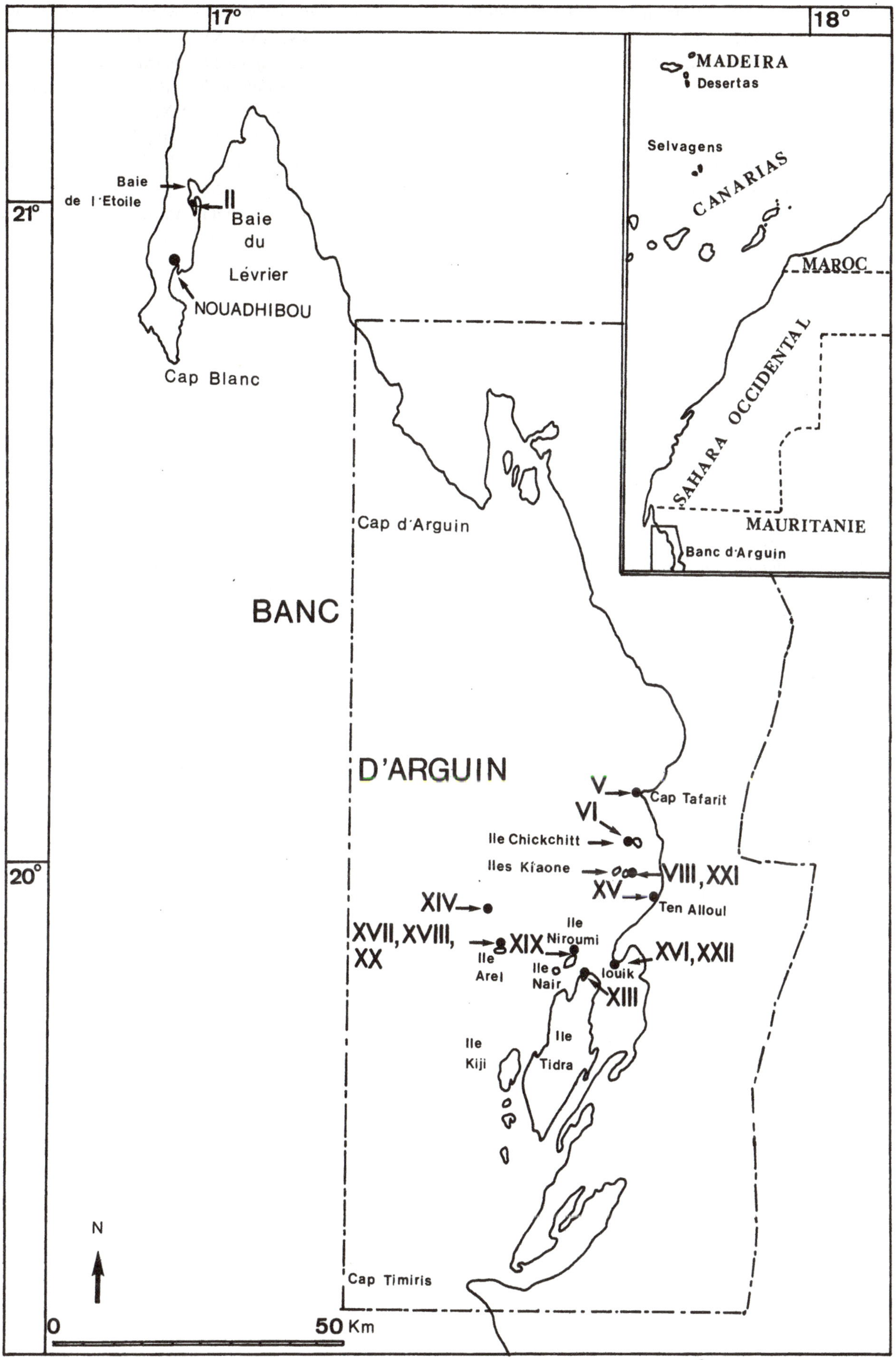

Fig. 1. Map to show collecting sites

* *Amphiroa beauvoisii* Lamouroux, VI: tetrasporangia; VIII, XVIII, XIX, XXI.
* *Amphiroa cryptarthrodia* Zanardini, VIII: tetrasporangia.
Amphiroa fragilissima (Linnaeus) Lamouroux, XX.
** *Antithamnion antillanum* Børgesen, XX.
Antithamnionella elegans (Berthold) Price *et* John, V.
* *Callithamnion byssoides* Arnott *ex* Harvey *in* Hooker, VIII: male and female organs; XIV: tetrasporangia.
Callithamnion sp., XX.
Centroceras clavulatum (C. Agardh *in* Kunth) Montagne *in* Durieu de Maisonneuve, VI.
Ceramium flaccidum (Kützing) Ardissone var. *byssoideum* Harvey, II, V, XIII, XVII: carposporophytes, tetrasporophytes; XVIII: tetrasporophytes; XX: males, tetrasporophytes.
* *Ceramium codii* (Richards) Mazoyer, XX.
* *Ceramium cornutum* Dangeard, XIII: carposporophytes, tetrasporophytes.
Ceramium diaphanum (Lightfoot) Roth, VI, XX: tetrasporophytes.
Ceramium (ledermannii?) Pilger, XVII, XX.
** *Ceramium leptozonum* Howe, XVII.
* *Ceramium tenerrimum* (Martens) Okamura var. *brevizonatum* Petersen, VIII, XX.
Ceramium sp., VIII.
Ceratodictyon variabile (Greville *ex* J. Agardh) R. E. Norris, XVII: tetrasporangia; XVIII.
Champia parvula (C. Agardh) Harvey, V, VI: tetrasporangia; XIII, XVII, XX.
Chondria dasyphylla (Woodward) C. Agardh, XX: tetrasporangia.
Chondria sp., XX: tetrasporangia.
Corallina (cubensis?) (Montagne) Kützing, XVII, XX.
** *Corallina granifera* Ellis *et* Solander, VI.
** *Cottoniella filamentosa* (Howe) Børgesen, XVII.
** *Cottoniella filamentosa* var. *fusiformis* (Børgesen) Cormaci, Furnari *et* Scammacca, XX.
** *Crouania franciscii* Cormaci, Furnari *et* Scammacca, VIII, XVII: carposporangia.
Cryptonemia seminervis (C. Agardh) J. Agardh, VIII.
* *Dasya baillouviana* (Gmelin) Montagne, XIII.
Dasya (caraibica?) Børgesen, XIII: tetrasporangia.
Dasya sp., XVII: carposporophytes, tetrasporophytes.
* *Dohrniella antillarum* (W. R. Taylor) Feldmann-Mazoyer, VIII.
* *Dudresnaya sp.*, XVI: tetrasporangia.
Erythrotrichia boryana (Montagne) Trevisan, VIII.
Erythrotrichia carnea (Dillwyn) J. Agardh, VI, VIII, XIII, XV, XVI, XVII, XVIII, XIX, XXI.
* *Erythrotrichia simplex* Dangeard, VIII.
* *Falkenbergia hillebrandii* (Bornet) Falkenberg "phase", VIII, XV, XVII, XXI.
Fosliella farinosa (Lamouroux) Howe *in* Britton *et* Millspaugh, VI, VIII, XIII, XV, XVII, XVIII, XX, XXI: conceptacles.
Gelidium crinale (Turner) Lamouroux *in* Bory, VI.
Gelidium pusillum (Stackhouse) Le Jolis, XXI.
Gigartina acicularis (Roth) Lamouroux, V, VI.
Griffithsia spp., VIII, XVIII.
Haliptilon subulatum (Ellis *et* Solander) Johansen, XXI: conceptacles.
* *Halymenia actinophysa* Howe, XIII: carposporophytes.
* *Halymenia floresia* (Clemente) C. Agardh, XIII: tetrasporophytes; XXII: carposporophytes.
Herposiphonia secunda (C. Agardh) Falkenberg, XX.
Herposiphonia secunda (C. Agardh) Falkenberg f. *tenella* (C. Agardh) Wynne, VI, XVI: males; XVIII: female, tetrasporophytes.
Heterosiphonia crispella (C. Agardh) Wynne, VIII, XIV: tetrasporangia; XVII, XIX, XXI.
* *Hydropuntia dentata* (J. Agardh) Wynne, VIII.
* *Hypnea arbuscula* Dangeard, XX: tetrasporangia.
Hypnea cervicornis J. Agardh, XVII, XVIII.
Hypnea musciformis (Wulfen *in* Jacquin) Lamouroux, VI, XIV, XX.
Jania adhaerens Lamouroux, VI, XVII, XVIII, XIX, XX, XXI.
* *Jania capillacea* Harvey, VI, XVII, XVIII, XIX, XX.
* *Jania capillacea* Harvey, VI, XVII, XVIII, XIX, XX.
Jania rubens (Linnaeus) Lamouroux, VIII.
Jania (verrucosa?) Lamouroux, XX.
* *Laurencia paniculata* J. Agardh, XVII.
Laurencia (papillosa?) (Forsskål) Greville, XVIII.
* *Laurencia perforata* Montagne, VI.
Melobesia membranacea (Esper) Lamouroux, VI, VIII, XXI.
** *Nitophyllum punctatum* (Stackhouse) Greville, VIII.
** *Pleonosporium borreri* (J. E. Smith) Nägeli *ex* Hauck, VIII.
Plocamium cartilagineum (Linnaeus) Dixon, V, VI, VIII, XXI.
* *Pneophyllum lejolisii* (Rosanoff) Chamberlain, VI, VIII, XXI.
** *Pneophyllum confervicola* (Kützing) Chamberlain f. *minutula* (Foslie) Chamberlain, VI, VIII.
** *Polysiphonia gorgoniae* Harvey, XX.

Polysiphonia spp. (4 pericentral cells), VI, XIV, XVI, XVII, XVIII, XIX, XX.

* *Rhodophyllis divaricata* (Stackhouse) Papenfuss, VIII.

Rhodymenia pseudopalmata (Lamouroux) Silva, VI, VIII.

Rytiphloea tinctoria (Clemente) C. Agardh, XIV, XVII, XX.

* *Sahlingia subintegra* (rosenvinge) Kornmann, VIII, XIII, XVIII, XXI.

Solieria tenera (J. Agardh) Wynne *et* W. R. Taylor, XIII, XVII, XIX, XX.

Spermothamnion (investiens?) (P. *et* H. Crouan) Vickers, XX.

** *Spermothamnion macromeres* Collins *et* Hervey, XX.

Spermothamnion (repens?) (Dillwyn) Rosenvinge, VIII.

Spermothamnion sp., XVI.

* *Spyridia clavata* Kützing, XX.

Spyridia filamentosa (Wulfen) Harvey *in* Hooker, VI, XIII, XIV, XVII, XVIII, XIX.

Spyridia hypnoides var. *typica* (Bory *ex* Belanger) Papenfuss, XV, XX.

Stylonema alsidii (Zanardini) Drew, II, V, VI, XIII, XIV, XVI, XVII, XVIII.

Taenioma sp., VIII.

Tiffaniella (capitata?) (Schousboe *ex* Bornet) Doty *et* Menez, XIII.

Titanoderma sp., VIII.

Conclusion

The tropical character of the surroundings of the Banc d'Arguin islands, well known at all the levels of animal kingdom (Sevrin-Reyssac 1983 b) seems confirmed by the affinities of the macroscopic marine algae, which, apart from the cosmopolitan ones, divide into warm temperate and pantropical species. The greater part of the algae newly recorded for Mauritania were first described from subtropical and tropical regions of the west Atlantic Ocean (from Brazil to Antilles).

Acknowledgements

This work was supported by grants of the Ministère français de l'Environnement, the Parc national de Port-Cros and the Parc national du Banc d'Arguin. Thanks are due to Prof. Ch. F. Boudouresque, P. Francour, G. Pergent and G. Vuignier for collecting the algae, and to M. Verlaque for assistance in identifying specimens.

References

Bodard, M. 1968. Les *Hypnea* au Sénégal (Hypnéacées, Gigartinales). *Bull. Inst. fond. Afrique Noire, 30 A*: 811–829.

Cuq, F. and P. Campredon. 1991. The coastal processes and littoral environment of the region of the Arguin Bank (Mauritania). *Wetlands Ecology and Management*. In press.

Dangeard, P. 1951. Sur les Gélidiacées de Dakar et de Port-Etienne. *Le botaniste 35*: 21–25.

Feldmann, G. 1964. Sur une nouvelle espèce iridescente de *Chondria* (Rhodophyceae, Rhodomelaceae). *Revue générale de Botanique 71*: 45–55.

Feldmann-Mazoyer, G. 1942. A propos de quelques *Spermothamnion* à polysporanges. *Bull. Soc. Hist. nat. Afrique du Nord 33*: 15–18.

Feldmann, J. 1937. Sur une nouvelle espèce de Laminariacée de Mauritanie *Ecklonia muratii* nov. sp. *Bull. Soc. Hist. nat. Afrique du Nord 28*: 325–327.

Gayral, P. 1960. Sur la présence au Maroc et à Dakar de *Levringia brasiliensis* (Montagne) B. Joly. *Revue Algologique 1*: 49–54.

Hariot, M. P. 1911. Algues de Mauritanie recueillies par M. Chudeau. *Bull. Soc. Bot. de France 58*: 438–445.

Lawson, G. W. and D. M. John. 1977. The marine flora of the Cap Blanc Peninsula. *Bot. J. Linnean Soc. 75*: 99–118.

Lawson, G. W. and D. M. John. 1982. The marine algae and coastal environment of Tropical West Africa. *Nova Hedwigia. 70*: 1–455.

Lawson, G. W. and D. M. John. 1987. The marine algae and coastal environment of Tropical West Africa (second edition). *Nova Hedwigia 93*: 1–415.

Price, J. M., D. M. John and G. W. Lawson. 1986. Seaweeds of the western coast of Tropical Africa and adjacent islands: a critical assessment. IV – Rhodophyta (Floridae). 1: Genera A–F. *Bull. Br. Mus. Nat. Hist., Bot. 15*: 1–122.

Price, J. M., D. M. John and G. W. Lawson. 1988. Seaweeds of the western coast of Tropical Africa and adjacent islands: a critical assessment. IV – Rhodophyta (Floridae). 2: Genera G. *Bull. Br. Mus. Nat. Hist., Bot. 18*: 195–273.

Seoane-Camba, J. 1960. Nota sobre algunas especies de Algas de la costa occidental africana (Sur de Cabo Blanco). *Inv. Pesq. 16*: 91–103.

Sevrin-Reyssac, J. 1980. Chlorophylle a et production primaire dans les eaux de la Baie du Lévrier et du Parc national du Banc d'Arguin. *Bull. Centre nat. Rech. océano. Pêches Mauritanie 9* (1): 56–65.

Sevrin-Reyssac, J. 1983 a. Quelques particularités de la chaîne alimentaire marine dans la région des îles du Banc d'Arguin (Parc national du Banc d'Arguin, Mauritanie). *Bull. Centre nat. Rech. océano. Pêches Mauritanie 11* (1): 41–52.

Sevrin-Reyssac, J. 1983 b. Affinité biogéographique de la région des îles du Banc d'Arguin (Parc national du Banc d'Arguin, Mauritanie). *Bull. Centre nat. Rech. océano. Pêches Mauritanie 11* (1): 53–56.

Sevrin-Reyssac, J. 1984. Conditions hydrologiques et phytoplancton dans la partie méridionale du Parc national du Banc d'Arguin (Mauritanie) en mars et avril 1983. *Bull. Centre nat. Rech. océano. Pêches Mauritanie 12* (1): 107–113.

Sourie, R. 1954. Contribution à l'étude écologique des côtes rocheuses du Sénégal. *Mém. IFAN*, Dakar: 38.

Botanica Marina
Vol. 34, pp. 201–210, 1991

Distribution and Thickness Patterns in Subtidal Encrusting Algae from Washington

M. N. Dethier, K. M. Paull* and M. M. Woodbury**

Friday Harbor Laboratories and Institute for Environmental Studies, University of Washington, Friday Harbor, Washington 98250, U.S.A.

* *Marine Science Center, Northeastern University, Nahant, Massachusetts 01908, U.S.A.*

** *West Indies Laboratory, Christiansted, St. Croix, Virgin Islands 00820, U.S.A.*

(Accepted 16 December 1990)

Abstract

Little is known about the ecology of subtidal fleshy (non-calcified) crustose algae, especially those in the northeastern Pacific. A survey of distributions, abundances, and thicknesses of these crusts was carried out along a depth gradient from −6 to −24 m at two sites in the San Juan Islands, Washington, USA. Encrusting algae dominate most space on the rock in these areas. The most abundant crusts are two species of *Peyssonnelia*, one of which previously had not been recorded north of California. Red crusts (fleshy and coralline) predominate and extend into deep water, while brown crusts are uncommon and inhabit shallower water zones. For all crusts surveyed, thickness decreases significantly with depth, probably as a result of the cost of supporting a thick thallus (whose lower tissues receive little light) under low ambient light. Of the common crust grazers (urchins, chitons, limpets), only one chiton species showed a significant decrease in abundance with depth. Urchins may contribute to some local patterns of crust distribution, for example the difference in abundances of coralline *vs.* fleshy crusts on horizontal and vertical surfaces. In laboratory feeding experiments, urchins consistently eat more fleshy than coralline crusts.

Introduction

Encrusting algae are a major component of rocky subtidal communities worldwide (*e.g.*, Adey and MacIntyre 1983, Littler and Littler 1984, Steneck 1986). However, the ecology of crustose algae is poorly understood, and even basic patterns of distribution and abundance are often unknown. Although crusts in the Atlantic grow deeper in the photic zone than any other algal form (reviewed in Vadas and Steneck 1988), no parallel work has been reported for Pacific species. In the northeastern Pacific, the distribution and abundance of intertidal fleshy (= non-calcified) species (Dethier 1987) and corallines (Paine 1984, Steneck and Paine 1986) have been described. But only a few studies exist on processes affecting distribution and abundance patterns of intertidal crusts (see below), and no ecological investigations have been reported for subtidal fleshy crusts from the northeast Pacific.

Distribution and abundance patterns of crustose algae can be affected by many factors such as competition, stress, and particulary herbivory. Dethier and Steneck (unpubl. observation) have studied the responses of a variety of fleshy and coralline crusts to physical stress and disturbance, and to herbivory. Many algae can increase their survival under intense herbivory by incorporation of a crustose phase or basal system in their life history (*e.g.*, Slocum 1980, Lubchenco 1980, Dethier 1981, Jara and Moreno 1983). However many herbivores, particularly sea urchins, alter crust abundance patterns by biting deeply into them and removing large amounts of biomass (Steneck 1982,

1986, Paine 1984). In addition, herbivory may alter crust competitive hierarchies (Paine 1984).

This study was designed to determine depth-related distribution and abundance patterns of subtidal fleshy crustose algae in the San Juan Islands, northeastern Pacific. Particular emphasis was given to the depth dependence of abundance and thickness of various red crusts. The potential role of herbivory by sea urchins on distribution and abundance patterns of crustose algae also was examined.

Study Sites and Methods

Field surveys

Surveys were conducted using SCUBA during July and August of 1986 and 1988 in the San Juan Islands, Washington (48°30′N, 123°W). Most sites were located in or near San Juan Channel, in areas of rocky to cobble substrata and moderate to high current. Detailed transects were surveyed at Point George on Shaw Island, a site with moderate wave exposure and currents, and on Turn Island, a site with low wave exposure but high currents. Because winter sampling was not attempted, we have no data on seasonal changes in crust distributions or reproductive status. All encrusting algae studied grow very slowly relative to other macroalgae (*e.g.*, Bertness *et al.* 1983, Steneck and Paine 1986, Dethier 1987) and few seem to undergo marked seasonal changes (for an exception see Dethier 1981). Thus patterns of abundance, as quantified here, probably change only slowly through time.

To quantify crust and grazer abundances, a 10 m long horizontal transect line was established at each of four depths: 6, 12, 18, and 24 m. No areas had overstory canopies of the kelp *Nereocystis*, but patches of the understory kelps *Laminaria* spp. and *Agarum* spp. were present at shallow depths. All 6 m deep areas at Turn Island had heavy kelp canopies; crust data are not included from this site and depth because of possible confounding canopy effects.

Percentage covers of fleshy and coralline crusts and of sessile invertebrates were estimated in 10 × 0.1 m^2 quadrats placed at 1 m intervals along each transect line. Using regularly spaced quadrats eliminated observer bias. Estimations were made visually with the aid of squares marked off in the quadrat, and counting the squares occupied by each taxon. This method has been found to be as repeatable (and much faster) than methods using randomly placed points (Dethier unpubl. observation). No between-person error exists since all percentage cover estimates were done by one investigator (KMP).

The pooled percentage cover of all species of red fleshy crusts was estimated in underwater surveys due to the difficulty of consistently distinguishing species *in situ*. To quantify abundance of individual species at each depth, collections were made (with hammer and chisel) of each clone of red fleshy crust that was crossed by the horizontal 10 m transect line. In the laboratory, samples were hand-sectioned to determine species (see below) and crust thickness. Relative frequency of each species in these samples was calculated and multiplied by the mean percentage cover of all red crusts (quadrat data) as an estimate of the actual percentage cover of individual species at each depth. For instance, if 18 of 39 crust chips collected along a line were *Peyssonnelia pacifica* Kylin and the mean coverage of all red crusts along that line was 16.4% (from 10 quadrats), then the relative frequency of that species was calculated to be 18/39 (16.4) = 7.6%. While this method clearly has sources of error, it is unlikely that it consistently misrepresented the abundance of any species, and it does provide a reasonable measure of relative crust abundances.

Divers observing subtidal crusts in this region frequently noted an apparent prevalence of fleshy crusts on vertical surfaces relative to horizontal ones (transects included all substratum orientations). To investigate this pattern, in August 1988 eight pairs of 0.1 m^2 quadrats were surveyed at 6–12 m depth, each pair with one horizontal surface and one vertical one within ~ 2 m. Surfaces to be surveyed were chosen haphazardly, depending on availability of areas of adequate size and correct orientation and proximity. Coralline crusts (lumped), fleshy crusts (lumped), and herbivores (by species) were counted within each.

The most common local sea urchin was the large red *Strongylocentrotus franciscanus* (A. Agassiz) (up to 14 cm test diameter). Since urchins were patchy in distribution, they were quantified in two ways: in 1986 surveys they were counted in a swath that extended 0.5 m on either side of each horizontal transect line; this number was divided by 10 to estimate the number per m^2. In 1988, they were counted in 1.0 m^2 quadrats placed contiguously along a vertical transect line crossing 6 m depth intervals. Other herbivores (limpets and chitons) were counted in the same 10 quadrats at each depth as the crusts.

Laboratory urchin feeding experiments

Laboratory experiments were conducted to determine if *Strongylocentrotus franciscanus* showed any consistent preferences among crust species as food. Vadas (1977) found that urchin food preferences as determined in the laboratory closely corresponded with their diets in nature, although the latter were affected by prey availability. Vadas did not test preferences among crusts as food sources. Six chips of rock of similar size, each covered with one crust species, were imbedded in Koppers Splash Zone Compound and placed in holes in a 25 × 25 cm square of plywood such that each crust was roughly flush with the wood surface; this provided a realistic foraging surface for the urchins. Two sets of crust species were tested: a set of intertidal species, collected in the low intertidal zone on the west side of San Juan Island (where red urchins are present in pools), and a set from the region of the subtidal transects. Both sets contained coralline and fleshy crust species. One urchin (5–8 cm test diam.) was restricted to each of 5 replicate plywood squares and allowed to graze for 5 to 12 days on its choice of crusts. All urchins had been held without food in a tank with running seawater for the previous 3–4 weeks. Crust areas were measured using a grid of 9 mm^2 squares both before and after exposure to urchins. Controls without urchins were run simultaneously, and showed no changes in crust areas.

Statistical procedures

Most data from the field surveys were transformed before statistical analyses to eliminate heterogeneity of variances. Appropriate transformations are given in the Results. Linear regressions were performed to examine crust thickness patterns and herbivore abundance patterns with depth, and their significance tested with ANOVAs. Differences in relative crust abundances on horizontal and vertical surfaces were tested with a two-factor ANOVA. Urchin food preferences were analyzed using Kruskal-Wallis one-way ANOVA (Zar 1974) on changes in crust area over the course of the experiment.

Organisms

Because of prior paucity of crust collections from the subtidal zone and the difficulty of crust taxonomy in general, many species found in the survey are poorly known. Crustose corallines were not identified to species except for the chips used in the feeding experiments; identifications of subtidal coralline chips are tentative, as this flora has not been thoroughly identified. Crustose fleshy reds were identified using Dethier (1987) and references therein, plus Norris and Hollenberg (1969), Hollenberg and Abbott (1965), and the references below. Crusts included *Peyssonnelia* (two species), *Hildenbrandia occidentalis* Setchell *in* Gardner, *Rhodophysema elegans* (Crouan *et* Crouan *ex* J. Agardh) Dixon, and several crusts that are tetrasporophytes of erect algae: 'Cruoria profunda' [in California, known to alternate with *Opuntiella californica* (Farlow) Kylin] (DeCew 1983), 'Haematocelis' sp. (alternating in California with *Schizymenia* and/or *Farlowia*) (DeCew and West 1981 a), and 'Cruoriopsis aestuarii' [alternating with *Gloiosiphonia californica* (Farlow) J. Agardh] (DeCew *et al.* 1981). In addition, a 'Petrocelis'-like crust was common at one station; no tetrasporangia were observed, so this could represent the alternate form of either a *Mastocarpus* sp. or a *Gymnogongrus* sp. (DeCew and West 1981 b). Finally, a thick undescribed crust, likely another tetrosporophyte, was found at some of the deeper stations. All these reds except *Peyssonnelia* spp. and *H. occidentalis* are grouped as 'infrequent red crusts' for analysis.

Brown crusts were relatively uncommon. Most samples were identified as *Ralfsia fungiformis* (Gunn.) Setchell *et* Gardner, *R. pacifica* Hollenberg, and *R. confusa* Hollenberg. The latter species has not been recorded north of Santa Monica, California (Abbott and Hollenberg 1976) except by Dethier (1987), but was one of the most frequently collected brown crusts at our study sites.

Herbivores counted in the surveys included red urchins, the chitons *Tonicella* spp. and small (< 1 cm) *Dendrochiton* spp., several limpet species (mostly *Acmaea mitra* Rathke, which is a coralline specialist: Padilla 1984), and the small snails *Margarites*, *Lirularia*, and *Homalopoma* spp. Large macroalgae included the kelps *Nereocystis luetkeana* (Mertens) Postels *et* Ruprecht and *Agarum* spp. Sessile invertebrates consisted of a variety of sponges, bryozoans, tunicates, and cup corals.

Results

Patterns of crust abundance and thickness

Crustose algae are a dominant community component of rocky substrata at all depths surveyed, from −6 to −24 m (Fig. 1). Total crust coverage was high, often exceeding 80%, and fleshy red crusts occupied up to 70% of the space in individual quadrats. A general trend of decreasing percentage cover with depth existed for all crust taxa except *Hildenbrandia occidentalis*, with similar trends seen at both study sites (Fig. 1).

The single most common species was a red fleshy crust not previously reported north of Monterey, California: *Peyssonnelia profunda* Hollenberg *et* Abbott. *Peyssonnelia profunda* is most readily distinguished from *P. pacifica* by its larger tetrasporangia (50 × 100 μm *vs.* 30 × 50 μm). In the absence of tetrasporangia, the presence of rhizoids strongly suggests *P. profunda*

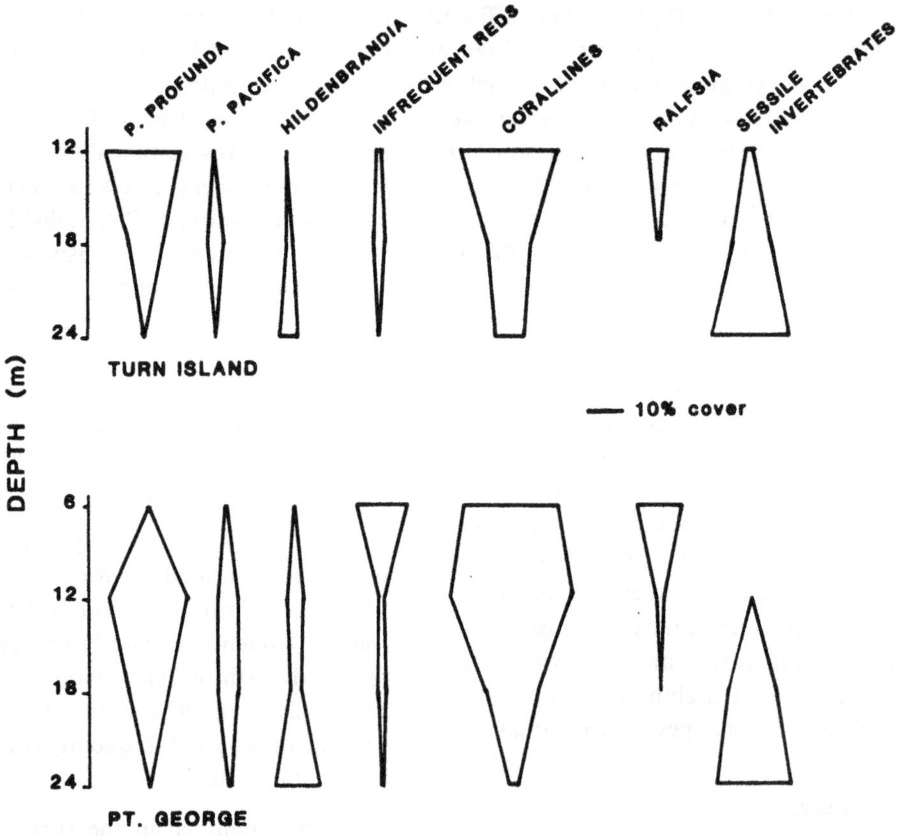

Fig. 1. Percentage cover of crusts and sessile invertebrates found subtidally at two sites. Each percentage is a mean from 10 quadrat samples. Error bars are not given because of the indirect calculation of cover values (see Methods).

(Hollenberg and Abbott 1965). All (N = 10) tetrasporic *P. profunda* samples we found had rhizoids while most (17/19) of the tetrasporic *P. pacifica* did not.

The cover of *Peyssonnelia pacifica* and *P. profunda* varied with depth. *Peyssonnelia profunda* was maximally abundant at 12 m, then decreased with depth (Fig. 1), whereas the rarer *P. pacifica* was nowhere common, but also disappeared in deeper water. Crustose coralline cover (all species pooled) also was highest at 12 m depth, then declined; individual coralline crust species patterns cannot be generated from our data. The 'infrequent red crusts' (see Organisms section) also decreased in abundance with depth, especially at Point George. *Ralfsia* spp. were found occasionally in shallow water, rarely at 18 m and never deeper. In contrast to the vertical distribution of all other species surveyed, *Hildenbrandia occidentalis* abundance increased with depth (Fig. 1).

Crust cross sections revealed that *Peyssonnelia pacifica*, *P. profunda*, and *Hildenbrandia occidentalis* became significantly thinner with increasing depth (Figs 2 and 3: ANOVAs for full linear regressions on log transformed data, p's < 0.001). Fewer data were available for the three other red crusts presented in Figure 3, but all showed similar trends. Only *Rhodophysema*, which is a thin crust (usually < 100 µm thick), showed no obvious depth-related pattern. Variation in crust thickness was high at all depths (Figs 2 and 3).

Quadrats sampled on neighboring horizontal and vertical surfaces (Fig. 4) showed that there is a significant relationship between crust type (fleshy *vs.* coralline) and rock orientation (two-factor ANOVA on log transformed data, $p = .0105$). Overall crust cover (combined coralline and fleshy) was not significantly different between the two orientations ($p = .09$), but the proportional cover of fleshy red crusts was higher and that of corallines was lower on vertical surfaces than on horizontal ones, confirming qualitative observations. The only other apparent biological difference between these surfaces is that the chiton *Tonicella lineata* (Wood) was more abundant on vertical surfaces (2.5 ± 0.8 s. e. *vs.* 1.0 ± 0.7 per 0.1 m^2, paired t-test, $p = .048$). No urchins were seen in any of these 16 small quadrats at the time of sampling.

Herbivory

The three most abundant herbivore groups found at the study sites capable of eating crusts were chitons, urchins, and limpets. The most common chiton, *Ton-*

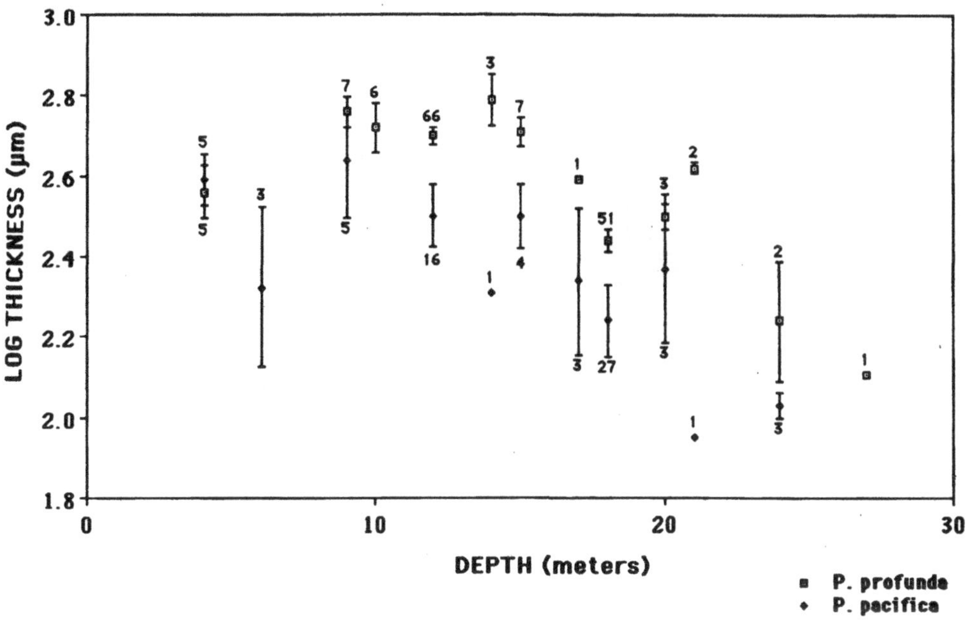

Fig. 2. Changes in thickness of two *Peyssonnelia* species with depth. Each point represents the mean thickness (one s. e.) for that species at that site and depth. Samples were taken from 8 sites and at slightly different depths, but each species was not sampled at all sites. N's for each sample are listed at the top. The regression line for *P. profunda* is log y = −.028x + 2.99 (r = −.51, N = 154), and for *P. pacifica* is log y = −.026x + 2.76 (r = −.40, N = 71).

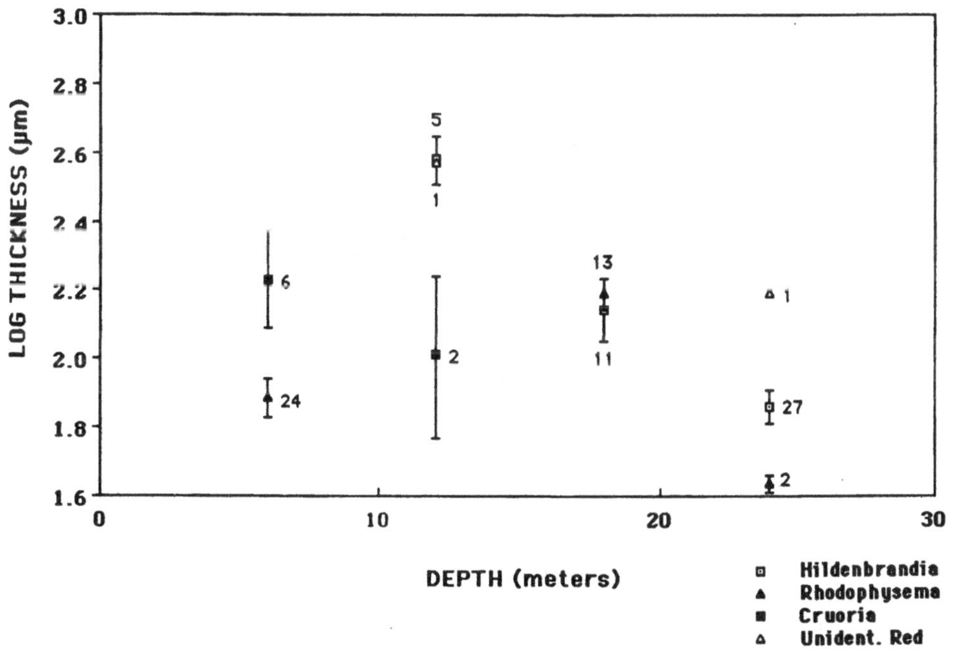

Fig. 3. Changes in thickness of four other red crusts with depth. Symbols as in Figure 2. The regression line for *Hildenbrandia occidentalis* is log y = −.05x + 3.22 (r = −.80, N = 42). Insufficient data were gathered to calculate meaningful regressions for the other three species.

icella, showed a weak (r = −.38) but significant (p = .0003, linear regression on square root transformed data) tendency to decrease in abundance with depth (Fig. 5), but no pattern with depth existed either for urchins or for the only common limpet, *Acmaea mitra*. All grazers occurred patchily (see error bars in Fig. 5). The only other common macroscopic grazers were small trochinid snails (archaeogastropods), which probably consume only microalgae (Steneck and Watling 1982).

While mean urchin densities were low (Fig. 5), they were abundant in patches, occurring at up to 11 per m^2 in the survey areas and even more densely in

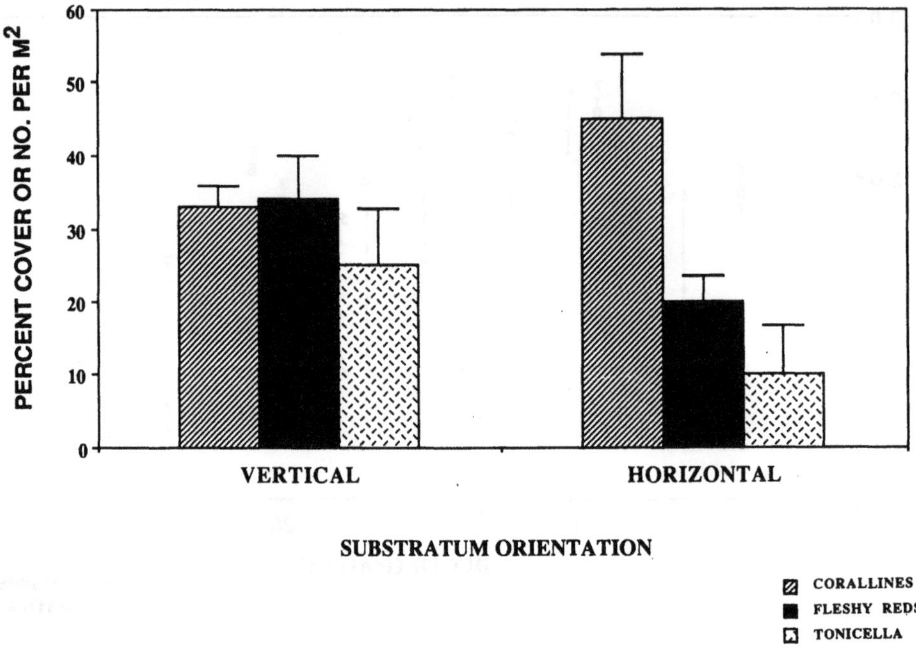

Fig. 4. Mean percent cover of fleshy and coralline crusts (all species lumped) and mean number of *Tonicella* per m^2 on vertical and horizontal surfaces. Error bars are 1 s.e., N = 8.

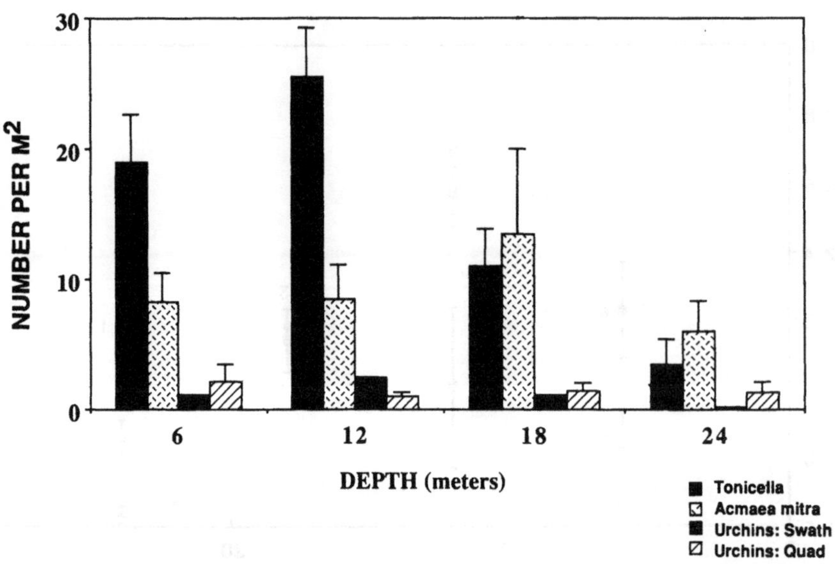

Fig. 5. Densities of herbivores in the horizontal transects (*Tonicella, Acmaea mitra*, and Urchins: Swath) and in cross-depth transects (Urchins: Quads) (see Methods). Number of quadrats with increasing depth = 30, 20, 20, and 20 for *Tonicella* and *Acmaea*, and = 10, 26, 15, and 6 respectively for urchins. Error bars = 1 s.e.

irregular patches elsewhere. Laboratory experiments showed that red urchins consume crusts, and exhibit marked crust-species preferences (Fig. 6A and B). ANOVAs on changes in crust area showed these differences among species to be highly significant (p values < 0.005). Overall, fleshy crusts were eaten much more readily than coralline species, in tests with both intertidal (Fig. 6A) and subtidal (Fig. 6B) crusts. In all but one case, corallines ranked lower as food sources than uncalcified species, despite highly variable thicknesses, cell sizes, and other morphological crust features. Thus calcification appears to provide a significant deterrent to urchin feeding.

Discussion

Although there is a large volume of literature exploring the patterns and causes of zonation of intertidal plants and animals, less is known about the processes

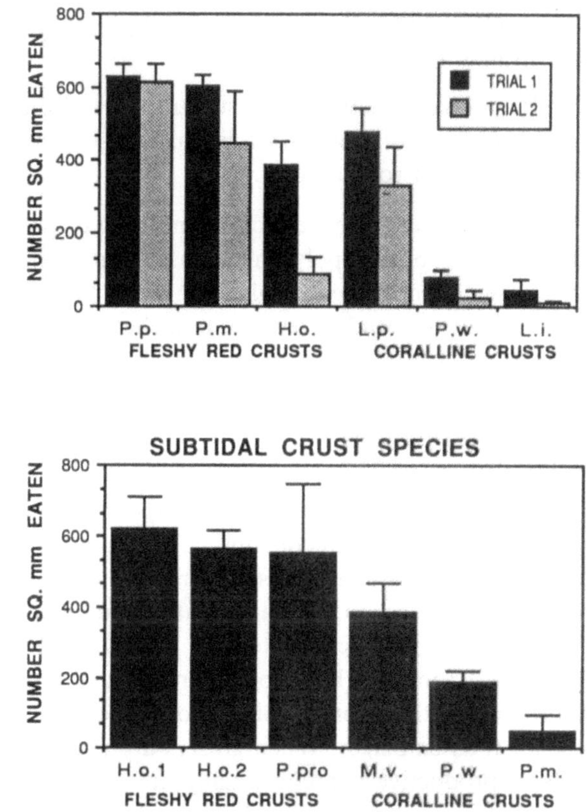

Fig. 6. Means (of 5 replicates) and 1 s. e. of the amounts of intertidal (6 A) and subtidal (6 B) crust species eaten by red urchins in the laboratory. Fleshy red crusts are represented by dark bars, corallines by light ones. A. Trial 1 (left bar) ran for 7 days, Trial 2 (right bar) for 5 days. P. p. = *Peyssonnelia pacifica*, P. m. = 'Petrocelis middendorffii', H. o. = *Hildenbrandia occidentalis*, L. p. = *Lithothamnion phymatodeum*, P. w. = *Pseudolithophyllum whidbeyense* (Foslie) Steneck *et* Paine, L. i. = *Lithophyllum impressum* Foslie. B. Amount eaten over 12 days. Two separate chips of *Hildenbrandia occidentalis* were put in each sampler so that equal areas of fleshy and coralline crusts were available. P. pro = *Peyssonnelia profunda*, M. v. = ?*Mesophyllum vancouveriense* (Foslie) Steneck *et* Paine, P. w. = ?*P. whidbeyense*, P. m. = ?*Pseudolithophyllum muricatum* Steneck *et* Paine.

behind zonation patterns of subtidal organisms. Macroalgal lower limits may be set by light, herbivory (especially by urchins), or a combination of these (Neushul 1965, Himmelman and Lavergne 1985, Witman 1985 and 1987, Vadas and Steneck 1988). Coralline crust morphology and species dominance with depth also are affected by herbivory (Steneck and Adey 1976, Steneck 1985), but much less is known about non-coralline crusts. The fact that the single most abundant crust species at the study sites, *Peyssonnelia profunda*, had not been reported north of central California (except for one record by Garbary *et al.* 1984) is indicative of the dearth of knowledge about these habitats.

Most of the subtidal encrusting organisms we studied had a depth-zonation pattern, showing some replacement with depth. *Ralfsia* spp., the only common brown crusts observed, were most abundant in shallow water and became less abundant quickly with depth. *Peyssonnelia* spp. and crustose corallines were most common at intermediate depths, then declined gradually. The decline of *Tonicella* with depth may be due to the decline of crustose corallines, which are its sole food (Padilla 1984). Only *Hildenbrandia occidentalis* increased with greater depths. In the intertidal zone, this species and its congener *H. rubra* grow slowly and are poor competitors, but are capable of persisting for long periods even when overgrown (suggesting a low metabolic rate), and of regrowing from small fragments when consumed by grazers (Dethier and Steneck, unpubl. observation). Similar qualities have been found in *H. rubra* in New England (Bertness *et al.* 1983) and in Australia (Underwood 1980), and in a subtidal red crust in New England (Sebens 1986). Slow growth but high persistence may prevent *H. occidentalis* from competing effectively in shallow water (where other crusts thrive), but may enable it to survive herbivory and temporary overgrowth in deeper water.

As algal crusts became less abundant in deeper water (Fig. 1), encrusting invertebrates increased. Competition from invertebrates may contribute to the decline of crusts, which are notoriously poor competitors for space (*e.g.*, Bertness *et al.* 1983, Menge *et al.* 1985, Sebens 1986, and references therein). We have no data on the processes responsible for the relative rarity of invertebrates in shallow zones, although in other regions grazing by omnivorous urchins restricts sessile invertebrates to deeper water (Vance 1979, Himmelman and Lavergne 1985, Briscoe and Sebens 1988). Urchins at our study sites showed no obvious zonation pattern with depth, but other predators such as fish, seastars, snails, and nudibranchs could be contributing to the absence of invertebrates in shallower waters. Foster (1975) suggested that at low light levels, sessile invertebrates have a competitive advantage over algae even if predators are present. Under higher light conditions, however, algae have the advantage regardless of grazer abundance, as long as some carnivores are present. Invertebrate larvae may have difficulty settling in shallower waters due both to algal whiplash and to various mobile invertebrates dislodging larvae or juveniles from the surfaces of smooth fleshy crusts.

For all crust species for which sufficient data were available (Figs 2 and 3), thickness decreased with depth. Variation in crust thickness was consistently high, probably due to local variation in herbivory, shading, microhabitats, and crust ages. Thin crustose thalli are nearly monolayered (and held perpendicular to incident light), as predicted for low light environments by models of functional geometry (Hay 1986). Such a form captures all incident light while minimizing self-shading and maximizing the number of cells in direct contact with the environment (for gas and nutrient exchange). In shallower, higher-light habitats, thicker thalli gain an advantage in being more resistant to overgrowth and to herbivores (Steneck 1986). It is unlikely that the deep crusts are thinner as a result of being intensively grazed, as there were fewer herbivores at the deeper sites (Fig. 5).

Urchin grazing in nearshore northeast Pacific sites is a major structuring parameter of subtidal communities (reviewed in VanBlaricom and Estes 1988). While the most obvious impact of urchin grazing is on erect macroalgal abundance (and perhaps evolution: Estes and Steinberg 1988), it is likely that they affect crust abundance as well. The virtual absence of 'Petrocelis' from subtidal regions may be attributable to its high preference as a food (Fig. 6 A: PM). In the absence of grazers, this common intertidal species (Dethier 1987) can thrive in subtidal habitats (Dethier and Steneck unpubl. observation). Urchins also might be responsible for the observed change in dominant crust species from horizontal to vertical surfaces. Overall crust abundance is almost identical in horizontal and vertical habitats, but fleshy reds are more (and corallines less) abundant on vertical faces. If red urchins forage more readily on horizontal than vertical surfaces as do their congeners *Strongylocentrotus droebachiensis* (O. F. Müller) (Sebens 1986), then fleshy crusts (their more preferred food) may survive better on vertical surfaces. In New England, Sebens (1986) found fleshy red crusts to be confined largely to vertical surfaces in the subtidal zone, whereas horizontal surfaces with urchins were dominated by corallines. On horizontal surfaces, some *Peyssonnelia* may persist owing to its relatively rapid recruitment and growth (Dethier and Steneck unpubl. observation) and to the patchiness of urchins. An alternative hypothesis is that fleshy crusts compete better under the lower light conditions of vertical surfaces, although in general corallines penetrate into darker waters farther than do fleshy crusts (Vadas and Steneck 1988).

In the laboratory, urchins ate one coralline species, *Lithothamnion phymatodeum* Foslie (Fig. 6 A), more readily than several of the fleshy crusts. This coralline may be easier to consume for two reasons:

1) its surface is quite rugose ('branched'), making it easier to bite, and

2) it often adheres poorly to the substratum and thus is easier to remove, sometimes because polychaetes bore into and burrow under it.

Sebens (1986) found the identical phenomena of poor adherence and boring polychaetes in *Lithothamnion glaciale* Kjellman in New England. Among the subtidal crusts at our study sites, one branched coralline (*Mesophyllum*) was eaten more readily than the smooth corallines, even though one of the smoother species was a thinner crust and thus easier to consume completely.

In conclusion, fleshy crustose algae, especially reds, are abundant components of subtidal assemblages at our study sites. Crust thicknesses are related to depth, probably in response to light, and relative abundances at each depth may be affected by herbivore food preferences. Ironically, herbivory is probably of great importance in the overall domination by crusts of some algal assemblages. When grazers are removed, crusts in productive habitats usually are overgrown by other algae and sessile animals (*e. g.*,, Dethier 1981, Steneck 1982, Jara and Moreno 1983, Bertness 1984, Levings and Garrity 1984, Morrison 1988). However, crusts are not always passive victims of overgrowth and herbivory, as some crusts retain space by inhibiting recruitment of invertebrates (Bertness *et al.* 1983, Breitburg 1984) and of other algae (Underwood 1980, Johnson and Mann 1986), or by inhabiting areas of high physical disturbance (*e. g.*, Daly and Mathieson 1977, Davis and Wilce 1987). Other crusts enhance recruitment of grazers and other organisms (Barnes and Gonor 1973, Morse *et al.* 1979, Rumrill and Cameron 1983, Sebens 1983). Species-specific responses to changes in the environment are virtually unknown; increased understanding of crust ecology may enable us to use crusts as indicators of environmental conditions, biotic and abiotic.

Acknowledgements

We wish to thank all our diving buddies, especially Paul Gabrielson and Steve Norton. R. S. Steneck generated ideas, and he, S. K. Guzik, and three anonymous reviewers improved the manuscript. Space and facilities at the Friday Harbor Laboratories were provided by Dr A. O. D. Willows, and research was supported by NSF grants OCE 8315136 and 8600262.

References

Abbott, I. A. and G. J. Hollenberg. 1976. *Marine Algae of California*. Stanford Univ. Press, CA, U. S. A.

Adey, W. H. and I. G. MacIntyre. 1973. Crustose coralline algae: a reevaluation in the geological sciences. *Geol. Soc. Am. Bull. 84*: 883–904.

Barnes, J. R. and J. J. Gonor. 1973. The larval settling response of the lined chiton *Tonicella lineata*. *Mar. Biol. 20*: 259–264.

Bertness, M. D. 1984. Habitat and community modification by an introduced herbivorous snail. *Ecology 65*: 370–381.

Bertness, M. D., P. O. Yund and A. F. Brown. 1983. Snail grazing and the abundance of algal crusts on a sheltered New England rocky beach. *J. exp. mar. Biol. Ecol. 71*: 147–164.

Breitburg, D. L. 1984. Residual effects of grazing: inhibition of competitor recruitment by encrusting coralline algae. *Ecology 65*: 1136–1143.

Briscoe, C. S. and K. P. Sebens. 1988. Omnivory in *Strongylocentrotus droebachiensis* (Müller) (Echinodermata: Echinoidea): predation on subtidal mussels. *J. exp. mar. Biol. Ecol. 115*: 1–24.

Daly, M. A. and A. C. Mathieson. 1977. The effects of sand movements on intertidal seaweeds and selected invertebrates at Bound Rock, New Hampshire USA. *Mar. Biol. 43*: 45–55.

Davis, A. N. and R. T. Wilce. 1987. Algal diversity in relation to physical disturbance: a mosaic of successional stages in a subtidal cobble habitat. *Mar. Ecol. Prog. Ser. 37*: 229–237.

DeCew, T. C. 1983. Culture studies in the Hildenbrandiales, Cryptonemiales, Gigartinales and Palmariales (Rhodophyta). PhD thesis, Univ. Calif., Berkeley, U. S. A.

DeCew, T. C. and J. A. West. 1981 a. Investigations on the life histories of three *Farlowia* species (Rhodophyta: Cryptonemiales, Dumontiaceae) from Pacific North America. *Phycologia 20*: 342–351.

DeCew, T. C. and J. A. West. 1981 b. Life histories in the Phyllophoraceae (Rhodophyta: Gigartinales) from the Pacific Coast of North America. I. *Gymnogongrus linearis* and *G. leptophyllus*. *J. Phycol. 17*: 240–250.

DeCew, T. C., J. A. West and E. K. Ganesan. 1981. The life histories and developmental morphology of two species of *Gloiosiphonia* (Rhodophyta: Cryptonemiales, Gloiosiphoniaceae) from the Pacific coast of North America. *Phycologia 20*: 415–423.

Dethier, M. N. 1981. Heteromorphic algal life histories: the seasonal pattern and response to herbivory of the brown crust, *Ralfsia californica*. *Oecologia 49*: 333–339.

Dethier, M. N. 1987. The distribution and reproductive phenology of intertidal fleshy crustose algae in Washington. *Can. J. Bot. 65*: 1838–1850.

Estes, J. A. and P. D. Steinberg. 1988. Predation, herbivory, and kelp evolution. *Paleobiology 14*: 19–36.

Foster, M. S. 1975. Regulation of algal community development in a *Macrocystis pyrifera* forest. *Mar. Biol. 32*: 331–342.

Garbary, D. J., G. I. Hansen and R. F. Scagel. 1984. Additions to the marine algae of Barkley Sound, Vancouver Island, British Columbia. *Syesis 17*: 43–45.

Hay, M. E. 1986. Functional geometry of seaweeds: ecological consequences of thallus layering and shape in contrasting environments. *In*: (T. J. Givnish, ed.) *On the Economy of Plant Form and Function*. Cambridge Univ. Press, Cambridge, MA, U. S. A. pp. 635–666.

Himmelman, J. H. and Y. Lavergne. 1985. Organization of rocky subtidal communities in the St. Lawrence Estuary. *Le Naturaliste Canadien 112*: 143–154.

Hollenberg, G. J. and I. A. Abbott. 1965. New species and new combinations of marine algae from the region of Monterey, California. *Can. J. Bot. 43*: 1177–1188.

Jara, H. F. and C. A. Moreno. 1983. Herbivory and structure in a midlittoral rocky community: a case in southern Chile. *Ecology 65*: 28–38.

Johnson, C. R. and K. H. Mann. 1986. The crustose coralline alga, *Phymatolithon* Foslie, inhibits the overgrowth of seaweeds without relying on herbivores. *J. exp. mar. Biol. Ecol. 96*: 127–146.

Littler, M. M. and D. S. Littler. 1984. Models of tropical reef biogenesis: the contribution of algae. *In*: (F. E. Round and D. J. Chapman, eds) *Progress in Phycological Research, Vol. 3*. Biopress Ltd., U. K., pp. 323–364.

Levings, S. C. and S. D. Garrity. 1984. Grazing patterns in *Siphonaria gigas* (Mollusca, Pulmonata) on the rocky Pacific coast of Panama. *Oecologia 64*: 152–159.

Lubchenco, J. 1980. Algal zonation in the New England rocky intertidal community: an experimental analysis. *Ecology 61*: 333–344.

Menge, B. A., J. Lubchenco and L. R. Ashkenas. 1985. Diversity, heterogeneity and consumer pressure in a tropical rocky intertidal community. *Oecologia 65*: 394–405.

Morse, D. E., N. Hooker, H. Duncan and L. Jensen. 1979. Gamma-aminobutyric acid, a neurotransmitter, induces planktonic abalone larvae to settle and begin metamorphosis. *Science 204*: 407–410.

Morrison, D. 1988. Comparing fish and urchin grazing in shallow and deeper coral reef algal communities. *Ecology 69*: 1367–1382.

Neushul, M. 1965. SCUBA diving studies of the vertical distribution of benthic marine plants. *Proc. 5th Mar. Biol. Assoc.*: 161–176.

Norris, R. E. and G. J. Hollenberg. 1969. Notes on marine algae of Washington and southern British Columbia. IV. *Syesis 2*: 115–119.

Padilla, D. K. 1984. The importance of form: differences in competitive ability, resistance to consumers and environmental stress in an assemblage of coralline algae. *J. exp. mar. Biol. Ecol. 79*: 105–128.

Paine, R. T. 1984. Ecological determinism in the competition for space. *Ecology 65*: 1339–1348.

Rumrill, S. S. and R. A. Cameron. 1983. Effects of gamma-aminobutyric acid on the settlement of larvae of the black chiton *Katharina tunicata*. *Mar. Biol. 72*: 243–247.

Sebens, K. P. 1983. Settlement and metamorphosis of a temperate soft-coral larva (*Alcyonium siderium* Verrill): induction by crustose algae. *Biol. Bull. 165*: 286–304.

Sebens, K. P. 1986. Spatial relationships among encrusting marine organisms in the New England subtidal zone. *Ecol. Monogr. 56*: 73–96.

Slocum, C. J. 1980. Differential susceptibility to grazers in two phases of an intertidal alga: advantages of heteromorphic generations. *J. exp. mar. Biol. Ecol. 46*: 99–110.

Steneck, R. S. 1982. A limpet-coralline alga association: adaptations and defenses between a selective herbivore and its prey. *Ecology 63*: 507–522.

Steneck, R. S. 1985. Adaptations of crustose coralline algae to herbivory: patterns in space and time. *In*: (D. F. Toomey and M. H. Nitecki, eds.) *Paleoalgology: Contemporary Research and Applications*. Springer Verlag, Berlin, pp. 352–366.

Steneck, R. S. 1986. The ecology of coralline algal crusts: convergent patterns and adaptive strategies. *Ann. Rev. Ecol. Syst. 17*: 273–303.

Steneck, R. S. and W. H. Adey. 1976. The role of the environment in control of morphology in *Lithophyllum congestum*, a Caribbean algal ridge builder. *Bot. Mar. 18*: 197–215.

Steneck, R. S. and L. Watling. 1982. Feeding capabilities and limitation of herbivorous molluscs: a functional group approach. *Mar. Biol. 68*: 299–319.

Steneck, R. S. and R. T. Paine. 1986. Ecological and taxonomic studies of shallow-water encrusting Corallinaceae (Rhodophyta) of the boreal northeastern Pacific. *Phycologia 25*: 221–240.

Underwood, A. J. 1980. The effects of grazing by gastropods and physical factors on the upper limits of distribution of intertidal macroalgae. *Oecologia 46*: 201–213.

Vadas, R. L. 1977. Preferential feeding: an optimization strategy in sea urchins. *Ecol. Monogr. 47*: 337–371.

Vadas, R. L. and R. S. Steneck. 1988. Zonation of deep water benthic algae in the Gulf of Maine. *J. Phycol. 24*: 338–346.

VanBlaricom, G. R. and J. A. Estes, eds. 1988. *The Community Ecology of Sea Otters*. Springer-Verlag, Berlin.

Vance, R. R. 1979. Effects of grazing by the sea urchin, *Centrostephanus coronatus*, on prey community composition. *Ecology 60*: 537–546.

Witman, J. D. 1985. Refuges, biological disturbance, and rocky subtidal community structure in New England. *Ecol. Monogr. 55*: 421–445.

Witman, J. D. 1987. Subtidal coexistence: storms, grazing, mutualism, and the zonation of kelps and mussels. *Ecol. Monogr. 57*: 167–187.

Zar, J. H. 1974. *Biostatistical Analysis*. Prentice-Hall, Englewood Cliffs, N. J., U. S. A.

Botanica Marina
Vol. 34, pp. 211–214, 1991

Screening of Brazilian Marine Algae for Hemagglutinins

I. Lima Ainouz and A. Holanda Sampaio*

Department of Biochemistry and Molecular Biology and **Laboratory of Marine Science (LABOMAR), Federal University of Ceará, Box 1065, 60001, Fortaleza, Ceará, Brazil*

(Accepted 19 January 1991)

Abstract

Twenty Brazilian marine algal species (sixteen red and four green) were analyzed for hemagglutinating activity using trypsin-treated erythrocytes of human A, B, 0 and AB groups, in addition to those of chicken, rabbit, sheep, goat and cow. Saline extracts from ten marine algae agglutinated rabbit erythrocytes, whereas seven, six, and four extracts agglutinated chicken, cow and sheep erythrocytes, respectively. Goat erythrocytes were not agglutinated by any of the algal extracts tested. The non-specific agglutination of human red blood cells was observed in extracts of two red algae (*Gracilaria ferox* and *Hypnea musciformis*) and one green alga (*Caulerpa cupressoides*).

Introduction

Compounds able to agglutinate red blood cells have been found in a great number of living organisms such as higher plants, vertebrates, invertebrates, microorganisms, and also viruses (Sharon and Lis 1989). These agglutinins are considered lectins, defined as proteins or glycoproteins which specifically recognize and bind some carbohydrates. As a consequence of their chemical properties they became a useful tool in several fields of biological research (immunology, cell biology, membrane structure and cancer research). Their occurrence in marine algae was first reported by Boyd *et al.* (1966) followed by Blunden *et al.* (1975), and Rogers *et al.* (1980) using only human erythrocytes when screening for hemagglutinins. Hori *et al.* (1981), Fabregas *et al.* (1984, 1985, 1986), and Muñoz *et al.* (1985, 1987) showed evidence of agglutinins in marine algae using untreated animal erythrocytes in addition to human A, B, 0 and AB groups. Rogers *et al.* (1980) and Hori *et al.* (1988) reported that hemagglutinating activity was more readily detected when erythrocytes were enzyme-treated.

These results also have shown that the rabbit erythrocytes are more sensitive to the agglutinins of marine algae than the human erythrocytes (Hori *et al.* 1981) especially when the cells are enzyme-treated (Hori *et al.* 1988).

The present work deals with the screening for hemagglutinins of red and green marine algae from Brazil against animal and human trypsinized erythrocytes. This is the first study on Brazilian algal agglutinins.

Material and Methods

Algal specimens

The red (16) and green (4) marine algal species listed in Table I were screened for hemagglutinating activity. They were collected on the NE Coast of Brazil (Fortaleza, State of Ceará) in October 1989. After collection they were packed in plastic bags and kept at −20 °C until used. The algal nomenclature used follows those of Taylor (1960) and Joly (1965).

Preparation of algal extracts

Samples of each alga were thawed and washed with distilled water, cut into small pieces and ground in a mortar with NaCl 0.85% (1 : 3, w/v). After filtration through cheese cloth the solution was centrifuged (9200 × g, 30 min, 4 °C) and the supernatant used for hemagglutination assays.

Table I. Hemagglutinating activity of marine algae. Titre of extracts with erythrocytes.

Species	Sheep	Rabbit	Chicken	Goat	Cow	A	B	0	AB
Chlorophyta									
Ulvales									
Ulvaceae									
Ulva lactuca Linnaeus	–	–	–	–	–	–	–	2^2	–
Siphonales									
Bryopsidaceae									
Bryopsis pennata Lamouroux	–	2^2	2^3	–	2^1	–	2^3	2^2	2^1
Caulerpaceae									
Caulerpa cupressoides (West) C. Agardh	2^1	2^5	2^4	–	2^2	2^1	2^1	2^1	2^1
Caulerpa mexicana (Sonder) J. Agardh	–	–	–	–	–	–	–	–	–
Rhodophyta									
Cryptonemiales									
Corallinaceae									
Corallina officinalis Linnaeus	–	–	–	–	–	–	–	–	–
Corallina subulata Ellis *et* Solander	–	–	–	–	–	–	–	–	–
Grateloupiaceae									
Cryptonemia crenulata J. Agardh	–	2^1	–	–	–	–	–	–	–
Girgartinales									
Hypneaceae									
Hypnea cervicornis J. Agardh	–	2^3	–	–	2^1	–	–	–	–
Hypnea musciformis (Wulfen) Lamouroux	2^2	2^7	–	–	2^5	2^2	2^2	2^5	2^2
Gracilariaceae									
Gracilaria cearensis Joly *et* Pinheiro	–	2^3	–	–	–	–	–	–	–
Gracilaria cervicornis (Turner) J. Agardh	–	2^1	2^3	–	–	–	–	–	–
Gracilaria cornea Maze *et* Schram	–	–	2^4	–	–	–	–	–	–
Gracilaria domigensis Sonder	–	–	–	–	–	–	–	–	–
Gracilaria ferox J. Agardh	2^2	2^7	2^6	–	2^4	2^4	2^4	2^4	2^4
Gracilaria sjoestedtii Kylin	2^2	–	2^8	–	–	–	–	–	–
Gelidiales									
Gelidiaceae									
Gelidiopsis gracilis (Kützing) Vickers	–	2^1	–	–	–	–	–	–	–
Ceramiales									
Rhodomelaceae									
Bryothamnion seaforthii (Turner) Kützing	–	2^8	2^9	–	2^8	–	–	–	–
Laurencia obtusa (Hudson) Lamouroux	–	–	–	–	–	–	–	–	–
Ceramiaceae									
Centroceras clavatum (C. Agardh) Montagne	–	–	–	–	–	–	–	–	–
Rhodymeniales									
Rhodymeniaceae									
Botryocladia occidentalis (Borgesen) Kylin	–	–	–	–	–	–	–	–	–

Hemagglutinating activity

Aliquots of 200 μL of the algal extracts were used to prepare serial two-fold dilutions made with NaCl (0.85%). Equal volumes of erythrocyte suspension were added to each tube, gently shaken, and incubated at 37 °C for 30 min. The mixture was left for 30 min at room temperature (28 °C), centrifuged (2000 × g, 30 sec), and the agglutination observed macroscopically. The titer was expressed as the reciprocal of the highest dilution showing positive results. The assays were carried out in triplicate and 0.85% NaCl solution plus the erythrocytes was used as control.

Erythrocytes

Red blood cells of rabbit, chicken, cow, sheep and goat (obtained by venous puncture) and human A, B, 0 and AB groups (obtained from the Center of Hematology of Ceará, Fortaleza, Ceará, Brazil) were used for hemagglutination tests, after treatment with trypsin. Each blood sample was collected into a pre-heparinised tube and washed three times with ten volumes of 0.85% NaCl. Saline containing trypsin (0.1 mg per ml) was added to the packed cells to give a 2% suspension and incubated at 25 °C for 60 min. After washing six times with cold saline, a 2% trypsin-treated erythrocyte suspension was prepared.

Results and Discussion

As shown in Table I three species of green algae out of four and ten species of red algae out of sixteen showed hemagglutinating activity against at least one of the trypsin-treated erythrocytes tested. Goat erythrocytes were not agglutinated by any of the algal extracts analyzed.

Chlorophyta

A saline extract of *Ulva lactuca* caused agglutination of 0 cells only. Blunden *et al.* (1975) found positive results for A, B and 0 human blood cells and later Chiles and Bird (1989) observed the agglutination of rabbit and sheep erythrocytes for the same species. *Ulva pertusa* Kjellman, *U. arasakii* Chihara and *U. conglobata* Kjellman were screened for agglutination by Hori and co-workers. They reported the agglutination of A and 0 cells, and rabbit, horse and duck erythrocytes for *U. arasakii* and also sheep, B and AB cells for *U. pertusa* (Hori *et al.* 1981). *Ulva conglobata* showed agglutination of sheep, B and 0 pronase-treated erythrocytes (Hori *et al.* 1988). The agglutination of untreated and treated 0 cells was therefore observed for all species of *Ulva* so far studied.

Bryopsis pennata was active with respect to rabbit, chicken, cow and human B, 0 and AB group erythrocytes. The agglutination of A, B and 0 papain-treated cells by extracts of *B. hypnoides* Lamouroux and *B. plumosa* (Hudson) Agardh was detected by Rogers *et al.* (1980). Hori *et al.* (1981) also found positive results for rabbit, horse, sheep, duck, A, B, 0 and AB erythrocytes by *B. hypnoides*. All species of *Bryopsis* tested by different authors caused agglutination of B and 0 cells and rabbit erythrocytes.

An extract of *Caulerpa cupressoides* was able to agglutinate all erythrocytes examined except those of goat but *C. mexicana* did not show any agglutinic activity. Recently Chiles and Bird (1989) reported the agglutination of rabbit and sheep erythrocytes by *C. paspaloides* (Bory) Greville and *C. sertularioides* (Gmelin) Howe. The latter was also able to agglutinate A, B, and 0 cells. The hemagglutinating activity was not detected before for other species of *Caulerpa* by Boyd *et al.* (1966), Rogers *et al.* (1980) and Hori *et al.* (1988).

Rhodophyta

The species of *Corallina* studied did not show hemagglutinic activity in agreement with the results of Blunden *et al.* (1975) and Hori *et al.* (1981) for *C. officinalis* and *C. pilulifera* Postels *et* Ruprecht.

Extracts of *Centroceras clavatum*, *Botryocladia occidentalis* and *Laurencia obtusa* were also inactive against any type of erythrocytes assayed. *Laurencia pinnafitida* (Hudson) Lamouroux and *L. undulata* Yamada were studied by Muñoz *et al.* (1987) and Hori *et al.* (1988) respectively, and similar results were found for rabbit, horse and sheep erythrocytes for both species.

Cryptonema crenulata and *Gelidiopsis gracilis* showed only weak agglutination of rabbit erythrocytes.

Bryothamnion seaforthii did not agglutinate human blood cells in spite of the high titers recorded for rabbit, cow and chicken erythrocytes.

Hypnea musciformis and *Gracilaria ferox* showed non-specific agglutination of human red blood cells and high agglutinic activity against rabbit erythrocytes. With the exception of *G. domingensis* all other species of *Gracilaria* were able to agglutinate at least one of the different erythrocytes analyzed. Hemagglutinins have been isolated and partially characterized from *Hypnea japonica* Tanaka (Hori *et al.* 1986) and *Gracilaria verrucosa* (Hudson) Papenfuss (Shiomi *et al.* 1981). The red algae have been shown to be the main source for isolation of hemagglutinins from marine algae.

The present results suggest the use of different types of erythrocytes for the detection of hemagglutinins in algal extracts even when the blood cells are enzyme-treated. Rabbit and chicken erythrocytes showed the greatest sensitivity to agglutination by these marine algal extracts. In general, no specificity was detected for human red blood cells.

Acknowledgements

The authors wish to thank Prof. Francisca Pinheiro Joventino for the identification of the algae. This research was supported by The National Research Council of Brazil (CNPq).

References

Blunden, G., D. L. Rogers and W. F. Farnham. 1975. Survey of British seaweeds for hemagglutinins. *Lloydia 38*: 162–168.

Boyd, W. C., L. R. Almodóvar and L. G. Boyd. 1966. Agglutinins in marine algae for human erythrocytes. *Transfusion* (Philadelphia) *6*: 82–83.

Chiles, T. C. and K. T. Bird. 1989. A comparative study of animal erythrocyte agglutinins from marine algae. *Comp. Biochem. Physiol. 94 B*: 107–111.

Fabregas, J., J. Llovo and A. Muñoz. 1985. Hemagglutinins in red sea weeds. *Bot. Mar. 28*: 517–520.

Fabregas, J., A. Muñoz and J. Llovo. 1986. Hemagglutinins in brown sea weeds. *J. Exp. Mar. Biol. Ecol. 97*: 213–219.

Fabregas, J., A. Muñoz, J. Llovo and J. Abalde. 1984. Agglutinins in marine red algae. *IRCS Med. Sci. 12*: 298–299.

Hori, K., K. Miyazawa, N. Fusetani, K. Hashimoto and K. Ito. 1986. Hypnins, low molecular weight peptidic agglutinins isolated from a marine alga, *Hypnea japonica. Biochim. Biophys. Acta 87*: 228–236.

Hori, K., K. Miyazawa and K. Ito. 1981. Hemagglutinins in marine algae. *Bull. Jpn. Soc. Sci. Fish. 47*: 793–798.

Hori, K., C. Oiwa, K. Miyazawa and K. Ito. 1988. Evidence for wide distribution of agglutinins in marine algae. *Bot. Mar. 31*: 133–138.

Joly, A. B. 1965. Flora marinha do litoral Norte do estado de São Paulo e regiões circunvizinhas. *Bolm. Fac. Filos. Cienc.*, Univ. S. Paulo, 294, *Bot. 21*: 1–393.

Muñoz, A., J. Llovo and J. Fabregas. 1985. Hemaglutininas de algas verdes. *Acta Cient. Compostelana 22*: 873–878.

Muñoz, A., J. Llovo and J. Fabregas. 1987. Different agglutinic activity of red marine algae against erythrocytes from several animal species. *Thalassas 1*: 87–89.

Rogers, D. J., G. Blunden, J. A. Topliss and M. G. Guiry. 1980. A survey of some marine organisms for hemagglutinins. *Bot. Mar. 23*: 569–577.

Sharon, N. and H. Lis. 1989. *Lectins*. Chapman and Hall Ltd. London. New York. 127 pp.

Shiomi, K., H. Yamanaka and T. Kikuchi. 1981. Purification and physicochemical properties of a hemagglutinin (GVA-1) in the red alga *Gracilaria verrucosa. Bull. Jpn. Soc. Sci. Fish. 47*: 1079–1084.

Taylor, W. R. 1960. *Marine Algae of the Eastern Tropical and Sub-Tropical Coast of Americas*. The University of Michigan Press. ix + 860 pp. Ann Arbor.

Botanica Marina
Vol. 34, pp. 215–220, 1991

Phycochemical Studies on *Dermonema abbottiae* (Nemaliales – Rhodophyta)

S. Afaq-Husain, M. Shameel*, K. Usmanghani**, M. Ahmad** and V. U. Ahmad***

PCSIR Laboratories Complex, Karachi-75280, Pakistan

* *Department of Botany, University of Karachi, Karachi-75270, Pakistan*
** *Department of Pharmacognosy, Faculty of Pharmacy University of Karachi, Karachi-75270, Pakistan*
*** *HEJ Research Institute of Chemistry, University of Karachi, Karachi-75270, Pakistan*

(Accepted 19 January 1991)

Abstract

A marine red alga, *Dermonema abbottiae* Afaq-Husain, Nizamuddin *et* Shameel, collected from the coast of Karachi, Pakistan, has been investigated for its sterol and fatty acid composition. The sterol profile, identified through MS and ^{1}H-NMR was shown to consist of cholesta-5,20,24- trien-3β-ol, desmosterol, cholesterol, brassicasterol, 24-methylene cholesterol, cholesta-5,25-dien-3β,24-diol, 24-methyl cholesterol, fucosterol and *nor* 31-cycloartanol. The fatty acids were analysed by GC-MS. The esterified fractions yielded methyl tridecylate, palmitate, margarate, stearate, behenate, myristoleate and oleate, and in non-methylated fractions pentadecylic, palmitic, margaric and stearic acids were found.

Introduction

Investigations on the chemical constituents of red seaweeds of Pakistan have been carried out by Bano *et al.* (1986, 1987, 1988) and Hayee-Memon *et al.* (1991). During a taxonomic study of the red algae of the coast of Pakistan *Dermonema abbottiae* Afaq-Husain, Nizamuddin *et* Shameel was described as a new species belonging to the Liagoraceae, Nemaliales (Afaq-Husain *et al.* 1991). It was selected for phycochemical studies primarily because no chemical work has so far been reported in the literature on any species of *Dermonema* and secondly to determine the taxonomic significance of liagosterol in the order Nemaliales as suggested by Fattorusso *et al.* (1975).

Dermonema abbottiae is also interesting from an ecological point of view because it occurs at the upper margin of mid-littoral zone (Afaq-Husain *et al.* 1991) and as such it remains exposed to the sun for long periods and receives a higher light intensity, even when submerged at high water, than other species of red seaweeds. Moreover, this plant is mucilaginous and not cartilaginous, but is tough in consistency and appeared promising for phycochemical studies.

Material and Methods

The plants of *Dermonema abbottiae* were collected from intertidal rocks between the upper and mid-littoral zones at the coasts of Buleji, Paradise point and Naugaza Mazar (2 km east of Cape Monze) near Karachi, Pakistan in June and July 1987. The algal samples were first cleaned with fresh water and then with distilled water to remove sand particles and other debris and finally dried in shade.

Sterol extraction

The dried material (700 g) was exhaustively extracted by percolation with $CHCl_3$ at room temperature in an aspirator. This procedure was repeated three times and the combined extract on evaporation under reduced pressure resulted in 12.2 g of thick dark reddish-

Table I. Sterols of *Dermonema abbottiae*.

Systematic name	Common name	Molecular formula	Mol. wt.	Mass and fragmentation pattern	Spectral data
Cholesta-5,20,24-trien-3β-ol		$C_{27}H_{42}O$ [1]	382	m/z 382 (M^+,$C_{27}H_{42}O$), 367 (M^+-CH_3), 364 (M^+-H_2O), 349 (M^+-CH_3-H_2O), 313 (M^+-C_5H_9), 283 (M^+-CH_3-C_6H_{11}-H), 271, 253, 239, 227, 213, 199, 185, 171, 159, 141, 129, 109, 95, 83, 81, 69, 57.	^{1}H-NMR (300 MHz, $CDCl_3$) ppm: 0.67 (3H,s,Me-18), 1.00 (3H,s,Me-19), 1.21 (18 H,m,H-1,2,4,7,11,12,15,22,23), 3.51 (1H,m,H-3β), 5.34 (2H,m,H-6,24), 5.15 (2H,s,H-21).
Cholesta-5,24,dien-3β-ol	Desmosterol = 24-dehydrocholesterol	$C_{27}H_{44}O$ [2]	384	m/z 384 (M^+,$C_{27}H_{44}O$), 369 (M^+-CH_3), 366 (M^+-H_2O), 351 (M^+-CH_3-H_2O), 300, 299 (M^+-CH_3-C_5H_9-H), 271 (M^+-C_8H_{15}, side chain -2H), 255 M^+-C_8H_{15}-H_2O), 253 M^+-side chain-2H-H_2O), 241 223, 207, 197, 185, 165, 149, 131, 117, 105.	^{1}H-NMR (300 MHz, $CDCl_3$) ppm: 0.67 3H,s,18-Me), 0.87 (6H,s,26,27-Me), 0.99 (3H,s,19-Me), 1.13 (3H,d,J = 7.0H Hz,21-Me), 3.51 (3β-H), 5.33 (2H,m,6,24-H).
Cholest-5-en-3β-ol	Cholesterol	$C_{27}H_{46}O$ [3]	386	m/z 386 (M^+,$C_{27}H_{46}O$), 371 (M^+-CH_3), 353 (M^+-CH_3-H_2O), 273 (M^+-C_8H_7-side chain), 252 (M^+-side chain -C_{16}-C_{17}), 231 (M^+- side chain-ring D cleavage), 229 (M^+-side chain-C_{16}-C_{17}-OH), 213 (M^+-side chain-H_2O-ring D cleavage), 121,107.	
24β-methylcholesta-5,22-dien-3β-ol	Brassicasterol = crinosterol	$C_{28}H_{46}O$ [4]	398	m/z 398 (M^+,$C_{28}H_{46}O$), 356 (M^+-C_3H_6), 340 (M^+-C_4H_{10}), 314 (M^+-C_6H_{12}), 300 (M^+-C_7H_{13}-H), 271 (M^+-C_8H_{15}-2H), 255 (M^+-C_9H_{17}-H_2O), 213, 161, 145, 121, 107, 93, 81.	^{1}H-NMR (400 MHz, $CDCl_3$) ppm: 0.67 (3H,s,18-Me), 0.99 (3H,s,19-Me), 1.24 (1H,d,J = 7 Hz, 21-Me), 3.51 (1H,m,3β-H), 5.35 (1H,d,J = 6.5 Hz, 6-H).
24-methylene cholest-5-en-3β-ol	24-methylene cholesterol = ostreasterol	$C_{28}H_{46}O$ [5]	398	m/z 398 (M^+, $C_{28}H_{46}O$), 365 (M^+-CH_3-H_2O), 314 (M^+-C_6H_{12}), 271 (M^+-C_9H_{17}-2H), 255 (M^+-C_9H_{17}-H_2O), 213 ($C_{12}H_{23}$-H_2O), 185 (M^+-$C_{14}H_{27}$-H_2O), 145, 121, 95, 81.	^{1}H-NMR (400 MHz, $CDCl_3$) ppm: 0.67 (3H,s,18-Me), 0.84 (3H,s,19-Me), 1.24 (3H,d,J = 7 Hz,21-Me), 1.60 (6H,m,26,27-Me), 3.15 (2H,s,28-H), 3.50 (1H,m,3β-H), 5.33 (1H,m,5-H).
Cholesta-5,25-dien-3β,24-diol		$C_{27}H_{44}O_2$ [6]	400	m/z 400 (M^+, $C_{27}H_{44}O_2$), 382 (M^+-H_2O), 367 (M^+-CH_3-H_2O), 349 (M^+-CH_3-2H_2O), 315 (M^+-C_6H_{11}-2H), 300, 271, 255, 213, 161, 145, 121, 107, 81.	^{1}H-NMR (300 MHz, $CDCl_3$) ppm: 0.67 (3H,s,Me-18), 0.99 (3H,s,Me-19), 1.24 (3H,d,J = 7 Hz,Me-21), 1.71 (3H,s,Me-27), 3.51 (2H,m,βH-3,24), 5.33 (1H,m,H-6).
24-methylcholest-5-en-3β-ol	24-methyl cholesterol	$C_{28}H_{48}O$ [7]	400	m/z 400 (M^+,$C_{28}H_{48}O$), 385 (M^+-CH_3), 382 (M^+-H_2O), 367 (M^+-CH_3-H_2O), 349,300,273 (M^+-side chain), 271, 255, 199, 145, 121, 81.	^{1}H-NMR (400 MHz, $CDCl_3$), ppm: 0.66 (3H,s,18-Me), 0.89 (3H,s,19-Me), 1.24 (1H,d,J = 7.0 Hz,21-Me), 1.60 (6H,m,26,27,-H), 3.48 (1H,m,3β-H), 5.33 (1H,m,5-H).
stigmasta-5,24(28)-dien-3β-ol	Fucosterol	$C_{29}H_{48}O$ [8]	412	m/z 412 (M^+,$C_{29}H_{48}O$), 397 (M^+-CH_3), 367 (M^+-C_2H_3-H_2O), 341 (M^+-CH_3-C_4H_7-H), 327 (M^+-CH_3-C_5H_9-H), 314 (M^+-C_7H_{14}), 299 (M^+-C_7H_{14}-CH_3), 271 (M^+-side chain-ring C+D cleavage), 255 (M^+-side chain-H_2O), 229 (M^+-side chain-ring D cleavage), 213, 183, 149, 121, 107, 83, 69.	^{1}H-NMR (300 MHz, $CDCl_3$) ppm: 0.84 (6H,s,18-Me,19-Me), 0.91 (3H,d,J = 7.0 Hz,21-Me), 1.35 (3H,J = 12.8 Hz,29-Me), 4.29 (3β-H), 5.15 (2H,m, 5,28-H).

Table I. Continued.

Systematic name	Common name	Molecular formula	Mol. wt.	Mass and fragmentation pattern	Spectral data
6 *Nor* 31,9β,19-cyclo-artanostan-3β-ol	*Nor* 31-cycloartanol	$C_{29}H_{50}O$ [9]	414	m/z 414 (M^+,$C_{29}H_{50}O$), 399 (M^+-CH_3), 396 (M^+-H_2O), 381 (M^+-CH_3-H_2O), 314 (M^+-$C_6H_{11}O$-H), 301 (C_8H_5-2H), 283 (C_8H_{15}-2H-H_2O), 271 (C_9H_{17}-H_2O), 213 ($C_{13}H_{25}$-2H-H_2O), 187 ($C_{15}H_{29}$-H_2O), 145, 133, 107, 95, 81, 69, 55.	^{1}H-NMR (300 MHz, $CDCl_3$) ppm: 0.67 (3H,s,18-Me), 0.84 – 1.00 (15H,m,21,26,27,28,29), 1.21 – 1.30 (2H,m,1,2,6,7,11,12,15,16,22,23,24), 2.40 (2H,s,19-H), 3.51 (1H,m,3β-H), 5.35 (1H,m,5-H).

Table II. Fatty acids of *Dermonema abbottiae* analysed as methyl esters

Systematic name	Common name	Molecular formula	Mol. wt.	Retention time (min.)	Relative % age	Significant ions in the mass spectra
A. Saturated fatty acid methyl esters:						
Methyl *n*-trideca-noate	Methyl tridecylate	$C_{14}H_{28}O_2$	228	17′21″	13.03	GC-MS m/z 228 (M^+, $C_{14}H_{28}O_2$, 100%), 211 (M^+-17, 3%), 199 (M^+-29, 10%), 185 (M^+-43, 54%), 171 (14%), 157 (5%), 143 (12%), 129 (55%), 115 (15%), 101 (8%), 87 (18%), 73 (88%).
Methyl *n*-hexa-decanoate	Methylpalmitate	$C_{17}H_{34}O_2$	270	19′21″	7.20	GC-MS m/z 270 (M^+, $C_{17}H_{34}O_2$, 100%), 239 (M^+-31, 33%), 227 (M^+-43, 43%), 213 (6%), 199 (13%), 185 (17%), 171 (13%), 157 (6%), 143 (55%), 129 (26%), 115 (10%), 101 (20%), 87 (90%).
Methyl *n*-hepta-decanoate	Methyl margarate	$C_{18}H_{36}O_2$	284	20′18″	28.65	GC-MS m/z 284 (M^+, $C_{18}H_{36}O_2$, 17%), 241 (M^+-43, 9%), 227 (M^+-57, 76%), 213 (90%), 199 (61%), 185 (73%), 171 (88%), 157 (80%), 143 (65%), 129 (74%), 115 (81%), 101 (85%), 87 (92%), 73 (100%).
Methyl *n*-octa-decanoate	Methyl stearate	$C_{19}H_{38}O_2$	298	21′47″	20.03	GC-MS m/z 298 (M^+, $C_{19}H_{38}O_2$, 5%), 267 (M^+-31, 2%), 255 (M^+-43, 3%), 241 (1%), 227 (1%), 213 (4%), 199 (2%), 185 (2%), 171 (2%), 157 (2%), 143 (3%), 129 (5%), 115 (3%), 101 (2%), 87 (8%), 73 (11%).
Methyl *n*-decosa-noate	Methyl behenate	$C_{23}H_{46}O_2$	354	22′10″	15.17	GC-MS m/z 354 (M^+, $C_{23}H_{46}O_2$, 2%), 280 (M^+-74, 7%), 266 (2%), 252 (1%), 238 (2%), 224 (1%), 210 (2%), 196 (2%), 168 (3%), 154 (5%), 140 (11%), 126 (12%), 112 (19%), 98 (42%), 84 (44%), 70 (50%).
B. Unsaturated fatty acid methyl esters:						
methyl-5,10,tridecanoate	Methyl myrist-oleate	$C_{15}H_{28}O_2$	240	16′36″	5.97	GC-MS m/z 240 (M^+, $C_{15}H_{28}O_2$, 82%), 211 (M^+-29, 11%), 197 (M^+-43, 14%), 183 (23%), 169 (28%), 155 (32%), 141 (38%), 127 (47%), 113 (64%), 94 (77%), 85 (80%), 71 (100%).
Methyl-9,10-octa-decenoate	Methyl oleate	$C_{19}H_{36}O_2$	296	21′40″	9.95	GC-MS m/z 296 (M^+, $C_{19}H_{36}O_2$, 11%), 278 (M^+-18, 6%), 264 (M^+-32, 4%), 222 (M^+-74, 3%), 208 (1%), 194 (3%), 180 (2%), 166 (2%), 152 (4%), 138 (6%), 124 (16%), 110 (9%), 96 (17%), 82 (26%), 68 (26%).

black residue. Again MeOH was added into the same aspirator containing the material after $CHCl_3$ extraction and kept for three weeks, and this procedure was also repeated three times. The combined MeOH extract was reduced under vacuum and gave 11.8 g of thick dark red syrupy liquid. Both the $CHCl_3$ and MeOH extracts were then partitioned between ethylacetate and distilled water. The ethylacetate layers, thus obtained, after evaporation under reduced pressure yielded dark reddish-black viscous residues which were then utilized for column chromatography over silica gel 60 (70–230 mesh, E. Merck).

Fatty acid extraction

The dried thalli of *D. abbottiae* (200 g) were extracted by soaking them in hexane-$CHCl_3$ (1:1 v/v) in an aspirator. This procedure was repeated three times and the combined solvent mixture was evaporated under reduced pressure and gave a thick syrupy dark-reddish material weighing 4.23 g. It was saponified by refluxing at 100 °C for 10 h with 10% KOH (25 ml) and then concentrated under vacuum. Water and Et_2O were added to the reaction mixture and this treatment was repeated three times. The aqueous alkaline layer was acidified with HCl and then extracted with Et_2O. The ether fraction was washed three times with H_2O and once with hydrogen carbonate to remove residual non-saponifiable impurities. The resultant Et_2O fraction was dried over anhydrous Na_2SO_4 and on evaporation of the Et_2O a residue (12 mg) was obtained.

Esterification

The fatty acid fraction so obtained was then esterified with diazomethane. To a quantity of 5.0 mg fatty acid fraction 1 ml of diazomethane was added and the reaction was left overnight in a deep-freeze. It was then evaporated under reduced pressure and the residue was dissolved in chloroform, and aliquots were directly injected into gas liquid chromatograph (GLC) and gas chromatograph-mass spectrometer (GC-MS). From the column chromatography a yellowish waxy fraction was obtained from MeOH extract through hexane- Et_2O (4:1 v/v) and this was not transmethylated but directly injected into the GLC and GC-MS.

Instrumentation

The methylated and non-methylated fatty acid fractions were analysed by gas liquid chromatography on a GC-Hewlett Packard gas chromatograph equipped with a model 11/73 DEC data system and a 1.2 m × 4 mm glass column packed with gas chrom Q (100–120 mesh, OV 101,1%). The column temperature was programmed between 70–240 °C with a rate of increase of 8 °C/min. The carrier gas (helium) flow rate was 32 ml/min, injector temperature was 250 °C.

Results and Discussion

Nine sterols were identified from the extract of *Dermonema abbottiae* by ^{1}H-NMR and mass spectrometry: cholesta-5,20,24-trien-2β-ol [1], desmosterol [2], cholesterol [3], brassicasterol [4], 24-methylene cholesterol [5], cholesta-5,25-dien-3β,24-diol [6], 24-methyl cholesterol [7], fucosterol [8] and *nor* 31-cycloartanol [9]. The details of their mass spectra and fragmentation patterns are presented in Table I. Cholesterol was present in large quantity, while the remaining ones were detected in small amounts. The major sterols of red algae are C_{27} containing compounds. A survey of the literature reveals that cholesterol commonly occurs in red algae as a major constituent with its biosynthetic precursor desmosterol also common (Goodwin 1974, Chardon-Loriaux *et al.*, Bano *et al.* 1987, 1988), whereas other sterols such as 22-dehydrocholesterol, brassicasterol, 5,25-dihydroxycholesterol, 25-hydroxy-24-methyl cholesterol, fucosterol, cholesta-5, 25-dien-3β, 24-diol and β-sitosterol have also been detected in small quantities in various red algae (Morisaki *et al.* 1976, Combaut *et al.* 1979, Kabore *et al.* 1983). However, 22-dehydrocholesterol has been found in relatively large

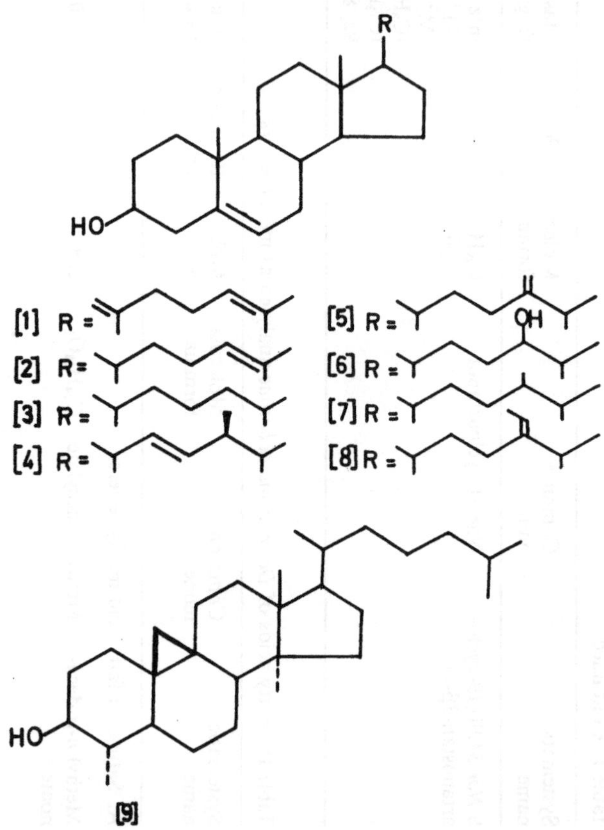

amounts in *Hypnea japonica* Tanaka (Tsuda *et al.* 1960) and *H. musciformis* (Wulf.) Lamour. (Meunier *et al.* 1970) and *Gracilaria foliifera* (Forssk.) Børg. (Hayee-Memon *et al.* 1991).

Fattorusso *et al.* (1975) found liagosterol in substantial quantities in *Liagora distenta* (Mert.) C. Ag. and *Scinaia furcellata* (Turn.) Bivona and suggested the possibility that liagosterol might be of some taxonomic significance to the order Nemaliales. This suggestion was confirmed when liagosterol was reported from *Asparagopsis armata* Harv. (Combaut *et al.* 1979), another member of this order, but it was not detected in *A. sandfordiana* (Bano *et al.* 1988). However, liagosterol has been detected in *Rhodymenia palmata* Grev. (Morisaki *et al.* 1976), a member of Rhodymeniales. This sterol is absent in *D. abbottiae* indicating that it is not of usual occurrence in all the members of Nemaliales.

Dermonema abbottiae resembles *Palmaria palmata* (L.) Kunt. (= *Rhodymenia palmata*) in its sterol contents. Noteworthy are desmosterol, cholesterol, brassicasterol, 24-methylene cholesterol, fucosterol and cycloartanol (Gibbons *et al.* 1967, Morisaki *et al.* 1976). Cholesta- 5,25-dien-3β, 24-diol detected in *D. abbottiae* have also been reported from the taxonomically closely related seaweeds, *Asparagopsis armata* – *Falkenbergia rufolanosa* (Harv.) Schmitz (Combaut *et al.* 1979) and *Rissoella verruculosa* (Berth.) J. Ag. (Kabore *et al.* 1983). It appears that *D. abbottiae* exhibits a complex sterol profile. In certain aspects it resembles closely related species and in others distantly related ones.

The fatty acid fraction of *D. abbottiae* consists of two portions, one was methylated with diazomethane and the other not. The GC-MS data of the methylated and non-methylated fatty acid fractions are presented in Tables II and III. The esterified fractions reveal the presence of methyl tridecylate, methyl palmitate, methyl margarate, methyl stearate and methyl behenate as saturated esters and methyl myristoleate and methyl oleate as unsaturated esters. The non-methylated fractions show the presence of pentadecylic, palmitic, margaric and stearic acids. The identity of the fatty acids was confirmed by comparison of the mass spectra of the unknown methylated and non-methylated fatty acids with those of the National Bureau of Standards' mass spectra library. The non-methylated fractions yielded only four fatty acids as compared to the methylated ones. It appears, therefore, better to methylate the fatty acid fraction.

The fatty acid composition of *D. abbottiae* exhibits a similarity with those of four Rhodophyta species from Australian waters studied by Johns *et al.* (1979). The

Table III. Fatty acids of *Dermonema abbottiae*

Systematic name	Common name	Molecular formula	Mol. wt.	Retention time (min.)	Relative % age	Significant ions in the mass spectra
n-Pentadecanoic acid	Pentadecylic acid	$C_{15}H_{30}O_2$	242	16′43″	13.97	GC-MS m/z 242 (M^+, $C_{15}H_{30}O_2$, 100%), 211 (M^+-31, 87%), 199 (M^+-43, 82%), 185 (M^+-57, 32%), 171 (12%), 157 (31%), 143 (77%), 129 (66%), 115 (22%), 101 (63%), 87 (75%), 73 (22%).
n-Hexadecanoic acid	Palmitic acid	$C_{16}H_{32}O_2$	256	17′36″	16.48	GC-MS m/z 256 (M^+, $C_{16}H_{32}O_2$, 10%), 213 (M^+-43, 6%), 199 (M^+-57, 24%), 185 (65%), 171 (38%), 157 (16%), 143 (42%), 129 (81%), 115 (55%), 101 (37%), 87 (91%), 73 (100%).
n-Heptadecanoic acid	Margaric acid	$C_{17}H_{34}O_2$	270	19′30″	24.37	GC-MS m/z 270 (M^+, $C_{17}H_{34}O_2$, 100%), 239 (M^+-31, 84%), 227 (M^+-43, 78%), 213 (M^+-57, 29%), 199 (68%), 185 (94%), 171 (80%), 157 (36%), 143 (75%), 129 (71%), 115 (65%), 101 (80%), 87 (92%), 73 (59%).
n-Octadecanoic acid	Stearic acid	$C_{18}H_{36}O_2$	284	20′31″	45.16	GC-MS m/z 284 (M^+, $C_{18}H_{36}O_2$, 5%), 253 (M^+-31, 1%), 241 (M^+-43, 9%), 227 (M^+-57, 53%), 213 (62%), 199 (78%), 185 (87%), 171 (90%), 157 (78%), 143 (73%), 129 (61%), 115 (74%), 101 (84%), 115 (74%), 101 (84%), 87 (86%), 73 (100%).

relative percentages of methylated and non-methylated fatty acids indicated that margaric and stearic acids are the major constituents, while palmitic acid occurs in small quantity (Tables II and III). Margaric acid has also been found as the major constituent of the fatty acid fraction of *Gracilaria foliifera* (Hayee-Memon *et al.* 1991). It is surprising that in the majority of the red seaweeds studied from Karachi, Pakistan palmitic acid has been found to be the major fatty acid component (Qasim 1986, Bano *et al.* 1987, Shameel 1990). Methyl myristoleate and methyl oleate are the unsaturated fatty acids detected, which are in much smaller quantity than the saturated fatty acids. This is in accordance with what has been observed in other marine red algae (Qasim 1986, Shameel 1990). It appears that in these algae oleic acid is the major unsaturated fatty acid, as has also been found in *D. abbottiae*.

References

Afaq-Husain, S., M. Nizamuddin and M. Shameel. 1991. The structure and reproduction of a new taxon *Dermonema abbottiae* (Nemaliales-Rhodophyta) from the coast of Pakistan. *Pak. J. Sci. Ind. Res. 34*: (in press).

Bano, S., N. Bano, V. U. Ahmad and M. Shameel. 1986. Isolation of natural products from a seaweed *Botryocladia leptopoda* (Rhodophyta) of Karachi coast. *In*: (R. Ahmad and A. San Pietro, eds): Prospects for biosaline research. *Proc. US-Pak. Biosal. Res. Worksh.*: 533–538.•

Bano, S., Shafiuddin, S. Perveen, N. Bano, V. U. Ahmad and M. Shameel. 1987. Chemical constituents of red algae *Acanthophora dendroides* and *Halymenia porphyroides*. *Pak. J. Bot. 19*: 253–257.

Bano, S., A. Hayee, V. U. Ahmad, W. Shaikh, K. Usmanghani and M. Shameel. 1988. Marine natural products. Part VIII. Steroids from a red alga *Asparagopsis sandfordiana*. *Polish J. Chem. 62*: 905–906.

Chardon-Loriaux, I., M. Morisaki and N. Ikekawa. 1976. Sterol profiles of red algae. *Phytochem. 15*: 723–725.

Combaut, G., Y. Bruneau, L. Codomier and J. Teste. 1979. Comparative sterol composition of the red alga *Asparagopsis armata* and its tetrasporophyte *Falkenbergia rufolanosa*. *J. Nat. Prod. 4*: 150–151.

Fattorusso, E., S. Magno, C. Santacroce, D. Sica, G. Impellizzeri, S. Mangiafico, G. Oriente, M. Piattelli and S. Sciuto. 1975. Constituents of red algae. II. Sterols of some red algae. *Phytochem. 14*: 1479–1582.

Gibbons, G. F., L. J. Goad and T. W. Goodwin. 1967. The sterols of some marine red algae. *Phytochem. 6*: 677–683.

Goodwin, T. W. 1974. Sterols. *In*: (W. D. P. Stewart, ed.): *Algal Physiology and Biochemistry*. Blackwells, Oxford, pp. 266–280.

Hayee-Memon, A., M. Shameel, M. Ahmad, V. U. Ahmad and K. Usmanghani. 1991. Phycochemical studies on *Gracilaria foliifera* (Gigartinales, Rhodophyta). *Bot. Mar. 34*: 107–111.

Johns, R. B., P. D. Nichols and G. J. Perry. 1979. Fatty acid composition of ten marine algae from Australian waters. *Phytochem. 18*: 799–802.

Kabore, S. A., G. Combaut, J.-P. Vidal, L. Codomier, J. Passet, J.-P. Girard and J.-C. Rossi. 1983. Sterols of the red alga *Rissoella verruculosa*. *Phytochem. 22*: 1239–1240.

Meunier, H., S. Zelenski and L. Worthen. 1970. Sterols in algae. *In*: (H. W. Youngken, ed.): Food and drugs from the sea. *Proc. Conf. Drugs Sea 2*: 319–325.

Morisaki, M., S. Kidooka and N. Ikekawa. 1976. Studies on steroids. XXXIX. Sterol profiles of red algae – (2). *Chem. Pharm. Bull. 24*: 3214–3216.

Qasim, R. 1986. Studies on fatty acid composition of eighteen species of seaweeds from the Karachi coast. *J. Chem. Soc. Pak. 8*: 223–230.

Shameel, M. 1990. Phycochemical studies on fatty acids from certain seaweeds. *Bot. Mar. 33*: 429–432.

Tsuda, K., K. Sakai, K. Tanabe and Y. Kishida. 1960. Steroid studies. XVI. Isolation of 22-dehydrocholesterol from *Hypnea japonica*. *J. Am. Chem. Soc. 82*: 1442–1443.

Botanica Marina
Vol. 34, pp. 221–227, 1991

The Marine Algae of Mar Piccolo, Taranto (Southern-Italy): A Re-assessment

E. Cecere, M. Cormaci* and G. Furnari*

Istituto Sperimentale Talassografico C.N.R., v. Roma, 3, 74100 Taranto, Italy
* *Istituto e Orto Botanico dell'Università, v. A. Longo, 19, 95125 Catania, Italy*

(Accepted 20 January 1991)

Abstract

A comparison has been carried out between the present marine flora in Mar Piccolo basin (Taranto, southern Italy) and that occurring in the same basin about 70 years ago, the latter being based on both literature data and herbarium collections.

The present flora shows a decrease in Phaeophyceae. Many species have disappeared; the newly settled ones are ubiquitous or very tolerant to sewage. The changes are a result of the increase of domestic and industrial pollution which has occurred during recent years.

Introduction

There are very few cases in which it is possible to verify the probable modifications of the algal flora in areas of limited extension after a period of many years. One of such published cases is the study conducted by Wilkinson and Tittley (1979) at three places in northern Scotland where floristic lists go back nearly 100 years. On the contrary, comparisons made after brief periods are more frequent both for areas which have been roughly modified directly by man (Munda 1967, Huvè *et al.* 1973, Riouall 1972, 1977) and protected areas continually surveyed such as marine parks (Marcot-Coqueugniot 1986). To make a comparison of the former type the Mar Piccolo in Taranto (southern Italy) is a suitable site where, about seventy years ago, floristic studies were carried out (Pierpaoli 1923), some of which were published later (Pierpaoli 1959, 1960). Recently, studies on phytobenthic communities have been carried out in the same area (Cecere *et al.* 1988) and a floristic list has been prepared. Therefore, it appeared interesting to compare the present data with those of Pierpaoli (1923, 1959, 1960) in order to show phytocenoses evolution, even if only from a floristic point of view. In fact, during recent years, the environmental conditions have been modified especially by the increase in the number of sewage outlets.

Description of the Area Studied

The basin, lying to the North of the town of Taranto (Mediterranean Sea, southern Italy) (Fig. 1), has a surface area of 20.72 km^2 and is divided by two promontories of land into two smaller basins, which are called the First and the Second Inlets. It is connected with Mar Grande basin through two channels, of which only one is important for the water exchange in the basin. Most of the hard substrata is artificial. The soft substrata, except for a few small gravelly areas, are sandy or clayey near the shore and muddy in the central zone of both inlets. The maximum depth is twelve meters in the center of the First Inlet and eight meters in the central part of the Second one.

The basin is characterized by the presence of about thirty submarine freshwater springs. The lowest water temperature (10 °C) occurs in February, the highest (28 °C at the surface and 26 °C at the bottom) in August. Salinity shows a mean value of 36‰ in the first 3 m. However, considerably lower values occur

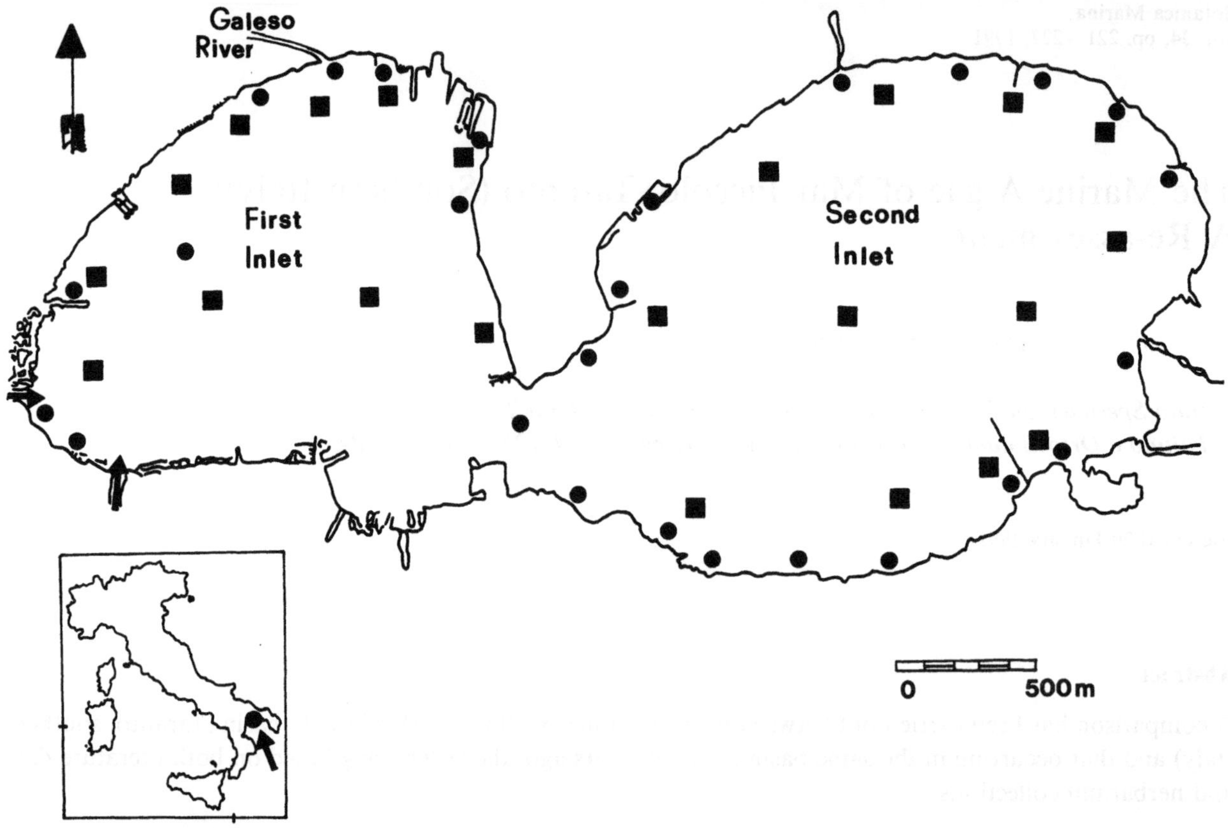

Fig. 1. Map of Mar Piccolo showing intertidal sampling stations (circles) and sublittoral ones (squares). The arrows indicate the entrance to the basin. The small picture shows the localization of Taranto.

near the outfall of the river Galeso and the freshwater springs. Many domestic sewage outlets flow into the basin.

Material and Methods

Intertidal collections were made at 26 stations and sublittorally (by SCUBA diving) at 22 stations, up to the maximum depth of the basin (Fig. 1), every two months in 1987 and 1988. The floristic list has been compared with the lists published by Pierpaoli (1923, 1959, 1960) and also by examining her herbarium specimens, held at the Marine Biology Station in Porto Cesareo (University of Lecce) and at the Istituto Talassografico in Taranto.

Results

One hundred and one taxa of algae have been identified, of which 61 are Rhodophyceae, 13 Phaeophyceae and 27 Chlorophyceae giving a total of 82 species, 1 subspecies, 1 form, 3 varieties and 14 specimens determined only to generic level. The list prepared from Pierpaoli's material lists 111 taxa of algae of which 65 are Rhodophyceae, 22 Phaeophyceae, 24 Chlorophyceae giving a total of 105 species, 2 forms,

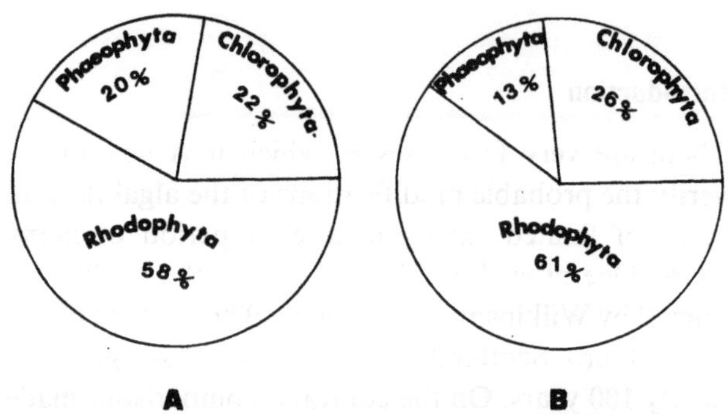

Fig. 2. Percentage of each Division in the floristic list by Pierpaoli (A) and in the present one (B).

3 varieties and 1 specimen determined only at generic level. By a quantitative comparison only, it is seen that the present flora is similar to that found by Pierpaoli, even if only 41 species are common to both the lists. Sixty taxa have been found only by the present authors and 70 were found only by Pierpaoli (Table I). Also the percentage incidence of the three Divisions in the two lists is not very different, with the exception of a decrease in the Phaeophyceae from 20% to 13% (Fig. 2). Nevertheless, the two floras differ from a qualitative point of view and the comparison becomes more significant if every species and its ecological meaning is considered. From this point

Table I. Floristic list of algae collected by both Pierpaoli (A) and the Authors (B). The species not published but found in Pierpaoli's herbarium are marked with an asterisk. In the reproductive phenology columns: T = Tetrasporophyte; M = male gametophyte; F = female gametophyte including carposporophyte; f = fertile, but without any other indications; n = no indications of the reproductive phenology; st = sterile; P = paraspores; G = gametangia; S = sporangia.

List of Species	A	B	Reproductive phenology		Names of species as reported by Pierpaoli
			A	B	
Rhodophyceae					
Acrochaetium savianum (Menegh.) Näg.	+	+	T	st	*Chantrasia saviana* Ardiss.
Alsidium helminthochorton (Latourr.) Kütz.	+	+	n	st	
Antithamnion cruciatum (C. Agardh) Näg.	+	+	M F T	st	
Callithamnion corymbosum (Sm.) Lyngb.	+	+	st	st	
Callithamnion granulatum (Ducluz.) C. Agardh	+	+	F T	st	
Chondria tenuissima (Good. *et* Woodw.) C. Agardh	+	+	n	T	
Corallina officinalis L.	+	+	f	T	
Fosliella farinosa (Lamour.) Howe	+	+	f	st	*Melobesia farinosa* Lamour.
Gelidium crinale (Turn.) Lamour.	+	+	T	T	
Gelidium pusillum (Stackh.) Le Jol.	+	+	T	st	
Gracilaria dura (C. Agardh) J. Agardh	+	+	F	M T	
Gracilaria verrucosa (Huds.) Papenf.	+	+	n	F T	*Gracilaria confervoides* Grev.
Herposiphonia secunda (C. Agardh) Ambronn f. *tenella* (C. Agardh) Wynne	+	+	n	T	*Herposiphonia tenella* Näg.
Hypnea musciformis (Wulf.) Lamour.	+	+	n	st	*
Polysiphonia variegata (C. Agardh) Zanard.	+	+	n	F T	*
Porphyra leucosticta Thur. *in* Le Jol.	+	+	n	F	*
Porphyra linearis Grev.	+	+	M	F	
Porphyra umbilicalis (L.) J. Agardh	+	+	M	F	*Porphyra laciniata* De Toni
Pterothamnion plumula (Ellis) Näg.	+	+	f	st	*Antithamnion plumula* (Ellis) Thur.
Rytiphloea tinctoria (Clemente) C. Agardh	+	+	f	st	
Solieria filiformis (Kütz.) Gabrielson	+	+	n	T	*Gracilaria confervoides* Grev.
Spyridia filamentosa (Wulf.) Harv. *in* Hook.	+	+	M F T	T	
Acrosorium venulosum (Zanard.) Kylin	+		F T		*Nitophyllum venulosum* Zanard.
Amphiroa rigida Lamour.	+		f		
Bangia atropurpurea (Roth) C. Agardh	+		n		*Bangia fuscopurpurea* Lyngb. and *Porphyra atropurpurea* De Toni
Callithamnion decompositum J. Agardh	+		n		*
Callithamnion hookeri (Dillwyn) S. F. Gray	+		T		*Callithamnion scopulorum* C. Agardh
Catenella caespitosa (With.) Irvine	+		M F		*Catenella opuntia* (Good. *et* Wood.) Grev.
Ceramium ciliatum (Ellis) Ducluz.	+		F		
Ceramium circinatum (Kütz.) J. Agardh	+		n		*
Ceramium diaphanum (Lightf.) Roth v. *strictum* (Kütz.) G. Mazoyer	+		F T		*Ceramium strictum* Grev. *et* Harv.
Ceramium fastigiatum (Roth) Harv.	+		M F T		
Ceramium radiculosum Grunow	+		F		
Ceramium rubrum (Huds.) C. Agardh v. *barbatum* (Kütz.) J. Agardh	+		F T		*Ceramium rubrum* (Huds.) C. Agardh f. *barbata* Hauck
Ceramium tenuissimum (Roth) Areschoug	+		F T		*Ceramium tenuissimum* (Lyngb.) J. Agardh
Chylocladia verticillata (Lightf.) Bliding	+		n		* *Chylocladia kaliformis* (Good. *et* Woodw.) Hook.
Contarinia squamariae (Menegh.) Denizot	+		n		*Rhyzophillis squamariae* Kütz.
Dasya baillouviana (Gmelin) Mont.	+		T		* *Dasya elegans* (Martens) Agardh
Dasya ocellata (Gratel.) Harv.	+		T		
Dasya punicea Menegh.	+		n		*
Dipterosiphonia rigens (Schousb.) Falkenb.	+		n		*
Erythrotrichia investiens (Zanard.) Born.	+		n		
Gelidium latifolium (Grev.) Born. *in* Born. *et* Thur.	+		n		
Grateloupia proteus Kütz.	+		n		
Griffithsia phyllamphora J. Agardh	+		st		*
Griffithsia tenuis C. Agardh	+		n		
Gymnothamnion elegans (Schousb. *ex* C. Agardh) J. Agardh	+		n		
Herposiphonia secunda (C. Agardh) Ambronn f. *secunda*	+		T		*Herposiphonia secunda* Näg.
Heterosiphonia crispella (C. Agardh) Wynne	+		n		*Heterosiphonia wurdemanni* Falkenb.
Jania rubens (L.) Lamour.	+		n		*Corallina rubens* L.
Lomentaria articulata (Huds.) Lyngb. v. *articulata*	+		n		*Lomentaria articulata* Lyngb.
Lomentaria clavellosa (Turn.) Gaillon	+		F T		*Chylocladia clevellosa* (Turn.) Grev.
Lomentaria uncinata Menegh.	+		T		*Chylocladia uncinata* Menegh.

Table I. Continued.

List of Species	A	B	Reproductive phenology A	Reproductive phenology B	Names of species as reported by Pierpaoli
Microcladia glandulosa (Solander) Grev.	+		F T		
Naccaria wiggii (Turn.) Endl.	+		n		
Nitophyllum punctatum (Stackh.) Grev.	+		T		*Nitophyllum punctatum* De Toni
Peyssonnelia squamaria (Gmelin) Decne.	+		n		
Pleonosporium borreri (Sm.) Näg. *ex* Hauck	+		F T		*Pleonosporium borreri* (Engl. Bot.) Näg.
Polysiphonia polyspora (C. Agardh) J. Agardh	+		F		*
Polysiphonia sertularioides (Gratel.) J. Agardh	+		F T		
Pterosiphonia pennata (Roth) Falkenb.	+		n		
Rhodochorton velutinum (Hauck) Hamel	+		n		*Chantransia minutissima* (Zanard.) Hauck
Rhodophyllis divaricata (Stackh.) Papenf.	+		n		*Rhodophyllis appendiculata* J. Agardh
Rhodymenia ligulata Zanard.	+		n		
Spermothamnion irregulare (J. Agardh) Ardiss.	+		n		
Acanthophora najadiformis (Del.) Papenf.		+		F	
Acrochaetium sp.		+		st	
Alsidium corallinum C. Agardh		+		T	
Antithamnion tenuissimum (Hauck) Schiffner		+		st	
Callithamnion sp.		+		st	
Callithamnion cordatum Boerg.		+		st	
Ceramium sp.		+		st	
Ceramium diaphanum v. *zostericulum* Thur.		+		P	
Ceramium flaccidum (Kütz.) Ardiss.		+		st	
Ceramium tenerrimum (Martens) Okam.		+		st	
Chondria mairei G. Feldm.		+		st	
Erythrotrichia sp.		+		st	
Feldmannophycus rayssiae (J. *et* G. Feldm.) Augier *et* Boudouresque		+		st	
Gastroclonium clavatum (Roth) Ardiss.		+		st	
Gelidium spathulatum (Kütz.) Born.		+		st	
Gigartina acicularis (Roth) Lamour.		+		st	
Gracilaria sp.		+		st	
Gracilaria bursapastoris (Gmelin) Silva		+		M F T	
Gracilaria longa Gargiulo, De Masi *et* Tripodi		+		F	
Grateloupia filicina (Wulf.) C. Agardh		+		T	
Griffithsia sp.		+		st	
Griffithsia cf. *aracnoidea* C. Agardh		+		st	
Griffithsia opuntioides J. Agardh		+		st	
Griffithsia schousboei Mont. *in* Webb		+		F	
Hypnea sp.		+		st	
Hypnea cervicornis J. Agardh		+		st	
Laurencia pinnatifida (Huds.) Lamour.		+		st	
Lomentaria articulata (Huds.) Lyngb. v. *linearis* Zanard.		+		T	
Lomentaria compressa Kylin		+		F T	
Lomentaria subdichotoma Ercegovic		+		st	
Polysiphonia sp.		+		st	
Polysiphonia subulata (Ducluz.) J. Agardh		+		F T	
Porphyra sp.		+		st	
Radicilingua reptans (Zanard.) Papenf.		+		st	
Radicilingua thysanorhizans (Holmes) Papenf.		+		st	
Rhodymenia pseudopalmata (Lamour.) Silva		+		T	
Stylonema alsidii (Zanard.) Drew		+		st	
Stylonema cornu-cervi Reinsch		+		st	
Titanoderma sp.		+		st	
Phaeophyceae					
Cystoseira barbata (Good. *et* Wood.) C. Agardh	+	+	G	G	*Cystoseira barbata* Agardh and *C. hoppii* Agardh
Cystoseira compressa (Esper) Gerloff *et* Nizamuddin	+	+	G	G	*Cystoseira abrotanifolia* J. Agardh
Dictyota dichotoma (Huds.) Lamour. v. *dichotoma*	+	+	n	F	* *Dictyota dichotoma* (Huds.) Lamour. f. *fimbriata*
Ectocarpus sp.	+	+	n	st	
Ectocarpus siliculosus (Dillwyn) Lyngb. v. *confervoides*	+	+	n	S	*Ectocarpus confervoides* Le Jol.
Petalonia fascia (Müll.) Kuntze	+	+	n	S	*Phyllitis fascia* (Fl. Dan.) Kütz.
Scytosiphon lomentaria (Lyngb.) Link	+	+	n	S	

Table I. Continued.

List of Species	A	B	Reproductive phenology A	Reproductive phenology B	Names of species as reported by Pierpaoli
Colpomenia sinuosa (Roth) Derbès *et* Solier	+		n		
Cutleria adspersa (Roth) De Not.	+		n		*Cutleria adspersa* De Toni
Cystoseira ercegovicii Giaccone	+		G		*Cystoseira discors* Agardh
Cystoseira sauvageauana Hamel	+		G		*Cystoseira selaginoides* Naccari
Cystoseira spinosa Sauv.	+		G		*Cystoseira erica-marina* Naccari
Dictyopteris polypodioides (Desf.) Lamour.	+		n		
Dilophus fasciola (Roth) Howe	+		n		*Dictyota fasciola* (Roth) Lamour.
Ectocarpus siliculosus (Dillwyn) Lyngb. v. *siliculosus*	+		n		*Ectocarpus siliculosus* Le Jol.
Feldmannia caespitula (J. Agardh) Knoeppfler-Péguy	+		S		*Ectocarpus paradoxus* Mont.
Halopteris filicina (Gratel.) Kütz.	+		st		
Hincksia sandriana (Zanard.) Silva	+		n		*Ectocarpus elegans* Menegh.
Padina pavonica (L.) Lamour.	+		n		*Padina pavonia* Lamour.
Punctaria latifolia Grev.	+		n		
Sphacelaria cirrosa (Roth) C. Agardh	+		f		*Sphacelaria hystrix* Suhr
Taonia atomaria (Woodw.) J. Agardh	+		f		
Asperococcus compressus Griffiths *ex* Hook.		+		st	
Dictyota cf. *indica* Sonder *in* Kütz.		+		st	
Dictyota dichotoma (Huds.) Lamour. v. *intricata* (C. Agardh) Grev.		+		st	
Dictyota linearis (C. Agardh) Grev.		+		st	
Stictyosiphon adriaticus Kütz.		+		st	
Stictyosiphon soriferus (Reinke) Rosenvinge		+		st	
Chlorophyceae					
Blidingia minima (Näg. *ex* Kütz.) Kylin	+	+	n	st	*Enteromorpha micrococca* Kütz f. *tortuosa* and E. *prolifera* J. Agardh v. *crispatissima*
Bryopsis plumosa (Huds.) C. Agardh	+	+	n	st	
Caulerpa prolifera (Forssk.) Lamour.	+	+	n	st	
Chaetomorpha linum (Müll.) Kütz.	+	+	n	st	*Chaetomorpha aerea* (Dillwyn) Kütz.; *C. breviarticulata* Hauck; *C. fibrosa*; *C. crassa* Kütz. and *C. tortuosa* Kütz.
Cladophora prolifera (Roth) Kütz.	+	+	n	st	*Cladophora prolifera* Kütz.
Cladophora rupestris (L.) Kütz.	+	+	n	st	*Cladophora rupestris* Kütz. f. *mediterranea* Kütz.
Cladophora sericea (Huds.) Kütz.	+	+	n	st	* *Cladophora gracilis* (Griff.) Kütz.
Enteromorpha compressa (L.) Grev.	+	+	n	st	
Enteromorpha flexuosa (Wulfen *ex* Roth) J. Agardh	+	+	n	st	*Enteromorpha flexuosa* J. Agardh
Enteromorpha intestinalis (L.) Link	+	+	n	st	*Enteromorpha intestinalis* Link and *E. intestinalis* (L.) Link f. *genuina* Hauck
Enteromorpha prolifera (Müll.) J. Agardh	+	+	n	st	*Enteromorpha prolifera* J. Agardh
Ulva rigida C. Agardh	+	+	n	st	*Ulva lactuca* L. f. *laciniata* J. Agardh; *U. lactuca* L. f. *cribosa* J. Agardh and *U. lactuca* (L.) Le Jol. f. *genuina* Hauck
Acetabularia acetabulum (L.) Silva	+		n		
Bryopsis cupressoides Kütz.	+		n		
Cladophora pellucida (Huds.) Kütz.	+		n		
Cladophora vagabunda (L.) van den Hoek	+		n		
Dasycladus vermicularis (Scopoli) Krasser	+		n		
Enteromorpha intestinalis v. *asexualis* Bliding f. *cornucopiae* Lyngb.	+		n		
Enteromorpha linza (L.) J. Agardh	+		n		
Halimeda tuna (Ellis *et* Solander) Lamour.	+		n		
Percursaria percursa (C. Agardh) Bory	+		n		
Udotea petiolata (Turra) Boerg.	+		n		
Ulothrix subflaccida Wille	+		n		
Valonia utricularis (Roth) C. Agardh	+		n		
Bryopsis sp.		+		st	
Bryopsis corymbosa J. Agardh		+		st	
Cladophora sp.		+		st	
Cladophora dalmatica Kütz.		+		st	
Cladophora glomerata (L.) Kütz.		+		st	
Cladophora hutchinsiae (Dillwyn) Kütz.		+		st	
Cladophora laetevirens (Dillwyn) Kütz.		+		st	

Table I. Continued.

List of Species	A	B	Reproductive phenology A	Reproductive phenology B	Names of species as reported by Pierpaoli
Cladophora ruchingeri (C. Agardh) Kütz.		+		st	
Codium decorticatum (Woodw.) Howe		+		st	
Enteromorpha flexuosa ssp. *linziformis* (Bliding) Bliding		+		st	
Ulva sp.		+		st	
Ulva curvata (Kütz.) De Toni		+		st	
Ulva olivascens Dang.		+		st	
Ulva rotundata Bliding		+		st	
Ulva scandinavica Bliding		+		st	

of view, the most important absences are those of *Cystoseira ercegovicii, C. sauvageauana* and *C. spinosa*. Many the species associated with them also disappeared; for example, those living as epiphytes: *Microcladia glandulosa, Pleonosporium borreri, Griffithsia tenuis, Ceramium circinatum, C. diaphanum, C. tenuissimum* and *Dasya baillouviana*; and those generally living in their shadow: *Peyssonelia squamaria, Contarinia squamariae, Rhodymenia ligulata, Acrosorium venulosum, Cutleria adspersa, Halopteris filicina, Halimeda tuna* and *Udotea petiolata*. On the contrary, the newly arrived species have little ecological meaning as they are ubiquitous species or very tolerant of high concentrations of sewage, such as *Grateloupia filicina, Codium decorticatum* and *Gigartina acicularis*.

Finally, it is interesting to point out that the number of fertile Rhodophyceae (23 out of 61, equal to 38%) found during the present survey is somewhat low compared to that deducible from Pierpaoli's papers (1923, 1959, 1960) (33 out of 65, equal to 51%) (Table I). Further, this last value is probably an underestimate, as Pierpaoli explicitly refers to the sterility state for only two species. Therefore, the remaining 30 species, for which Pierpaoli does not refer to the reproductive phenology can be supposed to be fertile, even although their fertility is not clearly referred to.

Discussion and Conclusion

It is difficult to know the environmental condition of an area only from a floristic list, without any vegetational studies. However, on the basis of the species found by Pierpaoli, it can be supposed that in the 1920's in Mar Piccolo there were environmental conditions similar, for example, to those occurring at present time in other analogous areas of Italy, such as the pond of Marsala (western Sicily) (Calvo *et al.* 1982). In fact, as in that pond at the present, Mar Piccolo which shows more marked lagoon peculiarities, did have meadows of *Cymodocea nodosa* (Ucria) Ascherson and *Zostera* sp. on soft bottoms, and isolated specimens of *C. ercegovicii, C. barbata*, and *C. spinosa*, on the rock. The presence of these species, certainly depended on a good environmental condition and Pierpaoli herself reported that the water in Mar Piccolo was "limpida e pulita" (clear and clean) (Pierpaoli, 1923: 6). On the contrary, the quality of the present flora, as well as the typology of the vegetation (Cecere *et al.* 1988), show a high degree of pollution of the basin, which now has turbid water, because of much sewage.

The present floristic comparison, carried out after over 60 years, indicates the importance of phytobenthos as a recorder of environmental changes. In Mar Piccolo these are mainly due to domestic pollution which operates slowly, and whose effects only become noticeable after many years. The present results are similar to those of Munda (1967) for an area in the South Netherlands and of Huvè *et al.* (1973) and Riouall (1972, 1977) for the ponds of Berre and Vaine (southern France) where, respectively, an increase of eutrophycation and a decrease of salinity occurred. Also in these areas the environmental alterations determined a remarkable change of floristic composition, especially from a qualitative point of view. In particular, a decrease of Phaeophyceae and of the most stenoecious Rhodophyceae, and a predominance of Ulvales and Cladophorales among the Chlorophyceae were observed.

References

Calvo, S., G. Giaccone and R. Ragonese. 1982. Tipologià della vegetazione sommersa dello Stagnone di Marsala (TP). *Naturalista sicil., S. IV, VI 2*: 187–196.

Cecere, E., M. Cormaci, G. Furnari, A. Tursi and O. Caciorgna. 1988. Phytocenoses in the Mar Piccolo in Taranto (Ionian Sea, Southern Italy): mesolittoral level and infralittoral fringe. *Rapp. Comm. int. Mer Médit. 31 (2)*: 3.

Huvè, H., A. Kiener and R. Riouall. 1973. Modifications de la flore et des populations ichtyologiques des étangs de Berre et de Vaine (Bouches-du-Rhone) en fonction des conditions hydrologiques crées par le déversament de la Durance. *Bull. Museum Hist. Nat. Marseille XXXIII*: 123–134.

Marcot-Coqueugniot, J. 1986. Le phytobenthos du port de Port-Cros (Var, Méditerranée): comparaison a sept années d'intervalle. *Sci. Rep. Port-Cros natl. Park, Fr. 12*: 73–92.

Munda, I. 1967. Changes in the algal vegetation of a port of the deltaic area in the southern Netherlands (Veerse Meer) after its closure. *Bot. Mar. 10*: 141–157.

Pierpaoli, I. 1923. Prima contribuzione allo studio delle alghe del Golfo di Taranto. *Rivista di Biologia 5 (6)*: 1–19.

Pierpaoli, I. 1959. L'epifitismo nelle alghe. Note sugli ambienti tarantino e anconetano. *Thalassia Jonica 2*: 46–51.

Pierpaoli, I. 1960. Microfotografie di alghe del Golfo di Taranto. *Thalassia Jonica 3*: 100–106.

Riouall, R. 1972. Contribution à l'étude de la flore des étangs de Berre et de Vaine (Bouches-du-Rhone). These Doctorat Spécialité. Aix-Marseille 2. pp. 528.

Riouall, R. 1977. Evolution de la flore benthique des étangs de Berre et de Vaine (Bouches-du-Rhone) de 1971 à 1975. *Bull. Museum Hist. Nat. Marseille XXXVII*: 141–147.

Wilkinson, M. and I. Tittley. 1979. The marine algae of Elie, Scotland: a re-assessment. *Bot. Mar. 22*: 249–256.

Botanica Marina
Vol. 34, pp. 229–239, 1991

Histochemical and Electron Microscope Studies of Ascospore Ontogeny in Three Marine Fungi with Cleistothecia: *Amylocarpus encephaloides* Currey, *Dryosphaera navigans* Koch *et* Jones and *Eiona tunicata* Kohlm.

R. G. Johnson, E. B. G. Jones and S. T. Moss

School of Biological Sciences, Portsmouth Polytechnic, King Henry Building, King Henry 1 Street, Portsmouth PO1 2DY, U.K.

(Accepted 20 January 1991)

Abstract

Ascospore appendage development in three species of cleistothecial marine Ascomycotina is described and each exhibits a different mode of appendage ontogeny: outgrowths in *Amylocarpus encephaloides*; fragmentation in *Eiona tunicata*; combination of outgrowths and fragmentation in *Dryosphaera navigans*. Ascospore walls differ in that in *D. navigans* there are three layers (exosporium, episporium, mesosporium), *E. tunicata* has an episporium and mesosporium while in *A. encephaloides* there is a single wall layer divided into three histochemically distinct zones. In *A. encephaloides* the mature wall is enrobed with a homogeneous, electron-dense amorphous layer through which the spine-like appendages project; the origin of this layer is speculated on. The taxonomic significance of these observations is discussed. *Eiona tunicata* and *A. encephaloides* are considered best placed in the Plectomycetes while *D. navigans* has no affinities with this group because of the morphology and structure of the hymenium and the structure of the ascus and the ascospore wall. This species is best assigned, for the present, to the Coronophorales.

Introduction

The three genera considered in this paper are all monotypic. *Amylocarpus* and *Eiona* are placed in the Eurotiaceae (Plectomycetes) because of their closed ascocarps with a pseudo-parenchymatous peridium, globose asci irregularly arranged within the centrum and ascospores which are passively released (Kohlmeyer and Kohlmeyer 1979). Little is known of ascocarp or ascospore ontogeny in *Dryosphaera navigans* Koch *et* Jones (= *Crinigera maritima* Schmidt) and this species has therefore been placed in the Ascomycotina *incertae sedis* by both Kohlmeyer (1986) and Hawksworth and Eriksson (1986). However, Koch and Jones (1989) assigned the fungus to the Ascomycotina, Parenchymatomycetidae, and the order Coronophorales (*sensu* Luttrell 1951) but with an uncertain taxonomic position at the familial level.

Ascospores of *Amylocarpus encephaloides* Currey are subglobose to ovoid, 8–16 µm diameter, one-celled, hyaline and possess 10–25 awl-shaped, radiating appendages (Jones and Moss 1978, Kohlmeyer and Kohlmeyer 1979) which Jones and Moss (1978) described as outgrowths of the spore wall.

Eiona tunicata Kohlm. has ellipsoid, 7.5–12 µm long × 3.5–5 µm wide, one-celled, hyaline ascospores with fibrillar appendages radiating from each pole (Jones and Moss 1978, Kohlmeyer and Kohlmeyer 1979). Fragmentation of an outer wall layer being the method of appendage ontogeny (Jones and Moss 1978).

In *Dryosphaera navigans* the ascospores are ellipsoid (7–) 10–13 µm long × 3.5–5 (–6.5) µm wide, hyaline, one-septate and possess subterminal, thread-like appendages (Kohlmeyer and Kohlmeyer 1979, Jones *et al.* 1980 as *Crinigera maritima*). Jones and Moss (1980) stated that in *C. maritima* the spore appendages are outgrowths of the spore, whereas Jones *et al.* (1980) showed that ascospores are surrounded by a sheath composed of two layers. The outer layer swells, becomes more diffuse and forms ball-like structures, while the strands of the inner layer are tightly coiled around the spore. These later uncoil to form sub-terminal radiating appendages to which the ball-like structure may remain attached.

Prior to the work of Jones and Moss (1978, 1980) and Jones *et al.* (1980) investigations of *A. encephaloides*, *E. tunicata* and *D. navigans* were confined to the light microscope. T. W. Johnson (1963) examined spore and appendage ontogeny in *A. encephaloides* but could not elucidate the mechanism of appendage formation. Transmission electron microscopy may elucidate more precisely the origins of the appendages in these cleistothecial forms and enable a comparison with other appendage types.

Material and Methods

Ascocarps of the fungi to be examined were collected on driftwood from various parts of the world but chiefly from the sand dunes of Løkken, Grønhøj and Blokhus, Denmark. Wood was incubated in sterile plastic boxes at 20 °C and examined (at regular intervals over a 10 month period) for ascocarps.

Material for light microscopy was examined with a Wild M20 microscope. Techniques and procedures for the examination of material for scanning electron microscopy were those described by Moss and Jones (1977) and material for the histochemical studies and transmission electron microscopy were those described by R. G. Johnson (1980). Fresh material was used for our investigations so that developmental stages could be examined. In the plates presented, it has only been possible to reproduce selective stages in the ascospore appendage ontogeny.

Results

Amylocarpus encephaloides Currey

The ascogenous system of *A. encephaloides* does not develop as a discrete hymenium but ramifies throughout the ascocarp centrum. Asci are produced as lateral branches at intervals along the septate ascogenous hyphae (Figs 4, 6). The ascogenous hyphae and asci possess an electron-transparent, single-layered wall (Figs 6–8) and each ascus is delimited from the ascogenous hypha by an electron-transparent cross-wall (Fig. 6) that has a central pore with associated Woronin bodies. The ascogenous hyphae have dense cytoplasm containing numerous tubular mitochondria and electron-dense, membrane-bounded, volutin-like bodies (Fig. 6). The volutin-like bodies are reduced or absent in the young developing ascus being replaced by oil droplets which are confined mainly to the base and tip of the young, diploid ascus (Fig. 6). The diploid nucleus occupies a central position in the ascus (Fig. 7).

As the diploid ascus becomes subglobuse (Fig. 5), the number of oil droplets increases and they become distributed throughout the cytoplasm (Fig. 7). In addition, the cytoplasmic glycogen and mitochondrial complement also increases, the former being closely associated with the oil droplets. Later, profiles of smooth endoplasmic reticulum accumulate in the cytoplasm concomitant with invaginations of the ascal plasma-membrane. These membranous proliferations may form double membrane profiles which coalesce to form the ascus vesicle and ultimately delimit the ascospore initials (Figs 8, 9). Each ascospore is initially irregular in shape and comprises dense cytoplasm containing tubular mitochondria, glycogen rosettes and a haploid nucleus (Figs 8–11). The most prominent components of the spore cytoplasm are the numerous, small oil droplets (Fig. 8). In contrast, the epiplasm lacks oil droplets but retains a large glycogen complement (Fig. 10). The endomembrane profiles of the smooth endoplasmic reticulum persist in the epiplasm and probably supply the membrane material required to complete delimitation of the ascospores (Fig. 9).

Maturation of the spore cytoplasm involves coalescence of the small oil droplets, an increase in glycogen deposits and the formation of stacks of smooth endoplasmic reticulum profiles adjacent to the delimiting membrane. The inner delimiting membrane becomes the spore plasma membrane while the outer delimiting membrane becomes the spore investing membrane. Between the two delimiting membranes is deposited a single, electron-transparent wall layer (Fig. 11). The epiplasm becomes less compact, more vesicular and contains less glycogen (Fig. 10).

Epiplasmic vesicles and membrane profiles become associated with the spore investing membrane to form the thick, electron-dense membrane complex (Fig. 12). During wall deposition, regions of the membrane complex evaginate and contain amorphous, appendage precursor material (Fig. 12). The appendages, each within a separate evagination of the membrane complex, become adpressed to the spore surface. At maturity the spore wall differentiates into three distinct zones: a thick (410 nm wide), electron-dense central zone between thinner (50–80 nm wide) electron-transparent zones. Histochemically, the outer, thin, electron-transparent zone stains intensely for neutral, Periodic Acid-Schiff (PAS) positive carbo-

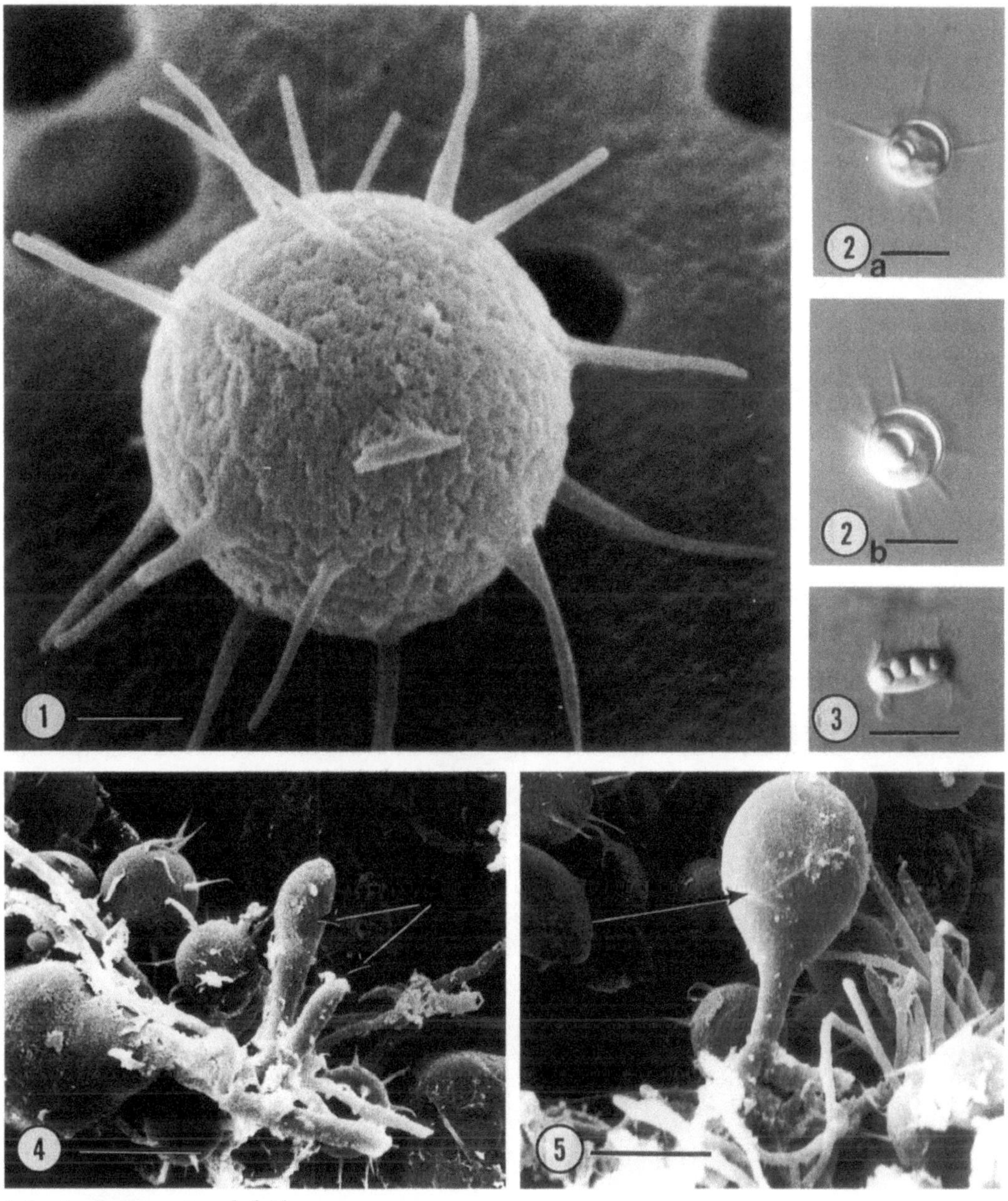

Figs 1, 2, 4, 5. *Amylocarpus encephaloides.*
Fig. 1. Scanning electron micrograph of ascospore with radiating appendages. Bar line = 2 μm. Fig. 2a, b. Light micrographs of ascospores. Bar line = 10 μm. Figs 4, 5. Immature asci (arrowed). Bar lines = 10 μm.

Fig. 3. *Eiona tunicata*, light micrograph of ascospore. Bar line = 10 μm.

hydrates, as do the appendages which are extensions of this wall zone (Table I). The thick, central, more electron-dense zone is not PAS positive but stains intensely with Bests' carmine and weakly with alcian blue at pH 2.5 and pH 0.5 (Table I); which indicates that this region is composed of acidic polysaccharides. Bests' carmine appears to be staining an unknown carbohydrate(s), which is not glycogen since diastase digestion does not remove the staining. The inner, electron-transparent region failed to stain with any histochemical test employed (Table I). Thus the ascospore wall of *A. encephaloides* comprises 3 chemically distinct zones.

As the ascospore wall and appendages mature, the oil droplets in the spore cytoplasm coalesce into a single large guttule which occupies the centre of the spore (Figs 2a, b, 13). The peripheral spore cytoplasm contains a single haploid nucleus, mitochondria, glycogen rosettes and granules, some smooth endoplasmic reticulum profiles and electron-dense membrane-bounded volutin bodies (Fig. 13). Histochemically, fully developed spores contain neutral PAS positive carbohydrates, particularly glycogen, proteins and RNA (Table I). Volutin bodies also stain for proteins, carbohydrates, RNA and polyphosphates.

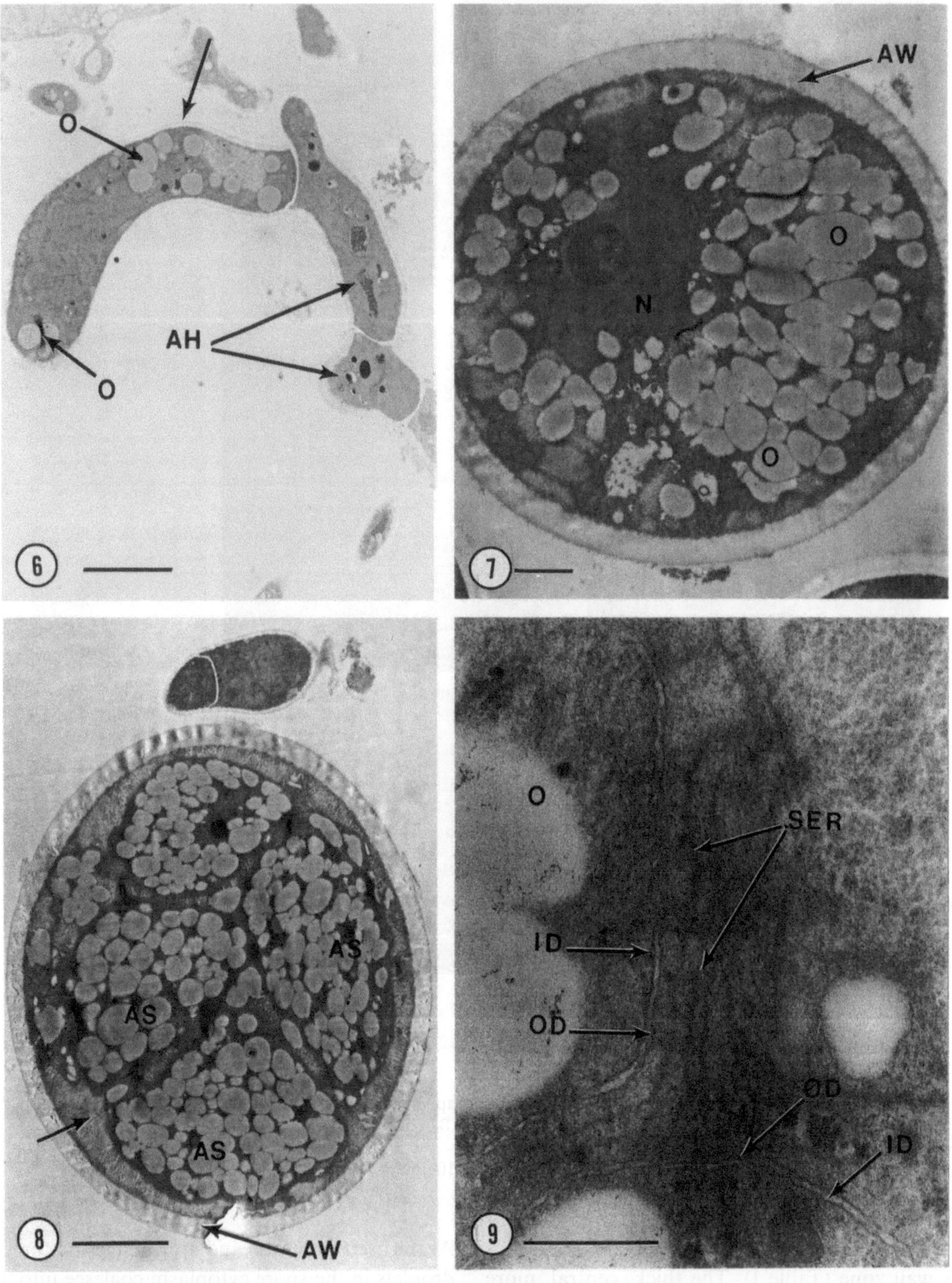

Figs 6–9. *Amylocarpus encephaloides*. All transmission electron micrographs.
Fig. 6. A diploid ascus (arrowed) developing as a lateral branch of the ascogenous hypha (AH), note oil deposits (O). Fig. 7. Transverse section of an ascus showing a diploid nucleus (N) surrounded by an accumulation of oil deposits (O); AW = ascus wall. Fig. 8. Transverse section of an ascus with 4, young, irregularly shaped, delimited ascospores (AS). AW = ascus wall, epiplasm arrowed. Fig. 9. Higher magnification of delimited ascospores showing spores separated from the epiplasm by a pair of unit membranes. ID = inner delimiting membrane, OD = outer delimiting membrane, O = oil droplet, SER = smooth endoplasmic reticulum. Bar lines: Figs 6, 8 = 5 µm; Fig. 7 = 2 µm; Fig. 9 = 0.5 µm.

Figs 10–13. *Amylocarpus encephaloides*. All transmission electron micrographs.
Fig. 10. Transverse section of an ascus showing the development of the spore wall (SW). Note the coalescing oil deposits (O) within the spore and that the epiplasm (EP) is poorly organized. AW = ascus wall. Fig. 11. Adjacent ascospores (AS) with

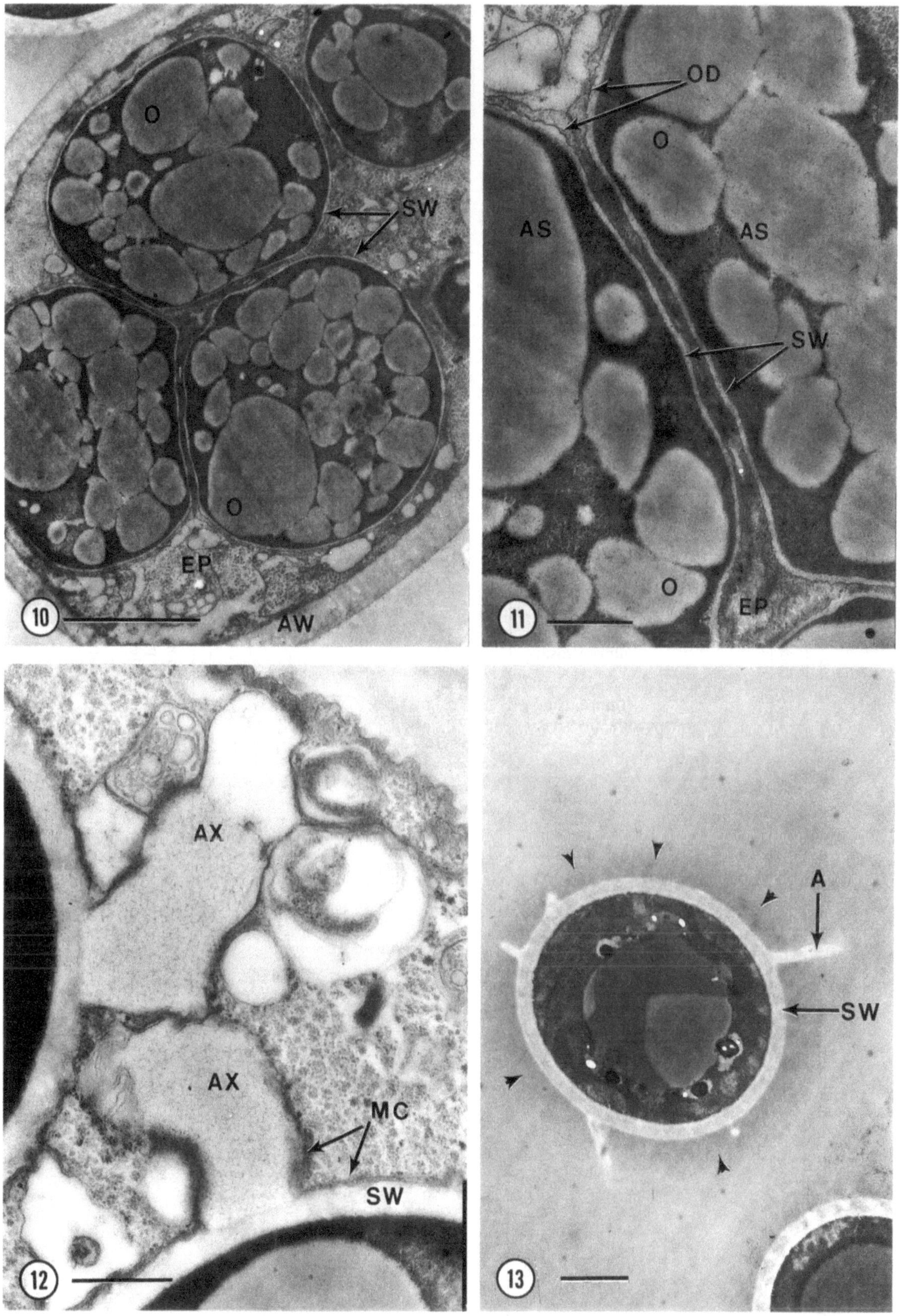

developing walls (SW). O = oil deposits, OD = outer delimiting membrane, EP = epiplasm. Fig. 12. Ascospore with thick walls (SW) surrounded by the membrane complex (MC). Note evaginations of the membrane complex containing appendage precursor material (AX). Fig. 13. Fully developed ascospore with extended appendages (A). The spore wall (SW) is tripartite with an increased electron-density of the central wall zone. An electron-dense 'halo' of amorphous material surrounds the spore surface (arrowheads). Bar lines: Fig. 10 = 5 µm; Figs 11, 12 = 1 µm; Fig. 13 = 2 µm.

Table I. Results of the histochemical tests on the mature ascospores of *Amylocarpus encephaloides*, *Eiona tunicata* and *Dryosphaera navigans*

		Amylocarpus encephaloides					*Eiona tunicata*					*Dryosphaera navigans*					
Histochemical Test	Chemical Component detected	Ascospore appendages	Ascospore wall	Ascospore ground cytoplasm	Volutin bodies	Oil guttules	Ascospore appendages (Episporium)	Mesosporium	Ascospore ground cytoplasm	Volutin bodies	Oil guttules	Ascospore appendages (Exosporium)	Episporium	Mesosporium	Ascospore ground cytoplasm	Volutin bodies	Oil guttules
Periodic acid-Schiff (PAS)	Neutral carbohydrate	+++	+++[1]	+++	+++	–	++	–	+++	++	–	+++	+++	–	+++	+++	–
Bests' Carmine	Glycogen	–	+++[2]	++	+/–	–	–	–	+++	–	–	–	–	–	+++	–	–
Alcian blue pH 2.5	Acidic carbohydrate	–	+[2]	–	–	–	+/–	–	–	–	–	++	++	–	–	–	–
Alcian blue pH 0.5	Sulphated mucins	–	+/–[2]	–	–	–	–	–	–	–	–	++	++	–	–	–	–
Dinitrofluorobenzene	Total protein	–	–	++	–	–	+/–	–	+	+	–	+/–	+/–	–	+	+	–
Acrolein-Schiff	Total protein	–	–	+++	+++	–	+	–	++	++	–	+	+	–	++	++	–
Fast green F. C. F.	Basic protein	–	–	–	–	–	–	–	–	–	–	–	–	–	–	–	–
Toluidine blue pH 6.0	Neutral/Acidic components	++[3]	–	+++[5]	+++[3]	–	++[3]	–	++[3]	++[3]	–	++[3]	++[3]	–	++[5]	++[3]	–
Toluidine blue pH 4.2	Acidic components	–	–	++[5]	+++[3]	–	–	–	+[5]	++[3]	–	+/–	+/–	–	+[5]	+[3]	–
Methyl green-Pyronin	D. N. A./R. N. A.	–	–	+++[6]	+++[6]	–	–	–	++[6]	++[6]	–	–	–	–	++[6]	++[6]	–
Feulgens	D. N. A.	–	–	–	–	–	–	–	–	–	–	–	–	–	–	–	–
Sudan black B	Total lipid	–	–	–	–	+++	–	–	–	–	+++	–	–	–	–	–	+++
Ebel's test	Polyphosphates	–	–	–	+++	–	–	–	–	++	–	–	–	–	–	++	–

+++ = Strongly positive
++ = Positive
+ = Weakly positive
+/– = Uncertain staining
– = Negative

1 = Only outer region of ascospore wall positive
2 = Only central region of ascospore wall positive
3 = Orthometachromatic blue staining only
4 = Metachromatic red staining only
5 = Orthometachromatic and metachromatic blue and red staining
6 = R. N. A. only detected

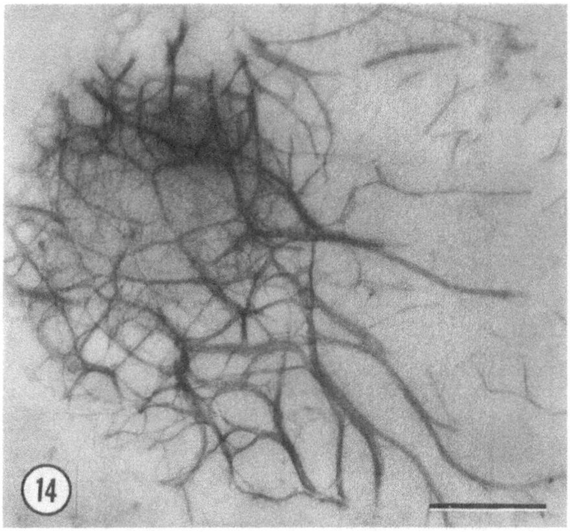

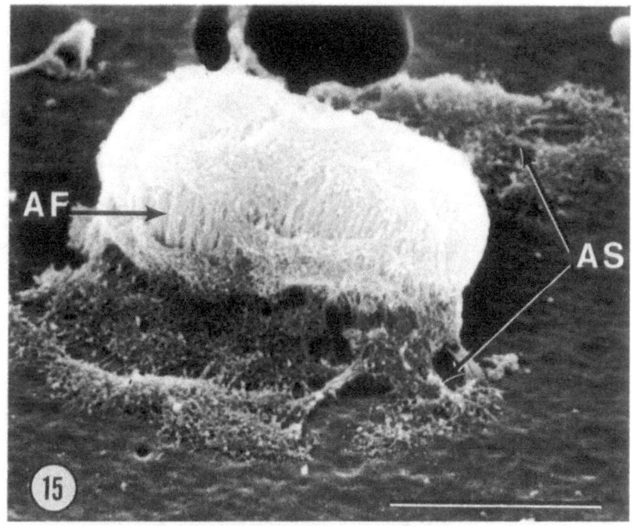

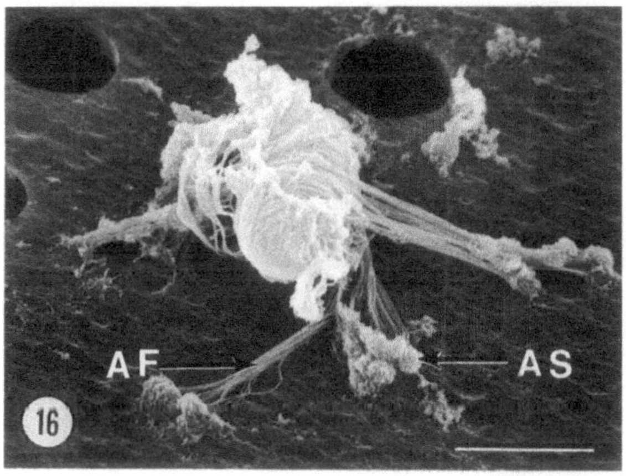

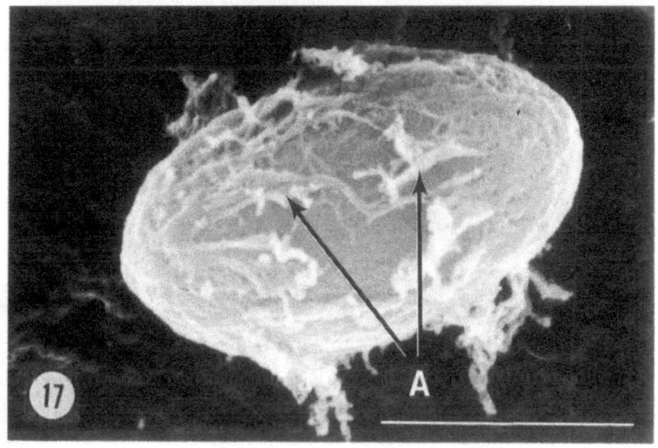

Fig. 14. *Eiona tunicata.* Transmission electron micrograph. Section through a mass of appendage fibres showing their random, branched structure. Bar line = 0.5 μm.

Figs 15, 16. *Dryosphaera navigans.* Scanning electron micrographs.
Fig. 15. Outer amorphous sheath (AS) of the exosporium attaching ascospore to surface, note the appendage fibres (AF) beneath the sheath. Fig. 16. Appendages fibres uncoiled with 'ball-like' remnants of the amorphous sheath (AS) at their tips. AF = appendage fibres. Bar lines = 0.5 μm.

Fig. 17. *Eiona tunicata.* Scanning electron micrograph. Early stages in appendage (A) extension, with appendage fibres beginning to separate at the mid region of the ascospore. Bar line = 5 μm.

Upon release into seawater, ascospores lose their delimiting membrane complex and the appendages become erect (Figs 1, 2a, b). Ascospores possess 10–25, tapering, spine-like appendages that vary in length (0.1–4 μm) and thickness (320–360 nm). The outer surface of the ascospore is coated with an amorphous material through which the spines project (Fig. 1). In section, this amorphous material is electron-dense and forms a thick homogeneous layer around the spore (Fig. 13). The origin of this material is unknown, but it probably develops after the appendages are formed since it was not observed around the spores while they were still within the outer delimiting membrane in the ascus. This amorphous material may be derived from either a thin, outer layer deposited around the spore which swells in seawater, or, from material released from or during the degradation and deliquescence of the ascus.

Eiona tunicata Kohlm.

Fully developed ascospores of *E. tunicata* are one-celled, ovoid with a subterminal appendage comprising radiating fibres at each pole (Figs 3, 17). Ascospores lack an exosporium, but possess an episporium and mesosporium. Fragmentation of the outer region of the episporium gives rise to disorganized, frequently branched, electron-dense fibres which separate at the midpoint of the ascospore and detach from the spore wall except at the spore poles (Fig. 17). Individual appendage fibres vary in length, thickness and degree of branching (Fig. 14).

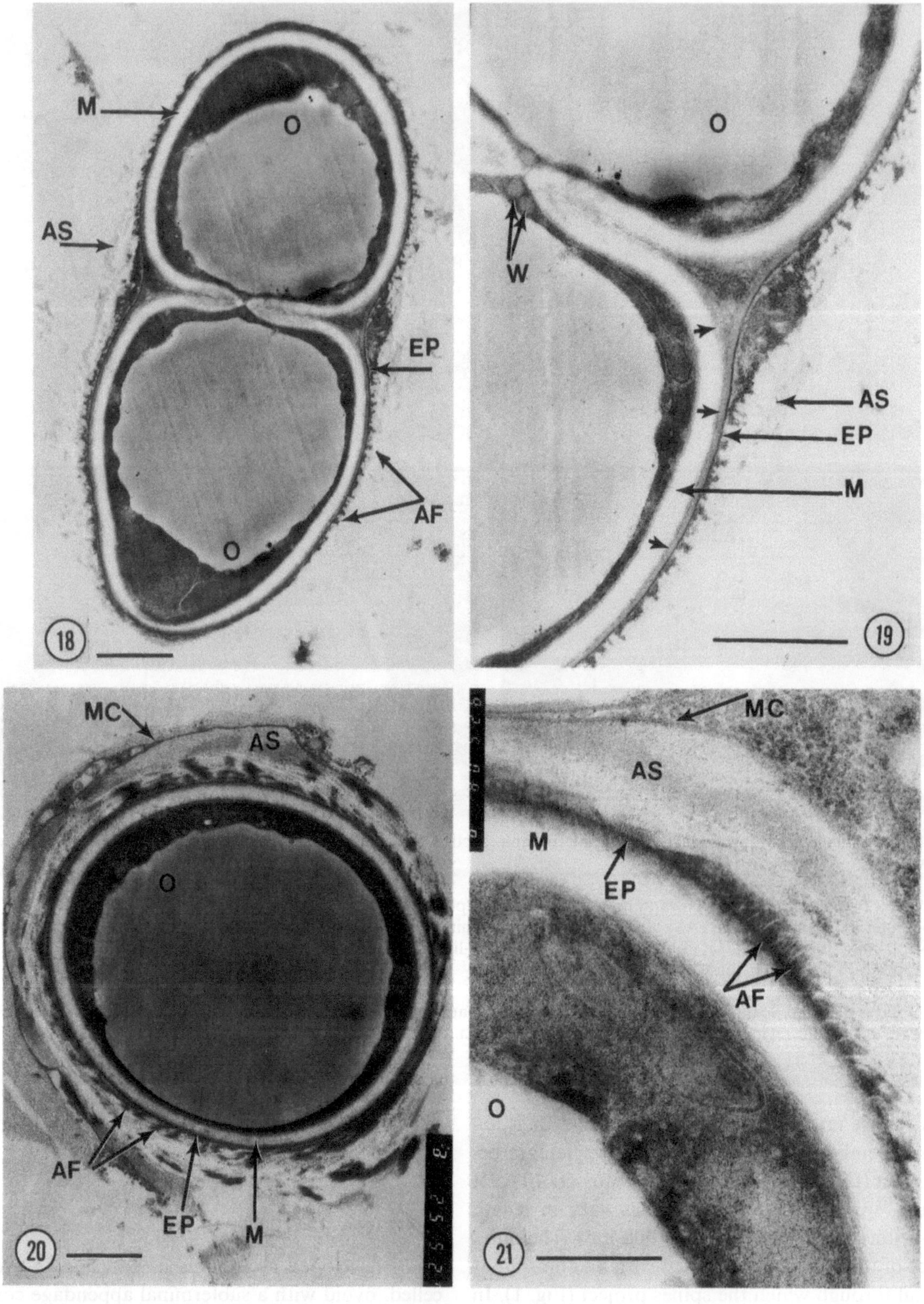

Figs 18–21. *Dryosphaera navigans*. All transmission electron micrographs.
Figs 18, 19. Low (18) and higher (19) magnifications of a longitudinal section of an ascospore in which the amorphous sheath (AS) is gradually lost with the unwinding of the appendage fibres (AF) from the spore wall. O = oil deposits, M = mesosporium, EP = episporium, W = Woronin bodies. Arrowheads indicate the outer, electron-dense zone of the mesosporium. Figs 20, 21. Transverse section of ascospore, the amorphous sheath (AS) covering the appendage fibres (AF) is bounded on its outer surface by the membrane complex (MC). Appendage fibres arise from the episporium (EP), M = mesosporium, O = oil deposits. Bar lines: Figs 18–20 = 1 μm; Fig. 21 = 0.5 μm.

Histochemically, the appendages, and hence the episporium, are composed of weakly acidic, PAS positive polysaccharides with protein (Table I). The electron-transparent mesosporium did not stain with any of the histochemical techniques utilized. In *E. tunicata* oil droplets are dispersed throughout the spore cytoplasm and the electron-dense, membrane-bounded volutin bodies are small. The ground, spore cytoplasm is dense and contains proteins, mitochondria, glycogen rosettes and other PAS positive carbohydrates (Table I).

Dryosphaera navigans Koch *et* Jones

Ascospores of *D. navigans* possess an inner mesosporium, a middle episporium and an outer exosporium (Figs 18–21). The mesosporium is an electron-transparent layer within which is contained an electron-dense zone on its outer surface (Figs 19, 21). A crosswall is formed from both zones of the mesosporium and has a central pore with associated Woronin bodies (Figs 18, 19). Histochemically, the mesosporium failed to stain with any of the histochemical techniques employed (Table I). Outside the mesosporium is an electron-dense episporium (Fig. 19) the chemical composition of which cannot be resolved with the staining procedures used (Table I).

Immediately after ascospore release from the ascus the appendage is adpressed to the episporium (Fig. 20). The appendages of *D. navigans* appear to be attached subterminally as seen under the light microscope (Jones *et al.* 1980). However, the exosporium or appendage comprises individual (64 nm wide), electron-dense fibres embedded in a less electron-dense, amorphous matrix. The latter structure is bounded by an electron-dense, membranous structure (Fig. 20) probably the membrane complex or an homologous structure. Appendage fibres originate from the outer surface of the electron-dense episporium and are orientated circumferentially around the ascospore (Figs 20, 21). After release from the ascus the amorphous matrix disperses to expose the appendage fibres (Figs 15, 16). The exposed fibres unwind progressively from the central region of the spore surface towards the spore poles. At the spore poles, appendage fibres may not unwind from the episporium, thereby giving the characteristic subterminal appendage (Jones *et al.* 1980, Koch and Jones 1989).

Histochemically, the spore appendage comprises highly acidic, PAS positive polysaccharides (Table I). Alcian blue staining at pH 0.5 indicates the presence of sulphated polysaccharides. However, it is uncertain whether the staining is due to the appendage fibres or the amorphous matrix of the exosporium: staining of old spores would have elucidated the staining reactions of components of the exosporium, but insufficient material was available.

Discussion

The results of histochemical and transmission electron microscopical studies of species of the marine cleistothecial genera *Amylocarpus*, *Eiona* and *Dryosphaera* demonstrate a diversity of structure (Table II).

Three topics are worthy of discussion:

1. Appendage ontogeny of *Amylocarpus encephaloides*, *Eiona tunicata* and *Dryosphaera navigans*

The three cleistothecial species examined illustrate different modes of appendage ontogeny: outgrowths in *A. encephaloides*, fragmentation in *E. tunicata* and a combination of outgrowths and fragmentation in *D. navigans* (Table II). Furthermore there is no structural or histochemical similarities of appendages between the species.

T. W. Johnson (1963) failed to elucidate the mechanism of appendage ontogeny in *A. encephaloides*. However, similar methods of appendage formation to those observed in this investigation were reported by Jones and Moss (1978) for *A. encephaloides* and *E. tunicata* and by Jones and Moss (1980) for *D. navigans*. Later, Jones *et al.* (1980) amended their interpretation and considered the appendage fibres of *D. navigans* to be formed by fragmentation of an outer wall layer. From the TEM studies in this investigation it is clear that the appendage fibres are outgrowths of the spore because they are individual structures attached directly to the episporium, but not derived from it.

2. Other ascospore and appendage characters of *Amylocarpus encephaloides*, *Eiona tunicata* and *Dryosphaera navigans*

Ascospores of *D. navigans* possess an exosporium, episporium and mesosporium whereas in *E. tunicata* the exosporium is absent. *Amylocarpus encephaloides* ascopores have a single wall layer divided into three histochemically distinct zones, which are also demarcated by variations in electron-density. Histochemically the three zones in *A. encephaloides* spore walls can be equated with the chemical composition of the exosporium, episporium and mesosporium of other marine fungi examined. The mesosporium in *D. navigans* is distinct from that of *E. tunicata* and *A. encephaloides* in that it contains an electron-dense zone on its outer surface.

Table II. Comparison of ascospore characters of *Amylocarpus encephaloides*, *Eiona tunicata* and *Dryosphaera navigans*

Ascospore character / Species	Spore shape	Membrane complex	Number of septa	Single spore wall layer of 3 histochemically distinct zones	Mesosporium	Bipartite mesosporium	Episporium	Exosporium	Episporial appendages	Exosporial appendages	Outgrowing spine appendages	Histochemical composition of the appendages: Acidic carbohydrates	Neutral carbohydrates	Protein
Amylocarpus encephaloides	Spherical-ovoid	+	0	+	−	−	−	−	−	−	+	−	+	−
Eiona tunicata	ellipsoid	−	0	−	+	−	+	−	+	−	−	+	+	+
Dryosphaera navigans	ellipsoid	+	1	−	+	+	+	+	−	+	−	+	+	+

\+ = character present − = character absent

Although appendage morphology of the three cleistothecial species is distinct (Table II), their morphology concurs with the descriptions and illustrations of Kohlmeyer (1960, 1968), T. W. Johnson (1963), Schmidt (1969, 1974), Koch (1974), Kohlmeyer and Kohlmeyer (1979), Jones and Moss (1978, 1980) and Jones *et al.* (1980).

Superficially the appendages of *D. navigans* and *E. tunicata* are similar in that they are attached subterminally and consist of fibres. However, in *E. tunicata* the appendage is derived from the outer region of the episporium, whereas in *D. navigans* it originates from the exosporium. The outer region of the episporium in *E. tunicata* may constitute an exosporium which is not readily distinguished from the episporium with the techniques employed. If this is the case then *E. tunicata* may be similar to the archetypal appendaged marine ascomycete hypothesized by Kirk (1976). The archetype, according to Kirk (1976), lacks ornamentations and an exosporium, but has a thicker spore wall than other marine fungi which may be due to an exosporium being imperceptibly fused with the episporium. *Eiona tunicata* is only slightly more advanced than this situation in having appendages composed of disorganized fibres.

Released ascospores of *A. encephaloides* have a homogeneous, electron-dense, amorphous layer around the spore wall through which the spine appendages project (Hyde *et al.* 1989). The origin of this amorphous layer is uncertain. However, it is absent around immature spores and a number of hypotheses can be forwarded to explain its origin. Firstly, the layer may be mucilage secreted by released spores prior to germination as described by Hyde *et al.* (1986) for *Halosphaeria appendiculata* Linder. Secondly, it may be formed by the swelling of an outer covering of the spore wall in seawater. Such sheaths have been described for *Leptosphaeria* spp., (Hyde and Jones 1989). The constant thickness and homogeneity of this outer covering suggests that its origins are sporal since an irregular structure would be formed if it was deposited from the disorganized, enucleated epiplasm. The composition and histochemistry of the spore cytoplasm in all cleistothecial species is similar.

3. Taxonomic significance of the observations

Plectomycetes are characterized by cleistothecial ascocarps which lack paraphyses but contain ascogenous hyphae formed at various levels. The asci are globose or broad and non-dehiscent; they deliquesce and liberate the ascospores within the ascocarp (Webster 1975, Talbot 1976). It is these centrum characteristics rather than the actual form of the ascocarp that are considered taxonimically important. In *Amylocarpus encephaloides* globose, non-dehiscent asci are found scattered, irregularly throughout the centrum, a character also noted by Kohlmeyer and Kohlmeyer (1979). A similar situation also occurs in *Eiona tunicata* (Kohlmeyer and Kohlmeyer 1979) thereby indicating the affinity of these two species with the Plectomycetes. *Dryosphaera navigans*, although having a cleistothecium, has been placed in the Ascomycotina *incertae sedis* (Kohlmeyer and Kohlmeyer 1979) because young developmental stages of ascosarps, asci and ascospores are unknown. The asci are clavate and appear to develop in a clump (Kohlmeyer and Kohlmeyer 1979) which suggests that an organised hymenium may be present which would not support the inclusion of *D. navigans* in the Plectomycetes.

Recent studies by Koch and Jones (1989) of three cleistothecial fungi have shown that the asci in *D. navigans* are formed in an irregular hymenium on the expanded upper part of a stem-like structure arising from the base of the ascocarp and stretching about 1/3 up into the centrum. Asci are thick-walled when young, especially towards the ascus tip and are persistent, but the ascus tip at maturity deliquesces to release ascospores into the ascocarp cavity. The cleistothecial ascocarp with branched appendages and thick-walled asci suggest affinities with the Erysiphales. Koch and Jones (1989) considered that on the basis of centrum morphology and the asci (thick-walled but not functionally bitunicate) *D. navigans* is better placed in the Coronophorales (Ascomycotina, Parenchymatomycetidae) with an uncertain taxonomic position at the familial level.

In terms of ascospore stucture *D. navigans* is the only cleistothecial species examined which has a crosswall and 3 spore wall layers with a distinct exosporium forming the appendage. These are features common to many of the marine pyrenomycetous species investigated (Jones *et al.* 1986, Jones and Moss 1987). *Amylocarpus encephaloides* and *E. tunicata* are single-celled with 1 and 2 wall layers respectively. Apart from spore cytoplasmic characters, ascospore and ascospore appendage ontogeny and morphology are different in the 3 cleistothecial species.

Acknowledgements

We thank G. Bremer, R. Healey and C. Derrick for photographic assistance; Dr J. Koch, Copenhagen and Dr G. Rees for assistance with the field work in Denmark and Dr G. Rees for Figure 3.

References

Hawksworth, D. L. and O. Eriksson. 1986. The names of accepted orders of ascomycetes. *Systema Ascomycetum 5*: 175–184.

Hyde, K. D., E. B. G. Jones and S. T. Moss. 1986. How do fungal spores attach to surfaces? *In*: (S. Barry, D. R. Houghton, G. C. Llewellyn and C. O'Rear, eds), *Biodeterioration 6*. Commonwealth Agricultural Bureaux and Biodeterioration Society, London. pp. 584–589.

Hyde, K. D. and E. B. G. Jones. 1989. Observations on ascospore morphology in marine fungi and their attachment to surfaces. *Bot. Mar. 32*: 205–218.

Hyde, K. D., S. T. Moss and E. B. G. Jones. 1989. Attachment studies in marine fungi. *Biofouling 1*: 287–298.

Johnson, T. W. 1963. Some aspects of morphology in marine Ascomycetes: *Amylocarpus* Currey, *Herpotrichiella* Petrak and *Torpedospora* Meyers. *Nova Hedwigia 6*: 157–168.

Johnson, R. G. 1980. Ultrastructure of ascospore appendage of marine Ascomycetes. *Bot. Mar. 23*: 501–527.

Jones, E. B. G., R. G. Johnson and S. T. Moss. 1986. Taxonomic studies of the Halosphaeriaceae – Philosophy and rationale for the selection of characters in the delineation of genera. *In*: (S. T. Moss, ed.) *The Biology of Marine Fungi*. Cambridge University Press, Cambridge. pp. 211–229.

Jones, E. B. G. and S. T. Moss. 1978. Ascospore appendages of marine Ascomycetes: an evaluation of appendages as taxonomic criteria. *Mar. Biol. 49*: 11–26.

Jones, E. B. G. and S. T. Moss. 1980. Further observations on the taxonomy of the Halosphaeriaceae. *Bot. Mar. 23*: 483–500.

Jones, E. B. G., S. T. Moss and J. Koch. 1980. Light and scanning electron microscope observations of the marine ascomycete *Crinigera maritima*. *Trans. Br. mycol. Soc. 74*: 625–631.

Jones, E. B. G. and S. T. Moss. 1987. Key and notes on genera of the Halosphaeriaceae examined at the ultrastructural level. *Systema Ascomycetum 6*: 179–200.

Kirk, P. W. 1976. Cytochemistry of marine fungal spores. *In*: (E. B. G. Jones, ed.) *Recent Advances in Aquatic Mycology*. Elek Science, London. pp. 177–192.

Koch, J. 1974. Marine fungi on driftwood from the west coast of Jutland, Denmark. *Friesià 10*: 209–250.

Koch, J. and E. B. G. Jones. 1989. The identity of *Crinigera maritima* Schmidt, and three new genera of marine cleistothecial ascomycetes. *Can. J. Bot. 67*: 1183–1197.

Kohlmeyer, J. 1960. Wood inhabiting marine fungi from the Pacific North West and California. *Nova Hedwigia 2*: 293–343.

Kohlmeyer, J. 1968. Dänische Meerespilze (Ascomycetes). *Ber. Dtsch. Bot. Ges. 81*: 53–61.

Kohlmeyer, J. 1986. Taxonomic studies of the marine Ascomycotina. *In*: (S. T. Moss, ed.) *The Biology of the Marine Fungi*. Cambridge University Press, Cambridge. pp. 199–210.

Kohlmeyer, J. and E. Kohlmeyer. 1979. *Marine Mycology: The Higher Fungi*. Academic Press, New York. 690 pp.

Luttrell, E. S. 1951. Taxonomy of the Pyrenomycetes. *University of Missouri Studies* 24: 1–120.

Moss, S. T. and E. B. G. Jones. 1977. Ascospore appendages of marine ascomycetes: *Halosphaeria mediosetigera*. *Trans. Br. mycol. Soc. 69*: 313–315.

Schmidt, I. 1969. *Corollospora intermedia*, nov. spec., *Carbosphaerella leptosphaerioides* nov. spec., und *Crinigera maritima*, nov. gen., nov. spec., 3 neue marine Pilzarten von der Ostseeküste. *Natur und Naturschutz in Mecklenburg 7*: 5–14.

Schmidt, I. 1974. Höhere Meerespilze der Ostsee. *Biologische Rundschau 12*: 96–112.

Talbot, P. H. B. 1976. *Principles of Fungal Taxonomy*. MacMillan, London. 274 pp.

Webster, J. 1975. *Introduction to Fungi*. Cambridge University Press, Cambridge. 424 pp.

Botanica Marina
Vol. 34, pp. 241 – 249, 1991

Trace Metal Content in Some Seaweeds from the Northern Adriatic

I. M. Munda and V. Hudnik

Biological Institute, Slovene Academy of Science and Arts and "Boris Kidrič" Institute of Chemistry, 61000 Ljubljana, Yugoslavia

(Accepted 9 February)

Abstract

Twenty-three common seaweeds from the eulittoral and upper sublittoral in the northern Adriatic (Piran and Rovinj, Istrian coast) were analysed for levels of trace metals (Mn, Zn, Cu, Cd, Pb, Co, Ni). Differences in the levels of metals between species and genera were considered as well as differences between seaweeds from the two areas. Seasonal variations were also observed in some species. Results indicated that Mn concentrations in algal tissue are the main distinguishing character between taxonomic groups as well as in seasonal- and habitat-conditioned variations.

Introduction

In spite of some experimental studies on heavy metal uptake in Adriatic benthic algae (Munda 1979, 1982, 1984, Kremer and Munda 1982) there is still a lack of information on their accumulation under varied field conditions.

The increasing concentration of heavy metal pollutants in estuarine and coastal waters has stimulated numerous studies in different areas on trace metal levels in seaweeds (*e.g.* Bryan 1969, Bryan and Hummerstone 1973, Hägerhäll 1973, Fuge and James 1973, 1974, Haug *et al.* 1974, Skipnes *et al.* 1975, Morris and Bale 1975, Eide *et al.* 1980, Rice and Lapointe 1981, Higgins and Mackey 1987 a, b). Brief information on Sb content in some species was given by Strohal *et al.* (1975). However, there is still a lack of data for Adriatic benthic algae.

Metals present in algae are a sensitive indicator of metal contamination in the aquatic environment. Due to high concentration factors seaweeds are important indicators in metal pollution studies (Phillips 1977).

The northern Adriatic is a heavily polluted shelf area which receives domestic sewage and industrial discharges from several rivers (*e.g.* Po, Adige, Tagliamento, Isonzo) and directly from urban effluents. Increases in the concentrations of heavy metals in this area were reported by Branica (1978).

Algal material for the present study was collected at Piran and Rovinj on the Istrian coast (see map – Fig. 1). River-born pollutants are transferred in the surface water along this coast by a regular circular convection current. The two sampling sites exhibit some differences in their salinity/temperature regimes. However, the tidal range is wider at Piran than at Rovinj. A further difference between the sampling areas is in their geological structure *viz.* Paleocene flysch at Piran and Cretaceous limestone around Rovinj. At Piran a heavy load of domestic sewage is the main source of pollution as well as industrial effluents (ship-building and chemical industry) from the neighbouring harbours of Isola, Koper and Trieste. Regarding concentrations of heavy metals, phosphates, nitrates and fecal coliform bacteria, the sources of pollution in the Gulf of Trieste have been dealt with by Tušnik *et al.* (1989). Analyses were carried out directly in outlets of domestic and industrial wastes, such as the fish-canning factory ("Delamaris") in Koper. Rovinj receives pollutants from a local fish-canning factory ("Mirna") and from domestic sewage, which has increased drastically during the last few years due to tourism and extended urbanisation of the area.

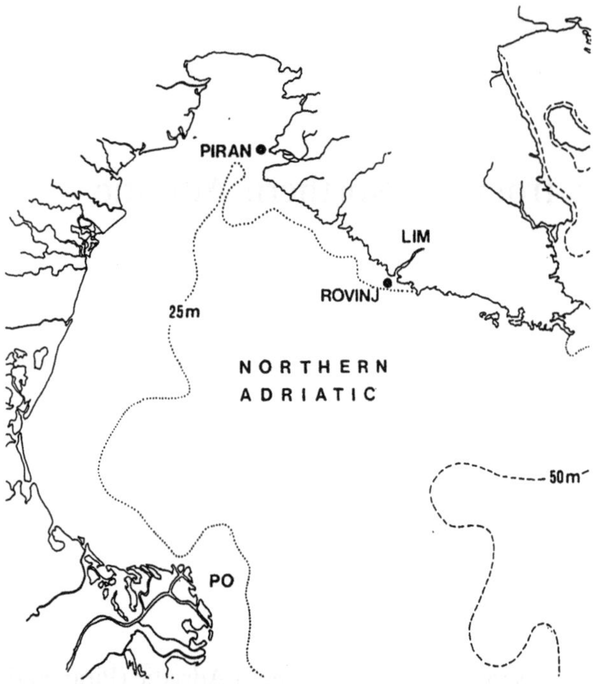

Fig. 1. Map of the northern Adriatic with sampling sites.

The concentrations of heavy metals (Zn, Mn, Cd, Cu, Co, Ni, Pb) in 23 common species of benthic algae, both perennial and seasonal, were measured as an indicator of metal pollution in the northern Adriatic shelf area. Differences in levels of metals between species and genera were considered as well as differences between samples from the two localities with different ecological characteristics. Attention was paid to the relation between the metals concentrated in algae and in two cases seasonal variations in metal levels were observed.

Material and Methods

Algal material was collected from the rocky substrate in the eulittoral and upper sublittoral to 7 m depth at Piran and at Rovinj. At Piran collections were carried out throughout the year, whereas at Rovinj collections were only made in April, May and November. Sampling sites at Rovinj were close to the harbour, in the relatively clean bay of Faborsa and in the Lim fjord (Limski kanal) which is strongly influenced by freshwater.

Five to ten replicates were taken for each species during sampling. The plants were rinsed with distilled water and dried at room temperature. The dry weight was than determined by further drying at 100 °C for 24 hours. The material was ground and subsamples were dissolved in a mixture of $HClO_4$ and HNO_3 (1 : 1) in sealed glass tubes at 140 °C. The trace metal content (Cd, Cu, Co, Mn, Ni, Pb) was determined by electrothermal (Perkin Elmer HGA 72) and flame atomic absorption spectrometry for Zn (AA 6 Techtron Varian). The results are means of three replicates and are expressed as $\mu g\ g^{-1}$ of dry weight. The precision and detection limits of the procedures applied are given in Table I.

Table I. The precision (s_r) and detection limits (D. L.) for determination of metals in seaweeds (results in $\mu g\ g^{-1}$).

Element	D. L.	S_r (%)
Cd	0.06	2
Co	3	5
Cu	2	3
Mn	0.2	2
Ni	3	5
Pb	2	4
Zn	6*	3

* – flame

Results

Salinity and temperature conditions and the levels of primary nutrients in the surface water at the time of sampling are given in Table II. Further extensive data on salinity and temperature conditions and the distribution of primary nutrients in the northern Adriatic were collected during cruises by the research vessel "Vila Velebita" (Škrivanić and Barić 1979).

Data on regional, seasonal and vertical distribution of trace metals (Zn, Cd, Pb, Cu) were collected for this area during the same cruises (Bubić *et al.* 1979). The overall average concentrations of Zn ranged from 11.8 $\mu g\ L^{-1}$ in August, over 5.5 $\mu g\ L^{-1}$ in February to a minimum of 2.3 $\mu g\ L^{-1}$ in May. Variations for Cd were from 0.11 $\mu g\ L^{-1}$ in August to 0.032 $\mu g\ L^{-1}$

Table II. Hydrographic conditions during sampling (0 m).

	Temperature	Salinity	Primary nutrients (μ g-atom. L^{-1}) PO_4-P	NO_3-N
Piran				
January	7.9	37.3	0.24	4.13
March	7.7	37.8	0.28	3.07
April	11.5	37.6	0.12	1.86
May	12.5	33.9	0.06	1.03
June	19.4	35.7	0.03	0.53
July	21.8	34.2	0.01	0.31
October	18.5	35.2	0.28	2.28
November	13.8	36.5	0.30	4.39
Rovinj				
April	11.9	37.2	0.48	9.55
May	14.5	36.8	0.28	4.55
November	13.8	36.5	0.12	8.22
Lim fjord (Limski kanal)				
April	19.7	10.8	–	–

Table III. Trace metal levels in the surface water at Rovinj (April–May).

Element Site	Zn	Mn	Co	Cd ($\mu g\ L^{-1}$)	Ni	Cu	Pb
Harbour	19.3	10	<1	0.03	<1	14.9	1.90
Bay of Faborsa	7.3	9	<1	0.07	<1	2.2	1.54
Lim fjord (brackish water)	5.7	10	<1	0.10	<1	0.9	0.70

in May. The two metals exhibited a similar trend of seasonal variations. Seasonal variations in Pb were slight (0.30 $\mu g\ L^{-1}$ in November to 0.65 $\mu g\ L^{-1}$ in August), whereas average concentrations of Cu varied between 1.4 $\mu g\ L^{-1}$ (February) and 0.57 $\mu g\ L^{-1}$ (November).

The total trace metal content in the surface water near Piran during algal sampling (April) was 60 $\mu g\ L^{-1}$ of Zn, 5 $\mu g\ L^{-1}$ of Mn, < 1.0 $\mu g\ L^{-1}$ of Co, 0.1 $\mu g\ L^{-1}$ of Cd, < 1.0 $\mu g\ L^{-1}$ of Ni, 0.8 $\mu g\ L^{-1}$ of Pb and 2.0 $\mu g\ L^{-1}$ of Cu. More recent measurements revealed a lower Zn content (14.0 $\mu g\ L^{-1}$) whereas the concentrations of other metals were on the same level.

The measurements of the trace metal levels at Rovinj in April–May are given in Table III.

Results presented in Table IV indicate that the two metals, essential in cell metabolism (Mn, Zn), are present in relatively high amounts in algal tissue. Although the concentration of Zn in seawater at Piran was higher than that of Mn, the concentration of Mn in most algal samples exceeded by two to four times that of Zn; in some cases even more. The toxic metals Cd, Cu and Pb were found in low concentrations in the material investigated, with exception of a few samples. Cobalt and nickel concentrations were low in algal tissue and below the detection limits of the method used.

Phaeophyceae (Fucophyceae)

Fucales (Table IV). In the perennial eulittoral seaweed *Fucus virsoides* the Mn content was relatively high and increased from winter towards summer, with a peak in August and a subsequent decline in autumn. Seasonal variations in Zn content were less pronounced and a slight increase was observed from spring towards autumn. Consequently, the relation between the two metals varied throughout the seasons.

In samples of *Fucus virsoides* from Rovinj the Mn content was lower than in those from Piran. Differences for Zn were less pronounced. In samples from Piran, Mn dominated over Zn, whereas in the Rovinj samples the concentrations of both metals were similar although in May the Zn content was slightly higher than that of Mn.

A similar Mn content was found in the eulittoral *Cystoseria compressa* and its Zn content was also at a similar level to that in *Fucus virsoides*.

In two sublittoral fucoids, *Cystoseira spicata* and *Cystoseira corniculata* the Mn content was lower than in the two eulittoral fucoids. This was so for both the Piran and the Rovinj samples. Differences from the eulittoral fucoids were less pronounced for the Zn content. It seems likely that the eulittoral fucoids accumulate more Mn and Zn than those from deeper water. In sublittoral fucoids the two metals were present in an approximately 1 : 1 ratio, whereas Mn dominated over Zn in the eulittoral samples.

Sphacelariales (Table IV). The only representative of this order, *Halopteris scoparia*, was collected at between 3 m and 7 m depth. It is dominant in the algal biomass of slightly polluted habitats. Samples from Rovinj revealed high concentrations of both Mn and Zn. In a sample collected in May the Mn content was twelve times as high as the Zn content.

Among representatives of the Dictyotales (Table IV) *Padina pavonica* was collected in winter, spring and mid-summer, *Dictyota dichotoma* in spring and *Dictyopteris membranacea* in June. Results revealed a high level of Mn in *Padina pavonica*. The accumulation of Zn was, on the other hand, at the same level as in the fucoids, except in January. A slight decline in Zn content from spring towards autumn was notable, whereas the Mn content varied irregularly with the seasons. This might relate to the appearance of successive generations. An isomorphic alternation of gametophytes and sporophytes has been shown for this species by Ramon and Friedmann (1965). In samples from Rovinj both Mn and Zn levels were lower than in the material from Piran.

The Mn content was lower in *Dictyota dichotoma* than in *Padina* and again lower in the Rovinj samples than in those from Piran. In *Dictyopteris membranacea* the concentrations of Mn and Zn were similar to those found in the spring samples of *Dictyota*. It seems likely, however, that among the Dictyotales *Padina pavonica* has the highest ability to accumulate Mn.

Among the Scytosiphonales (Table IV) the seasonal annual *Scytosiphon lomentaria* was collected both at Piran and at Rovinj. It forms erect thalli from March to May. The accumulation of Zn and Mn increases towards the end of the vegetative period. The Mn content was higher in samples from Piran than in those from Rovinj whereas the opposite was true for the Zn content. In the related *Colpomenia sinuosa* the Mn content was notably higher than the Zn content.

A spring sample of *Ectocarpus siliculosus* revealed that Mn was more than five times higher than Zn.

The so-called toxic metals Cd, Cu and Pb were present in most samples of brown algae at low concentrations. The concentration of Cu in algal tissue was about tenfold that of Cd, corresponding approximately to the relation of their levels in the sea (*cf.* Branica 1978). There where, however, only small scale variations in the concentrations of Cu, Cd and Pb in algal tissue, compared with variations in Mn and Zn. Seasonal variations, as observed in *Fucus virsoides* were inconspicuous. A minimum in Cd content in June coincided with a maximum in Pb content. The concentration of toxic metals was slightly lower in the *Fucus* samples from Rovinj than in those from Piran. Relatively high concentrations of Cd and Pb were found in the eulittoral *Cystoseira compressa*. An elevated Cd and relatively high Cu content was noted in the sublittoral *Cystoseira spicata* (Rovinj sample). With the exception of the samples just mentioned the level of toxic metals in fucoids was low.

Higher concentrations of toxic metals were found in representatives of the Dictyotales. In *Padina pavonica* from Piran the Cd content was high and decreased from winter towards summer. The relatively high Pb content showed the opposite trend. Copper concentration was high only in winter and spring. In samples of *Padina* from Rovinj, the content of all toxic metals was lower.

High levels of toxic metals (Cu, Pb) were found in *Dictyota dichotoma*. As in *Padina*, the concentrations of Pb and Cu were slightly lower in the material from Rovinj than in that from Piran. In *Dictyopteris membranacea* the Pb content was high. These results indicate that representatives of the Dictyotales have a greater ability to accumulate toxic metals than the Fucales.

A high concentration of Pb was also found in *Halopteris scoparia*. Exceptionally high concentrations of toxic metals were found in representatives of the Scytosiphonales (Table IV) collected at Rovinj. In *Scytosiphon lomentaria* from Piran both Cu and Pb concentrations showed a slight increase towards the end of the vegetative period of this species, whereas the Cd content tended to decrease. In the Rovinj material the highly elevated concentrations of Cu and Pb even increased in May (22 $\mu g\ g^{-1}$ of Cu and 68 $\mu g\ g^{-1}$ of Pb). In the related *Colpomenia sinuosa* from Rovinj the Cu and Pb concentrations were at the same level as in *Scytosiphon*.

It is noteworthy that the concentrations of Cu and Pb were not increased to such high levels in other brown algae, collected at the same time at Rovinj. In Rovinj harbour, where the samples were collected, a highly increased Cu content was notable, whereas the Pb content was only slightly elevated.

It is known, however, that annuals and short-lived perennials tend to accumulate more toxic metals than perennials which persist for several years. We might tentatively assume that representatives of the Scytosiphonales have an increased ability to accumulate Cu and Pb, with metal-binding proteins and organic acids possibly involved in their intensive uptake.

Rhodophyceae (Bangiophyceae)

Among the red algae investigated (Table IV), notable variations in trace metal concentrations were observed, relating to the taxonomic groups as well as influences of the habitats.

From the Bangiales only *Porphyra leucosticta* was analysed. It was collected in a brackish habitat (Lim fjord) near Rovinj and in it the levels of metals were different from the rest of the red algae tested. The Mn and Zn concentrations were low; the Zn concentration exceeded by 50% that of Mn. The Cu content was slightly elevated.

In the rest of the red algae both Zn and Mn were higher than in *Porphyra* and in all the cases the Mn concentration in the thalli was higher than that of Zn.

In the erect, articulate corallines, belonging to the order Cryptonemiales (*Jania rubens* and *Corallina officinalis*) the Mn content was relatively high and surpassed by two to four times the Zn content. This was true for the samples from Piran, where an increase from winter/spring towards summer was observed. The relatively high Pb content tended also to increase towards summer. As in some of the brown algae, the *Corallina* sample from Rovinj exhibited a lower Mn content than the Piran sample, while the Zn content was on the same level.

Gelidium pusillum and *Pterocladia capillacea*, as representatives of the Gelidiales, had high concentrations of Mn and Pb. They were at the same level in both species, indicating that taxonomic relationships can be of primary importance for the affinities of seaweeds towards dissolved metals. The two species were collected at different littoral levels and at differ-

Table IV. Trace metal content in Adriatic seaweeds.

Species	Locality	Month	Zn	Mn	Cd	Cu	Pb
			($\mu g\ g^{-1}$ dry weight)				
Phaeophyceae (Fucophyceae)							
Fucales							
Fucus virsoides (Don.) J. Ag.	Piran	January	54	132	0.5	5	3
		April	53	97	0.4	3	2
		June	40	222	0.3	3	6
		August	60	475	0.4	4	2
		October	74	285	0.5	5	2
	Rovinj	April	26	44	0.3	3	2
		May	33	28	0.2	3	2
Cystoseira compressa (Esper) Gerloff *et* Nizamuddin	Piran	April	50	139	1.5	4	6
Cystoseira spicata Ercegović	Rovinj	April	29	64	0.7	10	4
Cystoseira corniculata (Wulf.) Zanard. *ex* Hauck	Piran	July	43	40	0.3	4	4
	Rovinj	May	36	52			
Sphacelariales							
Halopteris scoparia (L.) Sauv.	Rovinj	April	15	186	0.1	6	8
		May	106	456	0.2	3	11
Dictyotales							
Dictyopteris membranacea (Stack.) Batt.	Rovinj	June	83	176	0.4	3	11
Dictyota dichotoma (Huds.) Lam.	Piran	April	48	157	0.9	5	6
		May	85	105	0.7	5	8
	Rovinj	April	45	80	0.9	3	5
Padina pavonica (L.) Thivy	Piran	January	105	275	1.1	8	8
		April	69	335	0.9	7	10
		June	52	223	0.6	5	10
		July	59	332	0.6	4	11
	Rovinj	April	36	100	0.4	2	3
Scytosiphonales							
Scytosiphon lomentaria (L.) Link.	Piran	March	29	42	0.4	2	3
		April	33	98	0.1	6	8
	Rovinj	April	57	21	0.4	9	22
		May	102	75	0.2	22	69
Colpomenia sinuosa Derb. *et* Sol.	Rovinj	April	54	215	0.7	22	53
Ectocarpus siliculosus (Dillw.) Lyngb.	Piran	March	28	169	0.3	4	9
Rhodophyceae (Bangiophyceae)							
Bangiales							
Porphyra leucosticta Thur.	Lim/Rovinj	April	23	15	0.1	6	2
Cryptonemiales							
Jania rubens L.	Piran	January	45	124	0.2	3	8
		August	60	195	0.3	4	15
Corallina officinalis L.	Piran	May	30	123	0.1	2	4
		June	42	154	0.2	3	7
	Rovinj	November	44	66	0.3	3	7
Gelidiales							
Gelidium pusillum (Stackh.) Le Jol.	Piran	January	60	214	0.2	15	15
Pterocladia capillacea (Gmel.) Born. *et* Thur.	Piran	November	73	217	0.5	5	15
Ceramiales							
Ceramium ciliatum (Ellis) Ducluz.	Rovinj	April	23	75	0.3	1	5
Laurencia obtusa (Huds.) Lam.	Rovinj	April	17	85	0.3	2	6
Chlorophyceae							
Ulvales							
Enteromorpha spp.	Piran	January	48	34	0.1	3	2
		June	42	215	0.3	6	6
		October	14	24	0.1	4	9
	Rovinj	April	31	41	0.4	3	6
		May	38	133	0.9	5	8
Ulva rigida (C. Ag.) Thur.	Rovinj	November	81	125	0.3	8	
Cladophorales							
Cladophora dalmatica Kütz.	Rovinj	April	61	126	0.3	2	5
Codiales							
Codium vermilara (Olivi) Delle Chiaje	Piran	May	11	82	0.1	3	2
	Rovinj	May	26	36	0.3	6	7
Codium effusum (Raf.) Delle Chiaje	Piran	August	38	349	0.2	2	9

ent seasons. *Gelidium pusillum* forms mats in the uppermost eulittoral, whereas *Pterocladia capillacea* is usually found in the sublittoral (between 3 m and 7 m depth). In both species the Pb content was high (15 $\mu g\ g^{-1}$ of dry weight) whereas the concentration of Cu was increased only in the eulittoral representative of the Gelidiales. A similarity in metal levels was likewise observed among representatives of the Ceramiales, *Ceramium ciliatum* and *Laurencia obtusa*. The Mn and Zn concentrations in both species were similar and the lowest amongst the red algae tested, with the exception of *Porphyra*. These two related species showed similarities in the levels of Cd, Cu and Pb.

All the red algae investigated showed a low Cd and relatively high Pb content. It was obvious that, like in the brown algae, the Mn content of the thalli is likely to be the main distinguishing character between the taxonomic groups.

Chlorophyceae

Only a few samples of green algae were examined (Table IV).

Ulvales. Mixed samples of several *Enteromorpha* species were collected in the eulittoral, where they form continuous mats. They occur throughout the year in several succeeding generations. Several species were present in such mixed populations, including *Enteromorpha intestinalis* (L.) Link., *E. prolifera* (O. F. Müll.) J. Ag., *E. clathrata* (Roth.) Grev. Pronounced variations in the Mn content were obvious. Concentrations of this metal were highest in late spring (May/June). They were again lower in the sample from Rovinj than in that from Piran at a comparable season. Variations in the Zn content were less conspicuous. Both Mn and Zn contents were notably lower in autumn, coinciding with an increased Pb content. In the samples from Rovinj the Cd content was higher than in those from Piran.

In *Ulva rigida* collected in November from Rovinj the Cu content was slightly elevated.

Cladophorales. *Cladophora dalmatica* collected in eulittoral depressions during spring exhibited similar concentrations of trace metals to *Ulva rigida*.

Codiales. The two *Codium* species, *Codium vermilara* and *Codium effusum* differed in their Mn content. In samples from Piran the Mn content was about four times higher in *Codium effusum* than in *C. vermilara*. In samples of the latter from Rovinj, the Mn content was even lower. On the contrary, samples of *C. vermilara* from Rovinj showed both higher Zn and Pb contents than those from Piran.

Discussion and Conclusions

The data presented above give information on the trace metal content in some northern Adriatic seaweeds, that are common in the vegetation of this polluted area.

Analyses of seaweeds are important for the detection of metals discharged into the environment. Small-scale gradients can be detected since seaweeds, in general, reflect the ambient dissolved metal pollutants. Brown algae, in particular the fucoids, are useful biomonitors (Hägerhäll 1973, Haug *et al.* 1974, Morris and Bale 1975, Foster 1976, Cullinane and Whelan 1982).

Trace metal uptake by seaweeds is regarded as a two stage process (Skipnes *et al.* 1975, Eide *et al.* 1980, Manley 1984). Firstly metals are transferred from seawater to the apparent free space (AFS) which includes the cell wall and intercellular spaces, by a rapid and reversible physico-chemical process. During the second stage they are transferred across the membranes by slower, regulated processes which require metabolic energy. The magnitude of the first-stage metal transfer is dependent on the concentration of metal ions in seawater and of diverse polyanions in the cell walls and intercellular spaces of marine algae. Algal polysaccharides, responsible for the first step of accumulation into the ASF, are alginic acid and diverse sulphated polysaccharides, containing fucose, in brown algae (Haug and Smidsröd 1967 a, Smidsröd and Haug 1968, Myklestad 1969) and sulphated galactans of the agar and carrageenan type in red algae. In green algae, *e.g. Enteromorpha*, the negatively charged polysaccharides of the cell wall are represented by sulphated glucuronoxylorhamnanes (Percival 1979). In most brown algae alginic acid is the dominant polyanion. It was proved that the uronic acid composition of alginates, which vary from species to species and even between the various tissues of the same plant, is responsible for the ion exchange properties (Haug and Smidsröd 1967 a, b, Smidsröd and Haug 1968). Alginic acid, rich in guluronic acid, has greater affinity for divalent cations than the mannuronic acid rich one.

Some authors (*e.g.* Higgins and Mackey 1987 a, Manley 1984) have determined the proportion of the AFS-bound cations in some Pacific brown algae [*Eklonia radiata* (C. Agardh) J. Agardh, *Macrocystis pyrifera* (L.) C. Agardh] and reported a high proportion of AFS bound Zn and Cd (90% and 60%), whereas it was less for Cu. If non-regulated processes of direct ion exchange play the major role in metal uptake these species are less appropriate as biomonitores of long-term metal pollution.

Studies of ionic selectivity of cell walls in Atlantic brown algae (fucoids) have proved that alginates have a low affinity for Zn though it is accumulated in high concentrations in brown algae (Bryan 1969, Haug *et al.* 1974, Skipnes *et al.* 1980). Long-term studies of ionic uptake under field conditions (Myklestad *et al.* 1978) have revealed that the uptake of Zn and Cd is regulated and hence metabolically mediated. The accumulation exhibited winter minima and summer maxima. The Pb content, which is, on the other hand, governed by an ion exchange process with the algal polysaccharides, showed no seasonal variations. It was, however, obvious that different metals showed different patterns of uptake in fucoid algae. Haug *et al.* 1974, Myklestad *et al.* 1978 and Eide *et al.* 1980 suggested a predominantly regulated uptake of Zn and Cd into Atlantic fucoids. There is, on the other hand, only sparse information of the mechanism of Mn uptake into marine algae. Morris and Bale (1975) showed that accumulation factors for Mn were not consistent over wider ranges of external concentrations, which indicated a regulated nature of its uptake into *Fucus vesiculosus* L.

For higher plants, however, ample evidence exists that the uptake of Mn is metabolically mediated (Moore 1972).

Considering long-term integrated changes of metal pollution in the marine environment, it became obvious, that they are reflected in the concentrations of those metals in marine algae, which were accumulated by a regulated uptake, resulting in strong binding. Zinc, Mn and also Cd concentrations in marine algae are the results of irreversible accumulation rather than an equilibrium between the plants and their surrounding water. Lead and Cu are accumulated apparently by different, merely physico-chemical mechanisms.

In Adriatic water the Zn content was higher than that of Mn at Piran and in the harbour of Rovinj, whereas Mn was slightly higher than Zn in the two unpolluted sites of the Rovinj area (Table III). In spite of similar metal levels or even a higher Zn than Mn content in seawater, the concentration of Mn in algal thalli usually exceeded by two for four times that of Zn.

The results presented in Table IV indicate that accumulation of trace metals varied between species from the same habitat. In several cases similarities in metal accumulation within the taxonomic groups were obvious.

In all the Adriatic material investigated here, variations in the Mn content were most pronounced. This relates to differences between the species as well as to seasonal- and habitat-conditioned variations. Variations in the Zn content were less pronounced and did not always follow the same trend as the Mn content. No definite trends were found for variations of the toxic metals (Cu, Pb, Cd).

Similarities between related species regarding their affinities for trace metals, first of all for Mn, were most obvious among Adriatic red algae, such as among the erect corallines or representatives of the orders Gelidiales and Ceramiales. Among the Gelidiales, elevated concentrations of the toxic metals Cu and Pb were characteristic. It was obvious, however, that *Porphyra leucosticta* deviated in the levels of most metals from the rest of the red algae investigated. Among the brown algae representatives of the Dictyotales and Sphacelariales tended to accumulate more of the toxic metals Cd, Cu and Pb than representatives of the Fucales. A high Mn level found in *Padina pavonica* is also noteworthy.

Unlike red and brown algae, *Codium* species revealed a pronounced difference in their Mn and Pb content, whereas in most of the Adriatic algae investigated, taxonomic relationships seemed connected with similar affinities for trace metals.

Differences between species regarding affinities for particular elements were noted by Black and Michell (1952). Recent investigations carried out by Mariani *et al.* (1985, 1990) revealed a structural basis for cation distribution in the cell walls of Adriatic brown, red and green algae. Differences between the species investigated are based on the localisation of the cation-binding sites in the apoplast. The basis for differences found between the Adriatic species investigated *viz.* the similarities within taxonomically related groups, might be due to a similar structure of the cell walls and the chemical characteristics of their negatively charged polysaccharides.

Further observed variations relate to the sampling sites. Differences in accumulation were found within the same species collected at Piran and at Rovinj. In general, a lower Mn level was found in the samples from Rovinj than in those of the same species from Piran (*e.g.* in *Fucus virsoides*, *Cystoseira corniculata*, *Dictyota dichotoma*, *Padina pavonica*, *Scytosiphon lomentaria*, *Corallina officinalis*, *Enteromorpha* spp., *Codium vermilara*). Differences in the Zn content were less conspicuous and in some cases Zn was even higher in the samples from Rovinj than in those of the same species from Piran (*e.g. Scytosiphon lomentaria*, *Codium vermilara*).

It is probable, however, that the combined effects of long-lasting pollution are responsible for these differences. Our analyses of seawater at Piran and Rovinj, carried out in spring, revealed a higher Zn content at Piran. These data are not appropriate to yield an explanation for differences in Mn accumulation in seaweeds from the two areas. Due to the vicinity of industrial effluents from the harbours of Isola, Koper and Trieste, Piran receives a more continuous supply with heavy metals than Rovinj. Furthermore, in the mixed outlets of industrial and domestic effluents which predominate at Rovinj the chelating effects of substances from domestic sewage must also be considered (*e.g.* amino acids). Organometallic chelates cannot be taken up by algae.

Ecological implications in heavy metal uptake were demonstrated by Bryan and Hummerstone (1973), Fuge and James (1973, 1974), Munda (1979, 1982, 1984), Kremer and Munda (1982). Differences between algal samples from the two areas might relate not only to different metal levels, but also to different ecological conditions, such as tidal range, temperature and salinity regimes, dissolved nutrients as well as different geological structures (flysch at Piran and limestone at Rovinj). The Mn content in seawater could also be effected by nutrient enrichment in the water and sediments (Hunt and Kelly 1988).

In connection with ecological implications of metal uptake it is also noteworthy that a lower Mn and Zn accumulation was found in sublittoral than in eulittoral fucoids (Table IV).

The extremely high Cu and Pb levels, found in *Scytosiphon lomentaria* from Rovinj could be due to metal pollution in the harbour where the plants were collected. Copper concentrations were elevated, while those of Pb were at a normal level, when water samples were taken. The ability to accumulate metals obviously increases towards the end of the vegetative period of this spring species.

Seasonal variations were studied in the eulittoral fucoid *Fucus virsoides* and revealed a peak in Mn content in August. Seasonal variations were less conspicuous for the rest of the metals measured. Among the green algae, seasonal variations in trace metal content were obvious in mixed samples of diverse *Enteromorpha* species, possibly related to the appearance and disappearance of succeeding generations.

Seasonal variations in trace metal content in marine algae might be related to growth rates and metabolic processes. As shown by Rice (1984), the concentrations of non-essential metals in algae decrease with increasing growth rate, while that of the essential micronutrient Mn increases. The summer peaks in Mn content in the Adriatic algae may agree with this finding. Also Myklestad *et al.* 1978 and Eide *et al.* 1980 reported that Zn and Cd are taken up quickly in summer and slowly in winter, while for Cu and Pb no seasonal variations were obvious. Hägerhäll (1973) found maxima in Zn, Cu and Cd accumulation in summer (July/August) in the brown alga *Ascophyllum nodosum* (L.) Le Jol. The author suggested that seasonal variations in metal uptake might be related to seasonal variations in metabolic rate which affects the active ion uptake.

Since Mn uptake is an active, energy requiring process, we might assume that its uptake is dependent on the metabolic state of the plants, which varies during the seasons and also in the different sampling sites (Piran, Rovinj) and depths.

We conclude that the Adriatic seaweeds tend to accumulate more Mn than Zn and that Mn levels in the thalli are the main distinguishing character between the algal samples investigated, both from ecological and taxonomic aspects. It was obvious, that plants have different requirements for Zn than for Mn (Collins 1981, Atri 1983). As observed by Collander (1941) the rate of Mn uptake varies considerably between species.

During previous experiments on the accumulation of single metals and metals pairs in *Fucus vesiculosus* (Munda and Hudnik 1986) it became obvious that Mn and Zn were accumulated to the same levels when both metals were added to the media in equal concentrations, which were much higher than those in natural habitats. It seems likely, however, that seaweeds have higher requirements for Mn than for Zn. When both metals were present in high concentrations, plants were saturated with Mn according to their metabolic requirements. Both essential metals, Mn and Zn, showed similar concentration factors. Under field conditions, on the other hand, the plants revealed an intensified Mn accumulation. At the low Mn concentrations in Adriatic water high concentration factors are needed to satisfy the metabolic requirements of the seaweeds.

References

Atri, F. R. 1983. *Schwermetalle* in *Wasserpflanzen*. 55. Fischer Verlag, Stuttgart – New York. 105 pp.

Black, W. A. P. and R. L. Michell. 1952. Trace elements in the common brown algae and in sea water. *J. mar. biol. Ass. U.K. 30*: 575 – 584.

Branica, M. 1978. Distribution of ionic Cu, Pb, Cd and Zn in the Adriatic water. *Thalasia Jugosl. 14*: 151 – 155.

Bryan, G. W. 1969. The absorption of zinc and other metals by the brown seaweed *Laminaria digitata*. *J. mar. biol. Ass. U.K. 49*: 225 – 243.

Bryan, G. W. and L. G. Hummerstone. 1973. Brown seaweed as an indicator of heavy metals in estuaries of south-west England. *J. mar. biol. Ass. U.K. 53*: 705 – 720.

Bubić, S., L. Sipos and M. Branica. 1979. Cruises of the Research Vessel "Vila Velebita" in the Kvarner Region of the Adriatic Sea. V. Electroanalytical determination of ionic zinc, cadmium, lead and copper in seawater. *Thalassia Jugosl. 15*: 89 – 111.

Collander, R. 1941. Selective absorption of cations by higher plants. *Plant. Physiol. 16*: 691 – 720.

Collins, J. C. 1981. Zinc. Effect of heavy metal pollution on plants. 1. Effects of trace metals on plant function. *In*: (N. W. Lepp, ed.) *Pollution monitoring series 5*: 165 – 170, London – New York.

Cullinane, J. P. and P. N. Whelan. 1982. Copper, cadmium and zinc in seaweeds from the south coast of Ireland. *Mar. Poll. Bull. 13*: 205 – 208.

Eide, L., S. Myklestad and S. Melsom. 1980. Long-term uptake and release of heavy metals by *Ascophyllum nodosum* (L.) Le Jol. (Phaephyceae) in situ. *Environ. Poll. Ser. A. Ecol. Biol. 23*: 19 – 28.

Foster, P. 1976. Trace metals in brown algae, a comparison of two analytical methods. *Chemosphere 3*: 151 – 154.

Fuge, R. and K. H. James. 1973. Trace metal concentrations in brown seaweeds. Cardigan Bay, Wales. *Marine Chem. 1*: 281 – 293.

Fuge, R. and K. J. James. 1974. Trace metal concentrations in *Fucus* from the Bristol Channel. *Mar. Poll. Bull. 5*: 9 – 15.

Haug, A. and O. Smidsröd. 1967 a. Strontium, calcium and magnesium in brown algae. *Nature 215*: 1167 – 1168.

Haug, A. and O. Smidsröd. 1967 b. Strontium-calcium selectivity of alginates. *Nature 215*: 757.

Haug, A., S. Melsom and S. Omang. 1974. Estimation of heavy metal pollution in two Norwegian fjord areas by analysis of the brown alga *Ascophyllum nodosum*. *Environ. Poll. 7*: 179 – 192.

Hägerhäll, B. 1973. Marine botanical-hydrographical trace element studies in the Öresund Area. *Bot. Mar. 16*: 53 – 64.

Higgins, H. W. and D. J. Mackey. 1987 A. Role of *Eklonia radiata* (C. Ag.) J. Agardh in determining trace metal availability in coastal waters. I. Total trace metals. *Aust. J. Mar. Freshw. Res. 38*: 307 – 315.

Higgins, H. W. and D. J. Mackey. 1987 B. Role of *Eklonia radiata* (C. Ag.) J. Agardh in determining trace metal availability in coastal waters. II. Trace metal speciation. *Aust. J. Mar. Freshw. Res. 38*: 317 – 328.

Hunt, C. D. and J. R. Kelly. 1988. Manganese cycling in coastal regions. Response to eutrofication. *Estuarine coastal and shelf science 26*: 527 – 559.

Kremer, B. P. and I. M. Munda. 1982. Ecophysiological studies of the Adriatic seaweed *Fucus virsoides*. *Mar. Ecol. 3* (1): 75 – 95.

Manley, S. L. 1984. Micronutrient uptake and translocation by *Macrocystis pyrifera* (Phaeophyta). *J. Phycol. 20*: 192 – 201.

Mariani, P., C. Tolomio and P. Braghetta. 1985. An ultrastructural approach to the adaptive role of the cell wall in the intertidal alga *Fucus virsoides*. *Protoplasma 128*: 208 – 217.

Mariani, P., C. Tolomio, B. Baldan and P. Braghetta. 1990. Cell ultrastructure and cation localization in some benthic marine algae. *Phycologia 29*: 253 – 262.

Moore, D. P. 1972. Mechanism of micronutrient uptake by plants. *In*: *Micronutrients in Agriculture*: Soil Sci. Soc. Amer. Inc., Madison. pp. 171 – 993.

Morris, A. W. and A. J. Bale. 1975. The accumulation of cadmium, copper, manganese and zinc by *Fucus vesiculosus* in the Bristol Channel. *Estuarine and coastal marine science 3*: 153 – 163.

Munda, I. M. 1979. Temperature dependence of Zn uptake in *Fucus virsoides* (Don.) J. Ag. and *Enteromorpha prolifera* (O. F. Müll.) J. Ag. from the Adriatic Sea. *Bot. Mar. 22*: 149 – 152.

Munda, I. M. 1982. The effect of different pollutants on benthic marine algae. *Rapp. Comm. Int. Mer. Médit.* VI. *Journ. Poll., Cannes*: 721 – 726.

Munda, I. M. 1984. Salinity dependent accumulation of Zn, Co and Mn in *Scytosiphon lomentaria* (Lyngb.) Link. and *Enteromorpha intestinalis* (L.) Link. from the Adriatic Sea. *Bot. Mar. 27*: 371 – 376.

Munda, I. M. and V. Hudnik. 1986. Growth response of *Fucus vesiculosus* to heavy metals, singly and in dual combinations, as related to accumulation. *Bot. Mar. 29*: 401 – 412.

Myklestad, S. 1979. Ion exchange properties of sulphated polysaccharides in brown algae. *Proc. Int. Seaweed Symp. 6*: 545 – 552.

Myklestad, S., I. Eide and S. Melsom. 1978. Exchange of heavy metals in *Ascophyllum nodosum* (L.) Le Jol. in situ by means of transplantion experiments. *Environ. Poll. 16*: 277 – 284.

Percival, E. 1979. The polysaccharides of green, red and brown seaweeds: their basic structure, biosynthesis and function. *Br. Phycol. J. 14*: 103 – 117.

Phillips, D. J. M. 1977. The use of biological indicator organisms to monitor trace metal pollution in marine and estuarine environments – a review. *Environ. Poll. 13*: 281 – 317.

Ramon, F. and I. Friedmann. 1965. The gametophyte of *Padina* in the Mediterranean. *Proc. Int. Seaweed Symp. 5*: 183 – 196.

Rice, D. L. 1984. A simple mass transport model for metal uptake by marine macroalgae growing at different rates. *J. Exp. Mar. Biol. Ecol. 82*: 175 – 182.

Rice, D. L. and B. E. Lapointe. 1981. Experimental and outdoor studies with *Ulva fasciata* Delile. II. Trace metal chemistry. *J. exp. mar. biol. ecol. 54*: 1 – 11.

Skipnes, D., T. Roland and A. Haug. 1975. Uptake of zinc and strontium by brown algae. *Physiol. Plant. 34*: 314 – 320.

Smidsröd, O. and A. Haug. 1968. Dependence upon uronic acid composition of some ion-exchange properties of alginates. *Acta Chem. Scand. 22*: 1989 – 1997.

Strohal, P., D. Huljev, S. Lulić and M. Picer. 1975. Antimony in coastal marine environment, North Adriatic. *Estuarine and coastal marine science 3*: 119 – 123.

Škrivanić, A. and A. Barić. 1979. Cruises of the research vessel "Vila Velebita" in the Kvarner Region of the Adriatic Sea. III. Hydrographic conditions. IV. Distribution of the primary nutrients. *Thalassia Jugosl. 15*: 35 – 60 and 61 – 88.

Tušnik, P., V. Turk and R. Planinc. 1989. Assessment of the level of pollution of the coastal sea in the eastern part of the Gulf of Trieste. *Biol. vestnik 37* (4): 47 – 64.

Botanica Marina
Vol. 34, pp. 251–260, 1991

The Nutritional Eco-physiology of *Chaetomorpha linum* and *Ulva rigida* in Peel Inlet, Western Australia

P. S. Lavery and A. J. McComb*

Dept of Botany, University of Western Australia, Nedlands, 6009, Australia

* *School of Biological and Environmental Sciences, Murdoch University, Murdoch, 6150, Australia*

(Accepted 9 February 1990)

Abstract

The uptake rates and critical tissue concentrations of nitrogen and phosphorus were determined for *Chaetomorpha linum* and *Ulva rigida*, the dominant algae in Peel Inlet, Western Australia. Both species had rate-saturating mechanisms of phosphate uptake described by Michaelis-Menten type functions; *C. linum* had the faster uptake rate (667 c.f. 272 µg P g dwt^{-1} h^{-1}) although *U. rigida* had a lower half-saturation value. Both species displayed linear relationships between ammonium uptake rates and substrate concentrations with *C. linum* having the greater slope (4.4 c.f. 1.7). *Chaetomorpha linum* also had a linear increase in uptake rate with increasing concentration of nitrate, but *U. rigida* showed rate-saturating kinetics; below 750 µg L^{-1}, *U. rigida* had the higher rate of uptake. *Ulva rigida* had critical tissue nitrogen and phosphorus concentrations of 20 and 0.25 mg g dwt^{-1} respectively. Corresponding concentrations for *C. linum* were 12 and 0.5 mg g dwt^{-1}. *Ulva* is frequently nitrogen limited during spring in Peel Inlet, reflecting the high nitrogen requirements of this plant compared to *Chaetomorpha* as well as the reduced ability of *Ulva* to store nutrients over winter. The growth rate of *Ulva* is negative in winter due to low salinity, temperature and irradience, and this results in negligible biomass at the time of highest inorganic nitrogen concentration. *Chaetomorpha linum* persists over winter allowing it to take advantage of elevated nutrient concentrations at the time; nutrients stored during winter support high growth rates in summer. The maximum tissue nitrogen levels recorded for *Ulva* were only 47% higher than the critical concentration while maximum tissue phosphorus concentration was almost 9 times the critical level. This seasonal limitation by nitrogen supports an earlier hypothesis that production in the system may be nitrogen limited in summer. It was concluded that differences in nutrient requirements and aquisition strategies afforded *C. linum* greater competitive potential for utilizing the seasonal and short-term increases in nutrient concentrations that occur in Peel Inlet.

Introduction

This is the second of two papers dealing with macroalgal nutrition in Peel Inlet, Western Australia. In the first (Lavery and McComb 1990), seasonal changes in tissue nutrient concentrations and the sources of macroalgal nutrition were examined. In this paper physiological aspects of the nutritional biology of *Chaetomorpha linum* Kütz. and *Ulva rigida* C. Agardh are compared, and the rates of nitrogen and phosphorus uptake and the critical tissue concentrations for nitrogen and phosphorus are reported.

The effects of eutrophication in promoting macroalgal growth have been well documented (Sawyer 1965, Waite and Mitchell 1972. Harlin and Thorne-Miller 1981, Kemp *et al.* 1983, Rosenberg 1985, Sfriso *et al.* 1987) and nutrient limitation has been proposed as a factor controlling macroalgal growth in Peel Inlet (Lukatelich and McComb 1986). In estuaries such as Peel Inlet, where nutrient input is highly seasonal, it is possible that species with the fastest rates of uptake have an advantage in utilizing periodic nutrient pulses. It has been speculated that this may account, in part, for the dominance of *Chaetomorpha linum* in that system (Lavery 1989, Lavery *et al.* 1991). Alternatively, uptake rates may be of little importance in determining the success of different species which have different nutrient requirements or strategies of nutrient utilization, such as luxury consumption or preferences for different ionic forms of the same nutrient. The physiological attributes of *U. rigida* and *C. linum* are examined in this paper, to determine the nutritional characteristics of these species and to assess any competitive advantages these may confer.

Material and Methods

Nutrient uptake rates

Uptake rates of phosphate, ammonium and nitrate + nitrite ('nitrate') were determined from nutrient depletion rates during short-term, 'batch culture' experiments. Due to logistical constraints it was not possible to conduct continuous flow experiments. Probyn and Chapman (1982), Gilbert and McCarthy (1984) and Harrison *et al.* (1989) point out that short-term, batch experiments tend to exaggerate normal uptake rates, particularly of ammonium. Alternatively, it can be argued that short-term experiments do not exaggerate uptake rates, but rather they disclose initial uptake rates which are often higher than rates determined over extended periods. Wallentinus (1984) suggests that short-term uptake rates should not be regarded as over-estimation but rather as an expression of the potential for an alga to utilize available nutrients. Since one of the primary aims of this work was to establish the potential rates of nutrient uptake for both species, the experimental design used was considered appropriate.

Collection

Algae were collected four to six weeks before use in experiments. Small, whole thalli of *Ulva rigida* and strands of *Chaetomorpha linum*, generally free of sediment, diatoms and other epiphytes, were collected from Boodalan Island in Peel Inlet (Fig. 1). Material was stored in estuary water and returned to the laboratory for preconditioning.

Preconditioning

Plants were preconditioned for four to six weeks at 25 °C and 250 μmol m^{-2} s^{-1} PAR (Daylight incandescent, 60 W bulbs and Cool White fluorescent 60 W tubes; 12/$\overline{12}$-L/D cycle). Enriched seawater medium (Table I) was used, omitting soil extract, since preliminary work showed that *C. linum* grew poorly in a number of artificial media. The medium was changed daily during preconditioning. A weight to volume ratio of approximately 1.0 g L^{-1} (fresh weight) was maintained throughout the experiment to reduce factors such as nutrient limitation and self-shading (Littler 1979). Plants used to determine phosphorus uptake rates were preconditioned with non-limiting concentrations of NH_4Cl (2 mg N L^{-1}) but no phosphorus, while those used in nitrogen uptake experiments were grown in non-limiting KH_2PO_4 (0.5 mg P L^{-1}) but no nitrogen.

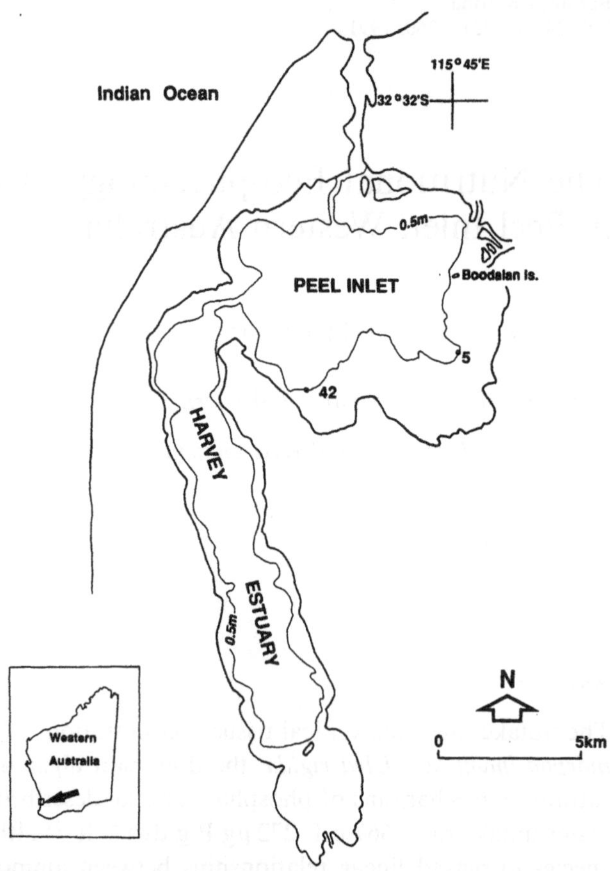

Fig. 1. The Peel-Harvey Estuarine System, showing its location in Western Australia and the sampling sites used in the study.

Table I. Enriched seawater medium used in salinity experiments. The recipe is a modification of SWM medium (McLachlan 1973).

Supplement	Concentration (L^{-1} filt. seawater)
Na NO_3	2 mg
KH_2PO_4	1 mg
Na_2SiO_3	24.4 mg
Fe. EDTA	0.692 mg
Trace metal solution*	TMS-1
Thiamine. HCl	100 μg
Biotin	0.5 μg
Cyanocobalamin	0.5 μg

* TMS-1 contains: 2.28 mMol Zn; 0.55 mMol Mn; 0.48 mMol Mo; 17,7 μMol Co; 19 μMol Cu; 4.32 mMol B; Chelate: Metal 2:1.

Experimental design

Rates of nitrogen and phosphorus uptake were determined over five treatment concentrations with non-limiting concentrations of all other nutrients in the medium. Phosphate-P treatment concentrations were in the range of approximately 0.05 – 1.20 mg L^{-1} as KH_2PO_4, with 2.0 mg N L^{-1} as NH_4NO_3. Nitrate and ammonium treatment concentrations were approximately 0.07 – 1.5 mg N L^{-1} as $NaNO_3$ and NH_4Cl respectively with 0.5 mg L^{-1} phosphate-P. Three replicate samples were used for each treatment.

Experiments were conducted in 2 L Ehrlenmeyer flasks, under the conditions described above, with a small pump (EKTA, Italy) circulating water through the flasks (2 L min^{-1}) to reduce boundary layer effects. Water was drawn from the bottom of the flask through a plastic tube (2 cm diameter) connected to the pump and returned by tube to the top of the flask. Inlet and outlet pipes were capped with perforated tubing to prevent damage to the algae. Dye tests revealed that the contents of each flask were completely mixed within 5 s. The medium was gently aerated with compressed, oil-free air. Plastic tubing was secured inside the flask and connected to an external syringe to withdraw water samples. All equipment was thoroughly washed and rinsed in de-ionized water.

Water samples were collected every 15 min for 4 h following the addition of algae to the treatment flasks. Samples were collected by syringe, filtered (GFC, Whatman No. 1), stored in polyethylene bags (Whirl-pac, NASCO, USA) and frozen until analysed. Controls were established as described above, but lacking macroalgal material, to estimate nutrient depletion due to microbial or other activity. Algal dry weights were determined after drying to constant weight at 70 °C.

Water samples were analysed for phosphate, ammonium or nitrate using standard methods (Lukatelich and McComb 1986), and the rates of depletion used to calculate rates of nutrient uptake. Regressions were performed on the first linear portion of the depletion curves and the negative of the slope taken as rate of uptake. Regressions were only accepted if statistically singificant ($\alpha = 0.05$) with a minimum of 4 data points.

The data were fitted to either linear or Michaelis-Menten type functions describing nutrient uptake. For the latter, the three linear transforms described by Dowd and Riggs (1965) were used to estimate V_{max} (maximum rate of uptake) and K_s (half saturation constant). The transformation which most accurately fitted the data was used.

In situ *nutrient uptake rates*

On a single occasion (November 1988), *in situ* rates of nutrient uptake were determined for plant material at Stations 5 and 42 in Peel Inlet (Fig. 1). The problems associated with measurement and interpretation of *in situ* nutrient uptake rates have been discussed by McCarthy (1981), and include inability to accurately define the physical and chemical environment and the resident population. For these reasons the assessment of *in situ* uptake rates in this study was intended only as a crude estimation. Phytoplankton and faunal contamination were reduced and controls established to account for these factors. Nitrogen and phosphorus concentrations were raised to easily detectable, but ecologically relevant concentrations.

Plastic bags were used as experimental vessels, allowing high light penetration and mixing of the internal water, through the action of water movement on the bags. Each was filled with 2 – 3 litres of treatment medium (0.48 μm filtered estuary water supplemented with phosphate and ammonium). Three treatment concentrations of each nutrient were used: 98, 170 and 240 μg L^{-1} phosphate-P (as KH_2PO_4) and 20, 180 and 220 μg L^{-1} ammonium-N (as NH_4Cl) with four replicates of each treatment. Whole thalli of *U. rigida* and clumps of *C. linum*, both of known fresh weight, were added to each treatment. Controls were established as described above but lacking algal material.

An initial sample of medium was collected from each bag for nutrient analysis, and the bags sealed, weighted and placed in the water column. They were retrieved and re-sampled after one hour.

Water samples were removed by syringe, and processed as described above. Algal material was returned to the laboratory for dry weight determination at 70 °C, material being dried to constant weight. Rates of uptake were calculated from the change in total nutrient content of each bag less the mean of any nutrient depletion in the control bags.

Critical tissue nutrient concentrations

Collection and medium

Plant material was collected from Peel Inlet and grown in filtered (0.45 μm) seawater medium. The medium was as described above, with either nitrogen or phosphorus additions. To ensure minimal nitrogen and phosphorus content of the base seawater, a single collection was made in mid-summer, when ambient nutrient levels were lowest (7 μg L^{-1} phosphate-P, 3 μg L^{-1} nitrate-N and 8 μg L^{-1} ammonium-N), stored in a dark cold-room, and used for all critical nutrient experiments.

Preconditioning

Preliminary experiments revealed that high nutrient concentrations (> 100 µg L^{-1} P and > 500 µg L^{-1} N) encouraged epiphytic contamination of *C. linum* by a small filamentous chlorophyte. Attempts at inducing sporulation to gain uni-algal cultures of *C. linum* were unsuccessful, as is often the case for free-floating algae (Norton and Mathieson 1983). Other options such as using selective culture media were of limited potential as host and epiphyte were chlorophytes and tended to respond in the same way to treatments. Surface sterilization was the most successful method for reducing epiphytic contamination, using a modification of the technique of Lawler *et al.* (1989). Material was dipped into 70% ethanol for 5 s, and held in de-ionized water for 5 min. It was returned to aquaria and dead or unhealthy strands removed after 7 days. The technique was effective in removing epiphytic contamination in about 30% of cases (Lavery 1989). Small strands of *C. linum* and disks of *U. rigida* (cut from the sub-marginal areas of large thalli) were preconditioned in seawater medium (Table I) at 25 °C and 250 µmol m^{-2} s^{-1} PAR (12/12 h-L/D) for 7 days prior to experimentation.

Experimental design

Determinations of critical tissue concentrations were conducted in two stages. In the initial stage 1 g samples of *C. linum* or single disks of *U. rigida* (approx. 0.1 g) were grown in 1 litre Ehrlenmeyer flasks at one of five nitrogen or phosphorus concentrations. All flasks were placed in growth cabinets, at the temperature and light levels described above, and gently mixed and aerated with oil-free compressed air. The flasks were randomly re-assorted in the cabinets each day to provide each with an average of any variation in the climate. Nitrogen treatments were in the range 0.0 – 1.5 mg L^{-1} added NH_4-N with 0.5 mg L^{-1} P as PO_4. Phosphorus treatments were in the range 0.0 – 0.25 mg L^{-1} added PO_4-P and 2.0 mg L^{-1} N as NH_4. Three replicates were established for each treatment. The medium was changed daily, and fresh weights of blotted material determined at the beginning and end of each 7 day period.

After 14 days the material was harvested back to the original weight (*U. rigida* material was re-cut into disks) and the three replicates returned to the treatment conditions. After a further 7 days the algae were weighed and the relative growth rates calculated from changes in biomass (Hunt 1978). Algal tissue was dried and analysed for total nitrogen or phosphorus, as described by Lukatelich and McComb (1986).

This two-stage approach had a number of advantages. Firstly, a period of less than three weeks was insufficient to establish detectable differences in growth rates between treatments. However, simply establishing replicates and allowing them to grow for more than three weeks resulted in nutrient limitation and self shading as the biomass increased. This problem was overcome by returning the treatments to a lower biomass for the final week and recording the growth rates over this period. Secondly, the high biomass of the first phase ensured that the tissue was not continuously bathed in high nutrient concentrations since even moderate nutrient uptake rates would rapidly lower the nutrient concentration in the medium; a range of tissue nutrient concentrations was still generated through the differences in nutrient loading to the treatments. This prevented the germination and rapid growth of epiphytic material in much the same way that Ryther *et al.* (1981) recommended pulse nutrient additions to suppress epiphytic growth on cultured seaweeds.

Observed growth rates were plotted against tissue nutrient concentrations, and the critical nutrient concentration estimated as that below which maximum growth rates were not maintained. An exponential function was used to describe the data.

Results

Nutrient uptake rates

The depletion of nutrients from the medium was linear over the period 15 – 75 minutes, and this part of the curve was used to estimate uptake rates (Fig. 2).

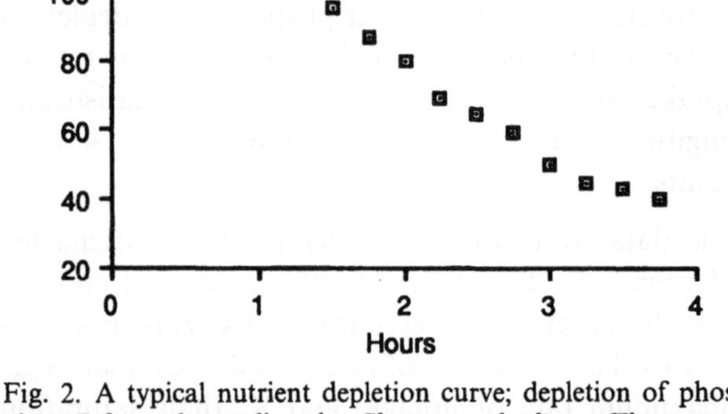

Fig. 2. A typical nutrient depletion curve; depletion of phosphate-P from the medium by *Chaetomorpha linum*. The rate of uptake was estimated from the linear regression of the first four points.

Chaetomorpha linum demonstrated greater phosphate and ammonium uptake rates than *Ulva rigida*, while the latter had higher nitrate uptake rates at low concentrations (Table II; Fig. 3).

Both species had rate-saturating mechanisms of phosphate uptake (Fig. 3) with the V_{max} for *C. linum* more than twice that of *U. rigida*. The half-saturation constant (K_s) of *C. linum* was almost three times that of *U. rigida*. The data were best estimated by the Michaelis-Menton-type functions described in Table II.

Both species displayed linear relationships between the rate of ammonium uptake and substrate concentration (Fig. 3). The slope was greater for *C. linum* (4.4) than *U. rigida* (1.73). Data conformed to the linear functions described in Table II.

The rate of uptake of nitrate by *U. rigida* was saturated as substrate concentrations increased and was described by Michaelis-Menten-type kinetics (Table II). This contrasted with the linear response of *C. linum* (Fig. 3; Table II). *Ulva rigida* exhibited greater rates of uptake up to a concentration of 750 µg N L^{-1}, while at higher concentrations *C. linum* showed a faster rate.

Considerable differences were found between the estimations of the Michaelis-Menten constants K_s and V_{max} using the three linear transforms outlined by Dowd and Riggs (1965). In all cases the S against S/V transformation yielded the greatest coefficient of regression, and most closely modelled the data.

The *in situ* rates of uptake of both nutrients varied almost linearly with substrate concentration and approximated the laboratory uptake rates described above (Fig. 4). For example, at 223 µg L^{-1} phosphate-P the expected uptake rates by *C. linum* and *U. rigida* were 235 and 160 µg g dwt^{-1} h^{-1} respectively, while the observed rates were 200 $\pm$ 12 and 227 $\pm$ 21 µg g dwt^{-1} h^{-1} respectively. The *in situ* uptake rate of ammonium-N at 203 µg L^{-1} by *U. rigida* (356 µg g dwt^{-1} h^{-1}) was similar to that recorded during laboratory estimations (440 µg g dwt^{-1} h^{-1}), however the *in situ* rate exhibited by *C. linum* (500 $\pm$ 66 µg g dwt^{-1} h^{-1}) was only about half of that expected from laboratory determinations (1048 µg g dwt^{-1} h^{-1}). There was no significant difference between the uptake rates of the two species for either ammonium or phosphate ($p = 0.152$ and 0.254 respectively), while substrate concentration had a significant influence ($p = 0.0001$ in both cases).

Critial tissue nutrient concentrations

The growth rate of *U. rigida* was saturated at tissue nitrogen and phosphorus concentrations (critical con-

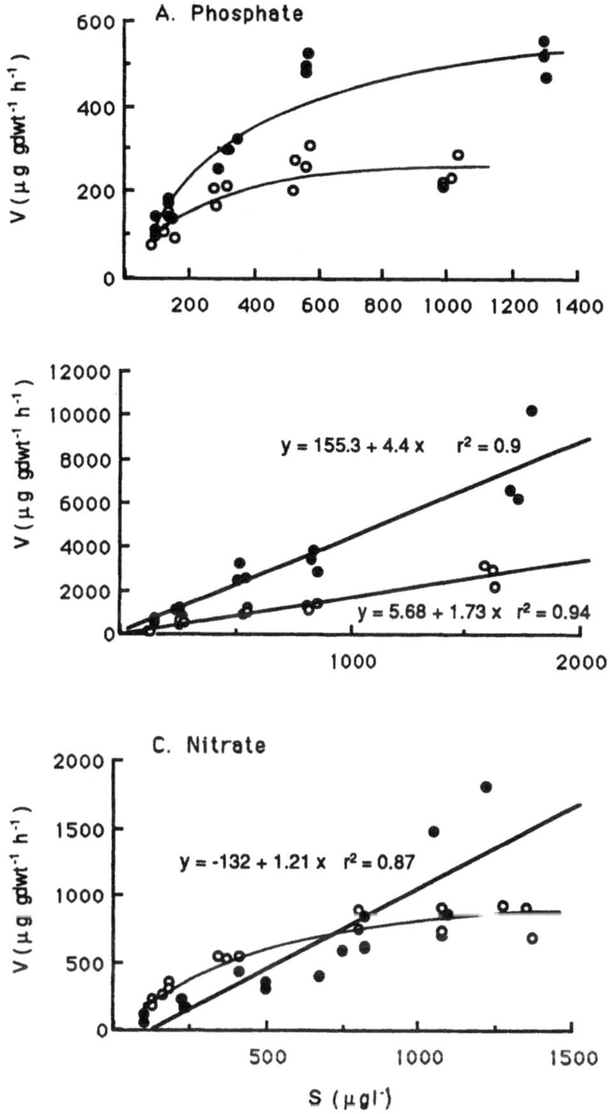

Fig. 3. Rate of (a) phosphate-P, (b) ammonium-N and (c) nitrate-N uptake by *U. rigida* (○) and *C. linum* (●) as a function of substrate concentration.

Table II. Uptake kinetics parameters of *C. linum* and *U. rigida* for phosphate, nitrate and ammonium from Michaelis-Menten (S against S/V) tranforms of raw data. Linear estimates are all significant ($p < .0005$ in all cases).

	C. linum		*U. rigida*	
	K_s (µg L^{-1})	V_{max} (µg g dwt^{-1} h^{-1})	K_s (µg L^{-1})	V_{max} (µg g dwt^{-1} h^{-1})
PO_4-P	321	667	113	272
NO_3-N	$V = -132 + 1.21\ S^*$		250 469	820 1193
NH_4-N	$V = 155 + 4.40\ S^*$		$V = 5.68 + 1.73\ S^*$	

* V = rate of nutrient uptake, and S = substrate concentration, from the linear regression of the data since no saturation kinetics were obtained.

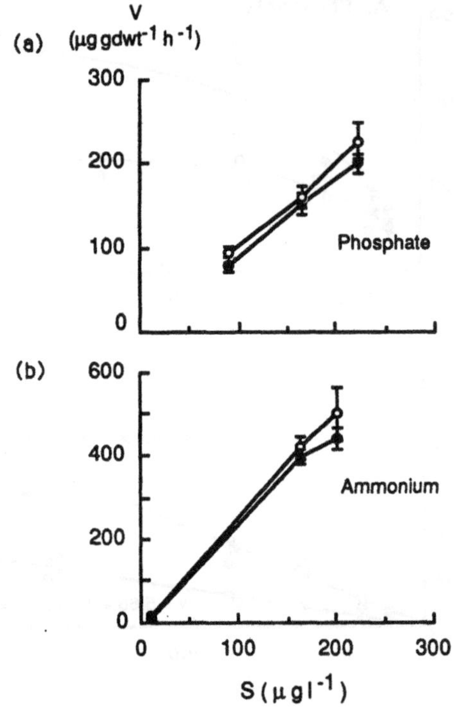

Fig. 4. Mean *in situ* uptake rates of (a) phosphate-P and (b) ammonium-N by *C. linum* (●) and *U. rigida* (o) at various substrate concentrations. n = 4 with bars representing the std error.

centrations) of 20 and 0.25 mg g^{-1} respectively (Fig. 5). Growth was most severely reduced by nitrogen deprivation; the zero nitrogen and phosphorus treatments had mean relative growth rates over the first week of 0.125 ± 0.004 and 0.207 ± 0.007 d^{-1} respectively. By the end of week three the zero phosphorus treatment had a relative growth rate 2.7 times that of the minus nitrogen treatment (0.024 compared to 0.009 d^{-1}).

Chaetomorpha linum had a lower nitrogen requirement than *Ulva rigida* but a higher phosphorus requirement. The growth rate of *C. linum* was saturated at tissue nitrogen and phosphorus concentrations (critical concentrations) of 12 and 0.5 mg g^{-1} respectively (Fig. 6). As for *U. rigida*, growth was most severely reduced by nitrogen deprivation; over week three the zero phosphorus treatment had a relative growth rate 2.27 times that of the minus nitrogen treatment (0.073 compared to 0.032 d^{-1}).

Discussion

Uptake rates

The observed nutrient uptake rates display a number of interesting features. Firstly, the differences in uptake rates were surprisingly large, given that the species are both simple, green algae; *Chaetomorpha linum* incorporated phosphate and ammonium at twice the

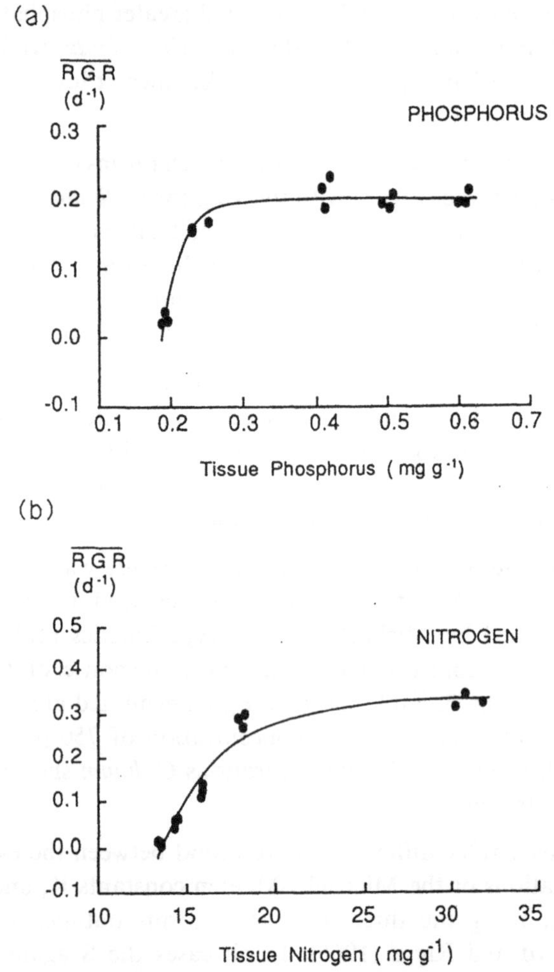

Fig. 5. Mean relative growth rates *U. rigida* as a function of (a) tissue phosphorus and (b) tissue nitrogen concentrations. Fitted curves are exponential approximations ($R^2 = 0.95$ in both cases).

rates of *Ulva rigida*. In the case of phosphate however, *U. rigida* had a lower K_s value indicating a greater affinity for phosphate at lower concentrations. The relative advantages of high V_{max} or low K_s will differ under various circumstances, with *C. linum* favoured during periods of short-term nutrient enrichment, such as discrete run-off events or high nutrient concentrations generated by sediment nutrient release. Interestingly, dense accumulations of *C. linum* have a demonstrated ability to induce sediment phosphorus flux through their effect of reducing oxygen levels in the overlying water, resulting in high P concentrations in the inter-algal water (Lavery and McComb 1991).

Species differences in uptake rates may be related to differences in morphology. Wallentinus (1984) observed that filamentous, delicately branched and monostromatic forms of algae generally exhibit the highest nutrient uptake rates as a result of their higher surface area to volume ratios and lack of internal and supportive tissue which cannot contribute to nutrient uptake. *Chaetomorpha linum*, an unbranched filamen-

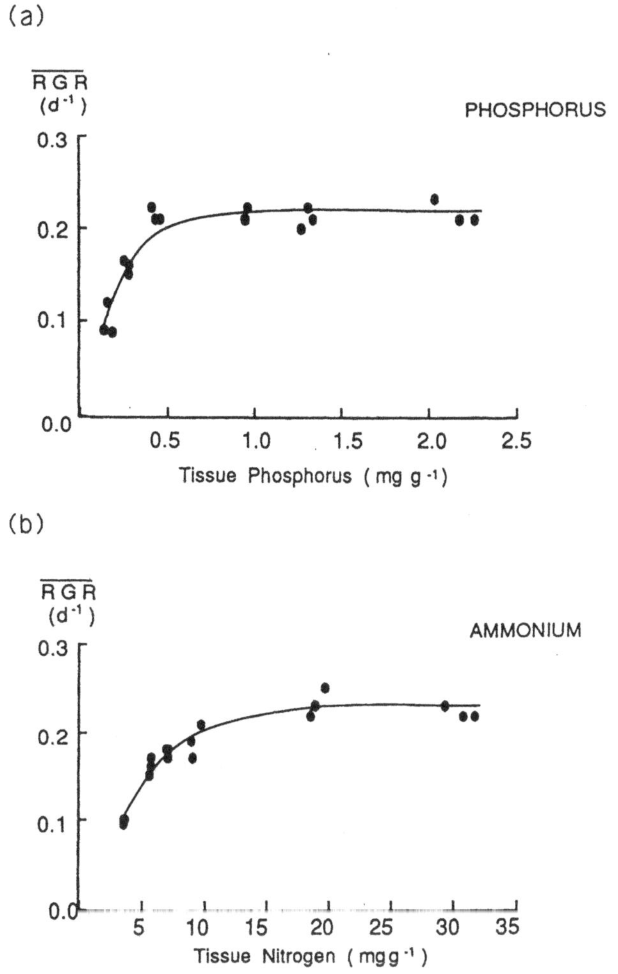

Fig. 6. Mean relative growth rates of *C. linum* as a function of (a) tissue phosphorus and (b) tissue nitrogen concentrations. Fitted curves are exponential approximations ($R^2 = 0.95$ in both cases).

tous-type alga, has a surface area : dry weight ratio 1.5 to 6 times that of the distromatic, membranous *U. rigida*, depending on which region of the *Ulva* thallus is measured (unpublished data). The area of thallus available for nutrient uptake is correspondingly higher for *C. linum* and this, apart from any physiological or biochemical considerations, could significantly enhance its rates of nutrient uptake.

Wallentinus (1984) lists a large range of published uptake rates of ammonium, nitrate and phosphate by macroalgae. Since that date nutrient uptake rates have been published by Probyn and McQuade (1985), Thomas and Harrison (1985), Thomas *et al.* (1987), Williams and Fisher (1985) and Duke *et al.* (1989). These literature values are generally lower than those reported here for *U. rigida* and *C. linum*, supporting the suggestion of Wallentinus (1984) that these simpler functional-form groups should have relatively high uptake rates. The K_s values for phosphate-P in this study are higher than many literature values, suggesting relatively poor efficiencies in incorporating nutrients at low concentrations. This may reflect localized adaptation to the relatively nutrient-rich estuarine environment, since most of the literature values relate to marine species.

Another interesting feature of the uptake experiments was the difference in nitrogen uptake mechanisms. Nitrate uptake by *U. rigida* conformed to a rate-saturating mechanism but for *C. linum* increased linearly with substrate concentration. A linear relationship was observed for ammonium uptake in both species. Linear relationships between S and V have been reported as components of di- or multi-phasic uptake mechanisms (D'Elia and DeBoer 1978, Haines and Wheeler 1978, MacFarlane and Smith 1982, Friedlander and Dawes 1985, Harrison *et al.* 1986, Thomas *et al.* 1987, Druehl *et al.* 1989). These multiphasic models of uptake generally consist of a high affinity, saturable uptake rate a low concentrations and a low affinity, non-saturable uptake mechanism, such as diffusion, at high concentrations. It is probable that the linear relationships between S and V observed in this study reflect similar di- or multiphasic uptake mechanisms. The high Y-intercept values for some of the regression suggest that the data conform to a multiphasic uptake mechanism rather than to simple linear functions which should theoretically approach the origin. Most examples of such mechanisms are for ammonium uptake and the evidence presented here represents one of only a few reported cases of a linear uptake phase for nitrate (Probyn and McQuade 1985, Harrison *et al.* 1986, Thomas *et al.* 1987, Druehl *et al.* 1989).

In all cases of saturable uptake rates, the S vs S/V linear transformation most accurately predicted V_{max} and K_s. This supports Dowd and Riggs' (1965) criticism of common use of Lineweaver-Burk transformations to estimate these constants. Nonetheless, the estimated values were usually higher than the raw data, or untransformed, uptake rates that were observed for a nutrient at a particular concentration. Thus, while the fitted curves accurately predict V over the experimental range of nutrient concentrations, extrapolation to higher substrate concentrations will reduce the accuracy; this should be considered when incorporating such data into models of nutrient uptake. On the other hand, the experimental nutrient levels used here cover the range of concentrations likely to be encountered *in situ*. Functions describing the algal responses at laboratory concentrations may reasonably reflect the response of algae in the field. Certainly the phosphate uptake rates observed *in situ* were reasonably similar to those expected from laboratory results. Differences between *in situ* and laboratory estimates may be expected because of differences in nutritional histories of the plants, and environmental factors which may limit growth or metabolic processes (*e.g.* Harlin and Craigie 1978, D'Elia and DeBoer 1978, Hanisak and Harlin 1978, Topinka 1978, Thomas and Turpin 1980, Lapointe and Tenore 1981, Thomas *et al.* 1987).

The ability of *C. linum* and *U. rigida* to utilize both nitrate and ammonium would be particularly advantageous in Peel Inlet. High ammonium concentrations persist inside algal banks throughout the year, while there is a strong winter influx of nitrate into the system (Lavery and McComb 1991). Up to concentrations of about 750 μg N L^{-1} *U. rigida* would have uptake rates for nitrate greater than those of *C. linum*. However, the biomass of *U. rigida* is lowest during winter, the period of maximum nitrate concentrations, due to light, salinity and temperature limitation (Lavery 1989) and so the capacity for luxury uptake may be limited. *Chaetomorpha*, on the other hand, demonstrates a higher rate of ammonium uptake at all concentrations tested and maintains high biomass over winter, the time of maximum ammonium concentrations. Thus, depending on the source available, either species would be able to take advantage of sudden high nitrogen concentrations although the persistence of *Chaetomorpha* biomass over winter would offer it an advantage in storing excess nutrient reserves.

Kautsky (1982) suggested that differences in the nutrient uptake parameters of species could determine the outcome of inter-specific competition between macroalgae at different nutrient concentrations. The potential of *Chaetomorpha* and *Ulva* to incorporate nitrogen was compared by assuming that the uptake rates and nitrate or ammonium were not affected by either the concentration of the other N-source or the nutritional history of the plants. With these assumptions the potential for nitrogen uptake could be calculated as the instantaneous rate of ammonium-plus-nitrate uptake at the ambient concentrations. In this case the high Y-intercept values in the linear regressions were omitted as they are considered spurious and would have over-estimated the rate of ammonium uptake by *Chaetomorpha* at low concentrations. Ambient concentrations for 1988 were available through the regular monitoring programme of the Centre for Water Research. As expected, *Chaetomorpha* had greater uptake rates of total inorganic nitrogen throughout the year despite the higher nitrate uptake rates of *Ulva* at lower concentrations (Fig. 7).

Cladophora montagneana Kütz., the previously dominant alga in Peel Inlet, is reported to have lower uptake rates but higher affinties for phosphate, ammonium and nitrate than the species in this study

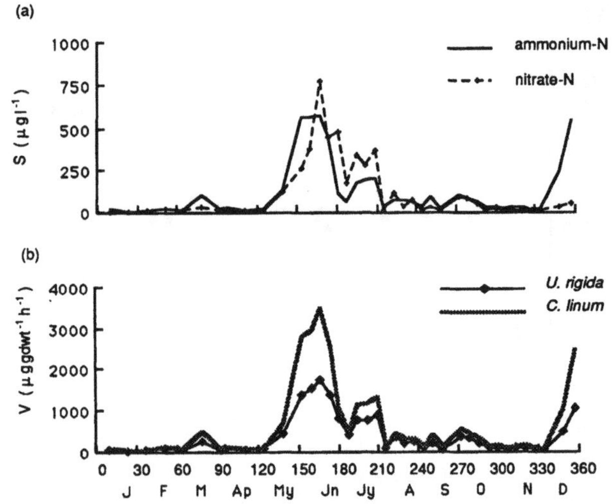

Fig. 7. The observed concentrations of ammonium-N and nitrate-N in at Station 7 in the centre of Peel Inlet (a), with (b) the calculated, uptake rates of these nutrients (NO_3 and NH_4) by *Ulva rigida* and *Chaetomorpha linum* based on rates determined in the laboratory.

(Gordon *et al.* 1981). The implications of the lower uptake rates for this species have been considered elsewhere (Lavery 1989, Lavery *et al.* 1991), and it suffices to say here that, at most nutrient concentrations, *Chaetomorpha* had faster rates of nutrient uptake than *Cladophora*. This supports the hypothesis that the nutrient uptake abilities of *Chaetomorpha* afforded it a competitive advantage over *Cladophora* and contributed to its replacement by *Chaetomorpha* as the dominant alga in the system.

It has been reported elsewhere (Lavery 1989, Lavery *et al.* 1991) that *Chaetomorpha* is seldom nutrient limited and that *Ulva* is frequently nitrogen limited during spring in Peel Inlet. The critical nutrient levels reported here, help to explain these differences. *Ulva* requires about half the tissue phosphorus of *Chaetomorpha*, while *Chaetomorpha* requires about half the tissue nitrogen of *Ulva*. The nitrogen limitation of *Ulva* over spring-summer reflects the high nitrogen requirements of this plant compared to *Chaetomorpha* as well as the reduced ability to store nutrients over winter. The maximum tissue nitrogen levels recorded for *Ulva* (29.4 mg g^{-1}) were only 47% higher than the critical concentration. This compares with a maximum tissue phosphorus concentration (2.22 mg g^{-1}) which was almost 9 times the critical level. This seasonal limitation by nitrogen supports the hypothesis of McComb *et al.* (1981) that production in the system may be nitrogen limited in summer, a conclusion based on tissue nutrient concentrations of *Cladophora* (Gordon *et al.* 1981).

The low phosphorus requirements reported here and the high inorganic phosphorus inputs to the system in winter may account for the lack of phosphorus limitation for both of these species. The relatively mild affect of phosphorus starvation on the plants compared to nitrogen in the critical concentration experiments also indicates a more effective phosphorus storage and recycling mechanism. In contrast, the phosphorus requirement of *Cladophora* (3.3 mg g^{-1}; Gordon *et al.* 1981) is considerably higher than either *Chaetomorpha* or *Ulva*, and *Cladophora* was often phosphorus limited, as indicated by tissue concentrations (McComb *et al.* 1981, Lavery *et al.* 1991).

The maintenance of *C. linum* biomass over winter would permit this plant to take full advantage of the elevated nutrient concentrations. This is reflected in the luxury consumption of both nitrogen and phosphorus in winter with concentrations about twice the critical concentration of nitrogen, and five times in the case of phosphorus. This appears to prevent nutrient limitation over summer. The occasions on which *Chaetomorpha* appeared to be phosphorus- and nitrogen-limited, were towards the end of summer, and followed winters of low nutrient loading. Nutrient limitation seems unlikely for the remainder of the year.

Acknowledgements

The authors gratefully acknowledge the assistance of staff from the Nutrient Analysis Laboratory at The University of Western Australia, who performed the nutrient analyses. Also to Mr I. McKernan and Dr R. J. Lukatelich, who assisted with field work. This work was undertaken at the Botany Department of The University of Western Australia. One of the authors (PSL) was in receipt of a Commonwealth Postgraduate Research Award.

References

D'elia, C. F. and J. A. DeBoer. 1978. Nutritional studies on two red algae. II. Kinetics of ammonium and nitrate uptake. *J. Phycol. 14*: 266–272.

Dowd, J. E. and D. S. Riggs. 1965. A comparison of estimates of Michaelis-Menten kinetic constants from various linear transformations. *J. Biol. Chem. 240*: 863–869.

Druehl, L. D., P. J. Harrison, K. E. Lloyd and P. A. Thompson. 1986. Phenotypic variation in nitrogen uptake by *Laminaria groenlandica* (Laminariales: Phaeophyta). *J. Exp. Mar. Biol. and Ecol. 127*: 155–165.

Duke, C. S., W. Litaker and J. Ramus. 1989. Effects of temperature, nitrogen supply, and tissue nitrogen on ammonium uptake rates of the Chlorophyte seaweeds *Ulva curvata* and *Codium decorticatum*. *J. Phycol. 25*: 113–120.

Friedlander, M. and C. J. Dawes. 1985. *In situ* uptake kinetics of ammonium and phosphate and chemical composition of the red seaweed *Gracilaria tikvahiae*. *J. Phycol. 21*: 448–453.

Gilbert, P. M. and J. J. McCarthy. 1984. Uptake and assimilation of ammonium and nitrate by phytoplankton: indices of nutritional status for natural assemblages. *J. Plankt. Res. 6*: 677–697.

Gordon, D. M., P. B. Birch and A. J. McComb. 1981. Effects of inorganic nitrogen on the growth of an estuarine *Cladophora* in culture. *Bot. Mar. 24*: 93–106.

Haines, K. C. and P. A. Wheeler. 1978. Ammonium and nitrate uptake by marine macrophytes *Hypnea musciformis* (Rhodophyta) and *Macrocystis pyrifera* (Phaeophyta). *J. Phycol. 14*: 319–324.

Hanisak, M. D. and M. M. Harlin. 1978. Uptake of inorganic nitrogen by *Codium fragile* ssp. *tomentosoides* (Chlorophyta). *J. Phycol. 14*: 450–454.

Harlin, M. M. and J. S. Craigie. 1978. Nitrate uptake in *Laminaria longicruris* (Phaeophyceae). *J. Phycol. 14*: 464–467.

Harlin, M. M. and B. Thorne-Miller. 1981. Nutrient enrichment of seagrass beds in a Rhode island coastal lagoon. *Mar. Biol. 65*: 221–229.

Harrison, P. J., L. D. Druehl, K. E. Lloyd and P. A. Thompson. 1986. Nitrogen uptake kinetics in three year classes of *Laminaria groenlandica* (Laminariales: Phaeophyta). *Mar. Biol. 93*: 29–35.

Harrison, P. J., J. S. Parslow and H. L. Conway. 1989. Determination of nutrient uptake kinetic parameters; a comparison of methods. *Mar. Ecol. Prog. Ser. 52*: 302–312.

Hunt, R. 1978. *Plant Growth Analysis*. The Institute of Biology's Studies in Biology No. 96. Edward Arnold, 67 pp.

Kautsky, L. 1982. Primary production and uptake of ammonium and phosphate by *Enteromorpha compressa* in an ammonium sulphate industry outlet area. *Aq. Bot. 12*: 23–40.

Kemp, W. M , R. R. Twilley, J. C. Stevenson, W. R. Boynton and J. C. Means. 1983. The decline of submerged vascular plants in upper Chesapeake Bay: Summary of results concerning possible causes. *Mar. Tech. Ser. 17*: 78–89.

Lapointe, B. E. and K. E. Tenore. 1981. Experimental outdoor studies with *Ulva fasciata* Delile. I: Interactions of light and nitrogen on nutrient uptake, growth and biochemical composition. *J. Exp. Mar. Biol. Ecol. 53*: 135–152.

Lavery, P. S. 1989. Factors controling the abundance and dominance of macroalgal species in eutrophic estuaries. Ph. D. Thesis, Dept. of Botany, The University of Western Australia. pp. 115, 165.

Lavery, P. S., R. J. Lukatelich and A. J. McComb. 1991. Changes in the biomass and species composition of macroalgae in a eutrophic estuary. *Est. Coast Shelf Sci.*: in press.

Lavery, P. S. and A. J. McComb. 1991. macroalgal-sediment nutrient interactions and their importance to macroalgal nutrition in a eutrophic estuary. *Est. Coast. Shelf Sci.*: in press.

Lawlor, H., M. Borowitzka and J. A. McComb. 1991. A rapid and cheap method of surface sterilization of *Ecklonia radiata* (Phaeophyta) for issue culture. *Bot. Mar. 34*: 261–264.

Littler, M. M. 1979. The effects of bottle volume, thallus weight, oxygen saturation levels and water movement on apparent photosynthetic rates in marine algae. *Aq. Bot. 7*: 21–34.

Lukatelich, R. J. and A. J. McComb. 1986. *Nutrient Recycling and the Growth of Phytoplankton and Macroalgae in The Peel-Harvey Estuarine System*. Report to the Waterways Commission, Western Australia. 73 pp.

MacFarlane, J. J. and F. A. Smith. 1982. Uptake of methylamine by *Ulva rigida*: Transport of cations and diffusion of free base. *J. Exp. Bot. 33*: 195–207.

McCarthy, J. J. 1981. Uptake of major nutrients by estuarine plants. *In*: (B. J. Neilson and L. E. Cronin, eds) *Estuaries and Nutrients*. Humana Press, New Jersey.

McComb, A. J., R. P. Atkins, P. B. Birch, D. M. Gordon and R. J. Lukatelich. 1981. Eutrophication in The Peel-Harvey Estuarine System, Western Australia. *In*: (B. J. Neilson and L. E. Cronin, eds) *Estuaries and Nutrients*. Humana Press, New Jersey.

McLachlan, J. 1973. Growth media-marine. *In*: (J. R. Stein, ed.) *Handbook of Phycological Methods. Culture Methods and Growth Measurements*. Cambridge University Press, Cambridge.

Norton, T. A. and A. C. Mathieson. 1983. The biology of unattached seaweeds. *In*: (F. E. Round and D. J. Chapman, eds) *Progress in Phycological Research*. Volume 2. Elsevier, Amsterdam, New York, Oxford.

Probyn, T. A. and A. R. O. Chapman. 1982. Nitrogen uptake characteristics of *Chordaria flagelliformis* (Phaeophyta) in batch mode and continuous mode experiments. *Mar. Biol. 71*: 129–133.

Probyn, T. A. and C. D. McQuaid. 1985. in situ measurements of nitrogenous nutrient uptake by kelp (*Ecklonia maxima*) and phytoplankton in a nitrate-rich upwelling environment. *Mar. Biol. 88*: 149–154.

Rosenberg, R. 1985. Eutrophication. The future marine coastal nuisance. *Mar. Poll. Bull. 16*: 227–231.

Ryther, J. H., N. Corwan, T. A. De Busk and L. D. Williams. 1981. Nitrogen uptake and storage by the red alga *Gracilaria tikvahiae*. *Aquaculture 26*: 107–115.

Sawyer, C. N. 1965. The sea-lettuce problem in Boston Harbor. *J. Water Poll. Cont. Fed. 37*: 1122–1133.

Sfriso, A., A. Marcomini and B. Pavoni. 1987. Relationships between macroalgal biomass and nutrient concentrations in a hypertrophic area of the Venice Lagoon. *Mar. Env. Res. 22*: 297–312.

Thomas, T. E. and P. J. Harrison. 1985. Effect of nitrogen supply on nitrogen uptake, accumulation and assimilation in *Porphyra perforata* (Rhodophyta). *Mar. Biol. 85*: 269–278.

Thomas, T. E. and D. H. Turpin. 1980. Desiccation enhanced nutrient uptake rates in the intertidal alga *Fucus distichus*. *Bot. Mar. 23*: 479–481.

Thomas, T. E., P. J. Harrison and D. H. Turpin. 1987. Adaptations of *Gracilaria pacifica* (Rhodophyta) to nitrogen procurement at different intertidal locations. *Mar. Biol. 93*: 569–580.

Topinka, J. A. 1978. Nitrogen uptake by *Fucus spirilis* (Phaeophyceae). *J. Phycol. 14*: 241–247.

Waite, T. and R. Mitchell. 1972. The effect of nutrient fertilization on the benthic alga *Ulva lactuca*. *Bot. Mar. 15*: 151–156.

Wallentinus, I. 1984. Comparisons of nutrient uptake rates for Baltic macroalgae with different thallus morphologies. *Mar. Biol. 80*: 215–225.

Williams, S. L. and T. R. Fischer. 1985. Kinetics of nitrogen-15 labelled ammonium uptake by *Caulerpa cupressoides* (Chlorophyta). *J. Phycol. 21*: 287–296.

Botanica Marina
Vol. 34, pp. 261–264, 1991

A Rapid and Inexpensive Method for Surface Sterilisation of *Ecklonia radiata* (Phaeophyta) for Tissue Culture

H. J. Lawlor, M. A. Borowitzka and J. A. McComb[1]

School of Biological and Environmental Sciences, Murdoch University, Murdoch, Western Australia, 6150, Australia

(Accepted 23 February 1991)

Abstract

Sections of the thallus of the brown alga *Ecklonia radiata* (C. Ag.) J. Ag. were surface sterilised for tissue culture by dipping in 70% ethanol for 30 s, followed by sterile deionised water for 30 min. A high percentage of aseptic viable explants could be obtained using this treatment, and growth *in vitro* was initiated.

Introduction

Tissue culture provides a useful technique for both basic research on algae and possible commercial exploitation of valuable products. Control over the growth conditions of cultures is improved when they are axenic and unialgal. Epiphytes (*i. e.* algae, bacteria, etc) are a major source of contamination in tissue cultures of macrophytes and many techniques have been used to remove them. For example, dragging the plant material through sterile seawater solidified with agar (0.8%) eliminated epiphytes from *Chondrus crispus* Stackhouse (Chen and Taylor 1978). Manual brushing, mechanical shaking in a suspension of abrasives, or ultrasound treatment were found to be effective for cleaning some red and brown macroalgae (Gibor *et al.* 1981). However, these methods are laborious, and the long term effect of disruptive procedures on the macroalgal tissues is unknown.

Most methods reported for surface sterilisation of macroalgae have been based on those used for higher plants. A saturated solution of calcium hypochlorite has been used to sterilise tissue from *Laminaria digitata* (Hudson) Lamouroux and *L. hyperborea* (Gunn.) Foslie (30–60 s treatment) (Fries 1980) and *Fucus spiralis* L. (2–5 min treatment) (Fries 1977, 1980), while a combination of 1.5% KI solution (10 min) and 70% ethanol (incubation time not given) sterilised tissue of *Laminaria japonica* Areschoug and *Undaria pinnatifida* (Harvey) Suringar (Yan 1984). Iodine was also the active ingredient in a sterilisation method for red algae using Jodopax (Fries 1963). A wide range of germicidal agents have also been tested by Gibor *et al.* (1981) on several species of red and brown macroalgae. Best results were obtained from a 5 min incubation in 1% Betadine which contains the active ingredient iodide. Stipe explants of *Laminaria saccharina* (L.) Lamouroux have also been sterilised with 1% Betadine or 1% sodium hypochlorite for 1–12 min (Lee 1985). For *L. angustata* Kjellman, an alcohol dip followed by flaming and the use of a 25 mm core, parallel to the long axis of the stipe, yielded a high percentage of sterile explants (Saga and Sakai 1983).

Antibiotics have also been used to sterilise algal cells (Fries 1963, Gibor *et al.* 1980, Saga and Sakai 1982, Polne-Fuller and Gibor 1982, Bradley *et al.* 1986, Fisher and Gibor 1987, Bradley *et al.* 1988). Generally explants were first surface sterilised, using one of the above methods, and then incubated on media to which the antibiotics were added. However, antibiotics are not always a satisfactory sterilising agent (Chen and Taylor 1978, Saga and Sakai 1983).

This report describes the establishment of *Ecklonia radiata* in axenic culture using a quick and inexpensive method for surface sterilisation.

[1] Author and address for reprint requests

Material and Methods

Mature *Ecklonia radiata* (C. Ag.) J. Ag. plants were collected at a depth of 1 m on limestone reefs off the coast near Perth, Western Australia. Plants with the fewest epiphytes were selected and were transported in plastic bags filled with air to minimise damage. Plants were stored at ambient seawater temperature and were sterilised and cultured within five hours of collection. In the laboratory, visible epiphytes were removed with a sterile scalpel and sections up to 6 cm long were cut from the holdfast, stipe, and lamina. These pieces were treated (85 per 250 mL screwcap jar) by submerging and shaking in the following solutions for the range of times shown in Table I: — ethanol (70% v/v in deionised water); potassium iodide, analytical grade (1.5% w/v in seawater); and sterile deionised water. Sections were then rinsed in two changes of sterile seawater and the cut surfaces of each piece were trimmed and discarded. Explants were cut into cubes about 4 mm each side and placed on solid medium in 9 cm Petri dishes sealed with parafilm (5–10 explants per dish). From stipes of sufficient diameter, a core comprising medullary and cortical tissues was removed parallel to the long axis, using a flame-sterilised 5 mm steel corer. The ends of the core were trimmed and discarded, and the remaining core was cut into 2 mm high discs.

The culture medium used was modified PES medium solidified with 7 gL^{-1} agar (Lawlor *et al.* 1988). A seawater based bacterial nutrient broth was used to detect the presence of contaminants (Gibor *et al.* 1981). Cultures were kept at 25 °C and continuous light (50 $\mu mole^{-2}$ s^{-1}). Sterility of the explants was determined by examination of the medium around and below the explant for bacterial and fungal growth, for at least 6 weeks.

Viability of the algal tissues was estimated after sterilisation, using hand sections stained with Evan's blue or neutral red (Millner *et al.* 1979). Fresh unsterilised tissue and tissue incubated for 10 min at 80 °C were used as the live and dead controls, respectively.

Results

Initially a solid sterility test medium based on seawater and bacterial nutrient broth (Gibor *et al.* 1981) was used to detect the presence of contaminants. However, for *Ecklonia radiata* this medium greatly reduced the induction and rate of callus growth. Since

Table I. Percentage of sterile explants of *E. radiata* following sterilisation and 6 weeks incubation on tissue culture medium. 100 explants were used per treatment. Viability was determined immediately after surface sterilisation and an asterisk indicates that treatment killed more than 20% of the tissue in all explants.

Treatment		Percentage sterile explants			
		Stipe		Holdfast	Lamina
(Time)		Not cored	Cored		
KI (1.5% w/v)					
10 s		0	0	0	0
15		0	0	0	0
30		0*	0*	0	0*
60		28*	31*	0*	25*
90		43*	40*	0*	25*
Ethanol (70% v/v)					
15 s		0	0	0	0
30		30	31	0	20
60		57	61	0	51*
90		63*	78*	10	60*
120		70*	100*	21	62*
Ethanol (70% v/v)	then KI (1.5% w/v)				
5 s	15 s	0	0	0	0
10	30	28*	30*	0*	25*
20	60	57*	80*	12*	58*
30	90	78*	100*	19*	80*
Ethanol (70% v/v)	then DI water				
30 s	10 min	60	64	0	61
30	20	69	71	10	65
30	30	75	85	20	81
30	60	82	85	20	80*
30	120	86*	100*	25*	85*
30	180	88*	100*	26	84*

any contamination was clearly visible on the solid tissue culture medium this step was eliminated.

Potassium iodide killed *Ecklonia radiata* tissue, whether used alone or following ethanol (Table I). Treatment for 30 to 60 s in ethanol resulted in live sterile explants. When the explants were treated with ethanol and then shaken in deionised water for 30 to 60 min the highest percentages of live sterile explants were obtained. Holdfasts gave the lowest percentage of sterile explants. Lamina tissues were killed during the 60 min freshwater treatment. The percentage of sterile stipe explants was increased by taking a core. When sterile explants were placed on PES medium, callus-like growth was apparent after 3 to 7 days (Lawlor *et al.* 1988). Cultures found to be clean after 6 weeks showed no contamination in later subcultures.

Discussion

Explants of *Ecklonia radiata* were successfully surface sterilised when treated for 30 s with 70% ethanol, followed by sterile deionised water. This may be due to the possibility that marine contaminants, especially marine bacteria, cannot survive in fresh water. Sterilisation periods for longer than one hour in deionised water damaged the algal cells. Although KI solution followed by ethanol has been used to sterilise *Laminaria japonica* and *Undaria pinnatifida* (Yan 1984); treatments sufficient to eliminate contamination in *E. radiata* killed the cells. Similarly, 70% ethanol alone, although a potent microbicide (Frobisher *et al.* 1974), also killed the algal cells.

Many factors influence the success of sterilisation of plants for tissue culture. The method of collection and transportation should not be underestimated; macroalgae which were damaged, allowed to become warm, or were stored before use, always had higher levels of contamination. Best results were obtained by using undamaged material, sterilised and cultured within five hours of collection.

The use of a scalpel to remove visible epiphytes is labour intensive, but effective for macroalgae such as *E. radiata*, causing little damage except to the outer layer of cells. Removal of a core of explant tissue is a further important step in the elimination of algal epiphytes, as well as surface bacterial and fungal contamination. In *E. radiata* this procedure is suited to the stipe and meristematic zone. As well as minimising contamination, this step eliminated the problem of leakage of phenolics from the epidermal and cortical cells into the culture medium, where these compounds may then interfere with growth (George and Sherrington 1984).

This paper provides an alternative approach for surface sterilisation of macroalgae. The method is now being used routinely for the tissue culture of *E. radiata*, and it has been applied successfully to species of *Sargassum* and *Cystophora*.

Acknowledgements

This study was supported by a grant from the Marine Sciences and Technology Grant Scheme.

References

Bradley, P. M., D. P. Cheney and N. Saga. 1986. A new one-step method for obtaining axenic cultures of algae. *Appl. Phyc. Forum 3*: 4–5.

Bradley, P. M., D. P. Cheney and N. Saga. 1988. One-step antibiotic disk method for obtaining axenic cultures of multicellular marine algae. *Plant Cell Tissue and Organ Culture 12*: 55–60.

Chen, L. C. M. and A. R. A. Taylor. 1978. Medullary tissue culture of the red alga *Chondrus crispus*. *Can. J. Bot. 56*: 883–886.

Fisher, D. D. and A. Gibor. 1987. Production of protoplasts from the brown alga *Sargassum muticum* (Yendo) Fensholt (Phaeophyta). *Phycologia 26*: 488–495.

Fries, L. 1963. On the cultivation of axenic red algae. *Physiol. Plant. 16*: 695–708.

Fries, L. 1977. Growth regulating effects of phenylacetic acid and p-hydroxyphenylacetic acid on *Fucus spiralis* L. (Phaeophyceae, Fucales) in axenic culture. *Phycologia 16*: 451–455.

Fries, L. 1980. Axenic tissue cultures for the sporophytes of *Laminaria digitata* and *Laminaria hyperborea* (Phaeophyta). *J. Phycol. 16*: 475–477.

Frobisher, M., R. D. Hinsdill, K. T. Crabtree and C. R. Goodheart. 1974. *Fundamentals of Microbiology*. Saunders, Philadelphia. pp. 850.

George, E. F. and P. D. Sherrington. 1984. *Plant Propagation by Tissue Culture*. Eastern Press, Reading. pp. 709.

Gibor, A., M. Polne-Fuller, M. Biniaminov and M. Neushul. 1981. Exploratory studies of vegetative propagation of marine algae; procedure for obtaining axenic tissues. *In*: (T. Levring, ed) *Proc. of Xth Internat. Seaweed Symposium*. Walter de Gruyter, Berlin. pp. 578–593.

Lawlor, H. J., J. A. McComb and M. A. Borowitzka. 1988. The development of filamentous and callus-like growth in axenic tissue cultures of *Ecklonia radiata* (Phaeophyta). *In*: (T. Stadler, J. Mollion, M. C. Verdus, Y. Karamanos, H. Horvan and D. Christian, eds) *Algal Biotechnology*. Elsevier Applied Science, London. pp. 139–150.

Lee, T. F. 1985. Aposporous gametophyte formation in stipe explants from *Laminaria saccharina* (Phaeophyta). *Bot. Mar. 28*: 179–185.

Millner, P. A., M. E. Callow and L. V. Evans, 1979. Preparation of protoplasts from the green alga *Enteromorpha intestinalis* (L.) Link. *Planta 147*: 174–177.

Polne-Fuller, M. and A. Gibor. 1984. Developmental studies in *Porphyra*. 1. Blade differentiation in *P. perforata* as expressed by morphology, enzymatic digestion and protoplast regeneration. *J. Phycol. 20*: 609–616.

Saga, N. and Y. Sakai. 1982. A new method for the culture of macroscopic algae, the one step selection method. *Jap. J. Phycol. 30*: 40–43.

Saga, N. and Y. Sakai. 1983. Axenic tissue culture and callus formation of the marine brown alga *Laminaria angustata*. *Bull. Jap. Soc. Sci. Fish. 49*: 1561–1563.

Yan, Z. M. 1984. Studies on tissue culture of *Laminaria japonica* and *Undaria pinnatifida*. *Hydrobiologica 115/116*: 314–316.

Botanica Marina
Vol. 34, pp. 265–269, 1991

A Seasonal Study of the Seagrass *Ruppia maritima* L. in Tampa Bay, Florida. Organic Constitutents and Tolerances to Salinity and Temperature

A. C. Lazar and C. J. Dawes

Department of Biology, University of South Florida, Tampa, Florida 33620. U.S.A.

(Accepted 2 March 1991)

Abstract

Two populations of the seagrass *Ruppia maritima* in Tampa Bay, Florida showed advanced seasonal growth cycles when compared with *Thalassia testudinum* and *Halodule wrightii.* Blade growth was evident in the winter corresponding to the highest levels of protein (20.7%) in that organ. High levels of soluble carbohydrate (26.7 to 52.0%) occurred in the rhizome in the spring and early summer months, when blade production was complete. Although the two sites in Tampa Bay differed with regard to exposure and salinity, both populations of *R. maritima* showed similar, broad tolerances to 9 combinations of salinity (0, 17.35, 35‰) and temperature (10, 20, 30 °C) in a factorial experiment measuring photosynthetic responses. The study suggests that *R. maritima* has a competitive advantage over other seagrasses by having broader tolerances to salinity and temperature and an earlier annual growth cycle. The question is raised why *Ruppia maritima* is found as a fringing seagrass or as a secondary component of seagrass beds in Tampa Bay.

Introduction

Ruppia maritima L. is a submerged marine angiosperm with broad tolerances to salinity and temperature, occurring from the southern part of Brazil (Koch and Seeliger 1988) to Newfoundland on the North Atlantic coast (Fernald and Wiegand 1914, Bourn 1935, McMillan 1974, Richardson 1980). *Ruppia* spp. occur in fresh water ditches (Richardson 1980, Wetzel and Penhale 1983) and will form submerged meadows in hypersaline ponds (110 to 132‰ salinity, Congdon and McComb 1981, Davis and Tomlinson 1974). Various populations can withstand temperature ranges of 0.5 to 35 °C (Wetzel and Penhale 1983).

Ruppia maritima also occurs in a wide variety of habitats along the west coast of Florida, forming extensive meadows at the mouths of rivers and on tidal flats as well as subtidal areas of high salinity (Dawes 1987). The species accounts for about 8.8% of the 5750 ha of seagrass meadows and usually occurs in shallow (< 0.2 m) regions in Tampa Bay, sometimes fringing other seagrass beds. Both annual and perennial intertidal to subtidal communities of *R. maritima* are found in Tampa Bay (Lewis *et al.* 1985). Thorne (1954) considered *Ruppia maritima* a freshwater species and others (Zieman 1982, Humm 1973) have suggested that it was not a true seagrass because, in contrast to seagrasses, it flowers at the water surface and is not limited to essentially marine situations.

The present study was designed to determine the seasonal tolerances of *Ruppia maritima* to 9 combinations of salinity and temperature in a factorial experiment measuring photosynthetic and respiratory responses and to compare the organic and energetic levels over a 12 month period. Two populations were selected in Tampa Bay, Florida so that tolerances could be compared. The goal was to compare the seasonal tolerances to salinity and temperature and allocation of proximate constituents and kilojoules of the blades, rhizome and roots with other seagrasses for the west coast of Florida.

Material and Methods

Ruppia maritima was sampled bimonthly at a northern site in Tampa Bay 55 km from the bay mouth near Courtney Campbell Causeway (CC: 27°57′57″N Lat., 82°33′12″W Long.) and at a southern site 22 km from the bay mouth at Piney Point (PP: 27°38′26″N Lat., 82°36′33″W Long.). The northern site was an intertidal estuarine flat situated 0.5 km from the mouth of a tidal creek where salinities were lower and fluctuated. The population at the southern site was subtidal, fringing mixed beds (2 to 4 m^2) of two other seagrasses, *Halodule wrightii* Ascherson and *Thalassia testudinum* Konig *et* Sims, about 0.1 km offshore, an area of relatively high salinity water.

Plants were randomly collected in the seagrass beds at each site by taking cores. Healthy, mature rhizome sections 6 cm long with 2 to 4 short shoots were selected after the growing tips and the first three short shoots were discarded. The sections were separated into roots, rhizomes and blades for determination of ash and organic constituents (protein, soluble carbohydrate, lipid) as described by Dawes and Lawrence (1983). Insoluble carbohydrate was determined by subtraction. Kilojoules were obtained by multiplying kilocaloric values for protein, lipid, and total (soluble and insoluble) carbohydrate (Brody 1964) by 4.184.

The factorial salinity/temperature experiments were carried out using a Gilson Respirometer (GRP Model 20) to measure photosynthetic and respiratory responses (μL O_2 m^{-2} s^{-1}). Four replicate reaction flasks containing 8 blade segments 1 cm long were used for each salinity treatment in the respirometer. Entire plants were acclimated in each of three salinities (0, 17.5, 35‰) for 3 days prior to the experiment based on the results of a preliminary salinity acclimation test. Blades were then cut into 1 cm sections and aspirated in the appropriate salinity seawater (0, 17.5, 35‰) until the sections no longer floated to remove gasses in the leaf parenchyma tissue (Hartman and Brown 1967, Bulthuis 1983, Kerr and Strither 1985). The blades were not acclimated to the three temperatures (10, 20, 30 °C) prior to the respirometer experiment as no changes were observed in responses based on preliminary temperature acclimation studies except at the highest temperature where plants showed deteriation after 3 days. The respirometer experiments were run under 200 μE m^{-2} s^{-1}, a non-saturating irradiance but well above the compensation point as determined by preliminary photosynthesis-irradiance curves. Ambient seawater was filtered and either diluted with deionized water to make 17.5‰ or concentrated by heating to 60 °C to make 35‰; deionized water was used for 9‰.

Results obtained from the temperature-salinity factorial experiment for the two populations were analysed as a 2 (populations) × 3 (temperature) × 3 (salinity) Complete Random Block Factorial Design in a SAS program (SAS Institute Inc. 1985) on an IBM 370 Computer. Normality of data and homogenity of variance were determined and no transformations were required.

Results

Water temperature at the two sites ranged from a low of 11 °C (CC: January 1986) to a high of 29 °C (PP: September 1985). The high salinity site had an average salinity of 30.7‰ (range 30 to 34‰) while the average salinity at the estuarine site was 25.7‰ (range 22 to 29‰). The lowest salinities at both sites occurred in the late summer and early autumn, a period of intense

Table I. Dry weight, organic constituents (percent of dry weight) and kilojoules (per gram dry weight) of the blades, rhizomes, and roots of *Ruppia maritima* from two populations in Tampa Bay, Florida. Means, standard deviations (±) and ranges are based on bimonthly collections over a period of 12 months (N = 6).

Organ	Dry Weight	Ash	Lipid	Protein	Carbohydrate	Kilojoules
Piney Point Population:						
Blades	13.1 (1.7)	29.9 (7.6)	2.2 (1.2)	13.8 (5.0)	19.4 (7.2)	13.4 (1.5)
	10.8–15.2	21.2–39.3	1.5–4.7	7.8–20.7	13.7–33.4	11.3–15.1
Rhizomes	15.4 (2.8)	31.7 (9.5)	2.0 (1.5)	7.6 (3.7)	29.6 (14.6)	12.6 (1.9)
	12.0–18.8	18.9–41.3	0.7–4.3	3.6–11.2	15.8–52.0	10.5–15.5
Roots	16.2 (4.0)	39.7 (10.8)	1.8 (0.6)	9.1 (4.0)	14.2 (4.3)	11.4 (1.9)
	12.3–24.0	26.9–52.0	1.1–2.8	3.5–14.5	9.5–22.1	9.2–13.4
Courtney Campbell Population:						
Blades	12.8 (1.6)	36.5 (3.4)	1.8 (0.6)	13.2 (5.0)	16.3 (5.3)	12.1 (0.9)
	11.4–15.7	31.7–41.1	1.0–2.7	6.5–20.7	11.9–25.9	11.3–13.4
Rhizomes	19.0 (2.4)	35.4 (13.3)	1.4 (0.7)	8.8 (3.9)	34.3 (14.1)	11.9 (2.0)
	14.8–21.3	19.0–55.6	0.9–2.4	3.1–13.8	11.8–48.1	8.8–14.7
Roots	26.4 (6.6)	38.8 (8.2)	4.2 (2.1)	11.1 (4.1)	17.0 (4.4)	12.1 (1.7)
	18.1–35.7	26.2–46.7	1.7–7.0	3.9–16.4	10.9–23.6	9.6–14.2

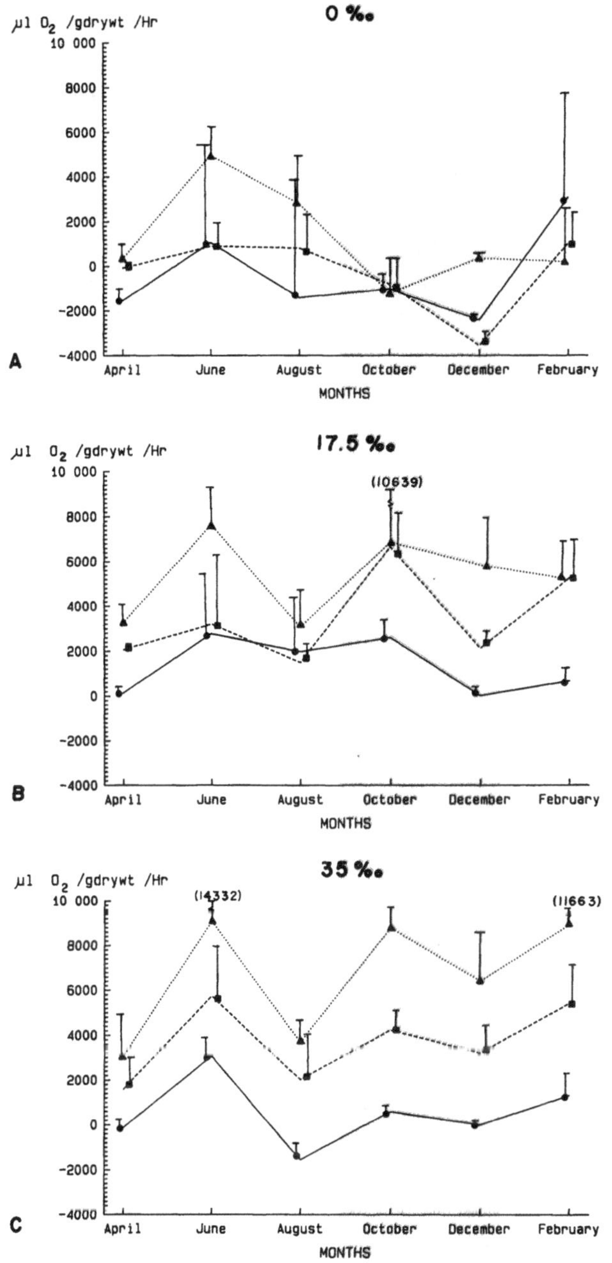

Fig. 1. Photosynthetic rates of *Ruppia maritima* blades collected bimonthly from Courtney Campbell in Tampa Bay, Florida expressed as µl O_2 gdwt^{-1} h^{-1}. Blades were exposed to 10 (bold line), 20 (dashed line) and 30 (dotted line) °C and 0‰ (Fig. 1A), 17.5‰ (Fig. 1B) and 30‰ (Fig. 1C) salinities. Vertical bars are ± 1 standard deviation.

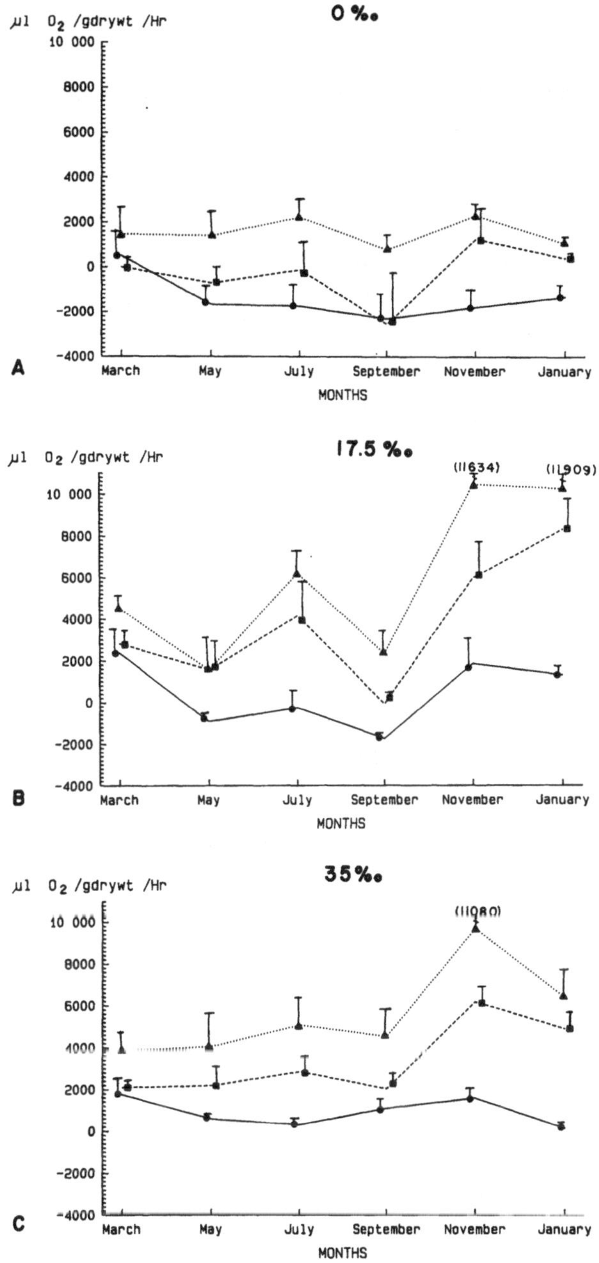

Fig. 2. Photosynthetic rates of *Ruppia maritima* blades collected bimonthly from Piney Point in Tampa Bay Florida expressed as µl O_2 gdwt^{-1} h^{-1}. See Figure 1 for explanation of symbols.

rains and highest water temperatures. Both populations showed new growth (rhizome and blade production) in January and February and die back of the blades in late summer and early autumn.

Both populations showed the same patterns with regard to organic constituents. Protein levels were highest in the blades (6.5 to 20.7%) and lowest in the rhizomes ((3.1 to 13.8%). Kilojoule levels were similar for blades, rhizomes, and roots and showed no seasonal shift. High levels for protein occurred in the blades during January and February (20.7% each population) and for soluble carbohydrate in the spring to summer months (estuarine population: April through August, 43 to 48%; high salinity population: July and September, 53 and 43,5%). No seasonal changes in dry weights or levels of lipid were evident in the blades, rhizomes, or roots (Table I). The lowest ash levels occurred in the late summer and early autumn (18.9 to 36.2%), a time of lowest salinities.

Blades of both populations did not show a uniform seasonal change in photosynthetic responses to temperature in the factorial experiment using three levels of salinity and temperature (Figs 1, 2). The blades from the high salinity population did show an increase in photosynthesis in the autumn when held in 17.5 and 35‰ salinities and exposed to 20 and 30 °C. Regardless of the time of year of collection, both populations showed significantly lower photosynthetic responses, shifting to respiration if held at 10 °C. The photosynthetic responses for plants held at 20 and 30 °C were not significantly different regardless of the salinity used.

Blades from both sites were able to tolerate deionized water during the 3 day acclimation period and the blade segments appeared healthy after testing in the respirometer. Photosynthetic responses of blades acclimated to 0‰ salinity were low (Fig. 1A, 2A) but evident throughout the 12 month period even for the plants from the high salinity site that were held at 30 °C (Fig. 2A). In contrast, the estuarine plants had a highly varied response at 30 °C and in 0‰ salinity, ranging from photosynthesis to respiration. Blades from both populations held at 10 or 20 °C in 0‰ showed little to no photosynthetic responses.

Both populations held in 17.5 and 35‰ salinities had similar photosynthetic responses which were not significantly different from one another. Both populations had low photosynthetic responses in August or September, a time when the plants in the field showed to blade damage and die back.

Discussion

The two populations of *Ruppia maritima* studied in Tampa Bay grow in distinct habitats that differ in exposure and salinity; one site being an estuarine tidal flat, the other being a submerged seagrass bed with stable, high salinities. In spite of this, both populations showed similar levels of organic constituents in their blades, roots, and rhizomes and similar tolerances to a factorial experiment using 9 combinations of salinities and temperatures.

Seasonal variations in levels of organic constituents are known for seagrasses. Soluble carbohydrate has been shown to be a storage product, being high in the summer in the rhizomes while protein is high in the spring in the blades during growth for *Thalassia testudinum* (Dawes and Lawrence 1979, Durako and Moffler 1985), *Halodule wrightii* and *Syringodium filiforme* Kützing (Dawes and Lawrence 1980, Dawes 1986), and in *Poisidonia oceanica* (L.) Delile (Pirc 1985). In the present study, the rhizomes of *Ruppia maritima* had the highest level of soluble carbohydrate in April through September and the blades had the highest level of protein in January and February. Thus, the highest level of protein corresponded with blade production and the highest level of soluble carbohydrate occurred just before blade die back.

Winter blade production by *Ruppia maritima* at both sites occurred during a period of lowest water temperatures, as also noted in Tampa Bay for *R. maritima* by Phillips (1960). It appears that *R. maritima* initiates growth in the winter, prior to the other seagrasses, stores carbohydrate in the summer and dies back in the autumn again before the other seagrasses. The species is usually considered to be a plant capable of colonizing a variety of habitats (Williams and McRoy 1982, Lewis *et al.* 1985), thus an advanced growth and die back cycle would enable *R. maritima* to develop at a time of year when other seagrasses are either dorminant or showing die back (Dawes 1986).

The broad tolerances of *Ruppia maritima* to salinity and temperature are well known based on its distribution and photosynthetic responses (Phillips 1960, Davis and Tomlinson 1974, Wetzel and Penhale 1983). The two populations in this study also showed similar broad tolerances whether the site is estuarine or high salinity seawater. Because of the wide variety of sites where the species grows, some researchers have suggested that *R. maritima* is not a 'true' seagrass (Humm 1973, Zieman 1982). The present study does not support this view because the blades from both populations showed highest photosynthetic responses in the higher two salinities although the plants were able to tolerate 0‰ salinity. The significant interaction between salinity and temperature based on photosynthetic responses for both populations (Table II) is in direct contrast to the 'phoma pattern' (Ritchie 1957) for estuarine fungi which show a lower tolerance to higher salinities if temperature also increases. This contrast further suggests that *R. maritima* is a euryhaline species which is adapted to oceanic habitats and can tolerate estuarine conditions.

In conclusion, *Ruppia maritima* shows a seasonally advanced growth cycle in Tampa Bay in relation to

Table II. Results of a randomized complete block ANOVA for the differences between the effect of salinity and temperature on photosynthesis in *Ruppia maritima* blades.

		F value
Courtney Campbell	Salinity	47.59**
	Temperature	55.30**
	Salinity X Temperature	5.89**
Pincey Point	Salinity	90.23**
	Temperature	126.16**
	Salinity X Temperature	4.15*

* = significant difference at the p < = .05 level
** = significant difference at the p < = .001 level
NS = No Significant difference

the other seagrasses as exhibited by organic constituent partitioning and this may allow it to colonize areas before other seagrasses. The broad tolerances of this species to salinity and temperature also give it a competitive advantage, yet it is usually restricted to shallow water sites, or forming fringe communities with other seagrasses. The limited occurrance of *R. maritima* in typical shallow seagrass beds may be due to other factors such as the requirement for flowering at the water surface, the effects of wave action, substrate composition, or a requirement for high irradiance levels.

References

Brody, S. 1964. *Bioenergetics and Growth.* Hafner, New York. 1023 pp.

Bourn, W. S. 1935. Sea-water tolerance of *Ruppia maritima* L. *Contrib. from the Boyce Thompson Inst. 7*: 249–255.

Bulthuis, D. A. 1983. Effects of temperature on the photosynthetsis-irradiance curve of the Australian seagrass, *Heterozostera tasmanica. Mar. Biol. Lett. 4*: 47–57.

Congdon, R. A. and A. J. McComb. 1981. The vegetation of the Blackwood River Estuary, south-west Australia. *J. Ecol. 69*: 1–16.

Davis, J. S. and P. B. Tomlinson. 1974. A new species of *Ruppia* in high salinity in Western Australia. *J. Arnold Arboretum 55*: 59–66.

Dawes, C. J. 1986. Seasonal proximate constitutents and caloric values in seagrasses and algae on the west coast of Florida. *J. Coastal Res. 2*: 25–32.

Dawes, C. J. 1987. The dynamic seagrasses of the Gulf of Mexico and Florida coasts. *In*: (M. J. Durako, R. C. Phillips, R. R. Lewis III, eds) *Proceedings of the Symposium on Subtropical-Tropical Seagrasses of the Southeastern U.S.* Fla. Dept. Nat. Res. Bureau Mar. Res., St. Petersburg, Florida. No. 42. pp. 25–38.

Dawes, C. J. and J. M. Lawrence. 1979. Effects of blade removal on the proximate composition of the rhizome of the seagrass *Thalassia testudinum* Banks *ex* Konig. *Aquat. Bot. 7*: 255–266.

Dawes, C. J. and J. M. Lawrence. 1980. Seasonal changes in the proximate constituents of the seagrass *Thalassia testudinum, Halodule wrightii,* and *Syringodium filiforme. Aquat. Bot. 8*: 371–380.

Dawes, C. J. and J. M. Lawrence. 1983. Proximate composition and caloric content of seagrasses. *Mar. Tech. Soc. J. 17*: 53–58.

Dawes, C. J., M. O. Hall and R. K. Riechert. 1985. Seasonal biomass and energy content in seagrass communities on the west coast of Florida. *J. Coastal Res. 1*: 255–262.

Durako, M. J. and M. D. Moffler. 1985. Spatial influences on temporal variations in leaf growth and chemical composition of *Thalassia testudinum* Banks *ex* Konig in Tampa Bay, Florida. *Gulf Res. Rep. 8*: 43–49.

Fernald, M. L. and K. M. Wiehand. 1914. The genus *Ruppia* in North America. *Rhodora 16*: 119–128.

Hartman, R. T. and D. L. Brown. 1967. Changes in the internal atmosphere of submerged vascular hydrophytes in relation to photosynthesis. *Ecology 48*: 252–258.

Humm, H. J. 1973. Section C. Seagrasses. *In*: (J. I. Jones, R. E. Ring, M. O. Rinkel and R. E. Smith, eds) *A Summary of the Knowledge of the Eastern Gulf of Mexico.* State Univ. System Florida, Institute of Oceanography, Tallahassee, Fl. pp. IIIC: 1–10.

Kerr, E. A. and S. Strither. 1985. Effects of irradiance, temperature, and salinity on photosynthesis of *Zostera muelleri. Aquat. Bot. 23*: 177–183.

Koch, E. W. and U. Seeliger. 1988. Germination ecology of two *Ruppia maritima* L. populations in southern Brazil. *Aquat. Bot. 31*: 321–327.

Lewis III, R. R., M. J. Durako, M. D. Moffler and R. C. Phillips. 1985. Seagrass meadows of Tampa Bay – A review. *In*: (S.-A. F. Treat, J. L. Simon, R. R. Lewis III and R. L. Whitman Jr., eds) *Tampa Bay Area Scientific Information Symposium.* Bellwether Press, Tampa BASIS P.O. Box 15759, Tampa, Fl. pp. 210–246.

McMillan, C. 1974. Salt tolerance of mangroves and submerged aquatic plants. *In*: (R. J. Reimold and W. H. Queen, eds) *Ecology of Halophytes.* Academic Press, New York. pp. 379–383.

Phillips, R. C. 1960. *Observations on the Ecology and Distribution of Florida Seagrasses.* Florida Board of Conservation, Tallahassee. Prof. Pap. Series 2. 72 pp.

Pirc, H. 1985. Growth dynamics in *Posidonia oceanica* (L.) Delile. 1. Seasonal changes of soluble carbohydrates, starch, free amino acids, nitrogen, and organic amines in different parts of the plant. *P.S.Z.N.I. Mar. Ecol. 6*: 141–165.

Richardson, F. D. 1980. Ecology of *Ruppia maritima* L. in New Hampshire (U.S.A.) tidal marshes. *Rhodora 82*: 403–440.

Ritchie, D. 1957. Salinity optima for marine fungi affected by temperature. *Amer. J. Bot. 44*: 870–874.

SAS Institute Inc. 1985. *SAS User's Guide: Statistics.* SAS Institute Inc., Cary N.C. Version 5 Edition. 956 pp.

Thorne, R. F. 1954. Flowering plants of the waters and shores of the Gulf of Mexico. *In*: (P. S. Galtsoff, ed.) *Gulf of Mexico, its Orgin, Waters, and Marine Life.* U.S. Fish and Wildlife Services, Washington D.C. Fisheries Bulletin 55. pp. 193–202.

Wetzel, R. L. and P. A. Penhale. 1983. Production ecology of seagrass communities in the lower Chesapeake Bay. *Mar. Tech. Soc. J. 17*: 22–31.

Williams, S. L. and C. P. McRoy. 1982. Seagrass productivity: the effect of light on carbon uptake. *Aquat. Bot. 12*: 321–344.

Zieman, J. C. 1982. *The Ecology of the Seagrasses of South Florida: A Community Profile.* U.S. Fish and Wildlife Services, Office of Biological Services, Washington, D.C. FWS/OBS-82/25. 158 pp.

Botanica Marina
Vol. 34, pp. 271 – 277, 1991

Fungi and Their Biomass in Detritus of the Seagrass *Thalassia hemprichii* (Ehrenberg) Ascherson

Veena Sathe and S. Raghukumar[1]

Biological Oceanography Division, National Institute of Oceanography, Dona Paula, Goa – 403 004, India

(Accepted 2 March 1991)

Abstract

Four different stages of detritus from the seagrass *Thalassia hemprichii* (Ehrenberg) Ascherson from two coral islands in the Lakshadweep Archipelago, Arabian Sea were examined for the presence of fungi. Direct microscopic examination of the detritus from Kalpeni Island revealed different kinds of hyphae and unicellular thraustochytrid fungi, particularly in later stages of decomposition. Highest values of biomass estimated by direct observations were $31\,840 \times 10^{-6}$ g mycelial fungi per g dry wt. detritus in detrital fragments from the sediment and 2320×10^{-6} g per g dry wt. of detritus for thraustochytrids in dark brown decaying leaves, corresponding to 3.2 and 0.23% per g dry wt. detritus respectively. The significance of these values which are higher than those reported for other submerged marine macrophyte detritus is discussed. Isolation of fungi in culture yielded thraustochytrids and terrestrial species of mycelial fungi from detritus.

Introduction

The transformation of macrophytic detritus through the mediation of microorganisms is now recognised to be an important process in the sea (Blum *et al.* 1988, Bowen 1987). Their activity results in the mineralization of detritus and a decrease in C/N ratio (Blum *et al.* 1988). The dead organic matter and the associated microorganisms form the base of the food webs for commercially important fishes and crustaceans (Odum 1971).

Most studies have indicated that bacterial biomass may form up to 1% of detritus (Cammen 1980, Findlay and Meyer 1984, Bowen 1988) but less is known of the biomass of fungi. Several studies have been made on the prevalence of fungi in decaying mangrove seedlings (Newell 1976) and mangrove leaves (Fell and Master 1973). Fungi appear to play a significant role in the decomposition of the intertidal salt marsh grass *Spartina alterniflora* Loisel (May 1974, Lee *et al.* 1980, Newell and Hicks 1982, Newell *et al.* 1986). Fungi have also been shown to be abundant in seagrasses such as *Posidonia oceanica* (L.) Delile, *Zostera* sp. and *Cymodocea nodosa* (Ucria) Aschers (Cuomo *et al.* 1985, 1988) and *Thalassia testudinum* Konig (Newell and Fell 1982). Cuomo *et al.* (1987) have presented evidence for an active role of ligno-cellulosic enzymes of marine fungi in *Posidonia oceanica* detritus. Inspite of these numerous reports, there have been problems in quantifying them, resulting in conflicting views on the biomass and importance of fungi in macrophyte detritus (Newell 1984).

Seagrasses have a very high productivity, with values up to 16 g C per day (Ott 1980). Consequently, they make an enormous contribution to the detrital material in shallow marine waters in temperate and tropical regions throughout the world (Kenworthy and Thayer 1984, Zieman *et al.* 1984). In a study on the relative abundance of fungi and bacteria in the detritus of seagrasses, Blum *et al.* (1988) concluded that fungi contribute a very low biomass. Newell and Fell (1982) and Newell *et al.* (1987) arrived at a similar conclusion. One possible reason for the reported low abundance of fungi would appear to be that these authors did not estimate fungal biomass in very late

[1] Corresponding author

stages of decomposition. Such stages characteristically harbour numerous marine species of fungi (Cuomo *et al.* 1985, 1988). Another possible reason is the lack of information on the biomass of the single-celled thraustochytrid fungi on such detritus. Thraustochytrids have been regularly isolated from submerged mangrove leaves (Fell and Master 1973, Newell 1976) and have been reported from a wide variety of substrates (Moss 1986, Raghukumar 1990). We have therefore carried out studies on the occurrence and prevalence of both mycelial and thraustochytrid fungi in the turtlegrass *Thalassia hemprichii* (Ehrenberg) Ascherson from the lagoons of two coral islands of the Lakshadweep archipelago during various stages of decomposition.

Material and Methods

The present study was carried out in the coral islands Kavaratti and Kalpeni of the Lakshadweep archipelago (10°30′ – 11.0°N Lat., 72.0° – 73.0°E Long.). The two islands are about 5 km long and 1 km wide, with a NE-SW orientation. The mean tidal range is about 2 m, the ambient water temperature being 28 to 30 °C. Large beds of seagrasses, including *Thalassia hemprichii* and *Cymodocea isoetifolia* (Ascherson) Graeb, mostly the former, are present below the subtidal level in the lagoons occurring on the western side of the island. The seagrass beds are interspersed with algal patches, mainly of *Gracilaria edulis* (Gmel.) Silva and *Laurencia* spp. (Qasim and Bhattathiri 1971, Jagtap 1977).

In order to study the fungi, green and decaying leaves of *Thalassia hemprichii* were collected from lagoons in the two islands immediately below low tide level and sorted into four stages of decomposition as follows: Stage 1 – green leaves; Stage 2 – light brown leaves; Stage 3 – easily disintegrating dark brown leaves and Stage 4 – broken down small fragments of leaves. In order to collect material of the last stage, sediment in the seagrass bed was stirred up so that the detrital fragments that had settled on the bottom came into suspension. These were then collected with a 250 μm sieve. The material was collected in sterile bags and analysed within 3 h of collection. The detrital material of stage 4 was compared microscopically with green leaves of the seagrass and based on anatomical similarity, namely absence of stomata, presence of intercellular spaces in mesophyll, presence of fibres and similarity of cell size and shape of the epidermis (Tomlinson 1980), ascertained as to their origin. Only such fragments were used for direct biomass estimation described here.

In order to isolate and enumerate fungi in various stages of detritus, 20 discs of 0.5 cm diameter and 0.0014 g average dry weight each, were cut from random locations on different leaves. Half of these discs were surface sterilized (Newell and Fell 1982), homogenised in a sterile mortar and pestle and taken in 10 mL sterile seawater. Aliquots of 0.1 mL were plated on cornmeal agar fortified with 10 000 units of sodium benzyl penicillin and 0.05 g streptomycin sulphate per 100 mL of medium. The same procedure was used for the remaining discs which were not surface sterilized. Material of Stage 4 could not be surface sterilised because of their small size and were therefore directly homogenized for plating. Isolations were carried out in sterile inoculation chambers with Petri dishes with nutrient medium left open for 1 min., used as controls. From the numbers of colonies isolated, the numbers of fungi per unit dry weight of detrital material was calculated.

Thraustochytrids were isolated and enumerated using the MPN pine pollen baiting method of Gaertner (1968). Serial dilutions of homogenized detritus-seawater suspensions given above were baited in 5 mL sterile seawater with sterile pine pollen. Five replicates of dilutions, 1.0, 0.5, 0.25 and 0.13 mL were used. Numbers of thraustochytrids per unit dry weight detritus were calculated using Thomas' formula (Anonymous 1981).

The preserved material was later used for direct observation of fungal hyphae and thraustochytrids, by directly immersing it in lactophenol-cotton blue stain for 48 h before mounting and observations. In order to confirm that the various kinds of filaments observed using the lactophenol-cotton blue staining did indeed belong to fungi, the material was subjected to 2 further treatments:

(1) A drastic bleaching and staining technique (Koske and Gemma 1989) used for observing mycorrhizal fungi involving boiling in 2.5% KOH, bleaching in 3% H_2O_2 in NH_4OH and staining in 0.01% gentian violet. Our observations have shown that cyanobacterial and actinomycete cell walls are destroyed by this treatment;

(2) Microchemical test for chitin done by hydrolyzing the material in 120% w/v KOH for 2 h at 120 °C and staining with potassium iodide and iodine solution in 1% H_2SO_4 (Gray 1973).

Fifty random microscope fields of the lactophenol-cotton blue stained material belonging to each stage were photographed under oil immersion. Measurements of the length and width of the fungal hyphae and diameter of thraustochytrids were made from each of the photomicrographs. From this, the bio-volumes for the total areas examined were estimated. Mycelial fungal biomass was determined by assuming a dry mass density of 0.2 g/cm^3 (Newell *et al.* 1986). Thraustochytrid biomass was determined by assuming a biomass of 21.97 pg C/cell of 51.0 μm^3 volume (Raghukumar *et al.* 1987) and by assuming that carbon formed 50% of the biomass. From the biomass values calculated for 50 microscope fields, biomass per g dry wt. leaf material was calculated using the dry wt./area factors of the leaf estimated independently.

Results

Microscopic examination of stained material revealed various kinds of hyphal structures mostly within the leaf material. No hyphae were detected in stage 1. Stages 2 to 4 revealed increasingly more hyphae, with the highest numbers in stages 3 and 4. Extremely fine intracellular hyphae (0.5 μm) traversing the walls of the cells were occasionally observed in stage 3 (Figs 1 and 2). More often, closely septate hyphae of 1.5 to 2.0 μm width were present within the leaf material

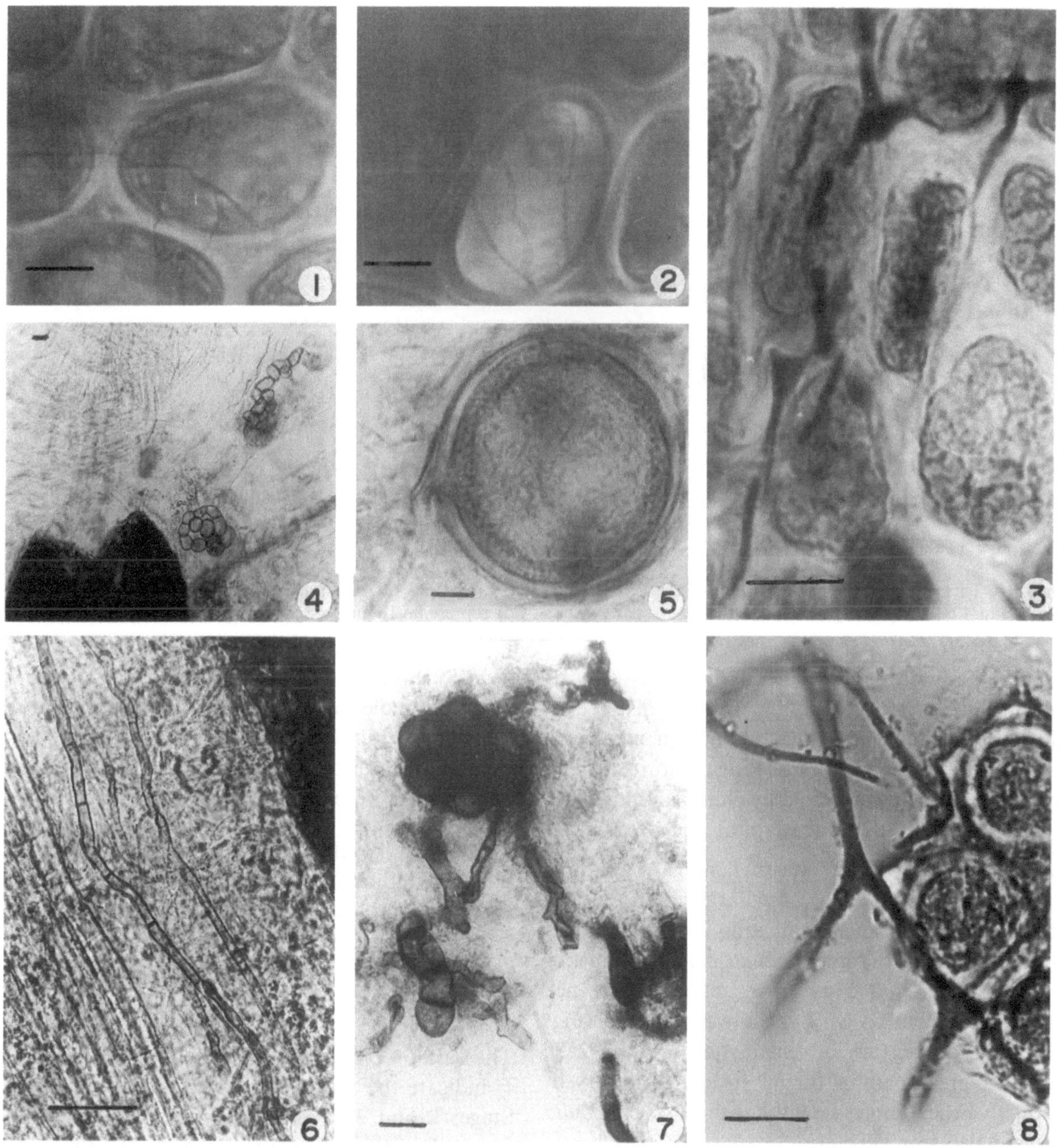

Figs 1 and 2. Fine intracellular hyphae of stage 3 detritus. Fig. 3. Hyphae with T-shaped branching in stage 4 detritus. Fig. 4. Ascocarp and aggregates of chlamydospore-like cells in stage 4 detritus. Fig. 5. Large, globose cells resembling chlamydospores in stage 4 detritus. Fig. 6. Closely septate hyphae running through the cells of detrital material. Fig. 7. Helical spores resembling those of *Cirrenalia* sp. in stage 4 detritus. Fig. 8. Detrital material of stage 4 stained for chitin revealing fungal hyphae with T-shaped branching. Bar represents 10 μm.

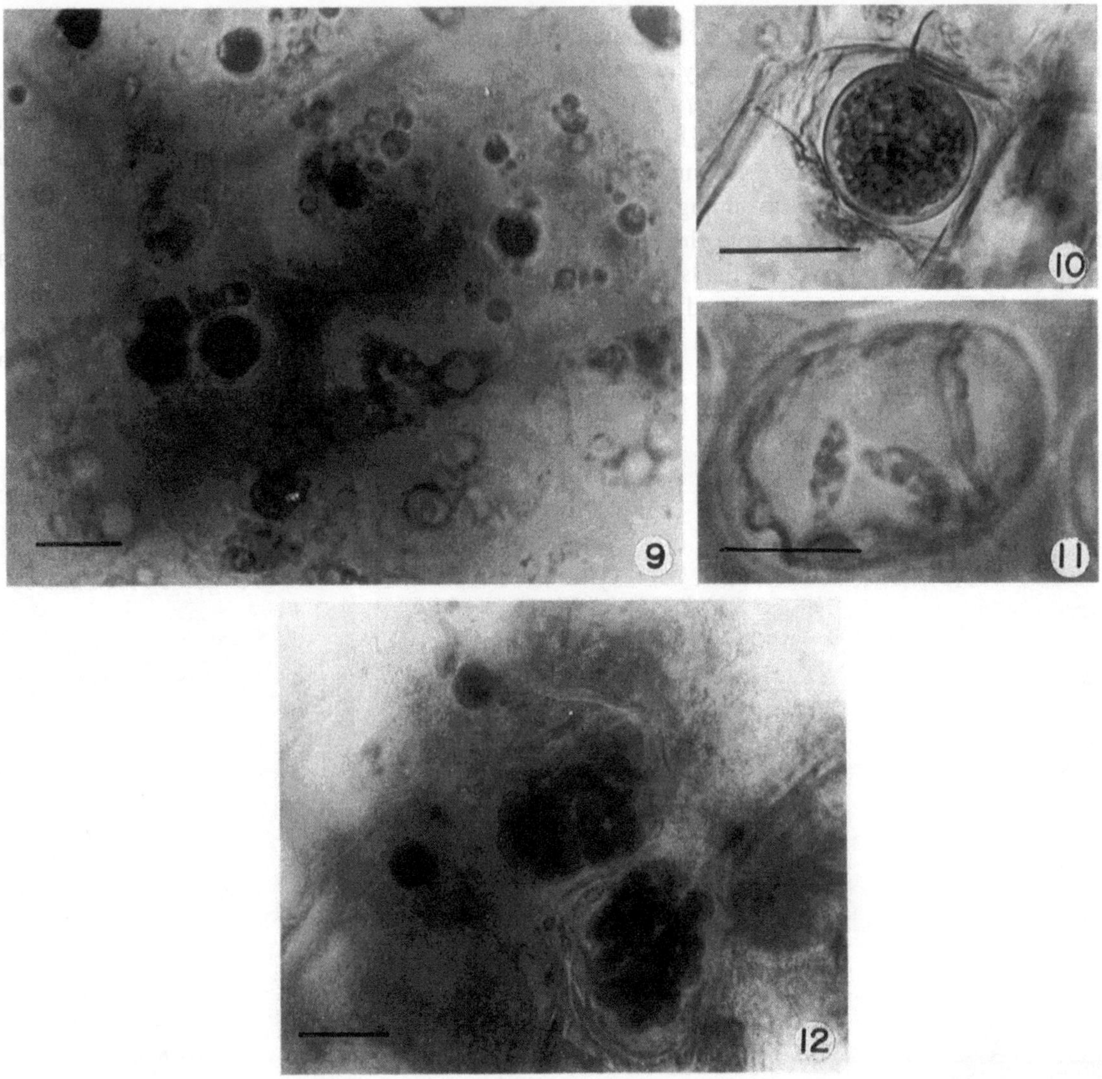

Figs 9, 10. Thraustochytrids on the surface of the leaf in stage 3. Fig. 11. Cells of *Labyrinthuloides minuta* inside the leaf cells of stage 3. Fig. 12. Thraustochytrids inside the leaf cells of stage 3. Bar represents 10 µm.

(Fig. 3). Many of these hyphae followed the cell wall contours, were intercellular and showed distinct T-shaped branching. The hyphae were frequently swollen at the site of branching. The fungal nature of these structures was also confirmed by chitin staining (Fig. 8). Such hyphae took up the typical violet stain.

Material of stage 4 often revealed fungal structures resembling ascocarps (Fig. 4). Large, globose cells measuring 130 µm and terminating a hypha, resembling the chlamydospores of *Endogone* sp. were frequently encountered (Fig. 5). An abundant presence of relatively thick brown hyphae were noticed (Fig. 6). Large aggregates of chlamydospore-like cells formed a great part of fungal structures observed. Distinctly helical spores resembling those of *Cirrenalia* sp. were also seen (Fig. 7).

Numerous thraustochytrids were observed on the surface of the detrital material of stage 3 (Figs 9, 10). Various stages of divisions of such cells could be clearly detected. The cells ranged from 3.5 to 11.0 µm in size, several of them containing zoospores. The distinct spindle shaped cells of *Labyrinthuloides minuta* (Watson *et* Raper) Perkins (Fig. 11) as well as clusters of thraustochytrid cells (Fig. 12) were present in the lumen of leaf cells.

The drastic bleaching and staining technique (Material and Methods) as well as the microchemical staining for chitin confirmed that the filaments observed in this study were fungal hyphae. The abundance of fungal hyphae observed using the chitin staining was the same as that using the lactophenol-cotton blue.

The data on bio-volume and biomass of fungi (Table I) indicate the absence of mycelial fungi in stage 1. Stages 2 and 3 revealed only very low biomass of fungi. On the contrary, stage 4 revealed 31 840 $\times 10^{-6}$ g fungus per g detritus. Thraustochytrids were detected with direct microscopy only in stage 3 where they contributed a dry wt. of 2320 $\times 10^{-6}$ g per g dry wt. of detritus.

Table I. Biomass of fungi in different stages of *Thalassia hemprichii* detritus from Kalpeni Island.

Stage of decomposition	Bio-volume (μm^3) per g dry wt. detritus	Dry weight biomass per g dry wt. detritus
1	0	0
2	8.44×10^4*	0.018×10
3	0.44×10* 2.7×10**	0.0009×10 2320×10
4	1.6×10*	$31\,840 \times 10$

* Mycelial fungi;
** Thraustochytrids

Cultural enumeration (Tables II and III) also revealed that mycelial fungi were most abundant in the last stage of decomposition. Very few fungi were observed in the other three stages. In samples from the Kalpeni lagoon, non-sporulating mycelial forms were most frequent in all stages whereas in Kavaratti, thraustochytrids dominated stage 4. Among mycelial fungi, only terrestrial species were isolated, even after surface sterilization of the detritus. Thraustochytrids were absent or low in the green leaves (stage 1) and stage 2. The commonly isolated species were *Labyrinthuloides minuta* and *Thraustochytrium motivum* Goldstein.

Discussion

While abundant fungal structures were not detected in stages 2 and 3, fungal fruiting bodies, conidia resembling those of the marine hyphomycete *Cirrenalia* sp. and masses of fungal tissue, as well as numerous hyphae were found in stage 4 of the detritus (Figs 4 to 7). These alone were considered for biomass estimations. Stage 3 contained fine filaments which could be considered filamentous cyanobacteria. However, such filaments did not occur on the surface but within the detrital substratum, in the lumen or cell walls. Endobiontic cyanobacteria are known within calcareous shells (Le Campion-Alsumard 1989), but we are not aware of instances where they have been reported within dead plant tissues. Therefore, such filaments probably belonged to fungi or actinomycetes. However, they were few and the overall biomass of stage 3 detritus was much less than that of stage 4. The variety of fungal structures detected (Figs 1 to 8) suggests that different fungal species were present.

Stage 4 detritus yielded a biomass of 0.032 g mycelial fungi per g dry wt. detritus, corresponding to 3% of the detrital material. Mycelial fungal biomass of values of up to 50% have been reported for standing dead tissues of the intertidal salt marsh grass *Spartina* (Newell *et al.* 1986). Submerged macrophytic detritus has not been sufficiently investigated for fungi. The biochemical indicator of higher mycelial fungi, ergosterol was found in amounts up to 60 µg per g dry wt. in submerged mangrove detritus (Newell *et al.* 1987). Using a conversion factor of 246 mg dry wt. fungus/mg ergosterol (Newell *et al.* 1986), this amounts to 1.48% g dry wt. fungi per g dry wt. detritus. Newell *et al.* (1987) reported negligible values for the seagrass *Thalassia testudinum*. Blum *et al.* (1988) also reported values of only 12×10^{-5} g per g detritus for the same substrate. The reasons for these

Table II. Species and numbers of fungi per g dry wt. in different stages of *Thalassia hemprichii* detritus in Kalpeni Island isolated after surface sterilization (+) and without (−).

Species	Stage 1 +	Stage 1 −	Stage 2 +	Stage 2 −	Stage 3 +	Stage 3 −	Stage 4
Acremonium sp.							1 420
Chaetomium sp.					704		
Graphium sp.							1 420
Humicola sp.							350
Penicillium sp.	704						1 060
Non-sporulating sp.	1408	1420		724	704	700	7 780
*Labyrinthuloides minuta**					560	1390	2 510
*Thraustochytrium motivum**							6 960
Other thraustochytrids*				280			
Total	2110	1420	0	1000	1970	2090	21 500

* Values calculated with MPN method.

Table III. Species and numbers of fungi per gram dry wt. in different stages of *Thalassia hemprichii* detritus in Kavaratti Island isolated after (+) and without (−) surface sterilization.

Species	Stage 1 +	Stage 1 −	Stage 2 +	Stage 2 −	Stage 3 +	Stage 3 −	Stage 4
Acremonium sp.			1340				
Aspergillus sp.	1450						
Humicola sp.							640
Thraustochytrids							48 750
Total	1450		1340				49 390

estimates which are lower than even our conservative estimate of 3% mycelial fungi in stage 4 detritus could be the following:

1. Very late stages of decomposition, such as the broken down small fragments of leaves corresponding to our stage 4 might not have been examined by Blum *et al.* (1988) and Newell *et al.* (1987). Miller and Jones (1982) have demonstrated that such late stages, where phenols and organic compounds have leached out support better growth of thraustochytrids in a decomposing macroalga. This might also be true with the mycelial fungi.

2. The physico-chemical characteristics of our study site might have favoured better fungal colonisation of detritus.

3. The Jones-Mollison technique (Newell and Hicks 1982) employed by Blum *et al.* (1988) might have yielded a lower biomass compared to the technique that we have used. The former technique, using homogenised material for direct detection of mycelial fungi might have serious handicaps (Newell *et al.* 1986). Mechanical disruption might make it difficult to recognize fungal structures. Fine hyphae (Figs 1 and 3) might go unnoticed. Intercellular hyphae pervading the cell wall might not get extracted during homogenisation (Newell and Hicks 1982). Large ascocarps and clumps of fungal structures might get disrupted beyond recognition. Non-homogenisation results in the retention of spatial configurations which makes the recognition of fungal structures easy.

Stage 4 detritus in the present study was devoid of cell contents and comprised only cell wall material of the leaf. It is likely that mycelial fungi which contributed a substantial biomass in this detritus would play an important part in lignocellulose degradation. The claim by Benner *et al.* (1984) that fungi may have an insignificant role in lignocellulose degradation in the sea might have to be evaluated in the case of *Thalassia* detritus and in other macrophytic detritus by studying late stages of decomposition and the fungi associated with them, a point also highlighted by Mouzouras *et al.* (1988).

Although thraustochytrids have been reported to be abundant in submerged plant material in the sea (Fell and Master 1973, Newell 1976), biomass estimations of these fungi have not been carried out. We have included the unmistakable spindle-shaped *Labyrinthuloides*-like cells and those within the detrital material in the cell lumens in our thraustochytrid counts. Thraustochytrids have been demonstrated within decomposing algal material (Miller and Jones 1983) and it is not likely that single-celled cyanobacteria occurred endobiotically in the decaying *Thalassia* leaves. However, our counts of thraustochytrids on the surface might have included such cyanobacteria and therefore our thraustochytrid biomass estimations may not be accurate. These fungi contributed 0.0023 g per g dry wt. detritus amounting to 0.2% in stage 3. Although this amount is not as high as mycelial fungi in stage 4, these values are much higher than even the total microbial biomass estimated for *Thalassia*, *Rhizophora*, *Halodule* and *Syringodium* by Blum *et al.* (1988). More reliable techniques for thraustochytrids are not presently available. Our biomass estimates were based on direct counts of only one focal plane of the substrate under the microscope. If it is presumed that the fungal hyphae and thraustochytrids occurred homogeneously within the detrital material, the biomass would be much higher than our conservative estimate. In view of these results, we believe that further critical studies and a reevaluation of the role and significance of fungal biomass in the nutrition of detritus feeders is warranted.

Cultural enumeration yielded the highest number of fungi in stage 4 detritus of Kalpeni Island, confirming the results of direct observations. Most such cultures were non-sporulating. Sporulating cultures of typical marine species were not isolated, apparently due to the inadequacy of the cultural techniques we have used. A number of terrestrial species were isolated even after surface sterilization of the samples. Although this procedure may not completely eliminate dormant fungal spores adhering to the surface of the samples, it nevertheless suggests the presence of such fungi in the form of active hyphae within the detritus. Although there is no direct evidence to date for this, such fungi have been regularly reported in marine macrophytic detritus (Newell 1976, Newell and Fell 1982), where they might indeed play a role.

Acknowledgements

This work was carried out under a project supported by the Department of Environment. One of us (V. S.) is grateful for a research fellowship from the Department of Environment, and to the Director, N. I. O., Dr A. H. Parulekar and Dr D. Chandramohan for support.

References

Anonymous. 1981. *Standard Methods for the Examination of Water and Wastewater.* American Public Health Association, Washington D.C. 1193 pp.

Benner, R., S. Y. Newell, A. F. Maccubbin and R. E. Hudson. 1984. Relative contribution of bacteria and fungi to rates of degradation of lignocellulosic detritus in salt marsh sediments. *Appl. Env. Microbiol.*: 36–40.

Blum, L. K., A. L. Mills, J. C. Ziemann and R. T. Ziemann. 1988. Abundance of bacteria and fungi in seagrasses and mangrove detritus. *Mar. Ecol. Prog. Ser. 42*: 73–78.

Bowen, S. H. 1987. Composition and nutritional value of detritus. *In*: (D. J. W. Moriarty and R. S. V. Pullin, eds) *Detritus and Microbial Ecology in Aquaculture.* International Center for Living Aquatic Resources Management, Manila, Phillippines.

Cammen, L. 1980. The significance of microbial carbon in the nutrition of the deposit feeding polychaete *Nereis succinea. Mar. Biol. 61*: 9–20.

Cuomo, V., F. Vanzanella, E. Fresi, F. Cinelli and L. Mazzella. 1985. Fungal flora of *Posidonia oceanica* and its ecological significance. *Trans. Br. mycol. Soc. 84*: 35–40.

Cuomo, V., S. Pegano, M. A. Pecorella and P. Parascandola. 1987. Evidence of the active role of ligno-cellulosic enzymes of marine fungi in degradation of *Posidonia oceanica* leaves. *Biochem. System. Ecol. 15*: 635–637.

Cuomo, V., E. B. Gareth Jones and S. Grasso. 1988. Occurrence and distribution of marine fungi along the coast of the Mediterranean Sea. *Prog. Oceanogr. 21*: 189–200.

Fell, J. W. and I. M. Master. 1973. Fungi associated with the degradation of mangrove (*Rhizophora mangle* L.) leaves in South Florida. *In*: (H. L. Stevenson and R. R. Colwell, eds) *Estuarine Microbial Ecology.* University of South Carolina Press, Columbia. pp. 455–465.

Findlay, S. and J. L. Meyer. 1984. Significance of bacterial biomass and production as an organic carbon source in lotic detrital systems. *Bull. Mar. Sci. 35*: 318–325.

Gaertner, A. 1968. Eine Methode des quantitativen Nachweises niederer mit Pollen köderbarer Pilze im Meerwasser und im Sediment. *Veröff. Inst. Meeresforsch. Bremerh. Sonderb. 3*: 75–92.

Gray, P. 1973 (ed). *The Encyclopaedia of Microscopy and Microtechnique.* Van Nostrand Reinhold Company, New York. 638 pp.

Jagtap, T. G. 1987. Distribution of algae, seagrasses and coral communities from Lakshadweep islands, eastern Arabian Sea. *Indian J. mar. Sci. 16*: 256–260.

Kenworthy, K. J. and G. W. Thayer. 1984. Production and decomposition of the roots and rhizomes of seagrasses, *Zostera marina* and *Thalassia testudinum* in temperate and subtropical marine ecosystems. *Bull. Mar. Sci. 35*: 364–369.

Koske, R. E. and J. N. Gemma. 1989. A modified procedure for staining roots to detect VA mycorrhizae. *Mycol. Res. 92*: 486–488.

Le Campion-Alsumard, T. 1989. Endolithic marine cyanobacteria. *In*: (G. Giraud and J. C. Thomas, eds) *Biological Potential of the Cyanobacteria.* Bull. Soc. Bot. Fr. Actual Bot. 136: pp. 99–112.

Lee, C., R. W. Howarth and B. L. Howes. 1980. Sterols in decomposing *Spartina alterniflora* and the use of ergosterol in estimating the contribution of fungi in detrital nitrogen. *Limnol. Oceanogr. 25*: 290–303.

May, M. S. 1974. Probable agents for the formation of detritus from the halophyte, *Spartina alterniflora. In*: (J. Reimold and W. H. Queen, eds) *Ecology of Halophytes.* Academic Press, New York. pp. 429–440.

Miller, J. D. and E. B. G. Jones. 1983. Observations on the association of thraustochytrid fungi with decaying seaweeds. *Bot. Mar. 26*: 345–351.

Mouzouras, R., E. B. G. Jones, R. Venkataswamy and D. M. Holt. 1988. Microbial decay of lignocellulose in the marine environment. *In*: (M.-F. Thompson, R. Sarojini and R. Nagabhushanam, eds) *Marine Biodeterioration.* Oxford and I. B. H. Publishing Co., Pvt. Ltd., New Delhi. pp. 345–356.

Moss, S. T. 1986. Biology and phylogeny of the Labyrinthulales and Thraustochytriales. *In*: (S. T. Moss, ed.) *The Biology of Marine Fungi.* Cambridge University Press, Cambridge. pp. 105–129.

Newell, S. Y. 1976. Mangrove fungi: The succession in the mycoflora of red mangrove (*Rhizophora mangle* L.) seedlings. *In*: (E. B. Gareth Jones, ed.) *Recent Advances in Aquatic Mycology.* Elek Science, London.

Newell, S. Y. 1984. Bacterial and fungal productivity in the marine environment: A contrastive overview. *In*: *Bacteriologie Marine.* Editions du CNRS, Paris.

Newell, S. Y. and J. W. Fell. 1982. Surface sterilization and the active mycoflora of leaves of a seagrass. *Bot. Mar. 25*: 339–346.

Newell, S. Y. and R. E. Hicks. 1982. Direct count estimates of fungi and biovolume in dead leaves of smooth cordgrass (*Spartina alterniflora* Loisel). *Oikos 39*: 261–268.

Newell, S. Y., R. D. Fallon and J. D. Miller. 1986. Measuring fungal biomass dynamics in standing dead leaves of a salt marsh vascular plant. *In*: (S. T. Moss, ed.) *The Biology of Marine Fungi.* Cambridge University Press, Cambridge. pp. 19–25.

Newell, S. Y., J. D. Miller and J. W. Fell. 1987. Rapid and pervasive occupation of fallen mangrove leaves by a marine zoosporic fungus. *Appl. Environ. Microbiol. 53*: 2464–2469.

Odum, E. P. 1971. *Fundamentals of Ecology.* W. B. Saunders Company, Philadelphia. 574 pp.

Ott, J. A. 1980. Growth and production in *Posidonia oceanica* (L.) Delile. *Mar. Ecol. 1*: 47–64.

Qasim, S. Z. and P. M. A. Bhattathiri. 1971. Primary production of a seagrass bed on Kavaratti atoll (Laccadives). *Hydrobiologia 38*: 29–38.

Raghukumar, S. 1990. Speculations on niches occupied by fungi in the sea with relation to bacteria. *Proc. Indian Acad. Sci. (Plant Sci.) 100*: 129–138.

Raghukumar, S., D. Chandramohan and N. Ramaiah. 1987. Contribution of the thraustochytrid *Corallochytrium limacisporum* Raghukumar to microbial biomass in coral reef lagoons. *Indian J. mar. Sci. 16*: 122–125.

Tomlinson, P. B. 1980. Leaf morphology and anatomy in seagrasses. *In*: (R. C. Phillips and C. P. McRoy, eds) *Handbook of Seagrass Biology.* Garland STPM Press, New York. pp. 7–28.

Zieman, J. C., S. A. Macko and A. L. Mills. 1984. Role of seagrasses and mangroves in estuarine foodwebs: Temporal and spatial changes in stable isotope composition and amino acid content during decomposition. *Bull. Mar. Sci. 35*: 380–392.

Botanica Marina
Vol. 34, pp. 279–284, 1991

Seasonal and Geographical Variations in the Antimicrobial Activity of Seaweeds from the Mangalore Coast of India

N. Vidyavathi and K. R. Sridhar

Department of Biosciences, Mangalore University, Mangalagangotri 574199, India

(Accepted 16 March 1991)

Abstract

Extracts of fourteen different species of seaweeds were tested against bacteria. The extracts of species from one location were more active than those from the same species from another location. Extracts of seaweeds from the post-monsoon season were much more active than from other seasons. The antimicrobial activity of seaweeds collected during the post-monsoon season was the highest in case of *Chaetomorpha antennina*, whereas it was least in *Hypnea valentiae*. Among the bacteria tested *Staphylococcus aureus* was most sensitive, and *Salmonella paratyphi* B was least sensitive. Fully grown *Gracilaria corticata* showed maximum activity against *S. aureus* compared to medium and young stages of growth. The middle part of the thalli of fully grown alga showed maximum activity compared with the terminal part and the basal part with holdfast.

Introduction

With increasing interest in seaweeds as a source of biologically active products of therapeutic potential investigations on antimicrobial activities are gaining more importance (Khaleafa *et al.* 1975, Aubert *et al.* 1979, Glombitza 1979, Naqvi *et al.* 1980, Blunden *et al.* 1981, Baker 1984, Moreau *et al.* 1984, 1988, Fenical and Paul 1984). There are few reports on the antimicrobial activity of Indian seaweeds (Naqvi *et al.* 1980, Srinivasarao and Parekh 1981, Sridhar *et al.* 1984, Padmakumar and Ayyakkannu 1986, Padmini Srinivasarao *et al.* 1986 a, 1986 b, 1988). The present investigation provides an antibacterial profile of extracts of common seaweeds from the west coast of India. Antibacterial activity of algal extracts from two different locations, in three seasons and extracted with different solvents are compared.

Material and Methods

1. Study area

Algae were collected from two representative stations viz., Someshwara (Station 1) and Surathkal (Station 2) (Fig. 1). Someshwara is a steep sandy shore, con-

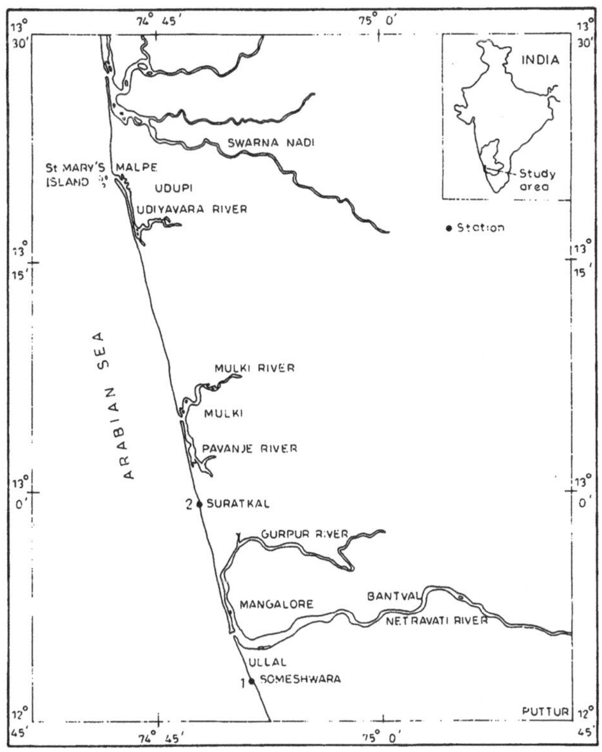

Fig. 1. Map of the investigation area with sampling stations.

sisting of many large rocks facing the full impact of the sea. At low tide several rocks are exposed and these are covered with a heavy growth of seaweed. Surathkal is a major shallow sandy shore with medium size rocks with cracks and crevices which harbour a considerable algal flora.

2. Collection and preservation

Healthy and well grown algae were collected from rocks and shells submerged under water during low tides. Plants with few epiphytes were selected and care was taken to remove the plants with their holdfast by using a sharp scalpel. They were separately cleaned on the spot with many changes of seawater in a trough to remove epiphytes, shells and other extraneous matter and were immediately transferred to separate polythene bags and placed on ice for return to the laboratory. A portion of each alga was again cleaned in seawater in the laboratory, blotted and stored in 4% formalin in seawater in small vials for identification. Voucher specimens were labelled and preserved in the laboratory. Some algae from the two sampling stations were collected at three seasons (post-monsoon, summer and monsoon) during 1988 – 1989 for detailed studies of seasonal effects on antimicrobial activity.

3. Processing and extraction

A. Cleaning

Each algal sample was thoroughly washed separately in sterile sea water in the laboratory to remove the remaining epiphytes and extraneous matter, followed by a rinse in distilled water to remove the salt on the surface of the samples. The water was drained off and the algae were spread on blotting paper and blotted to remove excess water. Whole plants with holdfast were selected for extraction.

B. Extraction

Twenty grams wet weight of the blotted samples were chopped into small pieces and ground thoroughly with a little of the respective solvents in a pestle and mortar.

i. Chloroform-methanol: The ground contents in the mortar were transferred to a soxhlet extractor and extracted with 100 mL of chloroform-methanol (2 : 1) over a low heat (55 °C) for 60 min. The extracted chloroform-methanol mixture was then evaporated to dryness at 55 °C. The dried extract was eluted with 5 mL of ethyl alcohol and stored in air-tight bottles in a refrigerator before bioassay (Ballantine *et al.* 1987).

ii. Acetone: The procedure for acetone extraction of seaweeds was similar to that of chloroform-methanol extraction. However, the dried extracts were eluted with 5 mL of acetone and preserved as described above.

4. Bioassay

The antimicrobial activity of the algal extracts was determined by the disk diffusion method (Cruickshank *et al.* 1975). Disks measuring 16.5 mm diameter were punched from Whatman No. 1 filter paper. Batches of 10 disks were dispensed to screw-capped tubes and sterilized by dry heat in hot-air oven at 140 °C for 1 h. The extract, measuring 700 µl, was pipetted aseptically into each bottle containing 10 disks. This allows 70 µl per disk, corresponding to 0.28 g wet weight of each alga. The disks with the extract were air-dried at 29 ± 1 °C. Disks prepared with ethanol or acetone served as controls.

A. Antibacterial assay

The antibacterial assay of the extracts was tested against *Bacillus subtilis*, *Eterobacter aerogenes*, *Escherichia coli*, *Proteus vulgaris*, *Salmonella paratyphi B*, *S. typhi*, *Serratia marcescens* and *Staphylococcus aureus*. These cultures were supplied by the National Chemical Laboratory, Poona, India. Sterile nutrient agar (NA) medium (Peptone, 5 g; sodium chloride, 5 g; beef extract, 1.5 g; yeast extract, 1.5 g; agar, 18 g and distilled water, 1000 mL; pH 7.4 ± 0.2) was poured into sterilized standard Petri plates as basal medium and allowed to solidify. Later 0.1 mL of inoculum in phosphate buffered saline (PBS) of 24 h cultures of bacteria grown on slants of NA medium was mixed with freshly sterilized NA medium at 45 °C and was overlayed on the basal medium and allowed to solidify. Four disks were placed aseptically above the seeded agar and pressed a little to facilitate proper diffusion. The petri plates inoculated with *Bacillus subtilis* and *Serratia marcescens* were incubated at 29 ± 1 °C and the remaining plates were incubated at 37 ± 1 °C. Control disks were also processed as described above. After 24 h all the plates except *Staphylococcus aureus* were scanned for growth of bacteria; plates with *S. aureus* were scanned after 48 h. Wherever visible inhibition was observed, the diameter of inhibition was measured using dial calipers.

Preliminary studies of *Gracilaria corticata* J. Ag. had shown pronounced antibacterial activity against *Staphylococcus aureus*, hence this seaweed was chosen to examine the antibacterial activity at different stages of growth as well as from different parts of the fully grown algal thallus. The three different growth stages were defined as young (2.5 cm), medium (6.5 cm) and fully grown (10 cm). For the study of different parts of the thallus the alga was separated into three parts *i.e.*, the basal region with holdfast (3 cm), middle (3 cm) and terminal (4 cm) portion. Twenty grams of blotted seaweed of each stage of growth and different parts were processed, extracted with acetone and tested for antibacterial activity against *S. aureus* as described above.

B. Antifungal assay

The antifungal activity of the extracts was tested against paddy seed pathogens viz., *Dichotomophthoropsis safeeullensis* Bhat, Rasheed *et* Sheeba, *Drechslera oryzae* (Breda de Haan) Subram. *et* Jain, *Trichoconiella padwickii* Ganguly and *Macrophomina phaseolina* (Tassi) Goid. These cultures were obtained through the courtesy of the Seed Technology Laboratory, Department of Applied Botany, University of Mysore, Mysore, India. Sterile potato dextrose agar (PDA) medium (Potato, infusion form, 200 g; dextrose, 20 g; agar, 18 g and distilled water, 1000 mL; pH 5.6 $\pm$ 0.2) was poured into sterilized Petri plates and allowed to solidify. Four disks of each extract were placed aseptically on the medium and pressed a little for proper diffusion. Five mm disks of five-day old test fungi grown on PDA in Petri plates were cut aseptically with cork borer and placed at 5 mm on either side of each paper disk containing extract. The plates were incubated at 29 $\pm$ 1 °C. The control disks were also processed as described above. After 48 h the plates were scanned for inhibition of fungal growth.

Results

1. Antibacterial activity

A. General observations: A total of 14 species of seaweeds viz., *Caulerpa taxifolia* (Vahl.) Ag., *Chaetomorpha antennina* (Bory.) Keutz., *Cladophora fascicularis* (Mert.) Kuetz., *Enteromorpha compressa* (L.) Grev., *Ulva lactuca* (L.) LeJolis, *Sargassum wightii* Grev., *Centroceras clavulatum* (C. Ag.) Mont., *Gelidium pusillum* (Stackh.) LeJolis, *Gigartina acicularis* (Wulff.) Lamour., *Gracilaria corticata* J. Ag., *Grateloupia filicina* (Wulff.) Ag., *G. lithophila* Boergs., *Hypnea musciformis* (Wulff.) Lamour. and *H. valentiae* (Turn.) Mont. were tested for antimicrobial activity. Figures 2 A and B show the details of the antibacterial activity of seaweeds of each station, season and solvent. The results are summarized in Table I. Irrespective of the location, post-monsoon samples showed more antibacterial activity compared to summer and monsoon samples. Chloroform-methanol extracts of post-monsoon samples were active against all eight bacteria compared to the other seasons, the acetone extracts of post-monsoon samples inhibited 50 per cent of test bacteria. *Staphylococcus aureus* was sensitive to the greatest number of extracts followed by *Proteus vulgaris*, whereas *Salmonella paratyphi* B was least sensitive. Chloroform-methanol extracts of seaweeds collected from Surathkal inhibited a greater number of bacteria compared to the extracts of seaweeds from Someshwara. *Chaetomorpha antennina*, *Gracilaria corticata* and *Grateloupia filicina* inhibited five test bacteria followed by *Enteromorpha compressa*, whereas the chloroform-methanol extracts of *Ulva lactuca* and *Sargassum wightii* did not inhibit any of the test bacteria. *Staphylococcus aureus* was most sensitive followed by *Proteus vulgaris*, *Enterobacter aerogenes* and *Bacillus subtilis*, whereas *Serratia marcescens* was not inhibited by any of the seaweed extracts.

B. Inhibition of *Staphylococcus aureus*: *Chaetomorpha antennina* showed good activity followed by *Gracilaria corticata*. Though the seaweeds of the post-monsoon showed good antibacterial activity (Table I), in the case of *S. aureus* the extracts of algae collected in the summer showed considerable activity. The chloroform-methanol extracts of *Gracilaria filicina* collected during the post-monsoon season showed the highest inhibition (28.3 mm diam.). However, *Ulva lactuca* collected from Someshwara in all seasons did not show any inhibitary activity to *S. aureus*.

C. Inhibition of other bacteria: *Chaetomorpha antennina*, *Centroceras clavulatum* and *Gracilaria corticata* from both Someshwara and Surathkal stations were active. The extracts of *Caulerpa teaxifolia*, *Cladophora fascicularis*, *Ulva lactuca*, *Sargassum wightii* and *Hypnea valentiae* from both stations did not show any inhibitory activity. The majority of seaweeds during post-monsoon season showed activity against *Bacillus subtilis*, *Enterobacter aerogenes* and *Escherichia coli*. The seaweeds such as *Cladophora fascicularis*, *Ulva lactuca*, *Sargassum wightii* and *Hypnea valentiae* did not show any inhibitary activity.

D. Activity of different growth stages/parts: Table II shows the details of activity on different stages of growth and different parts of *Gracilaria corticata* collected from Surathkal during the monsoon season on *Staphylococcus aureus*. The fully grown algae showed good activity, followed by algae with medium stage of growth. With regard to different parts, the middle part showed maximum activity followed by the terminal part, whereas the basal part with holdfast did not show any inhibition.

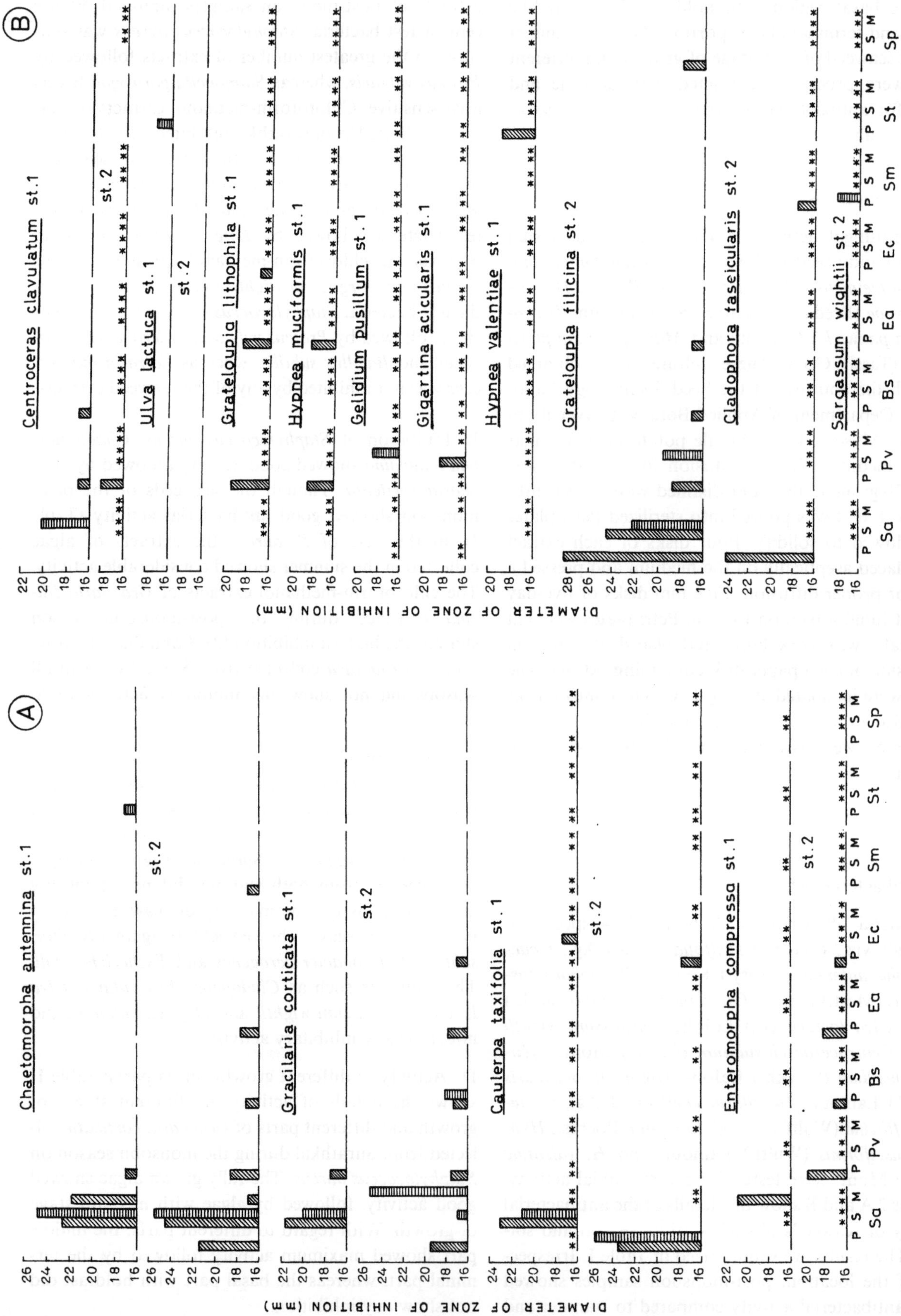

Fig. 2. Antibacterial activity of 14 seaweeds collected in different seasons from Mangalore coast of India. ▨, Chloroform-methanol extract; □, Acetone extract; *, Not screened due to nonavailability; P, Post-monsoon; S, Summer; M, Monsoon; Sa, *Staphylococcus aureus*; Pv, *Proteus vulgaris*; Bs, *Bacillus subtilis*; Ea, *Enterobacter aerogenes*; Ec, *Escherichia coli*; Sm, *Serratia marcescens*; St, *Salmonella typhi*; Sp, *S. paratyphi* B

Table I. Summary of activity of extracts of seaweeds on different bacteria.

Extracts	Seasons	Number of extracts tested	Number of extracts active on								Number of bacteria sensitive
			Sa	Pv	Bs	Ea	Ec	Sm	St	Sp	
Chloroform-methanol	Post-monsoon	17	5	10	5	6	4	2	1	1	8
	Summer	11	5	0	0	0	1	0	0	0	2
	monsoon	8	3	0	0	0	0	0	0	0	1
Acetone	Post-monsoon	17	1	0	1	0	0	1	2	0	4
	Summer	11	7	3	0	0	0	0	0	0	2
	monsoon	8	1	0	0	0	0	0	0	0	1
	Total	72	22	13	6	6	5	3	3	1	

Sa – *Staphylococcus aureus*
Pv – *Proteus vulgaris*
Bs – *Bacillus subtilis*
Ea – *Enterobacter aerogenes*
Ec – *Escherichia coli*
Sm – *Serratia marcescens*
St – *Salmonella typhi*
Sp – *S. paratyphi* B

2. Antifungal activity

To detect the antifungal activity of seaweeds, all the 72 extracts used for testing antibacterial activity (Table I) were also examined on fungal pathogens of paddy seeds. None of them showed inhibitory activity against these fungi.

Table II. Antibacterial activity of different growth stages and parts of acetone extracts of *Gracillaria corticata*[a] on *Staphylococcus aureus*

Growth stage/part	Length (cm)	Diameter of zone of inhibition (mm)
Fully grown	10.0	25.5
Medium size	6.5	23.7
Young stage	2.5	21.1
Basal part with holdfast	3.0	0.0
Middle part	3.0	26.4
Terminal part	4.0	23.1

[a] Collected from Surathkal (Station – 2) during monsoon season

Discussion

Variations in antibacterial activity were noticed from seaweeds of different geographical locations, habitats and seasons (Conover and Sieburth 1964, Henríquez *et al.* 1979, Ballantine *et al.* 1987), from different seaweed segments (Hornsey and Hide 1976) Al-Ogily and Knight-Jones 1977, Rosell and Srivastava 1987), extracts (Padmini Sreenivasarao *et al.* 1986 a) and fractions (Parekh *et al.* 1984, Padmini Sreenivasarao *et al.* 1987, 1988). However, Caccamese *et al.* (1979) found a similar GC-MS pattern and antibacterial activity of the active principle of *Laurencia decidua* Dawson collected from different habitats and seasons at Baja California. *Halimeda opuntia* (L.) Lamour. collected from many habitats and seasons at Puerto Rico did not show any marked seasonal variation and remained active against *Staphylococcus aureus* throughout the year, whereas against *Escherichia coli* two peaks were observed during October and November (Almodovar 1964). Srinivasarao and Parekh (1981) analysed *Enteromoprha intestinalis* (L.) Link and *Gracilaria corticata* collected from Gujarath Coast of India for antibacterial activity against *Staphylococcus aureus* and *Bacillus megaterium.* They were active throughout the year with a peak during the winter season (October – January). In the present study, post-monsoon (winter) samples of seaweeds were also more active than to summer and monsoon samples. The extracts of *Chaetomorpha antennina*, *Gracilaria corticata*, *Grateloupia filicina* and *Centroceras clavulatum* collected from Stations 1 and 2 were active only during summer against *Staphylococcus aureus*, whereas chloroform-methanol extracts of *Ulva lactuca* collected from Station 1 did not show any activity. Fully grown *Gracilaria corticata* as well as the middle part of the thalli in particular showed higher activity than other stages of growth and parts. These differences might be attributed to variable growth and segment dependent photosynthesis of the alga.

Station 2 (Surathkal) of the present study is shallow and calm compared with Station 1 (Someshwara) which is steep and more strongly affected by the tides. The overall antibacterial activity which was high during the post-monsoon season might be due to rapid growth, proliferation, photosynthetic activity, physical and chemical stresses as well as retention of the antibiotic principle. In fact, Khaleafa *et al.* (1975) indicated that antibacterial activity of seaweeds was often associated with photosynthesis. Ragan (1981) reviewed the literature on seaweed metabolites and concluded that most undergo diurnal and seasonal variations. Different populations of the same species of seaweed may accumulate significantly varied metabolites (Crews 1977, Mynderse and Faulkner 1978). The causes and the pattern of variation of chemical constituents of seaweeds under different conditions remain to be determined.

Acknowledgements

The authors thank Dr K. M. Kaveriappa, Department of Biosciences, Mangalore University for his constructive suggestions; Dr Sudheer A. Shetty, Department of Applied Botany, University of Mysore and D. J. Bhat, Department of Biosciences, Mangalore University for providing fungal cultures; Dr V. Chennubhotla, CMFRI, Cochin; Dr M. Mohan Joseph, College of Fisheries, Mangalore for the identification of seaweeds. The authors are grateful to Dr Felix Baerlocher, Department of Biology, Mount Allison University, Sackville, NB, Canada for correcting the manuscript and the editor for helpful comments. One of us (KRS) is thankful to the University Grants Commission, New Delhi, India for providing the financial support.

References

Almodovar, L. R. 1964. Ecological aspects of some antibiotic algae in Puerto Rico. *Bot. Mar. 6*: 143–146.

Al-Ogily, S. M. and E. W. Knight-Jones. 1977. Antifouling role of antibiotics produced by marine algae and bryozoans. *Nature 265*: 728–729.

Aubert, M., J. Aubert and M. Gauthier. 1979. Antibiotic substances from marine flora I. *In*: (Hoppe, H. A., T. Levring and Y. Tanaka, eds) *Marine Algae in Pharmaceutical Science*. Walter de Gruyter, Berlin. pp. 267–291.

Baker, J. T. 1984. Seaweeds in Pharmaceutical studies and applications. *Hydrobiologia 116/117*: 29–40.

Ballantine, D. L., W. H. Gerwick, S. M. Velez, E. Alexander and P. Guevara. 1987. Antibiotic activity of lipid soluble extracts from Caribbean marine algae. *Hydrobiologia 151/152*: 463–469.

Blunden, G., C. J. Barwell, K. J. Fidgen and K. Jewers. 1981. A survey of some British marine algae for anti-influenza virus activity. *Bot. Mar. 24*: 267–272.

Caccamese, S., P. L. Hager, K. L. Rinehart and R. B. Setzer. 1979. Characterization of *Laurencia* species by gas chromatography mass spectrometry. *Bot. Mar. 22*: 41–46.

Conover, J. T. and J. Mc. N. Sieburth. 1964. Effects of *Sargassum* distribution on its epibiota and antibacterial activty. *Bot. Mar. 6*: 147–157.

Crews, P. 1977. The breadth of monoterpene synthesis by marine red algae: potential difficulties in their application as taxonomic markers. *In*: (Faulkner, D. J. and W. H. Fenical, eds) *Marine Natural Products Chemistry*. Plenum Press, New York and London. pp. 211–223.

Cruickshank, R., J. P. Duguid, B. P. Morinion and R. H. A. Swain. 1975. *Medical Microbiology*, Vol. II, Churchill-Livingstone, New York.

Fenical, W. and V. J. Paul. 1984. Algae in medicine and pharmacology. *Hydrobiologia 116–117*: 135–140.

Glombitza, K. W. 1979. Antibiotics from algae. *In*: (Hoppe, H. A., T. Levring and Y. Tanaka, eds) *Marine Algae in Pharmaceutical Science*. Walter de Gruyter, Berlin. pp. 303–342.

Henríquez, P., A. Candia, R. Norambuena, M. Silva and R. Zemelman. 1979. Antibiotic properties of marine algae II. Screening of Chilean marine algae for antimicrobial activity. *Bot. Mar. 23*: 451–453.

Hornsey, I. S. and D. Hide. 1976. The production of antimicrobial compounds by British marine algae III. Distribution of antimicrobial activity within the algal thallus. *Br. Phycol. J. 11*: 175–181.

Khaleafa, A. F., A. M. Kharboush, A. Metwalli, A. F. Mohsen and A. Serwi. 1975. Antibiotic fungicidal action from extracts of some seaweeds. *Bot Mar. 18*: 163–165.

Moreau, J., D. Pesando and B. Caram. 1984. Antifungal and antibacterial screening of dictyotales from the French Mediterranean Coast. *Hydrobiologica 116/117*: 521–524.

Moreau, J., D. Pesando, P. Bernard, B. Caram and J. C. Pionnat. 1988. Seasonal variations in the production of antifungal substances by some dictyotales (brown algae) from the French Mediterranean Coast. *Hydrobiologia 162*: 157–162.

Mynderse, J. S. and D. J. Faulkner. 1978. Variations in the halogenated monoterpene metabolites of *Plocamium cartilagineum* and *P. violaceum*. *Phytochemistry 17*: 237–240.

Naqvi, S. W. A., S. S. Y. Kamat, L. Fernandes, C. V. G. Reddy, D. S. Bhakuni and B. N. Dhawan. 1980. Screening of some marine plant, from the Indian coast for biological activities. *Bot Mar. 24*: 51–55.

Padmakumar, K. and K. Ayyakkannu. 1986. Antimicrobial activity of some marine algae of Porto-Novo and Pondicherry water, East Coast of India. *Indian J. Mar. Sci. 15*: 187–188.

Padmini Srinivasarao, P., P. S. Rao and S. M. Karmarkar. 1986a. Antibacterial Substances from brown algae II. Efficiency of solvents in the evaluation of antibacterial substances from *Sargassum johnstonii*. *Bot. Mar. 29*: 503–504.

Padmini Srinivasarao, P., P. S. Rao and S. M. Karmarkar. 1986 b. Biological investigation on genus *Sargassum*. Antifungal activity of fractions of different species of *Sargassum*. *Phykos 25*: 6–11.

Padmini Srinivasarao, P., P. S. Rao and S. M. Karmarkar. 1987. Comparative study of antibacterial activity of fractions of *Sargassum johnstonii*. *Phykos 26*: 119–122.

Padmini Srinivasarao, P., P. S. Rao and S. M. Karmarkar. 1988. Antibacterial activity from Indian species of *Sargassum*. *Bot. Mar. 31*: 295–298.

Parekh, K. S., H. H. Parekh and P. S. Rao. 1984. Antibacterial activity of Indian seaweed. *Phykos. 23*: 216–221.

Ragan, M. A. 1981. Chemical constituents of seaweeds. *In*: (Lobban, C. S. and M. J. Wynne, eds) *The Biology of Seaweeds*. Blackwell Scientific Publications, London. pp. 589–626.

Rosell, K. G. and L. M. Srivastava. 1987. Fatty acids as antimicrobial substances in brown algae. *Hydrobiologia 151/152*: 471–475.

Sridhar, P., V. V. Lakshmi, H. Polasa, V. S. Reddy, P. C. H. Rao and G. Srimannarayana. 1984. Biological activity of some marine algal extracts. *Indian J. Mar. Sci. 13*: 90–91.

Srinivasarao, P. and K. S. Parekh. 1981. Antibacterial activity of Indian seaweed extracts. *Bot. Mar. 24*: 577–582.

Botanica Marina
Vol. 34, pp. 285–289, 1991

Decomposition of *Rhizophora mangle*, *Avicennia schaueriana* and *Laguncularia racemosa* Leaves in a Mangrove of Paranaguá Bay (Southeastern Brazil)

G. C. Sessegolo and P. C. Lana

Centro de Biologia Marinha, Universidade Federal do Paraná, 83200 Pontal do Sul, Paraná, Brazil

(Accepted 16 March 1991)

Abstract

The decomposition of senescent leaves of the mangrove species *Rhizophora mangle* L., *Avicennia schaueriana* Stapf *et* Leechman and *Laguncularia racemosa* Gaertn. was investigated at Baguaçu tidal creek (Paranaguá Bay, SE Brazil), under subtidal, intertidal and supratidal conditions. Overall decomposition rates were evaluated by weight loss over time in litter bags. Half-life for decomposition varied between 10.5 days for *Avicennia* leaves permanently immersed in the water column and 249 days for *Rhizophora* leaves under supratidal conditions. Instantaneous decomposition constants (k), that ranged from −0.003 (*Rhizophora*, supratidal) to −0.043 (*Avecennia*, immersed), were consistently higher with decreasing elevation of site location, indicating that submersion frequency is important in determining the rate of plant breakdown. Decomposition rates at the study site, much faster than many others reported for tropical or subtropical regions, were probably accelerated by the presence of an active grazing macrofauna.

Introduction

Mangroves have been traditionally considered as highly productive systems that support estuarine food webs and commercial fisheries through the export of organic matter. Despite the absence of adequate data to validate this generalization, there is compelling evidence that organic detritus is an important food source for many estuarine organisms. Plant material of low nutritional value needs to be transformed to 'protein-enriched' detritus as a pre-requisite to its ultimate incorporation into estuarine trophic webs. This transformation, mediated by decomposition processes, has been emphasized in remineralisation and detrital research. Decomposition involves mechanical fragmentation, through weathering and leaching, followed by colonization or direct utilization of particulate or dissolved organic matter by microorganisms (Camilleri and Ribi 1986). Decomposition of mangrove litter has been studied by a number of authors in world-wide scattered localities (Heald 1971, Lugo and Snedaker 1975, Albright 1976, Boomruang 1978, Cundell *et al.* 1979, Fell *et al.* 1980, Cintrón and Schaeffer-Novelli 1983). However, there have been very few studies of rates of litter decomposition in Brazilian mangroves (Adaime 1985, Panitz 1986, 1987, 1989, Schmidt 1988).

In this paper we measure *in situ* degradation rates of leaves of the mangrove species *Rhizophora mangle* L., *Avicennia schaueriana* Stapf *et* Leechman and *Laguncularia racemosa* Gaertn. in a subtropical mangrove of Paranaguá Bay (SE Brazil).

Material and Methods

The site of this investigation was Baguaçu tidal creek (25°33′03″S, 48°23′05″W) situated in Paranaguá Bay, a semi-enclosed estuarine system bounded by mangroves and saltmarshes (Fig. 1). The adjacent mangrove swamp extends for about 43 ha around the study site, and contains three mangrove species, *Rhizophora mangle*, *Laguncularia racemosa* and *Avicennia*

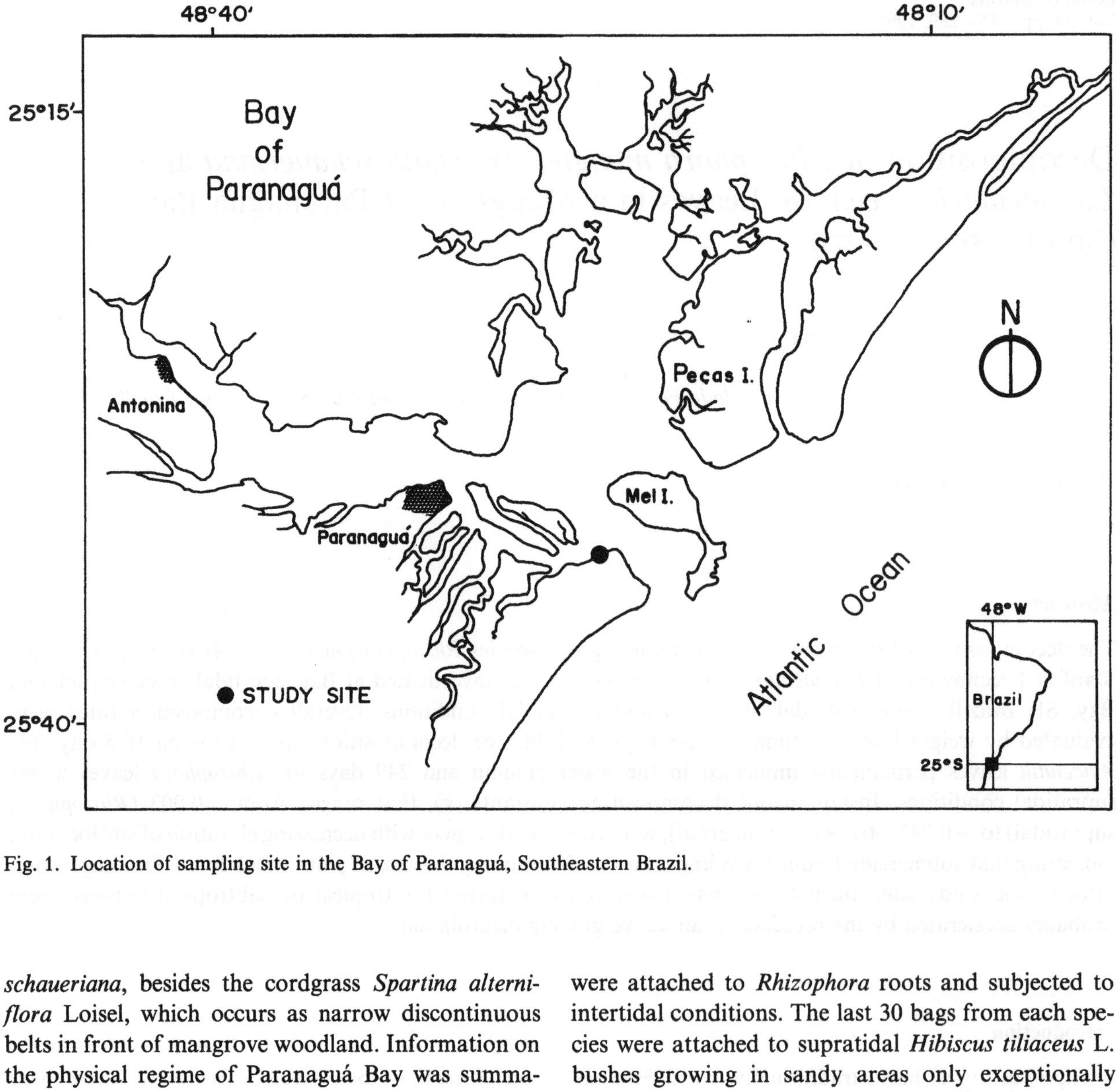

Fig. 1. Location of sampling site in the Bay of Paranaguá, Southeastern Brazil.

schaueriana, besides the cordgrass *Spartina alterniflora* Loisel, which occurs as narrow discontinuous belts in front of mangrove woodland. Information on the physical regime of Paranaguá Bay was summarized by Knoppers *et al.* (1987). Average salinity values at the study site are higher than 28‰; however, heavy precipitation, mainly in summer, can lead to a salinity decrease down to 20‰. Surface water temperature is around 20 °C in winter and 30 °C in summer. Local tides reach an almost semidiurnal pattern during maximum tidal amplitudes (*ca* 2 m).

Senescent mangrove leaves, yellow in colour and ready to fall, were picked from the trees on 3 November 1987. The leaves were divided into samples, each one consisting of 20 g of fresh weight from each of the three mangrove species. Each sample was placed in a separate nylon mesh bag (about 15 – 20 cm along each edge and 2.0 mm mesh size, large enough to permit entry of small invertebrates). Thirty bags from each species were attached to wooden stakes and exposed to subtidal conditions, at the tidal creek near the collecting site. Another 30 bags from each species were attached to *Rhizophora* roots and subjected to intertidal conditions. The last 30 bags from each species were attached to supratidal *Hibiscus tiliaceus* L. bushes growing in sandy areas only exceptionally flooded by saltwater. Average conversion factors for subsequent comparison with the dry weights of the field results were evaluated from 5 fresh samples of 20 g for each species, dried to constant weight at 80 – 90 °C.

Two bags from each species were retrieved from the three sites after 6, 13, 20 and 27 days, and then at monthly intervals for one year, until decomposition was complete. In the laboratory, the leaves were first searched for associated animals, the sediment was removed by gently rubbing the leaves in running tapwater and the samples were dried to constant weight at 80 – 90 °C.

Percentage weight loss was plotted against time, in order to evaluate decomposition half-time for each species under the three different exposure conditions. Instantaneous loss rates were calculated following an exponential model (Olson 1963).

Results

The factors for converting wet weight to dry weight were 0.48 for *Avicennia*, 0.36 for *Rhizophora* and 0.29 for *Laguncularia*. Loss rates, expressed by the constant k, are shown in Table I. Time courses for decomposition of the leaf material for the three species, under different exposure conditions, are shown in Figures 2, 3 and 4.

Table I. Decomposition constants (k) for the three leaf tissues under different exposure conditions.

Species	Subtidal	Inter-tidal	Supra-tidal
Avicennia schaueriana	−0.043	−0.019	−0.013
Rhizophora mangle	−0.015	−0.006	−0.003
Laguncularia racemosa	−0.016	−0.012	−0.008

Under subtidal conditions, leaf matter had a half-life of 10.5 days for *Avicennia*, 26 for *Laguncularia* and 36 for *Rhizophora*. At the intertidal site, *Avicennia* material had a half-life of 15 days, *Rhizophora*, of 118.5 days and *Laguncularia*, of 70.5 days. Leaf decomposition was much slower at the supratidal site, where *Avicennia* had a half-life of 55 days, *Rhizophora*, 249, and *Laguncularia*, 102 days.

The associated fauna were composed mainly of xanthid and porcellanid crabs, unidentified amphipods and oligochaetes and the polychaetes *Capitella* sp. and *Perinereis vancaurica* (Ehlers). Spats, juveniles and adults of the mangrove oyster *Crassostrea rhizophorae* (Guilding) were found attached to the subtidal bags. Even though the associated macrofauna were not systematically quantified, their bulk was frequently larger than that of the litter material, especially in the last months of the experiment.

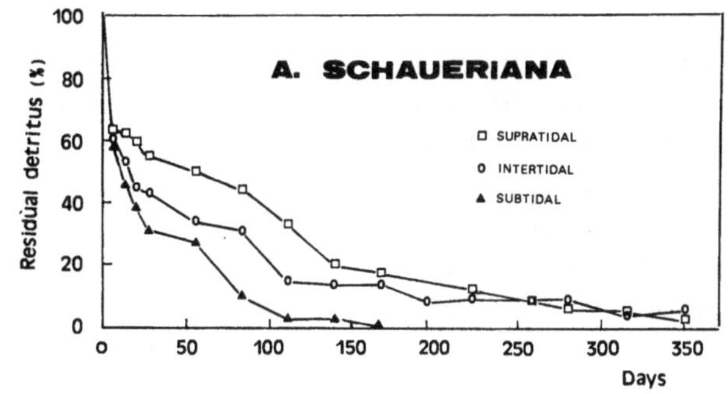

Fig. 2. Time course of leaf decomposition for *Avicennia schaueriana* under supratidal, intertidal and subtidal conditions.

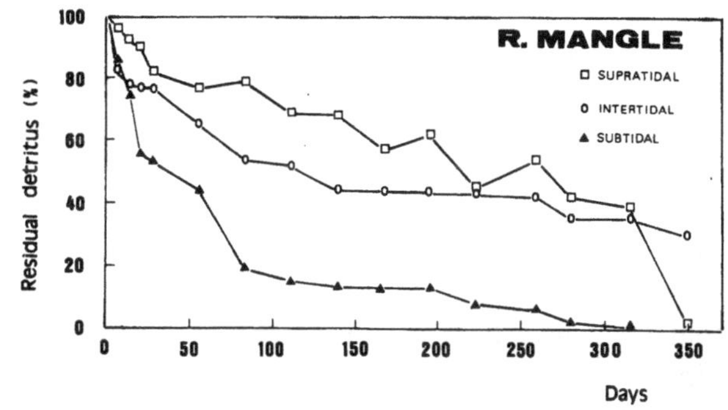

Fig. 3. Time course of leaf decomposition for *Rhizophora mangle* under supratidal, intertidal and subtidal conditions.

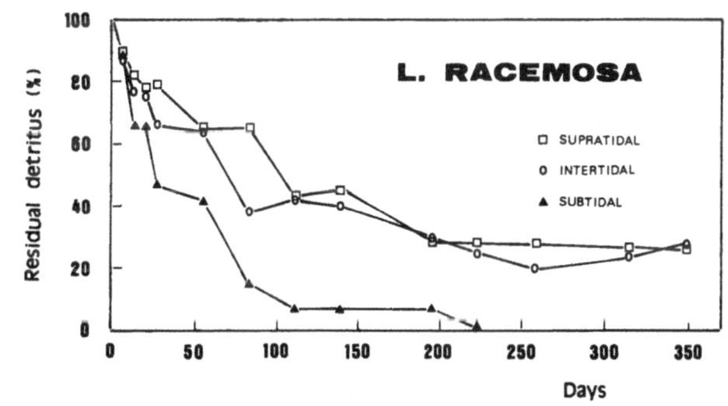

Fig. 4. Time course of leaf decomposition for *Laguncularia racemosa* under supratidal, intertidal and subtidal conditions.

Discussion

It has been suggested that decomposition dynamics in mangrove environments are primarily controled by temperature, humidity, soil pH, aeration and the nature of plant material, microbial populations and soil fauna (Lugo and Snedaker 1974).

The fairly rapid initial loss of weight in the early stages of the experiment for the three species under different exposure conditions was an expected feature, due to weathering and leaching out of plant material. A similar pattern has been widely reported (Boomruang 1978, Woodroffe 1982, Boto 1982).

Differences in the decomposition rates along a gradient of exposition have been also reported for other estuarine macrophytes (Long and Mason 1983). In our study, the decomposition rates were four to six times faster at the submersed site. This pattern was most probably related to:

(1) the higher levels of litter mechanical fragmentation and transport by tidal forces at the subtidal site and

(2) the development of subsurface anaerobic conditions at the supratidal site, where litter bags could be rapidly covered by fine sand with up to 5% of organic matter and less than 10% of water content (Lana, personal observation).

Anaerobic decomposition is known to be much slower than aerobic breakdown of organic matter (Boto 1982). This could explain, together with lower humidity that can retard the colonization by fungi, the much slower rates of plant breakdown at the supratidal site. In addition, anaerobiosis can be considered an intrinsic artifact of the litter-bag method, due to the crowding of leaves inside the bags. This may have happened at the intertidal and supratidal sites. However, anaerobiosis would be an unexpected condition in the well oxygenated waters of the tidal creek, where the dissolved oxygen content varies between 6.0 and 8.0 mL L^{-1} (Sessegolo, personal observation).

The consistently slower decomposition rates for *Rhizophora* leaves have been attributed to their relative thickness and higher tannin content (Boomruang 1978, Cundell *et al.* 1979, Adaime 1985, Benner and Hodson 1985). As pointed out by Benoit and Starkey (1968), the presence of tannin in mangrove leaves can probably delay the colonization by fungi and bacteria. González-Farias and Mee (1988) even hypothesized that there may be a negative feedback between the input of mangrove detritus to a tropical coastal lagoon and its degradation, such that as detritus input increases, detritus degradation rates decrease.

Our results are similar to those reported by Boomruang (1978) for *Rhizophora* and *Avicennia* leaves in a Thailandese mangrove but they differ widely from the decomposition rates reported by Adaime (1985) and Panitz (1986, 1987) for Brazilian subtropical mangroves located near our study site. It is difficult to make direct comparisons of half-life or instantaneous loss rates of leaf material evaluated by such studies, due to methodological differences. For instance, our conversion factors were much lower than those reported by Adaime (1985) for the same mangrove species. At the same time, it is difficult to assess the degree of submersion or exposure of litter bags in other studies, mainly in those submitted to intertidal conditions. The decomposition rates reported by Panitz (1986) for a mangrove located about 300 km south of our study site are up to three or four times slower. The rates reported by Adaime (1985) for a mangrove in Cananéia, about 100 km to the north, are either faster or slower than ours, depending on the species and the exposure site of the litter bags.

Some studies have implied that the associated macrofauna may play a significant role in the degradation rate of mangrove litter (Heald 1971, Boomruang 1978, Woodroffe 1982, Adaime 1985). This view is largely supported by our findings. Macrofaunal grazers, such as *Capitella* sp., which behaves locally as a true leaf-borer, may accelerate the decomposition of mangrove leaves. Adaime (1985) suggested that the activity of detritus-feeding invertebrates may be enhanced by higher salinities and higher summer temperatures, which would explain the faster decomposition rates recorded by her during summer. However, it was shown by Lana and Guiss (1991) that the density of local detritus-feeding populations is much higher in winter, at least in adjacent saltmarsh systems. This is a direct reflection of a higher availability of detritus in winter, due to concomitant higher mortality rates of the aboveground fractions of *Spartina alterniflora* (Lana *et al.* 1991). It is most probable that such a situation may also be expected in local mangrove environments. The effect of temperature could be in fact masked or supplanted by macrofaunal activity during winter and by the location of the fallen material within the mangrove.

Decomposition ***in situ*** may be responsible for only a small fraction of overall remineralisation at the study site, since most of litter material is carried away very quickly during each tidal cycle (Lana, personal observation). Hence, average decomposition rates for mangrove material in the estuarine waters of Paranaguá Bay may be even faster than those reported here for leaves exposed to intertidal or supratidal conditions.

Acknowledgements

This study was supported by the Brazilian National Research Council (CNPq) and by the Centro de Biologia Marinha (Universidade Federal do Paraná). Clarice Panitz and Yara Schaeffer-Novelli were most helpful in the planning of the research project. We wish to express special thanks to Elenise A. Bastos, Christoph B. Jaster, Nagib Machado da Veiga and to all the other colleagues who helped in field and laboratory work.

References

Adaime, R. R. 1985. Produção do bosque de mangue da gamboa Nóbrega (Cananéia, 25° Lat S-Brasil). Thesis Univ. S. Paulo, São Paulo, 305 pp.

Albright, L. J. 1976. *In situ* degradation of mangrove tissues. *New Zeal. Journ. Mar. Fresh. Res. 10*: 385–389.

Benner, R. and R. E. Hodson. 1985. Microbial degradation of the leachable and lignocellulosic components of leaves and wood from *Rhizophora mangle* in a tropical mangrove swamp. *Mar. Ecol. Progr. Ser. 23*: 221–230.

Benoit, R. E. and R. L. Starkey. 1968. Enzyme inactivation as a factor in the inhibition of decomposition of organic matter by tannins. *Soil. Sci. 105*: 203–208.

Boomruang, P. 1978. The degradation of mangrove leaves of *Rhizophora apiculata* (BI) and *Avicennia marina* (Forsk) Vierh. at Phuket Island, Thailand. *Res. Bull. Phuket Mar. Biol. Cent. 26*: 1–7.

Boto, K. G. 1982. Nutrient and organic fluxes in mangroves. *In*: (Clough, B. F., ed.) *Mangrove Ecosystems in Australia – Structure, Function and Management.* Proc. Australian National Mangrove Workshop 1979, Australian Institute of Marine Science and Australian National University Press. pp. 239–257.

Camilleri, J. C. and G. Ribi. 1986. Leaching of dissolved organic carbon (DOC) from dead leaves, formation of flakes from DOC, and feeding of flakes by crustaceans in mangroves. *Mar. Biol. 91*: 337–344.

Cintrón, G. and Y. Schaeffer-Novelli. 1983. *Introducion a la Ecologia del Manglar.* Unesco-Rostlac, Montevideo, 109 pp.

Cundell, A. M., M. S. Brown, R. Stanford and R. Mitchell. 1979. Microbial degradation of *Rhizophora mangle* L. leaves immersed in the sea. *Estuarine Coastal Mar. Sci. 9*: 281–286.

Fell, J. W., R. C. Cefalu, I. M. Master and A. S. Tallman. 1975. Microbial activities in the mangrove (*Rhizophora mangle*) leaf detrital system. *In*: (Walsh, G., S. C. S. Snedaker and H. Teas, eds) *Proc. International Symposium on the Biology and Management of Mangroves.* Gainesville, Institute of Food and Agricultural Sciences, University of Florida, vol. 2, pp. 661–679.

González-Farias, F. and L. D. Mee. 1988. Effect of mangrove humic-like substances on biodegradation rate of detritus. *J. Exp. Mar. Biol. Ecol. 119*: 1–13.

Heald, E. J. 1971. The production of organic detritus in a south Florida estuary. *Univ. Miami Sea Grant Tech. Bull. 6*: 110 pp.

Knoppers, B. A., F. P. Brandini and C. A. Thamm. 1987. Ecological studies in the Bay of Paranaguá. II. Some physical and chemical characteristics. *Neritica 2* (1): 1–36.

Lana, P. C. and C. Guiss. 1991. Influence of *Spartina alterniflora* on the structure and temporal variability of macrobenthic associations in a tidal flat of Paranaguá Bay (Southeastern Brazil). *Mar. Ecol. Progr. Ser.*: In press.

Lana, P. C., C. Guiss and S. T. Disaró. 1991. Seasonal variation of biomass and production dynamics for above- and belowground components of a *Spartina alterniflora* marsh in the euhaline sector of Paranaguá Bay (SE Brazil). *Est. Coast. Shelf Sci. 32*: 231–241.

Long, S. P. and C. F. Mason. 1983. *Saltmarsh Ecology.* Blackie & Son Ltd., Glasgow.

Lugo, A. E. and S. C. Snedaker. 1974. The ecology of mangroves. *A. Rev. Ecol. Syst. 5*: 39–64.

Lugo, A. E. & S. C. Snedaker. 1975. Properties of a mangrove forest in Southern Florida. *In*: (Walsh, G., S. C. Snedaker and H. Teas, eds) *Proc. International Symposium on the Biology and Management of Mangroves.* Gainesville. Institute of Food and Agricultural Sciences, University of Florida, v. 1. pp. 170–212.

Olson, J. S. 1963. Energy storage and the balance of producers and decomposers in ecological systems. *Ecology 44*: 322–331.

Panitz, C. M. N. 1986. Produção e decomposição de serapilheira no mangue do rio Itacorubi, Ilha de Santa Catarina, Florianópolis, Brasil (27°35′S–48°31′W). Thesis Univ. São Carlos, São Carlos, Brazil. 601 pp.

Panitz, C. M. N. 1987. Perfil funcional da gramínea *Spartina alterniflora* no mangue do rio Itacorubi, Ilha de Santa Catarina, Florianópolis, Brasil (27°35′S–48°31′W). *Simpósio Sobre Ecossistemas da Costa Sul e Sudeste Brasileira,* Vol. 1. Publicação da Academia de Ciências do Estado de São Paulo, 54: 100–116.

Panitz, C. M. N. 1989. *Spartina alterniflora* como fonte potencial de matéria orgânica no manguezal do rio Itacorubi, Ilha de Santa Catarina (Florianópolis, Brasil). *In*: (Poli, C. R. and P. A. M. Nascimento, eds) *Anais do II Seminário sobre Ciências do Mar.* Universidade Federal de Santa Catarina. pp. 107–118.

Robertson, A. I. 1986. Leaf-burying crabs: their influence on energy flow and export from mixed mangrove forests (*Rhizophora* spp.) in northeastern Australia. *J. Exp. Mar. Biol. Ecol. 102*: 237–248.

Schmidt, G. 1988. Manguezal de Cananéia. Concentração dos elementos químicos carbono, nitrogênio, cálcio, fósforo, manganês, ferro e cinzas em folhas de mangue e sua correlação com taxas de decomposição e a salinidade. Thesis Univ. S. Paulo, São Paulo, 370 pp.

Woodroffe, C. D. 1982. Litter production and decomposition in the New Zealand mangrove, *Avicennia marina* var. *resinifera. New Zeal. Journ. Mar. Fresh. Res. 16*: 179–188.

Botanica Marina
Vol. 34, pp. 291–301, 1991

Seasonality of Growth and Production of a Deep-water Population of *Halimeda tuna* (Chlorophyceae, Caulerpales) in the North-western Mediterranean

E. Ballesteros

Centre d'Estudis Avançats, CSIC, C. Sta. Bàrbara s/n. 17300 Blanes, Girona, Spain

(Accepted 16 March 1991)

Abstract

Growth of a deep-water population of *Halimeda tuna*, on a rocky bottom at 18 meters depth, has been recorded during an annual cycle at a site in the north-western Mediterranean Sea. Growth, as reflected in variations in the number of thalli and the production of new segments changed seasonally, being maximal in summer and minimal in winter. Growth appeared to be related mainly to temperature and irradiance whereas loss of segments seemed to be independent of environmental factors such as temperature, irradiance and nutrient concentration in the seawater. Physical disturbances, such as periods of high hydrodynamism (storms) and herbivory, may account for these losses. Seasonal changes in the epiphytic assemblage growing on the segments of *Halimeda* have also been demonstrated, with a maximum biomass and species richness in early summer. Production of *Halimeda tuna* has been estimated as 680 gdw m^{-2} y^{-1}, equivalent to 113.6 g organic C m^{-2} y^{-1} and to 314.3 g $CaCO_3$ m^{-2} y^{-1}, and the yearly P/B ratio amounted to 1.87 y^{-1}. These values are similar to other estimates of primary production and carbonate deposition for different deep-water species of *Halimeda* growing in tropical seas, so the contribution of *Halimeda* to productivity and sedimentology in Mediterranean waters may be far from negligible.

Introduction

Calcareous green algae of the genus *Halimeda* are widespread in tropical and subtropical seas (Hillis-Colinvaux 1980). Because of their wide bathymetric distribution (Littler *et al.* 1985), their high biomass (Wolanski *et al.* 1988) and coverage (Hillis-Colinvaux 1986 a) at different depths, and their high carbonate deposition (Neumann and Land 1975, Bach 1979, Jensen *et al.* 1985), *Halimeda* has been noted as a major contributor to the geomorphology of reef structures (Hillis-Colinvaux 1986 b) in tropical seas.

Coral reefs do not develop in the Mediterranean Sea, although the mean water temperature and nutrient concentrations in the seawater do approach values reported for the limits of coral reef distribution (Zabala and Ballesteros 1989). Nevertheless, the Mediterranean flora and fauna has much in common with tropical seas. This can be clearly observed in algal biogeography (Feldmann 1958, Alvarez *et al.* 1988). A number of genera with a pantropical distribution, such as *Acetabularia*, *Polyphysa*, *Valonia*, *Anadyomene*, *Udotea* and *Caulerpa* (Chlorophyta), *Zonaria* and *Sargassum* (Phaeophyta), and *Liagora*, *Galaxaura*, *Wrangelia* and *Amphiroa* (Rhodophyta) have Mediterranean representatives. However, there are only a small number of pantropical species inhabiting the Mediterranean Sea (Feldmann 1937), with *Halimeda tuna* being the most widespread one. In fact, *H. tuna* is a common and even abundant species in the submerged vegetation throughout the Mediterranean (Meinesz 1980).

In western Mediterranean waters, *H. tuna* occurs mainly in two quite different habitats: the photophilic and thermophilic communities of coastal lagoons and other sheltered places in the warmer waters of the central Mediterranean; and in the sciaphilic communities developing below 15 meters depth (Meinesz 1980, Boudouresque 1985).

The growth of *H. tuna* in Mediterranean deep-waters has already been noted by Drew (1969), but to date, there are no estimates of its primary production. Until the work of Meinesz (1980), dealing with the growth of *Halimeda tuna* at 5 meters depth on a vertical rocky wall from the French Mediterranean coast, it was thought that *Halimeda* only grew one segment per year (Feldmann 1937). Meinesz (1980) demonstrated the high turnover of *Halimeda* and emphasized the importance of rhizoids in the growth strategy of this species.

The objectives of this study were to quantify growth and production rates of a deep-water population in *H. tuna* and determine its variations during an annual cycle. Changes in the species composition and biomass of the epiphytic community growing on the thalli of *Halimeda* have also been recorded. The results provide an example of the growth and production of a deep-water community in the north-western Mediterranean Sea and serve as a reference to assess the importance of calcareous green algae to productivity and sedimentology processes in deep waters of warm-temperate seas.

Material and Methods

Underwater work was conducted on the rocky seabed around Tossa island (NE Spain, NW Mediterranean, 41°48′N – 2°26′E) (Fig. 1), using normal SCUBA techniques. Populations of *Halimeda tuna* were present from 12 – 15 meters to 30 – 35 meters depth. *Halimeda* was restricted to steeply inclined walls (70 – 90°) in shallow waters, but it became the dominant species of erect algae on sub-horizontal rocky substrate at a depth of 20 – 22 meters (Ballesteros, in press). Although the community was optimally developed between 22 and 32 meters, the primary sampling site was situated on an almost vertical wall at 18 meters depth, chosen in order to avoid long decompression times after the sampling dives.

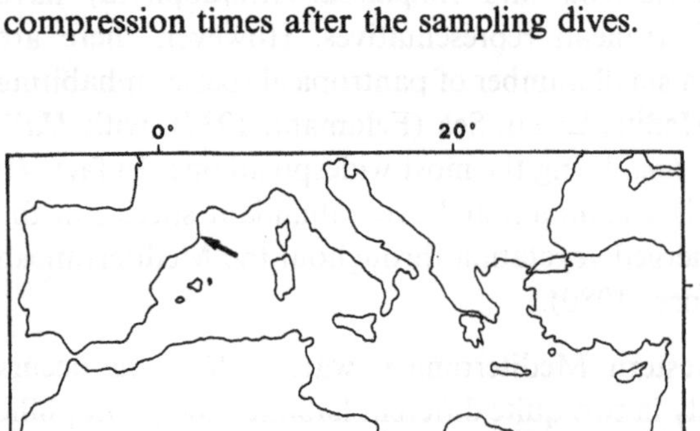

Fig. 1. Location of the sampling site (arrow) (Tossa de Mar, NW Mediterranean).

Samples were collected at approximately two-monthly intervals to determine the overall community structure and species composition. Growth was determined by marking all the erect thalli of *Halimeda* present in a permanent quadrat of 1000 cm^2 of sea bed. Individual thalli were identified with small numbered plastic tags firmly tied to the basal segment of each thallus by a short nylon thread. Distribution of *Halimeda* thalli inside the sampling area was also mapped by means of a plastic grid to account for possible losses of marks and to facilitate the identification of new thalli. Each thallus was schematically sketched *in situ* at approximately two-monthly intervals, with indications of the number of segments between each ramification and some notes about their appearance (*e.g.* basal segment, old segment, green segment) (Fig. 2). The different parts of the thallus were easily recognized when comparing the sketches of the same thallus taken at each time interval. New segments and branches were identified, and the number of segments lost could also be determined. After the final field measurements, all the plants were collected. Each thallus was carefully cleaned of epiphytes, dried at 100 °C for 24 hours and weighed. The regression between the number of segments and plant dry weight was calculated and used to estimate the average segment biomass. Subsamples of each thallus were combusted in a muffle furnace at 550 °C for six hours to obtain mean ash content (Brinkhuis 1985). Carbon content of dried segments was determined with a Carlo-Erba Nitrogen Analyzer 1500. Determination of average carbonate content was performed on plant material collected in June using a Bernard calcimeter (Siesser and Rogers 1971). Organic carbon content was estimated as the difference between total carbon and that contained in the carbonate component of the *Halimeda* plants. Production was calculated as the biomass of new segments produced at each time interval. Ratios of P/B were estimated as the quotient between production of a time interval and the initial biomass of the interval.

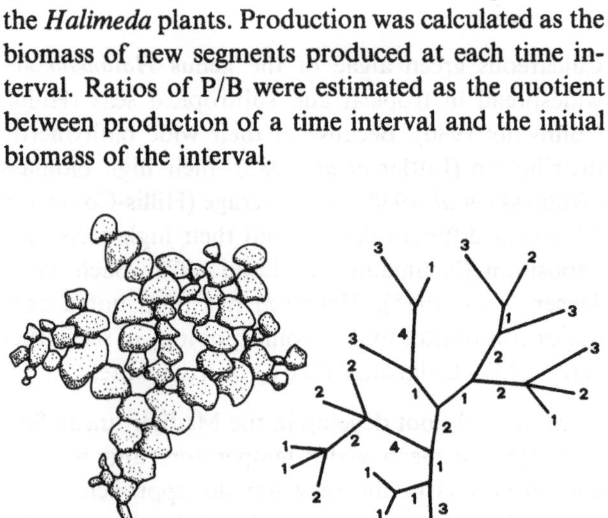

Fig. 2. Habit of a thallus of *Halimeda tuna* and its corresponding schematic sketch. Numbers in the sketch correspond to the number of segments between each ramification.

In order to relate growth of *Halimeda tuna* to the environmental factors usually limiting production of seaweeds, water transmittance, dissolved nutrient concentration and seawater temperature were measured at frequent intervals. Water transmittance was estimated from weekly Secchi disk measurements, and underwater irradiance was computed according to the method proposed by Weinberg and Cortel-Breeman (1978); irradiance at 23 meters depth has been calculated in this way in order to obtain an approximate value for light reaching the seabed where *Halimeda* grows. Temperature was monitored weekly at the sampling station at 18 meters depth and water samples for determination of dissolved nutrients were collected monthly at 18 meters depth. Nitrate and phosphate concentration in seawater was determined following methods recommended in Grasshoff *et al.* (1983).

Significance for correlation coefficients has been tested using a t-test (Sokal and Rohlf 1969).

Authorities for species are given in Tables I and II.

Results

The algal community

Halimeda tuna dominated the erect stratum and *Mesophyllum lichenoides* was the main component encrusting the substratum (Table I). In addition to *Halimeda*, *Udotea petiolata* was the only erect species with a significant biomass, whilst the identified species of the genus *Peyssonnelia* grew attached to the substratum. An unidentified melobesioid alga and *Lithophyllum expansum* were the other important encrusting species with high coverage.

Sixty one species of macroalgae were observed growing on *Halimeda* during the year (Table II). Small red algae dominated the epiphytic assemblage and, in this group, species of Ceramiaceae and Delesseriaceae had the greatest number of representatives. Brown algae, with the exception of *Sphacelaria cirrosa*, were rare and its appearance was highly seasonal. Green algae were only occasionally observed. The majority of epiphytic species had a spring–summer development and the number of species, and their overall biomass, were both greatest at the beginning of the summer (Table II).

Halimeda *growth and biomass*

The number of thalli of *Halimeda* was relatively constant during the year (Fig. 3) and some of them persisted for the whole annual cycle. The largest number of thalli were found in August and the maximum increase occurred during the June–August period.

Table I. Yearly average composition and species biomass of the *Halimeda tuna* community at Tossa island. Neither epiphytic species nor species with an average biomass lower than 0.1 gdw m^{-2} have been indicated.

	g dwt m^{-2}
Erect algae	
Halimeda tuna (Ellis *et* Solander) Lamouroux	243.7
Peyssonnelia harveyana J. Agardh	9.1
Udotea petiolata (Turra) Boergesen	9.0
Peyssonnelia bornetii Boudouresque *et* Denizot	6.6
Peyssonnelia squamaria (Gmelin) Decaisne	6.4
Peyssonnelia rubra (Greville) J. Agardh	1.5
Corallina elongata Ellis *et* Solander	1.0
Polysiphonia banyulensis Coppejans	0.4
Rodriguezella pinnata (Kützing) Schmitz	0.3
Halopteris filicina (Grateloup) Kützing	0.3
Peyssonnelia sp.	0.3
Predaea ollivieri J. Feldmann	0.2
Sphaerococcus coronopifolius Stackhouse	0.2
Dictyota dichotoma (Hudson) Lamouroux	0.2
Valonia macrophysa Kützing	0.1
Encrusting algae	
Mesophyllum lichenoides (Ellis) Lemoine	1081.7
Melobesiae (unidentified)	313.0
Lithophyllum expansum Philippi	104.6
Peyssonnelia rosa-marina Boudouresque *et* Denizot	63.3
Lithothamnion sp.	29.6
Lithophyllum incrustans Philippi	20.4
Contarinia peyssonneliaeformis Zanardini	1.6
Palmophyllum crassum (Naccari) Rabenhorst	0.4

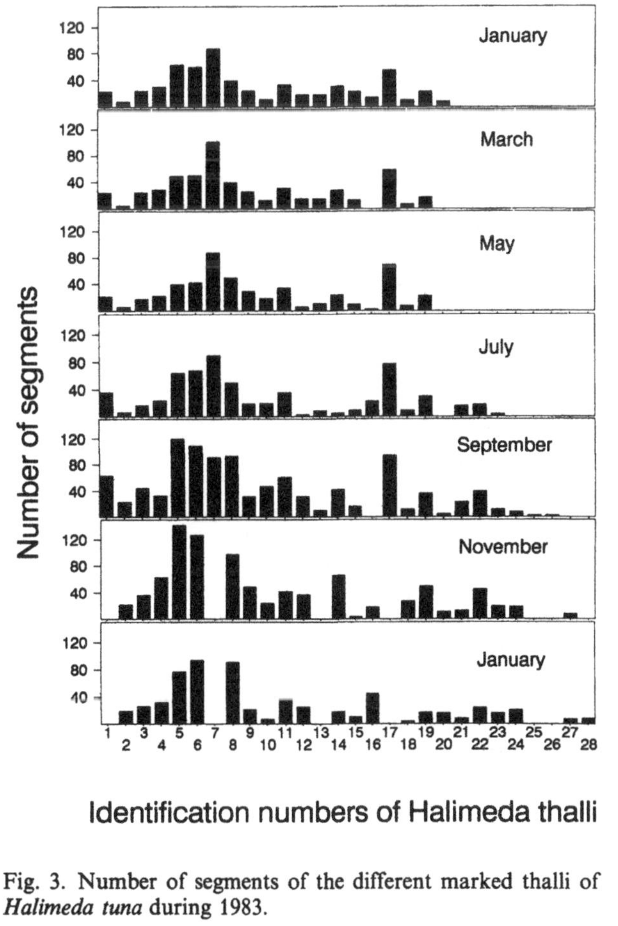

Fig. 3. Number of segments of the different marked thalli of *Halimeda tuna* during 1983.

Table II. Epiphytic species developing over the fronds of *Halimeda tuna* at different times of the year. Overall biomass is also indicated.

	Jan.	Mar.	Apr.	June	Aug.	Oct.
RHODOPHYTA						
Ceramiaceae						
Callithamnion decompositum J. Agardh	x	x	x	x	x	x
Callithamnion byssoides Arnott *ex* Harvey	x	x	x	x	x	x
Compsothamnion thuyoides (Smith) Schmitz	x	x	x	x	x	x
Ptilothamnion pluma (Dillwyn) Thuret	x	x	x	x	x	x
Spermothamnion johannis G. Feldmann	x	x	x	x	x	x
Ceramium codii (Richards) Mazoyer	x	x	x	x	x	x
Antithamnion heterocladum Funk	x	x	x	.	.	x
Antithamnionella spirographidis Schiffner	x	.	.	x	x	x
Seirospora sphaerospora J. Feldmann	.	.	x	x	x	x
Sphondylothamnium multifidum (Hudson) Nägeli	.	.	x	x	x	x
Pterothamnion plumula (Ellis) Nägeli	.	.	x	x	x	x
Ceramium diaphanum Roth v. *lophophorum* G. Feldmann	.	.	x	x	x	.
Ceramium cingulatum Weber van Bosse	.	.	x	x	x	x
Pterothamnion crispum (Ducluzeau) Nägeli	x	x	.	x	.	.
Wrangelia penicillata C. Agardh	x	.	.	x	x	.
Antithamnion tenuissimum (Hauck) Schiffner	.	.	.	x	x	x
Callithamniella tingitana (Bornet) G. Feldmann	.	x	x	.	.	.
Ceramium flaccidum (Kützing) Ardissone	.	.	x	x	.	.
Seirospora interrupta (Smith) Schmitz	.	.	.	x	x	.
Antithamnion cruciatum (C. Agardh) Nägeli	.	.	.	x	x	.
Crouania attenuata (C. Agardh) J. Agardh	.	.	.	x	x	.
Delesseriaceae						
Hypoglossum hypoglossoides (Stack.) Collins *et* Harvey	x	x	x	x	x	x
Apoglossum ruscifolium (Turner) J. Agardh	x	x	x	x	x	x
Myriogramme carnea (Rodríguez) Kylin	x	x	x	x	x	x
Erythroglossum sandrianum (Zanardini) Kylin	x	x	x	x	x	x
Nithophyllum micropunctatum Funk	x	.	.	x	x	x
Other groups						
Plocamium cartilagineum (Linné) Dixon	x	x	x	x	x	x
Falkenbergia rufolanosa (Harvey) Schmitz stage	x	x	x	x	x	x
Fosliella farinosa (Lamouroux) Howe	x	x	x	x	x	x
Gloiocladia furcata (C. Agardh) J. Agardh	x	x	x	x	x	x
Polysiphonia cf. *elongata* (Hudson) Sprengel	x	.	x	x	x	x
Erythrotrichia carnea (Dillwyn) J. Agardh	x	x	x	.	x	x
Rhodophyllis divaricata (Stackhouse) Papenfuss	x	x	x	x	.	.
Aphanocladia stichidiosa (Funk) Ardré	.	.	x	x	x	x
Champia parvula (C. Agardh) Harvey	.	.	x	x	x	.
Dasya hutchinsiae Harvey	.	.	x	x	x	.
Stylonema cornu-cervi Reinsch	.	.	.	.	x	x
Audouinella caespitosa (J. Agardh) Dixon	x	.	.	.	.	.
Stylonema alsidii (Zanardini) Drew	.	.	.	x	.	.
Lomentaria chylocladiella Funk	.	.	.	x	.	.
Eupogodon planus (C. Agardh) Kützing	.	.	.	.	.	x
PHAEOPHYTA						
Sphacelaria cirrosa (Roth) C. Agardh	x	x	x	x	x	x
Castagnea irregularis Sauvageau	.	.	x	x	.	.
Dictyota dichotoma (Hudson) Lamouroux	.	.	x	x	.	.
Giffordia mitchelliae (Harvey) Hamel	.	.	.	x	x	.
Castagnea cylindrica Sauvageau	.	.	.	x	.	.
Myrionema magnusii (Sauvageau) Loiseleur	.	.	.	x	.	.
Giraudia sphacelarioides Derbès *et* Solier	.	.	.	x	.	.
Nemacystus flexuosus (C. Agardh) Kylin	.	.	.	x	.	.
CHLOROPHYTA						
Cladophora albida (Hudson) Kützing	x	x	.	.	.	.
Enteromorpha multiramosa Bliding	.	.	.	x	x	.
TOTAL BIOMASS (g dwt m^{-2})	0.8	1.1	3.9	5.4	0.9	0.4
NUMBER OF SPECIES	25	21	31	44	35	28

Jan. = 13/1/83, Mar. = 6/3/83, Apr. = 29/4/83, June = 21/6/83, Aug. = 23/8/23, Oct. = 18/10/83.

The smallest number of thalli were observed in late March and the period of maximum loss occurred in September–October. There was great variability in the size of the thalli of *Halimeda* (Fig. 3). An increase in the number of segments in summer and a decrease in late autumn and winter was observed (Fig. 3). These variations were determined by changes in the number of thalli as well as by changes in the number of segments per thallus. In fact, inter-annual variation in the number of segments was a consequence of the seasonality in the appearance and disappearance of segments (Fig. 4).

Production of new segments was reduced during the period November–June, but it was high in summer. On the other hand, loss of segments was highest in autumn. The result of these two processes was a net growth of *Halimeda* between May and August and a decline of the population during the rest of the year (Fig. 4). Variance in segment production between the different thalli was noteworthy (Fig. 5) and can be partially explained by their different sizes. Large thalli produced more segments ($p < 0.001$) but seasonal differences and, perhaps, a great physiological heterogeneity among thalli may also account for the large variability observed.

Environmental factors

The inter-annual variation in temperature, irradiance and nutrient concentrations (Fig. 6) may affect *Halimeda* growth. A positive correlation between temperature and growth ($p < 0.01$) (Fig. 7a) indicates a strong dependence of growth on high temperatures. The relationship between irradiance and growth (Fig. 7b), although not statistically correlated ($p > 0.05$), indicates that growth was enhanced by relatively high irradiances. Therefore, temperature and light – in this order – were probably the main environmental factors affecting *Halimeda* growth. The inverse relationship between nitrate concentration in seawater and growth (Fig. 7c) is caused, in fact, by the inverse relationship between nutrient concentration and light and temperarature (Fig. 6). On the other hand, segment loss is not correlated ($p > 0.1$) with any environmental factor considered here and, because of its annual pattern (Fig. 4), it was probably related to periods of high disturbance (strong storms), which occurred in autumn, as is usual in north-western Mediterranean waters (Ballesteros 1989 a).

Productivity

The average carbon content of *H. tuna* (n = 12; mean ± S.D.) was 16.5 ± 1.4% of dry weight, mean organic content (ash free dwt) was 27.8 ± 6.0% of dry weight and carbonate content was 45.7 ± 11.7% of dry weight. Production of *Halimeda* in the sampled area ranged from 0.1 to 0.9 gdw m^{-2} d^{-1} (Table III) depending on the period of the year. Taking into account the average biomass of the *Halimeda* population (Table I), and assuming linearity between production and density of *Halimeda*, the production estimates would range between 0.5 and 3.7 gdw m^{-2} d^{-1}. Yearly production amounted to 165.8 gdw m^{-2}

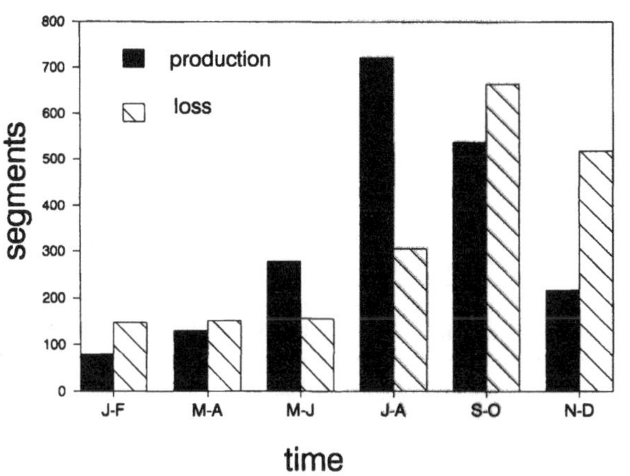

Fig. 4. Segment production and loss for the whole studied population of *Halimeda tuna* during 1983.

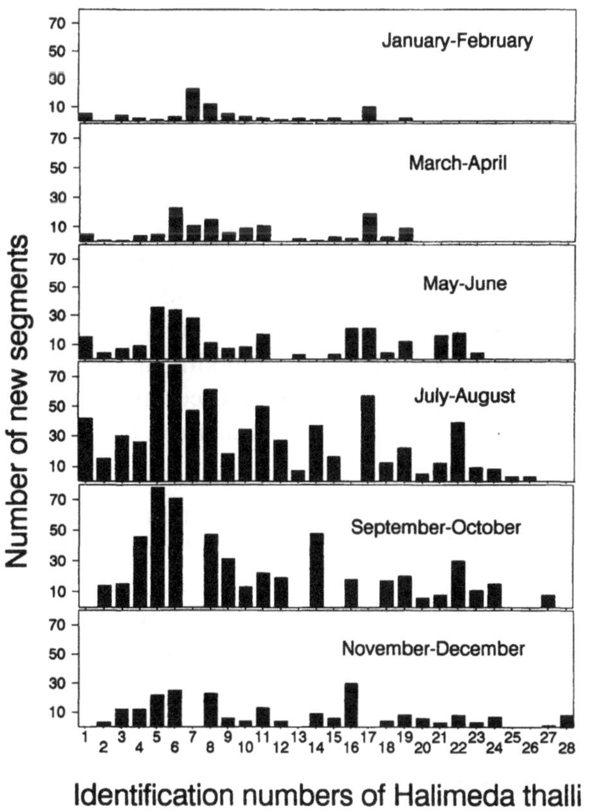

Fig. 5. Growth of new segments in the different marked thalli of *Halimeda tuna* during 1983.

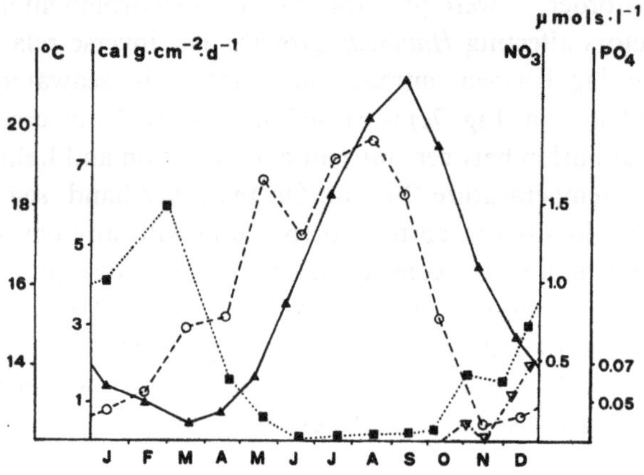

Fig. 6. Variation of temperature (°C) (solid line), irradiance (cal g cm^{-2} d^{-1}) (broken line), nitrate (dotted line) and phosphate (point and line line) concentration in seawater (both in micromols l^{-1}) measured adjacent to the community of *Halimeda tuna* off Tossa island during 1983.

y^{-1} (679.8 gdw m^{-2} y^{-1} referred to the overall population) equivalent to 46.1 g afdwt m^{-2} y^{-1} (191.0 g afdwt m^{-2} y^{-1} referred to the overall population) and 27.4 g organic C m^{-2} y^{-1} (113.6 g organic C m^{-2} y^{-1} referred to the overall population). Carbonate production was also very high (75.8 g $CaCO_3$ m^{-2} y^{-1}, 314.3 g $CaCO_3$ m^{-2} y^{-1} referred to the overall population). Daily P/B ratios ranged from 2.5 10^{-3} to 1.7 10^{-2} (Table III) and the yearly P/B ratio amounted to 1.87 y^{-1}, which indicates a high biomass turnover of *Halimeda*.

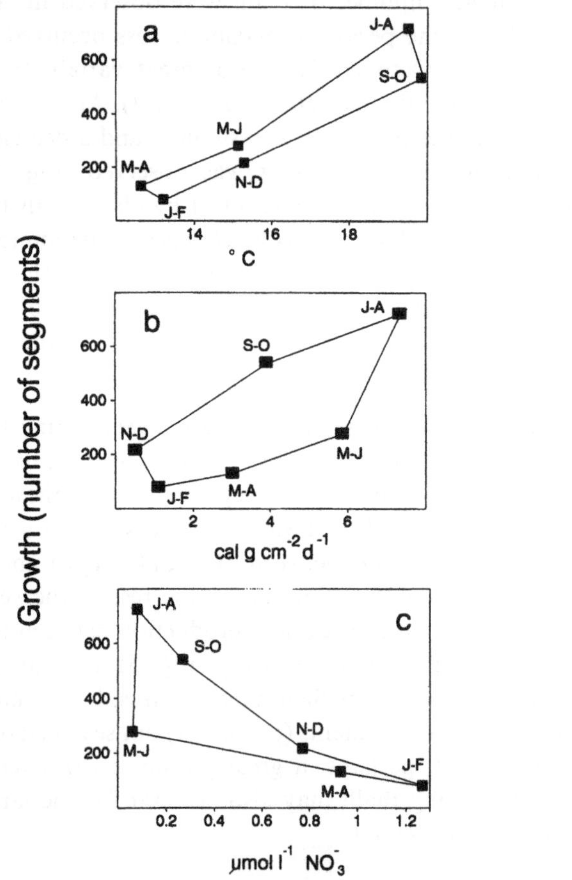

Fig. 7. Relationships between growth and temperature (a), between growth and irradiance (b) and between growth and nitrate concentration in seawater (c), with indication of the different periods of the year.

Discussion

The community of *Halimeda tuna* described here is a widespread deep-water community recently described by Ballesteros (1991) from the same geographic area. The dominance of *Halimeda* in Mediterranean deep-water communities developing at irradiances ranging from 0.34 to 7.59 cal g cm^{-2} d^{-1} (0.4 to 9.2% of surface irradiance) is noteworthy, although growth of *Halimeda* has been reported to occur at irradiances as low as 0.05% of surface irradiance in tropical seas (Littler *et al.* 1985, Hillis-Colinvaux 1986 a). Different physiological adaptations of *Halimeda* to dim light have been demonstrated (Solazzi and Tolomio 1976, Yokohama 1981) or suggested (Jensen *et al.* 1985). The encrusting layer of the community consists mainly of Corallinaceae and Peyssonneliaceae, two families with species adapted to dim light conditions (Larkum *et al.* 1967, Littler *et al.* 1986, Hooper 1985, Vadas and Steneck 1988).

Table III. Production and P/B ratios for the *Halimeda tuna* population growing inside the permanent quadrat extrapolated to m^2. Period of time, number of days, biomass at the end and at the beginning of the period, production, daily production and P/B ratio (x 10^{-3}) are indicated.

Period	days	Biomass		Production		daily P/B (d^{-1}) (10^{-3})
		final gdw m^{-2}	initial gdw m^{-2}	total gdw m^{-2}	daily gdw m^{-2} d^{-1}	
120183-060383	53	44.9	50.9	6.6	0.125	2.46
070383-290483	54	43.1	44.9	11.0	0.204	4.54
300483-210683	53	53.4	43.1	23.5	0.443	10.28
220683-230883	68	88.6	53.5	61.0	0.897	16.80
240883-281083	66	77.9	88.6	45.4	0.688	7.77
291083-120184	75	52.5	77.9	18.3	0.244	3.13

The epiphytic assemblage growing on *Halimeda* consists mainly of small sciaphilic species with opportunistic habits. Nevertheless, there are some photophilic species such as *Wrangelia penicillata*, *Crouania attenuata* and the members of the Phaeophyta. The species richness of the epiphytic assemblage is high, but slightly lower than other epiphytic assemblages of Mediterranean large algae (Rull and Gómez 1989) and seagrasses (Ballesteros 1987). The low epiphytic biomass can be attributed to low irradiances reaching the community, which probably prevents epiphytic growth in winter, and to the high segment turnover of *Halimeda*.

The average biomass of the *Halimeda* population (243.7 gdw m^{-2}) is higher than those found in shallow waters from tropical seas (Bach 1979, Hay 1981 a, Pairy 1988) but lower than that observed by Drew (1983), Drew and Abel (1988) and Wolanski *et al.* (1988) in deep-waters of the Great Barrier Reef.

The dynamics of the *Halimeda* population was quite similar to that described in another Mediterranean population by means of a demographic method (Meinesz 1980). This population was composed of small thalli and their turnover was very high: 172 new thalli appeared in 20×20 cm^2 of seabed area during one year. The smaller size and the higher turnover of this population can be attributed to a high degree of disturbance caused by its shallower distribution (only 5 meters depth). Nevertheless, Meinesz (1980) found a production of 652 new segments per 400 cm^2 per year, close to our estimates of 1962 segments per 1000 cm^2 per year. Meinesz (1980) also found that growth persisted throughout the year, without resting phases. However, although the period of highest growth (July – August) coincided in the two populations, the stand studied by Meinesz (1980) had the lowest growth in spring, not in winter. Meinesz (1980) also pointed out the behaviour of *Halimeda* rhizoidal systems as being similar to *Caulerpa* stolons. In our population, the different erect thalli were interconnected by a complex rhizoidal net growing below the encrusting thalli of *Mesophyllum lichenoides*. The development of these rhizoidal systems can be interpreted as a kind of vegetative growth to enhance *Halimeda* colonization and, protected as they are by the encrusting calcareous algae, they may serve as a spatial escape from predation; this strategy has to be added to other well-known defense mechanisms such as the high calcium carbonate content (Hillis-Colinvaux 1980), the aggregated growth form (Hay 1981 b), the production of toxic diterpenoid aldehydes (Paul and Fenical 1986), the nocturnal growth (Hay *et al.* 1988) and the chloroplast withdrawal during the night (Drew and Abel 1990).

Comparison of our data on production with results obtained by other authors is problematic because of different methods and units used. Some authors have evaluated the production of *Halimeda* by growth methods based on the collection or periodic recording of new segments. In terms of segment growth, Williams (1988) obtained a mean production of one segment per plant per day in Caribbean stands of *Halimeda incrassata* (Ellis) Lamouroux from deep-water, while Pairy (1988) found a yearly average of 3.3 new segments per plant per day in stands of *H. incrassata* growing in shallow waters from a Tahitian lagoon. Both values are higher than the 0.2 segments per plant per day that can be deduced from our data. In terms of biomass and carbonate production per areal unit, our values are intermediate for a selection of data from different species of *Halimeda* from tropical seas (Table IV). All these data emphasize

(1) the importance of *H. tuna* as a producer both of organic matter and calcium carbonate in Mediterranean deep phytal waters,

(2) the lower calcification rate of Mediterranean *H. tuna* compared with tropical *Halimeda* species and

(3) the high production rate considering the great depths, and the dim light, were *H. tuna* occurs.

Table IV. Production per unit area (total, organic carbon and calcium carbonate) reported for different species of *Halimeda*.

Species	Location	Depth m	Production			Reference
			Total gdw m^{-2} y^{-1}	Organic C g C m^{-2} y^{-1}	Carbonate g Ca CO_3 m^{-2} y^{-1}	
H. tuna	Mediterranean	18	680	114	314	this paper
H. opuntia + *H. copiosa*	Australia	5–8	1500	117	2227	Drew (1983)
H. opuntia f. *obovata*	Tahiti	1	–	60	1290	Pairy (1988)
H. opuntia	Florida	1–2	1250	75*	1088	Hudson (1985)
H. incrassata	Florida	3	41*	14*	10	Bach (1979)
H. incrassata	Tahiti	1	–	6	94	Pairy (1988)
H. incrassata	Bermuda	2	–	–	50	Wefer (1980)

* Calculated from author's published data.

Table V. Production rates (mg organic C gdw^{-1} h^{-1}) and calcification rates (mg inorganic C gdw^{-1} h^{-1}) recorded using different techniques for different species of *Halimeda*.

Species	Location	Organic Production mg C gdw^{-1} h^{-1}	Calcification Rates mg C gdw^{-1} h^{-1}	Method	Light µE m^{-2} s^{-1}	Reference
H. tuna	Mediterranean	0.13	0.04	Marking	38	this paper
H. tuna	Australia	0.11–0.35	0.14–0.50	^{14}C	24*	Borowitzka and Larkum (1976)
H. copiosa	Bahamas	0.04–0.05	0.06	O_2, ^{14}C	20	Jensen *et al.* (1985)
H. lacrimosa	Bahamas	0.10–0.12	0.06	O_2, ^{14}C	20	Jensen *et al.* (1985)
H. cryptica	Bahamas	0.24	0.16	O_2, ^{14}C	20	Jensen *et al.* (1985)
H. discoidea	Bahamas	0.09–0.11	0.09	O_2, ^{14}C	20	Jensen *et al.* (1985)
H. discoidea	Puerto Rico	–	0.20	^{45}Ca	1	Stark *et al.* (1969)
H. discoidea	Belize	0.20	–	O_2	500–2100	Littler *et al.* (1983)
H. dixoidea	Tahiti	0.32	0.17	O_2, pH-Alk	–	Pairy (1988)
H. monile	Belize	0.30	–	O_2	500–2100	Littler *et al.* (1983)
H. monile	Cuba	0.20	–	O_2	900*	Buesa (1977)
H. simulans	Cuba	0.37	–	O_2	900*	Buesa (1977)
H. incrassata	Tahiti	0.21	0.39	O_2, pH-Alk	*	Pairy (1988)
H. incrassata	Cuba	0.48	–	O_2	640**	Buesa (1977)
H. gracilis	Cuba	0.61	–	O_2	900**	Buesa (1977)
H. opuntia	Cuba	0.24	–	O_2	900**	Buesa (1977)
H. opuntia	Puerto Rico	–	0.26	^{45}Ca	1	Stark *et al.* (1969)
H. opuntia	Jamaica	–	0.07	^{45}Ca	–	Bohm and Goreau (1973)
H. opuntia	Tahiti	0.16	0.37	O_2, pH-Alk	–	Pairy (1988)

* Converted from lux. ** Converted from kj m^{-2} s^{-1}.

Some authors have estimated production of *Halimeda* by means of incubation techniques. Comparison of production rates obtained by growth or incubation methods have to be made with caution. Extrapolation of instantaneous values of photosynthesis to estimate annual production is complicated and, usually, inaccurate (Brinkhuis 1977) because the high variability of environmental factors and the physiological performances of the plants that change from hour to hour, from day to day and throughout the annual cycle (Lobban *et al.* 1985). A good example of interannual changes is provided in our study (see Fig. 6 and Table III). Also, sometimes there is no suitable information about the mean density of plants *in situ*. Nevertheless, conversion of production values obtained by growth methods to short periods of time can be used to make rough comparisons. If we take into account our original data, we can convert yearly estimates of organic carbon and calcium carbonate production per square meter to estimates of mean production per unit of biomass if the average standing crop and a net photosynthetic activity for 10 hours a day are considered. Such conversion gives an estimate of 0.13 mg organic C gdw^{-1} h^{-1} and 0.04 mg inorganic C gdw^{-1} h^{-1}. Values of *Halimeda* organic production from the literature range from 0.04 to 0.61 mg C gdw^{-1} h^{-1} and calcification rates ranges from 0.06 to 0.50 mg C gdw^{-1} h^{-1} (Table V). Production of organic carbon calculated from our data are very similar to those measured in different Caribbean deep-water species (Jensen *et al.* 1985) and they are in the lower range of the measures made in Australian *H. tuna* (Borowitzka and Larkum 1976), but they are lower than measures obtained for different *Halimeda* species at high irradiances (Buesa 1977, Littler *et al.* 1983). The low calcification rates estimated from our data (Table V) can be attributed to the low ash content (which consists mainly of calcium carbonate) of Mediterranean *H. tuna* compared with other deep-water species from tropical waters (Jensen *et al.* 1985). However, calcification rates of *H. tuna* in the Mediterranean exceed calcium carbonate deposition of encrusting coralline algae from temperate seas (Edyvean and Ford 1987, Delgado 1989) but are less than those of some articulated coralline algae (Pearse 1972, Borowitzka 1981) and *Padina pavonica* (L.) Thivy (Delgado 1989), a widespread brown alga in warm and tropical seas (Table VI).

The contribution of *Halimeda* stands to the production of organic matter and calcium carbonate in the Mediterranean can not be assessed accurately because the unknown extent of these *Halimeda* populations. Nevertheless, at a reduced scale, the primary production values of these *Halimeda* stands are close to the production estimated from deep-water algal assemblages from the Caribbean (Littler *et al.* 1986) and the temperate Atlantic (Peckol and Ramus 1988). Also, they are similar to measures made in the community of *Cystoseira zosteroides* (Turner) C. Agardh (Ballesteros 1990), another common deep-water algal assemblage in the north-western Mediterranean. Production of *Halimeda tuna* is only slightly lower than values found in communities developing in the lower

Table VI. Calcification rates (mg C gdw^{-1} h^{-1}) recorded for different species of calcareous algae.

Species	Location	Calcification Rates mg C gdw^{-1} h^{-1}	Methods	Light μE m^{-2} s^{-1}	Reference
Padina pavonica	Mediterranean	0.23	pH-Alk	2000	Delgado (1989)
Bossiella orbigniana	California	0.27	^{45}Ca	6*	Pearse (1972)
Amphiroa foliacea	Australia	0.05	^{14}C	600	Borowitzka (1981)
Corallina officinalis	Wales	0.02–0.03	^{14}C	100–300**	Pentecost (1978)
Lithophyllum expansum	Mediterranean	0.02	pH-Alk	350	Delgado (1989)
Lithophyllum incrustans	Wales	0.008***	morphological	–	Edyvean and Ford (1987)

* Converted from ergs cm^{-2} s^{-1}. ** Converted from klux. *** Calculated from author's published data.

infralittoral zone, at 10–15 meters depth, in the same geographic area (Ballesteros 1989 b). The maintenance of relatively high levels of production at increasing depths suggests a very efficient utilization of available light (Ballesteros 1989 a), and underscores the importance of deep-water algae in the contribution of the phytobenthos to the whole productivity of the sea.

The calcium carbonate production of *H. tuna*, on an area basis, (estimated as 314 gdw m^{-2} y^{-1}) is similar to that reported from maërl communities and for *Lithophyllum expansum* Philippi in the Mediterranean Sea (Delgado 1989, Canals *et al.* in press) (300–450 g $CaCO_3$ m^{-2} y^{-1}) and exceeds the carbonate deposition in *Posidonia oceanica* meadows (Romero 1988, Canals *et al.* 1988) (100–160 g $CaCO_3$ m^{-2} y^{-1}). As non-quantified carbonate deposition by encrusting corallines has to be added to *Halimeda* production, the contribution of the whole *Halimeda* community to sediment and coralline-reef formation (Bosence 1985) is very important. Although calcium carbonate production of the *H. tuna* community can not be compared to shallow, seaward portions of coral reefs dominated by corals and coralline algae, where 2000–4000 g $CaCO_3$ are deposited each year (Marsh 1970, Smith 1973, Littler 1973, Smith and Kinsey 1976), nor to the mounts of *H. opuntia* in shallow waters from the Great Barrier Reef (Drew 1983) or the Caribbean (Hudson 1985), it is similar to that reported from *Thalassia* beds (Smith and Atkinson 1983). It is also similar to the production of calcium carbonate estimated by Smith (1972) from the mainland shelf of Southern California and higher than psammophytic *H. incrassata* stands in shallow tropical lagoons (Bach 1979, Pairy 1988).

The production and productivity of *Halimeda tuna* shows great seasonal changes (Table III). Seasonality has been shown to occur in other species of *Halimeda* from tropical seas: *H. incrassata* (Wefer 1980, Pairy 1988) and *H. opuntia* (Hudson 1985, Pairy 1988). Temperature and light seem to be the main environmental factors regulating growth of *Halimeda tuna* in Mediterranean deep-waters. *Halimeda* is a genus with a pantropical distribution (Hillis-Colinvaux 1986 c), so relatively cold waters typical of winter months should affect *Halimeda* growth negatively. Low winter irradiances also should reduce the growth, and it is difficult to assess which factor, temperature or light, is the more critical, although temperature correlates better with growth (Figs 7a, b). Photosynthetic performances of *H. tuna* in relation to temperature and irradiance have to be experimentally assessed in order to understand growth dynamics.

A high translocation capacity has been indirectly demonstrated by Meinesz (1980) and Hay *et al.* (1988) in *Halimeda* species. It can be suggested that part of the growth of winter months could be maintained with products translocated from old segments created in summer. Thus, winter growth could not be attributed to primary production alone put, partially, to photosynthetate allocation from older parts of the plants. Such a mechanism would explain the shape of the relationship between irradiance and growth over the year (Fig. 7b) as growth from September to December could be partially sustained by relocation of photosynthetates. Rhizoidal connection between upright thalli could also play an important role in this relocation and would explain the great growth variability among the different thalli (Fig. 5) and the low correlation between growth and size. Experimental work is needed to resolve these questions.

The inverse relationship between nitrate concentration and growth is unusual and may be indirectly caused by the strong dependence of production on light and temperature. Thus, *Halimeda* is a paradigmatic example of a benthic algae whose production is light-dependent (Ballesteros 1989 a).

Halimeda has mainly evolved in tropical reefs, where a high competition for light and substrate and a high degree of herbivory do exist, so strong selection pressures have acted on the species of this genus to attain high levels of habitat adaptation. Therefore, it is necessary to assess the photosynthetic performances of *Halimeda tuna* at varying temperatures and irradiances, as well as the seasonal pattern of photosynthetates and nutrient allocation capability in order to understand the growth dynamics of *Halimeda tuna* in Mediterranean deep-waters.

Acknowledgements

The author thanks Dr O. Delgado and two anonymous reviewers for their detailed and constructive comments on an earlier version of this manuscript.

References

Alvarez, M., T. Gallardo, M. A. Ribera and A. Gómez. 1988. A reassessment of Northern Atlantic seaweed biogeography. *Phycologia 27*: 221–233.

Bach, S. D. 1979. Standing crop, growth and production of calcareous Siphonales (Chlorophyta) in a South Florida lagoon. *Bull. Mar. Sci. 29*: 191–201.

Ballesteros, E. 1987. Estructura i dinàmica del poblament algal de les fulles de *Posidonia oceanica* (L.) Delile als herbeis de Tossa de Mar. *Butll. Inst. Cat. Hist. Nat. 54*: 13–20.

Ballesteros, E. 1989 a. Production of seaweeds in North-Western Mediterranean marine communities: its relation with environmental factors. *Scient. Mar. 53*: 357–364.

Ballesteros, E. 1989 b. Estructura y dinámica de la communidad infralitoral de *Codium vermilara* (Olivi) Delle Chiaje de la Costa Brava (Mediterráneo Noroccidental). *Anal. Biol. 19*: 73–90.

Ballesteros, E. 1990. Structure and dynamics of the community of *Cystoseira zosteroides* (Turner) C. Agardh (Fucales, Phaeophyceae) in the North-Western Mediterranean. *Scient. Mar. 54*: 217–229.

Ballesteros, E. 1991. Structure of a deep-water community of *Halimeda tuna* (Chlorophyceae, Caulerpales) from the North-Western Mediterranean. *Collect. Bot. 19* (in press).

Böhm, E. L. and T. F. Goreau. 1973. Rates of turnover and net accretion of calcium and the role of calcium binding polysaccharides during calcification in the calcareous alga *Halimeda opuntia* (L.) Lamouroux. *Int. Rev. der ges. Hydrobiol. 58*: 723–740.

Borowitzka, M. A. 1981. Photosynthesis and calcification in the articulated coralline algae *Amphiroa anceps* and *Amphiroa foliacea. Mar. Biol. 62*: 17–23.

Borowitzka, M. A. and A. W. D. Larkum. 1976. Calcification in the green alga *Halimeda* II. The exchange of Ca^{++} and the occurrence of age gradients in calcification and photosynthesis. *J. Exp. Bot. 27*: 864–875.

Bosence, D. W. J. 1985. The "Coralligène" of the Mediterranean – a recent analogy for testing coralline algal limestones. *In*: (Toomey, D. F. and M. H. Niteck, eds) *Paleoalgology: Contemporary Research and Application*. Springer Verlag. Berlin. pp. 216–225.

Bourdouresque, C. F. 1985. Groupes écologiques d'algues marines et phytocoenoses benthiques en Méditerranée Occidentale: une revue. *Giorn. Bot. Ital.* 118: 7–42.

Brinkhuis, B. H. 1977. Comparisons of salt-marsh fucoid production estimated from three different indeces. *J. Phycol. 13*: 328–335.

Brinkhuis, B. H. 1985. Growth patterns and rates. *In*: (Littler, M. M. and D. S. Littler, eds) *Handbook of Phycological Methods: Ecological Field Methods. Macroalgae*. Cambridge University Press. Cambridge. pp. 461–477.

Buesa, R. J. 1977. Photosynthesis and respiration of some tropical marine plants. *Aquat. Bot. 3*: 203–216.

Canals, M., E. Ballesteros, J. Serra, B. Alonso and E. Catafau. 1988. The Pollensa Bay Carbonate Factory (Balearic Islands, NW Mediterranean Sea). *Rapp. Comm. int. Mer Medit. 31*: 298.

Canals, M., J. Acosta, E. Ballesteros, J. Serra, M. Zabala, A. Calafat, P. Herranz and J. L. Sanz. 1991. Carbonate particle production, transport and sedimentation, and seafloor morphology on the South Menorca Shelf and Canyon head (Balearic Islands, North-Western Mediterranean Sea). *Sedimentary Geology* (in press).

Delgado, O. 1989. Sistema carbónico-carbonatos en el Mediterráneo Noroccidental y relación con las principales comunidades fitobentónicas productoras de carbonatos. Ph. D. Thesis. Universitat de Barcelona.

Drew, E. A. 1969. Photosynthesis and growth of attached marine algae down to 130 meters in the Mediterranean. *Proceed. VI Int. Seaweed Symp. 6*: 151–159.

Drew, E. A. 1983. *Halimeda* biomass, growth rates and sediment generation on reefs in the Central Great Barrier Reef province. *Coral Reefs 2*: 101–110.

Drew, E. A. and K. M. Abel. 1988. Studies on *Halimeda*. I. The distribution and species composition of *Halimeda* meadows throughout the Great Barrier Reef Province. *Coral Reefs 6*: 195–205.

Drew, E. A. and K. M. Abel. 1990. Studies on *Halimeda*. III. A daily cycle of chloroplast migration within segments. *Bot. Mar. 33*: 31–45.

Edyvean, R. G. J. and H. Ford. 1987. Growth rates of *Lithophyllum incrustans* (Corallinales, Rhodophyta) from South West Wales. *Br. Phycol. J. 22*: 139–146.

Feldmann, J. 1937. *Recherches sur la Végétation Marine de la Méditerranée. La Côte des Albères*. Wolf. Rouen. 339 pp.

Feldmann, J. 1958. Origine et affinités du peuplement végétal benthique de la Méditerranée. *Rapp. Proc. Verb. Réun. CIESMM. 14*: 515–518.

Grashoff, K., M. Ehrhardt and K. Kremling. 1983. *Methods of Seawater Analysis*. Verlag Chemie. Weinheim.

Hay, M. E. 1981 a. Herbivory, algal distribution and the maintenance of between-habitat diversity on a tropical fringing reef. *Am. Nat. 118*: 520–540.

Hay, M. E. 1981 b. The functional morphology of turf-forming seaweeds: persistence in stressful marine habitats. *Ecology 62*: 739–750.

Hay, M. E., V. J. Paul, S. M. Lewis, K. Gustafson, J. Tucker and R. N. Trindell. 1988. Can tropical seaweeds reduce herbivory by growing at night? Diel patterns of growth, nitrogen content, herbivory and chemical versus morphological defenses. *Oecologia 75*: 233–245.

Hillis-Colinvaux, L. W. 1980. Ecology and taxonomy of *Halimeda*: primary producer of coral reefs. *Adv. Mar. Biol. 17*: 1–327.

Hillis-Colinvaux, L. W. 1986 a. Deep water populations of *Halimeda* in the economy of an atoll. *Bull. Mar. Sci. 38*: 155–169.

Hillis-Colinvaux, L. W. 1986 b. Historical perspectives on algae and reefs: have reefs been misnamed?. *Oceanus 29*: 43–48.

Hillis-Colinvaux, L. W. 1986 c. Distribution patterns of some Bryopsidales in the geologic past: their bearing on present distributions. *Bot. Mar. 29*: 271–277.

Hooper, R. G. 1985. Depth limits of the benthic algae of Newfoundland and the Grand Banks. *Br. Phycol. J. 20*: 185.

Hudson, J. H. 1985. Growth rate and carbonate production in *Halimeda opuntia*: Marquesas Keys, Florida. *In*: (Toomex, D. F. and M. H. Nitecki, eds) *Paleontology: Contemporary Research and Applications*. Springer Verlag, Berlin. pp. 257–263.

Jensen, P. R., R. A. Gibson, M. M. Littler and D. S. Littler. 1985. Photosynthesis and calcification in four deep-water *Halimeda* species (Chlorophyceae, Caulerpales). *Deep Sea Res. 32*: 451–464.

Larkum, A. W. D., E. A. Drew and R. N. Crosset. 1967. The vertical distribution of attached marine algae in Malta. *J. Ecol. 35*: 361–371.

Littler, M. M. 1973. The productivity of Hawaiian fringing-reef crustose corallinaceae and an experimental evaluation of production methodology. *Limnol. Oceanogr. 18*: 946–952.

Littler, M. M., D. S. Littler and P. R. Taylor. 1983. Evolutionary strategies in a tropical barrier-reef system: functional form groups of marine macroalgae. *J. Phycol. 19*: 229–237.

Littler, M. M., D. S. Littler, S. M. Blair and J. M. Norris. 1985. Deepest known plant life discovered on an uncharted seamount. *Science 227*: 57–59.

Littler, M. M., D. S. Littler, S. M. Blair and J. M. Norris. 1986. Deep water plant communities from an uncharted seamount off San Salvador Island, Bahamas: distribution, abundance and primary productivity. *Deep Sea Res. 33*: 881–892.

Lobban, C. S., P. J. Harrison and M. J. Duncan. 1985. *The Physiological Ecology of Seaweeds*. Cambridge University Press. Cambridge.

Marsh, J. A. 1970. Primary productivity of reef-building calcareous algae. *Ecology 51*: 255–263.

Meinesz, A. 1980. Contribution à l'étude des Caulerpales (Chlorophytes). Ph. D. Thesis. Université de Nice.

Neumann, A. C. and L. S. Land. 1975. Lime mud deposition and calcareous algae in the Bight of Abaco: a budget. *J. Sedim. Petrol. 45*: 763–786.

Payri, C. E. 1988. *Halimeda* contribution to organic and inorganic production in a Tahitian reef system. *Coral Reefs 2*. 251–262.

Paul, V. J. and W. Fenical. 1986. Chemical defense in tropical green algae, order Caulerpales. *Mar. Ecol. Progr. Ser. 34*: 157–169.

Pearse, V. B. 1972. Radioisotope study of calcification in the articulated coralline algae *Bossiella orbigniana*. *J. Phycol. 5*: 305–312.

Peckol, P. and J. Ramus. 1988. Abundances and physiological properties of deep-water seaweeds from Carolina outer continental shelf. *J. Exp. Mar. Biol. Ecol. 115*: 25–39.

Pentecost, A. 1978. Calcification and photosynthesis in *Corallina officinalis* L. using the $^{14}CO_2$ method. *Br. Phycol. J. 13*: 393–390.

Rull, J. and A. Gómez. 1989. Distribución de las algas epífitas sobre los ejemplares de *Cystoseira mediterranea* Sauvageau. *Anales Jardín Botánico 46*: 99–106.

Siesser, W. G. and J. Rogers. 1971. An investigation on the suitability of four methods used in routine carbonate analysis of marine sediments. *Deep Sea Res. 18*: 135–139.

Smith, S. V. 1972. Production of calcium carbonate on the mainland shelf of Southern California. *Limnol. Oceanogr. 17*: 28–41.

Smith, S. V. 1973. Carbon dioxide dynamics: a record of organic carbon production, respiration and calcification in the Enewetak reef flat community. *Limnol. Oceanogr. 18*: 106–120.

Smith, S. V. and P. W. Kinsey. 1976. Calcium carbonate production, coral reef growth and sea level change. *Science 194*: 937–939.

Smith, S. V. and M. Atkinson. 1983. Mass balance of carbon and phosphorus in Shark Bay, Western Australia. *Limnol. Oceanogr. 28*: 625–639.

Sokal, R. R. and F. J. Rohlf. 1969. *Biometry*. Freeman. New York. 859 pp.

Solazzi, A. and C. Tolomio. 1976. Effetti della profondita sull'apparato fotosintetico delle alghe. I. I pigmenti clorofilliani in *Halimeda tuna* Lam. (Chlorophyceae, Siphonales). *Mem. Biol. Mar. Oceanogr. 6*: 21–27.

Stark, L. M., L. Almodovar and R. W. Krauss. 1969. Factors affecting the rate of calcification of *Halimeda opuntia* (L.) Lamouroux and *Halimeda discoidea* Decaisne. *J. Phycol. 5*: 305–312.

Vadas, R. L. and R. S. Steneck. 1988. Zonation of deep-water benthic algae in the Gulf of Maine. *J. Phycol. 24*: 338–346.

Wefer, G. 1980. Carbonate production by algae *Halimeda*, *Penicillus* and *Padina*. *Nature 285*: 323–324.

Weinberg, S. and A. Cortel-Breeman. 1978. The estimation of the yearly cycle of submarine irradiance for ecological purposes. A methodological example based on data from Banyuls sur Mer. *Bijdragen tot de Dierkunde 48*: 35–44.

Williams, S. L. 1988. Disturbances and recovery of a deep-water Caribbean seagrass bed. *Mar. Ecol. Progr. Ser. 42*: 63–71.

Wolanski, E., E. Drew, K. M. Abel and J. O'Brien. 1988. Tidal jets, nutrient upwelling and their influence on the productivity of the alga *Halimeda* in the Ribbon Reefs, Great Barrier Reef. *Est. Coast. Shelf Sci. 26*: 169–201.

Yokohama, Y. 1981. Distribution of the green light absorbing pigments siphonoxanthin and siphonein in marine green algae. *Bot Mar. 24*: 637–640.

Zabala, M. and E. Ballesteros. 1989. Surface-dependent strategies and energy flux in benthic marine communities, or why corals do not exist in the Mediterranean? *Scient. Mar. 53*: 3–17.

Botanica Marina
Vol. 34, pp. 303–310, 1991

Ecology of *Chondrus crispus* Stackhouse on the Northern Coast of Spain

II. Reproduction

C. Fernández and M. P. Menéndez

Dpto. Biología de Organismos y Sistemas, Unidad de Ecología, Facultad de Biología, Universidad de Oviedo, Oviedo, Spain

(Accepted 2 April 1991)

Abstract

Patterns of reproduction in *Chondrus crispus* were studied in marginal populations growing on the north coast of Spain. In the population as a whole, fertile fronds present a low contribution in terms of number of fronds, but greater in biomass due to their larger size. Carposporic fronds show bimodal maxima and tetrasporic fronds a unimodal maxima.

Reproductive capacity was similar in both, gametophytes and tetrasporophytes, with an estimated number of spores/m^2/year of about 8×10^{10}. The greater number of fertile gametophytes is counterbalanced by high numbers of tetrasporangial sori and tetraspores. Some explanations for these variations are discussed.

Introduction

There have been a number of studies of reproduction in *Chondrus crispus* Stackhouse (see Chopin 1986, Pringle and Mathieson 1987, for bibliographies) but some problems still remain. One of these is the great variability in ploidy ratios, with dominance of haploid or diploid phases and a wide range of intermediate situations, due to physical environment or a differential reproductive capacity between phases (Bhattacharya 1985, Chopin *et al.* 1988, Lazo and McLachlan 1989, Mathieson 1989). Further information is also needed on the importance of reproduction by spores versus vegetative reproduction as this appears to decrease dramatically with the stability of the substratum (Lazo and McLachlan 1989).

The aim of the present study was to describe the phenology of the reproduction in two localities on the north coast of Spain, where isolated intertidal populations of *Chondrus crispus* occur (Fernández and Menéndez, in press). From the studies cited above we expected to find:

1) a bimodal pattern for reproductive gametophyte fronds and a unimodal one for sporophytes,

2) a different reproductive capacity between these phases of the life-history.

We used reproductive parameters such as density, biomass and size, and, as a new contribution, the phenology of spore production, based on the size of the reproductive structures.

Material and Methods

Samples were collected at two non-harvested sites on the Asturian coast (Santa Maria del Mar and Viodo, see Fig. 1). Both localities have similar substratum (rock platforms) but they differ in the degree of wave exposure, it being higher at Santa María del Mar than Viodo. Sampling was carried out at Santa María del Mar between June 1983 and June 1984, and at Viodo between June 1984 and June 1985. All collections were made monthly.

As the samples were located in homogeneous intertidal areas, with a gently slope and about 80% cover of *Chondrus crispus*, one haphazardly selected area of 3000 cm^2 was collected each month. This area satisfies the assumption of homogeneity for biomass values, number of fronds and size-development classes (Menéndez 1985).

All the reproductive fronds were taken from the sample area. Fertile gametophytes and tetrasporophytes were identified and their state of ripeness was determined. The ploidy composition of sterile fronds was carried out by the resorcinol-acetal test (Craigie and Leigh 1978). Subsequently, the lengths were measured, and the fronds were grouped into size classes of 3 cm intervals. All reproductive fronds belonged to the III size-development class as defined by McLachlan *et al.* (1988). Finally, the fronds were dried at 60 °C for 48 h and weighed.

The phenology of spore production was studied on the Santa María del Mar fronds. In each frond, the number of sporangial sori were counted. Up to ten reproductive fronds were haphazardly chosen from the fertile gametophytes and sporophytes. In both cases, ten carposporangial sori and ten tetrasporangial sori were measured in order to determine their volumes. Each sporangial sorus was considered as a revolution generated hyperelipsoid where:

$$Vm = \frac{\sum_{i=1}^{n} Vi}{n} = \frac{\sum_{i=1}^{n} 4/3\, a_i \cdot b_i \cdot c_i}{n}$$

Vm = average volume of the sporangia
a = major axis
b = minor axis
c = a transverse axis
n = number of observed sporangia

In the same way, ten carpospores and tetrasporangia were measured. Each one was considered as an ellipsoid inscribed in a square base prisma with a side equal to the minor axis of the ellipsoid and the height equal to the major axis:

$$Vm = \frac{\sum_{i=1}^{n} Vi}{n} = \frac{\sum_{i=1}^{n} a_i^2 \cdot b_i^2}{n}$$

Vm = average volume of the spore
a = minor axis
b = major axis
n = number of spores measured

This volume was divided by 4 to determine the tetraspore volume.

The Mann-Whitney U-test (Siegel 1956) was used to determine significant differences between means.

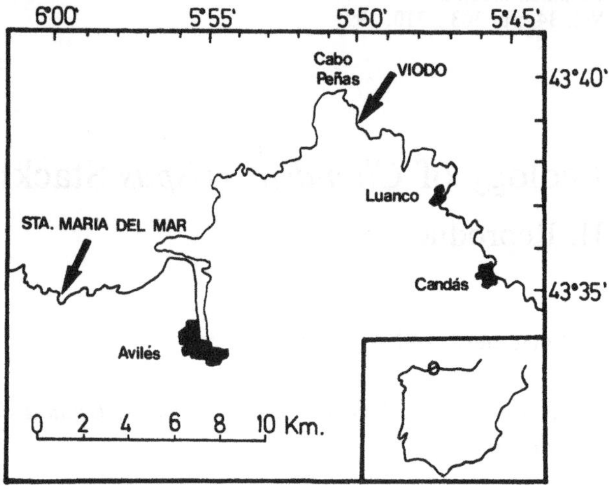

Fig. 1. Map showing study sites.

Results

Morphology of gametophytes and sporophytes

Throughout the study period we only found three male gametophytes and all were in August at Santa María del Mar. Female gametophytes carrying carposporangial sori were frequent throughout the year, and consisted of two different morphological types (Fig. 2). One had slim irregular branching and very conspicuous cystocarps (Fig. 2a). The other had wide fronds and less conspicuous cystocarps (Fig. 2b). Fertile tetrasporophytes were narrower and more ligulate, as pointed out by Chen and McLachlan (1972) and with tetrasporangial sori dispersed throughout the frond, starting where the frond begins to branch (Fig. 2c).

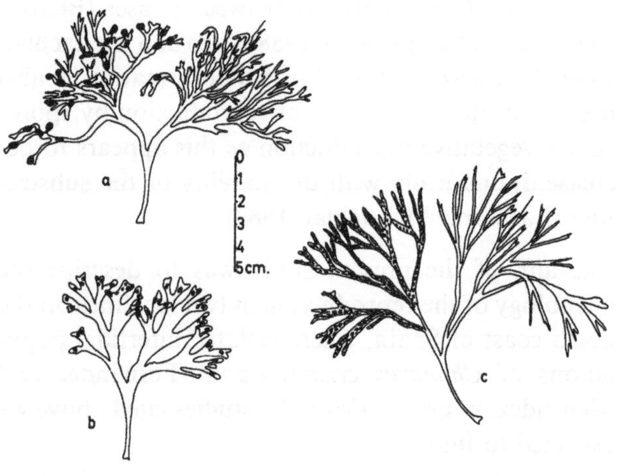

Fig. 2. Morphology of fertile fronds. a) Narrow female gametophyte b) Thick female gametophyte c) Tetrasporophyte.

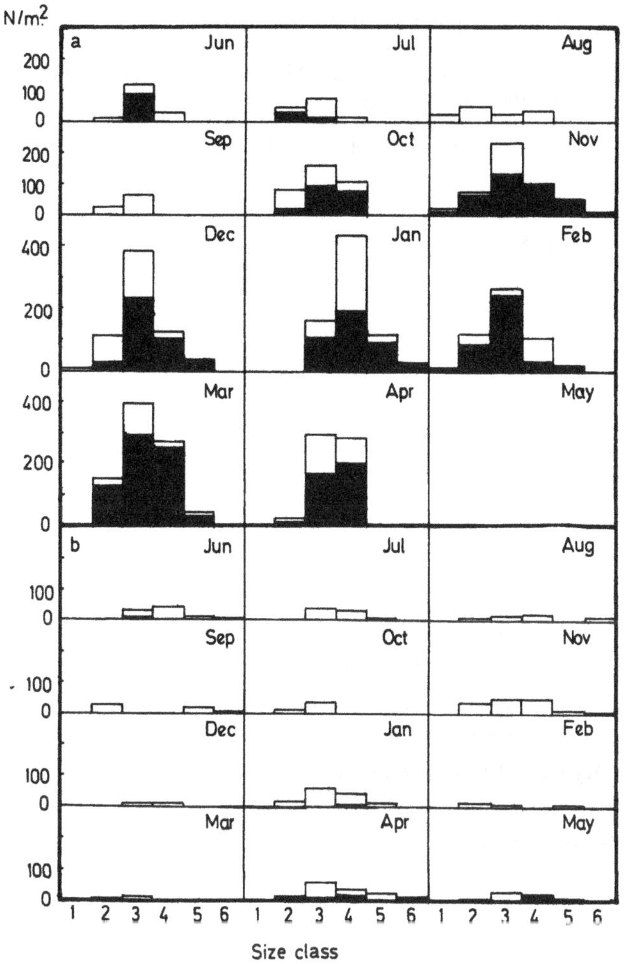

Fig. 3. Seasonal changes in the distribution of numbers of fertile fronds by size. Classes at 3 cm intervals a) Viodo b) Santa María del Mar. Black: tetrasporophytes White: female gametophytes.

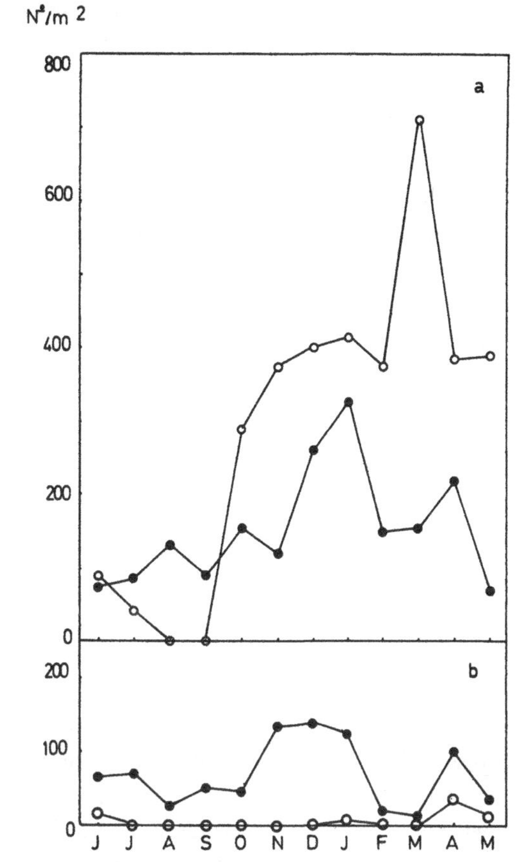

Fig. 4. Seasonal changes in density of fronds. a) Viodo b) Santa María del Mar. Open circles: tetrasporophytes Dark circles: female gametophytes.

Size distribution

The size range was wide for both, gametophytes and tetrasporophytes, at the two sites: for the gametophytes it was 6.5–19.9 cm at Santa María del Mar, and, 5.5–15.4 cm at Viodo. For tetrasporophytes the values were 6.9–20.6 cm and 5.7–19.5 cm, respectively. These mature fronds represented the largest found in the population (Fernández and Menéndez, in press). There were considerable differences between the site (Fig. 3); the reproductive population at Santa María des Mar was very small compared with that at Viodo. Another noteworthy fact is that there were fertile tetrasporophytes at Santa María del Mar, only during the months of January, April, May and June.

Number of fronds, biomass and sori rupture

Both, the number of fertile fronds and the biomass followed a marked seasonal pattern. The population density of tetrasporophytes showed a unimodal distribution with a maximum in winter or spring,

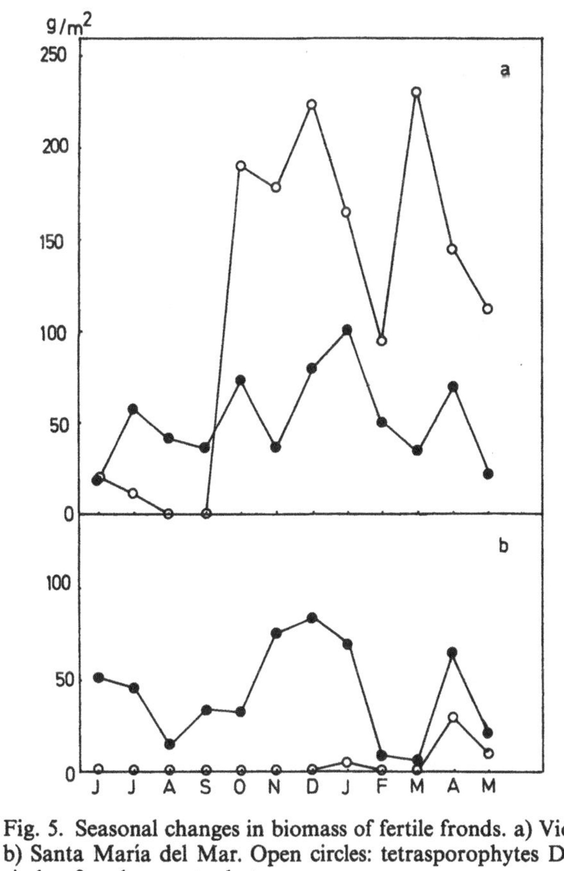

Fig. 5. Seasonal changes in biomass of fertile fronds. a) Viodo b) Santa María del Mar. Open circles: tetrasporophytes Dark circles: female gametophytes.

Table I. Density and biomass of fertile fronds, expressed as percentage of total population, at the two study sites.

		July	August	September	October	November	December	January	February	March	April	May	June
Santa María del Mar	% Density of whole population reproductive	1.5	0.3	1.7	1.0	5.1	1.5	3.6	0.4	0.4	3.4	1.1	1.0
	% Biomass of whole population reproductive	9.2	1.7	5.3	7.9	21.0	18.8	22.6	4.2	3.8	23.7	8.0	8.4
Video	% Density of whole population reproductive	6.5	2.7	3.1	16.2	13.3	22.1	29.1	12.9	17.1	10.8	?	2.4
	% Biomass of whole population reproductive	13.2	5.7	7.4	54.5	50.8	69.4	59.0	44.7	46.8	46.1	32.6	7.1

whereas gametophytes showed a bimodal distribution (Fig. 4) as has already been observed in the size distribution. The density was higher at Viodo, where tetrasporophytes dominated (Fernández and Menéndez, in press). The contribution of fertile fronds to total density (Table I) was minimal at Santa María del Mar, with values below 5%, whereas at Viodo the percentage reproductive was generally large, 10–30% from October to April.

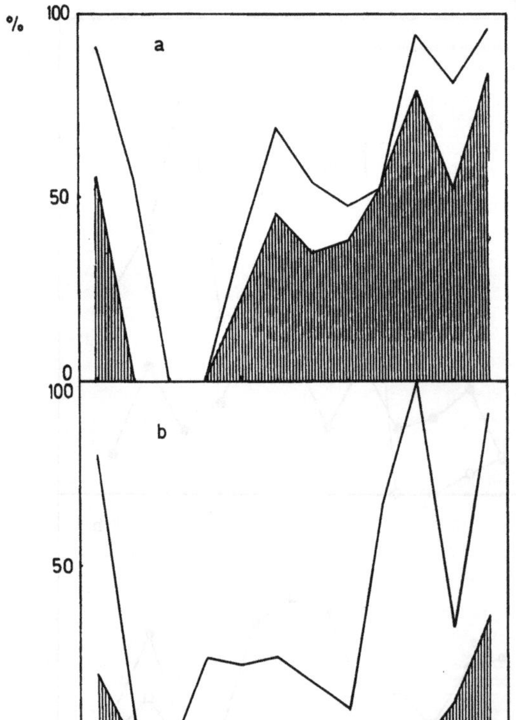

Fig. 6. Seasonal changes in percentage of broken sporangial sori. a) Viodo b) Santa María del Mar. ▥ Tetrasporophytes □ Female gametophytes.

The biomass of tetrasporophytes showed a bimodal distribution at Viodo (Fig. 5a) with maxima in winter and spring, whereas the effect was minimal at Santa María del Mar, where fertile tetrasporophytes were only occasionally present. Female gametophytes with cystocarps showed oscillations in spring and a maximum in winter (Fig. 5a and b). The contribution of reproductive fronds to total biomass of the population (Table I) was only significant at Santa María del Mar during the months of November, December, January and April, with values close to 20%. At Viodo during the period from October to April over 50% of the total biomass consisted of reproductive fronds.

Rupture of sori occurred from November to June at Viodo (Fig. 6a) but was restricted to the period February–June at Santa María del Mar (Fig. 6b) although the most intense period was from March to June.

Spore volume

Table II shows the mean volume and standard deviation for carpospores and tetraspores. Carpospores are bigger than tetraspores but significant differences in volume were observed during the year. For the carpospores, the largest mean volumes were recorded in spring and summer, and smallest in winter. The maximum tetraspore volume occurred in summer and minimum in spring.

Volume of sporangial sori

Figure 7 shows the mean annual variation in volume of the carposporangial sori, with a maximum in spring of 14.14 mm^3. Significant differences were observed between this season and the rest of the year (Table III). Tetrasporangial sori were smaller than carpo-

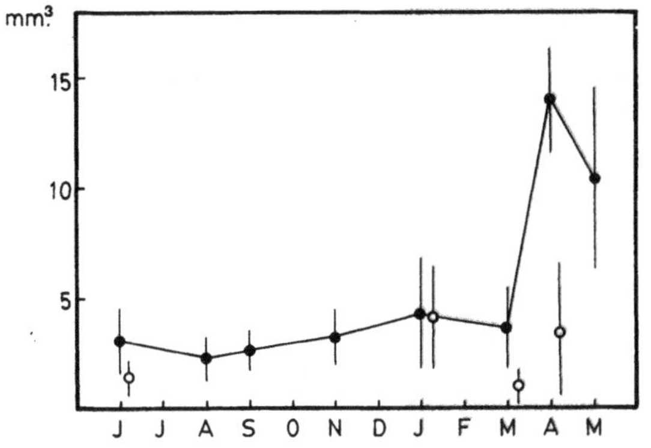

Fig. 7. Seasonal changes in volume of sporangial sori. Mean values ± S. D. Dark circles: carposporangial sori. Open circles: tetrasporangial sori.

sporangial sori, their maximum mean volume was 4.25 mm^3 in winter, and significant differences were observed between winter–spring and late spring–summer (Table IV).

Density of sporangial sori per frond

The number of carposporangial sori per frond did not show a regular pattern throughout the year because of the great variability of the data (Fig. 8). The only significant difference was in April (Table V). The number of tetrasporangial sori was constant throughout the fruiting period. All fertile fronds have plenty of sori, see Figure 2c, with an aproximate average number of 700 sori per frond.

Table II. Mann-Whitney U-test for differences between months and mean volume values ± S. D. for a) carpospores and b) tetraspores. All values in $mm^3 \times 10^6$.

a	June	August	January	April	May
June 5.89 ± 1.83	–				
August 7.08 ± 5.26	N. S.	–			
January 0.87 ± 0.33	**	N. S.	–		
April 9.83 ± 5.52	N. S.	N. S.	N. S.	–	
May 6.27 ± 4.68	N. S.	N. S.	*	*	–

b	June	January	April
June 9.47 ± 4.47	–		
January 4.42 ± 3.57	N. S.	–	
April 1.09 ± 0.25	*	**	–

* = P 0.05 ** = P 0.001.

Table IV. Mann-Whitney U-test for differences in mean values of tetrasporangial sori (mm^3) between months.

	June	January	April	May
June	–			
January	**	–		
April	N. S.	N. S.	–	
May	N. S.	***	*	–

* = P 0.05 ** = P 0.01 *** = P 0.001

Reproductive capacity

Frond reproductive capacity and density of spores are shown in Figure 9. Gametophytic fronds showed a maximum of around 2.5 $\times 10^8$ carpospores per frond from November to March (Fig. 9a). Tetrasporophytic fronds had the highest values in April, 23.5 $\times 10^8$ tetraspores per frond.

Total carpospores density per m^2 showed a maximum during November–January of 35–28 $\times 10^9$. The maximum for the tetraspores occurred in April and was 77.7 $\times 10^9$.

Table III. Mann-Whitney U-test for differences in mean values of carposporangial sori (mm^3) between months.

	June	August	September	November	January	March	April	May
June	–							
August	N. S.	–						
September	N. S.	N. S.	–					
November	N. S.	N. S.	N. S.	–				
January	N. S.	N. S.	N. S.	N. S.	–			
March	N. S.	N. S.	N. S.	N. S.	N. S.	–		
April	***	***	***	***	***	***	–	
May	***	***	***	***	***	***	N. S.	–

*** = P 0.001

Table V. Mann-Whitney U-test for differences in mean values of number of carposporangial sori per frond between months.

	June	August	September	November	January	February	March	April	May
June	–								
August	N.S.	–							
September	N.S.	**	–						
November	N.S.	*	N.S.	–					
January	N.S.	**	N.S.	N.S.	–				
February	N.S.	*	N.S.	N.S.	N.S.	–			
March	N.S.	*	N.S.	N.S.	N.S.	N.S.	–		
April	N.S.	*	N.S.	N.S.	N.S.	N.S.	N.S.	–	
May	N.S.	*	N.S.	N.S.	N.S.	N.S.	N.S.	N.S.	–

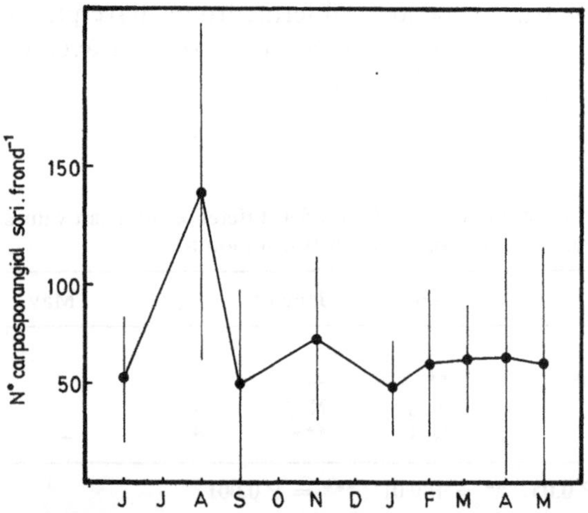

Fig. 8. Seasonal changes in the number of carposporangial sori per frond. Mean values ± S.D.

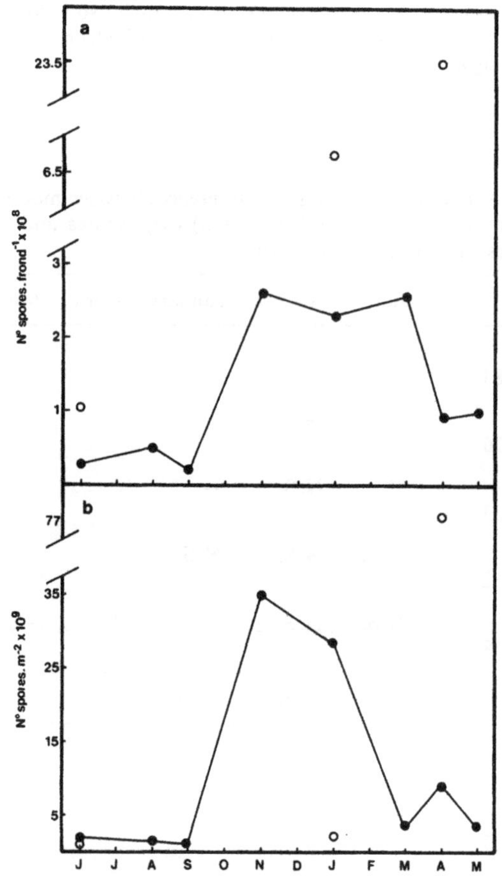

Fig. 9. Seasonal changes in frond reproductive capacity (a) and density of spores per m^2 (b). Dark circles: carpospores. Open circles: tetraspores.

Discussion

All the fertile individuals analyzed correspond to III size-development class as defined by McLachlan *et al.* (1988). The size ranges were similar in both localities. The fertile fronds being larger than the non-fertile ones (Fernández and Menéndez, in press). The minimum sizes of fertile fronds correspond with those found by Chen and McLachlan (1972) in laboratory cultures, for both tetrasporophytes and gametophytes. The percentage density of fertile fronds was low as would be expected in marginal populations, where a predominance of asexual reproduction has been reported (De Wreede and Klinger 1988) and, in populations growing on stable substratum (Chopin *et al.* 1988, Lazo and McLachlan 1989). However McLachlan *et al.* (1988) showed that studies of reproduction based on frond counts could give an underestimation of the potential capacity of plants. In contrast, the biomass of fertile fronds shows higher percentages in the whole population, according to the data of McLachlan *et al.* (1988).

The two study sites show great differences in the percentage of reproductive fronds with respect to the whole population. Tetrasporophytes are rare at Santa María del Mar, hence, there are fewer fertile fronds than at Viodo. Three reasons may explain this, firstly the different ploidy level of the populations (Fernández and Menéndez, in press); secondly, the exposure is higher at Santa María del Mar, and this may be a factor limiting the fertility period of tetrasporophytes, possibly due to their greater sensitivity (Chopin *et al.* 1988); and, thirdly, the different period of the observations, the two sites were sampled in different years.

Seasonality has a great influence on the annual pattern of denisty and biomass of mature fronds, with bi- or unimodal distribution of carposporophytes and tetrasporophytes, respectively. These maxima vary depending on the geographical situation, or, on a smaller scale, depending on the position within the vertical gradient of the coast, or the degree of wave exposure (Prince and Kingsbury 1973, Mathieson and Prince 1973, Mathieson and Burns 1975, Pybus 1977, Craigie and Pringle 1978, Chen and Taylor 1980, Tveter-Galagher *et al.* 1980, Mathieson 1982; Chopin *et al.* 1988, Lazo *et al.* 1989).

Seasonal patterns of ripeness of sori could be explained in the same way, but the maximum density of broken carposporangial sori occurs later than the maximum density of mature fronds. This agrees with observations reported by Chopin *et al.* (1988).

Although gametophytes and tetrasporophytes are described as isomorphic, there are differences in the terminal portions of the reproductive branches (Taylor and Chen 1973) and in the plant as a whole (Chen and McLachlan 1972). Tetrasporophytes tend to be narrower, ligulate and less branched than gametophytes (Craigie and Pringle 1978). We have found these characteristics in the fronds we have examined. No differences in size were found when sporophytic and gametophytic fronds were compared, and this again agrees with the observations of McLachlan *et al.* (1988). In the fronds examined here the gametophytes with wide fronds had less conspicuous cystocarps than those with narrow fronds. This variability could explain the seasonal differences in the carposporangial sori volume and the great variability in the density of carposoporangial sori per frond. The absence of males during this study is a frequent phenomenon, and has been reported by several authors (Mathieson and Prince 1973 includes an extensive bibliography). It is explained, at least partially, by the difficulty of identification (Taylor and Chen 1973) or, by their scarceness due to a critical gamete density (De Wreede and Klinger 1988).

Reproductive capacity, estimated as the number of spores/m^2/year or number of spores/m^2/month, was higher than in American intertidal populations (Bhattacharya 1985) and even more so than subtidal populations (Mathieson 1982). It shows a strong seasonal variation. These observations indicate that intertidal fronds are more reproductive than subtital ones.

Spore volume shows significative differences between months and phases. Minimal volume occurs in periods of maximal mature frond densities. Estimated carpospore volume was slightly larger than that for tetraspores, again agreeing with the observations of McLachlan *et al.* (1988). In spite of these differences, annual spore production was very similar in both phases. Two strategies are shown: a low and mantained carpospore production over a long period, and blooms of high tetraspore production at short times.

Chopin *et al.* (1988) suggested that a variation in spore production could partially explain the dominance of a phase. In our case, the similar composition (55% diploids *vs.* 45% haploids, Fernández and Menéndez, in press) of the whole population would be justified on the basis that no differences in frond survivorship has been noted (Bhattacharya 1985).

Reproduction control in *Chondrus crispus* is a complex phenomenon not strictly regulated by environmental factors and the absolute controlling factors remain unknown (McLachlan *et al.* 1988).

Acknowledgements

This work was supported by the Spanish Ministry of Science and Education (C. A. I. C. Y. T. Project N° 2636-83). We grateful to Dr J. McLachlan, Dr J. D. Pringle and a anonymous referee for helpful suggestions. Dr G. T. Boalch greatly improved the English of the manuscript.

References

Bhattacharya, D. 1985. The demography of fronds of *Chondrus crispus* Stackhouse. *J. Exp. Mar. Biol. Ecol. 91*: 217–231.

Chen, L. C.-M. and J. McLachlan. 1972. The life history of *Chondrus crispus* in culture. *Can. J. Bot. 50*: 1055–1060.

Chen, L. C.-M. and A. R. A. Taylor. 1980. Investigations of distinct strains of *Chondrus crispus* Stackh. 1. Field and laboratory observations. *Bot. Mar. 23*: 435–440.

Chopin, T. 1986. The red algae *Chondrus crispus* Stackhouse (Irish moss) and carrageenans -a review. *Can. Tech. Rep. Fish. Aquat. Sci. 1514*: v + 69 pp.

Chopin, T., J. D. Pringle and R. E. Temple. 1988. Reproductive capacity of dragraked and non-dragraked Irish moss (*Chondrus crispus* Stackhouse) beds in the southern Gulf of St. Lawrence. *Can. J. Fish. Aquat. Sci. 45(5)*: 758–766.

Craigie, J. S. and C. Leigh. 1978. Carrageenans and agars. *In*: (J. A. Hellebust and J. S. Craigie, eds) *Handbook of Phycological Methods, vol. II.* Cambridge Univ. Press. pp. 109–131.

Craigie, J. S. and J. D. Pringle. 1978. Spatial distribution of tetrasporophytes and gametophytes in four Maritime populations of *Chondrus crispus*. *Can. J. Bot. 56*: 2910–2914.

De Wreede, R. E. and T. Klinger. 1988. Reproductive strategies in algae. *In*: (J. Lovett Doust and L. Lovett Doust, eds) *Plant Reproductive Ecology*. Oxford Univ. Press, New York. pp. 267–284.

Fernández, C. and M. P. Menéndez. In press. Ecology of *Chondrus crispus* in the northern coast of Spain. I. Seasonal aspects. *Scient. Mar.*: In press.

Lazo, M. L. and J. McLachlan. 1989. Reproduction of *Chondrus crispus* Stackhouse (Rhodophyta, Gigartinales) in sublittoral Prince Edward Island, Canada. *Journal of Applied Phycology 1*: 359–365.

Lazo, M. L., M. Greenwell and J. McLachlan. 1989. Population structure of *Chondrus crispus* Stackhouse (Gigartinaceae, Rhodophyta) along the coast of Prince Edward Island, Canada. Distribution of gametophytic and sporophytic fronds. *J. Exp. Mar. Biol. Ecol. 126*: 45–58.

Mathieson, A. C. 1982. Reproductive phenology and sporeling ecology of *Chondrus crispus* Stackhouse. *In*: *Proc. of the Republic of China–U.S. Cooperative Science Seminar on Cultivation of Economic Algae*. Mangilao, Guam, Univ. Guam Lab. pp. 33–40.

Mathieson, A. C. 1989. Phenological patterns of northern New England seaweeds. *Bot. Mar. 32*: 419–438.

Mathieson, A. C. and R. L. Burns. 1975. Ecological studies of economic red algae. 5. Growth and reproductionof natural and harvested populations of *Chondrus crispus* Stackhouse in New Hampshire. *J. Exp. Mar. Biol. Ecol. 17*: 137–156.

Mathieson, A. C. and J. S. Prince. 1973. Ecology of *Chondrus crispus* Stackhouse. *In*: (M. J. Harvey and J. McLachlan, eds) *Chondrus crispus*. N. S. Institutc of Science, Halifax, Nova Scotia. pp. 53–79.

McLachlan, J. L., N. I. Lewis and M. L. Lazo. 1988. Biological considerations of *Chondrus crispus* Stackhouse (Rhodophyta, Gigartinaceae) in the Southern Gulf of St. Lawrence, Canada. *Gayana Bot. 45*: 29–54.

Menéndez, M. P. 1985. Biología de *Chondrus crispus* Stackhouse en la costa central asturiana. Tesis de Licenciatura, Univ. Oviedo. 98 pp.

Prince, J. S. and J. M. Kingsbury. 1973. The ecology of *Chondrus crispus* at Plymouth, Massachusetts. I. Ontogeny, vegetative anatomy, reproduction and life cycle. *Am. J. Bot. 60*: 956–963.

Pringle, J. D. and A. C. Mathieson. 1987. *Chondrus crispus* Stackhouse. *In*: (M. S. Doty, J. Caddy and B. Santelices, eds) *Case studies of seven commercial seaweeds resources. F. A. O. Fish. Tech. Paper 281*. pp. 49–122.

Pybus, C. 1977. The ecology of *Chondrus crispus* and *Gigartina stellata* (Rhosophyta) in Galway Bay. *J. Mar. Biol. Assoc. U. K. 57*: 609–628.

Siegel, S. 1956. *Non-parametric Statistics for the Behavioral Sciences*. McGraw-Hill Kogakusha Ltd., Tokyo. 311 pp.

Taylor, A. R. A. and L. C.-M. Chen. 1973. The biology of *Chondrus crispus* Stackhouse: Systematics, Morphology and Life-history. *In*: (M. J. Harvey and J. McLachlan, eds) *Chondrus crispus*. N. S. Institute of Science, Halifax, Nova Scotia). pp. 1–21.

Tveter-Gallagher, E. A. C. Mathieson and D. Cheney. 1980. Ecology and developmental morphology of male plants of *Chondrus crispus* (Gigartinales; Rhodophyta). *J. Phycol. 16*: 257–264.

Botanica Marina
Vol. 34, pp. 311–318, 1991

Helicascus kanaloanus, Helicascus nypae sp. nov. and *Salsuginea ramicola* gen. et sp. nov. from Intertidal Mangrove Wood

K. D. Hyde

Plant Pathology Branch, Queensland Department of Primary Industries, Mareeba Q. 4880, Australia

(Accepted 3 April 1991)

Abstract

The genus *Helicascus* is discussed and illustrated with interference light micrographs. One new species *Helicascus nypae*, is described from *Nypa* palm. A new genus *Salsuginea* (type species *S. ramicola*) is described from *Aegiceras corniculatum* to accommodate a taxon with spores similar to those of *Helicascus*, but which is otherwise morphologically dissimilar.

Introduction

The monotypic genus *Helicascus* Kohlm. was described from mangrove wood collected in Hawaii with *H. kanaloanus* Kohlm. as the type species (Kohlmeyer 1969). Recent collections of mangrove wood in South East Asia have resulted in several collections of *Helicascus*, some morphologically quite dissimilar to *H. kanaloanus* (Hyde 1988, 1989 a, b, c). In this paper, one new species of *Helicascus* from *Nypa* palm is described. A new genus *Salsuginea* is also proposed to accommodate fungi with ascospores that are similar to *Helicascus*, but which are otherwise morphologically quite dissimilar. All fungi are illustrated at the light microscope level.

Taxonomic part

The monotypic genus *Helicascus* Kohlm. was described from wood of dead prop roots and young plants of *Rhizophora mangle* L. (intertidal) collected in Hawaii by Kohlmeyer (1969). A description of *Helicascus* taken from Kohlmeyer (1969), Kohlmeyer and Kohlmeyer (1979) and material examined in this study, is given below.

Helicascus Kohlm. Can. J. Bot. (1969) 47: 1471.

Stromata with several locules, black, immersed in dead mangrove wood. Ascomata depressed ampulliform, horizontally arranged under a black pseudoclypeus, united radially into valsoid groups; ostioles connected into a common, central, periphysate porus. Hamathecium of numerous, persistent, anastomosing, delicate threads. Asci eight-spored, subcylindrical to oblong clavate, stipitate, fissitunicate, thick-walled, with an apical apparatus. Ascospores oboboidal, one-septate, constricted at the septum, brown; walls two-layered, covered by a mucilaginous sheath. Hosts: Dead intertidal mangrove trees (*Rhizophora, Nypa, Aegiceras, Kandelia*). Type species: *Helicascus kanaloanus* Kohlm.

Helicascus kanaloanus Kohlm. Can. J. Bot. (1969) 47: 1471. Figs 1–15

Stromata 0.6–0.78 mm high, 1.25–2.75 mm wide, lenticular, immersed, black, carbonaceous, enclosing 3–4(–5) loculi (Figs 1, 6, 8), covered by a pseudoclypeus. Ascomata 235–270 µm high, 440–800 µm long, depressed ampulliform (Fig. 1), united radially into valsoid groups (Fig. 6), horizontally arranged under a black pseudoclypeus (Fig. 8); ostioles 70–170 µm in diameter, uniting into a common central porus, stromatic tissue around the loculi composed of more or less isodiametric cells (Figs 2, 8), at the periphery composed of vertically arranged, longitudinal cells (Figs 5, 10); pseudoclypeus composed of host cells enclosed in black fungal hyphae (Figs 1, 9)

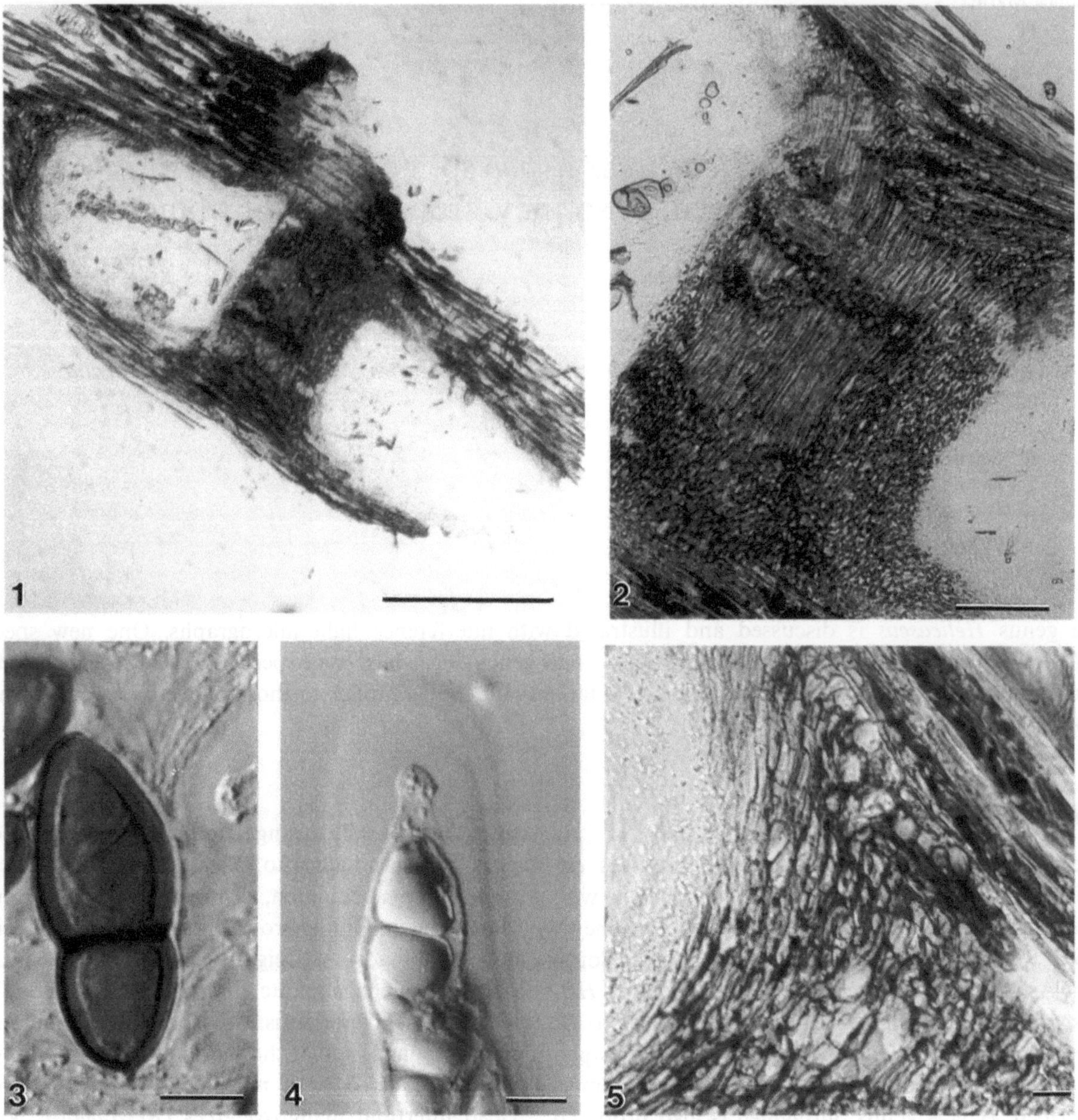

Figs 1–5. Interference light micrographs of *Helicascus kanaloanus* (Holotype NY 18591).
Fig. 1. Section through fruiting body showing depressed locules which unite into a common central porus. Fig. 2. Partition between loculi composed of brown, isodiametric and elongated cells. Fig. 3. Ascospore. Fig. 4. Ascus tip, showing ascus layers and ocular chamber. Fig. 5. Wedge of vertically arranged cells. Bar lines Fig. 1 = 500 μm, Fig. 2 = 100 μm, Figs 3–5 = 10 μm.

partitions between loculi formed of brown, isodiametric or elongated cells (Figs 2, 8). Neck 270–435 μm high, 255–300 μm in diameter; canal filled with periphyses (Fig. 7). Hamathecium, hyphae-like, numerous, persistent, filamentous, anastomosing. Asci 250–335 × 25–30 μm, 8-spored, subcylindrical, finally oblong clavate (Fig. 11), stipitate, bitunicate (fissitunicate), thick-walled, apically multilayered (Fig. 4), with an ocular chamber (Fig. 13); at maturity endoascus swelling in water and becoming coiled, finally stretching and pushing the ascus into the ostiolar canal. A faint ring was observed in asci from *Aegiceras corniculatum* (L.) Blanco, but was difficult to discern in other specimens. Ascospores 30–55 × 17–25 μm, uniseriate, obovoidal, unequally two-celled, highly constricted at the septum, dark brown, sometimes at one or both ends apiculate (Figs 3, 14–15) with a deliquescing sheath (Fig. 15). Saprobic on intertidal mangrove wood [*Rhizophora mangle L.*, *Aegiceras corniculatum*, *Kandelia candel* (L.) Druce]. Distribution: Hawaii, Brunei, Thailand.

Material examined: Hawaii, Oahu, Kaneohe Bay, Heeia Swamp, in wood of dead prop roots and young plants of *Rhizophora mangle*, 4 June 1968, J. Kohlm. (NY 18591: Holotype); Brunei, Kampong Tutong, in wood of dead intertidal buttress roots of *Kandelia candel*, December 1988, K. D. Hyde (KDH 1542,

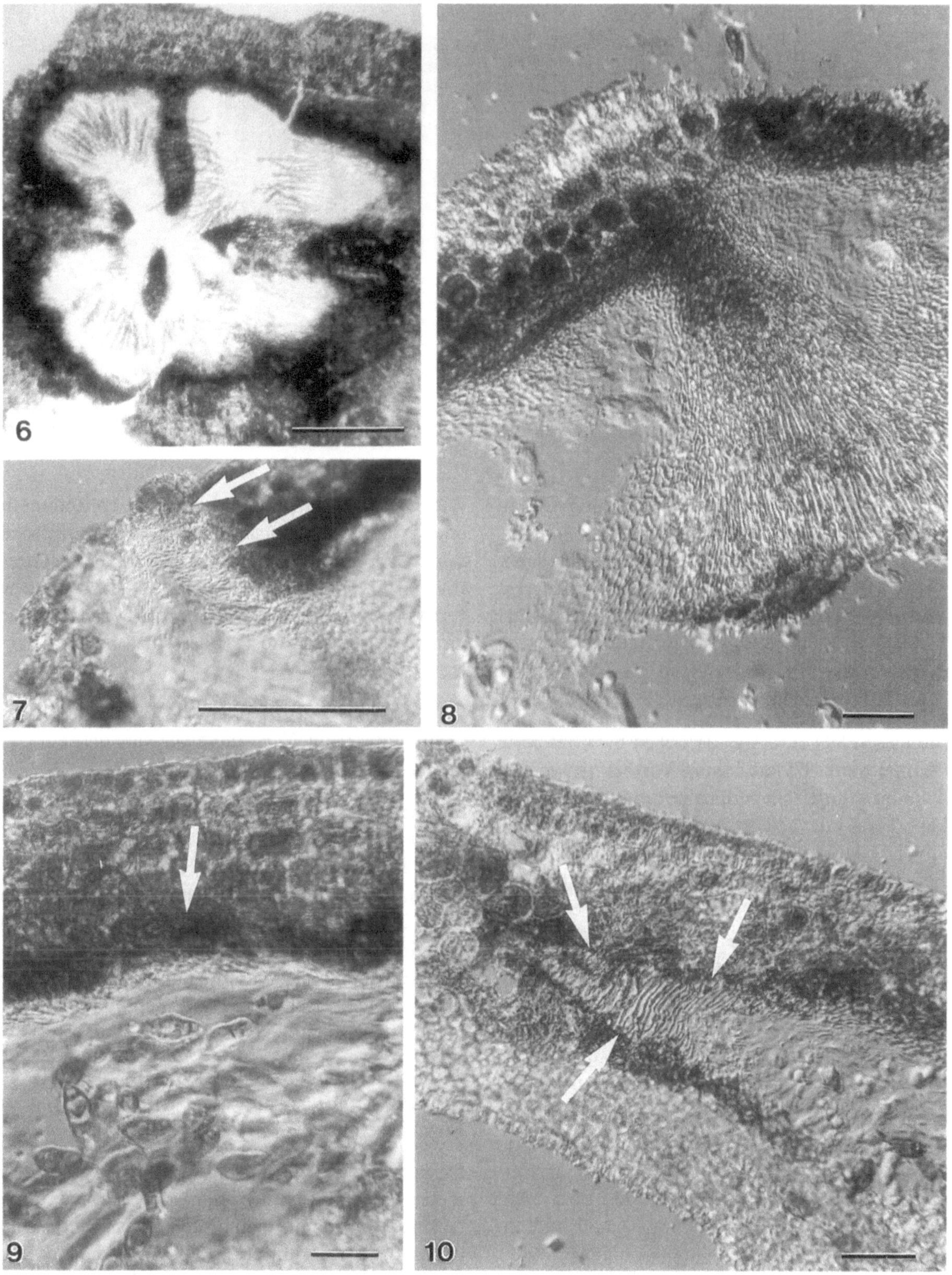

Figs 6–10. Interference light micrographs of *Helicascus kanaloanus* from *Kandelia candel* (BRIP 17104).
Fig. 6. Cross section through fruiting body illustrating the loculi and partitions. Fig. 7. Periphysate central ostiole (arrowed). Fig. 8. Section through fruiting body illustrating stroma and partition, the latter composed of vertically arranged elongate cells. Fig. 9. Section through stroma. Darkened fungal tissue (arrowed) grows amongst the host cells. Fig. 10. Margin between stroma and base of fruiting bodies. A wedge of vertically arranged elongate cells is present (arrowed). Bar lines. Fig. 6 = 500 μm, Figs 7–10 = 100 μm.

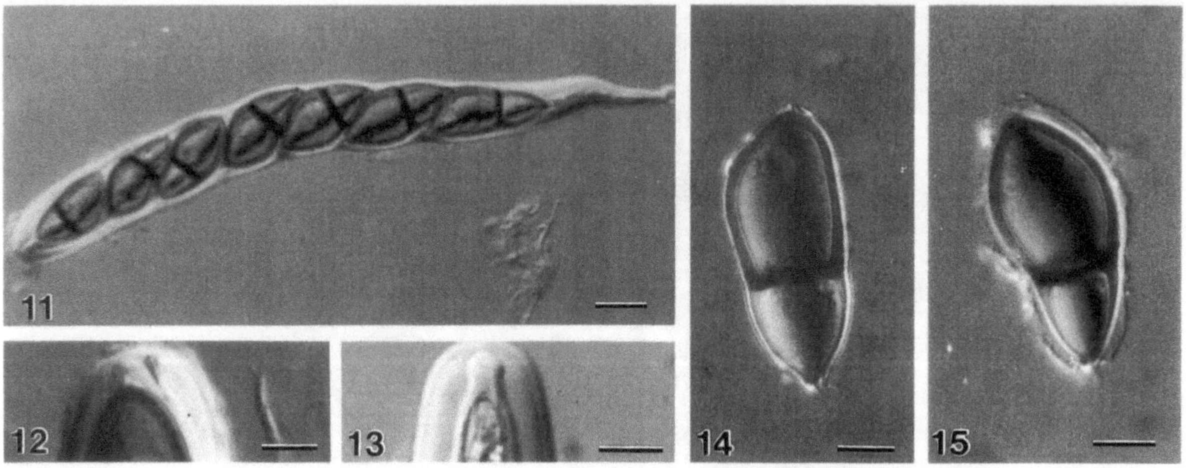

Figs 11–15. Interference light micrographs of *Helicascus kanaloanus* from *Kandelia candel* (BRIP 17104).
Figs 11–13. Asci. Note the ocular chamber. Figs 14–15. Ascospores. Note the mucilaginous sheath in Figure 15 and single apical germ pores. Bar lines = 10 µm.

1544, 1547, BRIP 17104); Brunei, Kampong Danau mangrove, on dead intertidal wood of *Rhizophora* sp., April 1986, K. D. Hyde (KDH 0786); Thailand, Phang Nga Bay, on dead intertidal roots of *Aegiceras corniculatum*, 10 January 1988, K. D. Hyde (KDH 0847–0849); Thailand, Ranong mangrove, on dead intertidal roots of *Aegiceras corniculatum*, October 1988, K. D. Hyde (KDH 1660, BRIP 17101).

The ascus of *Helicascus* was described by Kohlmeyer and Kohlmeyer (1979) as having 3 layers, the ectoascus forming a third, thin permeable outer layer around the base. I was only able to detect 2 layers and studies at the TEM level are needed to establish the nature of the ascus wall. The endoascus was coiled within the base, eventually stretching, splitting the ectoascus and becoming elongate.

Differences in ascospore size occurred between samples of this taxon taken from different hosts. In *Aegiceras*, ascospores were 30–45 × 17–25 µm, apical pores were small and remnants of the mucilaginous sheath were persistent. In *Kandelia* samples, ascospores were 33–55 × 18–25 µm, the apical pore was prominent, but usually only at one pole, and the sheath was persistent. In the type material of *Helicascus kanaloanus* ascospores were 36.5–48.5 × 18.0–22.5 µm, apical pores were small and the mucilaginous sheath deliquescing (Kohlmeyer 1969). Patil and Borse (1985) also reported this taxon from mangrove wood in India. Here ascospores were large 53–75 × 20–32 µm. Herbarium material of this collection was requested on several occasions, but this was not made available. It may later be necessary to describe more than one species of *Helicascus* from dicotyledonous mangrove hosts based on spore size and host. However, until larger numbers of samples from different hosts are examined and statistical analysis carried out I prefer to place all collections into one taxon. *Nypa fruticans* (Thunb.) Wurmb. also supported a species of *Helicascus*, but with ascospores of smaller dimension, with a persistent sheath and with verrucose spore walls. This fungus which is from a monocotyledonous host, is described below.

Helicascus nypae Hyde, sp. nov. Figs 16–20

Etymology: from *Nypa* the host.

Diagnosis

Stromata 260–390 µm alta, 750–1500 µm diametro, lenticulares, immersa, nigra, carbonacea, 3–4 loculares, ad centrum cum porus. Hamathecium filamentosum, numerosum, anastomoses, persistes. Asci 192–280 × 14–20 µm, 8-spori, pedunculati, subcylindrici, fissitunicati, cum camera apicali. Ascosporae 23–35 × 12–15 µm, uniseriatae, obovoideae, unaequaliter bicellularae, ad septum constrictae, brunneae, apiculatae, cum tunica gelatinosae.

Holotypus: Brunei, Tungit Api Api mangrove (salinity 14–28‰), on dead intertidal fronds of *Nypa fruticans*, December 1988, K. D. Hyde (BRIP 17103).

Stromata 260–390 µm high, 750–1500 µm in diameter, lenticular, immersed, black, carbonaceous, enclosing 3–4 loculi, loculi having a common central porus. Hamathecium 1.5–2.0 µm wide, numerous, persistent, anastomosing, in a gel. Asci 192–280 × 14–20 µm eight-spored, pedunculate, subcylindrical, fissitunicate, with an ocular chamber (Figs 19, 20). Ascospores 25–35 × 12–15 µm, uniseriate, obovoidal, unequally 2-celled, constricted at the septum, brown, sometimes at one or both ends apiculate, with a persistent mucilaginous sheath and a verrucose wall (Figs 16–18). Saprobic on intertidal *Nypa fruticans*. Distribution: South China Sea, Brunei.

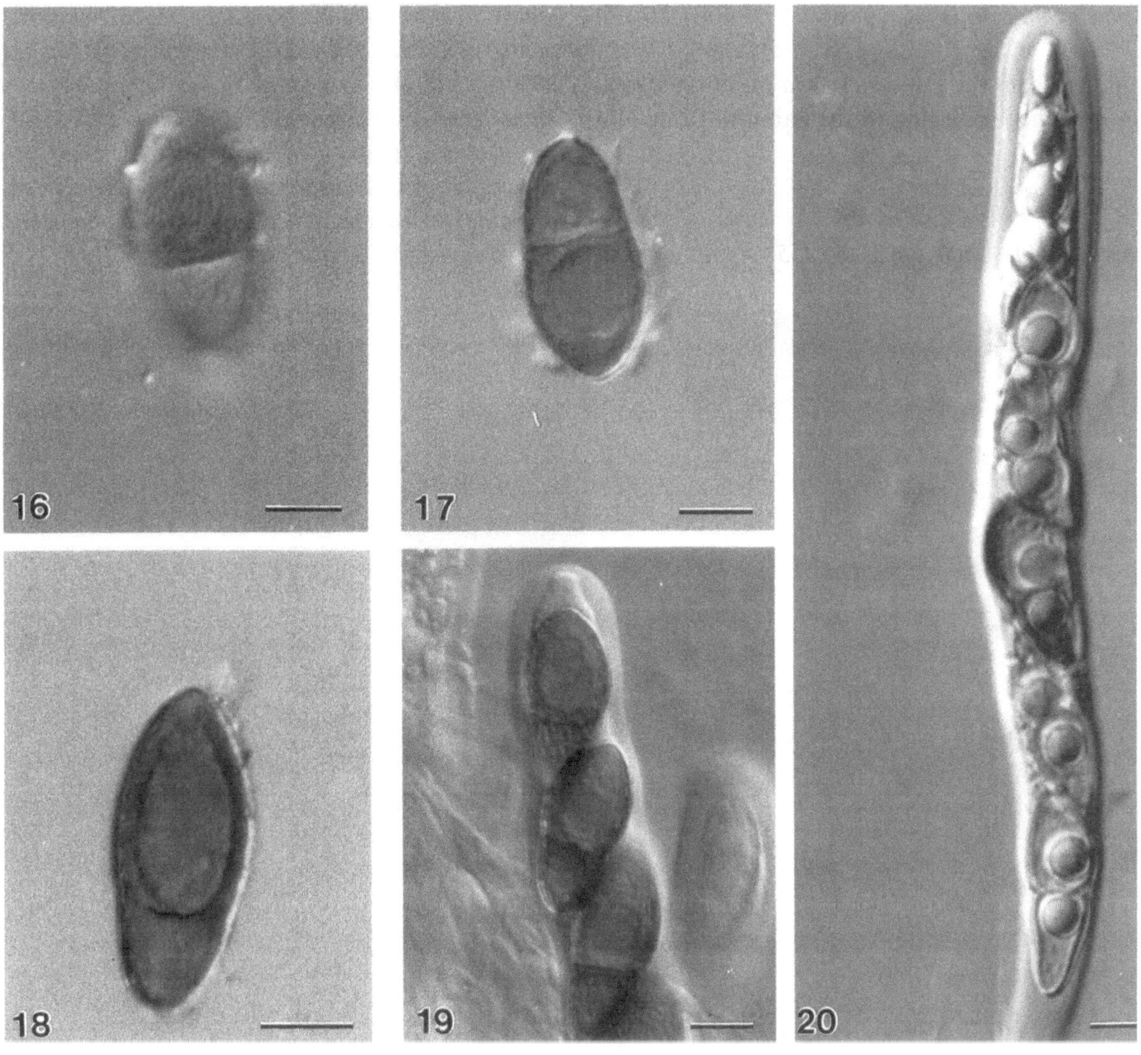

Figs 16–20. Interference light micrographs of *Helicascus nypae* (BRIP 17103).
Fig. 16–18. Ascospores. Note the mucilaginous sheath in Figures 16 and 17 and the verrucose wall in Figure 16. Figs 19, 20. Asci. Ascospores are uniseriate in Figure 20 and an ocular chamber can be seen at the apex in Figure 19. Bars lines = 10 µm.

Helicascus nypae differs from *H. kanaloanus* in the smaller ascus and ascospore dimensions, in having a persistent sheath to the ascospores (Fig. 17), in having distinctly verrucose ascospore walls (Figs 16, 19) and by a different host. *Helicascus nypae* has only been found on *Nypa* palm, while *H. kanaloanus* has been collected from dicotyledonous mangrove wood (Kohlmeyer 1969, Hyde and Jones 1988).

Salsuginea Hyde, gen. nov.

Etymology: from the Latin *salsugineus* meaning 'brackish', in relation to the brackish habitat in which this fungus develops.

Diagnosis

Ascomata immersa, fusiformes, conica vel subglobosa, clypeata, in sectione transversali fusiformes vel subglobosa, ostiolata, solitaria. Hamathecium tenuum et anastomoses. Asci 8-spori, fissitunicati, pedunculati, cum camera apicali. Ascosporae, unaequaliter, bicellularae, brunneae vel nigrae, cum porus gena, ad septum constrictae.

Typus generis: Salsuginea ramicola Hyde.

Ascomata immersed, fusiform, conical or subglobose, clypeate, appearing as a raised darkened area with a central ostiole and short neck, solitary. Peridium thin and hyaline at the base, of rounded cells at the sides, fusing with the clypeus at the top. Clypeus of dark fungal tissue. At the region between the side and base is a area of elongate parallel cells in the form of a wedge. Hamathecium thin, anastomosing, in a gel. Asci 8-spored, fissitunicate, pedunculate, with an ocular chamber and ring. Ascospores irregularly 2-celled, brown to black, obovoid, with hyaline apical germ pores, highly constricted at the septa. Type species: *Salsuginea ramicola* Hyde.

Salsuginea ramicola Hyde, sp. nov. Figs 21–34

Etymology: from the Latin *ramicola* meaning 'dwelling on branches', referring to the habitat of this fungus.

Diagnosis

Ascomata 1040–2600 µm alta, 455–1430 µm diametro, fusiformes, conica vel subglobosa, immersa, clypeata, ostiolata, solitaria. Peridium ad pagina superiore, cellularum rotundatum. Hamathecium tenuum, numerosum et anastomoses. Asci 440–512 × 29–34 µm, 8-spori, fissitunicati, pedunculati, cylindrici, cum camera apicali. Ascosporae 59–72 × 24–30 µm, uniseriatae, unaequaliter bicellularae, ad septum constrictae, brunneae vel nigrae, cum porus gena.

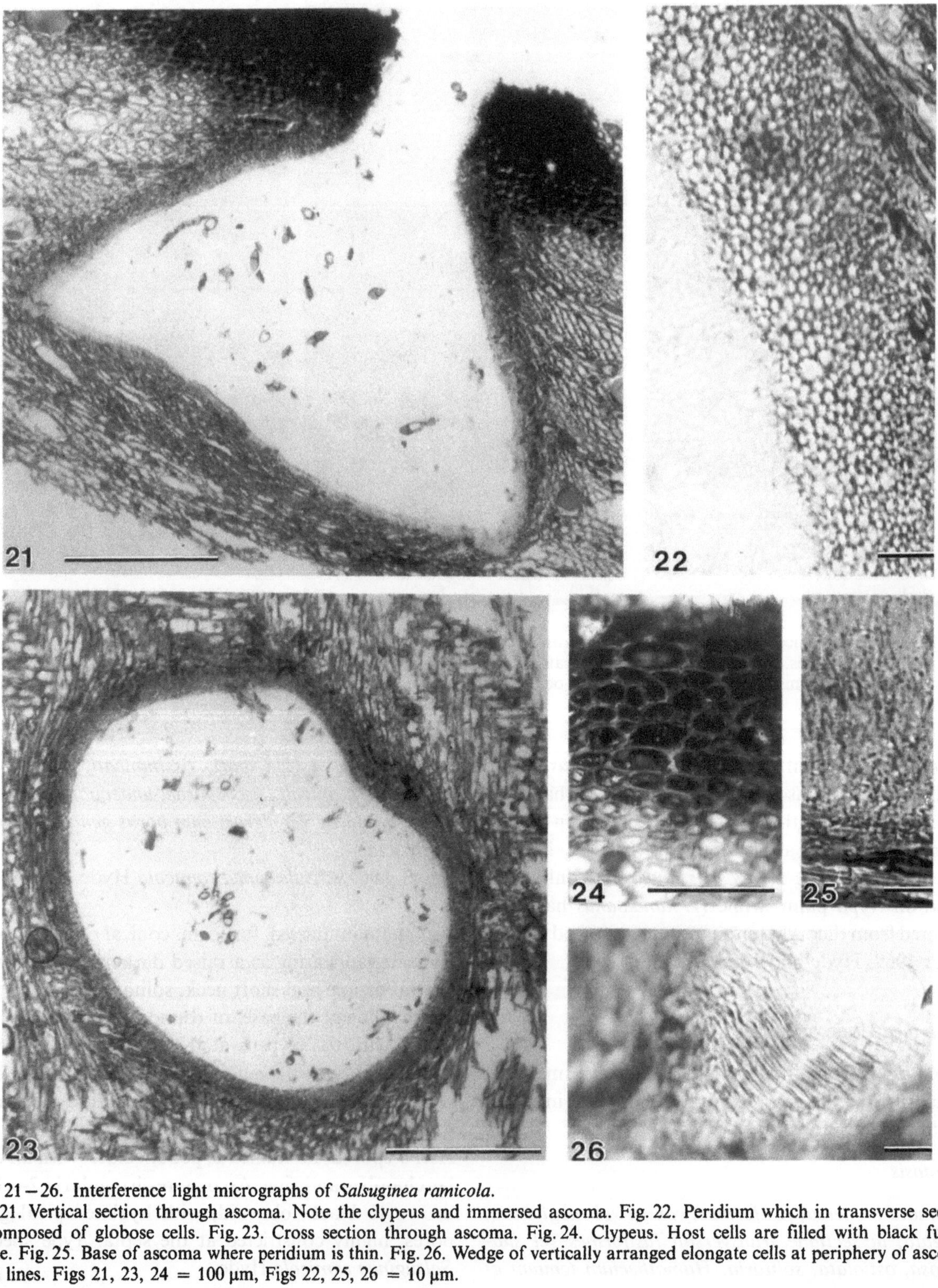

Figs 21–26. Interference light micrographs of *Salsuginea ramicola*.
Fig. 21. Vertical section through ascoma. Note the clypeus and immersed ascoma. Fig. 22. Peridium which in transverse section is composed of globose cells. Fig. 23. Cross section through ascoma. Fig. 24. Clypeus. Host cells are filled with black fungal tissue. Fig. 25. Base of ascoma where peridium is thin. Fig. 26. Wedge of vertically arranged elongate cells at periphery of ascoma. Bars lines. Figs 21, 23, 24 = 100 µm, Figs 22, 25, 26 = 10 µm.

Holotypus: Thailand, Ranong mangrove, *Aegiceras corniculatum*, October 1988, K. D. Hyde (BRIP 17102).

Ascomata 1040–2600 µm high, 455–1430 µm in diameter, fusiform, conical or subglobose, with or without a flattened base, immersed under a darkened clypeus (Fig. 21), appearing as a raised darkened area on the wood surface, with a central ostiole and short neck, solitary. Peridium up to 39 µm thick, of round cells (in cross section) at the sides, fusing at the top with the clypeus, thin at the base (Figs 21, 23, 25). Clypeus composed of host cells interdispersed with darkened fungal tissue and rounded melanin particles (Fig. 24). At the margin between the side and base is a wedge-shaped region of elongate parallel cells (Fig. 26). Hamathecium 0.9–1.8 µm in diameter, thin, numerous, anastomosing, in a gel (Figs 33, 34). Asci 440–512 × 29–34 µm, 8-spored, fissitunicate, cylindrical, pedunculate, with an apical apparatus consisting of a large distinctive ocular chamber and prominent ring (Figs 31–34). Ascospores 59–72 × 24–30 µm uniseriate, unequally 2-celled, obovoid, highly constricted at the septa, brown to black, smooth-walled, with hyaline apical germ pores, lacking a mucilaginous sheath (Figs 27–29). Saprobic on intertidal mangrove wood. Distribution: Brunei, North Sumatra, Thailand.

Other material examined: Tungit Api Api mangrove, dead intertidal wood of *Avicennia alba* Blume, December 1987, K. D. Hyde (KDH 0733–0735); *Xylocarpus granatum* König, July 1986, K. D. Hyde (KDH 0020), September 1985, K. D. Hyde (KDH 0169–0171), February 1986, K. D. Hyde (KDH 0384); Kampong Tutong mangrove, mangrove wood, September 1985, K. D. Hyde (KDH 0177); *Kandelia candel*, December 1988, K. D. Hyde (BRIP 17105); North Sumatra, Medan, *Xylocarpus granatum*, October 1986, K. D. Hyde (KDH 0256).

The genus *Salsuginea* differs from *Helicascus* mainly in the morphology of the fruiting body. In *Helicascus*

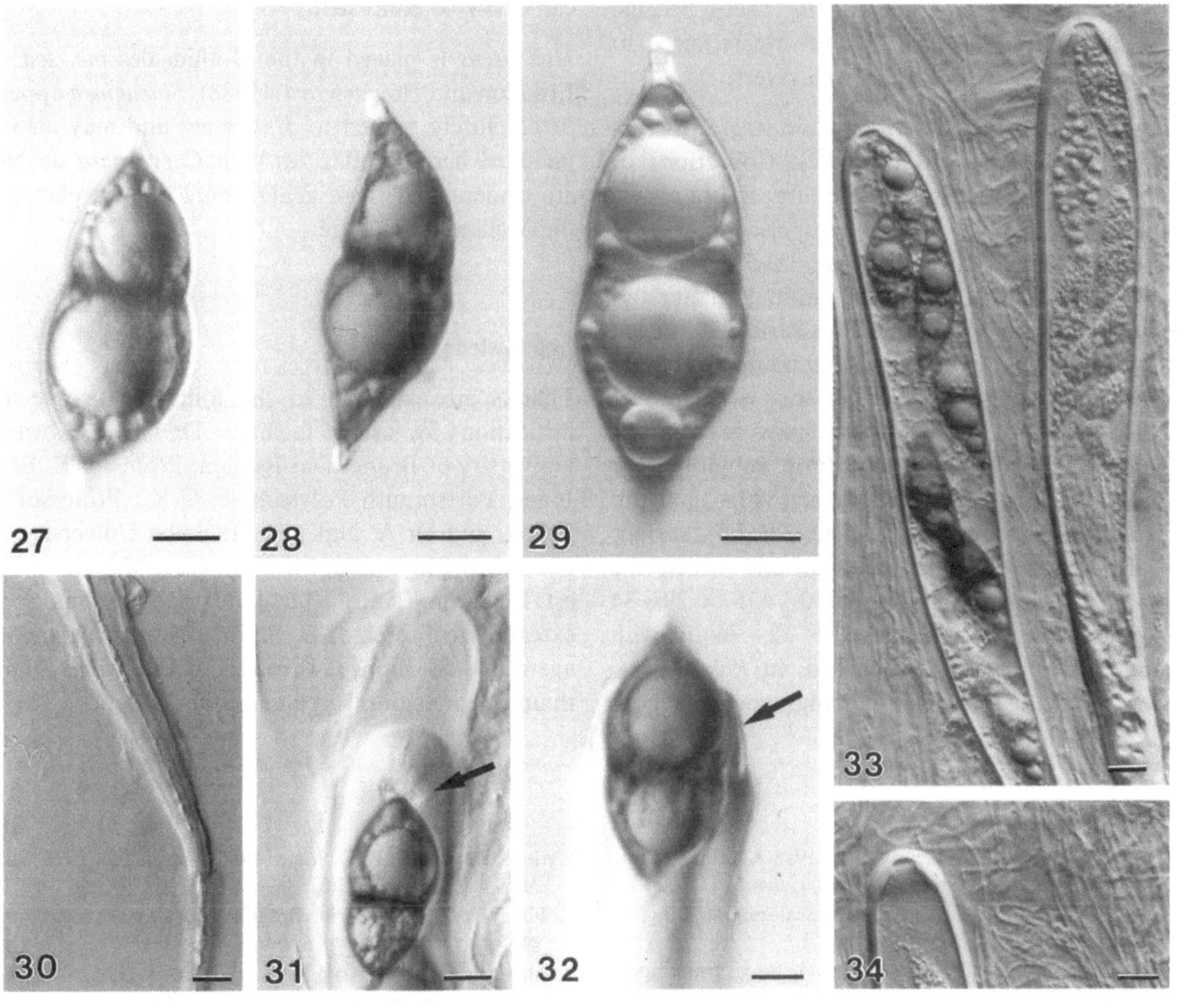

Figs 27–34. Interference light micrographs of *Salsuginea ramicola*.
Figs 27–29. Ascospores. Note the distinct apical pores. Fig. 30. Ascus. Note the fissitunicate dehiscence. Figs 31–32. Ascospores and asci. Note the ring structure (arrowed). Spores are released by fissitunicate dehiscence. Figs 33, 34. Asci and hamathecium. Note the distinct ocular chamber. Bars lines = 10 µm.

this is composed of a stroma in which 3–5 laterally depressed loculi develop with their necks fusing into a central porus. The sides and base are comprised of thin, hyaline, peridial cells, and elongate cells form the stroma partitioning the loculi (Figs 1, 2, 6, 8). In *Salsuginea* a single locule develops under a clypeus and thin stroma (Fig. 21). The side wall of the ascomata is composed of rounded cells (cross section) or textura intricata (in longitudinal section) and is thin at the base (Figs 21, 22, 25). The ring surrounding the ocular chamber of the ascus in *Salsuginea* is also distinctive (Figs 31, 32). Ascospores also differ, as in *Salsuginea* the apical pores are highly pronounced and there is never a mucilaginous sheath (Figs 27–29).

The asci of *Salsuginea* were similar to those in *Helicascus*, the number of wall layers being difficult to differentiate. The ectoascus of *Salsuginea* appeared to be comprised of a thin outer membrane and a thicker inner wall, while the endoascus was thick and coiled at the base. On dehiscence the endoascus stretched, uncoiled, split the ectoascus and the ascus became elongate. Studies at the TEM level are required to determine the number of ascus wall layers.

Salsuginea ramicola is described from *Aegiceras corniculatum* (Holotype: BRIP 17102). Collections of *Salsuginea* from other hosts varied in ascospore size and also morphology. In *Aegiceras* ascomata were fusiform or varied to subglobose. Ascomata were 1040–2600 μm high, 455–1430 μm in diameter and developed under a clypeus, the host surface darkened in a thin layer around several ascomata. Asci were 440–512 μm long, while ascospores were 59–72 × 24–30 μm with promnent germ pores at each end. In *Xylocarpus* ascomata were fusiform, subglobose or conoid, 715–1560 μm in diameter, 800–1300 μm high, and developed under a clypeus. The host surface was also darkened in the area around which the ascomata developed. Asci were 360–455 × 29–34 μm, while ascospores were 52–63 × 22–26 μm, with prominent germ pores at each end. In *Kandelia* ascomata were subglobose, developing under a clypeus with variable stromatic development laterally. Ascomata dimensions were 520–910 μm in diameter, 520–650 μm high. Asci were 212–320 μm long, while ascospores were 48–62 × 18–24 μm, with prominent germ pores. *Salsuginea ramicola* was also collected on *Avicennia marina*. Here ascomata were mostly fusiform, however, because of limited material measurements are not available. The differences in dimensions and morphology may later initiate the formation of several taxa based on dimensions and host. However, until larger numbers of samples are examined and statistical analysis carried out, I prefer to describe a single species.

There have been several records of *Helicascus* on mangrove wood in recent publications, many with a diverse spore range (Hyde 1988). The large ascospores in *Salsuginea* are very distinctive and collections named as *'Helicascus kanaloanus?'* in the publications of Hyde (1988, 1989 a, b, c) belong to *Salsuginea ramicola*. With other records it may be necessary for authors to reexamine their type material as these fungi can easily be confused.

Helicascus is placed in the Dothideales *inc. sed.* by Eriksson and Hawksworth (1988). *Salsuginea* appears to be closely related to *Helicascus* and may also be included here. Similarities with *Caryospora* de Not. are evident and these genera may best be placed in the Melanommataceae.

Acknowledgements

Thanks are extended to the following people and institutions for use of facilities: Dr David Edwards, University of Brunei Darussalam; Professor E. B. G. Jones, Portsmouth Polytechnic, U. K.; Professor K. Tubaki and Dr A. Nakagiri, Tsukuba University, Japan; Mr S. Havarand, Miss Prapasri, Mangrove Forest Research Centre, Thailand. Special thanks are extended to Professor E. B. G. Jones for his encouragement and support throughout this work. NY is thanked for loan of type material.

References

Eriksson, O. and D. L. Hawksworth. 1988. Outline of the ascomycetes – 1988. *Systema Ascomycetum 7*: 119–315.

Hyde, K. D. 1988. Studies on the tropical marine fungi of Brunei. *Bot. J. Linn. Soc. 98*: 135–151.

Hyde, K. D. 1989 a. Intertidal mangrove fungi from north Sumatra. *Can. J. Bot. 67*: 3078–3082.

Hyde, K. D. 1989 b. Ecology of tropical marine fungi. *Hydrobiologia 178*: 199–208.

Hyde, K. D. 1989 c. *Caryospora mangrovei* sp. nov. and notes on marine fungi from Thailand. *Trans. Mycol. Soc. Japan 30*: 333–341.

Hyde, K. D. and E. B. G. Jones. 1988. Marine mangrove fungi. *Marine Ecology* (Berlin) *104*: 15–33.

Kohlmeyer, J. 1969. Marine fungi of Hawaii including the new genus *Helicascus*. *Can. J. Bot. 47*: 1469–1487.

Kohlmeyer, J. and E. Kohlmeyer. 1979. *Marine Mycology. The Higher Fungi*. Academic Press, New York.

Patil, S. D. and B. D. Borse. 1985. Marine fungi from Maharastra (India) IV: Some Loculoascomycetes. *Trans. Mycol. Soc. Japan. 26*: 271–276.

Botanica Marina
Vol. 34, pp. 319–322, 1991

Particulate Organic Carbon, Nitrogen and Phosphorus Content in Roots, Rhizomes and Differently Aged Leaves of *Zostera noltii* Hornem. in Oosterschelde Estuary (Southwestern Netherlands)

J. L. Pérez-Lloréns, M. Muchtar*, F. X. Niell** and P. H. Nienhuis***

Departamento de Biología, Facultad de Ciencias del Mar, Universidad de Cádiz, 11510 Puerto Real, Cádiz, Spain

* *Centre for Oceanological Research and Development Alamat, Pasir Putih I, Ancol Timur, P.O.Box 580 DAK, Jakarta, 1101, Indonesia*

** *Departamento de Ecología, Facultad de Ciencias, Universidad de Málaga, Campus de Teatinos, 29071 Malaga, Spain*

*** *Delta Institute for Hydrobiological Research, Vierstraat 28, 4401 EA, Yerseke, The Netherlands*

(Accepted 7 April 1991)

Abstract

Total nitrogen, particulate organic carbon (POC) and phosphorus (orthophosphate, total phosphorus and hydrolyzable phosphate) have been measured in different-aged leaves, roots and rhizomes of *Zostera noltii* Hornem. plants growing in the Oosterschelde (Southwestern Netherlands) during the late spring. The highest nutrient amounts were detected in the leaves, especially in the youngest (the most actively growing parts), while lower contents were found in the underground parts: the roots and the rhizomes.

Introduction

Macrophytes possess the ability to accumulate nitrogen and phosphorus, accelerating the nutrient turnover in aquatic systems (Niell 1976, Brock *et al.* 1983). Studies on the elemental composition of macrophytes have been carried out by many investigators (Augier *et al.* 1982, Umebayashi and Sasaki 1985, Pellikaan and Nienhuis 1988, Pirc and Wollenweber 1988, Wahbeh 1988). These authors reported that the nutrient content of aquatic plants varies with species, plant part, time of the year, age of the plant and environment.

Because only a small part of the seagrass production is consumed by animals (Den Hartog 1980, Nienhuis and Groenendijk 1986), the bulk of the material enters the detritus pool, releasing part of the internal nutrients by leaching and microbial mineralization (Harrison and Mann 1975, Godshalk and Wetzel 1978, Pellikaan 1982, 1984, Velimirov 1987, Pellikaan and Nienhuis 1988), increasing, consequently, the energy flow in these environments. Hence, one of the main reasons for collecting data such as those reported here is to determine the nutrient content of the plant parts that enter the detritus pool (because aged leaves differ from the 'average'). The aim of this study was to collect this sort of information in a population of the seagrass *Zostera noltii* Hornem. from Oosterschelde during the early growing season. The research was carried out in June 1989 at the Delta Institute for Hydrobiological Research, Yerseke (The Netherlands).

Material and Methods

Selected plants (n = 100), collected at low tide, with healthy and complete leaves, were brought to the laboratory. Plants were generally free of visible epiphytic organisms; however, leaves were gently scraped with fingernails to remove all loosely attached material.

Leaves, cut from the underground parts, were separated according to age: the inner leaf was the youngest (leaf 1) and the outer (leaf 3) was the oldest. Underground parts were also separated into roots and rhizomes. Samples were oven dried at 70 °C for 2 d, pulverized and stored. Samples of the homogenate (1 – 1.5 mg dry weight) were taken for analysis of particulate organic carbon (POC) and nitrogen with a Carlo Erba C-N analyzer. Samples for phosphorus determination (6 mg dry weight) were analyzed with a Technicon auto-analyzer as two components: orthophosphate or molybdate-reactive phosphorus (Murphy and Riley 1962, modified by Hendriksen 1965) and total phosphorus (Armstrong and Tibbits 1968, Stanley and Richardson 1971). The difference between total phosphorus and orthophosphate is considered to be the hydrolyzable phosphate, bound to organic substances and polyphosphate (Golterman *et al.* 1978).

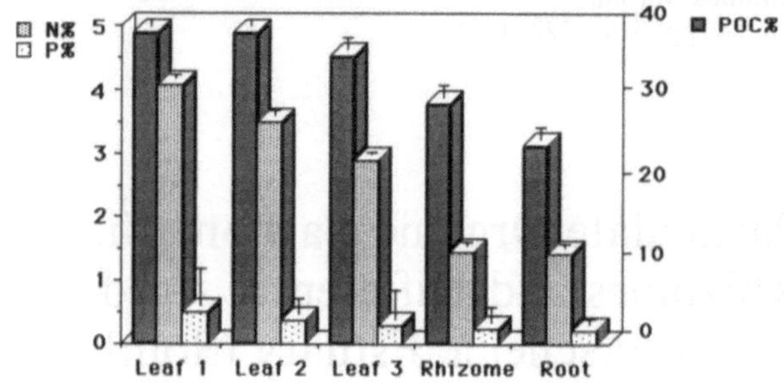

Fig. 1. Total nitrogen, total phosphorus and particulate organic carbon (POC) contents (as percentage of dry weight) detected in different-aged leaves (leaf 1 is the youngest and leaf 3 the oldest) and in roots and rhizomes of *Zostera noltii* plants. Each value is an average of three analyses. Vertical bars indicate S. D.

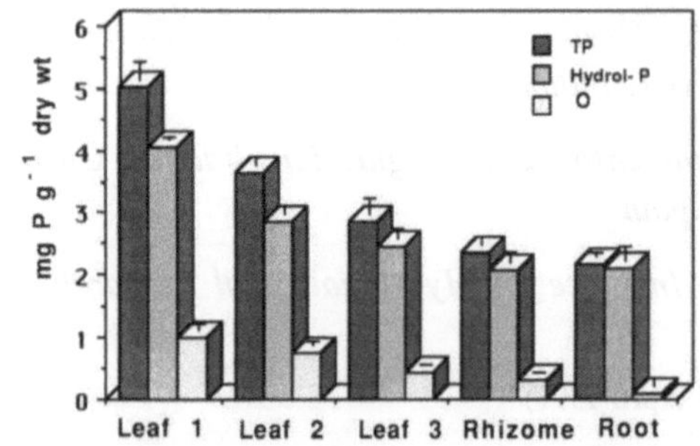

Fig. 2. Orthophosphate (O), total phosphorus (TP) and hydrolyzable phosphate (Hydrol-P) concentrations measured in leaves of different age (leaf 1 is the youngest and leaf 3 the oldest), roots and rhizomes of the seagrass *Zostera noltii.* Each value is an average of three analyes. Vertical bars indicate S. D.

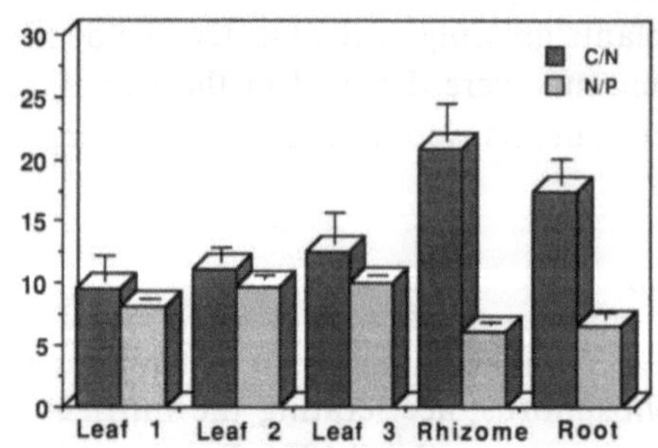

Fig. 3. C:N and N:P ratios estimated in leaves, roots and rhizomes of *Zostera noltii.* Each value is an average of three analyses. Vertical bars indicate S. D.

Results

Total nitrogen

Aerial parts contained on average 2.5 times more nitrogen than the underground parts (Fig. 1). The amount of nitrogen in the leaves varied depending on the age, from 4.0% in the youngest leaves to 2.9% in the oldest. No differences were found between rhizomes and roots.

Phosphorus content

The total phosphorus was on average 1.7 times higher in the aerial parts than in the roots and rhizomes (Figs 1, 2). The total phosphorus content in the leaves varied according to the age, with the highest values in the youngest leaves. No differences were detected between roots and rhizomes.

The distributional pattern of soluble phosphate was similar to that shown by total phosphorus. Leaves contained 3.8 times more soluble phosphate than roots or rhizomes, and the content varied depending on the age, reaching the highest values in the youngest leaves. Soluble-phosphate content detected in the rhizomes was higher than that in the roots (Fig. 3).

Particulate organic carbon (POC)

Leaves contained 1.5 times more POC than the underground parts. Differences of POC, in relation with the age, were detected in the leaves, especially in the oldest leaves which had the lowest content (Fig. 1). Differences were also found between roots and rhizomes, with more carbon in the rhizomes.

C:N:P ratio

Minimum C:N ratios were calculated for the aerial parts, increasing with the age of leaves. The highest ratio was found in the rhizome (Fig. 3). The highest N:P ratios were detected in the leaves, increasing slightly in the oldest leaves (Fig. 3). The C:N:P ratio increased with the age of the leaves: leaf 1, 78:8:1; leaf 2, 109:10:1; leaf 3, 126:10:1. The ratio recorded in underground parts were: 127:6:1 (rhizome) and 114:7:1 (roots).

Discussion

These results show that the nutrient content in *Z. noltii* depends, among other factors on the age of the leaves and on the part of the plant (leaves, roots and rhizomes). The highest nutrient levels were detected in the aerial parts, with the maximum content and the lowest C:N and N:P ratios in the youngest leaves (leaf 1) which are the most actively growing parts, acting as a natural sink for phosphorus and nitrogen (Stumpf 1952, Welsh and Denny 1979, Iizumi and Hattori 1982, Pérez-Lloréns and Niell 1989). Nutrient levels were always higher than the critical, limiting values of 0.75% for nitrogen and 0.07% for phosphorus (Gerloff 1975) reported for aquatic macrophytes. Assuming these critical values apply to *Z. noltii*, then these nutrients are not limiting for growth in Oosterschelde during the late spring, the initial growing season. According to the data of Vermaat *et al.* (1987), the total leaf growth of *Z. noltii* in The Netherlands showed maxima in spring and late summer. The high content of phosphorus and nitrogen and the low C:N ratio detected in the young leaves in late spring, before the main growing season, may be the result of active nutrient uptake, followed by a rapid protein synthesis before the fast growth, as Brock *et al.* (1983) pointed out for freshwater macrophytes.

The C:N:P ratio of *Z. noltii* deviates from the Redfield ratio. The ratio recorded for this seagrass is lower than those reported for benthic marine macroalgae and seagrasses (see Atkinson and Smith 1983). This lower ratio is additional evidence that nutrient availability causes no limitation of the specific growth rate. The ratio increases with the age of the leaves because these leaves have a large amount of structural carbon in relation to nitrogen and phosphorus. The C:N, N:P and C:N:P ratios for several seagrasses are shown in Table I. Leaves from *Z. noltii* populations growing in Oosterschelde show C:N ratios close to that reported for *Z. marina* L., and generally lower than those recorded for other seagrasses [*Posidonia oceanica* (L.) Delile, *Cymodocea nodosa* (Ucria) Aschers. and *Z. capricorni* Aschers.] which posses thicker leaves with large amounts of structural carbon as the main component of the cell walls (Tomlinson 1980). *Posidonia oceanica* and, to a lesser degree, *Cymodocea nodosa*, have thick rhizomes with low growth rates, which may live more than 10 years. These old rhizomes contain high amounts of carbon, and low contents of N and P, because these latter elements are mobilized to the active growing parts of the plant (Bristow and Whitcombe 1971, Iizumi and Hattori 1982). This explains the high C:N ratios. The rhizomes of the *Z. noltii* specimens dealt with in this paper were young, for this reason the C:N ratios are lower.

The C:N ratio may be used as a index of nutritive food quality (Niell and Mouriño 1981, Wahbeh 1988).

Table I. C:N, N:P and C:N:P ratios for several seagrasses.

Species	Plant part	C:N	N:P	C:N:P	References
Zostera marina	leaves	18	27	481:27:1	Birch (1975)
	leaves	12	10	120:10:1	Umebayashi (1989)
	leaves*	13	6	79: 6:1	Pellikaan & Nienhuis (1988)
	root + rhizomes*	11	15	170:15:1	Umebayashi (1989)
		19	3	57: 3:1	Pellikaan & Nienhuis (1988)
Zostera capricorni	leaves	21	17	349:17:1	Birch (1975)
	rhizomes	58	8	465: 8:1	Birch (1975)
Zostera noltii	leaves•	20– 40	–	–	Pirc & Wollenweber (1988)
	root + rhizomes•	35– 75	–	–	Pirc & Wollenweber (1988)
	leaves►	12	9	104: 9:1	This paper
	rhizomes	21	6	127: 6:1	This papar
	roots	16	7	114: 7:1	This paper
Posidonia oceanica	leaves	20– 35	–	–	Pirc & Wollenweber (1988)
	rhizomes•	25– 40	–	–	Pirc & Wollenweber (1988)
	roots•	50–100	–	–	Pirc & Wollenweber (1988)
Cymodocea nodosa	leaves•	20– 30	–	–	Pirc & Wollenweber (1988)
	rhizomes•	30–100	–	–	Pirc & Wollenweber (1988)
	roots•	70–115	–	–	Pirc & Wollenweber (1988)

* Mean values during the whole year (according to the authors, transformed data).
• Values range depending on the season.
► Mean values of three different aged leaves.

According to Russell-Hunter (1970), a ratio of 17:1 is adequate for good mammalian nutrition. Assuming that such ratio can be applied to marine animals (Wahbeh 1988) and to birds, *Z. noltii* leaves would constitute a good quality food for birds, since the ratio is below 13. This data is supported by Jacobs *et al.* (1981) who pointed out the importance of *Z. noltii* populations from The Dutch Wadden Sea (Netherlands) as a source of food for birds.

Acknowledgements

The authors thank Dr Paul G. Harrison, University of British Columbia, for criticism and comments on the manuscript. This work has been supported by the Spanish Commission for Science and Technology, Proyect N° PB-86-0677 and by fellowship of the 'Plan Nacional de Formación de Personal Investigador / 87. Plan Nacional'. The first author thanks the Director of the Delta Institute for Hydrobiological Research, Yerseke, for the working facilities.

References

Armstrong, F. A. and S. Tibbits. 1968. Photochemical combustion of organic matter in seawater for nitrogen, phosphorus and carbon determination. *J. Mar. Biol. Assoc. U.K. 48*: 143–152.

Atkinson, M. J. and S. V. Smith. 1983. C:N:P ratios of benthic marine plants. *Limnol. Oceanogr. 28*: 568–574.

Augier, H., H. Calvert, E. Wollaston and M. Santimone. 1982. A comparison of the C, H, N, protein and amino acids of *Posidonia australis* Hook. F. with that of *Posidonia oceanica* (L.) Delile and several other marine phanerogams. *Aquat. Bot. 12*: 69–80.

Birch, W. P. 1975. Some chemical and calorific properties of tropical marine angiosperms. *J. Appl. Ecol. 12*: 201–212.

Bristow, J. M. and M. Whitcombe. 1971. The role of roots in the nutrition of aquatic vascular plants. *Amer. J. Bot. 58*: 8–13.

Brock, Th. C. M., M. C. M. Bongaerts, G. J. M. A. Heijnen and J. H. F. G. Heijthuijsen. 1983. Nitrogen and phosphorus accumulation and cycling by *Nymphoides peltata* (Gmel.) O. Kuntze (Menyanthaceae). *Aquat. Bot. 17*: 189–214.

Den Hartog, C. 1980. Foreword. *In*: (R. C. Phillips and C. P. McRoy, eds) *Handbook of Seagrass Biology: An Ecosystem Perspective.* Garland STPM Press, New York. 353 pp.

Gerloff, G. C. 1975. *Nutritional Ecology of Nuisance Aquatic Plants.* Nat. Environt. Res. Center, Office of Research and Development, U. S. Environmental Protection Agency, Corvallis, OR. 78 pp.

Godshalk, G. L. and R. G. Wetzel. 1978. Decomposition of aquatic angiosperms. III. *Zostera marina* L. and a conceptual model of decomposition. *Aquat. Bot. 5*: 329–354.

Golterman, H. L., R. S. Clymo and M. H. M. Ohnstad. 1978. *Methods for Physical and Chemical Analysis of Fresh Waters.* IBP Handbook N° 8, 2nd edn. Blackwell, Oxford. 213 pp.

Harrison, P. G. and K. H. Mann. 1975. Detritus formation from eelgrass (*Zostera marina*). The relative effects of fragmentation, leaching and decay. *Limnol. Oceanogr. 5*: 924–934.

Hendriksen, A. 1965. An automated method for determining low level concentrations of phosphate in fresh and saline waters. *Analyst 90*: 29–34.

Iizumi, H. and A. Hattori. 1982. Growth and organic production of eelgrass (*Zostera marina* L.) in temperate waters of the Pacific coast of Japan. III. The kinetics of nitrogen uptake. *Aquat. Bot. 12*: 245–256.

Jacobs, R. P. W. M., C. Den Hartog, B. F. Braster and F. C. Carriere. 1981. Grazing of the seagrass *Zostera noltii* by birds at Terschelling (Dutch Wadden Sea). *Aquat. Bot. 10*: 241–259.

Murphy, J. and J. P. Riley. 1962. A modified single solution method for the determination of phosphate in natural waters. *Anal. Chim. Acta. 27*: 31–36.

Niell, F. X. 1976. C:N ratio in some marine macrophytes and its possible ecological significance. *Bot. Mar. 19*: 347–350.

Niell, F. X. and C. Mouriño. 1981. Contenido en carbono y nitrógeno en macrófitos intermareales. *Inves. Pesq. 45*: 178–210.

Nienhuis, P. H. and A. M. Groenendijk. 1986. Consumption of eelgrass (*Zostera marina*) by birds and invertebrates: an annual budget. *Mar. Ecol. Prog. Ser. 29*: 29–35.

Pellikaan, G. C. 1982. Decomposition processes of eelgrass (*Zostera marina*). *Hydrobiol. Bull. 16*: 83–92.

Pellikaan, G. C. 1984. Laboratory experiments on eelgrass (*Zostera marina*) decomposition. *Neth. J. Sea Res. 18*: 360–383.

Pellikaan, G. C. and P. H. Nienhuis. 1988. Nutrient uptake and release during growth and decomposition of eelgrass, *Zostera marina* L., and its effects on the nutrient dynamics of Lake Grevelingen. *Aquat. Bot. 30*: 189–214.

Pérez-Lloréns, J. L. and F. X. Niell. 1989. Emergence and submergence effects on the distributional pattern and exchange of phosphorus in the seagrass *Zostera noltii* Hornem. *Scient. Mar. 53*: 497–503.

Pirc, H. and B. Wollenweber. 1988. Seasonal changes in nitrogen, free amino-acids, and C:N ratio in Mediterranean seagrasses. *Mar. Ecol. 9*: 169–179.

Russell-Hunter, W. D. 1970. *Aquatic Productivity: An Introduction to Some Basic Aspects of Biological Oceanography and Limnology.* McMillan, London. 306 pp.

Stanley, G. H. and G. R. Richardson. 1971. The automation of the single reagent for total phosphorus. In: *Advances in Automated Analyses.* Technicon International Congress 1970, vol. I: 305–311.

Stumpf, P. K. 1952. Phosphate assimilation in higher plants. *In*: (W. D. McElroy and B. Glass, eds) *Phosphorus metabolism* vol. 2. Johns Hopkins, Baltimore. pp. 29–67.

Tomlinson, P. B. 1980. Leaf morphology and anatomy in seagrasses. *In*: (R. C. Phillips and C. P. McRoy, eds) *Handbook of Seagrass Biology: An Ecosystem Perspective.* Garland STPM Press, New York. 353 pp.

Umebayashi, O. and K. Sasaki. 1985. Distributional pattern of phosphorus, nitrogen and carbon in a leaf blade of eelgrass, *Zostera marina* L. *Bull. Tokai Reg. Fish. Res. Lab. 118*: 21–32.

Umebayashi, O. 1989. Eelgrass productivity on an intertidal flat of Central Japan. *Bull. Tokai Reg. Fish. Res. Lab. 127*: 17–30.

Velimirov, B. 1987. Organic matter derived from a seagrass meadow: origin, properties, and quality of particles. *Mar. Ecol. 8*: 143–173.

Vermaat, J. E., M. J. M. Hootsmans and P. H. Nienhuis. 1987. Seasonal dynamics and leaf growth of *Zostera noltii* Hornem., a perennial intertidal seagrass. *Aquat. Bot. 28*: 287–299.

Wahbeh, M. I. 1988. Seasonal distribution in the nutritional quality of different fractions of two seagrass species from Aqaba (Red Sea), Jordan. *Aquat. Bot. 32*: 383–392.

Welsh, R. P. H. and P. Denny. 1979. The translocation of ^{32}P in two submerged aquatic angiosperms species. *New Phytol. 82*: 645–656.

Botanica Marina
Vol. 34, pp. 323–325, 1991

The Transport of 3-Phosphoglycerate in Isolated Chloroplasts of the Green Seaweed *Bryopsis corticulans* Setch.

S. Yu and M. Pedersén

Department of Physiological Botany, University of Uppsala, Box 540, S-751 21 Uppsala, Sweden

(Accepted 7 April 1991)

Abstract

The transport of 3-phosphoglycerate (PGA) into isolated chloroplasts of the green seaweed *Bryopsis corticulans* was monitored as PGA-dependent O_2 evolution in a CO_2-free system. This O_2 evolution was inhibited completely by 10 mM pyrophosphate or 1.6 mM pyridoxal-5′-phosphate. In contrast to pyrophosphate, the inhibition by pyridoxal-5′-phosphate occurred only when it was added to the reaction system before the onset of light. When PGA-dependent O_2 evolution had already commenced the addition of pyridoxal-5′-phosphate had no inhibitory effect. These results suggest that the uptake of PGA by the chloroplasts is a carrier-mediated process.

Introduction

Work by Heldt, Flügge and others has established the existence of an orthophosphate (Pi) translocator in the inner membrane of spinach chloroplast envelope (Flügge and Heldt 1984, 1986, Heldt and Flügge 1986). The Pi-translocator facilitates a strict counter-exchange of Pi, triose phosphate (glyceraldehyde phosphate and dihydroxyacetone phosphate), and 3-phosphoglycerate (PGA) with high specificity. The main function of this translocator is to allow the export of photosynthetically fixed carbon in the form of triose phosphate from the chloroplast to the cytosol of the cell in exchange with PGA or inorganic phosphate. The triose phosphate can then be either used for the synthesis of sucrose or provide the cytosol with NAD(P) and ATP. The Pi-translocator isolated from spinach chloroplasts is a dimer with a molecular mass of about 61 000 (Flügge and Heldt 1984). A lysine and an arginine residue of the translocator are involved in the substrate binding. Pyridoxal-5′-phosphate (PLP) is known to be an irreversible inhibitor and pyrophosphate (PPi) a competitive inhibitor (Fliege *et al.* 1978, Flügge *et al.* 1980).

Although the appearance of the Pi-translocator during evolution has been suggested to have been a key event for the development of eukaryotic plants (Flügge and Heldt 1984), there is still very little known regarding the occurrence and distribution of this translocator in the plant kingdom (Robinson and Wiskich 1977, Rutter and Cobb 1983 a). Here we report a PGA-translocator in the chloroplasts of the marine green seaweed *Bryopsis corticulans* Setch.

Material and Methods

Bryopsis corticulans Setch. was collected in the intertidal zone in Taiping Bay, Quingdao (China) in the summer months. The thalli were washed in seawater to remove the epiphytes prior to the isolation of chloroplasts.

The thalli of *B. corticulans* were homogenized (1 : 1 w/v) in 10 mM tricine-NaOH buffer, pH 6.8, containing 500 mM NaCl. The homogenate was passed through two layers of cheesecloth and centrifuged at 200 × g to precipitate the cell debris. The chloroplasts were collected at a centrifugation of 1000 × g and washed twice by resuspension and recentrifugation. Oxygen evolution by the chloroplast fraction was measured at 25 °C using YSI Model 53 Biological Oxygen Monitor (Yellow Springs Instrument Co. Inc., Ohio) in a volume of 3.0 mL of 100 mM tricine-NaOH buffer, pH 7.0 containing 400 mM NaCl and the isolated chloroplasts corresponding to a chlorophyll concentration of 70–100 μg mL^{-1}. Then $NaHCO_3$, PGA or PLP was added in final concentrations of 10, 2 and 1.6 mM, respectively, as indicated. Irradiation was provided by a 400 W tungsten lamp (in a slide projector), which gave a irradiance of 680 μmol m^{-2} s^{-1} measured on the surface of the reaction vessel. A square glass container (4 cm thick) filled with water was used as a heat-filter between the projector and the reaction vessel. In the PGA-dependent O_2 evolution experiments, the buffer was adjusted to pH 2.6, heated and flushed with N_2 before use in order to expel the dissolved inorganic carbon. Chlorophyll was measured by the method of Arnon (1949).

Results and Discussion

Figure 1 shows the capability of the chloroplasts isolated from *B. corticulans* in carrying out PGA-dependent O_2 evolution (Trace B and C). This O_2 evolution became apparent only when endogenous dissolved inorganic carbon in the reaction mixture became exhausted. The O_2 evolution rates were quite similar whether PGA or $NaHCO_3$ was used as substrate (Trace A and B). The PGA-dependent O_2 evolution is an indication that PGA was able to enter the stroma of the isolated chloroplasts and join the Calvin cycle, where it will be converted to triose phosphate by utilizing NADPH and ATP derived from photosynthetic electron transfer (Heldt and Flügge 1986). When 1.6 mM PLP was added to the reaction mixture before the onset of light, PGA-dependent O_2 evolution was completely inhibited (Trace D). However, when the PGA-dependent oxygen evolution had already commenced, the addition of PLP, even in a two-fold concentration of 3.2 mM, did not have any inhibitory effect (Trace B). This may be explained by the protective effect of the substrate. That is, when in the light and in the presence of PGA, the PGA-translocator of *B. corticulans* chloroplasts may assume a different molecular conformation, which is induced by the proton gradient across the inner membrane of the chloroplast envelope due to the alkalization of the stroma as a result of photosynthesis (Flügge and Heldt 1986). This new conformation of the translocator molecule makes the lysine and arginine residues in the binding centre inaccessible to PLP. In the dark, such a protective effect was not offered by PGA due to the inability of the translocator to bind PGA as in the light. The substrate-protective effect occurrs also in the spinach Pi-translocator (Flügge and Heldt 1986). In contrast to the effect of PLP, the inhibitionof PGA-dependent O_2 evolution by PPi occurred both in light and dark (Trace C).

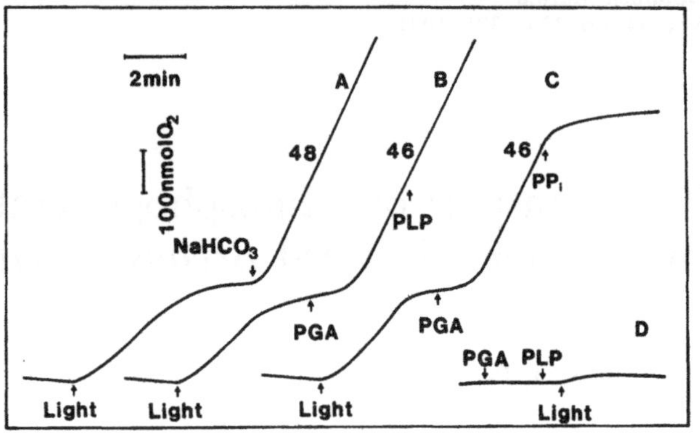

Fig. 1. Oxygen electrode traces of the isolated chloroplasts from *B. corticulans*. Traces A and B shows the same effectiveness of 10 mM $NaHCO_3$ and 2 mM PGA in the recovery of O_2 evolution when the endogenous inorganic carbon became exhausted. Traces C and D show the effects of 1.6 mM PLP and 10 mM PPi on the PGA-dependent O_2 evolution. Trace D shows the complete inhibition of 1.6 mM PLP on the O_2 evolution when it was added before the onset of light even in the presence of 2 mM of PGA. Numbers along the traces give rates of O_2 evolution in μmol O_2 mg^{-1} chl h^{-1}.

The uptake rate of PGA by chloroplasts of *B. corticulans* was slower compared to that by spinach chloroplasts (Walker *et al.* 1967) since more than one minute elapsed from the addition of PGA before O_2 evolution commenced (Trace B and C). Ten mM Pi showed no inhibition on the PGA-dependent O_2 evolution. This might indicate the PGA-translocator showed no affinity for Pi. In another experiment, phosphate in a concentration of 0, 0.4, 4, 10 and 20 mM was added with the O_2 evolution mixture for the isolated chloroplasts, the CO_2-dependent oxygen evolution rate was 35 μmol O_2 mg^{-1} chl h^{-1} for all these groups. These results differ from those obtained from higher plants (Edwards and Walker 1983), where the Pi-translocator shows high affinity towards both PGA and Pi, and therefore, the transport of the two chemicals is mutually inhibitory (Fliege *et al.* 1978). The chloroplasts isolated from higher plants usually have an optimal Pi concentration for maximal O_2 evolution (Leegood and Walker 1985). The uptake and accumulation of Pi in the algal chloroplasts deposited in the form of polyphosphate granules may involve a different mechanism not found in higher plants, and this polyphosphate pool may act as a phosphate buffer for sustained photosynthesis (Rutter and Cobb 1983 b).

In conclusion, isolated chloroplasts of *B. corticulans* were able to take up PGA as indicated by PGA-dependent O_2 evolution, and this O_2 evolution was inhibited by PPi and PLP, which suggests the existence of a PGA-translocator. This translocator, however, may differ from the the one reported in spinach, as the algal chloroplasts possessed a number of properties different from those of spinach.

References

Arnon, D. I. 1949. Copper enzyme in isolated chloroplasts. Polyphenoloxidase in *Beta vulgaris*. *Plant Physiol.* *24*: 1–15.

Edwards, G. and D. Walker. 1983. *In*: (G. Edwards and D. Walker, eds) *C_3, C_4: Mechanisms, and Cellular and Environmental Regulation, of Photosynthesis*. Blackwell Scientific Publications, Oxford. pp. 156–274.

Fliege, R., U. I. Flügge, K. Werdan and H. W. Heldt. 1978. Specific transport of inorganic phosphate, 3-phosphoglycerate and triosephosphates across the inner membrane of the envelope in the spinach chloroplasts. *Biochim. Biophys. Acta.* *502*: 232–247.

Flügge, U. I., M. Freisl and H. W. Heldt. 1980. Balance between metabolite accumulation and transport in relation to photosynthesis by isolated spinach chloroplasts. *Plant Physiol.* *65*: 574–577.

Flügge, U. I. and H. W. Heldt. 1984. The phosphate-triose phosphate-phosphoglycerate translocator of the chloroplast. *TIBS*. 530–533.

Flügge, U. I. and H. W. Heldt. 1986. Chloroplast phosphate-triose phosphate-phosphoglycerate translocator: Its identification, and reconstitution. *Methods Enzymol.* *125*: 716–730.

Heldt, H. W. and U. I. Flügge. 1986. Transport of metabolites across the chloroplast envelope. *Methods Enzymol.* *125*: 705–716.

Leegood, R. C. and D. A. Walker. 1985. Chloroplasts and protoplasts. *In*: (J. Coobs, D. O. Hall, S. P. Long and J. M. O. Scurlock, eds) *Techniques in Bioproductivity and Photosynthesis*. Second edition. Pergamon Press, Oxford. pp. 124–132.

Robinson, S. P. and J. T. Wiskich. 1977. Inhibition of CO_2 fixation by adenosine 5′-diphosphate and the role of phosphate transport in isolated pea chloroplasts. *Arch. Biochem. Biophys.* *184*: 546–554.

Rutter, J. C. and A. H. Cobb. 1983 a. Translocation of orthophosphate and glucose-6-phosphate in *Codium fragile* chloroplasts. *New Phytol.* *95*: 559–568.

Rutter, J. C. and A. H. Cobb. 1983 b. Photosynthesis by isolated *Codium fragile* chloroplasts of varying internal phosphate status. *New Phytol.* *95*: 549–557.

Walker, D. A., W. Cockburn and C. W. Baldry. 1967. Photosynthetic oxygen evolution by isolated chloroplasts in the presence of carbon cycle intermediates. *Nature* *216*: 597–599.

Botanica Marina
Vol. 34, pp. 327–340, 1991

Lissodinium Matzenauer, *emend.*, Based upon the Rediscovery of *L. schilleri* Matz., Another Member of the Family Podolampadaceae Lindemann (Dinophyceae)

M. C. Carbonell-Moore

Oregon State University, College of Oceanography, Oceanography Admn. Bldg. 104, Corvallis, Oregon 97331, U.S.A.

(Accepted 6 April 1991)

Abstract

The genus *Lissodinium* Matzenauer (Dinophyceae) is redefined, based upon a scanning electron microscope study of material collected in the Indian and Atlantic Oceans. These are cells very laterally compressed, an unusual feature within peridinioids. It is shown that Matzenauer might have misinterpreted the plate pattern and that *Blepharocysta schilleri* (Gaarder) Ballantine *in* Parke *et* Dixon (= *B. matzenaueri* Gaarder) and *B. compressa* Gaarder are indeed the same as *L. schilleri* Matz. (*Lissodinium*'s type species). In the case of *B. compressa* the plate pattern might have been partially described optically reversed by Gaarder. The relationships of *Lissodinium* with the two genera of the family Podolampadaceae Lindemann, *Blepharocysta* Ehrenberg and *Podolampas* Stein are discussed. It is proposed to include *Lissodinium* within this family.

Plate formula: Po X 3′ 1a 5″ 3C 5S 3‴ 3⁗.

Introduction

The main characteristics of the members of the family Podolampadaceae Lindemann (Dinophyceae) are the absence of both a depressed sulcus and a typical dinoflagellate cingulum. The presence of this structure within the family has been the subject of debate. Kofoid (1909 b), Lebour (1925), Schiller (1937), Nie (1939, 1942, 1945), Rampi (1941), Gaarder (1954) and Abé (1966) considered the well-developed plates in the hypotheca that may resemble the narrow girdle in other peridinioids, as postcingular plates. Balech (1954, 1963, 1988) has been the first to give these plates a cingular character, which allows the members of this family to be integrated into the Peridiniales (Sournia 1986). Due to the large development of the girdle plates, both the hypothecal plates and especially the apical plates, have been greatly reduced while the precingular series has been enlarged.

Two genera are traditionally included in the Podolampadaceae: *Blepharocysta* Ehrenberg, and *Podolampas* Stein. The former includes cells that are globular or ovoid, while the latter includes pyriform cells, which, to a certain extent, are dorso-ventrally compressed. The two genera maintain a remarkably uniform plate pattern. These are marine, generally warm tropical species although some can be occasionally found in temperate waters. One species of *Podolampas* has been described from Antarctica.

The genus *Lissodinium* was created by Matzenauer (1933) to describe some unusual cells that did not show a distinct girdle and were very difficult to orient due to an extreme lateral compression, a very unusual feature within peridinioids. Nevertheless, he concluded his specimens were dinoflagellates because they had numerous plates, intercalary bands and two large body openings. He named the type species *L. schilleri*. Later, Schiller (1937) included it in the family *Lissodiniaceae*.

The justification of having a separate genus for *Lissodinium* Matz. was rejected by Gaarder (1954) based on her observations of material collected by the "Michael Sars". She concluded that the organisms that she ascribed to *Blepharocysta* were the same as Matzenauer's. A closer look at her and Matzenauer's specimens, leads to the conclusion that these two cells do not look as identical as she claimed (Fig. 27; Fig. 25A and Figs 26A, B). In fact, as Matzenauer depicted his cells, fundamental differences are evident that Gaarder overlooked. For example, the anterior intercalary plate (1a) is dorsal in Gaarder's figures, while it is ventral in Matzenauer's (a_1). This characteristic would be unique in the order Peridiniales (Balech 1963). Gaarder included her cells in the genus *Blepharocysta*, and renamed *Lissodinium schilleri*, *Blepharocysta matzenaueri*. In the same study, Gaarder described a new species, *B. compressa*, which was similar to *L. schilleri*, but even more laterally compressed.

Nie (1945), in his description of the genus *Sinodinium* (for some researchers a synonym of *Palaeophalacroma* Schiller), emphasized the close relationship between the members of the family Podolampadaceae and the Lissodiniaceae. No new observations on *Lissodinium* were reported in his study, but he announced a new study dedicated to this genus, which apparently, was never published (Sournia 1986). He presented only redrawings of Matzenauer's species iconotype. This may mean that he did not have the opportunity to personally observe *Lissodinium* (Nie's studies have always included very detailed work, *e.g.* his papers of 1939 and 1942 on *Blepharocysta* and *Podolampas* respectively).

No subsequent reports, of any organisms resembling Matzenauer's or Gaarder's have been recounted since 1954, with the exception of Ballantine *in* Parke and Dixon (1964) and Steidinger and Williams (1970) who did not describe a tabulation pattern and could not make positive identification of their specimens.

Lissodinium has been generally accepted as a synonym of *Blepharocysta*, *e.g.* Ballantine *in* Parke and Dixon (1964), Sournia (1973), Taylor (1976), Tappan (1980) and Sournia (1986). Others such as Nie (1945), Balech (1963, 1988) and Taylor (1980) have kept it separate from *Blepharocysta*.

The present study intends to demonstrate:

1) that the way *Lissodinium schilleri* Matzenauer was depicted by its author was inaccurate and confused due to its unusual shape and,

2) that the species described by Gaarder as *B. compressa* was indeed the same *L. schilleri* without the intercalary bands as reported by Matzenauer for his species. The absence of intercalary bands makes Gaarder's species even more compressed than *L. schilleri*.

Finally, because of their similarities with *Blepharocysta* and *Podolampas* it is proposed that these lateral compressed cells be included within the genus *Lissodinium* as another member of the family Podolampadaceae.

Material and Methods

The cells of *Lissodinium schilleri* were isolated from phytoplankton samples collected both in the Atlantic (Dominican Republic) and in the Indian Ocean (Bay of Bengal), and then viewed under a scanning electron microscope using the technique described by Taylor (1972). This technique consists of isolating the cells with a micropipette under a light microscope (an inverted one is preferred), and then transferring the cells through several washes of distilled water to remove the salt and the formaldehyde. The series of transfers is easier to accomplish under a dissecting microscope. Once the cells are cleaned, they are transferred to an aluminum stub; as the cells dry on the stub, they adhere to its surface. Later the stub is coated with a metal under vacuum, usually a mixture of palladium and gold. Finally, the cells are ready to be observed and photographed (Polaroid) under the SEM. A Stereoscan 250, scanning electron microscope was used in this study, usually operated at 5 or 10 kV. The SEM enables one to observe the actual assembly of plates forming the cell as with the pieces of a jigsaw puzzle, in organisms with the shape of *L. schilleri*. It is very difficult to rotate such cells in an aquatic medium and to maintain them in a dorso-ventral or apical-antapical orientation for sufficient time to determine the plate patterns, but under the SEM detailed views were possible as many cells from both oceans were studied.

The plate nomenclature used is an adaptation of Balech's (1954, 1963 and 1988) which includes Kofoidean nomenclature (Kofoid 1909 a).

Results

Lissodinium Matzenauer, *emend.*

Description

Pebble shaped cells, very laterally compressed, with no distinct girdle. Dark protoplasm. Plate formula: Po X 3′ 1a 5″ 3C 5S 3‴ 3⁗.

Apical pore complex (APC) formed by plates Po and X, with a small spine on Po projected dorsally. The APC is surrounded by three plates conforming the apical region (Figs 1A, 7, 8, 11). Plate 1′ is very long, slender and it is slightly oblique with respect to the antero-posterior axis (Figs 1A, 2, 4). It does not touch the apical pore. It goes all the way from X to the ventral region, laying on S. a. (anterior sulcal). Plate

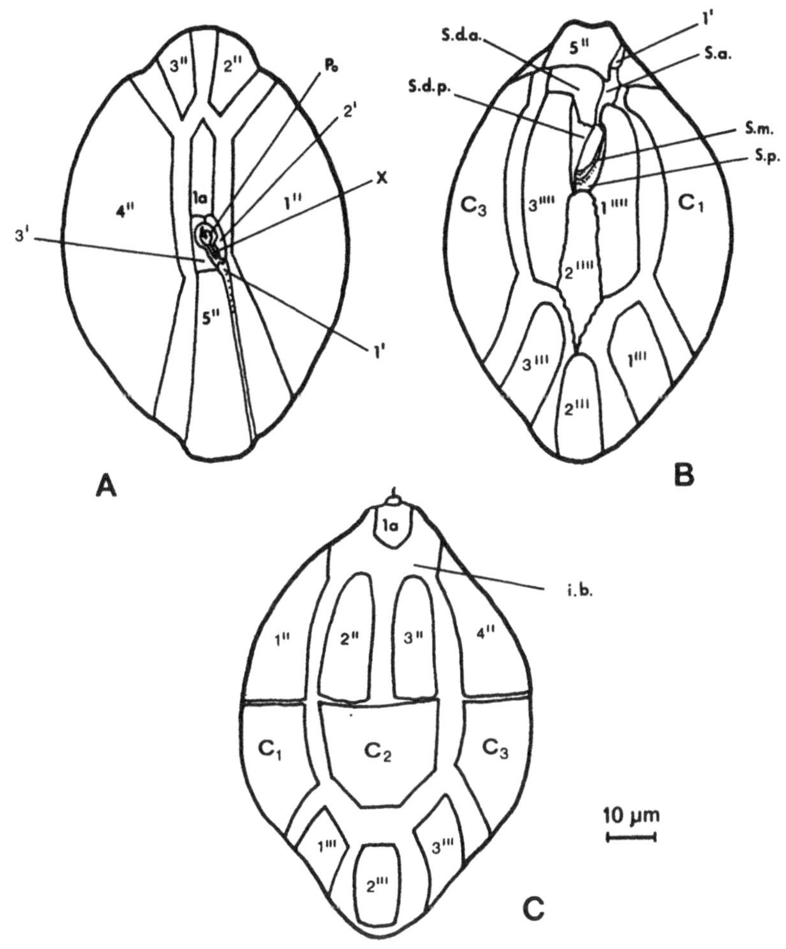

Fig. 1. Diagrams to illustrate tabulation patterns in *Lissodinium schilleri*.
Fig. 1A. Apical view. B. Antapical view. C. Dorsal view. i. b.: intercalary bands.

1′ separates the 1″ and 5″ plates and 2′ and 3′ are very small (Figs 1A, 7, 8, 11). They have a boomerang shape, dorsally embracing the apical pore while ventrally the X plate separates them from each other. The inner margins of 2′ and 3′ form a pronounced ridge (Figs 7, 8, 11). Dorsal to 2′ and 3′, there is a very elongated, pentagonal plate, the anterior intercalary 1a (Figs 1A, 2, 3, 7). Plates 1″ and 4″ in the epitheca (Figs 1A, 2, 3, 4, 14A) and C_1 and C_3 in the hypotheca (Figs 14A) are very well developed, forming almost all of both sides of the cell (Figs 12A, 13A, 13B). When observed from a lateral view, the other plates (part of 5″, part of 1a, 2″, part of C_2, 1‴, 1⁗ on the left side and part of 5″, part of 1a, 3″, part of C_2, 3‴, 3⁗ on the right side) seem to circumscribe a ring around these four large plates [1″ and C_1 on the left side (Figs 13A, 13B) and 4″ and C_3 on the right side (Fig. 12A)]. Both 2″ and 3″ are dorsal, one at each side of the cell (Figs 1C, 12B, 15). Plate 5″ is wider than the last two plates and lays on the S. d. a. (right anterior sulcal). Plate C_2 is six-sided. The three lower sides are much shorter than the upper sides and correspond to the sutures with each of the postcingular plates (Figs 1C, 12B, 15).

The sulcus is oval, big and has very reduced lists. There are five sulcal plates (Figs 17A, 17B). The anterior end of the sulcus is formed by S. d. a. which is hatchet shaped and by the S. a. which is small and narrows at its right end. The S. d. p. (right posterior sulcal) is broad anteriorly and forms the right side of the sulcus. The S. p. (posterior sulcal) has a J-shape and forms the left and the posterior sides of the sulcus. It has many rows of closely spaced round perforations that make the appearance of a screen mesh. Another small plate completes the sulcus, the S. m. (medial sulcal) which may contain a single row of round pores (Fig. 17A, 23).

The plates surrounding the sulcus (1⁗, 2⁗, 3⁗) (Figs 1B, 14B) are very elongated and have similar development. This series of plates is followed by another series (1‴, 2‴, 3‴) (Figs 1C, 12B, 15) which is almost in antapical position. The only postcingular plate that touches just one cingular plate (C_2) is 2‴. The other two postcingulars each touch two cingular plates.

Very well developed intercalary bands are present between almost all of the plates (*e.g.* Figs 1A–C, 2, 3, 4, 7, 8, 12A–B, 13A–B, 15). Neither the sutures

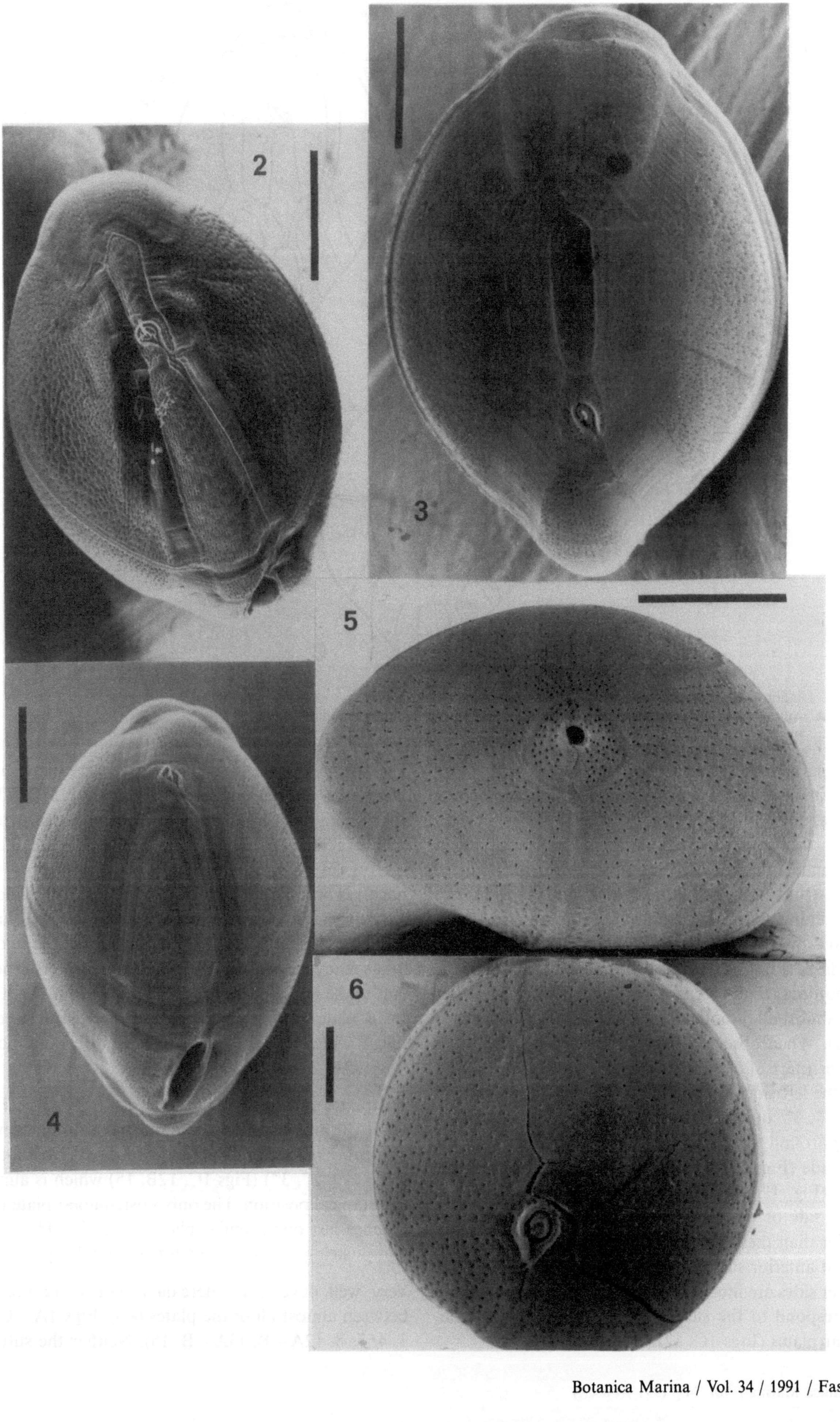
2
3
5
6
4

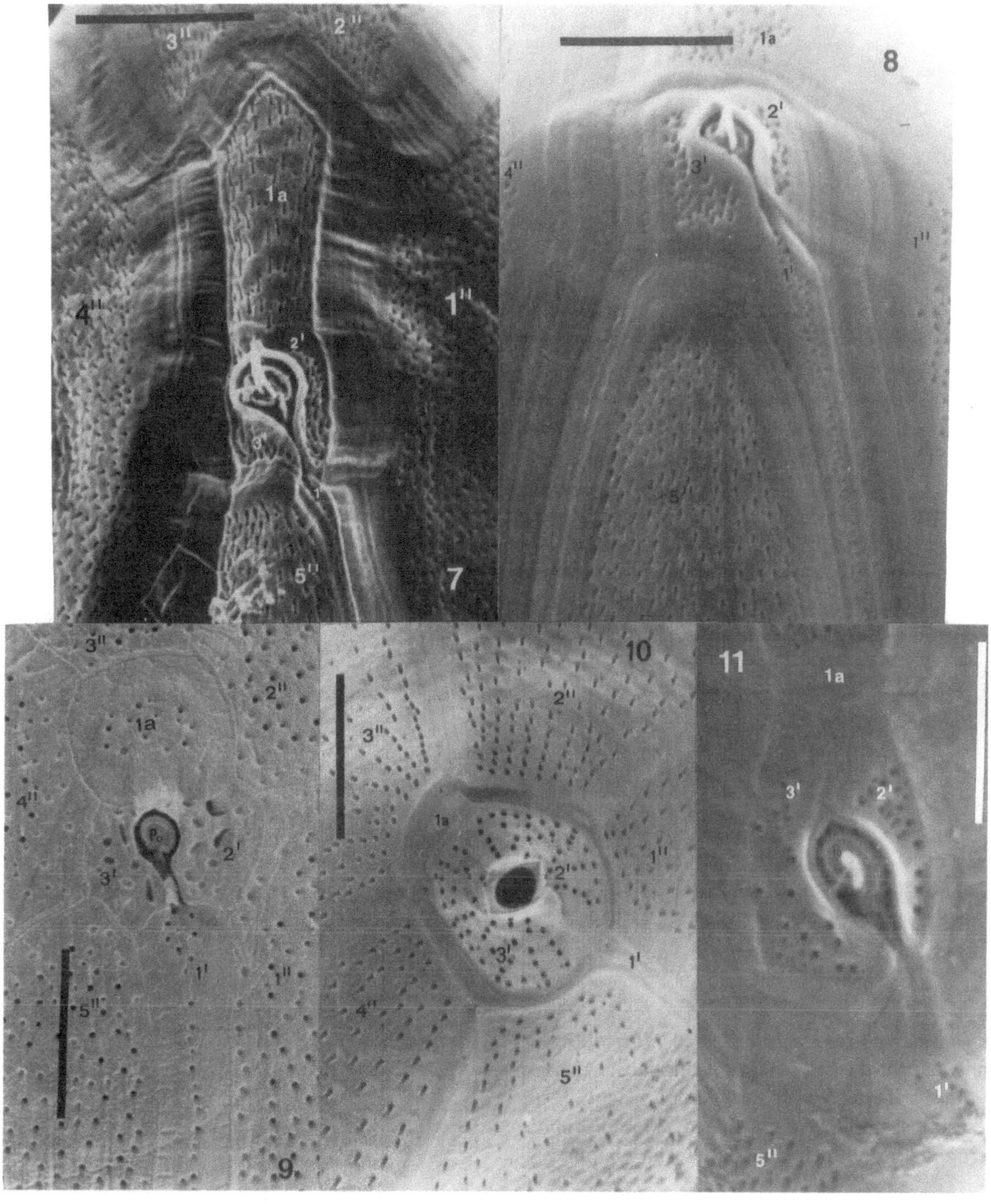

Figs 7–11. Apical zone views.
Fig. 7. *L. schilleri*. Detail of Figure 2. Fig. 8. *L. schilleri*. Detail of Figure 4. Fig. 9. *Blepharocysta* sp. Atlantic Ocean. Fig. 10. *Podolampas bipes*. Indian Ocean. Fig. 11. *L. schilleri*. Detail of Figure 3.
Figs 7–10: Scale bars = 10 µm. Fig. 11: Scale bar = 20 µm.

between the two big plates on each side of the cell (1″ and C_1; 4″ and C_3) (Figs 4, 12A, 13A–B, 14A) nor the sulcal plates (Figs 17A–B) have intercalary plates.

The surface of the plates presents pores of different sizes which may be round or elongated. Figure 16 shows the interior of some of the plates. No other features other than pores were found.

Figs 2–6. Apical views.
Fig. 2. *L. schilleri*. Atlantic Ocean. Fig. 3. *L. schilleri*. Indian Ocean. Fig. 4. *L. schilleri*. Indian Ocean. Different cell. Fig. 5. *Podolampas reticulata*. Indian Ocean. Fig. 6. *Blepharocysta splendor-maris*. Indian Ocean.
Fig. 2–5: Scale bars = 20 µm. Fig. 6: Scale bar = 10 µm.

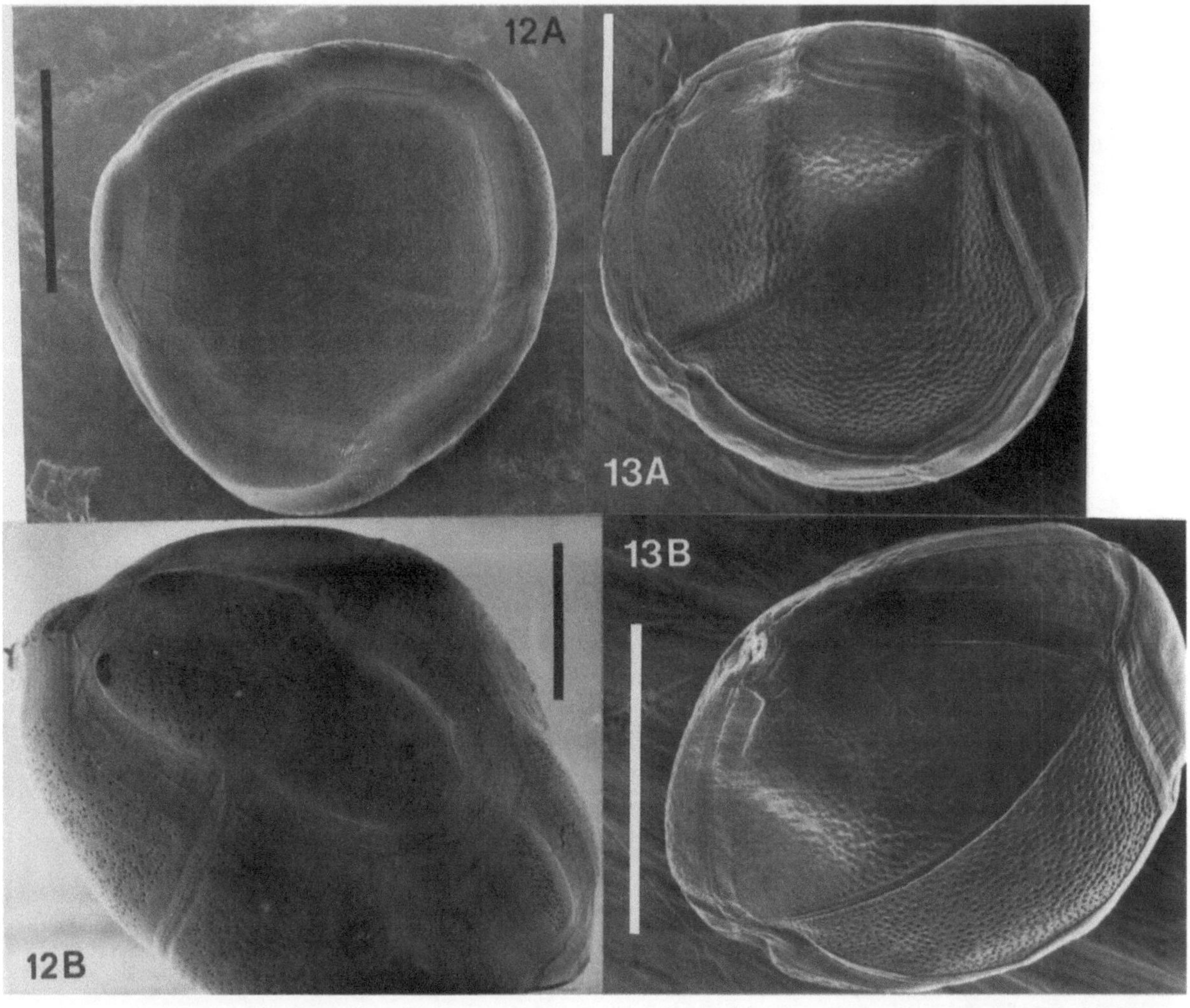

Figs 12–13. *L. schilleri.*
Fig. 12. A. Right lateral view. Same cell as in Figure 3. B. Dorsal view. Same cell as in Figure 3. Fig. 13. A. Left lateral view. Same cell as in Figure 2. B. Same cell as in Figure 2.
Figs 12A, 13B: Scale bars = 40 µm. Figs 12B, 13A: Scale bars = 20 µm.

Only megacytic cells were observed. The intercalary bands deform the cell making it difficult to measure the exact cell size. Width: 80–90 µm; height: 93 µm. These dimensions were calculated from the photomicrographs, and include the intercalary bands.

Type species: *L. schilleri* Matz., *emend.*
Matzenauer (1933): p. 482, fig. 77a–c. Schiller (1937): p. 481, fig. 553a–c.

Synonym: *Blepharocysta matzenaueri* Gaarder 1954, p. 6. *B. matzenaueri* f. *gibba* Gaarder 1954, p. 7, fig. 4d, e; 5a, b. *B. compressa* Gaarder 1954, p. 6, fig. 3a–e. *B. schilleri* (Matz.) Ballantine *in* Parke *et* Dixon 1964, p. 517. *B. schilleri* f. *gibba* (Gaarder) Ballantine *in* Parke *et* Dixon 1964, p. 517. *B. compressa* Gaarder? (*sic.*) Steindinger and Williams (1970): p. 43, fig. 8.

The cells observed in the present study were identified as *L. schilleri*, as it is discussed in the next section.

Discussion

Figure 25A is an enlarged lateral view of Matzenauer's iconotype of *L. schilleri*. Let's consider only the left side (Fig. 25B). Matzenauer oriented his figure in such way that both the ventral zone and the apical zone are on opposing sides of the cell. He improperly positioned the apical zone more or less on the equator of the cell instead of at the true apex. While the apical zone should be oriented 90 degrees in relation to the flagellar pore in all peridinioids, in Matzenauer's figure is oriented 180 degrees.

If we move the apical zone of Matzenauer's figure counterclockwise up to where the apical zone should be found in relation to the ventral zone, the suture of plates pr_2 and pr_3 moves dorsally (Fig. 25C). In this way, an intercalary plate (1a) can be seen (Fig. 25D). The a_1 plate, whose presence Matzenauer was not sure of, is in reality just the ventral sides of a_2 and a_3

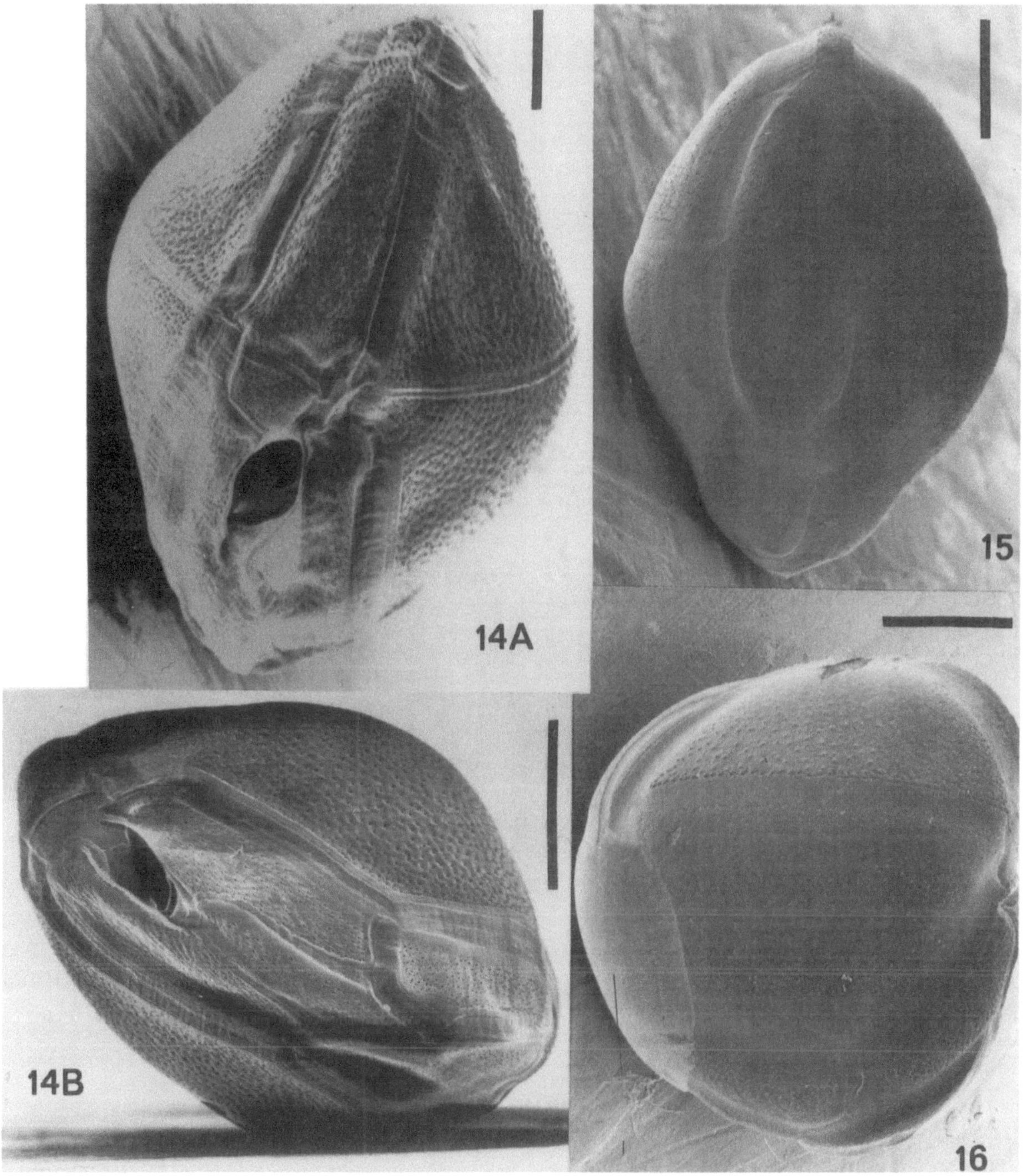

Figs 14–16. *L. schilleri*.
Fig. 14. A. Ventral view of Figure 2. B. Antapical view of the same cell. Fig. 15. Dorsal view of Figure 3. Fig. 16. Most of the left theca of a cell from the Atlantic Ocean.
Figs 14B, 15, 16: Scale bars = 20 µm. Fig. 14A: Scale bar = 10 µm.

(2′ and 3′ this study), that might have appeared to him as a different plate.
Likewise, if we move the suture between pr_2 and p_2 to the lower end of pr_3 (Fig. 25C) we can see that the cell is now divided by a regular line between the hypotheca and the epitheca. Moreover, if we elongate the plate series *e* and *p* as is found in lateral compressed cells, we then get a dorsal p_3 much smaller than Matzenauer's (Fig. 25E).

Finally, when we compare Matzenauer's transformed cell (Fig. 25E) with a cell from this study (Fig. 25F), strong similarities can be appreciated (using Matzenauer's nomenclature to simplify the comparison). It is possible to find a complete correspondence between Matzenauer's plates and the plates of the cells from this study. Figure 25G shows the same cell with the proper orientation (see Table I for plate nomenclature used in this study).

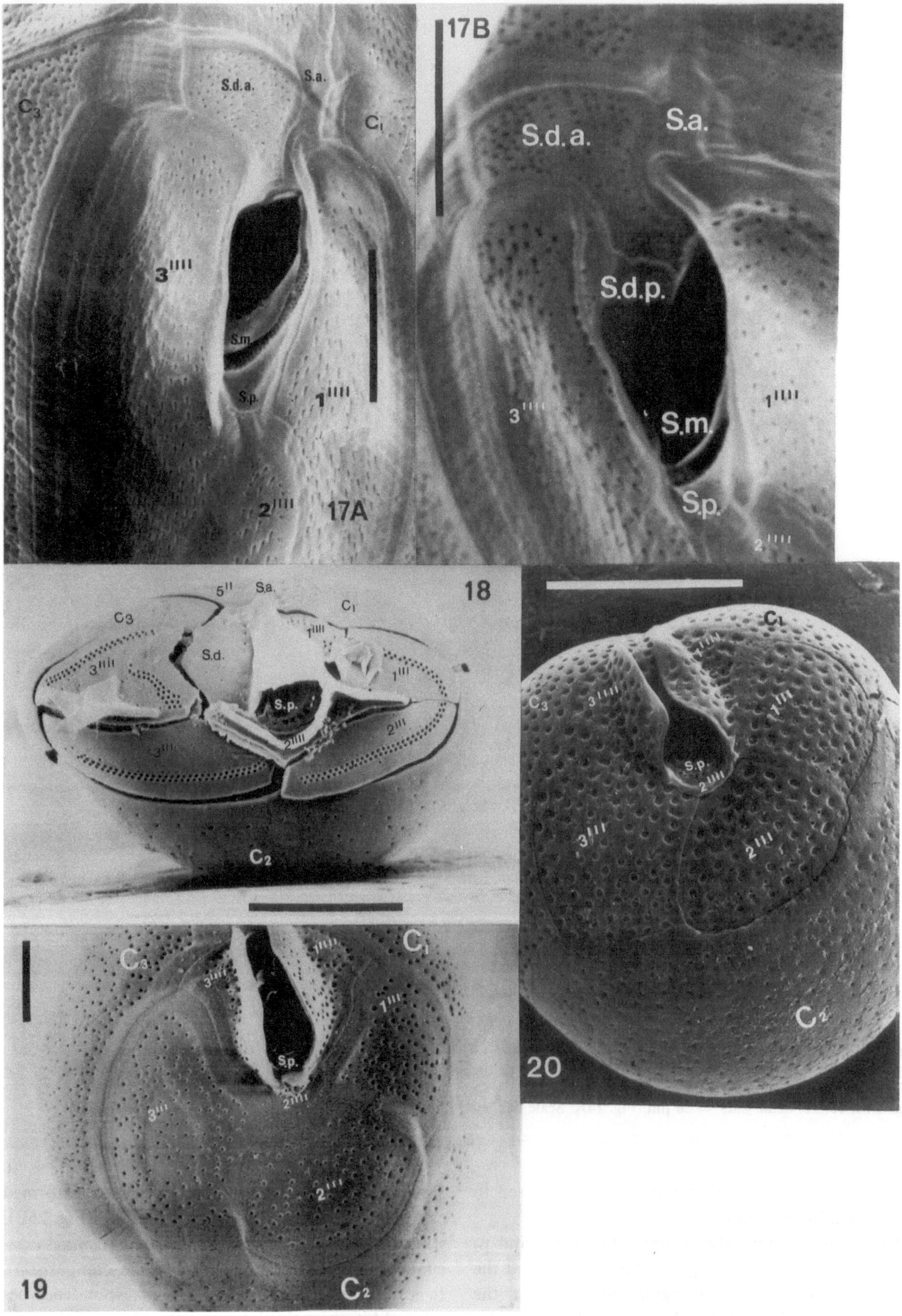

Figs 17–20.
Fig. 17. *L. schilleri*. A. Sulcal region, left side. Same cell as in Figure 2. B. Sulcal region, right side. Same cell. Fig. 18. *P. bipes*. Antapical view. Pacific Ocean. Fig. 19. *B. splendor-maris*. Antapical view. Different cell. Indian Ocean. Fig. 20. *B. paulsenii*. Antapical view. Atlantic Ocean.
Figs 17A–B, 19: Scale bars = 10 µm. Figs 18, 20: Scale bars = 20 µm.

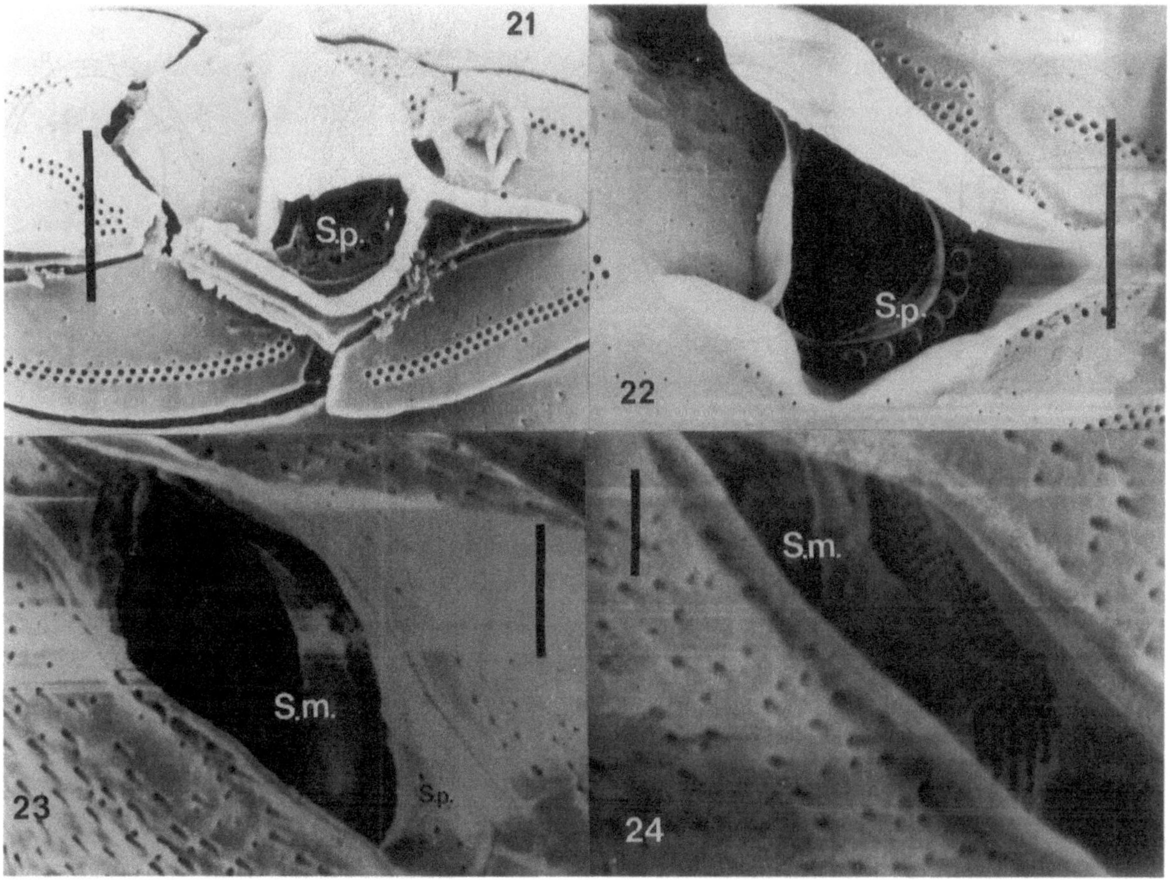

Figs 21–24.
Fig. 21. *Podolampas bipes*. Same cell as in Figure 18, closer view. Fig. 22. *P. bipes*. Another cell. Detail of the sulcal region. Indian Ocean. Fig. 23. *L. schilleri*. Detail of the sulcal region. Same cell as in Figure 2. Fig. 24. Detail of sulcal region of another cell. Atlantic Ocean.
Figs 21, 22: Scale bars = 10 μm. Fig. 23: Scale bar = 4 μm. Fig. 24: Scale bar = 2 μm.

Table I. Plate nomenclature discussed in the text.

Gaarder (1954) *B. compressa*	Gaarder (1954) *B. matzenaueri*	This Study *L. schilleri*	Matzenauer (1933) *L. schilleri*
1′	1′	1′	–
–	–	–	a_1
2′	2′	2′	a_2
3′	3′	3′	a_3
5″	5″	5″	pr_1
2″	1″	1″	pr_2
3″	2″	2″	pr_3
–	3″	3″	pr_4
4″	4″	4″	pr_5
1p	1p	1⁗	p_1
1‴ + 2⁗	1‴	C_1	p_2
2‴	2‴	C_2	p_3
3‴ + 4⁗	3‴	C_3	p_4
2p	2p	3⁗	p_5
1⁗	1⁗	2⁗	e_1
3‴	2⁗	1‴	e_2
–	3⁗	2‴	e_3
–	4⁗	3‴	e_4
l	*l*	S. a.	q_1
4‴	4‴	S. d. a.	q_2
1a	1a	1a	–
–	–	S. d. p.	–
–	–	S. p.	–
–	–	S. m.	–

Figures 26A and 26B represent the antapical and dorso-apical views from Matzenauer's original illustrations. Figures 26C and 26D show these same views where the nomenclature used in this study has been applied.

Matzenauer described *L. schilleri* from only one Indian Ocean sample and acknowledged his uncertainty of the actual orientation of his specimens. The delineation between the epitheca and the hypotheca was correctly determined by him when we consider between which plates this demarcation should go. Yet his descriptions and illustrations (Figs 25A, 26B) show an irregular division in such way that some of the epithecal plates are placed in the hypotheca (pr_3 and pr_4) causing him to be unable to relate his species to *Blepharocysta*. He used plasmolysis to clear the cells followed by Trypan blue to stain the plates. Matzenauer also observed megacytic cells as he described them presenting intercalary bands which contributed to the deformation of the cells. His misinterpretation of the plates may explain why cells exactly like his have never been reported elsewhere.

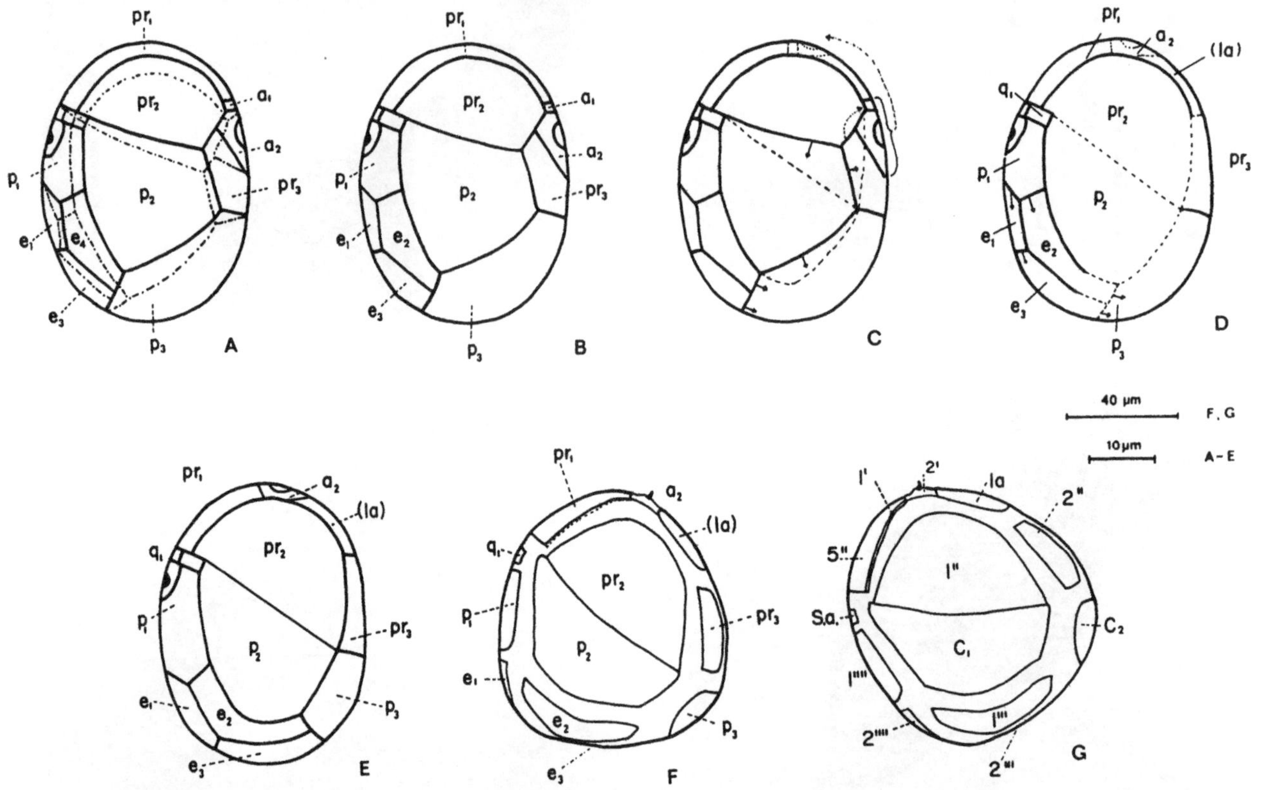

Fig. 25. Transformation of Matzenauer's iconotype of *L. schilleri* to a cell comparable with those of this study.
Fig. 25. *L. schilleri.* A. Iconotype, redrawn from Matzenauer (1933). Side view (left, solid lines; right, dashed lines, seen through the cell). B. Same as A. only left side, is shown. Note that e_4 should be e_2 (see originals of hypotheca, Figure 26A). C. Dashed lines show the proposed sutures and plates (except for that of a_1). Solid arrows indicate the direction where sutures are moved. Dashed-dotted arrows indicate the direction where plates are moved (see text). D. Dashed lines show the new position of the sutures. Some of Matzenauer's original sutures have been removed revealing the anterior intercalary (1a). E. Solid lines have replaced the dashed lines completing the transformation. F. View of the left side of a cell from this study. Matzenauer's nomenclature is used to facilitate the comparison. The 1′ is indicated by dashed lines between pr_1 and pr_2. G. Same. This study's nomenclature and correct orientation.

Gaarder (1954) renamed *L. schilleri*, *Blepharocysta matzenaueri* after she found some cells which she considered identical to *L. schilleri* and which she thought should belong to that genus. Although the valid name should have been *B. schilleri*, as noticed by Ballantine *in* Parke and Dixon (1964), the name of *B. matzenaueri* will be used throughout this paper because it is more closely related to Gaarder. Gaarder failed to notice that the dorsal anterior intercalary plate in her cells (Figs 27A, 27D) would be ventrally located in Matzenauer's (Fig. 25B), as a_1 would be the corresponding plate (Balech 1963). She considered that what Matzenauer called "die größere (Apikal) Öffung" was the dorsal intercalary plate and that his a_2 and a_3 around it were just intercalary bands. This conclusions is not correct. First, the intercalary bands between 1a and a_2 and a_3 are very small, smaller than the actual plates. Secondly, Gaarder believed that the apical zone in Matzenauer's figure was so tiny, that it should be found in the border of a_1. As it was shown before, the plates in the apical zone of Matzenauer's figure do correspond to plates of an apical zone, but in the wrong position in the cell. On the other hand, she failed to point out that the division between the epitheca and the hypotheca in his specimens was asymmetrical, while in hers it was symmetrical.

It may be said that the reason why both hypothecal series *e* and *p* (using Matzenauer's terminology) in Matzenauer's and Gaarder's cells look so small is because they tried to depict elongated plates in an oval or round body, where the actual cell is very laterally compressed. If we compare the sutures, we see that the ventral side of C_1 (Fig. 26G) (p_2 in Matzenauer's; Fig. 25E; 1‴ in Gaarder's; Fig. 27B) coincides precisely with the left side of 1⁗ (p_1 in Matzenauer's, 1p in Gaarder's).

Gaarder described the cells she called *B. matzenaueri* as being ellipsoidal. She stated they were difficult to clear and did not sketch them. Her illustrations correspond to a special form of these cells which she named *B. matzenaueri* f. *gibba*, in which 2‴ (her nomenclature) is extremely bulged. The cells observed

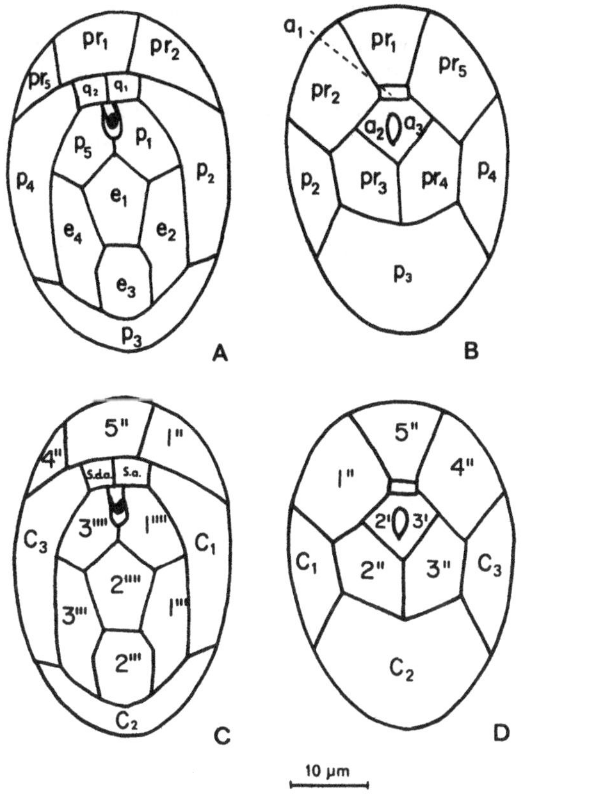

Fig. 26. *L. schilleri.* Redrawn from Matzenauer (1933).
Fig. 26. A. Antapical view. B. Dorso-apical view. C. Same as Figure 26A, using this study's nomenclature. D. Same as Figure 26B, using this study's nomenclature.

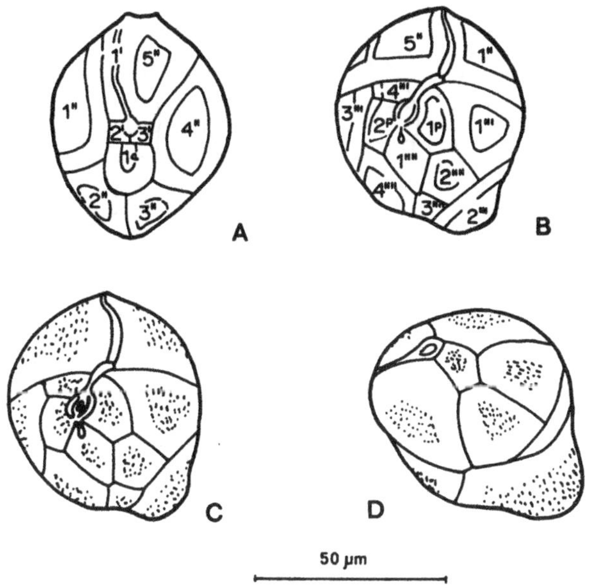

Fig. 27. *Blepharocysta matzenaueri* f. *gibba.* Redrawn from Gaarder (1954). A–D. Same cell.
Fig. 27. A. Apical view. B. Ventral view, showing partial left side. C. Same, nomenclature omitted. D. Dorsal view, showing partial left side.

in the present study closely resemble the ones depicted by Gaarder as *B. matzenaueri* f. *gibba.* Only after the original figures from Matzenauer are transformed as described above, can it be said that her cells correspond to *L. schilleri.*

Blepharocysta compressa as described and depicted by Gaarder (1954) (Fig. 28A–E) would be a bizarre *Blepharocysta* and should not be considered within this genus. The only cell Gaarder was able to clear and depict was probably a non-megacytic cell of *L. schilleri* (as redescribed here). This species may be the same *L. schilleri* (= *B. matzenaueri* Gaarder)but lacks intercalary bands which results in a much more laterally compressed cell. As in Matzenauer's case, there were some misinterpretations in the actual plate pattern. A non-megacytic cell of *L. schilleri* would be very flat, and the observation of apical, antapical, as well as dorsal and ventral views would be very difficult.

Lissodinium schilleri and *Blepharocysta compressa* have many similarities. They have the same shape and lateral compression, and the same four big lateral plates (1″, 4″; C_1 and C_3 this study) (Figs 12A, 13A–B, 28B–C). In both cells, 1′ is long and slender, laying on S. a. (*l*, Gaarder's nomenclature). The S. d. a. (4‴ of Gaarder's) has the same shape, size, position and relationship to the surrounding plates. The corresponding 1a, 3″ (considered as the left dorsal 2″ of this study) and 2‴ (C_2, this study) plates are very similar in both cells. Additionally, the ventral opening is oval and has very reduced flanges. Figures 28B and 28C suggest an apical region raised in the same say as in the individuals from this study (Fig. 12A).

Gaarder might have observed the epitheca of *B. compressa* optically reversed. The two dorsal plates of *L. schilleri* were interpreted as ventral and the only narrow ventral plate 5″ (this study), as dorsal, becoming 3″ (Gaarder's). Cases of optically reversed species descriptions are not uncommon in the literature. The epithecal plate pattern of *Gonyaulax tamarensis* [now *Alexandrium tamarense* (Lebour) Balech] was originally drawn by Lebour (1925) as a mirror image of the epitheca of other members of *Gonyaulax* (Taylor 1975). Likewise, Halim (1967) described optically reversed *Gonyaulax monilata* Howell and made it a new species, *Gessnerium mochimaensis* (Taylor 1979). In the apical view of the iconotype of *Amphidiniopsis kofoidii* Woloszynska there are sutures that are actually from the antapex (F. J. R. Taylor, personal communication).

Considering Figures 29A–C, Figure 29A corresponds to Gaarder's iconotype of *B. compressa.* Figure 29C is a cell from this study, where the intercalary bands have been removed in order to have a closer similarity. If we assume:

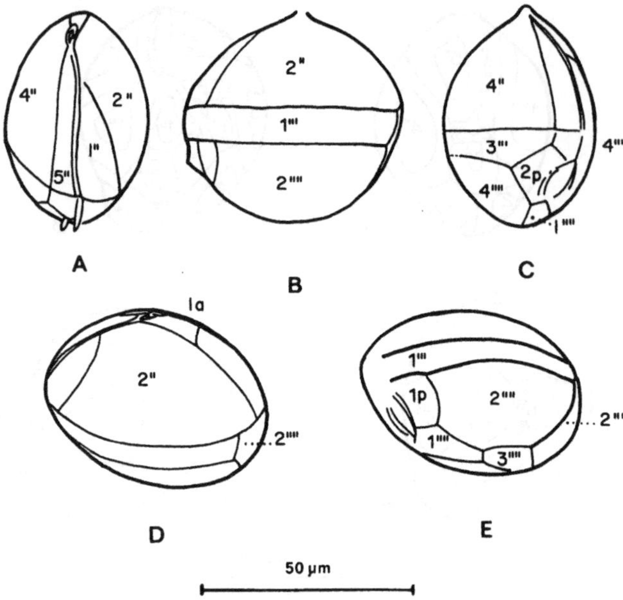

Fig. 28. *Blepharocysta compressa.* Redrawn from Gaarder (1954).
Fig. 28. Same cell. A. Ventral-apical view. B. Left lateral view. C. Part of the right side, and of the sulcal region. D. Left side. E. Partial view of the left side and antapical plates.

1) that Gaarder's plates 1″ and 5″ in the epitheca correspond to the dorsal plates when seen through the cell;

2) that the lower suture of her 1‴ is indeed the upper suture of 3‴ (this study) also seen through the cell and

3) that the left suture of her 2‴ (C_2) is also seen through the cell, when we remove these three sutures (Fig. 29B), we find that the two cells, Figures 29B and 29C are almost identical.

Also, compare the strong similarities between Figures 29B and 13B. Notice that in this study the right dorsal plate (3″ in this study) (Fig. 29C) cannot be seen as the cell is depicted, which is also the case in Gaarder's left view (Fig. 29B).

Cells of *B. compressa* and *B. matzenaueri* f. *gibba* were found by Gaarder in a sample from the same station. The latter species is the same as the former one (as considered in this study) with intercalary bands and a very bulgy C_2.

After Gaarder's original description of *B. compressa* the only literature report is the one published by Steidinger and Williams (1970) which neither included a tabulation pattern study nor a positive identification.

Neither Matzenauer nor Gaarder mentioned the small apical spine, which was not visible under the light microscope in the cells observed in this study. The small spine is also not seen in Steidinger and Williams' (1970), Figure 8 photomicrograph. The cells observed under the light microscope in the present study appeared exactly the same as the cell photographed by these authors and of approximately the same size. They reported it as 87 µm wide.

In general, the cells from the Indian Ocean seem to be larger than those from the Atlantic. The upper cell size range limits given earlier correspond to those of cells from the Indian Ocean. A similar case is observed in cells from other genera such as *Ceratocorys* Stein and *Goniodoma* Stein (personal observation). The apical plates in cells of *L. schilleri* from the Indian Ocean were considerably larger than those from the Atlantic, especially 1′. Also, the intercalary bands between the dorsal and ventral sutures of 2′ and 3′ were smaller in cells from the Atlantic (Fig. 7) than in cells from the Indian Ocean (Figs 8, 11).

Nie (1939, 1942, 1945) and Balech (1963, 1988) have widely discussed relationships between *Blepharocysta* and *Podolampas* and to some extent, with *Lissodinium.* The development of the plates in these genera is different and seems to depend upon the degree and type of the cell compression. In individuals such as *Blepharocysta* (Fig. 6), with almost no compression, the plate size varies. Almost each precingular plate is of a different size. In dorso-ventrally compressed individuals such as *Podolampas* (Fig. 5), the plates in the epitheca are of a more similar size. Plates 1″ and 5″ are pretty much of the same size, 2″ and 4″ are larger and oppose to each other. Plate 3″ is the widest of all, and occupies a significant part of the epithecal dorsal side. On the other hand, because of the lateral compression in *Lissodinium*, this relationship is different. Plates 1″ and 4″ are similar, 2″ and 3″ are also similar and 5″ is ventral and larger than the last two plates but much smaller than 1″ and 4″.

The anterior intercalary plate is also developed differently within the three genera. It is extremely reduced in *Podolampas* (Fig. 10), somewhat larger in *Blepharocysta* (Fig. 9) and more developed in *Lissodinium* (Fig. 7).

Despite the fact that *Lissodinium* and *Blepharocysta* have the same epithecal plate formula, some of the plates have considerably different relationships to each other. In *Lissodinium*, for example, the dorsal part of the epitheca is formed by two plates (2″ and 3″), while in *Blepharocysta* it is formed by just one (3″). The shape of the anterior intercalary plate in both genera is different. In *Lissodinium*, this plate is

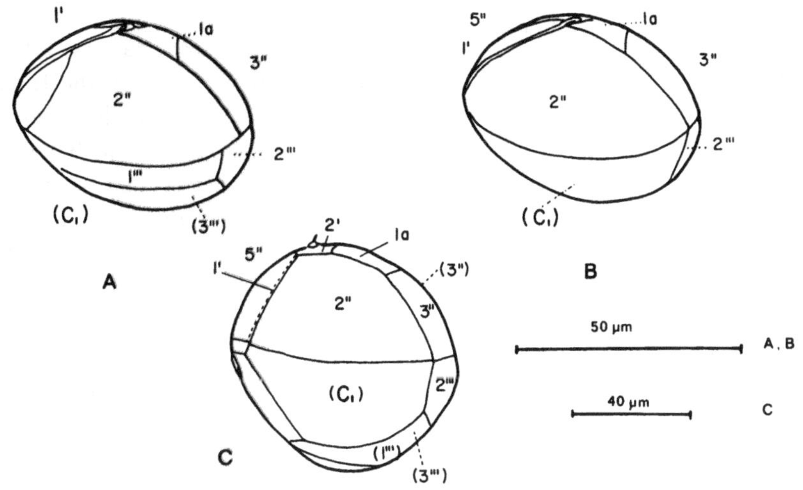

Fig. 29. Transformation of *B. compressa*'s iconotype to a cell comparable with those of this study. Plate designation in parenthesis indicates this study's nomenclature.
Fig. 29. A. Original from Gaarder. Left side. Note resemblances of outline, and especially 1a, 3″ and 2‴ with figure 13B. (3‴) is a plate form the right side, observed through the cell. B. The left suture of her 2‴ (this suture was observed through the cell), the left suture of her 2″ as well as the lower suture of her 1‴ have been removed. The left suture of her 2‴ has been replaced by the corresponding suture of this plate in the left side of the cell. C. *L. schilleri*. This study. Left side. Diagram where the intercalary bands have been removed and plates moved closer together. (3″) and (3‴) correspond to the right side of the cell.

pentagonal and very elongated. In *Blepharocysta* it is trapezoidal. In the former genus the intercalary plate has two sutures of more or less of the same size, with the two precingulars 2″ and 3″. In *Blepharocysta*, it has only one suture dorsally with 3″. The anterior margin of 1″ in *Lissodinium* has sutures with 2′ and 1a, while in *Blepharocysta*, only with 2′. The 4″ and 5″ anterior sutures are the same in both genera. The relationships within the epithecal plates in *Blepharocysta* spp. are the same (personal observation).

Within the cingular plate series it can be seen that in *Podolampas* (Fig. 18) and in *Blepharocysta* (Figs 19, 20), C_2 has the maximum development, while in *Lissodinium*, C_2 is the least developed due to the lateral compression (Figs 1C, 12B, 15).

In *Lissodinium* the right side of the sulcus is formed by the S. d. p. which is not present in *Blepharocysta* or *Podolampas*. In the latter two genera the right side of the sulcus is formed by a different plate, either by the right side of the S. p. or by the S. d. (right sulcal) respectively. The perforations in the S. p. in *Lissodinium* are similar to the rows of round pores found in the hypothecal plates of some *Podolampas* (Figs 17A, 18) and differ from the single row of large perforations present in the same plate in this genus (Figs 21, 22).

The development of the antapical plates is greater in *Lissodinium* (Fig. 1B, 14B), especially the 2⁗, than in *Blepharocysta* (Figs 19, 20) or in *Podolampas* (Fig. 18). This plate can be very small in both genera.

The main differences between *Lissodinium* and *Blepharocysta* can be summarized as follows:

1) Strong lateral compression in *Lissodinium*. Almost no compression in *Blepharocysta*.

2) The relationship among the epithecal plates in *Lissodinium* is different from that in *Blepharocysta*.

3) Intercalary bands are not present between the precingular and cingular plates in *Lissodinium* while they are present in *Blepharocysta*.

4) *Lissodinium* has five sulcal plates, while *Blepharocysta* has only four. The S. p. in *Blepharocysta* forms both the right and the left sides of the sulcus as well as the posterior end. In *Lissodinium* the right side of the sulcus is formed by a different plate, the S. d. p., which is not present in *Blepharocysta*.

5) The three plates of the antapical series have a similar development in *Lissodinium*, but are different in *Blepharocysta*. In *Blepharocysta*, 2⁗ is very small and has at least one row of round perforations similar to those found in the S. p. of *Podolampas*. In *Lissodinium* the 2⁗ is large and does not present such perforations.

6) *Lissodinium* also presents the small apical spine that can be found in *Podolampas* and that is absent in *Blepharocysta* spp.

The above discussion has demonstrated that cells from this study may be very well identified as *Lissodinium schilleri* and that *Blepharocysta matzenaueri* is indeed *L. schilleri*. The iconotype of *L. schilleri* was incorrectly depicted and the correct plate pattern is described in the present paper. The remarkable lateral compression and the other differences from *Blepharocysta* as pointed out earlier, justifies that these organisms should be kept in different genera.

Additionally, *Blepharocysta compressa* should be included as a synonym of *Lissodinium schilleri* because the epitheca of its iconotype appears to have been depicted optically reversed and some of the hypothecal sutures interpreted as forming a band of narrow plates that do not exist.

Finally, because of the similarities and close relationships of *Lissodinium* to *Blepharocysta* and *Podolampas*, it is proposed to include *Lissodinium* in the family Podolampadaceae and not in the Lissodiniaceae as was done by Schiller (1937).

Distribution

Lissodinium schilleri has a widespread distribution and has been reported from the Indian and the Atlantic Oceans. In the Indian it has been found in the Arabian Sea (Matzenauer) and in Bengal Bay (this study). In the Atlantic it has also been found in the Sargasso Sea (as *B. compressa* and *B. matzenaueri* f. *gibba*); west of the Canaries Islands (as *B. compressa*); off the coast of Morocco (as *B. matzenaueri* and *B. matzenaueri* f. *gibba*); northern coast of Spain (as *B. matzenaueri* f. *gibba*) (Gaarder 1954); British Islands (as *B. schilleri* and *B. schilleri* f. *gibba*) (Ballantine *in* Parke and Dixon 1964); Gulf of Mexico (as *B. compressa?*) (Steidinger and Williams 1970) and Dominican Republic, this study.

Acknowledgements

This study was made possible in part by the grant NSERC A-6137 to Dr F. J. R. Taylor. I am thankful to Max Taylor and to Laslo Veto and especially to my husband, Sandy.

References

Abé, T. H. 1966. The armoured Dinoflagellata: I. Podolampidae. *Publ. Seto mar. biol. Lab. 14*: 129–154.

Balech, E. 1954. Sur la tabulation de *Podolampas* et *Oxytosum*. *Rapp. Commun. 8e Congr. int. Bot., Sect. 17*: 114–116.

Balech, E. 1963. La familia Podolampaceae (Dinoflagellata). *Boln. Inst. Biol. mar., Mar del Plata 2*: 3–27, pl. 1–3.

Balech, E. 1988. *Los Dinoflagelados del Atlántico Sudoccidental.* Instituto Español de Oceanografia, Madrid. 310 pp. incl 88 pl.

Gaarder, K. R. 1954. Dinoflagellatae from the «Michael Sars» North Atlantic Deep-Sea Expedition 1910. *Rep. scient. Res. «Michael Sars» North.-Atl. Deep-Sea Exped.* 1910, *2*: 1–62 + 5 tables.

Halim, Y. 1967. Dinoflagellates of the South-East Caribbean Sea. *Int. Revue ges. Hydrobiol. 52*: 701–755.

Kofoid, C. A. 1909a. On *Peridinium steini* Jörgensen, with a note on the nomenclature of the skeleton of the Peridinidae. *Arch. Protistenk. 16*: 25–47, pl. 2.

Kofoid, C. A. 1909b. The morphology of the skeleton of *Podolampas*. *Arch. Protistenk. 16*: 48–61, pl. 3.

Lebour, M. V. 1925. *The Dinoflagellates of Northern Seas.* Marine Biological Association, Plymouth. pp. vii-250 incl pl. 1–35.

Matzenauer, L. 1933. Die Dinoflagellaten des Indischen Ozeans. (Mit Ausnahme der Gattung *Ceratium*.) *Bot. Arch. 35*: 437–510.

Nie, D. 1939. Dinoflagellata of the Hainan region. II. On the thecal morphology of *Blepharocysta*, with a description of a new species. *Contrib. biol. Lab. Sci. Soc. China*, Zool. Ser. *13*: 23–39.

Nie, D. 1942. Dinoflagellata of the Hainan region. IV. On the thecal morphology of *Podolampas* with description of species. *Sinensia 13*: 53–60.

Nie, D. 1945. Sinodiniidae, a new family of Peridiniida (Protozoa, Dinoflagellata). *Trans. amer. microsc. Soc. 64*: 196–202.

Parke, M. and P. S. Dixon. 1964. A revised check-list of British marine algae. *J. mar. biol. Ass. U.K. 44*: 499–542.

Rampi, L. 1941. Ricerche sul fitoplancton del Mare Ligure. 5. Le Podolampacee delle acque di Sanremo. *Annalis. Mus. civ. Stor. nat. G. Doria, Genova 61*: 141–152, pl. 5.

Schiller, J. 1937. Dinoflagellatae (Peridineae) in monographischer Behandlung. *In*: *«Rabenhorst's Kriptogamen-Flora» 10(3)* Teil 2(3/4): 321–590.

Sournia, A. 1973. Catalogue des espèces et taxons infraspécifiques de dinoflagellés marins actuels publiés depuis de la révision de J. Schiller. I. Dinoflagellés libres. *Beih. Nova. Hedw. 48*: xii-92.

Sournia, A. 1986. *Atlas du Phytoplankton Marin. Vol. I: Introduction, Cyanophycées, Dictyochophycées, Dinophycées et Raphidophycées.* CNRS, Paris. 219 pp.

Steidinger, K. A. and J. Williams. 1970. *Dinoflagellates. Mems. Hourglass Cruises* (Florida Dept. Natur. Resources, St. Petersburg) *2*: 1–251 incl pl. 1–45.

Tappan, H. 1980. *The Paleobiology of Plant Protists.* W. H. Freeman & Co., S. Francisco. 1028 pp.

Taylor, F. J. R. 1972. Application of the scanning electron microscope to the study of tropical microplankton. *J. mar. biol. Assoc. India 14*: 55–60.

Taylor, F. J. R. 1975. Taxonomic difficulties in red tide and paralytic shellfish poison studies: The "*tamarensis* complex" of *Gonyaulax*. *Environm. Lett. 9*: 103–119.

Taylor, F. J. R. 1976. Dinoflagellates from the International Indian Ocean Expedition. A Report on material collected by the R.V. «Anton Brun» 1963–1964. *Bibliotheca bot. 132*: 1–234, pl. 1–46.

Taylor, F. J. R. 1979. The toxigenic goniaulacoid dinoflagellates. *In*: (D. L. Taylor and H. H. Seliger, eds) *Toxic Dinoflagellate Blooms.* Elsevier/North Holland, New York. pp. 47–56.

Taylor, F. J. R. 1980. On dinoflagellate evolution. *Biosystems 13*: 65–108.

Botanica Marina
Vol. 34, pp. 341–358, 1991

Effects of Nutrient Enrichment on Microalgal Community Composition in a Coastal Shallow-water Sediment System: An Experimental Study

K. Sundbäck and P. Snoeijs*

Department of Marine Botany, University of Göteborg, Carl Skottsbergs Gata 22, S-41319 Göteborg, Sweden
* *Department of Ecological Botany, Uppsala University, Box 559, S-75122 Uppsala, Sweden*

(Accepted 7 April 1991

Abstract

An outdoor flow-through experimental set-up, was used to study the effect of increased concentrations of inorganic nitrogen (IN) and phosphorus (IP) on the microalgae of a sandy sediment. The compositional changes following nutrient addition were investigated in terms of the dominance of major taxonomic groups as well as the species composition of diatoms. Fresh samples were used to assess the viable biomass of major taxonomic and size groups, and cleaned diatom frustules were used for identification to species level. Nutrient addition to the water column stimulated the growth of filamentous cyanobacteria, diatoms and flagellates. Although statistically significant compositional changes with respect to both time and treatment were observed, the microflora was not subjected to any substantial shift in the dominance of major taxonomic groups. Instead, the major change occurred at the macroscopic level as a rapid growth of a filamentous green algal species. Diatoms which accounted for *ca* 50% of the microalgal biomass, were not outcompeted by any other algal group, probably because the Si-pool of the sediment counteracted secondary Si-limitation and because temperature was too low to allow cyanobacteria to fully dominate the biomass. Counts of both living cells and cleaned frustules showed that the diatom flora was dominated by cells in the size group 7–12 μm, however, depending on the counting method and the level of resolution, different compositional changes were noted. Biomass values of size groups of living diatom cells showed that N + P addition favoured larger cells. Frustule counts, which enable identification at the species level, revealed also that some small-sized species were enhanced by the nutrient addition. Canonical correspondence analysis (CCA) based on frustule counts, showed that the succession in both non-enriched and enriched containers was in the same direction, though faster and proceeding further with N + P addition. The CCA revealed that the P + N addition also favoured some small-sized taxa, such as *Nitzschia* and *Amphora* species within the size range 7–12 μm. Oxygen bubbles produced by the stimulated photosynthetic activity at the sediment surface appeared to selectively resuspend large motile diatom species, *e.g. Cylindrotheca*.

Introduction

During the last decade, eutrophication in coastal areas has worried scientists, the general public and local authorities in many countries, including Sweden. One of the symptoms of excessive eutrophication is an increased frequency of algal blooms, that may cause bottom hypoxia (Rosenberg *et al.* 1990). Oxygen deficiency does not only threaten sublittoral areas, but also shallow areas in sheltered bays. Such areas may be increasingly subjected to hypoxia because of the increasing accumulation of allochthonous organic matter such as drifting and decomposing macroalgae and detritus (Sundbäck *et al.* 1990). These symptoms, together with the fact that shallow-water sediment areas are important breeding and foraging areas for macrofauna and many commercial fish species, have drawn our attention to the impact of an increased nutrient load on the food webs of shallow-water sediment communities.

Cultural eutrophication of lakes has caused shifts in the dominance of the major taxonomic groups of phytoplankton, which in turn has induced associated changes in the biomass and composition of other trophic levels (Tilman *et al.* 1982). Parallel to these findings, the question has been raised whether an increasing eutrophication in the marine environment, implying changed ratios of nutrient supply, is currently changing the basis for the pelagic food web from a diatom based system towards a less edible flagellate system (Doering *et al.* 1989). In addition, investigations of nutrient effects on fresh-water periphyton have revealed shifts in the dominance of major taxonomic algal groups in benthic communities (Carrick *et al.* 1988). Less is known about the response of the sediment-associated microflora of shallow marine and brackish-water environments to the increasing load of nitrogen and phosphorus, especially in areas lacking regular tidal flushing of the habitat.

In temperate areas diatoms are important constituents of the microscopic primary producers in shallow-water sediments. Investigations from fresh-water benthic habitats have shown that periphyton diatoms respond to altered concentrations and ratios of inorganic nutrients by changed community composition (Pringle and Bowers 1984, Carrick and Lowe 1988). In the marine environment corresponding changes have been observed for salt-marsh diatom communities (Sullivan 1976, 1981).

This investigation is part of an experimental study, designed to investigate the effect of increased load of inorganic nitrogen and phosphorus on the structure and function of a marine shallow-water sandy sediment system on the west coast of Sweden (Nilsson *et al.* 1991). Since one of the main objectives of the experiment was to investigate the impact of elevated nutrient concentrations on the pathways of carbon between microscopic primary producers and consumers by measuring grazing rates using radioactive labels, we wanted to study in more detail the compositional changes of the microalgae of the sediment system. The changes were studied in terms of:

(1) possible shifts in the dominance of major taxonomic groups and

(2) possible shifts in the species composition of one of the dominant groups of microalgae, the diatoms.

Material and Methods

The experimental set-up

The effect of increased input of inorganic nitrogen (IN) and phosphorus (IP) in the water column above a sediment community, was studied using an outdoor flow-through system (Sundbäck *et al.* 1990) consisting of eight 30-litre containers placed in a floating frame at the Tjärnö Marine Biological Laboratory on the west coast of Sweden (N 58°52′, E 11°09′). This coastal area, in NE Skagerrak, is brackish and virtually non-tidal. Sandy sediment from a nearby shallow bay (0.2 m water depth) was collected in two layers that were treated separately: surface sediment layer (upper 1 cm) and lower sediment layer (1 to 10 cm depth). The sand was sieved (mesh size 0.5 mm) to remove macrofauna, homogenized and spread out in two layers (2 cm of surface sediment on top of a 8 cm thick bottom layer) in each experimental container (area 0.1 m^2). Surface seawater was filtered through 50 μm and 1 μm cotton-filter cartridges, and pumped from two 100-litre cisterns through the experimental containers using a multichannel peristaltic pump. The inflowing water entered the experimental containers 1 cm above the sediment surface and the flow-rate was 1.6 L/h corresponding to a turn-over time of 12 h. The height of the water column above the sediment in the containers was 18 cm. To allow the vertical chemocline in the sediment to stabilize, the containers were allowed to equilibrate for 10 days before the experiment was started.

From the third day of the experiment, the water in one of the two cisterns was enriched with approximately 85 μmol L^{-1} IN and 2.5 μmol L^{-1} IP, using $NaNO_3$ and K_2HPO_4, respectively, thus simulating nutrient levels and N/P ratios found in shallow coastal areas influenced by river or sewage-plant effluents (Granéli and Sundbäck 1985, Granéli *et al.* 1990). Four containers received this enriched water, corresponding to a mean nutrient input of 27.5 mmol IN and 0.74 mmol IP m^{-2} day^{-1}. Four other containers receiving non-enriched water from the other cistern served as controls. The approximate nutrient concentrations of the non-enriched water were: < 0.2 μM PO_4^{3-}, < 1 μM NO_3^- + NO_2^-, between 0.35 and 8.5 μM NH_4^+ and between 2.6 and 7.5 μM $Si(OH)_4$. The experiment was run for 4 weeks (12 August to 9 September 1988). The inner walls of the containers were kept clean from algal growth by carefully wiping the surfaces every 3–4 days with clean cloths rinsed in distilled water. Water temperature varied between 16.1 and 18.9 °C and salinity between 14 and 24‰ during the experiment.

Sampling

Sediment cores were taken with transparent PVC tubes designed for radiotracer measurements of primary and bacterial production and meiofaunal grazing rates in undisturbed sediment (*i.d.* 67.8 mm, length 150 mm; Jönsson in press). Samples were taken on days 1, 7, 14, 21 and 28. At each sampling, one core was taken from each container. The holes in the sediment surface in the containers caused by the core-sampling were filled with autoclaved sediment to prevent leakage of sulfide and other solutes from the anoxic layers. Subsamples for counts of living microalgae as well as diatom frustules were taken with a cut-off 2 mL plastic syringe (*i. d.* 9 mm) from the cores after the radiotracer measurements. Concentrations of NH_4^+, NO_3^- + NO_2^-, PO_4^{3-} and $Si(OH)_4$ in the inflowing and the overlying water of each experimental container was analysed as described in Nilsson *et al.* (1991).

Counts and biomass of living microalgae

For each sample used for counts of living microflora, 2 subsamples (*i. d.* 9 mm) from the top 5 mm were pooled. The samples were diluted to 6 mL with filtered seawater and treated in a sonication bath (Sonorex RK 100, 35 kHz) for 9 minutes. This time was selected by plotting number of living (= fluorescing) cells against sonication time. This treatment detached most of the epipsammic cells, but apparently did not break them, since motile cells could be seen in sonicated samples. The autotrophic cells were counted in a Bürker counting chamber using epifluorescence microscopy. Cells were grouped into size and shape classes, and then identified when possible to generic level. Generally cell dimensions of a minimum of 30 cells per size group were measured (for less frequent size groups, a minimum of 15 cells) and average cell volume and carbon content for each group were calculated using geometric formulae according to Edler (1977, 1979); the factor to convert cell volume to carbon varied between 0.085 and 0.11 (diatoms 0.085 – 0.089). Living cells were counted on four occasions (Days 1, 7, 14 and 28) in samples taken from all 8 containers (4 control and 4 enriched containers). To enable comparison between viable cells and cleaned frustule counts (see below), the absolute numbers of viable cells were converted to relative abundance values (RA, %) by size and shape group.

Analysis of cleaned diatom material

Permanent slides of diatoms were made from the same samples as those used for enumeration of the living microflora. Material from 6 containers (3 control and 3 enriched) were used. The sediment samples were centrifuged for 30 minutes at 1500 – 2000 rpm. Thereafter 0.5 mL H_2O_2 was added together with a pinch of potassium bichromate to destroy organic material and to detach the diatoms from the sand grains. The samples were washed and centrifuged three times with 10 mL double distilled water. Subsequently, the samples were shaken with 3 mL double distilled water, the sand grains were allowed to sink for 5 – 10 s and the diatom suspension was collected using a pipette; this procedure was repeated 6 times. Microscopic investigation confirmed that this treatment removed all diatom frustules from the sand grains. One drop of the diatom suspension was put on a cover slip that had been cleaned with 96% ethanol and left to dry in the air. Hyrax (refractive index 1.71) was used as the mounting medium. Diatom frustules were identified, counted and categorized into size groups by phase contrast microscopy using a 100 times oil immersion objective. Five hundred valves were counted along a transect across the middle of the drop. In total, 30 samples were analysed.

Scanning electron microscopy (SEM) was performed on a JEOL JSM-35 Scanning microscope at the Unit for Biological Structure Analysis (Department of Zoology, Uppsala University).

Species were identified using literature cited in Snoeijs (1989) and in addition, Brockmann (1950), Giffen (1966), Williams and Round (1987) and Lange-Bertalot and Krammer (1989). Nomenclature was updated according to the check-list of Hartley (1986) except for those taxa that were identified with the help of newer literature.

Data analysis of frustule counts

Community diversity and dominance were measured by species richness, the Shannon-Weaver index (log base = e) and Pielou's evenness index (Whittaker 1977, Patil and Tailllie 1979, Maarel 1989). The two quantitative indices were calculated from relative abundances.

Constrained ordination (Braak and Prentice 1988) by canonical correspondence analysis (CAA) implemented with the program CANOCO (Braak 1986), was applied to summarize variation in species composition in relation to time and environment. CANOCO gives a simultaneous ordination with sample and species scores, vectors indicating the direction and rates of change for quantitative predictor variables, and centroids for qualitative predictor variables (class variables). Eigenvalues measure the importance of successive ordination axes. Predictor variables can be used 'actively' (as constraints) in the ordination, or can be added 'passively' later. In both cases, CANOCO performs a multiple regression analysis to indicate which of the variables have significant effects on each axis, though the significance test is rigorous only for passive variables that were not used in the construction of the ordination. In the analyses class variables for date-treatment combinations were used as constraints. Nutrient concentrations ($NO_3 + NO_2$, NH_4, PO_4 and $Si(OH)_4$), number of days the experiment had been running (1, 7, 14, 21 and 28 days) and Shannon-Weaver diversity were used as passive quantitative variables. Two CCA analyses were made, one for all 101 identified taxa, and another for clusters of genera.

Statistical analyses

Data on biomass, number of living algae and relative abundances of frustule counts were tested with Taylor's power law (Green 1979) and were transformed by square root or the natural logarithm when necessary. Differences between dates and treatments were tested using Newman-Keuls multiple comparison test (NK) (Zar 1974) with differences accepted as significant when $p < 0.05$.

Results

Visual observations

Four days after the start of the treatment, the sediment surface in the enriched containers was more brownish in colour than in the control containers. In all 8 containers, oxygen bubbles were formed on the sediment surface during daytime. In the enriched containers, flakes of the sediment surface were lifted off by the rising oxygen bubbles towards the end of the experiment. On approximately Day 12, tufts of the filamentous green alga *Enteromorpha clathrata* (Roth) Greville were observed on the sediment in the enriched containers and by Day 28, the filaments started to float up to the water surface.

The relative abundances of different diatom taxa under each treatment are given in Table I.

Living autotrophic microflora

Biomass of major algal groups. The biomass obtained from cell counts of living autotrophic microalgae in the top 5 mm sediment was significantly stimulated by the nutrient addition (Fig. 1). After 2 weeks the microalgal biomass had increased from 113 to 455 mg C m^{-2} in the enriched containers and from 123 to 138 mg C m^{-2} in the control containers. The increase was significantly (NK) higher in the enriched containers. Although there was a trend that the microalgal biomass decreased in the enriched containers towards the end of the experiment, it was still twice as high as in the controls on Day 28 (349 as against 181 mg C m^{-2}).

Initially, diatoms accounted for 47% (= 56 mg C m^{-2}), cyanophytes for 42–47%, and various flagellates for 6–8% of the total living microalgal biomass. When considering the total biomass accounted for by organisms < 0.5 mm (= microflora + bacteria + meiofauna; ciliates and heterotrophic flagellates were not enumerated, but probably constituted a minor proportion; *cf.* Sundbäck *et al.* 1990), diatoms constituted 20% at the start and 13–15% at the end of the experiment. During the first week, the biomass of filamentous cyanobacteria (mainly *Oscillatoria* spp.) increased in both the control and enriched containers. Later their biomass decreased in the control containers and consequently the mean proportion of diatoms increased towards the end of the experiment, comprising on an average 55% on Day 28. In the nutrient-enriched containers, the increase of the biomass was due to both small-sized diatoms and filamentous cyanobacteria, and the relative proportions between these two main algal groups did not change during the course of the experiment. The final increase of the diatom biomass was 2.5-fold in the enriched containers. The proportion of flagellate biomass, in particular those within the size range 4–8 µm, increased significantly (NK) in the enriched containers, though they only comprised 15% of the total living microalgal biomass (Fig. 1).

Cell number of diatom size groups. The initial number of viable diatom cells was 7.5×10^5 cm^{-2} and the highest value was reached in enriched containers after

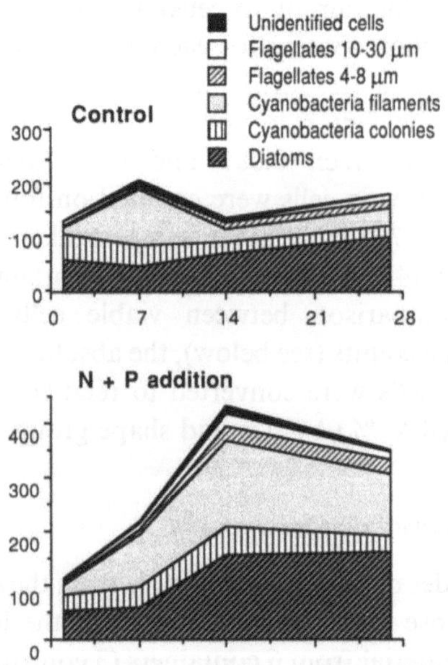

Fig. 1. Cumulative biomass of the major groups of living benthic microalgae in the top 5 mm of sediment in flow-through containers without (control) and with the addition of inorganic phosphorus and nitrogen to the overlying water (N + P addition). The curves show mean values of 4 replicate containers.

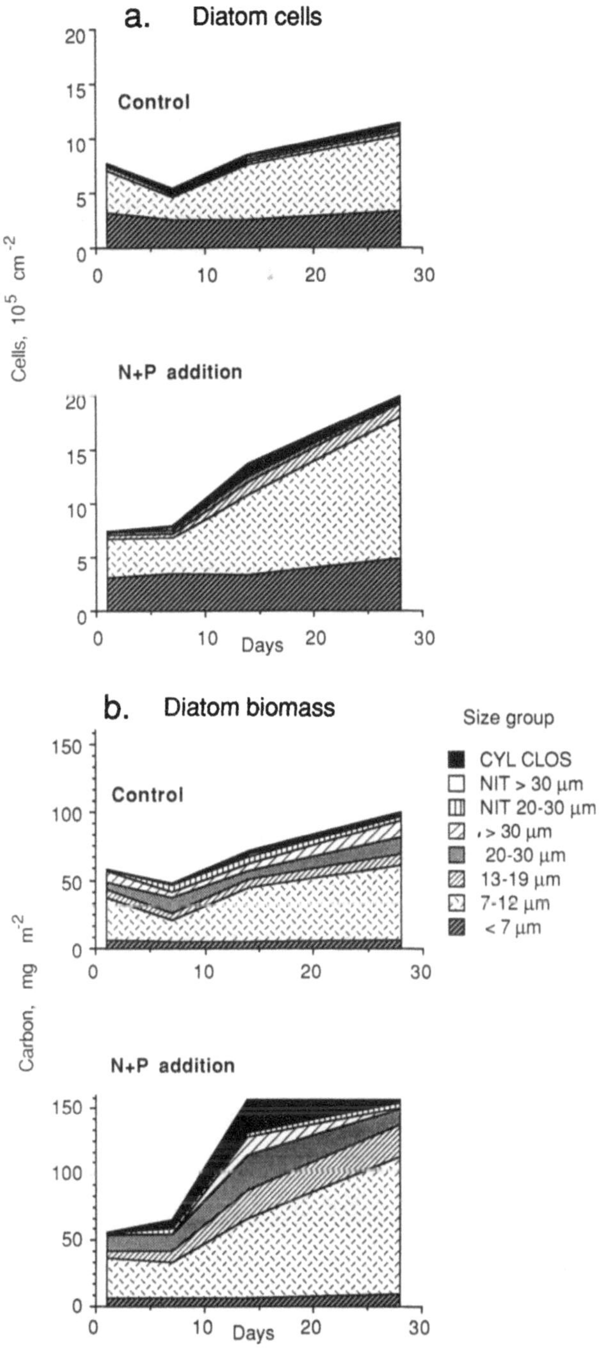

Fig. 2. Cumulative cell number (a) and biomass (b) of size groups of living diatoms in the top 5 mm of sediment. All diatoms are pennate forms; for size groups > 20 µm the genera *Nitzschia* (NIT) and *Cylindrotheca* (CYL) are shown separately. The curves show mean values of 4 replicate containers.

4 weeks (20×10^5 cells cm^{-2}) (Fig. 2a). Pennate taxa within the size group 7–12 µm accounted for most of the increase in enriched containers as well as in control containers, whereas the number of the smallest size group (< 7 µm) was unchanged (Fig. 2a). Cell numbers of the size group 13–19 µm increased significantly (NK) in enriched containers but not in control containers, and were significantly more abundant in enriched containers after 2 weeks (NK). The number of *Cylindrotheca closterium*, one of the few viable taxa identified at the species level, was significantly higher (NK) in the enriched containers after one week (Fig. 2a).

Biomass of diatom size groups. In terms of biomass (Fig. 2b), the smallest size group (< 7 µm) decreased and the larger size groups (> 12 µm) increased, but still the size group 7–12 µm dominated and accounted for the main biomass increase in enriched containers. The size group 13–19 µm increased significantly in enriched containers. This effect was more pronounced in terms of biomass (Fig. 2b) than in terms of cell numbers (Fig. 2a). In the enriched containers, the proportion of *Cylindrotheca closterium* increased from less than 1% to 18% of the total diatom biomass from Day 1 to Day 14 (significant increase with regard to both time and treatment). Furthermore, this diatom was significantly more abundant in enriched than in control containers (NK). At the end of the experiment, however, the biomass of diatom cells > 20 µm decreased in the nutrient enriched containers.

Relative abundance of viable cells and cleaned diatom frustules

Although the comparison of living cell counts and frustule counts was done on size-group level, the *Cylindrotheca closterium* and the genus *Nitzschia* in the size group > 20 µm were distinguished from the rest. On the whole, the proportions of the individual size groups agreed for the two counting techniques (Fig. 3), however, only the frustule counts revealed the increased RA of *Nitzschia* cells within the size range 7–12 µm. This increase was significantly (NK) greater in the enriched containers. The increase of *C. closterium* in the middle of the experiment was slightly less in the frustule counts as compared with the viable counts (Fig. 3).

Assuming that the RA values of frustules reflect RA values in the living diatom community (*cf.* Fig. 3), the results from the two counting methods were combined to illustrate quantitative changes of the 6 principal diatom genera (but including Fragilariaceae in one group) during the experiment (Fig. 4). At the end of the experiment all genera, except for *Cocconeis*, had increased their cell numbers in the control containers (= 'natural succession'); adding the nutrients resulted in more cells of all genera, largely following the pattern of 'natural succession' (Fig. 4a). Relative increases (= % increase from Day 1 to Day 28) (Fig. 4b) show that *Nitzschia* (+ 1300%) and *Amphora* (+ 450%) were the genera that profited most from the nutrient addition.

Table I. List of diatom taxa with mean relative abundance scores per treatment for each date, abbreviation codes and size groups. Separate size groups were used for *Nitzschia* (N) and *Cylindrotheca closterium* (CYL CLOS).

Diatom taxon	Code	Control					Nutrient load					Size group (μm)
		Date 0812	0818	0825	0901	0908	Date 0812	0818	0825	0901	0908	
Achnanthes bahusiensis (Grunow) Lange-Bertalot	ACH BAHU	2.8	1.5	1.5	4.6	1.4	1.5	2.0	1.7	3.2	1.6	13–19
Achnanthes delicatula (Kützing) Grunow	ACH DELI	0.3	0.5	0.3	0.1	0.1	0.9	0.1	0.1	0.2	0.5	13–19
Achnanthes cf. *delicatula* (Kützing) Grunow	ACH CFDE	10.4	9.2	8.0	4.7	5.5	9.5	8.6	6.9	5.0	6.2	7–12
Achnanthes cf. *delicatissima* Simonsen	ACH DECA	14.7	15.6	15.5	11.5	14.5	20.0	17.2	16.1	10.4	10.0	< 7
Achnanthes lemmermannii Hustedt	ACH LEMM	5.4	3.9	4.8	2.1	3.0	3.5	4.4	2.7	2.5	2.1	7–12
Achnanthes minutissima Kützing	ACH MINU	–	0.1	–	0.1	–	0.1	–	–	–	–	13–19
Achnanthes sp.	ACH SP	2.1	4.1	4.4	4.6	6.1	5.0	4.9	4.3	4.5	3.8	7–12
Amphora coffeaeformis (C. A. Agardh) Kützing	AMP COFF	0.5	1.1	0.5	1.3	0.7	0.6	0.5	0.7	0.8	0.9	20–30
Amphora cf. *coffeaeformis* (C. A. Agardh) Kützing	AMP CFCO	0.3	0.2	0.1	0.3	0.2	0.3	–	0.3	0.5	0.5	13–19
Amphora commutata Grunow	AMP COMM	–	0.1	–	–	–	0.1	–	–	–	–	>30
Amphora flebilis Simonsen	AMP FLEB	0.7	0.7	0.2	1.4	1.3	0.5	0.7	0.9	0.8	1.0	13–19
Amphora holsatica Hustedt	AMP HOLS	0.1	0.3	0.2	–	0.1	0.3	0.1	0.2	0.5	0.2	>30
Amphora laevis v. *laevissima* (Gregory) Cleve	AMP LAEV	–	–	–	0.1	–	–	–	0.1	–	–	>30
Amphora lineolata Ehrenberg	AMP LINE	–	–	–	–	–	–	–	0.1	–	–	>30
Amphora ostrearia de Brébisson	AMP OSTR	0.1	0.1	0.5	–	0.2	0.2	0.1	–	–	–	>30
Amphora ovalis (Kützing) Kützing	AMP OVAL	0.1	–	–	–	–	–	–	–	–	–	>30
Amphora staurophora Juhlin-Dannfeldt	AMP STAU	0.4	0.5	0.5	0.8	–	0.1	0.4	0.5	0.3	0.4	7–12
Amphora tenerrima Aleem *et* Hustedt	AMP TENE	0.7	0.9	0.5	1.2	0.7	0.5	0.8	0.5	1.1	1.2	7–12
Amphora veneta Kützing	AMP VENE	–	–	–	–	–	0.3	–	–	–	–	13–19
Amphora sp. 1	AMP SP 1	1.9	4.3	5.1	10.3	6.9	3.1	5.7	7.5	11.2	8.1	7–12
Amphora sp. 2	AMP SP 2	0.1	–	0.1	–	–	–	0.1	0.1	–	–	>30
Amphora sp. 3	AMP SP 3	–	–	–	–	–	0.1	–	–	–	–	>30
Amphora sp. 4	AMP SP 4	0.6	0.1	0.1	0.1	–	–	0.1	–	0.3	0.3	7–12
Anaulus balticus Simonsen	ANA BALT	–	0.1	–	–	0.1	–	0.1	–	–	–	7–12
Auliscus sculptus (W. Smith) Ralfs	AUL SCUL	0.1	–	–	–	–	–	–	0.1	–	–	>30
Berkeleya rutilans (Trentepohl) Grunow	BER RUTI	–	–	–	0.3	0.2	–	–	–	–	0.1	20–30
Catenula adhaerens (Mereschkowsky) Mereschkowsky	CAT ADHA	–	–	–	–	–	0.1	–	–	–	–	7–12
Chaetoceros cyst	CHA CYST	–	0.1	–	–	–	–	–	–	–	–	13–19
Cocconeis cf. *diminuta* Pantocsek	DOC DIMI	0.4	0.5	0.3	0.6	0.1	0.1	0.8	0.3	–	0.1	7–12
Cocconeis peltoides Hustedt	COC PELT	3.5	2.5	2.0	0.8	1.0	2.5	2.9	1.9	0.9	1.5	13–19
Cocconeis cf. *peltoides* Hustedt	COC CFPE	–	0.1	0.1	–	–	0.3	–	0.1	–	–	7–12
Cocconeis placentula Ehrenberg	COC PLAC	0.1	–	–	–	–	–	–	–	–	–	20–30
Cocconeis scutellum Ehrenberg	COC SCUT	–	–	–	–	0.1	–	–	–	0.1	–	20–30
Cocconeis stauroneiformis (W. Smith) Okuno	COC STAU	0.1	–	–	–	–	–	–	–	–	–	20–30
Cocconeis sp.	COC SP	0.1	–	0.1	–	0.1	–	0.2	–	–	–	7–12
Cyclotella atomus Hustedt	CYC ATOM	–	0.3	–	0.1	–	–	–	0.1	–	0.1	< 7
Cyclotella caspia Grunow	CYC CASP	0.1	0.3	0.1	–	–	0.2	–	–	–	0.1	< 7
Cylindrotheca closterium (Ehrenberg) Reimann *et* Lewin	CYL CLOS	0.1	1.3	1.8	2.5	0.3	0.3	1.2	3.7	0.9	0.3	CYL CLOS
Diatoma tenue C. A. Agardh	DIA TENU	0.1	–	–	–	–	–	–	–	–	–	13–19
Dimerogramma minor (Gregory) Ralfs	DIM MINO	0.3	0.7	0.5	0.1	0.3	0.3	0.6	0.1	0.1	0.1	20–30
Diploneis finnica (Ehrenberg) Cleve	DIP FINN	–	–	–	0.1	–	–	–	–	–	–	>30
Entomoneis paludosa v. *duplex* (Donkin) Czarnecki *et* Reinke	ENT PALU	–	0.1	–	–	0.5	0.1	–	0.2	0.3	0.1	20–30
Fragilaria vaucheriae (Kützing) Petersen	FRA VAUC	–	–	–	–	–	0.1	–	–	–	–	7–12
Fragilariaceae sp. 1	FRE SP 1	5.3	3.3	2.6	2.1	2.2	3.8	3.8	2.2	2.9	2.9	< 7
Fragilariaceae sp. 2	FRE SP 2	3.1	2.5	3.9	3.3	4.5	5.1	1.9	1.4	2.5	2.3	< 7
Fragilariaceae sp. 3	FRE SP 3	1.8	1.0	0.7	1.7	0.9	1.7	0.9	0.1	1.1	0.7	7–12
Fragilariaceae sp. 4	FRE SP 4	–	0.1	–	–	–	0.7	0.1	0.1	–	0.3	20–30

Table I. continued.

Fragilariaceae sp. 5	FRE SP 5	–	0.1	0.1	–	–	–	–	–	–	–	7–12
Fragilariaceae (rest group)	FRE SPP	0.1	0.1	0.5	0.5	0.3	0.1	0.1	–	0.2	0.1	7–12
Gomphonemopsis exigua (Kützing) Medlin	GOM EXIG	–	0.1	–	–	–	–	–	–	–	–	7–12
Hantzschia marina (Donkin) Grunow	HAN MARI	–	–	–	–	0.1	–	–	–	–	–	>30
Navicula abscondita Hustedt	NAV ABSC	0.3	0.4	0.7	0.5	0.7	0.3	1.1	0.5	0.7	0.1	13–19
Navicula amphipleuroides Hustedt	NAV AMPH	0.1	0.1	–	0.1	0.1	–	0.1	–	0.1	0.3	7–12
Navicula arenaria Donkin	NAV AREN	0.1	–	–	0.3	–	–	0.2	0.1	0.1	0.2	>30
Navicula atlantica (A. Schmidt) H. *et* M. Peragallo	NAV ATLA	–	–	–	0.1	–	0.1	–	0.1	–	–	>30
Navicula cf. *biskanteri* Hustedt	NAV BISK	2.5	2.3	3.1	0.8	1.8	2.3	2.5	1.9	0.4	2.4	7–12
Navicula cincta (Ehrenberg) Ralfs	NAV CINC	–	–	0.1	–	0.1	–	–	–	–	–	13–19
Navicula clamans Hustedt	NAV CLAM	0.2	0.1	–	–	0.3	0.3	–	0.1	–	0.1	13–19
Navicula cryptolyra Brockmann	NAV CRYP	14.7	12.1	9.5	6.9	11.6	13.4	8.1	7.5	5.8	7.4	7–12
Navicula digitoradiata (Gregory) Ralfs	NAV DIGI	–	–	–	0.1	–	–	–	–	–	–	>30
Navicula cf. *flanatica* Grunow	NAV FLAN	–	–	–	–	–	–	–	0.1	–	–	>30
Navicula forcipata Greville	NAV FORC	0.5	0.1	0.3	0.1	0.7	0.4	0.3	0.2	0.4	0.3	20–30
Navicula gregaria Donkin	NAV GREG	0.6	0.7	0.5	0.5	1.5	1.0	1.1	0.3	0.3	0.9	20–30
Navicula halophila (Grunow) Cleve	NAV HALO	–	–	–	0.5	–	–	–	–	–	–	13–19
Navicula heterovalvata Simonsen	NAV HETE	–	–	–	0.2	–	–	–	–	–	–	13–19
Navicula humerosa de Brébisson	NAV HUME	0.1	–	–	–	–	–	–	–	0.1	–	>30
Navicula cf. *jaernefeltii* Hustedt	NAV JAER	0.1	0.1	–	–	–	0.1	0.1	0.3	–	–	13–19
Navicula lanceolata (C. A. Agardh) Ehrenberg	NAV LANC	–	0.1	–	–	0.1	–	–	–	–	0.1	>30
Navicula cf. *perminuta* Grunow	NAV PERM	1.7	2.2	2.9	3.1	1.7	1.2	0.4	1.3	1.7	1.7	7–12
Navicula phyllepta Kützing	NAV PHYL	0.1	1.5	0.6	0.3	0.4	0.4	0.5	0.1	0.3	0.4	7–12
Navicula pygmaea Kützing	NAV PYGM	–	–	–	–	0.4	–	–	–	0.3	0.1	>30
Navicula rhynchocephala Kützing	NAV RHYN	–	–	–	–	–	–	–	–	0.1	–	>30
Navicula salinarum Grunow	NAV SANA	–	–	0.1	–	0.1	–	0.3	–	–	0.3	13–19
Navicula salinicola Hustedt	NAV SANI	2.3	1.1	1.2	0.7	0.9	1.2	0.9	0.5	1.1	1.4	7–12
Navicula subapiculata (Grunow) Hustedt	NAV SUBA	0.1	0.1	0.1	0.3	–	0.3	0.1	0.1	0.1	0.3	20–30
Navicula subinflata Grunow	NAV SUBI	0.3	0.5	0.4	0.2	0.1	0.6	0.6	0.3	0.2	0.3	>30
Navicula wuestii Simonsen	NAV WUES	–	–	0.2	–	–	–	0.3	0.1	–	0.1	7–12
Navicula sp. 1	NAV SP 1	12.1	9.1	8.3	7.9	5.9	8.6	8.9	8.3	7.0	6.9	< 7
Navicula sp. 2	NAV SP 2	–	–	–	–	0.1	–	–	–	–	–	13–19
Navicula sp. 3	NAV SP 3	–	–	–	0.1	–	–	–	–	–	–	7–12
Nitzschia aurariae Cholnoky	NIT AURA	1.9	3.9	6.6	11.1	10.6	1.4	6.7	13.9	20.3	21.2	N 7–12
Nitzschia cf. *aurariae* Cholnoky	NIT CFAU	3.0	4.1	6.9	4.6	7.6	3.5	5.2	9.9	8.3	7.4	N 7–12
Nitzschia dissipata (Kützing) Grunow	NIT DISS	0.4	0.5	0.3	0.3	0.5	0.2	0.3	0.1	–	0.4	N 20–30
Nitzschia frustulum (Kützing) Grunow	NIT FRUS	–	–	0.7	–	–	–	–	–	–	–	N 7–12
Nitzschia cf. *linearis* (C. A. Agardh) W. Smith	NIT LINE	–	0.1	–	–	–	–	–	–	–	–	N >30
Nitzschia littorea Grunow	NIT LITT	0.1	–	–	–	0.1	–	0.1	–	–	–	N >30
Nitzschia paleacea (Grunow) Grunow	NIT PALE	0.1	1.0	0.7	3.4	1.1	0.5	1.1	0.6	0.3	–	N 20–30
Nitzschia cf. *spathula* de Brébisson	NIT SPAT	–	0.1	–	0.1	0.1	0.1	–	–	–	–	N >30
Nitzschia thermaloides Hustedt	NIT THER	–	–	0.1	–	–	–	–	0.1	0.1	–	N >30
Opephora olsenii M. Møller	OPE OLSE	0.1	0.1	–	–	–	0.1	0.4	–	–	–	13–19
Opephora sp. 1	OPE SP 1	–	–	–	–	–	0.1	0.1	–	–	–	13–19
Opephora sp. 2	OPE SP 2	0.1	0.1	–	–	–	–	0.1	–	0.1	–	7–12
Pinnularia ambigua Cleve	PIN AMBI	0.1	0.1	0.5	0.1	0.4	0.1	0.1	–	–	–	>30
Pinnularia rectangulata (Gregory) Rabenhorst	PIN RECT	0.1	0.3	–	–	0.3	0.1	–	–	0.1	0.1	>30
Plagiogramma staurophorum (Gregory) Heiberg	PLA STAU	–	–	–	–	0.1	0.1	0.5	–	0.1	0.4	20–30
Pseudostaurosira sp.	PSE SP	2.1	2.4	0.7	0.9	1.1	1.5	1.5	0.3	1.1	1.2	7–12
Rhoicosphenia abbreviata (C. A. Agardh) Lange-Bertalot	RHO ABBR	–	–	–	–	0.1	–	–	–	–	–	>30
Stauroneis salina W. Smith	STA SALI	–	0.3	0.1	0.3	–	0.1	–	0.1	0.5	0.2	>30
Stauroneis tackei (Hustedt) Krammer *et* Lange-Bertalot	STA TACK	–	–	0.3	0.5	–	–	–	–	–	0.1	13–19
Striatella delicatula (Kützing) Grunow	STR DELI	0.1	–	–	0.5	–	–	–	0.1	–	–	13–19
Tropidoneis lepidoptera (Gregory) Cleve	TRO LEPI	–	0.1	0.1	–	0.1	–	–	–	–	–	>30

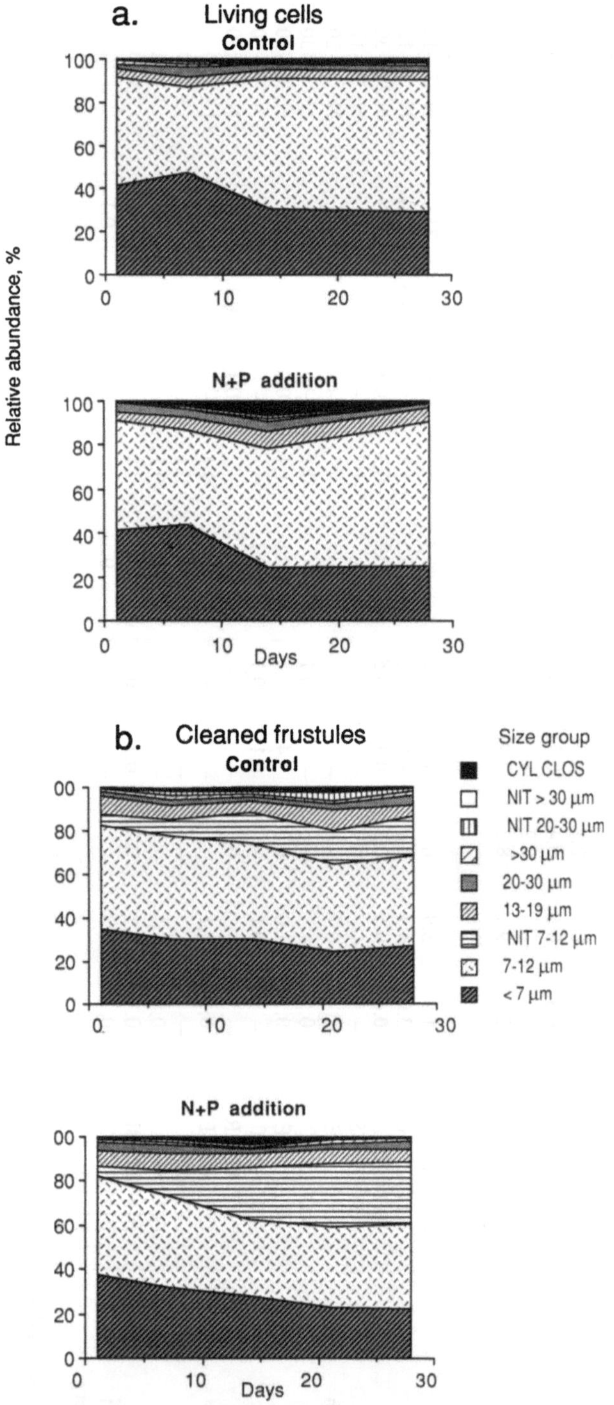

Fig. 3. Relative abundance of size groups of living cells (a) and frustules (b) of diatoms in the top 5 mm of sediment. All diatoms are pennate forms; the genera *Nitzschia* (NIT) and *Cylindrotheca* (CYL) are shown separately. Curves show mean values of 4 replicate containers for living diatoms and 3 replicate containers for frustules.

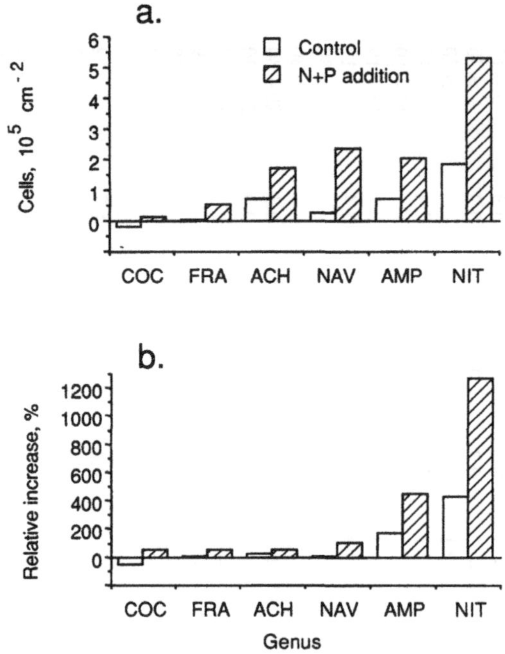

Fig. 4. Absolute (a) and relative increase (b) of the 6 principal diatom genera in the top 5 mm of sediment during the 4-week experiment (differences between first and last sampling date). Absolute cell number = relative abundance of frustules * total number of living cells of a genus. All values are means of 3 replicate containers. COC = *Cocconeis*, FRA = Fragilariaceae; ACH = *Achnanthes*, NAV = *Navicula*, AMP = *Amphora*; NIT = *Nitzschia*.

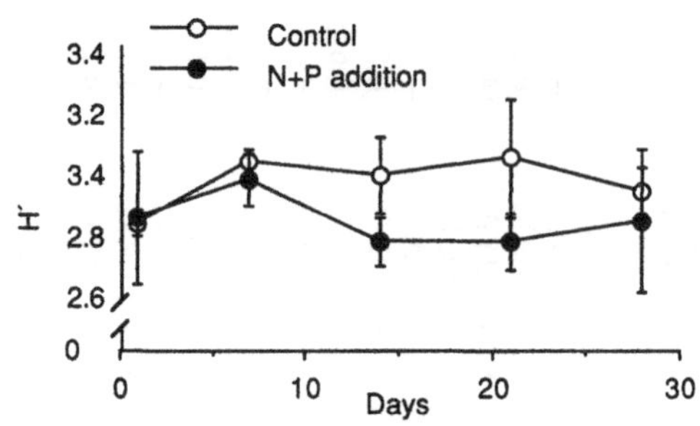

Fig. 5. Shannon-Weaver diversity (H′) based on relative abundance of diatom frustules in the top 5 mm of sediment. Curves show means of 3 replicate treatments ± SD.

Diversity and dominance

The diatom flora consisted mainly of small-sized epipsammic and epipelic species; 86% of the cells counted were less than 12 µm long. The dominating taxa (mean RA > 5%) were *Achnanthes* cf. *delicatissima* (mean RA 15%), *Navicula cryptolyra* (mean RA 10%), *Nitzschia aurariae* (mean RA 10%), *Navicula* sp. 1 (mean RA 8%), *Achnanthes* cf. *delicatula* (mean RA 8%), *Amphora* sp. 1 (mean RA 6%) and *Nitzschia* cf. *aurariae* (mean RA 6%) (Table I).

Shannon-Weaver diversity (H′) varied between 2.6 and 3.3 and depended about equally on species richness, ($r = 0.83$, $p < 0.001$) and evenness ($r = 0.77$, $p < 0.001$). After one week the H′-values were generally lower for the nutrient enriched containers, but this trend was not statistically significant (NK, $p > 0.05$) (Fig. 5).

CCA of cleaned diatom material

The data set yielded one dominant CCA axis (Fig. 6) with eigenvalue 0.10 (compared with 0.03, 0.02, and 0.02 for axes 2, 3 and 4, respectively). Nitrate + nitrite and phosphate concentrations, number of days and the Shannon-Weaver index gave significant ($p < 0.05$) coefficients in multiple regression analysis of ordination axis 1 on the environmental variables, whereas Si and NH_4 concentrations were not significant on any of the axes.

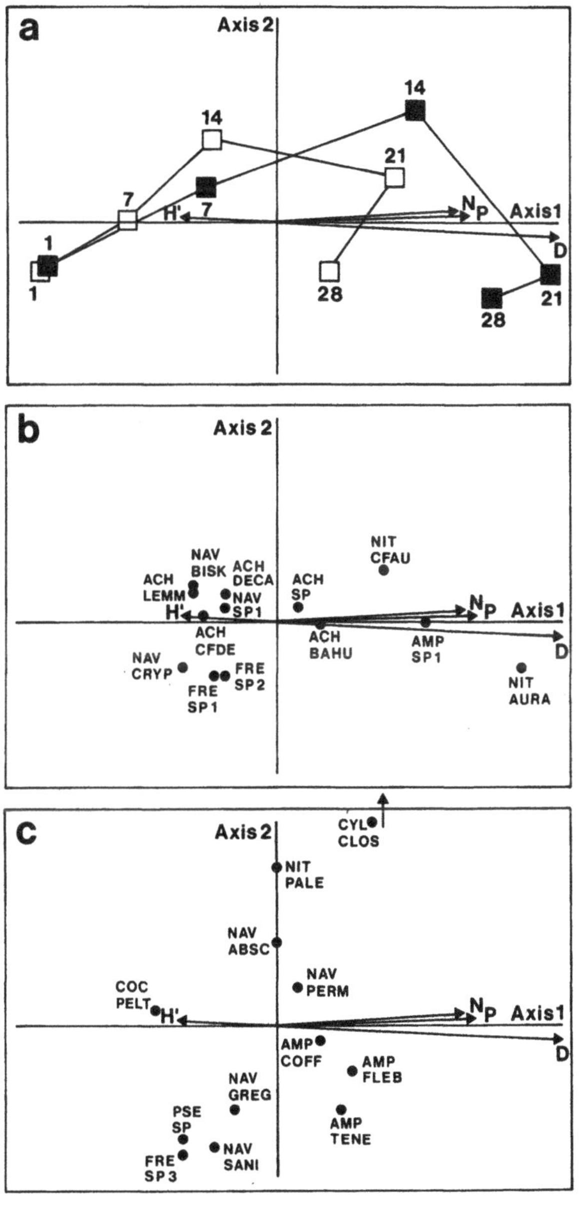

Fig. 6. CCA based on relative abundance (RA) of 101 diatom taxa with respect to number of days and treatment. (a) Ordination showing controids for day-treatment combinations (white squares = control, black squares = N + P addition). The arrows show the direction of variation in time (D for days), $NO_3 + NO_2$ concentration (N), PO_4 concentration (P) and Shannon-Weaver diversity (H′). (b) Taxon ordination for the 13 taxa with mean RA > 2.0%. (c) taxon ordination for the 12 taxa with mean RA 0.5–2.0%. For abbreviation of taxon names see Table I.

The squares in Figure 6a are centroids of samples taken from 3 replicate containers. They depict the changes in species composition with time in the control and the enriched containers. For Day 1, the centroids for both controls and enriched containers are very close to each other in the left part of the ordination. After nutrient addition, succession was faster in the enriched containers than in the controls, with the direction of the succession being from left to right along the main axis for both (Axis 1). In the last week of the experiment (between Days 21 and 28) succession was reversed for both control and enriched containers.

The vectors (shown by arrows) for environmental variables and community diversity (Fig. 6) are related to the correlations between each variable and Axis 1; they point in the direction of maximum increase of the variables. Those for the added nutrients (N and P) and number of days (D) all point to the right of the ordination in the direction of the succession, whereas the vector for diversity (H′) points in the opposite direction.

The taxon scores (Fig. 6b–c) summarize the distribution of the principal diatom taxa with respect to time and nutrient addition. Species succession during the experiment is shown from left to right. The taxa that were most favoured by nutrient addition, notably *Nitzschia aurariae* (Fig. 6b), are located in the far-right end of the ordination. No taxa with mean RA > 0.5% are found in the far-left part, thus indicating that no abundant diatom species were negatively affected by nutrient addition.

A separate CCA on the genus level (but including Fragilariaceae in one group) also produced one main axis of variation (Axis 1, eigenvalue = 0.07) (Fig. 7).

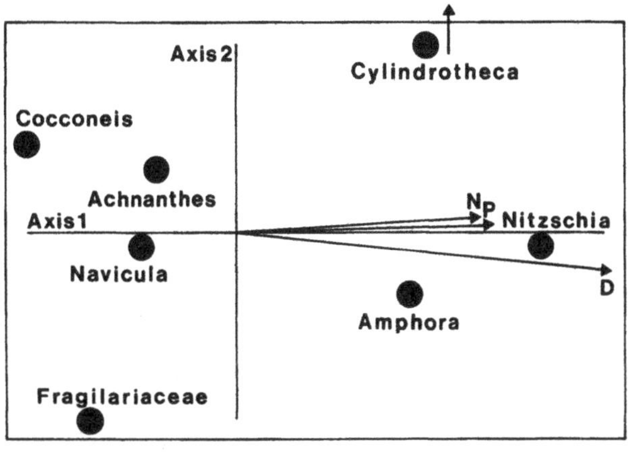

Fig. 7. Taxon ordination from CCA with respect to number of days and treatment. The scores of the seven dominant genera are indicated. The arrows show the direction of variation in time (D), $NO_3 + NO_2$ concentration (N) and PO_4 concentration (P).

Subsequent axes were of very little importance (eigenvalues < 0.01). The concentration of added nutrients (N and P) and number of days were significant in the ordination whereas Si and NH_4 concentrations were not. The pattern of the centroids (not shown) was similar to Fig. 6a. The taxon scores for the 7 dominating generic groups are shown: *Achnanthes* (mean RA 33%), *Navicula* (26%), *Nitzschia* (17%), *Amphora* (10%), Fragilariaceae including *Fragilaria, Pseudostaurosira* and *Staurosira* species (9%), *Cocconeis* (2%) and *Cylindrotheca* (1%). The successional pattern shows that the genera *Amphora, Nitzschia* and *Cylindrotheca* were favoured both with regard to time and P + N enrichment.

Notes on taxonomy

Most of the dominant diatom taxa encountered in this study were difficult to identify to species level even at a magnification of x1000 under oil immersion because of their small size and lack of visible structure on the valves. These are probably also the reasons why there are no or few references in the taxonomical literature about these species. Therefore we found it necessary to include some notes on taxonomy and scanning electron micrographs of the dominant taxa.

Achnanthes cf. *delicatula* (Kützing) Grunow (Fig. 8a–c)
Cells are broadly elliptical, sometimes with subrostrate ends, 6–10 × 4–5 µm, and with 10–14 striae in 10 µm on both valves. Sometimes on the rapheless valve there is a larger distance between the middle striae on one side of the valve (Fig. 8b–c). The hyaline area on the rapheless valve can be wider or narrower. The specimen in Figure 8c is similar to *Achnanthes* sp. of Kuylenstierna (1989; plate 35, fig. 349). This taxon is easy to distinguish from *A. delicatula* (Kützing) Grunow *sensu* Lange-Bertalot and Ruppel (1980). It may also be possible that several species are involved.

Achnanthes cf. *delicatissima* Simonsen (Fig. 8d)
Cells are elliptical with subrostrate ends, 6–8 × 3–4 µm, and no striae are visible in the light microscope (LM). Using SEM, the species was found to have 33–36 striae in 10 µm consisting of double rows of pores on the rapheless valve. This taxon appears to be very hyaline and is difficult to study not only by LM but also by SEM. The transmission electron micrograph in plate 36, figs 352–353 of Kuylenstierna (1989), shows an unidentified *Achnanthes* species which most likely is the same taxon and may be conspecific with *A. delicatissima* Simonsen.

Achnanthes sp.
Cells are narrowly elliptical, 6–9 × 3 µm, and with 16–18 striae in 10 µm on both valves. On the rapheless valve there is a larger distance between the middle striae on one side of the valve.

Amphora sp. 1 (Fig. 8e–g)
Valves are 8–15 × 2–3 µm, with no striae visible when observed by LM. Using SEM, the species was found to have 28–30 dorsal striae in 10 µm (double rows of pores), and *ca* 50 ventral striae in 10 µm. It resembles *Amphora coffeaeformis* v. *borealis* (Kützing) Cleve *sensu* Bérard-Therriault *et al.* (1986) in SEM but not when studied by LM. The number of striae is also quite different.

Cocconeis peltoides Hustedt (Fig. 9a–c)
Cells are broadly elliptical, 11–14 × 7–9 µm, with 15–18 striae in 10 µm on the rapheless valve, but in the LM no striae were visible on the raphe valve. Using SEM, the species was found to have ca. 38 striae in 10 µm on the raphe valve. This species is easy to recognize in the LM by the typical features of the rapheless valve. The raphe valve as described by Hustedt (1939) is incorrect as recognized by Simonsen (1987) from LM observations.

Fragilariaceae sp. 1 (Fig. 8h)
Cells are roundish, 2–3 µm in diameter and have *ca.* 22–24 striae in 10 µm. This species is identical with *Fragilaria* sp. A of Kuylenstierna (1989; plate 12, figs 125–126).

Navicula cryptolyra Brockmann (Fig. 9d–e)
Cells are broadly elliptical, 8–13 × 5–6 µm, and have 24–26 striae in 10 µm. The lateral areas are not visible in all frustules in the LM and this sometimes makes identification difficult as previously discussed by Håkansson and Stabell (1977).

Navicula cf. *perminuta* Grunow (Fig. 9f)
Cells are lanceolate with rounded ends, 8–12 × 3–3.5 µm, and have 18–20 striae in 10 µm. This taxon differs from the typical *N. perminuta* by the external raphe endings which in our specimens were sharply bent, while in *N. perminuta* their shape is more rounded (*cf.* plate 61, figs 767–771 in Kuylenstierna 1989).

Navicula sp. 1 (Fig. 9g–h)
Cells are broadly elliptical, 4–7 × 3–3.5 µm, with no striae which are visible in the LM, but the central nodule, the terminal nodules and the raphe canal are clearly distinct. Scanning electron micrographs show a large hyaline area and short marginal striae (*ca* 36 in 10 µm). In the LM the taxon resembles *Navicula nolens* Simonsen.

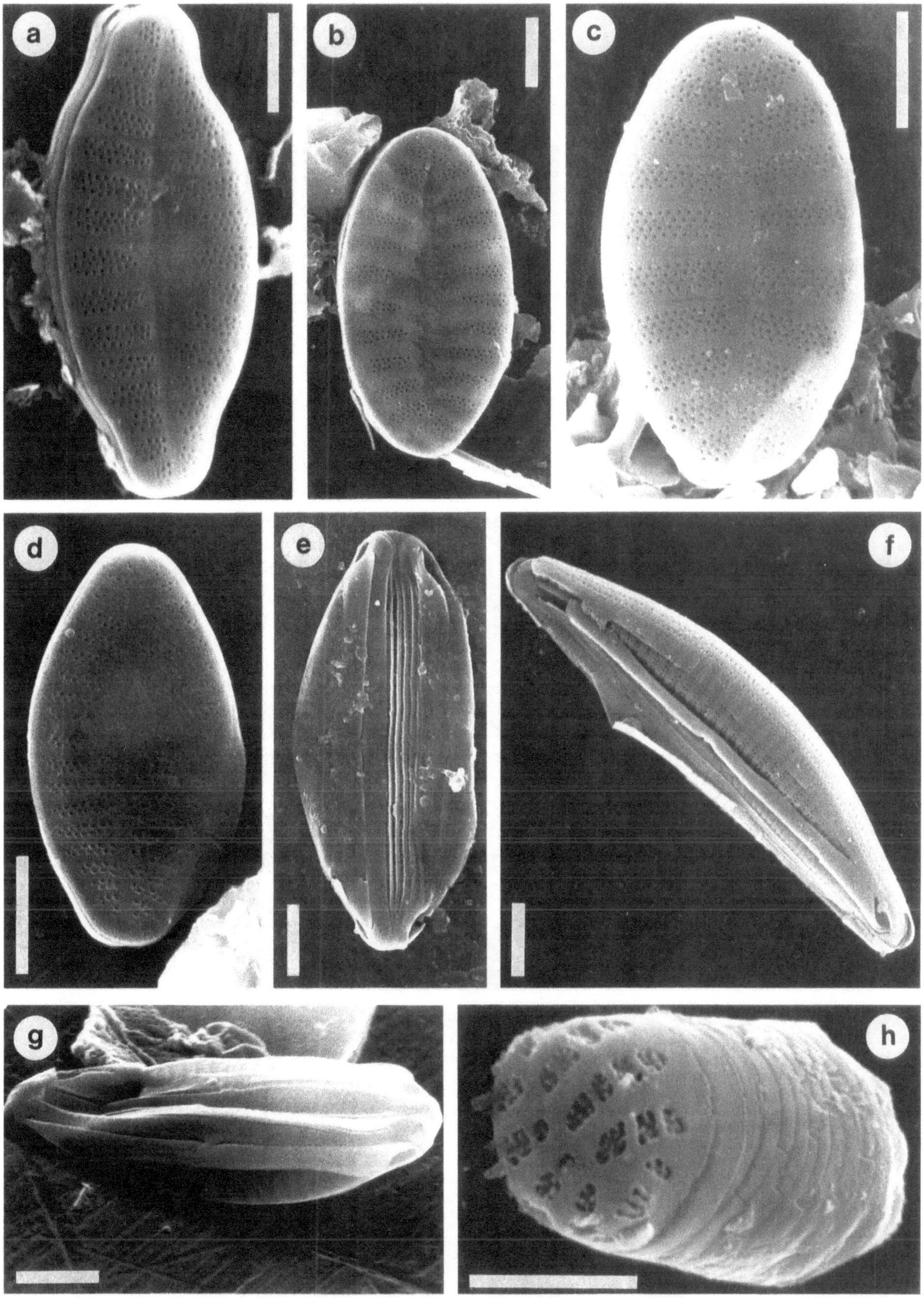

Fig. 8. Scanning electron micrographs of some dominant diatom taxa. Scale bar = 2 μm. a–c: *Achnanthes* cf. *delicatula*. d: *Achnanthes* cf. *delicatissima*. e–g: *Amphora* sp. 1. h: Fragilariaceae sp. 1.

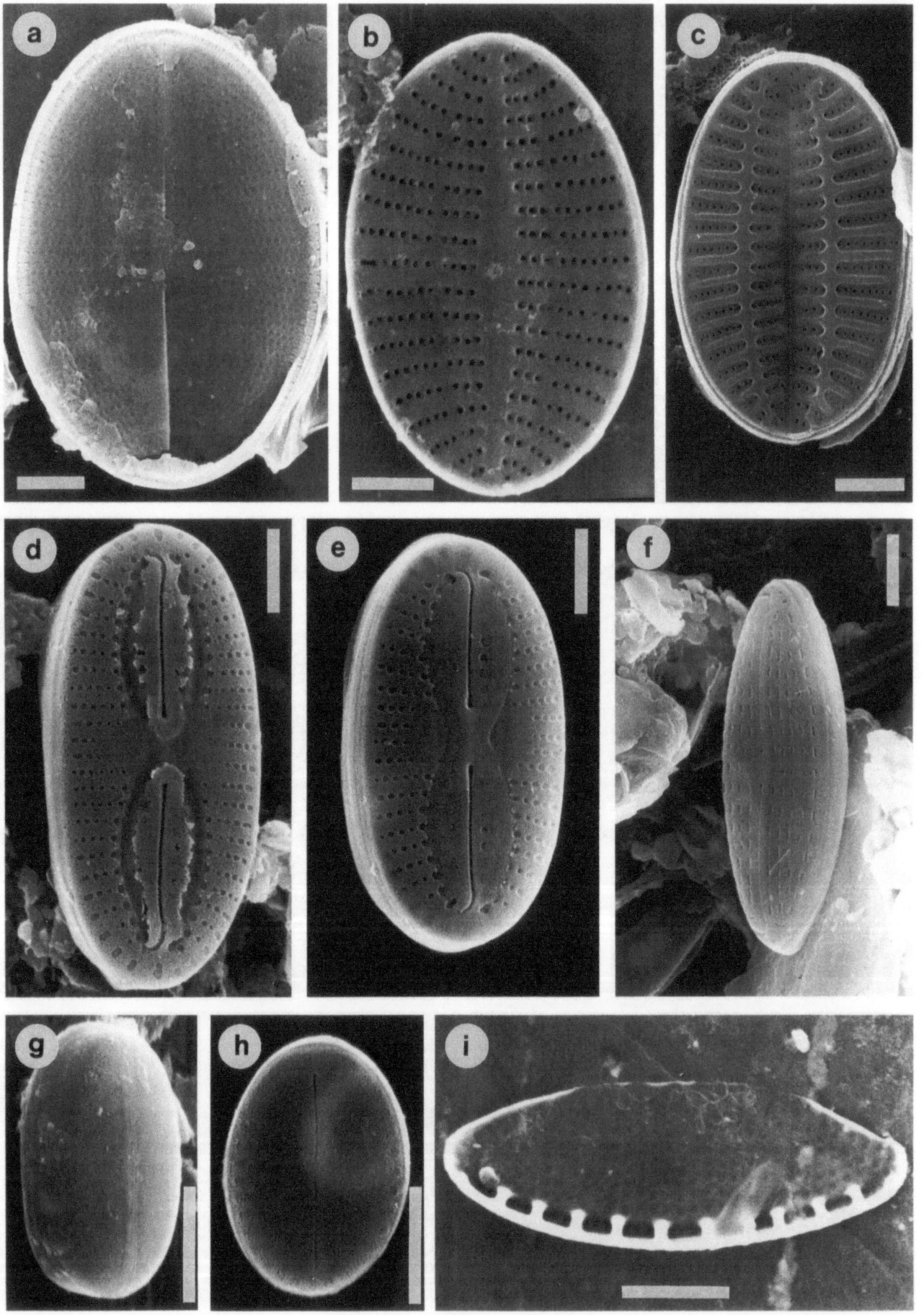

Fig. 9. Scanning electron micrographs of some dominant diatom taxa. Scale bar = 2 μm. a–c: *Cocconeis* peltoides. d–e: *Navicula cryptolyra* (somewhat eroded in d). f: *Navicula* cf. *perminuta*. g–h: *Navicula* sp. 1. i: *Nitzschia aurariae*.

Nitzschia aurariae Cholnoky (Fig. 9i)
Cells are linear-elliptical, 6–11 × 3–4 μm, with 14–18 fibulae and *ca* 45 striae in 10 μm. According to Krammer and Lange-Bertalot (1988), the apices of *N. aurariae* are never rostrate. In part of our frustules there was a tendency for subrostrate ends, and these are called *Nitzschia* cf. *aurariae*.

Discussion

Major algal groups

Nutrient addition to the water column resulted in enhanced microphytobenthic growth. This showed that the microalgae in the sediment were nutrient limited in the non-enriched containers and that nutrients were assimilated from the overlying water. This was also reflected as raised values of primary productivity and chlorophyll *a* content of the sediment (Nilsson *et al.* 1991). The microalgae were also capable, at least initially, to compete with fast-growing filamentous macroalgae for the nutrient pool of the overlying water (the phytoplankton was excluded from the containers by filtration).

One of the most obvious effects of the N + P-enrichment was the stimulated growth of some filamentous algae (the macroscopic green alga *Enteromorpha clathrata* and cyanobacteria). This agrees with previous observations that a proceeding eutrophication will favour opportunistic algal species such as fast growing filamentous macroalgae (Wallentinus 1984, Rosenberg *et al.* 1990). Carrick *et al.* (1988), who carried out experiments on periphyton in Lake Michigan using nutrient releasing substrata, found that N + P enrichment favoured algae other than diatoms, *e.g.* cyanobacteria and green algae, and they categorized these into a 'N + P guild'. It was typical for these filamentous taxa to modify physical aspects of the micro-environment (shading), a situation which was also likely to occur in our enriched containers. In addition, the chemical environment was probably influenced; Lee *et al.* (1975) observed higher concentrations of dissolved organic nitrogen in *Enteromorpha intestinalis* (L.) Link. communities and hypothetized that microscopic epiphytes with heterotrophic properties could be stimulated. Estuarine benthic diatoms have been shown to rapidly assimilate compounds such as amino acids (Admiraal *et al.* 1987).

The nutrient treatment in our experiment involved both a considerable increase of nutrient supply and a changed supply ratio. The concentration of IN increased on an average by a factor of 130 and IP by a factor of 40. No Si was supplied. Between 20 and 50% of this nutrient input dissappeared from the overlying water during the 4-week experiment (Nilsson *et al.* 1991). The nutrient uptake was a sum of algal assimilation, bacterial activity and adsorption processes in the sediment. According to the resource competition theory (Tilman *et al.* 1982), different relative supply rates of two or more limiting resources can be expected to result in (phytoplankton) communities dominated by different species or taxonomic groups. In our study, the N + P addition changed the mean Si : N : P mole ratio of the originally nutrient-poor surface water from 112 : 14 : 1 to 2.7 : 39 : 1. When the sediment microflora is considered, the changed Si : N : P ratio did not, despite the stimulated growth of cyanobacteria, lead to a dramatic shift in the dominance of the major taxonomic groups.

Carrick *et al.* (1988) interpreted that the shift in the algal community in Lake Michigan occurred because the N + P enrichment induced a Si-limitation of diatoms. Although Si was not added in our experiment, diatoms significantly increased their biomass. The optimal Si : N ratio of the medium for marine diatoms is considered to be approximately 1 : 1. The facts that the Si : N ratio in the overlying water of the enriched containers was usually < 0.1, and that the measured uptake from the water could cover between 20 and 90% of the theoretical Si-requirements for diatoms (Nilsson *et al.* 1991), suggest an additional Si-source, which is most probably the sediment itself. On the other hand, despite the generally higher concentrations of Si in sediments than in the overlying water, Si-limitation of sediment-inhabiting diatoms has been documented (Kelderman *et al.* 1988). Ultimately, Si may become a limiting factor in the presence of prolonged raised levels of N and P along the Swedish west coast.

From resource-limited competition experiments using natural phytoplankton communities comprising green algae, diatoms and cyanobacteria, it has been suggested that the nutrient ratios at which dominance shifts from one taxonomic group to another, are temperature dependent (Tilman *et al.* 1986). Our experiment was conducted in late summer, a season favourable for the growth of cyanobacteria in stagnant and shallow brackish-water benthic habitats (Snoeijs and Prentice 1989). Although the growth of cyanobacteria was significantly promoted by the N + P treatment, they did not outcompete the diatoms during the course of our 4-week experiment. One reason may be that the water temperature (16–19 °C) was not high enough to induce the shift. In a previous laboratory experiment with sublittoral marine sediment conducted at 14.5 °C, N + P fertilization also induced the growth of filamentous cyanobacteria, however, diatoms took over again at the end of the experiment (Sundbäck and Granéli 1988). Filamentous cyanobacteria are usually favoured by N-limitation because of their ability to fix N_2, which has also been found to occur in non-heterocystous species (Jones 1990). The high N/P ratio in the enriched containers would yield secondary P-limitation, and this may also have prevented the shift to a system dominated by cyanobacteria.

Although autotrophic flagellates were initially a minor constituent of the microflora, there was a trend towards increased biomass and this is in accordance with the hypothesis that flagellates become more important as eutrophication continues (Doering *et al.* 1989). Our analysis of the two groups labelled 'flagellates' is incomplete, since they include several taxonomic groups, such as green algae, dinoflagellates, chrysophytes, and probably also prymnesiophytes. Flagellates, both autotrophic and heterotrophic, constitute a rather overlooked functional group in marine sediments (Larsen 1985, Bak and Niewland 1989).

Measurable effects of the N + P treatment on the benthic microflora were observed after 2 weeks. A lag-time of a few weeks has also been observed in other experiments on sediment communities (Levinton 1985, Sundbäck *et al.* 1990). Thus our 4-week experiment probably only reflected the initial phase of the changes in the sediment system, which reacted rather slowly to environmental changes compared with both pelagial systems and non-sediment benthic systems.

The diatom community

Diversity. Shannon-Weaver diversity was rather low (2.6 – 3.3) compared with previously reported values for diatom communities of sandy substrates (3.2 – 4.6; Amspoker 1977, Sundbäck 1984) as well as for edaphic diatom communities (*ca* 4 – 5; Sullivan 1976, Cook and Whipple 1982). The sieving of the sediment may have decreased the initial diversity of the species pool. Algal diversity is expected to be highest at an intermediate level of disturbance, for example mechanical disturbance by grazing (Swamikannu and Hoagland 1989) or exposure to water movement (Sundbäck 1984). Thus, an additional reason for the rather low diversity may be the enclosure of the sediment in the experimental vessels. This enclosure would not only decrease the exposure to water movement but also exclude the immigration of macrofaunal grazers, as well as immigration of diatom species. Macrofaunal grazers were initially excluded from the containers and meiofaunal grazing pressure on autotrophic microalgae was rather low (Nilsson *et al.* 1991).

Diversity and evenness of the diatom community was only slightly negatively affected by the nutrient addition. Nutrient increase has been found to decrease diversity of benthic microalgal communities in fresh water (Carrick *et al.* 1988 and references therein), on salt marshes (Sullivan 1976), and estuarine diatom environments (Admiraal *et al.* 1989). On the other hand, Sullivan (1981) found that nitrogen enrichment could either decrease or increase H′ of edaphic diatom communities and Dam (1982) concluded that diversity indices appear to have no consistent relationship with the degree of water pollution.

Canonical correspondence analysis (CCA). The CCA showed that the succession in both non-enriched and enriched containers was in the same direction, though faster and proceeding further with N + P-addition. Reverse succession at the end of the experiment may be attributed to the loss of larger cells by oxygen bubble transportation. Certain diatom taxa were stimulated more by the nutrient addition than others. However, the CCA did not reveal any fundamental change in community composition that could be attributed to nutrient addition, and the question arises whether there are marine benthic diatom communities typical of eutrophic conditions. More information exists on the response to pollution of benthic microalgae from freshwater than from marine environments (*cf.* Wolf 1982). Using diatoms as freshwater indicator organisms in the 'classic' sense (*e.g.* Sládecek 1973) have been proven to be highly unreliable, whereas measures based on quantitative diatom species composition have been shown to be a useful tool in freshwater quality estimation (*e.g.* Lange Bertalot 1979). Kuhn *et al.* (1981) applied life-form strategies to interpret pollution effects on freshwater periphyton. Sullivan (1976, 1981) conducted *in situ* experiments on the response of edaphic diatom communities of salt marshes to fertilization by N and P. He observed compositional changes in the community, but did not succeed in grouping taxa with similar response patterns with the aid of multivariate methods, partly because the effect was shaded by light and temperature effects. Admiraal *et al.* (1984) and Jong and Admiraal (1984) made competition experiments with both isolated species and natural communities of estuarine diatoms. These workers found that selection mechanisms, such as inorganic carbon-limitation and allelopathy, are more important than inorganic nutrients on the nutrient-rich tidal flat.

In our experiment, the diatom taxa most favoured by the N + P addition were *Nitzschia* and *Amphora* species within the size range 7 – 12 μm. Despite their small size these species were important in terms of biomass. Of the 5 *Nitzschia* species listed by Wolf (1982) with a given code for nutrient content spectrum, 4 are categorized as eutrophic and 1 mesoeutrophic. Also others have reported that representatives of the genus *Nitzschia* increase because of fertilization (Sullivan 1981). Sundbäck (1984) found that the *N.* cf. *pusilla* Lange-Bertalot together with a small unidentified *Amphora* species increased as detritus accumulated in a sandy bay. The stimulation of *Amphora* sp. 1, *A. flebilis* and *A. tenerrima* by N + P-enrichment, agrees with the experience that small-sized *Amphora* species tend to rapidly colonize sand grains when trying to isolate and cultivate epipsammic species (Tufail 1987, Sundbäck unpublished). In the middle of the experiment the comparably large-sized *Cylindrotheca closterium* also increased significantly in the enriched containers. According to Admiraal (1977), *Nitzschia* (= *Cylindrotheca*) *closterium* tolerates high concentrations of inorganic N and P and laboratory experiments have shown that its capability to outcompete other tidal-flat diatom species increases with higher salinity and temperature (Jong and Admiraal 1984).

Size, morphology and motility of diatoms. The capability of macroscopic algae to compete for nutrients increases as the surface to volume ratio (S/V) of the species increases (Wallentinus 1984). This has also been shown for phytoplankton (for review see *e.g.* Fogg and Thake 1987). In our experiment the majority of the most abundant taxa were small-sized naviculoid forms, characterized as flat epipsammic diatoms occurring in cavities of sand grains (Sundbäck 1984). These species generally have a high S/V ratio and are capable of fast absorption of nutrients compared with large naviculoids and sigmoid cells (Baille 1987, Hudon and Legendre 1987). Carrick *et al.* (1988) coupled morphology and above all mobility of a diatom species to a certain nutritional guild. In our sandy sediment adnate and less motile epipsammic forms, such as *Achnanthes* species, were common initially. For the two diatom taxa most stimulated by the N + P enrichment, that is the two *Nitzschia* taxa, the morphology implies a lower S/V ratio than *e.g.* *Achnanthes* species, as well as the capability to move faster. Motility was likely to be an advantage when competing for light and nutrients, in particular as the microenvironment in the enriched containers was probably altered by the growth of *Enteromorpha* (*cf.* Hudon and Legendre 1987). The larger-sized, highly motile *Cylindrotheca closterium* cells benefitted from the N + P enrichment. On the other hand, these features facilitate rapid colonization and Hudon and Legendre (1987) suggest that this opportunistic strategy could present a positive trade off against their physical vulnerability.

When considering living cells and biomass, the fact that larger motile diatoms became initially more important in the enriched containers may also be related to the changed environment (changed water movement, shading) induced by the growth of filamentous green algae in combination with the decreased wave impact as a result of the experimental set-up. In addition, colonization space might have limited the growth of adnate epipsammic forms. This conclusion is supported by the significantly decreased chlorophyll *a* values and primary productivity on Day 28 (Nilsson *et al.* 1991). In particular cells of tychoplanktic species such as *Cylindrotheca closterium*, may be transported up from the sediment surface. In addition, the conditions of high oxygen content, high pH and inorganic C-limitation that are created when photosynthetic activity is high, may also function as a selective pressure on the microalgal community (Jong and Admiraal 1984).

Primary producers and higher trophic levels

The recorded changes in the biomass and composition of sediment-associated primary producers following N + P addition were coupled to changes both in functional variables (primary productivity, grazing pressure and nutrient flux) and structural variables, such as the composition of meiofauna. The meiofaunal biomass of, in particular, oligochaetes and copepods, increased significantly in enriched containers (Nilsson *et al.* 1991). As the autochthonous benthic primary production was the only possible source of increase of the organic carbon pool, the biomass increase of consumers must have been directly or indirectly (via bacteria and produced detritus) related to the stimulated algal growth. Since meiofauna only removed a small proportion of the microalgal biomass (*ca* 4–10% of the biomass per day; Nilsson *et al.* 1991), the total amount of microautotrophic biomass did not appear to limit meiofaunal growth. This suggests that the response of meiofauna may rather be connected to a qualitative change of the food sources rather than to the quantity of food. We do not presently know if and how this change is related to the compositional change of the primary producers. More information is needed on both the qualitative value of the food sources of benthic herbivores in sandy sediment (Carman and Thistle 1985), as well as on interactions between microscopic primary producers and bacteria (Hall and Fisher 1985).

Conclusions

Outdoor flow-through systems are useful tools for studying benthic microalgal communities both as a structural and functional component of shallow-water sediment systems. They mimic relatively well natural communities and enable replication and adequate control treatments to test and reveal patterns of change on the lower trophic levels in connection with environmental changes.

Increased concentration of IP and IN in the water column stimulated the growth of sediment-living microalgae. In particular, filamentous cyanobacteria and diatoms, as well as flagellates, were favoured. Although the N + P addition considerably lowered the Si/N ratio, diatoms were not outcompeted by any other microalgal group, probably because the Si-pool of the sediment prevented, at least initially, Si-limitation.

Despite the statistically significant compositional changes, the microflora was not subjected to any dramatic shift in the dominance of major taxonomic groups nor diatom species, but rather the major change occurred on the macroscopic level in the form of a rapid growth of the green alga *Enteromorpha clathrata*. Thus, in the long run enhanced growth of filamentous species can be expected to fundamentally change the character of bare sandy habitats and further contribute to the accumulation of drifting macroalgae and detritus in shallow areas.

Counts of both living cells and frustules showed that the diatom flora was dominated by cells within the size range 7–12 μm, however, depending on the method and level of resolution, different aspects of compositional changes were detected. Thus, both methods were necessary for the interpretation of the changes in composition.

The increasing abundance of small *Nitzschia* and *Amphora*-species with N + P enrichment supports previous findings that nutrient enrichment affects the composition of benthic diatom communities.

In stagnant conditions, a highly stimulated photosynthetic activity at the sediment surface may in itself function as a selective pressure on the microalgal community by selectively transporting away easily suspendible species.

Acknowledgements

Tjärnö Marine Biological Laboratory (Strömstad) provided excellent working facilities during the experiment. The Department of Marine Botany (Göteborg), and Department of Ecological Botany (Uppsala) provided facilities for subsequent analyses and data treatment. We thank Mats Kuylenstierna for valuable advice on identification of some diatom taxa and Gary Wife for help with SEM and preparing the photographs for publication. KS thanks her colleagues Benno Jönsson, Inger Lindström Swanberg and Per Nilsson for inspiring co-operation during the experiment. Financial support was provided by The Swedish Natural Science Research Council, The National Swedish Environmental Protection Agency and the Royal Society of Arts and Sciences in Gothenburg.

References

Admiraal, W. 1977. Tolerance of estuarine benthic diatoms to high concentrations of ammonia, nitrite ion, nitrate ion and ortophosphate. *Mar. Biol. 43*: 307–315.

Admiraal, W., H. Peletier and T. Brouwer. 1984. The seasonal succession patterns of diatom species on an intertidal mudflat: an experimental analysis. *Oikos 42*: 30–40.

Admiraal, W., C. Riaux-Gobin and R. W. P-M. Laane. 1987. Interactions of ammonium, nitrate, and D- and L-amino acids in the nitrogen assimilation of two species of estuarine benthic diatoms. *Mar. Ecol. Prog. Ser. 40*: 267–273.

Admiraal, W., E. D. de Ruyter van Steveninck and H. A. M. de Kruijf. 1989. Environmental stress in five ecosystems in the floodplain of the River Rhine. *The Science of the Total Environment 78*: 59–75.

Amspoker, M. C. 1977. The distribution of intertidal epipsammic diatoms on Scripps Beach, La Jolla, California, USA. *Bot. Mar. 20*: 227–232.

Bak, R. P. M. and G. Nieuwland. 1989. Seasonal fluctuations in benthic protozoan populations at different depths in marine sediments. *Neth. J. Sea Res. 24*: 37–44.

Baille, W. P. 1987. Diatom size distribution and community stratification in estuarine intertidal sediments. *Estuar. coast. Shelf Sci. 25*: 193–209.

Berárd-Therriault, L., A. Cardinal and M. Poulin. 1986. Les Diatomées (Bacillariophyceae) benthiques de substrats durs des eaux marines et saumâtres du Québec. 6. Naviculales: Cymbellaceae et Gomphonemaceae. *Naturaliste can. (Rev. Écol. Syst.) 113*: 405–429.

Braak, C. J. F. ter. 1986. Canonical correspondence analysis: a new eigenvector technique for multivariate direct gradient analysis. *Ecology 67*: 1167–1179.

Braak, C. J. F. ter and I. C. Prentice. 1988. A theory of gradient analysis. *Adv. Ecol. Res. 18*: 271–317.

Brockmann, C. 1950. Die Watt-Diatomeen der Schleswig-Holsteinischen Westküste. *Abh. Senckenb. naturf. Ges. 478*: 1–26.

Carman, K. R. and D. Thistle. 1985. Microbial food partioning by three species of benthic copepods. *Mar. Biol. 88*: 143–148.

Carrick, H. J. and R. L. Lowe. 1988. Response of Lake Michigan benthic algae to in situ enrichment with Si, N and P. *Can. J. Fish. Aquat. Sci. 45*: 271–279.

Carrick, H. J., R. L. Lowe and J. T. Rotenberry. 1988. Guilds of benthic algae along nutrient gradients: relationships to algal community diversity. *J. N. Am. Benthol. Soc. 7*: 117–128.

Cook, L. L. and S. A. Whipple. 1982. The distribution of edaphic diatoms along environmental gradients of a Louisiana salt marsh. *J. Phycol. 18*: 64–71.

Dam, H. van. 1982. On the use of measures of structure and diversity in applied diatom ecology. *Nova Hedwigia, Beih. 73*: 87–115.

Doering, P. H., C. A. Oviatt, L. L. Beatty, V. F. Banzon, R. Rice, S. P. Kelly, B. K. Sullivan and J. B. Frithsen. 1989. Structure and function in a model coastal ecosystem: silicon, the benthos and eutrophication. *Mar. Ecol. Prog. Ser. 52*: 287–299.

Edler, L. 1977. Phytoplankton and Primary Production in the Sound. Ph. D. thesis, Department of Marine Botany, University of Göteborg, pp. 82.

Edler, L. (ed.). 1979. Recommendations on methods for marine biological studies in the Baltic Sea. Phytoplankton and chlorophyll. *Baltic Mar. Biol. (BMB) Publ. No. 5*, p. 1–38.

Fogg, G. E. and B. Thake. 1987. *Algal Cultures and Phytoplankton Ecology*. The University of Wisconsin Press, Madison, pp. 269.

Giffen, M. H. 1966. Contributions to the diatom flora of South Africa. III. Diatoms of the marine littoral regions at Kidd's Beach near East London, Cape Province, South Africa. *Nova Hedwigia 13*: 245–293.

Granéli, E. and K. Sundbäck. 1985. The response of planktonic and microbenthic algal assemblages to nutrient enrichment in shallow coastal waters, southwest Sweden. *J. Exp. Mar. Biol. Ecol. 85*: 253–268.

Granéli, E., K. Wallström, U. Larsson, W. Granéli and R. Elmgren. 1990. Nutrient limitation of primary production in the Baltic Sea area. *Ambio 19*: 142–151.

Green, R. H. 1979. *Sampling Design and Statistical Methods for Environmental Biologists*. John Wiley and Sons, New York.

Håkansson, H. and B. Stabell. 1977. Identification of some small species of *Navicula*. *Bot. Not. 130*: 477–481.

Hall, S. L. and F. M. Fisher, Jr. 1985. Annual productivity and extracellular release of dissolved organic compounds by the epibenthic algal community of a brackish marsh. *J. Phycol. 21*: 277–281.

Hartley, B. 1986. A check-list of the freshwater, brackish and marine diatoms of the British Isles and ajoining coastal waters. *J. mar. biol. Ass. U. K. 66*: 531–610.

Hudon, C. and P. Legendre. 1987. The ecological implications of growth forms in epibenthic diatoms. *J. Phycol. 23*: 434–441.

Hustedt, F. 1939. Die Diatomeenflora des Küstengebietes der Nordsee vom Dollart bis zur Elbemündung. I. Die Diatomeenflora in den Sedimenten bei der unteren Ems sowie auf den Watten der Leybucht, des Memmert und bei der Insel Juist. *Abh. Naturwiss. Vereins Bremen 31*: 572–677.

Jones, K. 1990. Aerobic nitrogen fixation by *Lyngbya* sp., a marine tropical cyanobacterium. *Br. phycol. J. 25*: 287–289.

Jong, L. de and W. Admiraal. 1984. Competition between three estuarine benthic diatom species in mixed cultures. *Mar. Ecol. Prog. Ser. 18*: 269–275.

Jönsson, B. H. 1991. A ^{14}C-incubation technique for measuring microphytobenthic primary productivity in intact sediment cores. *Limnol. Oceanogr.*: In press.

Kelderman, P., H. J. Lindeboom and J. Klein. 1988. Light dependent sediment-water exchange of dissolved reactive phosphorus and silicon in a producing microflora mat. *Hydrobiologia 159*: 137–147.

Krammer, K. and H. Lange-Bertalot. 1988. Bacillariophyceae, 2. Teil: Bacillariaceae, Epithemiaceae, Surirellaceae. *Süßwasserflora von Mitteleuropa, Bd 2/2*. pp. 596.

Kuhn, D. L., J. L. Plafkin, J. Cairns Jr. and R. L. Lowe. 1981. Qualitative characterization of aquatic environments using diatom life-form strategies. *Trans. Am. Microsc. Soc. 100*: 165–182.

Kuylenstierna, M. 1989. Benthic Algal Vegetation in the Nordre Älv Estuary (Swedish west coast). Vol. 2. Plates. Dissertation, Dept. of Marine Botany, University of Göteborg, 72 plates.

Lange-Bertalot, H. 1979. Pollution tolerance of diatoms as a criterion for water quality estimation. *Nova Hedwigia, Beih. 64*: 285–304.

Lange-Bertalot, H. and M. Ruppel. 1980. Zur Revision taxonomisch problematischer, ökologisch jedoch wichtiger Sippen der Gattung *Achnanthes* Bory. *Arch. Hydrobiol., Suppl. 60*: 1–31.

Lange-Bertalot, H. and K. Krammer. 1989. *Achnanthes*, eine Monographie der Gattung. *Bibliotheca Diatomologica 18*. J. Cramer, Berlin. pp. 393.

Larsen, J. 1985. Algal studies of the Danish Wadden Sea II. A taxonomic study of psammobious dinoflagellates. *Opera Bot. 79*: 14–37.

Lee, J. J., M. E. McEnery, E. M. Kennedy and H. Rubin. 1975. A nutritional analysis of a sublittoral diatom assemblage epiphytic on *Enteromorpha* from a Long Island salt marsh. *J. Phycol. 11*: 14–49.

Levinton, J. S. 1985. Complex interactions of a deposit feeder with its resources: roles of density, a competitor, and detrital addition in the growth and survival of the mudsnail *Hydrobia totteni*. *Mar. Ecol. Prog. Ser. 22*: 31–40.

Maarel, E. van der. 1989. Species diversity in plant communities in relation to structure and dynamics. *In*: (H. G. During, M. G. A. Werger and G. H. Willems, eds) *Diversity and Pattern in Plant Communities*. S. P. B. Academic Publishing, The Hague. pp. 1–14.

Nilsson, P., B. Jönsson, I. Linström Swanberg and K. Sunbäck. 1991. Response of a marine shallow-water sediment system to an increased load of inorganic nutrients. *Mar. Ecol. Prog. Ser.*: In press.

Patil, G. P. and C. Taillie. 1979. An overview of diversity. *In*: (J. F. Grassie, G. P. Patil, W. Smith and C. Taillie, eds) *Ecological Diversity in Theory and Practice*. International Publishing House, Fairland. pp. 3–27.

Pringle, C. M. and A. Bowers. 1984. An in situ substratum fertilization technique: Diatom colonization on nutrient-enriched, sand substrata. *Can. J. Fish. Aquat. Sci. 41*: 1247–1251.

Rosenberg, R., R. Elmgren, S. Fleischer, P. Jonsson, G. Persson and H. Dahlin. 1990. Marine eutrophication. Case studies in Sweden. *Ambio 19*: 102–108.

Sládecék, V. 1973. System of water quality from the biological point of view. Ergebnisse der Limnologie, *Arch. für Hydrobiol. 7*: 1–218.

Simonsen, R. 1987. *Atlas and Catalogue of the Diatom Types of Friedrich Hustedt* (1–3). J. Cramer, Berlin–Stuttgart. 1: pp. 525, 2: Pl. 1–395, 3: Pl. 396–772.

Snoeijs, P. J. M. 1989. A check-list of the benthic diatoms at Forsmark (northern Baltic Sea). 1. Epilithic and epiphytic taxa. *Ann. Bot. Fennici 26*: 427–439.

Snoeijs, P. J. M. and I. C. Prentice. 1989. Effects of cooling water discharge on the structure and dynamics of epilithic algal communities in the northern Baltic. *Hydrobiologia 184*: 99–123.

Sullivan, M. J. 1976. Long-term effects of manipulating light intensity and nutrient enrichment on the structure of a salt marsh diatom community. *J. Phycol. 12*: 205–210.

Sullivan, M. J. 1981. Effects of canopy removal and nitrogen enrichment on a *Distichlis spicata*-edaphic diatom complex. *Estuar. coast. Shelf Sci. 13*: 119–129.

Sundbäck, K. 1984. Distribution of microbenthic chlorophyll-*a* and diatom species related to sediment characteristics. *Ophelia, Suppl. 3*: 229–246.

Sundbäck, K. and W. Granéli. 1988. Influence of microphytobenthos on the nutrient flux between sediment and water: a laboratory study. *Mar. Ecol. Prog. Ser. 43*: 63–69.

Sundbäck, K., B. Jönsson, P. Nilsson and I. Lindström. 1990. Impact of accumulating drifting macroalgae on a shallow-water sediment system: an experimental study. *Mar. Ecol. Prog. Ser. 58*: 261–274.

Swamikannu, X. and K. D. Hoagland. 1989. Effects of snail grazing on the diversity and structure of a periphyton community in a eutrophic pond. *Can. J. Fish. Aquat. Sci. 46*: 1698–1704.

Tilman, D., S. S. Kilham and P. Kilham. 1982. Phytoplankton community ecology: the role of limiting nutrients. *Ann. Rev. Ecol. System. 13*: 349–372.

Tilman, D., R. Kiesling, R. Sterner, S. S. Kilham and F. A. Johnson. 1986. Green, bluegreen and diatom algae: Taxonomic differences in competitive ability for phosphorus, silicon and nitrogen. *Arch. Hydrobiol. 106*: 473–485.

Tufail, A. 1987. Microbial communities colonising nutrient-enriched marine sediment. *Hydrobiologia 148*: 245–255.

Wallentinus, I. 1984. Comparison of nutrient uptake rates for Baltic macroalgae with different thallus morphologies. *Marine Biol. 80*: 215–225.

Whittaker, R. H. 1977. Evolution and measurement of species diversity. *Taxon 21*: 213–251.

Williams, D. M. and F. E. Round. 1987. Revision of the genus *Fragilaria. Diat. Res. 2*: 267–288.

Wolf, H. de. 1982. Methods of coding ecological data from diatoms for computer utilization. *Mededelingen Rijks Geologische Dienst. 36*: 95–98.

Zar, J. H. 1974. *Biostatistical Analysis*, Prentice-Hall, INC, Englewood Cliffs, H. J. pp. 620.

Botanica Marina
Vol. 34, pp. 359–364, 1991

The Effect of Light and Nutrient Conditions on Hair Cell Formation in *Spyridia filamentosa* (Wulfen) Harvey (Rhodophyta)

K. A. O'Connor and J. A. West

Department of Plant Biology, University of California, Berkeley, California 94720, U.S.A.

(Accepted 9 April 1991)

Abstract

The effects of light irradiance, specific nutrients, and overall enrichment level on hair cell production were determined for males, females and tetrasporophytes of *Spyridia filamentosa* (Wulfen) Harvey. At a 95% confidence level, males and females produced more hair cells in an enrichment medium without added nitrate and phosphate than in a standard enrichment medium. Females also formed more hair cells without added iron. Tetrasporophytes produced the same number of hair cells in the standard enrichment medium whether or not iron and nitrate were added. Tetrasporophytes produced fewer hair cells in the medium without added phosphate. Hairs formed without added phosphate were often short and abnormally shaped in both gametophytes and tetrasporophytes. Hair production in tetrasporophytes was dependent on light irradiance but not on the quality of light. Gametophytes, however, produced fewer hair cells in green light than in white light. Hair cell formation in male plants was affected more by overall enrichment concentration than it was in females and tetrasporophytes.

Introduction

Factors influencing hair cell formation and growth in a variety of green, red and brown algae have been investigated in a number of studies. Attention generally has been focused on hair cell formation under different light and nutrient conditions. Much of the literature has been strictly observational. For example, Nielsen (1972) noted that *Phaeophila tenuis* (Kylin) Nielsen and *P. dendroides* (Crouan) Batters formed hairs after a few days when transferred from an enriched sea water medium to an unenriched sea water medium, and that plants kept in the enriched medium did not form hairs. In addition, Nielsen (1979) noted that hairs appeared on *Acrochaete repens* Pringsheim as nutrients became depleted when the plants were grown in an enriched sea water medium. Yarish (1976) investigated the formation of setae (anucleate hair-like outgrowths of vegetative cells) in the Chlorophyta. He observed that *Bolbocoleon piliferum* Pringsheim, *Entocladia viridis* Reinke and *Entocladia flustrae* (Reinke) Taylor produced more setae when grown in low nitrate medium than in high nitrate medium. The same was true in low phosphate medium. For other Chaetophoracean algae examined, Yarish (1976) found that their cell growth was unrelated to phosphate and nitrate concentrations.

Few experimental studies on hair cell formation have been undertaken on the Rhodophyta (Whitton 1988). West (1971) noted that the abundance of hair cells in cultures of *Audouinella proskaueri* West was greater with increasing illumination. He noted that temperature was also a factor in hair cell production. More hair cells were formed at 15 °C than at 10 °C. Additionally, West observed that hair cell formation in *Audouinella* was not induced by nutrient depletion.

Material and Methods

The stock culture plants of *Spyridia filamentosa* were collected and maintained as described by West and Calumpong (1989). For the experimental studies,

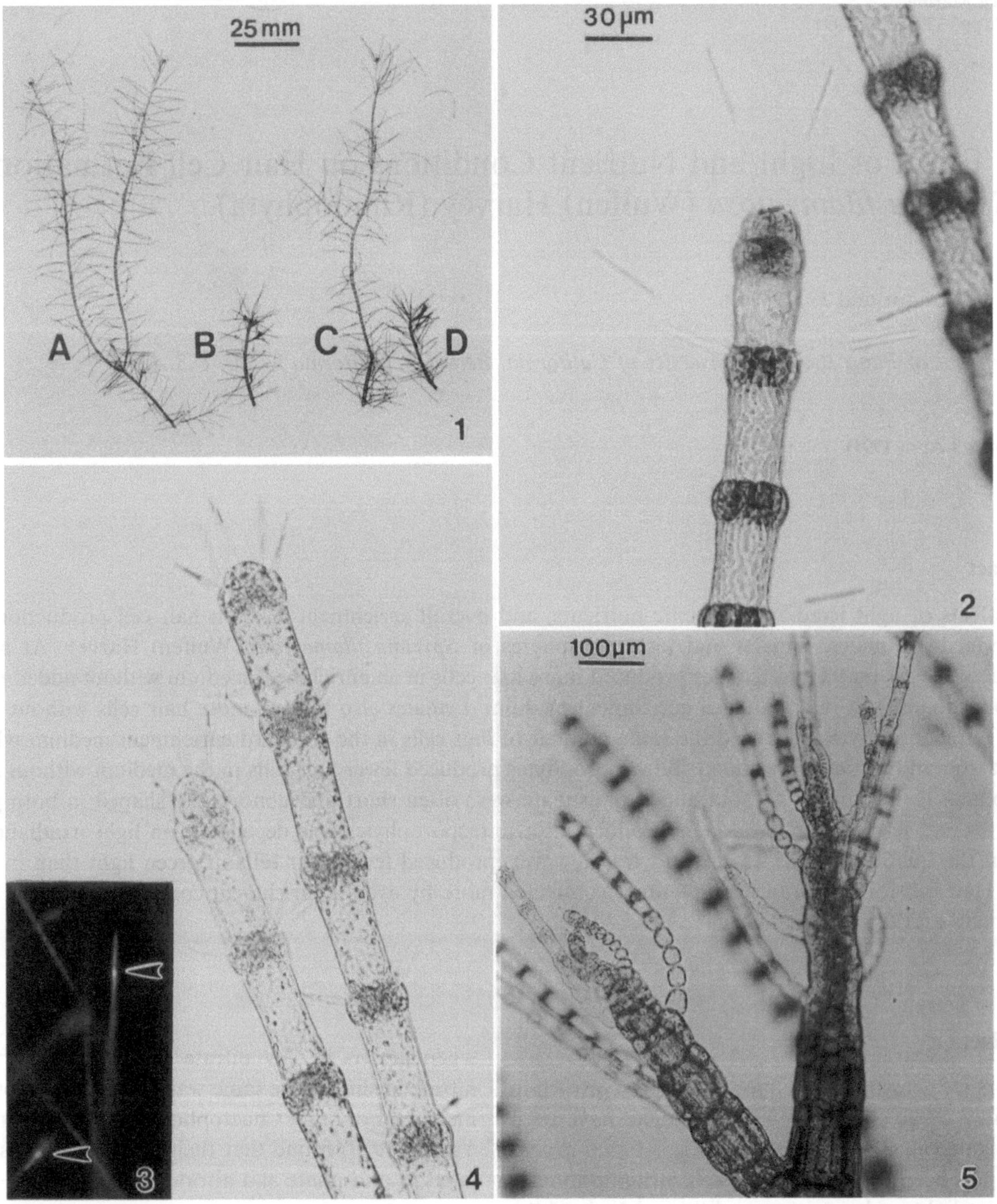

Fig. 1. Overall growth of plants in different light and enrichment strengths in standard PES medium. (A) Plant grown in 25 µmol $m^{-2} s^{-1}$ white light and PES/4. (B) Plant grown in 10 µmol $m^{-2} s^{-1}$ green light and PES/4. (C) Plant grown in 25 µmol $m^{-2} s^{-1}$ white light and PES/12. (D) Plant grown in 10 µmol $m^{-2} s^{-1}$ white light and PES/4. Fig. 2. Brachyblast apex of plant grown in PES/4 showing elongating hair cells. Fig. 3. Hair cells seen with DAPI fluorescence microscopy. The single nucleus in each is indicated by the arrowhead. Fig. 4. Brachyblast apices of plants grown in PES/4 enrichment medium lacking phosphate. Note stunted growth of hair cells. Fig. 5. Female grown in PES minus phosphate medium. Short, abnormal hairs are present but trichogynes are absent.

5 mm long apices were cut from lateral branches of the stock culture plants. These were transferred to 50 × 70 mm dishes containing 100 mL enriched sterile natural seawater. Seawater (35 parts per thousand) was obtained from the University of California Bodega Marine Laboratory pumping system after passing through a 12 µm filter. After storage for several weeks in 20 liter polyethylene carboys in a darkened room at 20–23 °C, the seawater was adjusted to 30 ppt with glass distilled water and sterilized in 4 L Pyrex flasks in a one atmosphere steam chamber for 30 min. Four different types of enrichment media were tested: Provasoli's enriched seawater (PES) medium (Starr and Zeikus 1987), a PES medium without added iron, PES without added nitrate, and PES without added phosphate. For each of the four media, four condi-

Table I. Concentrations of iron, phosphorus and nitrogen in natural seawater and enrichment sea water. (1) Lewin and Chen (1973). (2) Conomos, *et al.* (1979).

	Sea water	PES/4	PES/12
Fe (μM)	0.70[1]	1.8	0.60
P (μM)	1.0[2]	9.3	3.1
N (μM)	17[2]	160	55

tions were tested: 25 μmol m^{-2} s^{-1} white light and PES/7 (5 mL enrichment per liter seawater), 25 μmol m^{-2} s^{-1} white light and PES/12 (1.6 mL enrichment per liter seawater), 10 μmol m^{-2} s^{-1} white light and PES/4, and 10 μmol m^{-2} s^{-1} green light and PES/4. The iron, phosphorus and nitrogen levels in the natural seawater and enrichments are indicated in Table I. Tris buffer was eliminated from all media because it contributes to bacterial growth. Experiments were repeated once for males, females and tetrasporophytes. Plants were grown for two weeks under a 12 h daily photoperiod (12 : 12) in a 25 °C growth chamber illuminated by GE F20T12/CW cool white fluorescent light. Collars of black construction paper were wrapped around all culture dishes so plants received only epi-illumination. Neutral gray plastic mesh screening was placed over lids of culture dishes in the 10 μmol m^{-2} s^{-1} white light condition, and Roscolene #871 light green filters (see Fig. 6 for absorption spectrum) were placed over lids in the 10 μmol m^{-2} s^{-1} green light condition. Filters were obtained from Holzmueller, 2545 16th St., San Francisco, CA 94103. Photosynthetically active radiation (PAR) (380–700 nm) irradiance was determined with a QSI-140 integrating quantum irradiance meter (Biospherical Instruments, 4901 Marina Blvd., San Diego, CA 92117).

For each condition, hair cells from the 5 whorls of periaxial cells closest to the brachyblast apex were counted. Because hair cells are highly deciduous, the number of hair cells beyond the first whorl (dome) of periaxial cells decreases abruptly. Thus, only data obtained from the dome of periaxial cells on each

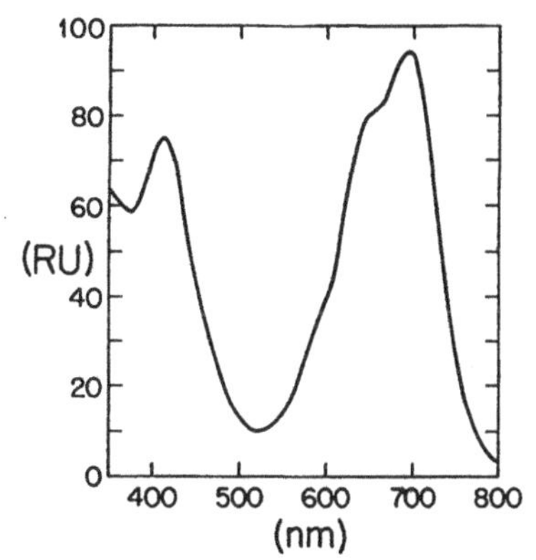

Fig. 6. Relative absorbance versus wavelength for green filter.

brachyblast is discussed in this paper. The procedure was repeated for ten brachyblasts from each of four plants in each growth condition. Counts were made only for fully elongate brachyblasts closest to the apex of the main axis. Results were analyzed using an ANOVA program in the Berkeley Interactive Statistical System (BLSS).

Results

For male and female plants, the mean number of hair cells increased when specific nutrients were removed from the standard PES medium (Fig. 7). In the female, the increase was significant for all nutrients tested: iron, phosphate, and nitrate. In the male, the increase was significant only for phosphate and nitrate. The tetrasporophyte showed a completely different response. It showed no significant difference in the mean number of hair cells in regular PES, PES minus iron, and PES minus nitrate. The mean number of hair cells on the tetrasporophyte decreased significantly in PES minus phosphate.

The morphology of *Spyridia* hair cells was normal in all the media except the PES minus phosphate. Many of the hairs grown in the absence of phosphate were irregular in outline and abnormally short (≈ 20–50 μm long compared with normal hair cells which are ≈ 250 μm long, see Figure 3), although they were produced in abundance (Fig. 4). Some females grown in PES minus phosphate had hairs but lacked trichogynes (Fig. 5).

There was no uniform pattern of response between the male, female, and tetrasporophyte for the different light and enrichment strength conditions (Fig. 8). Neither the female nor the tetrasporophyte showed any change in the mean number of hair cells from the PES/4 medium (Fig. 2) to the PES/12 medium. In contrast, the mean number of hair cells in the male decreased significantly in the PES/12 medium. The number of hair cells on all three life history phases decreased significantly from the 25 μmol m^{-2} s^{-1}, PES/4 condition to the 10 μmol m^{-2} s^{-1}, PES/4 condition. This decrease was most dramatic in the tetrasporophyte. The mean number of hair cells in the tetrasporophyte was the same in 10 μmol m^{-2} s^{-1} light regardless of whether the light was white or green. The male and female plants, however, produced fewer hair cells in 10 μmol m^{-2} s^{-1} green light than in 10 μmol m^{-2} s^{-1} white light.

In general, plants grew longer in 25 μmol m^{-2} s^{-1} light than in 10 μmol m^{-2} s^{-1} light (Fig. 1). Plants grown in 25 μmol m^{-2} s^{-1} and PES/12 were sometimes pale relative to those grown in 25 μmol m^{-2} s^{-1} and PES/4.

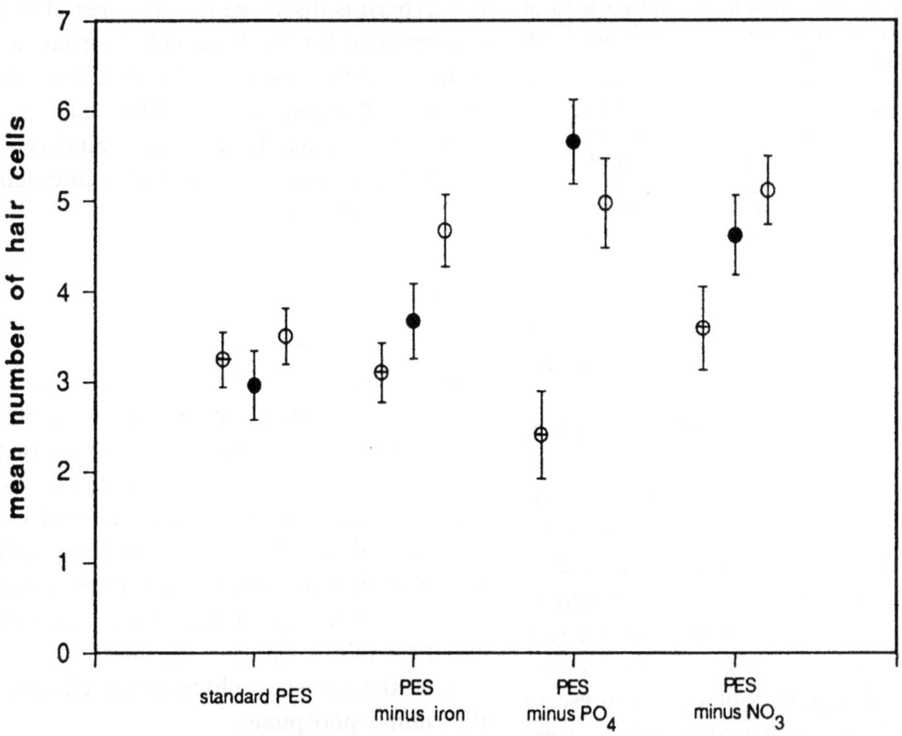

Fig. 7. Mean number of hair cells on first whorl of periaxial cells under different nutrient conditions. Bars represent 95% confidence intervals about the means. Note the difference in response between the male, female and tetrasphorophyte. ⊖ = tetrasporophyte. ● = male. ○ = female.

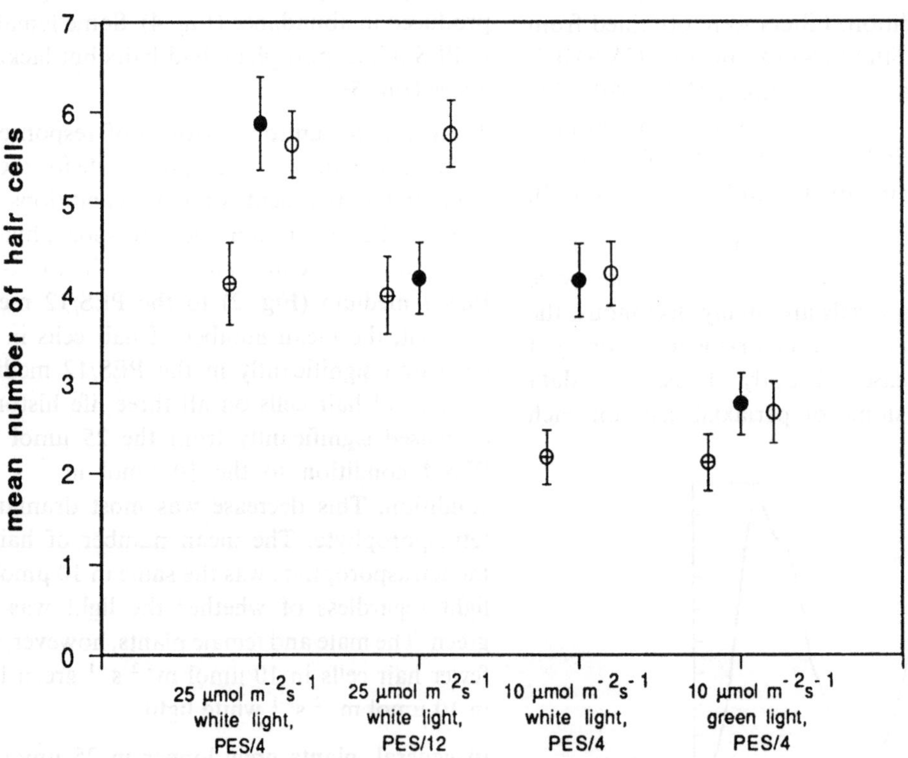

Fig. 8. Mean number of hair cells on first whorl of periaxial cells under different light conditions and different enrichment strengths. Bars represent 95% confidence intervals about the means. Note the lack of a single response pattern for the three phases. ⊖ = tetrasporophyte. ● = male. ○ = female.

Discussion

Our results show a complex relationship between hair cell production in *Spyridia filamentosa* and the difference light and nutrient conditions tested. One particularly interesting finding is that different factors induce hair cell production in males, females and tetrasporophytes. Other hair cell studies (Rueness, Mathisen and Tananger 1987, DeBoer and Whoriskey 1983, West 1971) have not examined possible differences in hair cell growth between different life history phases. It would be interesting to see if in other red algae the effect of growth conditions on hair cell production depends on the life history phase like in *S. filamentosa*.

The absence of added phosphate in the enrichment medium leads to an increase in hair production in males and females in *Spyridia filamentosa*. This is consistent with the studies of Gibson and Whitton (1987) and Yarish (1976) in which Chaetophoralean algae produced more hair cells when grown in media low in phosphate. It is important to note, however, that although both males and females of *S. filamentosa* produce more hair cells in the medium without added phosphate, this increase in number is probably not correlated with an increase in phosphorus absorption. Hairs produced without added phosphorus are one tenth to one fifth as long as hair cells produced in the other media. Thus gametophytes grown without added phosphate have less surface area for absorption than plants grown in standard PES. The role of phosphorus in hair cell formation is unclear, although phosphorus is clearly necessary in the production of morphologically normal hair cells.

Although the absence of either added nitrate or added phosphate induces hair cell formation in both males and females, a decrease in the total amount of nutrients does not. In fact, in the PES/12 medium, males form fewer hair cells than in the PES/4 medium. Females produce the same number of hair cells in both media. These results suggest that it may be the relative proportion of the different nutrients that is most significant in inducing hair cell formation in *Spyridia filamentosa*, not the overall amount of nutrients.

It is also important to note that although the results discussed are statistically significant, they may not be physiologically significant for the organism. Without doing uptake experiments for specific nutrients, it is impossible to determine whether a given increase in the mean number of hair cells per whorl of periaxial cells has any impact on nutrient absorption in the plant.

As in the study by Yarish (1976) in which the Chaetophoracean algae *Ochlochaete* and *Ectochaete* produced more hair cells at higher light intensities, tetrasporophytes of *Spyridia filamentosa* produce more hair cells in 25 μmol m^{-2} s^{-1} light than in 10 μmol m^{-2} s^{-1}. Light irradiance is more important than the quality of light. Just as many hair cells are produced in green light as in white light. Gametophytes, however, show a sensitivity to light quality like the brown algae *Dictyota* and *Scytosiphon* (Lüning 1981) by producing significantly fewer hair cells in green light than in white light.

Hair cell production in *Spyridia filamentosa* appears to be more strongly influenced by environmental factors in gametophytes than in tetrasporophytes. This may be linked to the difference in ploidy level between the diploid tetrasporophytes and the haploid gametophytes. However, not enough is known about the genetics, physiology, or ecology of *Spyridia* to speculate on why this may be true. If hair cells are involved in nutrient uptake in *S. filamentosa*, their role appears most crucial in gametophytes.

Acknowledgements

We appreciate the careful review of the manuscript by Dr David Ballantine.

References

Conomos, T. J., R. E. Smith, D. H. Peterson, S. W. Hayes and L. E. Schemel. 1979. Processes affecting seasonal distributions of water properties in the San Francisco Bay estuarine system. *In:* (T. J. Conomos, ed.) *San Francisco Bay, the Urbanized Estuary.* Pacific Division of the American Association for the Advancement of Science. Allen Press, Lawrence, Kansas. pp. 115–142.

DeBoer, J. A. and F. Whoriskey. 1983. Production and role of hyaline hairs in *Ceramium rubrum. Mar. Biol. 77*: 229–234.

Gibson, M. T. and B. A. Whitton. 1987. Hairs, phosphatase activity and environmental chemistry in *Stigeoclonium, Chaetophora* and *Draparnaldia* (Chaetophorales). *Br. Phycol. J. 22*: 11–22.

Lewin, J. and C. H. Chen. 1973. Changes in the concentration of soluble and particulate iron in seawater enclosed in containers. *Limnol. Oceanogr. 18*: 590–596.

Lüning, K. 1981. Photomorphogenesis of reproduction in marine macroalgae. *Ber. Dtsch. Bot. Ges. 94*: 401–417.

Nielsen, R. 1972. A study of the shell-boring marine algae around the Danish Island Laeso. *Bot. Tidsskr. 67*: 245–269.

Nielsen, R. 1979. Culture studies on the type species of *Acrochaete, Bolbocoleon* and *Entocladia* (Chaetophoraceae, Chlorophyceae). *Bot. Notiser 132*: 441–449.

Rueness, J., H. A. Mathisen and T. Tananger. 1987. Culture and field observations on *Gracilaria verrucosa* (Huds.) Papenf. (Rhodophyta) from Norway. *Bot. Mar. 30*: 267–276.

Starr, R. C. and J. Zeikus. 1987. UTEX – the culture of algae at the University of Texas at Austin. *J. Phycol. 23* (suppl.): 1–47.

West, J. A. 1971. Environmental control of hair and sporangial formation in the marine red alga *Acrochaetium proskaueri* sp. nov. *Proc. Int. Seaweed Symp. 7*: 377–384.

West, J. A. and H. P. Calumpong. 1989. On the reproductive biology of *Spyridia filamentosa* (Wulfen) Harvey (Rhodophyta) in culture. *Bot. Mar. 32*: 379–387.

Whitton, B. A. 1988. Hairs in eukaryotic algae. *In*: (F. E. Round, ed.) *Algae and the Aquatic Environment*. Biopress, Bristol. pp. 446–460.

Yarish, C. 1976. Polymorphism of selected marine Chaetophoraceae (Chlorophyta). *Br. Phycol. J. 11*: 29–38.

Botanica Marina
Vol. 34, pp. 365–367, 1991

Sulphonoglycolipid from the Green Alga *Enteromorpha flexuosa* (Wulf.) J. Ag.

A. K. Siddhanta, B. K. Ramavat, V. D. Chauhan, B. Achari*, P. K. Dutta* and S. C. Pakrashi*

Marine Algae Discipline, Central Salt & Marine Chemicals Research Institute, Bhavnagar 364002, India
* *Medicinal Chemistry Department, Indian Institute of Chemical Biology, Calcutta 700032, India*

(Accepted 4 May 1991)

Abstract

The methanolic extract of the marine green alga *Enteromorpha flexuosa* yielded 1,2-diacyl-3-0-(6′-sulpho-α-D-quinovopyranosyl)-glycerol, the acyl component consisting of 14:0 (1.7), 15:0 (0.9), 16:0 (57.7), 18:0 (2.3), 16:1 (9*Z*; 1.3), 18:1 (9*Z*; 26.7), 18:2 (9*E*:12*E*; 2.3) and 18:2 (9*Z*:12*Z*; 6.9%) fatty acids.

Introduction

The bioactivity of the marine green alga *Enteromorpha* is well documented (Chengkui and Junfu 1984, Naqvi *et al.* 1981, Rao and Parekh 1981 and Rao *et al.* 1982). In the course of our programme for isolating bioactive compounds from marine algae, a sulphonoglycolipid was encountered in the methanolic extract of *Enteromorpha flexuosa*. A number of sulphonoglycolipids have been obtained from marine sources (Fusetani and Hashimoto 1975, Ishizuka *et al.* 1973, Isono *et al.* 1967, Kitagawa *et al.* 1979 and Son 1990), and interest in them has increased in recent years because of their interesting biological activities (Fusetani and Hashimoto 1975, Isono *et al.* 1967). In this communication we wish to report the characterization of this compound.

Material and Methods

All solvents were redistilled prior to use. The alga *Enteromorpha flexuosa* (Wulf.) J. Ag. was collected from Okha Port (22°28′N, 69°05′E) on the west coast of India in January 1987. It was washed with tap water and cleaned to remove extraneous materials and epiphytes, and was then air-dried in the laboratory.

Optical rotation was measured on a JASCO DIP-360 polarimeter. Infrared (IR) spectrum was recorded on a Perkin-Elmer model 177 spectrophotometer and NMR spectra on a JEOL FX-100 machine with TMS as internal standard. Gas Chromatography-Mass Spectrometry (GC-MS) was performed on a Finnigan 4500 spectrometer interfaced with INCOS data system. A durabond fused silica (DB-1 bonded fused silica) capillary column (30 m × 0.25 mm) was used. Helium with a flow rate of 37–38 cm/sec was the carrier gas. The column oven temperature was kept at 80 °C for 2 minutes and then programmed at the rate of 5°/minute to 280 °C and was held at this temperature for 5 minutes. Temperature of the injection port was 280 °C.

For the isolation of the glycolipid, the alga *Enteromorpha flexuosa* (750 g dry weight) was repeatedly extracted with methanol (6 litres) at room temperature until the extract was colourless. The methanol extracts were concentrated under reduced pressure and the resulting concentrate was extracted with *n*-hexane (1.5 litres). The hexane-soluble fraction was subjected to repeated flash column chromatography on silica gel (230–400 mesh, 60 A°, Aldrich). The fractions 19 to 31 (eluted with the solvent chloroform: methanol-80: 20, saturated with water) were pooled and solvent removed under reduced pressure to afford amorphous **1** (120 mg), $[\alpha]_D$ + 42.2° ($CHCl_3$, *c* 1.24). IR: ν_{max}^{KBr} 3370, 1730, 1170 and 1035 cm^{-1}; ^1H NMR (100 MHz, $CDCl_3$, δ): 0.86, 0.96, 1.24, 2.04, 2.32, 2.80, 4.84 and 5.34 ppm. ^{13}C NMR (25 MHz, $CDCl_3$, 40 °C, δ): 14.0, 22.7, 24.9, 25.6, 27.3, 29.3, 29.8, 31.9, 34.2, 128.2, 128.3, 129.7, 129.9, 130.1, 173.9, 174.1 ppm.

Acetylation of 1: A solution of **1** (30 mg) in acetic anhydride (0.5 mL) and pyridine (0.2 mL) was heated on a water bath for 4 hours. The mixture was then cooled and the reagent removed on a rotary evaporator. The crude material was then chromatographed on a silica gel (2 g; 60 – 120 mesh) column to afford the triacetyl derivative of **1** (20 mg). ^1H NMR ($CDCl_3$, δ): 2.0 (3H, *s*,-$OCOCH_3$), 2.04 (6H, *s*, 2 × -$OCOCH_3$).

Hydrolysis of 1: A mixture of **1** (20 mg) with 2% aqueous methanolic NaOH (20 mL) was refluxed for 2 hours with the addition of a little benzene. The reaction mixture was cooled, water (20 mL) added and methanol removed on a water bath. It was then extracted with ether (100 mL). The aqueous layer was acidified with dilute HCl and re-extracted with ether (100 mL). The ethereal solution was washed with water, dried (Na_2SO_4) and then evaporated to afford the crude fatty acid mixture. This was then methylated by adding a 10% solution of acetyl chloride in methanol (10 mL) and kept overnight at room temperature. The mixture was partitioned between water and ether. The ether layer was washed with 10% $NaHCO_3$ solution, then with water and dried (Na_2SO_4). The crude methyl ester thus obtained after removal of the solvent was chromatographed over silica gel to afford pure methyl esters (10 mg).

Results and Discussion

The hexane fraction of the methanolic extract of *E. flexuosa* on repeated flash chromatography on silica gel yielded an amorphous compound (**1**); yield 0.02% dry weight; $[\alpha]_D$ + 42.2°. The IR spectrum of **1** showed the presence of hydroxyl group(s) (ν_{max}^{KBr} 3400 cm^{-1}), ester carbonyl (1730 cm^{-1}) and sulphonate function (1170 and 1035 cm^{-1}). The ^1H NMR spectrum showed peaks at δ 0.86 and 0.96 (*t*'s, terminal methyls), 1.24 [*br s*, -$(CH_2)_n$-], 2.04 (*m*, -CH_2-CH=CH-), 2.28 (*m*, -CH_2-COO-), 2.80 (*t*, -CH=CH-CH_2-CH=CH-), 4.84 (*br s*, anomeric proton of α-glycoside) and 5.34 (*m*, -*CH*=*CH*-) ppm indicating it to be a glycolipid (Kitagawa *et al.* 1979, Wood and Snyder 1967). The ^{13}C NMR spectrum of **1** in $CDCl_3$ showed all the peaks for saturated, mono- and di-unsaturated acyl moieties, though signals for sugar

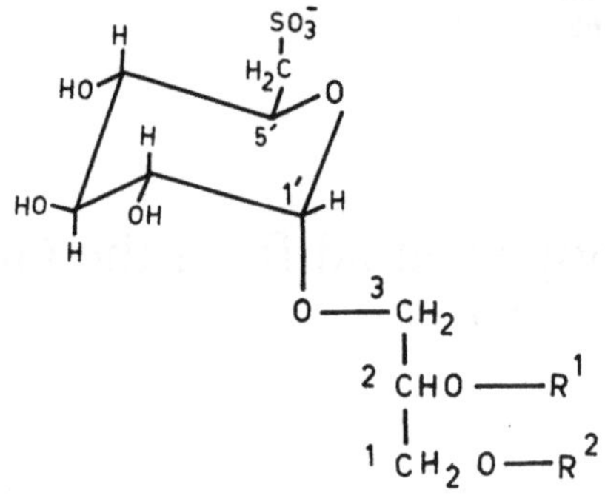

1

Fig. 1. **1**: 1,2-Diacyl-3-*O*-(6′-sulpho-α-D-quinovopyranosyl)-glycerol.
R^1, R^2 = Acyl groups consisting of 14:0 (1.7), 15:0 (0.9), 16:0 (57.7), 18:0 (2.3), 16:1 (9*Z*; 1.3), 18:1 (9*Z*; 26.7), 18:2 (9*E*:12*E*; 2.3) and 18:2 (9*Z*:12*Z*; 6.9%) fatty acids.

and glycerol parts were considerably broadened. When, however, the spectrum was measured in CD_3OD, such peaks could easily be distinguished. It is worthwhile to mention here that the broadening of the carbon signals belonging to the sugar and glycerol moieties of glycolipids, when measured in $CDCl_3$ has been ascribed (Johns *et al.* 1978) to a restriction of motion within parts of the molecule and the variation in the secondary structure of the lipid in different solvents. Among the additional peaks thus observed in CD_3OD, a characteristic signal at δ 54.3 ppm suggested the presence of -CH_2SO_3 group. The other peaks at δ 100.0 (C-1′, anomeric carbon), 74.9 (C-3′, C-4′), 73.2 (C-2′) and 69.7 (C-5′) ppm, were attributed to the sugar moiety and were in full agreement with the identification of sugar as sulphoquinovose (Kitagawa *et al.* 1979, Son 1990). The carbon resonances for diacyl glycerol moiety were assigned the chemical shifts at δ 64.3 (C-1), 71.7 (C-2) and 67.3 (C-3) ppm.

The diacyl nature of the glycerol moiety was confirmed by acetylation of **1** which yielded a triacetate only. To determine the fatty acid profile, **1** was hydrolysed with dilute NaOH to yield a mixture of fatty acids which was converted to methyl esters. Gas Chromatography-Mass Spectral (GC-MS) analysis revealed the presence of 14:0 (1.7), 15:0 (0.9), 16:0 (57.7), 18.0 (2.3), 16:1 (9*Z*; 1.3), 18:1 (9*Z*; 26.7), 18:2 (9*Z*:12*Z*; 6.9) and 18:2 (9*E*:12*E*; 2.3%) fatty acids. The presence of linoleic (18:2; 9*Z*:12*Z*) and linolelaidic (18:2; 9*E*:12*E*) acids as indicated by GC-MS, was in consonance with the proton NMR spectrum (δ 2.80 ppm for -CH=CH-CH_2-CH=CH-) of **1**. However, the predominance of Z over E isomer was evident from the ^{13}C NMR spectrum which contained, besides all the peaks expected of a saturated fatty acyl moiety, two additional peaks at δ 27.3 and 25.6 ppm, characteristic of vinylic and doubly vinylic methylene carbons of cisoid olefins (Johns *et al.* 1977). On the basis of the above evidence, the structure of **1** has been determined to be a 1,2-diacyl-3-*0*-(6'-sulpho-α-D-quinovopyranosyl)-glycerol derivative (acyl: as indicated in Fig. 1).

Acknowledgements

The authors wish to express their grateful thanks to Prof. M. M. Taqui Khan, Director, Central Salt & Marine Chemicals Research Institute, Bhavnagar, for his keen interest and help in this work.

The Department of Ocean Development, New Delhi, is gratefully acknowledged for funding. Thanks are due to Dr S. P. Dutta, Roswell Memorial Institute, New York, USA, for GC-MS analysis.

References

Chengkui, Z. (C. K. Tseng) and Z. Junfu (C. F. Chang). 1984. Chinese seaweeds in herbal medicine. *In*: (C. J. Bird and M. A. Ragan, eds). *Proceedings of the 11th International Seaweed Symposium.* Dr W. Junk Publishers, Dordrecht. pp. 152–154.

Fusetani, N. and Y. Hashimoto. 1975. Marine toxins. XLVIII. Structure of two water soluble hemolysins isolated from the green alga *Ulva pertusa. Agric. Biol. Chem. 39*: 2021–2025.

Ishizuka, I., M. Suzuki and T. Yamakawa. 1973. Isolation and characterization of a novel sulfoglycolipid, 'Seminolipid' from Boar testis and spermatozoa. *J. Biochem. 73*: 77–87.

Isono, Y., H. Mohri and Y. Nagai. 1967. Effect of egg sulfolipid on respiration of sea urchin. *Nature 214*: 1336–1338.

Johns, S. R., D. R. Leslie, R. I. Willing and D. G. Bishop. 1977. Studies on chloroplast membranes I. ^{13}C Chemical shifts and longitudinal relaxation times of carboxylic acids. *Aust. J. Chem. 30*: 813–822.

Johns, S. R., D. R. Leslie, R. I. Willing and D. G. Bishop. 1978. Studies on chloroplast membranes III. ^{13}C Chemical shifts and longitudinal relaxation times of 1,2-diacyl-3-(6-sulpho-O-quinovosyl)-sn-glycerol. *Aust. J. Chem. 31*: 65–72.

Kitagawa, I., Y. Hamamoto and M. Kobayashi. 1979. Sulfonoglycolipid from the sea urchin *Anthocidaris crassispina* A. Agassiz. *Chem. Pharm. Bull. 27*: 1934–1937.

Naqvi, S. W. A., Solimabi, S. Y. Kamat, L. Fernandes, C. V. G. Reddy, D. S. Bhakuni and B. N. Dhawan. 1981. Screening of some marine plants from the Indian coast for biological activity. *Bot. Mar. 24*: 51–55.

Rao, P. S. and K. S. Parekh. 1981. Antibacterial activity of Indian seaweed extracts. *Bot. Mar. 24*: 577–582.

Rao, P. S., K. S. Parekh, H. H. Parekh, H. M. Mody, B. A. Dave and S. B. Trivedi. 1982. Effect of seaweed extracts on *Mycobacterium. In*: (H. A. Hoppe and T. Levring, eds) *Marine Algae in Pharmaceutical Science 2.* Walter de Gruyter & Co., Berlin, New York. pp. 123–128.

Son, B. W. 1990. Glycolipids from *Gracilaria verrucosa. Phytochemistry 29*: 307–309.

Wood, R. and F. Snyder. 1967. Characterization and identification of glyceryl ether diesters present in tumor cells. *J. Lip. Res. 8*: 494–500.

Botanica Marina
Vol. 34, pp. 369–373, 1991

Effects of Seawater Phosphorus Concentration on Floridean Starch Content in *Agardhiella subulata* (C. Agardh) Kraft *et* Wynne (Rhodophyceae)

T. Chopin*,**, M. D. Hanisak** and F. E. Koehn**

* *Marine and Estuarine Research Group, Division of Sciences, University of New Brunswick, P.O. Box 5050, Saint John, N.B., E2L 4L5, Canada*
** *Harbor Branch Oceanographic Institution, 5600 Old Dixie Highway, Fort Pierce, Florida 34946, U.S.A.*

(Accepted 4 May 1991)

Abstract

Comparison of the carrageenan-selective CTAB extraction method with the non-selective ethanol extraction method for polysaccharides allowed determination of floridean starch content in unattached fronds of *Agardhiella subulata* cultivated for 4 weeks in tanks receiving seawater enriched with 53.5 μM nitrate and 0 to 20 μM phosphate (P_i) week $^{-1}$. The minimum floridean starch content was observed when carrageenan content was maximum, at an enrichment of 3 μM P_i week $^{-1}$. The alga appears to adjust a balance in the direction of the carbon flow into storage carbohydrate or structural carbohydrate pools in response to its physiological needs.

Introduction

In a previous paper the phycocolloids extracted from *Agardhiella subulata* (C. Agardh) Kraft *et* Wynne were identified as being mainly ι-carrageenan with less deviant ι-carrageenan and ν-carrageenan; maximum specific growth rate and carrageenan content were observed when seawater was enriched with 6 μM P_i week $^{-1}$ and 3 μM P_i week $^{-1}$, respectively (Chopin *et al.* 1990). While working on the polysaccharide extraction techniques, it appeared that the CTAB precipitation method was carrageenan-selective whereas the ethanol precipitation method yielded a mixture of carrageenan and floridean starch. Comparison of the two carrageenan extraction methods thus provided a way of determining floridean starch content in *A. subulata*.

Storage products of photosynthesis in the eukaryotic algae are carbohydrates, which are mainly glucose polymers (Craigie 1974, Kremer 1981). In the Florideophycidae the typical photosynthetic product is floridean starch, which is a branched glucan similar to amylopectin [$\alpha(1 \rightarrow 4)$ and $\alpha(1 \rightarrow 6)$ linkages] except for having a few $\alpha(1 \rightarrow 3)$ branch points (Peat *et al.* 1957, 1959 a, b). It is also different from land plant amylopectins in having average chain lengths of about 11–18 compared with 15 to more than 60 for land plant starches and in having a less granular organization which is extraplastidal (*i.e.*, in the cytoplasm) instead of intraplastidal (Meeuse 1962, Craigie 1974, Borowitzka 1978, Percival 1979, Hizukuri 1986).

While there have been a number of publications qualitatively studying and establishing the structure of floridean starch, quantitative studies on floridean starch in red algae remain limited (Sheath *et al.* 1979, Rosenberg and Ramus 1982, Macler 1986, Rotem *et al.* 1986, Knutsen and Grasdalen 1987). The present study investigates variations in floridean starch content when cultivating unattached plants of *A. subulata* at different P_i enrichments of seawater.

Material and Methods

Algae and culture conditions

The algal material of *A. subulata* used in this study was produced in earlier experiments whose culture conditions have been described (Chopin *et al.* 1990).

The same six P_i loadings — 0, 3, 6, 10, 15, and 20 μM P_i (as NaH_2PO_4) per week — were performed twice, in duplicate. As both experiments gave similar data, the results were combined.

Extraction and content of total polysaccharides

Samples were extracted and precipitated with two volumes of 95% ethanol, according to the technique of Stanley (1963) with some modifications (Chopin *et al.* 1987). A second extraction was performed, in the same way, from the pellet of the first one. Coagula were dried in a vacuum oven for 24 h at 60 °C and then weighed to determine the content or yield (= % DW).

Extraction and content of total carrageenans

Samples were extracted and carrageenans selectively precipitated with hexadecyltrimethylammonium bromide (CTAB) as described in Chopin *et al.* (1990). The coagula were also dried in a vacuum oven for 24 h at 60 °C and weighed.

Determination of neutral sugars by gas-liquid chromatography (GLC)

Aliquots from the above two types of coagulum were hydrolysed with 2 N HCl for 2 h at 100 °C, evaporated, and dissolved in distilled water (Elgastat UHQ). The hydrolysed polysaccharide solutions were used to prepare alditol-acetate derivatives according to the method of Sloneker (1972) and as described in Chopin *et al.* (1990).

^{13}C-nuclear magnetic resonance spectroscopy (^{13}C-NMR)

Proton composite pulse decoupled ^{13}C-NMR spectra were recorded on a Bruker AM360 sepctrometer (90.56 MHz) on samples in 1:1 D_2O/H_2O solution (30 mg mL^{-1}) at 80 °C using a 10 mm probe. A spectral width of 12.8 kHz was used with a flip angle of 45° (4.5 μs) and a total recycle time of 2.0 s. Spectra of 16K data points were accumulated. Chemical shifts (ppm) were measured relative to internal dimethyl sulfoxide (DMSO, 39.5 ppm) and are reported as values downfield from tetramethylsilane (TMS).

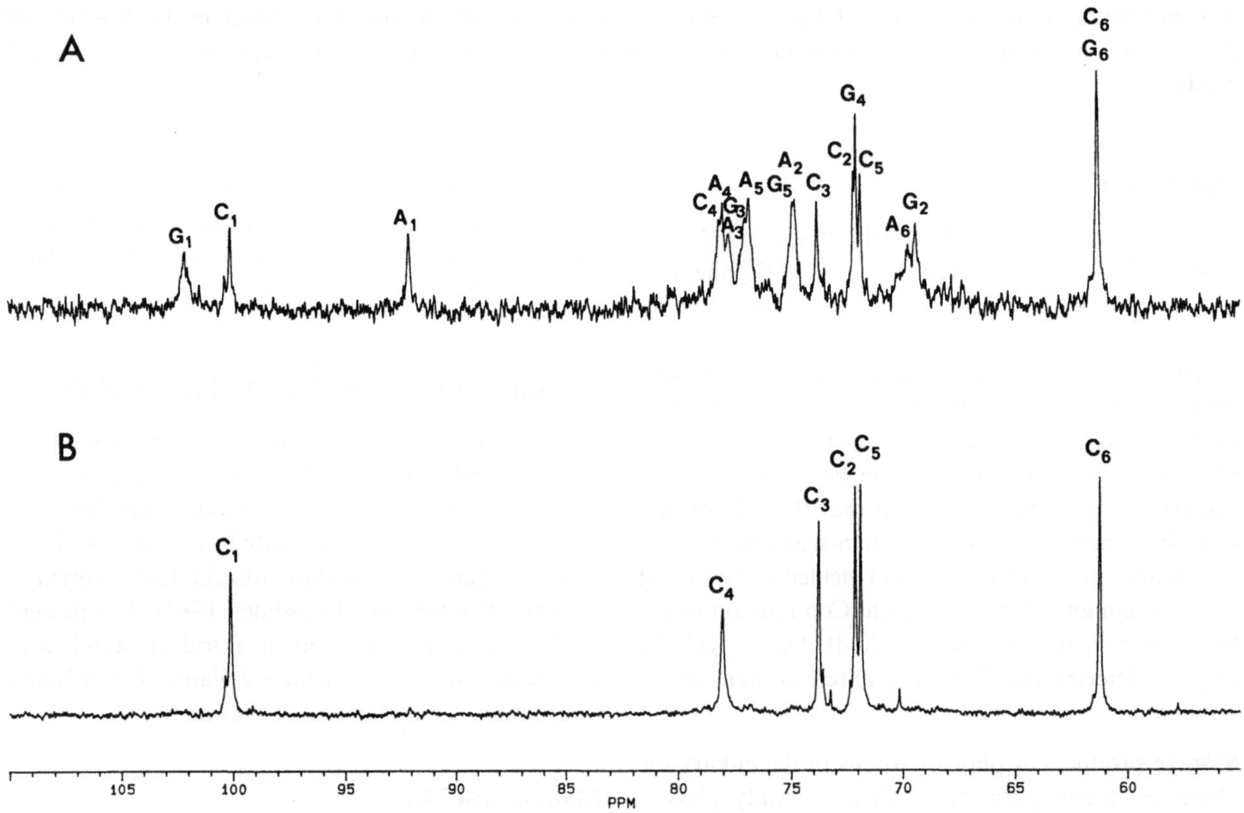

Fig. 1. ^{13}C-NMR spectra of the polysaccharides extracted from *A. subulata* by ethanol precipitation. A) first extraction, B) second extraction. Chemical shifts (ppm), measured relative to TMS, are assigned to carbon atoms of the repeating disaccharide of ι-carrageenan (G = 1,3-linked β-D-galactose-4-sulfate; A = 1,4-linked 3,6-anhydro-α-D-galactose-2-sulfate) and the repeating monosaccharide of floridean starch (C = glucose). The spectra were recorded at 80 °C with 22966 scans for A), and 19201 scans for B).

Results

^{13}C-NMR spectroscopy of samples precipitated with the ethanol method

The ^{13}C-NMR spectra of polysaccharides precipitated with ethanol from the first extract (Fig. 1A) showed 12 peaks with chemical shifts that were assigned to ι-carrageenan (Chopin *et al.* 1990), and 5 other peaks which could be assigned to carbon atoms of floridean starch (Dais and Perlin 1982, Bock *et al.* 1984, Lahaye *et al.* 1986, Knutsen and Grasdalen 1987). The sixth peak (C_6) assigned to floridean starch was not resolved from that of G_6 of the ι-carrageenan. No signals for 4-O methylgalactose and 6-O methylgalactose were observed, presumably because these compounds were present at too low a concentration to be detected at the signal-to-noise ratio of these experiments (lower than 10%).

The ^{13}C-NMR spectra of polysaccharides precipitated with ethanol from the second extract (Fig. 1B) clearly showed 6 major peaks with chemical shifts similar to those of floridean starch (Table I). The ι-carrageenan signals were barely noticeable at the signal-to-noise ratio of these experiments.

Table I. Chemical shifts (in ppm relative to TMS) of signals assigned to glucose of floridean starch in the ^{13}C-NMR spectrum of second extraction polysaccharides of *A. subulata* precipitated with ethanol.

C1	C2	C3	C4	C5	C6
100.2	72.2	73.8	78.1	71.9	61.3

^{13}C-NMR spectroscopy of samples precipitated with the CTAB method

A ^{13}C-NMR spectrum of the polysaccharides extracted from *A. subulata* with the CTAB method has been published (Chopin *et al.* 1990). It showed 12 major peaks characteristic of ι-carrageenan with no minor signals for floridean starch.

GLC analysis of polysaccharide hydrolyzates — CTAB vs. *ethanol methods*

In all GLC chromatograms from samples precipitated with the ethanol method a peak for glucose was recorded in addition to the peaks for galactose, 4-O-methylgalactose, 6-O-methylgalactose, and xylose (data not shown). In all GLC chromatograms from samples precipitated with the CTAB method only the last four peaks were recorded and no glucose was detected.

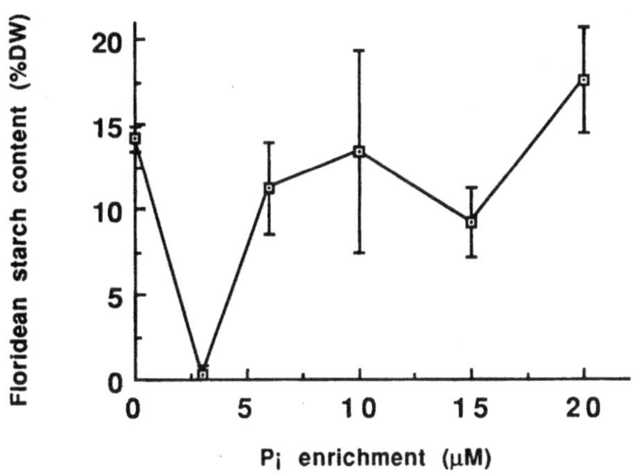

Fig. 2. Variations of floridean starch content (% DW) extracted from *A. subulata* cultured in different P_i enrichments. Values represent means (n = 4) ± 1 SD.

Floridean starch content

By subtracting the yield of polysaccharides from the CTAB precipitation (carrageenans only) from the yield of polysaccharides from the first and second ethanol extractions (carrageenans plus floridean starch), one could calculate the floridean starch content as a function of P_i enrichment (Fig. 2). The lowest content (0.33 ± 0.19% DW) was recorded for an enrichment of 3 µM P_i week $^{-1}$. When no P_i enrichment was provided, the floridean starch content was 14.16 ± 0.76% DW; P_i enrichments higher than 3 µM P_i week $^{-1}$ led to floridean starch contents fluctuating around an average value of 12.86 ± 3.61% DW.

Discussion

Polysaccharides precipitated with ethanol from the first extract were a mixture of mostly ι-carrageenan, with some floridean starch as shown by additional peaks to those of ι-carrageenan in the ^{13}C-NMR spectra. Polysaccharides from the second extract consisted almost entirely of floridean starch, with very little ι-carrageenan. A third extraction was processed but did not yield any quantifiable amount of polysaccharides. It was concluded that the two-step extraction method was exhaustive.

The chemical shifts of the additional peaks were similar to those found in *Gracilaria tenuistipitata* C. F. Chang *et* B. M. Xia and *G. eucheumoides* Harvey by Lahaye *et al.* (1986) and in *Furcellaria lumbricalis* (Hudson) Lamouroux by Knutsen and Grasdalen (1987), and were assigned to glucose of floridean starch. Moreover, all the samples with additional peaks also showed the presence of glucose, which is the repeating unit of floridean starch, in their GLC chromatograms. Conversely, both ^{13}C-NMR spectroscopy and GLC analysis of neutral sugars confirmed that samples precipitated with the CTAB method did not contain glucose. Hence, CTAB precipitation is superior to ethanol precipitation for the accurate measurement of carrageenan content because its carrageenan selectively eliminates the problem of co-precipitation with floridean starch.

When it was realized that the ethanol precipitated samples were contamined with floridean starch another attempt to extract these precipitated samples was made using CTAB precipitation. This step did not, however, allow the separation of floridean starch from carrageenans. Such a result was interpreted as due to an intimate association during ethanol precipitation which could not be disrupted by subsequent CTAB precipitation. Consequently, for an accurate measurement of carrageenan content, the powdered seaweed samples should be treated directly by the CTAB method. Thus, caution should be employed when reviewing carrageenan yields in publications based on ethanol precipitation as they probably are overestimates because of co-precipitation with floridean starch.

As all the storage and structural polysaccharides were exhaustively extracted after two consecutive extractions, the difference between the yield of polysaccharides from two consecutive ethanol precipitations and the yield of polysaccharides after CTAB precipitation should be a good estimation of the floridean starch content.

The lowest floridean starch content was recorded for an enrichment of 3 μM P_i week $^{-1}$, which was precisely the concentration at which the maximum carrageenan content was obtained (Chopin *et al.* 1990). At either lower or higher P_i concentrations the floridean starch content averaged 13.12 ± 3.18% DW. Studies on variations of floridean starch content according to variations of nutrient concentration are rare (Rosenberg and Ramus 1982, Macler 1986, Rotem *et al.* 1986, Knutsen and Grasdalen 1987). Kremer (1981) suggested that energy is generated at the expense of stored carbohydrates when ambient conditions do not allow photosynthetic surplus production of assimilates (respiratory pathway); Nagashima *et al.* (1971) showed that in the red alga *Serraticardia maxima* (Yendo) Johansen floridean starch and floridoside are utilized in dark respiration. Starch appears to be at the intersection of different metabolic pathways and available for energy production and biosynthesis of proteins, chloroplast and pigment synthesis, cell division, and cell wall formation (Craigie 1974). In *A. subulata* the floridean starch content dropped at the same time as the carrageenan content was reaching its maximal value. Therefore, it is possible that storage polysaccharides (floridean starch) are used for the synthesis of structural polysaccharides (carrageenans).

Knutsen and Grasdalen (1987) observed the highest floridean starch content in *Furcellaria lumbricalis* in late summer and the lowest content in winter. These authors related these fluctuations to variations in photosynthesis, growth conditions in the sea, and growth rate of the alga (maximal in spring – summer and minimal in winter). Rosenberg and Ramus (1982) recorded the highest levels of floridean starch in *Gracilaria foliifera* (Forsskål) Børgesen during spring and summer; moreover, during these two seasons, fluctuations in floridean starch content were inversely related to growth rate and soluble tissue nitrogen. Dawes *et al.* (1974) demonstrated that the highest total carbohydrate levels of several species of *Eucheuma* from Florida were recorded in the summer and the lowest in the late winter. During the active spring growth period, the percentage of total carbohydrate decreased, and the 3,6 anhydrogalactose/carbohydrate ratio showed a proportional increase of 3,6 anhydrogalactose, and, by inference, carrageenans. In *A. subulata*, variations of floridean starch content were not related to the specific growth rate as in previous studies. When the latter was at its highest value at 6 μM P_i week $^{-1}$ (Chopin *et al.* 1990), the former was just below its average value (11.20% DW). Thus, at maximal specific growth rate, the content of floridean starch was still significant; however, the carrageenan content was decreasing. Macler (1986) showed that in *Gelidium coulteri* Harvey the fixation rate and flow of photosynthate into floridoside, agar, floridean starch, and amino acids were dependent upon the nitrogen status of the tissue and light. Concerning phosphorus metabolism in *A. subulata*, it is hypothesized that a delicate balance, at a critical point in the metabolic pathways of nucleotide sugars prior to the synthesis of either floridean starch or carrageenans, is maintained to direct the flow of carbon towards either glucose and starch or galactose and carrageenans. Further studies, such as enzymatic and ^{14}C-labelling analysis, remain to be done in order to better understand the channelling mechanism of sugar residues between the different pathways.

Our observation that P_i may be an important regulator of the path of carbon flow in *A. subulata* should be followed up with studies on other species to determine its general occurrence amongst red algae. Interestingly, an increase in floridean starch was measured (J. S. Craigie, personal communication) in *Chondrus crispus* Stackhouse cultivated under 'Neish effect' conditions (Neish *et al.* 1977). It would also be desirable to determine if the N/P ratio is involved in this regulation, and if P_i enrichment in seawater affects the photosynthetic rate of carrageenophytes and, hence, their synthesis of storage and structural polysaccharides.

Moreover, by using the appropriate P_i enrichment in our culture system we produced plants with an increased carrageenan yield and, simultaneously, with a very low level of floridean starch. Such an enrichment strategy is of interest for the carrageenan industry, not only to maximize the yield of carrageenans but also to keep starch contamination of the carrageenans to a minimum.

Acknowledgements

This paper is Contribution No. 839 from the Harbor Branch Oceanographic Institution to whom T. Chopin is grateful for having received a postdoctoral fellowship. Thanks are extended to Dr J. S. Craigie for helpful discussion and useful criticism of the manuscript, and to J. Roy and J. Farrah for the preparation of the manuscript. This work was supported by NSERC grants OGP46376 nd EQP92706 to T. Chopin.

References

Bock, K., C. Pedersen and H. Pedersen. 1984. Carbon-13 nuclear magnetic resonance data for oligosaccharides. *In*: (S. R. Tripson and D. Horton, eds) *Advances in Carbohydrate Chemistry and Biochemistry*. Academic Press, New York. pp. 193–203.

Borowitzka, M. A. 1978. Plastid development and floridean starch grain formation during carposporogenesis in the coralline red alga *Lithothrix aspergillum* Gray. *Protoplasma 95*: 217–228.

Chopin, T., C. Bodeau-Bellion, J.-Y. Floc'h, E. Guittet and J.-Y. Lallemand. 1987. Seasonal study of carrageenan structures from female gametophytes of *Chondrus crispus* Stackhouse (Rhodophyta). *Hydrobiologia 151/152*: 535–539.

Chopin, T., M. D. Hanisak, F. E. Koehn, J. Mollion and S. Moreau. 1990. Studies on carrageenans and effects of seawater phosphorus concentration on carrageenan content and growth of *Agardhiella subulata* (C. Agardh) Kraft and Wynne (Rhodophyceae, Solieriaceae). *J. Appl. Phycol. 2*: 3–16.

Craigie, J. S. 1974. Storage products. *In*: (W. D. P. Stewart, ed.) *Algal Physiology and Biochemistry*. University of California Press, Berkeley. pp. 206–235.

Dais, P. and A. S. Perlin. 1982. High-field, ^{13}C-NMR spectroscopy of β-D-glucans, amylopectin, and glycogen. *Carbohydr. Res. 100*: 103–116.

Dawes, C. J., J. M. Lawrence, D. P. Cheney and A. C. Mathieson. 1974. Ecological studies of floridian *Eucheuma* (Rhodophyta, Gigartinales). III. Seasonal variation of carrageenan, total carbohydrate, protein, and lipid. *Bull. Mar. Sci. 24*: 286–299.

Hizukuri, S. 1986. Polymodal distribution of the chain lengths of amylopectins, and its significance. *Carbohydr. Res. 147*: 342–348.

Knutsen, S. H. and H. Grasdalen. 1987. Characterization of water-extractable polysaccharides from Norwegian *Furcellaria lumbricalis* (Huds.) Lamour. (Gigartinales, Rhodophyceae) by IR and NMR spectroscopy. *Bot. Mar. 30*: 497–505.

Kremer, B. P. 1981. Carbon metabolism. *In*: (C. S. Lobban and M. J. Wynne, eds) *The Biology of Seaweeds*. Blackwell Scientific Publications, Oxford. pp. 493–533.

Lahaye, M., C. Rochas and W. Yaphe. 1986. A new procedure for determining the heterogeneity of agar polymers in the cell walls of *Gracilaria* spp. (Gracilariaceae, Rhodophyta). *Can. J. Bot. 64*, 579–585.

Macler, B. A. 1986. Regulation of carbon flow by nitrogen and light in the red alga, *Gelidium coulteri*. *Plant Physiol. 82*: 136–141.

Meeuse, B. J. D. 1962. Storage products. *In*: (R. A. Lewin, ed.) *Physiology and Biochemistry of Algae*. Academic Press, New York. pp. 289–313.

Nagashima, H., S. Nakamura, K. Nisizawa and H. Terumitsu. 1971. Enzymic synthesis of floridean starch in a red alga, *Serraticardia maxima*. *Plant & Cell Physiol. 12*: 243–253.

Neish, A. C., P. F. Shacklock, C. H. Fox and F. J. Simpson. 1977. The cultivation of *Chondrus crispus*. Factors affecting growth under greenhouse conditions. *Can. J. Bot. 55*: 2263–2271.

Peat, S., J. R. Turvey and J. M. Evans. 1957. Isolation of nigerose from floridean starch. *Nature* (*London*) *179*: 261–262.

Peat, S., J. R. Turvey and J. M. Evans. 1959 a. The structure of floridean starch. Linkage analysis by partial acid hydrolysis. *J. Chem. Soc.*: 3223–3227.

Peat, S., J. R. Turvey and J. M. Evans. 1959 b. The structure of floridean starch. II. Enzymic hydrolysis and other studies. *J. Chem. Soc.*: 3341–3344.

Percival, E. 1979. The polysaccharides of green, red and brown seaweeds: their basic structure, biosynthesis and function. *Br. Phycol. J. 14*: 103–117.

Rosenberg, G. and J. Ramus. 1982. Ecological growth strategies in the seaweeds *Gracilaria foliifera* (Rhodophyceae) and *Ulva* sp. (Chlorophyceae): soluble nitrogen and reserve carbohydrates. *Mar. Biol. 66*: 251–259.

Rotem, A., N. Roth-Bejorano and S. Arad. 1986. Effect of controlled environmental conditions on starch and agar contents of *Gracilaria* sp. (Rhodophyceae). *J. Phycol. 22*: 117–121.

Sheath, R. G., J. A. Hellebust and T. Sawa. 1979. Floridean starch metabolism of *Porphyridium purpureum* (Rhodophyta) I. Changes during ageing of batch culture. *Phycologia 18*: 149–163.

Sloneker, J. H. 1972. Gas-lipid chromatography of alditol acetates. *In*: (R. L. Whistler and J. N. BeMiller, eds) *Methods in Carbohydrate Chemistry 6*. Academic Press, New York. pp. 20–24.

Stanley, N. F. 1963. Process for treating a polysaccharide of seaweeds of the Gigartinaceae and Solieriaceae families. U. S. Patent 3,094,517.

Botanica Marina
Vol. 34, pp. 375–381, 1991

Evidence for Ecotypic Divergence between *Phycodrys rubens* Populations from the Baltic Sea and North Sea

H. Rietema

Department of Marine Biology, Biological Centre, P.O. Box 14, 9750 AA Haren, The Netherlands

(Accepted 4 May 1991)

Abstract

Phycodrys rubens from the Baltic (Kiel Bight) kept in a salinity range grew and survived at all salinities from 33 down to 7.5‰ with maximum growth in the range from 23 down to 16.5‰. *Phycodrys rubens* from the North Sea, however, grew and survived at all salinities from 33 to 16.5‰. At 11‰ only some cells of the inoculum survived. Maximum growth occurred at salinities of 33 to 23‰. These differences in response pattern could be correlated rather well with salinities of their own natural habitats. Not only physiologically but also morphologically *Phycodrys rubens* from Helgoland and the Baltic differ from one another both in nature and in culture. These results suggest the existence of ecotypic variation between both populations.

Introduction

Since the last glaciation the region of the present Baltic Sea has had a dynamic hydrographic history. Periods with fresh and saline water alternated. The last incursion of sea water took place about 7500 yr BP. It is reasonable to assume that a marine algal flora was recruited from the North Sea with which it was also connected at that time. About 3000 yr BP, however, the salinity decreased and the Baltic got its present brackish character (Russell 1985). At present a gradual reduction in the number of marine algal species is associated with decreasing salinity along the axis of the Baltic Sea (Pankow 1971, Remane and Schlieper 1971, Christensen, Koch and Thomsen 1985). It is interesting that some typical marine (non-estuarine) algal species from the sublittoral zone of northern Atlantic seacoasts occur in low-salinity brackish parts of the Baltic Sea. One example is the foliaceous red alga *Phycodrys rubens* (L.) Batters (Ceramiales, Delesseriaceae), which is abundant in the sublittoral of the Kiel Bight. Long-term observations by investigators of the Institut für Meereskunde of the University of Kiel revealed that the relative abundance of this species has increased in this part of the Baltic during the last two decades and this could be related to increasing eutrophication of the marine environment and not to decreasing salinity (Breuer and Schramm 1988). Lowered salinity is apparently a more stable stress factor at the level where *Phycodrys* occurred in the Kiel Bight.

Survival of species exposed to a specific stress factor can be based on two different strategies:

1) tolerance or acclimation without changes in its original genome,

2) genetic adaptation attended with changes in its original genome.

As salinity has been, probably for a long period, the main stress factor in the Baltic, the influence of this factor on Baltic *Phycodrys rubens* was compared with that on material of this species from the North Sea, in order to obtain an answer to the question: which of the two survival strategies has been adopted by *Phycodrys* in the Baltic? The salinity responses of *Phycodrys rubens* could possibly also explain the distribution of this species in the Baltic. Generally temperature is considered as the most important factor delimiting seaweed species distribution (Hoek 1982a, b). Thermal tolerance requirement experiments in the laboratory almost always confirm this generalization (Breeman 1988). Therefore in addition to the salinity experiments thermal experiments were carried out. According to Breeman (1988), evidence points to the occurrence of thermal ecotypes within marine algal species. Russell (1988) recently produced some evidence for ecotypic, salinity-bound variation of some Baltic seaweeds.

Material and Methods

Phycodrys rubens plants were dredged from the bottom of the Kiel Bight (Western part of the Baltic Sea) at a depth of 5 to 10 meters in May 1988 and collected from the upper sublittoral of Helgoland in September 1988. Unialgal cultures were established using tip cuttings from field material at first drawn with a fine needle through agar. *Phycodrys* material from the Baltic and Helgoland were maintained in stock culture at 12 °C, long-day (16–$\bar{8}$ L/D) and dim light conditions in a Provasoli enriched seawater medium of 23‰ and 33‰ salinity, respectively. In preparation for experiments, fragments of both a Baltic and a North Sea strain were maintained for at least three months in seawater with a salinity of 33‰ S at 12 °C, long-day length (16–$\bar{8}$) conditions and a light intensity of 15 $\mu E\ m^{-2}\ s^{-1}$. Apical segments with a single apex obtained under these conditions and each about 2 mm long were dissected and transferred to different salinities. The range of salinities tested was from 33‰ through 23‰, 19‰, 16.5‰, 11‰ to 7.5‰ (at 12 °C and 15 $\mu E\ m^{-2}\ s^{-1}$). These different salinities were obtained by dilution of Provasoli enriched North Sea water with Provasoli enriched distilled water. Every fortnight the culture media were replaced. The influence of temperature on both strains was tested in culture cabinets with temperatures of 5, 10, 15, 20 and 25 °C, long-day length (16–$\bar{8}$) conditions and a light intensity of about 15 $\mu E\ m^{-2}\ s^{-1}$, and in a 0 °C and –2 °C water bath at long and short days, respectively. In this temperature experiment, tip cuttings from the North Sea strain were kept at 33‰ S and the Baltic strain at both 33 and 23‰ S.

For each experiment 4 or 5 tips were transferred to each experimental salinity or temperature condition; each tip was placed separately in a 55 mm wide sterile plastic Petridish ¾ full of medium and which was sealed with parafilm to prevent evaporation. The growth of individual tip cuttings could then be followed. Tip cuttings exposed to 0 °C and –2 °C were kept in culture tubes sealed with a cotton plug and in these tubes only survival was tested. Cuttings exposed to these very low temperatures were pretreated for 2 weeks at 5 °C.

In order to follow the growth of tip cuttings, drawings were made of each plant once a week on the same day and at the same time, using a dissecting microscope fitted with a camera lucida. These drawings were later analysed using a Hewlett Packard digitiser (model 9835A). The surface area and also the length of the main rib of each tip-derived plant were determined with this apparatus. From surface values relative growth rates, expressed as % increase per week, were calculated for each plant during the period of exponential growth as follows:

$$RG = 100 \frac{\ln S_2 - \ln S_1}{T_2 - T_1}$$

where S_1 = initial surface, S_2 = final surface, T_1 = initial time and T_2 = final time (*cf.* Cambridge *et al.* 1984). The values of S_1 and S_2 in this formula are S values of consecutive datum points, which were plotted on semilog paper and between it was possible to fit a long, straight line. From midrib lengths the linear growth rates were calculated.

Results

Phycodrys rubens plants from the North Sea and Baltic Sea differed conspicuously from one another in their ability to grow and survive in seawater with reduced salinities (Figs 1–4). Not only physiologically but also morphologically both strains were different at all salinities (Figs 3, 4, 6 and 7). Tip cuttings from the Baltic strain grew and survived in seawater with salinities variating from 33‰ down to 7.5‰.

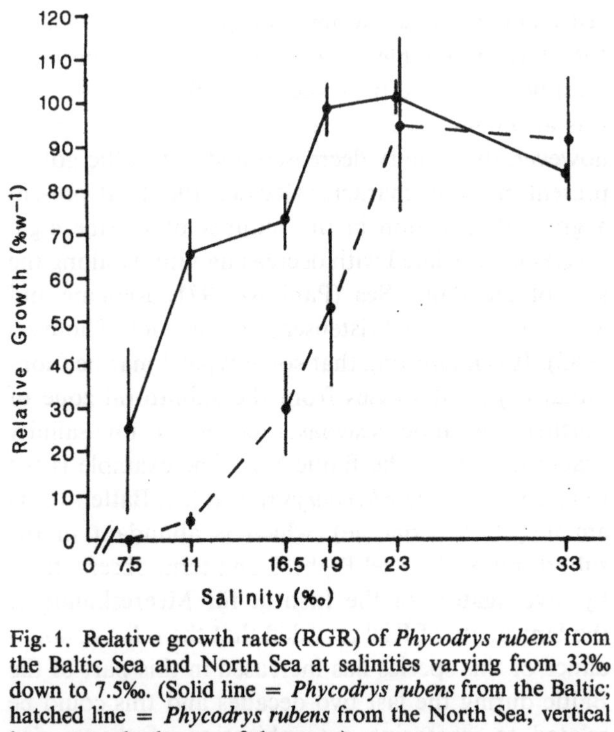

Fig. 1. Relative growth rates (RGR) of *Phycodrys rubens* from the Baltic Sea and North Sea at salinities varying from 33‰ down to 7.5‰. (Solid line = *Phycodrys rubens* from the Baltic; hatched line = *Phycodrys rubens* from the North Sea; vertical bar = standard error of the mean).

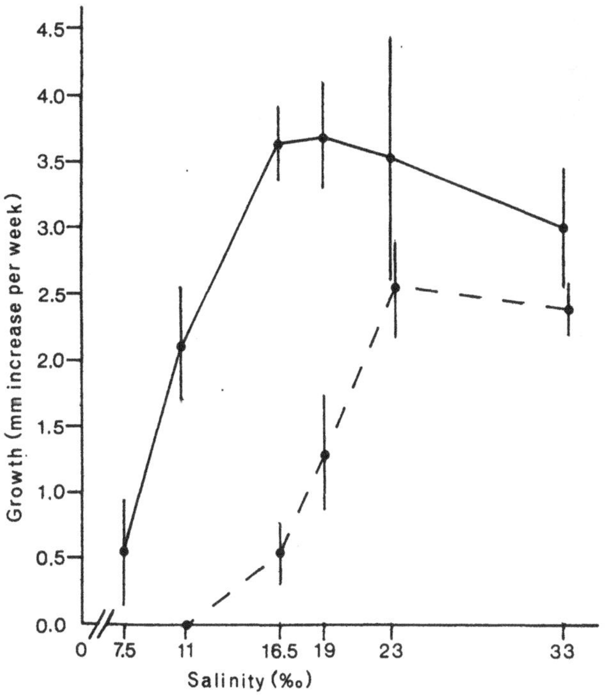

Fig. 2. Linear growth rates of *Phycodrys rubens* from the Baltic and North Sea at salinities from 33‰ down to 7.5‰. (Solid line = *Phycodrys* from the Baltic Sea; hatched line *Phycodrys* from the North Sea; vertical bar = standard error of the mean).

Maximum relative growth rates (RGR) occurred at 23‰ and 19‰ S and high RGR's were exhibited at 33, 16.5 and 11‰ S (Fig. 1). At 7.5‰ S the RGR was much reduced and growth only started up after a recovery phase during which considerable portions of the original inoculum died (Fig. 7c). Two of the five cuttings did not survive at all at 7.5‰ S. Absolute linear growth rates of the Baltic strain suggest a wider salinity range for maximal or nearly maximal growth (Fig. 2).

Tip cuttings from the North Sea strain, however, grew and survived in seawater with salinities from 33‰ down to 16.5‰ (Figs 1, 2 and 4). Both relative and absolute (linear) growth rates were maximal or nearly maximal at 33 and 23‰ S. These cuttings grew rather well at 19‰ S and only limited growth occurred at 16.5‰ S. No growth occurred at all at 11‰ and 7.5‰ S during the 8 weeks of observation. After transfer of cuttings kept for 8 weeks at 11‰ salinity into seawater with a salinity of 23‰, 2 of the 5 cuttings produced a new growing tip in thallus tissue which had given the visual impression of being completely dead. Obviously only a single cell or a small group of cells just survived a salinity of 11‰.

Tip cuttings of *Phycodrys* from both the North Sea and Baltic Sea grew and survived at temperatures from −2 °C up to 20 °C. At 25 °C the tip cuttings died. Maximum growth occurred at 10 and 15 °C (Figs 5 and 6). Also the growth of tip cuttings of the Baltic strain kept at different temperatures in seawater with a salinity of 23‰ slightly exceeded that at 33‰ S at all temperatures tested (Figs 4 and 6).

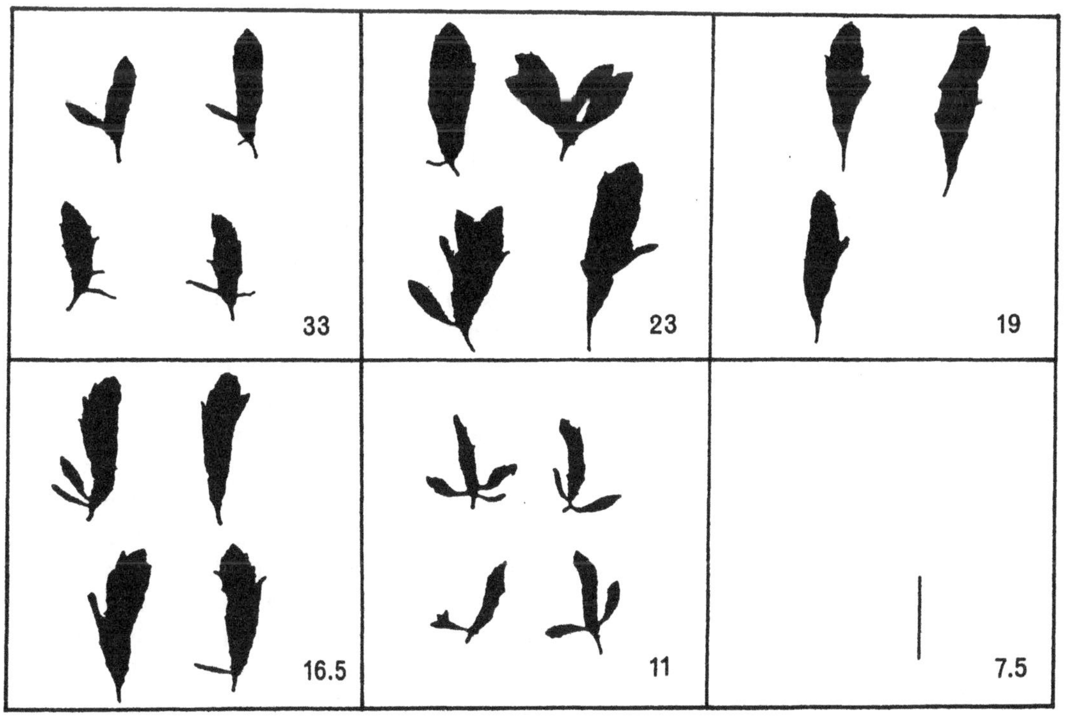

Fig. 3. Experimental plants of *Phycodrys rubens* from the Baltic Sea grown under salinity regimes varying from 33‰ down to 7.5‰. All plants are 7 weeks old and were derived from small about equal large tip cuttings. (Scale bar = 1 cm).

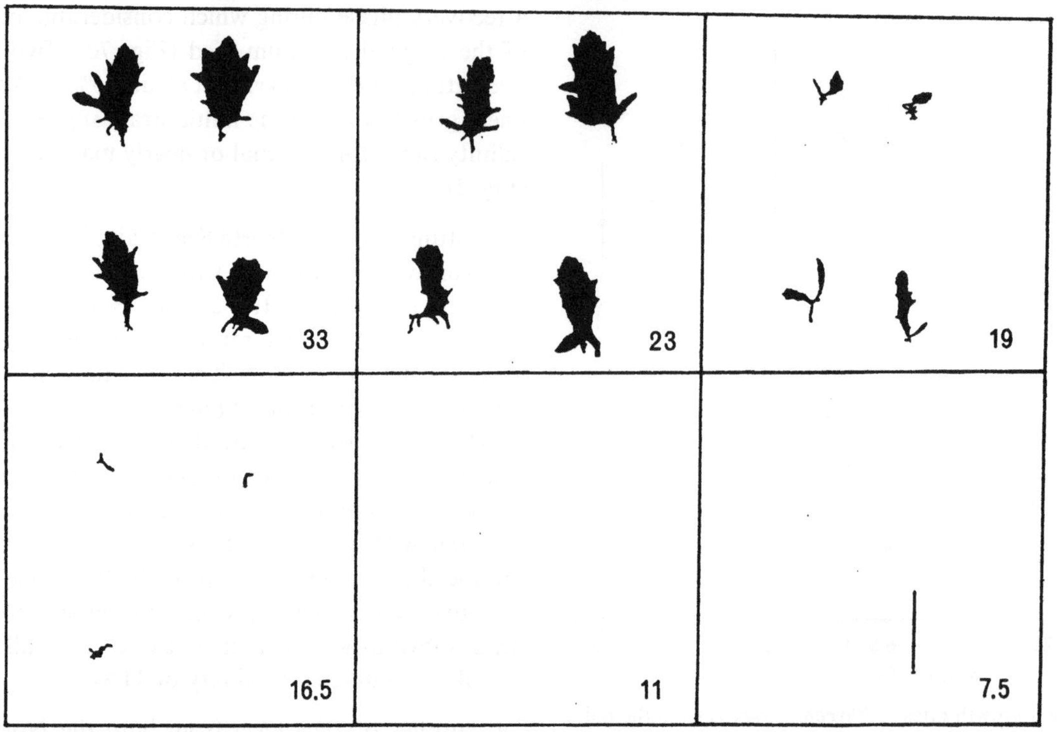

Fig. 4. Experimental plants of *Phycodrys rubens* from the North Sea grown under salinity regimes varying from 33‰ down to 7.5‰. All plants are 7 weeks old and were derived from small about equal large tip cuttings. (Scale bar = 1 cm).

Differences in morphology between Baltic and North Sea *Phycodrys* appeared distinctly from the length/width ratios of the foliaceous thalli of cultured plants (Figs 3, 4 and 7) and plants from nature (Fig. 8). For 15 6-weeks old Baltic and North Sea plantlets the average value amounted to 4.0 $\pm$ 0.43 and 2.4 $\pm$ 0.57, respectively. Differences in morphology also appeared in the margins of the thalli. The margin of Baltic plants were hardly creneated in contrast to North Sea plants (Figs 7 and 8). Obviously Baltic material had a tendency to form fewer new, distinct growing-points.

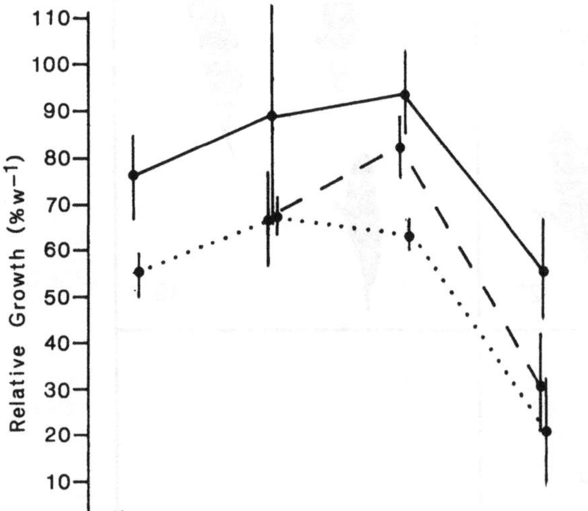

Fig. 5. Relative growth rates (RGR) of *Phycodrys rubens* from the Baltic Sea and North Sea at temperatures varying from 20 °C up to 5 °C. (Solid line = *Phycodrys* from the Baltic grown at 23‰ S; hatched line = *Phycodrys* from the Baltic grown at 33‰ S; dotted line = *Phycodrys* from the North Sea grown at 33‰; vertical bar = standard error of the mean).

Discussion

In culture, *Phycodrys rubens* from the Baltic exhibits a wider tolerance for reduced salinity than *Phycodrys rubens* from the North Sea, with optimum salinities (in the range 23–16.5‰) below that of seawater (33‰) and approximately those near Kiel (*ca* 14–20‰ S) (Smayda 1983). This suggest that the salinity preference is stable and genetically determined. Survival and limited growth of the Baltic *Phycodrys* at 7.5‰ S may explain the species innermost occurrence in the Baltic on the Island of Gotland where salinity is *ca* 7‰ (Pankow 1971). In contrast, North Sea *Phycodrys rubens* has a lower survival and growth limit at 16.5‰ and this cannot explain its geographic distribution in the Baltic. Apart from their different salinity tolerances and preferences *Phycodrys* from the Baltic and North Sea also differ morphologically from one another (Figs 3, 4, 6 and 7). All the above observations support the hypothesis of genetically determined or ecotypic divergence between North Sea and Baltic *Phycodrys rubens*. This contrast with

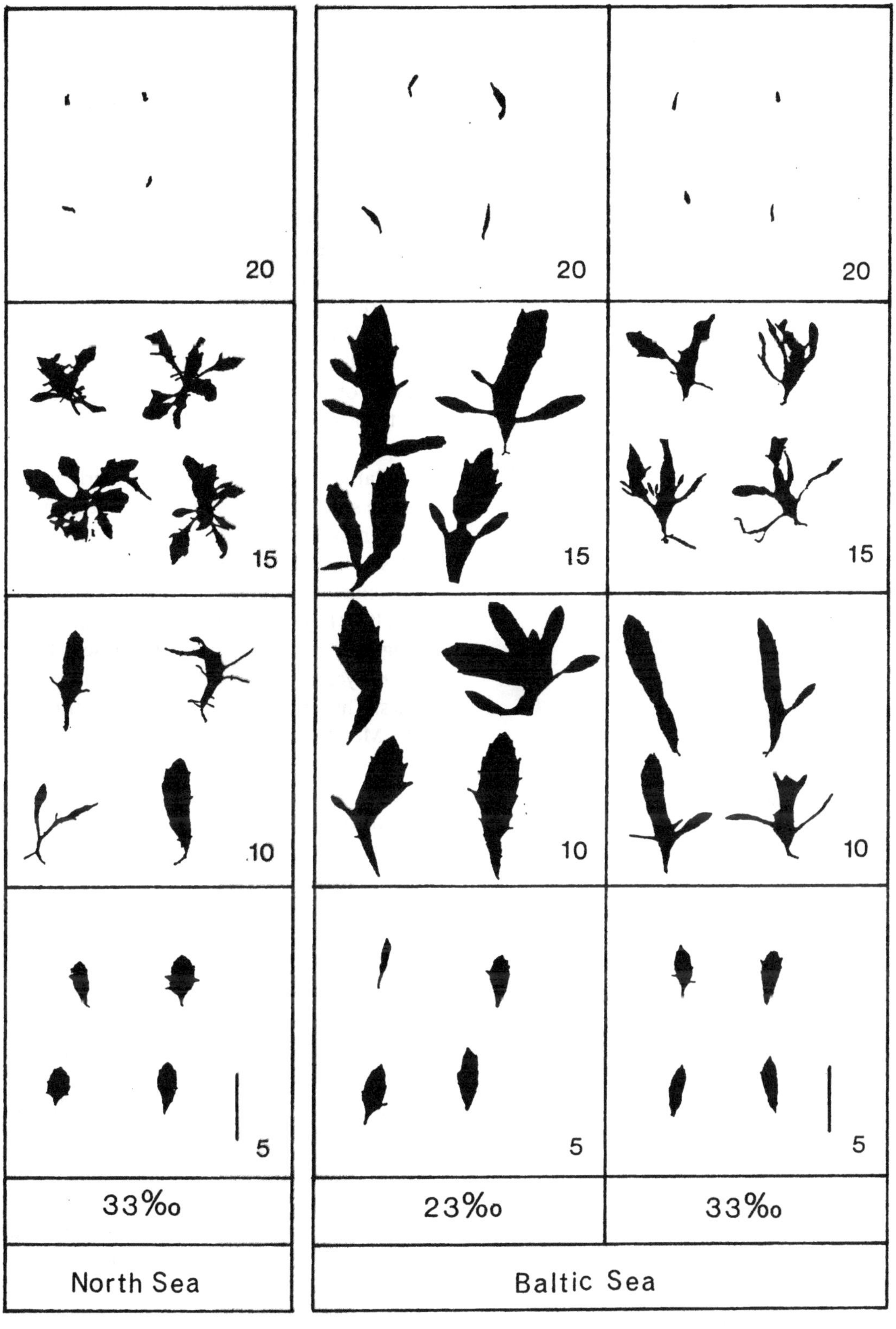

Fig. 6. Plants of *Phycodrys rubens* from the Baltic Sea and North Sea grown at salinities of 23 and 33‰ and 33‰ respectively. All plants are 8 weeks old and were grown from small, about equally sized tip cuttings. (Scale bar = 1 cm).

Schwenke's (1960) earlier conclusion that *Phycodrys rubens* from the Kiel Bight and Helgoland have the same range of salinity tolerance and that the Kiel Bight population lives near the species salinity limit. Schwenke's conclusion, however, was based on short term experiments in which the percentages of surviving cells after a 24 hours exposure to an abnormally lowered salinity (0–10‰ S) was used to determine differences in salinity tolerance. Russell (1988) used net photosynthetic oxygen production after 48 hours treatments with salinities of 0–102‰ to detect possible differences in salinity tolerance between Baltic

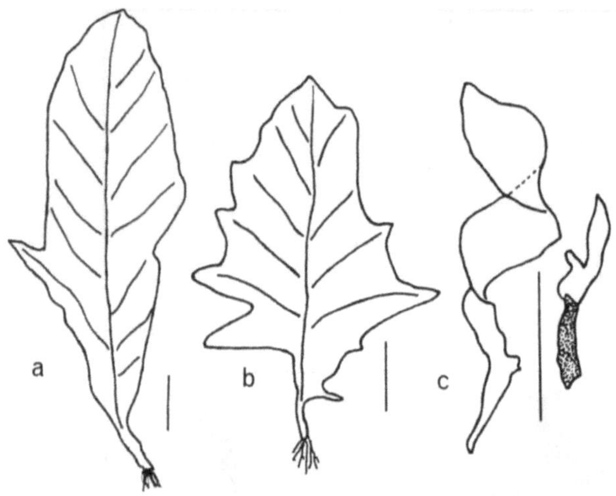

Fig. 7. Characteristic experimental plants of *Phycodrys rubens* from the Baltic Sea (a) and North Sea (b) grown for 6 weeks in salinities of 23‰ and of 2 *Phycodrys* plants from the Baltic grown for 6 weeks at 7.5‰ (c). (Scale bar = 2 mm).

Fig. 8. *Phycodrys rubens* plant from the Baltic collected in May 1988 in the Kiel Bight (a) and from the North Sea collected in May 1989 on Helgoland (b).

and extra-Baltic populations of the two brown algal species *Chorda filum* (L.) Stackh. and *Fucus vesiculosus* L. The maximum net O_2 production of plants from the Baltic and their extra-Baltic relatives occurred in salinities of their own natural habitat and he concluded therefore that they were adapted to prevailing salinity conditions with the reservation that the results of this type of short term experiment should be interpreted with caution. Rueness (1978) also found differences in salinity tolerance between Baltic and extra-Baltic populations of the red alga *Ceramium strictum* Harv. As in *Phycodrys rubens* the response pattern of both strains could be correlated with the salinities of their own natural habitats and he considered them as ecotypes. Baltic material of *Ceramium strictum*, differs in some morphological features from extra-Baltic material and was formerly considered a separate species [*Ceramium tenuicorne* (Kütz.) Waern] but its interfertility with the extra-Baltic material indicates conspecificity (Rueness 1978). In the past many endemic algal species were thought to exist in the Baltic (Lakowitz 1929) but it is unlikely that the geologically recent origin of the present Baltic in the last 3000 years has permitted the evolution of endemic species. The situation in the Baltic contrasts with that in the Mediterranean whose much older history (*ca* 5 million years) has probably permitted the origin of its numerous endemic algal species (*ca* 27% in the W. Mediterranean, *cf.* Prud'homme van Reine and van den Hoek 1990). Nonetheless it is interesting that the short geological history (thousands of years) of the Baltic has permitted the divergence of salinity ecotypes. Some of which are even morphologically different. This contrasts with the presumably long geologic periods (millions of years) required for the divergence of thermal ecotypes in seaweeds (Breeman 1988). In this context it is interesting that the Baltic strain and the North Sea strain of *Phycodrys rubens* had the same temperature responses, while their salinity responses were so different.

One common character of many algal species from the Baltic Sea is size reduction (Remane and Schlieper 1971). Also *Phycodrys* plants from the Baltic are generally very small. In culture, however, the absolute growth rates of the Baltic plants in their natural salinities were greater than those of *Phycodrys* from Helgoland (Figs 2, 3, 4 and 6). In their natural salinities the absolute growth rate of the Baltic *Phycodrys* amounted to about 3.7 mm/week and of North Sea *Phycodrys* to only 2.0 mm/week and this suggests that factors other than lowered salinity alone are responsible for size reduction. However, this difference in linear growth rate as observed in culture may be an experimental artefact and should be viewed with caution. To obtain different experimental salinities the

Provasoli-enriched North Sea water was diluted with Provasoli-enriched distilled water. It is quite conceivable that the dilution brought about the dilution of some adverse factor in the seawater by which the Baltic strain was consequently less affected at its lower salinity than the North Sea strain. In priniciple, the use of completely artificial sea-water is of course preferable, but this proved to be impossible as many algae failed to grow in it.

Acknowledgements

I gratefully acknowledge the use of working facilities and hospitality on the Institut für Meereskunde at Kiel. I would like to thank Mrs E. Kaminski for introducing me to the seaweeds from the Western Baltic Sea, Professor van den Hoek for his critical comments on the manuscript and Wiebe Kooistra for collecting the *Phycodrys* strain on Helgoland.

References

Breeman, A. M. 1988. Factors that keep seaweeds within their geographic boundaries; experimental and phenological evidence. *Helgoländer Meeresunters. 42*: 199–241.

Breuer, G. and W. Schramm. 1989. Changes in macroalgal vegetation of Kiel Bight (Western Baltic Sea) during the past 20 years. *Kiel. Meeresforsch. 6* (suppl.). (*Proc. Baltic Marine Biologists Symposium* 10): 241–255.

Cambridge, M., A. M. Breeman and C. van den Hoek. 1984. Temperature responses of some North Atlantic *Cladophora* species (Chlorophyceae) in relation to their geographic distribution. *Helgoländer Meeresunters. 38*: 349–363.

Christensen, T., C. Koch and H. A. Thomsen. 1985. *Distribution of Algae in Danish Salt and Brackish Waters*. Universtetsbogladen, København K.

Hoek, C. van den. 1982 a. Phytogeographic distribution groups of benthic marine algae in the North Atlantic Ocean. A review of experimental evidence from life history studies. *Helgoländer Meeresunters. 35*: 153–241.

Hoek, C. van den. 1982 b. The distribution of benthic marine algae in relation to the temperature regulation of their life histories. *Biol. J. Linn. Soc. 18*: 81–144.

Lakowitz, K. 1929. *Die Algenflora der gesamten Ostsee*. Danzig. pp. 1–474.

Pankow, H. 1971. *Algenflora der gesamten Ostsee* I. *Benthos*. VEB Fischer, Jena. pp. 1–419.

Prud'homme van Reine, W. F. and C. van den Hoek. 1990. Biogeography of Macronesian seaweeds. *Cour. Forsch.-Inst. Senckenberg 129*: 55–73.

Remane, A. and C. Schlieper. 1958. Die Biologie des Brackwassers. *In*: (A. Thienemann, ed.) *Die Binnengewässer*. E. Schweizerbart'sche Verlagsbuchhandlung, Stuttgart. pp. 1–348.

Rueness, J. 1978. Hybridization in red algae. *In*: (D. E. G. Irvine and J. H. Price, eds) *Modern Approaches to the Taxonomy of Red and Brown Algae*. Academic Press, London. pp. 247–262.

Russell, G. 1985. Recent evolutionary changes in the algae of the Baltic Sea. *Br. phycol. J. 20*: 87–104.

Russell, G. 1988. The seaweed flora of a young semi-enclosed sea: The Baltic. Salinity as a possible agent of flora divergence. *Helgoländer Meeresunters. 42*: 243–250.

Smayda, T. J. 1983. The phytoplankton of estuaries. *In*: (B. H. Ketchum, ed.) *Ecosystems of the World* 26. *Estuaries and Enclosed Seas*. Elsevier, Amsterdam. pp. 65–99.

Schwenke, H. 1960. Vergleichende Resistenzuntersuchungen an marinen Rotalgen aus Nord- und Ostsee. (Salzgehaltsresistenz). *Kieler Meeresforsch. 16*: 201–213.

Botanica Marina
Vol. 34, pp. 383–386, 1991

Assessment of Stocks of a Natural *Gracilaria* Population on Panay Island, Philippines

T. R. de Castro, N. G. Guanzon Jr. and Ma. R. J. Luhan

Aquaculture Department, Southeast Asian Fisheries Development Center, 5021 Tigbauan, Iloilo, Philippines

(Accepted 27 May 1991)

Abstract

Two peaks in biomass were recorded from natural beds of *Gracilaria* sp. at Leganes, Iloilo and Batan, Aklan study sites. The major peak occurred in February 1989 for both areas. The minor peak occurred in September 1988 at Batan and October 1988 at Leganes. Highest biomass at Ivisan, Capiz occurred in May 1988. Lowest biomass was recorded in June at Batan and December 1988 at Leganes. At Ivisan, no biomass was recorded from November 1988 to February 1989. Correlation analysis showed no relationship between biomass and temperature or pH at all study sites. However, salinity was negatively correlated with biomass at Leganes and Batan. Rainfall was inversely correlated with biomass. Based on salinity data gathered from the three study sites, *Gracilaria* sp. is euryhaline and can tolerate a wide range of salinity. The results show a marked seasonality in the biomass of *Gracilaria* sp.

Introduction

Gracilaria, a red algae, has long been used as human food in the Philippines and other parts of Asia. It is also one of the major sources of raw material for the manufacture of agar.

About 25 000 to 30 000 tons per year of dry *Gracilaria*, used in agar production, are harvested mainly from the wild (Santelices and Doty 1989). Although there is no available data on the export of *Gracilaria* from the Philippines, Japanese statistics give valuable information because Japan is a major export market for Philippine *Gracilaria*. In 1984 alone, Japan imported a total of 678 MT of *Gelidium* seaweed and 9462 MT of 'other seaweeds' most of which was *Gracilaria* (Armisen and Galatas 1987). Of the above total, 1470 MT of *Gracilaria* came from the Philippines.

The increasing demand and dwindling resources due to overharvesting of wildstocks have aroused interest in the culture of *Gracilaria*. Countries like Argentina (Mayer 1981, Borazo de Zaixzo 1987) and Chile (Black and Fonck 1981, Santelices *et al.* 1984, Poblete and Inostroza 1987) have good records on the management of standing stock in beds of *Gracilaria*. In Chile, resource-management programs for natural beds of *Gracilaria* have already started with the goal of conserving present stocks (Santelices *et al.* 1984, Poblete and Inostroza 1987). A study on the standing crop of *Gracilaria verrucosa* (Huds.) Papenfuss in Manila Bay (Trono and Azanza-Corrales 1981) is so far the only report from the Philippines. Thus, there is little information on the standing crop of *Gracilaria* in the Philippines especially in the Visayas where there are natural beds of *Gracilaria* which are being harvested throughout the year.

The study reported here was carried out to assess the standing stock in beds of *Gracilaria* sp. at three areas on Panay Island. The results will provide information on biomass seasonality. It will also provide baseline information for a proper management of existing resources as well as data for feasibility studies for large-scale cultivation.

Study Location and Methods

Three sampling sites (Fig. 1) were selected according to abundance and seasonality of *Gracilaria* sp.

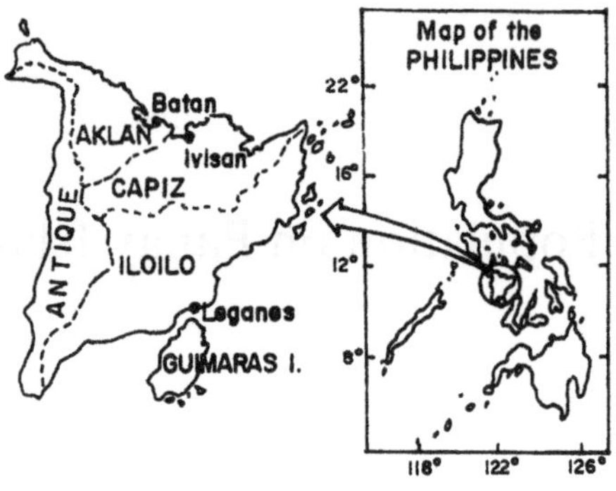

Fig. 1. Map of the Philippines indicating the location of the three study sites in Panay Island.

1. Leganes, Iloilo (122°36′N, 10°46′E). The site is located near the mouth of Gui-gui Creek. *Gracilaria* grows on a muddy substrate with water depth ranging from 0.3 to 1.5 m. The bed is approximately 4900 m^2. Effluents from nearby fishponds drain in this area before going out to the sea.

2. Palanas, Ivisan, Capiz (122°40′N, 11°32′E). The area is located in a cove with muddy bottom. The seaweed bed is approximately 19000 m^2 with water depth ranging from 0.3 to 1.0 m.

3. Batan, Aklan (122°30′N, 11°36′E). The site is located near the mouth of the river that flows into Batan Bay. The approximate area of the bed is 7900 m^2 with a water depth of 1–3 m.

The sampling device used was a 45 cm diameter ring (Trono and Butchan-Antalan 1987). The ring was thrown into the bed, and all the seaweed within the ring was collected and placed in one screen bag. Fifteen to 21 ramdom samples, depending upon the size of the bed, were collected monthly from each site from April 1988 to March 1989.

The samples were washed with fresh water to remove mud and associated organisms and weighed. The samples were then dried in an oven at 55–60 °C for 32–48 h. Standing crop biomass was expressed as dry weight of seaweed. Ecological parameters such as temperature, salinity and pH were measured at every sampling. Total rainfall at the sites for the duration of the study was obtained from the Philippine Atmospheric, Geothermal and Astronomical Services Administration (PAGASA).

Analysis of variance and Duncan's Multiple Range Test (DMRT) were used to compare the biomass for the different months at each study site. Correlation analysis was carried out to determine the relationship between biomass and temperature, salinity, pH, or total rainfall.

Results and Discussion

The genus *Gracilaria* abounds in the island of Panay but only a very few species form extensive beds. A survey by the authors on existing *Gracilaria* beds in Panay Island revealed that *Gracilaria* sp. (same as *Gracilaria* sp. 2, Trono *et al.* 1983) is found in commercial quantities. In some areas, it is harvested by townspeople and sold in the market for domestic consumption. This species was selected because of its abundance and its potential as an agar source. Preliminary screening of agar extracted from this species showed a high gel strength of over 500 g cm^{-2} (de Castro, unpublished data).

Two biomass peaks were recorded from beds at Leganes and Batan (Fig. 2). In Leganes, the major peak occurred in February 1989 (130.59 ± 8.39 g m^{-2} m^{-2}) and was significantly different from the other months ($P < 0.05$) except for January 1989 (120.17 ± 17.73 g m^{-2}). The minor peak was in October 1988 (64.77 ± 9.52 g m^{-2}) but was not significantly different from March 1989 ($P > 0.05$). At Batan, the major peak occurred in February 1989 (174.77 ± 13.47 g m^{-2}) and was significantly different from all the other months ($P < 0.05$). The minor peak occurred in September 1988 (95.01 ± 6.58 g m^{-2}) but was not significantly different from the months of August and March 1989 ($P > 0.05$). The higher biomass recorded at Ivisan occurred in May 1988 (42.32 ± 8.64 g m^{-2}) and was significantly different from the other months ($P < 0.05$). The lowest biomass recorded at Leganes was in December 1988 (13.55 ± 2.02 g m^{-2}) but was not significantly different from the months of June and November 1988 ($P > 0.05$). At Batan, the lowest biomass was in June 1988 (17.49 ± 2.98 g m^{-2}) and was not significantly different from the months of April, May, July, November and December 1988 ($P > 0.05$). At Ivisan, no biomass was recorded for the months of November 1988 to February 1989. The low values for standing crop in June 1988 for all sites

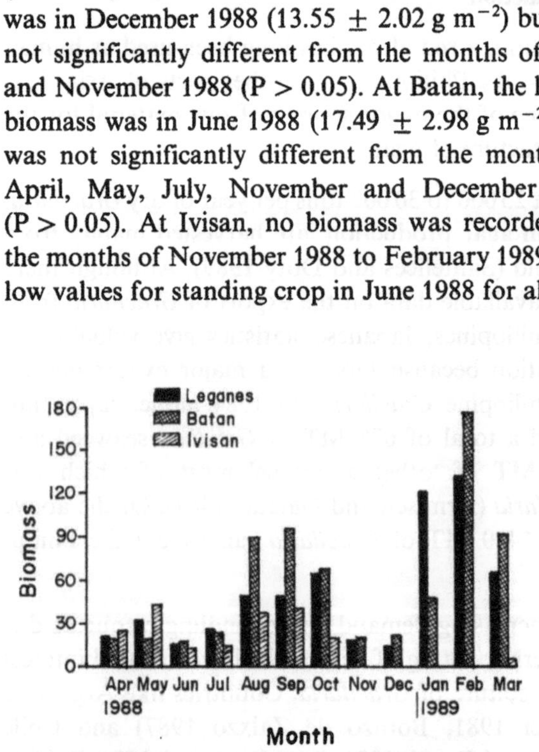

Fig. 2. Biomass (g m^{-2} dry weight) variations of *Gracilaria* sp. in the three study sites.

may be attributed to the shift from dry to wet season throughout the country (Fig. 3). This period is also characterized by strong wave action. Trono and Azanza-Corrales (1981) found that strong wave action during the rainy season can cause direct physical damage and uproot or shorten the thalli of *G. verrucosa* in Manila Bay. The above findings may also very well explain the low standing crop in June for all study sites. Correlation analysis showed that there is an inverse relationship between biomass and rainfall at all study sites ($P < 0.05$).

Several typhoons in late October until November 1988 greatly affected the standing crop at all study areas. At Ivisan, the whole area was covered with sand and silt so that no *Gracilaria* could be collected from November 1988 to February 1989. At Leganes and Batan, the standing crop was drastically lowered in November and December 1988. Borazo de Zaixso (1987), Brawley and Fei (1988), and Santelices *et al.* (1984) in their studies observed that the standing crop was greatly affected by losses due to storms. Santelices *et al.* (1984) observed that recovery, after storms, by *G. verrucosa* was unexpectedly fast due to the growth of erect thalli from portions buried in sand. Recovery after the storms was fastest at Leganes with the standing crop in January 1989 (120.17 ± 17.73 g m^{-2}) almost ten times that recorded in December 1988 (13.55 ± 2.02 g m^{-2}). The slowest recovery was at Ivisan, where visible growth was not seen until March 1989.

Sea temperatures ranged from 28–36 °C at Leganes, 27–33 °C at Batan and 28–34 °C at Ivisan (Fig. 3). Correlation analysis showed that temperature had no effect on the biomass at the three study sites ($P > 0.05$). This finding is similar to that of Trono and Azanza-Corrales (1981) for *G. verrucosa* in Manila Bay. Hurtado-Ponce and Umezaki (1987), however, showed that temperature had a significant effect on the growth of *G. verrucosa* from Manila Bay under controlled laboratory conditions with an optimum growth rate at 25–30 °C.

Salinity ranged from 19 to 33‰ at Leganes, 22 to 32‰ at Batan and 24 to 32‰ at Ivisan (Fig. 4). Correlation analysis showed that biomass is negatively correlated with salinity at Leganes and Batan but not at Ivisan ($P < 0.05$). Based on salinity data recorded from the three sites, *Gracilaria* sp. 2 is euryhaline because it can tolerate a wide range of salinity. Bird and McLachlan (1986) in their study of 17 species and strains of *Gracilaria* found that they were generally euryhaline under experimental conditions of long term exposure. They also indicated that maximum growth for the species they studied occurred at salinities of 15 to 38‰.

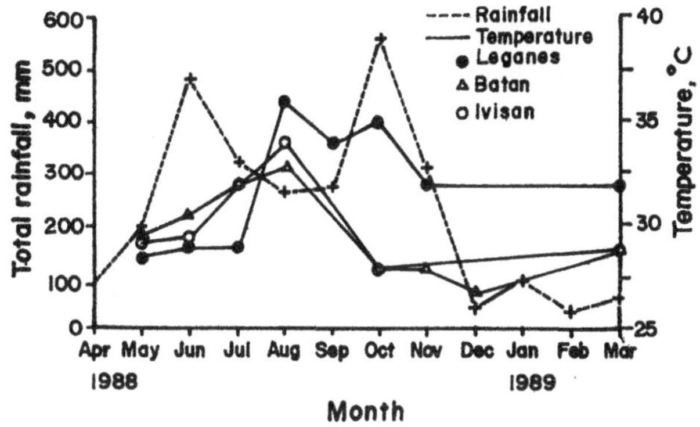

Fig. 3. Total rainfall and temperature readings in the three study sites.

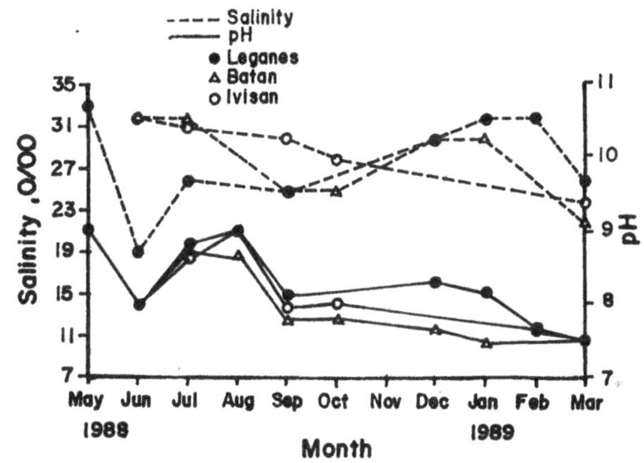

Fig. 4. Salinity and pH in the three study sites.

The pH levels ranged from 7.5–9.0 at Leganes, 7.4–8.7 at Batan and 7.5–9.0 at Ivisan (Fig. 4). Correlation analysis showed no relationship between biomass and pH at the three study sites ($P > 0.05$). This finding is similar to those of Trono and Azanza-Corrales (1981) for *G. verrucosa* in Manila Bay.

Associated plants found at the Batan, Aklan study site and observed throughout the year were *Acanthophora* and *Hypnea*. These grew luxuriantly from August to October. *Lyngbya* was observed entangled in *Gracilaria* at Leganes and Ivisan but only in small quantities.

This study shows that the standing stock of *Gracilaria* sp. was subject to a marked seasonality and this may also apply to any attempts to farm these species. It was also shown that the biomass varied from one study site to another. It is therefore important that proper management of the beds be instituted so that existing beds are not depleted. Proper management of the *Gracilaria* beds is also important as it will indicate how, when and how much of the bed can be harvested. Data on biomass seasonality will help determine the most suitable seasons for large-scale cultivation which in turn will give the highest yield.

Acknowledgements

We wish to thank Dr Gavino C. Trono Jr. for the advice and guidance he gave in the implimentation of this study and also for the critical review of this paper. Ms Larni Angeli Espada helped in the statistical analyses of the results.

References

Armisen, R. and F. Galatas. 1987. Production, properties and uses of agar. *In*: (D. J. McHugh, ed.) *Production and Utilization of Products from Commercial Seaweeds.* FAO Fish Tech. FAO Rome Pap. No. 288. pp. 1–49.

Bird, C. J. and J. McLachlan. 1986. The effect of salinity on distribution of species of *Gracilaria* Grev. (Rhodophyta: Gigartinales): An experimental assessment. *Bot. Mar. 29*: 231–328.

Black, H. J. and E. Fonck O'B. 1981. On the vegetation dynamics of *Gracilaria* spec. in Playa Changa, Coquimbo, Chile. *Proc. Int. Seaweed Symp. 10*: 223–228.

Boraso de Zaixso, A. L. 1987. *Gracilaria verrucosa* in Golfo Nuevo, Chubut, Argentina. Biological parameters and environmental factors. *Hydrobiologia. 151/152*: 239–244.

Brawley, S. H. and X. Fei. 1988. Ecological studies of *Gracilaria asiatica* and *Gracilaria lemaneiformis* in Zhanshan Bay, Qingdao. *Chin. J. Oceanol. Limnol. 6*: 22–34.

Hurtado-Ponce, A. Q. and I. Umezaki. 1987. Growth rate studies of *Gracilaria verrucosa* (Gigartinales, Rhodophyta). *Bot. Mar. 30*: 223–226.

Mayer, A. M. S. 1981. Studies on *Gracilaria* sp. in Bahia Arredondo, Chubut Province, Argentina. *Proc. Int. Seaweed Symp. 10*: 705–710.

Poblete, A. and I. Inostroza. 1987. Management of a *Gracilaria* natural bed in Lenga, Chile: A case study. *Hydrobiologia. 151/152*: 307–311.

Santelices, B. and M. S. Doty. 1989. A review of *Gracilaria* farming. *Aquaculture. 78*: 95–133.

Santelices, B., J. Vasquez, U. Ohme and E. Fonck. 1984. Managing wild crops of *Gracilaria* in Central Chile. *Hydrobiologia. 116/117*: 77–88.

Trono, G. C., Jr. and R. Azanza-Corrales. 1981. The seasonal variation in the biomass and reproductive states of *Gracilaria* in Manila Bay. *Proc. Int. Seaweed Symp. 10*: 743–748.

Trono, G. C. and T. Buchan-Antalan. 1987. Biomass production in *Acanthophora*-associated algal communities in Bacoor Bay, Cavite, Philippines. *Philipp. J. Sci.* Monograph No. 17. pp. 1–10.

Trono, G. C., R. Azanza-Corrales and D. Manuel. 1983. The genus *Gracilaria* (Gigartinales, Rhodophyta) in the Philippines. *Kalikasan, Philipp. J. Biol. 12*: 15–41.

Botanica Marina
Vol. 34, pp. 387–389, 1991

Change in the Anticoagulant Activity and Composition of a Fucan Sulfate from the Brown Seaweed *Ecklonia kurome* During Refrigerated Storage of the Fronds

T. Nishino and T. Nagumo

Department of Biophysics, School of Hygienic Sciences, Kitasato University, Kitasato 1-15-1, Sagamihara-shi, Kanagawa 228, Japan

(Accepted 28 May 1991)

Abstract

The fucan sulfate extracted from *Ecklonia* after it had been stored at −20 °C for 3 years differed from that extracted from fresh material. The anticoagulant activity of the extract from the stored material was only half that of a similar extract from fresh material.

Introduction

In a previous study, a potent anticoagulant fucan sulfate (C-II), which consisted of fucose (Fuc), galactose (Gal), glucuronic acid (GlcA) and sulfate in the molar ratio of 1.00 : 0.19 : 0.07 : 1.48 and had an average molecular weight of 21 000, was purified in 1986 from the brown seaweed *Ecklonia kurome* Okamura collected in the Japan sea in 1985 (Nishino *et al.* 1989). In 1988, for further study, we attempted to isolate fucan sulfate from the same seaweed fronds which had been stored at −20 °C until use. However, the fucan sulfate (C-II-88) obtained from the stored seaweed fronds differed very much from the fucan sulfate (C-II) isolated in 1986 with regard to the components and the anticoagulant activity, though both fucan sulfates were prepared from the same source. This fact suggests that the chemical and biological properties of the fucan sulfate may have changed during the period of storage of the seaweed fronds in a refrigerator. In order to confirm this suggestion, we attempted to isolate fucan sulfate from the seaweed fronds of *E. kurome* freshly collected in the Japan sea in 1989 and examined the properties of the carbohydrate. As expected, the properties of the newly obtained fucan sulfate (C-II-89) were similar to those of C-II but not of C-II-88.

We describe, herein, the compositional change of the fucan sulfate in the brown seaweed *E. kurome* during the period of preservation of the fronds in a refrigerator. The relationship between the anticoagulant activity and the primary structures of the fucan sulfates are also discussed.

Materials

The fucan sulfates, C-II, C-II-88 and C-II-89, were purified, according to the method described previously (Nishino *et al.* 1989), from three types of *E. kurome* frond;

(1) those fronds, which were freshly collected in 1985,

(2) those, which were the same fronds as (1) but stored at −20 °C until 1988, and

(3) those, which were freshly collected in 1989, respectively.

Methods

General methods were as described previously (Nishino *et al.* 1989). The constituents, the optical rotations, the infrared spectra, the molecular weights and the anticoagulant activities with respect to activated partial thromboplastin time (APTT), thrombin time (TT) and prothrombin time (PT) of each fucan sulfate were determined as described previously (Nishino *et al.* 1989). The absolute configurations of Gal in the fucan sulfate were determined by using galactose oxidase (Avigad *et al.* 1962 and Mourao and Perlin 1987). The hydrolyzates (Gal content, 0.4 mg) of the fucan sulfates were incubated with D-galactose oxidase (36 units, EC 1.1.3.9, from *Dactylium dendroides*, from Sigma) and peroxidase (16 units, EC 1.11.1.7, from horseradish, from Sigma) in 2 mL of 50 mM sodium acetate buffer, pH 7.0 for 2 h at 37 °C. Then, the reaction mixtures were boiled for 10 min to stop the reaction, and after the solvent had been removed by evaporation, the residues were trimethylsilylated with *N*-trimethylsilylimidazole-pyridine solution (TMSI-C, Gaskuro Kogyo Inc.). The amount of D-Gal oxidized with the enzyme was determined by measuring the Gal contents of the hydrolyzates before and after the enzyme-treatment by gas-liquid chromatography [3% SE-30 on Chromsorb W (3 mm × 2 m), at 125 °C] of the trimethylsilylated derivatives (internal standard; mannitol trimethylsilylate).

Table I: Properties of C-II, C-II-88 and C-II-89

Fucan sulfate	C-II[a]	C-II-88	C-II-89
Yield (%)[b]	0.8	1.1	0.8
$[\alpha]_D$ (c = 1.0, H_2O)	−129°	−125°	−98°
Molecular weight[c]	21 000	28 500	17 800
IR ν_{max}(KBr) cm^{-1}	1 240	1 238	1 240
	850	850	850
Components (mol. %)			
Fuc	36.6	29.4	42.6
Gal	6.8	19.7	4.7
Xyl	n. d.[d]	1.8	Trace
GlcA	2.4	1.0	0.8
sulfate	54.2	48.2	51.8
Anticoagulant activity (units · mg^{-1})[e]			
APTT	142	68	130
TT	11	13	19
PT	19	18	24

[a] Nishino *et al.* 1989.
[b] Expressed as a % of the dry weight of the seaweed fronds.
[c] Estimated by gel filtration on Sepharose CL-6B.
[d] Not detected.
[e] Expressed as units · mg^{-1} in relation to the activity of heparin (167 units · mg^{-1}).

Results and Discussion

The fucan sulfates, C-II, C-II-88 and C-II-89, each migrated as a single spot with the same mobilities on cellulose acetate membrane electrophoresis (200 V, 1 h) with 0.1 M zinc acetate, pH 6.6 and were eluted as a single carbohydrate peak from a Sepharose CL-6B column with 0.2 M sodium chloride.

The yields and physical, chemical and biological properties of the fucan sulfates are shown in Table I. Compared with C-II, the proportion of Gal in C-II-88 was significantly increased with decreases in the relative proportions of Fuc and sulfate. In addition to those results, C-II-88 was found to show a negative specific rotation as strong as C-II, though the former contained higher Gal and lower Fuc contents than the latter, suggesting that the absolute configuration of Gal in C-II-88 may be predominantly the L-form. Furthermore, Gal in the hydrolyzate of C-II was oxidized completely with galactose oxidase, indicating that all the Gal occurred as the D-isomer. On the other hand, about 80% of the total Gal in the hydrolyzate of C-II-88 was resistant to oxidation with the enzyme, indicating that most of the Gal in C-II-88 occurred as the L-isomer. The composition of C-II-89 was similar to that of C-II. Further tests showed that C-II and C-II-89 were fucan sulfates of the fucoidan type, whereas C-II-88 was of the galactofucan type. Although C-II and C-II-88 were obtained from similar seaweed fronds collected in the same period and at the same place but extracted at different times, their components differed. This result suggested that the component of C-II changed into that of C-II-88 during the period of storage of the seaweed fronds in a refrigerator, because the composition of C-II-89 from the fronds freshly collected was almost the same as that of C-II. In addition to this result, the molecular weight of C-II-88 was found to be much higher than that of C-II. This result indicated that the compositional change may not be caused by degradation of the polysaccharide with an enzyme such as glycosidase, but might be caused by an 'oxygenase', which converts L-Fuc into L-Gal, such as alkane 1-monooxygenase (EC 1.14.15.3, $RCH_3 \rightarrow RCH_2OH$) obtained from *Pseudomonas* (McKenna and Coon 1970). However, the mechanism remains to be clarified.

As for TT and PT, no significant difference in the activities of the fucan sulfates was observed. On the other hand, the anticoagulant activities of C-II-88 and C-II-89 were 48% and 92% of that of C-II with respect to APTT, respectively. This result suggested that the sugar components of fucan sulfates may have an influence on their anticoagulant activity because the activity of C-II-88 significantly decreased, as compared with those of C-II and C-II-89. In the fucan sulfates, C-II, C-II-88 and C-II-89, the respective molar ratios of sulfate groups to sugar residues were 1.18 : 1.00, 0.93 : 1.00 and 1.08 : 1.00. The slightly lower activity of C-II-89 than that of C-II may be due to the lower sulfate content of the former, because both fucan sulfates had similar sugar compositions.

It has been reported that the anticoagulant activity of heparin is dependent on the molecular weight of the polymer, indicating that the activity of fucan sulfates in the present study may also be related to their molecular weights (Laurent *et al.* 1978, Anderson *et al.* 1979 and Sache *et al.* 1982). In a previous study, we have reported that fucan sulfates with molecular weights ranging from about 10 000 to 30 000 showed similar anticoagulant activity with respect to APTT (Nishino *et al.* 1991). These results suggest that the higher contents of Fuc and sulfate and the lower content of another neutral sugar may be closely related to higher anticoagulant activity in fucan sulfates.

References

Anderson, L.-O., T. W. Holmer, E. A. Johnson and G. Soderstrom. 1979. Molecular weight dependency of the heparin potentiated inhibition of thrombin and activated factor X. Effect of heparin neutralization in plasma. *Thromb. Res. 15*: 531–541.

Avigad, G., D. Amaral and C. Horecker. 1962. The D-galactose oxidase of *Polyporus circinatus*. *J. Biol. Chem. 237*: 2736–2743.

Laurent, T. C., A. Tengbland, L. Thunberg, M. Höök and U. Lindahl. 1978. The molecular-weight-dependence of the anti-coagulant activity of heparin. *Biochem. J. 175*: 691–701.

McKenna, E. J. and M. J. Coon. 1970. Enzymatic ω-oxidation. IV Purification and properties of the ω-hydroxylase of *Pseudomonas oleovorans*. *J. Biol. Chem. 245*: 3882–3889.

Mourao, P. A. S. and A. S. Perlin. 1987. Structural features of sulfated glycans from the *Styela plicata* (Chordata-Tunicata). A unique occurrence of L-galactose in sulfated polysaccharides. *Eur. J. Biochem. 166*: 431–436.

Nishino, T., G. Yokoyama, K. Dobashi, M. Fujihara and T. Nagumo. 1989. Isolation, purification and characterization of fucose-containg sulfated polysaccharides from the brown seaweed *Ecklonia kurome* and their blood-anticoagulant activities. *Carbohydr. Res. 186*: 119–129.

Nishino, T., Y. Aizu and T. Nagumo. 1991. The relationship between the molecular weight and the anticoagulant activity of two types of fucan sulfates from the brown seaweed *Ecklonia kurome*. *Agric. Biol. Chem. 55*: 791–796.

Sache, E., M. Maillard, M. Bertrand, M. Maman, M. Kunz, J. Choay, J. Fareed and M. Messmore. 1982. Studies on a highly active anticoagulant fraction of high molecular weight isolated from porcine sodium heparin. *Thromb. Res. 25*: 443–458.

Botanica Marina
Vol. 34, pp. 391–401, 1991

The *Ascophyllum/Polysiphonia/Mycosphaerella* Symbiosis

II. Aspects of the Ecology and Distribution of *Polysiphonia lanosa* in Nova Scotia

D. J. Garbary, J. Burke and Tian Lining

Department of Biology, St Francis Xavier University, Antigonish, Nova Scotia, B2G 1C0, Canada

(Accepted 28 May 1991)

Abstract

The population ecology and distribution of *Polysiphonia lanosa* were studied at five sites along the Atlantic and Bay of Fundy coasts of Nova Scotia. *Polysiphonia lanosa* occupies a restricted portion of the *Ascophyllum nodosum* zone which is site dependent, and *P. lanosa* is not evenly distributed within its zone. Using five separate measures of *Polysiphonia* abundance, highest values at all sites occurred toward the lower end (but not at the lower limit) of the *A. nodosum* zone. The differential abundance and regular pattern of distribution of *P. lanosa* suggests that there is an optimal physical and/or competitive environment for the epiphyte in this part of the intertidal zone. The distribution pattern of *P. lanosa* on *A. nodosum* at all sites is highly contagious, as shown by high values for Morisita's index of dispersion ($I_d = 1.5$ to 8.3, significant at $p < 0.01$). At Tor Bay, at the two sampling heights where *P. lanosa* is most abundant, 31.7% and 14.5% of the *A. nodosum* fronds have no *P. lanosa*, and 10% of the *A. nodosum* fronds host 55.9% and 41.0% of the individuals of *P. lanosa*, respectively. Transplant studies suggest that the absence of *P. lanosa* in the southern Gulf of St Lawrence results neither from genetic features of the local *A. nodosum* populations, nor from an inability of of *P. lanosa* to survive environmental conditions in the southern Gulf. This absence may be accounted for by low *A. nodosum* densities in that part of the intertidal region where *P. lanosa* is most successful. The *Polysiphonia/Ascophyllum* symbiosis is explained using a hypothesis based on recruitment limitation.

Introduction

Polysiphonia lanosa (Linnaeus) Tandy is a marine red alga that grows almost exclusively on the fucoid *Ascophyllum nodosum* (Linnaeus) Le Jolis. It is occasionally reported on *Fucus* species (*e.g.* Cotton 1912, Feldmann 1954) and there is one report of *P. lanosa* on rock (Rattray 1885), although Dixon (1973) questioned if there was a complete absence of *Ascophyllum* in such cases. In addition, an ascomycete, *Mycosphaerella ascophylli* Cotton, is always found endophytic (parasitic?) in *A. nodosum* (Kohlmeyer and Kohlmeyer 1972, Garbary and Gautam 1989). Numerous observations have been made on various aspects of the symbiosis including morphological, ultrastructural, physiological and ecological studies of the symbionts and their interactions (*e.g.*, Webber 1967, Rawlence and Taylor 1970, 1972, Citharel 1972, Rawlence 1972, Penot 1974, Penot and Penot 1974, Harlin and Craigie 1975, Turner and Evans 1977, Fries 1979, 1980, Filion-Myklebust and Norton 1981, Higgins 1984, Lobban and Baxter 1983, Garbary and Gautam 1989, Pearson and Evans 1989, 1990, 1991), and Fries (1988) has shown that bacteria are also an important part of the symbiosis. Despite these studies, there is no explanation for the host specificity of *P. lanosa* and the biological basis for its interaction with *A. nodosum*.

Ascophyllum nodosum and *Polysiphonia lanosa* are widely distributed and abundant on temperate shores of the North Atlantic Ocean (Taylor 1957, South and Tittley 1986). The western Atlantic distributions of both *A. nodosum* and *P. lanosa* extend from New Jersey to the lower St Lawrence River, and Newfoundland. *Ascophyllum nodosum* extends further, and is also found in Hudson Bay, northern Labrador, James Bay, Hudson Strait, Baffin Island and Ellesmere Island (Taylor 1957, Lee 1980). Nova Scotia, with a diversity of oceanographic environments (*i. e.*, Atlantic coast, Bay of Fundy and Gulf of St Lawrence), is roughly in the middle part of the range of *P. lanosa*, and the species is locally very abundant. For this reason, Nova Scotia is suited for ecological studies of *P. lanosa*.

Although *P. lanosa* has a wide geographic distribution, and it is often quite abundant, there are few detailed autecological studies of this species. Several studies have dealt with the general ecology of *P. lanosa* from New England and eastern Canada as part of descriptive ecological studies of intertidal algae (*e. g.*, South and Hooper 1980, Mathieson *et al.* 1981 a, b, Thomas *et al.* 1983, Novaczek and McLachlan 1988). More detailed studies from New Brunswick by Lobban and Baxter (1983) described biomass distribution of *P. lanosa* along the primary shoots and lateral branches of *A. nodosum* with reference to year classes of the host. A further ecological study by Levin and Mathieson (1990) examined the abundance of *P. lanosa* in relation to exposure at three sites and two tidal heights (high and low).

As part a larger study of the *Ascophyllum*/*Mycosphaerella*/*Polysiphonia* symbiosis (see Garbary and Gautam 1989 for part I), this paper describes the distributional pattern of *P. lanosa* on *A. nodosum* and its intertidal and geographic distribution in Nova Scotia. The authors assumed that such studies would provide insight into the basis of host specificity of *P. lanosa* on *A. nodosum*.

Material and Methods

Mapping

The general distribution of *Polysiphonia lanosa* in Nova Scotia was established by reference to published literature, the Herbarium at the Institute for Marine Bioscience, National Research Council of Canada (NRCC), Halifax, N. S., and personal collections of the authors (specimens deposited at the Herbarium, Biology Department, St Francis Xavier University). These records and specimens were also used to establish the phenology and periodicity of *P. lanosa* in Nova Scotia (Table I).

Table I. The phenology and periodicity of *Polysiphonia lanosa* from Nova Scotia (+ = present, – = absent).

	J	F	M	A	M	J	J	A	S	O	N	D
Vegetative	+	+	+	+	+	+	+	+	+	+	+	+
Spermatangia	–	–	–	–	+	+	+	+	–	–	–	–
Carposporangia	–	–	–	–	–	–	+	+	+	+	–	–
Tetrasporangia	–	–	–	–	–	–	+	+	+	+	–	–

Study sites

Five study sites were chosen for studies of the intertidal distribution of *Polysiphonia lanosa*: Cape Canso and Tor Bay, Guysborough Co., Cranberry Cove, Halifax Co., Chebogue Point, Yarmouth Co. and Sandy Cove, Digby Co. (Fig. 1). The first three sites are along the Atlantic coast whereas the second two sites are in the Bay of Fundy and its approaches.

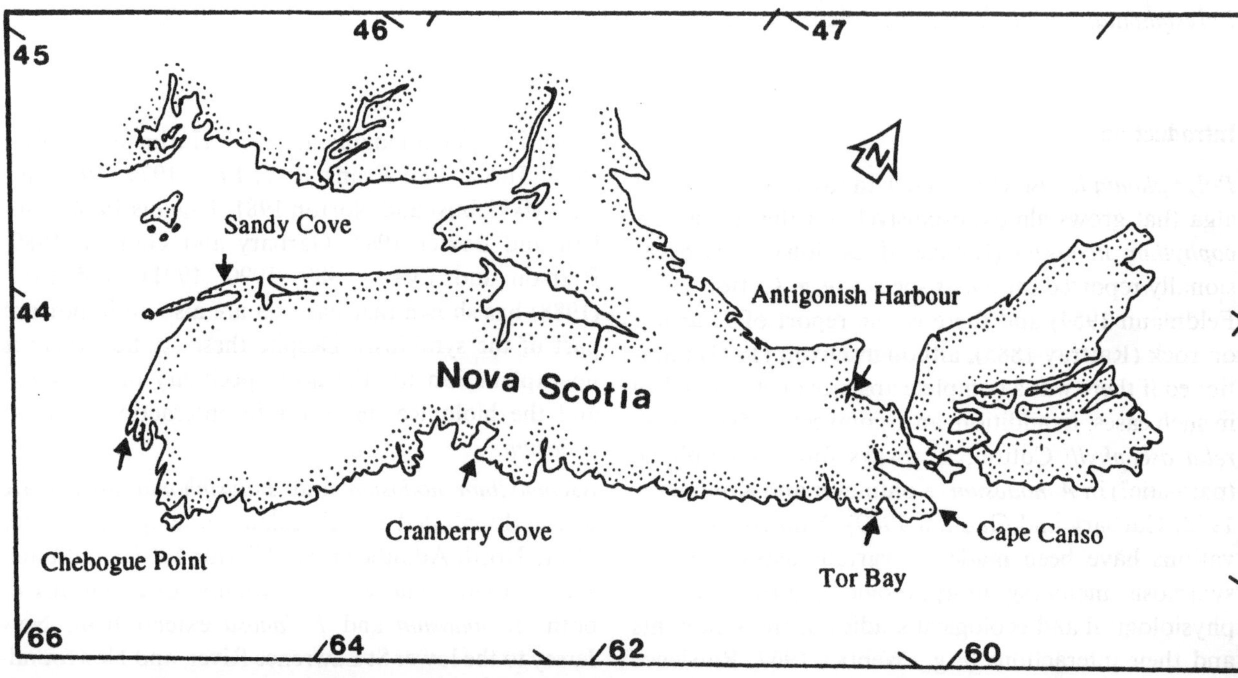

Fig. 1. Map of Nova Scotia indicating locations of study sites.

Sandy Cove and Cranberry Cove represent sites with continual rock slope (or very large boulders), whereas Chebogue Point, Tor Bay and Cape Canso are sites with large cobble or boulder substrata. All sites had a conspicuous *A. nodosum* zone through the mid-intertidal region where this species was the dominant macrophyte, and at all sites except Cape Canso, *A. nodosum* formed a virtually complete canopy layer over of the substratum with multiple layers of overlapping fronds.

Intertidal distribution

At each site, the vertical distribution limits of *A. nodosum* and *P. lanosa* were determined by means of a transect laid out perpendicular to the shore. Within its vertical range the distribution of *P. lanosa* was described at various sampling heights using a range of quantitative measures. Only four tidal heights were sampled at Cape Canso (because of the highly restricted vertical distribution of *P. lanosa*) and Cranberry Cove (because of the very small intertidal zone that was present). At the remaining sites six tidal heights were sampled for *P. lanosa.* In all cases, uppermost tidal height sampled was near the upper limit of distribution of *P. lanosa,* and lowermost height was near the lower limit of distribution for *P. lanosa* (Figs 2–3). In initial studies carried out during 1985 (all sites except Tor Bay), six 20 cm × 20 cm quadrats were evenly placed along a two-metre span perpendicular to the transect line, and up to five *A. nodosum* fronds were randomly chosen from separate plants in each quadrat based on the occurrence of the holdfast within the quadrat. In more detailed studies at Tor Bay, 50 cm × 50 cm quadrats were used, and up to ten fronds from separate holdfasts were collected from each quadrat. In all cases, fronds of *A. nodosum* were collected after shedding of receptacles.

Distribution of Polysiphonia lanosa *on* Ascophyllum nodosum

In the laboratory, the number of *P. lanosa* clumps, the length of the largest *P. lanosa* plant and total weight (fresh-damped) of *P. lanosa* on each *A. nodosum* frond were determined. The *P. lanosa* free *A. nodosum* frond was then weighed, and the maximum frond length was measured. In addition to basic descriptive statistics, one-way analysis of variance (AN-

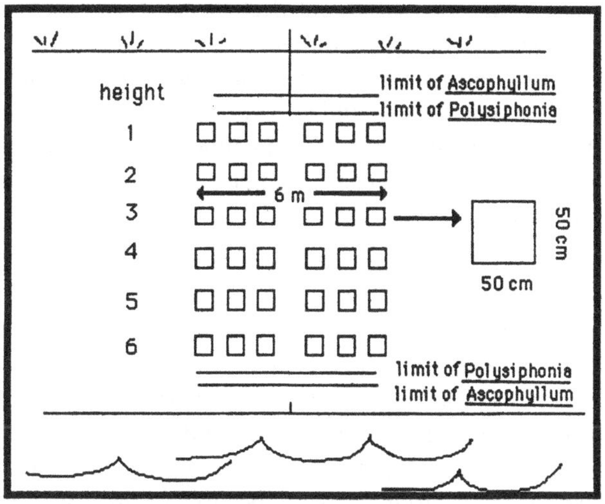

Fig. 2. Diagrammatic representation of the quadrat sampling used for *Polysiphonia lanosa.* Note that at Tor Bay larger quadrats (50 cm × 50 cm versus 20 cm × 20 cm) and greater number of fronds (10 versus 5) were used than at other sites.

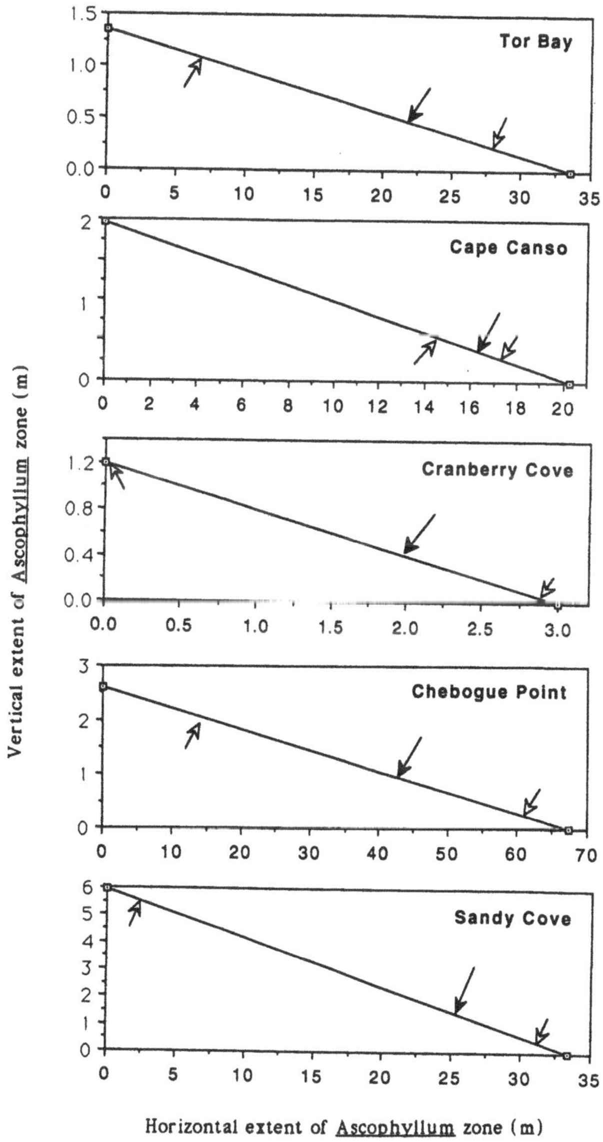

Fig. 3. Shore profiles of study sites indicating vertical and horizontal extents of *Ascophyllum nodosum* distribution. Vertical extent of *Polysiphonia lanosa* indicated with empty arrows; region of greatest abundance of *P. lanosa* indicated with solid arrows.

OVA) and Krustal-Wallis one-way non-parametric ANOVA (Goldman and Weinberg 1985) were performed to determine whether or not significant differences occurred in the measured parameters at the various heights. The Krustal-Wallis ANOVA was used because variances for various parameters were non-homogeneous according to the Bartlett test (Zar 1984). Statistical analyses were carried out using SYSTAT, version 5.0 (Wilkinson 1989), on a Macintosh SE/30 computer.

Morisita's index of dispersion (Brower and Zar 1984) was used to evaluate the distribution of numbers of *P. lanosa* plants on *A. nodosum* fronds at all tidal heights at Tor Bay, and at tidal heights with greatest numbers of *P. lanosa* plants at other sites. For this analysis, each *A. nodosum* frond was considered as a separate sample, and the extent to which *P. lanosa* has a random, uniform or aggregated distribution was determined and tested for significance. The index of distribution (I) is:

$$I_d = n(\Sigma X^2 - N)/N(N - 1)$$

where:

n = the number of *A. nodosum* fronds sampled at each vertical level;

N = the total number of *P. lanosa* clumps counted on all *A. nodosum* fronds;

ΣX^2 = the squares of the numbers of *P. lanosa* per *A. nodosum* frond, summed over all fronds.

The distribution can be generalized by comparing the calculated value of the index with the following values: as I_d approaches 1.0 the distribution is becoming more random; as I_d approaches 0 it is becoming more uniform; as I_d approaches n it is becoming more aggregated. Calculated dispersion values were assessed for departure from randomness using a Chi-square test.

Transplantation experiments

Two transplant experiments were carried out. The first experiment was performed to determine whether or not *P. lanosa* from the Atlantic coast could survive a different environmental regime in the southern Gulf of St. Lawrence. In one experiment, four rocks with over fifty attached *A. nodosum* fronds were moved from Tor Bay to the tidal rapids at the entrance to Captain's Pond, Antigonish Harbour on 26 July 1989. This site was selected because it has a relatively dense population of *A. nodosum* and remains ice-free throughout the winter. Host plants on these rocks bore from 15 to 350 macroscopically visible *P. lanosa* individuals or clumps. Rocks were retrieved on 1 October 1989 and 4 January 1990, and examined for survival of *P. lanosa.*

The second experiment was performed to determine whether or not fucoids from the southern Gulf of St Lawrence could become infected when exposed to *P. lanosa.* Ten rocks with 43 *Ascophyllum nodosum* fronds, and six rocks with 25 fronds of *Fucus vesiculosus* were transplanted from Antigonish Harbour to Tor Bay on 26 July 1989. Plants were placed in the intertidal zone where *Polysiphonia lanosa* was most abundant.

Results

Distribution of Polysiphonia lanosa *in Nova Scotia*

Polysiphonia lanosa is widely distributed in Nova Scotia including the outer reaches and primary portions of the Bay of Fundy, the Atlantic coast of mainland Nova Scotia and Cape Breton Island (Fig. 1). Relative to *A. nodosum*, however, which is found in most marine and brackish-water sites, *P. lanosa* has a restricted distribution in that it is apparently absent in the upper reaches of the Bay of Fundy, protected shores and inlets along the Atlantic coast, Bras d'Or Lake and Nova Scotian shores facing the Gulf of St Lawrence, all of which have greater or lesser populations of *A. nodosum.*

Seasonality and phenology (Table I)

Polysiphonia lanosa is present throughout the year, although it is reproductive only from May to September/October. Only unisexual plants were observed. *Polysiphonia* spermatangia were first seen in May, were common in June and declined in July. Carposporangia and tetrasporangia were mature by mid-July. In late July and throughout August, both kinds of sporangia were abundant. There was a decline in sporangial production in September, and by early October, very few carposporangia and tetrasporangia were present.

Numerous *P. lanosa* sporelings were found on *A. nodosum* in the late summer; these plants overwinter and presumably become reproductive the following summer. Overwintering plants have numerous broken branches and a dense covering of microalgal epiphytes in the spring. This epiphyte load, and the presence of brighter red (non-epiphytized) regenerating branches, suggest that little growth occurs during winter months. In the spring, some plants regenerated from single axes emerging from the *A. nodosum* thallus. Reproductive plants have survived at least one complete year. Although the potential life span of a single individual has not been resolved, the presence of large regenerating thalli in the early spring suggests that plants can survive at least two years.

Intertidal distribution

There was considerable variation in the percentage of the vertical range of *A. nodosum* that was occupied by *P. lanosa* (from 10–90%) (Fig. 3). At Cranberry Cove, *P. lanosa* was present virtually throughout the *A. nodosum* zone, whereas at Cape Canso *A. nodosum* occurred only in a narrow band about ¾ of the way down the *A. nodosum* zone. The primary study site at Tor Bay had 65% of the vertical extent of *A. nodosum* with epiphytic *P. lanosa*. At Cape Canso, where the *P. lanosa* was highly restricted (*i.e.* only 10% of the vertical extent of *A. nodosum*), it occurred at a tidal height equivalent to that with maximal growth of *P. lanosa* at other sites.

Quantification of *P. lanosa* abundance along its intertidal gradient showed a similar pattern at the five study sites (Table II). *Polysiphonia lanosa* tended to be less abundant near the top of its band, rose to peak about ¾ of the way towards the bottom of its distribution, and then declined toward its lower limit. This pattern was consistent regardless of the vertical extent of the *A. nodosum* zone or distance across the intertidal zone. The optimal growth region was at levels 4 or 5 (out of six) or level 3 (out of four) with all measures of growth (length of largest *Polysiphonia* clump per *Ascophyllum* frond, total mass of *Polysiphonia* per *Ascophyllum* frond, number of *Polysiphonia* clumps per *Ascophyllum* frond, mass or number of *Polysiphonia* per cm of *Ascophyllum*) being greatest in this region. The only exception to this pattern was at Cranberry Cove where the length of the largest *P. lanosa* was greatest at level 1 and then declined slightly. Although *P. lanosa* was present at

Table II. Five aspects of *Polysiphonia lanosa* growth or abundance, and length variation in fronds of *Ascophyllum nodosum* measured at the five study sites (mean ± s.e.). Note: the first three measures of *Polysiphonia* growth or abundance show values per *Ascophyllum* frond, whereas the second two show values per cm of *Ascophyllum* frond.

	Relative Tidal Height (high to low)					
Site	1	2	3	4	5	6
Polysiphonia growth and abundance measures						
TorBay						
Number per frond	0.60 ± 0.16	2.2 ± 0.81	7.02 ± 2.32	11.92 ± 2.82	21.00 ± 3.39	3.98 ± 1.16
Longest (cm) per frond	0.09 ± 0.02	0.18 ± 0.07	0.33 ± 0.07	0.90 ± 0.14	0.91 ± 0.11	0.31 ± 0.06
Mass (g) per frond	0.00 ± 0.00	0.03 ± 0.02	0.04 ± 0.03	0.17 ± 0.06	0.23 ± 0.12	0.02 ± 0.01
Number per cm	0.02 ± 0.00	0.06 ± 0.02	0.16 ± 0.04	0.21 ± 0.04	0.54 ± 0.07	0.08 ± 0.02
Mass (g) per cm	0.000 ± 0.00	0.001 ± 0.00	0.000 ± 0.00	0.004 ± 0.001	0.005 ± 0.002	0.000 ± 0.000
Ascophyllum length (cm)	21.3 ± 1.3	23.2 ± 1.5	32.0 ± 2.1	38.0 ± 2.7	32.0 ± 1.6	40.2 ± 2.3
Sandy Cove						
Number per frond	0.6 ± 0.3	1.5 ± 0.5	6.5 ± 1.0	7.2 ± 1.3	14.4 ± 2.4	3.8 ± 1.0
Longest (cm) per frond	33.2 ± 9.2	32.1 ± 6.8	44.5 ± 5.3	51.4 ± 3.9	38.6 ± 5.8	36.5 ± 7.8
Mass (g) per frond	0.11 ± 0.08	0.64 ± 0.30	2.30 ± 0.48	2.61 ± 0.05	2.86 ± 0.74	1.78 ± 0.68
Number per cm	0.01 ± 0.01	0.04 ± 0.01	0.15 ± 0.02	0.22 ± 0.03	0.36 ± 0.05	0.11 ± 0.03
Mass (g) per cm	0.002 ± 0.001	0.021 ± 0.009	0.050 ± 0.011	0.090 ± 0.018	0.071 ± 0.019	0.050 ± 0.019
Ascophyllum length (cm)	50.9 ± 2.8	38.7 ± 2.7	45.2 ± 2.8	33.4 ± 2.5	39.2 ± 1.6	42.2 ± 2.1
Chebogue Point						
Number per frond	14.5 ± 5.9	55.4 ± 17.2	25.6 ± 11.4	194.1 ± 62.3	19.2 ± 14.7	24.9 ± 9.6
Longest (cm) per frond	18.8 ± 3.5	25.4 ± 2.9	19.9 ± 3.5	27.6 ± 3.1	12.8 ± 5.2	21.3 ± 5.4
Mass (g) per frond	0.47 ± 0.20	2.79 ± 0.92	0.54 ± 0.37	5.71 ± 2.53	0.17 ± 0.15	1.53 ± 0.73
Number per cm	0.15 ± 0.06	0.54 ± 0.16	0.27 ± 0.12	1.54 ± 0.36	0.25 ± 0.16	0.27 ± 0.10
Mass (g) per cm	0.006 ± 0.003	0.026 ± 0.008	0.006 ± 0.004	0.040 ± 0.016	0.002 ± 0.002	0.018 ± 0.009
Ascophyllum length (cm)	74.7 ± 7.9	94.3 ± 6.0	62.0 ± 7.5	109.6 ± 10.9	59.7 ± 5.7	72.1 ± 7.7
Cape Canso						
Number per frond	5.6 ± 1.6	4.7 ± 1.5	18.6 ± 5.3	1.0 ± 0.9		
Longest (cm) per frond	3.6 ± 1.2	4.1 ± 1.5	15.3 ± 3.3	12.0 ± 11.0		
Mass (g) per frond	0.01 ± 0.01	0.02 ± 0.01	1.07 ± 0.40	0.04 ± 0.04		
Number per cm	0.28 ± 0.09	0.18 ± 0.05	0.51 ± 0.14	0.03 ± 0.02		
Mass (g) per cm	0.000 + 0.000	0.001 + 0.000	0.028 ± 0.011	0.001 ± 0.001		
Ascophyllum length (cm)	18.4 ± 2.1	22.3 ± 2.4	26.6 ± 3.4	21.4 ± 1.9		
Cranberry Cove						
Number per frond	9.5 ± 3.8	19.3 ± 3.6	24.0 ± 9.6	0.0 ± 0.0		
Longest (cm) per frond	26.7 ± 4.6	25.2 ± 2.5	22.3 ± 2.7	0.0 ± 0.0		
Mass (g) per frond	0.42 ± 0.22	0.79 ± 0.26	1.01 ± 0.54	0.0 ± 0.0		
Number per cm	0.09 ± 0.04	0.33 ± 0.16	0.37 ± 0.06	0.0 ± 0.0		
Mass (g) per cm	0.005 ± 0.002	0.011 ± 0.003	0.016 ± 0.009	0.0 ± 0.0		
Ascophyllum length (cm)	43.6 ± 5.4	65.4 ± 4.7	66.8 ± 4.4	35.5 ± 4.5		

height 4 at Cranberry Cove, its abundance was so low that it was not recovered by our sampling.

Within sites, mean frond length of *A. nodosum* is not constant through the portion of the intertidal zone sampled for *Polysiphonia lanosa* abundance, and at three sites these measures appeared correlated (Table II). Thus it is possible that some of our abundance measures (*i.e.* biomass and number of *Polysiphonia* per *Ascophyllum* frond) may be reflecting differences in *Ascophyllum* size, rather than real changes in *Polysiphonia* abundance. To account for this possibility, the data for *Polysiphonia* were also expressed as number of clumps and biomass per unit length of *Ascophyllum* frond. These data show a similar pattern of intertidal distribution as for average mass and number of clumps of *Polysiphonia* (Table II). All five measures of *Polysiphonia* abundance showed significant differences at different tidal levels ($p < 0.01$). The various aspects of *Polysiphonia* abundance and size were correlated with each other and with frond length and biomass of *Ascophyllum* (Tables III).

Table III. Correlation between number of clumps of *Polysiphonia lanosa* and mass of *P. lanosa* per frond of *Ascophyllum nodosum* with mass of *A. nodosum*, mass of *P. lanosa* per *Ascophyllum* frond, and the length of the largest *P. lanosa* plant per *Ascophyllum* frond. Data derived only from Tor Bay samples (in all cases $N = 360$ and $p < 0.01$).

	Number of clumps of *Polysiphonia* r	Mass of *Polysiphonia* r
Ascophyllum length	0.572	0.391
Ascophyllum mass	0.661	0.403
Mass of *Polysiphonia*	0.605	
Length of largest *Polysiphonia*	0.665	0.611

Table IV. The percentage of *Ascophyllum nodosum* fronds which did not host *Polysiphonia lanosa* and the percentage of *P. lanosa* individuals hosted by 10% of *A. nodosum* fronds. N = number of fronds sampled.

Height	N	% of *Ascophyllum* with no *Polysiphonia*	% of *Polysiphonia* on 10% *Ascophyllum*
1	61	72.1	61.1
2	58	77.6	76.3
3	60	46.7	59.9
4	60	31.7	55.9
5	60	14.5	41.0
6	65	46.1	63.5

Distribution pattern on Ascophyllum nodosum

Dispersion of *Polysiphonia lanosa*: The distribution of *P. lanosa* on *A. nodosum* is not random at any tidal level. This departure from randomness is shown graphically for levels 4 and 5 (Fig. 4) from Tor Bay, which have the greatest number of clumps. At these tidal levels the percentage frequency of *A. nodosum* is shown as a function of the number of *P. lanosa* individuals they host, and the highest frequencies are found for fronds with low numbers of clumps. These highly skewed distributions were even more apparent higher in the *P. lanosa* zone where over 70% of *A. nodosum* fronds had no *P. lanosa*, and up to 76.3% of *P. lanosa* clumps were found on only 10% of *A. nodosum* fronds (Table IV). Thus wherever *P. lanosa* occurs, it is concentrated on relatively few of the *A. nodosum* fronds.

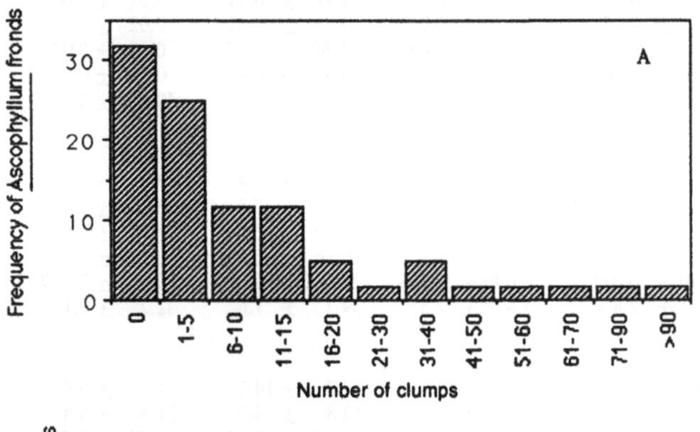

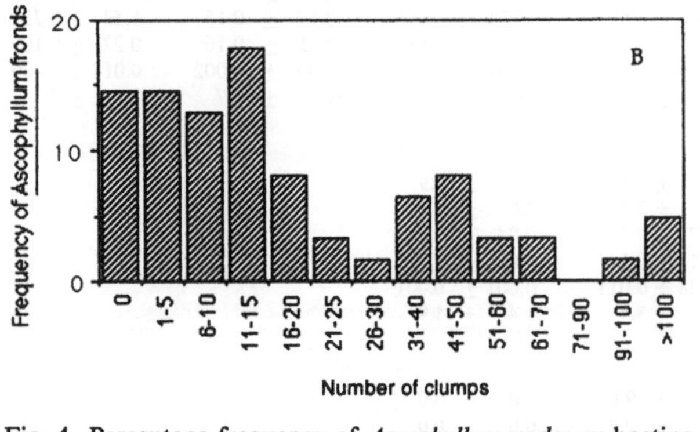

Fig. 4. Percentage frequency of *Ascophyllum nodosum* hosting different numbers of *Polysiphonia lanosa* individuals at Tor Bay: A) level 4 (N = 60, 11.9 ± 2.8, mean ± s.e.); B) Similar to A) but for level 5 (N = 60, 21.0 ± 3.4, mean ± s.e.). Note change of interval size for higher clump numbers.

Morisita's index of dispersion demonstrates that *P. lanosa* has a highly contagious spatial distribution on *A. nodosum*. This result was apparent for each level at Tor Bay (Table V) with the lowest I_d at level 5 where the largest, most dispersed (Table IV) *P. lanosa* population occurred. The contagious distribution was also observed from the populations at Cape Canso, Cranberry Cove, Sandy Cove, and Chebogue Point, where, because of smaller sample sizes, the index was computed only for the tidal height where *P. lanosa* was most abundant. Individual *A. nodosum* fronds vary in length and hence are not fully equivalent to

Table V. Morisita's index of dispersion (I_d) and the associated Chi^2 values calculated for the *Polysiphonia lanosa* populations at different tidal levels at Tor Bay and for tidal level of maximum abundance at other sites. n = number of *A. nodosum* fronds sampled. All Chi^2 values significant at $p < 0.01$.

Site	Height	n	I_d	Chi^2
Tor Bay	1	61	3.6	150.8
	2	58	8.3	988.9
	3	60	7.2	2636.9
	4	60	4.2	2360.1
	5	60	2.5	1939.5
	6	65	5.8	1204.9
Cape Canso	3	23	2.7	805.8
Cranberry Cove	3	20	4.0	1459.3
Sandy Cove	5	20	1.5	150.2
Chebogue Point	4	10	1.9	1799.2

a standard plot generally associated with Morisita's index of dispersion, thus introducing possible error into the analysis. However, the variation in size of *A. nodosum* fronds at each study site was relatively small (less than a doubling in mean size over tidal heights sampled) compared to the difference in mean number of *P. lanosa* individuals per frond where the minimal mean difference was over 13-fold (Sandy Cove).

Transplant studies

Transplantation of *Polysiphonia lanosa* to Antigonish Harbour: Two of the four rocks with the attached *A. nodosum* and *P. lanosa* were missing when examined on day 67. *Ascophyllum nodosum* fronds on one rock had 18 macroscopic *P. lanosa* individuals, three more than at the start of the experiment, suggesting growth of microscopic individuals. The one remaining rock was examined five months after transplantation, and a total of 240 *P. lanosa* plants was found (about 68% survival). *Ascophyllum nodosum* fronds adjacent to transplants were carefully examined for *P. lanosa*, but none were found.

Transplantation of fucoids from Antigonish Harbour to Tor Bay: After one month, no *P. lanosa* was found on the transplanted *A. nodosum* and some grazing by herbivores was apparent. After two months, three *P. lanosa* individuals, about 1 mm long, were found on a transplanted *A. nodosum* frond. No *P. lanosa* was found on transplanted *Fucus vesiculosus*. By this time, the herbivore damage was serious, and about half the fronds from each rock with *Ascophyllum* and *Fucus* plants had disappeared, and the remainder showed signs of heavy grazing. Previous transplant experiments indicated virtually complete removal of *Ascophyllum* (except for holdfast) after three months, and this experiment was terminated.

Discussion

Intertidal distribution

Generally, biological interactions (*e.g.*, Dayton 1971, 1975, Lubchenco 1980) and physical factors (*e.g.*, Schonbeck and Norton 1978, Druehl and Green 1982) are used to account for vertical distributions of intertidal algae. Upper distribution limits are largely considered to result from stresses associated with emersion and desiccation. In other words, physiological intolerance to physical factors such as desiccation during low tide is the primary causal factor setting the upper limits of algal distribution (Schonbeck and Norton 1978, Lubchenco 1980, Dring 1982, Underwood 1986). Some experiments have demonstrated that algae extend further toward shore in the intertidal zone where moisture conditions improved (see Underwood 1986). The lower limits of distribution in the intertidal zone are usually considered to be determined by interspecific competition (Dayton 1975, Lubchenco 1980, Schonbeck and Norton 1980, Dring 1982). Novaczek and McLachlan (1988) summarized much of this literature by stating that physical factors may determine the vertical extent of algae in the intertidal zone by exceeding limits of tolerance of individual species or by affecting competitive balances among species. When physical factors are not limiting, the boundaries of contiguous algal intertidal zones are mainly determined by competition and herbivory. It can be asked whether or not our observations on *Polysiphonia lanosa* are consistent with these models, or if other explanations are more satisfactory.

Paine (1979) found that the kelp, *Postelsia palmaeformis* Ruprecht, can be established at higher levels on the shore than it normally survives, and persist there for at least two natural generations. This suggests that increased desiccation at higher shore levels may not be the critical factor in determining the upper boundaries of these plants. An alternative explanation for the absence of algae at higher shore levels may be that the algae do not recruit in these areas. According to Dayton (1973) and Paine (1979) recruitment of *P. palmaeformis* at higher levels on the shore is difficult because of the limited power of spore dispersal. Also, because the submergence time per tide is shorter at higher tidal levels during each tidal cycle, the number of spores that float into upper-shore areas, and potentially attach to the substratum with each tidal cycle, is lower than would occur in the lower intertidal zone (Underwood 1986). Similarly, difficulties associated with recruitment, rather than physiological tolerance, may be important in explaining why *P. lanosa* is found in decreasing abundance towards the top of the *A. nodosum* zone.

At Tor Bay, the lower limit of *Polysiphonia lanosa* does not reach the lower limit of *Ascophyllum nodosum*, confirming previous observations (Mathieson *et al.* 1981 b, Thomas *et al.* 1983, Norvaczek and McLachlan 1988). Before assuming that biological interactions account for the lower boundary of *P. lanosa*, an interesting observation should be mentioned. According to Novaczek and McLachlan (1988), *P. lanosa* is usually absent in rock pools where *A. nodosum* is present, even though *P. lanosa* may be common on *A. nodosum* plants surrounding the rock pools. One might suggest, therefore, that a period of desiccation is critical for *P. lanosa*, and that *P. lanosa* may be intolerant of long periods of submersion, and therefore cannot reach the lower limit of *A. nodosum*. *Polysiphonia lanosa* may be similar to *Pelvetia canaliculata* (L.) Decaisne *et* Thuret which will not survive continuous immersion (Rugg and Norton 1987). Druehl and Green (1982) also showed that algal distributions correlated significantly with some aspects of their submersion/emersion histories, and they suggested that periodic desiccation plays an important role in determining vertical distributions of intertidal algae. Although differential herbivory as a causal factor cannot be ruled out, our observations on declining abundance of *P. lanosa* toward the lower limit of the *A. nodosum* zone, as well as those of Novaczek and McLachlan (1988) on the absence of *P. lanosa* in rock pools, support a similar interpretation for *P. lanosa*.

Polysiphonia lanosa is not evenly distributed within its vertical range, and all our measures of abundance showed significant differences along the tidal gradient. This difference in abundance of *P. lanosa* does not result from size variation of *A. nodosum*. Different environmental conditions prevail at different tidal levels and these may cause variation in abundance of *P. lanosa* which has a maximum development towards (but not at) the lower limit of the *A. nodosum* zone. Thus, regardless of the differences in vertical extent of the *A. nodosum* zone at the five sites (from *ca* 1 − 6 m), *P. lanosa* had the same distribution pattern in the intertidal zone at all sites.

Although we cannot explain the nature of the intertidal boundaries determining the upper and lower limits of *Polysiphonia lanosa*, it is clear that populations in Nova Scotia are not limited by the vertical distribution of the host (see Novaczek and McLachlan 1988 for information on *A. nodosum* zonation at numerous sites in Nova Scotia). One potential mechanism for limiting the intertidal distribution of *P. lanosa* is competitive exclusion caused by the occurrence of other epiphytes on *A. nodosum*. Accordingly, *Elachista fucicola* (Velley) Areschoug, which is also a common epiphyte on *A. nodosum* in Nova Scotia, may be occupying colonization sites. In general, however, *E. fucicola* was not a prominent epiphyte on *A. nodosum*, and most fronds were relatively free of macroscopic epiphytes (except *P. lanosa*).

Distribution pattern on Ascophyllum *thalli*

The contagious distribution *P. lanosa* may be characteristic of epiphyte populations in general. Kendrick and Hawkes (1988) reported that *Microcladia coulteri* (Harvey) Harvey has a highly contagious distribution on all three of its red algal host species, *Prionitis lanceolata* (Harvey) Harvey, *Iridaea cordata* (Turner) Bory and *Odonthalia floccosa* (Esper) Falkenberg. Kendrick and Hawkes (1988) suggested that epiphyte presence on a host thallus may enhance subsequent settlement and development of other individuals, thereby producing contagious distributions. Continued settlement of epiphyte spores on host thalli may be assisted by the proximity of reproductive epiphyte thalli. Hydrodynamic factors affecting the distribution of algae have been studied (Munteanu and Maley 1981, Sheath and Hambrook 1988), and continued settlement of epiphyte spores may be assisted by such physical factors. Pearson and Evans (1990) showed that spore settlement of *P. lanosa* is non-random on *A. nodosum*, and they suggested that hydrodynamic factors may be responsible.

Relationships between sites available for infection and the abundance of *Polysiphonia lanosa* were studied by Lobban and Baxter (1983), Burke (1986), Pearson and Evans (1989, 1990) and Levin and Mathieson (1990). These studies concurred that *P. lanosa* rarely occurred on branch segments, but almost exclusively in the axils of branches or sites of damage (*e.g.*, scars resulting from shedding of receptacles, damaged air bladders). Since most *Ascophyllum nodosum* plants have numerous potential sites, and these plants often occur in extremely dense stands, the contagious distribution of *P. lanosa* is almost surprising. Because of the high densities of *A. nodosum*, reproductive plants of *P. lanosa* from one host frond will contact other host fronds much of the time, thus leading to a more random epiphyte distribution. This apparent anomaly may be partly explained by the epidermis shedding of *A. nodosum* (Filion-Myklebust and Norton 1981). *Polysiphonia* produces abundant spores, and the fact that it is not more abundant suggests that it may have major dispersal restrictions that limit recruitment. The relationship between the timing of spore production and recruitment in *P. lanosa* is discussed below.

Transplantation of Polysiphonia lanosa *between different sites*

Previous investigations show that *P. lanosa* does not epiphytize all *A. nodosum* populations in Nova Scotia (McLachlan and Edelstein 1971, Bird *et al.* 1976, Burke 1986, Novaczek and McLachlan 1988). At Tor Bay transplanted *A. nodosum* (also *Fucus vesiculosus*) from Antigonish Harbour was more heavily grazed than were local populations. In addition, the *A. nodosum* fronds transplanted from Tor Bay to Antigonish Harbour were more prone to mussel colonization (2.69 versus 0.21 g mussel g^{-1} *A. nodosum*). These observations suggest possible differences in chemical composition among *A. nodosum* populations that might contribute towards differential occurrence of *P. lanosa*. However, the fact that some colonization of *P. lanosa* occurred on *A. nodosum* after transfer to Tor Bay, excludes the possibility that absence of *P. lanosa* in Antigonish Harbour is due to some some physical surface property, or chemical inhibition of *P. lanosa* by *A. nodosum* from the southern Gulf of St Lawrence. In addition, despite the fact that *A. nodosum* plants from these populations are morphologically different (unpublished observations), and may have fewer attachment sites (Levin and Mathieson 1990), this does not account for the absence of *P. lanosa* in Antigonish Harbour or the southern Gulf of St. Lawrence.

Phenology and recruitment of Polysiphonia lanosa

Spore maturation and release in *P. lanosa* occur primarily during July and August at Tor Bay, immediately following abscission of *A. nodosum* receptacles shed at this site during June and early July (Tian Lining 1990). Since the wound sites on *A. nodosum* are the primary locations for successful attachment of *P. lanosa*, and these are most abundantly formed by receptacle shedding (Lobban and Baxter 1983, Burke 1986, Levin and Mathieson 1990), the timing of spore release in *P. lanosa* is appropriate for recruitment. Similar timing of reproduction in these species in the eastern Atlantic was reported by Feldmann (1954).

Accounting for the geographic distribution of Polysiphonia lanosa *in Nova Scotia and Prince Edward Island*

Polysiphonia lanosa does not occur along the coasts of Nova Scotia in the Gulf of St Lawrence, nor is it present in adjacent Prince Edward Island (Bird *et al.* 1976, 1983, Lobban and Hanic 1984, Norvaczek and McLachlan 1988, personal observation) or Bras d'Or Lake in Cape Breton (McLachlan and Edelstein 1971, Burke 1986). The large seasonal temperature fluctuations in the southern Gulf of St Lawrence (relative to the Atlantic coast or Bay of Fundy) (Bird *et al.* 1976, Novaczek and McLachlan 1988), may partly account for the absence of *P. lanosa* in this region. Survival of *P. lanosa* transplants in Antigonish Harbour suggests, however, that the temperature regime is not the basis for its exclusion from the southern Gulf, although it is possible that spores or sporelings may be affected when larger plants are not. In addition, natural colonization of *A. nodosum* transplanted from Antigonish Harbour to Tor Bay suggests that the exclusion of *P. lanosa* from the southern Gulf does not have a genetic basis in an anti-epiphyte protection mechanism in local *A. nodosum* populations.

Along Atlantic and Bay of Fundy coasts where *P. lanosa* does occur, it is associated with dense populations of *A. nodosum*. In addition, it has a distinct intertidal distribution with a peak in abundance towards the lower portion of the *A. nodosum* zone. In the southern Gulf of St Lawrence intertidal *A. nodosum* populations are either extremely patchy, or consist of small, severely damaged plants in low densities, as a result of extreme ice scouring. Subtidal populations may be locally abundant in the southern Gulf region (Garbary, unpublished observations). We hypothesize that the absence of *P. lanosa* in the southern Gulf is explained largely by the absence of extensive *A. nodosum* populations in the part of the intertidal zone where *P. lanosa* is most successful. Without a large host population in appropriate microenvironments, *P. lanosa* may be simply unable to colonize.

An explanation of the Polysiphonia/Ascophyllum *symbiosis*

This study was carried out to further understanding of the symbiosis between *A. nodosum* and *P. lanosa*, in particular, the basis for the host-specificity that is usually present. It is important to stress at the outset that *P. lanosa* is not entirely host-specific, and that it may occur on various *Fucus* species, including *F. spiralis* L., *F. serratus* L. and *F. vesiculosus* L. (Cotton 1912, Feldmann 1954, Tian Lining 1990). Previously, the importance of having suitable settlement sites on *A. nodosum* for *P. lanosa* recruitment was stressed (Lobban and Baxter 1983, Burke 1986, Pearson and Evans 1989, 1990, Levin and Mathieson 1990), and we have noted a strong relationship betwen the formation of these sites (*e.g.* receptacle shedding) and spore maturation. Our study highlights the importance of position in the intertidal zone in limiting abundance. Even when these requirements are met, at tidal heights where *P. lanosa* is most abundant, about 30–55% of *A. nodosum* fronds may have five or fewer epiphytic individuals. This is low considering the hundreds of potential attachment sites per *Ascophyllum* plant (Burke 1986, Levin and Mathieson 1990).

Physiological studies have demonstrated that *Polysiphonia lanosa* is photosynthetically independent (Harlin and Craigie 1975, Turner and Evans 1977). Although transfer of fixed carbon may occur, it is generally not greater than can be accounted for by diffusion, and can be in either direction. Similar exchange was demonstrated for various ions (Penot 1974, Penot and Penot 1974). These physiological studies have been unable to demonstrate a basis for either host specificity or even the requirement of a host of any kind for *P. lanosa*. That *P. lanosa* requires a host can be concluded from the following observations:

1) it is always an epiphyte,

2) the rhizoid is highly specialized (Rawlence 1972, Rawlence and Taylor 1970, 1972),

3) sporelings, although readily obtained in unialgal culture from spores, eventually stop growing and die (Tian Lining and Garbary, unpublished observations), and

4) that secondary rhizoid production apparently does not occur unless brown algal exudates are present [*A. nodosum*, *Fucus* species and *Laminaria* all were effective (Pearson and Evans 1991)].

Based on the above information we suggest that *Polysiphonia lanosa* has a highly specialized niche in which tidal level (as well as other factors) is critical, in addition to the presence of a host with sufficient settlement sites to allow recruitment. Although there may be no physiological barrier to growth on *Fucus* species, fucoids other than *Ascophyllum nodosum* (except *F. vesiculosus*) usually do not grow in the correct environments at suitable tide levels, and rarely provide sufficient attachment sites at the appropriate time of year for recruitment to occur. Thus host specificity of *P. lanosa* on *A. nodosum* may be more apparent than real, and may be based on ecological factors which largely restrict colonization on to *Fucus* species, rather than a biochemical dependency that requires *A. nodosum*.

Acknowledgements

We thank Carolyn J. Bird for assistance with using the herbarium at NRCC. Lana Beth Barkhouse, Paul Fricker and Stephen Murphy provided valuable help with field work. Michael D. Guiry and two anonymous referees provided useful comments on the manuscript. This work was supported by a Natural Sciences and Engineering Research Council of Canada operating grant to D. G.

References

Bird, C. J., T. Edelstein and J. McLachlan. 1976. Investigations of the marine algae of Nova Scotia. XII. The flora of Pomquet Harbour. *Can. J. Bot. 54*: 2726–2737.

Bird, C. J., M. Greenwell and J. McLachlan. 1983. Benthic marine algal flora of the north shore of Prince Edward Island (Gulf of St. Lawrence), Canada. *Aquat. Bot. 16*: 315–335.

Brower, J. E. and J. H. Zar. 1984. *Field and Laboratory Methods for General Ecology*. Wm. C. Brown, Iowa.

Burke, J. M. 1986. Studies on the relationship between the obligate epiphyte *Polysiphonia lanosa* (Rhodophyta) and its host *Ascophyllum nodosum* (Phaeophyta). B. Sc. thesis, St. Francis Xavier University, Antigonish, Nova Scotia. 53 pp.

Citharel, J. 1972. Contribution à l'étude du métabolisme azoté des algues marines. Utilisation métabolique d'acide glutamique-^{14}C par *Ascophyllum nodosum* (Linne) Le Jolis et *Polysiphonia lanosa* (Linne) Tandy. *Bot. Mar. 15*: 157–161.

Cotton, A. D. 1912. Marine algae. *In*: (R. L. Praeger, ed.) A biological survey of Clare Island, (etc.). *Proc. R. Ir. Acad. 31*: 1–178.

Dayton, P. K. 1971. Competition, disturbance, and community organization: the provision and subsequent utilization of space in a rocky intertidal community. *Ecol. Monogr. 41*: 351–389.

Dayton, P. K. 1973. Dispersion, dispersal, and persistence of the annual intertidal alga, *Postelsia palmaeformis* Ruprecht. *Ecology 54*: 433–438.

Dayton, P. K. 1975. Experimental evaluation of ecological dominance in a rocky intertidal algal community. *Ecol. Monogr. 45*: 137–159.

Dixon, P. S. 1973. *Biology of the Rhodophyta*. Oliver and Boyd, Edinburgh.

Dring, M. J. 1982. *The Biology of Marine Plants*. Edward Arnold, London.

Druehl, L. D. and J. M. Green. 1982. Vertical distribution of intertidal seaweeds as related to patterns of submersion and emersion. *Mar. Ecol. Prog. Ser. 9*: 163–170.

Feldmann, J. 1954. *Inventaire de la Flore Marine de Roscoff. Supplément 6*. Travaux de la Station Biologique de Roscoff.

Filion-Myklebust, C. and T. A. Norton. 1981. Epidermis shedding in the brown seaweed *Ascophyllum nodosum* Le Jolis, and its ecological significance. *Mar. Biol. Let. 2*: 34–51.

Fries, L. 1988. *Ascophyllum nodosum* (Phaeophyta) in axenic culture and its response to the endophytic fungus *Mycosphaerella ascophylli* and epiphytic bacteria. *J. Phycol. 24*: 333–337.

Fries, N. 1979. Physiological characteristics of *Mycosphaerella ascophylli*, a fungal endophyte of the marine brown alga *Ascophyllum nodosum*. *Physiol. Plant. 45*: 117–121.

Fries, N. 1980. Chemical factors affecting ascospore germination and mycelial development in *Mycosphaerella ascophylli* Cotton. *Bot. Mar. 23*: 623–628.

Garbary, D. J. and A. Gautam. 1989. The *Ascophyllum, Polysiphonia, Mycosphaerella* symbiosis I. Population ecology of *Mycosphaerella* from Nova Scotia. *Bot. Mar. 32*: 181–186.

Goldman, R. N. and J. S. Weinberg. 1985. *Statistics, An Introduction*. Prentice-Hall, Englewood Cliffs, New Jersey.

Harlin, M. M. and J. S. Craigie. 1975. The distribution of photosynthate in *Ascophyllum nodosum* as it relates to epiphytic *Polysiphonia lanosa*. *J. Phycol. 11*: 109–113.

Higgins, J. M. 1984. Distribution, growth and differentiation of the endosymbiotic ascomycete fungus *Mycosphaerella ascophylli* Cotton isolated from the fucoid alga *Ascophyllum nodosum* (L.) Le Jolis growing in different intertidal environments. M. Sc. Thesis, University of Maine. 181 pp.

Kendrick, G. M. and M. W. Hawkes. 1988. The epiphyte *Microcladia coulteri* (Rhodophyta): changes in population structure with spatial and temporal variation in availability of host species. *Mar. Ecol. Prog. Ser. 43*: 79–86.

Kohlmeyer, J. and E. Kohlmeyer. 1972. Is *Ascophyllum nodosum* lichenized? *Bot. Mar. 15*: 109–112.

Lee, R. K. S. 1980. *A Catalogue of the Marine Algae of the Canadian Arctic*. National Museums of Canada, Ottawa Publications in Bot. No: 9 pp. 1–82.

Levin, P. S. and A. C. Mathieson. 1990. A host-epiphyte interaction: does it vary with wave exposure. *Proc. Northeast Algal Symposium, Woods Hole*, 1990, p. 18 (abstract only).

Lobban, C. S. and D. M. Baxter. 1983. Distribution of the red algal epiphyte *Polysiphonia lanosa* on its brown algal host *Ascophyllum nodosum* in the Bay of Fundy, Canada. *Bot. Mar. 26*: 533–538.

Lobban, C. S. and L. A. Hanic. 1984. Rocky shore zonation at North Rustico and Prim Point, Prince Edward Island. *Proc. N. S. Inst. Sci. 34*: 25–40.

Lubchenco, J. 1980. Algal zonation in the New England rocky intertidal community: an experimental analysis. *Ecology 6*: 333–344.

Mathieson, A. C., E. J. Hehre and N. B. Reynolds. 1981 a. Investigations of New England marine algae I: a floristic and descriptive ecological study of the marine algae at Jaffrey Point, New Hampshire, U. S. A. *Bot. Mar. 24*: 521–532

Mathieson, A. C., N. B. Reynolds and E. J. Hehre. 1981 b. Investigations of New England marine algae II: The species composition, distribution and zonation of seaweeds in the Great Bay estuary system and the adjacent open coast of New Hampshire. *Bot. Mar. 24*: 533–545.

McLachlan, J. and T. Edelstein. 1971. Investigations of the marine algae of Nova Scotia. IX. A preliminary survey of the flora of Bras d'Or Lake, Cape Breton Island. *Proc. N. S. Inst. Sci. 27*: 11–22.

Munteanu, N. and E. J. Maly. 1981. The effect of current on the distribution of diatoms settling on submerged glass slides. *Hydrobiologia 78*: 273–282.

Novaczek, I. and J. McLachlan. 1988 (1989). Investigations of marine algae of Nova Scotia XVII – Vertical and geographic distribution of marine algae on rocky shores of the Maritime Provinces. *Proc. N. S. Inst. Sci. 38*: 91–144.

Paine, R. T. 1979. Disaster, catastrophe, and local persistence of the sea palm *Postelsia palmaeformis*. *Science 205*: 685–687.

Pearson, G. A. and L. V. Evans. 1989. *Polysiphonia lanosa* – basiphyte interrelationships. *Br. Phycol. J. 24*: 310–311.

Pearson, G. A. and L. V. Evans. 1990. Settlement and survival of *Polysiphonia lanosa* (Ceramiales) spores on *Ascophyllum nodosum* and *Fucus vesiculosus* (Fucales). *J. Phycol. 26*: 597–603.

Pearson, G. A. and L. V. Evans. 1991. Stimulation of secondary rhizoid production in *Polysiphonia lanosa* by brown algal tissues and exudates. *Br. Phycol. J. 26*: 93–94.

Penot, M. 1974. Ionic exchange between the tissues of *Ascophyllum nodosum* (L.) Le Jolis and *Polysiphonia lanosa* (L.) Tandy. *Z. Pflanzephysiol. Bd. 73*: 125–131.

Penot, M. and M. Penot. 1974. Ions transport and exchange between *Ascophyllum nodosum* (L.) Le Jolis and some epiphytes. *Proc. Int. Seaweed Symp. 8*: 217–223.

Rattray, J. 1885. On some new cases of epiphytism among algae. *Trans. Bot. Soc. Edinburgh 16*: 209–219.

Rawlence, D. J. 1972. An ultrastructural study of the relationship between rhizoids of *Polysiphonia lanosa* (L.) Tandy (Rhodophyceae) and tissue of *Ascophyllum nodosum* (L.) Le Jolis (Phaeophyceae). *Phycologia 11*: 279–290.

Rawlence, D. J. and A. R. A. Taylor. 1970. The rhizoids of *Polysiphonia lanosa*. *Can. J. Bot. 48*: 607–611.

Rawlence, D. J. and A. R. A. Taylor. 1972. A light and electron microscopic study of rhizoid development in *Polysiphonia lanosa* (L.) Tandy. *J. Phycol. 8*: 15–24.

Rugg, D. A. and T. A. Norton. 1987. *Pelvetia canaliculata*, a high-shore seaweed that shuns the sea. *In*: (R. M. M. Crawford, ed.) *Plant Life in Aquatic and Amphibious Habitats*. Blackwell, Oxford. pp. 347–358.

Schonbeck, M. W. and T. A. Norton. 1978. Factors controlling the upper limits of fucoid algae on the shore. *J. Exp. Mar. Biol. Ecol. 31*: 303–313.

Schonbeck, M. W. and T. A. Norton. 1980. Factors controlling the lower limits of fucoid algae on the shore. *J. Exp. Mar. Biol. Ecol. 43*: 131–150.

South, G. R. and R. G. Hooper. 1980. A catalogue and atlas of the benthic marine algae of the Island of Newfoundland. *Memorial Univ. Nfld., Occ. Pap. Biol. 3*: 1–136.

South, G. R. and I. Tittley. 1986. *A Checklist and Distribution Index of the Benthic Marine Algae of the North Atlantic Ocean*. Huntsman Marine Laboratory and British Museum (Natural History), St. Andrews and London.

Sheath, R. G. and J. A. Hambrook. 1988. Mechanical adaptations to flow in freshwater red algae. *J. Phycol. 24*: 107–111.

Taylor, W. R. 1957. *Marine Algae of the Northeastern Coast of North America*. Univ. of Michigan Press, Ann Arbor.

Thomas, M. L. H., D. C. Arnold and A. R. A. Taylor. 1983. Rocky intertidal communities. *In*: (M. L. H. Thomas, ed.) *Marine and Coastal Systems of the Quoddy Region, New Brunswick*. Can. Spec. Publ. Fish. Aquat. Sci. Ottawa. pp. 35–73.

Tian Lining. 1990. Experimental studies on the *Ascophyllum nodosum, Polysiphonia lanosa* and *Mycosphaerella ascophylli* symbiosis. M Sc. thesis, St. Francis Xavier University. Antigonish, Nova Scotia. pp. 118.

Turner, C. H. C. and L. V. Evans. 1977. Physiological studies on the relationship between *Ascophyllum nodosum* and *Polysiphonia lanosa*. *New Phytol. 79*: 363–371.

Underwood, A. J. 1986. Physical factors and biological interactions: the necessity and nature of ecological experiments. *In*: (P. G. Moore and R. Seed, eds) *The Ecology of Rocky Coasts*. Columbia University Press. New York. pp. 372–390.

Webber, F. C. 1967. Observations on the structure, life history and biology of *Mycosphaerella ascophylli*. *Trans. Br. Mycol. Soc. 50*: 583–601.

Wilkinson, L. 1989. *SYSTAT: The System for Statistics*. SYSTAT, Inc., Evanston, IL.

Zar, J. H. 1984. *Biostatistical Analysis*. Prentice-Hall, Englewood Cliffs, New Jersey.

Botanica Marina
Vol. 34, pp. 403–410, 1991

Seasonal Patterns of Photosynthesis and Physiological Parameters and the Effects of Emersion in Littoral Seaweeds

G. J. Levitt and J. J. Bolton*

Sea Fisheries Research Institute, Private Bag X2, Rogge Bay 8012, Republic of South Africa

* *Botany Department, University of Cape Town, Private Bag, Rondebosch 7700, Republic of South Africa*

(Accepted 30 May 1991)

Abstract

Seasonal submersed P-I curves of five species of intertidal seaweeds of differing morphologies from different shore zones on a South African rocky shore have been measured. A single measurement of emersed photosynthetic rate of these seaweeds was performed for comparative purposes. The thin, sheet-like *Porphyra capensis*, occurring high on the shore, is the most photosynthetically productive species tested either when submersed or when emersed. Fleshy forms with flattened thalli are also highly photosynthetically productive, whilst the terete branched forms are least photosynthetically productive. A general seasonal pattern of spring maxima and winter minima in submersed photosynthesis is apparent. High calorific values and carbon content provide evidence that the possession of mucilage functions as a physiological mechanism to resist desiccation in the high shore species *Porphyra capensis*. Mid-shore species display similar photosynthesis and desiccation responses to emersion, whilst the low shore *Bifurcaria brassicaeformis* is less tolerant of emersion.

Introduction

There has been extensive study of primary production in intertidal macrophytes (*e.g.* Littler and Murray 1974, Mathieson and Norall 1975, Buesa 1977, Littler *et al.* 1979, Littler 1980, Oates 1986), as well as on the relationship between primary production and seaweed thallus morphology (Littler and Murray 1974, Littler 1980, 1981, Littler *et al.* 1983). Some studies have concentrated on seasonal photosynthesis (*e.g.* King and Schramm 1976, Brinkhuis 1977a, b), whilst other authors have investigated seaweed photosynthesis while emersed (*e.g.* Johnson *et al.* 1974, Brinkhuis *et al.* 1976, Dawes *et al.* 1978, Schonbeck and Norton 1978, 1979a, b, c), Quadir *et al.* 1979, Dring and Brown 1982). Oates (1986) investigated both the submersed and emersed contributions of photosynthesis in one algal species, but not on a seasonal basis. Studies of desiccation resistance of algae have been conducted (Jones and Norton 1979, Dromgoole 1980) and this has been related to photosynthesis (Brown 1987). Comparisons of emersed photosynthesis with submersed photosynthesis have been made by Johnson and Raven (1986) and Surif and Raven (1990).

This investigation combines seasonal measurements of submersed photosynthesis and a single measurement of emersed photosynthesis in a number of representative intertidal seaweeds of differing morphologies from a single shore with observations of shore zonation, functional form and other physiological parameters to gain a fuller picture of photosynthesis in an intertidal environment.

Material and Methods

Plants were collected at Dalebrook (34°07.4′S, 18°27.4′E) in False Bay which bounds the eastern shore of the Cape Peninsula, South Africa. The intertidal region consists of gently sloping ledges of Table Mountain sandstone exposed to moderate wave action. McQuaid (1985) documented 24 common species of macroalgae present on this shore. Five of the most abundant representatives of different functional plant forms with different intertidal distributions were chosen for this study. They were:

i) *Porphyra capensis* Kuetzing (Rhodophyta; thin, sheet-like form; littoral fringe);

ii) *Ulva rigida* C. Agardh (Chlorophyta; thin, sheet-like form; eulittoral);

iii) *Gigartina radula* (Esper) J. Agardh (Rhodophyta; thick cartilaginous form; eulittoral and shallow sublittoral);

iv) *Spachnidium rugosum* (L.) Greville (Phaeophyta; coarsely branched form; eulittoral);

v) *Bifurcaria brassicaeformis* (Kuetzing) Barton (Phaeophyta; coarsely branched form; low eulittoral).

Photosynthetic potential while submersed was measured in June, September, November 1984 and February, April, June, August, November 1985. Samples of each species were collected for determination of photosynthetic light curves using a Gilson respirometer (Gilson Medical Electronics Inc., Middleton, Wisconsin, USA). Tissue segments excised from branch tips or thallus edges (max. 0.1 g dw) were incubated at 15 °C (a temperature typical of the SW Cape — Anderson and Bolton 1985) in 15 mL of buffered (0.1 M $NaHCO_3$) nutrient enriched seawater solution (Provasoli 1968). This prevents N or C limitation of photosynthesis as, for example, inorganic C in normal seawater is not always saturating for photosynthesis (Surif and Raven 1989). Tissue segments were allowed to stabilise for three hours prior to incubation. Eight samples were incubated simultaneously with a control. Light was provided by incandescent white lights (General Electric reflector 2 × 100 W, 8 × 30 W). Segments were incubated after a one hour light equilibration period for six successive ten minute periods at each level of irradiance. Respiration in tissue segments was measured with six replicates of eight incubations after a one-hour dark equilibration period. Observations were converted to mg C fixed g dry wt^{-1} h^{-1}. Data were fitted to the hyperbolic tangent function $P = P_{max} \cdot \tanh(I/I_k)$ of Chalker (1980, 1981); where P = photosynthesis at irradiance I, P_{max} = maximum photosynthesis reached at irradiance I_k.

Measurement of photosynthesis of emersed seaweeds was carried out during September and October 1985. Seven whole plants of each species were collected from the middle of the shore zone they normally occupy at mid-tide, while still immersed. *Ulva rigida* plants were collected from rock pools. Plants were exposed to air at a time corresponding to their normal exposure during the tidal cycle so as not to disturb any innate rhythms, the fresh weight being recorded. They were then desiccated under incandescent lights at 1200 µE m^{-2} s^{-1} and 25 °C, a common ambient daytime temperature. Photosynthesis was measured 0.5, 1, 2, 3, 4,5, and 6 hours after emersion. Plants were weighed directly after measurement of photosynthesis, then dried to constant weight at 60 °C, a temperature which avoids the loss of volatile constituents (Westlake 1963). Percentage desiccation of the plants was calculated. Photosynthesis was measured using an ADC 225 Mk3 infra-red gas analyser (Analytical Development Company Ltd, Hoddesdon, England). A closed system was used because of simplicity and ease of use. Air was drawn from an incubation chamber enclosing the alga into the IRGA for analysis (in absolute mode) and then recycled back to the chamber. During photosynthesis CO_2 concentration declines until the compensation point of photosynthesis is reached. Measurement of the change in CO_2 over time is used to calculate the rate of photosynthetic CO_2 assimilation per unit mass. Atmospheric (reference) air acts as a control. Plants were placed so as to emulate as closely as possible the natural adaptation to resist water loss displayed by those species which fold over upon themselves as they desiccate (*e.g. Porphyra capensis*). Primary production of the emersed algae was calculated as follows (adapted from Sésták *et al.* 1971):

$$\text{mg C fixed g dw}^{-1}\,\text{h}^{-1} = \frac{(\text{ref} - \text{ana} \times 0.000002) \times \text{FR} \times 60 \times 12}{\text{DW} \times 44}$$

where:

ref = IRGA atmospheric CO_2 reference reading (ppm)

ana = experimental CO_2 photosynthetic reading (ppm)

FR = flow rate of air through IRGA (cm^3 min^{-1})

DW = dry weight of plant (g)

Five plants of each species were collected every month, freeze dried and analyzed for total percentage carbon and total percentage nitrogen by an Heraeus Universal Combustion Analyzer model CHN-RAPID using the method of Monar (1972). Freeze dried samples were analysed for percentage inorganic matter and calorific value using a DDR CP500 macro-bomb calorimeter.

All plants were non-juvenile, non-senescent plants, preference being given to the collection of healthy, minimally grazed individuals free of epiphytes.

Results

Immersed primary production

Values for P_{max} and I_k for each of the P-I curves obtained are shown in Figures 1 and 2. *Porphyra capensis* shows a high level of potential maximum production (Fig. 1), highest P_{max} levels being in late winter – early spring (August 1985 and September 1984) and lowest levels in autumn (April 1985).

Ulva rigida displays a late winter–early spring P_{max} peak (August 1985 and September 1984) and a late summer minimum (February 1985). *Porphyra capensis* and *Ulva rigida* show the highest net photosynthetic potential. *Gigartina radula* is the next most productive, with a P_{max} peak in spring (September 1984) and a summer low (November 1984). *Splachnidium rugosum* displayed a minimum P_{max} in summer (February 1985), but had no distinct seasonal maximum. The highest P_{max} for *Bifurcaria brassicaeformis* is distinct (April 1985), but the minimum is unclear.

Table I. Dark respiration rates of five littoral algae.

	Dark respiration rate (mg C g $dw^{-1} h^{-1}$ ± S.E.)	Percentage of P_{max}
Porphyra capensis	0.43 ± 0.06	10.4%
Ulva rigida	0.60 ± 0.11	13.2%
Gigartina radula	0.26 ± 0.04	12.1%
Splachnidium rugosum	0.37 ± 0.09	14.1%
Bifurcaria brassicaeformis	0.27 ± 0.08	16.0%

Splachnidium rugosum has the highest average I_k (Fig. 2), followed by *Gigartina radula* and *Porphyra capensis*. *Bifurcaria brassicaeformis* and *Ulva rigida* have

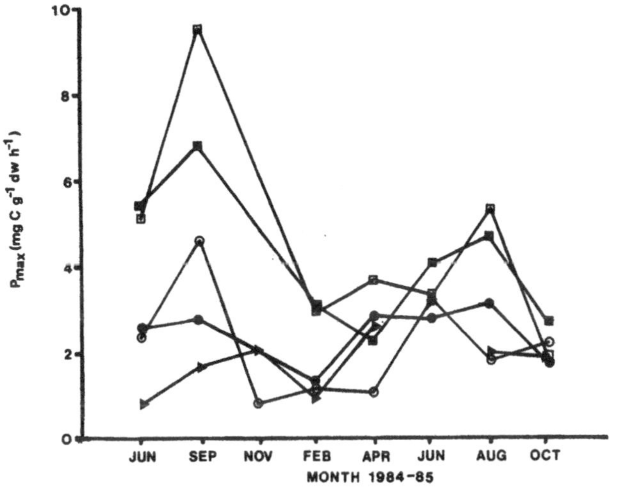

Fig. 1. Seasonal P_{max} values of five species of littoral seaweeds, ■ *P. capensis* □ *U. rigida* ○ *G. radula* ● *S. rugosum* ▶ *B. brassicaeformis*

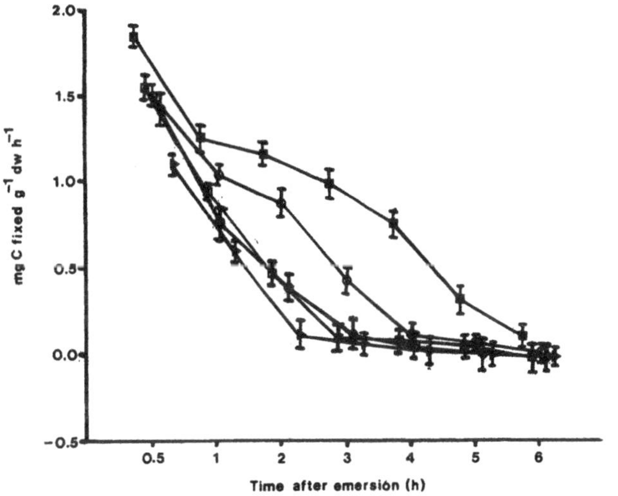

Fig. 3. Primary production versus time after emersion of five species of littoral seaweeds, ■ *P. capensis* □ *U. rigida* ○ *G. radula* ● *S. rugosum* ▶ *B. brassicaeformis*

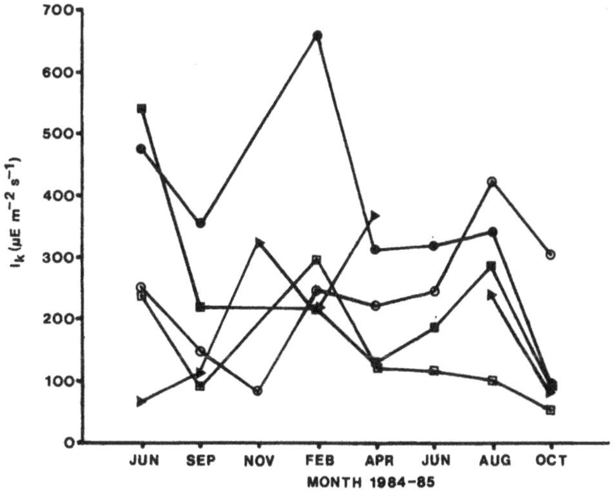

Fig. 2. Seasonal I_k values of five species of littoral seaweeds, ■ *P. capensis* □ *U. rigida* ○ *G. radula* ● *S. rugosum* ▶ *B. brassicaeformis*

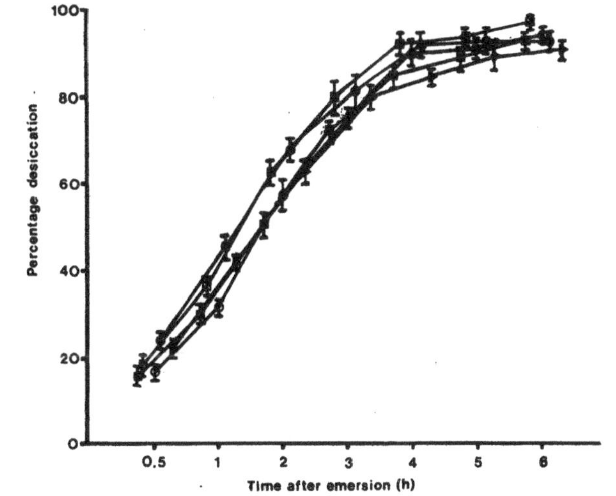

Fig. 4. Percentage desiccation versus time after emersion of five species of littoral seaweeds, ■ *P. capensis* □ *U. rigida* ○ *G. radula* ● *S. rugosum* ▶ *B. brassicaeformis*

the lowest I_k values. The I_k values are variable with no apparent seasonal pattern. Seasonally averaged dark respiration rates are shown in Table I. Species with the highest net production display the highest respiration rates (*Porphyra capensis* and *Ulva rigida*).

Emersed primary production

Figure 3 shows primary production at 0.5, 1, 2, 3, 4, 5 and 6 hours after emersion and Figure 4 the percentage desiccation. The rate of carbon fixation declines as time after emersion and percentage desicca-

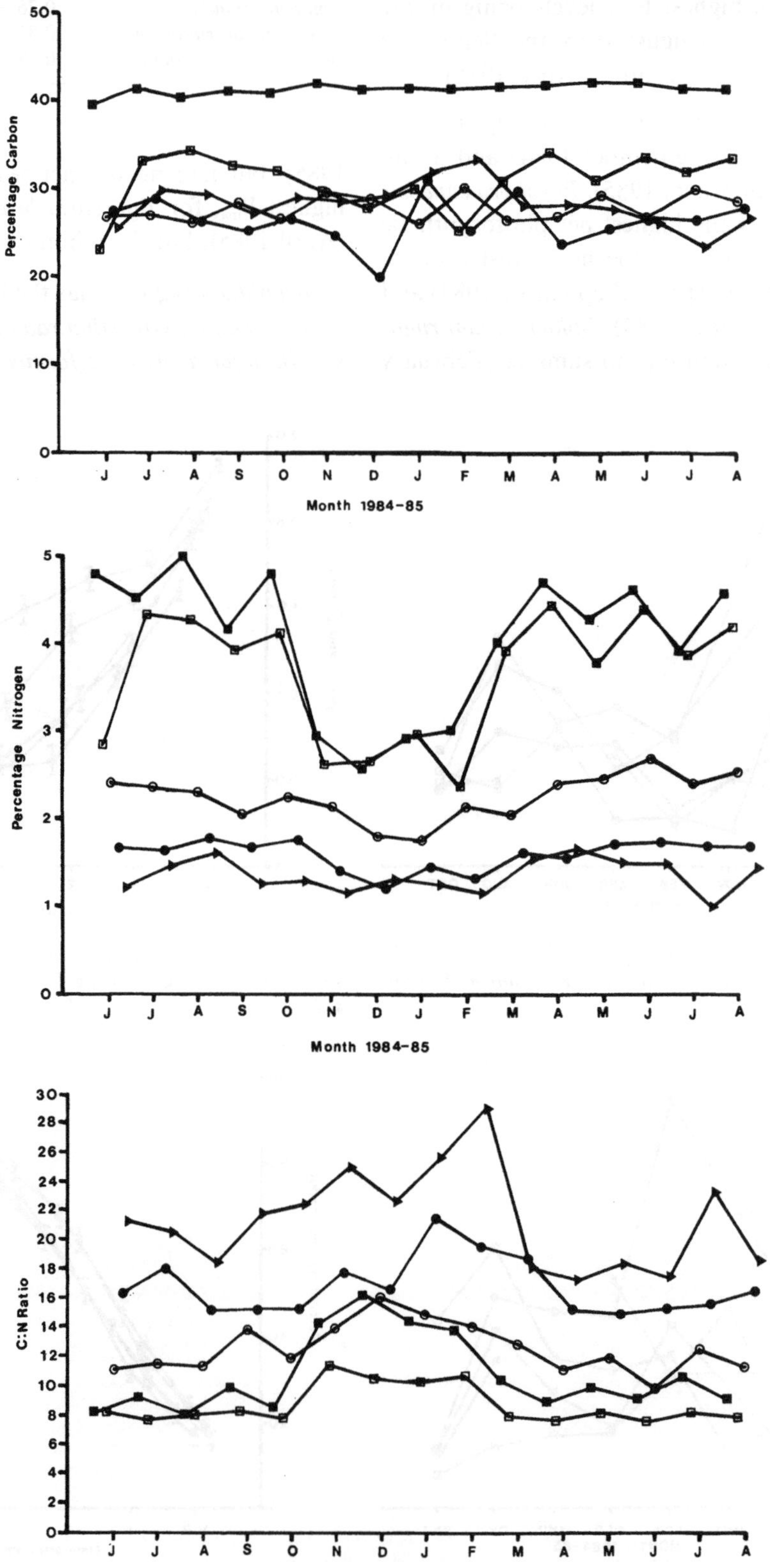

Fig. 5. Percentage carbon (a), percentage nitrogen (b), and C:N ratios (c) of five species of littoral seaweeds, ■ *P. capensis* □ *U. rigida* ○ *G. radula* ● *S. rugosum* ▶ *B. brassicaeformis*

tion increase. *Porphyra capensis* shows the highest levels of primary production, followed in decreasing order by *Gigartina radula, Ulva rigida, Splachnidium rugosum* and *Bifurcaria brassicaeformis*. The production of *Porphyra capensis* declines least rapidly as desiccation increases, a positive net photosynthetic rate being maintained throughout the experimental period. *Gigartina radula* is next most resistant to desiccation stress. *Ulva rigida* and *Splachnidium rugosum* have similar production versus desiccation curves, negative net production occurring after six hours. *Bifurcaria brassicaeformis* shows a pattern similar to *Ulva rigida* and *Splachnidium rugosum*, but displays a more rapid decline in production as desiccation increases in the early stages after emersion.

The general desiccation pattern is of an initial rapid rate of desiccation after emersion until a desiccation level of 80–90% is reached after four hours. It then increases slowly to *ca* 93% after six hours when net production is either negative or very close to zero.

Percentage carbon, percentage nitrogen and C : N ratios

One-way ANOVA was used to compare means between seasons (June–August = winter; September–November = spring; December–February = summer; March–May = autumn) for percentage carbon, percentage nitrogen and C : N ratios. Only *B. brassicaeformis* showed significant seasonal variation (at 5%) in carbon (Fig. 5a), with a summer maximum and winter minimum. All except *B. brassicaeformis* showed significant seasonal variation in nitrogen (Fig. 5b). Species with a seasonal nitrogen pattern had a summer minimum and a winter maximum, particularly evident in *P. capensis* and *U. rigida*. Seasonality in C : N ratio (Fig. 5c) was not significant for *S. rugosum*, but was for the other species (at 1%). A summer maximum and winter minimum was shown by those species with significant seasonal C : N ratios.

Energy content and inorganic matter

Table II shows the amount of inorganic matter as a percentage of total matter and calorific values on an ash-free dry weight basis. Four species have fairly similar amounts of inorganic matter (23.01–27.22%). *Porphyra capensis* has a low level of 8.40%. The same four species display similar calorific values (13.24–14.30 kJ g^{-1}), whilst *P. capensis* has a high energy content of 17.15 kJ g^{-1}.

Table II. Inorganic matter and energy content of five littoral algae.

	Inorganic matter (%)	kJ g^{-1} ash-free dry weight
Porphyra capensis	8.40	17.15
Ulva rigida	23.01	13.24
Gigartina radula	26.71	14.27
Splachnidium rugosum	27.22	14.30
Bifurcaria brassicaeformis	23.03	13.26

Discussion

Values of P_{max} for all five species are comparable with those observed in other studies (*e.g.* Buesa 1977, Wallentinus 1978, Merrill and Waaland 1979). Thin sheet-like forms (*P. capensis* and *U. rigida*) are the most productive, being better able to utilize light energy and absorbing nutrients. This is due to their large cells resulting in less internal cell-shading by the structural component (a small cell surface to volume ratio), their high thallus area to volume ratio due to their thin thalli (which presents the maximum thallus area to incident light), and because the entire thallus is composed of photosynthetic tissue (which permits the maximum exposure of photosynthetic pigment to incident light). Thick cartilaginous and coarsely branched forms have more structural non-photosynthetic components and their production rate per unit biomass is correspondingly less.

The late winter–spring seasonal photosynthetic maxima displayed by *P. capensis, U. rigida* and *G. radula* are comparable with values given in the literature. Brinkhuis (1977 a) observed a photosynthetic maximum for *Ascophyllum nodosum* (L.) Le Jol. in spring and a minimum during late summer–winter. In *Fucus vesiculosus* L. and *Ulva lactuca* L. Brinkhuis (1977 b) observed spring and summer photosynthetic maxima and winter minima. Littler *et al.* (1979) observed summer photosynthetic maxima in nine of 13 species and minima in winter–spring. At Dalebrook the season of lowest P_{max} was variable, ranging from late summer–mid-winter. Variable or extended minima were also observed by Brinkhuis (1977 a) for *Ascophyllum nodosum* as well as by Littler *et al.* (1979). The magnitude of primary production at Dalebrook is similar to values given in other studies and potentially much greater for the thin, sheet-like forms. These higher rates of production are consistent with what would be expected of opportunistic species common in temperate environments and can be compared with lower rates in stable, nutrient poor tropical environments (*e.g.* Buesa 1977).

The non-uniform pattern of seasonal P_{max} values may be due to differing biotic and abiotic factors. Factors controlling the seasonal potential rate of primary production may not be as obvious as total irradiance or photoperiod.

Species with high P_{max} values have lower respiration rates when expressed as a percentage of P_{max}. The high shore species (*e.g. Porphyra capensis*) having a lower rate of respiration than the low shore species (*e.g. Bifurcaria brassicaeformis*).

Biomass of *P. capensis*, *U. rigida* and *G. radula* at Dalebrook is low in winter and high in summer (McQuaid 1985). The summer biomass peak corresponds (with a lag phase for conversion of photosynthetic production to growth) to the high spring photosynthetic rate. The low winter biomass could be related to the broad autumn–winter period of minimum photosynthesis.

The value of I_k is indicative of a plant's physiological adaptation to its light environment (Chalker 1981). Seasonal variation in I_k values result from photoadaptive plant responses to varying irradiance. A higher I_k value is to be expected in summer and a lower I_k in winter (observed in sublittoral seaweeds, including *G. radula*, by Levitt and Bolton 1990). The lack of a seasonal pattern here may be explained by low light attenuation in the intertidal environment resulting in plants being photosynthetically light saturated during daylight irrespective of season.

These I_k variations could be a response to short term fluctuations in irradiance and not long term seasonal changes. Four species have average I_k values greater than 200 $\mu E\ m^{-2}\ s^{-1}$, indicating that they are sun-adapted. The lower average I_k of *U. rigida* may be due to it being a rock pool resident where it is subject to topographic shading and light attenuation by water. Adaptation to seasonal temperature changes is also important. Mathieson and Norall (1975) suggested that the optimal temperature and light requirements of seaweeds are adaptively adjusted to varying environmental regimes.

Chapman (1965) showed that the photosynthetic rate of emersed seaweeds is less than or equal to comparable submersed algae, whereas Johnson *et al.* (1974) found that low shore intertidal species had a lower photosynthetic rate while emersed than submersed while upper and middle shore plants had maximal photosynthetic rates when slightly desiccated. Quadir *et al.* (1979) reported a higher emersed photosynthetic rate for the high shore *Fucus distichus* L. and the reverse for the low shore *Iridaea cordata* (Turn.) Bory. Dring and Brown (1982) noted that photosynthesis in some emersed brown algae increased when initially desiccated, but decreased after further water loss. In our data, a lower rate of photosynthesis for emersed plants than P_{max} for submersed plants is apparent, the submersed environment being more favourable for achieving maximum photosynthetic rate. Due to high incident irradiance it is likely that P_{max} is often approached by intertidal algae while submersed. *Porphyra capensis* is well adapted to the upper shore as its photosynthetic rate is the highest of the desiccated plants. It is likely that *P. capensis* recovers more rapidly from a given degree of desiccation than the lower shore species, a phenomenon noted by Dring and Brown (1982) in *Pelvetia canaliculata* (L.) Dcne. *et* Thur., also a high shore species. Atmospheric boundary layer effects and the non-continuous availability of nutrients contribute towards reduced photosynthesis in emersed seaweeds (Brown 1987). Higher ambient temperatures also reduce photosynthesis in the intertidal. The maximum monthly average sea temperature in False Bay does not exceed 20 °C (Bolton 1986) while the surface temperature around exposed seaweeds can easily reach 30 °C or more. Respiration also increases as desiccation increases (Brinkuis *et al.* 1976), thus reducing net photosynthesis.

The broadly similar photosynthesis-desiccation responses shown by *G. radula*, *U. rigida* and *S. rugosum* is in keeping with their mid-shore habitat. *Gigartina radula* is slightly more productive and desiccates marginally slower. As a thick cartilaginous species, *C. radula* is more desiccation tolerant than the membranous *U. rigida*. *Bifurcaria brassicaeformis* is the least desiccation tolerant, growing in the lower eulittoral.

Most desiccation occurs in the first three to four hours after emersion. Quadir *et al.* (1979) observed almost identical rates of desiccation in three low, mid and high shore species (*Iridaea cordata*, *Ulva fenestrata* P. *et* R. and *Fucus distichus*). *Porphyra capensis* shows similar adaptations to *Fucus distichus* – as the thallus loses water the frond folds over on itself trapping moist air between the exposed part of the frond and the inner part close to the substratum, which then desiccates less rapidly. The outer exposed frond becomes hard and dry and prevents water loss in the unexposed part of the thallus.

Survival of prolonged exposure can be by avoiding tissue dehydration (*e.g. B. brassicaeformis* habitat) or by tolerating it. Schonbeck and Norton (1979 b) suggested that since the intertidal environment is characterised by predictable periodic immersion, a plant which loses water slowly may escape excessive dehydration at low tide. Hygroscopic polysaccharides in the form of mucus in the thallus may retard water loss. Conversely, the work of Jones and Norton (1979) suggests that littoral and sublittoral algae behave similarly with respect to water loss, which suggests the role of hygroscopic polysaccharides in the prevention of desiccation may not be as important as noted above. Support for the latter view also comes from the findings of Dromgoole (1980) that mucilage content does not contribute to species differences in desiccation resistance. Quadir *et al.* (1979) suggested that desiccation in *Fucus distichus* may be retarded by unsaturated fatty acids in the epidermal cells which act as a false cuticle.

The high calorific value displayed by *P. capensis* is most likely to be due the incorporation of carbohydrate (*i.e.* mucilage), thereby increasing the calorific value (Table II). *Ulva rigida* most likely uses its high photosynthetic rate for rapid growth rather than mucilage production. Himmelman and Carefoot (1975) showed an inverse relationship between calorific value and rate of growth in temperate seaweeds. A high carbon content and calorific value in *P. capensis* should result in a low inorganic content (carbon being incorporated in organic compounds) and the data reflects this.

The significant seasonal variation of C:N ratios in all except *B. brassicaeformis* results from significant seasonal variation in total nitrogen. Carter and Anderson (1986) also observed a summer minimum in nitrogen levels of the red intertidal seaweed *Gelidium pristoides* (Turner) Kuetzing. Higher C:N ratios in summer are due to the structural incorporation of carbon and indicate the maximum growth period which follows shortly after the maximum photosynthetic period. Reduction in thallus nitrogen due to increased photosynthesis would also result in higher C:N ratios. Species higher on the shore (*P. capensis, U. rigida* and *G. radula*) exhibit lower C:N ratios than the lower shore species (*S. rugosum* and *B. brassicaeformis*. This is probably coincidental because the more productive thin, sheet-like forms grow highest on the shore – C:N ratio and production rate show an inverse relationship (Niell 1976). Niell (1976) noted that Phaeophyta have higher C:N ratios than other divisions. The two representatives of the Phaeophyta (*S. rugosum* and *B. brassicaeformis*) show substantially higher C:N ratios. They are structurally more complex than the thin forms and a higher C:N ratio due to more carbon present in a more complex structure is to be expected.

References

Anderson, R. J. and Bolton, J. J. 1985. Suitability of the agarophyte *Suhria vittata* (L.) J. Ag. (Rhodophyta: Gelidiaceae) for mariculture: Geographical distribution, reproductive phenology and growth of sporelings in culture in relation to light and temperature. *S. Afr. J. mar. Sci. 3*: 169–178.

Bolton, J. J. 1986. Marine phytogeography of the Bengeula upwelling region on the west coast of Southern Africa: A temperature dependent approach. *Bot. Mar. 29*: 251–256.

Brinkhuis, B. H. 1977a. Seasonal variations in salt-marsh macroalgae photosynthesis. I. *Ascophyllum nodosum* ecad *scorpioides*. *Mar. Biol. 44*: 165–175.

Brinkhuis, B. H. 1977b. Seasonal variations in salt-marsh macroalgae photosynthesis. II. *Fucus vesiculosus* and *Ulva lactuca*. *Mar. Biol. 44*: 177–186.

Brinkhuis, B. H., N. R. Tempel and R. F. Jones. 1976. Photosynthesis and respiration of exposed salt-marsh fucoids. *Mar. Biol. 34*: 349–359.

Brown, M. T. 1987. Effect of desiccation on photosynthesis of intertidal algae from a southern New Zealand shore. *Bot. Mar. 30*: 121–127.

Buesa, R. J. 1977. Photosynthesis and respiration of some tropical marine plants. *Aquat. Bot. 3*: 203–216.

Carter, A. R. and R. J. Anderson. 1986. Seasonal growth and agar contents in *Gelidium pristoides* (Gelidiales, Rhodophyta) from Port Alfred, South Africa. *Bot. Mar. 29*: 117–123.

Chalker, B. E. 1980. Modelling light-saturation curves for photosynthesis: An exponential function. *J. theor. Biol. 84*: 205–215.

Chalker, B. E. 1981. Simulating light-saturation curves for photosynthesis and calcification by reef-building corals. *Mar. Biol. 63*: 135–141.

Chapman, V. J. 1965. The physiological ecology of some New Zealand seaweeds. *In*: (E. G. Young and J. L. McLachlan, eds) *Proc. 5th Int. Seaweed Symp.* Wiley and Sons, New York. pp. 29–54.

Dawes, C. J., R. E. Moon and M. A. Davis. 1978. The photosynthetic and respiratory rates and tolerances of benthic algae from a mangrove and salt marsh estuary: a comparative study. *Estuar. cstl mar. Sci. 6*: 175–185.

Dring, M. J. and F. A. Brown. 1982. Photosynthesis of intertidal brown algae during and after periods of emersion: a renewed search for physiological causes of zonation. *Mar. Ecol. Prog. Ser. 8*: 301–308.

Dromgoole, F. I. 1980. Desiccation resistance of intertidal and subtidal algae. *Bot. Mar. 23*: 149–159.

Himmelman, J. and T. H. Carefoot. 1975. Seasonal changes in the calorific value of three pacific coast seaweeds, and their significance to some marine invertebrate herbivores. *J. exp. mar. Biol. Ecol. 18*: 139–151.

Johnson, A. M. and J. A. Raven. 1986. The analysis of photosynthesis in air and water of *Ascophyllum nodosum* (L.) Le Jol. *Oecologia 69*: 288–295.

Johnson, W. S., A. Gigon, S. L. Gulmon and H. A. Mooney. 1974. Comparative photosynthetic capacities of intertidal algae under exposed and submerged conditions. *Ecology 55*: 450–453.

Jones, H. G. and T. A. Norton. 1979. Internal factors controlling the rate of evaporation from fronds of some intertidal algae. *New Phytol. 83*: 771–781.

King, R. J. and W. Schramm. 1976. Photosynthetic rates of benthic marine algae in relation to light intensity and seasonal variations. *Mar. Biol. 37*: 215–222.

Levitt, G. J. and J. J. Bolton. 1990. Seasonal primary production of understory Rhodophyta in an upwelling system. *J. Phycol. 26*: 214–220.

Littler, M. M. 1980. Morphological form and photosynthetic performances of marine macroalgae: Tests of a functional/form hypothesis. *Bot. Mar. 22*: 161–165.

Littler, M. M. 1981. The relationship between thallus form and the primary productivity of seaweeds. *In*: (G. E. Fogg and W. Eiffon Jones, eds) *Proc. 8th Int. Seaweed Symp.* The Marine Science Laboratories, Menai Bridge. pp. 398–403.

Littler, M. M. and S. N. Murray. 1974. The primary productivity of marine macrophytes from a rocky intertidal community. *Mar. Biol. 27*: 131–135.

Littler, M. M., S. N. Murray and K. E. Arnold. 1979. Seasonal variations in net photosynthetic performance and cover of intertidal seaweeds. *Aquat. Bot. 7*: 35–46.

Littler, M. M., D. S. Littler and P. R. Taylor. 1983. Evolutionary strategies in a tropical barrier reef system: functional-form groups of marine macroalgae. *J. Phycol. 19*: 229–237.

Mathieson, A. C. and T. L. Norall. 1975. Photosynthetic studies of *Chondrus crispus*. *Mar. Biol. 33*: 207–213.

McQuaid, C. D. 1985. Seasonal variation in biomass and zonation of nine intertidal algae in relation to changes in radiation, sea temperature and tidal regime. *Bot. Mar. 28*: 539–544.

Merrill, J. E. and J. R. Waaland. 1979. Photosynthesis and respiration in a fast growing strain of *Gigartina exasperata* (Harvey and Bailey). *J. exp. mar. Biol. Ecol. 39*: 281–290.

Monar, I. 1972. Analysenautomat zur simultanen Mikrobestimmung von C, H und N. *Mikrochim. Acta 6*: 784–806.

Niell, F. X. 1976. C : N ratio in some marine macrophytes and its possible ecological significance. *Bot. Mar. 19*: 347–350.

Oates, B. R. 1986. Components of photosynthesis in the intertidal saccate alga *Halosaccion americanum* (Rhodophyta, Palmariales). *J. Phycol. 22*: 217–223.

Quadir, A., P. J. Harrison and R. E. De Wreede. 1979. The effects of emergence and submergence on the photosynthesis and respiration of marine macrophytes. *Phycologia 18*: 83–88.

Schonbeck, M. W. and T. A. Norton. 1978. Factors controlling the upper limits of fucoid algae on the shore. *J. exp. mar. Biol. Ecol. 31*: 303–313.

Schonbeck, M. W. and T. A. Norton. 1979a. Drought-hardening in the upper-shore seaweeds *Fucus spiralis* and *Pelvatia canaliculata*. *J. Ecol. 67*: 687–696.

Schonbeck, M. W. and T. A. Norton. 1979b. An investigation of drought avoidance in intertidal fucoid algae. *Bot. Mar. 22*: 133–144.

Schonbeck, M. W. and T. A. Norton. 1979c. The effects of brief periodic submergence on intertidal fucoid algae. *Estuar. cstl mar. Sci. 8*: 205–211.

Sésták, Z., P. G. Jarvis and J. Catský. 1971. Criteria for the selection of suitable methods. *In*: (Z. Sésták, J. Catský and P. G. Jarvis, eds) *Plant Photosynthetic Production. Manual of Methods*. Junk, The Hague. pp. 1–48.

Surif, M. B. and J. A. Raven. 1989. Exogenous inorganic carbon sources for photosynthesis in seawater by members of the Fucales and the Laminariales (Phaeophyta): ecological and taxonomic implications. *Oecologia 78*: 97–105.

Surif, M. B. and J. A. Raven. 1990. Photosynthetic gas exchange under emersed conditions in eulittoral and normally submersed members of the Fucales and the Laminariales: interpretation in relation to C isotope ratio and N and water use efficiency. *Oecologia 82*: 68–80.

Wallentinus, I. 1978. Productivity studies on Baltic macroalgae. *Bot. Mar. 21*: 365–380.

Westlake, D. F. 1963. Comparisons of plant productivity. *Biol. Rev. 38*: 385–425.

Botanica Marina
Vol. 34, pp. 411–415, 1991

Formation of Cellulose Microfibrils on an Isolated Plasma Membrane of the Coenocytic Green Alga, *Boergesenia forbesii*

S. Mizuta[1] and T. Harada

Department of Biology, Faculty of Science, Kochi University, Akebono-cho Kochi 780, Japan

(Accepted 10 June 1991)

Abstract

The protoplasm isolated from the giant coenocytic green alga, *Boergesenia forbesii* (Harvey) Feldmann rapidly breaks down into many protoplasts in culture but maintains a sheet-like morphology under a condition of reduced Ca^{++}. The formation of cellulose microfibrils was demonstrated on the isolated plasma membrane from such a sheet.

Introduction

Cellulose microfibrils (crystallized cellulose, cellulose I) in plants are considered to be formed through the action of putative cellulose synthesizing complexes (terminal complexes) in the plasma membrane (Brown 1985). The correlation between the structure of several kinds of terminal complexes and microfibril assembly has been studied by the method of freeze-replication (Brown and Montezinos 1976, Giddings *et al.* 1980, Mueller and Brown 1980, Itoh and Brown 1984, Mizuta *et al.* 1989). But, for further clarification of the mechanisms of cellulose biogenesis, the plasma membrane and terminal complexes should be isolated and the assembly of microfibrils *in vitro* examined. To date, however, this has not been done owing to the considerable difficulty in isolating an intact plasma membrane and terminal complexes.

Boergesenia forbesii (Harvey) Feldmann is a marine unicellular coenocytic green alga consisting of a large central vacuole and a thin protoplasmic layer surrounded by the plasma membrane covered by cell walls. The thallus of *Boergesenia* grows to be quite large (approximately 5 cm in length) under both natural and culture conditions (Ishizawa *et al.* 1979) and many terminal complexes have been shown to be present within the plasma membrane (Itoh *et al.* 1984, Mizuta 1985). This organism should thus serve as appropriate material for isolating the intact big plasma membrane and terminal complexes embedded within the plasma membrane. *Boergesenia* protoplasm changes its shape from sheet to multi-rounded protoplasts by breaking down the plasma membrane when the cell is wounded mechanically in sea water but protoplast development can be disturbed by removal of Ca^{++} from the sea water (Ishizawa *et al.* 1979). It is thus possible to obtain large plasma membrane sheets that contain terminal complexes by removing tonoplasts and protoplasm from the protoplasmic sheets of cells by plasmolysis. In this study, cellulose microfibrils were found to form on the isolated plasma membrane of *Boergesenia.*

Material and Methods

Giant growing cells of *Boergesenia* were plasmolyzed in 0.7 M KCl for 15 min. After rupturing the cells, the protoplasm was pulled out gently from within the walls. The KCl solution was gradually replaced with Ca^{++}-reduced artificial sea water (Mizuta and Wada 1981) whose osmotic pressure had been modified with NaCl. Protoplasmic sheets were obtained by cutting the protoplasm open by micromanipulation. The sheets were floated on the surface of the medium layered on Formvar-coated grids with the tonoplast

[1] Correspondence and reprints.

surface facing upward, and membranes and organelles situated within them were removed by a stream of medium gently applied by a small syringe. Grids with a plasma membrane were then incubated for 0.5–2 h in the same medium at 25 °C, washed with redistilled water, dried, shadowed with platinum-carbon at 2×10^{-4} Pa and observed with a JEOL JEM 100U electron microscope.

Freeze-fracture was conducted by the method described previously (Mizuta and Okuda 1987).

Results and Discussion

Figure 1 shows protoplasts of *Boergesenia* developed from the protoplasm. Their formation was markedly inhibited by decreasing the concentration of Ca^{++} (Fig. 8). Following reduction to 1/10 (1.012 mM equiv.) of Ca^{++} from the normal concentration in artificial sea water (10.12 mM equiv.), relatively large masses of protoplasm were formed (Fig. 2), while by a 1/50 reduction of Ca^{++} (0.202 mM equiv.), large sheets (layers of protoplasm sandwiched between the

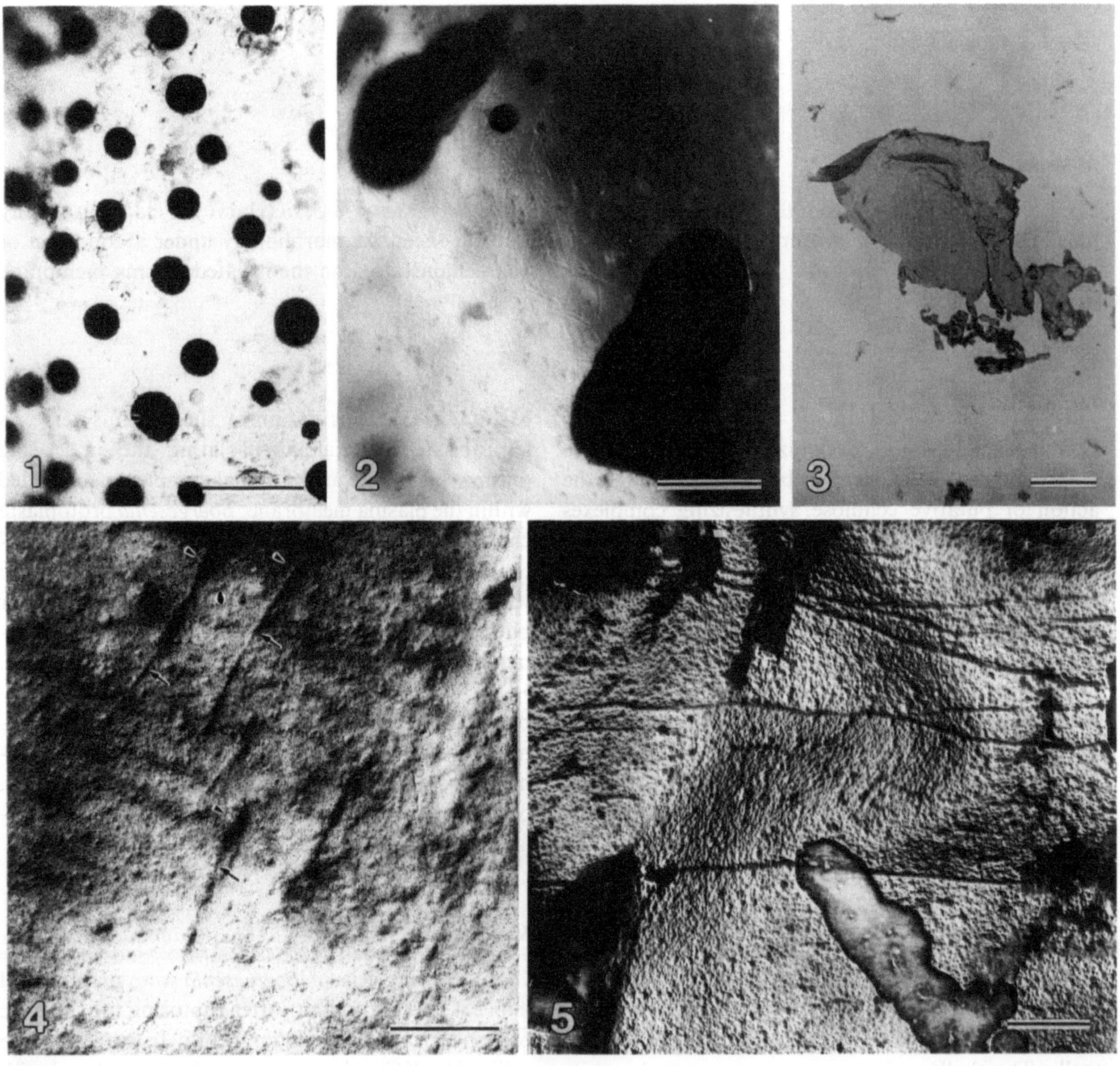

Figs 1–5.
Fig. 1. Protoplasts of *Boergesenia* developed from isolated protoplasm following 1 hour of culture. × 50, bar = 0.3 mm. Fig. 2. Masses of *Boergesenia* protoplasm. *Boergesenia* protoplasm were cultured in medium whose Ca^{++} concentration was reduced to 1/10 of that in artificial sea water (1.012 mM equiv.). × 50, bar = 0.3 mm. Fig. 3. Protoplasmic sheet of *Boergesenia* released from its thallus in Ca^{++}-free medium. × 10, bar = 1 mm. Fig. 4. Inside surface of the *Boergesenia* plasma membrane showing presumable terminal complexes (arrowheads) and nascent microfibrils (arrows) formed during culture for 30 min. × 33 500, bar = 0.5 µm. Fig. 5. Inside surface of the *Boergesenia* plasma membrane showing presumable nascent microfibrils formed during culture for 1 hour. × 25 000, bar = 0.5 µm.

plasma membranes and tonoplasts) were obtained (Fig. 3). This structure was found to remain for at least 12 and 24 hours in Ca^{++}-reduced (0.202 mM equiv.) and Ca^{++}-free artificial sea water, respectively.

Figure 4 shows an isolated plasma membrane, incubated for 30 minutes, dried and shadowed with platinum-carbon. The distribution of small membranous particles (10–15 nm in diameter) indicates that this sheet is a actual plasma membrane. The short filamentous structures (Fig. 4 arrows) were about 15–20 nm in width, which indicates that they correspond to the microfibrils in *Boergesenia* (Mizuta and Okuda 1987). Elliptical structures (Fig. 4, arrowheads) at the termini of the filaments were 90–100 nm in width. They may thus be considered as protrusions of the

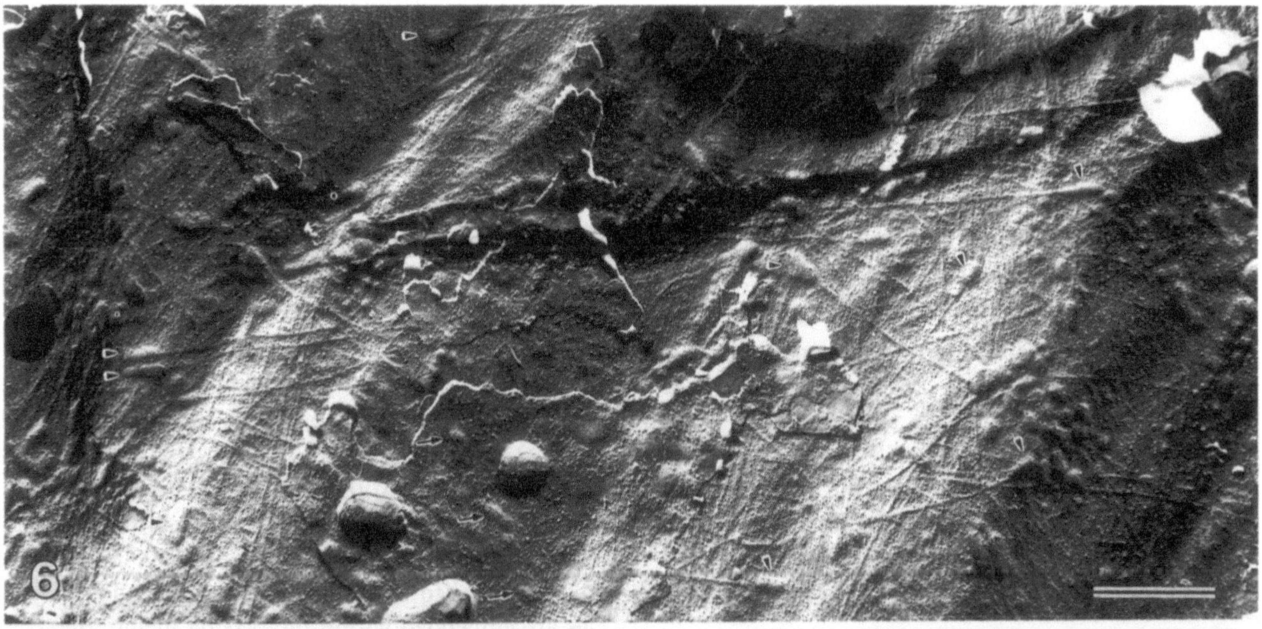

Figs 6, 7.
Fig. 6. Freeze-fractured E-face of the *Boergesenia* plasma membrane showing many elliptical terminal complexes (arrowheads) associated with microfibril impressions. Arrows show the terminal complexes observed on the PS-face. Aplanosporic cells of *Boergesenia* were fractured following 4 hours of culture at 25 °C. × 16000, bar = 1 µm. Fig. 7. Freeze-fractured E- and PS-faces of the *Boergesenia* plasma membrane showing terminal complexes covered and uncovered with plasma membrane leaflets. Small arrowheads, subunit particles of the terminal complexes visualized on the E-face; large arrowhead 1, elliptical terminal complex observed on the E-face; large arrowhead 2, elliptical terminal complex observed on the PS-face; large arrowhead 3, a part of an elliptical terminal complex showing that the subunit particles are existing inside it. × 55000, bar = 0.2 µm.

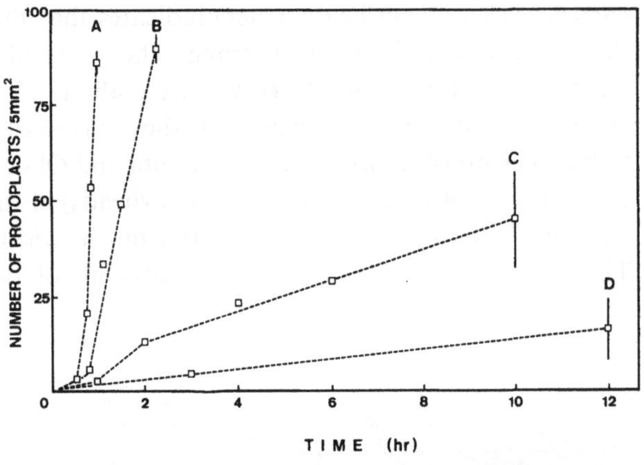

Fig. 8. Effect of Ca^{++} concentration on protoplasts formation from *Boergesenia* protoplasm. The protoplasm released from the *Boergesenia* thallus was cultured for 1–12 hours in reduced Ca^{++} concentrations (B–D). A, control (Ca^{++} concentration is 10.12 mM equiv.); B, 1/5 Ca^{++} (2.024 mM equiv.); C, 1/10 Ca^{++} (1.012 mM equiv.); D, 1/50 Ca^{++} (0.202 mM equiv.).

plasma membrane due to terminal complexes in *Boergesenia* situated under or within the plasma membrane. Many microfibril-like structures, 15–20 nm in width, were detected within 1 hour of culture (Fig. 5), but after 2 hours of culture, their number showed no change (not shown), which suggests that this constancy of the number of microfibrils appears to be due to the absence of a precursor of cellulose.

Terminal complexes in *Boergesenia* are comprised of linear multiple rows (approximately 50 nm in width) of particles (Mizuta and Okuda 1987) situated for the most part within the plasma membrane (Itoh and Brown 1984, Itoh *et al.* 1984, Brown 1985). However, elliptical structures similar to those observed on the isolated plasma membrane (Fig. 4) have also been observed on both the E- and P-faces of the freeze-fractured plasma membrane (Mizuta and Okuda 1987). Figure 6 shows a freeze-fractured plasma membrane of the E-face in the aplanosporic cell of *Boergesenia* cultured for 4 hours at 25 °C. This stage corresponds to the formation of randomly oriented microfibrils (Mizuta *et al.* 1985). Many elliptical structures (arrowheads) connected to microfibril impressions could be seen. No particles were present in the elliptical structures, approximately 100 nm in width. The structures were also on the PS-face but microfibril impressions connecting them to each other could not be seen (Fig. 6, arrows). It is assumed that particles of terminal complexes are present within elliptical structures but evidence for this is still vague. Therefore, further effort was made to fracture the elliptical structures.

Figure 7 shows the E- and PS-faces of the plasma membrane of the aplanosporic cell of *Boergesenia* cultured for 24 hours at 25 °C. Particulate structures of terminal complexes (small arrowheads) in this stage were relatively long and present on the E-face. Elliptical structures were noted on both E- (large arrowhead 1) and PS- (large arrowhead 2) faces. The portion of the structure on the PS-face indicated by the large arrowhead 3 in the figure suggests that the structure had been fractured, and that particles of the terminal complex are present within it. This evidence supports the possibility that terminal complex particles are present within elliptical structures.

As a result of drying of the isolated plasma membrane before shadowing, elliptical structures and their associated microfibril impressions could be discerned more clearly (Fig. 4) than when observed on the PS-face of a freeze-fractured plasma membrane (Fig. 6). Thus, elliptical structures in the isolated membrane would appear to be terminal complexes whose particles are embedded within the plasma membrane.

The present study on the formation of cellulose microfibrils on a released membrane system provides many possibilities for clarification of the mechanisms of cellulose assembly and its crystallization. Because of the presence of terminal complexes in isolated plasma membranes, it may be possible to isolate terminal complexes and crystallize cellulose from the terminal complexes *in vitro*. It would also be possible to examine the life and turnover of terminal complexes.

Acknowledgements

This work was supported in part by a grant, No. 01540565, from the Ministry of Education, Science and Culture of Japan.

References

Brown, R. M. Jr. 1985. Cellulose microfibril assembly and orientation: recent developments. *J. Cell Sci. Suppl. 2*: 13–32.

Brown, R. M. Jr. and D. Montezinos 1976. Cellulose microfibrils: Visualization of biosynthetic and orienting complexes in association with the plasma membrane. *Proc. Natl. Acad. Sci. USA 73*: 143–147.

Giddings, T. H., D. L. Brower and L. A. Staehelin. 1980. Visualization of particle complexes in the plasma membrane of *Micrasterias denticulata* associated with the formation of cellulose fibrils in the primary and secondary cell walls. *J. Cell Biol. 84*: 327–339.

Ishizawa, K., S. Enomoto and S. Wada. 1979. Germination and photo-induction of polarity in the spherical cells regenerated from protoplasmic fragments of *Boergesenia forbesii. Bot. Mag. Tokyo 92*: 173–186.

Itoh, T. and R. M. Brown, Jr. 1984. The assembly of cellulose microfibrils in *Valonia macrophysa* Kütz. *Planta 160*: 372–381.

Itoh, T., R. M. O'Neil and R. M. Brown, Jr. 1984. Interference of cell wall regeneration of *Boergesenia forbesii* protoplasts by Tinopal LPW, a fluorescent brightening agent. *Protoplasma 123*: 174–183.

Mizuta, S. 1985. Evidence for the regulation of the shift in cellulose microfibril orientation in freeze-fractured plasma membrane of *Boergesenia forbesii. Plant Cell Physiol. 26*: 53–62.

Mizuta, S. and S. Wada. 1981. Microfibrillar structure of growing cell wall in a coenocytic green alga, *Boergesenia forbesii. Bot. Mag. Tokyo 94*: 343–353.

Mizuta, S. and K. Okuda. 1987. *Boodlea* cell wall microfibril orientation unrelated to cortical microtubule arrangement. *Bot. Gaz. 148*: 297–307.

Mizuta, S., U. Kurogi, K. Okuda and R. M. Brown, Jr. 1989. Microfibrillar structure, cortical microtubule arrangement and the effect of amiprophos-methyl on microfibril orientation in the thallus cells of the filamentous green alga, *Chamaedoris orientalis. Ann. Bot. 64*: 383–394.

Mueller, S. C. and R. M. Brown, Jr. 1980. Evidence for an intramembrane component associated with a cellulose microfibril-synthesizing complex in higher plants. *J. Cell Biol. 84*: 315–326.

Botanica Marina
Vol. 34, pp. 417–424, 1991

Involvement of Cytoskeletal Microtubules in Microfibrillar Patterns in the Cell Walls of the Developing Coenocytic Green Alga, *Boodlea coacta*

S. Mizuta[1], S. Katoh, T. Harada, H. Yamada, K. Okuda and T. Morinaga

Department of Biology, Faculty of Science, Kochi University, Akebono-cho Kochi 780, Japan

(Accepted 10 June 1991)

Abstract

An examination was made of the features of cytoskeletal microtubules (MTs) in newly rounded cells of the coenocytic green alga, *Boodlea coacta* (Dickie) Murray *et* De Toni for comparison with newly formed microfibrillar textures. Fragments of *Boodlea* protoplasm, released into the culture medium from the mother cell by mechanical wounding, become protoplasts following 1 h of culture at 23 °C. These fragments contain many thallus-derived MTs that are initially long and curved, but subsequently disappear, their places being taken by numerous short MTs following completion of protoplast formation. Short MTs elongate in random directions in the cortical area of the cell. They increase in number reaching a maximum following 2 hours of culture. However, MTs decrease considerably in number over the next 2 hours of culture. After about 6 hours of culture, their number increases again and they are arranged parallel to each other. Cell walls begin to form after 2 hours of culture. Nascent microfibrils (MFs) appear as dots or randomly-oriented short lines of fluorescence following Fluostain (fluorescent agent that binds to cellulose) treatment. In thin sections, a lamella comprised of randomly-oriented MFs can be clearly seen within 2 and 3 hours but after additional culturing, microfibrils change their orientation from random to helicoid and then to a crossed fibrillar pattern following completion of the MT pattern of parallel orientation. The MF textures in cell walls of *Boodlea* thus appear to be related to MT cytoskeleton patterns.

Introduction

The cytoskeletal microtubules (MTs) in plants take the form of an interphase cortical array of MTs, the preprophase band of MTs, mitotic spindle fibers and phragmoplast MTs, and are closely associated with several dynamic events of cells (Lloyd *et al.* 1985). Microtubules distributed in the cortical area of cells are generally thought to regulate indirectly the angles of deposition of cellulose microfibrils (MFs). This concept was originally proposed by Ledbetter and Porter (1963) and is supported by the research of many investigators (Newcomb 1969, Hepler and Palevitz 1974, Gunning and Hardham 1982, Lloyd 1984). The strongest evidence for this concept is the parallelism of MTs and nascent MFs. Most studies on the correlation between MT and MF arrangements by using thin sections, tubulin immunofluorescence microscopy, and polarizing microscopy support this parallelism with minor exceptions (Newcomb 1969, Hepler and Palevitz 1974, Hardham *et al.* 1980, Gunning and Hardham 1982, Lloyd 1984). Parallelism is also important for explaining shape regulation of cells. In growing cylindrical cells, both MT and MFs are transversely arranged, and cells become spherical following treatment with antimicrotubule agents (see reviews by Green 1980, Lloyd 1984). In non-growing cells, whose walls are comprised of crossed fibrillar structures, MT/MF parallelism has also been demonstrated (Robinson and Quader 1982), and MFs exhibit abnormal arrangement following the disruption of MTs (Robinson *et al.* 1976, Robinson and Herzog 1977).

Microfibril orientation and cell shaping, however, appear not always to be in agreement with the above concept. In coenocytic siphonocladalean and clado-

[1] Correspondence and reprints

phoralean algae, which possess cross-polylamellated cell walls, transversely and obliquely arranged nascent MFs are not parallel to the cortical MT arrangement (Mizuta and Okuda 1987, Okuda and Mizuta 1987, Mizuta *et al.* 1989). The direction of growth of cylindrical cells in *Chaetomorpha* also is unrelated to cellulose MF orientation (Okuda and Mizuta 1987, Nishioka *et al.* 1990). Microtubules in these algae are always arranged longitudinally (Mizuta and Okuda 1987, Okuda and Mizuta 1987, Mizuta *et al.* 1989) and may possibly be related to maintaining cell growth in a longitudinal direction (Okuda *et al.* 1990).

These algae characteristically make protoplasts from isolated protoplasmic fragments (Ishizawa *et al.* 1979). The protoplasts rapidly start regenerating cell walls on their plasma membranes and develop both rhizoids and thalli on opposite sides (Ishizawa *et al.* 1979, Mizuta and Wada 1982, Mizuta 1985, Mizuta *et al.* 1985, Okuda and Mizuta 1985, Mizuta and

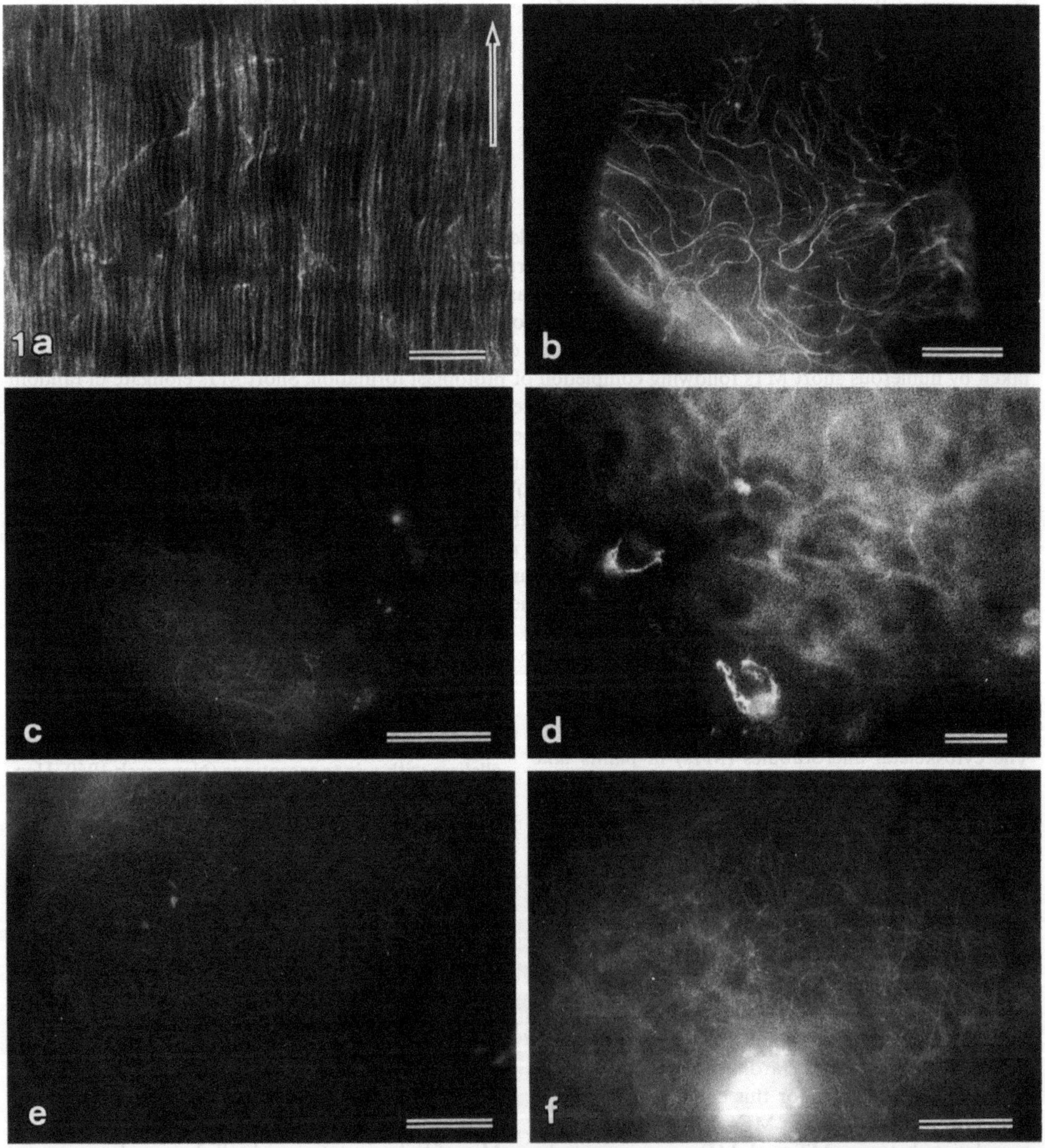

Fig. 1. a–l. Microtubules of the thallus (a), protoplasmic mass (b) and new round cells (c–l) in *Boodlea* stained with anti-tubulin antibody. Protoplasmic masses of *Boodlea* were cultured for 0.5 h (b), 1 h (c, d), 1.5 h (e), 2 h (f), 3 h (g), 4 h (h), 5 h (i), 6 h (j), 6.5 h (k) and 10 h (l) at 23 °C and then stained. An arrow in (a) indicates longitudinal direction of the thallus. Arrows

Okuda 1987). The protoplasts of *Boodlea* also regenerate cell walls consisting of crossed fibrillar textures following completion of the randomly oriented outer surface fibril layer (Okuda and Mizuta 1985, Mizuta 1987). It has yet to be clarified as to how the formation of this fibrillar pattern is controlled. In this paper, the dynamic features of the MT cytoskeleton during regeneration of cell walls in *Boodlea* are investigated and the action of MT cytoskeleton in generating the microfibrillar pattern is discussed.

Material and Methods

Cultures and observation of new cell walls with fluorescence microscopy

Boodlea coacta (Dickie) Murray *et* De Toni was cultured at 23 °C under a 14 h light (4 W m^2)/10 h dark cycle. The culture medium used was filtered sea water containing the ES-enrichment of Provasoli *et al.* (1957). The method of protoplast formation was basically that described previously (Mizuta 1985). For

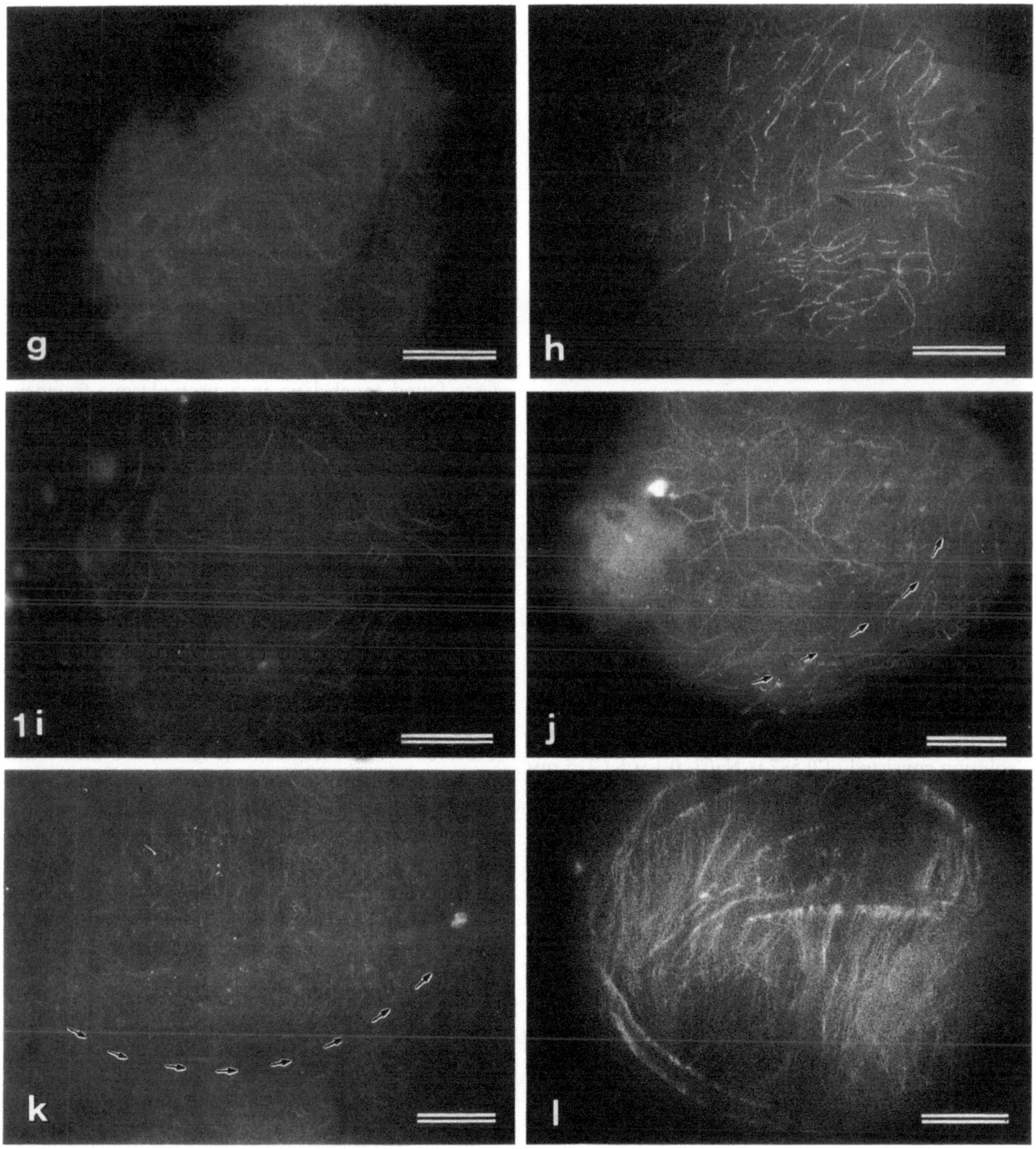

in (j) and (k) show the main orientation of MTs. a, × 1220, bar = 10 µm; b, × 660, bar = 20 µm; c, × 347, bar = 50 µm; d, × 1014, bar = 10 µm; e, × 688, bar = 20 µm; f, × 750, bar = 20 µm; g, × 736, bar = 20 µm; h, × 743, bar = 20 µm; i, × 750, bar = 20 µm; j, × 660, bar = 20 µm; k, × 660, bar = 20 µm; l, × 930, bar = 15 µm.

observation of MFs by fluorescence microscopy, cells cultured for various periods were incubated in culture medium containing 0.01% Fluostain I (Dojin Chemical Co. Kumamoto, Japan), a fluorescent brightener with molecular structure similar to that of Calcofluor White and excited by 365 nm UV. After 5 min incubation, the cells were placed on a glass slide, observed with a fluorescence microscope (Type BHS-RFK, Olympus Kogaku, Tokyo) and photographs were taken using Kodak Tri-X Pan film.

Thin sectioning

The cells were fixed in 50% sea water solution containing 4% glutaraldehyde and 0.8 M cacodylate buffer (pH 7.2) for 1 h at room temperature and then for 1 h at 2 °C. They were rinsed with 70% sea water and postfixed with 1% osmium tetroxide solution containing 0.8 M cacodylate buffer (pH 7.2). Dehydration of the fixed cells was conducted with acetone followed by embedding in Spurr resin. Thin sections (90–130 nm thick) obtained with a Reichert Om U2 ultramicrotome were picked up on Formvar-coated grids, incubated in saturated KOH-ethanol solution for 3 min to remove embedding resin and shadowed with platinum-carbon. A JEM 100U electron microscope (JEOL Ltd. Tokyo) was used for observation.

Replica preparation

The method for replica preparation was basically the same as described previously (Mizuta and Okuda 1987).

Immunofluorescence

The cells were fixed for 10 min at room temperature in 1.7% paraformaldehyde plus 0.4% glutaraldehyde in Pipes buffer solution (pH 6.9). The Pipes buffer solution consisted of 0.587 M NaCl, 0.1 M KCl, 5 mM $MgCl_2$, 5 mM ethylene glycol-bis (β-aminoethyl ether)-N, N, N′, N′-tetraacetic acid (EGTA) and 50 mM piperazine-N, N′-bis (ethanesulfonic acid) (Pipes). The fixed cells were rinsed three times with Pipes buffer solution, treated twice with another Pipes buffer solution containing 1% Triton X-100 and 1% Nonidet P-40 for 30 min to remove chlorophyll, and rinsed three times with the original Pipes buffer solution for 5 min. This was followed by treatment with phosphate-buffered saline (PBS: 0.137 M NaCl, 2.7 mM KCl, 1.5 mM KH_2PO_4, 8 mM Na_2PO_4, pH 7.3 with HCl) containing 0.5 mg/mL $NaBH_4$ for 5 min to remove free aldehydes and three rinses with PBS. Next, the cells were treated with the primary antibody by incubation in 400 μL of primary antibody solution containing rat monoclonal anti-yeast tubulin (Sera Labs., Crawley Down, England), diluted 1:100 with PBS, for 2–5 h in a dark chamber. After being rinsed three times with PBS, the cells were incubated in 400 μl of secondary antibody solution consisting of fluorescein isothiocyanate (FITC) labelled rabbit anti-rat immunoglobulin G (IgG) (Miles Science, Illinois, USA), diluted 1:150 with PBS, for 5 h at room temperature in a dark chamber. The cells, after being rinsed three times, were mounted on a glass slide with PBS containing 1% p-phenylenediamine and observed with a epifluorescence microscope (BHS-RFK, Olympus Kogaku, Tokyo) using a blue excitation system and absorption filters of G-520 and EY-455. To record optical results, Kodak Tri-X Pan film was used.

Frequency of MTs was determined with five cells in each culture period by counting the MTs in 10 μm squares of each cell.

Results

Cortical MTs in the *Boodlea* thallus run longitudinally parallel to each other (Fig. 1a). Each MT is extremely long with spacings of about 0.25 μm. Microtubules were found tangled up in a protoplasmic fragment released from thallus cells when cut (wounded) with scissors (Fig. 1b). Each of such fragments assumed a spherical shape (protoplasts) after 1 hour in culture after being wounded. Microtubule immunofluorescence in the protoplasts markedly decreased (Fig. 1c) with some local bright areas (Fig. 1c, d). However, a great number of short and randomly oriented MTs appeared instead (Fig. 1e) and the MTs increased in number reaching a maximum after 2 hours of culture (Fig. 1f). After 3 hours of culture, some of them elongated more, with concomitant decrease in number (Fig. 1g). During 4 and 5 hours of culture, MTs were observed as short fragments (Fig. 1h) and their number greatly decreased (Fig. 1i). But at 6 hours of culture, the number of MTs increased again and the MTs took on an ordered arrangement in localized regions (Fig. 1j, k, arrows). In most cases, they could be seen oriented unidirectionally after 10 hours of culture (Fig. 1l). The changes in MT concentration during culture are summarized in Figure 2.

The cell walls in a *Boodlea* thallus have a typical cross-polylamellated texture (Fig. 3a). Microfibrils in each fibril layer can be seen to run longitudinally and transversely when viewing the inside surface of the wall (Fig. 3b). At 2.5 hours of culture after wounding, many spots of randomly arranged fibrous fluorescence visualized with Fluostain I could be seen (Fig. 3c). These were considered to be nascent cellulose at sites of cellulose synthesizing complexes (dots in Fig.

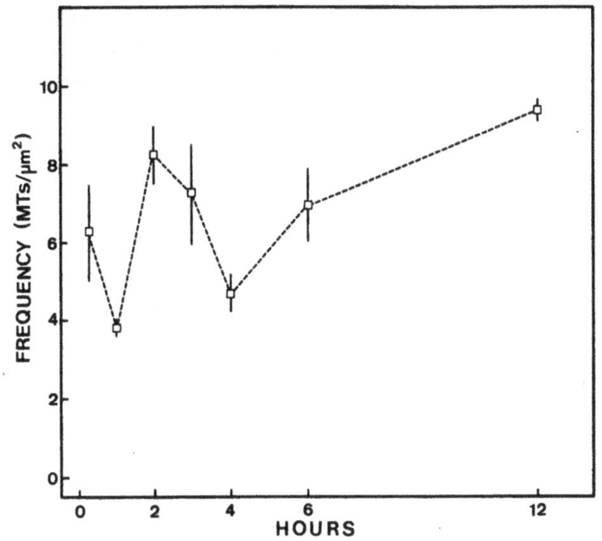

Fig. 2. Changes in concentration of microtubules during the culture of protoplasmic fragments for 12 hours. Each point indicates the mean value with standard deviation (vertical bar).

3d). A cross section following culture for 3 hours showed a wall layer comprised of randomly oriented MFs that had formed after 2–3 hours of culture (Fig. 3e). With additional culturing, however, the fluorescence of cellulose due to Fluostain I characteristically showed a helicoidal arrangement (Fig. 3f). This could also be seen in a cross section (Fig. 3g). The helicoidal cell wall continued to be generated with subsequent culturing (Fig. 3h). After about 10 hours, the arrangement of the fluorescence of cellulose changed from helicoid to a cross-pattern when the fluorescence was focused in the inside surface of the wall (Fig. 3i). This was also confirmed by images of cross sections at 24 hours of culture (Fig. 3j, k). The wall layers labeled R, H and C in Figure 3k exhibit random, helicoidal and crossed fibrillar layers, respectively.

Discussion

The microtubule cytoskeleton in *Boodlea* showed complex dynamic features in orientation and density during early protoplast development. The long convoluted MTs in protoplasmic fragments (Fig. 1b) may possibly be derive from those of the thallus since MTs are extremely long and their orientation is unusual. They virtually disappear when fragments develop into protoplasts. Their disappearance and the concomitant generation of randomly arranged short MTs indicates that MTs from thalli are possibly not used for the cytoskeleton of new cells, but rather a new MT cytoskeleton is formed instead. Microtubules in new cells were noted to change considerably in density and orientation during 10 hours of culture. That MT orientation becomes orderly during 6–10 hours of culture, accompanied by recovery of density, would indicate that the de-polymerized tubulin subunits may be re-polymerized into MTs oriented parallel to each other during this period. Likewise the decrease in MT density during 3–6 hours may be due to the making of longer re-oriented MTs from those that are randomly oriented.

The function of these cortical MTs may be to regulate the angles of nascent MF deposition (Robinson and Quader 1982, Gunning and Hardham 1982, Lloyd 1984). In cellulosic coenocytic green algae including *Boodlea*, these MTs do not appear to have this function, as is supported by the observation that newly formed spherical cells (aplanospores) lack the parallelism of MT and MFs (Itoh and Brown 1984, Okuda and Mizuta 1985, Mizuta and Okuda 1987, Okuda and Mizuta 1987, Mizuta *et al.* 1989). Hayano *et al.* (1988) found MTs of *Valonia* and *Boergesenia* aplanospores were oriented randomly throughout the development of cell walls, although orderly oriented MFs were noted consistently to form during this period. Thus, the relationship between the MT-cytoskeleton and microfibrillar pattern in cell walls remains unclear.

Should cell wall generation in *Boodlea* occur through discontinuous stages of randomly-oriented (stage 1), helicoidally-oriented (stage 2) and crossed (stage 3) MF formation, the regulation of MF orientation by the co-arrangement of MT and MFs would not appear very likely. Although both MT and MF orientation in stage 1 is random, the helicoidal orientation of MFs (stage 2) may no longer be considered to be a result of MT orientation (Neville 1985). Moreover, although MT orientation in stage 3 is unidirectional, a crossed-fibrillar pattern is produced, indicating the structural features of MT to be basically the same as in a thallus (Mizuta *et al.* 1989). The control of MF orientation due to MT/MF parallelism is thus not found in *Boodlea* new cells.

The patterns of the MT cytoskeleton of *Boodlea* in the present experiment may be associated with the regulation of microfibrillar patterns in cell walls. This is supported by the findings that the patterns of the microfibrillar texture in cell walls are reflected by MT cytoskeletal patterns since randomly oriented cellulose deposition was confirmed in 2.5 hours of culture following the appearance of dense randomly-oriented MTs in 2 hours (stage 1) and too, the helicoidal wall appears at low MT density in 3–6 hours and crossed fibrillar walls are formed following the appearance of parallel MTs. Although the effects of the disruption of MTs by antimicrotubule agents should be studied, the removal of MTs from thallus cells of *Chamaedoris*, a coenocytic green alga similar to *Boodlea*, indicates that longitudinally oriented MTs possibly regulate a certain system of alternating changes in MF orientation (Mizuta *et al.* 1989). The formation of randomly oriented MFs in stage 1 and those that are arranged in an orderly fashion in stage 3 would suggest that these MF orientations depend on the patterns of MT whose density increases prior to each of these two stages. The density of MTs in a helicoidal wall (stage 2), however, was noted to be quite low. Helicoidal MF orientation is considered to occur by spontaneous gradual change in MF orientation within liquid crystals of the cell wall matrix (Neville 1985).

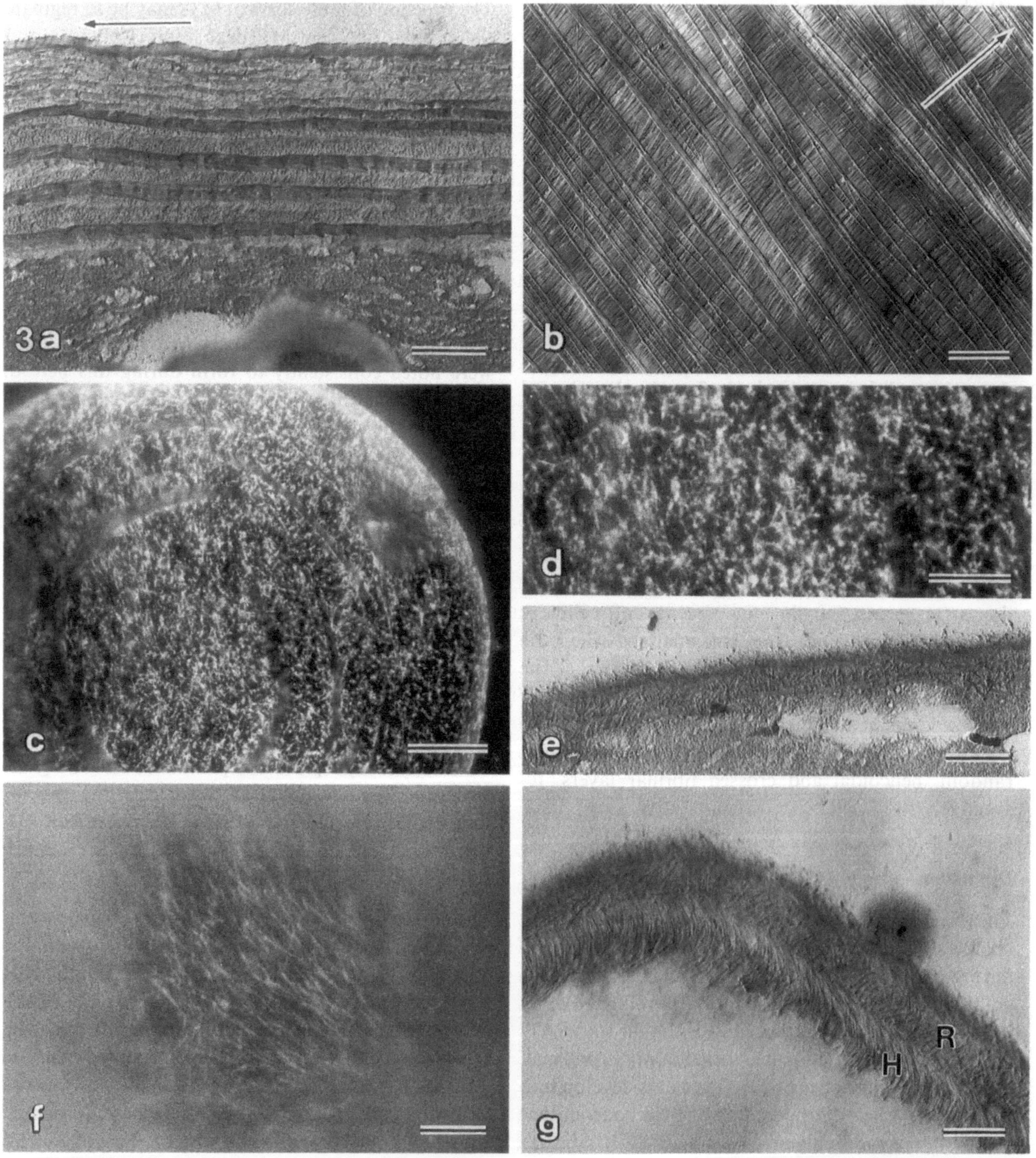

Fig. 3. Microfibrillar textures in *Boodlea* (a, b) and development of new cell walls (c–k) in the new round cells derived from the protoplasm. (a) Cross sectional view of the thallus wall. Arrow indicates longitudinal direction of the thallus. × 12000, bar = 1 μm. (b) Replica showing the inside surface of the thallus wall. Arrow indicates longitudinal direction of the thallus. × 5000, bar = 2 μm. (c) Nascent microfibrils developed on the new round cell cultured for 2.5 h. The microfibrils were stained with Fluostain I. × 643, bar = 20 μm. (d) Higher magnification of (c). Note the small fluorescent dots that are assumed to be nascent cellulose at sites of synthesizing complexes. × 1300, bar = 10 μm. (e) Cross section of the new cell showing the outermost, randomly oriented fibrillar layer. The cell was cultured for 3 h. × 16000, bar = 1 μm. (f) Fluorescence of developing microfibrils in the new cell

The helicoidal wall formation in stage 2 appears to occur in association with a decrease in MT density.

It is true that the alternating change of MF orientation in coenocytic green algae occurs in accordance with cell polarity, since MFs are formed as helices from pole to pole (Preston 1974, Mizuta 1987, Mizuta and Okuda 1987, Okuda *et al.* 1990). The fact that MTs in these algae are always arranged longitudinally and that when MTs are disrupted by anti-MT agents, both alternating MF formation and cell growth appear to be inhibited (Mizuta *et al.* 1989), suggests that MTs in these algae govern cell polarity. The formation of meridionally arranged MTs in the new cells of *Boodlea* could thus indicate the establishment of new cell polarity.

Changes in the patterns of the MT cytoskeleton and their relationship to microfibrillar texture in the early development of *Boodlea* new cells are diagrammatically shown in Figure 4. Microtubules in the thallus are ordinarily arranged longitudinally with very close spacing and MFs in the cell wall are alternately arranged longitudinally and transversely (A). Protoplasmic fragments from the thallus (B) possess many MTs that have possibly been derived from the thallus. These MTs disappear when fragments become spherical in shape (protoplasts) (C) and new MTs elongate in a random direction (D) assuming their highest density at 1 hour of culture (E). The formation of randomly oriented MFs starts (stage 1) subsequent to this. Although these decreases in density during 3–6 hours of culture, aligned MTs increase instead (F).

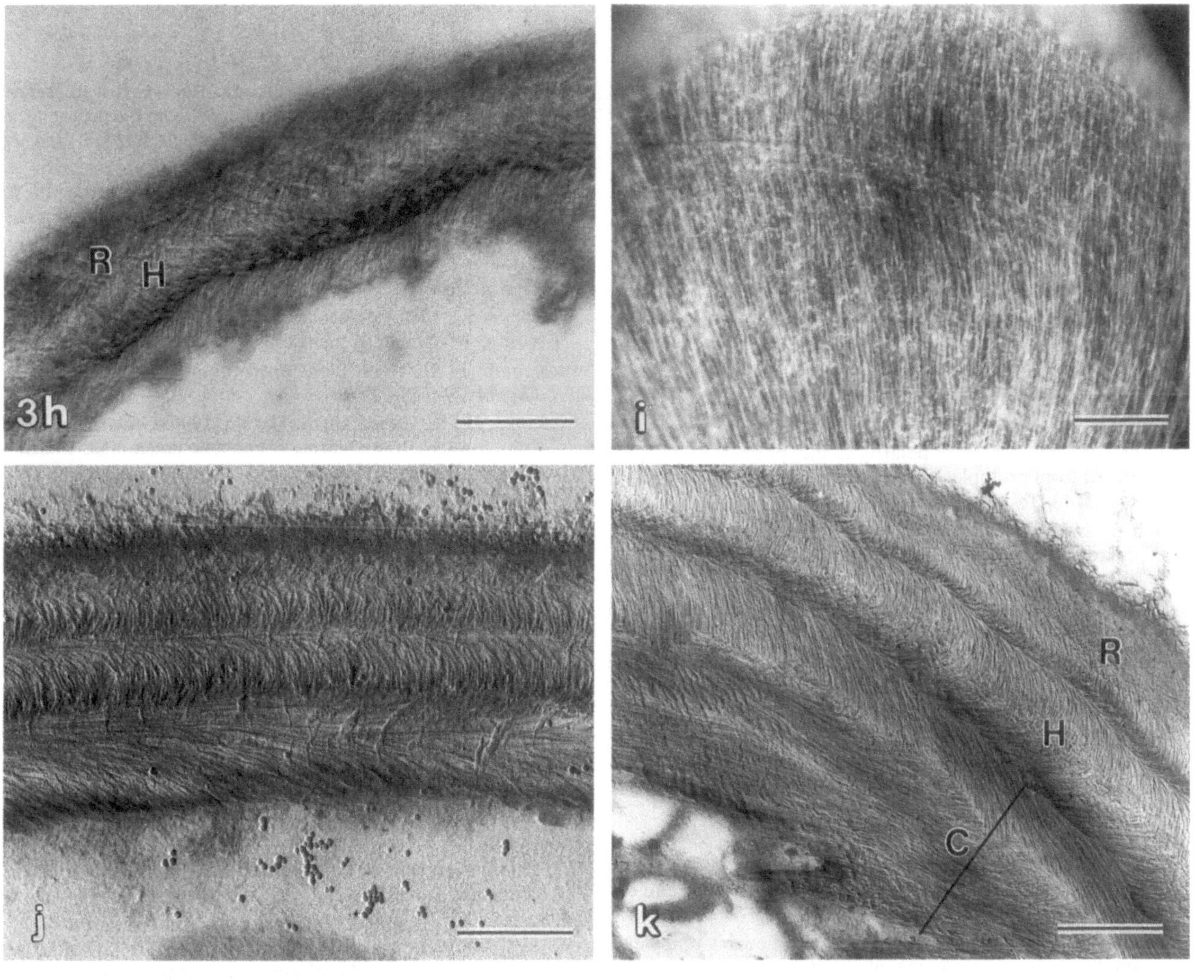

cultured for 4 h showing helicoidal orientation. Fluorescence was focused on the inside surface of the wall. × 1100, bar = 10 μm. (g) and (h) Cross sections of new walls formed after 4 h (g) and 6 h (h) of culture. A part of the helicoidal wall (H) can be seen inside the randomly oriented fibrillar layer (R). × 16200, bar = 1 μm. (i) Fluorescence of microfibrils in the new cell cultured for 10 h showing crossed fibrillar pattern. × 640, bar = 20 μm. (j) and (k) Cross sections of the new walls formed after 24 h of culture. R, H and C in (k) indicate wall layers consisting of random, helicoidal and crossed microfibrils, respectively. j, × 16000, bar = 1 μm; k, × 7500, bar = 2 μm.

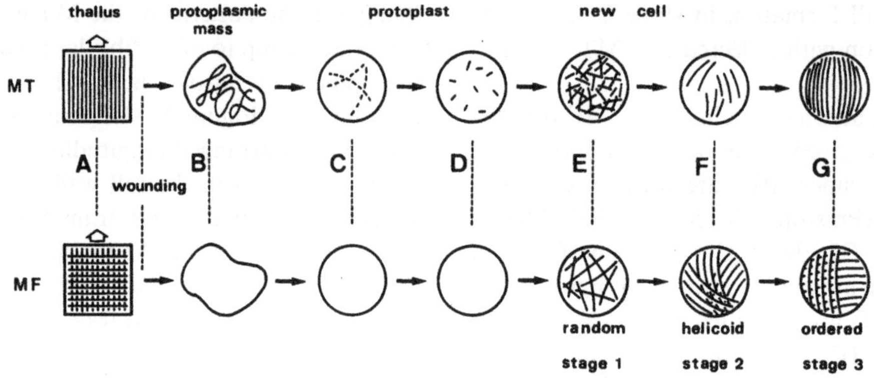

Fig. 4. Diagrammatical representation of changes in patterns of MT cytoskeleton and its relationship to the formation of microfibrillar texture in the *Boodlea* new cell. Arrows in thalli indicate the longitudinal direction of them.

During this period, MF orientation exhibits a helicoidal pattern (stage 2). With time, MTs become arranged meridionally with constant spacing (G) and crossed fibrillar orientation is resumed.

Acknowledgements

This work was supported in part by a grant, No. 01540565, from the ministry of Education, Science and Culture of Japan.

References

Green, P. B. 1980. Organogenesis – A biophysical view. *Ann. Rev. Plant Physiol. 31*: 51–82.

Gunning, B. E. S. and A. R. Hardham. 1982. Microtubules. *Ann. Rev. Plant Physiol. 33*: 651–698.

Hardham, A. R., P. B. Green and J. M. Lang. 1980. Reorganization of cortical microtubules and cellulose deposition during leaf formation in *Graptopetalum paraguayense. Planta 149*: 181–195.

Hayano, S., T. Itoh and R. M. Brown, Jr. 1988. Orientation of microtubules during regeneration of cell wall in selected giant marine algae. *Plant Cell Physiol. 29*: 785–793.

Hepler, P. K. and B. A. Palevitz. 1974. Microtubules and microfilaments. *Ann. Rev. Plant Physiol. 25*: 309–362.

Ishizawa, K., S. Enomoto and S. Wada. 1979. Germination and photoinduction of polarity in the spherical cells regenerated from protoplasmic fragments of *Boergesenia forbesii. Bot. Mag. Tokyo 92*: 173–186.

Itoh, T. and R. M. Brown, Jr. 1984. The assembly of cellulose microfibrils in *Valonia macrophysa* Kütz. *Planta 160*: 372–381.

Ledbetter, M. C. and K. R. Porter. 1963. A "microtubule" in plant cell fine structure. *J. Cell. Biol. 19*: 239–250.

Lloyd, C. W. 1984. Toward a dynamic helical model for the influence of microtubules on wall patterns in plants. *Int. Rev. Cytol. 86*: 1–51.

Lloyd, C. W., L. Clayton, P. J. Dowson, J. H. Doonan, J. S. Hulme, I. N. Roberts and B. Wells. 1985. The cytoskeleton underlying side walls and cross walls in plants: molecules and macromolecular assemblies. *J. Cell Sci. Suppl. 2*: 143–155.

Mizuta, S. 1987. Structure and generation of cell walls in cellulosic algae I. Microfibril arrangement patterns and cell wall formation. *Jpn J. Phycol. 35*: 45–59.

Mizuta, S. 1985. Assembly of cellulose synthesizing complexes on the plasma membrane of *Boodlea coacta. Plant Cell Physiol. 26*: 1443–1453.

Mizuta, S. and S. Wada. 1982. Effects of light and inhibitors on polylamellation and shift of microfibril orientation in *Boergensia* cell wall. *Plant Cell Physiol. 23*: 257–264.

Mizuta, S. and K. Okuda. 1987. *Boodlea* cell wall microfibril orientation unrelated to cortical microtubule arrangement. *Bot. Gaz.* 148: 297–307.

Mizuta, S., K. Sawada and K. Okuda. 1985. Cell wall regeneration of new spherical cells developed from the protoplasm of a coenocytic green alga, *Boergesenia forbesii. Jpn J. Phycol. 33*: 32–44.

Mizuta, S., U. Kurogi, K. Okuda and R. M. Brown, Jr. 1989. Microfibrillar structure, cortical microtubule arrangement and the effect of amiprophosmethyl on microfibril orientation in the thallus cells of the filamentous green alga, *Chamaedoris orientalis. Ann. Bot. 64*: 383–394.

Neville, A. C. 1985. Molecular and mechanical aspects of helicoid development in plant cell walls. *BioEssays 3*: 4–8.

Newcomb, E. H. 1969. Plant microtubules. *Ann. Rev. Plant. Physiol. 20*: 253–288.

Nishioka, H., K. Okuda and S. Mizuta. 1990. Growth and cell-shape control in the marine coenocytic green alga, *Chaetomorpha moniligera. Bot. Mar. 33*: 289–297.

Okuda, K. and S. Mizuta. 1987. Modification in cell shape unrelated to cortical microtubule orientation in growing thallus cells of *Chaetomorpha moniligera. Plant Cell Physiol. 28*: 461–473.

Okuda, K. and S. Mizuta. 1985. Analysis of cellulose microfibril arrangement patterns in the cell wall of new spherical cells regenerated from *Boodlea coacta. Jpn J. Phycol. 33*: 301–311.

Okuda, K., K. Matsuo and S. Mizuta. 1990. Characteristics of the deposition of microfibrils during formation of the polylamellate walls in the coenocytic green alga *Chamaedoris orientalis. Plant Cell Physiol. 31*: 357–364.

Preston, R. D. 1974. *The Physical Biology of Plant Cell Walls.* Chapman and Hall, London.

Provasoli, L., J. J. A. McLaughlin and M. R. Droop. 1957. The development of artificial media for marine algae. *Arch. Microbiol. 25*: 392–428.

Robinson, D. G. and H. Quader. 1982. The microtubule-microfibril syndrome. *In*: (C. W. Lloyd, ed.) *The Cytoskeleton in Plant Growth and Development*. Academic Press, London. pp. 109–125.

Robinson, D. G. and W. Herzog. 1977. Structure, synthesis and orientation of microfibrils II. A study of the action of microtubule inhibitors on microtubules and microfibril orientation in *Oocystis solitaria. Cytobiologie 15*: 463–474.

Robinson, D. G., I. Grimm and H. Sachs. 1976. Colchicine and microfibril orientation. *Protoplasma 89*: 375–380.

Botanica Marina
Vol. 34, pp. 425–428, 1991

Comparative Fatty Acid Composition of Five Species of *Dictyota* (Phaeophyta)

M. Shameel, W. Shaikh and R. Khan

Department of Botany, University of Karachi, Karachi-75270, Pakistan

(Accepted 10 June 1991)

Abstract

The fatty acid analyses of *Dictyota dichotoma* (Huds.) Lamour., *D. dumosa* Børg., *D. hauckiana* Nizam., *D. indica* Sond. *ex* Kütz. and *D. maxima* Zanard., found in Karachi coastal waters are reported. Ten saturated and four unsaturated fatty acids were identified by gas chromatography-mass spectrometry, of which only palmitic acid occurs in all of them, while pentadecylic, margaric, arachidic and oleic acids were found in four species, and others are of more selective distribution. *Dictyota indica* differs slightly in its fatty acid composition from other species, containing eicosatrienoic acid.

Introduction

Several species of *Dictyota* Lamouroux have been reported from the coast of Pakistan, out of which five species [*D. dichotoma* (Hudson) Lamouroux, *D. dumosa* Børgesen, *D. hauckiana* Nizamuddin, *D. indica* Sonder *ex* Kützing and *D. maxima* Zanardini] are conspicuous on sand-covered rocks and in rocky pools in mid- to sub-littoral areas (Shameel 1990 a). In such communities these species occur in abundance and are said to be a good source of food for invertebrates and herbivorous fishes (*e.g.* Teixeira *et al.* 1986).

Members of the order Dictyotales are a rich source of monocyclic, bicyclic and tricyclic diterpenes (Faulkner 1990). The thalli of *Dictyota* have recently received much attention, and approximately 90 diterpenes of different skeletal classes have been isolated from some 20 different species (Teixeira *et al.* 1986, Faulkner 1990, Ahmad *et al.* 1991). But apart from *D. dichotoma* (Johns *et al.* 1979) no investigations have so far been carried out on the fatty acids of any *Dictyota*. As fatty acids play an important role in the food chain of the marine environment, an assessment of these constituents appears to be a necessary prerequisite for their use in nutritional studies. In this communication an attempt has been made to extract and identify fatty acids of *Dictyota* of the abovementioned five species using gas-liquid chromatographic and mass spectrometric techniques.

Material and Methods

Large and healthy specimens of *Dictyota dichotoma*, *D. hauckiana* and *D. indica* were collected from mid- to sub-littoral rocks during the months of May to July 1987 and 1988 at Hawkes Bay and Buleji, and *D. dumosa* and *D. maxima* were picked up as drift material during June to July 1988 and 1989 from Sandspit, near Karachi, Pakistan. These species grow abundantly during May to July, therefore a multiple sampling employing 6–8 replicates collected during two consecutive years was carried out. The seaweed samples were first cleaned with tap water and later with distilled water to remove sand particles, animal castings and adhering detritus, and were then air-dried in the shade.

The dried seaweeds (1.4–1.8 kg) were milled and extracted three times by percolation at room temperature with $CHCl_3$: hexane (1 : 1 v/v) for 20 days. The combined extract so obtained was evaporated under vacuum, giving about 50 g of thick dark brownish residue. The procedures for saponification and esterification and the techniques for gas chromatography-mass spectrometry (GC-MS) were the same as described earlier (Shameel 1990 b).

Result and Discussion

The analyses of fatty acids of *Dictyota dichotoma* (1), *D. dumosa* (2), *D. hauckiana* (3), *D. indica* (4) and *D. maxima* (5) revealed the presence of 10 saturated and 4 unsaturated acids. The relative percentages of the total fatty acids of these species are presented in Table I. The total fatty acid pattern shows relatively high levels of saturated as compared to unsaturated components. Similar observations have also been made in other brown seaweeds (Qasim 1986, Shameel 1990 b). Palmitic acid (C16 : 0) was found to occur in the largest proportion in all the five species. This is not unusual, as hexadecanoic acid commonly occurs in the Dictyotales (Shaikh *et al.* 1990 a, b) and other brown algae (Bano *et al.* 1987).

Pentadecylic (C15 : 0), margaric (C17 : 0), arachidic (C20 : 0) and oleic (C18 : 1) acids are common in *Dictyota* of four species. The first one is absent in *D. maxima*, the second one in *D. dumosa* and the last one in *D. dichotoma* only. Oleic is the unsaturated fatty acid found in most of the brown seaweeds (Qasim 1986, Shameel 1990 b, Shaikh *et al.* 1991). Pentadecylic acid has also been reported from certain Dictyotales (Skaikh *et al.* 1990 a, b). However, the occurrence of margaric acid is unusual, as it has so far been only reported among brown algae from *Iyengaria stellata* (Børg.) Børg, a member of the Scytosiphonales (Usmanghani *et al.* 1987). Arachidic acid is fairly widely distributed in higher plants, but is rarely found in algae (Lie Ken Jie 1989).

Table I. Relative percentages of total fatty acids of *Dictyota* species: 1 = *D. dichotoma*, 2 = *D. dumosa*, 3 = *D. hauckiana*, 4 = *D. indica* and 5 = *D. maxima* (nd = not detected).

Fatty acid	Systematic Name (Common Name)	Formula (M^+)	Dictyota species (1)	(2)	(3)	(4)	(5)
I. Saturated Fatty Acids			95.73	73.97	79.48	61.03	85.24
C12 : 0	Dodecanoic acid (Lauric acid)	$C_{12}H_{24}O_2$ (200)	6.88	7.94	nd	nd	7.31
C13 : 0	Tridecanoic acid (Tridecylic acid)	$C_{13}H_{26}O_2$ (214)	nd	nd	nd	nd	2.84
C14 : 0	Tetradecanoic acid (Myristic acid)	$C_{14}H_{28}O_2$ (228)	nd	nd	6.27	6.53	6.11
C15 : 0	Pentadecanoic acid (Pentadecylic acid)	$C_{15}H_{30}O_2$ (242)	16.12	14.23	15.38	2.92	nd
C16 : 0	Hexadecanoic acid (Palmitic acid)	$C_{16}H_{32}O_2$ (256)	20.83	24.16	21.37	19.43	21.56
C17 : 0	Heptadecanoic acid (Margaric acid)	$C_{17}H_{34}O_2$ (270)	16.29	nd	18.31	21.15	17.43
C18 : 0	Octadecanoic acid (Stearic acid)	$C_{18}H_{36}O_2$ (284)	15.79	12.61	nd	nd	16.72
C19 : 0	Nonadecanoic acid (Nonadecylic acid)	$C_{19}H_{38}O_2$ (298)	nd	nd	18.15	nd	nd
C20 : 0	Eicosanoic acid (Arachidic acid)	$C_{20}H_{40}O_2$ (312)	19.82	11.32	nd	11.00	13.27
C22 : 0	Decosanoic acid (Behenic acid)	$C_{22}H_{44}O_2$ (340)	nd	3.71	nd	nd	nd
II. Unsaturated Fatty Acids			4.20	25.54	19.77	38.51	14.52
C12 : 1	Dodecenoic acid (Lauroleic acid)	$C_{12}H_{22}O_2$ (198)	4.20	nd	nd	nd	nd
C18 : 1	Octadecenoic acid (Oleic acid)	$C_{18}H_{34}O_2$ (282)	nd	22.07	19.77	20.29	12.35
C17 : 2	Heptadeca-9,12-dienoic acid	$C_{17}H_{30}O_2$ (266)	nd	3.47	nd	3.09	2.17
C20 : 3	Eicosa-8,11,14-trienoic acid	$C_{20}H_{34}O_2$ (306)	nd	nd	nd	15.13	nd

Lauric (C12:0), myristic (C14:0), stearic (C18:0) and heptadecadienoic (C17:2) acids are abundant in *Dictyota* of only three species and appear to have a restricted occurrence. The first one is absent in *D. hauckiana* and *D. indica*, the second one in *D. dichotoma* and *D. dumosa*, the third one in *D. hauckiana* and *D. indica*, and the last acid in *D. dichotoma* and *D. hauckiana*. Tridecylic acid (C13:0) is found only in *D. maxima*, nonadecylic acid (C19:0) only in *D. hauckiana*, behenic acid (C22:0) only in *D. dumosa*, and eicosatrienoic acid (C20:3) only in *D. indica*. This delineates a selective distribution of certain fatty acids within *Dictyota* species.

The saturated fatty acids identified in *Dictyota* species ranged from C12 to C22, in agreement with the observations of Johns *et al.* (1979). They reported 18 fatty acids from *D. dichotoma* on the basis of retention time measurements. The fatty acids reported in both the studies are pentadecylic (C15:0), palmitic (C16:0), margaric (C17:0), stearic (C18:0) and arachidic (C20:0) acids, while the detection of lauric (C12:0) and lauroleic (C12:1) acids is an addition to the previous information. Fresh-water algae generally lack fatty acids with more than 18 carbon atoms (Lie Ken Jie 1989). The occurrence of nonadecylic (C19:0), arachidic (C20:0) and behenic (C22:0) acids in *Dictyota* spp. suggests that the presence of higher acids may be an adaptation necessary for survival in the marine environment. The long chain fatty acids may be required to confer some distinctive physical properties upon the membranes in addition to serving as the precursors of prostanoids (Christie 1987).

In the composition of unsaturated fatty acids, the brown seaweeds usually show great diversity (Shameel 1990 b), but the *Dictyota* species investigated exhibited a very narrow range of diversity. Except for oleic acid, all the detected mono-, di- and triunsaturated acids are of restricted occurrence. Myristoleic, palmitoleic, linoleic and linolenic acids, which have been reported from several brown seaweeds of this region (Qasim 1986), were found to be absent in all five species of *Dictyota*. Although brown seaweeds are a good source of 20:4 and 20:5 fatty acids (Banaigs *et al.* 1984), they could not be detected in any of these *Dictyota* spp., which contain only 20:3 fatty acid. Johns *et al.* (1979) found high levels of 16:1ω5 (8%), an uncommon acid in algal lipids, with trace amounts of 18:1ω5 in *D. dichotoma* from Australian waters, but these acids could not be detected in *D. dichotoma* from Pakistan waters. It appears that regional differences may also occur within the same species.

Dictyota of all five species investigated differ from one another, although only slightly, in their fatty acid compositions. *Dictyota indica* appears to differ more in its fatty acid distribution than the other four species. It possesses comparatively less saturated acids and more unsaturated ones than others, as it revealed by the presence of mono-, di- and triunsaturated fatty acids. Oleic acid (C18:1) is found in other brown seaweeds studied from this area in comparatively large amounts (Qasim 1986, Shameel 1990 b). *Dictyota indica* also differed from other species in containing certain peculiar dolastane diterpenoids and amijiol (Ahmad *et al.* 1991). It is the only species of *Dictyota* showing the presence of eicosatrienoic acid (C20:3) and, although it is found there in appreciable amount, it has not yet been detected in any other brown alga from this region (Shameel 1990, Shaikh *et al.* 1990 a, b, 1991). *Dictyota* of the five species investigated do not show great anatomical variation, but differ remarkably from one another in size, general morphology and habitat ecology. These variations might cause slight differences in their fatty acid composition.

Acknowledgements

I gratefully acknowledge the valuable help rendered by my research colleage Prof. Dr K. Usmanghani, Department of Pharmacognosy, Faculty of Pharmacy, University of Karachi during the analysis and interpretation of GC-MS spectra.

References

Ahmad, V. U., S. Parveen, S. Bano, W. Shaikh and M. Shameel. 1991. Dolastane diterpenoids from the brown alga *Dictyota indica*. *Phytochem.* *30*: 1015–1018.

Banaigs, B., C. Francisco, E. Gonzalis and L. Codomier. 1984. Lipids from the brown alga *Cystoseira barbata*. *Phytochem.* *23*: 2951–2952.

Bano, S., S. Perveen, V. U. Ahmad, Shafiuddin, N. Bano and M. Shameel. 1987. Chemical constituents of *Endarachne binghamiae* (Scytosiphonales, Phaeophyta) from the Karachi coast. *Bot. Mar.* *30*: 371–372.

Christie, W. W. 1987. *High Performance Liquid Chromatography and Lipids*. Pergamon Press, Oxford, 272 pp.

Faulkner, D. J. 1990. Marine natural products. *Nat. Prod. Rep.* *7*: 269–309.

Johns, R. B., P. D. Nichols and G. J. Perry. 1979. Fatty acid composition of ten marine algae from Australian waters. *Phytochem.* *18*: 799–802.

Lie Ken Jie, M. S. F. 1989. Fatty acids and glycerides. *Nat. Prod. Rep.* *6*: 231–261.

Qasim, R. 1986. Studies on fatty acid composition of eighteen species of seaweeds from the Karachi coast. *J. Chem. Soc. Pak.* *8*: 223–230.

Shaikh, W., M. Shameel, A. Hayee-Memon, K. Usmanghani, S. Bano and V. U. Ahmad. 1990 a. Isolation and characterization of chemical constituents of *Stoechospermum marginatum* (Dictyotales, Phaeophyta) and their antimicrobial activity. *Pak. J. Pharm. Sci. 3* (2): 1–9.

Shaikh, W., M. Shameel, R. Khan, K. Usmanghani and V. U. Ahmad. 1990 b. Phycochemical studies on *Spatoglossum variabile* (Dictyotales, Phaeophyta). *Pak. J. Bot. 22*: 100–105.

Shaikh, W., M. Shameel, V. U. Ahmad and K. Usmanghani. 1991. Phycochemical studies on *Colpomenia sinuosa* (Scytosiphonales, Phaeophyta). *Bot. Mar. 34*: 77–79.

Shameel, M. 1990 a. *A Preliminary Check-List of Marine Algae from the Coast and Inshore Waters of Pakistan.* Seaweed Biol. Lab., MAHQ Biol. Res. Centre, Kar. Univ., Karachi. 56 pp.

Shameel, M. 1990 b. Phycochemical studies on fatty acids from certain seaweeds. *Bot. Mar. 33*: 429–432.

Teixeira, V. L., T. Tomassini, B. G. Fleury and A. Kelecom. 1986. Dolastane and secodolastane diterpenes from the marine brown alga, *Dictyota cervicornis. J. Nat. Prod. 49*: 570–575.

Usmanghani, K., M. Shameel, S. Siddiqui and S. A. Siddiqui. 1987. Studies on the fatty acids of *Iyengaria stellata* (Scytosiphonales, Phaeophyta). *Bot. Mar. 30*: 305–307.

Botanica Marina
Vol. 34, pp. 429–432, 1991

Anticoagulant, Fibrinolytic and Antiaggregant Activity of Carrageenans and Alginic Acid

K. C. Güven, Y. Özsoy[4] and O. N. Ulutin**

Institute of Marine Science and Geography, University of Istanbul, Vefa, Istanbul, Turkey

* *Department of Pharmaceutical Technology, Faculty of Pharmacy, University of Istanbul, Istanbul, Turkey*

** *Haemostatis and Thrombosis Research Center, Cerrahpaşa Medical Faculty, University of Istanbul, Istanbul, Turkey*

(Accepted 19 June 1991)

Abstract

Investigation of the effects of carrageenans and alginic acid on the coagulation and platelet aggregation systems was the aim of this study. The highest anticoagulant and fibrinolytic activities were found in λ-carrageenan and *Grateloupia dichotoma* carrageenan. Antiplatelet activity of carrageenans showed that ADP-induced aggregation was inhibited slightly by ι-carrageenan whereas λ- and κ-carrageenans were not effective. Ristocetin induced aggregation was inhibited by all the carrageenans tested. Alginic acid did not show any activity against coagulation, fibrinolytic and platelet aggregation systems.

Introduction

Anticoagulant activity has been investigated by many workers using such algal polysaccharides as, carrageenan (Elsner *et al.* 1937, Elsner 1938, Houck *et al.* 1957, Schimpf *et al.* 1969, Schwartz and Kellermayer 1969), its various types (Hawkins and Leonard 1962, 1963, Anderson and Duncan 1965, Kindness *et al.* 1979 a) and fucoidin (Springer *et al.* 1957).

Anticoagulant activity was demonstrated in a fraction of the extract of *Corallina rubens* L. (Güven *et al.* 1973/74) and *Pterocladia capillacea* (Gmel.) Born *et* Thur. (Güven *et al.* 1979) and also in extracts prepared from some species of Chlorophyta, Phaeophyta and Rhodophyta (Deacon-Smith *et al.* 1985 a). Fibrinolytic activity was found in *Corallina rubens* L. (Güven *et al.* 1973/74). Semisynthetic derivatives of algal polysaccharides such as laminarin sulphate (Hawkins and O'Neill 1955, Adams and Thorpe 1957) and sulphated alginic acid (Constantinides *et al.* 1954) also showed anticoagulant activities. The effect on the aggregation of human blood platelets has been studied for carrageenan (Vargaftig 1977, Kindness *et al.* 1979 b, 1980, McMillan *et al.* 1979) and also for algal extracts (Deacon-Smith *et al.* 1985 b). The biological activities of algal polysaccharides were summarized in an article by Güven *et al.* (1990).

In the work reported here the activities of different types of carrageenan and of alginic acid were studied on blood coagulation, fibrinolytic and platelet aggregation systems.

Material

The algal polysaccharides tested were: λ-, κ- and ι-carrageenan (Sigma, USA) and *Grateloupia dichotoma* J. Agardh carrageenan (Güven *et al.* 1984) and alginic acid (Roth, Germany). They were tested against: Factor X_a, Recalmix (Haemachem Inc. USA), fibrinogen (Kızılay, Turkey), antithrombin III, Factor X_a, substrate plasma, thrombin, thromboplastin, cephaloplastin, adenosine diphosphate (ADP) and ristocetin (Diagnostica Stago, France). Stock solutions used in the tests were: calcium chloride solution M/40, barbital buffer pH 7.8, borate-saline buffer pH 5.6.

The algal polysaccharides were dissolved in saline (0.9%) at 4 mg/mL and 1 mL was diluted with human plasma to 10 mL. Platelet aggregation tests were carried out in a Clinical Platelet Aggregometer (Malin Electronics, USA).

Methods

Tests were used for:

1. Anticoagulant activity

a) Heptest method (according to the manufacturer's procedure): 0.1 mL test solution was mixed with 0.1 mL Factor X_a. After 120 seconds 0.1 mL Recalmix solution was added and the clotting time recorded.

b) Hepaclot method (according to the manufacturer's procedure): 50 μL test solution was mixed with 50 μL antithrombin III solution. 100 μL Factor X_a, 100 μL substrate plasma and 100 μL calcium chloride solution were added in succession after 60, 90 and 30 seconds had elapsed. The clotting time was determined.

c) Thrombin Time (TT) (Pitney and Dacie 1953): 0.2 mL test solution and 0.2 mL thrombin solution (5 U/mL) were mixed and the time of appearance of a fibrin clot was noted.

d) Prothrombin Time (PT) (Quick 1945): 0.1 mL test solution was mixed with 0.1 mL thromboplastin and 0.1 mL calcium chloride solution and the time of fibrin clot appearance was recorded.

e) Activated Partial Thromboplastin Time (a-PTT) (Margolis 1957): 0.1 mL test solution and 0.1 mL cephaloplastin were mixed. After an incubation time of 3 min, 0.1 mL calcium chloride solution was added and the time of appearance of a fibrin clot was recorded.

All the tests were carried out in clotting tubes in a water bath at 37 °C.

2. Fibrinolytic activity

a) Fibrin Plate Method (Astrup and Müllertz 1952): 0.5 mL test solution and 9.5 mL acetic acid (0.16%) were mixed, centrifuged for 10 min at 2000 rpm and the supernatant was decanted. The substrate residue was dissolved in 0.25 mL borate-saline buffer and 30 μL of this solution was applied to a plate prepared with 1 mL of fibrinogen solution (1%), 9 mL of barbital buffer, 100 μL thrombin solution (20 U/mL) and 200 μL calcium chloride solution. The plates were kept at 37 °C for 24 h then the areas of lysis were measured.

b) Euglobulin Lysis Time (ELT) (Nordey *et al.* 1980): 0.5 mL test solution and 9.5 mL acetic acid (0.16%) were mixed and centrifuged for 10 min at 2000 rpm. The substrate residue was dissolved in 0.25 mL borate-saline buffer and kept at 37 °C in a water bath. Afterwards, 0.25 mL of calcium chloride solution was added and the time of formation of a clot was noted. The time of complete lysis of the clot was recorded.

3. Platelet aggregation (Born 1962)

The human blood (9 vol) was collected in trisodium citrate solution (1 vol) of 3.8% and centrifuged for 5 min at 1500 rpm. Then 0.40 mL of this plasma was mixed with 0.05 mL of polysaccharide solution and 0.05 mL ADP solution (5 mcg/mL) or 0.05 mL ristocetin solution (30 mg/mL). The platelet aggregation was measured in the aggrometer.

Results

1. Anticoagulant activity

The results are shown in Table I. In all the tests, namely Heptest, Hepaclot, Thrombin Time, Prothrombin Time and Activated Partial Thromboplastin Time methods, λ-carrageenan and *G. dichotoma* carrageenan showed high anticoagulant potency but the reaction was weak with ι- and κ-carrageenans with the exception that ι-carrageenan also prolonged a-PTT as λ-carrageenan did.

Table I. Anticoagulant activity of carrageenans and alginic acid. Each value is the mean of six experiments (time in sec.).

Samples \ Methods	Heptest	Hepaclot	TT	PT	a-PTT
λ-Carrageenan	300 >	49.75 ± 0.95	370.81 ± 6.41	360 >	180 >
κ-Carrageenan	35.85 ± 0.92	16.95 ± 1.15	50.71 ± 1.43	18.16 ± 0.70	173.50 ± 4.76
ι-Carrageenan	55.91 ± 0.88	21.70 ± 0.70	91.15 ± 4.53	44.62 ± 2.13	180 >
G. dichotoma Carrageenan	300 >	31.25 ± 1.45	342.51 ± 6.53	31.73 ± 2.48	180 >
Alginic acid	24.72 ± 0.20	20.70 ± 0.70	30.22 ± 2.08	15.33 ± 0.53	44.00 ± 5.07
Saline (Control)	23.30 ± 0.36	18.55 ± 0.35	29.58 ± 2.09	15.43 ± 0.51	41.66 ± 4.69

Table II. The fibrinolytic activity of carrageenans and alginic acid. Each value is the mean of six experiments.

Samples \ Methods	Fibrin Plate (mm^2)	ELT (min)
λ-Carrageenan	173.64 ± 1.94	no clot formation
κ-Carrageenan	129.76 ± 2.24	173 ± 5.04
ι-Carrageenan	156.42 ± 6.00	no clot formation
G. dichotoma Carrageenan	150.68 ± 4.66	no clot formation
Alginic acid	78.77 ± 0.15	120 ± 3.14
Saline (Control)	79.33 ± 3.14	120 ± 3.04

2. Fibrinolytic activity

The results are listed in Table II. In the fibrin plate method λ-carrageenan showed the highest activity, then came ι- and *G. dichotoma* carrageenans, and κ-carrageenan showed the least activity. In the ELT method, λ-, *G. dichotoma* and ι-carrageenans showed the highest and κ-carrageenan the lowest activity.

3. The effect on platelet aggregation system

The results from this experiment are shown in Table III. The ι-carrageenan inhibited platelet aggregation induced by ADP. There was no significant alteration with λ-, *G. dichotoma* or κ-carrageenans whereas ristocetin-induced aggregation was inhibited by all the carrageenan types.

Alginic acid had no effect on the aggregation system.

Discussion

The anticoagulant activity of carrageenan was first demonstrated by Elsner *et al.* (1937) who found that carrageenan was 7.5 times less active than heparin. Later, anticoagulant activity was also shown by carrageenan obtained from *Dellesseria sanguina* (Elsner 1938) and *Gigartina acicularis* (Houck *et al.* 1957). Among the carrageenan types tested, λ-carrageenan had the most activity (Anderson and Duncan 1965, Hawkins and Leonard 1962, 1963). This is confirmed by the findings of the present work in which the Heptest method was used in addition to the other methods.

Fibrinolytic activity was firstly demonstrated in a fraction of an extract of *Corallina rubens* L. by Güven *et al.* (1973/74). In the present work fibrinolytic activity of carrageenans was also studied first and it was found that λ-carrageenan had the strongest activity of the types tested. The effects of carrageenans on the aggregation system were studied previously and it was found that carrageenans synergistically enhanced the aggregation caused by ADP and inhibited thrombin-induced aggregation and that the most effective carrageenan was λ-type (Vargaftig 1977, Kindness *et al.* 1979 b). McMillan *et al.* (1979) showed that ι-carrageenan had the highest activity on aggregation induced by ADP or adrenaline. Kindness *et al.* (1980) showed that pretreatment of platelets with antithrombin III prevented subsequent aggregation initiated by carrageenan. In the present investigation we found that ι-carrageenan enhanced the inhibition of aggregation induced by ADP as had prevously been reported by Kindness *et al.* (1980). Deacon-Smith (1985 b) reported that the aggregation induced by ristocetin was completely inhibited only by *Laminaria saccharina* extract, although a number of other algae were tested. We found that ristocetin-induced aggregation was inhibited by all the carrageenans tested. We explain this phenomenon by the fact that carrageenan inhibited or blocked the receptor for von Willebrand Factor in the platelet membrane. *Grateloupia dichotoma* carrageenan, which was proven to be chemically λ-carrageenan (Güven and Güvener 1985 a, b), showed a similar activity to λ-carrageenan all the biological tests.

Alginic acid had no activity on the coagulation, fibrinolytic or aggregation system.

Table III. The effect of carrageenan on the ADP and ristocetin induced platelet aggregation. – No aggregation curve observed Each value is the mean of three experiments.

Samples \ Methods	ADP		Ristocetin	
	Angle	Maximum Amplitude	Angle	Maximum Amplitude
λ-Carrageenan	61.6 ± 1.69	9.36 ± 0.87	–	–
κ-Carrageenan	56.2 ± 8.86	7.58 ± 1.28	–	–
ι-Carrageenan	37.4 ± 2.87	3.70 ± 0.61	–	–
G. dichotoma Carrageenan	62.2 ± 5.50	10.1 ± 1.19	–	–
Alginic acid	66.0 ± 5.75	11.0 ± 2.63	70.8 ± 8.12	16.2 ± 3.97
Saline	68.0 ± 6.44	10.7 ± 1.49	64.5 ± 7.99	16.2 ± 3.53

References

Adams, S. S. and H. M. Thorpe. 1957. The anticoagulant activity and toxicity of laminarin sulphate K. *J. Pharm. Pharmacol. 9*: 459–463.

Anderson, W. and J. G. C. Duncan. 1965. The anticoagulant activity of carrageenan. *J. Pharm. Pharmacol. 17*: 647–654.

Astrup, T. and S. Müllertz. 1952. The fibrin plate method for estimating fibrinolytic activity. *Arch. Biochem. Biophys. 40*: 346–351.

Born, G. V. R. 1962. Aggregation of blood platelets by adenosine diphosphate and its reversal. *Nature 194*: 927–929.

Constantinides, P., A. Cairns and A. Werner. 1954. Antilipemic activity of sulfated polysaccharides. *Arch. Int. Pharmacodyn. 99*: 334–345.

Deacon-Smith, R. A., J. P. Lee-Potter and D. J. Rogers. 1985 a. Anticoagulant activity in extracts of British marine algae. *Bot. Mar. 28*: 333–338.

Deacon-Smith, R. A., J. P. Lee-Potter and D. J. Rogers. 1985 b. Platelet aggregation in the presence of extracts of British marine algae. *Med. Lab. Sci. 42*: 404–405.

Elsner, H., W. Broser and E. Bürgel. 1937, Über das Vorkommen von hochwirksamen, die Blutgerinnung hemmenden Stoffen in Rotalgen. *Z. Physiol. Chem. 246*: 244–247.

Elsner, H. 1938. Über das Vorkommen von hochwirksamen, die Blutgerinnung hemmenden Stoffen in Meeresalgen. II. *Z. Physiol. Chem. 252*: 196–200.

Güven, K. C., E. Güler, E. Aktin and H. Koyuncuoğlu. 1979. Studies on *Pterocladia capillacea* (Gmel.) Born et Thur. Part II. *In* (H. A. Hoppe, T. Levring and Y. Tanaka, eds) *Marine Algae in Pharmaceutical Science*. Walter de Gruyter, Berlin, New York. pp. 693–710.

Güven, K. C. and B. Güvener. 1985 a. Metachromatic identification of (iota-kappa-lambda) carrageenans. *Bot. Mar. 28*: 221–222.

Güven, K. C. and B. Güvener. 1985 b. A metachromatic method for the identification of alginic acid, agar and carrageenan. *Fette, Seifen, Anstrichmittel 87*: 172–176.

Güven, K. C., B. Güvener and E. Güler. 1990. Pharmacological activities of marine algae. *In* (I. Akatsuka, ed.) *Introduction to Applied Phycology*. SPB Academic Publishing, The Hague, The Netherlands. pp. 67–92.

Güven, K. C., B. Güvener and N. Güven. 1984. Succinic acid and Carrageenan from *Acanthophora delilei* Lamouroux and *Grateloupia dichotoma* J. Agardh. *Acta Pharm. Turc. 26*: 49–52.

Güven, K. C., Ş. B. Ulutin, E. Mutluay and O. N. Ulutin. 1973/74. Anticoagulant-antithrombin and fibrinolytic actions of extract of marine algae *Corallina rubens* L. *Haemostasis 2*: 260–268.

Hawkins, W. W. and A. N. O'Neill. 1955. The anticoagulant action in blood of sulphated derivatives of laminarin. *Can. J. Biochem. Physiol. 33*: 545–552.

Hawkins, W. W. and V. G. Leonard. 1962. Antipeptic and antithrombic properties of carrageenin. *J. Lab. Clin. Med. 60*: 641–648.

Hawkins, W. W. and V. G. Leonard. 1963. The antithrombic activity of carrageenin in human blood. *Can. J. Biochem. Physiol. 41*: 1325–1327.

Houck, J. C., R. K. Morris and E. J. Lazaro. 1957. Anticoagulant, lipemia clearing and other effects of anionic polysaccharides extracted from seaweed. *Proc. Soc. Exp. Biol. Med. 96*: 528–530.

Kindness, G., W. F. Long and F. B. Williamson. 1979 a. Enhancement of antithrombin III activity by carrageenans. *Thrombosis Research 15*: 49–60.

Kindness, G., W. F. Long, F. B. Williamson and J. Boyd. 1979 b. Effects of carrageenans on the aggregation of human blood platelets. *Thrombosis Research 15*: 3–15.

Kindness, G., F. B. Williamson and W. F. Long. 1980. Inhibition by antithrombin III of carrageenan- and xylan SP54-induced aggregation of human blood platelets. *Biochem. Soc. Trans. 8*: 84–85.

Margolis, J. 1967. Initiation of blood coagulation by glass and related surfaces. *J. Physiol. 137*: 95–109.

McMilan, R. M., D. E. McIntyre and J. L. Gordon. 1979. Stimulation of human platelets by carrageenans. *J. Pharm. Pharmacol. 31*: 148–152.

Nordey, H., P. Arnesen and H. C. Godal. 1980. The euglobin lysis time, a rapid and sensitive method for the assay of fibrinolytic activity after venos stanis. *Scand. J. Haemat. 25*: 401–406.

Pitney, W. R. and J. V. Dacie. 1953. A simple method of studying the generation of thrombin in recalcified plasma. *J. Clin. Path. 6*: 9–13.

Quick, A. J. 1945. On the quantitative estimation of prothrombin. *Amer. J. Clin. Path. 15*: 560–566.

Schimpf, Kl., J. Lenhard and G. Schaaf. 1969. The influence of the polysaccharide carrageenin on the clotting mechanism of the blood in the rabbit in vitro and in vivo. *Thromb. Diath. Haemorrh. 24*: 524–533.

Schwartz, H. J. and R. W. Kellermeyer. 1969. Carrageenan and delayed hypersensitivity II. Activation of Hageman factor by carrageenan and its possible significance. *Proc. Soc. Exp. Biol. Med. 132*: 1021–1024.

Springer, G. F., H. A. Würzel, G. M. McNeal, Jr., N. J. Ansell and M. F. Doughty. 1957. Isolation of anticoagulant fractions from crude fucoidin. *Proc. Soc. Exp. Biol. Med. 94*: 404–409.

Vargaftig, B. B. 1977. Carrageenan and thrombin trigger prostaglandin synthetase-independent aggregation of rabbit platelets: inhibition by phospholipase A_2 inhibitors. *J. Pharm. Pharmacol. 29*: 222–228.

Botanica Marina
Vol. 34, pp. 433–435, 1991

Mosquito Larvicidal Effect of Seaweed Extracts

T. Subramonia Thangam and K. Kathiresan

Centre of Advanced Study in Marine Biology, Parangipettai 608 502, Tamil Nadu, India

(Accepted 24 June 1991)

Abstract

Fifteen seaweeds, *Caulerpa peltata, C. racemosa, C. scalpelliformis, Chaetomorpha linum, Codium decorticatum, Dictyota dichotoma, Enteromorpha clathrata, E. intestinalis, Halimeda opuntia, Hypnea valentiae, Sargassum tenerrimum, S. wightii, Turbinaria conoides, T. oranata* and *Ulva lactuca* were extracted in acetone and evaluated for mosquito larvicidal activity against *Aedes aegypti* L. *Caulerpa scalpelliformis* and *Dictyota dichotoma* were found most effective with LC_{50} values of 53.70 and 61.65 mg L^{-1} respectively.

Introduction

Our earlier studies reported for the first time that seaweed extracts can kill mosquito larvae of *Anopheles stephensi* Liston, *Culex quinquefasciatus* Say and *C. tritaeniorhynchus* Giles (Subramonia Thangam and Kathiresan 1988 a, 1989). The present study evaluates the larvicidal activity of seaweed extracts against *Aedes aegypti* L., the vector of dengue and yellow fever.

Material and Methods

Fresh material of *Caulerpa peltata* Lamour., *C. racemosa* (Forssk.) Weber V. Bosse, *C. scalpelliformis* (R. Br.) Weber V. Bosse, *Chaetomorpha linum* (O. F. Muell.) Kuetz., *Codium decorticatum* (Woodward) Harvey, *Dictyota dichotoma* (Huds.) Lamour., *Enteromorpha clathrata* (Roth) J. Agardh, *E. intestinalis* (L.) Link, *Halimeda opuntia* f. *typica* Barton, *Hypnea valentiae* (Turn.) Mont., *Sargassum tenerrimum* J. Ag., *S. wightii* Greville, *Turbinaria conoides* Kuetz., *T. oranata* J. Ag., and *Ulva lactuca* L., were collected from southeast coast of India, shade-dried and powdered. The powdered samples (40 mesh size) were extracted in acetone using a Soxhlet apparatus. The extracts were concentrated using a rotary vacuum evaporator at 40–45 °C. Standard stock solutions of 1% (w/v) were prepared by dissolving the residues in acetone.

Test solutions were prepared at different concentrations ranging from 10 to 100 mg L^{-1}, by diluting the stock solution using unchlorinated, filtered tap water. Tween-80 was used as an emulsifier at the concentration of 0.02% (v/v) to the final test solutions. A control solution was prepared in tap water, using acetone and Tween-80 as used in the test solution of the highest concentration. Larval susceptibility tests were conducted according to the standard method (WHO/VBC 1975). Twenty five larvae of early fourth instar stage were released in 500 mL beakers containing 250 mL of test or control solutions and four replicates were run at a time. Mortality counts were made after 24 hours of the treatment. The LC_{50}, standard error, fiducial limits at 95% confidence and regression equation were calculated (Finney 1971).

The extracts of *Caulerpa scalpelliformis* and *Dictyota dichotoma* which were found effective were fractionated using silica gel (60–120 mesh) column chromatography. The column was eluted successively using methanol and acetone in different proportions. The different proportions of solvents (methanol : acetone) used were 0 : 10 (Fraction I), 0.5 : 9.5 (Fraction II), 1 : 9 (Fraction III), 2 : 8 (Fraction IV), 3 : 7 (Fraction V) and 10 : 0 (Fraction VI). The fractions were concentrated using a vacuum evaporator and the residues were dissolved in acetone to prepare 10% stock solutions. All the fractions were tested for larvicidal activity against *Aedes aegypti*.

Table I. Larvicidal activity of seaweed extracts against *Aedes aegypti*.

Alga	24 hours LC_{50} ± S. E. (mg L^{-1})	Fiducial limits (95%) (mg L^{-1})	Regression equation
Caulerpa scalpelliformis	53.70 ± 1.56	50.65 – 56.75	Y = 3.03x – 0.24
Dictyota dichotoma	61.65 ± 1.04	59.67 – 63.73	Y = 4.54x – 3.13
Enteromorpha clathrata	85.11 ± 1.56	82.04 – 88.16	Y = 4.17x – 3.05
Enteromorpha intestinalis	67.60 ± 1.49	64.69 – 70.51	Y = 3.22x – 0.90
Ulva lactuca	91.20 ± 1.32	88.61 – 93.79	Y = 5.88x – 6.52

Results

The results of the larval susceptibility to the seaweed extracts are shown in Table I. Extracts of *Caulerpa scalpelliformis* and *Dictyota dichotoma* were found most effective with LC_{50} values of 53.70 and 61.65 mg L^{-1} respectively. *Enteromorpha clathrata, E. intestinalis* and *Ulva lactuca* were also found effective with the LC_{50} values of 85.11, 67.60 and 91.20 mg L^{-1} respectively. The extracts of *Caulerpa peltata, C. racemosa, Chaetomorpha linum, Codium decorticatum, Halimeda opuntia, Hypnea valentiae, Sargassum tenerrimum, S. wightii, Turbinaria conoides* and *T. ornata* did not cause 50 per cent mortality of mosquito larvae even at 100 mg L^{-1}.

The fractionated acetone extracts showed differential larvicidal activity. Fraction IV of the *Caulerpa scalpelliformis* extract eluted in 2 : 8 methanol-acetone, exhibited the highest larvicidal activity with a LC_{50} value of 15.85 mg L^{-1} compared with that of 53.70 mg L^{-1} in the crude acetone extract. Fractions III and V were also found effective with LC_{50} values of 31.62 and 23.44 mg L^{-1} respectively. Among the six fractions eluted from the crude acetone extract of *Dictyota dichotoma*, fraction III eluted in 1 : 9 and fraction IV in 2 : 8 methanol-acetone were effective with LC_{50} values of 25.70 and 28.21 mg L^{-1} respectively. These values were lower than that of 61.65 mg L^{-1} of the crude acetone extract (Table II).

Discussion

Mosquito larvicidal activity of plant extracts has been reported earlier against *Aedes aegypti* with a LC_{50} value of 550 mg L^{-1} in acetone extract of *Oligochaeta ramosa* (Roxb.) Wagenitz (Saxena and Yadav 1983); 34.06 mg L^{-1} in petroleum ether extract of *Vinca rosea* L. (Kalyanasundaram and Das 1985) and 39.4 mg L^{-1} in petroleum ether extract of *Acorus calamus* L. (Sujatha *et al.* 1989). We have also earlier reported the mosquito larvicidal activity of marine plants with LC_{50} value of 17 mg L^{-1} in *Rhizophora apiculata* Blume stilt root against *Anopheles stephensi* (Subramonia Thangam and Kathiresan 1988 b), 22 mg L^{-1} in *Dictyota dichotoma* against *Culex quinquefasciatus* (Subramonia Thangam and Kathiresan 1988 a) and that of 46 mg L^{-1} in *Caulerpa scalpelliformis* against *Culex tritaeniorhynchus* (Subramonia Thangam and Kathiresan 1989). Compared with all these, the present study reports a low LC_{50} value of 15.85 mg L^{-1} in *Caulerpa scalpelliformis* (Fraction V) against *Aedes aegypti.* Hence *C. scalpelliformis* might possibly be used for mosquito control programmes. The solvent used for extraction may also be recycled with minimum loss. The abundant availability of seaweeds from the coastlines of India would make it economical for field use in mosquito control programmes.

Table II. Larvicidal activity of different fractions from column chromatographic separation of *Caulerpa scalpelliformis* and *Dictyota dichotoma* extracts.

Fraction	24 hours LC_{50} ± S. E. (mg L^{-1})	Fiducial limits (95%) (mg L^{-1})	Regression equation
C. scalpelliformis			
Crude extract	53.70 ± 1.56	50.65 – 56.75	Y = 3.03x – 0.24
Fraction III	31.62 ± 0.35	30.93 – 32.31	Y = 3.00x + 0.50
Fraction IV	15.85 ± 0.24	15.37 – 16.32	Y = 2.27x + 2.28
Fraction V	23.44 ± 0.35	22.75 – 24.13	Y = 1.92x + 2.37
D. dichotoma			
Crude extract	61.65 ± 1.04	59.67 – 63.73	Y = 4.54x – 3.13
Fraction III	25.70 ± 0.32	25.06 – 26.34	Y = 2.63x + 1.29
Fraction IV	28.21 ± 0.25	27.72 – 28.70	Y = 4.55x – 1.60

Fractions I, II, VI of *C. scalpelliformis* and Fractions I, II, V, VI of *D. dichotoma* did not cause LC_{50} even at 100 mg L^{-1}.

Acknowledgements

The authors are grateful to the Indian Council of Medical Research and Council of Scientific and Industrial Research, New Delhi for financial support and Prof. A. Subramanian, Director of this Centre for providing facilities.

References

Finney, D. J. 1971. *Probit Analysis.* Cambridge University Press, London, III edition. pp. 20–42.

Kalyanasundaram, M. and P. K. Das. 1985. Larvicidal and synergistic activity of plant extracts for mosquito control. *Indian J. Med. Res. 82*: 19–23.

Saxena, S. C. and R. S. Yadav. 1983. A new plant extract to suppress the population of yellow fever and dengue vector *Aedes aegypti* L. (Diptera : Culicidae). *Curr. Sci. 52*: 713–715.

Subramonia Thangam, T. and K. Kathiresan. 1988 a. Toxic effect of seaweed extracts on mosquito larvae. *Indian J. Med. Res. 88*: 35–37.

Subramonia Thangam, T. and K. Kathiresan. 1988 b. Toxic effect of mangrove plant extracts on mosquito larvae *Anopheles stephensi* L. *Curr. Sci. 57*: 914–915.

Subramonia Thangam, T. and K. Kathiresan. 1989. Larvicidal effect of marine plant extracts on mosquito *Culex tritaeniorhynchus. J. mar. biol. Ass. India 31*: 306–307.

Sujatha, C. H., V. Vasuki, T. Mariappan, M. Kalyanasundaram and P. K. Das. 1989. Evaluation of plant extracts for bio logical activity against mosquitoes. *Int. Pest Control 30*: 122–124.

WHO/VBC. 1975. Instructions for determining the susceptibility or resistance of mosquito larvae to insecticides. Mimeographed document (*Wld. Hlth. Org. Techn. Rep. Ser.*, 1970, No: 443 revised by WHO Expert Committee which met in Geneva, September 16–23, 1975), WHO/VBC/75. 583. pp. 1–5.

Botanica Marina
Vol. 34, pp. 437–445, 1991

Field and Culture Studies on the Life History of *Campylaephora crassa* (Okamura) Nakamura (Ceramiaceae, Rhodophyta)

S. M. Boo, S. Fredriksen*, J. Rueness* and I. K. Lee**

Department of Biology, Chungnam National University, Daejon 305-764, Korea

* *Department of Biology, Marine Botany, Oslo University, Oslo 3, Norway*
** *Department of Biology, Seoul National University, Seoul 151-742, Korea*

(Accepted 24 June 1991)

Abstract

The life history of an epiphytic species, *Campylaephora crassa* (Okamura) Nakamura, was investigated in the wild on the central eastern coast of Korea and under various laboratory conditions. Korean *C. crassa* plants showed no obvious seasonal differences in the occurrence of the reproductive organs, whereas their overall habit was variable. In culture the species completed a *Polysiphonia* type of life history without its host plants, but parasporangia occurred at a frequency of about 1% on young tetrasporophytes. The paraspores grew into tetrasporophytes, which later cycled a *Polysiphonia* type of life history. Crossability tests also documented that paraspores had the same nuclear phase as normal tetrasporophytes. We suggest that parasporangial plants of *C. crassa* may be mitotic diploids which show a temporary modification in the life history strategy. Maximal growth in the field occurred in May when the seawater temperature ranged from 14.9 to 18.2 °C, while in culture the optimum temperature for growth was between 16 and 20 °C under each photon flux density. Plants in the field had tetrasporangia and cystocarps throughout the year and the tetrasporophytes in culture formed tetrasporangia at most combinations of photon flux densities and temperatures. Both chlorophyll a and the phycobiliprotein content decreased at the higher irradiances.

Introduction

For the modern species concept, interpretation of life history is an important facet of experimental taxonomic studies because life histories are quite variable and some species may alter their pattern of growth in response to environmental differences. Life histories are very complex and diverse in the eukaryotic algae, of which the rhodophytes show the most complex infraspecific variation (Guiry 1987, Maggs 1988). To provide a better understanding of red algal species, therefore, developmental and environmental variations should be investigated at all stages of the life history in field and culture.

Campylaephora crassa (Okamura) Nakamura (1950) is a member of the tribe Ceramieae in the red algal family Ceramiaceae (Hommersand 1963). The species grows well in the infralittoral fringe and subtidal habitats of Korean coasts, where the plants have always been found to be epiphytic (Boo 1985). Its distribution is limited, ranging from Vladivostok to southern Japan (Okamura 1930, Funahashi 1966, Itono 1977). It resembles forms of taxa of the closely related genus *Ceramium* in habit, branching pattern and reproductive structures: *C. kondoi* Yendo from Japan (Okamura 1930), *C. rubrum* (Hudson) C. Agardh from England (Boo and Yoshida 1991) and *C. secundatum* Lyngbye from Denmark (Yendo 1918). However, *Campylaephora crassa* is distinguished by a thick cortex with rhizoidal cells (Okamura 1930), a discoid holdfast and large spores (Nakamura 1950).

Little published information is available on the life history of *Campylaephora crassa*, although the species is so variable in the field that it has been separated into five forms (Nakamura 1950, 1965, Itono 1977). The only culture studies of the species were carried out on spore development (Nakamura 1954) and furthermore there are no reports of field studies. Notoya (1979) investigated the life history in relation to temperatures and irradiances in *Campylaephora hypnaeoides* J. Agardh, the type of the genus *Campylaephora* (J. Agardh 1851). On the basis of floristic studies and Notoya's data, Hoek (1982) reported that the temperature responses of *C. hypnaeoides* seemed to be particularly adopted to the temperature conditions in its growing areas including the wide annual temperature fluctuations. There are no reports of field and culture studies on *C. japonica* Noda, the last taxon included in *Campylaephora*. In other genera of the Ceramieae, a *Polysiphonia* type of life history in culture has been reported in *Ceramium* (Edwards 1973), *Microcladia* (Gonzalez and Goff 1989 a), and *Herpochondria* (as *Microcladia*; Notoya and Saito 1980). Infraspecific variations in life history have been recognized in *Ceramium*, for example the cycling of parasporangial plants (Rueness 1973) and the absence of sexual plants in the field (Edwards 1973). On the other hand, Garbary *et al.* (1978) showed that the characteristic morphologies of some *Ceramium* species could differ under laboratory conditions. *Microcladia coulteri* Harvey and *M. californica* Farlow which were previously confused are separated into distinct species by life history, phenology and cross-inoculation experiments (Gonzalez and Goff 1989 a, b).

The present study characterized the life history of the epiphytic species *Campylaephora crassa* under various laboratory conditions. The phenology of a local population was described and an attempt was made to relate the observed field data to growth and pigment content in culture. A parasporangial life history in culture was also reported.

Material and Methods

Field observations were carried out at Sinnam (37°20′N, 129°3′E) on the eastern coast of Korea. This site was selected because of the great abundance of *Campylaephora crassa* plants throughout the year. The study area is moderately exposed to wave action and has a relatively small tidal range. Many plants were collected in the infralittoral fringe at monthly intervals from May 1989 to July 1990. All plants collected were examined for reproductive organs. For comparison, many plants were collected at Ganggu (36°3′N, 129°3′E), at about 190 km south of Sinnam, from March to May 1990 and at Choshi on the Pacific coast of the middle Honshu, Japan in January 1990.

The isolate used for culture experiments was collected in the infralittoral fringe at Anin (37°70′N, 128°8′E), at about 20 km north of Sinnam. A healthy tetrasporophyte growing on *Sargassum yezoense* (Yamada) Yoshida *et* T. Konno was isolated into unialgal culture and maintained as vegetative apices. Stock cultures were maintained in an enriched medium (IMR/2, Eppley *et al.* 1967) with 30‰ salinity at 12 °C under a 16:$\bar{8}$ h LD cycle. Photon flux density of about 20 μmol m^{-2} s^{-1} was given by fluorescent tubes (Philips TL 40 W/32). The clones were transferred to a culture room at 17 °C. During the experiments the culture medium was changed every 3 – 5 days.

To study the life history, *Campylaephora crassa* was followed through more than three generations starting from tetraspores. Three sporelings were grown in a polystyrene Petri dish containing 25 mL enriched seawater medium with three replicates. The culture was maintained under varying photon flux densities from 90 to 200 μmol m^{-2} s^{-1}. To determine the nuclear phase of plants originating from paraspores, crossability tests were carried out among the gametophytes of normal plants and the plants derived from, and plants with parasporangia. Three fertile female apices and one fertile male apex were cultured together in a Petri dish with three replicates.

To investigate growth and reproduction under laboratory conditions, a cross-gradient table was used in combinations of varying photon flux densities (20, 40, 100 and 250 μmol m^{-2} s^{-1}) and temperatures (16, 20, 24 and 28 °C). Five carposporelings, which were three days old after germination, were maintained in a dish with four replicates. The plants in two replicates were harvested after three weeks for determination of length and wet weight. The plants in the other replicates were examined at intervals of three days for the formation of tetrasporangia and the experiment was concluded after one month.

To relate the growth responses in laboratory conditions to the photosynthetic pigments, pigments were analyzed for all plants in the above mentioned culture. Two dishes at each combination were pooled with two replicates and, therefore, a total of 16 pigment extractions was taken at the four temperatures under each irradiance. According to Beer and Eshel (1985), fresh material (0.03 – 0.3 g) from each culture was ground in 5 mL 0.1 M phosphate buffer (pH = 6.8) and centrifuged. The absorbance was determined using a Shimadzu double beam UV-210 A at 445, 564, 592, 618 and 645 nm and the phycobilins quantified as mg pigment/g fresh weight. Chlorophyll a was extracted with 100% acetone from the solid material left after phycobiliprotein extraction, and the amount of chlorophyll a was determined (Jensen 1978).

Results

Phenology

Along the east coast of Korea, *Campylaephora crassa* plants were abundant and were always found to be epiphytic on *Prionitis divaricata* (Okamura) Kawaguchi, *Prionitis cornea* (Okamura) Dawson, *Sargassum* spp. and other macroalgae. In this central Korean population there were no obvious seasonal differences in the occurrence of a particular life history phase because both tetrasporic plants and carposporophyte-bearing females were collected every month, although males with spermatangia were not common (Table I). However, the plants were very variable during the study. They were secundly dichotomous when young (Fig. 1C) but irregularly pinnate when mature (Fig. 1B). The habit and branching pattern of Sinnam plants in March and May (Fig. 1A–B) were very

Table I. Phenology of *Campylaephora crassa* on the central eastern coast of Korea from May 1989 to July 1990.

	1989							1990				
	v	vi	vii	viii	ix	x	xii	ii	iii	iv	v	vii
Surface seawater temperature (°C)[1]	18.2	21.7	23.3	23.1	21.3	18.5	5.5	10.0	10.7	12.5	14.8	20.8
Cystocarp	+	+	+	+	+	+	+	+	+	+	+	+
Spermatangium	+	+	−	+	−	−	+	+	+	+	+	−
Tetrasporangium	+	+	+	+	+	+	+	+	+	+	+	+

[1] data from Anonymous, 1989, 1990.

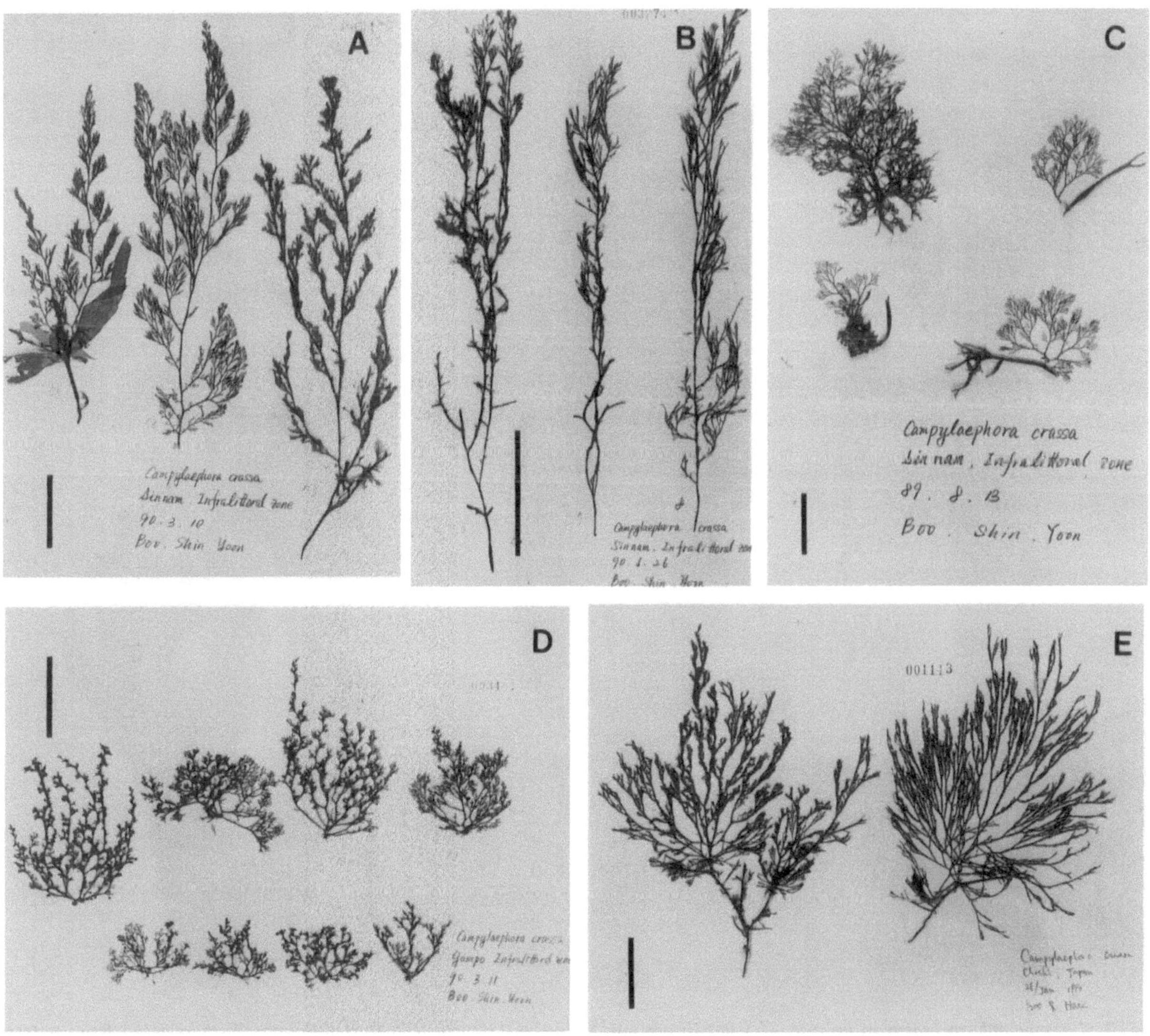

Fig. 1. *Campylaephora crassa* plants collected (A) at Sinnam, Korea in March (scale = 3.0 cm), (B) in May (scale = 3.5 cm), (C) in August (scale = 2.0 cm), (D) at Ganggu in March (scale = 4 cm) and (E) at Choshi, Japan in January (scale = 3 cm).

different to those of the plants in August (Fig. 1C), which were rather similar to the Ganggu plants in March (Fig. 1D). Maximum size was noted in May, when the largest plant was a carposporophyte-bearing female of 48 cm. In addition, plants in all phases of life history were collected at Ganggu and at Choshi. Japanese plants were likewise epiphytic on *P. corna* and their branching pattern was secundly dichotomous (Fig. 1E). The parasporangia observed in culture were not found in field-collected Korean and Japanese plants.

Life history in culture

In the absence of its host plants, *Campylaephora crassa* showed a typical *Polysiphonia* type of life history including the formation of male and female gametophytes from tetraspores, development of carposporophytes, spores from which gave rise to tetrasporophytes (Fig. 2A–F). The life cycle was completed in culture within 2–2.5 months. Male gametophytes became fertile (Fig. 2E) 15–20 days after germination of tetraspores (Fig. 2A–C) and after

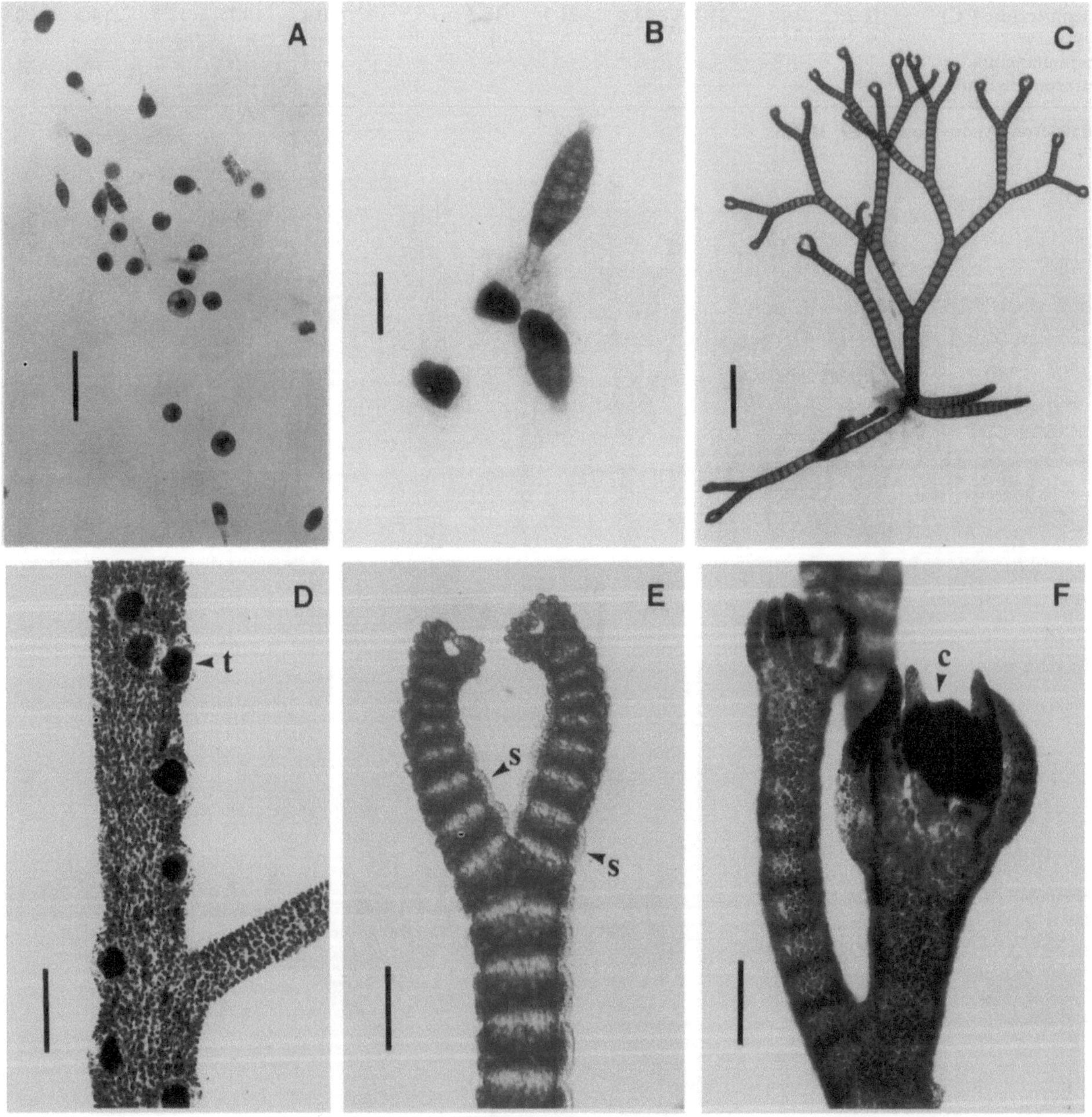

Fig. 2. *Campylaephora crassa* in culture, (A) tetrasporelings (scale = 130 µm), (B) four tetrasporelings released from a tetrasporangium (scale = 200 µm), (C) a young tetrasporophyte (scale = 1 mm), (D) a part with tetrasporangia (scale = 100 µm), (E) an apex with spermatangia (scale = 100 µm) and (F) an apex with cystocarps (scale = 100 µm). c: cystocarp, s: spermatangia, t: tetrasporangium.

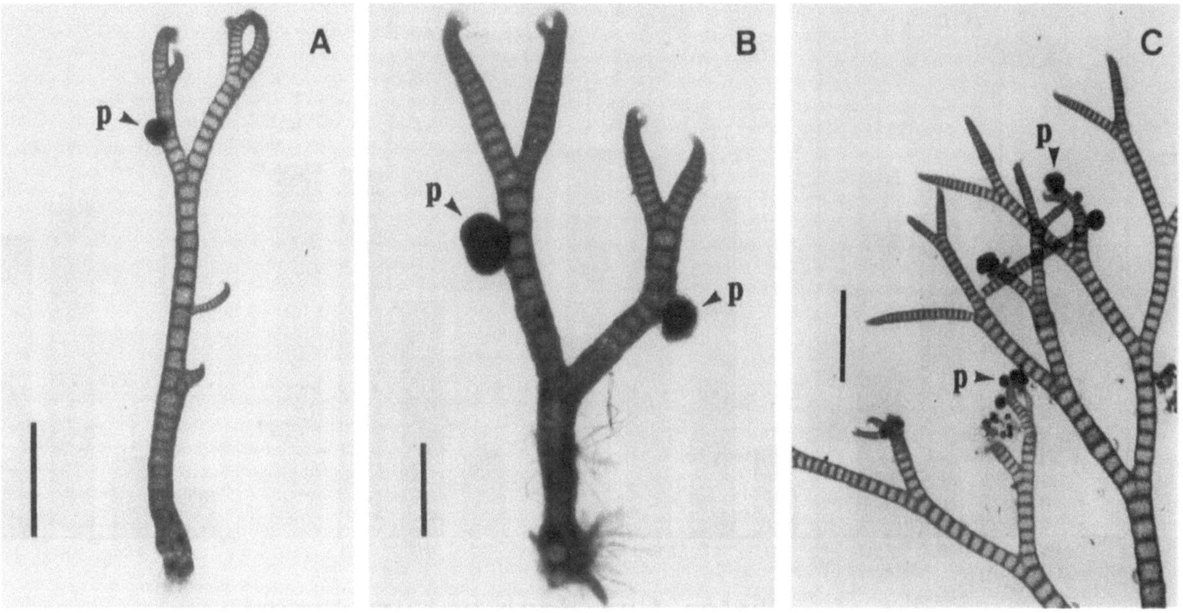

Fig. 3. Young plants of *Campylaephora crassa* in culture (scales = 100 µm) (A) with one parasporangial cluster, (B) with two parasporangial clusters and (C) with many parasporangial clusters. p: parasporangial cluster.

30–35 days cystocarps were observed (Fig. 2F). Carpospores germinated in a bipolar pattern, as did the tetraspores, and after 30–35 days mature tetrasporophytes were observed (Fig. 2D). The general habit and morphology of *C. crassa* in all phases of the life history was similar to young plants in the field but the irregular forms occurring in the field were not observed.

Parasporangial plants occurred at a frequency of about 1% on young tetraspophytes in culture. The parasporangial clusters (Fig. 3A–C) were observed on the young plants 21 days after germination at 12 °C under 40 µmol m^{-2} s^{-1} and 10 days after germination at 16 °C under 100 µmol m^{-2} s^{-1}. From each parasporangial cluster, seventeen to twenty spores were released. They germinated, in the same way as carpospores and tetraspores, and grew into tetrasporophytes. In addition, after the plants with parasporangial clusters released paraspores, they later also formed many tetrasporangia. Each tetrasporangium of the plants derived from paraspores and the plants with parasporangia released four tetraspores, which grew into two male and two female plants. Thus both the plants with parasporangia and the paraspores later returned to a *Polysiphonia* type of life history.

The culture strains used for the crossability tests were the strictly unisexual gametophytes of the normal plants, the plants with parasporangia and the plants derived from the paraspores. All the female plants used for crosses between strains produced many cystocarps after about 10 days and these released many carpospores (Table II). The developmental pattern of the progeny was similar to that of carpospores of within-strain crosses and a *Polysiphonia* type of life history was repeated.

Table II. Crossability tests in *Campylaephora crassa* between the gametophytes of normal plants, the parasporangial plants and the plants derived from paraspores.

Female \ Male	Normal plants	Parasporangial plants	Plants derived from paraspores
Normal plants	9+[1]	9+	9+
Parasporangial plants	9+	9+	9+
Plants derived from paraspores	9+	9+	9+

[1] Nine female plants were fertilized.

Growth and reproduction under laboratory conditions

In this experiment, results were similar whether plant length or wet weight was used for measuring growth, because length was correlated to wet weight ($r = 0.89$); $Y = -0.014 + 0.0046x$, where Y = wet weight and x = length. There was a significant decrease in plant length with increasing temperatures ($p < 0.05$, student t-test) at both 20 and 40 µmol m^{-2} s^{-1}, while no significant difference was found at 100 and 250 µmol m^{-2} s^{-1} ($P > 0.1$, student t-test) (Fig. 4). Within the range irradiances tested the growth

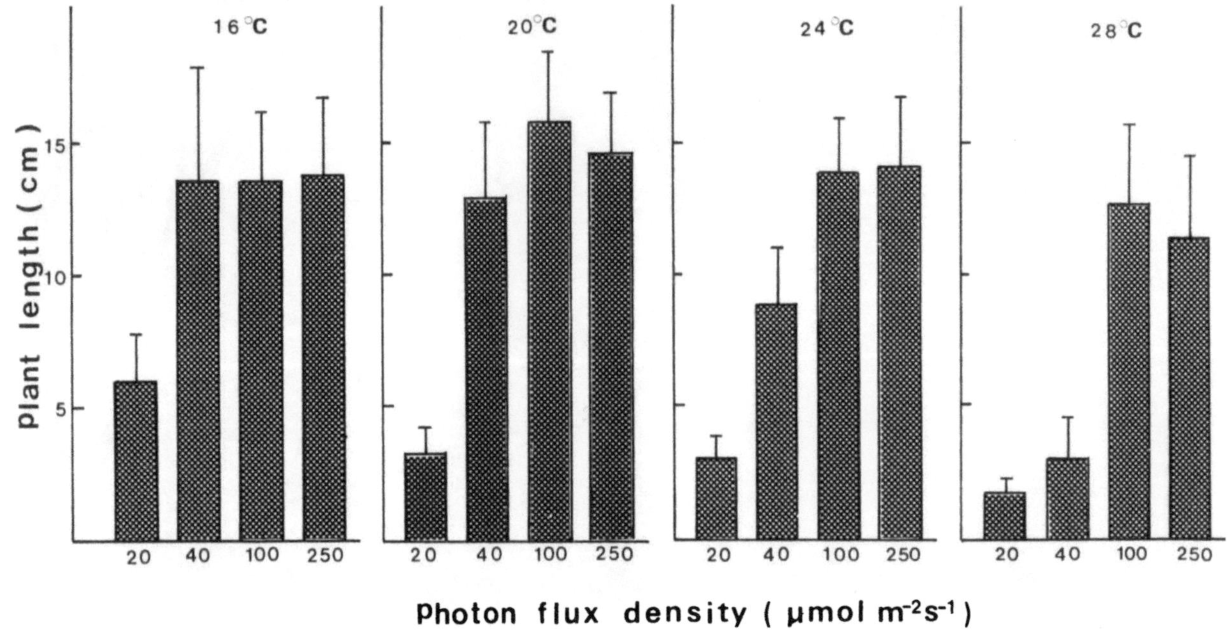

Fig. 4. Growth of *Campylaephora crassa* carposporelings in 16 combinations of PFDs and temperatures. Figures are mean values for ten plants at each combination (+SD).

responses also showed no significant differences between 16 and 20 °C ($p > 0.1$, student t-test). Growth was maximal under 100 and 250 μmol m^{-2} s^{-1} at 20 °C and also under 20 and 40 μmol m^{-2} s^{-1} at 16 °C. However, maximum growth, in terms of plant length, occurred at 20 °C under 100 μmol m^{-2} s^{-1} (Fig. 4), while by wet weight it was maximal at 16 °C under 250 μmol m^{-2} s^{-1} (data not shown).

The carposporelings became tetrasporic plants after 15 days under 100 μmol m^{-2} s^{-1} at 28 °C and under 40 μmol m^{-2} s^{-1} at 24 °C, respectively. The plants formed tetrasporangia at most combinations of photon flux densities and temperatures after one month, but they were not fertile under 20 and 40 μmol m^{-2} s^{-1} at the highest temperature (28 °C).

Photosynthetic pigments

The concentration of phycoerythrin, phycocyanin and chlorophyll a was determined after three weeks of growth. There was a significant reduction in the pig-

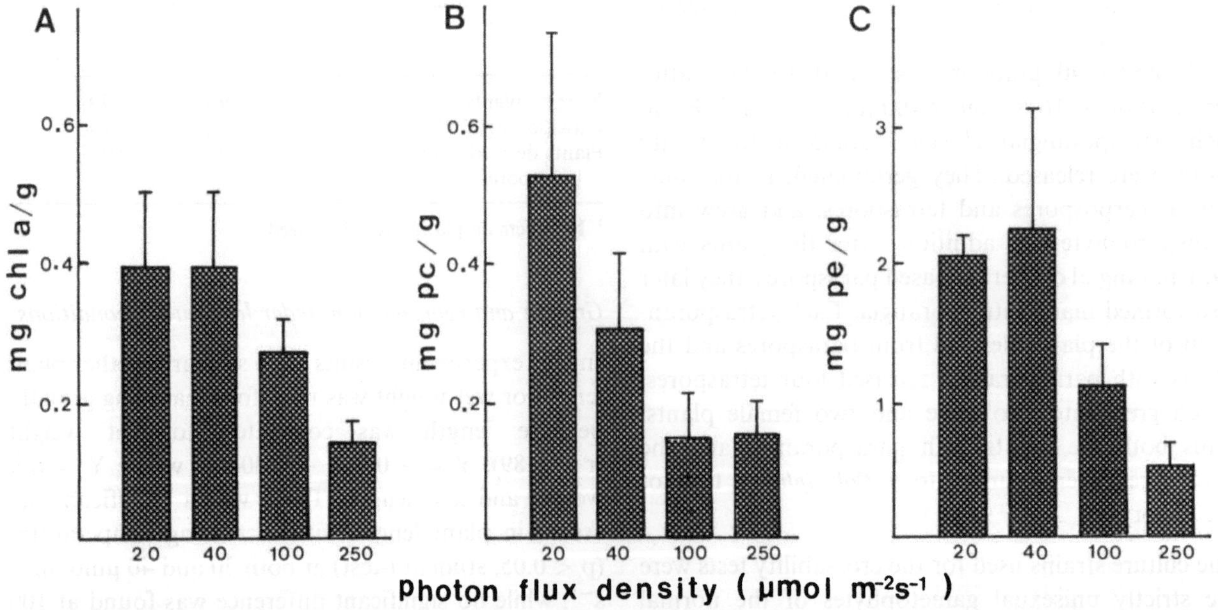

Fig. 5. Concentrations of (A) chlorophyll a, (B) phycocyanin and (C) phycoerythrin of *Campylaephora crassa* in 16 combinations of PFDs and temperatures. Figures are mean values for a total of 16 extractions at each irradiance (+SD).

ment content towards the higher light intensities ($p < 0.05$, student t-test) (Fig. 5). The high standard deviation in pigment content under 20 and 40 μmol m^{-2} s^{-1} is due to the poor growth of *C. crassa* under low light, hence little biomass (0.01 mg) was available for pigment extraction under these irradiances. The total concentration of pigments and the ratio of accessory pigments to chlorophyll a also tended to decrease with higher irradiance.

Discussion

Spatial and temporal phenotypic plasticities are major sources of taxonomic confusion especially in marine plants which frequently show different forms in different habitats (See Russell 1986). The discovery of new forms has often produced new taxonomic categories, for example new species within a genus range and new subspecies, varieties, forms or ecotypes in a species. In *Campylaephora crassa*, four forms were described on the Hokkaido coast (Nakamura 1950, 1965) and another form in Southern Japan (Itono 1977), but there have been no experimental taxonomic approaches, either in the field or in culture. In this study, Korean *C. crassa* plants showed no obvious seasonal differences in the occurrence of the reproductive organs. This needs an intensive quantitative analysis because it is unusual that in an area with a 15 °C seasonal temperature regime there is no ecological response in the species with regard to reproduction. The plants habit seems to be variable during this study. The young plants are distichous and dichotomous, similar to *C. crassa* f. *crassa* Nakamura or f. *cymosa* Nakamura. According to Article 26.1 of the Berlin code (Greuter *et al.* 1988), *Campylaephora crassa* f. *typica* Nakamura (1950) should be replaced by *C. crassa* f. *crassa*. In addition the aged plants are irregularly pinnate, similar to f. *borealis* Nakamura or f. *elongata* Nakamura. Our observations indicate that the forms *sensu* Nakamura (1950, 1965) may be interpreted as developmental or environmental variations. This agrees with Garbary *et al.* (1980) and Garbary (1989) who reported that some diagnostic features, *e.g.* branching pattern and degree of cortication differed in different life history stages and different environmental conditions in *Ceramium rubrum*, a related taxon to *Campylaephora crassa*. However, more work is required to determine if there is any genetic basis for these taxonomic categories.

Although *Campylaephora crassa* has always been found as an epiphyte in the field (Okamura 1930, Nakamura 1950, 1965, Funahashi 1966, Kang 1966, Boo 1985), Korean plants grow very well on plastic substrata in culture and completed a *Polysiphonia* type of life history, typical of the Ceramiaceae. Cultured plants were similar in habit and branching pattern to young plants in the field and showed no abnormal morphology. Therefore the epiphytism of *C. crassa* seems to be not a result of chemicals produced by its host plants but an aspect of physical anchorage, similar to that noted in *Microcladia coulteri* Harvey (Gonzalez and Goff 1989 a), a member of a related genus. Using cross-inoculation experiments, Gonzalez and Goff (1989 b) reported that the pattern of primary rhizoid development in *M. californica* Farlow and *M. coulteri* determined the mode of attachment and might influence the range of basiphytes suitable for each epiphyte. There are, however, other reports that some epiphytic species in the rhodophytes cannot complete their life history in the absence of their specific basiphytes (Gonzalez and Goff 1989 a, p. 54).

The definition, occurrence and fate of parasporangia in the Ceramiaceae have been reviewed by Maggs (1988). In *Plumaria elegans* (Bonnemaison) Schmitz, the plants from paraspores were found to be triploid (Drew 1939, Whittick 1977) and showed an apomeiotic life cycle in the field and culture (Rueness 1968, Whittick 1977). *Ceramium strictum* Harvey from Norway also formed parasporangia in the field and reproduced solely by them in culture (Rueness 1973). These findings were consistent with Drew's (1939) suggestion, that the triploids might have a wider geographical distribution than the haploid and the diploid plants. On the other hand, the paraspores can also be haploid because all progeny from paraspores of genetically male plants of *Callithamnion corymbosum* (Smith) Lyngbye were male (Hassinger-Huizinga 1952).

It is interesting that our plants in culture produced parasporangia on young tetrasporophytes. Furthermore the paraspores released grew into tetrasporophytes and returned to a *Polysiphonia* type of life history. In crossability tests, the plants derived from the paraspores were compatible with the normal plants and all progeny also showed a regular ceramiacean life history. These results indicate that the paraspores of *Campylaephora crassa* may be mitotic diploids and have the same nuclear phase as the plants with parasporangia, although we did not examine the nuclear phase of the paraspores by staining chromosomes or DNA quantification. We suggest that the formation of parasporangia in *Campylaephora crassa* is a temporary and accessory modification to the life history strategy.

According to Feldmann-Mazoyer (1941, p. 335), *Ceramium barbatum* Kützing (= *C. rubrum* var. *barbatum* Kützing) from France, which is very similar but allopatric to *Campylaephora crassa* (Boo 1985), formed parasporangia only on young plants in the Mediterranean Sea. This implies that Korean *C. crassa* plants may form parasporangia in some locations and that a further taxonomic study on the comparative morphology of, and hybridization between, the above two species is required.

Growth pattern varies with environmental factors such as temperature and irradiance in the Ceramieae. Temperature and irradiance responses have been investigated in *Ceramium* (Garbary 1979), *Campylaephora* (Notoya 1979) and *Herpochondria* (Notoya and Saito 1980). In *Campylaephora hypnaeoides* from Japan (Notoya 1979), growth was maximal at 20 °C under about 160 μmol m^{-2} s^{-1} (approximately 8000 lux) in gametophytes and at 20 °C under about 40 to 80 μmol m^{-2} s^{-1} (approximately 2000 to 4000 lux) in tetrasporophytes. Judging from figures 13 and 14 of Notoya (1979, p. 203–204), the growth responses seem to be more dependent on temperature than on light intensity. In our *C. crassa* plants, maximal growth occurred at 16 and 20 °C at all irradiances tested in culture. This agrees with the field data that the largest plants occurred in May when the surface water temperature ranged from 14.8 to 18.2 °C (Anonymous 1989, 1990). As seen in Figure 4, the growth of *C. crassa* depends less on light intensity at lower temperatures than at higher temperatures. This may be due to the fact that at high temperatures the respiration will be so high that there will be no photosynthetic surplus to allocate to growth. However, growth has shown to be light-limited in *Plocamium cartilagineum* (Linnaeus) Dixon, a subtidal undergrowth species (Kain 1987).

Reproduction is influenced by factors such as temperature, irradiance and daylength in some temperate species of Ceramiaceae (Edwards 1973, Fredriksen *et al.* 1990). Plants of *Campylaephora hypnaeoides* formed tetrasporangia and cystocarps only at 20 and 25 °C under the conditions tested in culture (Notoya 1979). Korean *C. crassa* plants in the field have tetrasporangia and cystocarps throughout the year, and tetrasporophytes in culture also formed tetrasporangia at most combinations of irradiances and temperatures. Our field and culture data indicate that environmental factors such as temperature and irradiance do not directly regulate the reproduction of *C. crassa*. Since reproduction in marine plants may depend on size or accumulation of stored reserves rather than on age, environmental and biological factors that promote the efficiency of photosynthesis can accelerate reproduction in marine plants.

Pigment concentration increases in plants growing in deep water (Saffo 1987). Although it has been studied in other rhodophytes (Saffo 1987, Fredriksen and Rueness 1989), little information is available on irradiance effect on pigment concentration in the Ceramiaceae. In the present experiments we found that both phycobiliprotein and chlorophyll a content decreased towards the higher photon flux densities. This could be observed in the bleaching of the plants. However, a high concentration of all pigments under low irradiances is an ecological strategy to absorb maximum light in deep water (Saffo 1987), and the phycobiliproteins may also protect against photooxidation of chlorophyll a during periods of high irradiance (Fredriksen and Rueness 1989).

Acknowledgements

This work was supported by a research grant from the Korean Science and Engineering Foundation to the first author (KOSEF 891-0409-011-2). The first author thanks Prof. T. Yoshida and Dr M. Masuda of Hokkaido University for hospitality and arrangements during his stay in Hokkaido. Many thanks go to Dr Y. Hara of Tsukuba University for his hospitality and escort in the field for a fantastic midnight sampling in Japan. Prof. M. Tatewaki kindly arranged collecting in Muroran, and Drs Y. Saito and A. Miura took me to sampling sites. Mrs W. K. Shin, H. S. Yoon, and Miss G. W. Park assisted in field sampling. Gratitude is extended to Prof. Y. J. Kim for criticizing the draft of this manuscript. The authors gratefully acknowledge Dr G. T. Boalch and two anonymous reviewers for their critical review of the manuscript.

References

Agardh, J. G. 1851. *Species genera et ordines algarum. Species genera et ordines algarum Floridearum.* 2(1). C. W. Gleerup. Lund. pp. 351.

Anonymous, 1989. *Annual Report of Oceanographic Observations. Vol. 36. 1987.* National Fisheries Research and Development Agency. Pusan. 468 pp.

Anonymous. 1990. *Annual Report of Oceanographic Observations. Vol. 37. 1988.* National Fisheries Research and Development Agency. Pusan. pp. 438.

Beer, S. and A. Eschel. 1985. Determining phycoerythrin and phycocyanin concentrations in aqueous crude extracts of red algae. *Aust. J. Mar. Freshw. Res. 36*: 785–792.

Boo, S. M. 1985. *A Systematic Study on Six Tribes of Ceramiaceae (Rhodophyta, Ceramiales) in Korea.* Seoul National University, Seoul. pp. 449. Pls 1-II. (in Korean).

Boo, S. M. and T. Yoshida. 1991. Typification of three names in *Campylaephora* (Rhodophyceae; Ceramiaceae). *Taxon 40*: 95–99.

Drew, K. M. 1939. An investigation of *Plumaria elegans* (Bonnem.) Schmitz. with special reference to tripolid plants bearing parasporangia. *Ann. Bot. N. S. 3*: 347–367.

Edwards, P. S. 1973. Life history studies of selected British *Ceramium species*. *J. Phycol. 9*: 181–184.

Eppley, R. W., R. W. Holmes and J. D. H. Strickland. 1967. Sinking rates of marine phytoplankton measured with a fluorometer. *J. Exp. Mar. Biol. Ecol. 1*: 191–208.

Feldmann-Mazoyer, G. 1941. *Recherches sur les Céramiacées de la Méditerranée Occidentale*. Imprimerie Minerva, Algiers. pp. 510. Pls. I–IV.

Fredriksen, S. and J. Rueness. 1989. Culture studies of *Gelidium latifolium* (Grev.) Born *et* Thur. (Rhodophyta) from Norway. Growth and nitrogen storage in response to varying photon flux density, temperature and nitrogen availability. *Bot. Mar. 32*: 539–546.

Fredriksen, S., S. M. Boo and J. Rueness. 1990. Photoperiodism and Rhodophyta) in culture. *Jpn J. Phycol. 38*: 43–49.

Funahashi, S. 1966. Marine algae from Vladivostok and its vicinity. *Jpn J. Phycol. 14*: 127–145 (in Japanese).

Garbary, D. 1979. Daylength and development in four species of Ceramiaceae (Rhodophyta). *Helgoländer wiss. Meeresunters. 32*: 213–227.

Garbary, D. 1989. Interoceanic hybridization in fully corticated *Ceramium* isolates (Rhodophyta) from Nova Scotia and Washington. *Korean J. Phycol. 3*: 89–93.

Garbary, D. J., D. Grund and J. McLachlan. 1978. The taxonomic status of *Ceramium rubrum* (Hudson) C. Ag. (Ceramiales, Rhodophyta) based on culture experiments. *Phycologia 17*: 85–94.

Garbary, D. J., D. Grund and J. McLachlan. 1980. Branching patterns and life history stages in *Ceramium rubrum* (Huds.) C. Ag. *Nova Hedwigia 33*: 249–260.

Gonzalez, M. A. and L. J. Goff. 1989 a. The red algal epiphytes *Microcladia coulteri* and *M. californica* (Rhodophyceae, Ceramiaceae). I. Taxonomy, life history and phenology. *J. Phycol. 25*: 545–558.

Gonzalez, M. A. and L. J. Goff. 1989 b. The red algal epiphytes *Microcladia coulteri* and *M. californica* (Rhodophyceae, Ceramiaceae). II. Basiphyte specificity. *J. Phycol. 25*: 558–567.

Greuter, W., H. M. Burdet, W. G. Chaloner, V. Demoulin, R. Grolle, D. L. Hawks-Worth, D. H. Nicolson, P. C. Silva, F. A. Stafleu, E. G. Voss and J. McNeill. 1988. International Code of Botanical Nomenclature. *Regum Veg. 118*: 1–328.

Guiry, M. D. 1987. The evolution of life history types in the Rhodophyta: an appraisal. *Cryptogamie, Algologie 8*: 1–12.

Hassinger-Huizinga, H. 1952. Generationswechsel und Geschlechtsbestimmung bei *Callithamnion corymbosum* (Sm.) Lyngb. *Arch. Protistenk. 98*: 91–125.

Hoek, C. van den. 1982. The distribution of benthic marine algae in relation to the temperature regulation of their life histories. *Biol. J. Linn. Soc. 18*: 81–144.

Hommersand, M. H. 1963. The morphology and classification of some Ceramiaceae and Rhodomelaceae. *Univ. California Publ. Bot. 35*: 165–366. Pls. 1–6.

Itono, S. 1977. Studies on the ceramiaceous algae (Rhodophyta) from the southern parts of Japan. *Biblioth. Phycol. 35*: 1–499. figs. 1–72.

Jensen, A. 1978. Chlorophylls and carotenoids. *In*: (J. A. Hellebust and J. S. Craigie, eds) *Physiological and Biochemical Methods. Handbook of Phycological Methods*. Cambridge Univ. Press, London. pp. 59–90.

Kain, J. M. 1987. Seasonal growth and photoinhibition in *Plocamium cartilagineum* (Rhodophyta) off the Isle of Man. *Phycologia 26*: 88–89.

Kang, J. W. 1966. On the geographical distribution of marine algae in Korea. *Bull. Pusan Fish. Coll. 9*: 1–125.

Maggs, C. A. 1988. Intraspecific life history variability in the Florideophycidae (Rhodophyta). *Bot. Mar. 31*: 465–490.

Nakamura, Y. 1950. New *Ceramium* and *Campylaephora* from Japan. *Sci. Pap. Inst. Algol. Res. Fac. Sci. Hokkaido Imp. Univ. 3*: 155–172.

Nakamura, Y. 1954. The structure and reproduction of the genera *Ceramium* and *Campylaephora* in Japan with special reference to criteria of classification. *Sci. Pap. Inst. Algol. Res., Fac. Sci. Hokkaido Imp. Univ. 4*: 15–62.

Nakamura, Y. 1965. Species of the genera *Ceramium* and *Campylaephora* especially those of northern Japan. *Sci. Pap. Inst. Algol. Res., Fac. Sci. Hokkaido Imp. Univ. 14*: 52–71.

Notoya, M. 1979. Life history of *Campylaephora hypnaeoides* J. Agardh (Ceramiaceae, Rhosophyta) in culture and environmental regulation of reproduction. *Jpn. J. Phycol. 27*: 201–204 (in Japanese).

Notoya, M. and Y. Saito. 1980. Life history, growth and environmental conditions for reproduction in *Microcladia elegans* Okamura (Ceramiales, Rhodophyta). *Jpn. J. Phycol. 28*: 191–195 (in Japanese).

Okamura, Y. 1930. *Icones of Japanese Algae*. Vol. 6. Kazamashobo, Tokyo. pp. 21–26. Pl. 269.

Rueness, J. 1968. Paraspores from *Plumaria elegans* (Bonnem.) Schmitz in culture. *Nytt Mag. Bot. 15*: 220–224.

Rueness, J. 1973. Culture and field observations on growth and reproduction of *Ceramium strictum* Harv. from the Oslofjord, Norway. *Norw. J. Bot. 20*: 61–65.

Russell, G. 1986. Variation and natural selection in marine macroalgae. *Oceanogr. Mar. Biol. Ann. Rev. 24*: 309–377.

Saffo, M. B. 1987. New light on seaweeds. *Bioscience 37*: 654–664.

Whittick, A. 1977. The reproductive ecology of *Plumaria elegans* (Bonnem.) Schmitz (Ceramiaceae; Rhodophyta) at its northern limits in the western Atlantic. *J. Exp. Mar. Biol. Exol. 29*: 223–230.

Yendo, K. 1918. Notes on algae new to Japan VIII. *Bot. Mag. Tokyo 32*: 79–81.

Botanica Marina
Vol. 34, pp. 447–453, 1991

Seasonal Variation in Growth, Reproduction and Nitrogen Content of *Fucus vesiculosus* L. in the Öresund, Southern Sweden

L. Carlson

Department of Marine Ecology, University of Lund, P. O. Box 124, S-22100 Lund, Sweden

(Accepted 29 June 1991)

Abstract

The autecology of *Fucus vesiculosus* at Barsebäck, Öresund, southern Sweden, was studied from May to December 1984. The length, growth rate, biomass increase, reproduction and nitrogen content were followed. *Fucus vesiculosus* at Barsebäck exhibited an average growth rate of 0.54 cm/week with the highest growth rates of 0.7–0.8 cm/week observed during June and August. The mean length of the vegetative tips increased almost linearly with time during May to September. At the end of the year the new growth constituted over 80% of the total plant biomass. The mean total length increase was 15.6 cm during the period studied. Two reproductive periods were observed during 1984. Mature receptacles were found during May to July and new receptacles were initiated in July on the same plants and matured during September and October.

The nitrogen content in *F. vesiculosus* varied both seasonally and between tissues of different ages within the plants. Variations were most pronounced in the new vegetative tips with the highest concentrations in winter; 1.71 mmol/g dw in December, and lowest during summer, about 0.5 mmol/g dw in July and August. Basal parts did not show any seasonal variation in nitrogen content. In November–December the vegetative tissue grown during the year contained almost 90% of the total nitrogen within the plant.

Introduction

Fucus vesiculosus L. is common along most of the Swedish coast, despite great environmental variations *e.g.* in salinity. It occurs in most localities with a suitable substrate in the Öresund, the sound between Sweden and Denmark, and is also the most abundant perennial alga in the near shore area here. The species exhibits great morphological variations which have also been observed in other parts of the Baltic (*e.g.* Müller-Stoll and Künzenbach 1956, von Wachenfeldt 1975, Waern 1952, Wallentinus 1979). *Fucus vesiculosus* also shows seasonal variation in both growth and reproduction (Knight and Parke 1950) as do other fucoid alga, *e.g. F. spiralis* L. (Niemeck and Mathieson 1976), and *F. distichus* (L.) Powell (Sideman and Mathieson 1983). The highest growth rates for most fucoids have been observed during late spring and early summer and they exhibit a single annual period of reproduction with time differences between species, except for *F. distichus* in New England which exhibited a bimodal pattern of reproduction (Sideman and Mathieson 1983).

Nitrogen in Scandinavian coastal waters fluctuates seasonally with highest concentrations during winter and early spring and minimum concentrations during summer (*e.g.* Granéli *et al.* 1986, Kornfeldt 1982, Strömgren 1986, Wallentinus 1984). Tissue nitrogen concentrations in various macroalgae have been shown to fluctuate in a similar way but with a time lag compared with the nitrogen content of the sea-water (Asare and Harlin 1983, Gangè *et al.* 1982, Harwick-Witman and Mathieson 1986, Kornfeldt 1982, Strömgren 1986, Wallentinus 1981). The algae accumulate nitrogen during the autumn and winter and utilize it when concentrations in the water are too low for the metabolic requirements and when other conditions for growth are favourable *e.g.* light and temperature (*e.g.* Gagnè *et al.* 1982). Nitrogen has been shown to be the element most frequently limiting the growth of marine macroalgae (*e.g.* Hanisak 1979, Topinka and Robbins 1976) as well as for phytoplankton (Granéli 1987, Granéli *et al.* 1986).

Fucus vesiculosus has been used as a bio-indicator of the release of radionuclides from Barsebäck nuclear power plant, situated 2.9 km south of the study area (Fig. 1) (Mattsson *et al.* 1980). To enable a better interpretation of the results from the bio-indicator studies the autecology of *F. vesiculosus* was investigated. The length, growth rate, biomass increase, and reproduction were studied during one vegetation period. The nitrogen content in different parts of the *Fucus* plants (receptacles, vegetative tips, 1-year-old, 2-year-old parts and basal parts) was also measured.

Study Area

Barsebäck is situated in the Öresund, southern Sweden (Fig. 1). The Öresund is one of the straits joining the brackish waters of the Baltic with the more saline waters of the Kattegat. The salinity in the area fluctuates, both in space and time, from about 8 to 25‰ in the near-shore water. Lowest salinity values are usually found in the upper water layer during the summer period, reflecting an outflow of water from the Baltic. Occasionally, during stormy weather, higher salinities are observed during summer which was the situation on June 18 (Fig. 3). The temperature variation is large with a minimum during January and February, when the near shore water is often frozen, to a maximum of about 20 °C in the summer months. The nuclear power plant does not affect the temperature at the study area. There is a marked variation in solar irradiance and day length with maxima in June and minima in December.

The coastal zone is characterized by shallow, virtually non-tidal waters and sandy bottoms with stones of varying sizes and distributions that contribute to the substrate. *Fucus vesiculosus* is found at most potentially suitable localities in varying numbers. At Barsebäck the maximum age of the *Fucus* plants was 3 to 4 years.

Material and Methods

This study was carried out during May to December 1984. The winter of 1984–1985 was severe and the near-shore waters were frozen which has not been the case for several years before this study (SMHI 1985). At the time of the ice break-up in April 1985 most of the *Fucus* plants had disappeared due to ice-scouring and the remaining plants were in too poor a condition to continue the investigation.

Samples of 10 plants of *F. vesiculosus* were collected once a month except in June and July when samples were taken twice monthly. The plants grew at a depth of about 0.5 m and plants were collected by hand, without viewing the plants, at a distance of one step.

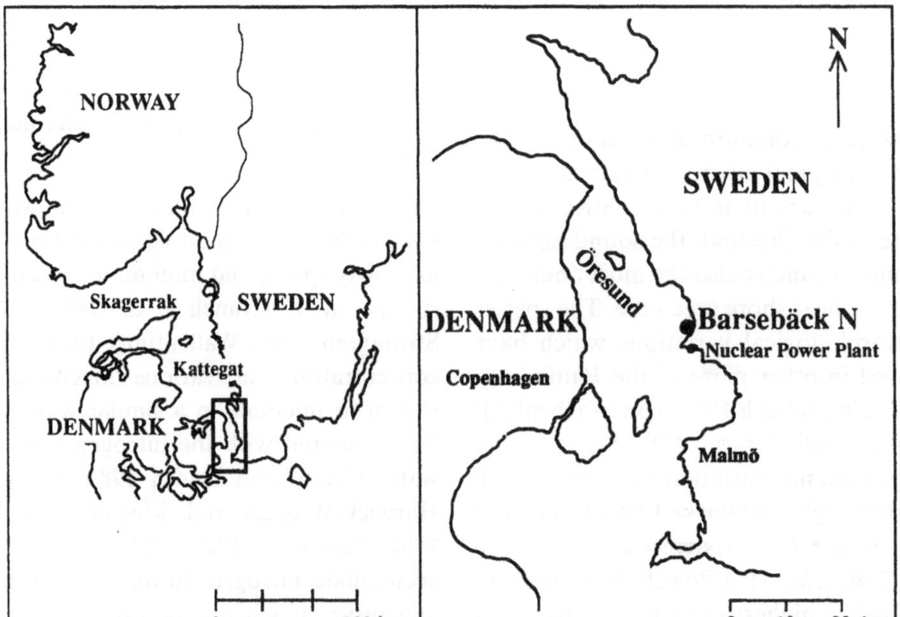

Fig. 1. Study area.

The length of the new fronds including vesicles, for both vegetative and fertile tips, were measured. A maximum of 20 vegetative and 20 fertile tips were studied in each plant. Growth rate was calculated as linear increase (cm/day) between the measurements. The *Fucus* plants were cleaned of epiphytes, rinsed in seawater and then dissected into different age parts (Fig. 2). At the last two sampling occasions the 1- and 2-year-old parts were pooled as it was difficult to separate them. The dry weight (dw) of each part was then determined after five days' drying at 65 °C.

The salinity and temperature of the surface water were measured on each sampling occasion.

The nitrogen content was measured as Kjeldahl nitrogen on sub-samples of the pooled samples of each of the different parts of *F. vesiculosus*. The nitrogen content of the whole plants was then calculated. The concentration of inorganic nitrogen in the ambient water was not measured. Nitrogen measurements in the Öresund performed in other studies have shown highest concentrations during winter followed by a rapid decrease in March and lowest concentrations, often close to the detection limit, during summer, April to August (*e.g.* Edler 1977, Granéli 1978).

The significance of differences between mean values was calculated by Student's t-test (Sokal and Rohlf 1981).

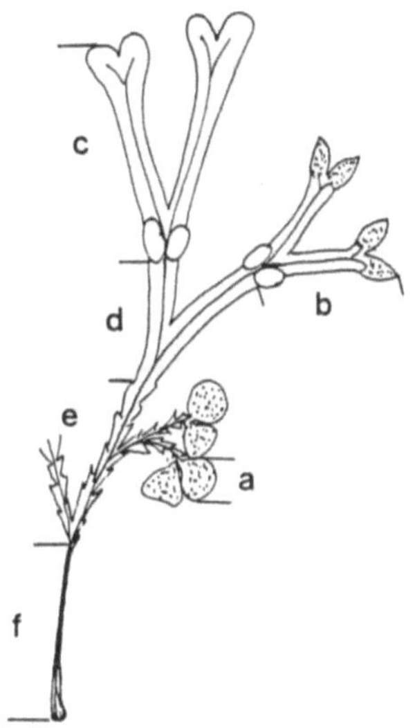

Fig. 2. Schematic figure of the different parts of *Fucus vesiculosus*, a) old receptacles, b) new fronds carrying receptacles, c) new vegetative fronds, d) 1-year-old vegetative tissue, e) 2-year-old vegetative tissue, f) basal parts.

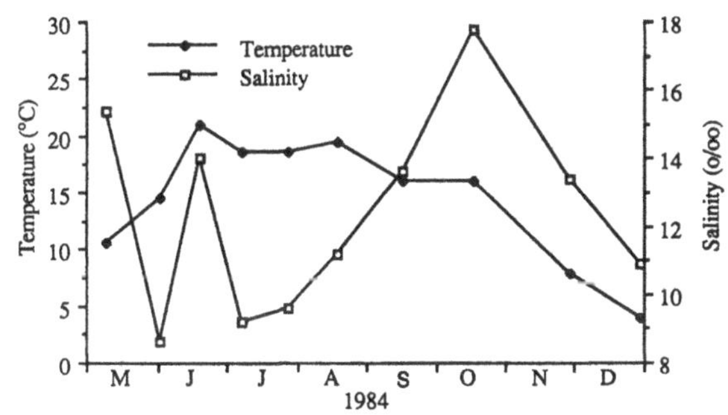

Fig. 3. Salinity (‰) and temperature (°C) of surface water at Barsebäck, 1984.

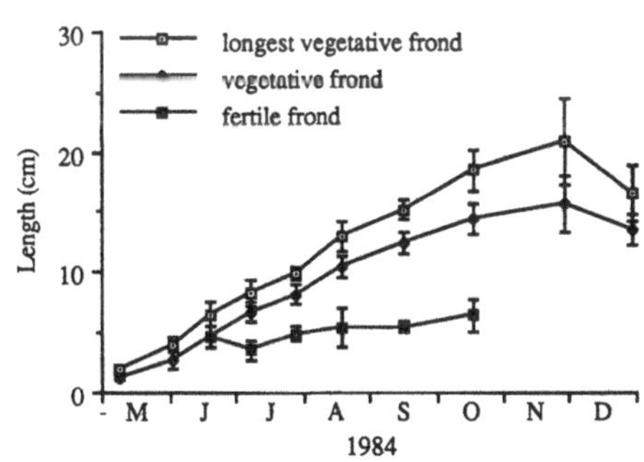

Fig. 4. Length (cm) of vegetative, longest vegetative and fertile tips including gas bladder of *Fucus vesiculosus*, vertical bars denote 95% confidence limit, (n = 10).

Results

The salinity varied between 8.6 and 17.8‰ with the highest salinities during spring and autumn due to mixing of surface water with the more saline bottom water during stormy weather (Fig. 3). The temperature varied between 4 °C in December and about 20 °C during the summer months, June to August (Fig. 3).

The elongation of the vegetative tips was almost linear with time during summer and early autumn, May–September $r^2 = 0.995$, even if the growth rate declined during July (not significant) (Figs 4, 5). In July several of the new tips started to initiate new receptacles and these tips were significantly shorter ($P < 0.001$) than the vegetative tips (Fig. 4). There was also a tendency for one of the fronds on each plant to grow faster than the others (Fig. 4). The mean length of the fastest-growing frond of the ten plants was significantly greater than the mean frond length of the plants during the investigation period ($P < 0.001$).

Individual variations in length between new vegetative tips within plants were larger than between means of the different plants on most occasions, the coefficients of variation (CV's) being 7.7–51.5 and 8.5–35.4, respectively. The coefficient of variation within plants varied seasonally with the highest values during May and June.

The growth rate of the vegetative tips was highest during late spring and summer (Fig. 5). During July the growth rate decreased but increased again during

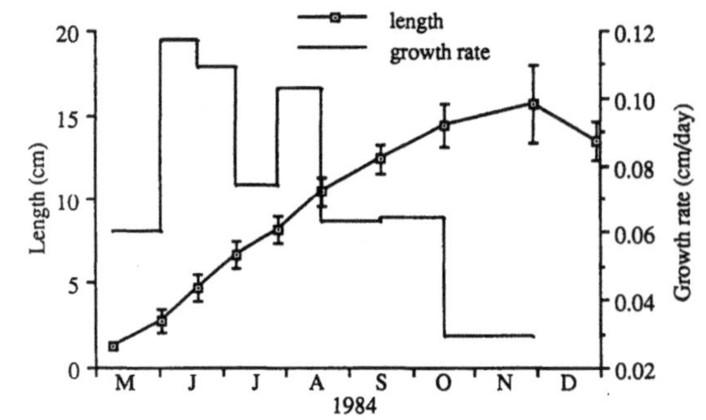

Fig. 5. Growth rate (cm/day) and length (cm) of vegetative tips of *Fucus vesiculosus*, vertical bars denote 95% confidence limit, (n = 10).

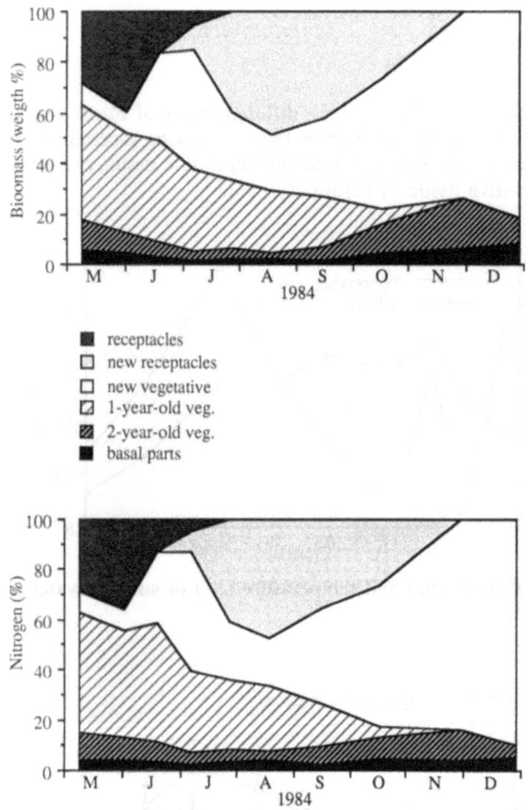

Fig. 6. a) Percentage biomass of the different parts of *Fucus vesiculosus* during a vegetation period, b) Percentage of Kjeldahl nitrogen in different parts of *Fucus vesiculosus*, receptacles, new vegetative fronds, 1-year-old vegetative tissue, 2-year-old vegetative tissue, basal parts, whole plant (n = 10).

August, whereafter the algae ceased to grow in November. The growth rate varied between 0.03 and 0.12 cm/day, average 0.08, corresponding to 0.21–0.84 cm/week with an average growth rate of 0.54 cm/week when measured as the mean length increase of new vegetative tips. The growth rate of the longest vegetative frond was higher, average 0.09 cm/day (0.05–0.14); 0.69 cm/week (0.35–0.98), but exhibited the same growth pattern except during the autumn when an increased growth rate was observed also during September–November.

Fucus vesiculosus at Barsebäck had two reproductive periods (Fig. 6a). The first was during spring with mature receptacles in stages 6 to 7 in the sequence given by Knight and Parke (1950), being present during May and June. The swollen receptacles released their gametes and during the end of June and the beginning of July most of the receptacles were completely depleted and began to decay. During July new receptacles were initiated on most of the plants and both new and old receptacles could be observed on the same individuals. In late November and December no receptacles were observed. Although the investigation ended in December due to the formation of ice, initiation of receptacles during February has been observed during winters with less severe ice conditions (Carlson unpubl.). Differences in the size of receptacles were observed with larger receptacles during spring than during summer and autumn (subjective observation).

Being a perennial, the growth of *F. vesiculosus* during the year changes the percentage biomass of tissues of different ages within the plant (Fig. 6a). The new growth of the year constituted almost 80% of the total biomass of a plant when the growth period was over. During July and August more than 50% of the new tips became fertile and were lost during the autumn. In October erosion of the wings from the previous years' growth began, followed by thickening of the midrib. At the end of November only wings on the vegetative parts formed during the same year were left.

The nitrogen content of *F. vesiculosus* varied seasonally and different parts of the *F. vesiculosus* plants exhibited different seasonal patterns (Fig. 7). The nitrogen content in the basal parts showed no seasonal variation. The nitrogen content of the other tissues decreased in all parts during late spring and started

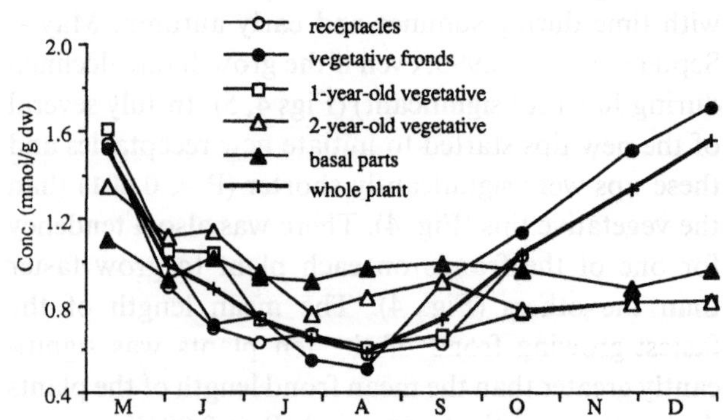

Fig. 7. Seasonal variation of Kjeldahl nitrogen (mmol N/g dw) in different parts of *Fucus vesiculosus*, receptacles, new vegetative fronds, 1-year-old vegetative tissue, 2-year-old vegetative tissue, basal parts, whole plant (n = 10).

to increase again in the new tips in late August. Comparing the nitrogen content in the differently aged tissues with their relative biomass showed that the older parts constituted less than 20% of the total and that the vegetative parts grown during the season constituted almost 90% of the total nitrogen content during winter (Fig. 6b).

Discussion

Length increase and growth rate (linear length increase per unit time) have been measured for different fucoids, mostly as the length of the whole plant from the holdfast to the end of the longest frond (*e.g.* Keser and Larson 1984, Knight and Parke 1950, Mathieson *et al.* 1976, Sideman and Mathieson 1983). A special method for measuring growth defined as the length increase of the outermost mm of apical apices has been developed by Strömgren (1977). In this study the length of new fronds, including the vesicles were measured. The *F. vesiculosus* plants were also used for radionuclide and nitrogen measurements and therefore growth measurements have been made on different plants each time. As there are differences in growth between apical fronds within plants as well as between plants, measurements on a random sample of single apices may give a more representative estimate of algal growth than recordings of whole plants, as pointed out by Strömgren (1986). Individual variations in growth between plants of *Ascophyllum nodosum* (L.) Le Jolis were observed by Strömgren (1981) but he found no correlation between the average growth rate of the different plants in the laboratory and the length of their respective apices. But there are also difficulties in growth measurements *in situ* on marked plants due to plant losses (Niemeck and Mathieson 1976, Sideman and Mathieson 1983, Thom 1983).

Measurement of the length of different fronds of the *Fucus* plants showed differences in length. At Barsebäck the vegetative fronds continued to grow while the fertile fronds almost ceased to grow after the initiation of receptacles. The observed growth of fertile tips was probably due to differences in time of receptable initiation. Knight and Parke (1950) observed that slow-growing fronds usually formed a reproductive fan forming receptacles while the other fronds continued to grow with one of the fronds always maintaining a peripheral lead. Müller-Stoll and Künzenbach (1956) also observed receptacle initiation on slow-growing fronds of *F. vesiculosus* at Hiddensee. Strömgren (1986) found no correlation between the growth of vegetative apices and the stage of maturation of the fertile apices of fucoids.

There are few papers in the literature concerning the autecology of *F. vesiculosus* from the Scandinavian waters. The growth rate of fucoids seems to vary both geographically and seasonally (see *e.g.* Knight and Parke 1950). Comparisons between growth studies are difficult as the measurements are usually made with different methods and during different time periods. This study covers only eight months, thus the lower growth rates during winter and spring are not included in the average growth rate. Measuring only the new fronds also does not reflect the shortening of the thallus which accompanies the thickening of the midrib (see *e.g.* Knight and Parke 1950). The highest growth rate for *F. vesiculosus* at Barsebäck was observed in June, as was recorded for *F. vesiculosus* from the Trondheimsfjord (Strömgren 1986), while Keser and Larson (1984) observed maximum growth in the autumn in Maine. Knight and Parke (1950) reported an average growth rate of *F. vesiculosus* in Britain of 0.48 cm/week, measured on whole plants during the whole year. *Fucus vesiculosus* from Barsebäck exhibited slightly higher growth rates with an average growth rate of 0.54 cm/week. Measuring only the longest vegetative fronds gave an average growth rate of 0.69 cm/week. The average yearly length increase of the *F. vesiculosus* at Barsebäck was 15.6 cm and for the longest vegetative tips the mean yearly increase was 20.8 cm. Measurements on the length of *F. vesiculosus* in the study area on a few occasions in the following years have given length values within the same range as those presented in the present study. At Askö in the Baltic Proper, a yearly length increase of 5–12 cm was reported for *F. vesiculosus* (Wallentinus 1979) and in the southern Baltic, Hiddensee, the average yearly length increase was 18 cm for the deep water form of *F. vesiculosus* (Müller-Stoll and Künzenbach 1956).

The relative biomass of the different parts of the thallus fluctuates seasonally and at the end of the growing period the new growth of the year constituted over 80% of the total plant biomass. Similar yearly variation in relative biomass was observed in the Askö area in the Baltic (Lamrén and Lamrén personal communication). In August, the new fertile tips accounted for about 70% of the total dry weight of the new tips and some plants had only a few vegetative tips, as was also observed by Knight and Parke (1950). The proportion of fertile tips can be as high as 80–95% of the total frondage in *F. vesiculosus* (Knight and Parke 1950). In *F. spiralis*, with ripe receptacles, the fertile fronds constituted more than 50% of the total fresh weight (Munda 1986).

Most fucoids exhibit one seasonal reproduction period with receptacles initiated at the beginning of the year or in the late autumn the preceding year and maturation during spring and early summer, but differences in reproductive patterns between fucoid species have been observed (Knight and Parke 1950, Müller-Stoll and Künzenbach 1956, Niemeck and Mathieson 1976, Sideman and Mathieson 1983, Strömgren 1986). In the Öresund area receptacles could be seen from February to October. During 1984 two reproductive periods were observed with new receptacles initiated in July on individuals that were fertile during spring. In July, plants with both mature and newly formed receptacles were found. Receptacles have also been found on *F. vesiculosus* at the same time of the year in the following years. Most other studies of the reproduction of *Fucus* species show, however, only one period of reproduction. One exception is the investigation by Sideman and Mathieson (1983) on *F. distichus* in New England. Most of the populations of *F. distichus* exhibited a bimodal reproductive periodicity but the individuals reproducing during each period were distinct from each other. Müller-Stoll and Künzenbach (1956) observed initiation of new receptacles in deep water forms of *F. vesiculosus* at Hiddensee in July but only on a few individual plants, in contrast to Barsebäck where receptacle initiation was common. The receptacles on *F. vesiculosus* at Hiddensee matured during the autumn and they were smaller than the receptacles initiated earlier, which was also observed at Barsebäck.

Nitrogen in perennial macroalgae has been found to follow the inorganic nitrogen concentration in the ambient water but often with a time lag indicating that nitrogen reserves which have been built up during the winter months were used for growth during April to June (Asare and Harlin 1983, Gagnè *et al.* 1982, Harwick-Witman and Mathieson 1986, Kornfeldt 1982, Strömgren 1986, Wallentinus 1981). In the present study nitrogen was measured in different parts of the alga and seasonal differences were observed between the parts. The metabolically most active tissue, new tips, showed the most pronounced variation. The same pattern was reported by Wallentinus (1981) for *F. vesiculosus* in the Askö area. Kremer and Munda (1982) also reported higher nitrogen concentrations in the growing apical tips compared with older tissue indicating a reduced participation of older tissue in active nitrogen metabolism. In this study it was also observed that the new parts grown during the year contained 80–90% of the nitrogen within the whole plants. Nitrogen increase during autumn was almost negligible in the older parts. The nitrogen concentrations in the basal parts did not vary during the year. Uptake rates of different parts of *F. vesiculosus* and *F. spiralis* showed highest uptake rates in the apical parts (Topinka 1978, Wallentinus 1984). Wallentinus (1984) found no significant differences in uptake rates between vegetative and young fertile parts of *F. vesiculosus*. At Barsebäck the increase in nitrogen concentration was slightly higher in the vegetative tips than in the fertile tips from August to October. The lowest uptake rate was measured for the basal parts by Wallentinus (1984) and the oldest part also showed lower productivity than the newer (Guterstam *et al.* 1978) showing that the oldest parts are not very metabolically active.

This investigation is part of a larger study of *F. vesiculosus* as a bio-indicator of radionuclides and the results indicate the problems encountered when comparing results from different seasons as growth, biomass of the different age tissues, and reproduction all exhibit seasonal variations (Carlson and Erlandsson 1991). A seasonal variation of zinc, has been observed in *Fucus*, with lowest concentration observed during the summer (Fuge and James 1974). Fuge and James (1974) suggested that the lower concentrations measured during summer could be an effect of dilution due to plant growth during summer. Similar variations in time and concentrations probably also exist for other elements. Fluctuations in concentration of several elements in algae are also a response to the ambient water concentrations, which is the primarly reason for using algae as indicators. To avoid possible dilution effects, *F. vesiculosus* for indicator studies should be collected during periods with low growth rate, *i.e.* during late autumn to spring. Alternatively samples should be taken at the same time of year in order to obtain comparable results and exclude effects due to the biology of the organism.

References

Asare, S. O. and M. Harlin. 1983. Seasonal fluctuations in tissue nitrogen for five species of perennial macroalgae in Rhode Island Sound. *J. Phycol. 19*: 154–157.

Carlson, L. and B. Erlandsson. 1991. Seasonal variation of radionuclides in *Fucus vesiculosus* L. from the Öresund, southern Sweden. *Environ. Pollut. 72*: In press.

Edler, L. 1977. Phytoplankton and primary production in the Sound. Ph. D. thesis. Dept. of Marine Botany, Gothenburg. pp. 82.

Fuge, R. and K. H. James. 1974. Trace metal concentrations in *Fucus* from the Bristol Channel. *Mar. Poll. Bull. 5*: 9–12.

Gagnè, J. A., K. H. Mann and A. R. O. Chapman. 1982. Seasonal patterns of growth and storage in *Laminaria longicruris* in relation to differing patterns of availability of nitrogen in the water. *Mar. Biol. 69*: 91–101.

Granéli, E. 1978. Algal assay of limiting nutrients for phytoplankton production in the Öresund. *Vatten 34*: 117–128.

Granéli, E. 1987. Nutrient limitation of phytoplankton biomass in a brackish water bay highly influenced by river discharge. *Estuar. Coast. and Shelf. Sci. 25*: 555–565.

Granéli, E., W. Granéli and L. Rydberg. 1986. Nutrient limitation at the ecosystem and the phytoplankton community level in the Laholm Bay, south-east Kattegat. *Ophelia. 26*: 181–194.

Guterstam, B., I. Wallentinus and R. Iturriaga. 1978. *In situ* primary production of *Fucus vesiculosus* and *Cladophora glomerata. Kieler Meeresforsch. Sonderheft 4*: 257–266.

Hanisak, M. D. 1979. Nitrogen limitation of *Codium fragile* ssp *tomentosoides* as determined by tissue analysis. *Mar. Biol. 50*: 333–337.

Hardwick-Witman, M. N. and A. C. Mathieson. 1986. Tissue nitrogen and carbon variations in New England estuarine *Ascophyllum nodosum* (L.) Le Jol. populations (Fucales, Phaeophyta). *Estuaries 9*: 43–48.

Keser, M. and B. R. Larson. 1984. Colonization and growth dynamics of three species of *Fucus. Mar. Ecol. Prog. Ser. 15*: 125–134.

Knight, M. and M. Parke. 1950. A biological study of *Fucus vesiculosus* L. and *Fucus serratus* L. *J. mar. biol. Ass. U.K. 24*: 439–515.

Kornfeldt, R.-A. 1982. Relation between nitrogen and phosphorus content of macroalgae and the waters of Northern Öresund. *Bot. Mar. 25*: 197–201.

Kremer, B. P. and I. M. Munda. 1982. Ecophysiological studies of the Adriatic seaweed, *Fucus virsoides. P. S. Z. N. I: Marine Ecology 3*: 75–93.

Mathieson, A. C., J. W. Shipman, J. R. O'Shea and R. C. Hasevlat. 1976. Seasonal growth and reproduction of estuarine fucoid algae in New England. *J. exp. mar. Biol. Ecol. 25*: 273–284.

Mattsson, S., R. Finck and M. Nilsson. 1980. Distribution of activation products from Barsebäck nuclear power plant (Sweden) in the marine environment. Temporal and spatial variations as established by seaweed. *Env. Poll. (Series B) 1*: 105–115.

Munda, I. M. 1986. Differences in heavy metal accumulation between vegetative parts of the thalli and receptacles in *Fucus spiralis* L. *Bot. Mar. 29*: 341–349.

Müller-Stoll, W. R. and R. Künzenbach. 1956. Über die standortbedingten Wuchsformen von *Fucus vesiculosus* in den Gewässern der Insel Hiddensee. *Arch. Protistenkunde 101*: 289–335.

Niemeck, R. A. and A. C. Mathieson. 1976. An ecological study of *Fucus spiralis* L. *J. exp. mar. Biol. Ecol. 24*: 33–48.

Sideman, E. J. and A. C. Mathieson. 1983. The growth, reproductive phenology and longevity of non-tide-pool *Fucus distichus* (L.) Powell in New England. *J. exp. mar. Biol. Ecol. 68*: 111–127.

SMHI. 1985. Väder och Vatten. *Monthly Report Swedish Meteorological and Hydrological Institute.* (In Swedish.)

Sokal, R. R. and F. J. Rohlf. 1981. *Biometry.* San Francisco: W. H. Freeman and Company. 859 pp.

Strömgren, T. 1977. Short-term effects of temperature upon the growth of intertidal Fucales. *J. exp. mar. Biol. Ecol. 29*: 181–195.

Strömgren, T. 1986. Annual variation in growth rate of perennial littoral algae from the west coast of Norway. *Aquat. Bot. 23*: 361–369.

Thom, R. M. 1983. Spatial and temporal patterns of *Fucus distichus* ssp *edentatus* (de la Pyl.) Pow. (Phaeophyceae: Fucales) in Central Puget Sound. *Bot. Mar. 26*: 471–486.

Topinka, J. A. 1978. Nitrogen uptake by *Fucus spiralis* (Phaeophyceae). *J. Phycol. 14*: 241–247.

Topinka, J. A. and J. V. Robbins. 1976. Effects of nitrate and ammonium enrichment on growth and nitrogen physiology in *Fucus spiralis. Limnol. Oceanogr. 21*: 659–664.

Wachenfeldt, T. von. 1975. The marine benthic algae and the environment in the Öresund I–III, Ph. D. thesis. Dept of Marine Botany, Lund. 328 pp.

Waern, M. 1952. Rocky-shore algae in the Öregrund archipelago. *Acta Phytogeogr. Suec. 30*: 1–298.

Wallentinus, I. 1979. Environmental influences on benthic macro-vegetation in the Trosa-Askö area, Northern Baltic Proper. II. The ecology of macroalgae and submerged phanerogams. *Contr. Askö Lab., Univ. Stockholm, Sweden 25*: 1–210.

Wallentinus, I. 1981. Chemical constituents of some Baltic macroalgae in relation to environmental conditions. In: (T. Levring, ed.) *Proc. of Xth International Seaweed Symposium.* de Gruyter, Berlin; New York. pp. 363–370.

Wallentinus, I. 1984. Comparisons of nutrient uptake rates for Baltic macroalgae with different thallus morphologies. *Mar. Biol. 80*: 215–225.

Botanica Marina
Vol. 34, pp. 455–457, 1991

Phlorotannins from the Brown Alga *Landsburgia quercifolia* (Hook. *fil. et* Harv.) Harv.

S-M. Li and K.-W. Glombitza[1]

Institut für Pharmazeutische Biologie, Universität Bonn, Nußallee 6, W-5300 Bonn 1, Fed. Rep. of Germany

(Accepted 1 July 1991)

Abstract

Using HPLC ten phlorotannins belonging to the group of fuhalols were isolated from an ethanolic extract of *Landsburgia quercifolia* as peracetates after acetylation. The structures of the isolated compounds were established by means of spectroscopic analysis, *i.e.* ^{1}H NMR spectra and direct comparison with spectra of authentic substances.

Similar to *Sargassum muticum*, the alga *Landsburgia quercifolia* contains exclusively phlorotannins of fuhalol-type.

Introduction

Brown algae often contain phlorotannins, oligomers or polymers of phloroglucinol (Ragan and Glombitza 1986).

As part of our investigations on phlorotannins in algae, we have studied phenolic constituents of the brown alga *Landsburgia quercifolia* (Cystoseiraceae) growing on coasts north of Banks Peninsula, New Zealand.

Czeczuga and Taylor (1987) reported on the detection of six carotenoids in acetone extract of *Landsburgia quercifolia.* The phenolic constituents have not yet been examined.

Material and Methods

The material examined was collected on 10 October 1987 at Great Barrier Island, New Zealand (voucher specimen deposited in the herbarium at University of Auckland, New Zealand, Department of Botany).

The ^{1}H NMR spectra (90 MHz) were recorded with a Bruker WH 90 using solvents as internal standards.

Extraction and isolation

Using the method described by Glombitza *et al.* (1978), 9.5 kg of frozen thallus of *Landsburgia quercifolia* (Hook. *fil. et* Harv.) Harv. were extracted with 20 L EtOH (96%). After concentration the aqueous solution was extracted with petrol, $CHCl_3$ and EtOAc, respectively. After evaporation of the solvent under reduced pressure, the residue of the EtOAc phase (50 g) was immediately acetylated with Ac_2O/pyridine (3 : 2, V/V). The crude acetylated phenols (50 g) dissolved in $CHCl_3$ were treated with a mixture of Et_2O and petrol (1 : 1). The polymeric phlorotannins were precipitated almost completely and removed by filtration. The yield was 17 g of low molecular weight oligomers. Thin layer chromatography (silica gel F_{254}, $CHCl_3/Me_2CO$, 9 : 1) of this sample showed eight spots which became red-violet on spraying with vanillin/H_2SO_4 (105 °C, 5–10 min). Portions of 500 mg were chromatographed on a silica gel (Merck, 230–400 mesh ASTM) column (4 × 14 cm) with a gradient: n-hexane, n-hexane/$CHCl_3$, $CHCl_3$, $CHCl_3$/Me_2CO and $CHCl_3$/MeOH. Five fractions were obtained; the first did not contain phenols. The last four fractions were further separated by HPLC (2 Knauer HPLC pumps, FR 30, gradient former) on LiChrosorb SI 60 (7 μm, 25 cm × 8 mm) with $CHCl_3$/

[1] Author to whom correspondence should be addressed.

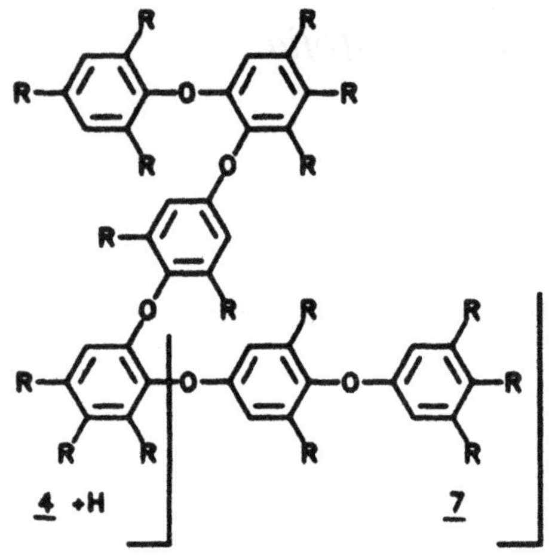

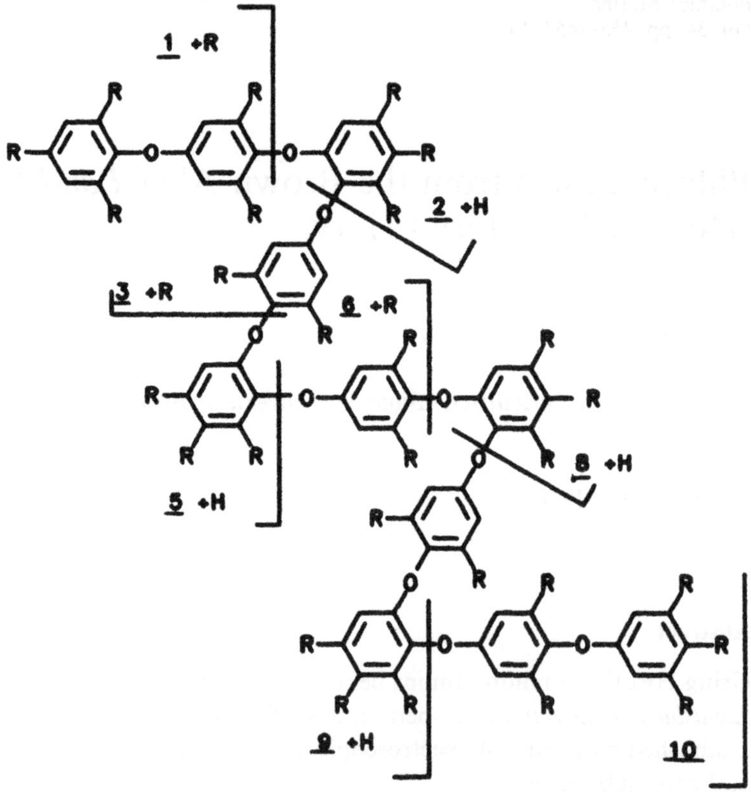

R=OAc

Isolated compounds

No.	peracetates of	yield (mg)	references of the first isolation
1	bifuhalol	2	Glombitza and Rösener, 1974
2	trifuhalol-A	5	Glombitza and Sattler, 1973
3	tetrafuhalol-A	5	Glombitza *et al.*, 1976
4	tetrafuhalol-B	5	Koch *et al.*, 1981
5	pentafuhalol-A	90	Glombitza *et al.*, 1976
6	hexafuhalol-A	1	Glombitza *et al.*, 1982
7	hexafuhalol-B	1.5	Glombitza and Li, 1991
8	heptafuhalol-A	12	Sattler *et al.*, 1977
9	nonafuhalol-A	62	Koch *et al.*, 1981
10	undecafuhalol-A	11	Koch *et al.*, 1981

EtOH gradients and detected at UV 275 nm (Knauer spectrometer and Shimadzu photodiode array UV-Vis detector).

Results and Discussion

The low molecular fraction achieved by the method described under Extraction and Isolation was pre-separated with liquid-vacuum-column-chromatography on silica gel. The fractions obtained were further purified on HPLC and compounds *1 – 10* (Figure 1) were isolated.

The structures of the compounds were elucidated by comparison with the spectra of authentic compounds. The ten substances contain two to eleven phloroglucinol units and belong exclusively to the group of fuhalols. Hexafuhalol-B octadecaacetate had been found in *Carpophyllum maschalocarpum* (Turn.) Grev. (Glombitza and Li 1991) and undecafuhalol-A octacosaacetate in *Bifurcaria bifurcata* Ross (Koch *et al.* 1981); the others have been isolated from various algae (Ragan and Glombitza 1986). With the exception of *4* and *7*, the compounds form a homologous series from bifuhalol consisting of two phloroglucinol units up to undecafuhalol-A eleven units.

Fuhalols are typical phlorotannins of the order Fucales. They have been found together with phloretholс, fucols and fucophlorethols in some genera (*Cystoseira*, *Bifurcaria*) (Glolbitza *et al.* 1976, Glombitza *et al.* 1978, Glombitza *et al.* 1981, Koch *et al.* 1981), alone (*Sargassum*) (Glombitza *et al.* 1982) or nearly alone (*Halidrys*) (Glombitza *et al.* 1980, Sattler *et al.* 1977).

In contrast to the other Cystoseiraceae (*e. g. Bifurcaria* and *Cystoseira*), which also contain other types of phlorotannins besides fuhalols, *Landsburgia quercifolia* contains only fuhalols. This is similar to *Sargassum muticum* (Yendo) Fensholt (Sargassaceae).

Acknowledgements

This research was supported by the Ministerium für Wissenschaft und Forschung, NRW. We thank the Central Analytic Laboratory of the Chemical Institutes, University of Bonn for the MS and NMR spectra.

References

Czeczuga, B. and F. J. Taylor. 1987. Carotenoid content in some species of the brown and red algae from the coastal area of New Zealand. *Biochem. Syst. Ecol. 15*: 5–8.

Glombitza, K.-W., M. Forster and W. F. Farnham. 1982. Antibiotics from algae – Part 25. Polyhydroxyphenyl ethers from the brown alga *Sargassum muticum* (Yendo) Fensholt, part II. *Bot. Mar. 25*: 449–453.

Glombitza, K.-W., C. Geisler and G. Eckhardt. 1980. Polyhydroxyphenyl ethers from Phaeophyceae *Halidrys siliquosa* (III). *Bot. Mar. 23*: 735–740.

Glombitza, K.-W. and S.-M. Li. 1991. Hydroxyphlorethols from the brown alga *Carpophyllum maschalocarpum. Phytochemistry 30*: 2471–2475.

Glombitza, K.-W. and H.-U. Rösener. 1974. Bifuhalol: ein Diphenyläther aus *Bifurcaria bifurcata. Phytochemistry 13*: 1245–1247.

Glombitza, K.-W., H.-U. Rösener and M. Koch. 1976. Polyhydroxyoligophenyle aus *Bifurcaria bifurcata. Phytochemistry 15*: 1279–1281.

Glombitza, K.-W. and E. Sattler. 1973. Trifuhalol, ein neuer Triphenyldiäther aus *Halidrys siliquosa* (1). *Tetrahedron letters 43*: 4277–4280.

Glombitza, K.-W., C. Schnabel and M. Koch. 1981. Antibiotica aus Algen, 27. Niedermolekulare Phlorotannine der Braunalga *Cystoseira baccata* (Gmelin) Silva, Teil II. *Arch. Pharm.* (Weinheim, Ger) *314*: 602–608.

Glombitza, K.-W., G. Wiedenfeld and G. Eckhardt. 1978. Antibiotica aus Algen, XX. Niedermolekulare Phlorotannine aus *Cystoseira baccata. Arch. Pharm.* (Weinheim, Ger) *311*: 393–399.

Koch, M., K.-W. Glombitza and H.-U. Rösener. 1981. Polyhydroxyphenyl ethers from *Bifurcaria bifurcata. Phytochemistry 20*: 1373–1379.

Ragan, M. A. and K.-W. Glombitza. 1986. Phlorotannins, brown algal polyphenols. *In*: (Round and Chapman, eds) *Progress in Phycological Research*, Vol. 4. Biopress Ltd, Bristol. pp. 129–241.

Sattler, E., K.-W. Glombitza, F. W. Wehrli and G. Eckhardt. 1977. Antibiotica aus Algen – XVI. Polyhydroxyphenyläther aus der Phaeophycee *Halidrys siliquosa. Tetrahedron 33*: 1239–1244.

Botanica Marina
Vol. 34, pp. 459–464, 1991

New Records of Marine Algae from Perú

J. A. West

Department of Plant Biology, University of California, Berkeley, California 94720, U.S.A.

(Accepted 2 July 1991)

Abstract

Bostrychia pinnata, *Caloglossa ogasawaraensis*, *C. stipitata* and *Halochlorococcum operculatum* are new records of marine algae from Perú that were collected from the mangroves of the Rio Tumbes estuary near Puerto Pizarro, Tumbes Province on 10 February 1990. All species were isolated into unialgal culture. *Caloglossa ogasawaraensis* males from Perú hybridized successfully with *C. ogasawaraensis* females from Ilha do Cardoso, São Paulo, Brazil producing viable tetrasporophytes in culture.

Introduction

The marine algae of Perú have received moderate interest since the first collections of Humboldt and Bonpland in 1802. Howe (1914), Dawson *et al.* (1964) and Acleto (1973) provided the most extensive floristic treatments. The Rio Tumbes estuary in Tumbes Province provides a biologically rich and diverse environment with many mangroves. These areas currently are semi-protected as part of a national park but the systematics and ecology of the marine algae there generally have been neglected. As in neighboring Ecuador, the Tumbes region is under development for shrimp mariculture with about 3000 hectares of ponds presently in production. Brief visits such as mine do not offer the long-term opportunities necessary for critical analysis of the biota and its ecological complexity. Marine algae new to Perú collected from the mangroves *Rhizophora* and *Avicennia* in several lateral channels of the main estuary of Rio Tumbes near Puerto Pizzaro (Fig. 13) on 10 February 1990 are listed.

Material and Methods

All species were collected and maintained alive in 15 mL plastic screw-cap centrifuge tubes with daily changes of sterile seawater. On return to Berkeley those were subdivided for herbarium specimens and for unialgal culture as described in West and Calumpong (1988). The culture isolates have been deposited with the Culture Collection of Algae, Department of Botany, University of Texas, Austin, TX 78713-7640, U. S. A. Specimens of each collection are deposited in the University of California Berkeley Herbarium (UC).

Results and Discussion

Division Rhodophyta

The red algae *Bostrychia*, *Caloglossa*, *Catenella* and *Stictosiphonia* are common components of the algal community that occurs on mangroves around the world. However, *Catenella* and *Stictosiphonia* are absent from the mangroves in Perú (Dawson *et al.* 1964, Acleto 1973).

Genus *Bostrychia* Montagne (Rhodomelaceae, Ceramiales)

In a recent monograph by King and Puttock (1989) *Bostrychia* was separated into two genera: *Bostrychia* with two tier cells per axial cell and *Stictosiphonia* with three to five tier cells per axial cell.

Bostrychia radicans (Montagne) Montagne. This was the only species of the genus recorded in Perú by earlier investigators, having been collected by Coker (Howe 1914) from Tumbes and by Dawson *et al.* (1964) from Talara. It is the most abundant red alga on *Rhizophora* prop roots in the Tumbes area.

Bostrychia pinnata J. Tanaka *et* Chihara. *Bostrychia pinnata* was described by Tanaka and Chihara (1984), based on specimens from Okinawa, Japan. Subsequently, it was observed in Australia by King and Puttock (1986) and West and Calumpong (unpubl.), in Indonesia by Tanaka and Chihara (1988) and in Singapore by West (1991). According to King and Puttock (1989), *B. pinnata* has been misidentified as *B. calliptera* (Montagne) Montagne in the Western Pacific. Uniform development of alternate lateral polysiphonous branches on every axial segment and the presence of peripherohaptera are features shared by both species, however, axial cortication and internal hyphae are present in *B. calliptera* and lacking in *B. pinnata*. The descriptions given in most works state that *B. calliptera* is corticated (*e.g.*, King and Puttock 1989, Cordeiro-Marino 1978, Schnetter and Bula Meyer 1982). The illustration of the type specimen of *B. calliptera* (Leprieur 355 and Montagne from French Guiana) in fig. 10c of King and Puttock (1989) clearly shows the cortication, however, the illustrations of *B. calliptera* collected from other localities show no cortication (*e.g.*, Cordeiro-Marino 1978, Schnetter and Bula Meyer 1982). By contrast, Itono (1986) clearly illustrates a well corticated *B. calliptera* from Ishigaki I., Okinawa, Japan. It appears that the exact morphological differences between *B. pinnata* and *B. calliptera* and their exact geographic distribution require reinvestigation. It is probable, too, that misidentifications have occurred. On *Rhizophora mangle* L. prop roots *B. pinnata* was mixed with *B. radicans*, *Caloglossa leprieurii* (Montagne) J. Agardh, *C. ogasawaraensis* Okamura and *C. stipitata* Post. All *B. pinnata* plants (Fig. 1) were non-reproductive, whereas *B. radicans* and *C. leprieurii* were frequently tetrasporic. One male and one female plant of *B. radicans* were also observed. *Bostrychia* was quite dry when collected, having been exposed to the air for eight hours during the low tide, and was also subjected to prolonged drying (1 – 2 h) in the laboratory without any apparent physiological damage. In culture, *B. pinnata* grows well but does not reproduce by spores. The morphology is typical, up to 1.5 cm long, with regular alternate branches (Fig. 2). About 5 mm from the main axis apex, lateral branch abscission occurs when one or two cells just above the base die (Fig. 3 arrowhead). Any water motion dislodges these diciduous vegetative propagules that regenerate new plants.

Genus *Caloglossa* J. Agardh (Delesseriaceae, Ceramiales)

Although Post (1943) treated this genus in great detail, the morphological variation of its many species make it difficult to identify species with certainty.

Caloglossa ogasawaraensis Okamura. The first record in Pacific South America of *C. ogasawaraensis* is a brief comment (footnote 2, p. 202) in Post (1943) concerning its presence in Ecuador although she provided no figures to document her record. In São Paulo, Brazil, Ugadim (1976) established the first report of *C. ogasawaraensis* on the Atlantic coast of South America. Braga *et al.* (1990) made a complete phenological study of this species at Ilha do Cardoso, São Paulo. These are all reports from the mangrove environments, but Soto (1982) observed it growing in abundance with aquatic bryophytes in a waterfall of the Rio El Salto near Heredia, Costa Rica. Soto's figures indicate that this material is similar to narrow forms of *Caloglossa leprieurii*. Schnetter and Bula Meyer (1982) list one collection of this species on wooden posts of a dock in Bahia de Buena Ventura Colombia.

At Puerto Pizarro *Caloglossa ogasawaraensis* is uncommon but intermixed with *C. leprieurii*, *C. stipitata*, *Bostrychia radicans* and *B. pinnata* on *Rhizophora* prop roots. The field-collected plants were narrow with many blades at the nodes.

The cultured plants (Fig. 4) are similar to field plants, having narrow vegetative blades with a midrib comprised of an axial cell and four pericentral cells (60 – 70 μm diam.). Alar cells are absent in the sector 1 – 3 mm above each node (Fig. 5), but above this area 3 – 5 alar cells develop on each side of the midrib widening the blade to 150 – 310 μm (Fig. 6). Each node initially forms the main axial blade, 1 or 2 lateral blades in 1 or 2 rhizoids, but secondary blades and rhizoids arise repeatedly near the bases of the first blades. The plant isolated into culture (3047) is a male that develops spermatangial sori on the widest sectors of the distal blades (Fig. 7). After spermatial discharge is completed, the sorus part of the blade dies and abscises. New secondary blades develop from pericentral cells and alar cells at the point of abscission on the blade (Fig. 8). Tetrasporophytes and female gametophytes were not observed in the field-collected plants. Crosses were successful with a 3047 male and a 3018 female from Ilha do Cardoso, São Paulo, Brazil producing viable carpospores and developing tetrasporophytes that also produced viable tetrasporangia.

Figs 1–6.
Fig. 1. *Bostrychia pinnata*. Field-collected plant. Fig. 2. *B. pinnata*. Apex of cultured plant with alternate branching from each axial cell. Fig. 3. *B. pinnata*. Lower axis of cultured plant with abscission of lateral branches caused by death of cells near base (arrow).
Fig. 4. *Caloglossa ogasawaraensis*. Habit of cultured plant with multiple blade and rhizoid formation from each node. Fig. 5. *C. ogasawaraensis*. Node with only pericentral cells formed by first 4–5 axial cells. Repeated formation of secondary blades from pericentral and marginal cells of the blades above the node. Three rhizoidal cells also derived from pericentral cells of the node. Use scale of Fig. 3. Fig. 6. *C. ogasawaraensis*. Second order blades with 4–5 alar cells. Use scale in Fig. 3.

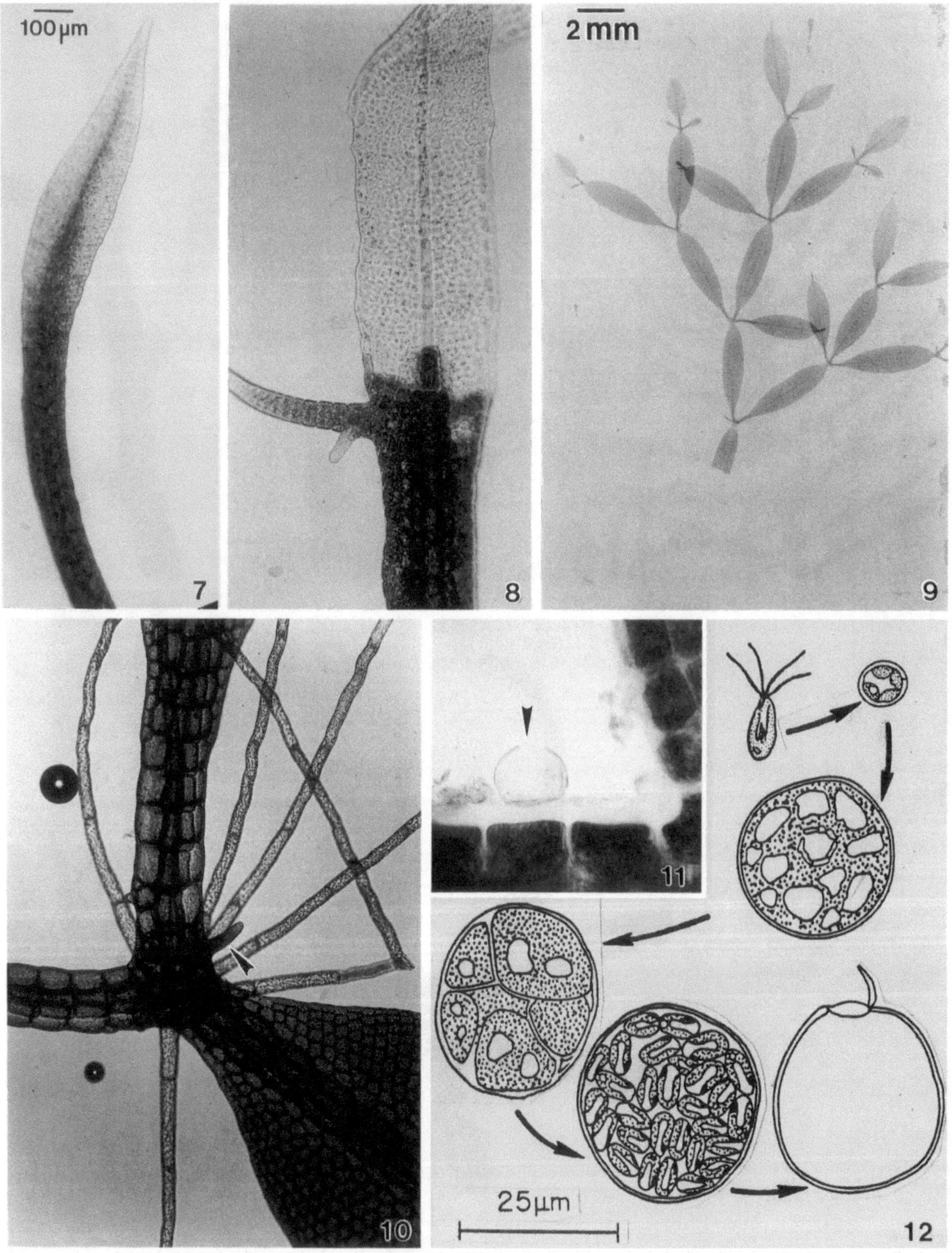

Figs 7–12.

Fig. 7. *C. ogasawaraensis.* Terminal blade with spermatangial sorus on both sides of axial cell. Fig. 8. *C. ogasawaraensis.* After spermatial release, sorus cells die and adventitous blades develop from marginal cells. Use scale in Fig. 3. Fig. 9. *C. stipitata.* Habit of cultured plant.

Fig. 10. *C. stipitata.* Node with 6 rhizoids, one developing lateral blade and one repressed lateral (arrowhead). Use scale in Fig. 7. Fig. 11. *Halochlorococcum operculatum.* Empty sporangial cell with discharge pore (arrowhead) on *B. radicans.* Fig. 12. *H. operculatum.* Zoospore, young cell, developing cell, developing sporangium, mature sporangium and empty sporangium with pore and operculum.

Caloglossa stipitata Post. Within an area of about 1 hectare three different collections were made of *Caloglossa* that had the characteristic stipitate nodes, wide internodes and rounded apices of *C. stipitata* and these were intermixed with *C. leprieurii*. One collection (3048) was isolated into culture and has continued vegetative growth in this stipitate form (Fig. 9). The distance between nodes is 4.0 – 4.5 mm and the widest part of each internodal blade is 970 – 1125 µm. Two opposite lateral blades and 5 – 7 rhizoids arise from the pericentral cells at each node, although sometimes the complete development of one or both laterals is suppressed (Fig. 10) so that a pseudo-dichotomous or linear unbranched appearance is evident. The first to sixth axial cells at each node usually develop only pericentral cells. This isolate has remained vegetative like the *C. stipitata* isolate (2992) from Singapore (West 1991). Post (1966) recorded *C. stipitata* in Florida, the only known citation for North America (Wynne 1986). For South America Schnetter and Bula Meyer (1982) recorded *C. beccarii* (Zanardini) De Toni and *C. stipitata*, two morphologically similar species, from several localities in Colombia. The apices of *C. beccarii* are considered to be gradually attenuate whereas those of *C. stipitata* are thought to be abruptly attenuate or rounded. Blade apex form appears in culture to be a feature dependent upon growth conditions, *i.e.*, during slow growth the apices become more rounded. King (1990) and others question whether or not *C. beccarii* and *C. stipitata* are taxa distinct from *C. leprieurii*. Until DNA-based molecular systematics and standard hybridization methodologies are applied this matter remains unresolved.

Division Chlorophyta

Genus *Halochlorococcum* Dangeard (Chlorococcaceae, Chlorococcales)

Halochlorococcum includes five species widely distributed in marine and estuarine habitats. They are either epiphytic, endophytic or free-living on various substrates (Kornmann and Sahling 1983, West *et al.* 1988).

Halochlorococcum operculatum Kornmann *et* Sahling. This alga, isolated as culture number (3050), was present as an epiphyte on the branches of a male *Bostrychia radicans* collection (3044). The vegetative cells were superficial on the *Bostrychia* branches in contrast to *H. porphyrae* (Setchell *et* Gardner) West and *H. moorei* (Gardner) Kornmann *et* Sahling which were embedded partially in the host (*Porphyra* and *Blidingia* respectively) extracellular matrix. This isolate grows vigorously in unialgal culture. Mature vegetative cells are spherical (21 – 25 µm diam.) when growing separately but become polygonal when aggregated. The parietal to central pyrenoid is 4.5 – 6.0 µm diam. and embedded in the parietal reticulate chloroplast. The larger cells become sporangia and produce about 32 zoospores that are quadriflagellate, elliptical (3.5 – 4.5 µm wide and 6 – 7 µm long) with a chloroplast with an anterio-lateral stigma. The spores are released through an apical circular pore (4.5 – 5.5 µm diam.) with an operculum that remains connected at one edge (Fig. 11). The empty sporangium has a wall about 2.0 – 3.0 µm thick. Figure 12 depicts most of the stages described above. Spores settle quickly within 10 – 20 minutes and become spherical, about 5.0 – 6.0 µm diam. The zoospores settle freely in aggregates on either the vessel bottom or on the side at the air-water interface nearest the light source, indicating that they are positively phototactic. This is the first record of *Halochlorococcum* in South America but it is probably more widespread.

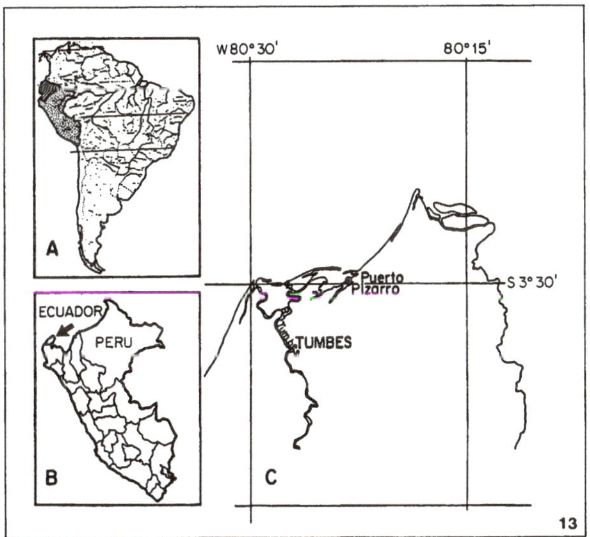

Fig. 13. A. Map of South America with Perú (stippled area) and Ecuador (lined area). B. Map of Perú showing Tumbes Province (arrow). C. Detail map of Tumbes River Estuary and Puerto Pizarro.

Acknowledgements

Partial travel funds were provided by the University of California Committee on Research. Valentin Santos and Francisco Rodriguez assisted with the collections. Dr Michael Wynne carefully reviewed the manuscript.

References

Acleto, C. 1973. Las algas marinas del Peru. *Boletin de la Sociedad Peruana de Botanica 6*: 1–164.

Braga, M., M. Fujii, N. Yokoya, V. de Eston, E. Plastino and M. Cordeiro-Marino. 1990. Macroalgal reproductive patterns in mangroves of Ilha do Cardoso, São Paulo, Brazil. *II Simpósio de Ecossistemas da Costa Sul e Sudeste Brasileira.* Estrutura, Funçáo e Manejo. Academia de Ciencias do Estado de São Paulo. Publicação ACIESP No. 71-2. pp. 209–217, 445–452.

Cordeiro-Marino, M. 1978. Rodoficeas bentonicas marinhas do Estado de Santa Catarina. *Rickia 7*: 1–243.

Dawson, E. Y., C. Acleto and N. Foldvik. 1964. The seaweeds of Perú. *Beihefte zur Nova Hedwigia 13*: 111 pp. + 81 pl.

Howe, M. A. 1914. The marine algae of Peru. *Memoirs Torrey Botanical Club 15*: 1–185.

Itono, H. 1986. New records of marine algae from the southern parts of Japan. *Japanese Journal of Phycology 34*: 74–82.

King, R. J. and C. F. Puttock. 1986. *Bostrychia pinnata* J. Tanaka *et* Chihara in Australia. *Bulletin of the National Science Museum. Tokyo*, Series B, *12*: 17–24.

King, R. J. and C. F. Puttock. 1989. Morphology and taxonomy of *Bostrychia* and *Stictosiphonia* (Rhodomelaceae/Rhodophyta) *Australian Systematic Botany 2*: 1–73.

King, R. J. 1990. Macroalgae associated with the mangrove vegetation of Papua New Guinea. *Bot. Mar. 33*: 55–62.

Kornmann, P. and P.-H. Sahling. 1983. Meeresalgen von Helgoland: Ergänzung. *Helgoländer Meeresuntersuchungen 36*: 1–65.

Post, E. 1943. Zur Morphologie und Ökologie von *Caloglossa. Archiv für Protistenkunde 96*: 123–220.

Post, E. 1966. *Caloglossa stipitata* in Florida. *Hydrobiologia 27*: 109–112.

Schnetter, R. and G. Bula Meyer. 1982. Marine Algen der Pazifikküste von Kolumbien, Chlorophyceae, Phaeophyceae, Rhodophyceae. *Bibliotheca Phycologica 60*: 1–287.

Soto, R. 1982. *Caloglossa ogasawaraensis* Okamura ["*ogassuarensis* Skuja"] (Rhodophyta, Ceramiales, Delesseriaceae) en Costa Rica. *Brenesia 19/20*: 251–253.

Tanaka, J. and M. Chihara. 1984. Taxonomic studies of Japanese mangrove macroalgae I. Genus *Bostrychia* (I) (Ceramiales, Rhodophyceae). *Bulletin of the National Science Museum Tokyo, Series B 10*: 169–176.

Tanaka, J. and M. Chihara. 1988. Macroalgae in Indonesian mangrove forests. *Bulletin of the National Science Museum Tokyo, Series B 14*: 93–106.

Ugadim, Y. 1976. Ceramiales (Rhodophyta) do litoral sul do estado de São Paulo, e do litoral sul do estado de Parana (Brasil). *Boletim do Botanica de Universidade São Paulo 4*: 133–172.

West, J. 1991. New algal records from the Singapore mangroves. *Gardens' Bulletin Singapore*; in press.

West, J., C. Smith and D. McBride. 1988. Observations on the marine unicellular endophyte *Chlorochytrium porphyrae* (Chlorophyceae). *Bot. Mar. 31*: 299–305.

West, J. and H. Calumpong. 1988. Mixed-phase reproduction of *Bostrychia* (Ceramiales, Rhodophyta) in culture. *I. B. tenella* (Lamouroux) J. Agardh. *Japanese Journal of Phycology 36*: 292–310.

Botanica Marina
Vol. 34, pp. 465–467, 1991

The Influence of Cytokinins on the Growth of *Macrocystis pyrifera*

R. de Nys, P. E. Jameson[1], and M. T. Brown

Botany Department, University of Otago, P.O. Box 56, Dunedin, New Zealand

(Accepted 1 July 1991)

Abstract

Discs, cut from blades of *Macrocystis pyrifera* (L.) C. Ag., were cultured in the presence of the cytokinins zeatin (Z) and isopentenyladenine (iP). Both cytokinins stimulated the growth of *M. pyrifera* discs as was shown by increases in disc diameter and dry weight/area. Maximum increase in disc diameter was promoted by similar concentrations of Z and iP (0.05 mg L^{-1}).

Introduction

While numerous authors have suggested that the cytokinins are involved in the regulation of algal growth (for review see Mooney and van Staden 1986), relatively few studies have been carried out on the role of cytokinins as endogenous growth regulators in the algae and in particular in the macroalgae.

We have shown that the cytokinin-like compounds in *Macrocystis pyrifera* (L.) C. Ag. vary on both a quantitative and qualitative basis within the plant and throughout the seasons (de Nys *et al.* 1990) and that this variation can be correlated with growth. To further test the hypothesis that the cytokinins act as endogenous growth regulators in this alga we have investigated the response of *M. pyrifera* to applied cytokinins.

Increased growth in brown algae in response to exogenous application of cytokinins has previously been demonstrated [*e.g.* see review by Mooney and van Staden (1986)]. The majority of these responses are to the synthetic cytokinin, kinetin, but Pedersén (1973) showed that addition of iP, which had been identified from seawater samples collected from the *Fucus-Ascophyllum* zone, could also increase the growth of *Ectocarpus confervoides* (Roth) Le Jolis and *Pylaiella littoralis* (L.) Kjellm. in artificial seawater.

We wished to test the response, in culture, of *M. pyrifera* to cytokinins which have been identified as part of the endogenous cytokinin complex within this seaweed. Zeatin has been identified by combined gas chromatography–mass spectrometry in *M. pyrifera*, while iP has been detected by radioimmunoassay (de Nys *et al.* 1990).

Consequently, in these experiments, we investigated the effect of the cytokinins zeatin (Z) and isopentenyladenine (iP) on the growth of *M. pyrifera*. The optimum concentration for growth was established and compared with endogenous levels previously measured within the alga (de Nys *et al.* 1990).

Material and Methods

Whole plants of *Macrocystis pyrifera* were collected from Otago Harbour, Dunedin, New Zealand in September and October 1986. Discs, 20 mm in diameter, were removed from the base of the first and second free vegetative blades below the apical scimitar. All cultures were incubated in an aerated one litre solution of synthetic seawater (Sigma sea salts, 40 g L^{-1}, Sigma Chemical Co.) enriched with Provasoli's medium (Provasoli 1968) and containing zeatin (Z) (experiments one and two) or isopentenyladenine (iP) (experiments one, two and three) at concentrations of 0.01, 0.05, 0.10, 0.50 mg L^{-1} (iP or Z) and 1.00 mg L^{-1} (zeatin only). Controls without the addition of

[1] To whom reprint requests should be addressed.

plant growth substances were run simultaneously. Three (experiment one) or six discs (experiments two and three) were cultured per flask at 10 °C under a 24 h light regime of 400–500 $\mu E\ m^{-2}\ s^{-1}$. Solutions were changed every three days and the disc diameter measured. At the termination of experiments one and two, the discs were oven dried (70 °C) and total dry weight per flask recorded. At the termination of experiment three, chlorophyll a, c, fucoxanthin and total pigment concentration were analysed using a modification of the extraction procedure of Seeley *et al.* (1972) which incorporated dimethylsulphoxide. Experiment one was terminated after six days. Discs were measured from experiment two after 9 and 12 days and experiment three after three, six, nine, 12 and 18 days. Initial disc diameter in all experiments was 20 mm.

Results were analysed by ANOVA, and Duncan's new multiple range test was used to determine the significance of differences in response to different cytokinin concentrations. An analysis of co-variance and an approximate Duncan's new multiple range test were used to determine whether the slopes of the regression of disc diameter against time were significantly different in experiment three. The significance of the relationship between dry weight/area and cytokinin concentration was determined using the combined data from experiments one and two and utilising a common regression which accounted for differences between the treatments. The control (zero cytokinin) was treated as 0.001 mg L^{-1} for the purpose of the regression. The analyses of variance, co-variance and regression were calculated with the aid of the general statistical programme Teddybear (Wilson 1975).

Results and Discussion

Stimulation of the growth of discs of *M. pyrifera* depended on concentration and cytokinin type.

Isopentenyladenine significantly stimulated growth in terms of disc diameter in all three experiments (Tables I and II). In experiment three an analysis of regression slopes (diameter/day, Table II) for the concentrations of iP showed significantly greater rates of growth at concentrations of 0.05 and 0.50 mg L^{-1}, both slopes being significantly different from each other with optimal growth at 0.05 mg L^{-1} iP. After 18 days growth, variation in the diameter of the discs at a single concentration became such that differences were no longer statistically significant. The decreased growth response observed in the 0.01 mg L^{-1} iP-supplemented medium was probably a consequence of a slower initial growth affecting all subsequent measurements at that concentration.

Table I. The effect of zeatin and isopentenyladenine on the growth increment of discs of *M. pyrifera*. Each disc was measured after six (experiment 1) or nine and 12 days (experiment 2) incubation. The values represent the average diameter of the three (experiment 1) or six discs (experiment 2) in each flask.

mg L^{-1}	Disc diameter (mm)		
	Experiment One	Experiment Two	
	day 6	day 9	day 12
Zeatin			
0.00	24.0^{abc}	22.3^{ab}	22.4^{a}
0.01	24.6^{ab}	21.8^{a}	*
0.05	25.3^{a}	25.8^{d}	28.3^{b}
0.10	23.0^{bc}	25.0^{cd}	22.3^{a}
0.50	22.8^{c}	23.8^{bc}	24.8^{ab}
1.00	24.5^{a}	22.3^{a}	21.3^{a}
Isopentenyladenine			
0.00	23.6^{a}	22.3^{a}	22.4^{a}
0.01	23.6^{a}	23.5^{ab}	25.5^{ab}
0.05	25.3^{ab}	26.2^{c}	30.7^{c}
0.10	26.0^{b}	24.6^{bc}	26.5^{b}
0.50	24.3^{ab}	23.0^{ab}	25.2^{ab}

* discs disintegrated

[a] Values, which are not separated by a horizontal line and which are follwed by the same letter are not significantly different from one another as assessed by Duncan's new multiple range test at $P \leq 0.05$.

Zeatin also significantly stimulated growth with the apparent optimal response at 0.05 mg L^{-1} in experiment one confirmed in experiment two (Table I). A decreased response to concentrations greater than 0.05 mg L^{-1} was noted in both experiments.

Maximum increases in disc diameter were promoted by similar concentrations of zeatin and iP (0.05 mg L^{-1}, Table I). This is in contrast to the observations of Kentzer *et al.* (1980) who showed that while maximum adventitious branching of thallus fragments of *Fucus vesiculosus* L. was promoted by a comparable level of zeatin (0.1 mg L^{-1}), maximum activity of iP was at a considerably lower concentration (0.001 mg L^{-1}). However, Pedersén (1973) reported that iP promoted maximum growth of both *Pylaiella littoralis* and *Ectocarpus confervoides* at a concentration of 0.04 mg L^{-1}.

Dry weights were recorded to ascertain if the observed increases in disc diameter could be attributed to true growth responses. The data from experiments 1 and 2 were analysed together. As there was a significant difference in the elevation of the regressions (the dry weights/area of the discs incubated in iP for 12 days being significantly lower than those incubated in iP for 6 days or Z for either 6 or 12 days) a common regression was used to eliminate the difference between experiments and emphasise the overall trend. This regression of weight per unit area against concentration shows a highly significant positive corre-

Table II. The effect of isopentenyladenine on the growth increment of discs of *Macrocystis pyrifera*. Each disc was measured every three days over an 18 day incubation period. The values represent the average diameter of the six discs in each flask.

mg L^{-1}	Experiment Three Disc diameter (mm) day 3	6	9	12	15	18	Regression slopes (diam/day)
Isopentenyladenine							
0.00	22.0[a]	24.8[a]	28.8[ab]	32.7[ab]	36.0[a]	40.0[ab]	1.14[ab]
0.01	22.6[ab]	25.5[a]	29.0[ab]	31.5[a]	33.8[a]	38.7[a]	1.01[a]
0.05	22.8[b]	26.8[b]	32.5[c]	37.8[c]	42.3[b]	46.2[b]	1.53[c]
0.10	22.2[a]	24.9[a]	28.3[a]	30.5[a]	33.3[a]	39.2[ab]	1.02[a]
0.50	23.0[b]	25.8[ab]	30.7[b]	35.8[bc]	41.7[b]	45.2[ab]	1.32[b]

[a] Values followed by the same letter within each column are not significantly different from one another as assessed by Duncan's new multiple range test ($p \leq 0.05$).

lation ($r = +0.7793$, $p < 0.001$) between dry weight/area and cytokinin concentration (Fig. 1). In the case of zeatin, a decrease in dry weight/area was observed at the highest concentration (1 mg L^{-1}).

After 18 days, discs cultured in the various iP supplemented media (experiment 3) showed no significant variation ($p > 0.05$) between treatments in either the chlorophyll a, c, fucoxanthin or total pigment concentration (data not presented).

The levels of cytokinins (0.05–0.50 mg L^{-1}) which significantly stimulated the growth of *M. pyrifera* in artificial culture corresponded to endogenous levels of cytokinin-like compounds detected in *M. pyrifera* (de Nys *et al.* 1990). For the purposes of making this comparison we have assumed that 1 g cellular fluid is approximately 1 mL.

The dry weight of the alga is approximately 10% of fresh weight (de Nys 1987) so that 9 g of the 10 g fresh weight extracted to determine the cytokinin content (de Nys *et al.* 1990) is cellular fluid. The concentration of endogenous cytokinins in the basal region of young blades harvested in October 1986 was *ca* 0.45 mg L^{-1} which corresponds with the concentrations of iP and Z which produced maximal levels of growth in the artificial growth media (*i.e.* 0.05–0.50 mg L^{-1}).

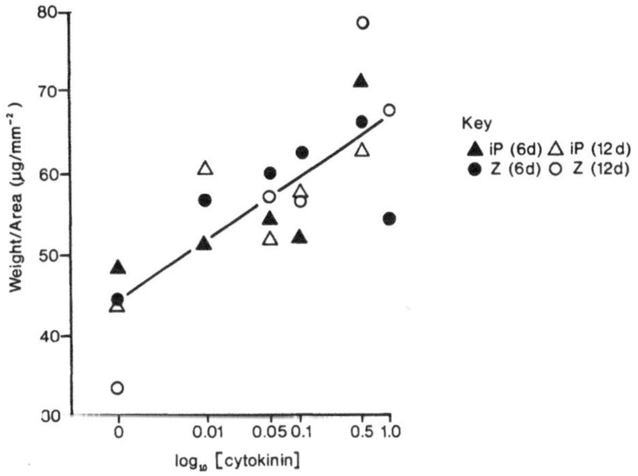

Fig. 1. The effect of increasing cytokinin concentration on the dry weight per unit area (µg mm^{-2}) of discs of *M. pyrifera* incubated in either isopentenyladenine (iP) or zeatin (Z) for 6 or 12 days. A common (within experiment) regression has been fitted to the data with the control (zero cytokinin) being treated as 0.001 mg L^{-1} for the purpose of the calculation. The regression ($y = 67.0 + 3.6x$) was highly significant ($R^2 = 59.3\%$, $P < 0.001$).

Acknowledgements

We wish to acknowledge the advice and assistance of Professor Peter Bannister especially with respect to the statistical analysis of results.

References

de Nys, R. 1987. An investigation of the cytokinins as endogenous growth regulators in *Macrocystis pyrifera*. M. Sc. Thesis. University of Otago, Dunedin, New Zealand.

de Nys, R., P. E. Jameson, N. Chin, M. T. Brown and K. J. Sanderson. 1990. The cytokinins as endogenous growth regulators in *Macrocystis pyrifera* (L.) C. Ag. (Phaeophyceae). *Bot. Mar. 33*: 467–475.

Kentzer, T., R. Synak, K. Burkiewicz and A. Banas. 1980. Cytokinin-like activity in sea water and *Fucus vesiculosis* L. *Biol. Plant 22*: 218–225.

Mooney, P. A. and J. van Staden. 1986. Algae and cytokinins. *J. Plant Physiol. 123*: 1–21.

Pedersén, M. 1973. Identification of a cytokinin, 6-(methyl-2-butenylamino) purine, in seawater and the effect of cytokinins on brown algae. *Physiol. Plant. 28*: 101–105.

Provasoli, L. 1968. Media and prospects for the cultivation of marine algae. *In*: (A. Watanabe, ed.) *Culture and Collection of Algae. Proceedings U.S.–Japan Conference, Hakone. September 1966. Japanese Society of Plant Physiologists*. pp. 63–75.

Seeley, G. R., M. J. Duncan and W. E. Vidaver. 1972. Preparative and analytical extraction of pigments from brown algae with dimethylsulphoxide. *Mar. Biol. 12*: 184–188.

Wilson, J. B. 1975. Teddybear – A statistical system. *N. Z. Stats. 10*: 36–42.

Botanica Marina
Vol. 34, pp. 469–473, 1991

The Effects of Aqueous Seaweed Sprays on the Growth of Maize

I. Jeannin, J.-C. Lescure* and J.-F. Morot-Gaudry*

Laboratoires Goëmar s. a., ZAC La Madeleine, BP 55, 35413 Saint-Malo Cedex, France

* *Laboratoire du Métabolisme et de la Nutrition des Plantes, I.N.R.A., 78026 Versailles Cedex, France*

(Accepted 4 July 1991)

Abstract

Seaweed extract (Goëmar GA 14), applied as a foliar spray increased the total fresh matter production of maize seedlings by 15 to 25% over the control. This was reflected in the increase of root and stem mass per plant. However, only minor differences were measured in stomatal resistance just after treatment and in net carbon assimilation rate two weeks after spraying. No other significant differences were observed in gaz exchanges characteristics. The physiological mechanisms responsible for the beneficial effects of seaweed concentrates on plants and the nature and mode of action of the compounds contained in these concentrates have not yet been elucidated.

Introduction

Seaweed extracts have been marketed for several years as fertilizers. The main seaweed extracts known to play a useful role in agriculture are: Maxicrop (Seaborn), Algifert (Marinure), Goëmar GA 14, Kelpak 66, Seaspray, Seasol, S. M. 3, Cytex and Seacrop 16. The diversity of commercial seaweed products has arisen not only from the use of different seaweed species, but also from the method of preparation, which varies according to the manufacturer.

Many different beneficial effects have been recorded for crops treated with seaweed extracts. As well as increased crop yields and improved quality of the products after harvest, seaweed extracts have been shown to ensure better seed germination, increase nutrient uptake, improve resistance of plants to unfavourable conditions of environment (frost, drought) and reduce incidence of fungal, insect and nematode attack (Jolivet *et al.* 1991). Seaweed extracts, typically used as a foliar spray, can increase the amount of limiting nutrient sufficiently to correct for a marginal deficiency (Abetz 1980).

Many reports on the effects of seaweed extracts in agriculture or horticulture have been based on commercial observations rather than properly conducted experiments. The major objective of this study was to determine the effects of the seaweed extract, Goëmar GA 14, on biomass production, photosynthesis and photosynthate partitioning patterns in young maize plants.

Material and Methods

Plant material

Maize (*Zea mays* L. cv DEA) seeds were germinated on sand for 7 d in a growth cabinet, then transplanted into pots of 1.3 L containing a 1 : 2 peat : argile mixture for an additional 5 weeks (until 11th leaf stage). The growth environment consisted of a dawn/day/dusk/night cycle of 20/25/22/17 °C, with dawn/day/dusk lengths of 2.5/7.5/6 h at 250/350/120 µmol $m^{-2}s^{-1}$ PAR respectively (cool-white fluorescent lamps, General Electric FG 6 PG 17/CW) and 70% RH. The nutrient solution (5.00 mM KNO_3, 3.00 mM $Ca(NO_3)_2$, 0.075 mM $Mg(NO_3)_2$, 0.65 mM $(NH_4)_2HPO_4$, 0.90 mM NH_4NO_3, 0.75 mM $MgSO_4$, and micronutrients, at pH 5.8) was automatically delivered at the rate of 100 mL every 30 min during the light period and 100 mL every 2 h during the night.

Treatments, sampling and assays

Three weeks after sowing, maize plants were separated according to their size into blocks each containing 16 to 20 uniform plants. Each block was sprayed with approximately 2 L of a treatment solution containing 0, 3.3 or 6.6 g seaweed cream GA 14 (*Ascophyllum nodosum* L. obtained from a patented process, Goëmar, Saint-Malo, France; Yvin *et al.*, 1989) per liter of demineralized water. After treatment the plants were randomly arranged in the growth chamber to minimize environmental differences.

Gas exchange measurements (CO_2 assimilation and transpiration rate, stomatal resistance) were begun the day following treatment and were conducted in the growth chamber using an infra-red gas analyser (LCA 2, Analytical Development Co., Hoddesdon-England). Thus, measurements were taken on days 1 (corresponding to emergence of the 5th leaf), 6 and 17 (appearance of the 9th leaf), between 9 and 12 a.m.

The plants were harvested 20 days after treatment and measurements of height, number of leaves and fresh weight of leaf, stalk (sheath + stem) and roots were recorded. Two experiments were conducted (Experiments I and II). Data represent the mean of 16 to 20 measurements.

Results

Seaweed extract (Goëmar GA 14) applied as a foliar spray at the concentration of 3.3 g/L, increased the total fresh matter production of maize seedlings by 15 to 25% over the control (Table I). The increase in biomass was manifested in both the roots and shoots, accompanied by a change in the shoot/root ratio favoring the roots. The increase in shoot mass was due to an increase in fresh stem weight per plant. The percentage of dry matter was proportionately similar in all three plant parts, suggesting that the increase in fresh weight is not attributable to a simple increase in water content. No noticeable effects were observed on either leaf number (data not shown) or plant height (Table I). Foliar application of 6.6 g/L seaweed extract had no beneficial effect on the growth of plants compared to the control (Table I). A third experiment done under slightly different conditions confirmed these results (data not shown).

A significant increase in stomatal resistance and consequently a decrease in transpiration rate was measured one day after treatment when 3.3 g/L Goëmar GA 14 was applied (Table II). Except for a decrease in the net carbon assimilation rate on day 17 after treatment, no other significant differences were observed in net carbon assimilation rate, transpiration and stomatal resistances when measured 1, 6 and 17 days after spraying.

Table I. Effects of GOEMAR seaweed extract on maize growth
1: Control; 2: seaweed extract (3.3 g/L); 3: seaweed extract (6.6 g/L).

		Height	Fresh Matter					Shoot/ Root
			Leaf	Stalk (stem + sheath)	Roots	Shoot	Total	
		cm	g	g	g	g	g	
Experiment 1								
1	mean	141.5	65.40 b	68.20 b	24.90 b	133.60 b	158.50 b	5.37
	std. dev	7.1	12.50	18.30	7.40	30.50	38.10	
2	mean	145.6	77.60 a	87.70 a	33.20 a	165.30 a	198.50 a	4.98
	std. dev	5.2	9.10	16.60	5.80	24.90	30.90	
			**	**	**	**	**	
Experiment II								
1	mean	135.19	55.96 b	56.24	22.47	112.20 b	134.67 b	4.99
	std. dev	9.87	9.06	12.01	4.89	20.40	24.68	
2	mean	136.94	64.31 a	64.91	26.09	129.22 a	155.31 a	4.95
	std. dev	10.15	11.90	14.97	5.53	26.20	31.02	
3	mean	134.75	57.31 b	58.66	23.19	115.97 ab	139.16 ab	5.00
	std. dev	8.11	13.16	16.37	5.25	29.04	34.00	
			*			*	*	

Significance level of variance analysis for $P < 0.05$ *, $P < 0.01$ **
a, b: classification by NEWMAN-KEULS test

Table II. Effects of GOEMAR seaweed extract on gas exchange characteristics
1: Control; 2: seaweed extract (3.3 g/L); 3: seaweed extract (6.6 g/L).

		Net Assimil. Rate µmol/m²/s		Transpiration Rate mmol/m²/s		Stomatal resistance 1/(mol/m²/s)		Height cm
Experiment I								
D + 1								
1	mean	4.80		3.03	a	9.15	b	74.00
	std. dev	0.95		0.48		2.15		3.27
2	mean	4.94		2.57	b	11.73	a	73.89
	std. dev	0.82		0.53		3.05		2.08
				***		***		
D + 6								
1	mean	5.06		4.27		4.81		94.27
	std. dev	1.61		0.49		1.33		2.86
2	mean	4.97		4.18		5.06		95.33
	std. dev	1.16		0.59		1.85		4.03
D + 17								
1	mean	6.96		4.79		3.72		131.50
	std. dev	1.09		0.56		1.20		5.45
2	mean	7.07		4.88		3.61		132.93
	std. dev	0.81		0.57		0.97		3.72
Experiment II								
D + 1								
1	mean	4.20		4.75	a	4.40	b	72.62
	std. dev	0.86		0.86		1.42		2.32
2	mean	4.38		4.24	b	5.46	a	72.31
	std. dev	0.98		0.63		1.39		3.16
3	mean	4.34		4.52	ab	4.80	ab	72.23
	std. dev	0.81		0.55		0.88		2.20
				*		**		
D + 6								
1	mean	5.98		4.87		3.38		90.46
	std. dev	1.23		0.44		0.84		3.36
2	mean	5.63		4.84		3.65		89.62
	std. dev	1.63		0.69		1.44		2.32
3	mean	6.18		5.05		3.29		90.23
	std. dev	0.95		0.64		1.37		1.34
D + 17								
1	mean	6.05	a	4.08		4.16		128.75
	std. dev	0.72		0.96		2.64		6.57
2	mean	6.18	a	4.19		4.67		130.25
	std. dev	0.91		0.69		6.22		9.19
3	mean	5.62	b	3.98		4.11		127.81
	std. dev	0.91		0.76		1.69		7.20
		*						

Significance level of variance analysis for $P < 0.05$ *, $P < 0.01$ **, $P < 0.001$ ***
a, b: classification by NEWMAN-KEULS test

Discussion and Conclusion

The foliar application of 3.3 g/L seaweed extract stimulated the growth of maize seedings favoring the accumulation of plant mass, particularly in roots and stems. The same proportionate increases in root and stem dry weight implies that the stimulation of fresh weight accumulation was not simply due to increased water content. Furthermore, only minor differences were observed in gas exchange characteristics after seaweed treatment, suggesting that there were no significant effects on the rate of carbon assimilation per unit of leaf area. The mass enhancement measured in these experiments could be due, not to a stimulation of the net carbon assimilation rate per unit of leaf area, but rather to a better sink capacity of stem and roots. It is possible, however, that a stimulation in the assimilation rate of carbon per unit of leaf area, not detectable by our method of measurement, could increase the organic matter accumulation in the plant after a longer period of growth.

Seaweed concentrates have already been shown to improve growth and increase the yield of a number of plants: potatoes (Blunden and Wildgoose 1977), wheat (Nelson and Van Staden 1984 a, Beckett and Van Staden 1989), sugar beet (Featonby-Smith and Van Staden 1983), and cucumbers (Nelson and Van Staden 1984 b). Application is most beneficial during the vegetative phases of growth (Finnie and Van Staden 1985). The most pronounced effects observed have been the development of a vigourous root system (Blunden and Wildgoose 1977) and enhanced stem diameter (Nelson and Van Staden 1984 a), consistent with the results presented in this paper.

Seaweed extracts, when applied as a foliar application, can contribute to the supply of plant nutrients. It is known that the leaf cuticle is partially permeable to water and dissolved solutes. Moreover, seaweed extracts can contain alginates or other agents which can reduce water surface tension and lead to the formation of a thin film on the leaf surface, thus providing larger surface contact, and the best conditions for optimal nutrient assimilation by the leaf (Mengel and Kirkby 1987).

It should be remembered that leaves are able to take up only relatively small quantities of nutrients in comparison with the plant's total demand. This is particularly true for macronutrients which are required by crops in very high amounts. Moreover, in compliance with manufacturer's recommendations, seaweed extracts are applied at very low concentrations, thus the total amount of plant nutrients supplied is very small indeed. At the very most, seaweed extracts may increase the amount of limiting nutrients to sufficiently correct for a marginal deficiency (Aitken and Senn 1965). On the basis that seaweed extracts supply very small amounts of elements when used as a foliar application, the observed benefits may be partially due to one or more of the organic compounds present (Abetz 1980). *Ascophyllum nodosum* used for the production of the Goëmar GA 14 extract can contain considerable quantities of phenolic compounds, laminaran, mannitol (Blunden *et al.* 1968, Booth 1969), fucoïdan, alginates and a wide range of organic growth regulators such as auxin (Augier 1976 a, Sanderson *et al.* 1987), gibberellins, substances with a cytokinin-like activity, cytokinins (Augier 1976 b, Blunden 1977, Tay *et al.* 1985), abscisic acid (Boyer and Dougherty 1988, Hirsch *et al.* 1989), and 1-amino-cyclopropane-1-carboxylic acid (Nelson and Van Staden 1985), a precursor of ethylene and numerous betaïnes (Abetz 1980, Blunden *et al.* 1981, Kingman and Moore 1982, Yvin *et al.* 1989).

These compounds have been suggested as the active ingredients which are partially responsible for the improved growth observed with seaweed products. Moreover, many of the beneficial effects of seaweed treatment on plants resemble those obtained following the application of plant regulators (Blunden and Wildgoose 1977). This has prompted speculation that this group of substances could be responsible for most of the improved performance of treated plants. Furthermore, it has been shown that, while plant growth can be stimulated by exposure to very low concentrations of some of the above listed substances, exposure to high concentrations will inhibit growth (Abetz 1980).

More research is needed to understand the physiological mechanisms responsible for the beneficial effects of seaweed concentrates on plants and to elucidate the nature and the mode of action of the compounds contained in these concentrates.

Acknowledgements

The authors are grateful to C. Bergmark for reading the manuscript and to S. Wuillème, R. Pailleux and B. Auclair for their technical assistance.

Recherche Agronomique/Agence Nationale de Valorisation de la Recherche/GOEMAR X8511403Q.

References

Abetz, P. 1980. Seaweed extracts: have they a place in Australian agriculture or horticulture? *J. aust. Inst. agric. Sci. 46*: 23–29.

Aitken, J. B. and T. L. Senn. 1965. Seaweed products as a fertilizer and soil conditioner for horticultural crops. *Bot. Mar. 8*: 144–148.

Augier, H. 1976a. Les hormones des algues. Etat actuel des connaissances. I. Recherche et tentative d'identification des auxines. *Bot. Mar. 19*: 127–143.

Augier, H. 1976b. Les hormones des algues. Etat actuel des connaissances. II. Recherches et tentative d'identification des gibbérellines, des cytokinines et de diverses autres substances de nature hormonale. *Bot. Mar. 19*: 245–254.

Beckett, R. P. and J. Van Staden. 1989. The effect of seaweed concentrate on the growth and yield of potassium stressed wheat. *Plant and Soil 116*: 29–36.

Blunden, G. 1977. Cytokinin activity of seaweed extracts. *In*: (D. J. Faulkner and W. H. Fenical, eds) *Marine Natural Products Chemistry*. Plenum Press, New York. pp. 337–344.

Blunden, G., S. B. Challen and D. L. Woods. 1968. Seaweed extracts as fertilizers. *J. Sci. Food Agric. 19*: 289–293.

Blunden, G., M. M. Barouni, S. M. Gordon, W. F. H. Mc Lean and D. J. Rogers. 1981. Extraction, purification and characterisation of Dragendorff-positive compounds from some British marine algae. *Bot. Mar. 24*: 451–456.

Blunden, G. and P. B. Wildgoose. 1977. The effects of aqueous seaweed extract and kinetin on potato yields. *J. Sci. Food Agric. 28*: 121–125.

Booth, E. 1969. The manufacture and properties of liquid seaweed extracts. *Proc. 6th Int. Seaweed symp.*, Santiago de Compostela, Spain. pp. 661–666.

Boyer, G. L. and S. S. Dougherty. 1988. Identification of abscisic acid in the seaweed *Ascophyllum nodosum*. *Phytochemistry 27*: 1521–1522.

Featonby-Smith, B. C. and J. Van Staden. 1983. The effect of seaweed concentrate and fertilizer on the growth of *Beta vulgaris*. *Z. Pflanzenphysiol. 112*: 155–162.

Finnie, J. F. and J. Van Staden. 1985. Effect of seaweed concentrate and applied hormones on *in vitro* cultured tomato roots. *J. Plant Physiol 120*: 215–222.

Hirsch, R., W. Hartung and H. Gimmler. 1989. Abscisic acid content of algae under stress. *Bot. Acta 102*: 326–334.

Jolivet, E., I. de Langlais-Jeannin and J. F. Morot-Gaudry. 1991. Les extraits d'algues marines: propriétés phytoactives et intérêt agronomique. *Année biologique*, Paris, France.

Kingman, A. R. and J. Moore. 1982. Isolation, purification and quantitation of several growth regulating substances in *Ascophyllum nodosum* (Phaeophyta). *Bot. Mar. 25*: 149–153.

Mengel, K. and E. A. Kirkby. 1987. Fertilizer application. *In*: (K. Mengel and E. A. Kirkby, eds) *Principles of Plant Nutrition*, 4th edition. International Potash Institute, Bern, Switzerland. pp. 328–329.

Nelson, W. R. and J. Van Staden. 1984a. The effect of seaweed concentrate on wheat culms. *J. Plant Physiol. 115*: 433–437.

Nelson, W. R. and J. Van Staden. 1984b. The effect of seaweed concentrate on growth of nutrient-stressed, greenhouse cucumbers. *HortScience 19*: 81–82.

Nelson, W. R. and J. Van Staden. 1985. 1-aminocyclopropane-1-carboxylic acid in seaweed concentrate. *Bot. Mar. 28*: 415–417.

Sanderson, K. J., P. E. Jameson and J. A. Zabkiewicz. 1987. Auxin in a seaweed extract: identification and quantitation of indole-3-acetic acid by gas chromatograph-mass spectrometry. *J. Plant Physiol. 129*: 363–367.

Tay, S. A. B., J. K. Macleod, L. M. S. Palni and D. S. Letham. 1985. Detection of cytokinins in a seaweed extract. *Phytochemistry 24*: 2611–2614.

Yvin, J.-C., R. Chabot and P. Savary. 1989. Les algues en agriculture. Influence sur les plantes cultivées et approche moléculaire des mécanismes d'action. *Perspect. agric. 134*: 74–80.

Botanica Marina
Vol. 34, pp. 475–482, 1991

Ulva lactuca Biofilters for Marine Fishpond Effluents

I. Ammonia Uptake Kinetics and Nitrogen Content

I. Cohen[1] and A. Neori[2]

Israel Oceanographic and Limnological Research, National Center for Mariculture, P.O. Box 1212, Eilat 88112, Israel

(Accepted 18 July 1991)

Abstract

The potential of the seaweed *Ulva lactuca* as a biofilter for effluents of intensive marine fishponds in arid regions has been studied in Eilat, Israel. One kg (wet wt) m^{-2} of *Ulva* in 1 m^2 tanks 60 cm deep, removed, during the day, over 90 percent of the ammonia from the fishpond effluents, at inflow fluxes up to 10 µmoles $L^{-1} h^{-1}$ (0.14 mole $m^{-2} d^{-1}$). At fluxes of 40 µmoles $L^{-1} h^{-1}$ (0.58 mole $m^{-2} d^{-1}$), nearly 40 percent were removed. Night time ammonia removal remained efficient at low ammonia fluxes. Ammonia uptake rate and percent nitrogen (N) in dry weight (DW) of the seaweed showed hyperbolic correlations with ammonia flux through the tanks. The data fitted the Michaelis-Menten kinetics, with a maximal uptake rate of 28 µmoles $L^{-1} h^{-1}$ (0.4 mole $m^{-2} d^{-1}$). Pulse ammonia uptake rate was approximately four times that of continuous uptake. Maximal percent N in DW of *U. lactuca*, when stocked at 1 kg m^{-2}, was 4.07. Increased stocking density (2, 4 and 6 kg m^{-2}) did not significantly affect ammonia uptake rate per tank. However, N-content was significantly higher, averaging 5.5 percent of DW. The data indicate that at low ammonia fluxes (below 10 µmoles $L^{-1} h^{-1}$) *U. lactuca* was N-limited and that at higher fluxes the algae became gradually light limited. Our results indicate that 10 m^2 of *Ulva* biofilter can remove from the effluents of our intensive grow-out ponds over 90 percent of the ammonia produced by 1 kg of daily feed ration or by approximately 75 kg of fish.

Introduction

Intensive culture of marine fishponds is being developed at the National Center for Mariculture in Eilat, Israel. The high densities of fish in the ponds cause problems of water quality which in turn limit the production (Krom *et al.* 1985 b). Ammonia is a major metabolite produced by the fish and is toxic to them at high levels. It is also a major component in aquaculture effluents, contributing to the eutrophication of aquatic environments. Removal of excess ammonia and other inorganic nutrients can improve water quality in the ponds, and at the same time reduce downstream eutrophication, which is particularly dangerous in the highly oligotrophic and enclosed Gulf of Eilat (Aqaba).

Seaweeds have been shown to reduce efficiently the level of seawater eutrophication in integrated mariculture (Ryther *et al.* 1975). In a series of studies they showed some seaweeds to have a high yield and nutrient uptake capacity and to be very adequate for cultivation and harvesting. These advantages could be augmented by the economic potential of many seaweeds in various industries. Motivated by these studies, Goldman (unpublished report) suggested the combination of colloid-rich seaweed culture with the fish culture effort in the arid south of Israel. Macroalgal 'nutrient traps' have thus become a component in the design of integrated desert mariculture in our institute (Gordin 1982).

[1] Current address: Department of Microbial Ecology, the Hebrew University of Jerusalem, Givat Ram, Jerusalem, Israel
[2] Corresponding author

Our current research on ways to incorporate seaweeds into intensive mariculture, considers the seaweed yield as a bonus in a water treatment system whose main products are stable water quality and clean effluents. This attitude (shared with Langton *et al.* 1977 and some others) is different than that in most other algal biotechnology research enterprises, in which the algae are the main or only product. Thus, we have chosen as the first candidate the local *Ulva lactuca* L. (Chlorophyta), not very valuable but sturdy, as our primary seaweed, rather than cope with the biological obstacles of culturing a more valuable but less robust plant.

Various species of the genus *Ulva* have been cultured in the laboratory and outdoors. They grew well in both artificial media and wastewater, showing a high capacity to take up nutrients (Duke *et al.* 1989 a, Fujita 1985, Guist and Humm 1976, Parker 1981, Rosenberg 1981). We have cultured *Ulva lactuca* successfully in Eilat for long periods under field conditions (unpublished and Vandermeulen and Gordin 1990) and its strong performance suggests that this species might indeed be a suitable biofilter for fishponds.

In the present study we quantify the operational factors – nutrient concentration, water exchange rate and stocking density – controlling nutrient uptake from effluents of intensive fishponds by *Ulva lactuca* under our field conditions and compare the results with other studies. This is an essential step for the design and operation of larger scale facilities.

Material and Methods

Thalli of *Ulva lactuca* L. (Vandermeulen and Gordin 1990), were collected from the Gulf of Eilat (Aqaba) and maintained outdoors in running seawater. Specimens have been deposited in the herbarium of the University of La Laguna, Tenerife, Spain (Dr Ricardo Haroun) and live and preserved samples have been maintained in our laboratory for later investigations.

All the experiments were conducted in 1 m^2, 600 L tanks (Vandermeulen 1989). The algae were grown unattached and kept suspended in the water column by air diffusers situated at the bottom of each tank. Fishpond effluents were pumped into a large holding tank (15 m^3) and then flowed by gravity to the experimental tanks. Temperatures were measured by minimum/maximum thermometers in each tank, but the temperature record is incomplete due to the breakdown of several thermometers. Light data were kindly provided by the Israeli Meteorological Service from a nearby facility.

In total five experiments are reported. At the beginning of each experiment the tanks were stocked at a fixed wet weight of centrifuged thalli and each treatment (in triplicates, except where indicated otherwise) was provided with a fixed flow-rate of fishpond effluents. The algae were harvested weekly, centrifuged in a converted washing machine and restocked at the original stocking density. The results of growth rate, yield and C : N ratio will be reported in the accompanying paper (Neori *et al.* 1991).

Samples for dry weight determination were oven dried (60 °C, 24 hours). Samples for CHN analyses (Perkin-Elmer Model 240) were rinsed with deionized water, dried and ground to a fine powder with a mortar and pestle. The rinsing step lead to a small (< 3%) loss of dry weight (DW), probably due to the salts in the seawater. Ammonia-N concentration (practically the only inorganic N dissolved in the fishpond water, see Porter *et al.* 1987 and Neori *et al.* 1989) was measured at the inflow and at the outflow water of each tank 2–3 times a week at 9:00 am. At this time the concentration was representative of daytime values as can be learned from the diel variation in ammonia concentration in the algal tanks and in the fishpond effluents reported later. The water samples were filtered (GF/F, Whatmann) and analyzed for total ammonia by a Technicon AA-2 autoanalyzer (Krom *et al.* 1985 a). The uptake rate was determined as $V = f(Si\text{-}So)$ where V = uptake rate (μmoles L^{-1} h^{-1}), f = water flow rate (number of water exchanges h^{-1}), Si = inflow ammonia-N concentration and So = ammonia-N concentration at the outflow (residual N). The Vmax and Ks were calculated from the linear S/V vs S plot (Atkins and Nimmo 1980). Overnight measurements at 3 h intervals were made twice during an experiment.

Statistical correlations and F-tests were made by the General Linear Model option of SAS. Since there were occasionally large differences in the inflowing concentration of ammonia between weeks, we have treated each week as a separate experimental unit and the data are presented accordingly. Data were averaged for each week in the Tables, but all the data were used in the analyses and plotted in the Figures.

To assess the importance of environmental differences between experiments we used in experiments 1, 2 and 3 a standard treatment with 1 kg of *Ulva lactuca* and 4 daily exchanges of ambient seawater. Inorganic nutrients, 2 mM of ammonia-N (as NH_4Cl) and 0.15 mM of P (as NaH_2PO_4), were supplied weekly, while stopping water exchange for 18 h (Vandermeulen and Gordin 1990). The rate of ammonia disappearance during the first 3 h of the nutrient supply was used to calculate maximal values of 'pulse' uptake rate.

Water exchange rate

Four experiments were conducted to determine the effects of exchange rate of fishpond effluents in the algal tanks on ammonia-N uptake by *Ulva lactuca.* Each tank was stocked with 1 kg of the seaweed. In the November 1987 experiment (exp. 1) water exchange rate was 4/day. In two experiments, in January, 1988 (exp. 2) and in February, 1988 (exp. 3) 4, 8 and 16 daily water exchanges were compared (in duplicates). In the last of the series, in January, 1989 (exp. 4), daily water exchange rates compared were 4, 8, and 12, and no CHN samples were taken.

Stocking density

The effect of high stocking densities of *Ulva lactuca* (above 1 kg per tank) on the rate of ammonia-N uptake at high ammonia-N fluxes was studied in May-June, 1988 (exp. 5). The tanks were stocked with 2, 4 and 6 kg of *U. lactuca* and maintained at 12 daily water exchanges of fishpond effluents.

Results

Water exchange rate

The water exchange rate experiments were all carried out in the winter to reduce the effect of the variation in environmental factors on the results. We also found highly insignificant correlation ($r^2 = 0.0115$, $P = 0.7$) between total daily irradiance and ammonia uptake in the combined data from the standard treatments of experiments 1, 2 and 3 (Table I).

Ammonia-N levels in the inflowing fishpond effluents varied markedly between experiments and, to a lesser extent, even between weeks in some experiments (Table II). The effect of ammonia-N concentration in the inflowing water can be seen by comparing treatments with similar water exchange rates between weeks and experiments. The effect of water exchange rate, although not independent of ammonia-N concentration, can be seen by comparing treatments within each week. Outflow concentration of ammonia-N correlated positively with both water exchange rate and inflowing ammonia-N concentration whereas percent uptake (removal efficiency) of ammonia-N by the seaweed correlated negatively with the same variables ($p < 0.05$). The percent uptake of ammonia-N by the seaweeds during day time was high at most levels of water flow rate and inflow ammonia-N concentration. Maximal removal efficiency, up to 96 percent, was obtained at low water exchange rates and ammonia-N concentrations of experiment 2. The removal efficiency dropped within each experiment with increasing water exchange rate. It also dropped between experiments with increasing ammonia concentration in the inflowing water. Thus, the removal efficiency dropped down to 39 percent at the highest water exchange rates and ammonia-N concentrations in experiment 4.

Nevertheless, V (the product of inflowing ammonia-N concentration, water exchange rate and the fraction of ammonia-N uptake in moles per tank, per liter or per unit area), increased both with increasing water exchange rate in each experiment and with increasing ammonia concentration in the inflowing water between experiments ($P < 0.05$). Most likely the value of V actually depends on the *in situ* concentration of nutrients in the tanks. The latter value is determined by three parameters, V itself, ammonia concentration in the inflowing water and water exchange rate. The two last parameters, multiplied by each other, produce ammonia-N flux as suggested by Fujita and Goldman (1985). They had found that the growth rate of *Gracilaria tikvahiae* McLachlan was predicted better by N-flux than by either flow rate or N-concentration alone. Nutrient flux reflects the interdependency of the effects of flow rate and nutrient concentration and describes better the availability of nutrients to the algae.

The dependence of V on ammonia-N flux in *Ulva lactuca* (Fig. 1a) shows a positive hyperbolic saturation, conforming with the Michaelis-Menten type kinetics. The values of Vmax and Ks which form the curve are 28 μmoles L^{-1} h^{-1} and 27.5 μmoles L^{-1} h^{-1}, respectively ($r^2 = 0.64$; $n = 206$). Complete ammonia removal at low ammonia fluxes is indicated by the initial slope (Vmax/Ks) of close to unity.

Plotting uptake rates as a function not of the ammonia-N flux but of the *in situ* ammonia-N concentration, or residual ammonia-N (Fig. 1b), shows also a positive hyperbolic correlation which fits the Michaelis-Menten kinetics (Vmax = 13.8 μmoles L^{-1} h^{-1}; Ks = 5.2 μM; $r^2 = 0.81$). The points are, however, much more scattered at the low concentration

Table I. Total daily irradiance and pulse 18 h ammoia-N uptake in the standard treatment of three uptake experiments with *Ulva lactuca* in 600 liter, 1 m^2 tanks.

Exp. No.	Week	Total irradiance (Watt h m^{-2} d^{-1})	Pulse ammoia uptake (μmoles L^{-1} per 18 h ± SE)
1	1	4645	654 ± 100
2	1	5160	393 ± 9
	2	3425	377 ± 21
3	1	6910	596 ± 44
	2	6625	334 ± 8

Table II. Experimental conditions and results in four ammonia uptake experiments with *Ulva lactuca* cultured in fishpond effluents in 600 L, 1 m^2 tanks (weekly means).

Exp. No.	Week	Water exchange rate (d^{-1})	Ammonia-N (μM ± SE)		Percent (± SE)	
			Inflow	Outflow	Uptake	N in DW
1	1	4	25 ± 4	2.8 ± 1.0	89 ± 2	2.40 ± 0.1
	2	4	31 ± 2	3.5 ± 1.0	91 ± 1	3.20 ± 0.2
2	1	4	10 ± 3	0.5 ± 0.4	96 ± 2	1.95 ± 0.15
		8		1.0 ± 0.5	89 ± 7	2.45 ± 0.25
		16		2.5 ± 0.4	76 ± 8	3.10 ± 0.1
	2	4	12 ± 2	0.5 ± 0.6	96 ± 4	1.55 ± 0.05
		8		1.1 ± 0.5	90 ± 5	2.15 ± 0.05
		16		1.9 ± 0.1	83 ± 2	3.25 ± 0.15
	3	4	14 ± 2	0.7 ± 0.4	95 ± 3	
		8		1.1 ± 0.5	91 ± 4	N.D.
		16		2.7 ± 0.6	80 ± 3	
3	1	4	27 ± 3	1.8 ± 0.5	93 ± 2	2.55 ± 0.05
		8		3.5 ± 0.1	87 ± 1	3.15 ± 0.15
		16		8.0 ± 0.7	69 ± 3	3.35 ± 0.45
	2	4	32 ± 4	3.0 ± 0.1	90 ± 3	2.85 ± 0.25
		8		5.7 ± 1.8	82 ± 6	3.80 ± 0.3
		16		12.7 ± 2.4	61 ± 10	3.70 ± 0.01
	3	4	48 ± 10	4.7 ± 0.5	89 ± 3	2.90 ± 0.10
		8		8.9 ± 3.3	81 ± 4	3.85 ± 0.05
		16		12.5 ± 2.1	71 ± 10	4.15 ± 0.35
4	1	4	77 ± 5	16.0 ± 7.0	79 ± 10	
		8		33.8 ± 7.3	56 ± 10	N.D.
		12		41.8 ± 2.4	47 ± 11	
	2	4	71 ± 4	25.2 ± 5.8	68 ± 7	
		8		35.5 ± 6.2	51 ± 7	N.D.
		12		43.1 ± 6.5	39 ± 6	
	3	4	78 ± 2	32.8 ± 6.3	62 ± 9	
		8		45.1 ± 3.6	49 ± 6	N.D.
		12		53.0 ± 0.2	39 ± 1	

N.D. = Not Determined

range and the Vmax is lower because the effect of water exchange rate is not fully appraised in this kind of plot.

The values of Vmax in the continuously-fed treatments are low by comparison with the standard treatment, where the algae were pulse-fed after a period of nutrient starvation. Maximal uptake rate for the pulse fed algae, grouped for all experiments, was 110 ± 9.6 μmoles L^{-1} h^{-1} for the first interval (3 h) of measurements. Even this value is possibly an underestimate, as maximal uptake rates by starved algae can take place within the first 15–30 min of nutrient supply (Thomas and Harrison 1987).

At night the seaweeds kept low the ammonia concentration in the outflowing water (*i.e.*, high night time removal efficiency) when ammonia-N flux was low, as illustrated in Figure 2 and also in Vandermeulen and Gordin (1990). However, at higher ammonia-N fluxes (achieved in this case by higher water exchange rates) removal efficiency decreased markedly. At fluxes of approximately 25 μmoles L^{-1} h^{-1} (the highest exchange rate in Fig. 2) ammonia concentration in the outflowing water late at night approached that in the inflowing water, meaning that most of the inflowing ammonia was not taken up.

Nitrogen content in seaweed tissue can be expected to depend on their nutritional status (Guist and Humm 1976, Bird *et al.* 1981, Duke *et al.* 1989 b, DeBusk *et al.* 1986). Indeed, like the ammonia uptake rate, the dependencies of percent nitrogen content in the dry plant tissue on N-flux and residual N (Fig. 3a and Fig. 3b, respectively) also show Michaelis-Menten kinetics. The values of the linear regression correlation coefficient (r^2) between S/%N and %N were 0.95 (n = 36) for both. Maximal calculated percent N in plant tissue (Nmax) for these experiments was 4.07, similar to the value attained in DeBusk *et al.* (1986).

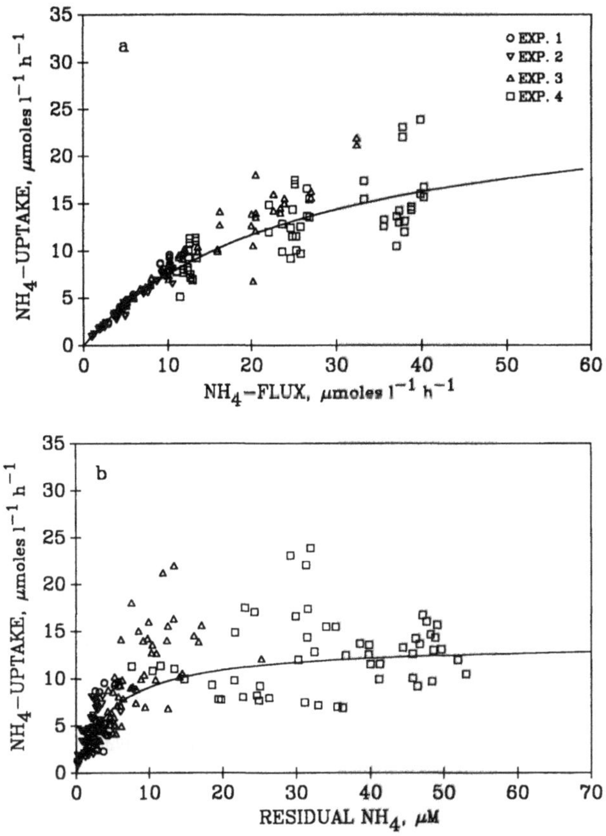

Fig. 1. Ammonia-uptake rate by *Ulva lactuca* in four water exchange rate experiments, with Michaelis-Menten curves fitted to the entire data sets: a, as a function of ammonia-N flux (Vmax = 28 µmoles L^{-1} h^{-1} and Ks = 27.5 µmoles L^{-1} h^{-1}); b, as a function of residual ammonia-N (Vmax = 13.8 µmoles L^{-1} h^{-1} and Ks = 5.2 µmoles L^{-1}).

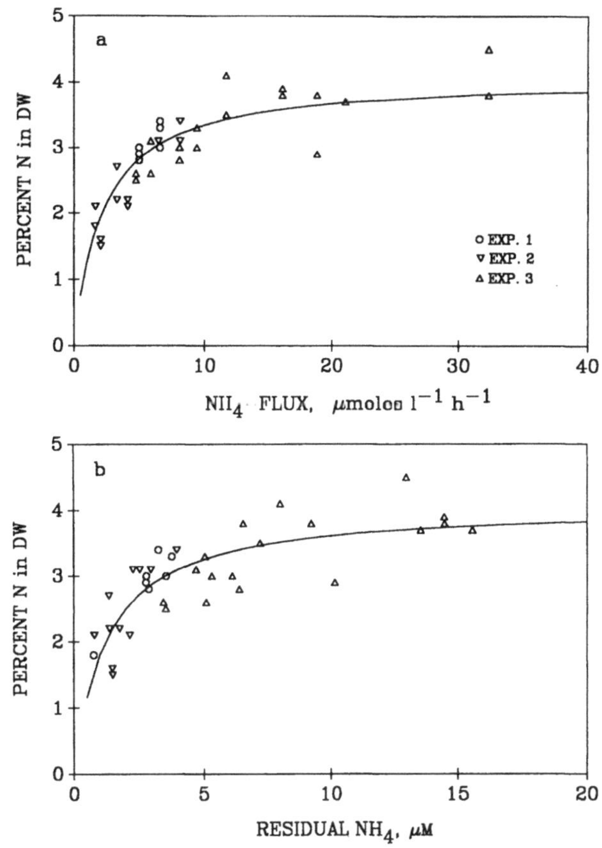

Fig. 3. Nitrogen content (percent of DW) of *Ulva lactuca* grown in fishpond effluents in three ammonia uptake experiments, with Michaelis-Menten curves fitted to the entire data sets: a, as a function of ammonia-N flux (Nmax = 4.07 percent and Ks = 2.2 µmoles L^{-1} h^{-1}); b, as a function of residual ammonia-N (Nmax = 4.02 percent and Ks = 1.24 µmoles L^{-1}).

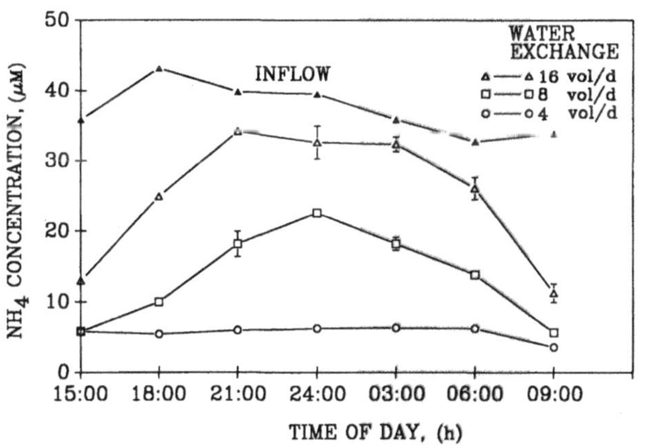

Fig. 2. A typical time-course of ammonia-N concentration in the water flowing into and out of the *Ulva lactuca* tanks during night time at three water exchange rates. SE bars do not appear when they are smaller than the symbol.

Effect of high stocking density

Ammonia uptake did not vary significantly between treatments with densities of 2, 4 and 6 kg of *Ulva lactuca* per m^2 (Table III). Only once, in the second week of this experiemnt, was the ammonia-N uptake in one treatment significantly higher ($p < 0.05$) than in another. Percent N ranged between 4.6–5.7%, higher than Nmax for the experiments with 1 kg m^{-2} of *U. lactuca*. Unfortunately, we did not include a 1 kg m^{-2} treatment in this experiment, which was carried out in the spring. Although not statistically significant, a definite trend of increased precent N with increased stocking density is evident in Table III.

Discussion

A number of studies have examined the ammonia uptake kinetics in several species of the genus *Ulva* (Table IV), most of them under laboratory conditions. Such results have indicated the great potential of the genus to treat wastewater, as has also been indicated in preliminary studies in our institute (Vandermeulen and Gordin 1990). We have obtained similar results in the present study, in fishpond effluents and under our field conditions, as close as possible to the practical application. In such experiments it is less easy to separate the effect of all environmental variables and, moreover, data could be location specific. Nevertheless, our uptake results do fall within the values ex-

Table III. Experimental conditions and results in a stocking density experiment with *Ulva lactuca* cultured in fishpond effluents in 600 L, 1 m^2 tanks with 12 water exchanges d^{-1} (weekly means).

Week	Stocking density ($kg\ m^{-2}$)	Ammonia-N (μM ± SE)		Percent ± SE	
		Inflow	Outflow	Uptake	N in DW
1	2	35 ± 1	13.6 ± 1.5	61 ± 4	4.9 ± 0.3
	4		13.9 ± 0.8	60 ± 2	5.3 ± 0.2
	6		13.2 ± 1.4	62 ± 4	5.4 ± 0.1
2	2	36 ± 3	17.2 ± 5.8	51 ± 16	5.4 ± 0.2
	4		19.4 ± 5.8	44 ± 14	5.7 ± 0.1
	6		12.8 ± 3.5	64 ± 6	5.7 ± 0.1
1	2	76 ± 25	20.7 ± 9.9	74 ± 3	4.6 ± 0.2
	4		24.0 ± 9.7	69 ± 3	4.8 ± 0.2
	6		23.3 ± 7.9	69 ± 3	5.1 ± 0.2

Table IV. Kinetic parameters of N uptake by *Ulva* species.

Species	Vmax ($\mu mol\ g^{-1}\ h^{-1}$)	Ks (μM)	Vmax/Ks		Reference
U. curvata	250	14	17.9		Rosenberg (1981)[a]*
U. lactuca	138	41	3.5	N-sufficient	Fujita (1985)*
U. lactuca	252	15	17.1	N-starved	Fujita (1985)*
U. curvata			9.2		Rosenberg & Ramus (1984)*
U. curvata			4–27.5		Duke *et al.* (1989 a)**
U. lactuca	50	5.2	9.5	N-sufficient	this study[a,c]**
U. lactuca	100	27.5		N-sufficient	this study[b,c]**
U. lactuca	390			N-starved	this study[d]**

[a] as a function of residual-ammonia-N_4
[b] as a function of ammonia-N flux, Ks in $\mu mol\ g^{-1}\ h^{-1}$
[c] continuous fishpond effluents
[d] inorganic nutrient pulse
* under laboratory conditions
** under field conditions

pected from the literature and certainly have a general application. We also point out that the water flow rate experiments were made during the winter season, the least favorable for *Ulva lactuca* growth in our system (Vandermeulen and Gordin 1990). The yield of the standard treatment in November was 58 percent of that in June (unpublished). Thus, the ammonia-N uptake results reported here are quantitatively at the low end of the yearly cycle.

In wastewater treatment, the seaweeds will likely be exposed to nutrients continuously. However, many of the nutrient uptake values in the literature are for nutrient pulses. In our results, uptake rate of ammonia in pulse treatments is approximately four fold that of continuous-flow treatments. This phenomenon of rapid nutrient uptake is typical of N-starved or N-depleted seaweeds. Probyn and Chapman (1982), using field collected *Chordaria flagelliformis* (O. F. Müll) C. Ag., fitted their results to the Michaelis-Menten expression. Comparing these results to the kinetic parameters determined in continuous-flow laboratory cultures, they concluded that the short-term experiments tended to over-estimate the Ks and Vmax by an order of magnitude. Similar results were reported by Rosenberg *et al.* (1984), Fujita (1985) and D'Elia and DeBoer (1978). Ryther *et al.* (1981/1982) also reported rapid ammonia uptake by N-limited algae which was 4–6 times higher than the long-term values. Thus, N-starved seaweeds can take up large quantities of ammonia in a short time. Unlike the continuous values, however, pulse rates cannot be sustained for more than a few hours, and thus are less important in relation to our application of effluent treatment.

The saturation kinetics as a function of N-flux shown by both ammonia-N uptake rate and percent tissue N indicate that at low N-fluxes (below 10 $\mu moles\ L^{-1}\ h^{-1}$ or 0.14 $moles\ m^{-1}\ d^{-1}$) *Ulva lactuca* was N-limited. At higher N-fluxes another factor became growth limiting, possibly light as shown below and in the accompanying paper (Neori *et al.* 1991).

Increased stocking density did not lead to nutrient limitation as the algae were grown at high ammonia fluxes. Ammonia-N uptake rate remained the same for all treatments in the stocking density experiment. However, percent N in algal tissue here was higher than for the 1 kg m^{-2} experiments and a positive correlation was suggested between percent N and stocking density. Neori *et al.* (1991) found stocking density did not influence percent carbon in the same experiment. Thus the carbon limitation of the algae, suggested for this experimental system by Vandermeulen and Gordin (1990), is not supported by our results. The seasonal difference between spring and winter is also unlikely to account for the observed high percent tissue N in the stocking density experiment compared with the water exchange rate experiments. Actually, if the effects were only seasonal the values of percent tissue N should have decreased as a result of the better temperature and light conditions in the spring and not increased, as observed here. Duke *et al.* (1986 and 1989 a) showed in studies with *Ulva curvata* (Kütz.) de Toni and *Codium decorticatum* (Woodw.) Howe, that improved light and temperature conditions decreased percent tissue N in the seaweeds. These findings can be reconciled with the increased percent N in our spring experiment, based on the study of Vymazal (1987) on optimal stocking density. The super-optimal stocking densities (2, 4 and 6 kg m^{-2}) in our experiment No. 5 must have led to reduced mean light availability in the tanks, in spite of it taking place in the spring, relative to the optimally-stocked tanks of the winter experiments 1–4. Vandermeulen and Gordin (1990), using the very same tanks, indeed reported a reduction in photon flux density at the bottom of 89% and 100% for tanks with 1 kg and 4 kg of *Ulva*, respectively.

In conclussion, the robust seaweed *Ulva lactuca* could continuously and efficiently strip nearly all the ammonia from fishpond effluents at low nutrient fluxes (below 10 μmoles L^{-1} h^{-1} or 0.14 moles m^{-2} d^{-1}). The rate of ammonia uptake increased at higher ammonia fluxes to 28 μmoles L^{-1} h^{-1} (or 0.4 moles m^{-2} d^{-1}), at the cost of reduced removal efficiency, especially at night. Nitrogen content of the seaweeds increased with increased ammonia flux and with increased stocking density. Ammonia uptake rate per tank was insensitive to algal stocking density in the range of 1 to 6 kg m^{-2}.

Ulva biofilters can operate both as recirculating units of intensive fishponds and as units for polishing effluents before their release. The data in this paper allow the design of pilot-size *Ulva* biofilters for intensive seawater fishponds. Our calculations show that 10 m^2 of the seaweed biofilter can take up completely and in one stage the ammonia produced in the fishpond from 1 kg of daily feed. This quantity feeds approximately 75 kg of *Sparus aurata* L., the fish grown in our grow-out ponds. A multi-stage biofilter system of similar size (10 m^2), designed to use the higher uptake capacity of the seaweed at high ammonia fluxes, can treat the effluents of even twice that weight of fish.

Acknowledgements

We are grateful to A. Davidson for chemical analyses, to H. Voet for help with statistical analyses and to the staff of the NCM for their assistance. H. Gordin provided encouragement and facilitated this research. M. Friedlander and C. Dawes critically reviewed earlier versions of this manuscript. This research was supported by the Israeli Ministry for Energy and Infrastructure Mariculture Project, The Arava Research and Development Fund, BARD research grant No. I-1634-89 and grant No. 3035189 from the Israeli Ministry for Science and Technology.

References

Atkins, G. L. and I. A. Nimmo. 1980. Current trends in the estimation of Michaelis-Menten parameters. *Anal. Biochem.* *104*: 1–9.

Bird, K. T., M. D. Hanisak and J. H. Ryther. 1981. Chemical quality and production of agars extracted from *Gracilaria tikvahiae* grown in different nitrogen enrichment conditions. *Bot. Mar. 14*: 441–444.

DeBusk, T. A., M. Blakeslee and J. H. Ryther. 1986. Studies on the outdoor cultivation of *Ulva lactuca* L. *Bot. Mar. 24*: 381–386.

D'Elia, D. F. and J. A. DeBoer. 1978. Nutritional studies of two red algae. II. kinetics of ammonium and nitrate uptake. *J. Phycol. 14*: 266–272.

Duke, C. S., B. E. Lapointe and J. Ramus. 1986. Effects of light on growth, RuBPase activity and chemical composition of *Ulva* species (Chlorophyta). *J. Phycol. 22*: 362–370.

Duke, C. S., W. Litaker and J. Ramus. 1989 a. Effects of the temperature, nitrogen supply and tissue nitrogen on ammonium uptake rates of the Chlorophyte seaweeds *Ulva curvata* and *Codium decorticatum*. *J. Phycol. 25*: 113–120.

Duke, C. S., W. Litaker and J. Ramus. 1989 b. Effect of temperature on nitrogen limited growth rate and chemical composition of *Ulva curvata* (Ulvales: Chlorophyta). *Mar. Biol. 100*: 143–150.

Fujita, R. M. 1985. The role of nitrogen status in regulating transient ammonium uptake and nitrogen storage by macroalgae. *J. Exp. Mar. Biol. Ecol. 92*: 283–301.

Fujita, R. M. and J. C. Goldman. 1985. Nutrient flux and growth of the red alga *Gracilaria tikvahiae* McLachlan (Rhodophyta). *Bot. Mar. 28*: 265–268.

Gordin, H. 1982. Aquaculture: potential developemnt. *In*: (P. G. Brewer, ed) *Oceanography: Present and Future*. Springer-Verlag, New York. pp. 347–361.

Guist, G. G. Jr. and H. J. Humm. 1976. Effects of Sewage effluents on growth of *Ulva lactuca*. *Biol. Sci. 4*: 267–271.

Krom, M. D., S. Grayer and A. Davidson. 1985a. An automated method of ammonia determination for use in mariculture. *Aquaculture 44*: 153–160.

Krom, M. D., C. B. Porter and H. Gordin. 1985b. Causes of fish mortalities in semi-intensively operated seawater ponds in Eilat, Israel. *Aquaculture 49*: 159–177.

Langton, R. W., K. C. Haines and R. E. Lyon. 1977. Ammonia-nitrogen produced by the bivalve mollusc *Tapes japonica* and its recovery by the red seaweed *Hypnea musciformis* in a tropical mariculture system. *Helgoländer wiss. Meeresunters. 30*: 217–229.

Neori, A., I. Cohen and H. Gordin. 1991. *Ulva lactuca* biofilters for marine fishpond effluents. II. growth rate, yield and C : N ratio. *Bot. Mar. 34*: 483–489.

Neori, A., M. D. Krom, I. Cohen and H. Gordin. 1989. Water quality conditions and particulate chlorophyll *a* of new intensive seawater fishponds in Eilat, Israel: daily and diel variations. *Aquaculture 80*: 63–78.

Parker, H. S. 1981. Influence of relative water motion on the growth, ammonium uptake and carbon composition of *Ulva lactuca* (Chlorophyta). *Mar. Biol. 63*: 309–318.

Porter, C. B., M. D. Krom, M. G. Robbins, L. Brickell and A. Davidson. 1987. Ammonia excretion and total N budget for gilthead seabream (*Sparus aurata*) and its effect on water quality conditions. *Aquaculture 66*: 287–297.

Probyn, T. A. and A. R. O. Chapman. 1982. Nitrogen uptake characteristics of *Chordaria flagelliformis* (Phaeophyta) in batch mode and continuous mode experiments. *Mar. Biol. 71*: 129–133.

Rosenberg, G. 1981. Ecological growth strategies in the red seaweed *Gracilaria foliifera* (Rhodophyceae). Ph. D. thesis, Yale University, pp. 151.

Rosenberg, G., T. A. Probyn and K. H. Mann. 1984. Nutrient uptake and growth kinetics in brown seaweeds: response to continuous and single additions of ammonium. *J. Exp. Mar. Biol. Ecol. 80*: 125–146.

Ryther, J. H., J. C. Goldman, C. E. Gifford, J. E. Huguenin, A. S. Wing, J. P. Clarner, L. D. Williams and B. E. Lapointe. 1975. Physical models of integrated waste recycling marine polyculture systems. *Aquaculture 5*: 163–177.

Ryther, J. H., N. Corwin, T. A. DeBusk and L. D. Williams. 1981/1982. Nitrogen uptake and storage by the red alga *Gracilaria tikvahiae* (McLachlan, 1979). *Aquaculture 26*: 107–115.

Thomas, T. E. and P. J. Harrison. 1987. Rapid ammonium uptake and nitrogen interactions in five intertidal seaweeds grown under field conditions. *J. Exp. Mar. Biol. Ecol. 107*: 1–8.

Vandermeulen, H. 1989. A low-maintenance tank for mass culture of seaweeds. *Aqua Eng. 8*: 67–71.

Vandermeulen, H. and H. Gordin. 1990. Ammonium uptake using *Ulva* (Chlorophyta) in intensive fishpond systems: mass culture and treatment of effluent. *J. Appl. Phycol. 2*: 363–374.

Vymazal, J. 1987. Ammonium uptake and biomass interaction in *Chladophora glomerata* (Chlorophyta). *Br. Phycol. J. 22*: 163–167.

Botanica Marina
Vol. 34, pp. 483–489, 1991

Ulva lactuca Biofilters for Marine Fishpond Effluents

II. Growth Rate, Yield and C:N Ratio

A. Neori[1], I. Cohen[2] and H. Gordin

Israel Oceanographic and Limnological Research, National Center for Mariculture, P.O.Box 1212 Eilat 88112, Israel

(Accepted 18 July 1991)

Abstract

The seaweed *Ulva lactuca* L. was cultured in effluents from intensive marine fishponds in Eilat, Israel. The specific growth rate and the yield of *U. lactuca* in nutrient-rich fishpond effluents were higher than in pulse-enriched fresh seawater. Both yield and specific growth rate of the seaweed showed positive hyperbolic correlations with N-flux which fitted the Michaelis-Menten kinetics. Maximal dry weight (DW) yield was 55 g m^{-2} d^{-1} and the maximal specific growth rate was 18% d^{-1}. The C : N ratio in plant tissue showed a negative hyperbolic correlation with N-flux, dropping from over 15 to approximately 7 (w : w) when grown at low and high N-fluxes, respectively. As the stocking density increased up to 6 kg m^{-2} under N-sufficient conditions, both specific growth rate and yield decreased significantly, as did mostly also the C : N ratio. The data suggest the seaweed to be N-limited at low N-fluxes and light limited at high N fluxes, especially at high stocking densities. The high yields obtained show that fishpond effluents supply *U. lactuca* with all its required nutrients including inorganic carbon. For high yield and high nitrogen content, *U. lactuca* should be kept at densities of 1–2 kg m^{-2} and at ammonia fluxes of about 0.5 moles m^{-2} d^{-1}. Nitrogen content can be increased further, but at reduced yield, by increasing the stocking density to 4 or 6 kg m^{-2}.

Introduction

As described in an earlier paper (Cohen and Neori 1991), the use of seaweed biofilters for mariculture effluents is a promising proposition (see also Langton *et al.* 1977). In numerous studies seaweeds, in particular *Ulva lactuca* L., (De Busk *et al.* 1986, Parker 1981) and other species of the genus, showed a high growth rate and nutrient uptake capacity, ease of cultivation and harvesting and potential economic value. In our research the seaweeds make up biofilters whose main product is clean effluent, whereas the algal yield is a by-product. High yields of consistently good seaweed quality, such as high N-content, can improve the attractiveness of such biofilters for treating mariculture effluents.

Previous studies under field conditions in Eilat showed that *Ulva lactuca* can grow free of contamination for long periods (Vandermeulen and Gordin 1990, Neori and Cohen, unpublished data). Yield in fresh seawater with weekly inorganic nutrient enrichment, measured over 8 months, averaged 200 g fresh wt m^{-2} d^{-1} [30 g dry weight (DW) m^{-2} d^{-1}]. This value equals the results for intensive seaweed cultivation in the literature (*e.g.*, DeBusk *et l.* 1986 and more references therein) and is much higher than for the best known field crops (Lapointe *et al.* 1976). In the present study we report the effect of water exchange rate, of ammonia-N concentration in the inflow and of the algal stocking density on specific growth rate, yield and C : N ratio for *U. lactuca* grown in fishpond effluents under field conditions.

[1] Corresponding author
[2] Current address: Department of Microbial Ecology, the Hebrew University of Jerusalem, Givat Ram, Jerusalem, Israel

Material and Methods

The source of experimental material, the experimental system and most of the procedures have been reported in the preceding paper (Cohen and Neori 1991). The specific growth rate (μ), as percentage increase in fresh weight production was calculated according to the equation $\mu = 100 \, Ln(W_t/W_0)/t$ where W_0 is the initial biomass and W_t is the biomass at day t (D'Elia and DeBoer 1978). Percent ash was determined from weighed DW samples ashed in a muffle furnace at 400 °C.

The response of growth rate, yield and C : N ratio of *Ulva lactuca* to fishpond effluent and to pulse-feeding in fresh seawater was compared in November, 1987 (exp. 1) by three treatments:

1. A standard treatment, with 1 kg m^{-2} of the seaweed, 4 daily exchanges of fresh seawater and one weekly (pulse) nutrient addition as described in the preceding paper (Cohen and Neori 1991).

2. The same as (1) but using nutrient-rich fishpond effluents and not fresh seawater. This treatment was aimed to determine possible interactions between the two forms of nutrient supply.

3. The same as (2) but without the weekly pulse nutrient addition, that is, the only nutrient source was the fishpond effluents.

The effect of exchange rate of fishpond effluents on the growth rate, yield and C : N ratio of *Ulva lactuca* was determined in experiments 2, 3 and 4 using also the results of treatment 3 from experiment 1 (no data on C : N ratio for exp. 4). The effect of stocking densities above 1 kg m^{-2} on *U. lactuca* growth rate, yield and C : N ratio was examined in experiment 5.

The Michaelis-Menten kinetic parameters of maximal yield (Ymax), maximal specific growth rate (μ_{max}), minimal C : N ratio (C : Nmin) and their respective values of Ks were calculated from the linear S/V vs S plots (Atkins and Nimmo 1980).

Results

No significant correlation ($r^2 = 0.071$; $p = 0.24$) was found between cumulative weekly irradiance and *Ulva lactuca* yield in experiments 1–4 (Table I). We thus presume that light (and temperture, which is tightly coupled to it) was not a major determining factor in the results of these experiments.

Yield and growth rate of *Ulva lactuca* in fishpond effluents of experiment 1, treatments no 2 and 3, were significantly higher ($P < 0.05$) than in the standard treatment of seawater and inorganic nutrient addition (Table II). The C : N ratios (w : w) for the fishpond effluents treatments were significantly lower ($p < 0.05$) than for the standard treatment, suggesting possible N-limitation in the latter. The addition of a nutrient pulse to fishpond effluents, treatment 2, did not effect the yields relative to treatment 3 but did reduce significantly the C : N ratio. As reported by Parker (1981), the differences in the C : N ratio were due mostly to changes in percent N and not in percent C.

Fishpond effluents also increased specific growth rate and yield of *Ulva lactuca* significantly ($P < 0.05$) relative to the standard treatment in the three water exchange rate experiments 2, 3 and 4 (Tables III, IV and V, respectively). All fishpond effluent treatments

Table I. Average daily irradiance and weekly DW yield of *Ulva lactuca* grown in the standard treatment of three ammonia uptake experiments in 600 liter, 1 m^2 tanks.

Exp. No.	Week	Irradiance ± SE (watt h m^{-2} d^{-1})	DW yield ± SE (g m^{-2} wk^{-1})
1	1	4950 ± 50	200 ± 9.6
	2	4700 ± 330	171 ± 3.8
2	1	3970 ± 1740	128 ± 8.3
	2	3530 ± 1375	168 ± 21.0
	3	4770 ± 770	97 ± 7.1
3	1	5670 ± 1050	121 ± 21.0
	2	6290 ± 1100	285 ± 13.0
	3	6995 ± 245	148 ± 16.0

Table II. Effect of nutrient source on the growth rate, yield and chemical composition of *Ulva lactuca* in experiment 1. See text for more details.

Week	Inflow ammonia concentration (μM ± SE)	Treatment No.	Water source	Inorganic nutrient addition	Growth rate (% d^{-1} ± SE)	Yield (g m^{-2} wk^{-1} ± SE) Fresh	Yield Dry	Percent C in DW ± SE	C : N Ratio ± SE (w : w)
1	1 ± 1	1	sea water	+	12.4 ± 0.38	1388 ± 45	200 ± 13	23.6 ± 0.07	12.2 ± 0.03
	25 ± 4	2	fish ponds	+	13.6 ± 0.60	1590 ± 78	247 ± 22	24.6 ± 0.52	7.8 ± 0.02
	25 ± 4	3	fish ponds	–	14.6 ± 0.39	1770 ± 30	259 ± 8	24.7 ± 0.45	8.5 ± 0.10
2	1 ± 1	1	sea water	+	10.1 ± 0.14	1027 ± 15	171 ± 4	28.9 ± 0.49	12.2 ± 0.20
	31 ± 2	2	fish ponds	+	12.6 ± 0.29	1415 ± 35	237 ± 10	28.2 ± 2.58	7.5 ± 0.10
	31 ± 2	3	fish ponds	–	11.6 ± 0.25	1260 ± 19	221 ± 5	26.4 ± 1.47	8.2 ± 0.03

Table III. Effect of fishpond water exchange rate on growth rate, yield and chemical composition of *Ulva lactuca* in experiment 2 (low inflow ammonia concentrations).

Week	Inflow ammonia conc. (μM ± SE)	Water exchange rate (d^{-1})	Growth rate (% d^{-1})	Yield (g m^{-2} wk^{-1} ± SE)		Percent C in DW ISE	C : N Ratio ± SE (w : w)
				Fresh	Dry		
1		standard	7.4	678 ± 25	128 ± 6	28.5 ± 0.61	10.0 ± 0.3
	10 ± 3	4	8.9	860 ± 8	158 ± 7	26.6 ± 1.05	13.7 ± 0.5
		8	10.0	1017 ± 23	186 ± 7	26.5 ± 2.10	10.8 ± 0.2
		16	10.0	1010 ± 40	190 ± 13	26.0 ± 0.15	8.4 ± 0.05
2		standard	9.1	892 ± 57	168 ± 15	26.6 ± 0.83	10.6 ± 0.2
	12 ± 2	4	10.4	1070 ± 10	195 ± 3	23.8 ± 0.30	15.4 ± 0.3
		8	9.6	957 ± 27	179 ± 5	24.1 ± 0.50	11.2 ± 0.03
		16	10.8	1136 ± 31	217 ± 3	26.6 ± 0.85	8.2 ± 0.1
3		standard	7.0	643 ± 125	97 ± 6	N. D.	N.D.
	14 ± 2	4	10.3	1050 ± 80	185 ± 27		
		8	8.4	800 ± 80	155 ± 8		
		16	10.2	1045 ± 55	195 ± 2		

Table IV. Effect of fishpond water exchange rate on growth rate, yield and chemical composition of *Ulva lactuca* in experiment 3 (medium ammonia concentrations).

Week	Inflow ammonia conc. (μM ± SE)	Water exchange rate (d^{-1})	Growth rate (% d^{-1})	Yield (g m^{-2} wk^{-1} ± SE)		Percent C in DW ± SE	C : N ratio ± SE (w : w)
				Fresh	Dry		
1	27 ± 3	standard	7.6	700 ± 100	121 ± 21	24.6 ± 0.3	11.9 ± 0.03
		4	13.4	1550 ± 50	261 ± 9	24.6 ± 0.2	9.7 ± 0.1
		8	13.1	1500 ± 100	275 ± 1	24.8 ± 1.8	7.7 ± 0.2
		16	11.8	1290[a] ± 10	233[a] ± 8	25.2 ± 2.7	7.5 ± 0.2
2	32 ± 4	standard	14.1	1675 ± 75	285[b] ± 13	25.9 ± 0.8	11.3 ± 0.1
		4	13.5	1575 ± 175	268[b] ± 30	25.2 ± 1.1	8.7 ± 0.4
		8	15.8	2025 ± 75	344[b] ± 13	27.7 ± 1.6	7.3 ± 0.1
		16	15.5	1966 ± 50	334[b] ± 9	26.8 ± 0.1	7.2 ± 0.03
3	48 ± 10	standard	9.3	915 ± 135	147 ± 16	23.8 ± 0.1	17.6 ± 0.7
		4	15.1	1875 ± 75	252 ± 36	29.4 ± 1.0	9.0 ± 0.1
		8	16.5	2175 ± 75	352 ± 3	29.0 ± 0.5	7.5 ± 0.04
		16	16.8	2250 ± 0	381 ± 16	26.2 ± 0.5	7.1 ± 0.2

[a] Washout of fragments observed. Data not included in the statistic analyses.
[b] Estimates (17% DW). Numbers not included in the statistical analyses.

decreased C : N ratio below its value in the standard treatment in experiment 3 (Table IV), but in experiment 2 (Table III) this was observed only in the highest water exchange rate treatment. This difference we attribute to the difference in inflow ammonia-N concentration, which was higher in experiment 3 than in experiment 2. The algae in the latter experiment were probably severly N-limited in the 4 and 8 d^{-1} water exchange rate treatments.

Higher values of both water exchange rate and ammonia-N concentration in the inflow correlated significantly ($p < 0.03$) with higher specific growth rate and yield in experiments 3 and 4 but not in experiment 2, in which ammonia levels were low. Water exchange rate was inversely correlated with C : N ratio ($p < 0.02$) (Tables III – IV). A comparison of the effect of inflow ammonia-N concentration on the C : N ratio at similar water exchange rates between experiments 2 and 3 also shows a consistent inverse correlation. These results conform with those of Parker (1981) for *Ulva lactuca* and of DeBoer *et al.* (1978) for red algae.

The two parameters, water exchange rate and inflow ammonia concentration, can be multiplied by each other to produce the unified continuous parameter of ammonia-N flux through the tanks as suggested by Fujita and Goldman (1985). This parameter correlated positively with the specific growth rate and yield for *Ulva lactuca* and negatively with its C : N ratio (Figs 1 – 3, respectively). The three correlations appear to follow saturation type curves which fit the Michaelis-Menten kinetics. Maximal specific growth rate (μ_{max}) was 18.6% d^{-1} with a half saturation constant (Ks) of 326 mmoles m^{-2} wk^{-1}. Maximal

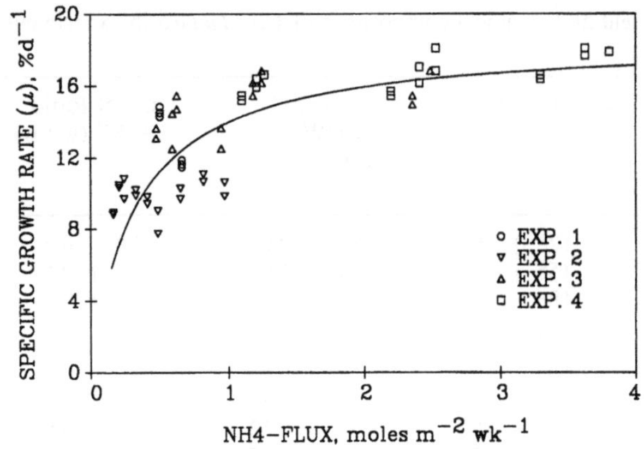

Fig. 1. Specific growth rate (μ) of *Ulva lactuca* in experiments 1–4 as a function of ammonia-N flux. A Michaelis-Menten curve has been fitted to the data (see text) with Vmax and Ks of 18.6 and 0.326, respectively.

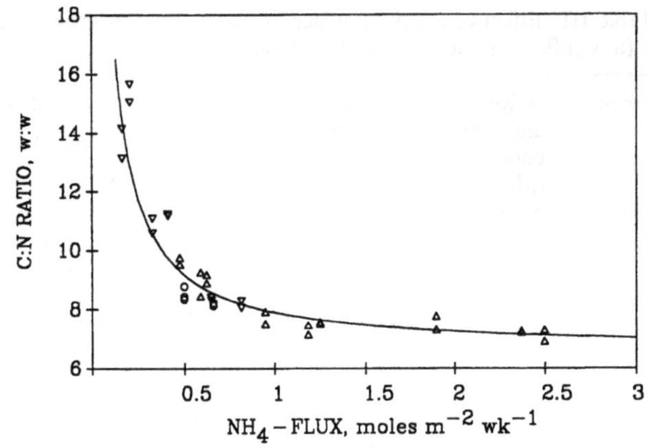

Fig. 3. C : N ratio of *Ulva lactuca* in experiments 1–3 as a function of ammonia-N flux. A Michaelis-Menten curve has been fitted to the data (see text) with Vmin and Ks of 6.6 and 0.195, respectively. Symbols as in Figure 1.

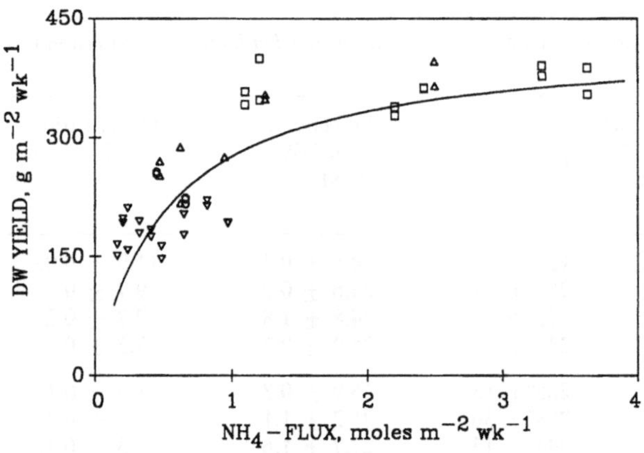

Fig.2. Dry yield of *Ulva lactuca* in experiments 1–4 as a function of ammonia-N flux. A Michaelis-Menten curve has been fitted to the data (see text) with Vmax and Ks of 422 and 0.53, respectively. Symbols as in Figure 1.

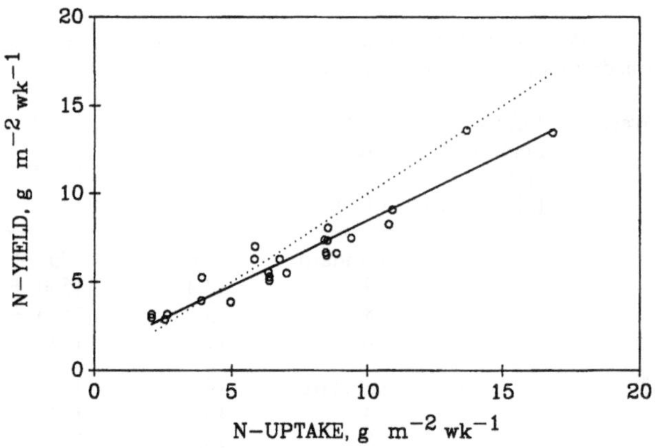

Fig. 4. Nitrogen production of *Ulva lactuca* grown in fishpond effluents in 600 L tanks at a stocking density of 1 kg m^{-2} in experiments 1–3 as a function of ammonia-N uptake. Symbols, data; Solid line, linear regression of the plotted data; Dotted line, a slope of unity.

Table V. Effect of fishpond water exchange rate on growth rate, yield and chemical composition of *Ulva lactuca* in experiment 4 (high ammonia concentrations).

Week	Inflow ammonia conc. (μM ± SE)	Water exchange rate (d^{-1})	Growth rate (% d^{-1})	Yield ($g\ m^{-2}\ wk^{-1} \pm SE$) Fresh	Dry
1	77 ± 5	4	15.3	1925 ± 25	350 ± 8
		8	15.6	1975 ± 25	334 ± 6
		12	16.5	2175 ± 25	385 ± 7
2	71 ± 4	4	16.2	2100 ± 50	374 ± 26
		8	16.6	2200 ± 100	362 ± 10
		12	17.9	2500 ± 50	372 ± 17
3	78 ± 2	4	16.6	2200 ± 0	374[a] ± 0
		8	17.5	2400 ± 150	408[a] ± 26
		12	17.9	2500 ± 0	425[a] ± 0

[a] Estimates (17% DW). Numbers not included in the statistical analyses.

calculated DW yield was 422 g m^{-2} wk^{-1} (*ca* 60 g m^{-2} d^{-1}) with a Ks of 530 mmoles ammonia-N m^{-2} wk^{-1}. The maximal DW yield (including ash, which for some of the algal samples constituted 30–40% of DW) actually obtained in our experiments was only 380 g m^{-2} wk^{-1} or 55 g m^{-2} d^{-1}. However, it has to be considered that small algal fragments and spores were lost from the tanks, especially at the high water exchange treatments. The C : N ratio decreased hyperbolically with increasing ammonia-N flux. The minimal calculated value of C : N ratio was 6.6 (Ks = 195 mmoles m^{-2} wk^{-1}).

Organic nitrogen yield showed a positive linear correlation ($r^2 = 0.9$; n = 28) with ammonia uptake, measured as the difference between influent and effluent ammonia concentrations multiplied by the water exchange (Cohen and Neori 1991) (Fig. 4). The slope of 0.74 (g N-yield per 1 g ammonia-N uptake)

Table VI. Effect of stocking density on growth, yield and chemical composition of *Ulva lactuca* in experiemnt 5 (sufficient nitrogen and 12 water exchanges d^{-1}).

Week	Stocking density (kg m^{-2})	Growth rate (% d^{-1})	Yield (g m^{-2} wk^{-1} ± SE)		Percent C in DW ± SE	C : N ratio ± SE (w : w)
			Fresh	Dry		
1	2	9.9	1987 ± 80	766 ± 105	30.1 ± 3.2	6.2 ± 0.3
	4	3.7	1183 ± 94	407 ± 65	33.8 ± 2.5	6.4 ± 0.2
	6	0.6	246 ± 72	63 ± 29	35.1 ± 0.7	6.5 ± 0.1
2	2	9.1	1783 ± 164	310 ± 30	36.3 ± 1.9	6.7 ± 0.1
	4	2.3	715 ± 52	137 ± 12	37.2 ± 0.7	6.5 ± 0.1
	6	2.1	940 ± 115	186 ± 20	35.7 ± 0.7	6.3 ± 0.1
3	2	8.7	1668 ± 44	275 ± 14	28.3 ± 0.6	6.2 ± 0.03
	4	4.0	1310 ± 250	253 ± 42	28.1 ± 0.7	5.9 ± 0.04
	6	2.5	1140 ± 31	231 ± 31	28.5 ± 0.3	5.6 ± 0.1

was significantly different from a slope of 1, which would represent 100% uptake (t test, $p < 0.01$). Carbon content in the plant tissue (Tables II–IV) was not correlated with treatments tested in each experiment, ranging between 23 to 29% of DW.

Increasing stocking density above 1 kg m^{-2} decreased growth rate and yield (Table VI). This agrees with several studies, which had shown the optimal stocking density of *Ulva lactuca* to be near 1 kg m^{-2} (*e.g.*, Ryther *et al.* 1984). The results of DW determination of the first week of this experiment were apparently faulty and therefore we refer to the fresh weight data. The differences between the yields obtained at the three stocking densities were very significant for the first week of the experiment and progressively decreased at the second and the third week. The values of the specific growth rate (μ) decreased with increasing stocking density. Even in the 2 kg m^{-2} treatment (Table VI) μ was still lower compared with 1 kg m^{-2} treatments in this study (Tables II–V). Carbon content in plant tissue was similar for the three stocking densities within each week. The variation between the weeks ranged between 28 to 37% of dry wt. These values are higher than those for the stocking density of 1 kg m^{-2} in the other experiments. However, the values of C : N ratio were all lower in this experiment than in the 1 kg m^{-2} experiments and ranged between 5.6 to 6.7. In the second and third week the C : N ratio decreased consistently with the increasing stocking density, indicating a trend of inverse relationship between them. This complements the observation made in this experiment by Cohen and Neori (1991) that percent N increased consistently at higher stocking densities.

Discussion

Fishpond effluents undoubtedly support excellent growth of *Ulva lactuca* under our field conditions. In the first experiment, the treatment using only fishpond effluents resulted in higher yield (by up to 38%) than in the standard treatment with only pulse nutrient addition. Although the treatment combining fishpond effluents and pulse nutrient addition did not have a consistent effect on the seaweed growth rate, yield and percent C, it did reduce its C : N ratio. This indicates that the algae accumulated some of the pulse-added nitrogen, but this was not sufficient to generate more growth. Our results confirm the exceptional suitability of *Ulva* for intensive culture in different types of nutrient-loaded water (*e.g.*, Ryther *et al.* 1984, Langton *et al.* 1977, Guist and Humm 1976).

Growth rate, yield and C : N ratio showed saturation kinetics in their correlations with ammonia-N flux. It corroborates our earlier findings that the seaweeds were N-limited at low N-fluxes (up to 10–15 μmoles l^{-1} h^{-1}, 1–1.5 moles m^{-2} $week^{-1}$). Above a certain N-flux the growth curve of the seaweeds leveled off and the C : N ratio in the plant tissue decreased without a concomitant decrease in percent C. These results indicate that at high N-fluxes the seaweeds might have become light-limited, as shown by Duke *et al.* (1986, 1989).

Maximal DW yield of *Ulva lactuca* obtained in N-saturation conditions was 55 g DW m^{-2} d^{-1}, above 30 g m^{-2} d^{-1} ash free. This value is higher than the theoretical maximum light-limited daily DW yield of 25 g m^{-2} d^{-1}, averaged annually for algal photosynthetic production in our region (Avron 1989). In addition, the experiments with *U. lactuca* took place during the winter months, in which yields are lowest under Eilat conditions (Cohen and Neori, unpublished). This explains the low specific growth rates obtained in our experiments relative to the maximal outdoor values reported in the literature (*e.g.*, Duke *et al.* 1986). Thus, the annual average ash free yield of *U. lactuca* in our tanks is likely to be above 30 g DW m^{-2} d^{-1}, which is remarkable considering we did not enrich the cultures with CO_2. Such high yield can be explained by several testable hypotheses, such as the following:

a) Uptake of organic nutrients, *e.g.*, free amino acids or other organic compounds by the seaweeds (Fries 1973, Gemmill *et al.* 1974);

b) A greater photosynthetic efficiency in our tanks, thanks to optimal light conditions, as compared to the efficiency anticipated by Avron (1989) in his calculation.

The supply of inorganic carbon to the system may have been bolstered by heterotrophic breakdown of dissolved organic substances, excreted amply by the fish. We have shown such breakdown processes to be intensive in our ponds (Krom and Neori 1989). Photosynthetic depletion of inorganic C increases the pH in the water. Indeed, afternoon pH values measured in the water of the standard treatments (which were fed clean seawater) were usually higher than in the effluent treatments, sometimes by over a whole pH unit (unpublished). Thus, fishpond effluents may provide to the seaweed also inorganic carbon, diminishing the need for CO_2 enrichment.

The saturation of ammonia-N uptake (Cohen and Neori 1991) occurred at N-flux of 24 mmoles m^{-2} h^{-1} while maximum growth rate and yield reported here were already obtained at N-fluxes of about 10 mmoles m^{-2} h^{-1}. The results of Duke *et al.* (1986, 1989) indeed showed that N uptake in *Ulva* is less limited by light than growth. Also, washout of algal fragments and spores, which was observed but not measured, must have contributed to the loss of some yield especially at high ammonia fluxes, which were brought about by high water exchange rates. This loss could account for the deficiency of N yield compared to N uptake in Figure 4.

We did not get higher yields by increasing the stocking density of *Ulva lactuca* above 1 kg m^{-2}. Indeed, the optimal density of 0.8 kg m^{-2} had been reported for *U. faciata* Delile, (Lapointe and Tenore 1981) and for *U. lactuca* (DeBusk *et al.* 1986), and an optimum of 1 kg m^{-2} was reported by Ryther *et al.* (1984) for *U. lactuca.* The fact that the increase in percent tissue nitrogen with increasing stocking densities (Cohen and Neori 1991) was not concomitant with a decrease in percent tissue carbon indicates that the algae probably did not become carbon limited but light limited (Duke *et al.* 1986, 1989). The progressive decline of the differences in yields between stocking densities during the three weeks of experiment 5 suggests that the algae might have been adapting to the light-limited conditions. Longer experiments may lead to better results with the high stocking densities. Within the ranges tested, this parameter is not critical to the optimal operation of the seaweed biofilter but it can be used to modify percent nitrogen in the yield.

Acknowledgements

We are grateful to H. Voet for help with statistical analyses and to S. Peled and the others at the NCM for their assistance. M. Friedlander and C. Dawes critically reviewed earlier versions of this manuscript. This research was supported by the Israeli Ministry for Energy and Infrastructure Mariculture Project, The Arava Research and Development Fund, BARD research grant No. I-1634-89 and grant No. 3035189 from the Israeli Ministry for Science and Technology.

References

Atkins, G. L. and I. A. Nimmo. 1980. Current trends in the estimation of Michaelis-Menten parameters. *Anal. Biochem. 104*: 1–9.

Avron, M. 1989. Efficiency of biosolar energy conversion by aquatic photosynthetic organisms. *In*: (Y. Cohen and E. Rosenberg, eds) *Microbial Mats, Physiological Ecology of Benthic Microbial Communities*. Amer. Soc. Microbiol., Washington, DC. pp. 387–389.

Cohen, I. and A. Neori. 1991. *Ulva lactuca* biofilters for marine fishpond effluents. I. Ammonia uptake kinetics and nitrogen content. *Bot. Mar. 34*: 475–482.

DeBusk, T. A., M. Blakeslee and J. H. Ryther. 1986. Studies on the outdoor cultivation of *Ulva lactuca* L. *Bot. Mar. 24*: 381–386.

DeBoer, J. A., H. J. Guigly, T. L. Israel and C. F. D'Elia. 1978. Nutritional studies of two red algae. I. Growth rate as a function of nitrogen source and concentration. *J. Phycol. 14*: 261–266.

D'Elia, C. F. and J. A. DeBoer. 1978. Nutritional studies of two red algae. II. kinetics of ammonium and nitrate uptake. *J. Phycol. 14*: 266–272.

Duke, C. S., B. E. Lapointe and J. Ramus. 1986. Effects of light on growth, RuBPase activity and chemical composition of *Ulva* species (Chlorophyta). *J. Phycol. 22*: 362–370.

Duke, C. S., W. Litaker and J. Ramus. 1989. Effects of the temperature, nitrogen supply and tissue nitrogen on ammonium uptake rates of the Chlorophyte seaweeds *Ulva curvata* and *Codium decorticatum. J. Phycol. 25*: 113–120.

Fries, E. 1973. Requirements of organic substances in seaweeds. *Bot. Mar. 16*: 19–31.

Fujita, R. M. and J. C. Goldman. 1985. Nutrient flux and growth of the red alga *Gracilaria tikvahiae* McLachlan (Rhodophyta). *Bot. Mar. 28*: 265–268.

Gemmill, E. R. and R. A. Galloway. 1974. Photoassimilation of ^{14}C-acetate by *Ulva lactuca. J. Phycol. 10*: 359–366.

Guist, G. G. Jr. and H. J. Humm. 1976. Effects of Sewage effluents on growth of *Ulva lactuca*. *Biol. Sci. 4*: 267–271.

Krom, M. D. and A. Neori. 1989. A total nutrient budget for an experimental fishpond with circularly moving seawater. *Aquaculture 83*: 345–358.

Langton, R. W., K. C. Haines and R. E. Lyon. 1977. Ammonia-nitrogen produced by the bivalve mollusc *Tapes japonica* and its recovery by the red seaweed *Hypnea musciformis* in a tropical mariculture system. *Helgoländer wiss. Meeresunters. 30*: 217–229.

Lapointe, B. E. and K. R. Tenore. 1981. Experimental outdoor studies with *Ulva fasciata* Delile. I. Interaction of light and nitrogen on nutrient uptake, growth and biochemical composition. *J. exp. mar. Biol. Ecol. 53*: 135–152.

Lapointe, B. E., L. D. Williams, J. C. Goldman and J. H. Ryther. 1976. The mass outdoor culture of macroscopic marine algae. *Aquaculture 8*: 9–21.

Parker, H. S. 1981. Influence of relative water motion on the growth, ammonium uptake and carbon composition of *Ulva lactuca* (Chlorophyta). *Mar. Biol. 63*: 309–318.

Ryther, J. H., T. A. DeBusk and M. Blakeslee. 1984. Cultivation and conversion of marine macroalgae. *Final Subcontract Report to SERI, the U. S. Dept. of Energy, Publication SERI/STR-231-2360 DE 84004522.* 81 pp. Available through National Technical Information Service, U. S. Dept. of Commerce, Springfield, Virginia.

Vandermeulen, H. and H. Gordin. 1990. Ammonium uptake using *Ulva* (Chlorophyta) in intensive fishpond systems: mass culture and treatment of effluent. *J. Appl. Phycol. 2*: 363–372.

Botanica Marina
Vol. 34, pp. 491–499, 1991

Antithamnionella floccosa (Rhodophyta) in Culture: Distribution, Life History and Chromosome Number

T. Jacobsen, J. Rueness and A. Athanasiadis*

Section for Marine Botany, Department of Biology, University of Oslo, P.O. Box 1069 Blindern, 0316 Oslo 3, Norway

* *Department of Marine Botany, University of Göteborg, Carl Skottsbergs Gata 22, S-41319 Göteborg, Sweden*

(Accepted 22 July 1991)

Abstract

Antithamnionella floccosa from Norway was grown under different combinations of temperature and salinity, and a *Polysiphonia* type of life history was demonstrated for the first time in culture. The alga survived temperatures from −1 °C to 15 °C and salinities from 18‰ to 32‰. The life cycle was completed in 8 months at 8 °C under a 16 : 8 light : dark cycle, but not under continuous light at the same temperature. At 14 °C and with a 16 : 8 light : dark cycle, gametangia were produced, but no carposporophyte development occurred. The results suggest that the southeastern boundary of distribution in Scotland and southern Norway is set by a lethal summer temperature of 17 °C. The northern distribution limit is probably a reproduction limit near a summer maximum of 6 °C, or due to infavourable ice and light conditions during winter. Distribution data from the western Atlantic coast indicate a broader temperature tolerance than on European coasts. This might be explained in terms of ecotypic differentiation. A haploid chromosome number of n = 31–33 was determined in apical cells of gametophytes.

Introduction

The geographical distribution of benthic marine algae is often correlated with their temperature tolerances and requirements (Breeman 1988, Cambridge *et al.* 1984, 1987, Hoek 1982 a, b, Yarish *et al.* 1984, 1986). Temperature may be limiting for survival, or for growth and reproduction (Hoek 1982 a, b). A comparison of the temperature responses, distribution data and summer and winter ocean isotherms, for a given species indicates the season at which the temperature is limiting (Breeman 1988). Other factors such as competition, salinity, daylength and substrate may also affect distribution, especially on a local scale (Bird and McLachlan 1986, Breeman 1988). Since ecotypic differentiation in response to temperature may influence the distribution range (Breeman 1988, Yarish *et al.* 1986), more than one isolate is often needed to explain the distribution pattern of a species. The isolates need not necessarily come from boundary populations (Breeman 1988).

In this study, the temperature and salinity tolerances and requirements of *Antithamnionella floccosa* (Müller) Whittick have been experimentally tested. This is a sublittoral, northern amphiatlantic species which is found from Scotland to the Murman Coast in the eastern Atlantic, and from Massachusetts to Labrador and Greenland in the western Atlantic (Fig. 1). It has also been recorded from Iceland, the Faroes, the Shetland Isles and Bear Island. According to the distribution data, the northern boundary of the species in the eastern Atlantic is near the 0 °C–1 °C February isotherm. This corresponds with a lethal temperature between −1 °C and 0 °C (Hoek 1982 a, b). However, in the western Atlantic *A. floccosa* is found further north than the −2 °C February isotherm, indicating that the species is not limited by a lethal water temperature in the north. The northern boundary also seems to follow the 8 °C August isotherm (5 °C August isotherm in the western Atlantic). This corresponds to a minimum summer temperature

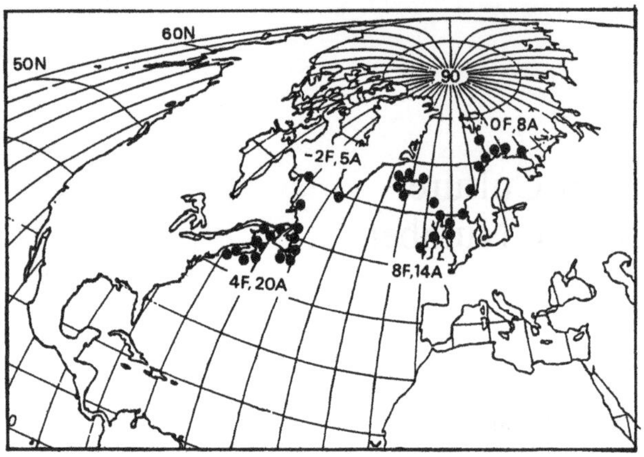

Fig. 1. The distribution of *Antithamnionella floccosa*, according to the following sources: Athanasiadis 1990, Batters 1902, Børgesen 1902, Edelstein *et al.* 1970, Harvey 1844, 1853, Jaasund 1965, Jonsson 1901, Juliussen 1979, Kleen 1874, Mathieson and Penniman 1986, Munda 1976, 1978, 1980, 1987, Pedersen 1976, Pybus 1977, Rosenvinge 1898, Sivertsen 1981, South and Hooper 1972, 1980, South *et al.* 1988, Taylor 1957, Tittley *et al.* 1982, Zinova 1912, Rueness pers. comm. and Athanasiadis and Rueness pers. comm. The ocean isotherms for August (A) and February (F) are shown.

of 6 °C for growth and reproduction (3 °C in the western Atlantic). The southern boundary of the species in the eastern Atlantic follows the the 14 °C August isotherm, which corresponds with a lethal summer temperature of 17 °C. To some extent, it also follows the 8 °C and the 6 °C February isotherm. If the alga is limited by the 8 °C February isotherm (corresponding with a maximum temperature of 9 °C for growth and reproduction), one would expect to find *A. floccosa* along the coasts of England. If the alga is limited by the 6 °C February isotherm as indicated by its distribution in Norway (corresponding with a maximum temperature of 7 °C for growth and reproduction), one would not expect the alga to grow in Scotland and the Faroes. In both cases one would expect the alga to grow in the Skagerrak. It therefore seems likely that the species is limited by a lethal summer temperature in the eastern Atlantic. The southern boundary in the western Atlantic may be either a lethal limit near the 20 °C August isotherm (corresponding with a lethal summer temperature of 23 °C) or a growth and reproduction limit at the 4 °C February isotherm (corresponding with a maximum temperature of 5 °C for growth and reproduction). In this study, some of these hypotheses about the distribution limits of *A. floccosa* have been tested experimentally.

Whittick (1980) described the life history of the species, based on field collections from Newfoundland, and suggested that it was of the regular *Polysiphonia* type. In this study, which is the first culture study of *A. floccosa*, the life history has been completed in culture.

Material and Methods

The material used in this study originated from a tetrasporophyte collected at Edøy (63°17′N, 8°15′E) on the west coast of Norway. The plant, growing epiphytically on *Laminaria hyperborea* (Gunn.) Foslie, was collected by K. Sivertsen in 1979 and isolated into unialgal culture by J. Rueness. Stock cultures have been kept as gametophytes derived from the original tetraspores and kept in a 30‰ half strength IMR medium (Eppley *et al.* 1967) at 8 °C, 8 : $\overline{16}$ light : dark cycle. Excised apical portions were used as the inoculum for the present study.

For completion of the life cycle, fertile male and female plants were placed in the same 25 mL polystyrene Petri dish and grown at 8 °C, 8 : $\overline{16}$ (L : D); 8 °C, 24 : $\overline{0}$ (L : D) and 14 °C, 16 : $\overline{8}$ (L : D). A photon irradiance of 50 μmol m^{-2} s^{-1} was provided by Philips fluorescent tubes (TL E 32W/33 and TL E 40W/33) and measured by a QSL-100 Laboratory Quantum Scalar Irradiance meter measuring photosynthetically active radiation (PAR). The medium was changed and the plants examined each week. As the carposporophyte developed the male and most of the female plants were removed. Carpospores were isolated and grown in separate Petri dishes. Tetrasporelings were also isolated into separate Petri dishes as they were liberated from the tetrasporophyte.

A cross-gradient growth table as described by Yarish *et al.* (1979) was used to study survival, growth and fertility under different combinations of temperature and salinity. Three to five apices (2–5 mm) from vegetative male plants were placed in each Petri dish. The plants were acclimated to exerimental conditions by progressively exposing them towards higher or lower temperatures and salinities, with two days at each 5 °C or 5‰ interval.

As a measure of growth, the plants were photographed or weighed, or the axial cells in the main axis and major side branches of unlimited growth were counted. The plants were dried at 50 °C for two days prior to weighing.

Figs 2–10. Life history of *Antithamnionella floccosa*.
Fig. 2. Spermatangia on the adaxial side of whorl branches. Scale bar = 30 μm. Fig. 3. A 4-celled carpogonial branch (CB) on a reduced, 2-celled whorl branch. Scale bar = 30 μm. Fig. 4. Early stage of a carposporophyte. Several involucre branches (IB)

have developed. Scale bar = 50 µm. Fig. 5. Plant bearing mature carposporophytes. Scale bar = 400 µm. Figs. 6 and 7. Pedicellate, cruciately divided tetrasporangia. Scale bars = 200 µm. and 30 µm, respectively. Fig. 8. Monoecious plant with carpogonial branches (CB) and immature spermatangia (S). Scale bar = 30 µm. Fig. 9. Creeping, sparsely branched thallus of plant grown in suboptimal conditions. Scale bar = 400 µm. Fig. 10. The typical branching form with pairs of sub-equal, opposite whorl branches and alternately arranged branches of unlimited growth. Scale bar = 200 µm.

Algae with no pigments were considered dead. In doubtful cases, the algae were placed in optimum conditions for 1–2 weeks to test for viability.

For cytological investigations, material was fixed in picric acid (Dammann 1930) and nuclei stained in chrome-haematoxylin (Melander and Wingstrand 1953).

Statistics

Weight data were analysed with one-way and two-way ANOVA (analysis of variance). Studentized range (Q-test) was used for comparing treatments (Parker 1979). The non-normal data were log transformed prior to analysis and a 0.05 significance level was used for all tests.

Results

Reproduction and life-history in culture

The life cycle of *Antithamnionella floccosa* was completed at 8 °C, 8 : $\overline{16}$ (L : D), but not under the other two culture regimes [8 °C, 24 : $\overline{0}$ (L : D) and 14 °C, 16 : $\overline{8}$ (L:D)]. Spermatangia and carpogonia developed at 14 °C, 16 : $\overline{8}$ (L : D), but no carposporophyte development occurred.

Spermatangia (Fig. 2) developed 1–2 weeks prior to carpogonia (Fig. 3). One to two weeks after the first carpogonial branches were observed, young carposporophytes could be seen (Figs 4, 5). Mature carpospores were shed 2–5 weeks later and started to germinate. About 4 months after germination, the tetrasporophytes were fertile (Fig. 6). The pedicellate, cruciately divided tetrasporangia measured 41–51 µm × 70 µm (Figs 6, 7). Tetraspores developed into reproductive gametophytes 3½–4 months after germination, thus completing the life cycle in 8 months.

Antithamnionella floccosa is dioecious with male and female structures on separate plants. No deviations from this pattern have been reported previously. However, the isolate from Edøy developed spermatangial clusters on fertile female plants (Fig. 8). The spermatangia developed later than the carpogonial branches, and were most abundant in the lower parts of the plant. Monoecious plants were observed at both 8 °C, 8 : $\overline{16}$ (L : D) and 14 °C, 16 : $\overline{8}$ (L : D), all those observed had initially been female. The spermatangia were non-functional since carposporophytes only developed when placed together with male plants.

Plants grown under suboptimal conditions often produced a creeping, sparsely branched thallus (Fig. 9). On transfer to more favourable conditions, the plants developed a typical branching form (Fig. 10). Lateral branches that were shed from the main thallus produced rhizoids at their bases, and developed into fully grown plants, showing that the species can be vegetatively propagated.

Some of the carpogonial branches (5.5%, n = 429) were found in whorl branches consisting of three or more cells. The rest were found on two-celled branches, which are described as typical for the species.

Growth and survival under different culture conditions

The effects of temperature and salinity on growth are shown in Figures 11, 12 and 13.

Antithamnionella floccosa survived temperatures from −1 °C to 15 °C, and salinities from 18‰ to 32‰. The alga grew between 3 °C–15 °C and 23‰–32‰. It survived −1 °C, 0 °C and 3 °C but showed little or no growth. On the formation of ice in the Petri-dish (~ −1.9 °C), the alga died. The alga grew poorly at 18‰, and died within 3 weeks at 18 °C and 20 °C,

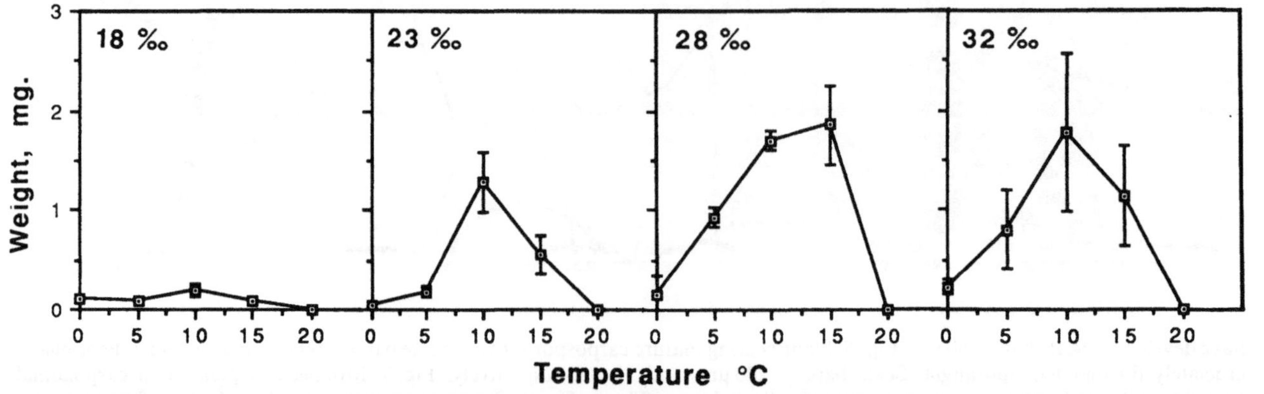

Fig. 11. Yield of *Antithamnionella floccosa* at various combinations of temperature and salinity after 6½ weeks in culture. Light regime was 50 µmol $m^{-2}s^{-1}$, 16 : 8 (L : D). Error bars represent SD of 5 separate measurements.

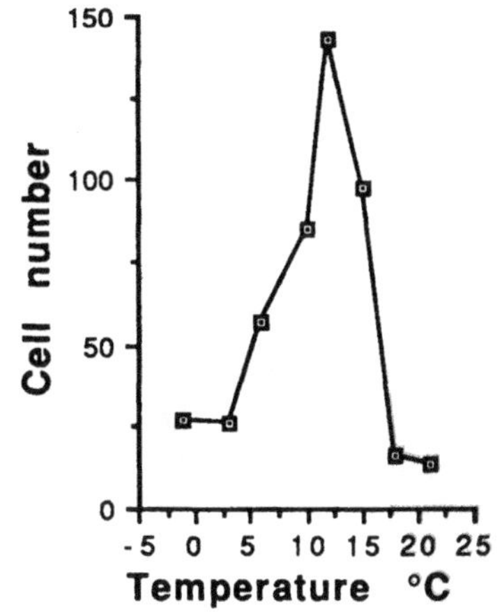

Fig. 12. Number of axial cells in both the main axis and the major side branches of unlimited growth of *Antithamnionella floccosa*. Inoculum was apices of 20 ± 3 cells. Measurements made after 21 days at 30‰, 40 µmol m^{-2} s^{-1}, 16 : 8 (L : D).

and within 12 days at 0‰, 6‰ and 11‰. Maximum growth took place at 12 °C (Figs 12, 13). At 23‰ there was no significant differnce between algae grown at 0 °C and 5 °C ($Q_{0.05}$ = 3.54), but at the highest salinities tested (28‰ and 32‰), significant differences in growth between 0 °C and 5 °C were recorded ($Q_{0.05}$ = 14.82 and 4.94, respectively) (Fig. 11). At 23‰, the algae grown at 10 °C were significantly larger than those grown at 15 °C ($Q_{0.05}$ = 9.50). At 28‰ and 32‰ there were no significant differences in growth between the two temperatures ($Q_{0.05}$ = 2.07 and 3.40, respectively). At 18‰ growth was poor at all the temperatures tested and no significant differences were seen ($F_{0.05}$ = 0.60).

At 5 °C and 10 °C the male plants became fertile at 23‰ and 28‰, while at 0 °C the algae became fertile at 28‰ and 32‰. At 15 °C no plants were fertile. The plants became fertile after approximately three weeks in culture.

Cytology

Chromosome counts were made in late prophase stages of mitosis in apical cells of a male gametophyte. Individual chromosomes were distinguished as deeply staining globular bodies spread over the whole nuclear area of the cell. Chromosomes were counted at several planes of focus (Fig. 14) and these showed a haploid number of 31 to 33 chromosomes.

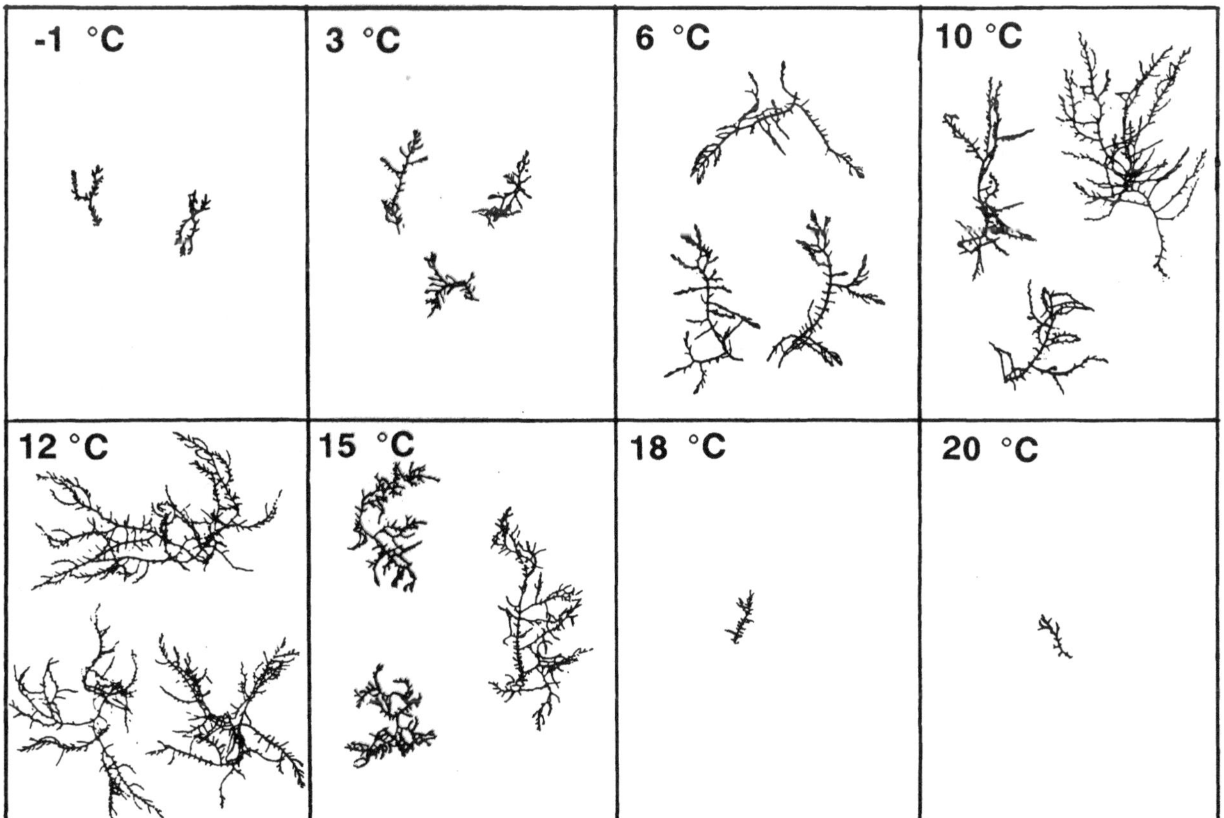

Fig. 13. Growth responses of *Antithamnionella floccosa* after 29 days under varying temperature regimes, and 40 µmol m^{-2} s^{-1}, 16 : 8 (L : D), 30‰. Apical cuttings (2–5 mm) were used as inoculum. "Squares" are 12 × 9 mm.

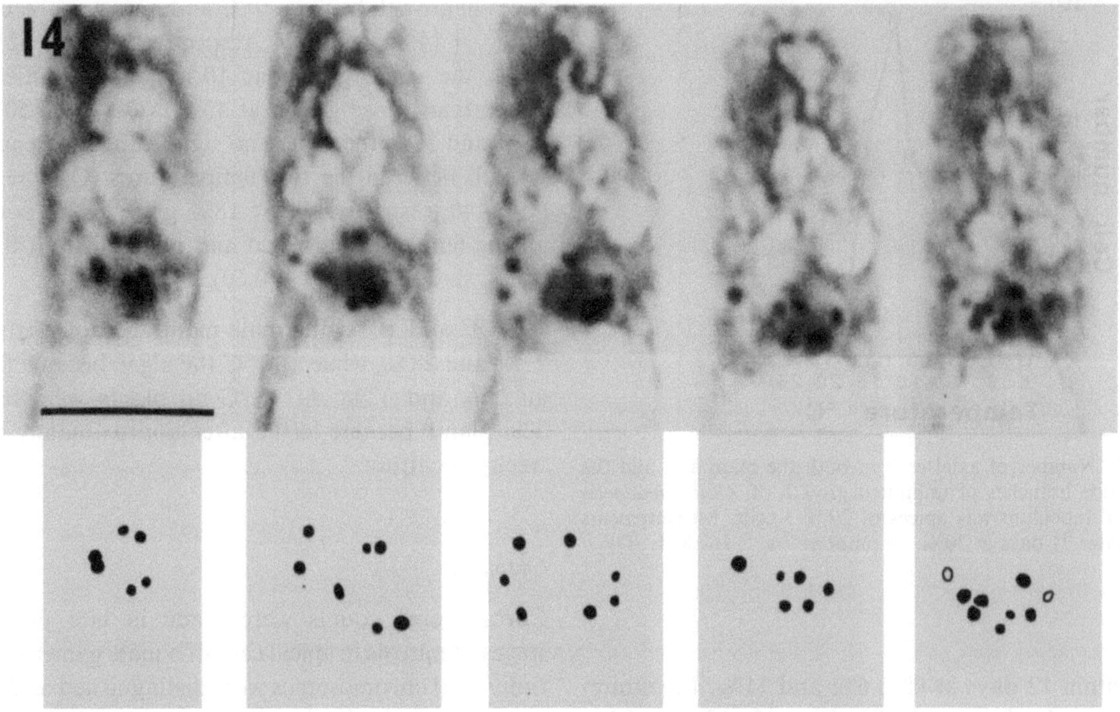

Fig. 14. *Antithamnionella floccosa.* Late prophase in an apical cell. Five focal planes showing 6, 7, 7, 6, and 7 chromosomes (Chromosomes indicated by ring are visible in two successive planes). Scale bar = 10 µm.

Discussion

Northern limit

The results suggest that the lethal minimum temperature for *Antithamnionella floccosa* is between −1 °C and −2 °C, *i.e.* near freezing point. Along the coast of Norway, the mean February temperature both at the surface and at a depth of 4 m is above 1 °C (Sætre 1973, Cambridge *et al.* 1987). The extreme temperatures are somewhat lower, and ice occasionally forms. This may explain the absence of the species from the littoral zone. In the Canadian Arctic and on southern Greenland, the species survives in deep, ice-free waters (Rosenvinge 1898, South and Hooper 1972). The alga has been found at 25 m in Newfoundland (South and Hooper 1980) and at 10 m on Bear Island (Athanasiadis 1990). On the basis of this, one would also expect to find *A. floccosa* in deep water in the southern part of Spitzbergen. There may be several reasons for its absence here. The substrates is loose and there is very little light in the winter (Spitzbergen is situated 20–30 °C north of southern Greenland). Another possibility is that the alga has passed unrecorded in Spitzbergen. Further north on both Spitzbergen and Greenland, the alga is probably limited by a combination of ice and insufficient light.

Antithamnionella floccosa grew at both 3 °C and 5 °C, and it is therefore unlikely to have a growth limit of 6 °C in the north east Atlantic Ocean (near the 8 °C August isotherm). Fertile gametophytes were observed at temperatures below 6 °C, but the plant might still have a reproductive limit of 6 °C since other parts of the life cycle may be limiting for distribution, *e.g.* the development of carposporophytes. According to Zinova (1912), carposporophytes are found in Vostotsjnaja Litsa (Murman Coast) only in June and July when the temperature is above 6 °C (about 8–9 °C). A reproductive limit of 6 °C might explain the absence of the species from Spitzbergen, where the water does not reach this temperature. Attempts to complete the life cycle at temperatures below 6 °C may clarify this problem. If the western Atlantic population has similar temperature requirements for sexual reproduction, it may reproduce vegetatively in Labrador and Greenland. The western Atlantic population may also be of another ecotype, with a reproductive limit of 3 °C (near the 5 °C August isotherm). Ecotypic differentiation between populations from the eastern and western Atlantic has previously been demonstrated for the two amphiatlantic species *Callithamnion tetragonum* (With.) Agardh and *Lomentaria orcadensis* (Harvey) Collins *ex* Taylor (Yarish *et al.* 1986).

Southern limit

The alga died at 18 °C, which supports the hypothesis that the upper lethal limit for *A. floccosa* is near the 14 °C August isotherm in Scotland and western coast of Norway. The western Atlantic population seems to be of another ecotype since it has a southern boundary corresponding to a lethal limit of 23 °C. Another possibility is that it grows in deeper waters where the temperature is less extreme. Other species, like *Lomentaria articulata* (Hudson) Lyngbye, are found submerged near their southern limit of distribution (Yarish *et al.* 1986). If the west Atlantic population has a southern growth and reproductive limit at the 4 °C February isotherm, it is unlikely also to have a reproductive limit at the 5 °C August isotherm in the north. This would give too small a temperature range for reproduction.

In order to understand the distribution completely and solve some of the questions raised here, similar experiments should be made using western Atlantic isolates.

Salinity

Antithamnionella floccosa barely survived 18‰ and died at 11‰. This alone does not exclude the species from the littoral zone, or limit its geographical distribution.

Two-way variance tests (ANOVA) showed interaction between temperature and salinity, which means that the response to temperature depends on the salinity. *Antithamnionella floccosa* grew best at 10 – 15 °C and 28 – 32‰. Edwards (1979) found that *Callithamnion hookeri* (Dillw.) Gray grew best at low salinities when the temperature was also low, but no such interaction was observed for *A. floccosa*.

Reproduction

Completion of the life cycle at 8 °C, 8 : $\overline{16}$ (L : D) indicates that the alga reproduces in autumn/winter at its southern limit of distribution. At the northern limit of distribution, it seems to reproduce in summer (Zinova 1912). Whittick *et al.* (1989) found the same phenomenon among several Rhodophycean species in Newfoundland. The plants which reproduced in summer had their northern limit in the area, those which reproduced in winter had their southern limit near Newfoundland. In Newfoundland, at the centre of distribution of *A. floccosa*, fertile plants have been found in all months except July – September (Whittick *et al.* 1989). This indicates that *A. floccosa* is more dependent on temperature than on daylength for the completion of its life-cycle, and will reproduce when the temperature is right.

The life cycle is of the *Polysiphonia* type as described by Whittick (1980). Dioecious plants which become monoecious are known from other members of the genus, such as *Antithamnionella spirographides* Shiffner (Drew 1955). In culture, *A. floccoca* developed carposporophytes only in the proximity of male plants. This might indicate that self-fertilization is rare or does not occur in this species. However, if monoecious plants are self-fertile, this may be an important mechanism for maintaining the population at the edge of its range where the alga is scarce. Vegetative propagation by fragmentation is probably also important near the distribution boundaries.

Cytology

A haploid chromosome number of approximately 31 to 33 corresponds to the haploid number in other species of *Antithamnionella* (32 – 40; Magne 1986, Rao 1960), *Antithamnion* (*ca* 32; Rueness and Rueness 1973), and *Scagelia* (30 – 40; Athanasiadis and Rueness 1992, but is different from that of *Pterothamnion* species, which have a haploid compliment of 24 ± 2 and c. 23 chromosomes (Athanasiadis 1985, Magne 1964). While these investigations suggest the presence of two separate groups (*Pterothamnion vs Antithamnion-Antithamnionella-Scagelia*) with basic numbers of (22)23 – 24(26) and (30)32(40) respectively, a recent report of c. 24 chromosomes in *Antithamnion cruciatum* (C. Agardh) Näg. (Kapraun 1989) suggests that the second group should be regarded as a complex of polyploid-aneuploid derivatives. This view is also supported by previous reports of n = 46 ± 4 in *Antithamnion heterocladum* Funk (Athanasiadis 1983), 2n = ca. 50 in *Scagelia pylaisaei* (Mont.) Wynne (see Cole 1990), and n = 55 – 65 in *Antithamnion cruciatum* (Athanasiadis 1986). The latter species has also been recorded with a polyploid number of 85 – 110 towards the northern limit of its range (Whittick and Hooper 1977).

References

Athanasiadis, A. 1983. The life history of *Antithamnion heterocladum* (Rhodophyta, Ceramiales) in culture. *Bot. Mar. 26*, 153 – 157.

Athanasiadis, A. 1985. The taxonomic recognition of *Pterothamnion crispum* (Rhodophyta, Ceramiales), with a survey of the carposporophyte position in genera of the Antithamnieae. *Br. phycol. J. 20*: 381 – 389.

Athanasiadis, A. 1986. A comparative study of *Antithamnion tenuissimum* and three varieties of *A. cruciatum*, including var. *scandinavicum* var. nov. (Rhodophyceae). *Nord. J. Bot. 6*: 703 – 709.

Athanasiadis, A. 1990. Evolutionary biogeography of the North Atlantic Antithamnioid algae. *In*: (D. J. Garbary and G. R. South, eds) *Evolutionary Biogeography of the Marine Algae of the North Atlantic*. NATO ASI Series, Vol. G 22. Springer-Verlag, Berlin, Heidelberg. pp. 219 – 240.

Athanasiadis, A. and J. Rueness. 1992. Biosystematic studies in the genus *Scagelia* (Rhodophyta, Ceramiales) from Scandinavia: Genetic variation, life histories and chromosome numbers. *Phycologia 31* (in press).

Batters, E. A. L. 1902. A catalogue of the British marine algae. *J. Bot., Lond. (Suppl.), 40*: 1–107.

Bird, C. J. and J. McLachlan. 1986. The effect of salinity on distribution of species of *Gracilaria* Grev. (Rhodophyta, Gigartinales): An experimental assessment. *Bot. Mar. 29*: 231–238.

Breeman, A. M. 1988. Relative importance of temperature and other factors in determining geographic boundaries of seaweeds: experimental and phenological evidence. *Helgoländer Meeresunters. 42*: 199–241.

Børgesen, F. 1902. *Botany of the Faeröes Based on Danish Investigations. Pt. II.* Ernest Boyesen, Copenhagen. pp. 339–532.

Cambridge, M. L., A. M. Breeman, S. Kraak and C. van den Hoek. 1987. Temperature responses of tropical to warm temperature *Cladophora* species in relation to their distribution in the North Atlantic Ocean. *Helgoländer Meeresunters. 41*: 329–354.

Cambridge, M., A. M. Breeman, R. van Oosterwijk and C. van den Hoek. 1984. Temperature responses of some North Atlantic *Cladophora* species (Chlorophyceae) in relation to their geographic distribution. *Helgoländer Meeresunters. 38*: 349–363.

Cole, K. M. 1990. Chromosomes. *In*: (K. M. Cole and R. G. Sheath, eds) *Biology of Red Algae*. Cambridge University Press. pp. 73–101.

Dammann, H. 1930. Entwicklungsgeschichtliche und zytologische Untersuchungen an Helgoländer Meeresalgen. *Helgoländer Meeresunters. 18*: 1–37.

Drew, K. M. 1955. Sequences of sexual and asexual phases in *Antithamnionella spirographidis* Schiffner. *Nature. Lond. 175*: 813–814.

Edelstein, T., L. Chen and J. McLachlan. 1970. Investigations of the marine algae of Nova Scotia. VIII. The flora of Digby Neck Peninsula, Bay of Fundy. *Can. J. Bot. 48*: 621–629.

Edwards, E. 1979. A cultural assessment of the distribution of *Callithamnion hookeri* (Dillw.) S. F. Gray (Rhodophyta, Ceramiales) in nature. *Phycologia 18*: 251–263.

Eppley, R. W., R. W. Holmes and J. D. H. Strickland. 1967. Sinking rates of marine phytoplankton measured with a fluorometer. *J. Exp. Mar. Biol. Ecol. 1*: 191–208.

Harvey, W. H. 1844. Description of a new British species of *Callithamnion*. *Ann. Nat. Hist. 14*: 186–187.

Harvey, W. H. 1853. Contributions to a history of the marine algae of North America. *Bibl. phycol. 20*. Reprint 1976.

Hoek, C. van den. 1982 a. Phytogeographic distribution groups of benthic marine algae in the North Atlantic Ocean. A review of experimental evidence from life history studies. *Helgoländer Meeresunters. 35*: 153–214.

Hoek, C. van den. 1982 b. The distribution of benthic marine algae in relation to the temperature regulation of their life histories. *Biol. J. Linn. Soc. 18*: 81–144.

Jaasund, E. 1965. Aspects of the marine algal vegetation of north Norway. *Bot. Gothoburg. 4*: 1–147.

Jónsson, H. 1901. The algae of Iceland 1. Rhodophyceae. *Bot. Tidsskr. 24*: 127–155.

Juliussen, R. 1979. Den marine benthosvegetasjonen i ytre del av Ofotfjorden. Cand. real. thesis. University of Tromsø. pp. 108.

Kapraun, D. F. 1989. Karyological investigations of chromosome variation patterns associated with speciation in some Rhodophyta. *In*: (R. Y. George and A. W. Hulbert, eds) *Carolina Coastal Oceanography Symposium* 1987. pp. 65–76.

Kleen, E. 1874. Om Nordlandens högre hafsalger. *K. svenska Vetensk-Akad. Handl. 41 (9)*: 1–46.

Magne, F. 1964. Recherches caryologiques chez les Floridées (Rhodophycées). *Cahiers de Biologie Marine 5*: 461–671.

Magne, F. 1986. Anomalies du développement chez *Antithamnionella sarniensis* (Rhodophyceae, Ceramiaceae). II: nature des individus issus des tétraspores. *Cryptogamie: Algologie 7*: 215–229.

Mathieson, A. C. and C. A. Penniman. 1986. Species composition and seasonality of New England seaweeds along an open coastal-estuarine gradient. *Bot. Mar. 29*: 161–176.

Melander, Y. and K. Wingstrand. 1953. Gomori's haematoxylin as a chromosome stain. *Stain Technology 28*: 217–223.

Munda, I. 1976. The distribution of *Halosaccion ramentaceum* (L.) J. Ag. Associations in the Icelandic coastal area and their hydrographically conditioned variations. *Bot. Mar. 19*: 179–182.

Munda, I. 1978. Survey of the benthic algal vegetation of the Dyrafjördur, Northwest Iceland. *Nova Hedwigia 29*: 281–403.

Munda, I. 1980. Survey of the bentic algal vegetation of the Borgarfjordur, Southwest Iceland. *Nova Hedwigia 32*: 855–927.

Munda, I. 1987. Characteristic features of the benthic algal vegetation along the Snaefellsnes Peninsula (Southwest Iceland). *Nova Hedwigia 44*: 339–448.

Parker, R. E. 1979. *Introductory Statistics for Biology*. 2nd. ed. London. pp. 122.

Pedersen, P. M. 1976. Marine, benthic algae from southernmost Greenland. *Meddr Grønland 199 (3)*: 1–79.

Pybus, C. 1977. A List of algae found on artificial substrata in Galway Bay. *Ir. Nat. J. 19*: 72–75.

Rao, C. S. P. 1960. Cytology of red algae. *In*: *Proceedings of the Symposium on Algology*. Kachroo, New Dehli. pp. 37–45.

Rosenvinge, L. K. 1898. Om algevegetationen ved Grønlands kyster. Oversigt over Grønlands havalger. *Meddr Grønland 20*: 1–125.

Rueness, J. and M. Rueness. 1973. Life history and nuclear phases of *Antithamnion tenuissimum*, with special references to plants bearing tetrasporangia and spermatangia. *Norw. J. Bot. 20*: 205–210.

Sætre, R. 1973. Temperatur og saltholdighetsnormaler for overflatelaget i norske kystfarvann. *Fiskets Gang 59*: 166–172.

Sivertsen, K. 1981. Algevegetasjonen i Frøyfjorden, Sør-Trøndelag. Cand. real. thesis. University of Oslo.

South, G. R. and R. G. Hooper. 1972. Additions to the marine algal flora of insular Newfoundland II. With remarks on some species new for southern Labrador. *Naturaliste can. 99*: 263–270.

South, G. R. and R. G. Hooper. 1980. A catalogue and atlas of the benthic marine algae of the Island of Newfoundland. *Mem. Univ. Nfld. Occ. Pap. Biol. 3*: 1–136.

South, G. R., I. Tittley, W. F. Farnham and D. W. Keats. 1988. A survey of the benthic marine algae of southwestern New Brunswick, Canada. *Rhodora 90*: 419–451.

Taylor, W. R. 1957. *Marine Algae of the Northeastern Coast of North America*. 2nd. ed. Univ. of Michigan Press, Ann Arbor, Michigan. 509 pp.

Tittley, I., W. F. Farnham and P. W. G. Gray. 1982. Seaweeds of the Faroes II: Sheltered fjords and sounds. *Bull. Br. Mus. nat. Hist. Botany series 10*: 133–151.

Whittick, A. 1980. *Antithamnionella floccosa* (O. F. Müll.) nov. comb.: a taxonomic re-appraisal of *Antithamnion floccosum* (O. F. Müll.) Kleen (Rhodophyta: Ceramiales). *Phycologia. 19*: 74–79.

Whittick, A. and R. G. Hooper. 1977. The reproduction and phenology of *Antithamnion cruciatum* (Rhodophyta: Ceramiaceae) in insular Newfoundland. *Can. J. Bot. 55*: 520–524.

Whittick, A., R. G. Hooper and G. R. South. 1989. Latitude, distribution and phenology: reproductive strategies in some Newfoundland seaweeds. *Bot. Mar. 32*: 407–417.

Yarish, C., A. M. Breeman and C. van den Hoek. 1984. Temperature, light, and photoperiod responses of some Northeast American and West European endemic rhodophytes in relation to their geographic distribution. *Helgoländer Meeresunters. 38*: 273–304.

Yarish, C., A. M. Breeman and C. van den Hoek. 1986. Survival strategies and temperatures responses of seaweeds belonging to different biogeographic distribution groups. *Bot. Mar. 29*: 215–230.

Yarish, C., K. W. Lee and P. Edwards. 1979. An improved apparatus for the culture of algae under varying regimes of temperature and light intensity. *Bot. Mar. 22*: 395–397.

Zinova, A. D. 1912. Vodorosli Murmana, 1. [Algae Murmanicae, Chlorophyceae et Rhodophyceae]. *Trav. Soc. Nat. St. Pétersb. 43*: 171–343.

Botanica Marina
Vol. 34, pp. 501–504, 1991

Fatty Acid Composition of Nine Green Seaweeds

M. Shameel and R. Khan

Department of Botany, University of Karachi, Karachi-75270, Pakistan

(Accepted 21 August 1991)

Abstract

A variety of saturated and unsaturated fatty acids has been isolated and identified from 3 members of the Ulvales: *Ulva anandii* Amjad *et* Shameel, *U. lactuca* L. and *U. taeniata* (Setch.) Setch. *et* Gard., 4 of the Caulerpales: *Caulerpa faridii* Nizam., *C. manorensis* Nizam., *C. qureshii* Nizam. and *C. veravalensis* Thivy *et* Chauh. and 2 of the Codiales: *Codium dwarkense* Børg. and *Co. flabellatum* Silva *et* Nizam. by gas chromatography-mass spectrometry. The seaweeds were collected from the coast of Karachi, Pakistan. Saturated fatty acids were found in much greater quantity (81.9–92.8%) than unsaturated ones, palmitic acid was the major component (76.8–87.9%). Oleic was the most commonly occurring unsaturated fatty acid (2.5–8.9%), linolenic was the only triunsaturated acid present. Members of the Codiales differed more than those of the other two orders.

Introduction

The fatty acids occurring in algal lipids are generally straight chain molecules containing an even number of carbon atoms. The metabolic behaviour of algal lipids is largely determined by the structure of their component fatty acids. Much work has been carried out on the fatty acid composition of brown (Shaikh *et al.* 1990, Shameel *et al.* 1991) and red seaweeds of this area (Afaq-Husain *et al.* 1991, Hayee-Memon *et al.* 1991). Apart from two general works (Shameel and Khan 1989, 1990) no detailed investigation has been conducted on the fatty acids of green seaweeds of Karachi. In this communication an attempt has been made to extract and identify the fatty acids from nine species of benthic green algae, collected from the coast of Karachi, Pakistan.

Material and Methods

Three seaweeds of the order Ulvales *i.e. Ulva anandii* Amjad *et* Shameel, *U. lactuca* Linnaeus and *U. taeniata* (Setchell) Setchell *et* Gardner, 4 members of Caulerpales *i.e. Caulerpa faridii* Nizamuddin, *C. manorensis* Nizamuddin, *C. qureshii* Nizamuddin and *C. veravalensis* Thivy *et* Chauhan and 2 species of Codiales *i.e. Codium dwarkense* Børgesen and *Co. flabellatum* Silva *et* Nizamuddin were collected partly from mid-littoral to upper sub-littoral rocks and pools at the rocky ledge and partly from the sandy bay of Buleji, near Karachi during November 1989 and February 1990. They were thoroughly cleaned to remove sand particles and animal castings, dried in the shade and finely milled. About 1 kg of the dried seaweed was extracted three times by percolation at room temperature with $CHCl_3$: hexane (1 : 1 v/v) for 20 days. The extract so obtained, was evaporated under vacuum, and yielded about 50 g of a thick brownish green residue. The procedures for saponification and esterification and the techniques for gas chromatography-mass spectrometry were the same as described earlier (Shameel 1990).

Results and Discussion

The relative percentages of the individual fatty acids isolated and identified are presented in Table I. The saturated fatty acids (S) were present in much greater quantity (81.9–92.8%) than the unsaturated ones.

Table I. Relative percentages of the fatty acids isolated from green seaweeds. 1 = *Ulva anandii*, 2 = *U. lactuca*, 3 = *U. taeniata*, 4 = *Caulerpa faridii*, 5 = *C. manorensis*, 6 = *C. qureshii*, 7 = *C. veravalensis*, 8 = *Codium dwarkense*, 9 = *Co. flabellatum* (S = Saturated fatty acids, M = Monounsaturated fatty acids, D = Diunsaturated fatty acids, T = Triunsaturated fatty acids).

Fatty acid	Ulvales			Caulerpales				Codiales	
	(1)	(2)	(3)	(4)	(5)	(6)	(7)	(8)	(9)
S	87.37	88.46	89.83	81.87	92.84	84.35	83.75	88.26	88.06
C8:0	–	–	–	–	–	–	–	0.49	0.48
C9:0	–	–	–	–	–	–	0.29	–	–
C11:0	1.42	1.51	–	–	–	–	–	0.98	–
C12:0	–	0.72	–	2.68	0.97	2.32	–	–	0.87
C13:0	–	–	–	–	1.35	–	1.35	–	–
C14:0	0.63	0.53	0.72	0.25	–	0.43	0.52	0.47	0.45
C15:0	0.55	–	0.60	–	0.78	–	0.61	–	–
C16:0	82.87	83.43	84.75	76.83	87.91	78.72	80.26	83.79	83.59
C17:0	–	–	–	–	–	0.51	–	–	–
C18:0	1.90	2.27	2.39	2.11	1.83	2.37	0.72	2.53	2.67
C22:0	–	–	1.37	–	–	–	–	–	–
M	10.57	9.27	9.83	16.24	4.81	13.66	9.63	11.69	11.90
C11:1	1.13	1.23	–	2.29	–	1.55	–	–	–
C12:1	–	1.51	2.62	1.40	0.87	1.19	–	2.77	2.72
C13:1	–	–	–	–	–	–	2.23	–	–
C14:1	0.29	–	–	–	–	–	–	–	–
C16:1	–	–	–	3.76	–	3.48	1.49	–	1.75
C18:1	8.03	6.32	7.21	4.57	2.46	4.17	4.08	8.92	7.43
C19:1	–	–	–	–	0.51	–	–	–	–
C20:1	1.12	0.21	–	4.22	0.97	3.27	1.83	–	–
D	2.03	2.23	0.29	1.83	2.32	1.94	4.02	–	0
C17:2	–	–	–	–	–	–	0.43	–	–
C18:2	2.03	2.23	0.29	1.83	2.32	1.94	2.47	–	–
C20:2	–	–	–	–	–	–	1.12	–	–
T	0	0	0	0	0	0	0.57	0	0
C18:3	–	–	–	–	–	–	0.57	–	–

Amongst the saturated acids palmitic (C16:0) and stearic (C18:0) acids were common to all the species investigated, while myristic acid (C14:0) was present in all the species except *Caulerpa manorensis*. Palmitic acid was the most abundant (76.8–87.9%) of all the acids isolated. It has also been reported as the major fatty acid in the other groups of seaweeds (Qasim 1986, Shameel 1990, Shameel *et al.* 1991). Myristic and stearic acids have been detected in several green microalgae (Klenk *et al.* 1963, Hinchcliffe and Riley 1972).

The saturated fatty acids (S) identified ranged from C8:0 to C22:0, with C10:0, C19:0, C20:0 and C21:0 missing. Behenic acid (C22:0) was only found in *Ulva taeniata* in very small quantities (1.4%). Green algae generally lack fatty acids with more than 18 carbon atoms (Wood 1988). Unlike higher plants, the saturated acids of green seaweeds are low in molecular weight (Gunstone 1987). In the species investigated lauric (C12:0) and pentadecylic (C15:0) acids were found in 4–5 species, while caprylic (C8:0), undecylic (C11:0) and tridecylic (C13:0) acids could be detected only in 2–3 species and they appear to have a restricted occurrence. The saturated fatty acids, caprylic (C8:0), pelargonic (C9:0), myristic (C14:0), pentadecylic (C15:0) and margaric (C17:0) acids were found in traces (0.25–0.78%). This appears to be a characteristic feature of green seaweeds as these acids are found in large amounts in other groups of seaweeds (Qasim 1986, Shameel 1990). Pelargonic (C9:0), margaric (C17:0) and behenic (C22:0) acids were found in *Caulerpa veravalensis*, *C. qureshii* and *Ulva taeniata* respectively and appear to be exceptional in green seaweeds. Arachidic acid (C20:0), which is fairly widely distributed in higher plants (Lie Ken Jie 1989) and to some extent in other groups of algae (Shameel *et al.* 1991), was not detected in green seaweeds. It was also not detected in several groups of marine microalgae (Klenk *et al.* 1963, Shaw 1966, Menzel and Wild 1989).

Amongst unsaturated fatty acids (Table I) the monounsaturated acids (M) were most abundant (4.8–13.7%). Although unsaturated acids were present in small amounts, oleic acid (C18:1) was common in all the species investigated and found in the largest quantity (2.5–8.9%). It has also been reported to occur in largest proportion among ethenoids *i.e.* unsaturated acids, in other epilithic algae (Qasim 1986, Shameel 1990). The monounsaturated fatty acids ranged from C11:1 to C20:1, with C15:1 and C17:1 lacking. Apart from oleic acid all the monoethenoids are of restricted occurrence. Decylacrylic (C13:1), myristoleic (C14:1) and nonadecylenic (C19:1) acids were found in *Caulerpa veravalensis*, *Ulva anandii* and *Caulerpa manorensis* respectively and probably are exceptional in green seaweeds. The remaining fatty acids were present in 4–7 species.

Although no fatty acid with more than one double bond was common to all the species the most common was the essential fatty acid, linoleic (C18:2), which was present in 0.3–2.5%. It was also frequently found in other groups of seaweeds (Qasim 1986, Shameel 1990). Diunsaturated fatty acids (D) were found in very small quantities (0.3–4.0%). In brown and red seaweeds these acids are usually present in an amount greater than this (Shameel 1990). Heptadecadienoic (C17:2) and eicosadienoic (C20:2) acids were only found in *Caulerpa veravalensis* and no other seaweed. Linolenic acid (C18:3) was the only triunsaturated fatty acid (T) found and was present only in this species and no other alga, while no tetraunsaturated fatty acid could be detected. Moreover, C18 is the only fatty acid, which occurred with one, two and three double bonds. Generally green seaweeds are good sources of 18:3 and 18:4 fatty acids (Hinchcliffe and Riley 1972, Takagi *et al.* 1985, Menzel and Wild 1989). Although arachidic acid (C20:0) was absent, its mono- and diunsaturated homologs were present in the benthic green algae. Eicosapentaenoic acid (C20:5ω3) has been detected in several microalgae (Ben-Amotz *et al.* 1985, Nichols *et al.* 1989).

Fatty acid composition also revealed some interesting results from a chemotaxonomic point of view (Table I). The three species of *Ulva* investigated differed slightly in their fatty acid composition, but *U. taeniata* differed more than the other two species. The range of its saturated fatty acids (S) was C14:0 to C22:0, while the other two exhibited a range of C11:0 and C18:0. Its diunsaturated fatty acids (D) ranged between C12:1 and C18:1, while others showed a larger range of C11:1 and C20:1. *Ulva anandii* closely resembles *U. indica* Anand in its fatty acid composition (Shameel and Khan 1990), indicating that they may be conspecific. Similarly, *U. taeniata* closely resembles *U. fasciata* Delile in the composition of its fatty acids (Shameel and Khan 1990), it may also suggest that they are conspecific, as the former has been considered as *U. fasciata* f. *taeniata* by some phycologists (*e.g.* Saifullah and Nizamuddin 1977). The allied genus *Enteromorpha* also exhibits specific differences in its fatty acid composition (Klenk *et al.* 1963, Shaw 1966).

The four species of *Caulerpa* investigated differed in their fatty acid composition. Coppejans and Meinesz (1988) discussed the similarities of *C. manorensis* with *C. qureshii* and *C. veravalensis* and considered the last two as conspecific. However, their fatty acid composition indicates that they are distinct species. *Caulerpa veravalensis* differs from other species in possessing pelargonic (C9:0), decylacrylic (C13:1), heptadecadienoic (C17:2), eicosadienoic (C20:2) and linolenic (C18:3) acid. It further differs from others in having large amounts of di- and triunsaturated fatty acids. *Caulerpa qureshii* differs from *C. veravalensis* in having C12:0, C17:0, C11:1 and C12:1 and in lacking C9:0, C13:0, C15:0, C13:1, C17:2, C20:2 and C18:3 fatty acids. *Caulerpa manorensis* differs from other species in having nonadecylenic acid (C19:1), and the largest amount of saturated and smallest amount of monounsaturated fatty acids. Hence, these species are delineated further. *Caulerpa veravalensis* exhibits the largest range of fatty acids, as compared with the other 3 species investigated in this paper as well as other species investigated previously (Shameel and Khan 1989).

The two species of *Codium* resembled one another to a great extent in the pattern of fatty acid composition (Table I). However, *Codium flabellatum* differed from *Codium dwarkense* in having lauric (C12:0) and palmitoleic (C16:1) acids and in lacking undecylic acid (C11:0). Generally the three groups of green seaweeds exhibited more or less the same pattern of fatty acid composition. The Ulvales and Caulerpales differed slightly from one another, but the Codiales differed more than the other two orders. No monounsaturated fatty acid beyond C18:1 and no di- and triunsaturated acids could be detected in its members. These acids were also absent in *Codium iyengarii* Børg. (Shameel 1990), although such acids have been detected in *Codium fragile* (Sur.) Har. (Klenk *et al.* 1963). These orders also differ slighlty in the distribution of fatty acids from the marine members of Chlorococcales and Eustigmatales (Menzel and Wild 1989) as well as other groups of microalgae (Ben-Amotz *et al.* 1985). This indicates that one may find differences, though slight, among various orders of green seaweeds, regarding the pattern of their fatty acid composition.

References

Afaq-Husain, S., M. Shameel, K. Usmanghani, M. Ahmad and V. U. Ahmad. 1991. Phycochemical studies on *Dermonema abbottiae* (Nemaliales-Rhodophyta). *Bot. Mar. 34*: 215–220.

Ben-Amotz, A., T. G. Tornabene and W. H. Thomas. 1985. Chemical profile of selected species of microalgae with emphasis on lipids. *J. Phycol. 21*: 72–81.

Coppejans, E. and A. Meinesz. 1988. Marine algae of Papua New Guinea (Mandang Prov.). 1. Caulerpaceae (Chlorophyta-Caulerpales). *Blumea 33*: 181–196.

Gunstone, F. D. 1987. Fatty acids and glycerides. *Nat. Prod. Rep. 4*: 95–128.

Hayee-Memon, A., M. Shameel, M. Ahmad, V. U. Ahmad and K. Usmanghani. 1991. Phycochemical studies on *Gracilaria foliifera* (Gigartinales, Rhodophyta). *Bot. Mar. 34*: 107–111.

Hinchcliffe, P. R. and J. P. Riley. 1972. The effect of diet on the component fatty acid composition of *Artemia salina*. *J. mar. biol. Ass. U.K. 52*: 203–211.

Klenk, E., W. Kniprath, D. Eberhagen and H. D. Koof. 1963. Über die Ungesättigten Fettsäuren der Fettstoffe von Süßwasser- und Meeresalgen. *Z. Physiol. Chem. 334*: 44.

Lie Ken Jie, M. S. F. 1989. Fatty acids and glycerides. *Nat. Prod. Rep. 6*: 231–261.

Menzel, K. and A. Wild. 1989. Fatty acid composition in the lipids of some marine Chlorococcales and Eustigmatales. *Z. Naturforsch.* C: *Biosci. 44*: 743–748.

Nichols, P. D., D. J. Holdsworth, J. K. Volkman, M. Daintith and S. Allanson. 1989. High incorporation of essential fatty acids by the rotifer *Brachionus plicatilis* fed on the prymnesiophyte alga *Pavlova lutheri*. *Aust. J. Mar. Freshwat. Res. 40*: 645–656.

Qasim, R. 1986. Studies on fatty acid composition of eighteen species of seaweeds from the Karachi coast. *J. Chem. Soc. Pak. 8*: 223–230.

Saifullah, S. M. and M. Nizamuddin. 1977. Studies on the marine algae from Pakistan: Ulvales. *Bot. Mar. 20*: 521–536.

Shaikh, W., M. Shameel, A. Hayee-Memon, K. Usmanghani, S. Bano and V. U. Ahmad. 1990. Isolation and characterization of chemical constituents of *Stoechospermum marginatum* (Dictyotales, Phaeophyta) and their antimicrobial activity. *Pak. J. Pharm. Sci. 3* (2): 1–9.

Shameel, M. 1990. Phycochemical studies on fatty acids from certain seaweeds. *Bot. Mar. 33*: 429–432.

Shameel, M. and R. Khan. 1989. Phycochemical studies on fatty acid composition in *Caulerpa* (Bryopsidophyceae). *Pak. J. Bot. 21*: 343–346.

Shameel, M. and R. Khan. 1990. Fatty acid composition of *Ulva* (Chlorophyceae). *Pak. J. Bot. 22*: 39–42.

Shameel, M., W. Shaikh and R. Khan. 1991. Comparative composition of fatty acids of five species of *Dictyota* (Phaeophyta). *Bot. Mar. 34*: 425–428.

Shaw, R. 1966. Polyunsaturated fatty acids of microorganisms. *Adv. Lipid Res. 4*: 107–174.

Takagi, T., M. Asahi and Y. Itabashi. 1985. Fatty acid composition of twelve algae from Japanese waters. *Yukagaku 34*: 1008–1012.

Wood, B. J. B. 1988. Fatty acids and lipids in algae. *In*: (C. Ratledge and S. G. Wilkinson, eds): *Microbial Lipids*, vol. 1, Acad. Press, N. Y. pp. 807–867.

Botanica Marina
Vol. 34, pp. 505–511, 1991

Relationships between the Movements of Chloroplasts and Cytoskeletons in Diatoms

S.-T. Chen and C.-W. Li[1]

Institute of Life Science, National Tsing Hua University, Hsinchu, Taiwan

(Accepted 22 August 1991)

Abstract

Movements of the chloroplasts were studied in two marine centric diatoms: in *Ditylum brightwellii* the chloroplasts move outward and inward during the light-dark cycle, and in *Odontella regia* the chlorplasts are stationary in the cytoplasm. Physical relationships between chloroplasts and cytoskeleton were determined by immunofluorescence microscopy and electron microscopy, and the possible involvement of the cytoskeleton in the movements of chloroplasts was investigated using cytoskeleton-inhibitors. The results show that microtubules and microfilaments surround the chloroplasts of *D. brightwellii* and are involved in their movements. Neither microtubules nor microfilaments have a close relationship with the chloroplasts of *O. regia*.

Introduction

Light-mediated movements of chloroplasts have been observed in several diatoms (Senn 1919, Kiefer 1973, Anderson and Sweeney 1978). The chloroplasts of *Biddulphia* sp. are usually spread through the whole cell, but in both high light intensities and darkness, they aggregate around the centrally located nucleus (Senn 1919). The chloroplasts of *Ditylum brightwellii* (West) Grün. are distributed peripherally in the light but aggregate around the centrally located nucleus in the dark (Anderson and Sweeney 1978). In *Lauderia borealis* Gran, however, the chloroplasts are evenly distributed around the periphery of the cell in the dark or under moderate light, while upon exposure to intense light they move to the valvar ends of the cell, forming two aggregates of equal size (Kiefer 1973). No detailed analysis of the mechanism of the movement is available. Anderson and Sweeney (1978) attributed the diurnal chloroplast movement in *D. brightwellii* to the expansion and contraction of cell vacuoles, in response to net ion fluxes during the light-dark cycle. However, it is well known that intracellular cytoskeletons are frequently involved in the movement of organelles in many biological systems (Wagner 1979, Kamiya 1981, Haupt 1982, Beckerle and Porter 1983, Menzel and Schliwa 1986 a, b). The purpose of our study was to elucidate whether the major cytoskeletons, microtubules and microfilaments, have any role in the positioning of the chloroplasts in diatoms. *Ditylum brightwellii* and *Odontella regia* Schultze were chosen to investigate this question: chloroplasts in *O. regia* are stationary in the cytoplasm, thus providing a basis for comparison. Physical relationships between chloroplasts and cytoskeletons were determined by immunofluorescence microscopy and electron microscopy, and the possible involvement of the cytoskeleton in the movements of chloroplasts were determined using cytoskeleton-inhibitors.

[1] Correspondence and reprints: Chia-Wei Li, Institute of Life Science, National Tsing Hua University, Hsinchu, Taiwan.

Material and Methods

Cell culture

A unialgal culture of *Ditylum brightwellii* was isolated from nearshore water at La Jolla, California in May 1980. A unialgal culture of *Odontella regia* was obtained from Dr Y.-L. L. Chen, National Sun Yat-Sen University, Taiwan. Both cultures were maintained in F/2 medium (McLachlan 1973) under cool-white fluorescent light (55–60 microEinsteins m^{-2} sec^{-1}) at 20–25 °C in an 8 h light: $\overline{16}$ h dark cycle.

Immunofluorescence microscopy

Cells were fixed in 1% glutaraldehyde for 30 min and immediately extracted with 0.1% Triton X-100 in PBS, pH 7.4, plus 0.12% Azide for 12 h at 20 °C. Extracted cells were reduced in 1 mg mL^{-1} sodium borohydride in 1 : 1 PBS: methanol for 20 min. After PBS rinses the cells were incubated in primary antiserum (containing rabbit polyclonal antibodies against sea urchin tubulin or rabbit polyclonal antibodies against chicken actin, all obtained from Polysciences) for 22 h at 20 °C, then extensively rinsed in PBS containing 0.05% Tween 20. The cells were incubated for 7 h in fluorescein-conjugated goat anti-rabbit IgG (Cappel Laboratories), rinsed in PBS, and stored in an anti-bleach solution (containing 0.1% p-phenylenediamine, 90% glycerol, and 0.1 M Tris). The labelled cells were either stored at −80 °C or directly examined under a Nikon epi-fluorescence microscope with a blue excitation filter.

Electron microscopy

For transmission electron microscopy, cells were fixed in 1% glutaraldehyde in F/2 medium for 30 min at room temperature, then post-fixed in 0.2% OsO_4 for 30 min. For better cytoskeletal preservation, the fixed cells were incubated in 0.25% tannic acid for 30 min before dehydration (Fujiwara and Linck 1982, Seagull and Heath 1979). The cells were filter-dehydrated in a continuous ethanol series and flat embedded in Spurr's resin. Sections were cut with a diamond knife, stained with uranyl acetate and lead citrate, and examined in a Hitachi H-600 electron microscope.

For scanning electron microscopy, the fixed cells were dehydrated in an ethanol series, then critical point dried. The dried cells were slightly damaged using a tiny glass needle under a stereomicroscope, coated with gold in a sputter coater, and observed in a Hitachi S-570 scanning electron microscope.

Pharmacological treatment

Colchicine and β-lumicolchicine were dissolved in F/2 medium. Cytochalasin B was prepared as stock solutions in dimethyl sulfoxide (DMSO) and subsequently diluted with F/2 medium to yield the appropriate concentration. Incubations were carried out in 20 mL aliquots containing about 1×10^4 cells. At intervals, samples of cells were removed from the incubation culture for observation under an Olympus BH-2 differential interference contrast microscope. The remaining cells were processed for immunocytochemistry as described above. A control without inhibitors was run for each experiment. Recovery from the various experimental treatments was checked at 2 h intervals. Concentrations of the inhibitors used were: for *D. brightwellii*, colchicine (1 or 5 μg mL^{-1}), β-lumicolchicine (1 or 5 μg mL^{-1}), and cytochalasin B (0.1 or 1 μg mL^{-1}); for *O. regia*, colchicine (2 or 20 μg mL^{-1}), β-lumicolchicine (2 or 20 μg mL^{-1}), and cytochalasin B (0.1 or 2 μg mL^{-1}).

Results

Movement of chloroplasts

At the end of the $\overline{16}$ h dark period most of the cytoplasm and organelles in *D. brightwellii* aggregated to form a central mass, which appeared to be suspended by a few radiating cytoplasmic strands (Fig. 1). During the light period, the number of cytoplasmic strands increased, chloroplasts moved outward along the strand (Fig. 2) at about 0.1–0.3 μm min^{-1}, the cytoplasm gradually formed a thin layer at the periphery of the cell, so that finally the chloroplasts were distributed evenly throughout the peripheral cytoplasm (Fig. 3); This took place in less than 4 h of light. Returned to the dark, the chloroplasts and peripheral cytoplasm gradually moved inwards at a slightly slower speed than for the outward movement. Chloroplasts of cells maintained in either continuous light or continuous dark showed no rhythmic movements.

In *O. regia*, the chloroplasts, which were evenly distributed in the peripheral cytoplasm (Fig. 4), did not move during the light-dark cycle, nor was any distinct cytoplasmic streaming observed.

Distribution of the cytoskeleton in normal culture condition

In *D. brightwellii*, the immunofluorescent images of tubulins and actins, which revealed the distribution of microtubules and microfilaments, corresponded well to the distribution of chloroplasts and their associated cytoplasmic strands, during both the light and dark period (Figs 5–8). In *O. regia*, the tubulin- and actin-staining showed that no microtubules or microfilaments were associated with the chloroplasts (Figs 9, 10). This was not due to poor counter-reactivity between the antibodies used and the cytoskele-

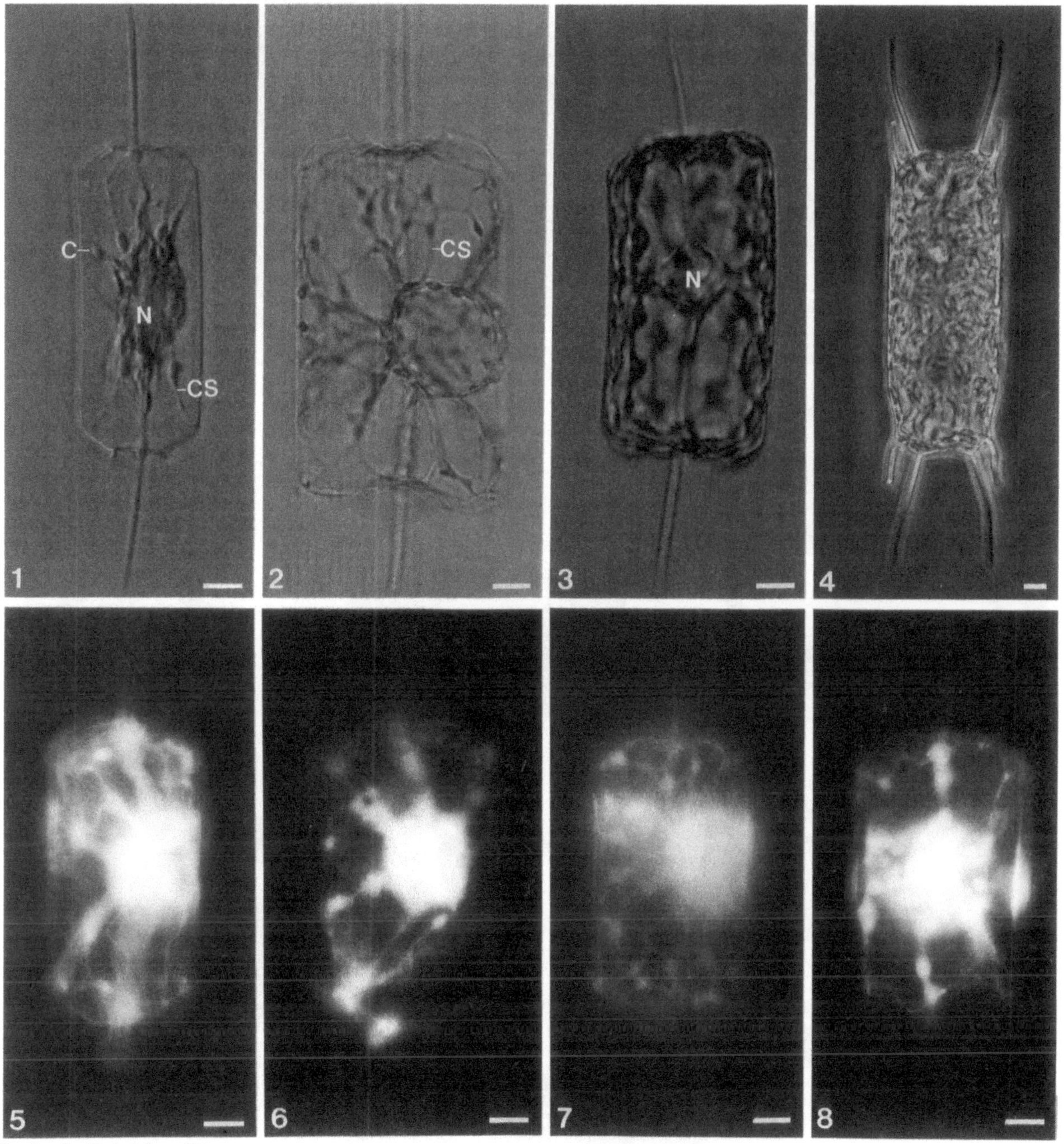

Figs 1–3. *D. brightwellii.*
Fig. 1. A cell at the end of the dark period. Fig. 2. A cell after 4 h in the light. Fig. 3. A cell at the end of the light period.

Fig.4. A cell of *O. regia* with evenly distributed and motionless chloroplasts.

Figs 5–8. Immuno-stained cells of *D. brightwellii* under fluorescence micriscopy.
Fig. 5. A cell at the end of the light period, showing the distribution of microtubules. Fig. 6. A cell at the end of the dark period, showing the distribution of microtubules. Fig. 7. A cell at the end of the light period, showing the distribution of microfilaments. Fig. 8. A cell at the end of the dark period, showing the distribution of micrifilaments. C, chloroplast; CS, cytoplasmic strand; N, nucleus. Scale bar: Figs 1–8, 10 µm.

ton in this organism, since both the microtubule-organizing center (Fig. 9) and the contractile ring (Fig. 10) could be easily recognized by immunofluorescent staining.

It is intriguing that the chloroplasts of the immunostained cells of *D. brightwellii* still emit light, even when the autofluorescence has been eliminated by extensive detergent pretreatment and a green barrier filter. This phenomenon was not observed in *O. regia* and suggests that the chloroplasts in *D. brightwellii* might be entangled by a network or basket of microtubules and microfilaments.

In *D. brightwellii*, thin sections revealed that abundant microtubules and microfilaments, 24 nm and 7 nm in

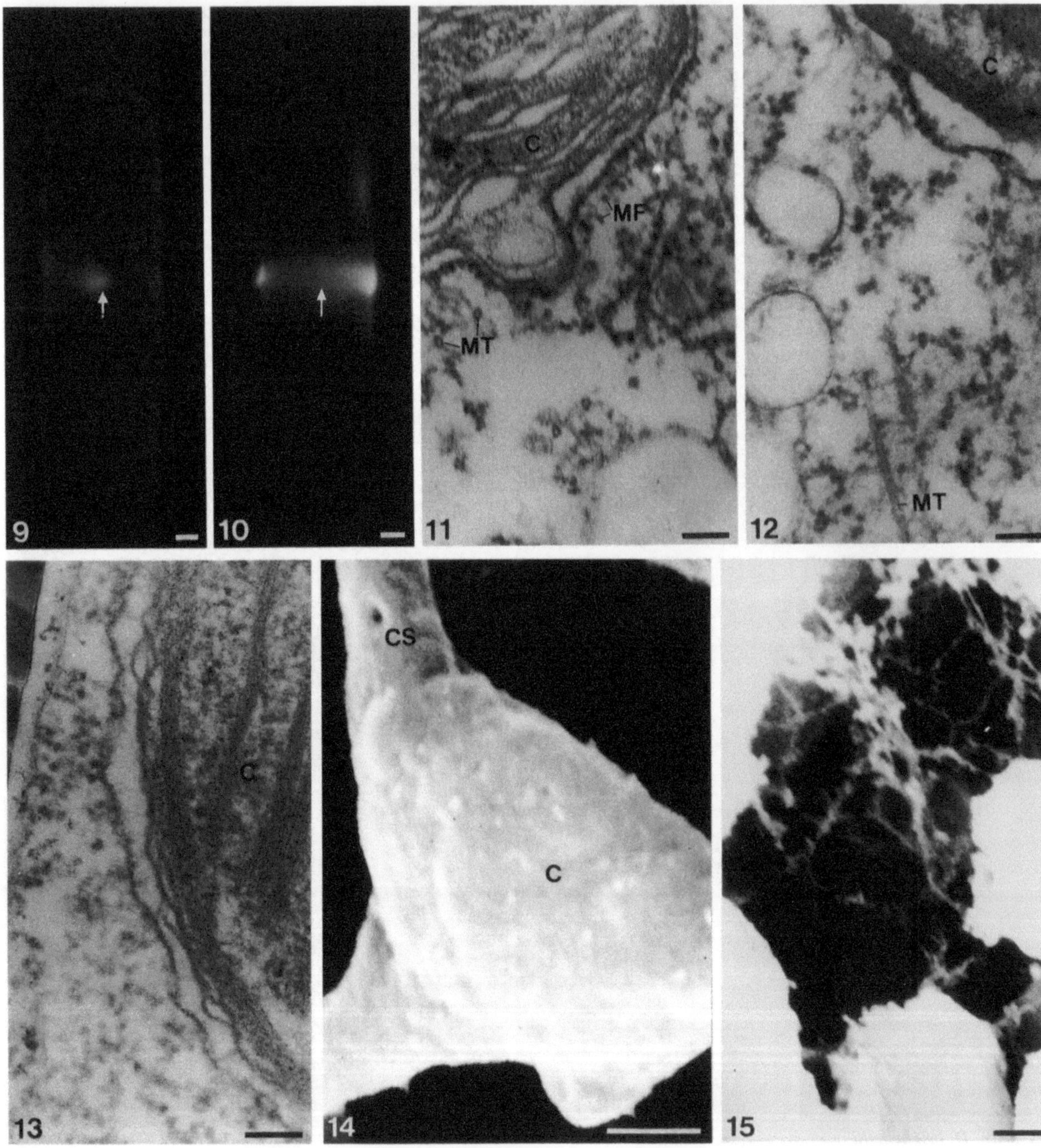

Figs 9, 10. Immuno-stained cells of *O. regia* under fluorescence microscopy.
Fig. 9. An interphase cell, showing the microtubule-organizing center (arrow). Fig. 10. An interphase cell, showing the contractile ring (arrow).

Figs 11, 12. Sections of *D. brightwellii* at the end of the light period in transmission electron microscopy, showing abundant microtubules (MT) and microfilaments (MF) in the vicinity of the chloroplast.

Fig. 13. A section of *O. regia* at the end of the light period in transmission electron microscopy, showing no cytoskeletons in the vicinity of the chloroplast.

Figs 14, 15. Crush-opened cells of *D. brightwellii* in scanning electron microscopy.
Fig. 14. A cytoplasmic strand with a chloroplast (C) inside. Fig. 15. A damaged cytoplasmic strand, showing a network of cytoskeletons and a chloroplast (C) which might have been dislocated but originally entangled by the network. Scale bar: Figs 9–10, 10 μm; Figs 11–13, 0.1 μm; Figs 13–14, 0.5 μm.

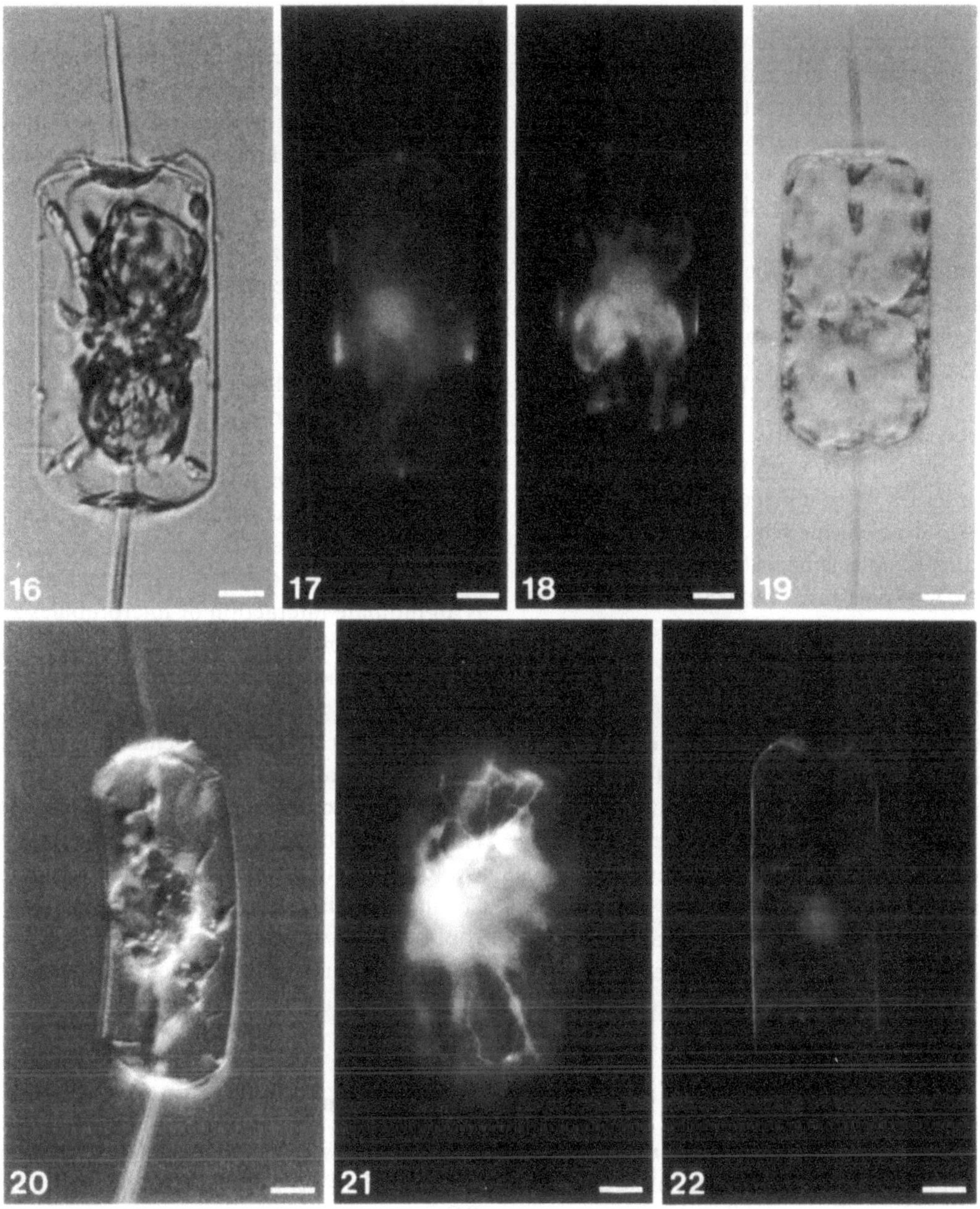

Figs 16–22. *D. brightwellii.*
Fig. 16. A cell treated with 5 μg mL^{-1} colchicine 1 h before the end of the dark period, showing that the chloroplast cannot move away from the central mass even when the cell has been in the light for 6 h. Fig. 17. Immuno-staining of a cell treated as in Fig. 16, showing degradation of the microtubule system. Fig. 18. Immuno-staining of a cell treated as in Fig. 16, showing degradation of the microfilament system. Fig. 19. A cell treated with 5 μg mL^{-1} β-lumicolchicine 1 h before the end of the dark period, showing the light-mediated chloroplast movement is not affected. Fig. 20. A cell treated with 0.5 μg mL^{-1} cytochalasin B 1 h before the end of the dark period, showing the chloroplast cannot move away from the central mass even although the cell has been in the light for 6 h. Fig. 21. Immuno-staining of a cell treated as in Fig. 20, showing the intact microtubule system. Fig. 22. Immuno-staining of a cell treated as in Fig. 20, showing degradation of the microfilament system. Scale bar: Figs 16–22, 10 μm.

diameter respectively, were present in the vicinity of the chloroplasts (Figs 11, 12). It was common to observe 40–50 cross sections of microtubules within 2 μm of each chloroplast in thin sections. In *O. regia*, it was difficult to find any microtubules or microfilaments near the chloroplasts in thin sections (Fig. 13). Cytoplasmic strands with enclosed chloroplasts could be observed in critical point-dried cells of *D. brightwellii*, which had been opened with a fine glass needle (Fig. 14). A filamentous network which might surround a chloroplast was also observed (Fig. 15) in the partially damaged cytoplasm.

Pharmacological treatment

At 1 μg mL^{-1} colchicine, chloroplast movements in *D. brightwellii* were little altered, while at 5 μg mL^{-1} movements were severely affected. Chloroplasts of cells treated with 5 μg mL^{-1} colchicine one h before the end of the dark period, could not move away from the central mass during the light period (Fig. 16). Chloroplasts of cells treated with 5 μg mL^{-1} colchicine during the light period, moved inwards in 3 h under continuous illumination and had a similar configuration to cells during the dark period. Immunofluorescent staining showed that both microtubules and microfilaments, which were abundant in the cytoplasmic strand and entangle the chloroplast in normal cells, were clearly broken down in 5 μg mL^{-1} colchicine treated cells (Figs 17, 18). When transferred into a colchicine-free medium, treated cells recovered the light-mediated chloroplast movement in 6 h. In a control, cells were treated with lumicolchicine (a mixture of colchicine isomers derived by UV radiation of colchicine which can mimic the cytotoxic actions of colchicine, except the binding with tubulins). At 1 or 5 μg mL^{-1} lumicolchicine, the distribution of chloroplasts in *D. brightwellii* was not affected (Fig. 19), and the chloroplasts moved normally during the light-dark cycle.

With 0.1 μg mL^{-1} cytochalasin B, chloroplast movements in *D. brightwellii* were partially affected, while at 0.5 μg mL^{-1} the treated cells exhibited a similar response to the 5 μg mL^{-1} colchicine treatment (Fig. 20). Immunofluorescent images of tubulins and actins in 0.5 μg mL^{-1} cytochalasin B treated cells showed that the network of microtubules is intact (Fig. 21), while the network of microfilaments is seriously damaged (Fig. 22). On transfer into a cytochalasin B-free medium, the treated cells recovered light-mediated chloroplast movements in 4 h.

At 20 μg mL^{-1} colchicine or 2 μg mL^{-1} cytochalasin B, the distribution of chloroplasts in *O. regia* was not affected, even after 12 h of incubation, although both mitosis and cytokinesis were inhibited.

Discussion

Chloroplast movement in higher plants and algae is generally considered to occur along bundles of microfilaments and to be driven by actin-myosin interaction (for a review, see Haupt 1982), with the exception of a few green algae, *Ulva* (Britz 1984), *Caulerpa* (Kuroda and Manabe 1984, Manabe and Kuroda 1984), and *Bryopsis* (Menzel and Schliwa 1986 a, b). In *Ulva*, cytochalasin B does not affect the circadian chloroplast movement, but colchicine seriously inhibits it (Britz 1976). In *Caulerpa*, ultrastructural observations and preliminary pharmacological studies point to a mechanism of chloroplast movement involving microtubules and perhaps a dynein-like molecular motor (Kuroda and Manabe 1984, Manabe and Kuroda 1984). In *Bryopsis*, detailed analysis of the spatial relationship between the chloroplasts and cytoskeletons plus pharmacological studies indicate that there is an intimate interaction of microtubules and microfilaments maintaining the spatial organization of the cytoplasm and supporting chloroplast movement (Menzel and Schliwa 1986 a, b). From the present study it seems clear that the chloroplasts in *D. brightwellii* are entangled within a network of microfilaments and microtubules, and that the light-mediated chloroplast motility is driven in association with these two cytoskeletons. In contrast, microtubules and microfilaments have no close relationship with the chloroplasts in *O. regia* which do not move. The similarity of the mechanism of chloroplast movement in *Ditylum* and *Bryopsis* seems to be an example of convergent evolution.

Acknowledgements

This study was supported by grant NSC 78-02030B007-14 from the National Science Council, ROC.

References

Anderson, L. M. J. and B. M. Sweeney. 1978. Role of inorganic ions in controlling sedimentation rate of a marine centric diatom *Ditylum brightwellii*. *J. Phycol.* *14*: 204–214.

Beckerle, M. C. and K. R. Porter. 1983. Analysis of the role of microtubules and actin in erythrophore intracellular motility. *J. Cell Biol.* *96*: 354–362.

Britz, S. J. 1976. Studies on a Circadian Rhythm of Chloroplast Movement in *Ulva*. PhD thesis. Harvard Univ., Cambridge, Mass.

Fujiwara, K. and R. W. Linck. 1982. The use of tannic acid in microtubule research. *Methods in Cell Biol.* *24*: 217–233.

Haupt, W. 1982. Light-mediated movement of chloroplasts. *Ann. Rev. Plant Physiol.* *33*: 205–233.

Kamiya, N. 1981. Physical and chemical basis of cytoplasmic streaming. *Ann. Rev. Plant Physiol.* *32*: 205–236.

Kiefer, D. A. 1973. Chlorophyll a fluorescence in marine centric diatoms: responses of chloroplasts to light and nutrient stress. *Mar. Biol.* *23*: 39–46.

Kuroda, K. and E. Manabe. 1984. Microtubules associated cytoplasmic streaming in *Caulerpa*. *Proc. Japan Acad. Ser. B.* *59*: 131–134.

Manabe, E. and K. Kuroda. 1984. Ulltrastructural basis of the microtubule-associated cytoplasmic streaming in *Caulerpa*. *Proc. Japan Acad. Ser. B. 60*: 118–121.

McLachlan, J. 1973. Growth media-marine. *In*: (J. R. Stein, ed) *Handbook of Phycological Methods, Culture Methods and Growth Measurements*. University Press, Cambridge. pp. 25–51.

Menzel, D. and M. Schliwa. 1986 a. Motility in the siphonous green alga *Bryopsis*. I. Spatial organization of the cytoskeleton and organelle movements. *Eur. J. Cell Biol. 40*: 275–285.

Menzel, D. and M. Schliwa. 1986 b. Motility in the siphonous green alga *Bryopsis*. II. Chloroplast movement requires organized arrays of both microtubules and actin filaments. *Eur. J. Cell Biol. 40*: 286–295.

Seagull, R. W. and I. B. Heath. 1979. The effects of tannic acid on the *in vivo* preservation of microfilaments. *Eur. J. Cell Biol. 20*: 184–188.

Senn, G. 1919. Weitere Untersuchungen über Gestalts- und Lageveränderungen der Chromatophoren. IV. und V. *Z. Bot. 11*: 81–139.

Wagner, G. 1979. Actomyosin as a basic mechanism of movement in animals and plants. *In*: (W. Haupt and M. E. Feinleib, eds) *Physiology of Movements*. Springer-Verlag, Berlin, Heidelberg, New York.

Botanica Marina
Vol. 34, pp. 513–520, 1991

Scinaia pseudo-moniliformis sp. nov. (Galaxauraceae, Rhodophyta) from the Sea of Japan[1]

M. Kajimura

Marine Biological Station, Shimane University, Kamo, Saigo, Oki-gun, 685 Japan

(Accepted 22 August 1991)

Abstract

A new deep-water moniliform species of *Scinaia* is reported from the Sea of Japan.

Thalli of *Scinaia pseudo-moniliformis* Kajimura, sp. nov. are up to 9.5 cm in height, dichotomously branched up to seven times and dioecious. The spermatangial sori are of the modified apical cap-type, consisting of oblong segments of 2–13 × 1.5–4 µm, with 1–3 hypodermal cell layers. Utricles of 3–5 × 1–2 µm distally and 26–33 × 10–26 µm basally are simple and without a solid basal segment. Lateral sterile branches occur on hypogynous cells, are 1–3 in number and consist of 1–4 cells. Mature gonimoblasts consist of a pseudoparenchymatous inner part and filamentous outer part. Mature carposporangia are 5–7 × 7–13 µm. The lower gonimoblast cells, carpogonium, hypogynous cell and basal cell of carpogonial branch form a fusion cell.

Scinaia pseudo-moniliformis Kajimura sp. nov. is most closely related to *Scinaia moniliformis* J. Agardh in appearance, however reduction of the lateral sterile branches on the hypogynous cell is the main difference between them.

Introduction

The genus *Scinaia* was established by Bivona-Bernardi (1822). Huisman (1985) included both *Pseudoscinaia* Setchell and *Pseudogloiophloea* Levring in the genus which at present includes some 34 species.

Repeated collections of the deep-water flora in the Oki Islands since 1966 have yielded many new species of marine algae, four of which belong to *Scinaia* and one of the four recently collected is described in the present paper.

Four of the known species of the genus *Scinaia boergesenii* Tseng (Tseng 1941), *Scinaia hormoides* Setchell (Setehell 1914), *Scinaia howensis* Huisman (Huisman 1986) and *Scinaia moniliformis* J. Agardh (J. Agardh 1884) belong to so called "*Scinaia moniliformis* group" (Huisman 1986) which have regularly constricted and moniliform thalli and *Scinaia pseudo-moniliformis* sp. nov. also belongs to this group.

Evidence that *Scinaia moniliformis* may have evolved from *Scinaia pseudo-moniliformis* sp. nov. is also discussed and a synoptic and dichotomous key to the five species belonging to "*Scinaia moniliformis* group" is presented.

Seven species of *Scinaia* are reliably known from Japan, namely, *Scinaia cottonii* Setchell (Setchell 1914; middle to southern part of the Pacific coast; rare), *Scinaia flabellata* Kajimura (Kajimura 1988; the Oki Islands; rare), *Scinaia hormoides* Setchell (middle part of the Pacific coast; rare), *Scinaia japonica* Setchell (Setchell 1914; Honshu and Kyushu; common), *Scinaia okiensis* Kajimura (Kajimura 1988; the Oki Islands; common), *Scinaia pseudo-moniliformis* sp. nov. (the Oki Islands, rare) and *Scinaia tokidae* Kajimura (Kajimura 1988; the Oki Islands; common), and a synoptic and dichotomous key to these seven Japanese species of *Scinaia* is also produced.

[1] Contribution No. 51 from Oki Marine Biological Station, Shimane University.

Material and Methods

Sixty-seven mature specimens were examined for this study of *Scinaia pseudo-moniliformis* sp. nov. (23 male plants, 44 female) and all were collected by dredge (Kajimura 1981, 1987) from depths of 35 to 40 m off Tsudo in the Oki Islands in 1983, 1988 and 1989.

Most of the specimens are deposited in the Phycological Herbarium of the National Science Museum, Tokyo (TNS).

Both fresh and formalin preserved material was examined. Small pieces of a frond were stained with a 1% aqueous solution of aniline blue acidified with acetic acid (9 : 1 v/v) for 10–40 minutes and sectioned with a freezing microtome. Sections were mounted in a 50% aqueous solution of rice syrup acidified with acetic acid (33 : 1 v/v). The holotype specimen of *Scinaia japonica* [Herbarium of the University of California (UC) No. 90835] and syntype specimens of *Scinaia moniliformis* [Botanical Museum, Lunds University (LD) Herb. Agardh Nos. 32207 and 32208] were also examined.

Observations and Discussion

Scinaia pseudo-moniliformis sp. nov.

Thalli moniliformis, usque ad 9.5 cm lati, dichotome aut raro trichotome ramosi uaque ad 7plo in 3 dimensionibus. Segmenta oblonga, 2–13 × 1.5–4 mm. Epidermis ex utriculis et cellulis elongatis assimilantibus constans. Utriculi 3–5 × 1–2 µm in parte distali thalli et 26–33 × 10–26 µm in segmento basali; chromatophora irregulatim formata et minus pigmentum praebentia quam cellulae assimulantes epidermales et hypodermales. Regio hypodermica e 1–3 stratis cellularum aut sphaericarum aut pyriformium usque ad clavatas, valde pigmentosarum, hypodermalium constans. Chromatophora irregulatim vittiformia, parietalia in cellulis et hypodermalibus et epidermalibus.

Plantae dioeciae. Sori in spermatangiis solum in dimidio superno unius aut plurium segmentorum successivorum distaliumque. Sori evolventes spermatangiorum utriculos et degenerantes et plus minusve pyriformes includunt; illi sori spermatangiis superficialibus tam dense tecti ut utriculi videri a superficie non possint. Sori spermatangiorum maturi sine utriculis. Cellulae 2–3 matricales spermatangiorum plus minusve elongatae in cellulis corticalibus formatae; cellulae corticales parvae et subrotundae, e cellulis hypodermalibus exorientes. Omnis cellula matricalis spermatangiorum spermatangia 2–4 ad extremitates efferens. Monosporangia in cellulis assimulantibus in epidermide aliquando singulatim et terminaliter formata.

Rami carpogoniales 3-cellulares maturitate; carpogonium trichogyne usque ad 30 µm longa instructum. Cellula hypogyna cellulas primas ramorum lateralium steriliumque abscidens antequam initia gonimoblasti formata. Cellula hypogyna initia 1–3 ramorum lateralia steriliaque abscidens; omnia cellula aliquando 1–3plo longistrorsum aut transverse dividens; itaque rami laterales sterilesque 1–4-cellulares. Cellula basalis rami carpogonialis cellulas 4 primas filamentorum pericarpialium faciens.

Post fecundationem 1–3 initia gonimoblasti ex carpogonio fecundato terminaliter vel subterminaliter vel oblique vel lateraliter exorientia. Pars exterior gonimoblasti perfecte evoluti ex filamentis liberalibus consistens; pars interior parenchymata et ex cellulis dense compactis, subrotundis, parietes crassos habentibus, consistens. Omnes aut plures cellulae superiores filamentorum liberorum in carposporangia 5–7 × 1–13 µm maturitate factae. Aliquot cellulae inferiores gonimoblasti maturi, una cum carpogonio et cellula hypogyna et cellula basali rami carpogonialis cellulam-coalescentem formantes. Cystocarpia matura pyriformia usque ad sphaerica, distaliter contracta ad partem cerviciformem ad ostiolum unicum ducentem; pericarpium crassum. Cystocarpia 120–260 µm diametro, in medulla inclusa et per thallum omnino dispersa.

Holotype collection: 40 m depth off Tsudo, the Oki Islands (36°09′N, 133°14′E) on October 4, 1989.

Holotype (female) OS 10077-A, isotype (male) OS 10077-B, both deposited in the Phycological Herbarium of the National Science Museum, Tokyo (TNS).

Additional specimens examined: The following additional collections are all from the type locality.

1) OS 10059, female (1 specimen), at 40 m depth, M. Kajimura, 17 October 1983.
2) OS 10060, female (9 specimens), at 40 m depth, M. Kajimura, 19 October 1988.
3) OS 10061, spermatangial (1 specimen), at 40 m depth, M. Kajimura, 19 October 1988.
4) OS 10062, spermatangial (1 specimen), at 40 m depth, M. Kajimura, 20 October 1988.
5) OS 10064, female (3 specimens), at 40 m depth, M. Kajimura, 21 October 1988.
6) OS 10067, female (1 specimen), at 35 m depth, M. Kajimura, 8 August 1989.
7) OS 10068, spermatangial (3 specimens) and female (1 specimen), at 35 m depth, M. Kajimura, 10 August 1989.
8) OS 10069, spermatangial (3 specimens), at 35 m depth, M. Kajimura, 17 August 1989.
9) OS 10070, spermatangial (3 specimens), at 35 m depth, M. Kajimura, 4 September 1989.

10) OS 10071, spermatangial (1 specimen) and female (1 specimen), at 35 m depth, M. Kajimura, 5 September 1989.
11) OS 10072, spermatangial (1 specimen) and female (1 specimen), at 35 m depth, M. Kajimura, 7 September 1989.
12) OS 10073, spermatangial (1 specimen) and female (2 specimens), at 35 m depth, M. Kajimura, 11 September 1989.
13) OS 10074, spermatangial (2 specimens) and female (1 specimen), at 35 m depth, M. Kajimura, 12 September 1989.
14) OS 10075, spermatangial (2 specimens) and female (2 specimens), at 35 m depth, M. Kajimura, 14 September 1989.
15) OS 10077, spermatangial (3 specimens) and female (10 specimens), at 40 m depth, M. Kajimura, 4 October 1989.
16) OS 10078, female (1 specimen), at 40 m depth, M. Kajimura, 5 October 1989.
17) OS 10079, female (3 specimens), at 40 m depth, M. Kajimura, 12 October 1989.
18) OS 10080, spermatangial (1 specimen) and female (1 specimen), at 40 m depth, M. Kajimura, 25 October 1989.
19) OS 10081, female (1 specimen), at 40 m depth, M. Kajimura, 30 October 1989.

Distribution: Known only from the type locality.

This species is named for its similarity to *Scinaia moniliformis* in appearance of the thallus.

Vegetative structure

Thalli are moniliform, up to 9.5 cm in height, branched di- or more rarely trichotomously up to seven times in three dimensions (Figs 1–4). Holdfasts are discoid, axils narrow. Segments are normally oblong and more or less bullet-shaped, 2–13 × 1.5–4 mm in size. The epidermis consists of utricles and elongate assimilatory cells (Figs 5, 7). Utricles are 3–5 × 1–2 µm in size in distal part of thallus and 26–33 × 10–26 µm in the basal segment, and have irregularly shaped chromatophores (Fig. 6) with less pigmentation than assimilatory cells in epidermis and hypodermal cells. The hypodermal region consists of one to three layers of spherical or pear-shaped to clavate, densely pigmented, hypodermal cells and colorless rhizoidal filaments. Internally plants are typical of classical *Scinaia*. The medulla consists of a central axis composed of dichotomous thick-walled filaments and loosely and radially disposed dichotomous filaments extending from the central axis to hypodermis. The central axis is visible from outside of the plant. Chromatophores are irregularly band-shaped and are parietal in both hypodermal and epidermal cells.

Figs 1, 2. *Scinaia pseudo-moniliformis* Kajimura, sp. nov. Fig. 1. Habit of female holotype specimen in liquid preserved condition. Fig. 2. Habit of male isotype specimen in liquid preserved condition.

3

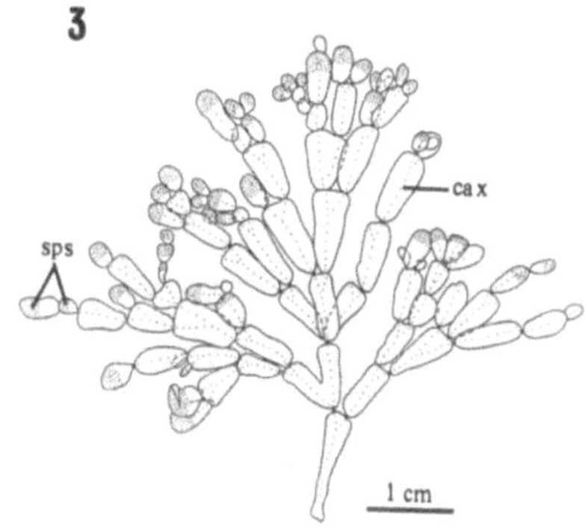

4

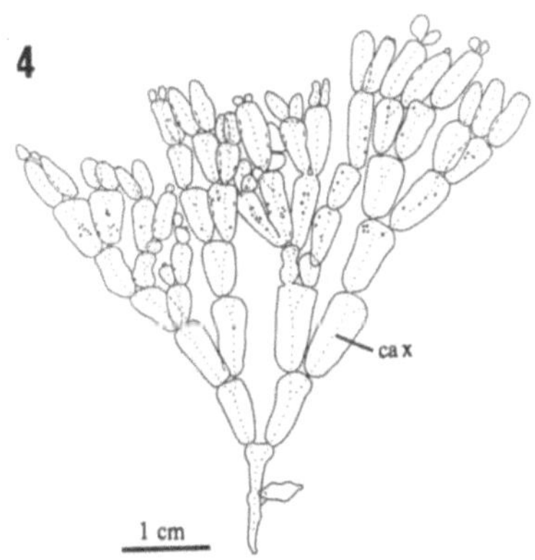

Figs 3, 4. *Scinaia pseudo-moniliformis* Kajimura, sp. nov. Fig. 3. Habit of male isotype specimen in liquid condition. Fig. 4. Habit of a liquid preserved female specimen (OS 10067) collected at a depth of 35 m off Tsudo, the Oki Islands 8 August, 1989.

Abbreviations used in Figures

bca, basal cell of carpogonial branch; ca, carposporangium; cax, central axis; cbi, carpogonial branch initial; cp, carpogonium; fgf, free gonimoblast filament; fu, fusion cell; fug, fused part of gonimoblast; gf, gonimoblast filament; gi1, gi2, gi3, gonimoblast initial 1, 2, 3; hy, hypogynous cell; hyc, hypodermal cell; mf, medullary filament; msp, monosporangium; o, ostiole; per, pericarpial cell; rhf, rhizoidal filament; sp, spermatangium; spm, spermatangial mother cell; sps, spermatangial sorus; stb, lateral sterile branch of hypogynous cell; tr, trichogyne; u, utricle; ycb, young carpogonial branch.

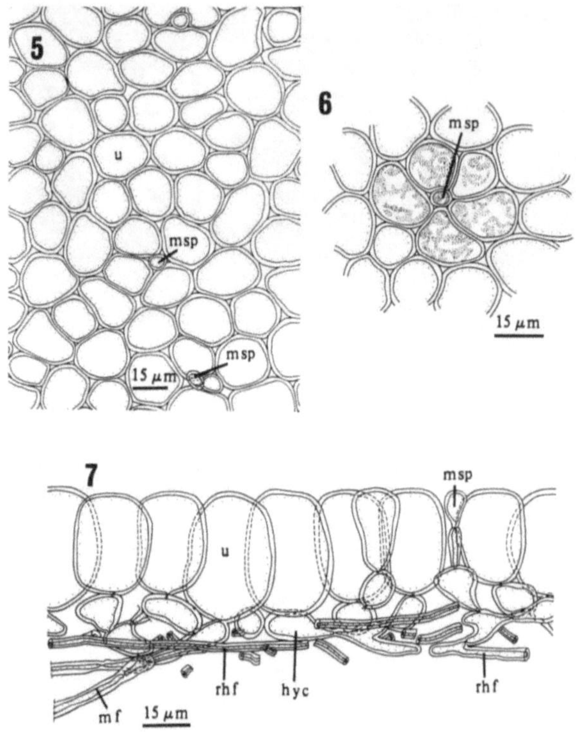

Figs 5–7. *Scinaia pseudo-moniliformis* Kajimura, sp. nov. Fig. 5. Surface view of thallus. Fig. 6. Surface view of thallus showing chromatophores in utricles and a monosporangium. Fig. 7. Cross section of cortex showing utricles, monosporangium, hypodermal cells, rhizoidal filaments and medullary filaments.

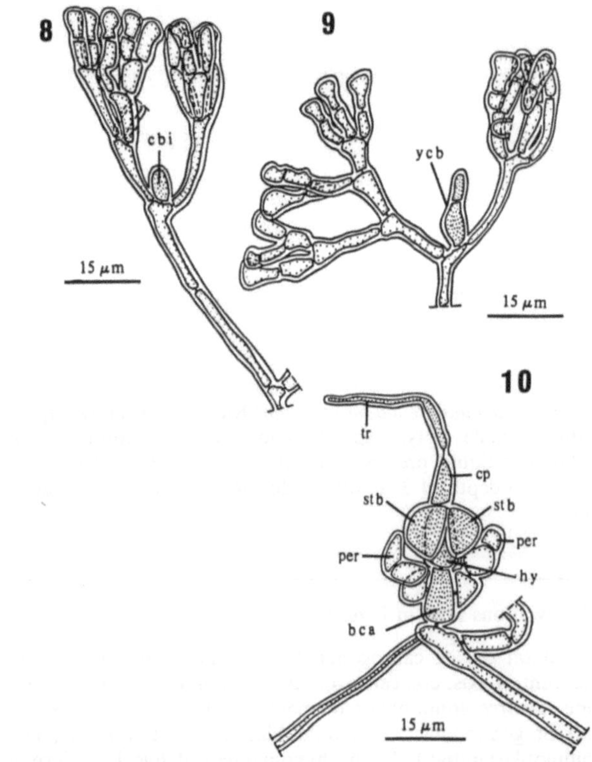

Figs 8–10. *Scinaia pseudo-moniliformis* Kajimura, sp. nov. Fig. 8. Carpogonial branch initial. Fig. 9. Two-celled young carpogonial branch. Fig. 10. Three-celled mature carpogonial branch with elongate trichogyne formed terminally on the carpogonium.

Reproductive structure

Spermatangia and monosporangia

Plants are dioecious (Figs 1–4). Spermatangial sori are restricted to approximately the upper half of one to several successive distal segments (Figs 1, 3, 22). Developing spermatangial sori include degenerating, more or less pear-shaped utricles (Fig. 20), however the spermatangial sori are usually so densely covered with superficial spermatangia that utricles are not evident in surface view. Fully mature spermatangial sori completely lack utricles (Figs 21, 23). More or less elongate spermatangial mother cells are, two to three in number, formed on roundish small cortical cells arising from hypodermal cells (Figs 20, 21). Each spermatangial mother cell (produces two to four spermatangia) terminally to subterminally (Figs 20, 21).

Monosporangia are sometimes formed singly and terminally on the assimilatory cells in the epidermis and each has a chromatophore (Figs 5–7).

Carpogonial branch and cystocarp

Carpogonial branches are three-celled at maturity, the carpogonium being provided with a trichogyne up to 30 µm in length (Fig. 10). The hypogynous cell starts to cut off initial cells of lateral sterile branches prior to the formation of gonimoblast initials (Fig. 10). The hypogynous cell cuts off one to three lateral sterile branch initials each of which may further divide once to three times longitudinally or transversely (Figs 10–17), consequently producing one to four-celled lateral sterile branches. The basal cell of the carpogonial branch produces four initial cells of pericarpial filaments.

After fertilization, one to three gonimoblast initials arise terminally, subterminally, obliquely or laterally from the fertilized carpogonium (Figs 11–16). Young gonimoblast filaments are all free (Fig. 17). Young gonimoblast filaments are all free (Fig. 17). Outer part of the fully developed gonimoblast consist of free filaments, however the inner part of it is pseudoparenchymatous and consists of densely compacted, roundish, thick-walled cells (Fig. 18). All or the upper several cells of the free filaments become carposporangia of 5–7 × 7–13 µm at mature size (Figs 18, 19). Some lower cells of the mature gonimoblast, the carpogonium, the hypogynous cell and the basal cell of the carpogonial branch form a fusion cell leaving sterile branch cells distinct (Figs 17, 18). Mature cystocarps with a thick pericarp consist of interwoven filaments and are pear-shaped to spherical, tapering distally to a cervix-like portion leading to a single ostiole (Fig. 18). Cystocarps, 120–260 µm in diameter, are embedded in the medulla, and scattered throughout the thallus (Figs 2, 4).

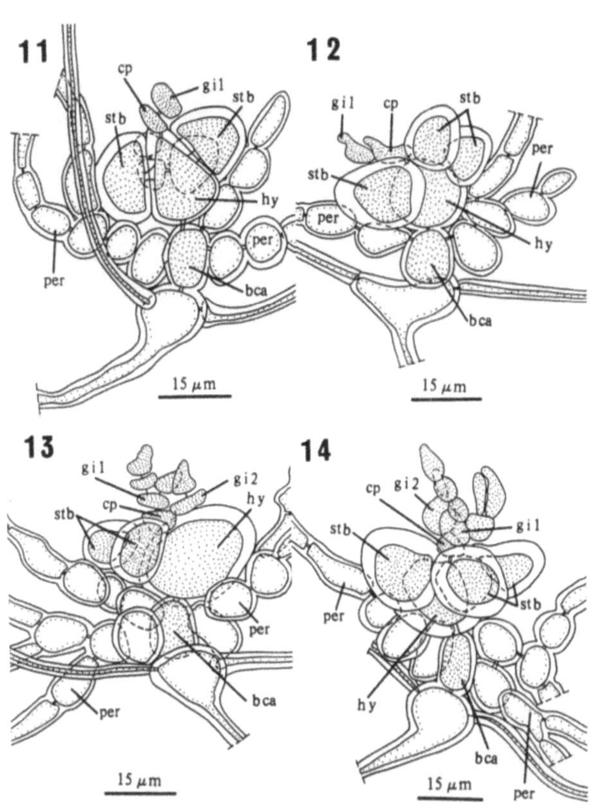

Figs 11–14. *Scinaia pseudo-moniliformis* Kajimura, sp. nov. Fig. 11. First gonimoblast initial arising obliquely from carpogonium; the hypogynous cell bearing one four-celled and one one-celled, lateral sterile branches. Fig. 12. First gonimoblast initial arising laterally from the carpogonium, the hypogynous cell bearing one two-celled and one-celled, lateral sterile branches. Fig. 13. Young gonimoblast with two gonimoblast initials arising terminally and subterminally from the carpogonium, the hypogynous cell bearing a single two-celled lateral sterile branch. Fig. 14. Young gonimoblast with two gonimoblast initials arising terminally and subterminally from the carpogonium, the hypogynous cell bearing one two-celled and two one-celled sterile lateral branches.

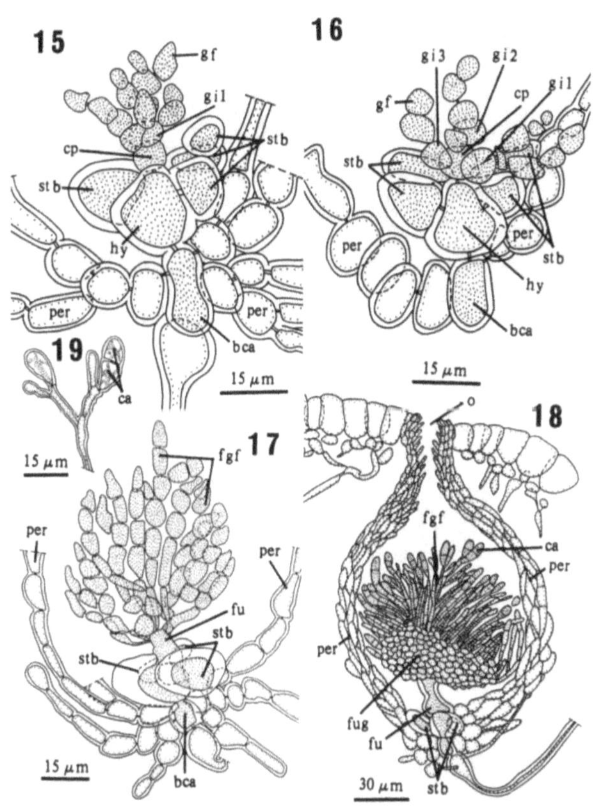

Figs 15–19. *Scinaia pseudo-moniliformis* Kajimura, sp. nov. Fig. 15. Young gonimoblast with a single gonimoblast initial arising terminally from the carpogonium, the hypogynous cell bearing one three-celled and one one-celled lateral sterile branches. Fig. 16. Developing gonimoblast with three gonimoblast initials arising laterally and terminally from the carpogonium, the hypogynous cell bearing two two-celled lateral sterile branches. Fig. 17. Longitudinal section of developing cystocarp showing distinct fusion cell and free gonimoblast filaments. Fig. 18. Longitudinal section of mature cystocarp showing fusion cell and fully develop gonimoblast consisting of free filaments in upper half and pseudo-parenchyma in lower half. Fig. 19. Upper part of a mature free gonimoblast filament showing three carposporangia.

Remarks

Spermatangial sori of *Scinaia pseudo-moniliformis* sp. nov. are considered to be modifications of the known apical cap-type and distinct not only from the typical apical cap-type described for *Scinaia tokidae*, but also from the intercalary belt-type of *Scinaia okiensis* as well as from the irregular type produced by *Scinaia flabellata*, *Scinaia japonica* and *Scinaia cottonii*.

Of the known species of the genus, four belonging to the "*Scinaia moniliformis* group" (*Scinaia boergesenii*, *Scinaia hormoides*, *Scinaia howensis* and *Scinaia moniliformis*) are compared with *Scinaia pseudo-moniliformis* below and in Table I. *Scinaia pseudo-moniliformis* sp. nov. is related to *Scinaia boergesenii* in characters such as shape and structure of the basal segment of thallus, height of thallus, frequency of dichotomous branching of thallus and structure of mature gonimoblast, but the former is distinct from the latter in characters such as the shape of the segments of thallus, number of hypodermal cell layer, size of utricles, size of carposporangia and size of segments of thallus as shown in Table I. *Scinaia pseudo-moniliformis* sp. nov. is related to *Scinaia hormoides* in characters such as the frequency of dichotomous branching of the thallus, number of hypodermal cell layer and height of the thallus, but the former is distinct from the latter in characters such as the shape of the segments of thallus, the size of utricles, sexuality, shape and structure of the basal segment of thallus and size of segments of thallus as shown in Table I. *Scinaia pseudo-moniliformis* sp. nov. is related to *Scinaia howensis* in characters such as height of thallus, number of hypodermal cell layer and sexuality, but the former is distinct from the latter in characters such as the frequency of dichotomous branching of thallus, shape

20

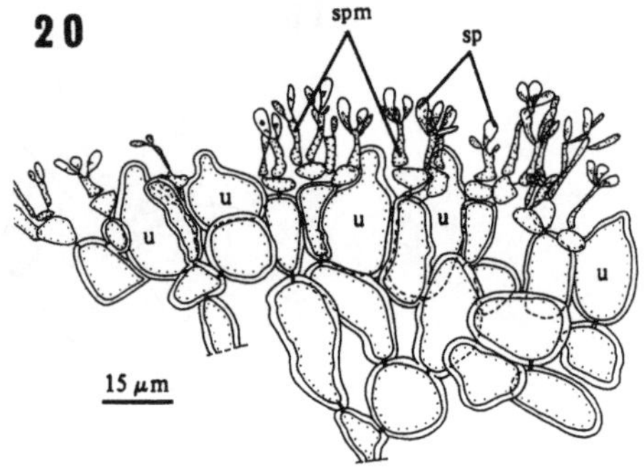

21

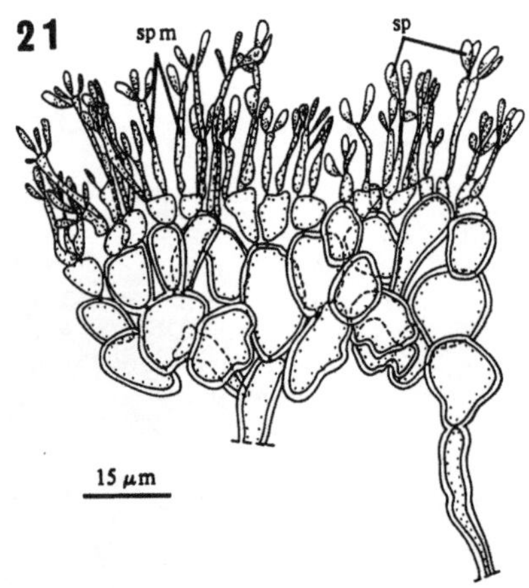

Figs 20, 21. *Scinaia pseudo-moniliformis* Kajimura, sp. nov. Fig. 20. Cross section of developing spermatangial sorus containing degenerating pear-shaped utricles. Fig. 21. Cross section of the fully mature spermatangial sorus lacking utricles.

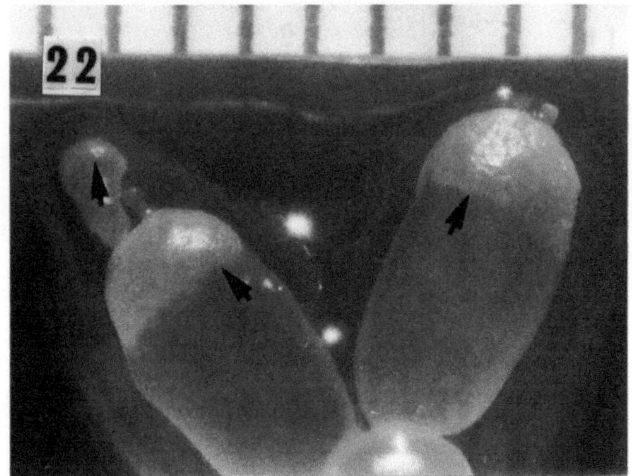

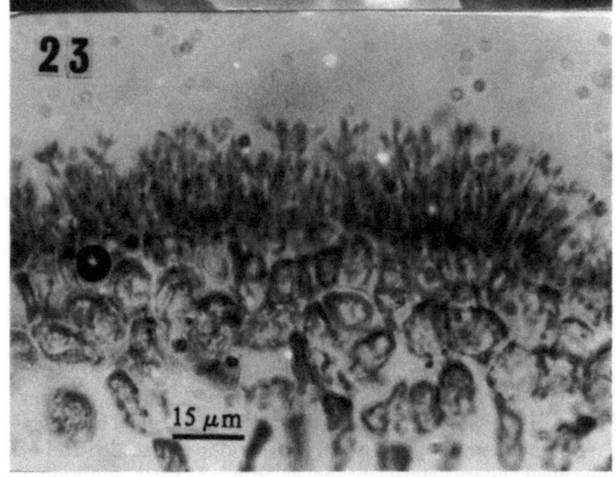

Figs 22, 23. *Scinaia pseudo-moniliformis* Kajimura sp. nov. Fig. 22. Spermatangial sori of the modified apical cap-type (arrows) on distal thallus segments (one section of the rule: 1 mm). Fig. 23. Cross section of the fully mature spermatangial sous.

of segments of thallus, size of segments of thallus, size of utricles, number of lateral sterile branches on the hypogynous cell, cell number of lateral sterile branches on the hypogynous cell, size of carposporangia, component of fusion cell and shape and structure of the basal segment of thallus as shown in Table I. *Scinaia pseudo-moniliformis* sp. nov. is related to *Scinaia moniliformis* in characters such as the shape of segments of thallus and shape and structure of the basal segment, sexuality and spermatangial sori but the former is distinct from the latter in characters such as the size of segments of thallus, number of hypodermal cell layer, size of utricles, number of lateral sterile branches on the hypogynous cell and cell number of lateral sterile branches on the hypogynous cell (Huisman 1986), number of gonimoblast initials, structure of mature gonimoblast, component of fusion cell, height of thallus and frequency of dichotomous branching of thallus as shown in Table I. *Scinaia pseudo-moniliformis* sp. nov. is most closely related to *Scinaia moniliformis* in appearance, however reduction of lateral sterile branches on hypogynous cells (Huisman 1986) is a difference between those two species as shown in Table I and the latter species is considered to have evolved from the former.

To facilitate determination of the five moniliform species of *Scinaia* discussed above, a synoptic and dichotomous key has been produced.

Key to Species of *Scinaia moniliformis* Group

1. Segments of thallus regularly oblong 2
1. Segments of thallus more or less subglobular or globular to obovate 3
   2. Utricles 3 – 5 × 1 – 2 µm distally and 26 – 33 × 10 – 26 µm basally; lateral sterile branches on hypogynous cell 1 – 3 in number and 1 – 4-celled; gonimoblast initials 1 – 3 in number; inner part of mature gonimoblast pseudoparenchymatous and outer part of it filamentous; lower gonimoblast cells, carpogonium, hypogynous cell and basal cell of carpogonial branch forming a fusion cell. . *S. pseudo-moniliformis*

Table I. Comparison of *Scinaia pseudo-moniliformis* sp. nov. and its allied species.

Species	*S. pseudo-moniliformis* sp. nov.	*S. moniliformis*[1]	*S. boergesenii*[2]	*S. howensis*[3]	*S. hormoides*[4]
Size of utricles	3–5 × 1–2 μm distally, 26–33 × 10–26 μm basally	25–30 × 17–23 μm distally, 35–50 × 20–30 μm basally	25–35 × 15–32 μm	25–35 × 15–20 μm distally, 45–65 × 30–40 μm basally	24–25 × 20–22 μm
Structure of gonimoblast	Inner part pseudoparenchymatous, outer part filamentous	Mostly filamentous	Inner part pseudoparenchymatous, outer part filamentous	Mostly filamentous?	?
Component of fusion cell	Lower gonimoblast cells, carpogonium, hypogynous cell and basal cell of carpogonial branch	Free gonimoblast filaments, carpogonium and hypogynous cell	?	Carpogonium, hypogynous cell and basal cell of carpogonial branch	?
Size of carposporangia	5–7 × 7–13 μm	5–8 × 10–15 μm	6–10 × 8–20 μm	12 × 20 μm	?
Number of gonimoblast initials	1–3	3 or 4	?	?	?
Cell number of lateral sterile branches	1–4	One one-celled and one two-celled	?	One one-celled and one two-celled	?
Number of lateral sterile branches	1–3	2	?	2	?
Number of hypodermal cell layer	1–3	1–2	1–2	1–3	1–3
Sexuality	Dioecious	Dioecious	Dioecious?	Dioecious	Monoecious
Spermatangial sori	Modified apical cap-type	Modified apical cap-type	?	?	In stellate clusters
Basal segment of thallus	Simple, not solid	Simple, solid?	Simple, not solid	Y-shaped, solid, constricted	Y-shaped, solid, unconstricted
Shape of segments	Regularly oblong	Regularly oblong	Subglobular, obovate or oblong	Subglobular, obovate or ovoid	Globular to obovate, at times oblong
Size of segments	2–13 × 1.5–4 mm	10–15 (–17) × 4–5 mm	2–12 × 2–6 mm	3–6 (–8) × 1–3 mm	3–10 × 3 mm
Branching of thallus	Up to 7 times	Up to 11 times	7–8 times	Up to 13 times	7–8 times
Height of thallus (cm)	Up to 9.5	Up to 22	Ca. 7	Up to 8	4–6

[1]Based on J. Agardh (1884), Setchell (1914), Huisman (1986) and personal communication with Dr Kraft; [2]based on Tseng (1941, 1984); [3]Huisman (1986); [4]Setchell (1914), Huisman (1986).

2. Utricles 25–30 × 17–23 µm distally and 35–50 × 20–30 µm basally; lateral sterile branches on hypogynous cell regularly 2 in number and consist of one 1-celled and one 2-celled; gonimoblast initials are 3 or 4 in number; mature gonimoblast mostly filamentous; free gonimoblast filaments, carpogonium and hypogynous cell forming a fusion cell ..... *S. moniliformis*
3. Basal segment of thallus simple and not solid ... . *S. boergesenii*
3. Basal segment of thallus Y-shaped and solid .. 4
4. Basal segment of thallus constricted .... *S. howensis*
4. Basal segment of thallus unconstricted .... *S. hormoides*

Seven species of *Scinaia* have been reported from the vicinity of Japan to date as mentioned above, including the four endemic species (*Scinaia tokidae*, *Scinaia okiensis*, *Scinaia flabellata* and *Scinaia pseudo-moniliformis* sp. nov.) all of which have been collected from the Oki Islands (Kajimura 1988 and present study). On the other hand, Japanese records of *Scinaia moniliformis* (Okamura 1930, 1936) are being rejected because its reproductive structure has not been reported and it may be identical with *Scinaia pseudo-moniliformis* sp. nov.

To facilitate their determination, a synoptic and dichotomous key is produced.

Key to Japanese Species of *Scinaia*

1. Thallus cylindrical and not regularly moniliform 2
1. Thallus regularly moniliform or flattened, not cylindrical 5
2. Thallus branching in two dimensions .... *S. flabellata*
2. Thallus branching in three dimensions 3
3. Plant monoecious, spermatangial sori irregularly shaped .... *S. japonica*
3. Plant dioecious, spermatangial sori terminal cap-type or intercalary belt-type 4
4. Spermatangial sori apical cap-type .... *S. tokidae*
4. Spermatangial sori intercalary belt-type .... *S. okiensis*
5. Thallus flattened and branching in two dimensions, plant monoecious, spermatangial sori irregularly shaped .... *S. cottonii*
5. Thallus moniliform and branching in three dimensions 6
6. Basal segment of thallus Y-shaped and solid monoecious .... *S. hormoides*
6. Basal segment of thallus simple and not solid dioecious .... *S. pseudo-moniliformis*

Acknowledgements

The present writer wishes to acknowledge his indebtedness to Dr P. C. Silva, Herbarium, University of California, Berkeley for the loan of the holotype specimen of *Scinaia japonica* and helpful suggestions. He also wishes to acknowledge his indebtedness to Dr P. Lassen of the Botanical Museum, Lunds University (LD) for the loan of the syntype specimens of *Scinaia moniliformis*. He would like to further acknowledge Dr G. T. Kraft, School of Botany, University of Melbourne, Dr G. T. Boalch, the Laboratory, Plymouth for their many helpful suggestions and critical reading of the manuscript, and Mrs A. Shipman for production of the Latin diagnosis.

References

Agardh, J. G. 1884. Till algernes systematik. Nya bidrag. VII Florideae. *Lunds Univ. Årsskr. 21*: 1–117.

Bivona-Bernardi, A. 1822. *Scinaia*, Algarum Marinum Novum Genus. *L'Iride 1*: 232–234.

Husman, J. M. 1985. The *Scinaia* assemblage (Galaxauraceae, Rhodophyta): a re-appraisal. *Phycologia 24*: 403–418.

Huisman, J. M. 1986. The red algal genus *Scinaia* (Galaxauraceae, Nemaliales) from Australia. *Phycologia 25*: 271–296.

Kajimura, M. 1981. *Streptophyllopsis* a new genus of Laminariaceae, Phaeophyta, from Japan. *Mem. Fac. Sci., Shimane Univ. 15*: 75–87.

Kajimura, M. 1987. Deep-water flora of benthic marine algae in the Oki Islands, Sea of Japan. *Bot. Mar. 30*: 373–385.

Kajimura, M. 1988. Three new deep-water species of *Scinaia* (Galaxauraceae, Rhodophyta) from the Sea of Japan. *Bot. Mar. 31*: 175–185.

Okamura, K. 1930. *Icones of Japanese algae. 6*. Kazamashobō, Tokyo. pp. 101. (English pages).

Okamura, K. 1936. *Nippon Kaisōshi* (*Description of the Japanese algae*). Uchida-rōkakuho, Tokyo. pp. 964.

Setchell, W. A. 1914. The *Scinaia* assemblage. *Univ. Calif. Publ. Bot. 6*: 79–152.

Tseng, C. K. 1941. Studies of the Chaetangiaceae of China. *Bull. Fan. Mem. Inst. Biol., Bot. Ser. 11*: 83–118.

Botanica Marina
Vol. 34, pp. 521–526, 1991

Chaetoceros bermejensis sp. nov., a New Planktonic Diatom from the Gulf of California

D. U. Hernández-Becerril[1]

University of Bristol, Department of Botany, Woodland Road, Bristol BS8 1UG, U.K.

(Accepted 22 August 1991)

Abstract

A new species of the diatom genus *Chaetoceros* Ehr., *Ch. bermejensis* sp. nov., is described and illustrated by light and electron microscopy, from net plankton samples from the Gulf of California. It is included within the subgenus Hyalochaete Gran, section Cylindrica Ostenfeld. The main characteristics of *Ch. bermejensis* are: its relative fragility, the shape of the chains and cells, the small depressions at the apical areas of the valves, and the presence of numerous small, slit-like rimoportulae at the centre of the terminal valves only. Comparisons with related species are made; the closest related species seems to be *Ch. weissflogii* Schütt. Questions about the presence of more than one rimoportula per valve are raised.

Introduction

During the analysis of net plankton samples from the Gulf of California and coasts of Baja California, Mexico, many species of the diatom genus *Chaetoceros* Ehr. were encountered, some of them being studied by electron microscopy for the first time and others being new records for the Gulf of California (Hernández-Becerril, 1989).

In this paper a new *Chaetoceros* species, belonging to the subgenus Hyalochaete, is described and illustrated on the basis of light (LM), scanning (SEM) and transmission (TEM) electron microscopy. Its most remarkable feature is the presence of numerous labiate processes (rimoportulae) on the valve face of the terminal valves in the chain. The species was neither very common nor abundant in the Gulf of California, but is fairly conspicuous because of its relatively large size, chain form with small apertures and relative fragility.

[1] Present address: Centro de Investigaciones de Quintana Roo, Depto. de Ecología Acuática, Apdo. postal 424, Chetumal, Q.R. 77000 México.

Material and Methods

Material used was preserved marine net plankton samples collected from the Gulf of California and coasts of Baja California; samples were obtained from five cruises during 1982–1986 (Hernández-Becerill 1989).

Observations were carried out under light microscopy (Olympus CH), once the material had been rinsed. After preliminary LM observations and measurements, specimens were isolated with a micropipette and put on to coverslips, air-dried, mounted on an aluminium stub, and then coated with gold for study by scanning electron microscopy (Phillips 501, at 10–12 kv). As this new *Chaetoceros* species is delicate and lightly silicified, the critical point drying method was used to avoid the collapse of valves (Medlin 1978), after rinsing and dehydration of the material. Transmission electron microscopy (TEM) (JEOL 1200 EX) observations were also made, using cleaned material. The cleaning method followed Simonsen (1974) and Hasle (1978).

The terminology adopted follows that proposed by Anonymous (1975) and Ross *et al.* (1979); terminology specific to *Chaetoceros* is taken from Rines and

Hargraves (1988). However, the following modifications are made here: the term 'aperture' is preferred over 'foramen' (Rines and Hargraves 1988), because foramen refers to a structure of the valve of other diatoms, while the aperture is the space between sibling cells in a chain; the central hyaline field is termed the annulus (Stosch 1977); measurements were made of the apical axis (typically the width in the girdle view), pervalvar axis, and the aperture (all in µm).

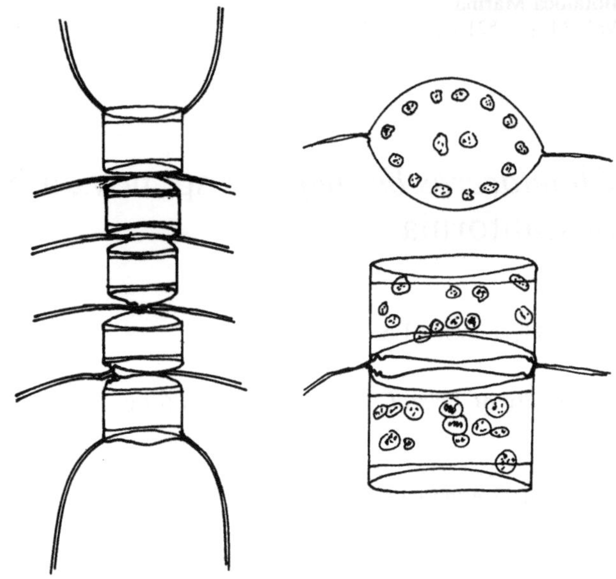

Fig. 1. *Chaetoceros bermejensis* sp. nov. View of a whole chain, apical view, and two sibling cells with chromatophores, showing the small depressions from where the intercalary setae arise.

Observations

Genus *Chaetoceros* Ehrenberg
Subgenus Hyalochaete Gran
Section Cylindrica Ostenfeld

Chaetoceros bermejensis sp. nov.
Synonyms: *Chaetoceros teres*? *sensu* Okamura, 1907, p. 94, pl. 4, figs 53, 54.
'Chaetoceros californianus' Hernández-Becerril, 1989, p. 69, pls 60, 70, *non Chaetoceros californicus* Grunow *in* Van Heurck

Diagnosis

Cellulae in catenas plus minusve rectas, leniter siliceae. Cellula aspectu cincturae, rectangularis, axe pervalvari 30–45µm longo, axe apicali 37–40 µm longo; anguli cellulae graciles, cellulas contiguas non tangens. Aperturae inter cellulas parvae, 3–5 µm. Limbus altus, lineae cincturae rectae. Cellula, aspectu valvae, subcircularis ad parum elliptica. Setae tenues, longae, ad basim connatae, fortuito divergentes. Ad apices valvae et prope bases setarum depressiones duae parvae. Ad centrum cellularum terminalium solarum 30–40 rimoportulae parvae. Multa chromatophora parva discoidea in utraque cellula, in setis carentes.

Locus typicus: 25°47′N, 109°35′W, 11 Dec., 1984.
Iconotype: Figs 1, 2.
Holotype: To be deposited in the diatom collection of the Instituto de Ciencias del Mar y Limnología (MEXU-D), Cd. de México, Mexico.
Isotype: to be deposited in the diatom collection of the British Museum, Natural History (BM), London, England.

Description in LM (Figs 1–3): The cells are joined in straight or slightly twisted, delicate chains of variable length. Cells are rectangular in girdle view with the corners more or less rounded and without contact between sibling cells, although the aperture is narrow. The valve face is flat in the usual girdle view, but convex when viewed from an apex. The mantle is low. Cells are subcircular to elliptical in valve view. The setae are rather thin and long, and arise from small depressions of the poles of adjacent valves, fuse immediately and then diverge at varied angles. Terminal setae differ from intercalary setae only in orientation, curving towards the chain axis. Numerous, small and round chromatophores occur in the cells. Measurements: 37–40 µm apical axis, 30–45 µm pervalvar axis, 3–5 µm aperture. Habitat: Planktonic, Type location; Gulf of California. The epithet of this new species is due to the Gulf of California's old name: Bermejo Sea ('Mar Bermejo').

Description in SEM and TEM: Again, in EM the cells appear fragile and delicate (Figs 4, 5). The valve face may be convex or nearly flat (Fig. 4), and the valve mantle is relatively low. The valves are very lightly silicified and perforated by small poroids that are more numerous on the valve face, becoming fewer close to the base of the setae (Figs 7–9). A weak pattern of costae radiates from the centre of the valve (Fig. 9). There are numerous (30–40) rimoportulae at the centre of terminal valves only. The rimoportulae are very small and slit-like (Figs 10, 11). Each opens at the centre of a small external depression, the slits being randomly orientated (Figs 12, 13). Near the base of the setae, at the apical margins of the valve are two depressions, each to one side of the seta base (Figs 5, 6, 8). The bases of the setae are quite short and spine less (Figs 6, 8). Distally the setae bear spiral rows of spines and have rows of pores also running in an apparently spiral pattern (Figs 12, 14). The setae are circular in cross section throughout. The morphology of terminal and intercalary setae is similar.

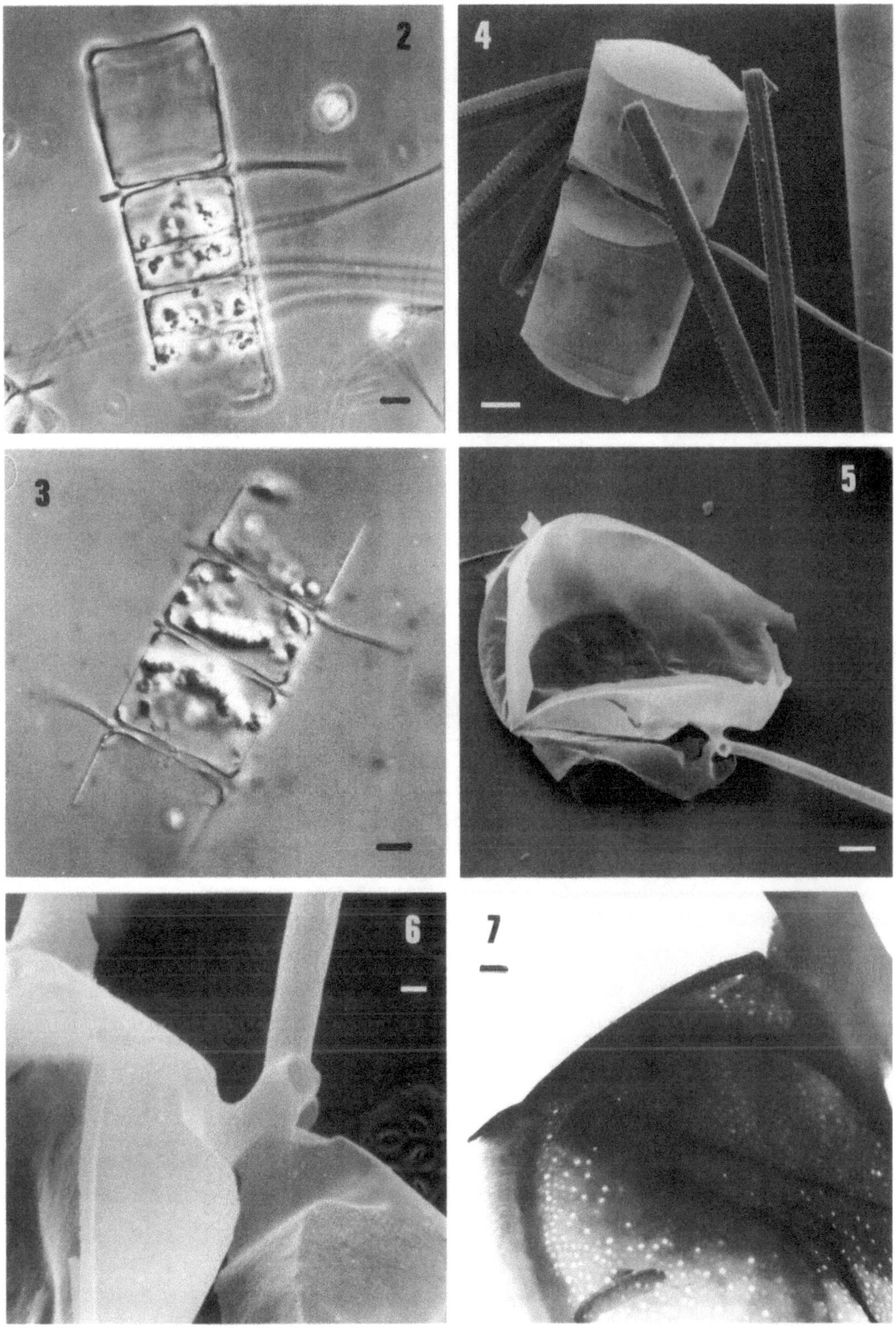

Figs 2–7. *Chaetoceros bermejensis* sp. nov.
Fig. 2. Part of a typical chain showing a large terminal cell (the upper one), LM: Fig. 3. Part of another chain showing the intercalary cells and the numerous chromatophores, LM. Fig. 4. A short chain, SEM (after critical point drying). Fig. 5. Two sibling valves, SEM (no critical point drying). Fig. 6. Detail of the base of the setae and depressions at apical areas, SEM. Fig. 7. Detail of the valve, TEM. Scale bars = 10 µm, Figs 2–4; = 5 µm, Fig. 5; = 1 µm, Figs 6, 7.

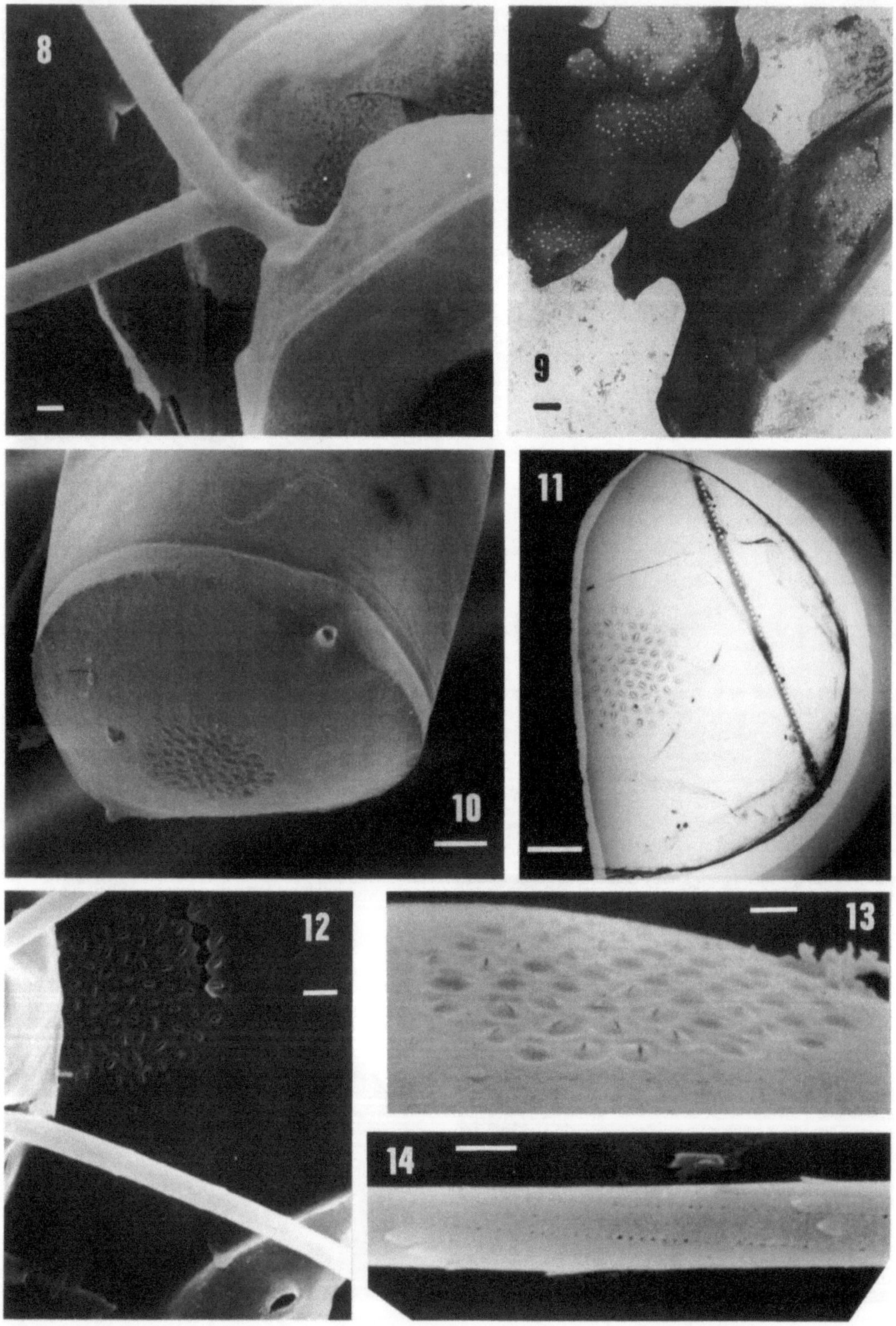

Figs 8–14. *Chaetoceros bermejensis* sp. nov.
Fig. 8. Sibling valves showing the depressions at apical areas, SEM. Fig. 9. Sibling valves with broken setae, TEM. Fig. 10. Terminal valve with several small rimoportulae, SEM: Fig. 11. Another view of the rimoportulae, TEM. Fig. 12. Valve face showing the rimoportulae and part of a seta, SEM. Fig. 13. Higher magnification of the rimoportulae at a terminal valve, SEM. Fig. 14. Detail of the middle part of a seta, SEM. Scale bars = 5 µm, Figs 10, 11; = 2 µm, Fig. 12; = 1 µm, Figs 8, 9, 13, 14.

Discussion

Morphology of Chaetoceros bermejenisis

Early studies of *Chaetoceros* species using electron microscopy (basically TEM) (Okuno 1951, Helmcke and Krieger 1953, Hendey *et al.* 1954) showed what Desikachary (1956) later summarized: the cell wall is simply perforated by small poroids with some 'vein-like thickenings' (costa pattern) in thin-walled species (mainly the subgenus Hyalochaete). Okuno (1956) contributed observations of the 'central area' (central hyaline field or annulus), and the costae radiating from this field, features usually found in species of the subgenus Hyalochaete.

Evensen and Hasle (1975) confirmed previous observations on *Chaetoceros* and added comments on the structure and position of the processes (rimoportulae) in every valve for species of the subgenus Phaeoceros, and in terminal valves for species of the subgenus Hyalochaete.

The new species describes here, *Ch. bermejensis*, shows a morphology typical of species of the subgenus Hyalochaete, as its wall is thinly silicified, perforated by small poroids with a pattern of costae. However, the central hyaline field (annulus) is not observed in intercalary valves. The structure of the setae differs from that of the valves (Evensen and Hasle 1975), but seta structure is the same throughout, *i.e.* terminal and intercalary setae.

The most striking feature of *Ch. bermejensis* is, however, the presence of numerous rimoportulae in the terminal valves. As stated above, each rimoportula lies within its own annulus (small external depressions), rather than all of the rimoportulae lying within a common annulus. No previous report has described this character to be constant, although a similar finding has recently been noted for *Chaetoceros coarctatus* Lauder (Hernández-Becerril 1989), which has several small rimoportulae in every valve in a chain and belongs in the subgenus Phaeoceros, and also Rines and Hargraves (1990) documented some few terminal valves with multiple processes within a Hyalochaete taxon, *Ch. compressus* var. *hirtisetus* Rines *et* Hargraves. Therefore, several interesting questions, which can not be answered at the moment are raised: Why is this character shared by species from different subgenera? What is the significance of having more than one rimoportula per valve?

As regards the first point, this particular character found in taxa belonging to different subgenera may indicate parallel evolution. It also brings to discussion the usefullness of the present classification system that divides the genus *Chaetoceros* into two subgenera, with subdivision into a number of sections, and the need for further studies that contribute to clarify its taxonomy.

The second question is an old one, which refers to the function of the rimoportula in *Chaetoceros* (Evensen and Hasle 1975), and whether or not it is the same in all the species. It is possible that the function may include differential permeability to certain chemicals, as is thought for other centric diatoms (Fryxell and Hasle 1974), and this fact may be increased by having more than one rimoportula; this is especially important in tropical, oceanic conditions. Yet another idea concerns the evolution of the raphe (Hargraves and Rines, personal communication).

Comparison with related species

Chaetoceros bermejensis has been placed in the section Cylindrica Ost., subgenus Hyalochaete, because its valves are nearly circular, the apertures narrow, chromatophores numerous, and the terminal setae are no thicker than the other setae. Within this section are included *Ch. lauderi* Ralfs, *Ch. weissflogii* Schütt, and *Ch. teres* Cleve. Positive identification of these three species is difficult and has caused some confusion. For instance, the morphology of the resting spores is the diagnostic character separating *Ch. lauderi* from *Ch. teres* (Hargraves 1979). *Chaetoceros weissflogii* resembles *Ch. lauderi*, especially in its resting spores, although the general appearance of the intact chain is different; *Ch. weissflogii* has been included as a synonym of *Ch. lauderi* (*e.g.* Allen and Cupp 1935). *Chaetoceros teres* is the only one of these three species that has been recorded from the Gulf of California (Hernández-Becerril 1987).

Other species considered within this section are two tropical ones: *Chaetoceros buceros* Karsten and *Ch. bacteriastroides* Karsten. There seems to be no clear relationship between *Ch. bermejensis* and these two species. On the other hand, in an earlier contribution (Hernández-Becerril 1989) it was proposed that *Ch. bacteriastroides* be placed in a new section, Bacteriastrodea, due to its distinctive characters. The only feature in common between *Ch. bacteriastroides* and the new species, is the morphology of the setae (Fryxell 1978, Hernández-Becerril 1989).

Chaetoceros bermejensis is most closely related to *Ch. weissflogii.* The form of the chain is similar in both species (following the original description and drawings by Schütt, 1895, fig. 17b, *non* fig. 17a), but resting spores have not been observed in *Ch. bermejensis.* Cells of *Ch. weissflogii* are joined in a different way to those of *Ch. bermejensis*, and the shape of the cells, especially the terminal ones differ in the two species, in *Ch. bermejensis* the apical axis is larger than the pervalvar axis. In addition, the apertures are narrower, without a characteristic shape in *Ch. bermejensis*, while the original description of *Ch. weissflogii* makes no mention of the small depressions close to the bases of the setae. Illustrations of *Ch. weissflogii* by Peragallo and Peragallo (1879–1908) and Ikari (1926) differ from Schütt's (1895) original drawings. Meunier (1913) illustrated a species more like Schütt's (1895), but this still strongly resembles *Ch. lauderi*. Thus, confusion surrounds the precise identity of *Ch. weissflogii*.

Acknowledgements

Many thanks and appreciation are due to Dr R. M. Crawford for his guidance, and also to Prof. F. E. Round for his advice and comments. Mr R. Ross kindly translated the diagnosis into Latin. Many comments made by Dr P. E. Hargraves and Dr J. E. B. Rines considerably improved this paper, as well as others made by Dr E. Cox. Technical support was provided by Mr R. Porter, Mr S. Martin, and Mr T. Colborn. Material was provided by CIB, CICIMAR and ICML. Financial support was given by CONACYT (Mexican Council for Science and Technology).

References

Allen, W. E. and E. E. Cupp. 1935. Plankton diatoms of the Java Sea. *Ann. Jardin Bot. Buitenzorg 44*: 101–224.

Anonymous. 1975. Proposals for a standarization of diatom terminology and diagnoses. *Nova Hedwigia, Beih. 53*: 323–354.

Desikachary, T. V. 1956. Electron microscope studies on diatoms. *J. Royal Microsc. Soc., Ser. III 76*: 9–36.

Evensen, D. L. and G. R. Hasle. 1975. The morphology of some *Chaetoceros* (Bacillariophyceae) species as seen in the electron microscope. *Nova Hedwigia, Beih. 53*: 153–174.

Fryxell, G. A. 1978. Chain-forming diatoms: Three species of Chaetoceraceae. *J. Phycol. 14*: 62–71.

Fryxell, G. A. and G. R. Hasle. 1974. Coscinodiscineae: some consistent patterns in diatom morphology. *Nova Hedwigia, Beih. 45*: 69–96.

Hargraves, P. E. 1979. Studies on marine plankton diatoms. IV. Morphology of *Chaetoceros* resting spores. *Nova Hedwigia, Beih. 64*: 99–120.

Hasle, G. R. 1978. Diatoms. *In*: (A. Sournia, ed.) *Phytoplankton Manual*. UNESCO. Paris. pp. 136–142.

Helmcke, J.-G. and W. Krieger. 1953. *Diatomeenschalen im elektronenmikroskopischen Bild*. Teil 1. J. Cramer, Weinheim. 20 pp., pls. 1–102.

Hendey, N. I., D. H. Cushing and G. W. Ripley. 1954. Electron microscope studies on diatoms. *J. Roy. Microsc. Soc., ser. III 74*: 22–34.

Hernández-Becerril, D. U. 1987. A checklist of planktonic diatoms and dinoflagellates from the Gulf of California. *Nova Hedwigia 45* (1–2): 237–261.

Hernández-Becerril, D. U. 1989. The morphology and taxonomy of species of the diatom genera *Asteromphalus* and *Chaetoceros*. Ph. D. Thesis. Univ. of Bristol, England. 138 pp., 109 pls.

Ikari, J. 1926. On some *Chaetoceras* of Japan. I. *Bot. Mag., Tokyo 40* (478): 517–534.

Medlin, L. K. 1978. The use of critical point drying for cleaned diatom valves. *Bacillaria 1*: 169–172.

Meunier, A. 1913. Microplancton de la Mer Flamande, 1ere. partie. Le genre *Chaetoceros* Ehr. *Mem. Mus. Roy. Hist. Nat., Belgique 7* (2): 1–58, pls. 1–7.

Okamura, K. 1907. Some *Chaetoceras* and *Peragallia* of Japan. *Bot. Mag., Tokyo 21* (244): 89–106, pls. 3, 4.

Okuno, H. 1951. Electron microscopical study on antarctic diatoms (1). *J. Jap. Bot. 26* (10): 305–310.

Okuno, H. 1956. Electron microscopical study on fine structures of diatom frustules. XIV. observations on the genus *Chaetoceros. Bot. Mag., Tokyo 69* (14): 182–200.

Peragallo, H. and M. Peragallo. 1897–1908. *Diatomeés marines de France et districts maritimes voisins*. M. J. Tempere. Grez-sur-loin. 491 pp., 137 pls.

Rines, J. E. B. and P. E. Hargraves. 1988. The *Chaetoceros* Ehrenberg (Bacillariophyceae) flora of Narragansett Bay, Rhode Island, U. S. A. *Bibliotheca Phycologica 79*, 196 pp.

Rines, J. E. B. and P. E. Hargraves. 1990. Morphology and taxonomy of *Chaetoceros compressus* Lauder var. *hirtisetus* var. nova, with preliminary consideration of closely related taxa. *Diatom Res. 5*: 113–127.

Ross, R., E. J. Cox, N. I. Karayeva, D. G. Mann, T. B. B. Paddock, R. Simonsen and P. A. Sims. 1979. An amended terminology for the siliceous components of the diatom cell. *Nova Hedwigia, Beih. 64*: 513–533.

Schütt, F. 1895. Arten von *Chaetoceras* and *Peragallia*. Ein Beitrag zur Hochseeflora. *Ber. Deutsch. Bot. Ges. 13*: 35–48, pls. IV, V.

Simonsen, R. 1974. The diatom plankton of the Indian Ocean expedition of R. V. "Meteor" 1964–1965. *"Meteor" Forsch., Reih. D. 19*: 1–66, 41 pls.

Stosch, H. A. von. 1977. Observations on *Bellerochea* and *Streptotheca*, including descriptions of three new planktonic diatom species. *Nova Hedwigia, Beih. 54*: 113–166.

Botanica Marina
Vol. 34, pp. 527–536, 1991

Contribution to the Seaweed Flora of the Canary Islands: New Records of Florideophyceae

M. Sansón, J. Reyes and J. Afonso-Carrillo

Department of Vegetal Biology (Botany), University of La Laguna, 38271 La Laguna, Canary Islands, Spain

(Accepted 3 October 1991)

Abstract

Seven species of Florideophyceae are reported from the Canary Islands for the first time: *Predaea huismanii* Kraft, *Platysiphonia caribaea* Ballantine *et* Wynne, *Jania longifurca* Zanardini, *Spermothamnion flabellatum* Bornet, *Callithamnion decompositum* J. Agardh, *Helminthocladia agardhiana* Dixon and *Rissoella verruculosa* (Bertolini) J. Agardh. Data concerning ecological conditions, geographical distribution, morphology, anatomy and phenology are presented. Observations on *Platysiphonia miniata* (C. Agardh) Børgesen which has previously only been reported once from the Canary Islands, are also given.

Introduction

Afonso-Carrillo and Gil-Rodríguez (1982) have shown that the Canary Islands have a seaweed flora comprised of species with different biogeographical distribution patterns: Atlantic-American, Atlantic-European, Mediterranean and Atlantic-African (Feldmann 1946).

The first general account of the marine algae of the Canary Islands was given by Børgesen (1925–30), who recorded 369 species. These records were mainly from the intertidal zone of two localities (Puerto Orotava in Tenerife and near Las Palmas in Gran Canaria) during the winter of 1920–21. Later, Gil-Rodríguez and Afonso-Carrillo (1980) published a checklist increasing the number of benthic marine algae to 434 known species. Further studies (Afonso-Carrillo *et al.* 1984, Gil-Rodríguez *et al.* 1985, Viera-Rodríguez *et al.* 1987, Afonso-Carrillo 1988, 1989, Afonso-Carrillo and Sansón 1989) have increased the known flora to 461 species (93 Chlorophyta, 79 Phaeophyta, 264 Rhodophyta and 22 Cyanophyta). The present work adds seven species of Florideophyceae to the flora.

Material and Methods

Specimens were collected between 1985 and 1990 at several localities in Tenerife and Lanzarote and were deposited at TFC (Departamento de Biología Vegetal, Botánica, Universidad de La Laguna, Canary Islands). Permanent slides for anatomical studies were prepared from specimens preserved in 4% formalin in sea water, sectioned by hand with a razor blade and mounted in 20% aqueous 'Karo' dextrose. Camera lucida drawings and photomicrographs were obtained with the use of a Zeiss microscope. *Jania longifurca* Zanardini specimens were studied by scanning electron microscopy procedures as outlined in Afonso-Carrillo *et al.* (1984 b). Selected dried fragments were sectioned and rinsed under distilled water. After air drying, fragments were coated with gold and viewed with a Hitachi S-450 Stereoscan Microscope.

Results and Discussion

1. *Predaea huismanii* Kraft (Figs 1–5)

The genus *Predaea* (Nemastomaceae) was described by De Toni (1936) and includes generally gelatinous species distinguished by features including habit, shape of outer cortical cells, presence or absence of gland cells, site of origin of the gonimoblast, number of carpogonial branch cells and number of nutritive cells on cells adjacent to auxiliary cells. According to Verlaque (1990), thirteen species of *Predaea* are known from tropical, subtropical and warm temper-

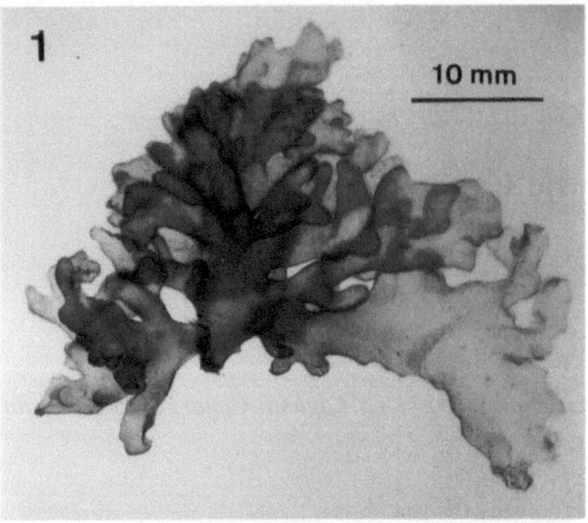

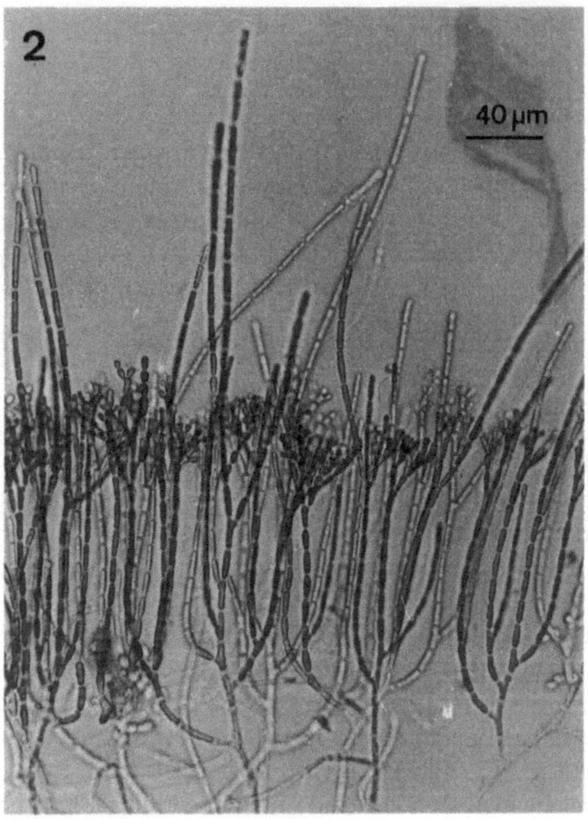

Figs 1–2. ***Predaea huismanii*** Kraft.
Fig. 1. Habit. Fig. 2. Transverse section of the plant showing exserted assimilatory cortical filaments.

ate regions of the Northern and Southern Hemispheres. According to Kajimura (1987), only three of these species have been reported from the Atlantic Ocean: *Predaea feldmanii* Børgesen from St. Helena I., North Carolina, Venezuela and Ghana; *P. weldii*

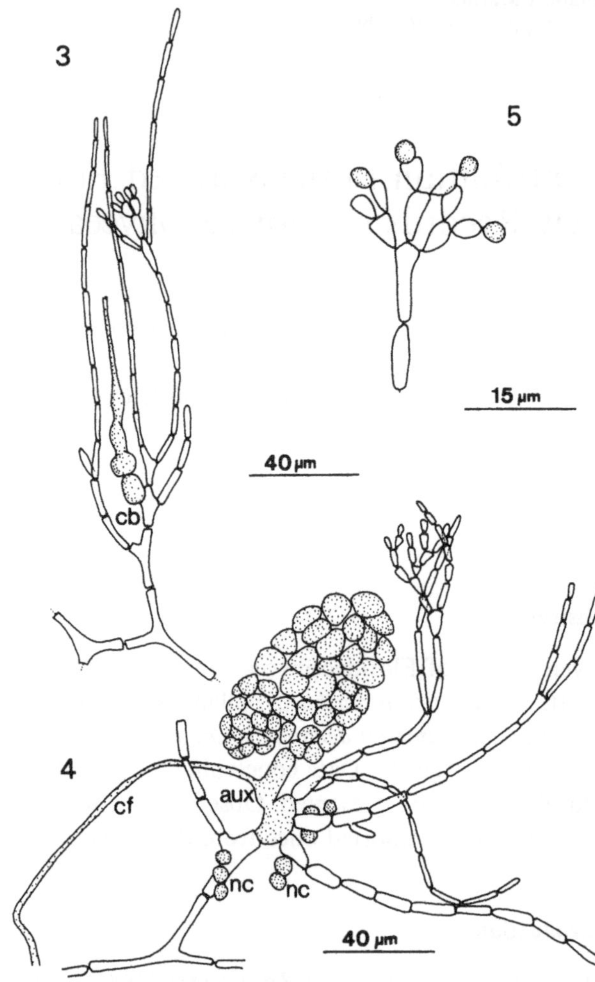

Figs 3–5. *Predaea huismanii* Kraft.
Fig. 3. Cortical filaments showing a 3-celled carpogonial branch (cb). Fig. 4. Mature carposporophyte. Note the small chains of nutritive cells (nc) and the lateral point of fusion of the connecting filament (cf) to the auxiliary cell (aux). Fig. 5. Spermatangia terminal on distichous pinnae.

Kraft *et* Abbott from Hawaii, Australia and Venezuela and *P. masonii* De Toni from Clarion I., Pacific and Atlantic Mexico, North Carolina and Ghana.

Plants collected in the Canary Islands (Fig. 1) have assimilatory filaments which are variable in length and exserted (Fig. 2), resulting in the hairy texture of the thallus. The Canarian plants agree with the description of *P. huismanii* Kraft (Kraft 1984) from Australia, the only known locality for this species. Plants are characterized by

(a) 3-celled carpogonial branches,

Figs 6–12.
Figs 6–8. *Platysiphonia caribaea* Ballantine *et* Wynne.
Fig. 6. Habit. Fig. 7. Detail of a multicellular rhizoid and the origin of an erect branch. Note the origin of the rhizoid from flanking cells and the endogenous origin of the branch. Fig. 8. Apical region of an erect branch.
Figs 9–12. *Platysiphonia miniata* (C. Agardh) Børgesen.
Fig. 9. General aspect of erect branches showing numerous tetrasporangial stichidia. Fig. 10. Detail of unicellular rhizoids. Fig. 11. Lateral branch arising from the distal end of a central cell. Fig. 12. Apical region of an erect branch.

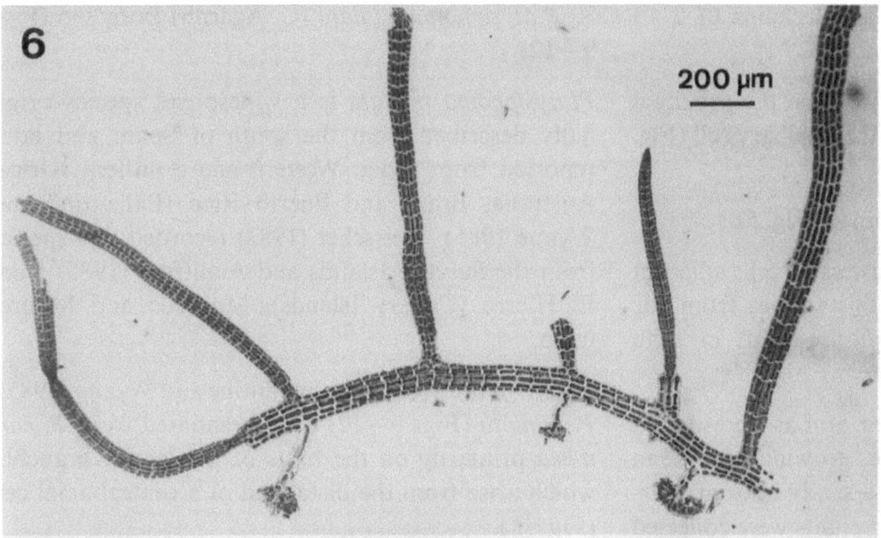
6
200 µm

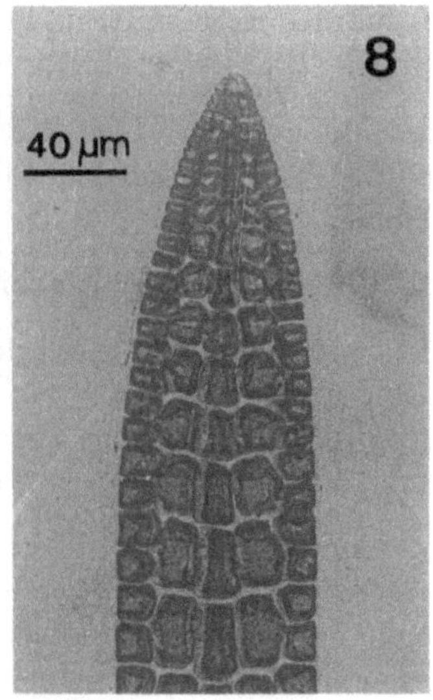
8
40 µm

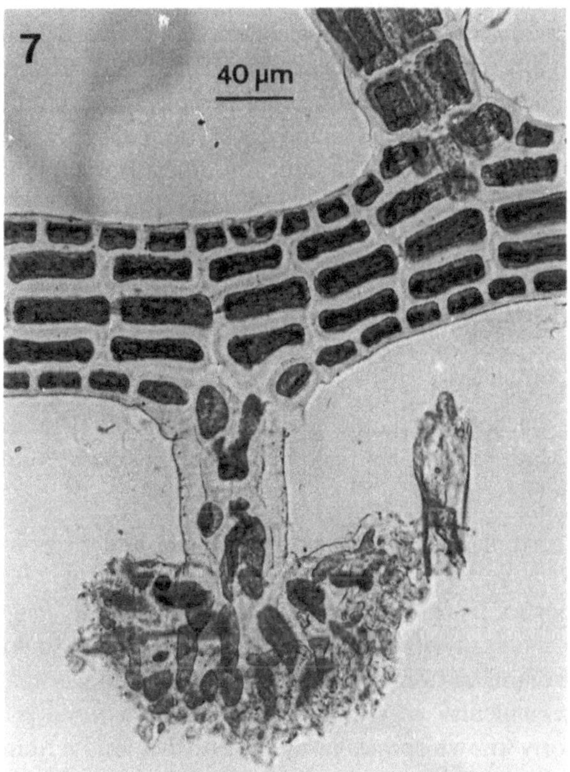
7
40 µm

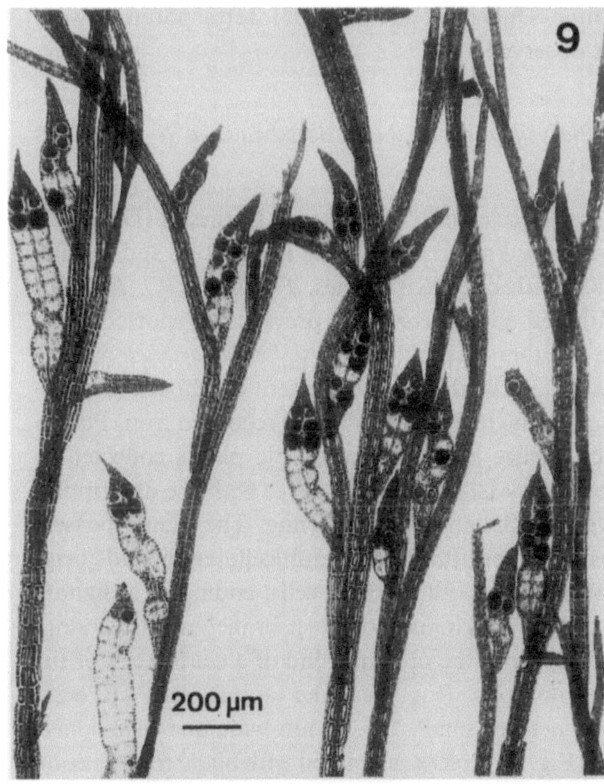
9
200 µm

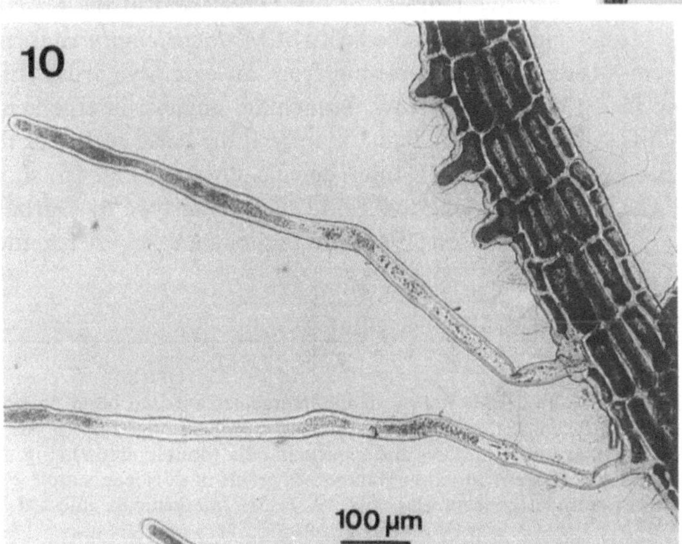
10
100 µm

11
40 µm

12
40 µm

(b) nutritive cells single or in small chains of 2–3 cells (Fig. 4),

(c) up to 2 clusters of nutritive cells on the proximal and two distal cells adjacent to the auxiliary cell (Fig. 4) and

(d) distichous spermatangial pinnae (Fig. 5).

The presence of nutritive cell clusters on cells adjacent to the auxiliary cell separate this species from the rather similar *Nemastoma confusum* Kraft *et* John (Kraft and John 1976).

Predaea huismanii was collected at Las Teresitas, a sheltered locality in E Tenerife, growing at 1–3 m depth, on an anchor rope over a sandy bottom. Carposporangial and spermatangial plants were collected in June–July [TFC (Phyc) 5703]. Tetrasporangia were not observed.

2. *Platysiphonia caribaea* Ballantine *et* Wynne (Figs 6–8)

Eleven species of *Platysiphonia* Børgesen (Delesseriaceae) are known world wide (Ballantine and Wynne 1985) with only two species, *P. miniata* (C. Agardh) Børgesen and *P. caribaea*, previously reported from the Northern Atlantic. *Platysiphonia caribaea* was recently described from the Western Caribbean (Ballantine and Wynne 1985) and only recorded from Puerto Rico, Belize and Mexico. Sterile plants collected in the Canary Islands (Fig. 6) agree with the description given by Ballantine and Wynne. The plants possess postrate axes attached by multicellular rhizoids originating from both flanking cells, ending in a digitate disc (Fig. 7). Unramified erect branches arise endogenously from the proximal end of a central cell of the postrate axes (Fig. 7). The apical cells of erect branches are small, 14–15 µm wide and hemispherical (Fig. 8). Apices associated with blade regeneration are evident (Fig. 6).

Platysiphonia caribaea was collected at Orzola, a semiexposed locality at NNE of Lanzarote, growing in tide pools mixed with other caespitose species [TFC (Phyc) 5704]. Previously it is only known from deep reef environments in the Caribbean (Ballantine and Wynne 1985).

3. *Platysiphonia miniata* (C. Agardh) Børgesen (Figs 9–12)

Platysiphonia miniata is a widespread species originally described from the south of Spain and later reported from India, Western and Southern Africa, Australia, Brazil and Puerto Rico (Ballantine and Wynne 1985). Weisscher (1983) recorded this species from the Salvage Islands and Audiffred (1985) from El Hierro (Canary Islands), Morocco and Mauretania.

Based on the account in Ballantine and Wynne (1985), *P. miniata* (Figs 9–12) is differentiated from *P. caribaea* primarily on the basis of the lateral branches which arise from the distal end of a central axial cell (Fig. 11).

Platysiphonia miniata was collected at Punta del Hidalgo, a semiexposed locality in NE Tenerife, growing in tide pools mixed with other caespitose Rhodomelaceae species as an epiphyte of *Galaxaura rugosa* (Ellis *et* Solander) Lamouroux. Tetrasporic plants were collected in January [TFC (Phyc) 5705].

4. *Jania longifurca* Zanardini (Figs 13–19)

Jania longifurca (Corallinaceae) has been frequently reported from the Mediterranean Sea (Hamel and Lemoine 1953, Giaccone 1978, Athanasiadis 1987), the European Atlantic coast, from Roscoff (France) to Mauretania (Lemoine 1924, Ardré 1970) and Madeira (Levring 1974).

According to Lemoine (1924), Hamel and Lemoine (1953) and Bressan (1974), among species of *Jania* reported from the Eastern North Atlantic, *J. longifurca* is easily distinguished by intergenicula up to 400 µm diam. and genicular cells considerably longer than intergenicular cells. Nevertheless, *J. longifurca* is a poorly known species in need of further study. Some vegetative features of plants collected in the Canary Islands were studied with SEM. *Jania longifurca* forms erect branches raising from an extensive crust (Fig. 13), with narrow branching angles, intergenicula 200–400 µm diam., mainly at the basal region of the plant (Fig. 14). Intergenicula surfaces (Fig. 15) show elongate poligonal cells called *Jania*-type by Garbary and Johansen (1982) and trichocyte bases are rounded

Figs 13–19. *Jania longifurca* Zanardini.
Fig. 13. Habit of a group of plants arising from a basal crust. Fig. 14. SEM. Aspect of the intergenicula at the basal portion. Fig. 15. SEM. Surface of intergenicula showing the *Jania*-type surface and the trichocytes (arrows) *Corallina*-type. Fig. 16. SEM. Longitudinal fracture of a branch showing intergenicular medullary cells (arrow) and genicular cells (double arrow). Fig. 17. SEM. Intergenicular transverse fracture showing thin walled medullary cells containing numerous grains of floridean starch. Fig. 18. SEM. Surface view of terminal branches showing tetrasporangial conceptacles. Fig. 19. SEM. Intergenicula infected by *Choreonema thuretii*.

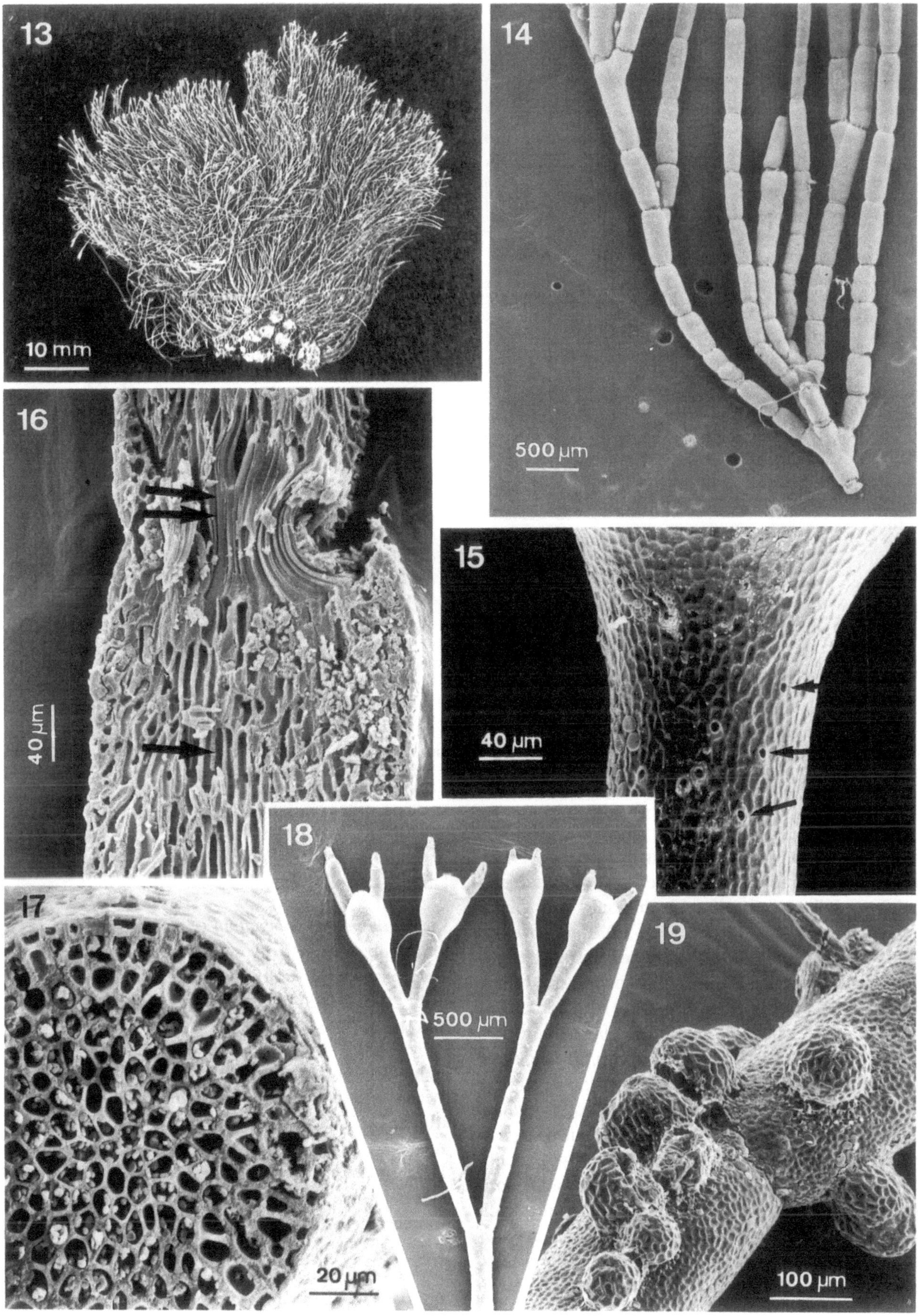
13
10 mm
14
500 µm
16
40 µm
15
40 µm
18
500 µm
17
20 µm
19
100 µm

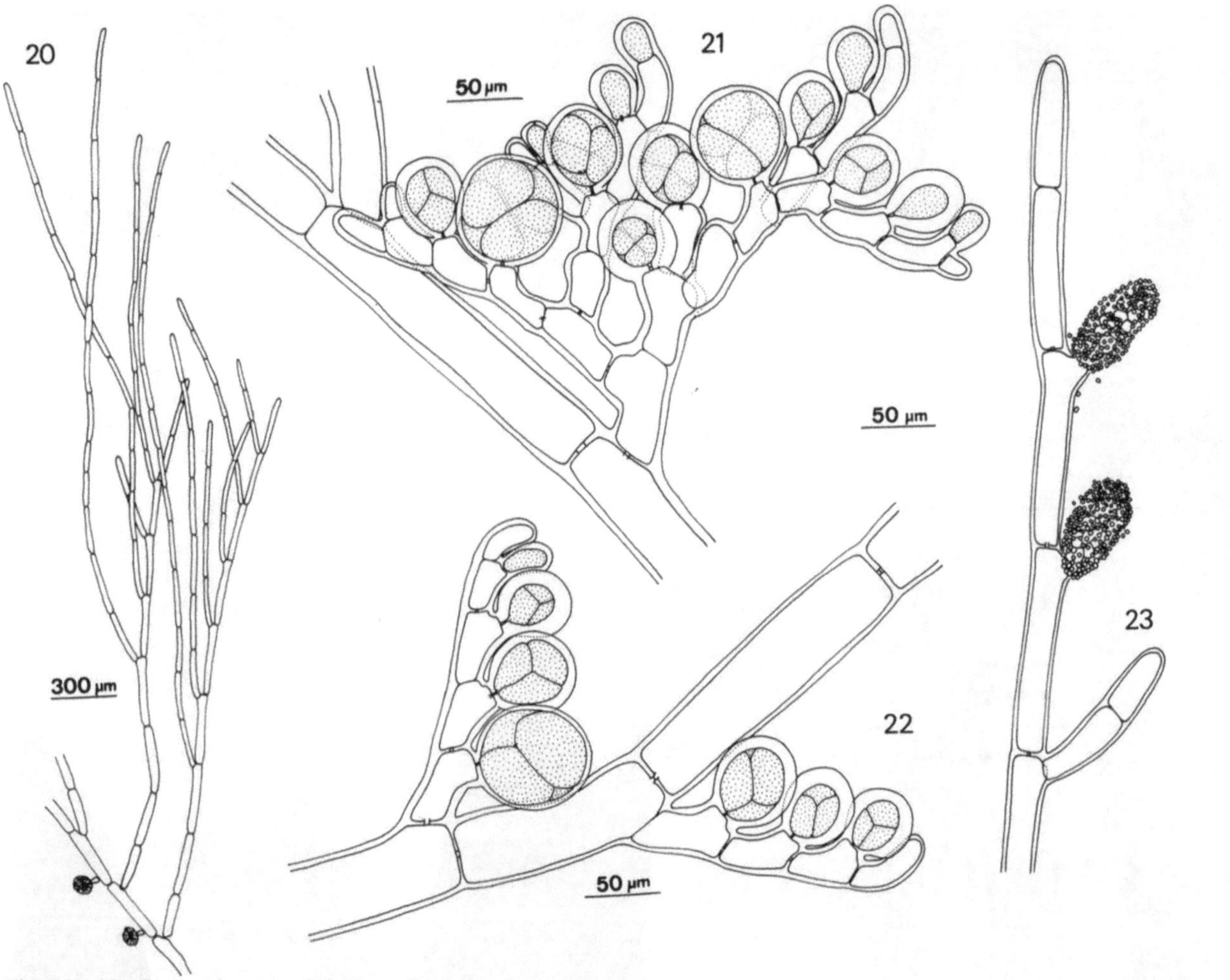

Figs 20–23. *Spermothamnion flabellatum* Bornet.
Fig. 20. Habit. Figs 21–22. Tetrasporangia placed on the adaxial side of special branches. Note the older tetrasporangia at the proximal portion of the branches. Fig. 23. Spermatangial heads.

as in *Corallina*-type (Garbary and Johansen 1982). Longitudinal fractures of branches show intergenicular medullary cells, 40–60 µm long, and longer genicular cells, 100–120 µm (Fig. 16). Intergenicular transverse fractures show thin walled medullary cells containing numerous floridean starch grains and thick walled cortical cells (Fig. 17). Only tetrasporangial conceptacles, 360–480 µm outer diam., were observed (Fig. 18). Among *Jania* species, *J. longifurca* may be characterized as possessing several vegetative features previously reported as typical of *Corallina*: genicular cells relatively much longer than intergenicular cells, extensive crusts and *Corallina*-type trichocytes (Johansen and Silva 1978, Garbary and Johansen 1982). Nevertheless, other vegetative features such as strictly dichotomous branching, intergenicula surfaces of the *Jania*-type and infection by *Choreonema thuretii* (Bornet) Schmitz (Fig. 19) are typical of *Jania*.

Jania longifurca was collected in tide pools at Playa de los Patos, an exposed locality in N Tenerife [TFC (Phyc) 5610].

5. *Spermothamnion flabellatum* Bornet *in* Bornet *et* Thuret (Figs 21–23)

Spermothamnion flabellatum (Ceramiaceae) has been commonly reported from the Mediterranean Sea (Feldmann-Mazoyer 1941, Athanasiadis 1987). Gallardo *et al.* (1985) have recently reported it from the Atlantic coast of Spain as well.

Plants examined agree with previous accounts of the species. Thalli reach up to 0.7 cm in height and grow attached to utricles of *Codium* by short rhizoids ending in a digitate disc. The rhizoids originate at the distal end of the postrate axial cells (Fig. 20). Tetrasporangia are placed adaxially on specialized short branches or occasionally are terminal on short stalks (Fig. 21). The older tetrasporangia were observed towards the branch base (Fig.22), in contrast to that observed by Feldmann-Mazoyer (1941). Spermatangial heads are elongated and sessile on the adaxial side of erect branches (Fig. 23). Female plants were not observed.

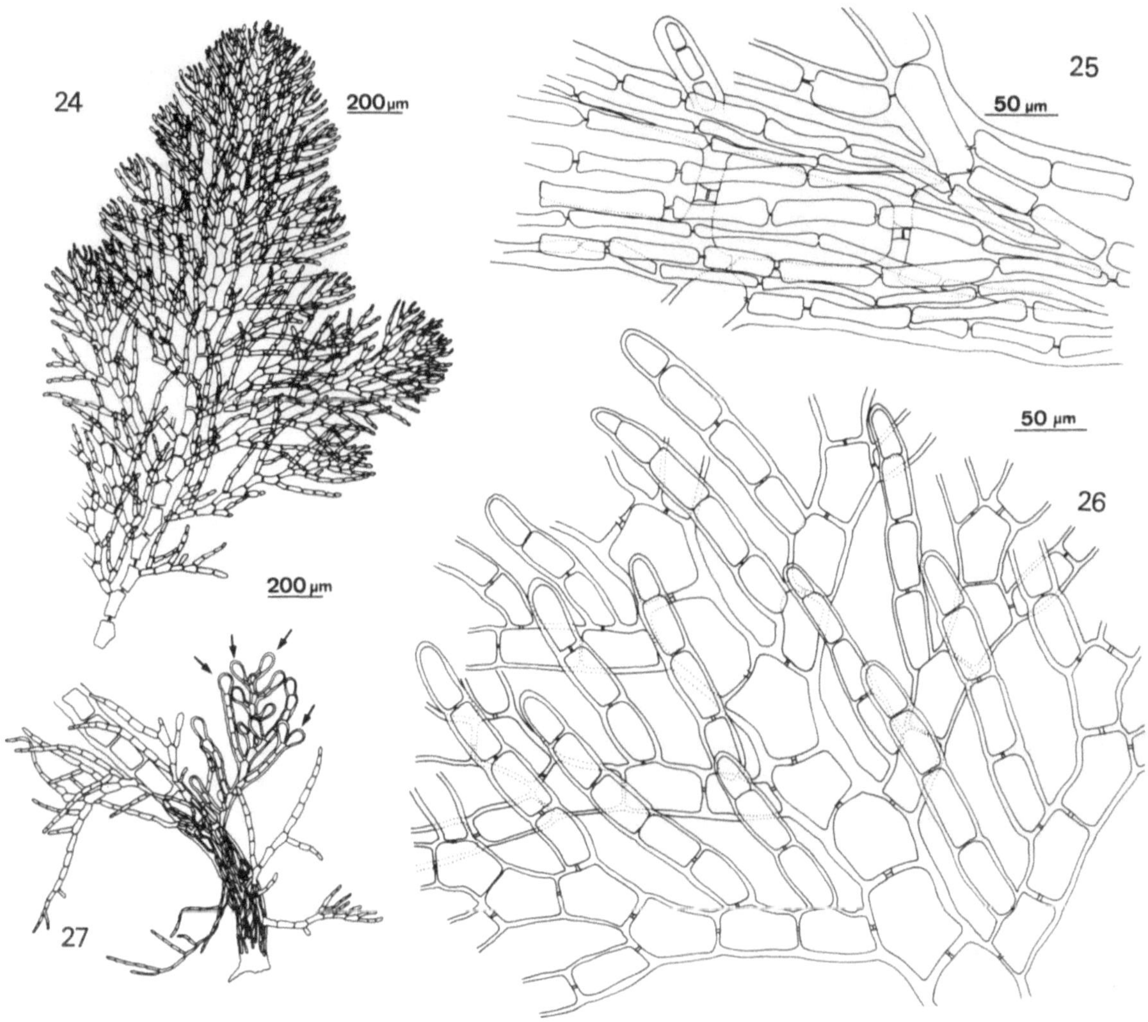

Figs 24–27. *Callithamnion decompositum* J. Agardh.
Fig. 24. Habit. Fig. 25. Dense basal cortication. Fig. 26. Detail of the branching showing consecutive adaxial branching by two adjacent branchlets. Fig. 27. Inflated hyaline cells (arrows) grouped on a special lateral branch.

Spermothamnion flabellatum was collected at Orzola, Lanzarote, growing on *Codium bursa* (L.) C. Agardh in tide pools [TFC (Phyc) 5705].

6. *Callithamnion decompositum* J. Agardh (Figs 24–26)

Callithamnion decompositum (Ceramiaceae) has been recorded from the Eastern North Atlantic from the Faroes to Senegal (Dixon and Price 1981). Although the range of morphological expression of *C. decompositum* is not clearly elucidated yet, the specimens from the Canary Islands (Fig. 24) agree with the accounts of Halos (1965), as *Aglaothamnion*, and Dixon and Price (1981). It differs from *C. tripinnatum* C. Agardh in its greater cell dimensions, dense basal cortication (Fig. 25) and consecutive adaxial branching on the basal-most two cells of lateral branches (Fig. 26). Apical organization is always complanate, a feature that clearly segregates it from *C. hookeri* (Dillwyn) S. F. Gray (see Dixon and Price 1981). Terminal inflated pear-shaped hyaline cells, up to 85 µm long and 45 µm wide, which are grouped on special lateral branches (Fig. 27) are very commonly produced. They were observed both in carposporangial and tetrasporangial plants and appear different from parasporangia reported by Halos (1965, Figs 2E, 3C).

Callithamnion decompositum was collected at Puerto de la Cruz, N Tenerife, growing in an exposed tide pool. Tetrasporangial, carposporangial and spermatangial plants were collected in May–June [TFC (Phyc) 5707].

7. *Helminthocladia agardhiana* Dixon (Figs 28–31)

Dixon (1962) proposed the name *H. agardhiana* Dixon (Helminthocladiaceae) as a substitute for *H. hudsonii* (C. Agardh) J. Agardh because the specimen collected by J. Agardh from Tangier was not conspecific with that to which the basionym *Mesogloia hud-*

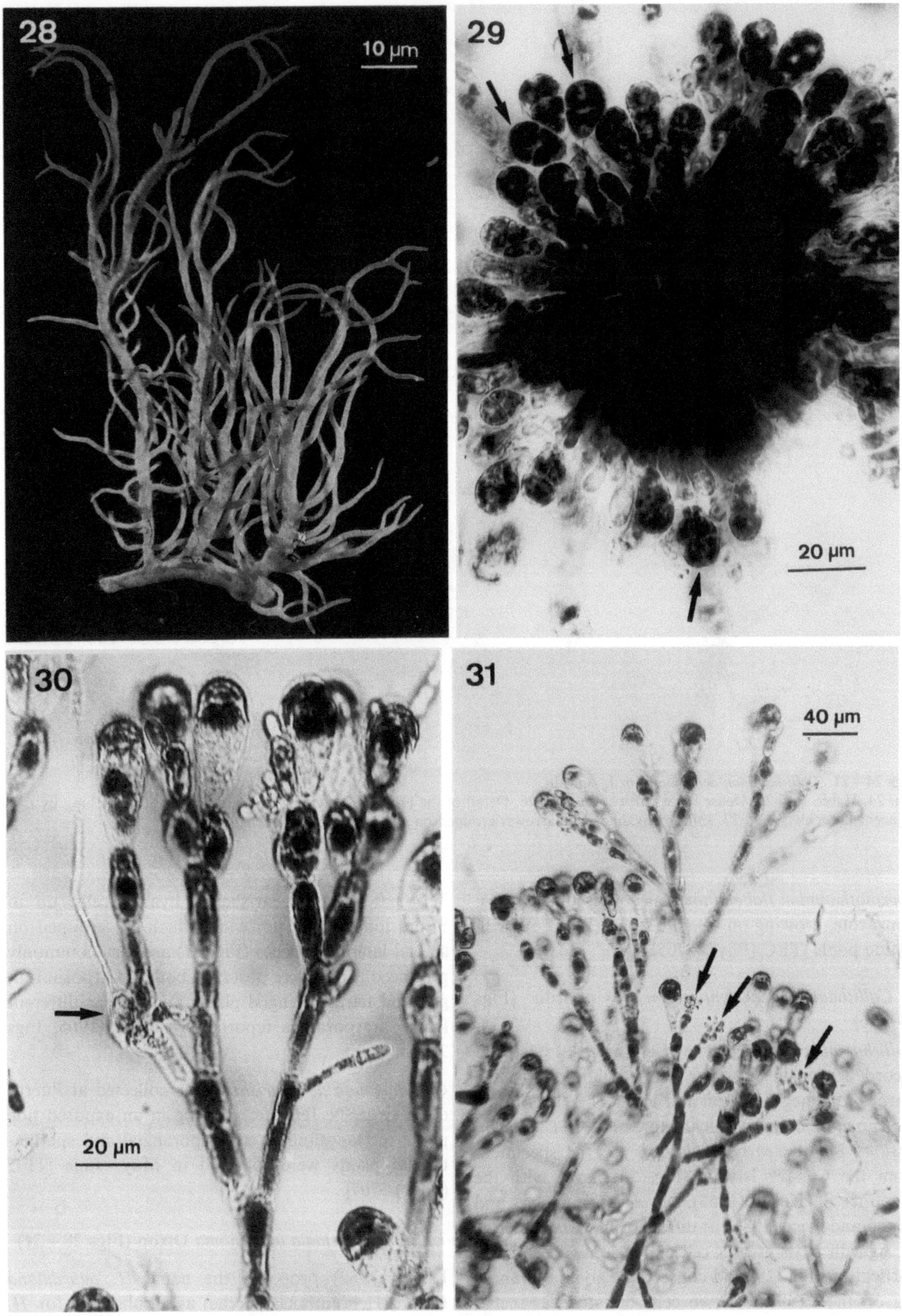

Figs 28–31. *Helminthocladia agardhiana* Dixon.
Fig. 28. Habit. Fig. 29. Aspect of carposporophyte showing carpotetrasporangia (arrows). Fig. 30. Cortical filaments with a carpogonial branch (arrow). Fig. 31. Spermatangial groups terminal on cortical filaments (arrows).

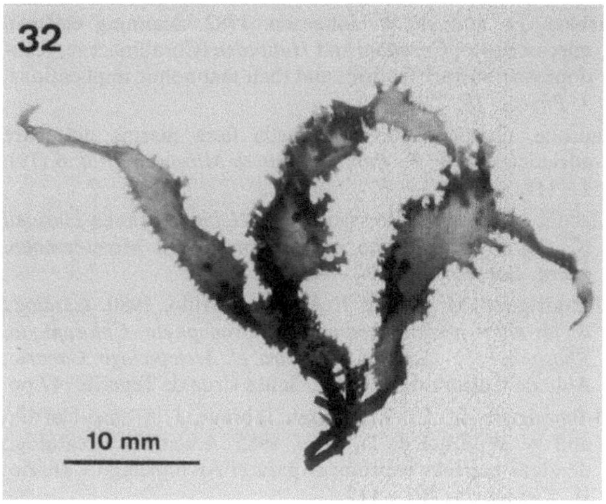

Fig. 32. *Rissoella verruculosa* (Bertolini) J. Agardh. Habit.

sonii C. Agardh was applied. *Helminthocladia agardhiana* is restricted to Morocco (Dangeard 1949), the Mediterranean Sea (Giaccone and Bressan 1973) and the Atlantic coasts of Spain (Gallardo *et al.* 1985).

The Canarian material (Fig. 28) is in good agreement with previous accounts of this taxon (Feldmann 1939; Bodard 1971; Dixon 1962). According to Bodard (1971), amongst the species of *Helminthocladia* only *H. agardhiana* and *H. senegalensis* Bodard have carpotetrasporangia (Fig. 29). In *H. agardhiana*, gonimoblasts lack involucral cells (Fig. 29), a feature that cleary segregates this species from *H. senegalensis.*

Helminthocladia agardhiana was collected at El Médano, a sheltered locality in S. Tenerife, growing epilithic at 1–3 m depth, ocassionally epiphyte on *Cymodocea nodosa* (Ucria) Ascherson. Cystocarpic (Fig. 30), carpotetrasporic (Fig. 29) and spermatangial plants (Fig. 31) were collected in May–June [TFC (Phyc) 5708].

8. *Rissoella verruculosa* (Bertolini) J. Agardh (Fig. 32).

Rissoella verruculosa (Rissoellaceae) has only previously been reported from the Western Mediterranean Sea (Feldmann 1937, Gallardo *et al.* 1985).

The Canarian material agrees with Kylin (1930). Plants possess cartilaginous and flattened erect fronds which are twisted and proliferous from the margins (Fig. 32). Plants arise from a short stipe and are attached to the sustratum by a small crust. Globose cystocarps with an obvious pore protrude externally from the cortex. Spermatangial and tetrasporangial plants were not observed.

Rissoella verruculosa was collected at El Guincho, an exposed locality in N Tenerife, growing on basaltic rocks in the upper littoral zone below the *Chthamalus stellatus* band. Cystocarpic plants were collected in May–August [TFC (Phyc) 5709].

Conclusion

As a result of their geographic position, the Canary Islands are subject to settlement of spores from seaweed species with different biogeographical affinities. *Predaea huismanii*, recently described by Kraft (1984) from Rottnest Island, is only previously known from Western Australia. *Platysiphonia caribaea*, described by Ballantine and Wynne (1985) has only previously been reported from the Western Caribbean. *Rissoella verruculosa* has previously been considered as an endemic species of the Mediterranean Sea (Feldmann 1937).

The remainder of the species newly reported from the Canary Islands (*Jania longifurca, Spermothamnion flabellatum, Callithamnion decompositum* and *Helminthocladia agardhiana*) have been previously recorded from the Eastern Atlantic and their presence in the Canary Islands is not unexpected.

Acknowledgements

Thanks are due to A. Padrón for providing technical assistance with the electron microscopy and two anonymous reviewers for their critical reading of the manuscript, comments and suggestions. The work was supported by grant PB 89-0601 from the Spanish DGICYT.

References

Afonso-Carrillo, J. 1988. Structure and reproduction of *Spongites wildpretii* sp. nov. (Corallinaceae, Rhodophyta) from the Canary Islands, with observations and comments on *Spongites absimile* comb. nov. *Br. phycol. J. 23*: 89–102.

Afonso-Carrillo, J. 1989. Morphology, anatomy and vegetative reproduction of *Fosliella paschalis* (Corallinaceae, Rhodophyta). *Phycologia 28*: 331–341.

Afonso-Carrillo, J. and M. C. Gil-Rodríguez. 1982. Aspectos biogeográficos de la flora ficológica marina de las Islas Canarias. *Actas II Simp. Ibér. Estud. Bentos Mar. 3*: 41–48.

Afonso-Carrillo, J., M. C. Gil-Rodríguez, R. Haroun Tabraue, M. Villena Balsa and W. Wildpret de la Torre. 1984a. Adiciones y correcciones al catálogo de algas marinas bentónicas para el Archipiélago Canario. *Vieraea 13*: 27–49.

Afonso-Carrillo, J., M. C. Gil-Rodríguez, R. Haroun-Tabraue and W. Wildpret. 1984 b. Algunos aspectos de la aplicación de la microscopía electrónica de barrido al estudio de las algas Corallinaceae (Rhodophyta). *Cah. Biol. mar. 25*: 427–433.

Afonso-Carrillo, J. and M. Sansón. 1989. *Clave ilustrada para la determinación de los macrófitos marinos bentónicos de las Islas Canarias*. Dep. Biología Vegetal, Univ. La Laguna. 55 pp.

Ardré, F. 1970. Contribution à l'étude des algues marines du Portugal. I – La Flore. *Portugaliae Acta Biol. B. 10* (1–4): 1–423, 56 pl.

Athanasiadis, A. 1987. A survey of the seaweeds of the Aegean Sea with taxonomic studies on species of the tribe Antithamnieae (Rhodophyta). Thesis, Department of Marine Botany, Univ. of Gothenburg. vii + 709 pp.

Audiffred, P. A. J. 1985. Marine algae of El Hierro (Canary Islands). *Vieraea 14*: 157–183.

Ballantine, D. L. and M. J. Wynne. 1985. *Platysiphonia* and *Apoglossum* (Delesseriaceae, Rhodophyta) in the tropical western Atlantic. *Phycologia 24*: 459–465.

Bodard, M. 1971. Étude morphologique et cytologique d'*Helminthocladia senegalensis*(Rhodophycées), Némalionale nouvelle à carpotétraspores et à cycle haplodiplophasique. *Phycologia 10*: 361–374.

Børgesen, F. 1925–30. The marine algae of Canary Islands especially from Teneriffe and Gran Canaria. I – Chlorophyceae, II – Phaeophyceae, III – Rhodophyceae, part. 1, 2, 3. *K. Danske Vidensk. Selsk. Biol. Medd. 5* (5): 1–123 (1925); *6* (2): 1–112 (1926); *6* (6): 1–97 (1927); *8* (1): 1–97 (1929); *9* (1): 1–159 (1930).

Bornet, E. and G. Thuret. 1867. Notes algologiques. Reprinted in *Bibliotheca phycologica 9*: 196 pp + 50 pl.

Bressan, G. 1974. Rodoficee calcaree dei mari Italiani. *Boll. Soc. adriat. Sci. nat. 59*: 1–132.

Dangeard, P. 1949. Les algues marines de la côte occidentale du Maroc. *Le Botaniste 34*: 89–189.

De Toni, G. 1936. *Noterelle di nomenclature algologica. VII. Primo elenco di Floridee omonime*. Brecia.

Dixon, P. S. 1962. Taxonomic and nomenclatural notes on the Florideae, II. *Bot. Notiser 115*: 245–260.

Dixon, P. S. and J. H. Price. 1981. The genus *Callithamnion* (Rhodophyta, Ceramiaceae) in the British Isles. *Bull. Br. Mus. nat. Hist. (Bot.) 9* (2): 99–141.

Feldmann, J. 1937. Recherches sur la Végétation Marine de la Méditerranée. La cote des Albères. *Revue algol. 10*: 1–339.

Feldmann, J. 1939. Une Némalionale à carpotétraspores: *Helminthocladia hudsonii. Bull. Hist. Nat. Afr. Nord 30*: 87–97.

Feldmann, J. 1946. La Flore Marine des Iles Atlantides. *Mém. Soc. Biogéogr. 8*: 395–435.

Feldmann-Mazoyer, G. 1941. *Recherches sur les Céramiacées de la Méditerranee occidentale*. Alger. 510 pp.

Gallardo, T., A. Gómez Garreta, M. A. Ribera, M. Alvarez and F. Conde. 1985. A preliminary checklist of Iberian benthic marine algae. *Real Jardín Botánico. Madrid.* 83 pp.

Garbary, D. and H. W. Johansen. 1982. Scanning electron microscopy of *Corallina* and *Haliptilon* (Corallinaceae, Rhodophyta): surface features and their taxonomic implications. *J. Phycol. 18*: 211–219.

Giaccone, G. 1978. Revisione della flora marina del mare adriatico. *W. W. F., Parco marino di Miramare, Ital. 6* (19): 1–118.

Giaccone, G. and G. Bressan. 1973. *Helminthocladia hudsonii* (C. Ag.) J. Ag.: prima segnalazione per il Mediterraneo. *Giorn. Bot. Ital. 107*: 37–42.

Gil-Rodríguez, M. C. and J. Afonso-Carrillo. 1980. *Catálogo de las algas marinas bentónicas (Cyanophyta, Chlorophyta, Phaeophyta y Rhodophyta) para el Archipiélago Canario.* Aula de Cultura de Tenerife, Santa Cruz de Tenerife. 47 pp.

Gil-Rodríguez, M. C., R. Haroun Tabraue, J. Afonso-Carrillo and W. Wildpret de la Torre. 1985. Adiciones al catálogo de algas marinas bentónicas para el Archipiélago Canario. II. *Vieraea 15*: 101–112.

Halos, M. T. 1965. Sur trois Callithamniées des environs de Roscoff. *Cah. Biol. Mar. 6*: 117–134.

Hamel, G. and P. Lemoine. 1953. Corallinacées de France et d'Afrique du Nord. *Archs Mus. natn. Hist. Nat., Paris, Ser. 7, 1*: 15–136.

Johansen, H. W. and P. C. Silva. 1978. *Janieae* and *Lithotricheae*: two new tribes of articulated Corallinaceae (Rhodophyta). *Phycologia 17*: 413–417.

Kajimura, M. 1987. Two new species of *Predaea* (Nemastomaceae, Rhodophyta) from the Sea of Japan. *Phycologia 26*: 419–428.

Kraft, G. T. 1984. The red algal genus *Predaea* (Nemastomaceae, Gigartinales) in Australia. *Phycologia 23*: 3–20.

Kraft, G. T. and J. A. Abbott. 1971. *Predaea weldii*, a new species of Rhodophyta from Hawaii, with an evaluation of the genus. *J. Phycol. 7*: 194–202.

Kraft, G. T. and D. M. John. 1976. The morphology and ecology of *Nemastoma* and *Predaea* species (Nemastomataceae, Rhodophyta) from Ghana. *Br. phycol. J. 11*: 331–344.

Kylin, H. 1930. Über die Entwicklungsgeschichte der Florideen. *Lunds Univ. Årsskr. NF, 2, Bd 26* (6): 1–104.

Lemoine, P. 1924. Corallinacées du Maroc. I. *Bull. Soc. Sc. Nat. Maroc. IV* (5–6): 113–134.

Levring, T. 1974. The marine algae of the archipelago of Madeira. *Bol. Mus. Mun. Funchal 28* (125): 1–111.

Viera-Rodríguez, M. A., P. A. J. Audiffred, M. C. Gil-Rodríguez, W. F. Prud'homme van Reine and J. Afonso-Carrillo. 1987. Adiciones al catálogo de algas marinas bentónicas para el Archipiélago Canario. III. *Vieraea 17*: 227–235.

Villena Balsa, M., J. Afonso-Carrillo and W. Wildpret de la Torre. 1987. Morfología, estructura y reproducción de una especie epífita del género *Jania* (Corallinaceae, Rhodophyta) de las Islas Canarias. *Vierraea 17*: 19–42.

Weisscher, F. C. M. 1983. Marine algae from Selvagem Pequena (Salvage Islands) (CANCAP-Project Contributions n° 19). *Bol. Mus. Mun. Funchal 35* (152): 41–80.

Yoshida, T. 1980. Two new species of red algae from the west coast of Kyushu, Japan. *Jap. J. Phycol. 28*: 69–74.

Botanica Marina
Vol. 34, pp. 537–539, 1991

Mosquito Larvicidal Activity of Marine Plant Extracts with Synthetic Insecticides

T. Subramonia Thangam and K. Kathiresan

Centre of Advanced Study in Marine Biology, Annamalai University, Parangipettai 608502, Tamil Nadu, India

(Accepted 4 October 1991)

Abstract

Two seaweeds, *Caulerpa scalpelliformis*, *Dictyota dichotoma* and one mangrove sample, stilt root of *Rhizophora apiculata* were extracted in acetone and were combined with synthetic insecticides, DDT, BHC and malathion and evaluated for mosquito larvicidal activity against *Aedes aegypti*. The extracts showed synergism with the insecticides. The stilt root of *Rhizophora apiculata* exhibited highest synergistic activity with BHC.

Introduction

The synthetic insecticides, DDT, BHC and malathion are commonly used in India in mosquito control programmes. However, these chemicals are well known to cause environmental pollution. To overcome this, natural insecticides of plant origin have been studied in recent years. In this context, marine plant extracts have been reported for the first time in our laboratory to have mosquito larvicidal activity (Subramonia Thangam and Kathiresan 1988 a, b, 1989, 1991 a, b). Terrestrial plant extracts, when combined with synthetic insecticides, have been reported to reduce the required insecticide dose and hence also reduce pollution. Such reports for marine plants are not available. The present study was carried out to evaluate the synergistic mosquito larvicidal activity of extracts of *Caulerpa scalpelliformis*, *Dictyota dichotoma* and *Rhizophora apiculata* — stilt root in combination with DDT, BHC and malathion against *Aedes aegypti* L., the vector of dengue and yellow fever.

Material and Methods

The seaweeds, *Caulerpa scalpelliformis* (R. Br.) Weber V. Bosse and *Dictyota dichotoma* (Huds.) Lamour., were collected from the southeast coast of India. The stilt roots of *Rhizophora apiculata* Blume were collected from the Pichavaram mangrove forest on the southeast coast of India. They all were washed thoroughly in tap water, shade-dried and powdered to 40-mesh size. The powdered samples were extracted in acetone using a Soxhlet apparatus. The extracts were concentrated using a vacuum evaporator. Standard stock solutions of 1% (w/v) were prepared by dissolving the residues in acetone. Larval susceptibility tests of the synthetic insecticides were carried out following the standard method (WHO/VBC 1975). Stock solutions of the insecticides, DDT, BHC and malathion (Technical grade) were prepared at the concentrations of 100 mg L^{-1} (w/v) by dissolving them in acetone. Twelve different concentrations of the insecticides, 0.06, 0.08, 0.10, 0.12, 0.14, 0.16, 0.18, 0.20, 0.22, 0.24, 0.26 and 0.28 mg L^{-1} were prepared by diluting the stock solutions using unchlorinated, filtered tap water. Tween-80, at the concentration of 0.02% (v/v), was used as an emulsifier. A control solution was prepared in tap water using acetone and Tween-80 as used in the test solution at the highest concentration. Twenty five larvae of early fourth instar stage were released in 500 mL beakers containing 250 mL of the test or control solutions and four replicates were run at a time. Mortality counts were made after 24 hours of the treatment. The LC_{50}, standard error, fiducial limits at 95% confidence and regression equation were calculated (Finney 1971).

Another set of the insecticide solutions were also prepared, at above mentioned concentrations. The acetone extracts of *Caulerpa scalpelliformis*, *Dictyota dichotoma* and *Rhizophora apiculata* – stilt root were added to the 250 mL test solutions of the synthetic insecticide at 1 or 5 mg L^{-1}. The larval susceptibility tests were carried out as mentioned above and the synergistic factor (SF) of the marine plant extracts was calculated using the formula of Kalyanasundaram and Babu (1982):

$$SF = \frac{LC_{50} \text{ of insecticide with plant extract}}{LC_{50} \text{ of insecticide}}$$

A value of less than one, confirms synergism; the smaller the value, the higher the synergistic activity.

Results

The effect of synthetic insecticides with or without the marine plant extracts on mosquito larvae is shown in Table I. Plant extracts in general, exhibited synergism with the insecticides against the larvae of *Aedes aegypti*. The stilt root extract of *Rhizophora apiculata* showed synergistic effect at both 1 and 5 mg L^{-1}. The LC_{50} of DDT, BHC and malathion was 0.155, 0.209 and 0.158 mg L^{-1} respectively. When the stilt root extract of *R. apiculata* at 1 mg L^{-1} was added to the DDT, BHC and malathion, the LC_{50} of the insecticides was reduced to 0.151, 0.169 and 0.126 mg L^{-1} and the synergistic factor was 0.97, 0.81 and 0.80 respectively. When the extract of *R. apiculata* was added to DDT, BHC and malathion at 5 mg L^{-1}, the synergistic factor was 0.77, 0.68 and 0.71 respectively. The extracts of *Caulerpa scalpelliformis* and *Dictyota dichotoma* showed synergistic activity with the insecticides at 5 mg L^{-1} but not at 1 mg L^{-1}. The synergistic factor of *C. scalpelliformis* extract with DDT, BHC and malathion at 5 mg L^{-1} was 0.89, 0.71 and 0.84 respectively. The synergistic factor of *D. dichotoma* extract with DDT, BHC and malathion was 0.97, 0.76 and 0.85 respectively.

Among the plant extracts studied, the stilt root extract of *Rhizophora apiculata* was found to be the most effective synergist with all three insecticides. All three plant extracts showed maximum synergism with BHC. The stilt root extract of *R. apiculata* at 5 mg L^{-1} was the most effective with a synergistic factor of 0.68.

Discussion

In the present study, marine plant extracts were found to act synergistically with synthetic chemicals in causing larval mortality of *Aedes aegypti* (Table I). Similar synergistic activity was observed in extracts of terrestrial plants, *Ocimum sanctum* L. and *Azadirachta indica* Adr. Juss. with phenthoate and *O. basilicum* L. and *A. indica* Adr. Juss. with fenthion (Kalyanasundaram and Babu 1982); *Leucas aspera* (Willd.) Link

Table I. Synergistic activity of marine plant extracts with synthetic insecticides on the larvae of *Aedes aegypti*

Insecticide	24 hours $LC_{50} \pm$ S. E. (mg L^{-1})	Fiducial limits (95%) (mg L^{-1})	Regression equation	Synergistic factor
DDT	0.155 ± 0.0007	0.153 – 0.157	Y = 7.692x + 11.228	–
DDT + *R. apiculata***	0.151 ± 0.0008	0.150 – 0.152	Y = 6.667x + 10.474	0.97
DDT + *R. apiculata**	0.120 ± 0.0007	0.119 – 0.121	Y = 6.667x + 11.139	0.77
DDT + *C. scalpelliformis**	0.138 ± 0.0009	0.136 – 0.140	Y = 4.760x + 9.094	0.89
DDT + *D. dichotoma**	0.151 ± 0.0007	0.150 – 0.152	Y = 7.692x + 11.315	0.97
CD at 1% level of significance	0.003			0.013
BHC	0.209 ± 0.0007	0.207 – 0.211	Y = 11.111x + 12.554	–
BHC + *R. apiculata***	0.169 ± 0.0007	0.168 – 0.170	Y = 10.000x + 12.721	0.81
BHC + *R. apiculata**	0.141 ± 0.0007	0.140 – 0.142	Y = 8.330x + 12.087	0.68
BHC + *C. scalpelliformis**	0.148 ± 0.0007	0.147 – 0.149	Y = 7.692x + 11.382	0.71
BHC + *D. dichotoma**	0.158 ± 0.0007	0.157 – 0.159	Y = 9.090x + 12.284	0.76
CD at 1% level of significance	0.003			0.046
Malathion	0.158 ± 0.0007	0.157 – 0.159	Y = 9.090x + 12.284	–
Mal + *R. apiculata***	0.126 ± 0.0007	0.125 – 0.127	Y = 7.143x + 11.426	0.80
Mal + *R. apiculata**	0.112 ± 0.0006	0.111 – 0.113	Y = 7.143x + 11.791	0.71
Mal + *C. scalpelliformis**	0.132 ± 0.0007	0.131 – 0.133	Y = 7.407x + 11.514	0.84
Mal + *D. dichotoma**	0.135 ± 0.0007	0.134 – 0.136	Y = 7.692x + 11.689	0.85
CD at 1% level of significance	0.008			0.003

** – at the concentration of 1 mg L^{-1}; * – at the concentration of 5 mg L^{-1}.

with phenthoate and fenthion and *Pedalium murex* L. with fenthion (Kalyanasundaram and Das 1985); *Bougainvillea glabra* Choisy with DDT, BHC and malathion (Subramonia Thangam and Kathiresan 1990). As compared with *B. glabra*, the synergistic activity of the marine plants was low with DDT, BHC and high with malathion.

This synergism may be due to the plant extracts inhibiting some factors such as detoxifying enzymes in mosquito larvae, which can act against the synthetic chemicals. This is supported by the report of Vasantharaj David and Kumaraswami (1978) that the plant alkaloids such as pyrethrins, cinerins and jasmoline inhibited the action of detoxifying enzymes, which were formed in insects against the toxicity of synthetic insecticides.

It is clear from the study that the acetone extracts of *Rhizophora apiculata* – stilt root, *Caulerpa scalpelliformis*, *Dictyota dichotoma* can be combined with synthetic insecticides to reduce the consumption of the insecticides which would considerably reduce the aquatic pollution caused by using synthetic insecticides in mosquito control programmes.

Acknowledgements

The authors are grateful to the Indian Council of Medical Research and Council of Scientific and Industrial Research, New Delhi for financial support, Prof. A. Subramanian, Director of this Centre and the authorities of Annamalai University for providing facilities.

References

Finney, D. J. 1971. *Probit Analysis.* Cambridge University Press, London, III edition, pp. 20–42.

Kalyanasundaram, M. and C. J. Babu. 1982. Biologically active plant extracts as mosquito larvicides. *Indian J. Med. Res. 76*: 102–106.

Kalyanasundaram, M. and P. K. Das. 1985. Larvicidal and synergistic activity of plant extracts for mosquito control. *Indian J. Med. Res. 82*: 19–23.

Subramonia Thangam, T. and K. Kathiresan. 1988 a. Toxic effect of seaweed extracts on mosquito larvae. *Indian J. Med. Res. 88*: 35–37.

Subramonia Thangam, T. and K. Kathiresan. 1988 b. Toxic effect of mangrove plant extracts on mosquito larvae *Anopheles stephensi* L. *Curr. Sci. 57*: 914–915.

Subramonia Thangam, T. and K. Kathiresan. 1989. Larvicidal effect of marine plant extracts on mosquito *Culex tritaeniorhynchus*. *J. Mar. Bio. Ass. India 31*: 306–307.

Subramonia Thangam, T. and K. Kathiresan. 1990. Synergistic effects of insecticides with plant extracts on mosquito larvae. *Tropical Biomedicine 7*: 135–137.

Subramonia Thangam, T. and K. Kathiresan. 1991 a. Mosquito larvicidal activity of mangrove plant extracts against *Aedes aegypti*. *Int. Pest Control* (in press).

Subramonia Thangam, T. and K. Kathiresan. 1991 b. Mosquito larvicidal effect of seaweed extracts. *Bot. Mar. 34*: 433–435.

WHO/VBC. 1975. Instructions for determining the susceptibility or resistance of mosquito larvae to insecticides. Mimeographed document (Wld. Hlth. Org. Tech. Rep. Ser., 1970, No. 443 revised by WHO Expert Committee, which met in Geneva, September 16–23, 1975), WHO/VBC/75. 583, pp. 1–5.

Botanica Marina
Vol. 34, pp. 541–552, 1991

Temperature Responses and Distribution of Macroalgae Belonging to the Warm-temperate Mediterranean-Atlantic Distribution Group

S. Orfanidis[1]

Biologische Anstalt Helgoland (Zentrale), Notkestr. 31, D-2000 Hamburg 52, Fed. Rep. of Germany

(Accepted 5 October 1991)

Abstract

The thermal requirements and limits of tolerance of four macroalgae belonging to the warm-temperate Mediterranean-Atlantic distribution group (*Kuckuckia spinosa*, *Choristocarpus tenellus*, *Cystoseira barbata* and *Falkenbergia rufolanosa*) were experimentally determined and correlated to their geographic boundaries in the North Atlantic Ocean. "Sufficient" growth (> 20% of maximum rate) for *C. tenellus*, *C. barbata* and *F. rufolanosa* proceeded at 10–20 °C, and that of *K. spinosa* at 10–25 °C. Plurilocular sporangia in *K. spinosa* were produced at a temperature range of 15–25 °C. *Cystoseira barbata* was the only species tested that tolerated two months at −1 °C. The lowest temperature survived by *K. spinosa* and *C. tenellus* was 1 °C, and by *F. rufolanosa* was 5 °C. The upper temperature limit of *C. tenellus* was between 20 and 25 °C and of *F. rufolanosa* 27 °C; *K. spinosa* and *C. barbata* survived up to 29 and 30 °C respectively. The nature of the northern and southern boundaries of these species, where they are clear from the geographical and temperature limits, is heterogenous. *Falkenbergia rufolanosa* and *C. tenellus* survived a temperature range < 20 °C between their upper and lower lethal limits; whereas *K. spinosa* and *C. barbata* survived a range > 20 °C, making these latter species potential inhabiters of the eastern American temperature region.

Introduction

The experimental approach in seaweed biogeography has made possible the formulation and testing of hypotheses relating to their geographic distribution. It is generally accepted that temperature is the most important factor determining the distribution of the marine algae on an uninterrupted rocky coastline (Lüning 1990). Hence, most hypotheses (see Breeman 1988) have dealt with the nature of geographical boundaries by relating the actual distribution limits with the temperature limits determined from laboratory studies.

Two types of temperature boundaries were stressed by Hutchins (1947) and Hoek (1982 a, b) for marine organisms:

(1) growth and/or reproduction boundaries (summer temperatures may be too cold for warm-adapted species to grow and/or reproduce at the polar end of their distribution; winter temperatures may be too warm for cold-adapted species to grow and/or reproduce at the equatorial end of their distribution) and

(2) lethal boundaries (winter temperatures may be lethal for warm-adapted species at the polar end of their distribution; summer temperatures may be lethal for cold-adapted species at the equatorial end of their distribution).

The known temperature responses of species belonging to the warm-temperate Mediterranean-Atlantic distribution group could give a general picture for the characteristic temperature features of Mediterranean algae (see Breeman 1988). However, the data were based on a small number of species from the rich flora of the region (Hoek 1975, Lüning 1990) and

[1] Present address: Botanical Institute, Department of Biology, University of Thessaloniki, 54006 Thessaloniki, Greece.

interesting variations or additions to the basic scheme are still to be expected particularly since experimental evidence from macroalgae endemic to the Mediterranean is still lacking.

Experimental biogeographical studies need to be based on reliable taxonomic records in order to avoid misrepresentation of the geographic distribution. However, according to McLachlan and Bird (1984) such studies can also provide a suitable test for a more accurate interpretation of a species and its possible ecotypic differentiation.

The objective of the present study was to investigate the temperature responses of four macroalgae (*Kuckuckia spinosa*, *Choristocarpus tenellus*, *Cystoseira barbata* and *Falkenbergia rufolanosa*) belonging to the warm-temperate Mediterranean-Atlantic group and to relate the February and August isotherms coinciding with the geographical boundaries of these species in the North Atlantic with the temperatures limiting their survival, growth and reproduction.

Material and Methods

Unialgal cultures

The species utilised in this study were isolated as spores or fragments of plants collected near Thessaloniki, Greece, (Thermaikos Gulf; 40.38°N, 22.58°E), near Naples, Italy, (Porto Mergellina; 40.50°N, 14.15°E) and near Rade de Villefranche-sur-Mer, France, (43.42°N, 7.18°E). Collection site and isolator are listed in Table I. A culture of *Cystoseira barbata* was established by transporting young zygotes in an insulated cooling container to K. Lüning's laboratory in Hamburg. The cloned unialgal cultures isolated by D. G. Müller were sent to Hamburg in a similar way.

Unialgal cultures were propagated at 15 ± 2 °C in sterile crystallising dishes (90 and 110 mm in diameter) and 2 or 3 L beakers aerated with membrane-filtered air (Sartorius cellulose nitrate filter, pore size 0.2 µm), depending on the size of the plants. Standard growth medium, heat-sterilised von Stosch's Enriched Seawater (VSES) (Guiry and Cunningham 1984) was used. This had a salinity of 32‰ and was changed every second week. A photon fluence rate of 20–40 $\mu mol\ m^{-2}\ s^{-1}$ fluorescent light, 16 h light per day, was provided by cool-white fluorescent tubes (Osram L 40 W/25 S).

Temperature tolerance tests

Temperature tolerance experiments were performed by incubating plants in a temperature range of −1 to 33 °C. The upper temperature tolerance was determined by incubating for 2 weeks, while the lower temperature tolerance by incubating for 2, 4 and 8 weeks. Exceptions are noted in the Results. Temperature-controlled water baths (Haake D8-V water bath-cryostat units; Haake, Karlsruhe, Germany) were used for temperatures > 0 °C (all ± 0.1 °C) and a constant temperature room for −1 °C (± 0.5 °C). Glycerin was added to the 0, 1, and 2 °C water baths to prevent freezing. Standard light conditions were cool-white fluorescent light, and a photon fluence rate of 30 $\mu mol\ m^{-2}\ s^{-1}$ and 12 h light per day.

Sterile glass tubes (20 mL) and screw-top bottles (250 and 500 mL) containing the test organisms were placed in the temperature-controlled system at the desired conditions, without prior adaptation. The experimental individuals of *Cystoseira barbata* were tested in aerated 2 L beakers. The medium in each culture was changed three times a week for an upper tolerance test and once a week for a lower tolerance test. The new medium was first equilabrated to the experimental temperatures.

After the treatments the algae were transferred into new sterile glass tubes and screw-top bottles filled with VSES. Plants treated for lower temperature tol-

Table I. Species, locality, date of collection, habitat and isolator.

Culture number	Species	Locality and date of collection	Habitat	Isolator
2087	*Asparagopsis armata* Harvey tetrasporic phase = *Falkenbergia rufolanosa* (Harvey) Schmitz (Nemaliales, Rhodophyta)	Villefranche-sur-Mer, France 1 September 1980	Sublittoral zone 20 m depth	D. G. Müller
2207	*Kuckuckia spinosa* (Kützing) Kuckuck (Ectocarpales, Phaeophyceae)	Naples, Italy 24 March 1975	Unknown	D. G. Müller
2262	*Choristocarpus tenellus* (Kützing) Zanardini (Sphacelariales, Phaeophyceae)	Villefranche-sur-Mer, France 1 September 1980	Sublittoral zone 20 m depth	D. G. Müller
2511	*Cystoseira barbata* C. Agardh (Fucales, Phaeophyta)	Thessaloniki, Greece 5 January 1987	Upper sublittoral zone	S. Orfanidis

erance were post-cultivated at 15 ± 2 °C and for upper temperature tolerance at 20 ± 1 °C (both cases: 20 μmol m^{-2} s^{-1}, 16 h hours light per day). Four weeks later, the viability of the plants was investigated visually and microscopically to determine whether plants had survived and exhibited normal coloration, regenerated from injured thallus, or were dead. In cases where there was any doubt about the viability of the treated material, the post-cultivation was prolonged for another 4 weeks, with renewal of VSES. The upper survival test for each species was performed at least twice. The lower survival test was run once and samples were taken after 2, 4 and 8 weeks.

Relative growth rate

Transparent PVC tanks (80 × 30 cm, 30 cm high), filled with distilled water, were used as temperature-controlled baths for the growth experiments at 0, 5, 10, 15, 20, 25 and 30 °C (all ± 0.2 °C). Standard light conditions were cool-white fluorescent light, 60–80 μmol m^{-2} s^{-1} (water surface), and 16 h light per day. Fragments of *Kuckuckia spinosa* and *Choristocarpus tenellus* were allowed to settle on plastic petri dishes for 3 days. Eight thalli were then drawn with a camera lucida and measured; for *K. spinosa* only the upright thalli were measured. The petri dishes were placed in 2 L beakers filled with VSES, and immersed in the PVC baths. A preliminary growth experiment with *K. spinosa* had shown the formation of reproductive organs within 7 days at 25 °C followed by a cessation of growth. Thus the above experiments were monitored over a period of 5 days at temperatures between 15 to 30 °C and over a period of 7 days at temperatures between 0 to 10 °C. The thalli of *Cystoseira barbata* (n = 5–6) were first marked and weighted and then put into aerated 2 L beakers. Growth was determined as fresh weight after 2 weeks cultivation time. Fresh weight was determined after removing excess water with blotting-paper. The culture medium was changed at the end of the first week. The growth of *Falkenbergia rufolanosa* was estimated visually.

Growth was measured as relative growth rate (RGR) calculated as follows:

$$RGR = (\ln N_t - \ln N_o)/t \times 7 \text{ (week}^{-1})$$

[N_o = initial length or fresh weight; n_t = length or fresh weight after t days; t = experimental time (days)]

Relative growth rates that exceeded 20% and 80% of the maximum value were defined as 'sufficient' growth and 'optimum' growth, respectively. The parameter 'sufficient' growth is appropriate in the context of geographic boundaries.

Reproduction

Experiments on reproduction were performed in culture rooms adjusted to temperatures of 5, 10, 15 °C (all ± 2 °C) and 20, 25 °C (all ± 1 °C). A photon fluence rate of 30–35 μmol m^{-2} s^{-1} was provided by cool-white fluorescent tubes (Osram L 40 W/25 S), 16 h light per day. At 15 °C, a special light cabinet was also used (Lüning 1980) providing 8 h light per day. For all the experiments young immature plants were placed in crystallising dishes (90 mm in diameter) and in aerated 2 L beakers. The medium was changed weekly at 15 to 25 °C, and every second week at 5 to 10 °C. The plants were examined microscopically every second week and the presence or absence of sporangia was noted.

Distribution and temperature records

Records used in the maps of geographical distribution were derived from literature sources cited in the legend of Figure 1. The oceanic isotherms used in this study were adopted from Gorskov World Ocean Atlas (1985). More detailed local temperature records, taken from various sources, are given below.

Results

Kuckuckia spinosa grew optimally between 15 to 25 °C and sufficiently at 10 to 25 °C (Fig. 2). The lowest survival temperature was −1 °C for 2 weeks, 1 °C for 4 weeks and 3 °C for 8 weeks incubation time; death occurred at −1 and 0 °C after 8 weeks incubation time. In individuals incubated for 4 weeks at −1 and 0 °C and for 8 weeks at 1 and 2 °C, chilling injury appeared but the injured thallus could regenerate (Table II). In the upper temperature tolerance experiments individuals survived at 28 °C but not at 30 °C or above. Plants treated at 29 °C could regenerate from a few living cells (Table III). The life history consisted of immediate replication of the same phase through plurispores. After 2 weeks of cultivation of *K. spinosa*, mature and empty plurilocular sporangia were detected at 20 and 25 °C. The number of sporangia per plant was higher at 25 °C than at 20 °C. At 15 °C the first immature sporangia were observed after 2 weeks and mature and empty sporangia two weeks later under both daylengths. Formation of a few sporangia was observed at 10 °C after 8 to 12 weeks of cultivation. At 5 °C the plants were able to grow vegetatively but they remained sterile even after 12 weeks.

Choristocarpus tenellus grew optimally at 20 °C and sufficiently at 10 to 20 °C (Fig. 2). The lower temperature tolerance was tested over a period of 56 days

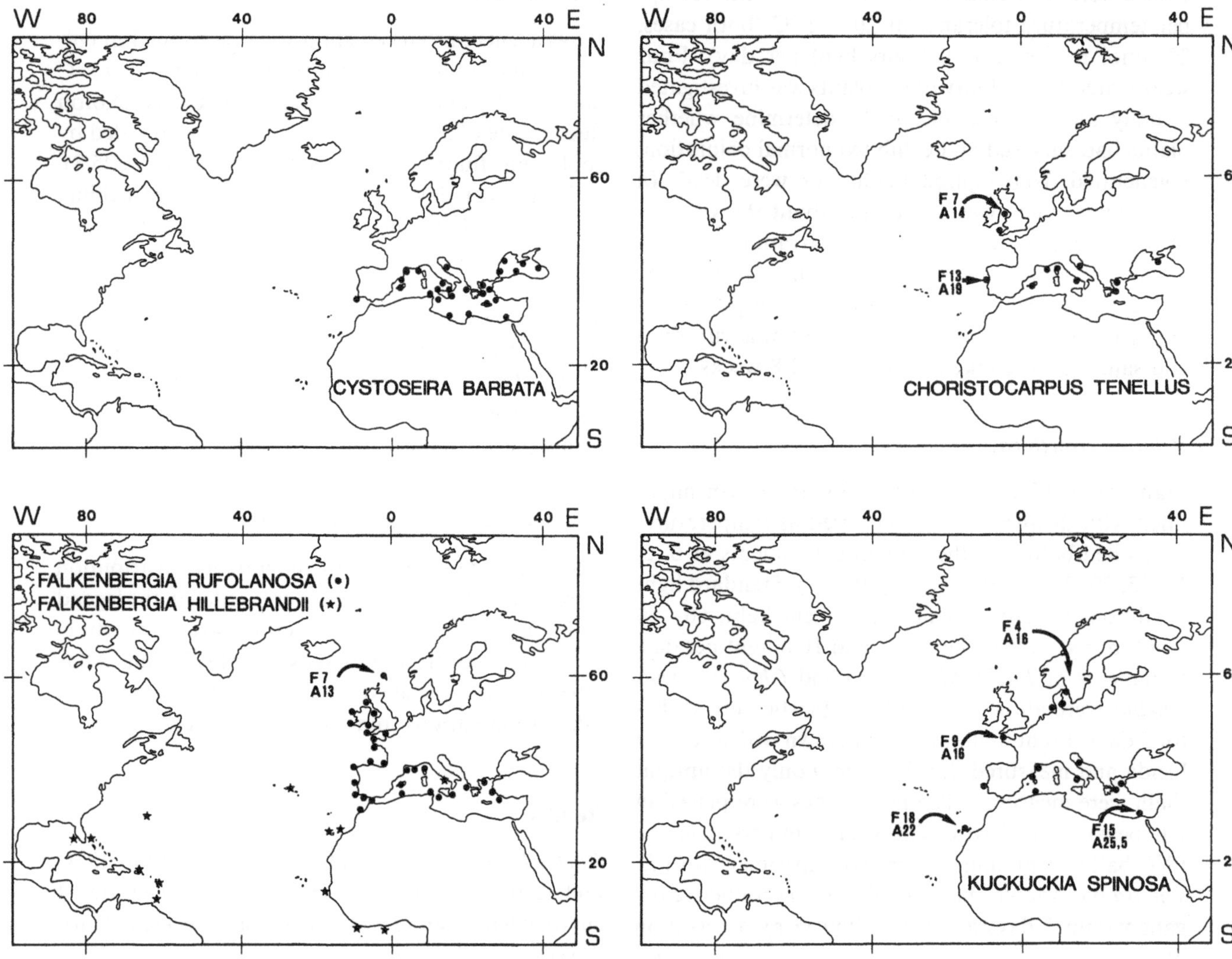

Fig. 1. Distribution of four species in the North Atlantic Ocean. Maps show the geographic distribution and the ocean isotherms for February (F) and August (A) at the boundaries. *Kuckuckia spinosa* [includes *K. crinigera* (Kuckuck) Hamel; see Cardinal (1964) and *K. kylinii* Cardinal; see Pedersen 1989]. Sources: eastern N Atlantic (Kuckuck 1958, South and Tittley 1986), W Sweden (Kylin 1947), N France (Feldmann and Magne 1964), Portugal (Ardré 1970), Mediterranean (Furnari 1984), S France (Boudouresque *et al.* 1984), Aegean Sea (Diapoulis and Haritonidis 1987). *Choristocarpus tenellus*. Sources: British Isles (Hiscock and Maggs 1984), Atlantic coasts of Iberian peninsula (Gallardo *et al.* 1985), Mediterranean (Coppejans and Boudouresque 1976), Balearic Islands (Ribera and Gómez 1985), Corsica (Coppejans 1981), Aegean Sea (Athanasiadis 1987, Diapoulis and Haritonidis 1987), Black Sea (Zinova 1967). *Cystoseira barbata*. Sources: Portugal (Ardré 1970), Mediterranean (Furnari 1984), Balearic Islands (Ribera and Gómez 1985), S France (Belsher *et al.* 1976, Boudouresque *et al.* 1984), Tunisia (Meñez and Mathieson 1981), Libya (Nizamuddin *et al.* 1978), Egypt (Nasr *et al.* 1967), W Greece (Schnetter and Schnetter 1981), Aegean Sea (Athanasiadis 1987), Black Sea (Zinova 1967). *Falkenbergia rufolanosa* (●) (○ = record from drift material) and *F. hillebrandii* (★). Sources: Atlantic (Feldmann and Feldmann 1942), British Isles (Norton 1986), Netherlands (Stegenga and Mol 1983), N France (Feldmann 1954), W France (Hoek and Donze 1966, Bouxin and Dizerbo 1971, L'Hardy-Halos 1972), N Spain (Weber-Peukert and Schnetter 1982), NW Spain (Donze 1968, Perez-Cirera 1975), Portugal (Ardré 1970), SW Spain (Seoane-Camba 1965), Balearic Islands (Ribera and Gómez 1984), S France (Belsher *et al.* 1976, Boudouresque *et al.* 1984), Corsica (Boudouresque and Perret 1977), S Italy (Funk 1927, Cinelli *et al.* 1979, Cormaci and Furnari 1979), Tunisia (Meñez and Mathieson 1981), W Greece (Schnetter and Schnetter 1981), Aegean Sea (Athanasiadis 1987, Diapoulis *et al.* 1986), Marocco (Gayral 1958), Azores (Schmidt 1931), Canaries (Johnston 1969, Lawson and Norton 1971), Senegal (Sourie 1954), Liberia (De May *et al.* 1977), Gold Coast (Lawson 1956), W Florida (Dawes 1974), E Florida (Juett *et al.* 1976).

(Table IV). Plants survived 1 day incubation time at −1 °C; 14 days at 0 °C; 28 days at 1 °C and all 56 days at 2 °C. Death occurred after 3 days at −1 °C and after 35 days at 0 °C. Chilling injury and regeneration appeared in plants incubated for 2 days at −1 °C, 21 to 28 days at 0 °C and 35 to 56 days at 1 °C. Individuals investigated for their upper temperature tolerance survived at 26 °C but not at 28 °C or above. At 27 °C they became reversibly injured (Table III). Plants incubated in the 25 °C culture room began to bleach after 4 weeks and died within 8 weeks (Table III). Plants remained sterile under all test conditions.

Table II. Lower temperature tolerance of the investigated species. Total exposure time to experimental temperatures was 8 weeks. Results evaluated after 4 or 8 weeks post-cultivation at 15 °C, 20 μmol photoms m^{-2} s^{-1}, 16 h light per day, +++: plants alive; R: regeneration from injured thallus; ---: plants dead; nt. = not tested.

Species	Experimental time (weeks)	Experimental temperatures (°C)						
		−1	0	1	2	3	4	5
Falkenbergia rufolanosa	2	---	---	+++	+++	+++	+++	+++
	4	---	---	R	R	+++	+++	+++
	8	---	---	---	---	---	---	+++
Kuckuckia spinosa	2	+++	+++	+++	+++	+++	+++	+++
	4	R	R	+++	+++	+++	+++	+++
	8	---	---	R	R	+++	+++	+++
Cystoseira barbata	2	R	nt.	nt.	nt.	nt.	nt.	+++
8-months-old plants	4	R	nt.	nt.	nt.	nt.	nt.	+++
	8	R	R	R	+++	+++	+++	+++

Table III. Upper temperature tolerance of the investigated species. Standard exposure time to experimental temperatures was 2 weeks. In cases (*) the exposure time was 8 weeks at 25 ± 1 °C, 30−35 μmol photons m^{-2} s^{-1}, 16 h light per day, and in case (**) was 8 weeks at 30 ± 0.2 °C, 60−80 μmol photons m^{-2} s^{-1}, 16 h light per day. Results were evaluated after 4 or 8 weeks post-cultivation at 20 °C, 20 μmol photons m^{-2} s^{-1}, 16 h light per day (see Table II for explanation of symbols).

Species	Experimental temperatures (°C)								
	25	26	27	28	29	30	31	32	33
Choristocarpus tenellus	+++	+++	R	---	---	---			
	+++	+++	R	---	---	---			
	---*								
Falkenbergia rufolanosa	+++	+++	R	---	---	---			
	+++	+++	---	---	---				
	+++	+++	R	---	---				
	R*								
Kuckuckia spinosa		+++	+++	+++	R	---	---	---	
		+++	+++	+++	R	---	---		
Cystoseira barbata									
2-months-old plants			R	R	R				
			R	R	R	R	---	---	---
5−8-months-old plants			R	R	R	R	R	---	---
			R	R	R	R	R	---	---
10-months-old plants						R**			

Two-month-old plants of *Cystoseira barbata* grew optimally at 15 °C and sufficiently at 10 to 25 °C (Fig. 2). The lower temperature tolerance was tested on 8-month-old plants which survived for 8 weeks at 3 °C. Eight weeks cooling at −1 to 2 °C caused chilling injury on the primary laterals and the main axis; regeneration occurred from the main axis (Table II). The upper temperature tolerance test was carried out on 2-month-old and 5- to 8-month-old plants at 27 to 33 °C. Regeneration, mostly from the main axis, was observed up to 31 °C from the older plants and up to 30 °C from the younger ones (Table III). A 10-month-old plant (*ca* 10 cm high) lost most of its primary laterals after 8 weeks at 30 °C and the main axis grew dark but was still able to regenerate (Table III). Plants cultivated for 16 months in aerated 3 L beakers at 15 °C (20−25 μmol m^{-2} s^{-1}, 16 h light per day) reached a size of *ca* 18 cm (main axis). Plants remained sterile under all test conditions.

The lowest survival temperatures of *Falkenbergia rufolanosa* were 1 °C for 2 weeks, 3 °C for 4 weeks and 5 °C for 8 weeks incubation time; death occurred at −1 to 0 °C after 2 weeks and at 4 °C after 8 weeks. Chilling injury and regeneration appeared in plants incubated for 4 weeks at 1 and 2 °C (Table II). In the upper temperature tolerance test the plants survived incubation at 26 °C but not at 28 °C or above. At 27 °C they became reversibly injured (Table III). Plants incubated in the 25 °C culture room began to bleach within 4 to 8 weeks incubation time but they were able to regenerate (Table III). At 10−20 °C, plants grew well without producing sporangia for at least 12 weeks.

Upper temperature tolerance tests were run at least twice. Usually identical values were obtained in the first two runs. If not, the difference was no more than 1 °C. Two difficulties should be emphasised:

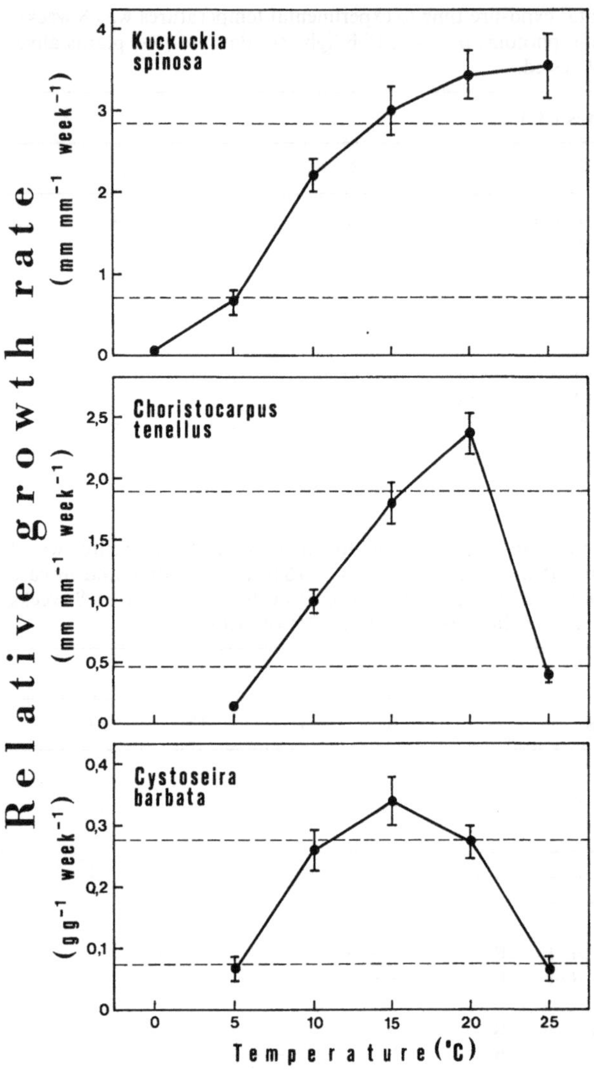

Fig. 2. Mean relative growth rates (RGR) of *Kuckuckia spinosa* (n = 8), *Choristocarpus tenellus* (n = 8) and *Cystoseira barbata* (n = 5–6) measured at 5 °C temperature intervals (vertical bars = 95% confidence intervals). Standard light conditions: 60–80 μmol m^{-2} s^{-1}, 16 h light per day. Broken lines = 20% and 80% of maximum RGR, defining "sufficient" growth and "optimum" growth respectively.

(1) the non quantitative technique employed, *i. e.* the quantity of experimental plant material could have varied from one experiment to an other;

(2) the difficulty of separating young and old tissue in smaller algae. Younger thalli of *Cystoseira barbata* are more sensitive to heating than older ones (Table III; see also Lüning 1984).

Discussion

Methodological aspects

Exposure time is fundamental to the determination of temperature tolerance (Levitt 1980, Larcher and Bauer 1981). In this study an exposure time of 2 weeks was used to determine the upper temperature tolerance, which was found to be long enough for the red alga *Polyneura hilliae* (Grev.) Kylin (Yarish *et al.* 1987), and nearly long enough in the case of *Enteromorpha linza* (Linn.) J. Ag. For *E. linza* the highest tolerated temperature decreased by 1 °C between 2 and 6 weeks incubation time (Orfanidis 1990). *Choristocarpus tenellus* was tested over a period of 56 days in order to determine the adequate exposure time for lower temperature tolerance (Table IV). The lowest survival limit for this alga was constant from an incubation time of 35 days onwards. However, *Colpomenia peregrina* (Sauvag.) Hamel and *Gigartina teedii* (Roth) Lamour. reach their constant lowest survival limit after 42 and 56 days cultivation respectively (Orfanidis 1990). In view of this result, it was decided to use 8 weeks exposure for all species.

In recent studies (Cambridge *et al.* 1984, 1987, 1990, Yarish *et al.* 1984, 1986) a temperature tolerance-dependent on daylength regime was documented: survival at lower temperatures was enhanced in short-day conditions for all tested species. In the present study 12 h light per day was used but a shorter day-length might have been more relevant to approximate the northern lethal boundaries. However, the northern lethal boundary of *Lomentaria articulata* (Hudson) Lyngbye, a warm-temperate Mediterranean-Atlantic alga, might occur in spring when low temperatures coincide with long-days and high irradiances (Yarish *et al.* 1986). The southern lethal boundaries are met under long-day conditions. *Choristocarpus tenellus* and *Falkenbergia rufolanosa* cultivated in long-days were more easily injured at lower temperatures than if incubated under 12 h light per day (Table III).

In intercalary growing species, like *Kuckuckia spinosa*, one might expect a logarithmic growth pattern (Kurogi and Yoshida 1966, Geesink 1973). In species growing mainly at the apex, like *Cystoseira barbata*, a linear elongation of the main axis and branches was observed (Bird *et al.* 1977, Guiry *et al.* 1987, see also Kain 1987) but the total increase in cell number (*i. e.* fresh weight) was exponential (Edwards 1977). Fragments of *Choristocarpus tenellus* grew mostly in two opposite directions and followed also a logarithmic pattern (Orfanidis 1990).

Biogeographic hypotheses and temperature responses

Kuckuckia spinosa is an eulittoral to sublittoral epiphytic alga with an *Ectocarpus*-like morphology. The northernmost population of *K. spinosa* which has been consistently collected by Feldmann and Magne (1964) is from Roscoff, Brittany. In the North Sea and western Baltic it seems to be a rare species; it has been observed mostly in crude cultures growing together with other algae (Kuckuck 1958, Kylin 1947). The distribution data (Fig. 1) suggest that the northern boundary is a growth and/or reproduction boundary at the 16 °C August isotherm (corresponding with a minimum mean summer temperature of 14 °C permitting growth and/or reproduction). The fact that *K. spinosa*, at its northern range, behaves as a summer annual (Feldmann and Magne 1964) supports this hypothesis. A northern lethal boundary is not expected because the winter temperatures in the North Sea and W. Baltic are several degrees lower than those on the open European Atlantic coasts. The southern records in the Atlantic and Mediterranean (Fig. 1) suggest either a southern lethal boundary at the 25.5 °C August isotherm (temperature record from Alexandria, Egypt from Nasr and Aleem 1948; corresponding with a maximum summer temperature of 28.5 °C) or a southern growth and/or reproduction boundary at the 18 °C February isotherm (corresponding with a maximum mean winter temperature of 20 °C permitting growth and/or reproduction).

According to the thermal requirements for the growth of *Kuckuckia spinosa* (Figs 2, 3) one would expect a northern growth boundary near the 10 °C August isotherm and an occurrence on coasts of the British Isles and NW Norway. Perhaps the northern boundary is governed by the lowest temperature permitting sufficient reproduction (10 to 15 °C, daylength independent); this could be necessary for the maintenance of a population in nature. The northern boundary is thus a reproduction boundary at the 16 °C August isotherm (Table V) and not a growth boundary at the 10 °C August isotherm as proposed by Pedersen (1989). In his study, an isolate from Lanzarote, Canaries, was used which exhibited very similar temperature responses to the Villefranche-sur-Mer isolate treated here. The available experimental data (Figs 2, 3, Table III) indicate a southern lethal boundary at the 25.5 °C August isotherm in the Mediterranean (Table V). According to this view, the species could exist in the Azores and Madeira and could extend farther south to the West African coast.

Choristocarpus tenellus is a rare deep-water species with, however, a more or less continuous occurrence from the Black Sea and the Mediterranean to the British Isles. The northernmost recorded populations (Isle of Man and North Cornwall, Britain; Fig. 1) suggest that low winter temperatures could prevent *C. tenellus* from penetrating the North Sea, because summer temperatures in that area are several degrees higher than those experienced along the west coasts of the British Isles. Otherwise, the distribution gives little information on the nature of the boundary along the west coasts of the British Isles. The southernmost record in the Atlantic Ocean (Fig. 1) suggests a lethal southern boundary at the 19 °C August isotherm (corresponding with a maximum summer temperatures of 22 °C).

The temperature responses of *Choristocarpus tenellus* (Figs 2, 3, Tables III, IV) do not explain its absence from NW Scotland and NW Norway. One would expect a northern growth boundary near the 10 °C August isotherm, because the winter temperatures on these coasts are within the survival range. Possibly, reproduction requires higher temperatures than growth, as is the case in *Kuckuckia spinosa*. Evidently, further investigation on the life history of this species is required to solve the problem of the northward limit. However, several species belonging to the warm-temperate Mediterranean-Atlantic distribution group appear not to be able to reach their potential distribution limit (Breeman 1988). The occurrence of *C.*

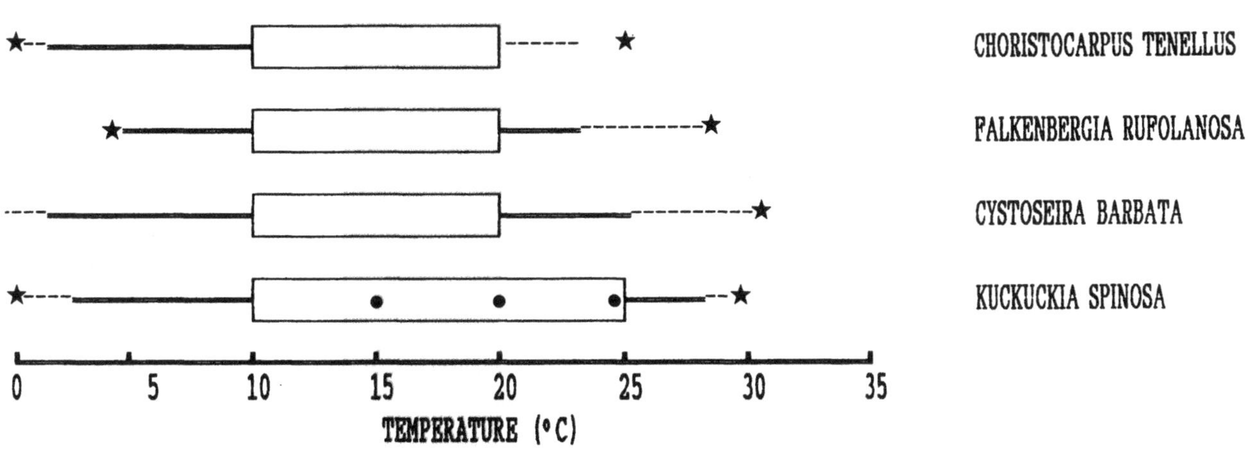

Fig. 3. Temperature intervals limiting growth, reproduction and survival of the investigated macroalgae. The symbols in the diagram are defined as (▬▬) survival without injury in the interval (------) regeneration from injured thallus during the post-cultivation in the interval; (★) lethal temperature; (□) sufficient growth in the interval; (•) asexual reproduction.

Table IV. Lower temperature tolerance of *Choristocarpus tenellus* over 56 days incubation (see Table II for explanation of symbols).

Time (days)	Experimental temperatures (°C)						
	−1	0	1	2	3	4	5
1	+++	+++	+++	+++	+++	+++	+++
2	R	+++	+++	+++	+++	+++	+++
3	−−−	+++	+++	+++	+++	+++	+++
4	−−−	+++	+++	+++	+++	+++	+++
5	−−−	+++	+++	+++	+++	+++	+++
6	−−−	+++	+++	+++	+++	+++	+++
7	−−−	+++	+++	+++	+++	+++	+++
10	−−−	+++	+++	+++	+++	+++	+++
14	−−−	+++	+++	+++	+++	+++	+++
21	−−−	R	+++	+++	+++	+++	+++
28	−−−	R	+++	+++	+++	+++	+++
35	−−−	−−−	R	+++	+++	+++	+++
42	−−−	−−−	R	+++	+++	+++	+++
49	−−−	−−−	R	+++	+++	+++	+++
56	−−−	−−−	R	+++	+++	+++	+++

Table V. Survival between northern and southern lethal isotherms (based on experimental data) and actual poleward and equatorward biogeographical boundaries.

Species	Experimental interval of survival (°C)	Northern-Southern lethal isotherms (corres. to exper.)	Difference (°C)	Poleward biogeographical boundary	Equatorward biogeographical boundary
Choristocarpus tenellus	1−(20−25)	2−22	18	7 °C February, winter lethal or 14 °C August, summer growth/reprod.	19 °C August, summer lethal
Falkenbergia rufolanosa	5−27	6−24	18	7 °C February, winter lethal	possibly 24 °C August, summer lethal
Kuckuckia spinosa	1−29	2−26	24	16 °C August, summer reproduction	25.5 °C August, summer lethal
Cystoseira barbata	(−1)−30	(−1)−27	28	?	?

tenellus in the northern Black Sea but not in the North Sea could be due to the clear water of the Black Sea (Zenkevitch 1963). This would allow growth in deeper waters preventing the extinction of the population during cold winters. However, another explanation might be the occurrence of a temperature ecotype in the Black Sea. On the basis of the experimental data (Figs 2, 3, Table III) the southern boundary is a lethal one at the 19 °C August isotherm (Table V).

The distribution of *Cystoseira barbata* is restricted to the Mediterranean and Black Seas and the southern coasts of Portugal (Fig. 1). It grows in the upper sublittoral zone. Its occurrence in the eastern Mediterranean Sea and in the Black Sea (extreme local temperatures −1,4 to 29 °C from Zenkevitch 1963) indicates a survival potential of *ca* −1 to 29 °C. Therefore, one would expect a wider distribution of the species along the North Atlantic coasts. This hypothesis is supported by the fact that *C. barbata* is a perennial species flourishing during the winter months in the Gulf of Naples (Funk 1927) and in the Gulf of Thessaloniki (pers. observation).

The experimental data obtained for *Cystoseira barbata* (Figs 2, 3, Tables II, III) explain its occurrence in the Mediterranean and the Black Sea, but there is no apparent reason for its restriction to south Portugal (see below). This alga was the only species in the present work that survived −1 °C. Similar results have been obtained for two other species belonging to this distribution group. The sporophytes of *Laminaria hyperborea* (Gunnerus) Foslie survived at −1.5 °C (Lüning 1986) and the gametophytes and tetrasporophytes of *Lomentaria articulata* survived at 0 °C (Yarish *et al.* 1986).

Falkenbergia rufolanosa is the tetrasporic phase of *Asparagopsis armata* (Feldmann 1965, Oza 1977) which first appeared in Europe at the beginning of this century. Since its first collection on the European Atlantic coasts, *A. armata* has spread, together with its *Falkenbergia* phase, northwards up to the Shetland Isles, apparently by vegetative means (Irvine *et al.* 1975). The tetrasporic phase has been recorded in the Atlantic in the eulittoral and sublittoral zones, where it is epiphytic, rarely epilithic, and often free floating

(Dixon and Irvine 1977, Gayral 1958). In the Mediterranean Sea, *F. rufolanosa* is restricted to the low sublittoral zone (Cinelli *et al.* 1979). The distribution (Fig. 1) suggests that the northern boundary is a lethal one at the 7 °C February isotherm (corresponding to a 6 °C minimum winter temperature). If a growth and/or reproduction boundary occurred at the 14 °C August isotherm, *F. rufolanosa* would be expected at Helgoland and on the Danish coasts, where the summer temperatures rise above 13 °C. The record from the Netherlands (Stegenga and Mol 1983) does not indicate a penetration into the North Sea because it is based on drift material. The southern distribution limit cannot be determined from the available records pending an accurate taxonomic discrimination (see Dixon 1964, Dixon and Irvine 1977) between *A. armata\F. rufolanosa* and *A. taxiformis\F. hillebrandii* which is widely distributed in tropical regions.

On the basis of the experimental data (Fig. 3, Tables II, III) the northern boundary of *Falkenbergia rufolanosa* is a lethal one at the 7 °C February isotherm (Table V). Apparently this species is unable to spread northwards to Bergen, Norway, because the 4 °C minimum winter temperature (5 °C February isotherm) in this region causes lethal damage (Fig. 3, Table II). Independently of the nature of the southern boundary, the 24 °C August isotherm (Table V) should prevent the distribution to tropical waters and the southernmost distribution limit of the eulittoral populations should occur in the Canaries. The presence of this species in Macaronesian islands has recently been confirmed (Prud'homme van Reine, pers. comm.). In the Mediterranean Sea the southern boundary on the eastern Sicilian coasts is located at 26 °C August isotherm (Gorshkov 1985), which corresponds with slightly lower summer temperatures in the sublittoral habitat of this species.

Extreme winter temperatures around 0 °C probably prevent the distribution of *Choristocarpus tenellus* and *Falkenbergia rufolanosa* and cause rarity of *Kuckuckia spinosa* in the southern North Sea (Fig. 3, Tables II, IV).

The species studied (Table V) can survive between their northern and southern lethal boundaries (corresponding to monthly isotherms) the following temperature spans: *Choristocarpus tenellus* 18 °C, *Falkenbergia rufolanosa* 18 °C, *Kuckuckia spinosa* 24 °C and *Cystoseira barbata* 28 °C. The relatively narrow survival limits of *C. tenellus* and *F. rufolanosa* are in accordance with the experimental data of other species falling within the warm-temperate Mediterranean-Atlantic group (Breeman 1988). This restricts their distribution to eastern Atlantic coasts. A species may succesfully inhabit the temperate coasts of eastern N. America if it tolerates a minimum temperature span of 20 °C (Hoek 1982 a, b). The other two species belonging to this distribution group (*K. spinosa* and *C. barbata*) might potentially disperse to the eastern American coasts but they probably lack dispersal mechanisms or they have not had sufficient time. Such eurythermal temperature characteristics were also found by Yarish *et al.* (1986, 1987) in the red alga *Lomentaria articulata* (survived at least 0 to 27 °C).

Biogeography of the genus Cystoseira *in the Mediterranean*

Cystoseira is phylogenetically an old genus with a tropical to warm temperate distribution in both hemispheres. Fossil records from the Miocene (see Parker and Dawson 1969) certify its presence in the ancient equatorial ocean (Tethys Sea). The Mediterranean Sea as a remnant of this ocean was occupied by tropical marine biota until it was completely closed at the end of the Miocene (Messinian salinity crisis). In the Miocene, tropical planktonic foraminifera extended northward at least to Gibraltar (Cifelli 1976). After the Messinian crisis (5 Ma) some of the late Miocene biota survived the oceanographic changes at the end of Miocene, and returned to the Mediterranean to establish the early Pliocene biota (Raffi and Marasti 1982). At that time the Mediterranean Sea was a warm subtropical or marginally tropical sea (Raffi *et al.* 1985) consequently losing its coral reefs (Lüning 1990), the inhabitors of the sublittoral zone in most tropical regions. From the current distribution of *Cystoseira* spp. it may be inferred that those conditions could not be limiting for an ancestor of this genus. Empty niche space and the climate changes in the late Pliocene and the Pleistocene may have promoted speciation. The survival capacity at tropical temperatures and the extension of the survival range to the lower end of the temperature scale, *e.g. C. barbata* (Fig. 3), makes these algae ideal inhabitants of the Mediterranean region. In addition, the large eggs of the Cystoseiraceae indicate a storage capacity which may be seen as an evolutionary advance to invade sublittoral regions (Clayton 1984). The restriction of *Laminaria* spp. to deeper, colder waters in this region discloses their inability to compete for the same ecological niche. Today a great number of the *Cystoseira* spp. are endemic in the Mediterranean Sea indicating that the rate of speciation in this genus is greater than the dispersal rate (Hoek 1975). Evidently, *C. barbata* has not yet reached its potential northern limit. This species needs temperatures of at least 10 °C for adequate growth. It is interesting to note that during the last glaciation the northern growth limit would have been situated in southern Portugal, near the present northernmost record on the Atlantic coast.

Maps of geographical distribution

Gaps in the species' distribution could be attributed to lack of suitable substratum or to undersampling. Coasts with limited rocky substrate occur from northern France to Denmark, in eastern England between Flamborough Head in Yorkshire and Herne Bay in Kent, and from the estuary of the Loire (W France) to the Basque coasts (Lüning 1990). In addition, the whole eastern part as well as some other parts of the Mediterranean (*e.g.* Aegean Sea) have been poorly investigated, in particular in the low sublittoral zone. Many species also appear to be absent from the west coast of Africa in spite of potentially suitable temperature regimes. Again, absence of rocky substrates or undersampling may be the explanation.

Conclusion

The attempt to identify the temperature processes that limit the distribution of a marine benthic alga may be an invaluable tool in understanding the nature of its geographical boundaries, and in testing its taxonomy. Several species appear unable to reach their potential limits of tolerance indicating the role of other ecological factors or historical events determining the present distribution.

The temperature responses of *Choristocarpus tenellus* are irrelevant in explaining its northern distribution boundary; its southern boundary is set by an August lethal limit. *Falkenbergia rufolanosa* has its northern boundary at a February lethal limit. The temperature range tolerated by this species may limit its distribution in tropical waters and distinguish it from the morphologically similar *F. hillebrandii.* The latitudinal distribution of *Kuckuckia spinosa* is restricted by an August reproduction limit in the north and by an August lethal limit in the south. Its longitudinaly distribution is probably limited by the absence of long range dispersal mechanisms. *Cystoseira barbata* is a species still expanding its distribution.

Acknowledgements

This work has supported by the German Academic Exchange Service (DAAD) and is based on my Ph. D. study in the Department of Biology, University of Thessaloniki. I am very grateful to Prof. Dr K. Lüning for advice and for his hospitality and working facilities during my stay in his laboratory. I thank Prof. Dr K. Lüning, Dr H. Kirkman and Dr I. tom Dieck for critical review of an earlier draft. I thank Prof. Dr D. G. Müller for the isolates of *Kuckuckia spinosa, Choristocarpus tenellus* and *Falkenbergia rufolanosa.* Thanks are also due to Prof. Dr Haritonidis and Prof. Dr Tsekos for help as supervisors and members of the advisory committee for my Ph. D. work.

References

Ardré, F. 1970. Contribution à l'étude des algues marines du Portugal. I. La flore. *Portugaliae Acta Biol. (B) 10*: 137–555.

Athanasiadis, A. 1987. A survey of the seaweeds of the Aegean Sea with taxonomic studies on species of the tribe Antithamnieae (Rhodophyta). Goterna, Kungälv. pp. viii + 174.

Belsher, T., H. Augier, C. F. Boudouresque and E. Coppejans. 1976. Inventaire des algues marines benthiques de la rade et des îles d'Hyères (Méditerranée France). *Trav. sci. Parc nation. Port-Cros 2*: 39–89.

Bird, C. J., T. Edelstein and J. McLachlan. 1977. Studies on *Gracilaria.* Experimental observations on growth and reproduction in Pomquet Harbour, Nova Scotia. *Naturaliste can. 104*: 245–255.

Boudouresque, C. F. and N. Perret. 1977. Inventaire de la flore marine de Corse (Méditerranée): Rhodophyceae, Phaeophyceae, Chlorophyceae et Bryopsidophyceae. *Bibliotheca Phycologia 25*: 1–170.

Boudouresque, C. F., M. Perret-Boudouresque and M. Knoepffler-Peguy. 1984. Inventaire des algues marines benthiques dans les Pyrénées-Orientales (Méditerranée, France). *Vie et Milieu 34*: 41–59.

Bouxin, H. and A. H. Dizerbo. 1971. Les algues de l'Archipel des Glenan. *Botanica Rhedonica, sér. A. 10*: 201–226.

Breeman, A. M. 1988. Relative importance of temperature and other factors in determining geographic boundaries of seaweeds: experimental and phenological evidence. *Helgoländer Meeresunters. 42*: 199–241.

Cambridge, M. L., A. M. Breeman and C. van den Hoek. 1990. Temperature responses limting the geographical distribution of two temperate species of *Cladophora* (Cladophorales: Chlorophyta) in the North Atlantic Ocean. *Phycologia 29*: 74–85.

Cambridge, M. L., A. M. Breeman, S. Kraak and C. van den Hoek. 1987. Temperature responses of tropical to warm temperate *Cladophora* species in relation to their distribution in the North Atlantic Ocean. *Helgoländer Meeresunters. 41*: 329–354.

Cambridge, M. L., A. M. Breeman, R. van Oosterwijk and C. van den Hoek. 1984. Temperature responses of some North Atlantic *Cladophora* species (Chlorophyceae) in relation to their geographic distribution. *Helgoländer Meeresunters. 38*: 349–363.

Cardinal, A. 1964. Étude sur les Ectocarpacées de la Manche. *Nova Hedwigia 15*: 1–86.

Cifelli, R. 1976. Evolution of ocean climate and the record of planktonic foraminifera. *Nature,* London 264: 431–432.

Cinelli, F., E. Fresi, L. Mazzella and M. P. Ponticelli. 1979. Deep algal vegetation of the western Mediterranean. *Giorn. Bot. Ital. 113*: 173–188.

Clayton, M. N. 1984. Evolution of the Phaeophyta with particular reference to the Fucales. *Progr. Phycol. Res. 3*: 11–46.

Coppejans, E. 1981. Végetation marine de la Corse (Méditerranée). 4. Documents pour la flore des algues. *Biol. Jaarb. 49*: 76–97.

Coppejans, E. and C. F. Boudouresque. 1976. Présence de *Choristocarpus tenellus* (Kützing) Zanardini a Port-Cros. *Trav. sci. Parc nation. Port-Cros 2*: 195–197.

Cormaci, M. and G. Furnari. 1979. Flora algale marina della Sicilia orientale: Rhodophyceae, Phaeophyceae e Chlorophyceae. *Informatore Botanico Italiano 11*: 221–250.

Dawes, C. J. 1974. *Marine Algae of the West Coast of Florida*. University of Miami Press, Coral Gables, Florida. pp. 201.

De May, D., D. M. John and G. W. Lawson. 1977. A contribution to the littoral ecology of Liberia. *Bot. Mar. 20*: 41–46.

Diapoulis, A., Haritonidis, S. and Th. Koussouris. 1986. Spring benthic flora of Rhodos Island, Greece. *Thalassographica. 9*: 49–57.

Diapoulis, A. and S. Haritonidis. 1987. A qualitative and quantitative study of the marine algae in the Saronikos Gulf (Greece). *Marine Ecology 8*: 175–189.

Dixon, P. S. 1964. *Asparagopsis* in Europe. *Nature, London, 204*: 902.

Dixon, P. S. and L. M. Irvine. 1977. *Seaweeds of the British Isles. Vol. 1. Rhodophyta.Part 1. Introduction, Nemaliales, Gigartinales*. British Museum (Natural History), London. pp. 252.

Donze, M. 1968. The algal vegetation of the Ria de Arosa (NW Spain). *Blumea 16*: 159–183.

Edwards, P. 1977. An analysis of the pattern and rate of cell division, and morphogenesis of sporelings of *Callithamnion hookeri* (Dillw.) S. F. Gray (Rhodophyta, Ceramiales). *Phycologia 16*: 189–196.

Feldmann, J. 1954. Inventaire de la flore marine de Roscoff. *Trav. Stat. Biol. Roscoff 6 (Suppl.)*: 1–152.

Feldmann, J. and G. Feldmann. 1942. Recherches sur les Bonnemaisoniacées et leur alternance de générations. *Ann. Sci. Nat., Ser. 11, Botan. 3*: 75–175.

Feldmann, J. and F. Magne. 1964. Additions à l'inventaire de la flore marine de Roscoff. Algues, chambignons, lichens. *Editions de la Station Biologique de Roscoff*: 1–28.

Feldmann, G. 1965. Le development des tetraspores de *Falkenbergia rufolanosa* et le cycle des Bonnemaisoniales. *Rev. Gen. Bot. 72*: 621–626.

Funk, G. 1927. Die Algenvegetation des Golfs von Neapel. Nach neueren ökologischen Untersuchungen. *Pubbl. Staz. Zool. Napoli 7 (Suppl.)*: 1–507.

Furnari, G. 1984. The benthic marine algae of southern Italy. Floristic and geobotanic considerations. *Webbia 38*: 349–369.

Gallardo, T., A. Gómez Garreta, M. A. Ribera, M. Alvarez and F. Conde. 1985. *A Preliminary Checklist of Iberian Benthic Marine Algae*. Real Jardin Botanico, Madrid.

Gayral, P. 1958. *Algues de la côte Atlantique Marocaine*. Société des sciences naturelles et physiques du Maroc, Rabat. pp. 523.

Geesink, R. 1973. Experimental investigations on marine and freshwater *Bangia* (Rhodophyta) from the Netherlands. *J. exp. mar. Biol. Ecol. 11*: 239–247.

Gorshkov, S. G. (ed.) 1985. *World Ocean Atlas. Vol. 2: Atlantic and Indian Oceans*. Pergamon Press, Oxford.

Guiry, M. D. and E. M. Cunningham. 1984. Photoperiodic and temperature responses in the reproduction of north-eastern Atlantic *Gigatina acicularis* (Rhodophyta: Gigartinales). *Phycologia 23*: 357–367.

Guiry, M. D., G. Tripodi and K. Lüning. 1987. Biosystematics, genetics and upper temperature tolerance of *Gigartina teedii* (Rhodophyta) from the Atlantic and Mediterranean. *Helgoländer Meeresunters. 41*: 283–295.

Hiscock, S. and C. A. Maggs. 1984. Notes on the distribution and ecology of some new and interesting seaweeds from south-west Britain. *Br. phycol. J. 19*: 73–87.

Hoek, C. van den. 1975. Phytogeographic provinces along the coasts of the northern Atlantic Ocean. *Phycologia 14*: 317–330.

Hoek, C. van den. 1982 a. The distribution of benthic marine algae in relation to the temperature regulation of their life histories. *Biol. J. Linn. Soc. 18*: 81–144.

Hoek, C. van den. 1982 b. Phytogeographic distribution groups of benthic marine algae in the North Atlantic Ocean. A review of experimental evidence from life history studies. *Helgoländer Meeresunters. 35*: 153–214.

Hoek, C. van den and M. Donze. 1966. The algal vegetation of the rocky Côte Basque (SW France). *Bull. Cent. Etud. Rech. sci. Biarritz 6*: 289–319.

Hutchins, L. W. 1947. The bases for temperature zonation in geographical distribution. *Ecol. Monogr. 17*: 325–335.

Irvine, D. E. G., M. D. Guiry, I. Tittley and G. Russell. 1975. New and interesting marine algae from the Shetland Isles. *Br. phycol. J. 10*: 57–71.

Johnston, C. S. 1969. The ecological distribution and primary production of macrophyte marine algae in the eastern Canaries. *Int. Rev. ges. Hydrobiol. 54*: 473–490.

Juett, L., C. J. Miller, S. J. Moure and E. S. Ford. 1976. Summer marine algae at Vero Beach, Florida. *Fla Scient. 39*: 76–80.

Kain, J. M. 1987. Seasonal growth and photoinhibition in *Plocamium cartilagineum* (Rhodophyta) off the Isle of Man. *Phycologia 26*: 88–99.

Kuckuck, P. (Kornmann, P. (ed.)) 1958. Ectocarpaceen-Studien V. *Kuckuckia, Feldmannia. Helgoländer Meeresunters. 6*: 171–192.

Kurogi, M. and T. Yoshida. 1966. Experimental cultivation of *Porhyra umbilicalis* (L). Kütz. in Matsushima Bay, North East Honshu, Japan. *Bull. Tohoku reg. Fish. Res. Lab. 26*: 91–107.

Kylin, H. 1947. Die Phaeophyceen der schwedischen Westküste. *Lunds Univ. Arsskr., N.F., Adv. 2, 43 (4)*: 1–99.

Larcher, W. and H. Bauer. 1981. Ecological significance of resistance to low temperature. *In*: (O. L. Lange, P. S. Nobel, C. B. Osmond and H. Ziegler, eds) *Encyclopedia of Plant Physiology, New Series, Vol. 12 A*. Springer-Verlag, Berlin. pp. 403–437.

Lawson, G. W. 1956. Rocky shore zonation on the Gold Coast. *J. Ecol. 44*: 153–170.

Lawson, G. W. and T. A. Norton. 1971. Some observations on littoral and sublittoral zonation at Teneriffe (Canary Isles). *Bot. Mar. 14*: 116–120.

Levitt, J. 1980. *Responses of Plants to Environmental Stresses. Vol. I. Chilling, Freezing, and High Temperature Stresses*. Academic Press, New York. p. 497.

L'Hardy-Halos, M. T. 1972. Recherches en scaphandre autonome sur le peuplement végétal de l'infralittoral rocheux: La Baie de Morlaix (Nord-Finistère). *Bull. Soc. scient. Bretagne 47*: 177–192.

Lüning, K. 1980. Control of algal life-history by daylength and temperature. *In*: (J. H. Price, D. E. G. Irvine and W. F. Farnham, eds) *The Shore Environment, Vol. 2: Ecosystems*. Academic Press, London. pp. 915–945.

Lüning, K. 1984. Temperature tolerance and biogeography of seaweeds: the marine algal flora of Helgoland (North Sea) as an example. *Helgoländer Meeresunters. 38*: 305–317.

Lüning, K. 1986. New frond formation in *Laminaria hyperborea* (Phaeophyta): a photoperiodic response. *Br. phycol. J. 21*: 269–273.

Lüning, K. 1990. *Seaweed Biogeography and Ecophysiology*. John Wiley & Sond, New York.

McLachlan, J. and C. J. Bird. 1984. Geographical and experimental assessment of the distribution of *Gracilaria* species (Rhodophyta: Gigartinales) in relation to temperature. *Helgoländer Meeresunters. 38*: 319–334.

Meñez, E. G. and A. C. Mathieson. 1981. The marine algae of Tunisia. *Smithson. Contr. mar. Sci. 10*: 1–59.

Nasr, A. H. and A. A. Aleem. 1948. Ecological studies of some marine algae from Alexandria. *Hydrobiologia 1*: 251–281.

Nasr, A. H., I. A. Bekheet and R. K. Ibrahim. 1967. An electrochromatographic investigation of the amino acid pattern of some marine algae from Alexandria. *Hydrobiologia 29*: 80–92.

Nizamuddin, M., J. A. West and E. G. Meñez. 1978. A list of marine algae from Libya. *Bot. Mar. 22*: 465–476.

Norton, T. A. 1986. *Provisional Atlas of British Seaweeds.* British Phycological Society and the Biological Research Centre, Port Erin. pp. 164.

Orfanidis, S. 1990. Temperature and photoperiodic responses of several mediterranean macroalgae in relation to their distribution in the North Atlantic Ocean. Ph. D. thesis, University of Thessaloniki.

Oza, R. M. 1977. Culture studies on induction of tetraspores and their subsequent development in the red alga *Falkenbergia rufolanosa* (Harvey) Schmitz. *Bot. Mar. 20*: 29–32.

Parker, B. C. and E. Y. Dawson. 1969. Non-calcareous marine algae from California Miocene deposits. *Nova Hedwigia 10*: 273–295.

Pedersen, P. M. 1989. Studies on *Kuckuckia spinosa* (Fucophyceae, Sorocarpaceae): life history, temperature gradient experiments, and synonymy. *Nord. J. Bot. 9*: 443–447.

Perez-Cirera, J. L. 1975. Catalogo floristico de las algas bentonicas de la Ria de Corme y Lage, NO. de Espana. *Anal. Inst. Bot. Cavanilles 32*: 5–87.

Raffi, S. and R. Marasti. 1982. The Mediterranean bioprovince from the Pliocene to the Recent: observations and hypotheses based on the evolution of the taxonomic diversity of molluscs. *In*: (E. Montanaro Gallitelli, ed.) *Plaeontology, Essential of Historical Geology (Proc. Intern. Meeting, Venice). S. T. E. M. Mucchi, Modena.* pp. 151–177.

Raffi, S., S. M. Stanley and R. Marasti. 1985. Biogeographic patterns and Plio-Pleistocene extinction of Bivalvia in the Mediterranean and southern North Sea. *Paleobiology 11*: 368–388.

Ribera Siguán, M. A. and A. Gómez Carreta. 1984–1985. Checklist of the benthic marine flora of the Balearic Islands. I. Rhodophyceae. Collect. Bot. 15, 377–406. II. Phaeophyceae, Chlorophyceae. *Collect. Bot. 16*: 25–41.

Schmidt, O. C. 1931. Die marine Vegetation der Azoren, in ihren Grundzügen dargestellt. *Bibliotheca Bot. 102*: 1–116.

Schnetter, R. and M. L. Schnetter. 1981. Marine Benthosalgen von Kephallinia (Jonische Inseln). *Bibliotheca Phycol. 51*: 111–152.

Seoane-Camba, J. A. 1965. Estudios sobre las algas bentonicas en la costa sur de la Peninsula Ibérica (litral de Cadiz). *Inv. Pesq. 29*: 3–216.

Sourie, R. 1954. Contribution à l'étude écologique des côtes rocheuses du Sénégal. *Mém. Inst. fr. Afr. noire 38*: 1–342.

South, G. R. and I. Tittley. 1986. *A Checklist and Distributional Index of the Benthic Marine Algae of the North Atlantic Ocean.* British Museum (Natural History), London. pp. 76.

Stegenga, H. and I. Mol. 1983. *Flora van de Nederlandse Zeewieren.* Koninklijke Nederlandse Natuurhistorische Vereniging, Amsterdam. pp. 263.

Weber-Peukert, G. and R. Schnetter. 1982. Floristische und ökologische Untersuchungen über benthische Meeresalgengesellschaften der Küste Asturiens (Spanien). 1. Ein Beitrag zur Planzengeographie der iberischen Nordküste. *Nova Hedwigia 36*: 65–80.

Yarish, C., A. M. Breeman and C. van den Hoek. 1984. Temperature, light and photoperiod responses of some Northeast American and West European endemic rhodophytes in relation to their geographic distribution. *Helgoländer Meeresunters. 38*: 273–304.

Yarish, C., A. M. Breeman and C. van den Hoek. 1986. Survival strategies and temperature responses of seaweeds belonging to different distribution groups. *Bot. Mar. 29*: 215–230.

Yarish, C., H. Kirkman and K. Lüning. 1987. Lethal exposure times and preconditioning to upper temperature limits of some temperate North Atlantic red algae. *Helgoländer Meeresunters. 41*: 328–327.

Zenkevitch, L. 1963. *Biology of the Seas of the U. S. S. R.* George Allen and Unwin, London. pp. 955.

Zinova, A. D. 1967. *"Identifcation of Green, Brown and Red algae in the Southern Seas of U. S. S. R."*. Izdatel'stvo Akademii Nauka, Moskva, Leningrad [in Russian]. pp. 398.

Botanica Marina
Vol. 34, pp. 553–560, 1991

Polysaccharides of Algae

44. Investigation of Sulfated Galactan from *Laurencia nipponica* Yamada (Rhodophyta, Rhodomelaceae) Using Partial Reductive Hydrolysis

A. I. Usov and M. Ya. Elashvili

N. D. Zelinsky Institute of Organic Chemistry, Academy of Sciences of the USSR, Leninsky Prospect 47, Moscow B-334, USSR

(Accepted 5 October 1991)

Abstract

A new approach to partial depolymerization of polysaccharides containing acid-labile glycosidic bonds, namely, reductive hydrolysis in the presence of borane-4-methylmorpholine complex, was applied to the structural analysis of a complex agar-like sulfated galactan from the red seaweed *Laurencia nipponica.* Agarobiitol (**1**) and its 2^2-sulfate (**3**) were shown to be the main components of the hydrolyzate. Also isolated were 2^1-O-methyl-agarobiitol (**2**), two disulfates of the reduced tetrasaccharide of the porphyran type (**4** and **6**), one reduced branched pentasaccharide disulfate containing xylose residue (**5**), and a high-molecular mass fraction devoid of 3,6-anhydrogalactose residues. Structures of these fragments were elucidated using compositional analysis, secondary ion mass spectrometry (SIMS), and ^{13}C-NMR spectroscopy.

Introduction

Isolation and preliminary characterization of sulfated galactans from four *Laurencia* species were described in our recent publication (Usov *et al.* 1989). It was shown by chemical methods and ^{13}C-NMR spectroscopy that these polysaccharides belong to the agar group, being agarose derivatives with a rather high content of alkali-stable sulfate groups and with a reduced amount of 3,6-anhydro-L-galactose residues. Some features of ^{13}C-NMR spectra of galactans from *L. nipponica* Yamada, *L. obtusa* (Huds.) Lamour. and *L. papillosa* (Forsk.) Grev., particularly the complex character of the anomeric region consisting of at least eight resonances, could not be satisfactorily explained from the data available. The present paper is devoted to the more detailed investigation of one of these galactans which was isolated from the sublittoral form of the Pacific red seaweed *Laurencia nipponica.*

Reductive hydrolysis, *e.g.* action of acid in the presence of borane-4-methylmorpholine complex (Garegg *et al.* 1988), has been introduced in the practice of structural analysis of red seaweed galactans in our laboratory (Usov and Elashvili 1989, 1991) and independently by Stevenson and Furneaux (1991). It was shown that complete reductive hydrolysis of polysaccharides containing 3,6-anhydrogalactose derivatives results in quantitative reduction of these acid-labile monosaccharides into the corresponding alditols, whereas other monosaccharides remain unreduced (Usov and Elashvili 1989, 1991). This procedure permits the simultaneous estimation of 2-O-methyl-3,6-anhydrogalactose, 3,6-anhydrogalactose and all the other monosaccharides in one sample of hydrolyzate using gas liquid chromatography (GLC). Milder hydrolysis in the presence of the above mentioned reducing agent results in selective splitting of 3,6-anhydrogalactose glycosidic bonds giving rise to a series of oligosaccharides with a 3,6-anhydrogalactitol residue at each 'reducing' end; sulfate groups are substantially retained. These oligosaccharides are stable and more convenient for chromatographic separation than the products of partial acid hydrolysis or methanolysis. Their structures, including the position of sulfate groups, may be elucidated using ^{13}C-NMR spectroscopy (Usov and Elashvili 1989, 1991). The present work is an example of application of reductive hydrolysis to the structural analysis of a rather complex representative of agar-like polysaccharides.

Material and Methods

Polysaccharide preparation

Laurencia nipponica (sublittoral form) pretreatments, hot water extraction and isolation of sulfated galactan by Cetavlon precipitation followed by transformation into the sodium salt, were carried out according to Usov *et al.* (1989).

Chromatography

The GLC of acetylated alditols and aldononitriles was performed on a Hewlett-Packard 5890A chromatograph with a flameionization detector, Ultra-1 capillary column, and 3393A integrator at 175–290 °C (10°/min).

The high performance liquid chromatography (HPLC) was carried out with a Gilson 303 pump and Knauer differential refractometer. Analytical ion-exchange chromatography was performed on a 25 × 0.4 cm column of DEA-Silasorb 7 µm (Diagnosticum, USSR) using 0.08 M NaCl as the eluant at a flow rate of 0.75 mL/min. A 25 × 1 cm semi-preparative column with the same ion-exchanger and eluant at a flow rate of 3 mL/min was used for isolation of sulfated oligosaccharides. Reversed-phase chromatography was carried out on a 25 × 1 cm column of Silasorb C-18 7 µm (Diagnosticum, USSR) in water at a flow rate of 2.7 mL/min.

A 87 × 2.6 cm column of Toyo-pearl TSK HW-40S (V_o 135 mL), Gilson peristaltic pump and Knauer differential refractometer were used for gel chromatography with distilled water as the eluant at a rate of 0.75 mL/min at room temperature. Desalting was performed on a 30 × 2.6 cm column of Sephadex G-10 in water.

NMR spectroscopy and mass spectrometry

The ^{1}H-NMR spectra were obtained with a Bruker WM-250 spectrometer for oligosaccharide solutions in D_2O at 30 °C. The ^{13}C-NMR spectra were recorded with a Bruker AM-300 spectrometer for the same solutions using methanol (50.15 ppm) as an internal standard. An attached proton test (APT) technique (Patt and Shoolery 1982) was used for assignment of C-1 and C-6 signals of 3,6-anhydrogalactitol residues and C-6 of other hexose residues. Signals in the ^{13}C-NMR spectrum of agarobiitol were assigned using selective heteronuclear {^{1}H-^{13}C} double resonance technique.

Mass spectra were obtained with a Hitachi M-80A mass spectrometer using SIMS technique and positive ion mode. About 1–10 µg of a sample were dissolved in 1 µL water and then placed on the probe tip, previously covered with 1 µL glycerol containing traces of aqueous NaCl solution. Sodium chloride was added to enhance the formation of $(M + Na)^+$ ionms, *cf.* Suzuki and Harada (1985). Bombardment was carried out by Xe^+ ions at 8 keV.

Complete reductive hydrolysis

First borane-4-methylmorpholine complex (Aldrich, 30–40 mg) and then 2 M CF_3COOH (1 mL), containing inositol (0.9 mg/mL), were added to a weighed sample of oligo- or polysaccharide (0.5–5.0 mg). The mixture was heated in a stoppered flask for 8 h at 100 °C, cooled and the acid was evaporated *in vacuo* with 3 × 10 mL of ethanol. A 6% solution of hydroxylamine hydrochloride in anhydrous pyridine (1 mL) was added to the residue, the mixture was heated at 100 °C for 30 min, cooled, then Ac_2O (1 mL) was added and the mixture was heated at 199 °C for 1 h. The resulting solution was evaporated *in vacuo* with 3 × 10 mL of toluene, chloroform (5 mL) and water (5 mL) were added to the residue and mixed. The $CHCl_3$-layer was separated, filtered, concentrated and used for GLC.

The exact quantities of methyl β-D-xylopyranoside (Reanal, Hungary), methyl α-D-galactopyranoside (Reanal, Hungary), and methyl 3,6-anhydro-α-D-galactopyranoside (synthesized according to Lewis *et al.* 1963) separately or in mixtures of different proportions were treated as above to obtain D-xylononitrile, D-galactononitrile, and 3,6-anhydrogalactitol acetates. The molar response factors relative to inositol acetate were found to be 0.41 ± 0.01, 0.49 ± 0.01, and 0.66 ± 0.03, respectively. These values were used to calculate the corresponding monosaccharide content in oligo- and polysaccharides. The value 0.49 was used also for mono-O-methylgalactoses, and 0.66 for 2-O-methyl-3,6-anhydrogalactose. Complete hydrolysis for the determination of monosaccharide composition of substances devoid of 3,6-anhydrogalactose was carried out as described above, but without the addition of borane-4-methylmorpholine complex.

Partial reductive hydrolysis

Sulfated galactan (200 mg) was dissolved in water (15 mL), the solution was heated to 60 °C, borane-4-methylmorpholine complex (1.35 g) was then added followed by 2 M CF_3COOH (5 mL). The mixture was incubated at 65 °C for 8 h, the acid was evaporated with water, the residue was dissolved in small amount of water and chromatographed on a column with TSK HW-40S. A fraction eluting at 135–199 mL contained high-molecular weight hydrolysis products (HMF) and a fraction eluting at 200–295 mL contained oligosaccharides. The oligosaccharide mixture was separated by ion-exchange chromatography (Fig. 1) to isolate charged oligosaccharides **3**–**6** and a mixture of neutral oligosaccharides. The latter was desalted on a Sephadex G-10 column and chromatographed on Silasorb C-18 to obtain individual neutral oligosaccharides **1** and **2**. The yields, composition, and some characteristics of substances isolated are given in Table I.

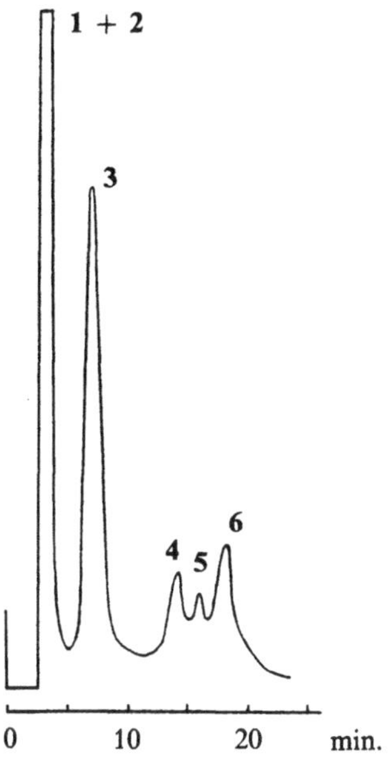

Fig. 1. Separation of neutral and charged oligosaccharides on a 25 × 1 cm column of DEA-Silasorb using 0.08 M NaCl as eluant.

Alkaline modification

Acid oligosaccharides (about 10 mg) were dissolved in M NaOH (2 mL), the solutions were heated at 80 °C for 2 h, neutralized with AcOH and desalted on Sephadex G-10. The modified tetrasaccharides **4**-M and **6**-M, containing one sulfate per molecule, were purified by chromatography on a 25 × 1 cm colum of DEA-Silasorb. Oligosaccharide **5** was recovered after alkaline treatment unchanged.

Fraction HMF (50 mg) was dissolved in M NaOH (10 mL), $NaBH_4$ (30 mg) was added and the mixture was heated at 80 °C for 6 h, cooled, neutralized with AcOH, desalted on TSK HW-40S and lyophilized to give an alkali-modified fraction HMF-M, yield 86%; its composition is given in Table I.

Table I. Yields and some characteristics of partial reductive hydrolysis products of the galactan from *Laurencia nipponica*.

Substance	Yield %	HPLC retention time, min		SIMS, m/z	Monosaccharide composition (molar proportion relative to galactose)				
		Silasorb C-18	DEA-silasorb, 25 × 1 cm colum		2-O-methyl-3,6-anhydro-galactitol	3,6-anhydro-galactitol	xylose	6-O-methyl-galactose	galactose
1	20	6.0	3.5	349 (M+Na)	–	0.93	–	–	1
2	1.5	9.0	3.5	341 (M+H) 363 (M+Na) 385 (M+2Na-H)	0.89	tr.	–	tr.	1
3	30	–	6.5	451 (M+Na) 473 (M+2Na-H)	tr.	0.89	–	–	1
4	2.5	–	13.5	877 (M+Na) 899 (M+2Na-H)	0.06	0.33	–	0.06	1
5	1.5	–	16.0	1009 (M+Na) 1031 (M+2Na-H)	0.06	0.30	0.31	0.05	1
6	4	–	18.5	877 (M+Na) 899 (M+2Na-H)	0.07	0.31	–	0.05	1
HMF[a])	10	–	–	–	0.03	0.07	0.31	0.20	1
4-M	–	–	5.0	757 (M+Na)	0.08[c])	0.88[c])	–	0.08	1
6-M	–	–	5.0	757 (M+Na)	0.10[c])	0.90[c])	–	0.06	1
HMF-M[b])	–	–	–	–	0.03[c])	0.19[c])	0.34	0.19	1

[a]) SO_3Na content 30%.
[b]) SO_3Na content 16%.
[c]) Complete reductive hydrolysis data.

Results and Discussion

Sulfated galactan from a sublittoral form of *Laurencia nipponica* was isolated as described by Usov *et al.* (1989) with a yield of 23.3%. It contained 14.9% SO_3Na, 19.05% of 3,6-anhydrogalactose and 51.35% of other monosaccharides, mainly galactose with some xylose and mono-O-methyl-galactoses. Repeated analysis of the polysaccharide, using the complete reductive hydrolysis procedure, gave the following result: it contains 51 residues of 3,6-anhydrogalactose, 10 residues of xylose, 7 residues of 2-O-methyl-3,6-anhydrogalactose, 2 residues of 6-O-methyl-galactose and one residue of 3-O-methyl-galactose per every 100 galactose residues.

Partial reductive hydrolysis resulted in splitting of 3,6-anhydrosugar glycosidic bonds and reduction of the corresponding fragments. After chromatography of the hydrolyzate on TSK HW-40S a high-molecular fraction (HMF, 10%) and a mixture of neutral and charged oligosaccharides were obtained. Ion-exchange chromatography of the oligosaccharide mixture gave rise to a fraction of neutral oligosaccharides and to sulfated oligosaccharides **3**–**6** (see Fig. 1). Individual oligosaccharides **1** and **2** were further isolated by reversed-phase chromatography of the neutral fraction. The yields and chromatographic mobilities for oligosaccharides are given in Table I.

Degree of polymerization as well as degree of sulfation for charged oligosaccharides were determined by secondary ion mass spectrometry. Intense peaks of $[M + Na]^+$ ions, often accompanied by peaks of $[M + H]^+$ and $[M + 2Na - H]^+$ ions, were observed in mass spectra of all the oligosaccharides **1**–**6**. These data were supplemented with the results of monosaccharide composition analysis by GLC after complete hydrolysis (Table I). It should be noted that acid oligosaccharides **4**–**6**, isolated by ion-exchange chromatography, are not strictly individual compounds, but contain some components differing from the main substance in one methyl group at O–6 of galactose or O–2 of 3,6-anhydrogalactitol residues. However, the presence of a methyl group in neutral disaccharide **2** permits its facile separation from **1** by reversed-phase chromatography. Taking into account the comparatively low level of methylation of the parent polysaccharide, we have only elucidated structures of the main components of the acid oligosaccharides **4**–**6**.

As expected for the agar-like structure of the parent galactan, oligosaccharide **1** proved to be 3,6-anhydro-4-O-β-D-galactopyranosyl-L-galactitol (agarobiitol, see Fig. 2). Its ^{1}H- and ^{13}C-NMR spectra (Tables II and III) coincided completely with the corresponding spectra of an authentic specimen obtained after partial reductive hydrolysis of agar (Usov and Elashvili 1989, 1991).

Oligosaccharide **2** obtained in a much smaller amount was identified as 2^1-O-methyl-agarobiitol. The position of the O-methyl group in 3,6-anhydro-L-galactitol residue followed from the composition analysis and from the ^{13}C-NMR spectrum, taking into account the known α- and β-effects of methylation. The ^{1}H-NMR spectrum of **2** was also completely interpreted using homonuclear double resonance and two-dimensional technique (Tables II and III).

Acid oligosaccharide **3** was the main product of partial reductive hydrolysis of the parent galactan. Its composition corresponded to agarobiitol monosulfate. The position of sulfate at C-2 of β-D-galactopyranose residue was determined by comparison of ^{13}C-NMR spectra of **3** and agarobiitol **1**. These spectra differed one from another in accordance with the known α- and β-effects of sulfation (see Usov 1984, Contreras *et al.* 1988).

One of two sulfate groups was split off upon alkaline treatment of the acid oligosaccharide **4** giving rise to a modified oligosaccharide **4**-M. According to chromatographic mobility, composition, and mass spectrum, this compound was an agarotetraitol monosulfate. The position of sulfate group at C-2 of the galactopyranose residue nearest to the 'reducing' end of the oligosaccharide was elucidated from comparison of ^{13}C-NMR spectra of **4**-M and agarotetraitol (see Usov and Elashvili 1991). These spectra coincided in most of the signals but differed in the positions of resonances belonging to G2r (α-effect of sulfation +6.9 ppm), G3r and G1r (β-effects of sulfation −1.7 and −1.5 ppm respectively), as well as A4r and A1nr (influence of the remote sulfate group, +0.8 and +0.1 ppm respectively).

The structure of **4**-M and the method of its formation indicated that the starting tetrasaccharide **4** contained an L-galactose 6-sulfate residue. Indeed, ^{13}C-NMR spectrum of **4**, which was interpreted using the known spectral data on the porphyran-type oligosaccharides (Morrice *et al.* 1983, Lahaye *et al.* 1985, 1989), corresponded completely to the structure suggested.

Acid oligosaccharide **6** was also partially desulfated upon alkaline treatment with the formation of a 3,6-anhydro-L-galactose residue. The tetrasaccharide monosulfate **6**-M obtained did not differ from **4**-M in chromatographic mobility, composition, and mass spectrum, but it was evident from the ^{13}C-NMR spectrum that this compound was an agarotetraitol derivative with a sulfate group at C-2 of the terminal non-reducing galactose residue. The signal for this carbon was shifted downfield by 9.0 ppm (α-effect of sulfa-

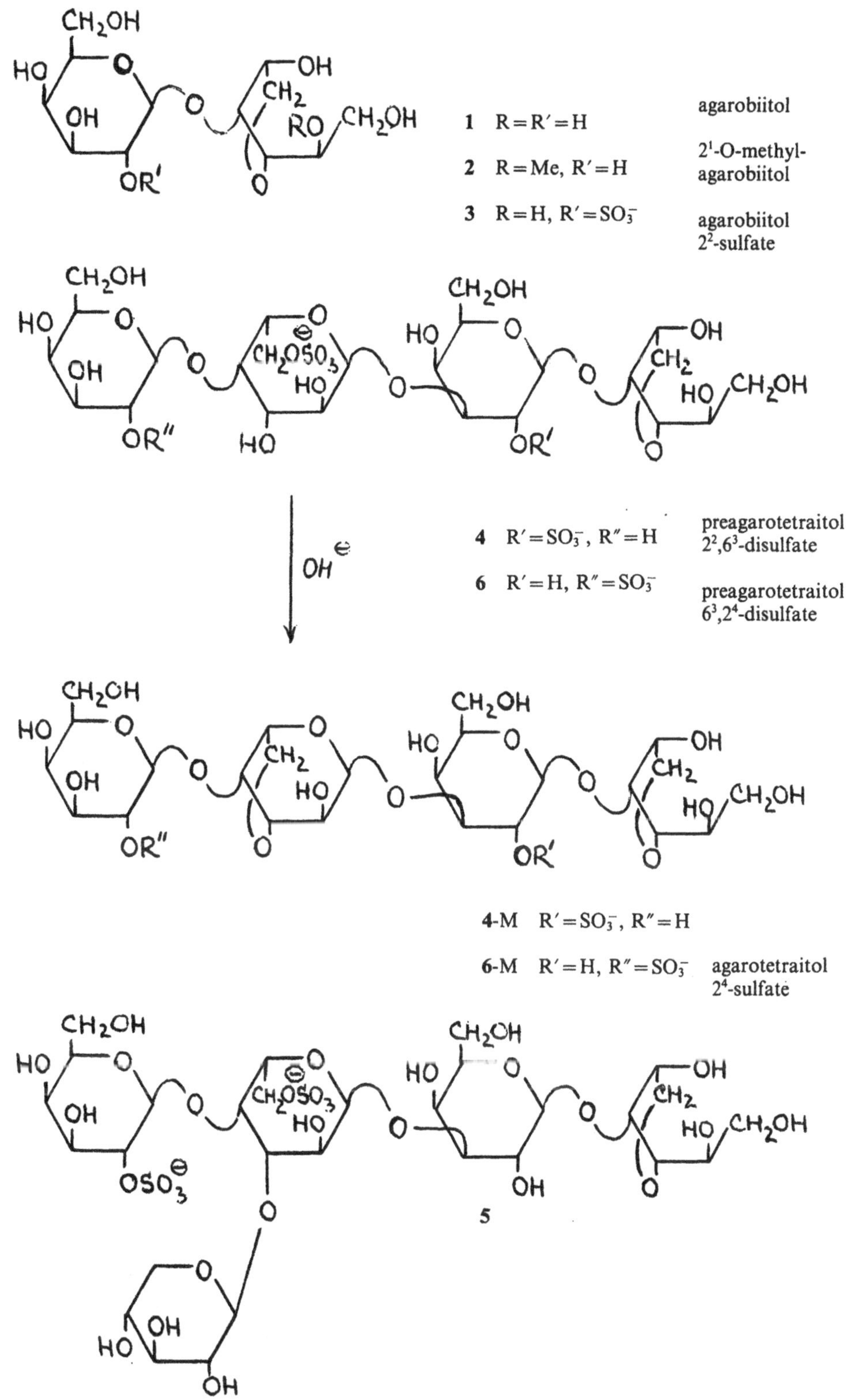

Fig. 2. Structures of the reduced oligosaccharides **1**–**6**.

tion) as compared to the spectrum of agarotetraitol (Usov and Elashvili 1991); the corresponding β-effects were detected on the signals of G3nr (−0.7 ppm) and G1nr (−1.5 ppm). The effect of remote sulfation was observed also in the signals of A4nr (+0.6 ppm), A3nr (−0.3 ppm), and A1nr (+0.1 ppm). According to this evidence, the structure **6** was attributed to the starting oligosaccharide. Its ^{13}C-NMR spectrum, as in the case of **4**, was completely interpreted in accordance with the known spectral data for the porphyran-type oligosaccharides (Morrice *et al.* 1983, Lahaye *et al.* 1985, 1989).

Table II. Chemical shifts (in ppm from TMS) of signals in ^{13}C-NMR spectra of agarobiitol **1**, its 2'-O-methyl derivative **2**, and acid oligosaccharides **3**–**6**, **4**-M, and **6**-M.

Caron atom[a])	Oligosaccharide							
	1	**2**	**3**	**4**	**4**-M	**5**	**6**	**6**-M
G1nr	102.9	102.8	101.1	103.7	102.7	101.4	101.4	101.2
G1r				101.2	101.2	102.6	102.6	102.7
G2nr	71.3	71.3	79.4	71.7	71.7	79.7	79.5	79.3
G2r				77.1	77.4	70.6	70.7	70.5
G3nr	73.4	73.4	72.4	73.2	73.2	72.8	72.7	72.5
G3r				81.3	80.8	81.2	80.8	82.5
G4nr	69.3	69.3	69.3	69.8	69.2	69.4	69.6	69.3
G4r				69.4	68.9	69.2	69.3	68.9
G5nr,r	75.9	76.1	75.8	75.9	75.9	76.0	76.1	75.9
						75.8		
G6nr,r	61.7	61.7	61.5	61.8	61.5	61.9	61.9	61.6
				61.6	61.3	61.4	61.5	
A1nr				102.0	98.7	101.4	101.4	98.7
A1r	63.5	60.6	63.5	63.5	63.5	63.5	63.5	63.5
A2nr				69.2	69.8	67.3	69.3	70.0
A2r	71.6	81.6	71.6	71.7	71.6	71.6	71.6	71.6
A3nr				70.5	80.0	73.5	70.7	79.9
A3r	84.2	83.9	84.4	84.2	84.1	84.0	84.2	84.2
A4nr				78.3	78.2	76.2	78.7	78.1
A4r	86.1	86.0	86.8	86.9	86.8	86.1	86.1	86.1
A5nr				69.8	75.3	70.2	70.3	75.6
A5r	75.9	75.9	75.8	75.9	75.9	75.8	75.9	75.9
A6nr				67.9	69.6	67.5	67.6	69.6
A6r	73.6	73.7	73.7	73.8	73.7	73.6	73.6	73.6
X1						101.0		
X2						73.8		
X3						75.9		
X4						69.8		
X5						65.6		
OMe		58.9						

[a]) designation analogous to the nomenclature used by Rochas *et al.* (1986) and Lahaye *et al.* (1989): G – carbon atoms of D-galactose residues, A – of 3,6-anhydro-L-galactose, 3,6-anhydro-L-galactitol or L-galactose 6-sulfate residues, X – of D-xylose residue; r – residue nearest to the 'reducing' end, nr – to the non-reducing end of the oligosaccharide.

According to monosaccharide composition analysis and mass spectral data, acid oligosaccharide **5** was a pentasaccharide disulfate containing three galactose, one xylose, and one 3,6-anhydrogalactitol residues. Sulfate groups were resistant to alkali and the substance was recovered unchanged after alkaline treatment. The ^{13}C-NMR spectrum of **5** was similar in many respects to the spectrum of oligosaccharide **6** but contained additional signals belonging to the terminal non-reducing xylose residue. Its pyranose form and anomeric β-configuration were proved by characteristic positions of C-2–C-5 resonances (Kovač *et al.* 1980). A large coupling constant for the H-1 signal in the ^{1}H-NMR spectrum of **5** was additional evidence for β-configuration of the xyloside bond (*cf.* Altona and Haasnoot 1980). Other resonances in the ^{13}C-NMR spectrum of **5** were assigned by comparison with the spectrum of **6**. The upfield position of the C-1 signal of the β-D-xylopyranose residue (101.0 ppm) corresponded to its attachment to O–3 of L-galactose residue (*cf.* Shashkov *et al.* 1988). It was evident from the spectrum that the latter residue is sulfated at C-6, and hence the alkaline stability of the oligosaccharide **5** may be considered as additional chemical evidence that position 3 of this L-galactose residue is xylosylated.

The high-molecular mass fraction (HMF) obtained after partial reductive hydrolysis differed considerably from the parent polysaccharide in its composition: it had no 3,6-anhydrogalactose and was enriched in xylose, methylated galactoses, and sulfate. According to the 3,6-anhydrogalactitol content, HMF was a mixture of reduced oligosaccharides with an average degree of polymerization of about 10. All the signals characteristic of the ^{13}C-NMR spectra of **4**–**6** were found in the spectrum of HMF together with methyl group resonances at 58.9 and 57.3 ppm belonging to 6-O-methyl- and 3-O-methyl-galactose residues. These mono-O-methyl-galactoses were also identified by GLC after complete acid hydrolysis of HMF. Alkaline treatment of HMF resulted in elimination of about one half of the sulfate present with concomitant formation of the corresponding amount of 3,6-

Table III. Several proton resonances in ^{1}H-NMR spectra of agarobiitol **1**, its 2^1-O-methyl derivative **2**, and acid oligosaccharides **3** and **5**.

Oligosaccharide	1		2		3		5	
Residue[a])	Gnr	Ar	Gnr	Ar	Gnr	Ar	Xyl	Anr
H-atom	Chemical shifts in ppm							
H1	4.560	3.65–3.78	4.543	3.83	4.758	–	4.665	5.325
H1	–	–	–	3.72	–	–	–	–
H2	3.510	3.93	3.48	3.55	4.250	4.100	3.400	4.115
H3	3.660	3.90	3.64	3.92	3.865	3.945	3.500	4.319
H4	3.96	4.33	3.90	4.28	4.010	4.325	3.658	4.560
H5ax	–	–	–	–	–	–	3.322	–
H5eq	–	–	–	–	–	–	4.003	–
H5	–	5.00	3.65–3.8	4.37	–	4.455	–	–
H6	–	3.99	3.65–3.8	3.94	–	4.025	–	–
Ht	–	3.86	3.65–3.8	3.84	–	3.870	–	–
	Spin-spin coupling constants, Hz							
1, 1	–	–	–	12.5	–	–	–	–
1, 2	7.7	–	8.0	4.5	8.0	–	7.5	4.0
1, 2	–	–	–	5.5	–	–	–	–
2, 3	10.0	7.0	10.0	6.5	10.0	4.0	9.0	11.0
3, 4	3.5	2.0	3.5	1.5	3.5	1.0	9.0	3.5
4, 5eq	–	–	–	–	–	–	1.5	–
4, 5	–	–	–	–	–	–	7.5	–
5ax, 5eq	0.5	4.2	1.0	4.0	1.0	4.5	–	1.0
5, 6	–	–	–	–	–	–	11.5	–
5, 6	–	4.5	–	4.5	–	4.5	–	–
6, 6	–	2.5	–	2.5	–	2.5	–	–
	–	10.0	–	12.5	–	10.0	–	–

[a]) Designation as in Table II.

anhydro-L-galactose residues. The ^{13}C-NMR spectrum of the modified fraction HMF-M contained two signals at 98.7 and 98.6 ppm corresponding to C-1 of 3,6-anhydro L galactose residues linked to different neighbours, namely, to β-D-galactopyranose 2-sulfate or to non-sulfated β-D-galactopyranose, respectively.

Summarizing all the evidence obtained, the conclusion may be inferred that the sulfated galactan from *Laurencia nipponica* has a very complex structure of the agar type. Its carbohydrate chain is a hybrid of agarose and porphyran with many sulfate, O-methyl groups, and β-D-xylopyranose residues as substituents. Most of sulfate groups occupy positions 2 of 3-O-linked β-D-galactopyranose residues. This structural feature gives rise to some downfield shift of the anomeric signal of the neighbouring 3,6-anhydro-α-L-galactopyranose residues resulting in characteristic complication of the anomeric spectral region. On the basis of the amount of products isolated after partial reductive hydrolysis, the proportion of different disaccharide units may be calculated for the parent galactan molecule giving the following results: per every 29 agarobiose units there are two 2^1-O-methylagarobiose, 43 agarobiose 2^2-sulfate, two single preagarobiose t^1-sulfate (as in **4**), one single xylosyl-preagarobiose 6^1-sulfate (as in **5**), three single preagarobiose t^1,2^2-disulfate (as in **6**), and 13 contiguous disaccharide units of the porphyran type as in HMF. Additional non-regularity is introduced in the polysaccharide molecule by the presence of some 6-O-methyl- and 3-O-methylgalactose residues.

Small amounts of xylose are often observed in red seaweed galactans, but data on the nature and exact position of these residues are scarce. In one of our earliest investigation of the structure of a oligosaccharide from *Laingia pacifica* Yamada evidence was obtained that xylose residues may occupy position 2 of 3,6-anhydro-L-galactose and position 6 of D-galactose residues (Kochetkov *et al.* 1973). Xylosyl-agarobiose having β-D-xylopyranosyl residue at position 4 of β-D-galactopyranose and O-mthyl group of position 2 of 3,6-anhydro-L-galactose residue was isolated by Hirase *et al.* (1982) from *Laurencia undulata* Yamada polysaccharide. In the present work, in spite of the rather low xylose content in the starting polysaccharide, pentasaccharide **5** was isolated and its structure was tentatively elucidated according to its alkaline stability and ^{13}C-NMR spectral data. The same position of xylosyl residues attached to the backbone at O−3 of 4-linked L-galactose 6-sulfate residues was found recently in the xylogalactan sulfate from *Chondria macrocarpa* Hooker *fil. et* Harvey (Furneaux and Stevenson 1990). Both *Laurencia* and *Chondria* belong to the family Rhodomelaceae, thus the finding of similarly arranged xylosyl branches in these agar-like polysaccharides is worthy of note. It is interesting that the presence of a xylose residue at position 3 of L-galactose 6-sulfate residues of porphyran-like polysaccharide fragments should prevent their transformation into agarose-like fragments during the galactan biosynthesis.

Conclusion

Our work demonstrates that partial reductive hydrolysis is very useful for structural analysis of complex agar-like polysaccharides. Practically complete splitting of 3,6-anhydro-L-galactose glycosidic bonds takes place under the hydrolysis conditions used, regardless of the position of other substitutes, giving rise to reduced oligosaccharides where sulfate groups and xylose residues are retained. This splitting is more extensive than enzymatic hydrolysis with β-agarase. The main hydrolysis products, namely, agarobiitol derivatives, as well as minor substituents tetrasaccharides of the porphyran type, having α-L-galactose 6-sulfate residues, and high-molecular fragments devoid of 3,6-anhydro-L-galactose may be easily isolated, affording important information on the structure and degree of regularity of the parent polysaccharide.

Acknowledgements

The authors wish to thank Dr A. S. Shashkov for his advice in NMR spectral analysis and Dr V. L. Sadovskaya for mass spectrometry experiments.

References

Altona, C. and C. A. G. Haasnoot. 1980. Prediction of anti and gauche vicinal proton-coupling constants in carbohydrates: a simple additivity rule for pyranose rings. *Org. Magn. Reson. 13*: 417−429.

Contreras, R. R., J. P. Kamerling, J. Breg and J. F. G. Vliegenthart. 1988. ^{1}H- and ^{13}C-n. m. r. spectroscopy of synthetic monosulfated methyl α-D-galactopyranosides. *Carbohydr. Res. 179*: 411−418.

Furneaux, R. H. and T. T. Stevenson. 1990. The xylogalactan sulfate from *Chondria macrocarpa* (Ceramiales, Rhodophyta). *Hydrobiologia 204/205:* 615−620.

Garegg, P. J., B. Lindberg, P. Konradsson and I. Kvarnström. 1988. Hydrolysis of glycosides under reducing conditions. *Carbohydr. Res. 176*: 145−148.

Hirase, S., K. Watanabe, R. Takano and J. Tamura. 1982. Structural features of the sulfated polysaccharide isolated from the red seaweed *Laurencia undulata. XIth Int. Carbohydr. Symp.*, Abstracts. Vancouver, Canada, p. III-12.

Kochetkov, N. K., A. I. Usov, L. I. Miroshnikova and O. S. Chizhov. 1973. Polysaccharides of algae. XII. Partial hydrolysis of polysaccharide from *Laingia pacifica* Yamada. *Zh. obshch. khim. 43*: 1832−1839 (in Russian).

Kovač, P., J. Hirsch, A. S. Shashkov, A. I. Usov and S. V. Yarotsky. 1980. ^{13}C-N. m. r. spectra of xylooligosaccharides and their application to the elucidation of xylan structures. *Carbohydr. Res. 85*: 177−185.

Lahaye, M., W. Yaphe and C. Rochas. 1985. ^{13}C-N. m. r.-spectral analysis of sulfated and desulfated polysaccharides of the agar type. *Carbohydr. Res. 143*: 240−245.

Lahaye, M., W. Yaphe, M. T. Phan Viet and C. Rochas. 1989. ^{13}C-N. M. R. spectroscopic investigation of methylated and charged agarose oligosaccharides and polysaccharides. *Carbohydr. Res. 190*: 249−265.

Lewis, B. A., F. Smith and A. M. Stephen. 1963. 2,5- and 3,6-anhydrosugars and their derivatives. *Methods Carbohydr. Chem. 2*: 172−188.

Morrice, L. M., M. W. McLean, W. F. Long and F. B. Williamson. 1983. Porphyran primary structure. An investigation using β-agarase I from *Pseudomonas atlantica* and ^{13}C-NMR spectroscopy. *Eur. J. Biochem. 133*: 673−684.

Patt, S. L. and J. N. Shoolery. 1982. Attached proton test for carbon-13 NMR. *J. Magn. Reson. 46*: 535−539.

Rochas, C., M. Lahaye, W. Yaphe and M. T. Phan Viet. 1986. ^{13}C-N. M. R.-spectroscopic investigation of agarose oligomers. *Carbohydr. Res. 148*: 199−207.

Shashkov, A. S., G. M. Lipkind, Yu. A. Knirel and N. K. Kochetkov. 1988. Stereochemical factors determining the effects of glycosylation on the carbon-13 chemical shifts in carbohydrates. *Magn. Reson. Chem. 26*: 735−747.

Stevenson, T. T. and R. H. Furneaux. 1991. Chemical methods for the analysis of sulphated galactans from red algae. *Carbohydr. Res. 210*: 277−298.

Suzuki, M. and K.-I. Harada. 1985. Matrix effects in liquid SIMS of naturally occurring oligosaccharides and structural characterization of aminoglycoside antibiotics. *In*: (A. L. Burlingame and N. Castagnoli, Jr., eds) *Mass Spectrometry in the Health and Life Sciences*. Elsevier Science Publishers B. V., Amsterdam. pp. 149−170.

Usov, A. I. 1984. NMR spectroscopy of red seaweed polysaccharides: agars, carrageenans, and xylans. *Bot. Mar. 27*: 189−202.

Usov, A. I. and M. Ya. Elashvili. 1989. Specific fragmentation of red algal galactans under reducing conditions. *F. E. C. S. Fifth Int. Conf. Chem. Biotechnol. Active Nat. Prod., Conf. Proc.* Bulgarian Acad. Sci., Sofia. Vol. 2, 346−350.

Usov, A. I. and M. Ya. Elashvili. 1991. Quantitative determination of 3,6-anhydrogalactose derivatives and partial fragmentation of the red algal galactans under reductive hydrolysis conditions. *Bioorg. Khim. 17*: 839−848 (in Russian).

Usov, A. I., E. G. Ivanova and M. Ya. Elashvili. 1989. Polysacharides of algae. 41. Characterization of water-soluble polysaccharides from several representatives of the genus *Laurencia* (Ceramiales, Rhodophyta). *Bioorg. Khim. 15*: 1259−1267 (in Russian).

Botanica Marina
Vol. 34, pp. 561–573, 1991

Studies on the Genera *Gyrosigma* and *Pleurosigma* (Bacillariophyceae)

The Typus Generis of *Pleurosigma*, some Presumed Varieties and Imitative Species

F. A. S. Sterrenburg

Westerstraat 47, 1655LC, Sijbekarspel, The Netherlands

(Accepted 7 October 1991)

Abstract

The taxon generally referred to as *Pleurosigma angulatum* (Quekett) W. Smith was studied with the light- and electron microscope in large populations of various origins and fossil materials. There is no morphological discontinuity between *Pleurosigma quadratum* W. Sm., *Pleurosigma angulatum* sensu W. Smith and *Pleurosigma aestuarii* (Bréb. *ex* Kütz.) W. Smith and the species description of *Pleurosigma angulatum* is extended to account for this continuum. Taxa identified in recent publications as *Pleurosigma angulatum* were found to correspond to *Pleurosigma rhombeum* (Grunow *in* Cleve *et* Grunow) H. Peragallo and a hitherto undescribed species, *Pleurosigma stidolphii* n. sp. The highly complex nomenclatural problems are discussed.

Introduction

Pleurosigma W. Smith is listed as a nomen genericum conservandum in the International Code of Botanical Nomenclature with *P. angulatum* (Quekett) W. Smith as its type. Discrepancies in the data given for *Navicula angulata* by Quekett (1848) and for *Pleurosigma angulatum* by W. Smith (1852) raise the question of whether these taxa are indeed identical. Quekett illustrated a slender lanceolate diatom unlike *Pleurosigma angulatum* sensu Smith, who stressed that the outline is always more or less angular and called Quekett's taxon 'var. B'. Modern publications still mention an angular outline as diagnostic for *Pleurosigma angulatum*, thus excluding Quekett's original taxon.

Is smoothly rounded or angular shape a character for this species or its varieties, and if so, which?

In Quekett's illustration the striae are drawn as 16.5 oblique and 17.5 transverse in 10 µm, too coarse to be representative for Smith's taxon. Quekett was the first to specify the angle of intersection of the oblique striae, which creates a complex situation: the stria angle ('relative to the sides') is given as '51 deg. in some specimens, nearly 60 deg. in others' and the illustrations show a stria intersection angle of 51 deg. Stria angle was found to be a stable parameter in this genus in Sterrenburg 1991 and the findings then indicated that 51 deg. falls outside the range of variation for *Pleurosigma angulatum* sensu W. Smith. Can a stria angle ranging from 51 to 60 deg. – as still given in more recent publications – apply to this species? If not, there are only two options:

1) If Quekett's data are considered valid, his *Nav. angulata* description included two taxa: '51 deg.' and '60 deg.', while his illustrations show a (non-angular, '51 deg.') taxon other than Smith's.

2) If Quekett's angular measurements are not considered valid and are to be disregarded in his diagnosis, the remaining data (stria fineness, outline) also differ from Smith's taxon.

Hendey (1964) states that the decussate stria pattern changes to perpendicular arrangement at the poles. This had never been described or drawn in the previous century of observations. Since then, specimens without this pattern have still been identified as *Pleurosigma angulatum* in publications. Data in Van der Werff and Huls (1957–1974) suggest that this feature is inconstant: '*Pleurosigma angulatum* mostly shows a uni- or bipolar perpendicular pattern'.

Can this species display non-perpendicular stria pattern at one or both poles?

There are evidently some long-standing questions for this typus generis and to clarify these, the goal was a comparative study of Quekett and Smith specimens, of wild populations of different geographic provenance and of taxa ranked as this species in recent publications or regarded as varieties (see Van Landingham 1968–1979).

Material and Methods

A search in the collections of the Royal College of Surgeons, the Natural History Museum and the Quekett Microscopical Club (all in London) failed to retrieve a Quekett type slide or sample. Quekett's slides in the collection of the RCS are labelled *'Pl. angulatum'* in his hand and thus postdate Smith's publication. The oldest slide found is BM 23671, W. Smith, other Smith slides examined were BM 23672 and BM 23666. For further data on Quekett's taxon see Sterrenburg 1990 b.

About 50 samples from British, Dutch and German coasts, the Arabian Gulf, New Zealand, Mediterranean, Caribbean, W. Spain, Patagonia and W. Africa plus Holocene samples from the collections of the Geological Survey of The Netherlands, were included for comparison. Slides of the following materials have been deposited in the Natural History Museum, London and in the collections of the Academy of Natural Sciences, Philadelphia, Pa., USA:

– Texel, Nature Reserve 'De Schorren', the Netherlands, 21 Sept. 1989, sample Sterrenburg #280, (BM 81609)

– Toko Mouth, South Island, New Zealand, 14 Jan. 1983, Stidolph collection SRS FW-B 146, (BM 81610)

– Otago Harbour, New Zealand, coll. A. J. Doig, Oct. 1961, Stidolph collection. (BM 81608)

– Bahia Fosforescente, Puerto Rico, 20-3-1976, leg. J. Nelson Navarro, (BM 81611)

Specimens were examined in naphrax and often in air with a variety of light microscopic equipment. For further methods and definitions see Sterrenburg 1991. All data given refer to personal findings, not to quotes from the literature unless this is specifically mentioned.

Observations

In Smith's slides and all other populations of *Pleurosigma angulatum* sensu Smith, the following features were found to be stable: colour in darkfield (mounted in naphrax and in 'dry mount', where the colours are identical although their intensity varies), type of axial costa path, shape and orientation of central and terminal areas. The raphe angle varies from +8 to +13 deg., for the stria angle the range is from 57 to 62 deg., never approaching 51 deg. In all populations seen, including Smith's slides, a bipolar terminal perpendicular pattern was consistently present.

Stidolph (1980) reported that *Pleurosigma angulatum* populations in the Southern hemisphere did not show the terminal perpendicular pattern, but amended this subsequently (Stidolph 1988). Two New Zealand samples supplied by that author showed the following:

– Toko Mouth: This contains several *Peurosigma* species; one of these is compatible with *Pleurosigma angulatum* sensu Smith including the terminal perpendicular stria pattern.

– Otago Harbour: this contains the taxon referred to as *Pleurosigma angulatum* in Stidolph 1980 (pl. 12, fig. 4). Although it shows the angular outline usually associated with *Pleurosigma angulatum*, the LM revealed major differences from this and other described species. The taxon is dexcribed as *Pleurosigma stidolphii* n. sp. under 'Species descriptions' (Figs 38–46).

Navarro (1982) also identified a taxon as *Pleurosigma angulatum* but the illustrations indicated the absence of a terminal perpendicular pattern and additional differences. Examination of the material (Bahia Fosforescente) showed that the taxon was incompatible with *Pleurosigma angulatum*, but matched *Pleurosigma rhombeum* Grun., examined in isotype slides; see under 'Species Descriptions' (Figs 35–37, 47–56).

Smoothly rounded forms can be seen in any population, sometimes as the dominant form; they are also present in Smith's slides of *Pleurosigma quadratum* (BM 23666) and of *Pleurosigma aestuarii.*

Presumed varieties

Pleurosigma angulatum and *Pleurosigma aestuarii* have been ranked as varieties of *Pleurosigma angulatum* in some publications, as individual species in others. The Texel sample (#280), a rich gathering of all three, with specimens fully conforming to those in Smith's BM slides for each, also showed a profusion of intermediates. The findings were as follow:

- there is a continuum in size range, already evident from the data given by Smith
- there is a continuum in outline; the 'pinched' apices regarded as a feature of *aestuarii* can equally occur in specimens conforming to *angulatum* as regards size (Figs 5, 8)

- the width-length ratio in a single population can range from about 1:2.5 (elliptic or rhombic) to over 1:8 (linear lanceolate) without associated differences
- the degree of silicification of the valves was seen to decrease in parallel with the shift from *quadratum* to *aestuarii.* While the color (in the sense of hue) in darkfield remains stable, there is a corresponding very gradual shift in its intensity: from intense deep blue (*quadratum*) via deep blue (*angulatum*) to faint deep blue (*aestuarii*)
- one feature is constant in all three: a bipolar perpendicular terminal stria pattern, which is a very rare feature in the genus as a whole
- stria angle and fineness, raphe angle and axial costa path show no discontinuity for these three taxa
- in agreement with earlier studies (Sterrenburg 1990 a, 1991) specimens that are otherwise identical (size, outline, striation, raphe and stria angle, axial costa path) show marked differences in the shape of the central raphe fissures (Figs 18, 20–22)

A selection of forms from a single slide of this sample is shown in Figures 1–22. Additional similarly flourishing populations from this and other North Sea coast locations yielded identical findings. The database amounts to hundreds of valves.

SEM findings

Like the light microscope, the scanning electron microscope (SEM) does not furnish a criterion for separating these taxa. The findings are illustrated for typical *Pleurosigma quadratum*, *angulatum* and *aestuarii* specimens on a single stub in Figures 23–34.

In all three taxa the central raphe fissures are variable, illustrated for *Pleurosigma angulatum* specimens in Figures 19, 23, 24. No specific difference could be established for the three taxa in question.

The abundance of the organism in the Texel population permitted a study on the variability of the structures called 'central bars' (Schrader 1973). Specimens conforming to *Pleurosigma aestuarii* on a single stub were selected for identical outline, size and stria fineness. The 'central bars' constituted the single most variable structure in these specimens and Figures 32–34 show some of the many forms observed. No specific difference in this respect could be established for the three taxa in question.

Discussion

An examination of previous descriptions of *Pleurosigma aestuarii*, *Pleurosigma angulatum* and *Pleurosigma quadratum* reveals that differentiation of these taxa has been based on size and outline. The observations on massive living populations showed a continuum in both respects. Large (presumably *Pleurosigma quadratum*) individuals may be smoothly rounded, while small (presumably *Pleurosigma aestuarii*) individuals can be angular, with all possible intermediates. Reference to angular outline as a criterion, also in recent publications (*e.g.* Krammer and Lange-Bertalot 1986) may have been due to the fact that W. Smith specified angular outline for a species that is described by the epithet 'angulatum'. In fact, the majority of specimens in Smith's slides – as well as in other populations examined – has a non-angular outline. The non-angular outline in Quekett's figures and the total absence of any reference in the text of Quekett 1848 to 'angle' other than the angle of intersection of the striae indicate that the epithet was chosen for the stria pattern (which was a new discovery then), not the valve shape. Undulate valves have been seen in several populations for both *angulatum* and *aestuarii* forms (Figs 5, 9, 10, 12) and the individuality of *Pleurosigma angulatum var. undulatum* Grunow *in* Cleve *et* Grunow is rejected. Hendey (1964) concluded that outline is an artificial criterion for distinction of *quadratum* as a variety. Neither is it specific for *aestuarii*: the 'pinched' apices also occur in presumably *angulatum* specimens (Figs 5, 8).

It has been pointed out (Sterrenburg 1991) that traditionally emphasized criteria such as outline need not *per se* distinguish individual taxa. The features described there as reliable LM criteria (*e.g.* stria angle; raphe angle and path of the axial costa; apical details) show no discontinuity for the taxa in question either. Studies with the SEM did not reveal other interspecific differences. The central bars were found to be highly variable in several of the 'states' defined and discussed in Cardinal *et al.* (1989). By themselves, they need not be more reliable taxonomic criteria than some others in use. When the total complex of morphological features of the central nodule is taken into account, broadly different definitions can sometimes emerge permitting distinction of species, as shown here for *Pleurosigma rhombeum* and *Pleurosigma stidolphii.* Cardinal *et al.* (1989) refer to the illustration of *Pleurosigma angulatum* in Navarro (1982), but the taxon in question was *Pleurosigma rhombeum*, as demonstrated in the present paper.

Thus, no morphological criteria could be found for differentiation between *Pleurosigma quadratum*, *Pleurosigma angulatum* and *Pleurosigma aestuarii*. Observations indicate that the 'Schorren' habitat supports flourishing populations throughout the year and these contain all three taxa plus intermediates. I have observed similar conditions in samples from other locations on the Dutch and British coast and they also apply to Smith's slides. The findings would all be compatible with the vegetative cycle of a single organism; *quadratum* as early stage, *angulatum* representing mid-range and *aestuarii* as late stage.

Some samples may be almost pure *aestuarii*. While this may have suggested the taxon's individuality, it does not necessarily imply it. The phenomenon would again be compatible with the hypothesis of a biological continuum between *quadratum*, *angulatum* and *aestuarii*: it can be interpreted as an end-stage of an old population where zygote production has not occurred. This is supported by the fact that other gatherings may yield an abundance of large (and thus presumably early) forms: sampling took place shortly after a phase of zygote production. Thus interpreted, the 'Schorren' population (where the vast majority of individuals was observed to be alive) is about intermediate: while there is a good stock of large forms, medium-sized ones are more frequent and the smaller (*aestuarii*) forms are the most abundant.

The species description of *Pleurosigma angulatum* must be extended to include three taxa hitherto considered individual entities, see under 'Species Descriptions'. Terminal bipolar perpendicular stria pattern is considered specific as no exception has been found. The width/length ratio found for *Pleurosigma angulatum* in Hendey 1964 was 1:3.3 to 1:4.7, a value greatly extended by the present findings.

As regards the 'pinching' seen in the *aestuarii* (and *angulatum*) forms: the valves of the *angulatum* forms that show this are always of a slender type with narrow apices. It is proposed that the somewhat rostrate apices are not caused by constriction (pinching) of the valve outline, but by the opposite. The width of the valve at the site where 'pinching' occurs is fairly constant at circa 6 μm. This is slightly more than the lateral dimension of the termination of the axial rib, which consists of the solid silica plate carrying the helictoglossa on the inside and the terminal raphe fissure on the outside. The axial rib is the first structure to become silicified on morphogenesis of the valve (see Mann 1982). If there would be a minimum size for the helictoglossa and the terminal raphe fissure, the valve cannot become narrower than a slightly greater minimum value (*ca* 6 μm) in order to accommodate them. Instead of being 'pinched', the margins cannot further converge towards the apex but have to run parallel — which is indeed the aspect presented. 'Pinching' has not been seen for the broader forms with wider apices.

The question of the possible nomenclatural consequences of the findings presented here raises formidable problems — which the present paper does not intend to tackle. Should such an attempt be made, the rules of the ICBN would have to be applied to the following situation:

— no original material has been recovered of *Navicula angulata* Quekett, the basionym of *Pleurosigma angulatum* (Quekett) W. Smith and listed in the ICBN as the type of the genus *Pleurosigma* (nom. cons.);

— the protologue was based on two species, only one of which was illustrated. Taxa with an intersection angle near 50 deg. have indeed been described from the North Sea, *e. g. Pleurosigma latiusculum* Peragallo;

— the identity of the species illustrated in the protologue is uncertain but it certainly was not *Pleurosigma angulatum* sensu W. Smith and subsequent authors, while the non-illustrated species may have been;

— the species definition of *Pleurosigma angulatum* sensu W. Smith must be extended to include *Pleurosigma aestuarii* as well as *Pleurosigma quadratum*.

Despite its early date and the fact that Quekett's book is not a study on the diatoms at all, its illustrations of diatoms were far ahead of anything produced on the Continent — even much later. British microscope optics were superior then. Also, the angular measurements mentioned by Quekett were quite feasible: in his time the larger British microscopes featured accurately graduated rotating stages. I have used British instruments of that period for angular measurements and accuracy was consistently better than 1 deg. overall. In other words, Quekett's data are not to be taken lightly.

Figs 1–14. *Pleurosigma angulatum* from the reference population (sample Sterrenburg #280). Bar = 10 μm.
Figs 1–4. 'Quadratum' forms, angular and more rounded. Figs 5–9. 'Angulatum' forms. Note pinching in figs 5 and 8. Undulate margins in 5 and 9. Figs 10–14. 'Aestuarii' forms. Note undulate margins in 10 and 12.

Figs 1–14, 35, 38–40, 47–48: × 500; other LM figs × 1500.

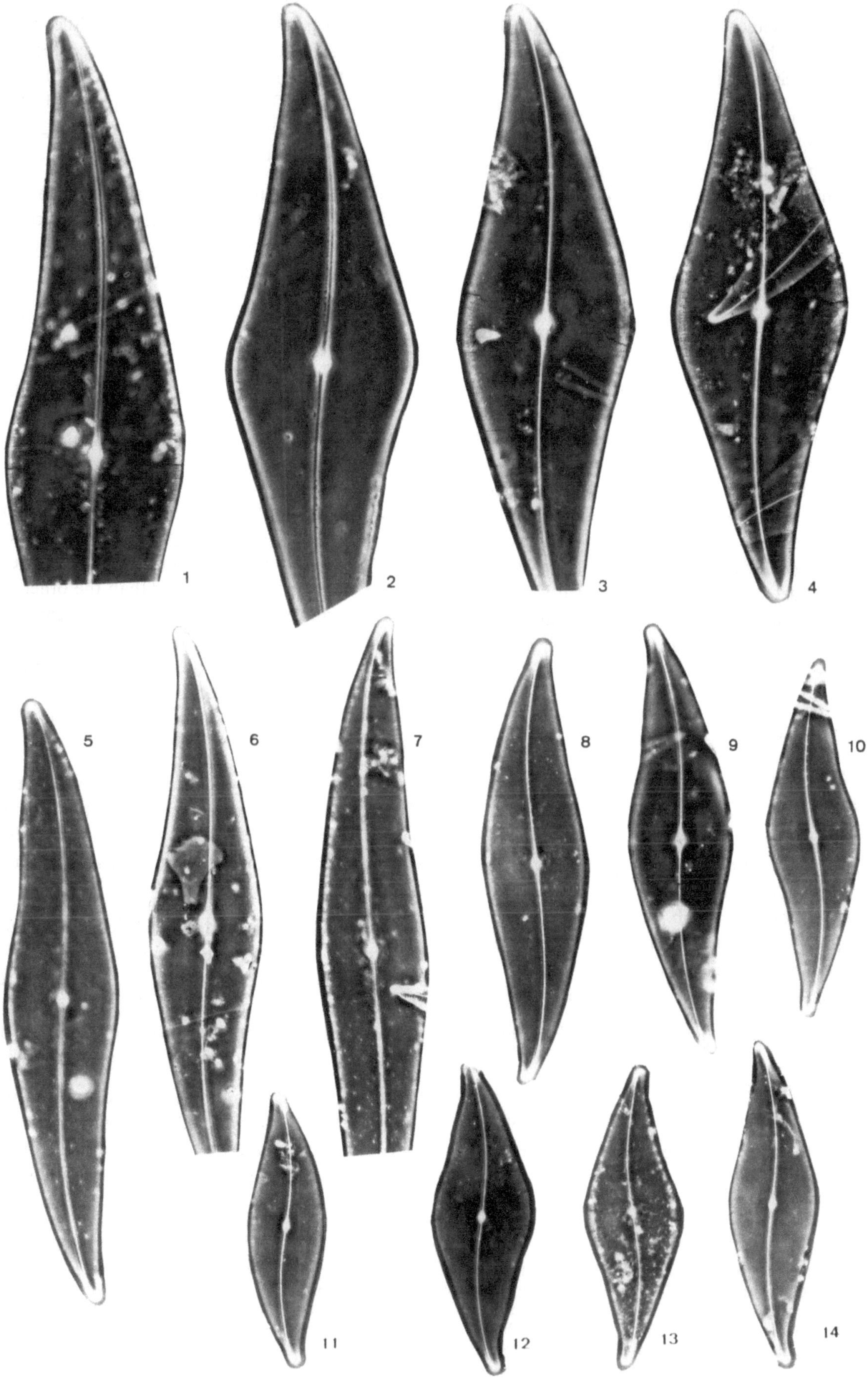

Bars for all LM figures equal 10 µm, SEM figures: scales indicated.

The striae intersection angle for *Pleurosigma angulatum* given by Reimer (in Patrick and Reimer 1966) (51 to 60 deg.) should not be regarded as more than a quote of Quekett's data (Reimer, pers. comm.).

Although Peragallo (1890–1891) stated that *Pleurosigma angulatum* is 'trop connu pour avoir besoin d'une description', this was optimistic. Species superficially similar to *Pleurosigma angulatum* will all require re-examination on the basis of their respective type material. A future study will specifically focus on some imitative taxa not showing terminal perpendicular pattern.

Pleurosigma angulatum has been mentioned in floristic lists from the tropics. However, in large numbers of tropical samples, *Pleurosigma angulatum* conforming to Smith's slides has not been observed by Hendey (pers. comm.), van den Heuvel (pers. comm.) or by the present author. For the Northern Hemisphere, I cannot report a more southerly location than the coast of the Iberian peninsula. For the Southern Hemisphere I have seen the organism in samples from Argentina (leg. Martha Ferrario) as well as New Zealand (leg. Stuart R. Stidolph). An appeal is made to colleagues for samples considered to contain this organism from the tropics and additional Southern Hemisphere locations.

Species Descriptions

Pleurosigma angulatum sensu W. Smith, emend. Sterrenburg Figs 2–34

Synonyms: *Navicula angulata* Quekett *pro parte quoad typum*, *Navicula aestuarii* Bréb. *ex*. Kütz., *Pleurosigma aestuarii* (Bréb. *ex* Kütz.) W. Smith, *Pleurosigma angulatum* var. *aestuarii* (Bréb. *ex* Kütz.) Van Heurck, *Pleurosigma quadratum* W. Smith, *Pleurosigma angulatum* var. *quadratum* (W. Smith) Van Heurck, *Pleurosigma angulatum* var. *undulatum* Grun.

Slides examined: see under 'Material and Methods'.

Extended species description

Valves sigmoid, from linear-lanceolate to elliptic/rhombic, with acute to obtuse ends. The valvar outline consists of convex and more or less concave segments, the latter can tend to convex in the smallest forms. Larger forms slightly vaulted in the middle, more so at the ends, smaller valves tend to become flatter centrally, still vaulted at the ends, frequently with a 'pinched' aspect. Length 80–300 μm, width 26–65 μm for the full range of forms; width-length ratio from about 1:2.5 to over 1:8.

Colour: yellowish to ochre in brightfield in air. In darkfield (resin and 'dry' mounts) intense deep blue ('quadratum') gradually shifting to deep blue ('angulatum') and faint deep blue ('aestuarii').

Axial costa: single curvature, raphe angle +8 to +13 deg., almost median and straight for the central fifth of its length, gradually curving and becoming moderately eccentric until at the distal end it suddenly approaches the periphery of the valve outline. The 'quadratum' form generally has longer, the 'aestuarii' form shorter straight sections.

The central raphe fissures – not always discernible in the LM – are mostly oppositely deflected to the concavity of their axial costa branch. One fissure may be more deflected and/or longer than the other. They are very variable, however, and deflection can also be homolateral (Figs 18, 22, 23) without associated differences. This is yet another example of the great variation in raphe fissure deflection pattern described for several other *Gyrosigma* and *Pleurosigma* spp. (Sterrenburg 1990 a, 1991).

Terminal areas sideways deflected, dilated in the form of narrow funnels. Axial area narrow, but raphe distinctly visible. Central areas small, oval to slightly rhombic, often slightly asymmetric, its longitudinal axis not rotated.

Oblique striae crossing at 57–62 deg., terminally in 2 perpendicular systems. Transverse *ca* 18–22, oblique striae *ca* 17–21 in 10 μm, terminally equally fine for the transverse, *ca* 20–24 in 10 μm for the longitudinal striae. Smaller forms tend to have finer and at any rate less contrasted striae.

The 'central bars' (Schrader 1973) are highly variable; a general description is as follows: symmetric or asymmetric, crescent-shaped; as more silica is deposited at this location it may spread sideways, making the bars

Figs 15–28. *Pleurosigma angulatum* from the reference population (sample Sterrenburg #280). Bar = 10 μm.
Fig. 15. Apex of tpyical 'aestuarii' form. Fig. 16. Apex of typical 'angulatum' form. Fig. 17. Apex of typical 'quadratum' form. Figs 18–22. Central raphe fissures of typical 'angulatum' forms, schematic diagrams drawn to the left of the photomicrographs. Figs 19, 20 and 21 show apparently most common shapes, 18 and 22 show examples of homolateral deflection. Fig. 23. An example of homolateral deflection in a 'quadratum' form, where the shape of 19 also appears to be the most usual. Fig. 24. Apparently most usual shape in 'aestuarii' forms, corresponding to 19. Other shapes also occur. Fig. 25. Apex of 'angulatum' form, outside. For 'quadratum' and 'aestuarii' forms the aspect is identical. Figs 26–28. Apex inside; 26: 'angulatum', 27: 'quadratum', 28: 'aestuarii'. No visible differences.

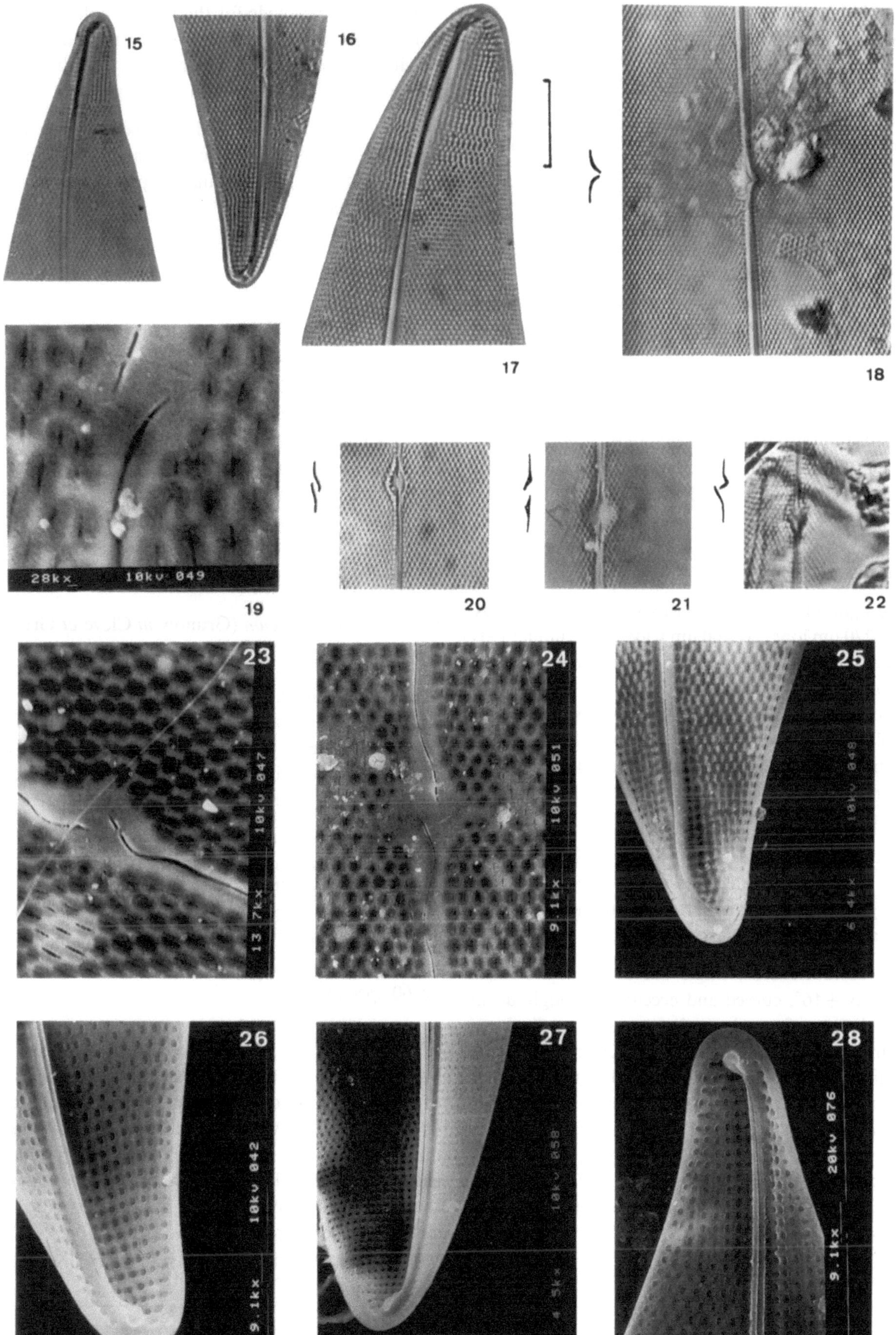
15
16
17
18
19
20
21
22
23
24
25
26
27
28
28kx 10kv 049
13 7kx 10kv 047
9 1kx 10kv 051
9 1kx 10kv 042
9 1kx 20kv 076

somewhat angular or broadening their outline by irregular lateral extensions.

Typical habitat: marine litoral mudflats.

Pleurosigma stidolphii Sterrenburg n. sp. Figs 38–46

Diagnosis

Valva sigmoidea, lanceolata ad rhombica apicibus acutis 100–270 μm longa, 30–60 μm lata. Raphe monocurvata, angulo +12° ad +16°. Area centralis parva, rotunda, symmetrica, non rotata. Striae transversae 23–27, obliquae 22–25 in 10 μm, in angulo 65° ad 70°. In aquis marinis.

Materials examined: Otago Harbour, coll. A. J. Doig, Oct. 1961; additional New Zealand populations. Holotype in author's collection, syntypes deposited in the collections at London (BM 81608) and Philadelphia Pa., USA.

Description

Valve sigmoid, lanceolate to rhombic with acute ends, 100–270 μm long, 30–60 μm wide. Valve moderately to markedly vaulted, raphe situated on a ridge. At the ends the valve slopes very gently to the concave side of the axial costa, to the convex side it drops steeply. Vaulting decreases in later vegetative stages, but the raphe remains elevated. Valvar outline consists of concave and concave to straight (tending to convex in later stages) segments.

Colour in air pale yellowish. In darkfield (resin) the axial area is highlighted, the valve is deep blue with silvery blue patches, the latter gradually diminish in later stages.

Axial costa elevated, single curvature, raphe angle +12° to +16°, curved and eccentric throughout, at some distance from the ends becoming so displaced towards its convexity that it is almost peripheral. Terminal fissures sideways deflected; central fissures long and projecting far into the central area. They approach each other so closely that often the ends cannot be resolved in the LM. Opposite deflection to the concavity of the corresponding half of the axial costa appears to be the usual form, but the shape shows marked variation.

Axial area rather wide for the genus, raphe distinctly visible. Central area small, oval to irregularly rhombic, its longitudinal axis not rotated. Terminal areas narrow, slightly dilated on the concavity.

Oblique stria angle 65–70 deg. ('Speciosi' sensu Peragallo 1890–1891). Transverse striae 23–27, oblique 22–25 in 10 μm. No terminal perpendicular pattern. Marine/estuarine species.

This description is based on 50 valves in the type and other populations.

Taxonomic comments

This species differs from *Pleurosigma angulatum* in raphe angle and path of axial costa, stria angle, vaulting and absence of terminal perpendicular pattern. The decrease in vaulting and gradual shift in the intensity of colour in darkfield for later vegetative stages are analogous to the findings in *Pleurosigma angulatum*. Despite some similarity, the morphology is essentially different from *Pleurosigma rhombeum* Grun., see below.

Pleurosigma rhombeum (Grunow *in* Cleve *et* Grunow 1880) H. Peragallo Figs 35–37, 47–56

Material examined: isotype slide 'Apia, Samoa' in the Kinker collection; slide Samoa BM 55261; Bahia Fosforescente, Puerto Rico (leg. Navarro, BM 81611), additional personally collected samples from the Caribbean and W. coast of Sulawesi (Indonesia).

As both *Pleurosigma stidolphii* and the taxon described as *Pleurosigma angulatum* in Navarro (1982) showed some similarity to the data for *Pleurosigma rhombeum* Grun. as given in Peragallo 1890–1891, a study was made of the latter species. This permitted an extended and corrected species description based on 60 specimens.

Description

Valve sigmoid, rhombic to broad lanceolate consisting of convex and concave segments, 150–280 μm long, 45–67 μm wide. Valve vaulted and with a longitudinal fold on the concavity of the axial rib, raphe elevated, valve falls off steeply on the convexity of the axial costa. Colour ochre in air, deep blue with silvery patches in resin and darkfield.

Figs 29–34. *Pleurosigma angulatum* from the reference population (sample Sterrenburg #280). bar = 10 μm.
Fig. 29. Central bars in 'quadratum'. Other forms also present, see 31–34. Fig. 30. Central bars in 'angulatum'. Other forms also present, see 31–34. Figs 31–34. Some of the shapes of the central bars in specimens conforming to 'aestuarii' and selected for identical size, outline and striation; 31: symmetric, 32: asymmetric, 33: angular, 34: irregular. Similar shapes are seen in 'quadratum' and 'angulatum' forms.

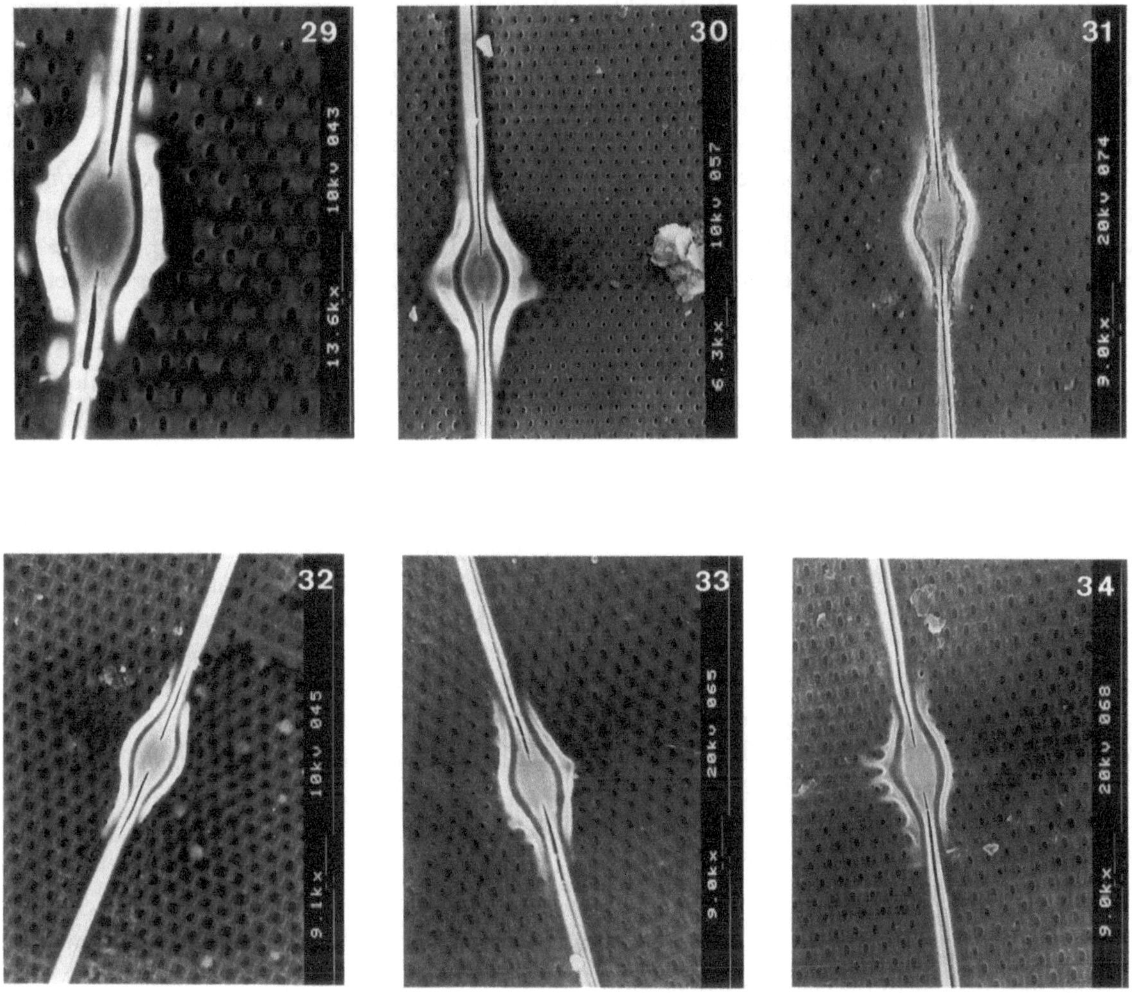

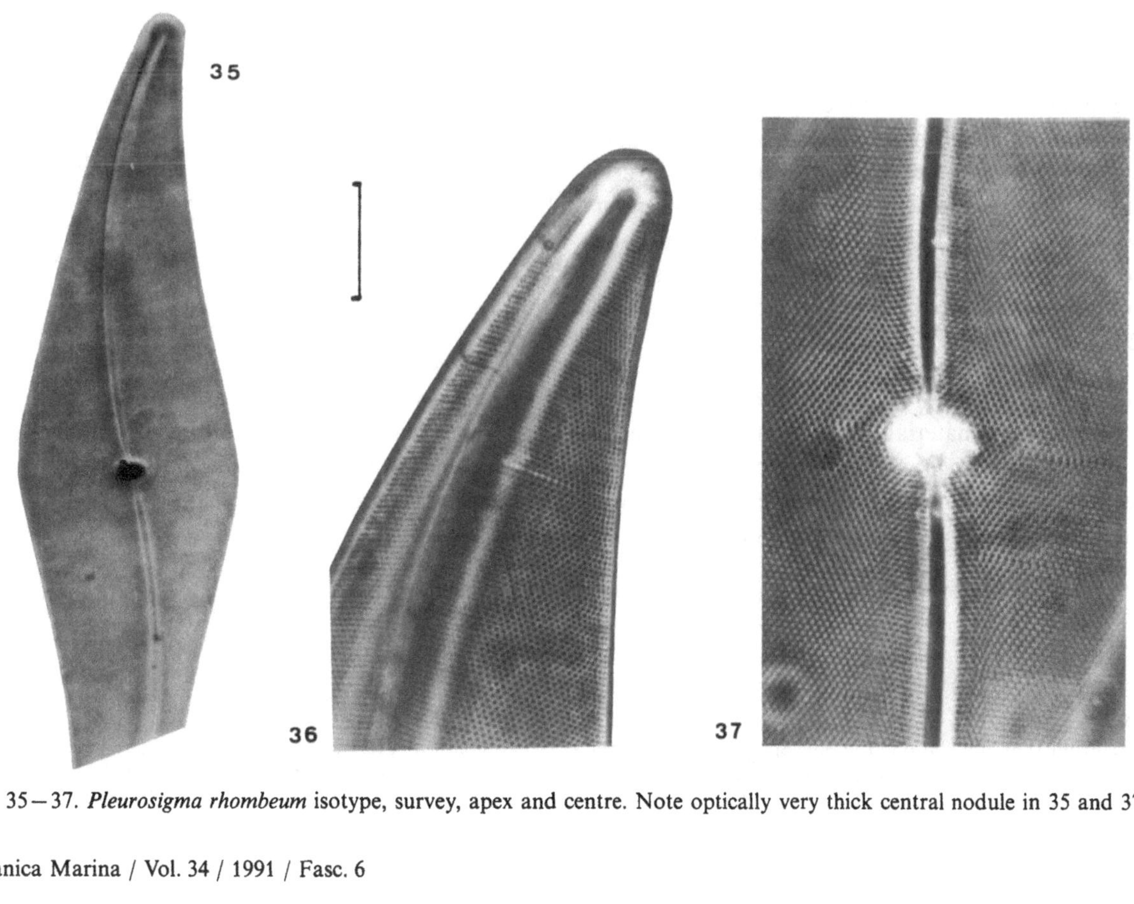

Figs 35–37. *Pleurosigma rhombeum* isotype, survey, apex and centre. Note optically very thick central nodule in 35 and 37.

Axial costa curved throughout, markedly displaced to the convexity towards the ends, raphe angle +18 to +22 deg. Axial area wide, raphe clearly visible, terminal fissures long and sideways curved. Central fissures clear, terminations droplet shaped, deflection often heterolateral to the concavity of the corresponding half of the axial costa but unstable. Central area transversely elliptic, sometimes slightly asymmetric, its longitudinal axis is not rotated. Terminal areas not dilated, practically symmetric.

Stria angle 66–74 deg. ('Speciosi' sensu Peragallo 1890–1891), transverse somewhat curved, 19–22 in 10 μm, oblique 17–20 in 10 μm. No terminal perpendicular pattern. In the LM, the internal central nodule presents as a marked laterally extended thickening of the valve.

Taxonomic comments

In some respects *e.g.* stria angle, raphe angle, outline and colour, this species resembles *Pleurosigma stidolphii*, which was initially equated to it. The morphological difference that excludes conspecificity and is clearly visible in the LM (figs 35, 37) is that *Pleurosigma rhombeum* has a thick internal central nodule as also seen in *Pleurosigma normanii*, for instance (Sterrenburg 1990 a). This is absent in *Pleurosigma stidolphii* and is reflected in a wholly different internal and external morphology of the central valve portion. Also, *Pleurosigma stidolphii* does not have the droplet-shaped external central raphe fissure ends.

Neither in the BM slide nor in the Kinker collection isotype slide was there a single specimen with the coarse striation given in Peragallo 1890–1891 and Cleve (1894) – and specifically mentioned as a criterion for differentiation from *Pleurosigma angulatum* in the latter. Foged (1975) identified a specimen as *Pleurosigma rhombeum* and specified 16 transverse and oblique striae, but the illustration shows *ca* 18–20 transverse and oblique striae.

Serious confusion may result from the fact that the Apia, Samoa type material contains another superficially similar *Pleurosigma* sp.! In principle this could create exactly the same sort of problems as the two different species contained in Quekett's protologue of *Navicula angulata*, especially as the protologue of *Pleurosigma rhombeum* is not sufficiently precise to separate the two. I have not been able to definitely equate this second taxon to a known species on the basis of type studies and will leave it as 'sp. indet.' but it differs specifically from *Pleurosigma rhombeum* in having a small round central area without thickening of the internal central nodule, absence of longitudinal fold in the valve and of the droplet-shaped central raphe fissure ends, much less eccentric axial costa with smaller raphe angle and shorter and less deflected apical raphe fissures. The frustule outline is also more smoothly rounded and broadly elliptical, but whether this permits distinction is dubious. It should be stressed that this species has indeed been confused with *Pleurosigma rhombeum* in practice, as is evidenced by the fact that one of the 4 specimens in the Kinker collection isotype slide (a selected mount) is not *Pleurosigma rhombeum*, but the other species! As it is much less rhombic than the other specimens, it is less representative of Grunow's description.

The taxon identified as *Pleurosigma angulatum* in Navarro 1982 matches *Pleurosigma rhombeum* in all respects, especially as regards raphe and stria angle, thickened internal central nodule, droplet-shaped central raphe fissure ends and longitudinal fold in the distal valve. Only a small number of valves is present in the samples seen and these give the impression of being rather early forms (robustly silicified, somewhat blunt apices like *'Pleurosigma quadratum'*). The Samoa type population does not present a wide morphological range either: the size range is fairly small and the aspect of the valves is rather monotonous. The small difference in apical width between the Caribbean and Samoan specimens is fully comparable to the variation illustrated in the present paper for *Pleurosigma angulatum*. The Caribbean specimens are therefore presented as illustrative of the SEM aspect of *Pleurosigma rhombeum* (Figs 47–45). The LM findings (*e.g.* raphe fissures, internal central nodule) are fully confirmed. Another difference between *Pleurosigma rhombeum* and *Pleurosigma stidolphii* is that the former does not have the bisected areolae (obex areolae, Stidolph 1988) of the latter, but whether bisected areolae are a stable character in this genus requires further study.

Figs 38–46. *Pleurosigma stidolphii* Sterrenburg.
Figs 38. From type population. Fig. 40. Specimen from other N. Z. population. Fig. 41. Apex, note absence of perpendicular pattern. Fig. 42. Centre and apparently usual raphe fissures. Fig. 43. Centre inside, note absence of inner elevated central nodule and bisected 'obex areolae'. Fig. 44. Centre outside, somewhat different from 42. Fig. 45. Apex inside. Fig. 46. Apex outside.

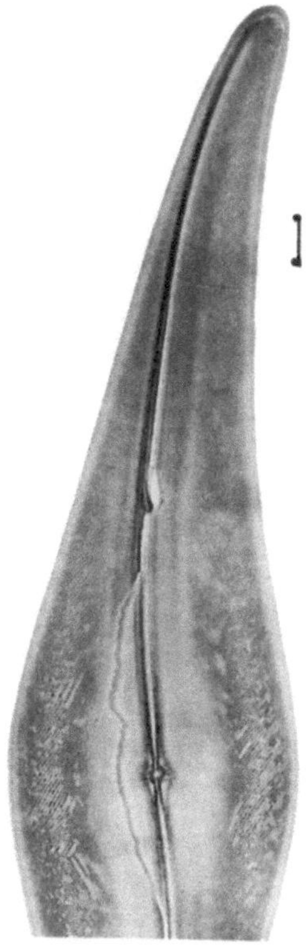

38

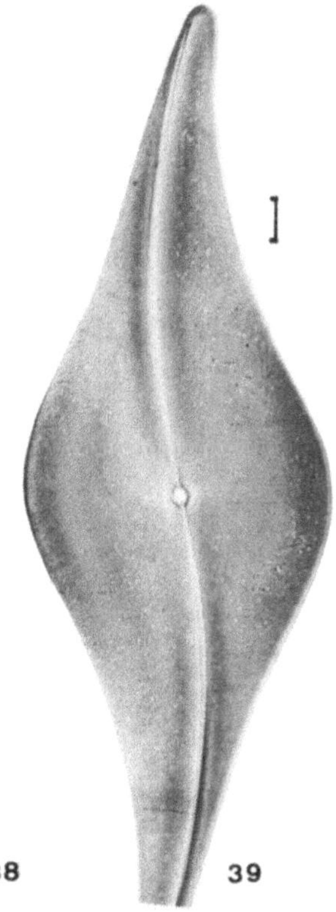

39

40

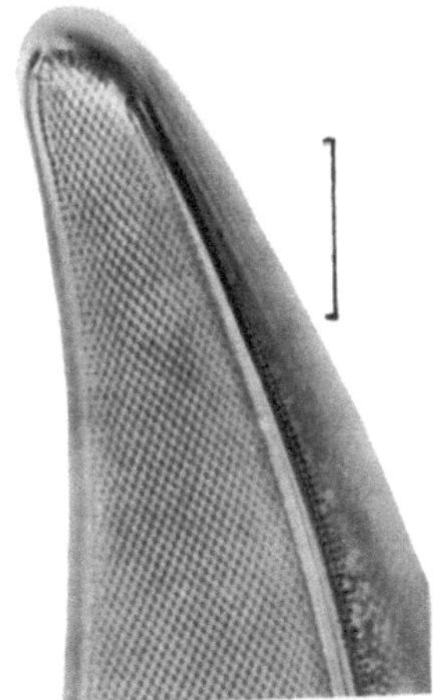

41

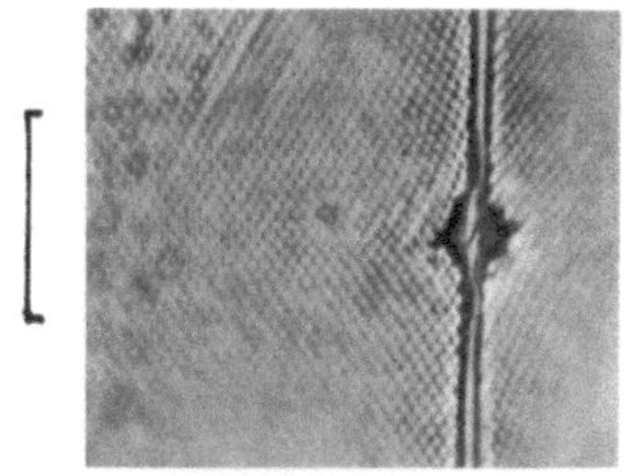

42

43
4µM 20KV 66 032 S

44
4µM 20KV 67 010 S

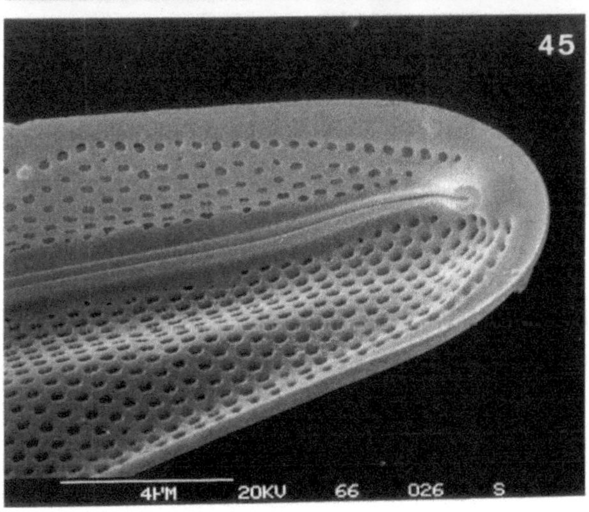
45
4µM 20KV 66 026 S

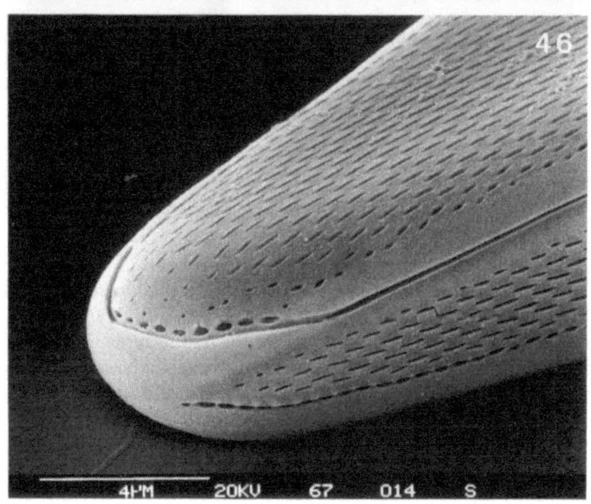
46
4µM 20KV 67 014 S

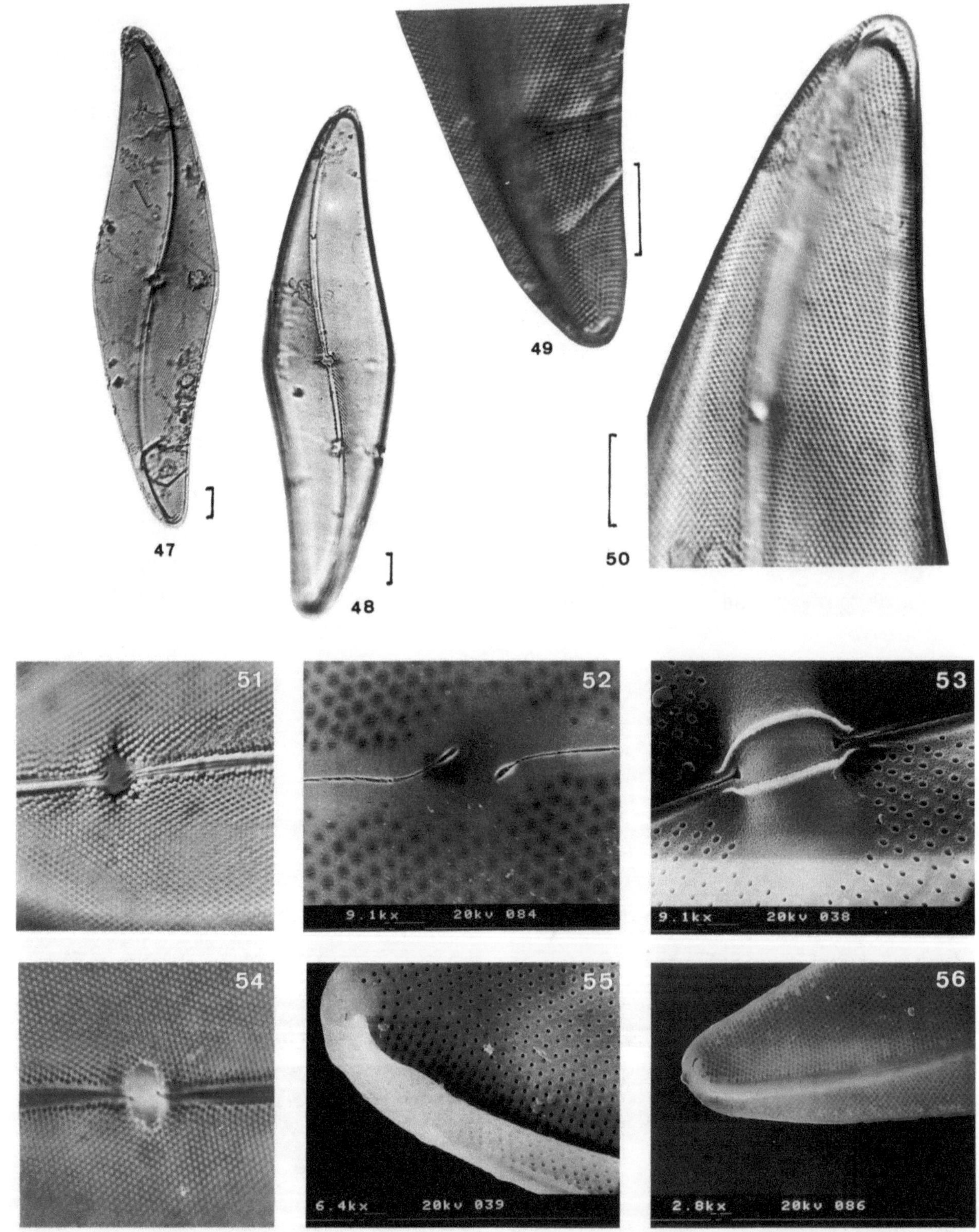

Figs 47–56. *Pleurosigma rhombeum*, Caribbean.
Fig. 47, 48. Representative specimens. Note thick central nodules (shadows). Fig. 49. Apex, inside. Note type of vaulting ('fold'). Fig. 50. Apex, outside. Note absence of perpendicular pattern. Fig. 51. Centre, outside. Focus to show thick central area (lateral shadowing). Fig. 52. Centre, outside. Raphe fissures as in 54. Fig. 53. Centre, inside. Note elevated central nodule and triangular raphe terminations. Fig. 54. Centre, outside. Focused on raphe fissure ends to show droplet-shaped termination. (Isotype slide). Fig. 55. Apex, inside. Compare with 49 for vaulting. Fig. 56. Apex, outside, compare with 50.

Acknowledgements

I wish to acknowledge the valuable help of:

Miss E. Allen and Mrs M. Gibbs of the Hunterian Museum, the Royal College of Surgeons of England (London) for access to and assistance with the Quekett collection, Mr H. S. Henderson and Mr C. G. Hussey (Natural History Museum) for efforts in a further search for Quekett's slides; Patricia Sims, David Williams and Barrie Paddock, Natural History Museum, for assistance during examination of types; Theo Lissenberg and Hein de Wolf, Geological Survey of the Netherlands, Haarlem, for help with SEM studies and fossil materials; Dr André Cardinal, GIROQ, Quebec, for discussions on our respective findings; Dr Martha Ferrario, Dr J. Nelson Navarro and Hans van den Heuvel for interesting samples; Norman Ingram Hendey for data on his observations; and Mr K. Boot, supervisor of De Schorren Nature Reserve, for his permission to collect in this protected area. For the SEM photomicrographs of Pleurosigma stidolphii I am indebted to D. Hopcroft and R. Bennett (DSIR Electron Microscope Laboratory, Palmerston North, New Zealand). Stuart R. Stidolph is owed a debt for valiant support of my efforts in relation to these organisms by the exchange of samples, views and data.

The editor and referees are thanked for their valuable suggestions.

References

Cardinal, A., M. Poulin and L. Berard-Therriault. 1989. New criteria for species characterization in the genera Donkinia, Gyrosigma and Pleurosigma (Naviculaceae, Bacillariophyceae). *Phycologia 28*: 15–27.

Cleve, P. T. and A. Grunow. 1880. Beiträge zur Kenntniss der Arctischen Diatomeen. *Kongliga Svenska Vetenskabs-Akademiens Handlingar 17*, *2*: 1–121.

Cleve, P. T. 1894. Synopsis of the naviculoid diatoms, I. *Kongliga Svenska Vetenskaps-Akademiens Handlingar*, n. f. 26, 2: 1–194.

Hendey, N. I. 1964. *An Introductory Account of the Smaller Algae of British Coastal Waters. V. Bacillariophyceae (Diatoms)*, pp. 317, pl. 45, London.

Mann, D. G. 1982. An ontogenetic approach to diatom systematics. (D. G. Mann, ed.) *7th Int. Diatom Symposium*, Koenigstein. pp. 113–141.

Navarro, J. N. 1982. A survey of the marine diatoms of Puerto Rico. V. Suborder Raphidineae: families Achnanthaceae and Naviculaceae (excluding *Navicula* and *Mastogloia*). *Bot. Mar. 25*: 321–338.

Patrick, R. and Ch. Reimer. 1966. The diatoms of the United States, exclusive of Alaska and Hawaii. 1. *Monograph of the Academy of Natural Sciences of Philadelphia 13*, p. 688.

Peragallo, H. 1890–1891. Monographie du genre Pleurosigma et des genres alliés. *Le Diatomiste 1*: 1–35.

Quekett, J. 1848. *A Practical Treatise on the Use of the Microscope*, 438–439. Pl. 9, Figs 241. London.

Schrader, H.-J. 1973. Types of raphe structures in the diatoms. *Nova Hedwigia Beiheft 45*: 195–230.

Smith, W. 1852. Notes on the Diatomaceae, with descriptions of British species included in the genus Pleurosigma. *Ann. Mag. nat. Hist. 9*, *2*: 1–12.

Sterrenburg, F. A. S. 1990 a. Studies on the genera Gyrosigma and Pleurosigma (Bacillariophyceae). A new phenomenon: co-existence of dissimilar raphe structures in populations of several species. *In*: (M. Ricard, ed.) *Œvrage dediée à la mémoire du Professeur Henry Germain (1903–1989)*. Koenigstein. pp. 235–242.

Sterrenburg, F. A. S. 1990 b. The quest for Quekett. In search of *Nav. angulata* Quekett 1848. *Microscopy 36*: 468–474.

Sterrenburg, F. A. S. 1991. Studies on the genera Gyrosigma and Pleurosigma (Bacillariophyceae). Light microscopical features for taxonomy. *Diatom Research 6*: In press.

Stidolph, S. R. 1980. A record of some coastal marine diatoms from Porirua Harbour, North Island, New Zealand. *New Zealand Journal of Botany 18*: 379–403.

Stidolph, S. R. 1982. Observations and remarks on morphology and taxonomy of the diatom genera Gyrosigma Hassall and Pleurosigma W. Smith. *Nova Hedwigia 47*: 377–388.

Van der Werff, A. and H. Huls. 1957–1974. *Diatomeeënflora van Nederland.* Abcoude/Den Haag.

Van Landingham, S. 1968–1979. *Catalogue of the Fossil and Recent Genera and Species of Diatoms and their Synonyms*, Vols. 1–8, Lehre/Vaduz.

Botanica Marina
Vol. 34, pp. 575–587, 1991

Species of the Toxigenic Dinoflagellate Genus *Alexandrium* in Southeastern Australian Waters

G. M. Hallegraeff, C. J. Bolch, S. I. Blackburn and Y. Oshima*

CSIRO Division of Fisheries, Marine Laboratories, GPO Box 1538, Hobart, Tasmania 7001, Australia
* *Faculty of Agriculture, Tohoku University, 1-1 Tsutsumidori, Amamiyamachi, Sendai, Miyagi 981, Japan*

(Accepted 15 October 1991)

Abstract

Five species of the toxigenic, marine dinoflagellate genus *Alexandrium* are reported from south-eastern Australian waters: *A. minutum* Halim, *A. catenella* (Whedon *et* Kofoid) Balech, *A. tamarense* (Lebour) Balech, *A. affine* (Fukuyo *et* Inoue) Balech and a new species to be described as *A. margalefi* in Balech's monograph on the genus *Alexandrium*. Production of paralytic shellfish poisons was confirmed for *A. minutum* and *A. catenella*, but isolates of *A. margalefi*, *A. tamarense* and *A. affine* were non-toxic. All the species produce smooth-walled, mucilaginous resting cysts but these ranged in shape from spherical (*A. margalefi*, *A. affine*), hemispherical (*A. minutum*) to cylindrical with rounded ends (*A. catenella*, *A. tamarense*). Considerable variation in the cell shape of wild and cultured populations is documented for *A. margalefi*, *A. catenella* and *A. affine*. However, details of shape of the first apical plate (with or without ventral pore), the apical pore complex (with anterior attachment pore) and the posterior sulcal plate (with posterior attachment pore) proved to be conservative taxonomic characters. Historic reports of PSP toxins (Batemans Bay 1935) and of a chain-forming gonyaulacoid dinoflagellate (*Gonyaulax conjuncta* Wood) in New South Wales waters are reviewed. The species *A. margalefi*, *A. tamarense* and *A. affine* are new records for the Australian region.

Introduction

Dinoflagellates that produce paralytic shellfish poisons (PSP) are receiving increasing attention due to the public health risk they pose and the economic impacts they have on aquaculture developments. Many of the causative organisms are small gonyaulacoid species with four apical plates, no intercalary plates and no conspicuous ornamentation; they produce smooth-walled, mucilaginous resting cysts. For many years, the best known representatives were known as *Gonyaulax tamarensis* Lebour and *Gonyaulax catenella* Whedon *et* Kofoid. These species clearly do not belong to the genus *Gonyaulax* Diesing, and have been variously allocated to *Alexandrium* Halim, *Gessnerium* Halim, *Goniodoma* Stein, *Protogonyaulax* Taylor or *Pyrodinium* Plate. Following a re-description of *A. minutum* Halim from the type locality (Balech 1989), a consensus was reached at the Fourth International Conference on Toxic Marine Phytoplankton, Lund, May 1989, on using the name *Alexandrium* (as amended by Balech 1990 a) for species such as *Alexandrium* (= *Gonyaulax*) *tamarense* (Lebour) Balech and *Alexandrium* (= *Gonyaulax*) *catenella* (Whedon *et* Kofoid) Balech.

While these species are well-known from temperate waters of Europe, North America and Japan, until recently no *Alexandrium* were known from Australian waters. The Australian medical literature has a single account (Le Messurier 1935) of wild mussels collected in February 1935 near Batemans Bay (New South Wales) that produced typical PSP symptoms in mice; however, the causative organism was not identified. Wood's (1954) taxonomic survey of dinoflagellates from the Australian region includes a single record of a chain-forming species, *Gonyaulax conjuncta* Wood, collected in 1945 from Port Hacking, New South

Wales; however, its description is inadequate. The first modern reports of the genus *Alexandrium* from Australian waters are those of Hallegraeff *et al.* (1988), who recorded the chain-forming species *Alexandrium catenella* from Port Phillip Bay, Melbourne, and the red water species *Alexandrium minutum* from Port River, Adelaide. The production of PSP toxins by these species was confirmed by both mouse-bioassays and high-performance liquid chromatography. This finding stimulated a more extensive survey for *Alexandrium* species in south-eastern Australian waters and for their benthic resting cysts in sediments (Bolch and Hallegraeff 1990). In the present work, we extend our observations on toxin profiles, resting cysts and morphological characteristics of wild and cultured populations of *A. catenella* and *A. minutum*. In addition, we record the presence in south-eastern Australian waters of non-toxic populations of *A. margalefi* Balech, *A. tamarense* and *A. affine* (Fukuyo et Inoue) Balech.

Material and Methods

The materials used are listed in Table I and sample locations are indicated in Figure 1. Plankton samples were collected with a 20 µm mesh plankton net, first examined in the living state and then preserved with formaldehyde, glutaraldehyde or Lugol's iodine. Dinoflagellate thecae were examined and photographed with a Zeiss Axioplan light microscope using bright field, phase contrast or differential interference illumination. Thecae were also stained with calcofluor white M2R (Sigma Chem. Co.) and examined using epifluorescent illumination (Zeiss filter set 02) (Fritz and Triemer 1985). Selected samples were dehydrated in increasing concentrations of acetone, collected on Nuclepore filters (2 µm) and critical point dried from liquid CO_2. Portions of these filters were then mounted on aluminium stubs, sputter coated with platinum and examined with a Philips 515 scanning electron microscope (SEM). Surface sediments containing cysts were collected with a modified Craib corer, sonicated to dislodge detritus particles, screened through a 90 µm sieve to remove larger particles and then concentrated onto a 20 µm sieve (see Bolch and Hallegraeff 1990). Individual living dinoflagellate cells and cysts were isolated by micropipette under a Nikon Diaphot inverted microscope and placed into tissue culture wells containing 2 mL of modified GPM medium (see Bolch *et al.* 1991). Cultures were incubated at 17.5 °C at a light intensity of 80 $\mu Em^{-2}s^{-1}$ (12 h light : $\overline{12}$ h dark). Successful cultures were transferred to 150 mL Ehrlenmeyer flasks for further study and

Table I. Materials used in the present investigation.

Species and locality	Date	Plankton (p) or cysts (c)	Toxin analyses	Cultures established (CSIRO strain number; isolator)
Alexandrium minutum				
Port River, Adelaide	27 Oct 1987	p	toxic	AMAD 06 (J. Cannon/S. Blackburn)
Port River, Adelaide	8 Sept 1988	c	not tested	see Bolch *et al.* 1991
Port River, Adelaide	11 Nov 1988	p	toxic	AMAD 15 (J. Cannon/S. Blackburn)
Alexandrium margalefi				
River Derwent, Tasmania	9 Dec 1988	p	non-toxic	AMGDE 01 (C. Bolch)
River Derwent, Tasmania	18 Nov 1987	p	not tested	–
Murdunna, Tasmania	4 Dec 1989	p	not tested	–
Recherche Bay	31 May 1990	c	not tested	–
Hastings Bay	3 July 1990	c	not tested	–
Alexandrium catenella				
Hobsons Bay, Melbourne	April 1986	p	not tested	–
Williamstown, Melbourne	10 Mar 1988	p	toxic	ACPP 09 (Ø. Moestrup)
Port Phillip Bay, Melbourne	3 Mar 1988	p/c	toxic	ACPP 02 (S. Blackburn)
Port Phillip Bay, Melbourne	8 Jan 1988	p	not tested	–
Port Phillip Bay, Melbourne	7 May 1990	p	not tested	ACPP 11 (S. Blackburn)
Batemans Bay	10 Jan 1991	p	not tested	–
Alexandrium tamarense				
Port Phillip Bay, Melbourne	Oct 1988	p	not tested	–
Port MacDonnell, S. Australia	Nov 1988	p/c	not tested	–
Devonport, Tasmania	24 Sept 1987	p/c	not tested	–
Bell Bay, Tasmania	12 Dec 1987		non-toxic	ATBB 01 (C. Bolch)
Alexandrium affine				
Bell Bay, Tasmania	10 Sept 1987	p/c	non-toxic	AABB 01 (C. Bolch)

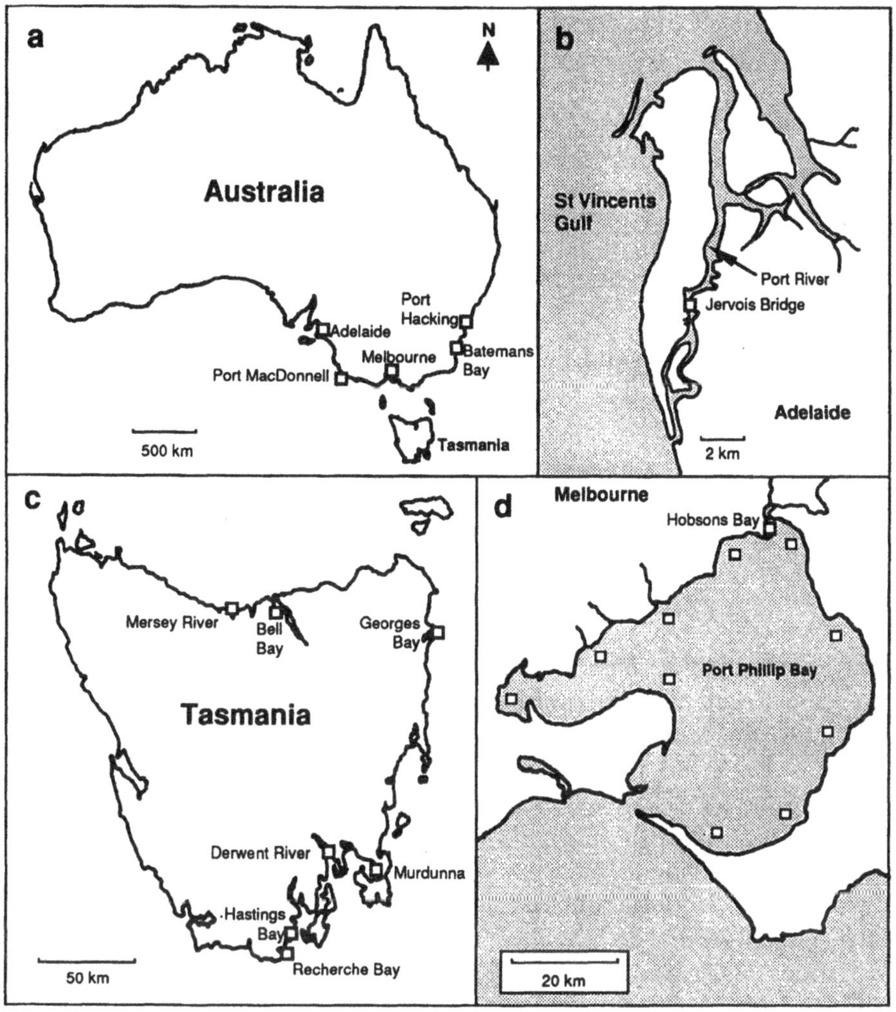

Fig. 1. Map of Australia (a) with details of sampling locations in Port River, Adelaide (b), Tasmania (c) and Port Phillip Bay, Melbourne (d).

toxin analysis, and have been deposited in the CSIRO Culture Collection of Microalgae. High Performance Liquid Chromatographic analyses for PSP toxins were performed as described by Oshima *et al.* (1989 b).

Results and Discussions

Five species of the dinoflagellate genus *Alexandrium* are reported from south-eastern Australian waters. We here describe their morphological characteristics from both wild and cultured material, their distribution, benthic resting cysts and toxicology.

Alexandrium minutum Halim 1960 Figs 2–9
[Synonym: *Alexandrium ibericum* Balech 1985]

Bolch *et al.* 1991, figs 1–7; Halim 1960, figs 1a–j; Hallegraeff *et al.* 1988, figs 7–12.

Description

This small, subspherical dinoflagellate (Fig. 3), 22–26 μm long, 19–23 μm wide, occurs as single cells or rarely in pairs. The first apical plate (1′) and the apical pore complex (P_o) are often in direct contact (Figs 2, 7), except in cells in which the margins of plates 2′ and 4′ obscure this connection (Fig. 5). In some cells the plate 1′ can be drawn out into a thread-like extension connecting it to P_o (A. Rees, unpublished observations). In natural field samples (Port River, October 1987), the 1′-P_o contact varied from direct connection (24% of cells), contact via a thread-like connection (22%) to contact obscured by plate margins (54%). Cultured cells, on the other hand, showed a higher proportion of cells with direct contact between 1′ and P_o (Bolch *et al.* 1991; A. Rees, unpublished observations). The apical pore plate (P_o) is large, with a comma-shaped pore (Fig. 5). The first apical plate has a conspicuous small ventral pore on

Table II. Toxin composition (mol %) of *Alexandrium* species in south-east Australian waters.

Toxin fraction	*Alexandrium minutum*		*Alexandrium catenella*	
	isolate AMAD 06	isolate AMAD 15	isolate ACPP 02	isolate ACPP 09
Saxitoxin	–	–	–	–
neosaxitoxin	–	–	0.4	1.0
decarbamoyl saxitoxin	–	–	0.1	0.1
gonyautoxin-1	45.2	13.5	3.3	3.9
gonyautoxin-2	15.7	9.3	–	0.1
gonyautoxin-3	10.8	7.8	0.2	0.4
gonyautoxin-4	28.3	69.4	18.0	26.2
gonyautoxin-5	–	–	4.0	1.9
gonyautoxin-6	–	–	57.3	30.4
decarbamoyl gonyautoxin-2	–	–	0.1	–
decarbamoyl gonyautoxin-3	–	–	–	0.1
fraction C-1	–	–	0.6	1.0
fractin C-2	–	–	11.3	21.2
fraction C-3	–	–	0.5	2.3
fraction C-4	–	–	4.1	11.4

its right anterior margin (Figs 2, 7). Anterior attachment pores (in P_o) and posterior attachment pores (in S_p) have only been observed in actively dividing cultures. Unlike in *A. catenella* and *A. tamarense*, in *A. minutum* these pores were typically absent in field samples (Figs 5, 6). The 6th precingular plate is characteristically narrow (Figs 2, 4), the anterior sulcal plate (S_a) is narrow and the posterior sulcal plate (S_p) is almost rectangular (Fig. 6). Australian cells always had smooth thecal plates in contrast to literature reports on material from Italy in which the hypotheca may show a distinct reticulate ornamentation (Montresor *et al.* 1990). This species is very similar to *A. lusitanicum* Balech from Portugal, which has a wider S_a plate (Balech 1985), and to *A. andersonii* Balech from North America, which has a differently shaped 6″ plate (Balech 1990 b).

As part of its life-cycle, this dinoflagellate produces a colourless, mucoid resting cyst, roughly hemispherical in shape, circular (24–29 μm diamter) when seen from above and reniform (15–19 μm high) in lateral view (Figs 8, 9). This cyst was first described from surface sediments from Port River, Adelaide (Bolch *et al.* 1991). It has also been reported from Vilaine Bay, France (E. Erard, pers. comm.) and from Spain (Blanco 1989, *Alexandrium* sp. 2, figs 7, 8).

Distribution

Within the Australian region, this species is only known from the Port River area near metropolitan Adelaide, South Australia (Fig. 1b). First recognised in October 1986 (Hallegraeff *et al.* 1988), this dinoflagellate now produces annually recurrent red water blooms (up to 10^8 cells l^{-1}) in the period September–November (Cannon 1990). This taxon has also been recorded from Egypt, Greece, France, Spain, Portugal, Ireland, Italy, Yugoslavia, Turkey and has been found once off New York, on the east coast of America (Balech, pers. comm.).

Toxicology

Wild mussels from the Port River have been found to contain up to 2700 μg saxitoxin equivalent/100 g shellfish meat (Hallegraeff *et al.* 1988) but fortunately no commercial shellfish farms are located in the affected area. The distinctive toxin chemistry of this dinoflagellate (gonyautoxins −1, −2, −3 and −4; Table II) was first described from cultures established from South Australian waters (Oshima *et al.* 1989 a). Similar toxin profiles (predominantly gonyautoxins −2 and −3) have been found in French mussels and oysters contaminated with *A. minutum* (Nezan and Ledoux 1989).

Figs 2–9. *Alexandrium minutum* from Port River, Adelaide.
Fig. 2. SEM. Wild cell in ventral view, showing the shape of the first apical plate (1′) with ventral pore (arrow) and the narrow 6th precingular plate (6″). Fig. 3. LM. Cultured cell showing cell contents. Fig. 4. LM. Cultured cell in ventral view, showing plate tabulation. Fig. 5. SEM. Detail of the apical pore plate (P_o) with comma-shaped apical pore. The connection between P_o and 1′ is obscured by the margins of the 2′ and 4′ plates. Wild cell. Fig. 6. SEM. Antapical view of the sulcal region showing the rectangular posterior sulcal plate (S_p). Wild cell. Fig. 7. LM. Squashed epithecal plates, showing direct contact between P_o and 1′ Cultured cell. Micrography by Dr A. J. J. Rees. Fig. 8. LM. Cyst from Port River sediments, seen from above. Fig. 9. LM. The same cyst as in Fig. 8, seen from the side. All scale bars 10 μm; except Fig. 6 (5 μm), Fig. 5 (1 μm).

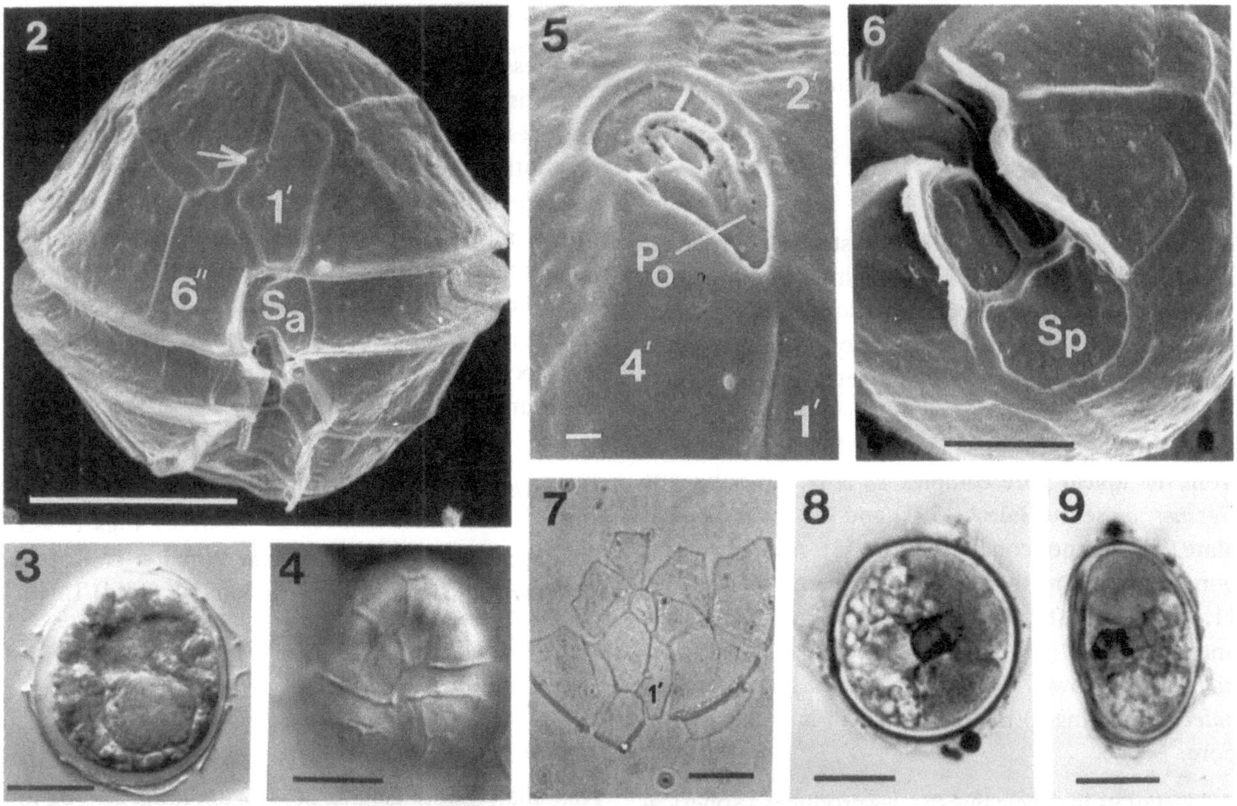

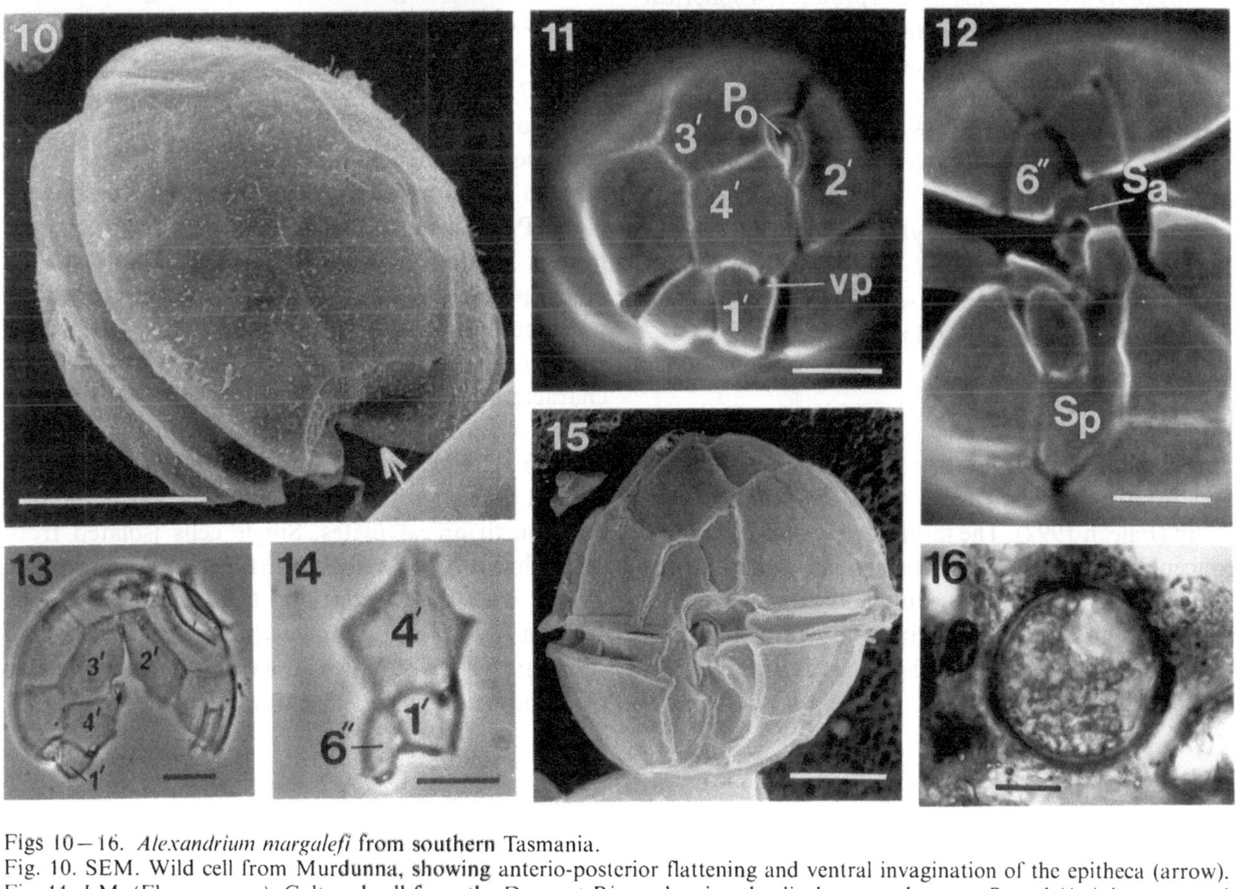

Figs 10–16. *Alexandrium margalefi* from southern Tasmania.
Fig. 10. SEM. Wild cell from Murdunna, showing anterio-posterior flattening and ventral invagination of the epitheca (arrow). Fig. 11. LM. (Fluorescence). Cultured cell from the Derwent River, showing the displacement between P_o and 1′. A large ventral pore (v.p.) is present in the anterior left corner of the quadrangular first apical plate. Calcofluor stained. Fig. 12. LM. (Fluorescence). Detail of the sulcal plates, showing the oblique anterior border of the posterior sulcal plate (S_p). Cultured cell. Calcofluor stained. Fig. 13. LM. Squashed epithecal plates of a cultured cell. The second apical plate (2′) does not contact 1′ Fig. 14. LM. Detached epithecal plates of a cultured cell, showing the quadrangular 1′ and narrow 6th precingular plate (6″). Fig. 15. SEM. Large, cultured cell with subspherical shape and no invagination. Fig. 16. LM. Spherical cyst from Hastings Bay sediments. All scale bars 10 μm; except Fig. 14 (5 μm).

Alexandrium sp. Figs 10–16
(to be described as a new species *Alexandrium margalefi* in Balech's monograph on the genus *Alexandrium*; Balech 1992, figs 10–32).

Description

Wild cells (Fig. 10) are anterio-posteriorly flattened, 30–35 μm wide, 22–26 μm high, consisting of a broadly rounded hypotheca and a subconical epitheca with a distinct ventral invagination. Cultured cells tend to be larger (35–40 μm diameter) and subspherical in shape (Fig. 15). The distinctive short, quadrangular first apical plate (1′) is completely displaced from the apical pore complex P_o (Fig. 11; subgenus *Gessnerium* sensu Balech 1991) and the second apical plate (2′) does not contact 1′ (Fig. 13). A large ventral pore is present in the anterior left corner of 1′ (Figs 11, 14). The posterior sulcal plate (S_p) has an oblique anterior border (Fig. 12). This species is related to *A. monilatum* (Howell) Balech, *A. foedum* Balech and *A. balechi* (Steidinger) Balech, which all have no ventral pore.

As part of its life cycle this species produces a spherical mucoid cyst, 28–34 μm diameter, with clear globular contents except for a yellow accumulation body. This cyst type has been found in Hastings Bay and Recherche Bay, southern Tasmania (Fig. 16) and in Galicia, Spain (Blanco 1989, fig. 9, *Alexandrium* sp. 3). In both cases, cysts have been successfully germinated into *A. margalefi* cultures. The breach-like archeopyle of *A. margalefi* cysts is distinct from that of *A. monilatum* cysts, which have an attached operculum (Bolch *et al.* 1991).

Distribution

This is the first record of this species since its description from cultured material from the Ria of Vigo, Spain (Balech 1992). The motile cell was found in low concentrations in southern Tasmanian waters, in the River Derwent and off Murdunna; the cyst was found in Recherche Bay and Hastings Bay (Fig. 1c). This species has also been identified from Bass Strait (Dr D. Hill, pers. comm.).

Toxicology

Cultures of this species proved negative for PSP toxins. It is possible, however, that this dinoflagellate produces haemolytic toxins similar to those reported for another member of the subgenus *Gessnerium*, *A. monilatum* (Erker *et al.* 1985). In southern Tasmanian waters, mortality of juvenile cage-reared salmonids was associated with a mixed bloom of *Gymnodinium mikimotoi* Miyake *et* Kominami (= *G. nagasakiense* Takayama *et* Adachi), *G.* sp. (to be described as a new species *pulchellum* Larsen) and *A. margalefi* (Murdunna, December 1989).

Alexandrium catenella (Whedon *et* Kofoid) Balech 1985 Figs 17–32
[Synonyms: *Gonyaulax catenella* Whedon *et* Kofoid, *Protogonyaulax catenella* (Whedon *et* Kofoid) Taylor]

Bolch and Hallegraeff 1990, fig. 8; Fukuyo 1985, figs 2h–n; Hallegraeff *et al.* 1988, figs 1–2.

Description

This small dinoflagellate, 20–22 μm long, 25–32 μm wide, occurs as single cells (Figs 25, 26) or more often in characteristic short chains of 2, 4 or 8 cells long (Figs 17, 18, 24, 32). Cells in chains tend to be anterio-posteriorly compressed (Fig. 19) while single cells are often isodiametrical in shape (Fig. 26). The first apical plate (1′) is connected to the triangular apical pore plate (P_o) (Figs 20, 27, 28). An anterior attachment pore is located to the right of the fishhook-shaped apical pore (Figs 20, 21, 29) and a posterior attachment pore is located in the right half of the sulcal posterior plate (Fig. 22). Both pores are transitory characters associated with chain formation; they are absent when single cells are produced in culture (Figs 30, 31). Unlike *A. minutum* and *A. tamarense*, no ventral pore is present on the suture between the first and fourth apical plates. Single cells isolated from Port Phillip Bay, Melbourne (Fig. 25) corresponded closely with the description of *A. fundyense* Balech 1985, a species only known thus far from the cold subarctic waters of North America (type locality: Bay of Fundy). However, in culture these cells formed

Figs 17–32. *Alexandrium catenella* from Port Phillip Bay, Melbourne.
Fig. 17. LM. Four-celled chain of anterio-posteriorly flattened cells. Field sample from Hobsons Bay, April 1986. Fig. 18. SEM. Two-celled chain from Hobsons Bay. Fig. 19. LM. Anterio-posteriorly flattened theca in lateral view. Hobsons Bay. Fig. 29. LM. Apical view of epithecal plates, showing direct contact between P_o and 1′ Hobsons Bay. Fig. 21. SEM. Detail of the triangular apical pore plate (P_o) with fishhook-shaped apical pore and anterior attachment pore (aap). Hobsons Bay. Fig. 22. SEM. Detail of the posterior sulcal plate (S_P) with posterior attachment pore (pap). Hobsons Bay. Fig. 23. LM. Cylindrical cyst with rounded ends from Port Phillip Bay sediments. Fig. 24. LM. Four-celled chain, cultured from a plankton sample collected in May 1990.

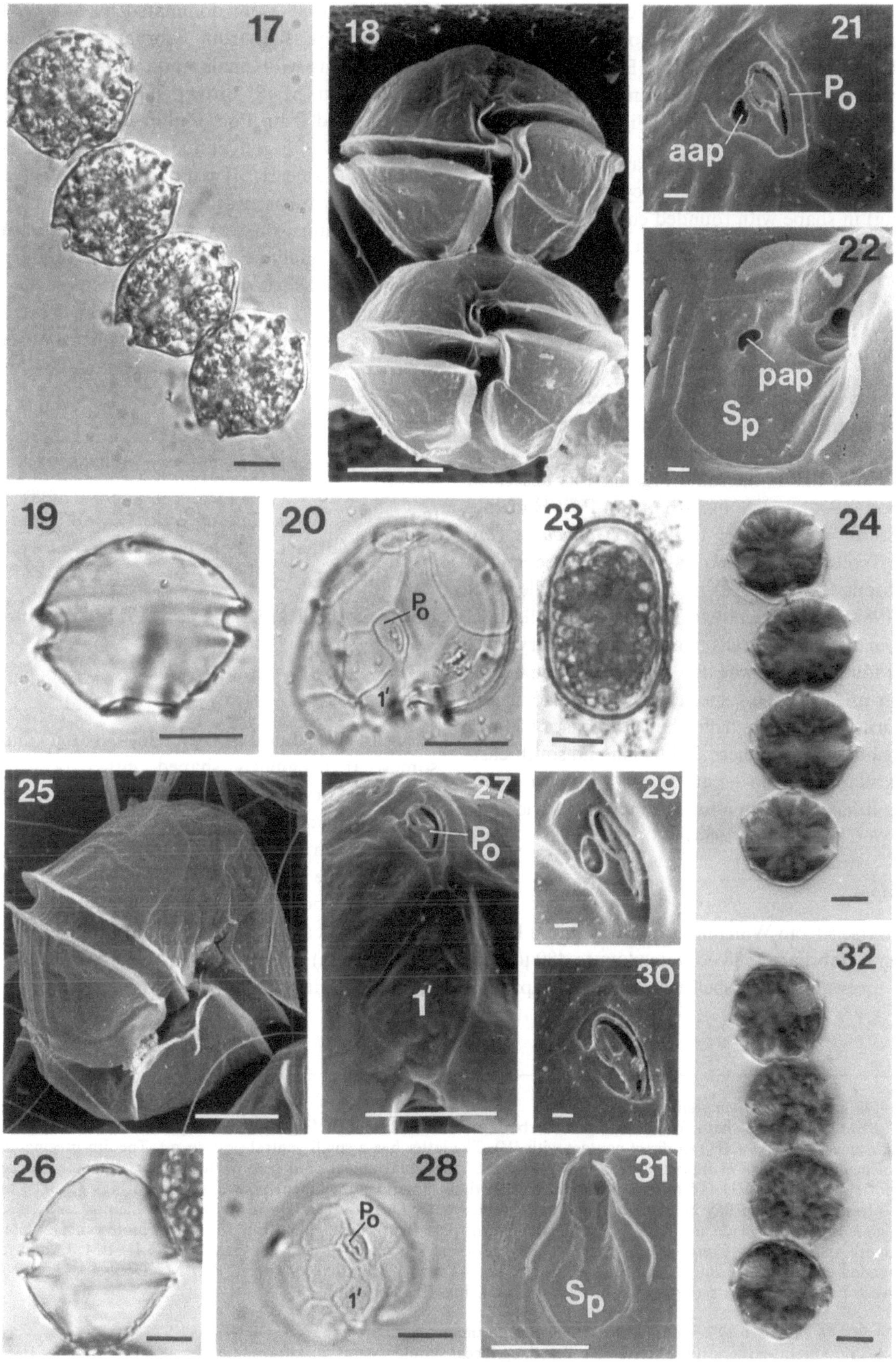

Fig. 25 (SEM), 26 (LM). Single isodiametrical cells from a plankton sample collected in January 1988. Fig. 27 (SEM), 28 (LM). Detail of the connection between P_o and 1′ (without ventral pore); January 1988. Fig. 29. SEM. Detail of the triangular apical pore plate with elliptical anterior attachment pore; January 1988. Fig. 30. SEM. Detail of a triangular apical pore plate without anterior attachment pore; plankton, January 1988. Fig. 31. SEM. Detail of posterior sulcal plate (S_p) without posterior attachment pore; plankton, January 1988. Fig. 32. LM. Four-celled chain cultured from a single cell, isolated from Williamstown in March 1988. All scale bars 10 μm, except Fig. 27 (5 μm), Figs 21, 22, 29, 30 (1 μm).

chains of up to 8 cells long and cell characteristics gradually intergraded with those of typical *A. catenella* (type locality: San Francisco Bay). In Port Phillip Bay, this dinoflagellate only rarely formed chains but pairs of cells were common (Fig. 18).

As part of its life cycle, this dinoflagellate produces a colourless, mucoid cyst, roughly cylindrical (35–40 μm long) in shape with rounded ends (Fig. 23). This cyst type has been found in Port Phillip Bay sediments, but germination attempts have been unsuccessful.

Distribution

Within the Australian region, this species is known thus far from Port Phillip Bay, Melbourne and from Batemans Bay, New South Wales (Table I). It is possible that this is the same organism as described by Wood (1954; p. 258, fig. 161) from Port Hacking, New South Wales, under the name *Gonyaulax conjuncta.* Typical A. *catenella* cells were first recognised in Hobsons Bay (Fig. 1d) in April 1986 (Hallegraeff *et al.* 1988). Subsequently, a bloom (3.10^6 cells L^{-1}) of *A. fundyense* type cells occurred in January 1988 and gradually gave way to *A. catenella* type cells (Moestrup 1989). This species is widespread in the Northern Hemisphere, from the west coast of North America between southern California and south-eastern Alaska, and from Japan. In the Southern Hemisphere, it has also been observed in South Africa and southern Chile (Taylor 1984).

Toxicology

Wild mussels from Williamstown, Port Phillip Bay in January–February 1988 contained up to 480 μg saxitoxin equivalent/100 g shellfish meat. Toxin profiles in January 1988 were dominated by C_1-C_4 toxins while toxin profiles from February 1988 contained C_1-C_4 toxins with small amounts of gonyautoxin-2 and saxitoxin (Mr R. Brown, unpublished data). Cultures isolated from Port Phillip Bay in March 1988 produced C_2, C_4, gonyautoxins −4, −6 and −1, 3, 5 (traces) (Table II). It is not yet clear whether this variation in toxin profiles reflects the presence of different strains (or even different species), but electrophoretic analyses of isozymes (Blackburn and Ward, unpublished data) also suggest genetic heterogeneity of *A. catenella* cultures from Port Phillip Bay. It is likely that this same organism caused PSP toxins in wild mussels from Batemans Bay in February 1935 (Le Messurier 1935).

Alexandrium tamarense (Lebour) Balech 1985
Figs 33–42

[Synonyms: *Gonyaulax tamarensis* Lebour, *Gonyaulax excavata* (Braarud) Balech, *Protogonyaulax tamarensis* (Lebour) Taylor]

Bolch and Hallegraeff 1990, fig. 9; Fukuyo 1985, figs 2a–g.

Description

Medium-size dinoflagellate, 26–38 μm diameter, somewhat irregularly shaped, either occurring as single cells or in pairs (Fig. 33). The first apical plate (1′) has a small ventral pore and is directly connected to the large triangular apical pore plate (P_o) (Figs 34, 40). In some cells a circular to ellipsoidal anterior attachment pore is present to the right of the fishhook-shaped apical pore (Fig. 35). The posterior sulcal plate (S_p) is longer than wide with a V-shaped ventral margin and sometimes has a posterior attach-

Figs 33–42. *Alexandrium tamarense.*
Fig. 33. SEM. Two-celled chain from Devonport harbour, northern Tasmania. Fig. 34. SEM. Apical view of cell showing the triangular apical pore plate (P_o) in direct contact with (1′). The latter has a small ventral pore (v. p.). The 6th precingular plate (6″) is of medium width; Devonport. Fig. 35. SEM. Detail of triangular apical pore plate with anterior attachment pore (aap); Devonport. Fig. 36. SEM. Detail of an apical pore plate without anterior attachment pore; Devonport. Figs 37, 38. LM. Cells collected from Port Phillip Bay in October 1988 (micrographs by Dr D. Hill). Fig. 39. SEM. Detail of the posterior sulcal plate (S_p) which is longer than wide and has a posterior attachment pore (pap); Devonport. Fig. 40. LM. (Fluorescence). Detail of P_o and 1′ (with ventral pore). Cultured cell from Bell Bay, northern Tasmania. Calcofluor stained. Fig. 41. LM. Cylindrical cyst with rounded ends from Devonport harbour sediments. Fig. 42. LM. Cylindrical cyst from Port MacDonnell plankton bloom, South Australia. All scale bars 10 μm (5 μm), Fig. 40 (2 μm), Figs 35, 36 (1 μm).

Figs 43–50. *Alexandrium affine* cultured from Bell Bay, Tasmania.
Fig. 43. SEM. Anterio-posteriorly compressed cell with epithecal shoulders and thick, porulate thecal plates. Fig. 44. SEM. Detail of the subrectangular apical pore plate (P_o), which is in direct contact with 1′. The latter has a very small ventral pore (v. p.). Fig. 45. LM. Two-celled chain, with the cells being longer than wide. Fig. 46. SEM. Antapical view of deeply indented sulcus with the left list (arrow) being higher than the right list. Fig. 47. LM. Detached epithecal plates showing the shape of 1′ (with ventral pore) and 6″ Fig. 48. LM. Detail of detached apical pore plate with large round anterior attachment pore, located to the dorsal right of the fishhook-shaped apical pore. Fig. 49. LM. Detached posterior sulcal plate (S_p) with posterior attachment pore (pap), located near the middle of the right margin. Fig. 50. LM. Spherical cyst from Bell Bay sediments, which germinated to produce the cells in Figs 43, 44. All scale bars 10 μm, except Figs 44, 48, 49 (5 μm).

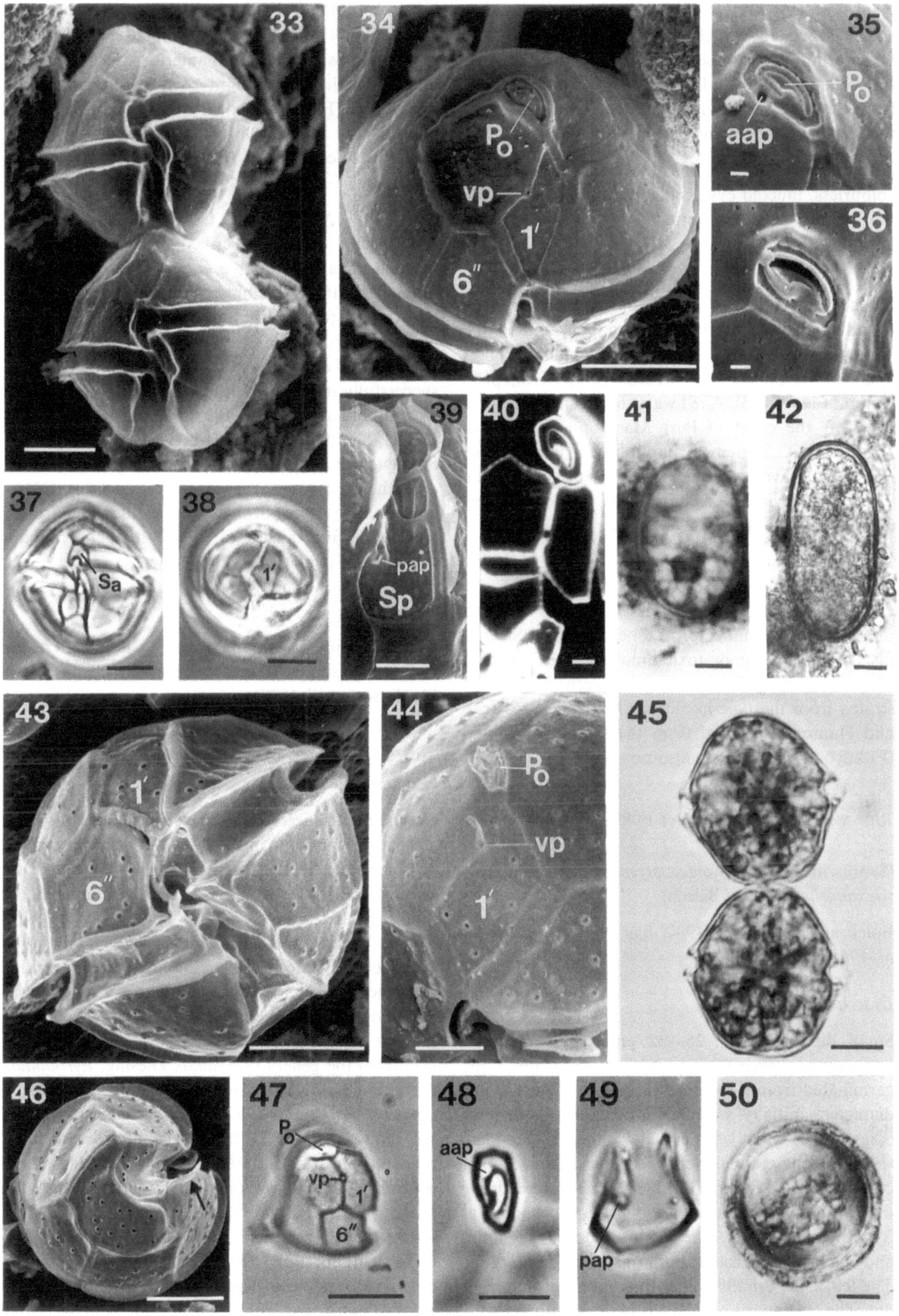
33
34
Po
vp
1′
6″
35
Po
aap
36
39
pap
Sp
40
41
42
37
Sa
38
1′
43
1′
6″
44
Po
vp
1′
45
46
47
Po
vp
1′
6″
48
aap
49
pap
50

ment pore (Fig. 39). Both pores are transitory characters associated with chain formation (Fig. 36). The 6th precingular plate is of medium width. The Australian material is indistinguishable from material from the type locality in the Tamar Estuary, southwest England (Moestrup and Hansen 1988).

As part of its life cycle, this dinoflagellate produces a colourless, mucoid cyst, roughly cylindrical in shape (35 – 40 µm long) with rounded ends (Figs 41, 42).

Distribution

Within the Australian region, this species has been found in the Mersey River, Devonport and in the Tamar River, Bell Bay, in northern Tasmania (Fig. 1) and in Port Phillip Bay, Melbourne (D. Hill, pers. comm.; Figs 37, 38). A red water bloom in November 1988 off the coast of Port MacDonnell, southern Australia (D. Steffensen, unpublished data) is also thought to have been caused by this species (Fig. 1). This dinoflagellate is common in Japan, North America and Europe (Taylor 1984).

Toxicology

Cultures established from plankton samples collected off Bell Bay, Tasmania, proved negative for PSP toxins. This is a common toxic dinoflagellate species in North America, Europe and northern Japan, but strains from the type locality in England (Moestrup and Hansen 1988) and from the Gulf of Thailand (Fukuyo *et al.* 1988) are also non-toxic.

Alexandrium affine (Inoue *et* Fukuyo) Balech 1985
Figs 43 – 50

[Synonyms: *Protogonyaulax affinis* Inoue *et* Fukuyo; *Alexandrium fukuyoi* Balech]

Bolch and Hallegraeff 1990, figs 10a – e; Fukuyo *et al.* 1985, figs 24 – 29.

Description

Small dinoflagellates, 25 – 32 µm diameter, which have comparatively thick, porulate thecal plates. Cells germinated from a spherical resting cyst (28 – 30 µm diameter) initially were slightly anterio-posteriorly compressed with well-defined epithecal shoulders (Fig. 43) but gradually changed into chain-forming cultures (2 – 4 cells long) in which the cells were longer than wide, with a rounded epitheca and trapezoidal hypotheca (Fig. 45). The sulcus is deeply indented and can have high lists at both right and left margins. The left list is usually higher than the right one (Fig. 46). The first apical plate (1′) is in direct contact with P_o (Fig. 44) and has a small ventral pore in the anterior third of the suture between the 1′ and 4′ plates (Fig. 47). The apical pore plate is sub-rectangular and has a large round anterior attachment pore located to the dorsal right of the fishhook-shaped apical pore (Fig. 48). A posterior attachment pore is present near the middle of the right margin of the posterior sulcal plate S_p (Fig. 49).

Distribution

This species was first described from the saline Lake Saroma in Japan (Fukuyo *et al.* 1985) where it produces red water containing chain-forming dinoflagellates of up to 16 cells long. Within the Australian region, the spherical resting cyst (Fig. 50) was isolated from Bell Bay, sediments, northern Tasmania, and successfully germinated into chain-forming cultures. This dinoflagellate is also known from Korea (Lee 1990), the Philippines (J. MacLean, pers. comm.), Spain (since 1985; Fraga, pers. comm.) and Brazil (Balech, pers. comm.). Genetic studies using DNA-sequencing have confirmed that the Australian isolate has close affinities with *A. affine* cultures from Spain (C. Scholin, pers. comm.).

Toxicology

Cultures from Tasmania proved to be negative for PSP toxins, similar to evidence from Spanish cultures (Anderson *et al.* 1989).

Conclusions

Nowhere is the need for accurate species identification more important than in the case of toxic species which are of public health or economic significance. Taxonomic difficulties with the toxigenic, gonyaulacoid dinoflagellates have been considerable (Taylor 1975). The re-description of *Alexandrium minutum* from the type locality (Balech 1989) has only recently clarified the generic concepts within this group. *Protogonyaulax* Taylor and *Gessnerium* Halim are now recognised as synonyms of *Alexandrium* Halim, and its delineation from the genera *Gonyaulax* Diesing, *Pyrodinium* Plate and *Goniodoma* Stein has been clarified (Steidinger 1990, Balech 1990 a). Species concepts within the genus *Alexandrium* have also only recently been refined (Balech 1985, 1992). At present, the taxonomy is primarily based on morphological criteria while the range of variability of 'conservative' characters in vegetative cells still needs to be tested. Until the 1980s most workers only distinguished between single-celled, *A. 'tamarense'*-like species (with or without ventral pore) and chain-forming *A. 'catenella'*-like species. Balech (1985, 1992) now recognises 30 *Alex-*

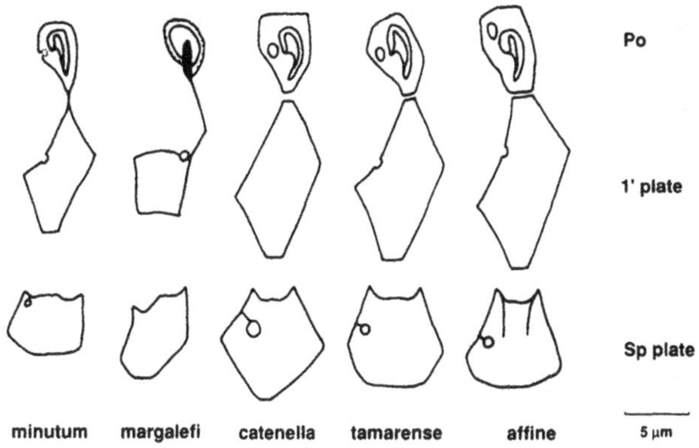

Fig. 51. Details of plate tabulation of five Australian *Alexandrium* species. Summary of shape of the apical pore complex, P_o, (with anterior attachment pore), its connection with the first apical plate (with or without ventral pore) and the shape of the posterior sulcal plate (with posterior attachment pore).

andrium species, based on cell shape and size, chain formation, shape and position of the P_o plate, shape and position of (attachment) pores in the P_o and S_p plates, presence and size of the ventral pore, shape and displacement of the first apical plate (1′), and shape and size of the 6th precingular (6″) and anterior (S_a) and posterior (S_p) sulcal plates.

The present work on cultured and wild populations of *A. catenella* and *A. margalefi* demonstrates considerable variability in cell shape. Single-celled populations from Port Phillip Bay (Fig. 25; initially identified as *A. fundyense* Balech) could only be confirmed as *A. catenella* after they produced 4–8 celled chains in culture (Fig. 32). Similarly, anterio-posteriorly compressed cells from southern Tasmanian waters (Fig. 10) could only be confirmed as *A. margalefi* after they started to produce isodiametrical cells in culture (Fig. 15; cultured cells were also used in the type description!). A careful study of the variability of wild and cultured cells of type locality material of *A. tamarense* (Lebour) Balech and *A. excavatum* (Braarud) Balech *et* Tangen (only distinguished by the presence of epithecal shoulders) has now also indicated their synonymity (Moestrup and Hansen 1988). In contrast to cell shape, in this study the shape of the first apical plate (with or without ventral pore), the apical pore complex (with anterior attachment pore) and the posterior sulcal plate (with posterior attachment pore) proved to be conservative taxonomic characters (Fig. 51).

The presence of a small ventral pore at the margins of 1′ and 4′, at times, may be difficult to recognise by standard light microscopy (in the case of very small pores; Fig. 7) or scanning electron microscopy (when the pores fill in; Fig. 44). However, when using a combination of both techniques and/or plate staining (Fritz and Triemer 1985), this appears to be a conservative taxonomic character (Table III). The degree of displacement of P_o and 1′ is a variable character in *A. minutum* (Hallegraeff *et al.* 1988, Balech 1989) but is more constant in other species (*e.g. A. margalefi*, no contact, Fig. 11; *A. catenella*, always in contact, Fig. 27). The presence or absence of an anterior attachment pore in P_o and a posterior attachment pore in S_p is a variable character related to chain formation (Figs 35, 36). However, if present, their position and shape appear to be conservative taxonomic characters (*e.g. A. catenella*, Figs 21, 22;

Table III. Summary of *Alexandrium* species in south-eastern Australian waters and their diagnostic characteristics. Compare Figure 51.

Species	Cell size	Chain formation	P_o/1′ contact	Ventral pore in 1′ plate	Resting cyst	PSP toxins	Distribution in Australia
A. minutum	22–26 µm long, 19–23 µm wide	Single cells; rarely in pairs	Direct contact or thread-like connection	+	hemispherical	+	Port River, Adelaide
A. margalefi	30–35 µm wide, 22–26 µm high	Single cells; rarely in pairs	Completely displaced	+	spherical	–	Southern Tasmania, Bass Strait
A. catenella	20–22 µm long, 25–32 µm wide	Chains of 2, 4 or 8 cells long	Direct contact	–	cylindrical with rounded ends	+	Port Phillip Bay, Batemans Bay
A. tamarense	26–38 µm diameter	Single cells; rarely in pairs	Direct contact	+	cylindrical with rounded ends	–	Northern Tasmania, Port Phillip Bay, Port MacDonnell
A. affine	25–32 µm diameter	Chains of 2, 4 (up to 16) cells long	Direct contact	+	spherical	–	Northern Tasmania

A. affine, Figs 47, 49; summarized in Fig. 51). Physiological condition of cells can influence cell shape and plate rim overgrowth and can cause the pores in S_p and in 1′ to vary in position from close to the plate margin to a position well away from the margin, but still connected by a thread-like connection (Fig. 22).

The shape of the resting cyst (hypnozygote) (spherical in *A. margalefi* and *A. affine*; hemispherical in *A. minutum*; ellipsoidal with rounded ends in *A. catenella and A. tamarense*) can also be of value in species distinction (Table III). In the case of toxic species, toxin profiles have only limited diagnostic value; *A. minutum* appears to have a distinctive dominance of gonyautoxins in isolates from Australia, Spain and France, but *A. catenella* shows considerable variability in toxin profiles (both gonyautoxins, saxitoxins, C-toxins) dependent upon geographic origin (Table II; Oshima *et al.* 1990).

Starting in the 1970s, the increased size and speed of cargo vessels discharging sea water ballast in Australian ports has raised concerns about the possibility that toxic dinoflagellates, not endemic to Australian waters, can be accidentally introduced via cyst stages contained in ship's ballast tanks. In a survey of 80 cargo vessels entering Australian ports, 40% contained viable dinoflagellate cysts and 6% carried the cysts of toxic *Alexandrium* species; toxic cultures of both *Alexandrium catenella* and *A. tamarense* have been established from this source (Hallegraeff and Bolch 1991). Ships containing *A. catenella* cysts thus have been found to arrive in Port Phillip Bay, Melbourne, where plankton blooms of this dinoflagellate have been observed since 1986 (Hallegraeff *et al.* 1988, Hallegraeff and Bolch 1992). Unfortunately, species of *Alexandrium* are easily overlooked in routine plankton surveys, and as their benthic resting cysts do not fossilize, it is virtually impossible to prove that introduction has occurred. Early reports of PSP toxins in mussels from Batemans Bay (Le Messurier 1935, most likely due to *A. catenella*) and of a similar chain-forming dinoflagellate in Port Hacking, Sydney (Wood 1954) indicate that at least some of these species may be endemic to the Australian region. However, plankton and cyst surveys in Port River, Adelaide, in 1983 failed to detect *A. minutum* (Hallegraeff, unpublished data) in an area which since 1986 has produced annually recurrent red water blooms (Cannon 1990). This has led to speculation that *A. minutum* could be an introduced species and genetic studies using DNA sequencing have now confirmed a close affinity between Australian and Spanish isolates of this species (C. Scholin, pers. comm.). The apparent disjunct global distribution of estuarine dinoflagellates such as *A. minutum* (mainly confined to South Australia and Southern Europe), *A. margalefi* (Tasmania, Spain) and *Gymnodinium catenatum* (Tasmania, Japan, Spain, Mexico: see Blackburn *et al.* 1989) also supports the suggestion that these cyst-forming species may have been transported across oceanic boundaries, either in ships' ballast water or associated with transfer of shellfish stocks.

Acknowledgements

We are indebted to Prof. E. Balech (Argentina) for introducing us to *Alexandrium* taxonomy and for helpful comments and discussions. The following persons kindly provided us with field samples: Dr G. Arnott (Marine Science Labs., Queenscliff, Victoria); Mr R. Brown (Tasmanian Department of Sea Fisheries); Ms J. Cannon (University of Adelaide); Dr D. Hill (University of Melbourne); Dr Ø. Moestrup (Institut for Sporeplanter, Copenhagen); Dr G. Parry (Marine Science Labs., Queenscliff) and Dr D. Steffensen (Engineering and Water Supply Dept., S. A.). Mr R. Brown assisted with the HPLC toxin analyses, and Dr A. J. Rees provided expert assistance with light and electron microscopy. Mr C. A. Scholin (Woods Hole Oceanographic Institution) collaborated in elucidating the genetic affinities of our Australian *Alexandrium* cultures. This study was partially supported by grant 89/39 from the Fishing Industry Research Development Council and a grant from the Ministry of Education, Science and Culture of Japan (International Scientific Research Program).

References

Anderson, D. M., J. J. Sullivan and B. Reguera. 1989. Paralytic shellfish poisoning in Northwest Spain: The toxicity of the dinoflagellate *Gymnodinium catenatum*. *Toxicon 27*: 665–674.

Balech, E. 1985. The genus *Alexandrium* or *Gonyaulax* of the tamarensis group. *In*: (D. M. Anderson, A. W. White and D. G. Baden, eds) *Toxic Dinoflagellates*, Elsevier, N. Y. pp. 33–38.

Balech, E. 1989. Redescription of *Alexandrium minutum* Halim (Dinophyceae), type species of the genus *Alexandrium*. *Phycologia 28*: 206–211.

Balech, E. 1990 a. A short diagnostic description of *Alexandrium*. *In*: (E. Graneli, B. Sundström, L. Edler and D. Anderson, eds) *Toxic Marine Phytoplankton*, Elsevier Science Publishing Co., New York. p. 77.

Balech, E. 1990 b. Four new dinoflagellates. *Helgoländer Meeresunters. 44*: 397–396.

Balech, E. 1992. *The Genus* Alexandrium *Halim (Dinoflagellata)*. Sherkin Island Marine Station, Cork, special publication (in press).

Blackburn, S. I., G. M. Hallegraeff and C. J. Bolch. 1989. Vegetative reproduction and sexual life cycle of the toxic dinoflagellate *Gymnodinium catenatum* from Tasmania, Australia. *J. Phycol. 25*: 577–590.

Blanco, J. 1989. Quistes de dinoflagelados de las costas de Galicia. I. Dinoflagelados gonyaulacoides. *Scient. Mar. 53*: 785–796.

Bolch, C. J. and G. M. Hallegraeff. 1990. Dinoflagellate cysts in Recent marine sediments from Tasmania, Australia. *Bot. Mar. 33*: 173–192.

Bolch, C. J., S. I. Blackburn, J. A. Cannon and G. M. Hallegraeff. 1991. The resting cyst of the red tide dinoflagellate *Alexandrium minutum* Halim (Dinophyceae). *Phycologia 30*: 215–219.

Cannon, J. A. 1990. Development and dispersal of red-tides in the Port River, South Australia. *In*: (E. Graneli, B. Sundström, L. Edler and D. Anderson, eds) *Toxic Marine Phytoplankton*, Elsevier, New York. pp. 110–115.

Erker, E. F., L. J. Slaughter, E. L. Bass, J. Pinion and J. Wutoh. 1985. Acute toxic effects in mice of an extract from the marine algae *Gonyaulax monilata*. *Toxicon 23*: 761–767.

Fritz, L. and R. E. Triemer. 1985. A rapid, simple technique utilising Calcofluor white M2R for the visualisation of dinoflagellate thecal plates. *J. Phycol. 21*: 662–664.

Fukuyo, Y. 1985. Morphology of *Protogonyaulax tamarensis* (Lebour) Taylor and *Protogonyaulax catenella* (Whedon and Kofoid) Taylor from Japanese waters. *Bull. Mar. Sci. 37*: 529–537.

Fukuyo, Y., K. Yoshida and H. Inoue. 1985. *Protogonyaulax* in Japanese coastal waters. *In*: (D. M. Anderson, A. W. White and D. G. Baden, eds) *Toxic Dinoflagellates*, Elsevier Science Publishing Co., New York. pp. 27–32.

Fukuyo, Y., P. Pholpunthin and K. Yoshida. 1988. *Protogonyaulax* (Dinophyceae) in the Gulf of Thailand. *Bull. Plankton Soc. Japan 35*: 9–20.

Halim, Y. 1960. *Alexandrium minutum* nov. g. nov. sp. dinoflagellé provocant des 'eaux rouges'. *Vie et Milieu 11*: 102–105.

Hallegraeff, G. M. and C. J. Bolch. 1991. Transport of toxic dinoflagellate cysts via ships' ballast water. *Mar. Poll. Bull. 22*: 27–30.

Hallegraeff, G. M. and C. J. Bolch. 1992. Transport of diatom and dinoflagellate resting spores in ships' ballast water: Implications for plankton biogeography and aquaculture. *J. Plankton Res.* (in press).

Hallegraff, G. M., D. A. Steffensen and R. Wetherbee. 1988. Three estuarine Australian dinoflagellates that can produce paralytic shellfish toxins. *J. Plankton Res. 10*: 533–541.

Lee, S. G. 1990. Identification of three dinoflagellate species belonging to the genus *Alexandrium* occurring in Chinhae Bay. *Bull. Nat. Fish. Res. Dev. Agency 44*: 1–8.

Le Messurier, D. J. 1935. A survey of mussels on a portion of the Australian coast. *Med. J. Aust. 1*: 490–492.

Moestrup, Ø. 1989. A toxic bloom of *Alexandrium* in Port Phillip Bay, during the austral summer, 1988. *Red Tide Newsletter 2(1)*: 3.

Moestrup, Ø. and P. J. Hansen. 1988. On the occurrence of the potentially toxic dinoflagellates *Alexandrium tamarense* (= *Gonyaulax excavata*) and *A. ostenfeldii* in Danish and Faroese waters. *Ophelia 28*: 195–213.

Montresor, M., D. Marino, A. Zingone and G. Dafnis. 1990. Three *Alexandrium* species from coastal Tyrrhenian waters (Mediterranean Sea). *In*: (E. Graneli, B. Sundström, L. Edler and D. Anderson, eds) *Toxic Marine Phytoplankton*, Elsevier Science Publishing Co., New York. pp. 82–87.

Nezan, E. and M. Ledoux. 1989. Red Tide by *Alexandrium minutum* in France. *Red Tide Newsletter 2(1)*: 2–3.

Oshima, Y., M. Hirota, T. Yasumoto, G. M. Hallegraeff, S. I. Blackburn and D. A. Steffensen. 1989 a. Production of paralytic shellfish toxins by the dinoflagellate *Alexandrium minutum* Halim from Australia. *Bull. Japan Soc. Sci. Fish. 55*: 925.

Oshima, Y., K. Sugino and T. Yasumoto. 1989 b. Latest advances in HPLC analysis of paralytic shellfish toxins. *In*: (S. Natori, K. Hashimoto and Y. Ueno, eds) *Mycotoxins and Phycotoxins '88*, Elsevier Science Publishers, New York. pp. 319–326.

Oshima, Y., K. Sugino, H. Itakura, M. Hirota and T. Yasumoto. 1990. Comparative studies on paralytic shellfish toxic profile of dinoflagellates and bivalves. *In*: (E. Graneli, B. Sundström, L. Edler and D. Anderson, eds) *Toxic Marine Phytoplankton*, Elsevier Science Publishing Co., New York. pp. 391–396.

Steidinger, K. A. 1990. Species of the *tamarensis/catenella* group of *Gonyaulax* and the fucoxanthin derivative-containing gymnodinioids. *In*: (E. Graneli, B. Sundström, L. Edler and D. Anderson, eds) *Toxic Marine Phytoplankton*, Elsevier Science Publishing Co., New York. pp. 11–16.

Taylor, F. J. R. 1975. Taxonomic difficulties in red tide and paralytic shellfish poison studies: the "*tamarensis*" complex of *Gonyaulax*. *Environmental Letters 9*: 103–119.

Taylor, F. J. R. 1984. Toxic dinoflagellates: taxonomic and biogeographic aspects with emphasis on *Protogonyaulax*. *In*: (E. P. Ragelis, ed.) *Seafood Toxins*, American Chemical Society Symposium Series 262. pp. 77–97.

Wood, E. J. F. 1954. Dinoflagellates in the Australian region. *Aust. J. Mar. Freshw. Res. 5*: 171–351.

Botanica Marina
Vol. 34, pp. 589–591, 1991

On the Reported Presence of Vaterite and Aragonite in Coccoliths of *Emiliania huxleyi*

J. R. Young, J. M. Didymus*, and S. Mann*

Palaeontology Department, The Natural History Museum, London SW7 5BD, U.K.
* *School of Chemistry, University of Bath, Bath, BA2 7AY, U.K.*

(Accepted 15 October 1991)

Abstract

X-ray diffraction analysis of coccoliths of the prymnesiophyte alga *Emiliania huxleyi* have been undertaken in order to test anomalous reports of phases other than calcite occurring in these coccoliths. Only calcite was found in our studies and we conclude that the earlier results were spurious.

Introduction

The coccolithophore *Emiliania huxleyi* (Lohmann) Hay *et* Mohler is an extremely abundant member of the marine phytoplankton and is probably responsible for more calcification than any other single species. *Emiliania huxleyi* is also very easy to maintain in laboratory cultures, and so has proven a fruitful organism for the investigation of biomineralization processes (Willbur and Watabe 1963, Klaveness 1976, Westbroek *et al.* 1984, 1989, Mann and Sparks 1988). This work has resulted in an appreciation that although *E. huxleyi* is a single-celled organism the coccolithogenesis process is sophisticated with precise regulation of both nucleation and crystal growth. One well known result appears to be in discord with this model of a precise, highly evolved biomineralization process: Wilbur and Watabe (1963) investigated the mineralogical composition of *E. huyleyi* coccoliths by means of X-ray diffraction (XRD) analysis of powdered bulk samples. They reported that, under standard culture conditions, the coccoliths were entirely formed of calcite but that under conditions of nitrogen depletion three polymorphs of calcium carbonate were present – calcite, vaterite, and aragonite.

This apparent tendency for vaterite and aragonite to substitute for calcite would have considerable and interesting implications for the biomineralization process. As one aspect of current work on the process, we attempted to reproduce the original results.

Material and Methods

Culture work was carried out at the University of Leiden Geobiochemistry Unit. *Emiliania huxleyi* Strain L was used for this work since it produces coccoliths morphologically similar to those of the strains used by Wilbur and Watabe (1963). Strain L was originally isolated from the North Atlantic and was kindly provided by Dr J. Green (Plymouth Marine Laboratory). Batch culturing was used in 250 mL flasks, at 200 °C, with twice daily agitation, a 16–8 hour light–dark cycle, and illumination of *ca* 3000 lux. Natural sea water media were used, with addition of trace elements, vitamins and variable quantities of major nutrients (nitrogen as KNO_3 and phosphate as K_2HPO_4). These conditions are similar to those of Wilbur and Watabe (1963). Precise reproduction of conditions is impossible, since natural sea water was used in both cases.

Cultures were harvested after one week – during the late logarithmic/early stationary phase of culture growth. The coccoliths were isolated by centrifugation and, cleaned with NaOCl and Triton-X detergent, and washed in distilled water buffered with 0.05 M $NaHCO_3$.

XRD powder analysis was carried out using a Philips X-ray powder diffractometer with Cu Kalpha (1.5405 angstroms) radiation. Samples were prepared by freeze-drying beforehand.

Table I. Comparison of our data with that of Wilbur and Watabe (1963).
N – number of repeat experiments. HPO_4^-, NO_3^- concentration of these nutrient ions in the culture media. N. B. In natural sea water typical values are: HPO_4^- 0.1 – 1 μm; NO_3^- 1 – 10 μm.

Culture medium	N	HPO_4	NO_3	Calcite	Vaterite	Aragonite
Our results:						
Eplley's f/50	3	1 μm	10 μm	100%	/	/
low nitrate	1	50 μm	10 μm	100%	/	/
Eppley's f	1	50 μm	500 μm	100%	/	/
Wilbur and Watabe:						
Nitrate deficient	4	72 μm	400 μm	15 – 30%	0 – 70%	18 – 60%
Guillard's f	1	72 μm	1800 μm	100%	/	trace

Results

Table I summarises our results, and those of Wilbur and Watabe (1963). All the coccolith material we analysed by X-ray diffraction was entirely composed of calcite, typical results are given in Table II. No traces of aragonite or vaterite were detected. In addition we have carried out numerous electron diffraction studies of individual coccolith elements at all stages of development and found no evidence of polymorphs other than calcite.

Wilbur and Watabe (1963) suggested that aragonite and vaterite were produced in 'nitrogen deficient' media. Their nitrogen deficient medium however had very high absolute nitrate levels and only slight nitrate deficiency relative to phosphate. Our tested material included samples from media of similar composition and with much lower relative and absolute nitrate levels.

Table II. Experimental XRD line data for one sample, K1, low nitrate. Standard line values from Joint Committee for Powder Diffraction Standards (1974).

Observed lines	Standard lines	Face indices (hkl)
3.854	3.852	012
3.034	3.030	104
2.838	2.834	006
2.495	2.495	110
2.282	2.284	113
2.090	2.094	202
1.923	1.926	024
1.910	1.907	018
1.872	1.873	116
1.624	1.626	211
1.603	1.604	122
	1.582	
1.525	1.525	214

Discussion and Conclusion

The occurrence of aragonite and vaterite reported by Wilbur and Watabe (1963) is anomalous in the light of subsequent analyses, and our attempts to reproduce it were unsuccessful. It seems likely that the original result was spurious; introduced by some form of contamination, chemical reaction during preparation, or instrumental error. This conclusion fits well with suggestions that crystal nucleation is regulated by an organic matrix. It is likely that an assembly of macromolecules constitutes the nucleation matrix and possesses definite arrays of binding sites that recognise structural (geometrical and stereochemical) similarities in specific calcite faces (Mann and Sparks 1988). A possible candidate for the matrix is a group of complex acidic polysaccharides which are closely associated with completed coccoliths and with coccolith growth (de Vrind-de Jong *et al.* 1986, van Emburg *et al.* 1986). The plane on which nucleation occurs appears to be precisely regulated so it is difficult to envisage polymorphs with different crystal structure being incorporated in the same structure without disruption of the regularity of coccolith form.

Acknowledgements

We are grateful for the interest and assitance of Peter Westbroek, Kees Linschooten, Robert Kempers and other members of the Leiden geobiochemistry research group. We also thank Alan Carver of Bath University for help with the XRD measurements.

References

Emburg, P. R. van, E. W. de Jong and W. T. Daems. 1986. Immunochemical localization of a polysaccharide from biomineral structures (coccoliths) of Emiliania huxleyi. *J. Ultrastructure Molecular Structure Res. 94*: 246–259.

Joint Committee on Powder Diffraction Standards. 1974. Card. 24–27: Calcite.

Klaveness, D. 1976. *Emiliania huxleyi* (Lohmann) Hay and Mohler. III. Mineral deposition and the origin of the matrix during coccolith formation. *Protistologica 12*: 217–224.

Mann, S. and N. H. C. Sparks. 1988. Single crystalline nature of coccolith elements of the marine alga *Emiliania huxleyi* as determined by electron diffraction and high resolution transmission electron microscopy. *Procs. R. Soc. London B/ 234*: 441–453.

Vrind-de Jong E. W. de, A. H. Borman, R. Thierry, P. Westbroek, M. Gruter and J. P. Kamerling. 1986. Calcification in the coccoliths *Emiliania huxleyi* and *Pleurochrysis carterae*. II. Biochemical aspects. *In* (B. S. C. Leadbeater and R. Riding, eds) *Biomineralization in Lower Plants and Animals*. Systematics Assoc. Spec. Vol. 30. pp. 207–217.

Westbroek, P., E. W. de Jong, P. van der Wal, A. H. Borman, J. P. M. de Vrind, D. Kok, W. C. de Bruijn and S. B. Parker. 1984. Mechanism of calcification in the marine alga *Emiliania huxleyi. Phil. Trans. R. Soc. Lond. B/304*: 435–444.

Westbroek, P., J. R. Young and K. Linschooten. 1989. Coccolith production (biomineralization) in the marine alga *Emiliania huxleyi. J. Protozool. 36*: 368–373.

Botanica Marina
Vol. 34, p. 593, 1991

Acknowledgement

The editors and the publishers wish to thank the following colleagues for their kind assistance in acting as referees for the journal this year.

W. Admiraal, Bilthoven, The Netherlands
P. O. Ang Jr, Vancouver, Canada
K. E. Apt, Stanford, CA, U.S.A.
E. Balech, Necochea, Argentina
D. L. Ballantine, Mayaguez, PR, U.S.A.
C. J. Bird, Halifax, Canada
K. T. Bird, Fort Pierce, FL, U.S.A.
G. Blunden, Portsmouth, U.K.
C. F. Boudouresque, Marseille, France
P. F. Cannon, Kew, Surrey
A. Cardinal, Québec, Canada
F. Colijn, The Hague, The Netherlands
M. Cormaci, Catania, Italy
J. S. Craigie, Halifax, Canada
C. J. Dawes, Tampa, FL. U.S.A.
J. D. Dodge, Egham, U.K.
G. Drebes, List, Germany
J. Fábregas, Santiago, Spain
J. Y. Floc'h, Brest, France
G. A. Fryxell, College Station, TX, U.S.A.
N. D. Hallam, Clayton, Australia
G. M. Hallegraeff, Hobart, Australia
P. E. Hargraves, Kingston, RI, U.S.A.
M. M. Harlin, Kingston, RI, U.S.A.
P.G. Harrison, Vancouver, Canada
C. den Hartog, Nijmegen, The Netherlands
S. J. Hawkins, Port Erin, U.K.
B. R. Heimdal, Bergen, Norway
B. Hickel, Plön, Germany
C. van den Hoek, Haren, The Netherlands
A. Q. Hurtado-Ponce, Iloilo City, Philippines
J. Jewett-Smith, Port Aranses, TX, U.S.A.
J. M. Jones, Port Erin, U.K.
B. Kloareg, Roscoff, France
G. T. Kraft, Parkville, Australia
J. W. La Claire II, Austin, TX, U.S.A.
M. Lahaye, Nantes, France
W. J. Langston, Plymouth, U.K.
M. Lazarri, Buenos Aires, Argentina
M. L. Lazo, Vancouver, Canada
I. K. Lee, Seoul, Korea
J. E. Lyngby, Horsholm, Denmark
C. A. Maggs, Belfast, Ireland
F. L. Magne, Paris, France
A. C. Mathieson, Durham, NH, U.S.A.
A. J. McComb, Nedlands, Australia
J. McLachlan, Halifax, Canada
A. Meinesz, Nice, France
O. Moestrup, Copenhagen, Denmark
H. P. Molitoris, Regensburg, Germany
I. Munda, Ljubljana, Jugoslavia
T. A. Norton, Port Erin, U.K.
M. Pedersén, Uppsala, Sweden
M. Polne-Fuller, Santa Barbara, CA, U.S.A.
W.F. Prud'homme van Reine, Leiden, The Netherlands
J. Ramus, Beaufort, NC, U.S.A.
J. Raven, Dundee, U.K.
D. C. Reed, Santa Barbara, CA, U.S.A.
D. J. Rogers, Portsmouth, U.K.
R. Ross, Bishops Frome, U.K.
J. Rucness, Oslo, Norway
G. Russell, Liverpool, U.K.
B. Santelices, Santiago, Chile
R.B. Searles, Durham, NC, U.S.A.
A. Sournia, Paris, France
J. van Staden, Pietermaritzburg, Rep. of South Africa
K. Sundbäck, Göteborg, Sweden
A. R. A. Taylor, Fredericton, Canada
F. J. Taylor, Leigh, New Zealand
I. Tsekos, Thessaloniki, Greece
H. Vandermeulen, Rexdale, Canada
M. Verlaque, Marseille, France
T. J. Ward, Hobart, Australia
J. A. West, Berkeley, CA, U.S.A.
M. Whitfield, Plymouth, U.K.
A. Whittick, St John's, Canada
J. M. Wolfe, Houghton, N. Y. U.S.A.
H. B. S. Womersley, Adelaide, Australia
M. J. Wynne, Ann Arbor, MI, U.S.A.

Bei Fragen zur Produktsicherheit wenden Sie sich bitte an:
If you have any questions regarding product safety,
please contact:

Walter de Gruyter GmbH
Genthiner Straße 13
10785 Berlin
productsafety@degruyterbrill.com